Prefixes for Powers of 10[†]

Multiple	Prefix	Abbreviation
10^{24}	yotta	Y
10^{21}	zetta	Z
10^{18}	exa	E
10^{15}	peta	P
10^{12}	tera	T
10^{9}	**giga**	**G**
10^{6}	**mega**	**M**
10^{3}	**kilo**	**k**
10^{2}	hecto	h
10^{1}	deka	da
10^{-1}	deci	d
10^{-2}	**centi**	**c**
10^{-3}	**milli**	**m**
10^{-6}	**micro**	**μ**
10^{-9}	**nano**	**n**
10^{-12}	**pico**	**p**
10^{-15}	femto	f
10^{-18}	atto	a
10^{-21}	zepto	z
10^{-24}	yocto	y

† Commonly used prefixes are in bold. All prefixes are pronounced with the accent on the first syllable.

The Greek Alphabet

Alpha	A	α	Nu	N	ν
Beta	B	β	Xi	Ξ	ξ
Gamma	Γ	γ	Omicron	O	o
Delta	Δ	δ	Pi	Π	π
Epsilon	E	ϵ, ε	Rho	P	ρ
Zeta	Z	ζ	Sigma	Σ	σ
Eta	H	η	Tau	T	τ
Theta	Θ	θ	Upsilon	Y	υ
Iota	I	ι	Phi	Φ	ϕ
Kappa	K	κ	Chi	X	χ
Lambda	Λ	λ	Psi	Ψ	ψ
Mu	M	μ	Omega	Ω	ω

Terrestrial and Astronomical Data[†]

Acceleration of gravity at earth's surface	g	$9.81 \text{ m/s}^2 = 32.2 \text{ ft/s}^2$
Radius of earth	R_E	$6370 \text{ km} = 3960 \text{ mi}$
Mass of earth	M_E	$5.98 \times 10^{24} \text{ kg}$
Mass of sun		$1.99 \times 10^{30} \text{ kg}$
Mass of moon		$7.36 \times 10^{22} \text{ kg}$
Escape speed at earth's surface		$11.2 \text{ km/s} = 6.95 \text{ mi/s}$
Standard temperature and pressure (STP)		$0°C = 273.15 \text{ K}$ $1 \text{ atm} = 101.3 \text{ kPa}$
Earth–moon distance[‡]		$3.84 \times 10^{8} \text{ m} = 2.39 \times 10^{5} \text{ mi}$
Earth–sun distance (mean)[‡]		$1.50 \times 10^{11} \text{ m} = 9.30 \times 10^{7} \text{ mi}$
Speed of sound in dry air (at STP)		331 m/s
Speed of sound in dry air (20°C, 1 atm)		343 m/s
Density of air (STP)		1.29 kg/m^3
Density of water (4°C, 1 atm)		1000 kg/m^3
Heat of fusion of water (0°C, 1 atm)	L_f	333.5 kJ/kg
Heat of vaporization of water (100°C, 1 atm)	L_v	2.257 MJ/kg

† Additional data on the solar system can be found in Appendix B and at http://nssdc.gsfc.nasa.gov/planetary/planetfact.html.
‡ Center to center.

Mathematical Symbols

$=$	is equal to
$\equiv$	is defined by
$\neq$	is not equal to
$\approx$	is approximately equal to
$\sim$	is of the order of
$\propto$	is proportional to
$>$	is greater than
$\geq$	is greater than or equal to
$>>$	is much greater than
$<$	is less than
$\leq$	is less than or equal to
$<<$	is much less than
Δx	change in x
dx	differential change in x
$|x|$	absolute value of x
$|\vec{v}|$	magnitude of $\vec{v}$
$n!$	$n(n-1)(n-2)\dots1$
Σ	sum
$\lim$	limit
$\Delta t \to 0$	Δt approaches zero
$\dfrac{dx}{dt}$	derivative of x with respect to t
$\dfrac{\partial x}{\partial t}$	partial derivative of x with respect to t
$\displaystyle\int_{x_1}^{x_2} f(x)dx$	definite integral
$= F(x)\Big|_{x_1}^{x_2} = F(x_2) - F(x_1)$	

Abbreviations for Units

A	ampere	H	henry	nm	nanometer (10^{-9} m)
Å	angstrom (10^{-10} m)	h	hour	pt	pint
atm	atmosphere	Hz	hertz	qt	quart
Btu	British thermal unit	in	inch	rev	revolution
Bq	becquerel	J	joule	R	roentgen
C	coulomb	K	kelvin	Sv	seivert
°C	degree Celsius	kg	kilogram	s	second
cal	calorie	km	kilometer	T	tesla
Ci	curie	keV	kilo-electron volt	u	unified mass unit
cm	centimeter	lb	pound	V	volt
dyn	dyne	L	liter	W	watt
eV	electron volt	m	meter	Wb	weber
°F	degree Fahrenheit	MeV	mega-electron volt	y	year
fm	femtometer, fermi (10^{-15} m)	Mm	megameter (10^6 m)	yd	yard
ft	foot	mi	mile	μm	micrometer (10^{-6} m)
Gm	gigameter (10^9 m)	min	minute	μs	microsecond
G	gauss	mm	millimeter	μC	microcoulomb
Gy	gray	ms	millisecond	Ω	ohm
g	gram	N	newton		

Some Conversion Factors

Length

1 m = 39.37 in = 3.281 ft = 1.094 yd

1 m = 10^{15} fm = 10^{10} Å = 10^9 nm

1 km = 0.6215 mi

1 mi = 5280 ft = 1.609 km

1 lightyear = 1 $c\cdot$y = 9.461×10^{15} m

1 in = 2.540 cm

Volume

1 L = 10^3 cm^3 = 10^{-3} m^3 = 1.057 qt

Time

1 h = 3600 s = 3.6 ks

1 y = 365.24 d = 3.156×10^7 s

Speed

1 km/h = 0.278 m/s = 0.6215 mi/h

1 ft/s = 0.3048 m/s = 0.6818 mi/h

Angle–angular speed

1 rev = 2π rad = 360°

1 rad = 57.30°

1 rev/min = 0.1047 rad/s

Force–pressure

1 N = 10^5 dyn = 0.2248 lb

1 lb = 4.448 N

1 atm = 101.3 kPa = 1.013 bar = 76.00 cmHg = 14.70 lb/in^2

Mass

1 u = [(10^{-3} mol^{-1})/N_A] kg = 1.661×10^{-27} kg

1 tonne = 10^3 kg = 1 Mg

1 slug = 14.59 kg

1 kg weighs about 2.205 lb

Energy–power

1 J = 10^7 erg = 0.7373 ft$\cdot$lb = 9.869×10^{-3} L$\cdot$atm

1 kW$\cdot$h = 3.6 MJ

1 cal = 4.184 J = 4.129×10^{-2} L$\cdot$atm

1 L$\cdot$atm = 101.325 J = 24.22 cal

1 eV = 1.602×10^{-19} J

1 Btu = 778 ft$\cdot$lb = 252 cal = 1054 J

1 horsepower = 550 ft$\cdot$lb/s = 746 W

Thermal conductivity

1 W/(m$\cdot$K) = 6.938 Btu$\cdot$in/(h$\cdot$ft$^2\cdot$F°)

Magnetic field

1 T = 10^4 G

Viscosity

1 Pa$\cdot$s = 10 poise

fifth edition

PHYSICS

FOR SCIENTISTS AND ENGINEERS

Volume 1
Mechanics, Oscillations and
Waves; Thermodynamics

W. H. Freeman and Company
New York

Publisher:	Susan Finnemore Brennan
Senior Development Editor:	Kathleen Civetta/Jennifer Van Hove
Assistant Editors:	Rebecca Pearce/Amanda McCorquodale/Eileen McGinnis
Marketing Manager:	Mark Santee
Project Editors:	Georgia L. Hadler/Cathy Townsend,
	PreMediaONE, A Black Dot Group Company
Cover and Text Designers:	Marcia Cohen/Blake Logan
Illustrations:	Network Graphics/PreMediaONE,
	A Black Dot Group Company
Photo Editors:	Patricia Marx/Dena Betz
Production Manager:	Julia DeRosa
Media and Supplements Editors:	Brian Donnellan
Composition:	PreMediaONE, A Black Dot Group Company
Manufacturing:	RR Donnelley & Sons Company

Cover image: Digital Vision

Library of Congress Cataloging-in-Publication Data
Physics for Scientists and Engineers. - 5th ed.

    p. cm.
    By Paul A. Tipler and Gene Mosca
    Includes index.
    ISBN: 0-7167-0809-4 (Vol. 1 Hardback Ch. 1-20, R)
    ISBN: 0-7167-0900-7 (Vol. 1A Softcover Ch. 1-13, R)
    ISBN: 0-7167-0903-1 (Vol. 1B Softcover Ch. 14-20)
    ISBN: 0-7167-0810-8 (Vol. 2 Hardback Ch. 21-41)
    ISBN: 0-7167-0902-3 (Vol. 2A Softcover Ch. 21-25)
    ISBN: 0-7167-0901-5 (Vol. 2B Softcover Ch. 26-33)
    ISBN: 0-7167-0906-6 (Vol. 2C Softcover Ch. 34-41)
    ISBN: 0-7167-8339-8 (Standard Hardback Ch. 1-33, R)
    ISBN: 0-7167-4389-2 (Extended Hardback Ch. 1-41)

Printed in the United States of America

First printing 2003

PT: For Claudia

GM: For Vivian

CONTENTS IN BRIEF

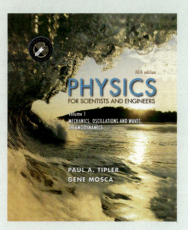

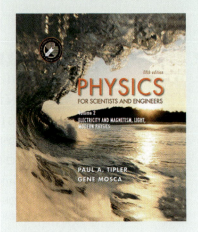

VOLUME 2

CONTENTS

CHAPTER 10

CONSERVATION OF ANGULAR MOMENTUM / 309

CHAPTER R

SPECIAL RELATIVITY / R-1

CHAPTER 11

GRAVITY / 339

CHAPTER 12 *

STATIC EQUILIBRIUM AND ELASTICITY / 370

PART III **THERMODYNAMICS /532**

APPENDIX A

SI UNITS AND CONVERSION FACTORS / AP-1

APPENDIX B

NUMERICAL DATA / AP-3

APPENDIX C

PERIODIC TABLE OF ELEMENTS / AP-6

APPENDIX D

REVIEW OF MATHEMATICS / AP-8

VOLUME 2

PART IV ELECTRICITY AND MAGNETISM /651

CHAPTER 21

THE ELECTRIC FIELD I: DISCRETE CHARGE DISTRIBUTIONS / 651

CHAPTER 22

THE ELECTRIC FIELD II: CONTINUOUS CHARGE DISTRIBUTIONS / 682

PART VI MODERN PHYSICS: QUANTUM MECHANICS, RELATIVITY, AND THE STRUCTURE OF MATTER

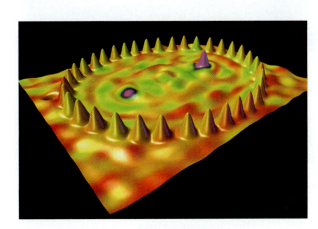

CHAPTER 36

ATOMS

CHAPTER 37

MOLECULES

CHAPTER 38

SOLIDS AND THE THEORY OF CONDUCTION

PREFACE

We are exceptionally pleased to present the fifth edition of *Physics for Scientists and Engineers*. Over the course of this revision, we have built upon the strengths of the fourth edition so that the new text is an even more reliable, engaging and motivating learning tool for the calculus-based introductory physics course. With the help of reviewers and the many users of the fourth edition we have carefully scrutinized and refined every aspect of the book, with an eye toward improving student comprehension and success. Our goals included helping students to increase their problem-solving ability, making the text more accessible and fun to read, and keeping the text flexible for the instructor.

Examples

One of the most important ways we've addressed our goals was to add some new features to the side-by-side worked examples that were introduced in the fourth edition. These examples juxtapose the problem-solving steps with the necessary equations so that it's easier for students to watch the problem unfold.

The side-by-side format for the worked examples came from a student suggestion; we've just added a few finishing touches:

• After each problem statement, students are asked to *Picture the Problem*. Here, the problem is analyzed both conceptually and visually, with students frequently directed to draw a free-body diagram. Each step of the solution is then presented with a written statement in the left-hand column and the corresponding mathematical equations in the right-hand column.

• *Remarks* at the end of the example point out the importance or relevance of the example, or suggest a different way to approach it.

• NEW *Plausibility Checks* remind students to check their results for mathematical accuracy, and for reasonableness as well.

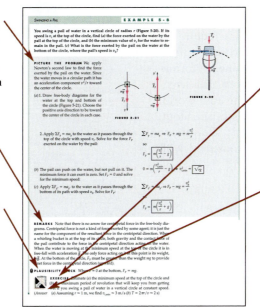

• An *Exercise* often follows the solution of the example, allowing students to check their understanding by solving a similar problem without help. Answers are included with the Exercise to provide immediate feedback and alternative solutions.

• NEW *Master the Concept Exercises* appear at least once in each chapter and help build students' problem-solving skills online.

Every example has been scrutinized, with additional steps added wherever an assumption might have been made, new Remarks included, and new follow-up exercises, free-body diagrams added where appropriate. The answers are now boxed to make them easier to find. Our new features include the Plausibility Check, which offers quick tests that help students learn to evaluate their answers with logic. We've also added interactive Master the Concept exercises to help students work through key problems. The exercises follow examples in the textbook and are marked with a Master the Concept icon that directs students to our Web site. There, the exercise is set up with algorithmically generated variables and students work the problem with step-by-step guidance and immediate feedback.

This edition also includes two types of specialized examples that provide unique problem-solving opportunities for students. The Try it Yourself examples prompt students to take an active role in solving the problem, and the Put It in Context examples approximate the real life scenarios they might encounter as scientists.

Try It Yourself examples

Like the regular worked example, these use the side-by-side format, but here the Picture the Problem section is sometimes missing, and the descriptions in the left-hand column are more terse. These examples take students step-by-step through the solution without doing the math for them. Students find it helpful to cover the right-hand column and attempt to perform the calculations on their own before looking at the equations. In this way, students can think through the steps as they fill in the answers.

New Put It in Context examples

Each chapter now identifies at least one worked example as "context rich." These examples may include information not needed to solve the problem, or may require the student to find additional information in tables or to draw from experience or previously obtained information. Context-rich examples reflect the way that scientists and engineers solve problems in the real world. Laura McCullough of the University of Wisconsin, Stout, and Thomas Foster of Southern Illinois University, Edwardsville, initiated this feature and consulted with us in creating many of these examples.

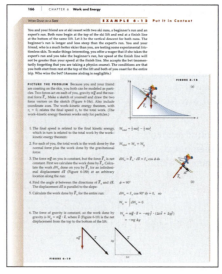

Practice Problems

Care has been taken to improve the quality and clarity of the end-of-chapter problems. About twenty percent of the 4,500 problems are new, written by Charles Adler of St. Mary's College of Maryland. Conceptual problems have been grouped together at the beginning of each problem set, and a new category of Estimation and Approximation problems have been added to encourage students to think more like scientists or engineers. Answers to odd-numbered problems appear at the back of the text. Solutions to approximately twenty-five percent of the problems appear in the newly revised Student Solutions Manual. This was written by David Mills of the College of the Redwoods to provide detailed solutions and to mirror the popular side-by-side format of the textbook examples.

About 1,100 of the text's problems are included in the new iSOLVE homework service. These problems can be accessed at www.whfreeman.com/tipler5e. About a third of the iSOLVE problems are Checkpoint Problems, which ask students to note the key principles and equations they're using and indicate their confidence level.

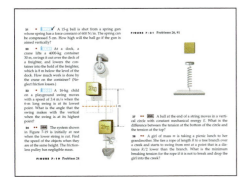

Each problem is marked with:

• a series of one, two, or three bullets, which identify its level of difficulty

• a SSM icon if the answer is in the Student Solutions Manual

• an iSOLVE icon if the problem is part of the isolve homework service and a iSOLVE ✔ icon if the problem is a Checkpoint problem.

Features

This new edition of *Physics* has a number of textual features that make the book a valuable teaching tool. Key aspects of the last edition have been revised, and some new features have been added to make the book more engaging, inviting and up to date.

New chapter-opening pedagogy

•Each chapter now begins with a photograph and a question that is answered in a worked example within the chapter. These draw students into the material and provide motivation for problem solving.

•Chapter outlines list the major section headings, giving students a "road map" to the chapter.

•Chapter goal statements highlight the main ideas of the chapter.

PART I **MECHANICS**

Motion in One Dimension

CHAPTER **2**

MOTION IN ONE DIMENSION IS MOTION ALONG A STRAIGHT LINE, LIKE THAT OF A CAR ON A STRAIGHT ROAD. THIS DRIVER ENCOUNTERS STOPLIGHTS AND DIFFERENT SPEED LIMITS ON HER COMMUTE ALONG A STRAIGHT HIGHWAY TO SCHOOL.

? How can she estimate her arrival time? (See Example 2-2.)

2-1 Displacement, Velocity, and Speed
2-2 Acceleration
2-3 Motion With Constant Acceleration
2-4 Integration

Content improvements

Chapter R, an optional "mini" chapter in Volume 1, brief enough to be covered in a lecture or two, allows instructors to include this popular modern topic early in the course. The chapter avoids the abstraction associated with the Lorentz transformations and focuses on the basic concepts of length contraction, time dilation, and simultaneity, using thought experiments involving meter sticks and light clocks. The relation between relativistic momentum and relativistic energy is also developed.

Quantum Theory: Chapters 17, "Wave-Particle Duality and Quantum Physics," and 27, "The Microscopic Theory of Electrical Conduction" of the fourth edition have been moved to their more traditional location in Volume II of the fifth edition as Chapters 37 and 38. Should instructors wish to include these chapters earlier in the course, both chapters are available on the web at www.whfreeman.com/tipler5e.

Changes in Approach: Dozens of smaller, yet significant improvements in content have been made throughout the book. For example:

- Motion-diagrams are introduced in Section 3-3 and used to estimate the direction of the acceleration vector using the definition of acceleration.

- In Section 4-4, frictional forces are now introduced qualitatively, allowing for free-body diagrams that include frictional forces. A quantitative treatment of frictional forces appears in Section 5-1.

- Section 4-7 introduces problems with two or more objects. Selecting a separate set of coordinate axes for each object is a robust problem-solving practice when using Newton's laws with systems consisting of two or more objects. The value of this practice is revealed in the example where Steve is sliding down the glacier while Paul has already fallen over its edge.

- In Section 8-8, "Systems With Variable Mass," the basic equation of motion for an object with continuously varying mass (the rocket equation) is developed using an object that is acquiring mass—like an open boxcar in the rain—rather than one that is losing mass—like a rocket spewing exhaust gasses. This approach facilitates both the development of the basic equation of motion and the application of it to certain situations.

- In Chapter 9, "Rotation," there is a new section that provides problem-solving guidelines for applying Newton's Second Law to rotation.

- In Section 13-3 the discussion of buoyancy now includes the buoyant force on objects supported by a submerged surface.

- In Chapter 18, work-energy relations are expressed in terms of the work done on the system. The first law of thermodynamics is now expressed in terms of the work done on the system also. (The Educational Testing Service has adopted the convention that the work term in first law of thermodynamics be the work done on the system. This will be adhered to on all Advanced Placement physics exams.)

More engineering and biological applications

Additional applications emphasize the relevance of physics to students' experiences, further studies, and future careers.

New focus on common pitfalls

Topics that commonly cause confusion are identified with a new 🔴 icon where the difficulty is addressed. For example, in Section 3-4 the icon is used to identify the discussion pointing out that the horizontal and vertical motions are independent in projectile motion.

For instructor and student convenience, the fifth edition of *Physics for Scientists and Engineers* is available in five paperback volumes—

Vol. 1A Mechanics (Ch. 1-13, plus a mini-chapter on relativity, Ch. R) 0-7167-0900-7

Vol. 1B Oscillations & Waves; Thermodynamics (Ch. 14-20) 0-7167-0903-1

Vol. 2A Electricity (Ch. 21-25) 0-7167-0902-3

Vol. 2B Electrodynamics, Light (Ch. 26-33) 0-7167-0901-5

Vol. 2C Elementary Modern Physics (Ch. 34-41) 0-7167-0906-6

or in four hardcover versions—

Vol. 1 Mechanics, Oscillations and Waves; Thermodynamics (Ch. 1-20, R) 0-7167-0809-4

Vol. 2 Electricity, Magnetism, Light & Modern Physics (Ch. 21-41) 0-7167-0810-8

Standard Version (Vol 1A-2B) 0-7167-8339-8

Extended Version (Vol 1A-2C) 0-7167-4389-2

New design and improved illustrations

The book has a warmer, more colorful look. Each piece of art has been carefully considered and many have been revised to increase clarity. Approximately 245 new figures have been added, including many new free-body diagrams within the worked examples. New photos bring to life the many real-world applications of physics.

Optional sections

The book was designed to allow professors to be flexible by designating certain sections "optional." These sections are marked with an *, and professors who choose to skip this section can do so knowing that their students won't be missing any material they will need in later chapters.

Summary

End of chapter summaries are organized with important topics on the left and relevant remarks and equations on the right. Here the key equations from the chapter appear together for easy reference.

Exploring essays

Students are invited to examine interesting extensions of the chapter concepts in Exploring sections, which are now found on the Web. These short pieces relate the chapter concepts to everything from the weather to transducers.

Media and Print Supplements

The supplements package has been updated and improved in response to reviewer suggestions and those from users of the fourth edition.

For the Student:

Student Solutions Manual: *Vol. 1, 0-7167-8333-9; Vol. 2, 0-7167-8334-7.* The new manual prepared by David Mills of College of the Redwoods, Charles Adler of St. Mary's College of Maryland, Ed Whittaker of Stevens Institute of Technology, George Zober of Yough Senior High School and Patricia Zober of Ringgold High School provides solutions for about twenty-five percent of the problems in the textbook, using the same side-by-side format and level of detail as the textbook's worked examples.

Study Guide: *Vol. 1, 0-7167-8332-0; Vol. 2, 0-7167-8331-2.* Prepared by Gene Mosca of the United States Naval Academy and Todd Ruskell of Colorado School of Mines, the Study Guide describes the key ideas and potential pitfalls of each chapter, and also includes true and false questions that test essential definitions and relations, questions and answers that require qualitative reasoning, and problems and solutions.

Student Web Site: Robin Jordan of Florida Atlantic University has put together a site designed to make studying and testing easier for both students and professors. The Web site includes:

- **On-line quizzing:** Multiple choice quizzes are available for each chapter. Students will receive immediate feedback, and the quiz results are collected for the instructor in a grade book.

- **iSOLVE homework service**: *0-7167-5802-4.* About one-fourth of the book's end-of-chapter problems, 1,100 altogether, are available on-line in W.H. Freeman's iSOLVE homework service. This service will offer each student a different version of every problem similar to CAPA and WebAssign, and the iSOLVE problems will be marked with an icon in the textbook. Homework scores can be collected in a grade book. Students may purchase access to iSOLVE for three semesters at a time.

• **iSOLVE Checkpoint problems:** A third of our iSOLVE questions are Checkpoint problems, which prompt students to describe how they arrived at their answer and to indicate their confidence level. All student responses will be gathered and included in the instructor's grade book report. Rolf Enger of the U.S. Air Force Academy inspired the development of Checkpoints to help professors gauge their student's understanding of the material.

• **Master the Concept exercises:** For each chapter, one or more exercises from the book will be available on-line so students can practice working the problem with randomized variables and step-by-step guidance. The on-line exercise will walk the student slowly through the problem-solving process and use interactive animations, simulations, video, and other graphic aids to help students visualize the problem. Teachers can collect grade book information on their progress. These premium examples are called out in the book with a Master the Concept icon.

Homework services: In addition to the iSOLVE network, there are three other homework services that are compatible with this textbook. End of chapter problems are available in WebAssign as well as CAPA: A Computer-Assisted Personalized Approach. A list of all the fifth edition problems included in WebAssign and CAPA is posted on the instructor's section of the *Physics* Web site. Our text is also compatible with the University of Texas Interactive Homework Service.

> **The iSolve homework service is available at**
> www.whfreeman.com/tipler5e
>
> **For more information about WebAssign, CAPA or UTX homework services, find their Web sites at:**
> http://webassign.net/info
> http://www.pa.msu.edu/educ/CAPA/
> http://hw.utexas.edu/hw.html

For the Instructor:

Instructor's Resource CD-ROM: *0-7167-9839-5*. This multi-faceted resource will give instructors the tools to make their own Web sites and presentations. The CD contains illustrations from the text in .jpg format, Powerpoint Lecture Slides for each chapter of the book, Lab Demonstration Videos, and Applied Physics videos in QuickTime format, and Presentation Manager Pro v.2.0, as well as all of the solutions to the end-of-chapter problems in editable Microsoft Word format.

Instructor's Resource Manual: The updated IRM contains Classroom Demonstrations for each chapter, a film and video guide with suggestions for each chapter, links to valuable Web sites, and links to free sources for Physlets, animations, and other teaching tools. This manual will be available on the book's Web site at www.whfreeman.com/tipler5e.

Instructor's Solutions Manual: *Vol. 1, 0-7167-9640-6; Vol. 2, 0-7167-9639-2*. This guide contains fully worked solutions for all of the problems in the textbook, using the side-by-side format wherever possible. It is available in print and is also included in editable Word files on the Instructor's CD-ROM.

Test Bank: *In print, 0-7167-9652-X; CD-ROM, 0-7167-9653-8*. Prepared by Mark Riley of Florida State University and David Mills of College of the Redwoods, this set of more than 4,000 multiple choice questions is available both in print and on a CD-ROM for Windows and Macintosh users. All questions refer to specific sections in the book. The CD-ROM version of the Test Bank makes it easy to add, edit and re-sequence questions to suit your needs.

Transparencies: *0-7167-9664-3*. Approximately 150 full color acetates of figures and tables from the text are included, with type enlarged for projection.

Acknowledgments

We are grateful to the many instructors, students, colleagues, and friends who have contributed to this, and to earlier editions.

Charles Adler of St. Mary's College of Maryland authored the excellent new problems. David Mills of the College of the Redwoods extensively revised the solutions manual. Robin Jordan of Florida Atlantic University created the innovative Master the Concept exercises and iSOLVE Checkpoint problems. Laura McCullough of the University of Wisconsin, Stout, and Thomas Foster of Southern Illinois University, Edwardsville, drawing from their background in Physics Education Research, were instrumental in providing context-rich examples in every chapter as well as our new Estimation and Approximation problems. We received invaluable help in accuracy checking of text and problems from professors:

Karamjeet Arya,
San Jose State University

Michael Crivello,
San Diego Mesa College

David Faust,
Mt. Hood Community College

Jerome Licini,
Lehigh University

Dan Lucas,
University of Wisconsin, Madison

Jeannette Myers,
Clemson University

Marian Peters,
Appalachian State University

Paul Quinn,
Kutztown University

Michael G. Strauss,
University of Oklahoma

George Zober,
Yough Senior High School

Patricia Zober,
Ringgold High School

Many instructors and students have provided extensive and helpful reviews of one or more chapters. They have each made a fundamental contribution to the quality of this revision, and deserve our gratitude. We would like to thank the following reviewers:

Edward Adelson,
The Ohio State University

Todd Averett,
The College of William and Mary

Yildirim M. Aktas,
University of North Carolina at Charlotte

Karamjeet Arya,
San Jose State University

Alison Baski,
Virginia Commonwealth University

Gary Stephen Blanpied,
University of South Carolina

Ronald Brown,
California Polytechnic State University

Robert Coakley,
University of Southern Maine

Robert Coleman,
Emory University

Andrew Cornelius,
University of Nevada at Las Vegas

Peter P. Crooker,
University of Hawaii

N. John DiNardo,
Drexel University

William Ellis,
University of Technology - Sydney

John W. Farley,
University of Nevada at Las Vegas

David Flammer,
Colorado School of Mines

Tom Furtak,
Colorado School of Mines

Patrick C. Gibbons,
Washington University

John B. Gruber,
San Jose State University

Christopher Gould,
University of Southern California

Phuoc Ha,
Creighton University

Theresa Peggy Hartsell,
Clark College

James W. Johnson,
Tallahassee Community College

Thomas O. Krause,
Towson University

Donald C. Larson,
Drexel University

Paul L. Lee,
California State University, Northridge

Peter M. Levy,
New York University

Jerome Licini,
Lehigh University

Edward McCliment,
University of Iowa

Robert R. Marchini,
The University of Memphis

Pete E.C. Markowitz,
Florida International University

Fernando Medina,
Florida Atlantic University

Laura McCullough,
University of Wisconsin at Stout

John W. Norbury,
University of Wisconsin at Milwaukee

Melvyn Jay Oremland,
Pace University

Antonio Pagnamenta,
University of Illinois at Chicago

John Parsons,
Columbia University

Dinko Pocanic,
University of Virginia

Bernard G. Pope,
Michigan State University

Yong-Zhong Qian,
University of Minnesota

Ajit S. Rupaal,
Western Washington University

Todd G. Ruskell,
Colorado School of Mines

Mesgun Sebhatu,
Winthrop University

Marllin L. Simon,
Auburn University

Zbigniew M. Stadnik,
University of Ottawa

G. R. Stewart,
University of Florida

Michael G. Strauss,
University of Oklahoma

Chin-Che Tin,
Auburn University

Stephen Weppner,
Eckerd College

Suzanne E. Willis,
Northern Illinois University

Ron Zammit,
California Polytechnic State University

Problems/solutions reviewers

Lay Nam Chang,
Virginia Polytechnic Institute

Mark W. Coffey,
Colorado School of Mines

Brent A. Corbin,
UCLA

Alan Cresswell,
Shippensburg University

Ricardo S. Decca,
Indiana University-Purdue University

Michael Dubson,
University of Colorado at Boulder

David Faust,
Mount Hood Community College

Philip Fraundorf,
University of Missouri, Saint Louis

Clint Harper,
Moorpark College

Kristi R.G. Hendrickson,
University of Puget Sound

Michael Hildreth,
University of Notre Dame

David Ingram,
Ohio University

James J. Kolata,
University of Notre Dame

Eric Lane,
University of Tennessee, Chattanooga

Jerome Licini,
Lehigh University

Daniel Marlow,
Princeton University

Laura McCullough,
University of Wisconsin at Stout

Carl Mungan,
United States Naval Academy

Jeffry S. Olafsen,
University of Kansas

Robert Pompi,
The State University of New York
at Binghamton

R. J. Rollefson,
Wesleyan University

Andrew Scherbakov,
Georgia Institute of Technology

Bruce A. Schumm,
University of California, Santa Cruz

Dan Styer,
Oberlin College

Jeffrey Sundquist,
Palm Beach Community College - South

Cyrus Taylor,
Case Western Reserve University

Fulin Zuo,
University of Miami

Study Guide & Test Bank reviewers

Anthony J. Buffa,
California Polytechnic State University

Mirela S. Fetea,
University of Richmond

James Garner,
University of North Florida

Tina Harriott,
Mount Saint Vincent, Canada

Roger King,
City College of San Francisco

John A. McClelland,
University of Richmond

Chun Fu Su,
Mississippi State University

John A. Underwood,
Austin Community College

Media reviewers

Mick Arnett,
Kirkwood Community College

Colonel Rolf Enger,
U.S. Air Force Academy

John W. Farley,
The University of Nevada at Las Vegas

David Ingram,
The Ohio State University

Shawn Jackson,
The University of Tulsa

Dan MacIsaac,
Northern Arizona University

Peter E.C. Markowitz,
Florida International University

Dean Zollman,
Kansas State University

Media focus group participants

Edwin R. Jones,
University of South Carolina

William C. Kerr,
Wake Forest University

Taha Mzoughi,
Mississippi State University

Charles Niederriter,
Gustavus Adolphus College

Cindy Schwarz,
Vassar College

Dave Smith,
University of the Virgin Islands

D.J. Wagner,
Grove City College

George Watson,
University of Delaware

Frank Wolfs,
University of Rochester

We also remain indebted to the reviewers of past editions. We would therefore like to thank the following reviewers, who provided immeasurable support as we developed the fourth edition:

Michael Arnett,
Iowa State University

William Bassichis,
Texas A&M

Joel C. Berlinghieri,
The Citadel

Frank Blatt,
Michigan State University

John E. Byrne,
Gonzaga University

Wayne Carr,
Stevens Institute of Technology

George Cassidy,
University of Utah

I.V. Chivets,
Trinity College, University of Dublin

Harry T. Chu,
University of Akron

Jeff Culbert,
London, Ontario

Paul Debevec,
University of Illinois

Robert W. Detenbeck,
University of Vermont

Bruce Doak,
Arizona State University

John Elliott,
University of Manchester, England

James Garland,
Retired

Ian Gatland,
Georgia Institute of Technology

Ron Gautreau,
New Jersey Institute of Technology

David Gavenda,
University of Texas at Austin

Newton Greenberg,
SUNY Binghamton

Huidong Guo,
Columbia University

Richard Haracz,
Drexel University

Michael Harris,
University of Washington

Randy Harris,
University of California at Davis

Dieter Hartmann,
Clemson University

Robert Hollebeek,
University of Pennsylvania

Madya Jalil,
University of Malaya

Monwhea Jeng,
University of California – Santa Barbara

Ilon Joseph,
Columbia University

David Kaplan,
University of California – Santa Barbara

John Kidder,
Dartmouth College

Boris Korsunsky,
Northfield Mt. Hermon School

Andrew Lang (graduate student),
University of Missouri

David Lange,
University of California – Santa Barbara

Isaac Leichter,
Jerusalem College of Technology

William Lichten,
Yale University

Robert Lieberman,
Cornell University

Fred Lipschultz,
University of Connecticut

Graeme Luke,
Columbia University

Howard McAllister,
University of Hawaii

M. Howard Miles,
Washington State University

Matthew Moelter,
University of Puget Sound

Eugene Mosca,
United States Naval Academy

Aileen O'Donughue,
St. Lawrence University

Jack Ord,
University of Waterloo

Richard Packard,
University of California

George W. Parker,
North Carolina State University

Edward Pollack,
University of Connecticut

John M. Pratte,
Clayton College and State University

Brooke Pridmore,
Clayton State College

David Roberts,
Brandeis University

Lyle D. Roelofs,
Haverford College

Larry Rowan,
University of North Carolina
at Chapel Hill

Lewis H. Ryder,
University of Kent, Canterbury

Bernd Schuttler,
University of Georgia

Cindy Schwarz,
Vassar College

Murray Scureman,
Amdahl Corporation

Scott Sinawi,
Columbia University

Wesley H. Smith,
University of Wisconsin

Kevork Spartalian,
University of Vermont

Kaare Stegavik,
University of Trondheim, Norway

Jay D. Strieb,
Villanova University

Martin Tiersten,
City College of New York

Oscar Vilches,
University of Washington

Fred Watts,
College of Charleston

John Weinstein,
University of Mississippi

David Gordon,
Wilson, MIT

David Winter,
Columbia University

Frank L.H. Wolfe,
University of Rochester

Roy C. Wood,
New Mexico State University

Yuriy Zhestkov,
Columbia University

Of course, our work is never done. We hope to continue to receive comments and suggestions from our readers so that we can improve the text and correct any errors. If you believe you have found an error, or have any other comments, suggestions, or questions, send us a note at asktipler@whfreeman.com. We will incorporate corrections into the text during subsequent reprinting.

Finally, we would like to thank our friends at W. H. Freeman and Company for their help and encouragement. Susan Brennan, Kathleen Civetta, Georgia Lee Hadler, Julia DeRosa, Margaret Comaskey, Dena Betz, Rebecca Pearce, Brian Donnellan, Jennifer Van Hove, Patricia Marx, and Mark Santee were extremely generous with their creativity and hard work at every stage of the process. We are also grateful for the contributions of Cathy Townsend and Denise Kadlubowski at PreMediaONE and the help of our colleagues Larry Tankersley, John Ertel, Steve Montgomery, and Don Treacy.

Paul Tipler
Alameda, California

Gene Mosca
Annapolis, Maryland

ABOUT THE AUTHORS

PAUL A TIPLER

Paul Tipler was born in the small farming town of Antigo, Wisconsin, in 1933. He graduated from high school in Oshkosh, Wisconsin, where his father was superintendent of the Public Schools. He received his B.S. from Purdue University in 1955 and his Ph.D. at the University of Illinois in 1962, where he studied the structure of nuclei. He taught for one year at Wesleyan University in Connecticut while writing his thesis, then moved to Oakland University in Michigan, where he was one of the original members of the Physics department, playing a major role in developing the physics curriculum. During the next 20 years, he taught nearly all the physics courses and wrote the first and second editions of his widely used textbooks *Modern Physics* (1969, 1978) and *Physics* (1976, 1982). In 1982, he moved to Berkeley, California, where he now resides, and where he wrote *College Physics* (1987) and the third edition of *Physics* (1991). In addition to physics, his interests include music, hiking, and camping, and he is an accomplished jazz pianist and poker player.

GENE MOSCA

Gene Mosca was born in New York City and grew up on Shelter Island, New York. His undergraduate studies were at Villanova University and his graduate studies were at the University of Michigan and the University of Vermont, where he received his Ph.D. in 1974. He taught at Southampton High School, the University of South Dakota, and Emporia State University. Since 1986 Gene has been teaching at the U.S. Naval Academy. There he coordinated the core physics course for 16 semesters, and instituted numerous enhancements to both the laboratory and classroom. Proclaimed by Paul Tipler as, "the best reviewer I ever had," Mosca authored the popular Study Guide for the third and fourth editions of the text.

Systems of Measurement

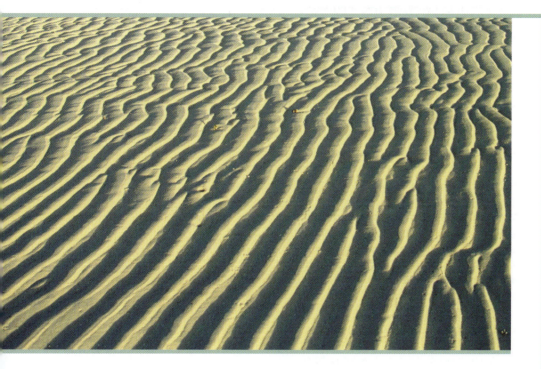

THE NUMBER OF GRAINS OF SAND ON A BEACH MAY BE TOO GREAT TO COUNT, BUT WE CAN ESTIMATE THE NUMBER BY USING REASONABLE ASSUMPTIONS AND SIMPLE CALCULATIONS.

? **How many grains of sand are on your favorite beach? (See Example 1-6.)**

We have always been curious about the world around us. Since the beginnings of recorded thought, we have sought ways to impose order on the bewildering diversity of events that we observe. Science is a process of searching for fundamental and universal principles that govern causes and effects in the universe. The process of science is the building, testing, and connecting of falsifiable models in an effort to describe, explain, and predict reality. The process involves hypotheses, repeatable experiments and observations, and new hypotheses. The prime criteria for determining the value of a scientific model is its simplicity and its usefulness in correctly making predictions or explaining observations concerning a broad spectrum of phenomena.

Today we think of science as divided into separate fields, although this division occurred only in the last century or so. The separation of complex systems into smaller categories that can be more easily studied is one of the great successes of science in general. Biology, for example, is the study of living organisms. Chemistry deals with the interaction of elements and compounds. Geology is the study of the earth. Astronomy is the study of the solar system, the stars and galaxies, and the universe as a whole. Physics is the science of matter and energy, space and time. It includes the principles that govern the motion of particles and waves, the interactions of particles, and the properties of molecules, atoms, and atomic nuclei, as well as larger-scale systems such as gases, liquids, and solids.

Some consider physics the most fundamental science because its principles supply the foundation of the other scientific fields.

Physics is the science of the exotic and the science of everyday life. At the exotic extreme, black holes boggle the imagination. In everyday life, engineers, musicians, architects, chemists, biologists, doctors, and many others routinely command such subjects as heat transfer, fluid flow, sound waves, radioactivity, and stresses in buildings or bones to perform their daily work. Countless questions about our world can be answered with a basic knowledge of physics. Why must a helicopter have two rotors? Why do astronauts float in space? Why do moving clocks run slow? Why does sound travel around corners while light appears to travel in straight lines? Why does an oboe sound different from a flute? How do CD players work? Why is there no hydrogen in the atmosphere? Why do metal objects feel colder than wood objects at the same temperature? Why is copper an electrical conductor while wood is an insulator? Why is lithium, with its three electrons, extremely reactive, whereas helium, with two electrons, is chemically inert?

➤ **In this chapter, we will begin to prepare ourselves to answer some of these questions by examining the units and dimensions used. Any time a measurement is made, the accuracy of the measurement should be stated. If a fuel-gauge reading is 100 gallons, it does not mean that there are exactly 100 gallons of fuel left. So, what exactly does it mean, and how do I express it?**

Classical and Modern Physics

The earliest recorded efforts to systematically assemble knowledge concerning motion came from ancient Greece. In Aristotle's (384–322 B.C.) system of natural philosophy, explanations of physical phenomena were deduced from assumptions about the world, rather than derived from experimentation. For example, it was a fundamental assumption that every substance had a "natural place" in the universe. Motion was thought to be the result of a substance trying to reach its natural place. Because of the agreement between the deductions of Aristotelian physics and the motions observed throughout the physical universe, and the lack of experimentation that could overturn the ancient physical ideas, the Greek view was accepted for nearly two thousand years. It was the Italian scientist Galileo Galilei (1564–1642), whose brilliant experiments on motion established once and for all the absolute necessity of experimentation in physics, who initiated the disintegration of Aristotelian physics. Within a hundred years, Isaac Newton had generalized the results of Galileo's experiments into his three spectacularly successful laws of motion, and the reign of the natural philosophy of Aristotle was over.

Experimentation during the next two hundred years brought a flood of discoveries—and raised a flood of new questions. These inspired the development of new models to explain them. By the end of the nineteenth century, Newton's laws for the motions of mechanical systems had been joined by equally impressive laws from James Maxwell, James Joule, Sadi Carnot, and others to describe electromagnetism and thermodynamics. The subjects that occupied physical scientists through the end of the nineteenth century—mechanics, light, heat, sound, electricity and magnetism—are usually referred to as *classical physics*. Because classical physics is what we need to understand the macroscopic world we live in, it dominates Parts I through V of this text.

The remarkable success of classical physics led many scientists to believe that the description of the physical universe was complete. However, the discoveries of X rays by Wilhelm Röntgen in 1895, and of radioactivity by Antoine Becquerel and Marie and Pierre Curie in the next few years seemed to be outside the framework of classical physics. The theory of special relativity proposed by Albert Einstein in 1905 contradicted the ideas of space and time of Galileo and Newton. In the same year, Einstein suggested that light energy is quantized; that is, that light comes in discrete packets rather than being wavelike and continuous as was thought in classical physics. The generalization of this insight to the quantization of all types of energy is a central idea of quantum mechanics, one that has many amazing and important consequences. The application of special relativity, and particularly quantum theory, to such microscopic systems as atoms, molecules, and nuclei, which has led to a detailed understanding of solids, liquids, and gases, is often referred to as *modern physics*. Modern physics is the subject of Part VI of this text.

While classical physics is the main subject of this book, from time to time in the earlier parts of the text we will note the relationship between classical and modern physics. For example, when we discuss velocity in Chapter 2, we will take a moment to consider velocities near the speed of light and briefly cross over to the relativistic universe first imagined by Einstein. After discussing the conservation of energy in Chapter 7, we will discuss the quantization of energy and Einstein's famous relation between mass and energy, $E = mc^2$. Just a few chapters later, in Chapter R, we will study the nature of space and time as revealed by Einstein in 1903.

1-1 Units

We all know of things that cannot be measured—the beauty of a flower, or of a Bach fugue. As certain as our knowledge of these things may be, we readily admit that this knowledge is not science. The ability not only to define but also to measure is a requisite of science, and in physics, more than in any other field of knowledge, the precise definition of terms and the accurate measurement of quantities have led to great discoveries. We begin with some preliminary chores, such as establishing some basic definitions and learning about units and how to deal with them in equations. The fun comes later.

Measurement of any quantity involves comparison with some precisely defined unit value of the quantity. For example, to measure the distance between two points, we need a standard unit, such as a meter. The statement that a certain distance is 25 meters means that it is 25 times the length of the unit meter. That is, a standard meterstick fits into that distance 25 times. It is important to include the unit, in this case meters, along with the number 25 in expressing this distance, because there are other units of distance such as kilometers or miles that are in common use. To say that a distance is 25 is meaningless. The magnitude of any physical quantity must include both a number and a unit.

The International System of Units

A small number of fundamental units are sufficient to express all physical quantities. Many of the quantities that we shall be studying, such as velocity, force, momentum, work, energy, and power can be expressed in terms of three fundamental measures: length, time, and mass. The choice of standard units for these fundamental quantities determines a system of units. The system used universally in the scientific community is called SI (for *Système International*). The standard SI unit for length is the meter, the standard unit of time is the second, and the standard unit of mass is the kilogram. Complete definitions of the SI units are given in Appendix B.

Length The standard unit of length, the **meter** (abbreviated m), was originally defined by two scratches on a bar made of a platinum-iridium alloy kept at the International Bureau of Weights and Measures in Sèvres, France. This length was chosen so that the distance between the equator and the North Pole along the meridian through Paris would be 10 million meters (Figure 1-1). The meter is

FIGURE 1-1 The meter was originally chosen so that the distance from the equator to the North Pole along the meridian through Paris would be 10^7 m.

now defined in terms of the speed of light—the meter is the distance light travels through empty space in 1/299,729,458 second. (This makes the speed of light exactly 299,792,458 m/s.)

EXERCISE What is the circumference of the earth in meters? (*Answer* About 4×10^7 m)

Time The unit of time, the **second** (s), was originally defined in terms of the rotation of the earth and was equal to (1/60)(1/60)(1/24) of the mean solar day. The second is now defined in terms of a characteristic frequency associated with the cesium atom. All atoms, after absorbing energy, emit light with wavelengths and frequencies characteristic of the particular element. There is a set of wavelengths and frequencies for each element, with a particular frequency and wavelength associated with each energy transition within the atom. As far as we know, these frequencies remain constant. The second is defined so that the frequency of the light from a certain transition in cesium is exactly 9,192,631,770 cycles per second. With these definitions, the fundamental units of length and time are accessible to laboratories throughout the world.

(a)

Mass The unit of mass, the **kilogram** (kg), which equals 1000 grams (g), is defined to be the mass of a standard body, also kept at Sèvres. A duplicate of the standard 1-kg body is kept at the National Institute of Standards and Technology (NIST) in Gaithersburg, Maryland. We shall discuss the concept of mass in detail in Chapter 4, where we will see that the weight of an object at a given point on earth is proportional to its mass. Thus the masses of objects of ordinary size can be compared by weighing them.

In our study of thermodynamics and electricity, we shall need three more fundamental physical units: one for temperature, the kelvin (K) (formerly the degree Kelvin); one for the amount of a substance, the mole (mol); and one for electrical current, the ampere (A). There is another fundamental unit, the candela (cd) for luminous intensity, which we shall have no occasion to use in this book. These seven fundamental units, the meter (m), second (s), kilogram (kg), kelvin (K), ampere (A), mole (mol), and candela (cd), constitute the international system of units or SI units.

The unit of every physical quantity can be expressed in terms of the fundamental SI units. Some frequently used combinations are given special names. For example, the SI unit of force, $kg \cdot m/s^2$ is called a newton (N). Similarly, the SI unit of power, $1 \; kg \cdot m^2/s^3 = N \cdot m/s$, is called a watt (W). When a unit like the newton or the watt is someone's name, it is written starting with a lowercase letter. Abbreviations for such units start with uppercase letters.

Prefixes for common multiples and submultiples of SI units are listed in Table 1-1. These multiples are all powers of 10. Such a system is called a decimal system. The decimal system based on the meter is called the metric system. The prefixes can be applied to any SI unit; for example, 0.001 second is 1 millisecond (ms); 1,000,000 watts is 1 megawatt (MW).

(b)

(a) Water clock used to measure time intervals in the thirteenth century. (b) Cesium fountain clock with developers Jefferts & Meekhof.

TABLE 1-1

Prefixes for Powers of 10†

Multiple	Prefix	Abbreviation
10^{18}	exa	E
10^{15}	peta	P
10^{12}	tera	T
10^{9}	giga	G
10^{6}	mega	M
10^{3}	kilo	k
10^{2}	hecto	h
10^{1}	deka	da
10^{-1}	deci	d
10^{-2}	centi	c
10^{-3}	milli	m
10^{-6}	micro	μ
10^{-9}	nano	n
10^{-12}	pico	p
10^{-15}	femto	f
10^{-18}	atto	a

$\dagger$ The prefixes hecto (h), deka (da), and deci (d) are not multiples of 10^3 or 10^{-3} and are rarely used. The other prefix that is not a multiple of 10^3 or 10^{-3} is centi (c). The prefixes frequently used in this book are printed in red. Note that all prefix abbreviations for multiples 10^6 and higher are uppercase letters; all others are lowercase letters.

Other Systems of Units

Another decimal system still in use but gradually being replaced by SI units is the cgs system, based on the *centimeter*, *gram*, and *second*. The centimeter is defined as 0.01 m. The gram is now defined as 0.001 kg. Originally the gram was defined as the mass of one cubic centimeter (cm^3) of water at 4°C. The kilogram is then the mass of 1 liter (1000 cm^3) of water.

In another system of units, the U.S. customary system, a unit of force, the pound, is chosen to be a fundamental unit. In this system, the unit of mass is then defined in terms of the fundamental unit of force. The pound is defined in terms of the gravitational attraction of the earth at a particular place for a standard body. The fundamental unit of length in this system is the foot and the unit of time is the second, which has the same definition as the SI unit. The foot is defined as exactly one-third of a yard, which is now defined in terms of the meter:

$$1 \text{ yd} = 0.9144 \text{ m} \qquad\qquad 1\text{-}1$$

$$1 \text{ ft} = \tfrac{1}{3}\text{yd} = 0.3048 \text{ m} \qquad\qquad 1\text{-}2$$

making the inch exactly 2.54 cm. This scheme is not a decimal system. It is less convenient than the SI or other decimal systems because common multiples of the unit are not powers of 10. For example, 1 yd = 3 ft and 1 ft = 12 in. We will see in Chapter 4 that mass is a better choice for a fundamental unit than force because mass is an intrinsic property of an object, independent of its location. Relations between the U.S. customary system and SI units are given in Appendix A.

1-2 Conversion of Units

All physical quantities contain both a number and a unit. When such quantities are added, subtracted, multiplied, or divided in an algebraic equation, the unit can be treated like any other algebraic quantity. For example, suppose you want to find the distance traveled in 3 hours (h) by a car moving at a constant rate of 80 kilometers per hour (km/h). The distance is the product of the speed v and the time t:

$$x = vt = \frac{80 \text{ km}}{\cancel{h}} \times 3 \, \cancel{h} = 240 \text{ km}$$

We cancel the unit of time, the hours, just as we would any algebraic quantity to obtain the distance in the proper unit of length, the kilometer. This method of treating units makes it easy to convert from one unit of distance to another. Suppose we want to convert our answer of 240 km to miles (mi). Using the fact that 1 mi = 1.61 km, we divide each side of this equality by 1.61 km to obtain

$$\frac{1 \text{ mi}}{1.61 \text{ km}} = 1$$

Because any quantity can be multiplied by 1 without changing its value, we can now change 240 km to miles by multiplying by the factor (1 mi)/(1.61 km):

$$240 \text{ km} = 240 \, \cancel{\text{km}} \times \frac{1 \text{ mi}}{1.61 \, \cancel{\text{km}}} = 149 \text{ mi}$$

The factor (1 mi)/(1.61 km) is called a **conversion factor.** All conversion factors have a value of 1 and are used to convert a quantity expressed in one unit of measure into its equivalent in another unit of measure. By writing out the units explicitly and canceling them, we do not need to think about whether we multiply by 1.61 or divide by 1.61 to change kilometers to miles, because the units tell us whether we have chosen the correct or incorrect factor.

(*a*) Laser beam from the Macdonald Observatory used to measure the distance to the moon. The distance can be measured within a few centimeters by measuring the time required for the beam to go to the moon and back after reflecting off a mirror (*b*) placed on the moon by the Apollo 14 astronauts.

(a)

(b)

USING CONVERSION FACTORS   **EXAMPLE 1-1**

Your employer sends you on a trip to a foreign country where the road signs give distances in kilometers and the automobile speedometers are calibrated in kilometers per hour. If you drive 90 km/h, how fast are you going in meters per second and in miles per hour?

PICTURE THE PROBLEM We use the facts that 1000 m = 1 km, 60 s = 1 min, and 60 min = 1 h to convert to meters per second. The quantity 90 km/h is multiplied by a set of conversion factors each having the value 1, so the value of the speed is not changed. To convert to miles per hour, we use the conversion factor (1 mi)/(1.61 km) = 1.

1. Multiply 90 km/h by a set of conversion factors that convert km to m and h to s:

$$\frac{90\ \cancel{km}}{\cancel{h}} \times \frac{1000\ m}{1\ \cancel{km}} \times \frac{1\ \cancel{h}}{60\ \cancel{min}} \times \frac{1\ \cancel{min}}{60\ s} = \boxed{25\ m/s}$$

2. Multiply 90 km/h by 1 mi/1.61 km:

$$\frac{90\ \cancel{km}}{h} \times \frac{1\ mi}{1.61\ \cancel{km}} = \boxed{55.9\ mi/h}$$

EXERCISE What is the equivalent of 65 mi/h in m/s? (*Answer* 29.1 m/s)

1-3 Dimensions of Physical Quantities

The area of a surface is found by multiplying one length by another. For example, the area of a rectangle of sides 2 m and 3 m is $A = (2\ m)(3\ m) = 6\ m^2$. The units of this area are square meters. Because area is the product of two lengths, it is said to have the dimensions of length times length, or length squared, often written L^2. The idea of dimensions is easily extended to other nongeometric quantities. For example, speed is said to have the dimensions of length divided by time, or L/T. The dimensions of other quantities such as force or energy are written in terms of the fundamental quantities of length, time, and mass. Adding two physical quantities makes sense only if the quantities have the same dimensions. For example, we cannot add an area to a speed to obtain a meaningful sum. If we have an equation like

$$A = B + C$$

the quantities A, B, and C must all have the same dimensions. Addition of B and C also requires that these quantities be in the same units. For example, if B is an area of 500 in.^{2} and C is 4 ft^2, we must either convert B into square feet or C into square inches in order to find the sum of the two areas.

We can often find mistakes in a calculation by checking the dimensions or units of the quantities in our result. Suppose, for example, that we mistakenly use the formula $A = 2\pi r$ for the area of a circle. We can see immediately that this cannot be correct because $2\pi r$ has dimensions of length whereas area must have dimensions of length squared. Dimensional consistency is a necessary but not a sufficient condition for an equation to be correct. An equation can have the correct dimensions in each term without describing any physical situation. Table 1-2 gives the dimensions of some quantities we will encounter in physics.

TABLE 1-2

Dimensions of Physical Quantities

Quantity	Symbol	Dimension
Area	A	L^2
Volume	V	L^3
Speed	v	L/T
Acceleration	a	L/T^2
Force	F	ML/T^2
Pressure (F/A)	p	M/LT^2
Density (M/V)	ρ	M/L^3
Energy	E	ML^2/T^2
Power (E/T)	P	ML^2/T^3

EXAMPLE 1 - 2

The pressure in a fluid in motion depends on its density ρ and its speed v. Find a simple combination of density and speed that gives the correct dimensions of pressure.

PICTURE THE PROBLEM We note from Table 1-2 that both pressure and density have units of mass in the numerator, whereas speed does not contain M. We therefore divide the units of pressure by those of density and inspect the result.

1. Divide the units of pressure by those of density:

$$\frac{[p]}{[\rho]} = \frac{M(L/T^2)}{M/L^3} = \frac{L^2}{T^2}$$

2. By inspection, we note that the result has dimensions of v^2. The dimensions of pressure are thus the same as the dimensions of density times speed squared:

$$[p] = [\rho][v^2] = \frac{M}{L^3}\left(\frac{L}{T}\right)^2 = \boxed{\frac{M}{LT^2}}$$

REMARKS When we study fluids in motion in Chapter 13, we will see from Bernoulli's law that for a fluid moving at a constant height, $p + \frac{1}{2}\rho v^2$ is constant where p is the pressure in the fluid.

1-4 Scientific Notation

Handling very large or very small numbers is simplified by using scientific notation. In this notation, the number is written as a product of a number between 1 and 10 and a power of 10, such as 10^2 ($= 100$) or 10^3 ($= 1000$). For example, the number 12,000,000 is written 1.2×10^7; the distance from the earth to the sun, about 150,000,000,000 m, is written 1.5×10^{11} m. The number 11 in 10^{11} is called the **exponent**. For numbers smaller than 1, the exponent is negative. For example, $0.1 = 10^{-1}$, and $0.000\,1 = 10^{-4}$. The diameter of a virus, which is about 0.00000001 m, is written 1×10^{-8} m.

When numbers in scientific notation are multiplied, the exponents are added; when divided, the exponents are subtracted. These rules can be seen from some simple examples:

$$10^2 \times 10^3 = 100 \times 1{,}000 = 100{,}000 = 10^5$$

Similarly,

$$\frac{10^2}{10^3} = \frac{100}{1000} = \frac{1}{10} = 10^{2-3} = 10^{-1}$$

In this notation, 10^0 is defined to be 1. To see why, suppose we divide 1000 by itself. We have

$$\frac{1000}{1000} = \frac{10^3}{10^3} = 10^{3-3} = 10^0 = 1$$

EXAMPLE 1 - 3

In 12 g of carbon, there are $N_A = 6.02 \times 10^{23}$ carbon atoms (Avogadro's number). If you could count 1 atom per second, how long would it take to count the atoms in 1 g of carbon? Express your answer in years.

PICTURE THE PROBLEM We need to find the total number of atoms to be counted, N, and then use the fact that the number counted equals the counting rate R multiplied by the time t.

1. The time is related to the total number of atoms N, and the rate of counting $R = 1$ atom/s:

$$t = \frac{N}{R}$$

2. Find the number of carbon atoms in 1 gram:

$$N = \frac{6.02 \times 10^{23} \text{ atoms}}{12 \text{ g}} \times 1 \text{ g} = 5.02 \times 10^{22} \text{ atoms}$$

3. Calculate the number of seconds it takes to count these at 1 per second:

$$t = \frac{N}{R} = \frac{5.02 \times 10^{22} \text{ atoms}}{1 \text{ atom/s}} = 5.02 \times 10^{22} \text{ s}$$

4. Calculate the number of seconds in a year:

$$n = \frac{365 \text{ d}}{1 \text{ y}} \times \frac{24 \text{ h}}{1 \text{ d}} \times \frac{3600 \text{ s}}{1 \text{ h}} = 3.15 \times 10^7 \text{ s/y}$$

5. Use the conversion factor 3.15×10^7 s/y (a handy quantity to remember) to convert the answer in step 3 to years:

$$t = 5.02 \times 10^{22} \text{ s} \times \frac{1 \text{ y}}{3.15 \times 10^7 \text{ s}}$$

$$= \frac{5.02}{3.15} \times 10^{22-7} \text{ y} = \boxed{1.59 \times 10^{15} \text{ y}}$$

REMARKS The time required is about 100,000 times the age of the universe.

EXERCISE If you divided the task so that each person counted different atoms, how long would it take for 5 billion (5×10^9) people to count the atoms in 1 g of carbon? (*Answer* 3.19×10^5 y)

How Much Water?   **EXAMPLE   1 - 4**

A liter (L) is the volume of a cube that is 10 cm by 10 cm by 10 cm. If you drink 1 liter of water, how much volume in cubic centimeters and in cubic meters would it occupy in your stomach?

PICTURE THE PROBLEM The volume of a cube of side ℓ is $V = \ell^3$. The volume in cubic centimeters is found directly from $\ell = 10$ cm. To find the volume in cubic meters, convert cm^3 to m^3 using the conversion factor $1 \text{ cm} = 10^{-2}$ m.

1. Calculate the volume in cm^3:

$$V = \ell^3 = (10 \text{ cm})^3 = 10^3 \text{ cm}^3$$

2. Convert to m^3:

$$10^3 \text{ cm}^3 = 10^3 \text{ cm}^3 \times \left(\frac{10^{-2} \text{ m}}{1 \text{ cm}}\right)^3 = 10^3 \text{ cm}^3 \times \left(\frac{10^{-6} \text{ m}^3}{1 \text{ cm}^3}\right) = \boxed{10^{-3} \text{ m}^3}$$

REMARKS Note that the conversion factor (which equals 1) can be raised to the third power without changing its value, enabling us to cancel units.

Care is required when adding or subtracting numbers written in scientific notation when their exponents don't match. Consider, for example,

$$(1.200 \times 10^2) + (8 \times 10^{-1}) = 120.0 + 0.8 = 120.8$$

To find the sum without converting both numbers into ordinary decimal form, it is sufficient to rewrite either of the numbers so that its power of 10 is the same as that of the other. For example, we can find the sum by writing $1.200 \times 10^2 = 1200 \times 10^{-1}$ and then adding:

$$(1200 \times 10^{-1}) + (8 \times 10^{-1}) = 1208 \times 10^{-1} = 120.8$$

When the exponents are very different, one of the numbers is much smaller than the other. The smaller number can often be neglected in addition or subtraction.

For example,

$$(2 \times 10^6) + (9 \times 10^{-3}) = 2{,}000{,}000 + 0.009$$
$$= 2{,}000{,}000.009 \approx 2 \times 10^6$$

where the symbol $\approx$ means "is approximately equal to."

When raising a power to another power, the exponents are multiplied. For example,

$$(10^2)^4 = 10^2 \times 10^2 \times 10^2 \times 10^2 = 10^8$$

1-5 Significant Figures and Order of Magnitude

Many of the numbers in science are the result of measurement and are therefore known only to within some degree of experimental uncertainty. The magnitude of the uncertainty depends on the skill of the experimenter and the apparatus used, and often can only be estimated. A rough indication of the uncertainty in a measurement is inferred by the number of digits used. For example, if we say that a table is 2.50 m long, we are often saying that its length is between 2.495 m and 2.505 m. That is, we know the length to about ± 0.005 m $= \pm 0.5$ cm. If we used a meterstick with millimeter markings and measured the table length carefully, we might estimate that we could measure the length to ± 0.5 mm rather than ± 0.5 cm. We would indicate this precision when giving the length by using four digits, such as 2.503 m. A reliably known digit (other than a zero used to locate the decimal point) is called a **significant figure.** The number 2.50 has three significant figures; 2.503 m has four. The number 0.001 03 has three significant figures. (The first three zeroes are not significant figures but merely locate the decimal point.) In scientific notation, the number 0.001 03 is written 1.03×10^{-3}. A common student error is to carry more digits than the certainty of measurement warrants. Suppose, for example, that you measure the area of a circular playing field by pacing off the radius and using the formula for the area of a circle, $A = \pi r^2$. If you estimate the radius to be 8 m and use a 10-digit calculator to compute the area, you obtain $\pi (8 \text{ m})^2 = 201.0619298 \text{ m}^2$. The digits after the decimal point give a false indication of the accuracy with which you know the area. If you found the radius by pacing, you might expect that your measurement was accurate to only about 0.5 m. That is, the radius could be as great as 8.5 m or as small as 7.5 m. If the radius is 8.5 m, the area is $\pi (8.5 \text{ m})^2 = 226.9800692 \text{ m}^2$, whereas if it is 7.5 m, the area is $\pi (7.5 \text{ m})^2 = 176.714587 \text{ m}^2$. There is a general rule to guide you when combining several numbers in multiplication or division:

> The number of significant figures in the result of multiplication or division is no greater than the least number of significant figures in any of the factors.

In the previous example, the radius is known to only one significant figure, so the area is also known only to one significant figure. It should be written as $2 \times 10^2 \text{ m}^2$, which says that the area is probably between 150 m² and 250 m².

The precision of the sum or difference of two measurements is only as good as the precision of the least precise of the two measurements. A general rule is:

> The result of addition or subtraction of two numbers has no significant figures beyond the last decimal place where both of the original numbers had significant figures.

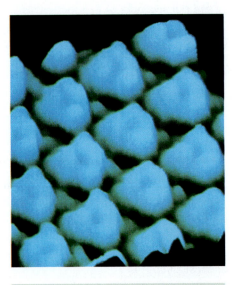

Benzene molecules of the order of 10^{-10} m in diameter as seen in a scanning electron microscope.

Chromosomes measuring on the order of 10^{-6} m across as seen in a scanning electron microscope.

SIGNIFICANT FIGURES

EXAMPLE 1-5

Find the sum of 1.040 and 0.21342.

PICTURE THE PROBLEM The first number, 1.040, has only three significant figures beyond the decimal point, whereas the second, 0.21342 has five. According to the rule stated above, the sum can have only three significant figures beyond the decimal point.

Sum the numbers, keeping only three digits beyond the decimal point:

$$1.040 + 0.21342 = \boxed{1.253}$$

EXERCISE Apply the appropriate rule for significant figures to calculate the following: (a) 1.58×0.03, (b) $1.4 + 2.53$, (c) $2.34 \times 10^2 + 4.93$ (*Answer* (a) 0.05, (b) 3.9, (c) 2.39×10^2)

Most examples and exercises in this book will be done with data to three (or sometimes four) significant figures, but occasionally we will say, for example, that a table top is 3 ft by 8 ft rather than taking the time and space to say it is 3.00 ft by 8.00 ft. Any data you see in an example or exercise can be assumed to be known to three significant figures unless otherwise indicated. The same assumption holds for the data in the end-of-chapter problems. In doing rough calculations or comparisons, we sometimes round off a number to the nearest power of 10. Such a number is called an **order of magnitude.** For example, the height of a small insect, say an ant, might be 8×10^{-4} m $\approx 10^{-3}$ m. We would say that the order of magnitude of the height of an ant is 10^{-3} m. Similarly, though the height of most people is about 2 m, we might round that off and say that the order of magnitude of the height of a person is 10^0 m. By this we do not mean to imply that a typical height is really 1 m but that it is closer to 1 m than to 10 m or to $10^{-1} = 0.1$ m. We might say that a typical human being is three orders of magnitude taller than a typical ant, meaning that the ratio of heights is about 1000 to 1. An order of magnitude does not provide any digits that are reliably known.

Distances familiar in our everyday world. The height of the woman is of the order of 10^0 m and that of the mountain is of the order of 10^4 m.

Earth, with a diameter of the order of 10^7 m, as seen from space.

The diameter of the Andromeda galaxy is of the order of 10^{21} m.

TABLE 1-3

The Universe by Orders of Magnitude

Size or Distance	(m)	Mass	(kg)	Time Interval	(s)
Proton	10^{-15}	Electron	10^{-30}	Time for light to cross nucleus	10^{-23}
Atom	10^{-10}	Proton	10^{-27}	Period of visible light radiation	10^{-15}
Virus	10^{-7}	Amino acid	10^{-25}	Period of microwaves	10^{-10}
Giant amoeba	10^{-4}	Hemoglobin	10^{-22}	Half-life of muon	10^{-6}
Walnut	10^{-2}	Flu virus	10^{-19}	Period of highest audible sound	10^{-4}
Human being	10^{0}	Giant amoeba	10^{-8}	Period of human heartbeat	10^{0}
Highest mountain	10^{4}	Raindrop	10^{-6}	Half-life of free neutron	10^{3}
Earth	10^{7}	Ant	10^{-4}	Period of earth's rotation	10^{5}
Sun	10^{9}	Human being	10^{2}	Period of earth's revolution around sun	10^{7}
Distance from earth to sun	10^{11}	Saturn V rocket	10^{6}	Lifetime of human being	10^{9}
Solar system	10^{13}	Pyramid	10^{10}	Half-life of plutonium-239	10^{12}
Distance to nearest star	10^{16}	Earth	10^{24}	Lifetime of mountain range	10^{15}
Milky Way galaxy	10^{21}	Sun	10^{30}	Age of earth	10^{17}
Visible universe	10^{26}	Milky Way galaxy	10^{41}	Age of universe	10^{18}
		Universe	10^{52}		

It may be thought of as having no significant figures. Table 1-3 gives some typical order-of-magnitude values for a variety of sizes, masses, and time intervals encountered in physics.

In many cases the order of magnitude of a quantity can be estimated using reasonable assumptions and simple calculations. The physicist Enrico Fermi was a master at using cunning order-of-magnitude estimations to generate answers for questions that seemed impossible to calculate because of lack of information. Problems like these are often called **Fermi questions.** The following is an example of a Fermi question.

EXPLORING

How many piano tuners are there in Chicago? Find out this, and more, at www.whfreeman.com/tipler5e.

BURNING RUBBER

E X A M P L E   1 - 6

What thickness of rubber tread is worn off the tire of an automobile as it travels 1 km (0.6 mi)?

PICTURE THE PROBLEM We assume the tread thickness of a new tire is 1 cm. This may be off by a factor of two or so, but 1 mm is certainly too small and 10 cm is too large. Since tires have to be replaced after about 60,000 km (about 37,000 mi), we will assume that the tread is completely worn off after 60,000 km. In other words, the rate of wear is 1 cm of tire per 60,000 km of travel.

Use 1 cm wear per 60,000 km travel to compute the thickness worn after 1 km of travel:

$$\frac{1 \text{ cm wear}}{60,000 \text{ km travel}} = \frac{1.7 \times 10^{-5} \text{ cm wear}}{1 \text{ km travel}}$$

$$\boxed{\approx 0.2 \text{ } \mu\text{m wear per km of travel}}$$

EXERCISE How many grains of sand are on a 0.50 km stretch of beach that is 100 m wide? *Hint: Assume that the sand is 3 m deep. Estimate that the diameter of one grain of sand is 1.00 mm. (Answer* $\approx 2 \times 10^{14}$)

SUMMARY

The fundamental units in the SI system are the meter (m), the second (s), the kilogram (kg), the kelvin (K), the ampere (A), the mole (mol), and the candela (cd). The unit(s) of every physical quantity can be expressed in terms of these fundamental units.

Topic	Relevant Equations and Remarks
1. Units	The magnitude of physical quantities (for example, length, time, force, and energy) are expressed as a number times a unit.
Fundamental units	The fundamental units in the SI system (short for *Système International*) are the meter (m), the second (s), the kilogram (kg), the kelvin (K), the ampere (A), the mole (mol), and the candela (cd). The unit(s) of every physical quantity can be expressed in terms of these fundamental units.
Units in equations	Units in equations are treated just like any other algebraic quantity.
Conversion	Conversion factors, which are always equal to 1, provide a convenient method for converting from one kind of unit to another.
2. Dimensions	The two sides of an equation must have the same dimensions.
3. Scientific Notation	For convenience, very small and very large numbers are generally written as a factor times a power of 10.
4. Exponents	
Multiplication	When multiplying two numbers, the exponents are added.
Division	When dividing two numbers, the exponents are subtracted.
Raising to a power	When a number containing an exponent is itself raised to a power, the exponents are multiplied.
5. Significant Figures	
Multiplication and division	The number of significant figures in the result of multiplication or division is *no greater than* the least number of significant figures in any of the numbers.
Addition and subtraction	The result of addition or subtraction of two numbers has no significant figures beyond the last decimal place where both of the original numbers had significant figures.
6. Order of Magnitude	A number rounded to the nearest power of 10 is called an order of magnitude. The order of magnitude of a quantity can often be estimated using reasonable assumptions and simple calculations.

- • Single-concept, single-step, relatively easy
- •• Intermediate-level, may require synthesis of concepts
- ••• Challenging
- SSM Solution is in the *Student Solutions Manual*
- iSOLVE Problems available on iSOLVE online homework service
- iSOLVE ✓ These "Checkpoint" online homework service problems ask students additional questions about their confidence level, and how they arrived at their answer

In a few problems, you are given more data than you actually need; in a few other problems, you are required to supply data from your general knowledge, outside sources, or informed estimates.

Conceptual Problems

1 • SSM iSOLVE Which of the following is *not* one of the fundamental physical quantities in the SI system? (*a*) Mass. (*b*) Length. (*c*) Force. (*d*) Time. (*e*) All of the above are fundamental physical quantities.

2 • iSOLVE In doing a calculation, you end up with m/s in the numerator and m/s^2 in the denominator. What are your final units? (*a*) m^2/s^3. (*b*) 1/s. (*c*) s^3/m^2. (*d*) s. (*e*) m/s.

3 • iSOLVE The prefix giga means (*a*) 10^3, (*b*) 10^6, (*c*) 10^9, (*d*) 10^{12}, (*e*) 10^{15}.

4 • iSOLVE The prefix mega means (*a*) 10^{-9}, (*b*) 10^{-6}, (*c*) 10^{-3}, (*d*) 10^6, (*e*) 10^9.

5 • SSM iSOLVE The prefix pico means (*a*) 10^{-12}, (*b*) 10^{-6}, (*c*) 10^{-3}, (*d*) 10^6, (*e*) 10^9.

6 • iSOLVE The number 0.0005130 has _____ significant figures. (*a*) one, (*b*) three, (*c*) four, (*d*) seven, (*e*) eight.

7 • iSOLVE The number 23.0040 has _____ significant figures. (*a*) two, (*b*) three, (*c*) four, (*d*) five, (*e*) six.

8 • What are the advantages and disadvantages of using the length of your arm for a standard length?

9 • True or false:

(*a*) Two quantities must have the same dimensions in order to be added.
(*b*) Two quantities must have the same dimensions in order to be multiplied.
(*c*) All conversion factors have the value 1.

Estimation and Approximation

10 •• SSM
The angle subtended by the moon's diameter at a point on the earth is about 0.524° (Figure 1-2). Use this and the fact that the moon is about 384 Mm away to find the

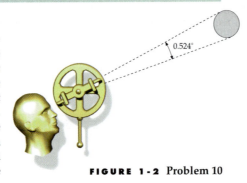

FIGURE 1-2 Problem 10

diameter of the moon. (The angle θ subtended by the moon is approximately D/r_m, where D is the diameter of the moon and r_m is the distance to the moon.)

11 •• SSM iSOLVE ✓ The sun has a mass of 1.99×10^{30} kg and is composed mostly of hydrogen, with only a small fraction being heavier elements. The hydrogen atom has a mass of 1.67×10^{-27} kg. Estimate the number of hydrogen atoms in the sun.

12 •• Most soft drinks are sold in aluminum cans. The mass of a typical can is about 0.018 kg. (*a*) Estimate the number of aluminum cans used in the United States in one year. (*b*) Estimate the total mass of aluminum in a year's consumption from these cans. (*c*) If aluminum returns $1/kg at a recycling center, how much is a year's accumulation of aluminum cans worth?

13 •• In his essay "There's plenty of room at the bottom," Richard Feynman proposed writing the entire *Encyclopaedia Brittanica* on the head of a pin. (*a*) Estimate the size of the letters needed if we assume that a pinhead is 1/16 in across (the value that Feynman used). (*b*) If the atomic spacing in a metal is about 0.5 nm (5×10^{-10} m), about how many atoms across is each letter?

14 •• SSM (*a*) Estimate the number of gallons of gasoline used per day by automobiles in the United States and the total amount of money spent on it. (*b*) If 19.4 gal of gasoline can be made from one barrel of crude oil, estimate the total number of barrels of oil imported into the United States per year to make gasoline. How many barrels per day is this?

15 •• There is an environmental debate over the use of cloth versus disposable diapers. (*a*) If we assume that between birth and 2.5 y of age, a child uses 3 diapers per day, estimate the total number of disposable diapers used in the United States per year. (*b*) Estimate the total landfill volume due to these diapers, assuming that 1000 kg of waste fills about 1 m^3 of landfill volume. (*c*) How many square miles of landfill area at an average height of 10 m is needed for the disposal of diapers each year?

16 ••• Each binary digit is termed a bit. A series of bits grouped together is called a word. An 8-bit word is called a byte. Suppose a computer hard disk has a capacity of 20 gigabytes. (*a*) How many bits can be stored on the disk? (*b*) Estimate the number of typical books that can be stored on the disk assuming each character requires one 8-bit word.

17 •• **SSM** Estimate the yearly toll revenue of the George Washington Bridge in New York. At last glance, the toll is $6 to go into New York from New Jersey; going from New York into New Jersey is free. There are a total of 14 lanes.

Units

18 • Express the following quantities using the prefixes listed in Table 1-1 and the abbreviations listed on page EP-1; for example, 10,000 meters = 10 km. (a) 1,000,000 watts, (b) 0.002 gram, (c) 3×10^{-6} meter, (d) 30,000 seconds.

19 • Write each of the following without using prefixes: (a) 40 μW, (b) 4 ns, (c) 3 MW, (d) 25 km.

20 • **SSM** Write the following (which are not SI units) without using abbreviations. For example, 10^3 meters = 1 kilometer: (a) 10^{-12} boo, (b) 10^9 low, (c) 10^{-6} phone, (d) 10^{-18} boy, (e) 10^6 phone, (f) 10^{-9} goat, (g) 10^{12} bull.

21 •• **SOLVE** In the following equations, the distance x is in meters, the time t is in seconds, and the velocity v is in meters per second. What are the SI units of the constants C_1 and C_2? (a) $x = C_1 + C_2 t$, (b) $x = \frac{1}{2} C_1 t^2$, (c) $v^2 = 2 C_1 x$, (d) $x = C_1 \cos C_2 t$, (e) $v^2 = 2 C_1 v - (C_2 x)^2$.

22 •• **SOLVE** If x is in feet, t is in seconds, and v is in feet per second, what are the units of the constants C_1 and C_2 in each part of Problem 21?

Conversion of Units

23 • From the original definition of the meter in terms of the distance from the equator to the North Pole, find in meters (a) the circumference of the earth and (b) the radius of the earth. (c) Convert your answers for (a) and (b) from meters into miles.

24 • **SOLVE** The speed of sound in air is 340 m/s. What is the speed of a supersonic plane that travels at twice the speed of sound? Give your answer in kilometers per hour and miles per hour.

25 • **SSM** **SOLVE** A basketball player is 6 ft $10\frac{1}{2}$ in. tall. What is his height in centimeters?

26 • Complete the following: (a) 100 km/h = _____ mi/h, (b) 60 cm = _____ in., (c) 100 yd = _____ m.

27 • The main span of the Golden Gate Bridge is 4200 ft. Express this distance in kilometers.

28 • **SSM** Find the conversion factor to convert from miles per hour into kilometers per hour.

29 • Complete the following: (a) 1.296×10^5 km/h² = _____ km/(h·s), (b) 1.296×10^5 km/h² = _____ m/s², (c) 60 mi/h = _____ ft/s, (d) 60 mi/h = _____ m/s.

30 • There are 1.057 quarts in a liter and 4 quarts in a gallon. (a) How many liters are there in a gallon? (b) A barrel equals 42 gallons. How many cubic meters are there in a barrel?

31 • **SOLVE** There are 640 acres in a square mile. How many square meters are there in one acre?

32 •• **SOLVE** A right circular cylinder has a diameter of 6.8 in and a height of 2 ft. What is the volume of the cylinder in (a) cubic feet, (b) cubic meters, (c) liters?

33 •• **SSM** In the following, x is in meters, t is in seconds, v is in meters per second, and the acceleration a is in meters per second squared. Find the SI units of each combination: (a) v^2/x, (b) $\sqrt{x/a}$, (c) $\frac{1}{2} a t^2$.

Dimensions of Physical Quantities

34 • What are the dimensions of the constants in each part of Problem 21?

35 •• The law of radioactive decay is $N(t) = N_0 e^{-\lambda t}$, where N_0 is the number of radioactive nuclei at $t = 0$, $N(t)$ is the number remaining at time t, and λ is a quantity known as the decay constant. What is the dimension of λ?

36 •• **SSM** The SI unit of force, the kilogram-meter per second squared (kg·m/s²) is called the newton (N). Find the dimensions and the SI units of the constant G in Newton's law of gravitation $F = G m_1 m_2 / r^2$.

37 •• An object on the end of a string moves in a circle. The force exerted by the string has units of ML/T^2 and depends on the mass of the object, its speed, and the radius of the circle. What combination of these variables gives the correct dimensions?

38 •• **SOLVE**✓ Show that the product of mass, acceleration, and speed has the dimensions of power.

39 •• The momentum of an object is the product of its velocity and mass. Show that momentum has the dimensions of force multiplied by time.

40 •• What combination of force and one other physical quantity has the dimensions of power?

41 •• **SSM** **SOLVE**✓ When an object falls through air, there is a drag force that depends on the product of the surface area of the object and the square of its velocity, that is, $F_{air} = CAv^2$, where C is a constant. Determine the dimensions of C.

42 •• Kepler's third law relates the period of a planet to its orbital radius r, the constant G in Newton's law of gravitation ($F = G m_1 m_2 / r^2$), and the mass of the sun M_s. What combination of these factors gives the correct dimensions for the period of a planet?

Scientific Notation and Significant Figures

43 • **SSM** Express as a decimal number without using powers of 10 notation: (a) 3×10^4, (b) 6.2×10^{-3}, (c) 4×10^{-6}, (d) 2.17×10^5.

44 • Write the following in scientific notation: (a) 3.1 GW = _____ W, (b) 10 pm = _____ m, (c) 2.3 fs = _____ s, (d) 4 μs = _____ s.

45 • **SOLVE** Calculate the following, round off to the correct number of significant figures, and express your result in scientific notation: (a) $(1.14)(9.99 \times 10^4)$, (b) $(2.78 \times 10^{-8}) - (5.31 \times 10^{-9})$, (c) $12\pi/(4.56 \times 10^{-3})$, (d) $27.6 + (5.99 \times 10^2)$.

46 • Calculate the following, round off to the correct number of significant figures, and express your result in scientific notation: (a) $(200.9)(569.3)$, (b) $(0.000000513)(62.3 \times 10^7)$, (c) $28{,}401 + (5.78 \times 10^4)$, (d) $63.25/(4.17 \times 10^{-3})$.

47 • **SSM** **iSOLVE** A cell membrane has a thickness of about 7 nm. How many cell membranes would it take to make a stack 1 in high?

48 • Calculate the following, round off to the correct number of significant figures, and express your result in scientific notation: (a) $(2.00 \times 10^4)(6.10 \times 10^{-2})$, (b) $(3.141592)(4.00 \times 10^5)$, (c) $(2.32 \times 10^3)/(1.16 \times 10^8)$, (d) $(5.14 \times 10^3) + (2.78 \times 10^2)$, (e) $(1.99 \times 10^2) + (9.99 \times 10^{-5})$.

49 • **SSM** Perform the following calculations and round off the answers to the correct number of significant figures: (a) $3.141592654 \times (23.2)^2$, (b) $2 \times 3.141592654 \times 0.76$, (c) $4/(3\pi) \times (1.1)^3$, (d) $(2.0)^5/3.141592654$.

General Problems

50 • On many of the roads in Canada the speed limit is 100 km/h. What is the speed limit in miles per hour?

51 • **SSM** If you could count $1 per second, how many years would it take to count 1 billion dollars (1 billion = 10^9)?

52 • Sometimes a conversion factor can be derived from the knowledge of a constant in two different systems. (a) The speed of light in vacuum is 186,000 mi/s = 3×10^8 m/s. Use this fact to find the number of kilometers in a mile. (b) The weight of 1 ft³ of water is 62.4 lb. Use this and the fact that 1 cm³ of water has a mass of 1 g to find the weight in pounds of a 1-kg mass.

53 •• **iSOLVE** The mass of one uranium atom is 4.0×10^{-26} kg. How many uranium atoms are there in 8 g of pure uranium?

54 •• **iSOLVE✓** During a thunderstorm, a total of 1.4 in. of rain falls. How much water falls on one acre of land? (1 mi² = 640 acres.)

55 •• An iron nucleus has a radius of 5.4×10^{-15} m and a mass of 9.3×10^{-26} kg. (a) What is its mass per unit volume in kg/m³? (b) If the earth had the same mass per unit volume, what would be its radius? (The mass of the earth is 5.98×10^{24} kg.)

56 •• Evaluate the following expressions: (a) $(5.6 \times 10^{-5})(0.0000075)/(2.4 \times 10^{-12})$, (b) $(14.2)(6.4 \times 10^7)(8.2 \times 10^{-9}) - 4.06$, (c) $(6.1 \times 10^{-6})^2(3.6 \times 10^4)^3/(3.6 \times 10^{-11})^{1/2}$, (d) $(0.000064)^{1/3}/[(12.8 \times 10^{-3})(490 \times 10^{-1})^{1/2}]$.

57 •• **SSM** The astronomical unit (AU) is defined in terms of the distance from the earth to the sun, namely 1.496×10^{11} m. The parsec is the radius of a circle for which a central angle of 1 s intercepts an arc of length 1 AU. The light-year is the distance that light travels in 1 y. (a) How many parsecs are there in one astronomical unit? (b) How many meters are in a parsec? (c) How many meters in a light-year? (d) How many astronomical units in a light-year? (e) How many light-years in a parsec?

58 •• If the average density of the universe is at least 6×10^{-27} kg/m³, then the universe will eventually stop expanding and begin contracting. (a) How many electrons are needed in a cubic meter to produce the critical density? (b) How many protons per cubic meter would produce the critical density? ($m_e = 9.11 \times 10^{-31}$ kg; $m_p = 1.67 \times 10^{-27}$ kg.)

59 •• **SSM** The Super-Kamiokande neutrino detector in Japan is a large transparent cylinder filled with ultrapure water. The height of the cylinder is 41.4 m and the diameter is 39.3 m. Calculate the mass of the water in the cylinder. Does this match the claim posted on the official Super-K Web site that the detector uses 50,000 tons of water? The density of water is 1,000 kg/m³.

60 ••• The table below gives experimental results for a measurement of the period of motion T of an object of mass m suspended on a spring versus the mass of the object. These data are consistent with a simple equation expressing T as a function of m of the form $T = Cm^n$, where C and n are constants and n is not necessarily an integer. (a) Find n and C. (There are several ways to do this. One is to guess the value of n and check by plotting T versus m^n on graph paper. If your guess is right, the plot will be a straight line. Another is to plot log T versus log m. The slope of the straight line on this plot is n.) (b) Which data points deviate the most from a straight-line plot of T versus m^n?

Mass m, kg	0.10	0.20	0.40	0.50	0.75	1.00	1.50
Period T, s	0.56	0.83	1.05	1.28	1.55	1.75	2.22

61 ••• The table below gives the period T and orbit radius r for the motions of four satellites orbiting a dense, heavy asteroid. (a) These data can be fitted by the formula $T = Cr^n$. Find C and n. (b) A fifth satellite is discovered to have a period of 6.20 y. Find the radius for the orbit of this satellite, which fits the same formula.

Period T, y	0.44	1.61	3.88	7.89
Radius r, Gm	0.088	0.208	0.374	0.600

62 ••• **SSM** The period T of a simple pendulum depends on the length L of the pendulum and the acceleration of gravity g (dimensions L/T^2). (a) Find a simple combination of L and g that has the dimensions of time. (b) Check the dependence of the period T on the length L by measuring the period (time for a complete swing back and forth) of a pendulum for two different values of L. (c) The correct formula relating T to L and g involves a constant that is a multiple of π, and cannot be obtained by the dimensional analysis of Part (a). It can be found by experiment as in Part (b) if g is known. Using the value $g = 9.81$ m/s² and your experimental results from Part (b), find the formula relating T to L and g.

63 ••• **iSOLVE✓** The weight of the earth's atmosphere pushes down on the surface of the earth with a force of 14.7 lb for each square inch of the earth's surface. What is the weight in pounds of the earth's atmosphere? (The radius of the earth is about 6370 km.)

Motion in One Dimension

MOTION IN ONE DIMENSION IS MOTION ALONG A STRAIGHT LINE, LIKE THAT OF A CAR ON A STRAIGHT ROAD. THIS DRIVER ENCOUNTERS STOPLIGHTS AND DIFFERENT SPEED LIMITS ON HER COMMUTE ALONG A STRAIGHT HIGHWAY TO SCHOOL.

? **How can she estimate her arrival time? (See Example 2-2.)**

We begin our study of the physical universe by examining objects in motion. The study of motion, whose measurement, more than 400 years ago gave birth to physics, is called **kinematics.**
➤ In this chapter, we start with the simplest case of kinematics, the motion of a particle along a straight line, like the motion of a car moving along a flat, straight, narrow road. A particle is an object whose position can be described by a single point. Anything can be considered to be a particle—a molecule, a person, or a galaxy—as long as we can reasonably ignore its internal structure.

2-1 Displacement, Velocity, and Speed

Figure 2-1 shows a car at position x_i at time t_i and at position x_f at a later time t_f. The change in the car's position, called the **displacement,** is given by $x_f - x_i$. We use the Greek letter Δ (uppercase delta) to indicate the change in a quantity; thus, the change in x is written Δx:

$$\Delta x = x_f - x_i \qquad\qquad 2\text{-}1$$

DEFINITION—DISPLACEMENT

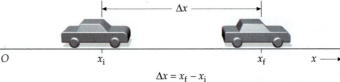

$$\Delta x = x_f - x_i$$

The notation Δx (read "delta x") stands for a single quantity, the change in x. It is not a product of Δ and x any more than $\cos \theta$ is a product of $\cos$ and θ. By convention, the change in a quantity is always its final value minus its initial value.

Velocity is the rate at which the position changes. The **average velocity** of the particle is defined as the ratio of the displacement Δx to the time interval $\Delta t = t_f - t_i$:

$$v_{av} = \frac{\Delta x}{\Delta t} = \frac{x_f - x_i}{t_f - t_i} \qquad\qquad 2\text{-}2$$

DEFINITION—AVERAGE VELOCITY

FIGURE 2-1 A car moving in a straight line. A coordinate axis consists of a line along the path of the car. A point on this line is chosen to be the origin O. Other points on it are assigned a number x, the value of x being proportional to its distance from O. Points to the right of O are positive as shown, and points to the left are negative. When the car travels from point x_i to point x_f, its displacement is $\Delta x = x_f - x_i$.

Displacement and average velocity may be positive or negative. A positive value indicates motion in the positive x direction. The SI unit of velocity is meters per second (m/s), and the U.S. customary unit is feet per second (ft/s).

DISPLACEMENT AND VELOCITY OF A COMET   **EXAMPLE 2-1**

A comet moving toward the sun is first seen at $x_i = 3.0 \times 10^{12}$ m relative to the sun (see Figure 2-2). Exactly one year later, it is seen at $x_f = 2.1 \times 10^{12}$ m. Find its displacement and average velocity.

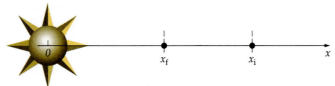

FIGURE 2-2

PICTURE THE PROBLEM Comets move in elliptical orbits around the sun. Here we consider just the distance from the sun as if the comet moved in one dimension. We are given x_i and x_f. If we choose $t_i = 0$, then $t_f = 1$ y $= 3.16 \times 10^7$ s. The average velocity is $\Delta x / \Delta t$.

1. The displacement is found from its definition:
$$\Delta x = x_f - x_i = (2.1 \times 10^{12} \text{ m}) - (3.0 \times 10^{12} \text{ m})$$
$$= -9 \times 10^{11} \text{ m}$$

2. The average velocity is the displacement divided by the time interval:
$$v_{av} = \frac{\Delta x}{\Delta t} = \frac{-9 \times 10^{11} \text{ m}}{3.16 \times 10^7 \text{ s}}$$
$$= -2.85 \times 10^4 \text{ m/s} = \boxed{-28.5 \text{ km/s}}$$

REMARKS Both displacement and average velocity are negative, because the comet moved toward smaller values of x. Note that the units, m for Δx, and m/s or km/s for v_{av}, are essential parts of the answers. It is meaningless to say "the displacement is -9×10^{11}" or "the average velocity of a particle is -28.5."

EXERCISE A jet plane leaves the gate in Detroit at 2:15 P.M. Its average velocity is 500 km/h for the trip to Chicago, which is 483 km away. When does it arrive at the gate in Chicago? (*Answer*   3:13 P.M. Detroit time, which is actually 2:13 P.M. Chicago time)

DRIVING TO SCHOOL   **EXAMPLE 2 - 2** **Put It in Context**

It normally takes you 10 min to travel 5 mi to school along a straight road. You leave home 15 min before class begins. Delays caused by a broken traffic light slow down traffic to 20 mph for the first 2 mi of the trip. Will you be late for class?

PICTURE THE PROBLEM You need to find the total time that it will take you to travel to class. To do so, you must find the time $\Delta t_{2\,mi}$ that you will be driving at 20 mph, and the time $\Delta t_{3\,mi}$ for the remainder of the trip, during which you are driving at your usual speed.

1. The total time equals the time to travel the first 2 mi plus the time to travel the remaining 3 mi:

$$\Delta t_{tot} = \Delta t_{2\,mi} + \Delta t_{3\,mi}$$

2. Using $\Delta x = v_{av}\Delta t$, solve for the time taken to travel 2 mi at 20 mi/h:

$$\Delta t_{2\,mi} = \frac{\Delta x}{v_{av}} = \frac{2\;mi}{20\;mi/h} = 0.1\;h = 6\;min$$

3. Using $\Delta x = v_{av}\Delta t$, relate the time to travel the last 3 mi at the usual speed:

$$\Delta t_{3\,mi} = \frac{\Delta x}{v_{av}} = \frac{3\;mi}{v_{usual}}$$

4. Using $\Delta x = v_{av}\Delta t$, solve for the v_{usual}, the speed needed for you to travel the 5 mi in 10 min:

$$v_{usual} = \frac{\Delta x_{tot}}{\Delta t_{usual}} = \frac{5\;mi}{10\;min} = 0.5\;mi/min$$

5. Solve for $t_{3\,mi}$:

$$\Delta t_{3\,mi} = \frac{3\;mi}{0.5\;mi/min} = 6\;min$$

6. Solve for the total time:

$$\Delta t_{tot} = \Delta t_{2\,mi} + \Delta t_{3\,mi} = 12\;min$$

7. The trip takes 12 min with the delay, compared to the usual 10 minutes. Because you wisely allowed yourself 15 min for the trip, *you will not be late for class.*

The **average speed** of a particle is the ratio of the total distance traveled to the total time from start to finish:

$$\text{Average speed} = \frac{\text{total distance}}{\text{total time}} = \frac{s}{t} \qquad\qquad 2\text{-}3$$

Since the total distance and total time are both always positive, the average speed is always positive.

AVERAGE SPEED IN A FOOT RACE   **EXAMPLE 2 - 3**

You run 100 m in 12 s then turn around and jog 50 m back toward the starting point in 30 s (see Figure 2-3). Calculate (*a*) your average speed, and (*b*) your average velocity for the total trip.

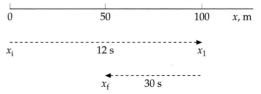

FIGURE 2-3

PICTURE THE PROBLEM We use the definitions of average speed and average velocity, noting that average *speed* is the total *distance* divided by Δt, whereas the average *velocity* is the *net displacement* divided by Δt:

(*a*) 1. Your average speed equals the total distance divided by the total time:

$$\text{Average speed} = \frac{s}{t}$$

2. Calculate the total distance traveled and the total time:

$$s = s_1 + s_2 = 100\;m + 50\;m = 150\;m$$

$$t = 12\;s + 30\;s = 42\;s$$

3. Use s and t to find your average speed:

$$\text{Average speed} = \frac{150 \text{ m}}{42 \text{ s}} = \boxed{3.57 \text{ m/s}}$$

(b) 1. Your average velocity is the ratio of the net displacement Δx to the time interval Δt:

$$v_{av} = \frac{\Delta x}{\Delta t}$$

2. Your net displacement is $x_f - x_i$, where $x_i = 0$ is your initial position and $x_f = 50$ m is your final position:

$$\Delta x = x_f - x_i = 50 \text{ m} - 0 = 50 \text{ m}$$

3. Use Δx and Δt to find your average velocity:

$$v_{av} = \frac{\Delta x}{\Delta t} = \frac{50 \text{ m}}{42 \text{ s}} = \boxed{1.19 \text{ m/s}}$$

PLAUSIBILITY CHECK The world record for a 100-m race is just under 10 s, so 10 m/s is about the maximum possible speed obtainable. The result of 3.57 m/s for the average speed in (*a*) is reasonable, given that you merely jogged for one-third of the distance. If you had obtained 35.7 m/s for the average speed, that would have been a clue that something was wrong with the calculation.

REMARKS Note that your average speed is greater than your average velocity because the total distance traveled is greater than the total displacement. Also, note that the net displacement is the sum of the individual displacements. That is, $\Delta x = \Delta x_1 + \Delta x_2 = (100 \text{ m}) + (-50 \text{ m}) = 50 \text{ m}$, which is the Part (*b*), step 2 result.

A TRAIN-HOPPING BIRD   **EXAMPLE 2-4**

Two trains 75 km apart approach each other on parallel tracks, each moving at 15 km/h. A bird flies back and forth between the trains at 20 km/h until the trains pass each other. How far does the bird fly?

PICTURE THE PROBLEM This problem seems difficult at first, but viewed in the right way it is actually quite simple. We approach it by first writing an equation for the quantity to be found, the total distance s flown by the bird.

1. The total distance traveled by the bird equals its speed times the time:

$$s = (\text{average speed})_{bird} \times t$$
$$= (\text{speed})_{av \ bird} \times t$$

2. The time that the bird is in the air is the time taken for the trains to meet. The sum of the distances traveled by the two trains is $D = 75$ km. Find the time it will take the two trains to travel a total distance D:

$$s_1 + s_2 = (\text{speed})_{av \ 1} \times t + (\text{speed})_{av \ 2} \times t = D$$

so

$$t = \frac{D}{(\text{speed})_{av \ 1} + (\text{speed})_{av \ 2}}$$

3. The total distance traveled by the bird is therefore:

$$s = (\text{speed})_{av \ bird} \times t$$
$$= (\text{speed})_{av \ bird} \frac{D}{(\text{speed})_{av \ 1} + (\text{speed})_{av \ 2}}$$
$$= 20 \text{ km/h} \frac{75 \text{ km}}{15 \text{ km/h} + 15 \text{ km/h}}$$
$$= \boxed{50 \text{ km}}$$

REMARKS Some try to solve this problem by finding and summing the distances flown by the bird each time it moves from one train to the other. This makes a relatively easy problem quite difficult. It is important to develop a thoughtful, systematic approach to solving problems. Begin by writing an equation for the unknown quantity in terms of other quantities. Then proceed by determining the values for each of the other quantities in the equation.

Figure 2-4 depicts average velocity graphically. A straight line connects points P_1 and P_2 and forms the hypotenuse of the triangle having sides Δx and Δt. The ratio $\Delta x / \Delta t$ is the line's **slope,** which gives us a geometric interpretation of average velocity:

> The average velocity is the slope of the straight line connecting the points (t_1, x_1) and (t_2, x_2).

GEOMETRIC INTERPRETATION OF AVERAGE VELOCITY

Generally, average velocity depends on the time interval on which it is based. In Figure 2-4, for example, the smaller time interval indicated by t_2' and P_2' gives a larger average velocity, as shown by the greater steepness of the line connecting points P_1 and P_2'.

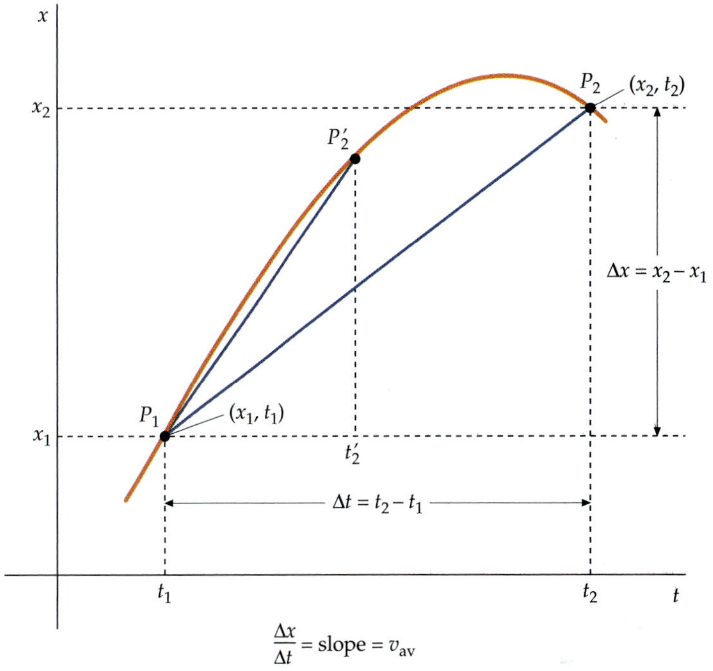

Instantaneous Velocity

On first consideration, defining the velocity of a particle at a single instant seems impossible. At a given instant, a particle is at a single point. If it is at a single point, how can it be moving? If it is not moving, how can it have a velocity? This age-old paradox is resolved when we realize that observing and defining motion requires that we look at the position of the object at more than one time. Consider Figure 2-5. As we consider successively shorter time intervals beginning at t_1, the average velocity for the interval approaches the slope of the tangent at t_1. We define the slope of this tangent as the **instantaneous velocity** at t_1. This tangent is the limit of the ratio $\Delta x / \Delta t$ as Δt, and therefore Δx, approaches zero. So we can say:

> The instantaneous velocity is the limit of the ratio $\Delta x / \Delta t$ as Δt approaches zero.
>
> $$v(t) = \lim_{\Delta t \to 0} \frac{\Delta x}{\Delta t}$$
>
> = slope of the line tangent to the
> x-versus-t curve[†]   2-4

DEFINITION—INSTANTANEOUS VELOCITY

This limit is called the **derivative** of x with respect to t. In the usual calculus notation, the derivative is written dx/dt

$$v(t) = \lim_{\Delta t \to 0} \frac{\Delta x}{\Delta t} = \frac{dx}{dt} \qquad 2\text{-}5$$

A line's slope may be positive, negative, or zero; consequently, instantaneous velocity (in one-dimensional motion) may be positive (x increasing), negative (x decreasing), or zero (no motion). The magnitude of the instantaneous velocity is the **instantaneous speed.**

$$\frac{\Delta x}{\Delta t} = \text{slope} = v_{av}$$

FIGURE 2-4 Graph of x versus t for a particle moving in one dimension. Each point on the curve represents the location x at a particular time t. We have drawn a straight line between positions P_1 and P_2. The displacement $\Delta x = x_2 - x_1$ and the time interval $\Delta t = t_2 - t_1$ between these points are indicated. The straight line between P_1 and P_2 is the hypotenuse of the triangle having sides Δx and Δt, and the ratio $\Delta x / \Delta t$ is its slope. In geometric terms, the slope is a measure of the line's steepness.

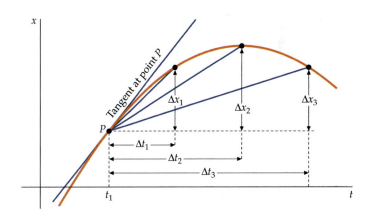

FIGURE 2-5 Graph of x versus t. Note the sequence of successively smaller time intervals, $\Delta t_1, \Delta t_2, \Delta t_3, \ldots$. The average velocity of each interval is the slope of the straight line for that interval. As the time intervals become smaller, these slopes approach the slope of the tangent to the curve at point t_1. The slope of this line is defined as the instantaneous velocity at time t_1.

[†] The slope of the line tangent to a curve is often referred to more simply as the "slope of the curve."

E X A M P L E   2 - 5   Try It Yourself

The position of a particle as a function of time is given by the curve shown in Figure 2-6. Find the instantaneous velocity at time $t = 2$ s. When is the velocity greatest? When is it zero? Is it ever negative?

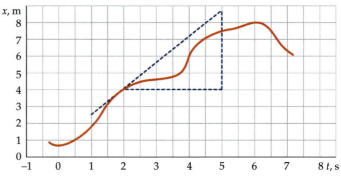

PICTURE THE PROBLEM In Figure 2-6, we have sketched the line tangent to the curve at $t = 2$ s. The tangent line's slope is the instantaneous velocity of the particle at the given time. You can use this figure to measure the slope of the tangent line.

FIGURE 2-6

Cover the column to the right and try these on your own before looking at the answers.

Steps

1. Find the values x_1 and x_2 on the tangent line at times $t_1 = 2$ s and $t_2 = 5$ s.

2. Compute the slope of the tangent line from these values. This slope equals the instantaneous velocity at $t = 2$ s.

3. From the figure, the slope (and therefore velocity) is greatest at about $t = 4$ s. The slope and velocity are zero at $t = 0$ and $t = 6$ s and are negative before 0 and after 6 s.

Answers

$x_1 \approx 4$ m, $x_2 \approx 8.5$ m

$v = \text{slope} \approx \dfrac{8.5 \text{ m} - 4 \text{ m}}{5 \text{ s} - 2 \text{ s}} = \boxed{1.5 \text{ m/s}}$

EXERCISE What is the average velocity of this particle between $t = 2$ s and $t = 5$ s? (*Answer* 1.17 m/s)

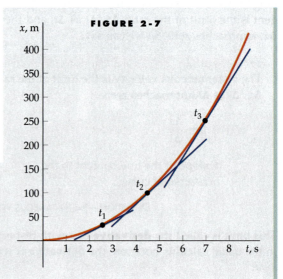

FIGURE 2-7

A STONE DROPPED FROM A CLIFF   **E X A M P L E   2 - 6**

The position of a stone dropped from a cliff is described approximately by $x = 5t^2$, where x is in meters measured downward from the original position at $t = 0$, and t is in seconds. Find the velocity at any time t. (We omit explicit indication of units to simplify the notation.)

PICTURE THE PROBLEM We can compute the velocity at some time t by computing the derivative dx/dt directly from the definition in Equation 2-4. The corresponding curve giving x versus t is shown in Figure 2-7. Tangent lines are drawn at times t_1, t_2, and t_3. The slopes of these tangent lines increase steadily, indicating that the instantaneous velocity increases steadily with time.

1. By definition the instantaneous velocity is:

$$v(t) = \lim_{\Delta t \to 0} \frac{\Delta x}{\Delta t} = \lim_{\Delta t \to 0} \frac{x(t + \Delta t) - x(t)}{\Delta t}$$

2. We compute the displacement Δx from the position function $x(t)$:

$$x(t) = 5t^2$$

3. At a later time $t + \Delta t$, the position is $x(t + \Delta t)$, given by:

$$x(t + \Delta t) = 5(t + \Delta t)^2 = 5\left[t^2 + 2t\Delta t + (\Delta t)^2\right]$$
$$= 5t^2 + 10t\Delta t + 5(\Delta t)^2$$

4. The displacement for this time interval is thus:

$$\Delta x = x(t + \Delta t) - x(t)$$
$$= \left[5t^2 + 10t\Delta t + 5(\Delta t)^2\right] - 5t^2$$
$$= 10t\Delta t + 5(\Delta t)^2$$

5. Divide Δx by Δt to find the average velocity for this time interval:

$$v_{av} = \frac{\Delta x}{\Delta t} = \frac{10t\Delta t + 5(\Delta t)^2}{\Delta t} = 10t + 5\Delta t$$

6. As we consider shorter and shorter time intervals, Δt approaches zero and the second term $5\Delta t$ approaches zero, though the first term, $10t$, remains unchanged:

$$v(t) = \lim_{\Delta t \to 0} \frac{\Delta x}{\Delta t} = \lim_{\Delta t \to 0} (10t + 5\Delta t) = \boxed{10t}$$

REMARKS If we had set $\Delta t = 0$ in steps 4 and 5, the displacement would be $\Delta x = 0$, in which case the ratio $\Delta x / \Delta t$ would be undefined. Instead, we leave Δt as a variable until the final step, when the limit $\Delta t \to 0$ is well defined.

To find derivatives quickly, we use rules based on the limiting process above (see Appendix Table A-4). A particularly useful rule is

$$\text{If } x = Ct^n, \quad \text{then} \quad \frac{dx}{dt} = Cnt^{n-1} \qquad\qquad 2\text{-}6$$

where C and n are any constants. Using this rule in Example 2-6, we have $x = 5t^2$, and $v = dx/dt = 10t$, in agreement with our previous results.

Relative Velocity

If you are sitting in an airplane moving with a velocity of 500 mi/h toward the east, your velocity is also 500 mi/h toward the east. However, 500 mi/h toward the east might be your velocity relative to the surface of the earth, or it might be your velocity relative to the air outside the airplane. (If the plane is flying in a jet stream, these two velocities would be very different.) Furthermore, your velocity is zero relative to the airplane itself. To specify the velocity of a particle you must also specify the **frame of reference.** In this discussion three different frames of reference are specified, the surface of the earth, the air outside the airplane, and the airplane itself.

Midair refueling. Each plane is nearly at rest relative to the other, though both are moving with very large velocities relative to the earth.

A frame of reference is an extended object whose parts are at rest relative to each other.

DEFINITION—FRAME OF REFERENCE

To make position measurements we use coordinate axes that are attached to reference frames. For a horizontal coordinate axis attached to the plane your position remains constant, at least it does if you remain in your seat. However, for a horizontal coordinate axis attached to the surface of the earth, and for a horizontal coordinate axis attached to a balloon floating in the air outside the plane, your position keeps changing. If you have trouble imagining a coordinate axis attached to the air outside the plane, instead imagine a coordinate axis attached to a balloon that is suspended in the air. The air and the balloon are at rest relative to each other, so together they form a single reference frame.

If a particle moves with velocity v_{pA} relative to reference frame A, which is in turn moving with velocity v_{AB} relative to reference frame B, the velocity of the particle relative to B is

$$v_{pB} = v_{pA} + v_{AB} \qquad\qquad 2\text{-}7a$$

For example, if you swim in a river parallel to the direction of the flow, your velocity relative to the shore v_{ys} equals your velocity relative to the water v_{yw} plus the velocity of the water relative to the shore v_{ws}:

$$v_{ys} = v_{yw} + v_{ws}$$

If you are swimming upstream at 2 m/s relative to the water, and the water is moving at 1.2 m/s relative to the shore, then your velocity relative to the shore is $v_{ys} = -2$ m/s $+ 1.2$ m/s $= -0.8$ m/s, where we have chosen downstream to be the positive direction.

A great surprise of twentieth-century physics was the discovery that Equation 2-7a is only an approximation. When we study the theory of relativity in Chapter 39, we will see that the exact expression for relative velocities is

$$v_{pB} = \frac{v_{pA} + v_{AB}}{1 + v_{pA}v_{AB}/c^2} \qquad\qquad 2\text{-}7b$$

where $c = 3 \times 10^8$ m/s is the speed of light in a vacuum. In all everyday cases with macroscopic objects, v_{pA} and v_{AB} are both so much smaller than c that Equations 2-7a and b give essentially the same result, but for high-speed objects, such as electrons or distant galaxies, the difference between these two equations becomes significant. Equation 2-7b has the interesting property that if $v_{pA} = c$, then v_{pB} also equals c, which is a core tenet of relativity, namely that the speed of light is the same in all reference frames.

EXERCISE Using Equation 2-7b, substitute c for v_{pA} and solve for v_{pB}, verifying that Equation 2-7b is in agreement with the result "the speed of light is the same in all reference frames."

2-2 Acceleration

Acceleration is the rate of change of the instantaneous velocity. When you step on your car's accelerator, for example, you expect to change your velocity. The **average acceleration** for a particular time interval $\Delta t = t_2 - t_1$ is defined as the ratio $\Delta v/\Delta t$, where $\Delta v = v_2 - v_1$:

$$a_{av} = \frac{\Delta v}{\Delta t} \qquad\qquad 2\text{-}8$$

DEFINITION—AVERAGE ACCELERATION

Acceleration has dimensions of length divided by time squared. The SI unit is meters per second squared, m/s². (In Equation 2-8, if the numerator is in m/s and the denominator is in s, then $\Delta v/\Delta t$ is in units of (m/s)/s. Multiplying the

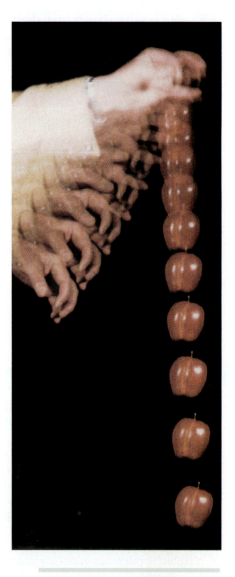

A falling apple captured by strobe photography at 60 flashes per second. The acceleration of the apple is indicated by the widening spaces between the images.

numerator and the denominator by 1 s, we find the units of $\Delta v/\Delta t$ to equal m/s^2.) We can write Equation 2-8 as $\Delta v = a_{av}\Delta t$. Thus, if a particle at rest accelerates at 5.1 m/s^2, its velocity after 1 s is 5.1 m/s, its velocity after 2 s is 10.2 m/s, and so on. **Instantaneous acceleration** is the limit of the ratio $\Delta v/\Delta t$ as Δt approaches zero. On a plot of velocity versus time, the instantaneous acceleration at time t is the slope of the line tangent to the curve at that time:

$$a = \lim_{\Delta t \to 0} \frac{\Delta v}{\Delta t}$$

$$= \text{slope of the line tangent to the } v\text{-versus-}t \text{ curve} \qquad 2\text{-}9$$

DEFINITION—INSTANTANEOUS ACCELERATION

Thus, acceleration is the derivative of velocity with respect to time, dv/dt. Since velocity is the derivative of the position x with respect to t, acceleration is the second derivative of x with respect to t, d^2x/dt^2. We can see the reason for this notation when we write the acceleration as dv/dt and replace v with dx/dt:

$$a = \frac{dv}{dt} = \frac{d(dx/dt)}{dt} = \frac{d^2x}{dt^2} \qquad 2\text{-}10$$

If acceleration remains zero, there is no change in velocity over time—velocity is constant. In this case, the curve of x versus t is a straight line. If acceleration is nonzero and constant, as in Example 2-9, then velocity varies linearly with time and x varies quadratically with time.

A Fast Cat

EXAMPLE 2-7

A cheetah can accelerate from 0 to 96 km/h (60 mi/h) in 2 s, whereas a Corvette requires 4.5 s. Compute the average accelerations for the cheetah and Corvette and compare them with the free-fall acceleration due to gravity, $g = 9.81$ m/s^2.

1. Find the average acceleration from the information given: cat $a_{av} = \dfrac{\Delta v}{\Delta t} = \dfrac{96 \text{ km/h} - 0}{2 \text{ s}}$

$$= 48 \text{ km/(h·s)}$$

car $a_{av} = \dfrac{\Delta v}{\Delta t} = \dfrac{96 \text{ km/h} - 0}{4.5 \text{ s}}$

$$= 21.3 \text{ km/(h·s)}$$

2. Convert to m/s² using 1 h = 3600 s = 3.6 ks:

cat $\dfrac{48 \text{ km}}{\text{h·s}} \times \dfrac{1 \text{ h}}{3.6 \text{ ks}} = 13.3 \text{ m/s}^2$

car $\dfrac{21.3 \text{ km}}{\text{h·s}} \times \dfrac{1 \text{ h}}{3.6 \text{ ks}} = 5.92 \text{ m/s}^2$

3. To compare the result with the acceleration due to gravity, multiply each by the conversion factor $1g/9.81 \text{ m/s}^2$:

cat $13.3 \text{ m/s}^2 \times \dfrac{1g}{9.81 \text{ m/s}^2} = \boxed{1.36g}$

car $5.92 \text{ m/s}^2 \times \dfrac{1g}{9.81 \text{ m/s}^2} = \boxed{0.60g}$

REMARKS Note that by expressing the time in kiloseconds in step 2, the kilo prefixes in km and ks cancel.

EXERCISE A car is traveling at 45 km/h at time $t = 0$. It accelerates at a constant rate of 10 km/(h·s). (*a*) How fast is it traveling at $t = 2$ s? (*b*) At what time is the car traveling at 70 km/h?

(*Answer* (*a*) 65 km/h (*b*) 2.5 s)

EXERCISE IN DIMENSIONAL ANALYSIS If a car starts from rest at $x = 0$ with constant acceleration a, its velocity v depends on a and the distance traveled x. Which of the following equations has the correct dimensions and therefore could be a possible equation relating x, a, and v?

(*a*) $v = 2ax$ (*c*) $v = 2ax^2$
(*b*) $v^2 = 2a/x$ (*d*) $v^2 = 2ax$

(*Answer* Only (*d*) has the same dimensions on both sides of the equation. Although we cannot obtain the exact equation from dimensional analysis, we can often obtain the functional dependence.)

VELOCITY AND ACCELERATION AS FUNCTIONS OF TIME   **EXAMPLE 2-8**

The position of a particle is given by $x = Ct^3$, where C is a constant having units of m/s³. Find the velocity and acceleration as functions of time.

1. We find the velocity by applying $dx/dt = Cnt^{n-1}$ (Equation 2-6):

$x = Ct^3$

$v = \dfrac{dx}{dt} = \boxed{3Ct^2}$

2. The time derivative of velocity gives the acceleration:

$a = \dfrac{dv}{dt} = \boxed{6Ct}$

PLAUSIBILITY CHECK We can check the units of our answers. For velocity, $[v] = [C][t^2] = (\text{m/s}^3)(\text{s}^2) = \text{m/s}$. For acceleration, $[a] = [C][t] = (\text{m/s}^3)(\text{s}) = \text{m/s}^2$.

2-3 Motion With Constant Acceleration

The motion of a particle that has constant acceleration is common in nature. For example, near the earth's surface all unsupported objects fall vertically with constant acceleration (provided air resistance is negligible). If a particle has a constant acceleration a, it follows that the average acceleration for any time interval is also a. Thus,

$$a_{av} = \frac{\Delta v}{\Delta t} = a \qquad\qquad 2\text{-}11$$

If the velocity is v_0 at time $t = 0$, and v at some later time t, the corresponding acceleration is

$$a = \frac{\Delta v}{\Delta t} = \frac{v - v_0}{t - 0} = \frac{v - v_0}{t}$$

Rearranging yields v as a function of time:

$$v = v_0 + at \qquad\qquad 2\text{-}12$$

CONSTANT ACCELERATION, v VERSUS t

This is the equation for a straight line in a v-versus-t plot (Figure 2-8). The line's slope is the acceleration a and its v intercept is the initial velocity v_0.
 The displacement $\Delta x = x - x_0$ in the time interval $\Delta t = t - 0$ is

$$\Delta x = v_{av}\Delta t = v_{av}t \qquad\qquad 2\text{-}13$$

For constant acceleration, the velocity varies linearly with time, and the average velocity is the mean value of the initial and final velocities. (This relation holds only if the acceleration is constant.) If v_0 is the initial velocity and v is the final velocity, the average velocity is

$$v_{av} = \tfrac{1}{2}(v_0 + v) \qquad\qquad 2\text{-}14$$

CONSTANT ACCELERATION, v_{av}

The displacement is then

$$\Delta x = x - x_0 = v_{av}t = \tfrac{1}{2}(v_0 + v)t \qquad\qquad 2\text{-}15$$

We can eliminate v by substituting $v = v_0 + at$ from Equation 2-12:

$$\Delta x = \tfrac{1}{2}(v_0 + v)t = \tfrac{1}{2}(v_0 + v_0 + at)t = v_0 t + \tfrac{1}{2}at^2$$

The displacement is thus

$$\Delta x = x - x_0 = v_0 t + \tfrac{1}{2}at^2 \qquad\qquad 2\text{-}16$$

CONSTANT ACCELERATION, $x(t)$

The first term on the right, $v_0 t$, is the displacement that would occur if a were zero, and the second term, $\tfrac{1}{2}at^2$, is the additional displacement due to the constant acceleration.

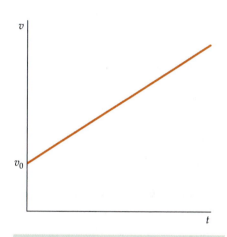

FIGURE 2-8 Graph of velocity versus time for constant acceleration.

"It goes from zero to 60 in about 3 seconds."
© Sydney Harris

Let's eliminate t from Equations 2-12 and 2-14 and find a relation between Δx, a, v, and v_0. From Equation 2-12, $t = (v - v_0)/a$. Substituting this into Equation 2-14,

$$\Delta x = v_{av}t = \frac{1}{2}(v_0 + v)t = \frac{1}{2}(v_0 + v)\left(\frac{v - v_0}{a}\right) = \frac{v^2 - v_0^2}{2a}$$

or

$$v^2 = v_0^2 + 2a\Delta x \qquad\qquad 2\text{-}17$$

CONSTANT ACCELERATION

Equation 2-17 is useful, for example, if we want to find the final velocity of a ball dropped from rest at some height x and we are not interested in the time the fall takes.

Problems with One Object

Many practical problems deal with objects in free-fall, that is, falling freely under the influence of gravity only. All objects in free-fall with the same initial velocity move identically. As shown in Figure 2-9, a feather and an apple, simultaneously released from rest in a large vacuum chamber, fall with identical motions. They have the same acceleration. The magnitude of this acceleration, designated by g, has the approximate value

$$g \approx 9.81 \text{ m/s}^2 = 32.2 \text{ ft/s}^2$$

Because g is the *magnitude* of the acceleration, it is always positive. If downward is the positive direction, then the free-fall acceleration is $a = g$; if upward is positive, then $a = -g$.

FIGURE 2-9

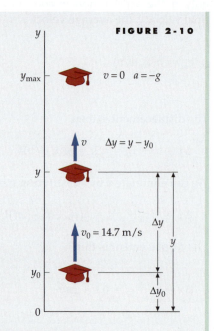

FIGURE 2-10

THE FLYING CAP

EXAMPLE 2-9

Upon graduation, a joyful physics student throws her cap straight upward with an initial speed of 14.7 m/s. Given that its acceleration is 9.81 m/s² downward (we neglect air resistance), (*a*) how long does it take to reach its highest point? (*b*) What is the distance to the highest point? (*c*) Assuming the cap is caught at the same height from which it was released, what is the total time the cap is in flight?

PICTURE THE PROBLEM When the cap is at its highest point, its instantaneous velocity is zero. **Thus, we translate the phrase "at its highest point" into the mathematical condition $v = 0$.**

(*a*) 1. Make a sketch of the cap in its initial position and again at its highest point. Include a coordinate axis and label the origin and the two positions of the cap.

2. The time is related to the velocity and acceleration: $v = v_0 + at$

3. To find the time at which the cap reaches its greatest height, set $v = 0$ and solve for t: $t = \dfrac{0 - v_0}{a} = \dfrac{-14.7 \text{ m/s}}{-9.81 \text{ m/s}^2} = \boxed{1.50 \text{ s}}$

(b) We can find the displacement from the time t and the average velocity:

$$\Delta y = v_{av}t = \tfrac{1}{2}(v_0 + v)t$$

$$= \tfrac{1}{2}(14.7 \text{ m/s} + 0)(1.50 \text{ s}) = \boxed{11.0 \text{ m}}$$

(c) 1. Set $\Delta y = 0$ in Equation 2-16 and solve for t:

$$\Delta y = v_0 t + \tfrac{1}{2}at^2$$

$$0 = (v_0 + \tfrac{1}{2}at)t$$

2. There are two solutions for t if $\Delta y = 0$. The first corresponds to the time at which the cap is released, the second to the time at which the cap is caught:

$$t = 0 \qquad \text{(first solution)}$$

$$t = -\frac{2v_0}{a} = -\frac{2(14.7 \text{ m/s})}{-9.81 \text{ m/s}^2} = \boxed{3 \text{ s}} \qquad \text{(second solution)}$$

REMARKS The $t = 3$ s solution also follows from a symmetry in the system—it takes the same time for the cap to fall from its greatest height as to rise to that height (see Figure 2-11). In reality the cap will not have a constant acceleration because air resistance has a significant effect on a light object like a cap. If air resistance is not negligible, the fall time will exceed the rise time.

EXERCISE Find $y_{max} - y_0$ using (a) Equation 2-15 and (b) Equation 2-16. (c) Find the velocity of the cap when it returns to its starting point. (*Answer* (a) and (b) $y_{max} - y_0 = 11.0$ m (c) -14.7 m/s; notice that the final speed is the same as the initial speed)

EXERCISE What is the velocity of the cap at the following points in time? (a) 0.1 s before it reaches its highest point; (b) 0.1 s after it reaches its highest point. (c) Compute $\Delta v/\Delta t$ for this 0.2-s-long time interval. (*Answer* (a) $+0.981$ m/s (b) -0.981 m/s (c) $[(-0.981 \text{ m/s}) - (+0.981 \text{ m/s})]/(0.2 \text{ s}) = -9.81$ m/s^2)

EXERCISE A car starting from rest gains speed at a constant rate of 8 m/s^2. (a) How fast is it going after 10 s? (b) How far has it gone after 10 s? (c) What is its average velocity for the interval $t = 0$ to $t = 10$ s? (*Answer* (a) 80 m/s (b) 400 m (c) 40 m/s)

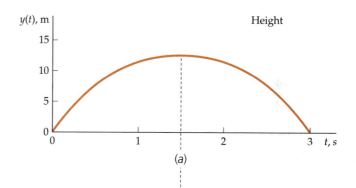

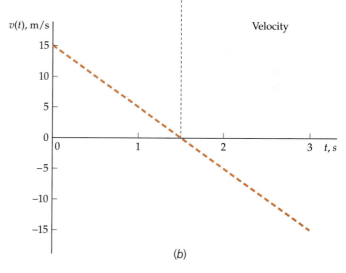

FIGURE 2-11

The next example concerns a car's **stopping distance**—how far it travels while coming to a halt.

A Car's Stopping Distance

EXAMPLE 2-10

On a highway at night you see a stalled vehicle and brake your car to a stop with an acceleration of magnitude 5 m/s^2. (An acceleration that reduces the speed is often called a *deceleration*.) What is the car's stopping distance if its initial speed is (a) 15 m/s (about 34 mi/h) or (b) 30 m/s?

PICTURE THE PROBLEM If we choose the direction of motion to be positive, the stopping distance and the initial velocity are positive, but the acceleration is negative. Thus, the initial velocity is $v_0 = 15$ m/s, the final velocity is $v = 0$, and the acceleration is $a = -5$ m/s². We seek the distance traveled, Δx. We do not need to know the time it takes for the car to stop, so Equation 2-17 is the most convenient formula to use.

(a) Using Equation 2-17 with $v = 0$, calculate the displacement Δx:

$$v^2 = v_0^2 + 2a\Delta x$$

so

$$\Delta x = \frac{v^2 - v_0^2}{2a} = \frac{0^2 - v_0^2}{2a} = -\frac{v_0^2}{2a}$$

$$= -\frac{(15 \text{ m/s})^2}{2(-5 \text{ m/s}^2)} = \boxed{22.5 \text{ m}}$$

(b) From the previous step we see that if $v = 0$, then $\Delta x = -v_0^2/(2a)$. Thus, Δx is proportional to the square of the initial speed. Using this observation and the Part (a) result, find the stopping distance for an initial speed equal to twice that in Part (a):

$$\Delta x = 2^2(22.5 \text{ m}) = \boxed{90 \text{ m}}$$

REMARKS The answer to (b) can also be gotten by directly substituting the initial speed of 30 m/s into the expression for Δx obtained in Part (a). Ninety meters is a considerable distance, roughly the length of a football field. Changing v_0 by a factor of 2 changes the stopping distance by a factor $2^2 = 4$ (see Figure 2-12). The practical implication of this squared dependence is that even modest increases in speed cause significant increases in stopping distance.

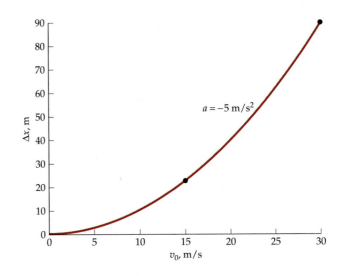

FIGURE 2-12 Stopping distance as a function of the initial velocity. The curve shows the case for Example 2-10, where the acceleration is $a = -5.0$ m/s²; the points shown on the red curve are the solutions to parts (a) and (b).

STOPPING DISTANCE **EXAMPLE 2-11** Try It Yourself

In Example 2-10, (a) how much time does it take for the car to stop if its initial velocity is 30 m/s? (b) How far does the car travel in the last second?

PICTURE THE PROBLEM (a) Except for the numbers, this is the same as Part (a) of Example 2-9. Use the same procedure shown in Example 2-10. (b) Since the speed decreases by 5 m/s each second, the velocity 1 s before the car stops must be 5 m/s. Find the average velocity during the last second and use that to find the distance traveled.

Cover the column to the right and try these on your own before looking at the answers.

Steps	Answers
(a) Find the total stopping time t.	$t = 6$ s

(b) 1. Find the average velocity during the last second. $v_{av} =$ $\boxed{2.5 \text{ m/s}}$

2. Compute the distance traveled from $\Delta x = v_{av}\Delta t$. $\Delta x_1 = v_{av}\Delta t =$ $\boxed{2.5 \text{ m}}$

REMARKS If Part (b) asked for the average velocity during the last 1.3 seconds (rather than during the last second), we can find the initial velocity v_1 during this interval using $\Delta v = a\Delta t$ (Equation 2-11).

Sometimes valuable insight can be gained about the motion of an object by assuming that the constant-acceleration formulas still apply even when the acceleration is not constant. This is the case in the following example.

THE CRASH TEST   **E X A M P L E   2 - 1 2**

In a crash test, a car traveling 100 km/h (about 62 mi/h) hits an immovable concrete wall. What is its acceleration?

PICTURE THE PROBLEM In this example, it is not appropriate to treat the car as a particle because different parts of the vehicle will have different accelerations as the car crumples to a halt. Moreover, the accelerations are not constant. However, we *can* treat a bolt in the center of the car as a particle. We need additional information to solve this problem—either the stopping distance or the time to stop. We can estimate the stopping distance using common sense. Upon impact, the center of the car will certainly move forward less than half the length of the car. We'll choose 0.75 m as a reasonable estimate of the stopping distance. Since the problem neither asks for nor provides the time we will use the relation $v^2 = v_0^2 + 2a\Delta x$.

1. Using $v^2 = v_0^2 + 2a\Delta x$, solve for the acceleration:

$$v^2 = v_0^2 + 2a\Delta x$$

so

$$a = \frac{v^2 - v_0^2}{2\Delta x} = \frac{0^2 - (100 \text{ km/h})^2}{2(0.75 \text{ m})}$$

2. Convert the velocity from km/h to m/s. In one hour there are 60^2 s $= 3.6$ ks:

$$(100 \text{ km/h}) \times \left(\frac{1 \text{ h}}{3.6 \text{ ks}}\right) = 27.8 \text{ m/s}$$

3. Complete the calculation of the acceleration:

$$a = \frac{0^2 - (100 \text{ km/h})^2}{2(0.75 \text{ m})} = -\frac{(27.8 \text{ m/s})^2}{1.5 \text{ s}}$$

$$= -514 \text{ m/s}^2 \approx \boxed{-500 \text{ m/s}^2}$$

REMARKS Note that the magnitude of this acceleration is greater than $50g$. This estimate is what the magnitude of the acceleration would be both if the displacement of the center of the car were actually 0.75 m and if the acceleration were constant.

EXAMPLE 2-13 Try It Yourself

An electron in a cathode-ray tube accelerates from rest with a constant acceleration of 5.33×10^{12} m/s^2 for 0.15 μs (1 μs = 10^{-6} s). The electron then drifts with constant velocity for 0.2 μs. Finally, it comes to rest with an acceleration of -2.67×10^{13} m/s^2. How far does the electron travel?

PICTURE THE PROBLEM The equations for constant acceleration do not apply to this problem directly because the acceleration of the electron varies with time. Divide the electron's motion into three intervals, each with a different constant acceleration and use the final position and velocity for one interval as the initial conditions for the next interval. Choose the origin to be at the electron's starting position, and the positive direction to be the direction of motion.

Cover the column to the right and try these on your own before looking at the answers.

Steps	Answers
1. Find the displacement and final velocity for the first 0.15-μs interval.	$\Delta x_1 = 6.00$ cm, $v_1 = 8.00 \times 10^5$ m/s
2. Use this final velocity as the constant velocity to find the displacement while it drifts at constant velocity.	$\Delta x_2 = 16$ cm
3. Use this same velocity as the initial velocity and Equation 2-17 with $v = 0$ to find the displacement for the third interval, in which the electron slows down.	$\Delta x_3 = 1.20$ cm
4. Add the displacements found in steps 1, 2, and 3 to find the total displacement.	$\Delta x = \Delta x_1 + \Delta x_2 + \Delta x_3$ $= 6.00$ cm $+ 16$ cm $+ 1.20$ cm $= \boxed{23.2 \text{ cm}}$

REMARKS In an X-ray machine electrons are accelerated from a hot wire to a metal target. They crash into it, coming abruptly to rest. As a result, the target emits X rays characteristic of the target metal.

(*left*) The two-mile-long linear accelerator at Stanford University, used to accelerate electrons and positrons in a straight line to nearly the speed of light. (*right*) Cross-section of the accelerator's electron beam as shown on a video monitor.

EXAMPLE 2-14 Try It Yourself

John climbs a tree to get a better view of the speaker at an outdoor graduation ceremony. Unfortunately, he leaves his binoculars behind. Marsha throws them up to John, but her strength is greater than her accuracy. The binoculars pass John's outstretched hand after 0.69 s and again 1.68 s later. How high is John?

PICTURE THE PROBLEM There are two unknowns in this problem, John's height h and the initial velocity of the binoculars v_0. We know that $y = h$ at $t_1 = 0.69$ s and $y = h$ at $t_2 = 0.69$ s $+ 1.68$ s $= 2.37$ s. Expressing h as a function of time t gives us two equations from which the two unknowns can be determined.

Cover the column to the right and try these on your own before looking at the answers.

Steps	Answers
1. Using $\Delta y = v_0 t + \frac{1}{2}at^2$, equate y for times t_1 and t_2, noting that $y = h$, and $a = -g$ in each case.	$h = v_0 t_1 - \frac{1}{2}g t_1^2$ and $h = v_0 t_2 - \frac{1}{2}g t_2^2$
2. Eliminate v_0 from these two equations and solve for h in terms of the times t_1 and t_2. This can be done by solving the first equation for v_0 and substituting into the second equation.	$h = \left(\dfrac{h + \frac{1}{2}g t_1^2}{t_1}\right)t_2 - \frac{1}{2}g t_2^2$ so $h = \boxed{8.02 \text{ m}}$

REMARKS We have two unknowns, h and v_0, but are given two times, t_1 and t_2, so we can write two equations and solve them for either or both of the two unknowns.

EXERCISE Find the initial velocity of the binoculars and the velocity of the binoculars as they pass John on the way down. (*Answer* $v_0 = 15.0$ m/s; $v_2 = -8.24$ m/s)

Problems with Two Objects

We now give some examples of problems involving two objects moving with constant acceleration.

CATCHING A SPEEDING CAR **EXAMPLE 2-15**

A car is speeding at 25 m/s (~90 km/h; ~56 mi/h) in a school zone. A police car starts from rest just as the speeder passes and accelerates at a constant rate of 5 m/s². (*a*) When does the police car catch the speeding car? (*b*) How fast is the police car traveling when it catches up with the speeder?

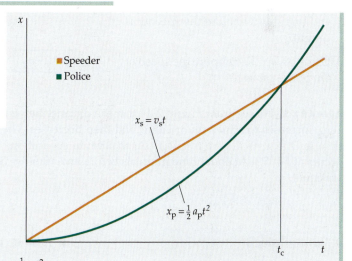

PICTURE THE PROBLEM To determine when the two cars will be at the same position, we write the positions of the speeder x_s and of the police car x_p as functions of time and solve for the time t when $x_s = x_p$.

(*a*) 1. Write the position functions for the speeder and the police car:

$$x_s = v_s t \quad \text{and} \quad x_p = \tfrac{1}{2}a_p t^2$$

FIGURE 2-13 The two curves depict the positions of the speeder and the police car. They have the same position at $t = 0$ and again at $t = t_c$.

2. Set $x_s = x_p$ and solve for the time t_c, for $t_c > 0$:

$$v_s t_c = \tfrac{1}{2}a_p t_c^2 \Rightarrow v_s = \tfrac{1}{2}a_p t_c \quad t_c \neq 0$$

$$t_c = \frac{2v_s}{a_p} = \frac{2(25 \text{ m/s})}{5 \text{ m/s}^2} = \boxed{10 \text{ s}}$$

(*b*) 1. The velocity of the police car is given by $v = v_0 + a t_c$, with $v_0 = 0$:

$$v_p = a_p t_c = (5 \text{ m/s}^2)(10 \text{ s}) = \boxed{50 \text{ m/s}}$$

REMARKS Notice that the final speed of the police car in (*b*) is exactly twice that of the speeder. Since the two cars covered the same distance in the same time, they must have had the same average speed. The speeder's average speed, of course, is 25 m/s. For the police car to start from rest and have an average speed of 25 m/s, it must reach a final speed of 50 m/s.

EXERCISE How far have the cars traveled when the police car catches the speeder? (*Answer* 250 m)

THE POLICE CAR   **E X A M P L E   2 - 1 6**   **Try It Yourself**

How fast is the police car in Example 2-15 traveling when it is 25 m behind the speeding car?

PICTURE THE PROBLEM The speed is given by $v_p = at_1$, where t_1 is the time at which $D = x_s - x_p = 25$ m.

Cover the column to the right and try these on your own before looking at the answers.

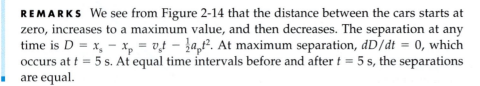

FIGURE 2-14

Steps	Answers
1. Sketch an x-versus-t curve showing the positions of the two cars at time t_1 (Figure 2-14).	
2. Using the equations for x_p and x_s from Example 2-15, solve for t_1 when $x_s - x_p = 25$ m. We expect two solutions, one shortly after the start time and one shortly before the speeder is caught.	$t_1 = (5 \pm \sqrt{15})$ s
3. Use $v_{p1} = a_p t_1$ to compute the speed of the police car at $t = t_1$.	$v_{p1} = \boxed{5.64 \text{ m/s}}$ and $\boxed{44.4 \text{ m/s}}$

REMARKS We see from Figure 2-14 that the distance between the cars starts at zero, increases to a maximum value, and then decreases. The separation at any time is $D = x_s - x_p = v_s t - \frac{1}{2} a_p t^2$. At maximum separation, $dD/dt = 0$, which occurs at $t = 5$ s. At equal time intervals before and after $t = 5$ s, the separations are equal.

A MOVING ELEVATOR   **E X A M P L E   2 - 1 7**

While standing in an elevator, you see a screw fall from the ceiling. The ceiling is 3 m above the floor. How long does it take for the screw to hit the floor if the elevator is moving upward and gaining speed at a constant rate of $a_f = 4.0$ m/s²?

PICTURE THE PROBLEM Write the position as a function of time for both the screw, y_s, and the floor, y_f. When the screw hits the floor $y_s = y_f$. Choose the origin to be the initial position of the floor, and designate "upward" as the positive direction.

1. Make a sketch of the elevator and the screw as shown in Figure 2-15. Include a coordinate axis indicating the positions of the screw and the floor:

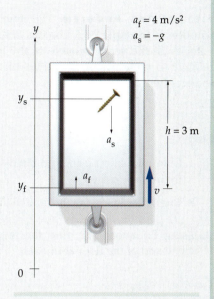

2. Write the position function for the elevator floor and for the screw:

$$y_f - y_{0f} = v_{0f}t + \tfrac{1}{2}a_f t^2$$

$$y_s - y_{0s} = v_{0s}t + \tfrac{1}{2}a_s t^2$$

3. At $t = t_1$ the screw hits the floor. At this time these positions are equal:

$$y_s = y_f$$

$$y_{0s} + v_{0s}t_1 + \tfrac{1}{2}a_s t_1^2 = y_{0f} + v_{0f}t_1 + \tfrac{1}{2}a_f t_1^2$$

4. At $t = 0$, the floor and the screw have the same velocity. Use this to simplify the step 3 result:

$$v_{0f} = v_{0s}$$

so

$$y_{0s} + v_{0s}t_1 + \tfrac{1}{2}a_s t_1^2 = y_{0f} + v_{0s}t_1 + \tfrac{1}{2}a_f t_1^2$$

$$y_{0s} + \tfrac{1}{2}a_s t_1^2 = y_{0f} + \tfrac{1}{2}a_f t_1^2$$

5. Use the given information to further simplify:

$$y_{0f} = 0, \quad a_f = 4.0 \text{ m/s}^2$$

$$y_{0s} = h = 3 \text{ m}, \quad a_s = -g$$

so

$$h - \tfrac{1}{2}g t_1^2 = 0 + \tfrac{1}{2}a_f t_1^2$$

or

$$h = \tfrac{1}{2}(g + a_f)t_1^2$$

6. Solve for the time:

$$t_1 = \sqrt{\frac{2h}{g + a_f}} = \sqrt{\frac{2(3 \text{ m})}{9.81 \text{ m/s}^2 + 4.0 \text{ m/s}^2}}$$

$$= \boxed{0.659 \text{ s}}$$

FIGURE 2-15 The coordinate axis is fixed to the building.

REMARKS The time of fall depends on the acceleration of the elevator, but not on its velocity. There is an "effective gravity" $g' = g + a_f$ in the frame of reference of the elevator. In the case (presumably hypothetical) in which the elevator itself is in free-fall, that is, $a_f = -g'$, the time of fall becomes infinite and the screw appears "weightless."

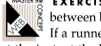 **EXERCISE** The speed of a good base runner is 9.5 m/s. The distance between bases is 26 m, and the pitcher is about 18.5 m from home plate. If a runner on first base edges 2 m off the base and then begins running at the instant the ball leaves the pitchers's hand, what is the likelihood that the runner will steal second base safely?

THE MOVING ELEVATOR **EXAMPLE 2-18 Try It Yourself**

Consider the elevator and screw in Example 2-17. Assume the velocity of the elevator is 16 m/s upward when the screw separates from the ceiling. (*a*) How far does the elevator rise while the screw is falling? How far does the screw fall? (*b*) What is the velocity of the screw and the velocity of the elevator at impact? (*c*) What is the velocity of the screw relative to the floor at impact?

PICTURE THE PROBLEM The time of flight of the screw is obtained in the solution of Example 2-17. Use this time to solve parts (a) and (b). For part (c), the velocity of the screw relative to the building equals the velocity of the screw relative to the elevator floor plus the velocity of the elevator floor relative to the building.

Cover the column to the right and try these on your own before looking at the answers.

Steps	Answers
(a) 1. Using Equation 2-16, find the distance the floor rises in time t_1.	$\Delta y_f = v_{f0}t_1 + \frac{1}{2}a_f t_1^2 = \boxed{11.4 \text{ m}}$
2. The screw starts out 3 m above the floor.	$\Delta y_s = \boxed{8.42 \text{ m}}$
(b) Using Equation 2-12, find the impact velocity of the screw and of the floor at impact.	$v = v_0 + at$, so
	$v_s = v_{s0} - gt_1 = \boxed{9.53 \text{ m/s}}$
	$v_f = v_{f0} + a_f t_1 = \boxed{18.6 \text{ m/s}}$
(c) Using Equation 2-7a, find the velocity of the screw relative to the elevator floor.	$v_{sb} = v_{sf} + v_{fb}$
	so
	$v_{sf} = v_{sb} - v_{fb} = 9.53 \text{ m/s} - 18.6 \text{ m/s}$
	$= \boxed{-9.10 \text{ m/s}}$

REMARKS The screw strikes the floor 8.4 m above its position when it leaves the ceiling. Relative to the building it is still rising when it strikes the floor. Note that at impact the velocity of the screw relative to the building is positive.

2-4 Integration

To find the velocity from a given acceleration, we note that the velocity is the function $v(t)$ whose derivative is the acceleration $a(t)$:

$$\frac{dv(t)}{dt} = a(t)$$

If the acceleration is constant, the velocity is that function of time which, when differentiated, equals this constant. One such function is

$$v = at, \qquad a = \text{constant}$$

More generally, we can add any constant to at without changing the time derivative. Calling this constant c, we have

$$v = at + c$$

When $t = 0$, $v = c$. Thus, c is the initial velocity v_0.

Similarly, the position function $x(t)$ is that function whose derivative is the velocity:

$$\frac{dx}{dt} = v = v_0 + at$$

We can treat each term separately. The function whose derivative is a constant v_0 is $v_0 t$ plus any constant. The function whose derivative is at is $\frac{1}{2}at^2$ plus any constant. Writing x_0 for the combined arbitrary constants, we have

$$x = x_0 + v_0 t + \tfrac{1}{2}at^2$$

When $t = 0$, $x = x_0$. Thus, x_0 is the initial position.

Whenever we find a function from its derivative, we must include an arbitrary constant in the general function. Since we go through the integration process twice to find $x(t)$ from the acceleration, two constants arise. These constants are usually determined from the velocity and position at some given time, which is usually chosen to be $t = 0$. They are therefore called the **initial conditions.** A common problem, called the **initial-value problem,** takes the form "given $a(t)$ and the initial values of x and v, find $x(t)$." This problem is particularly important in physics because the acceleration of a particle is determined by the forces acting on it. Thus, if we know the forces acting on a particle and the position and velocity of the particle at some particular time, we can find its position at all other times.

A function $F(t)$ whose derivative (with respect to t) equals the function $f(t)$ is called the **antiderivative** of $f(t)$. Finding the antiderivative of a function is related to the problem of finding the area under a curve. Consider motion with a constant velocity v_0. The change in position Δx during an interval Δt is

$$\Delta x = v_0 \Delta t$$

This is the area under the v-versus-t curve (Figure 2-16). If v_0 is negative, both the displacement and the area under the curve are negative. Normally we think of area as a quantity that cannot be negative. However, in this context that is not the case. In this case the "area under the curve" (the area between the curve and the time axis) is a negative quantity.

The geometric interpretation of the displacement as the area under the v-versus-t curve is true not only for constant velocity, but it is true in general, as illustrated in Figure 2-17. To show this we first divide the time interval into numerous small intervals, Δt_1, Δt_2, and so on. Then we draw a set of rectangles as shown. The area of the rectangle corresponding to the ith time interval Δt_i (shaded in the figure) is $v_i \Delta t_i$, which is approximately equal to the displacement Δx_i during the interval Δt_i. The sum of the rectangular areas is therefore approximately the sum of the displacements during the time intervals and is approximately equal to the total displacement from time t_1 to t_2. Mathematically, we write this as

$$\Delta x \approx \sum_i v_i \Delta t_i$$

where the Greek letter Σ (uppercase sigma) stands for sum. We can make the approximation as accurate as we wish by putting enough rectangles under the curve, each rectangle having a sufficiently small value for Δt. For the limit of smaller and smaller time intervals (and more and more rectangles), the resulting sum approaches the area under the curve, which in turn equals the displacement. The limit as Δt approaches zero (and the number of rectangles approaches infinity) is called an **integral** and is written

$$\Delta x = x(t_2) - x(t_1) = \lim_{\Delta t \to 0}\left(\sum_i v_i \Delta t_i\right) = \int_{t_1}^{t_2} v\, dt \qquad 2\text{-}18$$

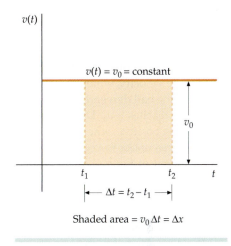

FIGURE 2-16 The displacement for the interval Δt equals the area under the velocity-versus-time curve for that interval. For $v(t) = v_0 =$ constant, the displacement equals the area of the shaded rectangle.

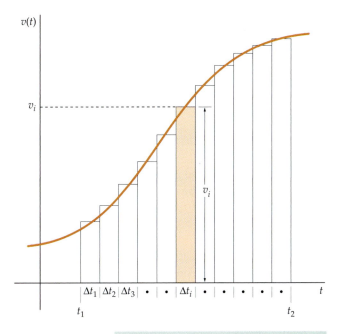

FIGURE 2-17 Graph of a general $v(t)$-versus-t curve. The total displacement from t_1 to t_2 is the area under the curve for this interval, which can be approximated by summing the areas of the rectangles.

It is helpful to think of the integral sign $\int$ as an elongated S indicating a sum. The limits t_1 and t_2 indicate the initial and final values of the variable t. The displacement is thus the area under the v-versus-t curve. Figure 2-18 demonstrates that the average velocity has a simple geometric interpretation in terms of the "area under a curve."

To illustrate that the displacement equals the area under the v-versus-t curve, consider what happens when you throw a golf ball straight up. The ball rises a meter or so, reverses direction, and then descends, gaining speed until you catch it. Assuming air resistance is negligible, the velocity of the ball is given by $v = v_0 + at$ (Equation 2-12), where the up direction is taken as positive and $a = -g$. Figure 2-19 is a plot of this velocity during the time that the ball is in free-fall. As shown, the velocity, which is initially positive, equals zero half way through the flight. As the ball descends, the velocity remains negative and, just before the ball is caught, reaches $-v_0$. During the rising portion of the motion, the area under the curve is positive, whereas during the descending portion it is negative. Thus, the total area under the curve for the entire flight is zero. It is easy to see that the displacement of the ball is also zero. Because the ball is thrown from the same place at which it is caught, the change in position is zero. Therefore, the displacement and the area under the v-versus-t curve are equal because they both are zero.

The process of computing an integral is called **integration**. In Equation 2-18, v is the derivative of x, and x is the antiderivative of v. This is an example of the fundamental theorem of calculus, whose formulation in the seventeenth century greatly accelerated the mathematical development of physics.

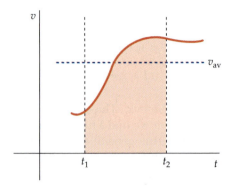

FIGURE 2-18 The displacement Δx during the time interval $\Delta t = t_2 - t_1$ is equal to the area of the shaded region. We know from the definition of average velocity that $\Delta x = v_{av} \Delta t$. This is just the area of a rectangle of height v_{av} and width Δt. Thus, the rectangular area $v_{av} \Delta t$ and the area under the v-versus-t curve must be equal.

$$\text{If } f(t) = \frac{dF(t)}{dt}, \quad \text{then} \quad F(t_2) - F(t_1) = \int_{t_1}^{t_2} f(t)\, dt \qquad \text{2-19}$$

FUNDAMENTAL THEOREM OF CALCULUS

The antiderivative of a function is also called the indefinite integral of the function and is written without limits on the integral sign, as in

$$x = \int v\, dt$$

Finding the function x from the derivative v (that is, finding the antiderivative) is also called integration. For example, if $v = v_0$, a constant, then

$$x = \int v_0\, dt = v_0 t + x_0$$

where x_0 is the arbitrary constant of integration. We can find a general rule for the integration of a power of t from Equation 2-6, which gives the general rule for the derivative of a power. The result is

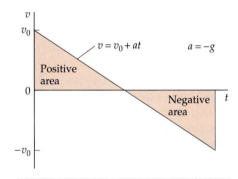

FIGURE 2-19 A v-versus-t curve for a golf ball thrown straight up that is caught by the thrower. The area under the curve is positive for the rising portion of the motion and negative for the descending portion. The area under the curve for the entire flight is zero.

$$\int t^n\, dt = \frac{t^{n+1}}{n+1} + C, \qquad n \neq -1 \qquad \text{2-20}$$

where C is an arbitrary constant. This can be checked by differentiating the right side using the rule of Equation 2-6. (For the special case $n = -1$, $\int t^{-1}\, dt = \ln t + C$, where $\ln t$ is the natural logarithm of t.)

The change in velocity for some time interval can similarly be interpreted as the area under the a-versus-t curve for that interval. This is written

$$\Delta v = \lim_{\Delta t \to 0} \left(\sum_i a_i \Delta t_i \right) = \int_{t_1}^{t_2} a\, dt \qquad \text{2-21}$$

We can now derive the constant-acceleration equations by computing the indefinite integrals of the acceleration and velocity. If a is constant, we have

$$v = \int a\, dt = a \int dt = v_0 + at \qquad\qquad 2\text{-}22$$

where we have expressed a times the constant of integration as v_0. Integrating again, and writing x_0 for the constant of integration, gives

$$x = \int (v_0 + at)dt = x_0 + v_0 t + \tfrac{1}{2}at^2 \qquad\qquad 2\text{-}23$$

It is instructive to derive Equations 2-22 and 2-23 using definite integrals instead of indefinite ones. For constant acceleration, Equation 2-21, with $t_1 = 0$, gives

$$v(t_2) - v(0) = a \int_0^{t_2} dt = a(t_2 - 0)$$

where the time t_2 is arbitrary. Because it is arbitrary, we can set $t_2 = t$ to obtain

$$v = v_0 + at$$

where $v = v(t)$ and $v_0 = v(0)$. To derive Equation 2-23, we substitute $v_0 + at$ for v in Equation 2-18 with $t_1 = 0$. This gives

$$x(t_2) - x(0) = \int_0^{t_2} (v_0 + at)dt$$

This integral is equal to the area under the v-versus-t curve (Figure 2-20). Evaluating the integral and solving for x gives

$$x(t_2) - x(0) = \int_0^{t_2} (v_0 + at)dt = v_0 t + \tfrac{1}{2}at^2 \Big|_0^{t_2} = v_0 t_2 + \tfrac{1}{2}at_2^2$$

where t_2 is arbitrary. Setting $t_2 = t$ we obtain

$$x = x_0 + v_0 t + \tfrac{1}{2}at^2$$

where $x = x(t)$ and $x_0 = x(0)$.

Having derived the constant-acceleration kinematic equations without any reference to average velocity, we can now show that the average velocity for the special case of constant acceleration is the mean value between the initial and final velocities as given by Equation 2-14. Let v_0 be the initial velocity at $t = 0$, and v be the final velocity at time t. According to the definition of average velocity, the displacement is

$$\Delta x = v_{av}\Delta t = v_{av}(t - 0) = v_{av}t \qquad\qquad 2\text{-}24$$

Also, from Equation 2-23, we have

$$\Delta x = v_0 t + \tfrac{1}{2}at^2$$

We eliminate the acceleration using $a = (v - v_0)/t$ from Equation 2-12. This gives

$$\Delta x = v_0 t + \frac{1}{2}\left(\frac{v - v_0}{t}\right)t^2 = v_0 t + \frac{1}{2}vt - \frac{1}{2}v_0 t = \frac{1}{2}(v + v_0)t \qquad\qquad 2\text{-}25$$

Comparing this with $\Delta x = v_{av}t$ (Equation 2-24), we have

$$v_{av} = \tfrac{1}{2}(v_0 + v_f)$$

which is Equation 2-14.

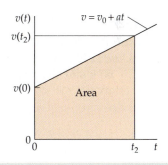

FIGURE 2-20 The area under the v-versus-t curve equals the displacement $\Delta x = x(t_2) - x(0)$.

The average velocity can be visualized using a v-versus-t curve (Figure 2-21). The displacement Δx equals the area under this curve. However, the average velocity is the area under the curve $v = v_{av}$ for the same time interval. Thus, the height of the $v = v_{av}$ curve is such that the areas under the two curves are equal. This implies that the areas of the two gray-shaded triangles are equal and that $v_{av} = \frac{1}{2}(v_1 + v_2)$.

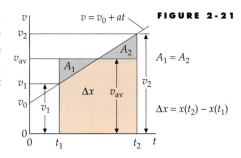

FIGURE 2-21

$v = v_0 + at$

$A_1 = A_2$

$\Delta x = x(t_2) - x(t_1)$

A COASTING BOAT   **EXAMPLE 2-19**

A Shelter Island ferryboat moves with constant velocity $v_0 = 8$ m/s for 60 s. It then shuts off its engines and coasts. Its coasting velocity is given by $v = v_0 t_1^2/t^2$, where $t_1 = 60$ s. What is the displacement of the boat for the interval $0 < t < \infty$?

PICTURE THE PROBLEM This velocity function is shown in Figure 2-22. The total displacement is calculated as the sum of the displacement Δx_1 during the interval $0 < t < t_1 = 60$ s and the displacement Δx_2 during the interval $t_1 < t < \infty$.

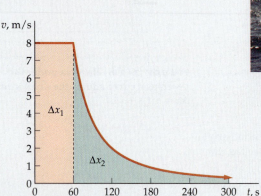

FIGURE 2-22

1. The velocity of the boat is constant during the first 60 s; thus the displacement is simply the velocity times the elapsed time:

$$\Delta x_1 = v_0 \Delta t = v_0 t_1 = (8 \text{ m/s})(60 \text{ s}) = 480 \text{ m}$$

2. The remaining displacement is given by the integral of the velocity from $t = t_1$ to $t = \infty$. We use Equation 2-18 to calculate the integral:

$$\Delta x_2 = \int_{t_1}^{\infty} v \, dt = \int_{t_1}^{\infty} \frac{v_0 t_1^2}{t^2} \, dt = v_0 t_1^2 \int_{t_1}^{\infty} t^{-2} \, dt = v_0 t_1^2 \left. \frac{t^{-1}}{-1} \right|_{t_1}^{\infty}$$

$$= -v_0 t_1^2 \left(\frac{1}{\infty} - \frac{1}{t_1} \right) = v_0 t_1 = (8 \text{ m/s})(60 \text{ s}) = 480 \text{ m}$$

3. The total displacement is the sum of the displacements found above:

$$\Delta x = \Delta x_1 + \Delta x_2 = 480 \text{ m} + 480 \text{ m} = \boxed{960 \text{ m}}$$

REMARKS Note that the area under the v-versus-t curve is finite. Thus, even though the boat never stops moving, it travels only a finite distance. A better representation of the velocity of a coasting boat might be the exponentially decreasing function $v = v_0 e^{-b(t-t_1)}$, where b is a positive constant. In that case, the boat would also coast a finite distance in the interval $t_1 \leq t \leq \infty$.

Displacement, velocity, and acceleration are important *defined* kinematics quantities.

Topic	Relevant Equations and Remarks	
1. Displacement	$\Delta x = x_2 - x_1$	2-1
Graphical interpretation	Displacement is the area under the v-versus-t curve.	
2. Velocity		
Average velocity	$v_{av} = \dfrac{\Delta x}{\Delta t}$	2-2
Instantaneous velocity	$v(t) = \lim\limits_{\Delta t \to 0} \dfrac{\Delta x}{\Delta t} = \dfrac{dx}{dt}$	2-5
Graphical interpretation	The instantaneous velocity is represented graphically as the slope of the x-versus-t curve.	
Relative velocity	If a particle moves with velocity v_{pA} relative to reference frame A, which is in turn moving with velocity v_{AB} relative to a second reference frame B, the velocity of the particle relative to B is $$v_{pB} = v_{pA} + v_{AB}$$	2-7
3. Speed		
Average speed	Average speed $= \dfrac{\text{total distance}}{\text{total time}} = \dfrac{s}{t}$	2-3
Instantaneous speed	Instantaneous speed is the magnitude of the instantaneous velocity.	
4. Acceleration		
Average acceleration	$a_{av} = \dfrac{\Delta v}{\Delta t}$	2-8
Instantaneous acceleration	$a = \dfrac{dv}{dt} = \dfrac{d^2 x}{dt^2}$	2-10
Graphical interpretation	The instantaneous acceleration is represented graphically as the slope of the v-versus-t curve.	
Acceleration due to gravity	The acceleration of an object near the surface of the earth in free-fall under the influence of gravity is directed downward and has the magnitude $$g = 9.81 \text{ m/s}^2 = 32.2 \text{ ft/s}^2$$	
5. Displacement and velocity as integrals	Displacement is represented graphically as the area under the v-versus-t curve. This area is the integral of v over time from some initial time t_1 to some final time t_2 and is written $$\Delta x = \lim_{\Delta t \to 0} \sum_i v_i \Delta t_i = \int_{t_1}^{t_2} v \, dt$$	2-18

Similarly, change in velocity is represented graphically as the area under the *a*-versus-*t* curve:

$$\Delta v = \lim_{\Delta t \to 0} \sum_i a_i \Delta t_i = \int_{t_1}^{t_2} a \, dt \qquad \text{2-21}$$

Velocity	$v = v_0 + at$	**2-12**
Displacement in terms of v_{av}	$\Delta x = x - x_0 = v_{av}t = \frac{1}{2}(v_0 + v)t$	**2-15**
Displacement in terms of a	$\Delta x = x - x_0 = v_0 t + \frac{1}{2}at^2$	**2-16**
v in terms of a and Δx	$v^2 = v_0^2 + 2a \, \Delta x$	**2-17**

PROBLEMS

- Single-concept, single-step, relatively easy
- • Intermediate-level, may require synthesis of concepts
- • • Challenging
- **SSM** Solution is in the *Student Solutions Manual*
- **iSOLVE** Problems available on iSOLVE online homework service
- **iSOLVE** ✓ These "Checkpoint" online homework service problems ask students additional questions about their confidence level, and how they arrived at their answer

In a few problems, you are given more data than you actually need; in a few other problems, you are required to supply data from your general knowledge, outside sources, or informed estimates.

For all problems, use $g = 9.81$ m/s² for the acceleration due to gravity and neglect friction and air resistance unless instructed to do otherwise.

Conceptual Problems

1 • What is the average velocity over the "round trip" of an object that is launched straight up from the ground and falls straight back down to the ground?

2 • **SSM** An object thrown straight up falls back to the ground. Its time of flight is T, its maximum height is H, and its height at release is negligible. Its average speed for the entire flight is (a) H/T, (b) 0, (c) $H/(2T)$, (d) $2H/T$.

3 • **iSOLVE** To avoid falling too fast during a landing, an airplane must maintain a minimum airspeed (the speed of the plane relative to the air). However, the slower the ground speed (speed relative to the ground) during a landing, the safer the landing. Is it safer for an airplane to land with the wind or against the wind?

4 • Give an example of one-dimensional motion where (a) the velocity is positive and acceleration is negative, and (b) the velocity is negative and the acceleration is positive.

5 • **SSM** Stand in the center of a large room. Call movement to your right "positive," and movement to your left "negative." Walk across the room along a straight line in such a way that, after getting started, your velocity is negative but your acceleration is positive. (a) Is your displacement initially positive or negative? Explain. (b) Describe how you vary your speed as you walk. (c) Sketch a graph of v versus t for your motion.

6 • True/false; explain: The displacement *always* equals the product of the average velocity and the time interval.

7 • True/false; explain:
(a) For the velocity to remain constant, the acceleration *must* remain zero.
(b) For the speed to remain constant, the acceleration *must* remain zero.

8 • • **SSM** Draw careful graphs of the position and velocity and acceleration over the time period $0 \leq t \leq 25$ s for a cart that
(a) moves away from the origin at a slow and steady (constant) velocity for the first 5 s;
(b) moves away at a medium-fast, steady (constant) velocity for the next 5 s;
(c) stands still for the next 5 s;
(d) moves toward the origin at a slow and steady (constant) velocity for the next 5 s;
(e) stands still for the last 5 s.

9 • True/false; explain: The average velocity *always* equals one-half the sum of initial plus final velocities.

10 • Identical twin brothers standing on a horizontal bridge each throw a rock straight down into the water below. They throw rocks at exactly the same time, but one hits the water before the other. How can this occur if the rocks have the same starting time?

11 •• **SSM** Dr. Josiah S. Carberry stands at the top of the Sears Tower in Chicago. Wanting to emulate Galileo, and ignoring the safety of the pedestrians below, he drops a bowling ball from the top of the tower. One second later, he drops a second bowling ball. While the balls are in the air, does their separation (*a*) increase over time, (*b*) decrease, or (*c*) stay the same? Ignore any effects that air resistance may have.

12 •• Which of the position-versus-time curves in Figure 2-23 best shows the motion of an object with constant positive acceleration?

FIGURE 2-23 Problem 12

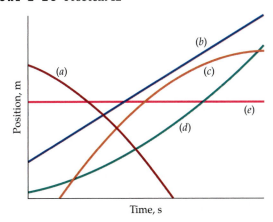

13 • **SSM** Which of the velocity-versus-time curves in Figure 2-24 best describes the motion of an object with constant positive acceleration?

FIGURE 2-24 Problem 13

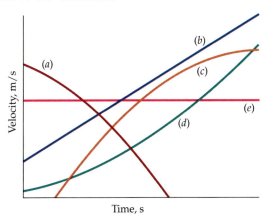

14 • Does the following statement make sense? "The average velocity of the car at 9 A.M. was 60 km/h."

15 • **SSM** Is it possible for the average velocity of an object to be zero during some interval, even though its average velocity for the first half of the interval is not zero? Explain.

16 • The diagram in Figure 2-25 tracks the path of an object moving in a straight line along the *x* axis. Assuming that the object is at the origin ($x_o = 0$) at $t_o = 0$, which point on the position-versus-time graph represents the instant the object is farthest from the origin? (*a*) A (*b*) B (*c*) C (*d*) D (*e*) E.

FIGURE 2-25 Problem 16

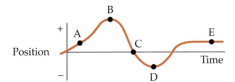

17 • **iSOLVE** If the instantaneous velocity does not change, will the average velocities for different intervals differ?

18 • If $v_{av} = 0$ for some time interval Δt, must the instantaneous velocity v be zero at some point in the interval? Support your answer by sketching a possible *x*-versus-*t* curve that has $\Delta x = 0$ for some interval Δt.

19 •• An object moves along a line as shown in Figure 2-26. At which point or points is its speed at a minimum? (*a*) A and E. (*b*) B, D, and E. (*c*) C only. (*d*) E only. (*e*) None of these is correct.

FIGURE 2-26 Problem 19

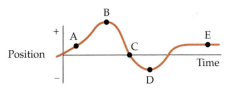

20 •• **SSM** **iSOLVE** For each of the four graphs of *x* versus *t* in Figure 2-27, answer the following questions. (*a*) Is the velocity at time t_2 greater than, less than, or equal to the velocity at time t_1? (*b*) Is the speed at time t_2 greater than, less than, or equal to the speed at time t_1?

FIGURE 2-27 Problem 20

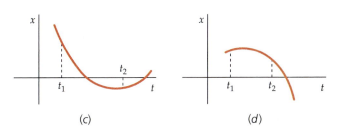

21 • True/false; explain:

(*a*) If the acceleration remains zero, the body cannot be moving.
(*b*) If the acceleration remains zero, the *x*-versus-*t* curve must be a straight line.

22 • Is it possible for a body to simultaneously have zero velocity and nonzero acceleration?

23 • `SOLVE` A ball is thrown straight up. What is the velocity of the ball at the top of its flight? What is its acceleration at that point?

24 • Find the average speed over the "round trip" of an object that is launched straight up from the ground, reaches a height H, and falls straight back down to the ground, hitting it after T seconds have elapsed. Express this in terms of the initial launch speed v_0.

25 • A bowling ball is thrown upward. While it is in flight, its acceleration is (a) decreasing, (b) constant, (c) zero, (d) increasing.

26 • At $t = 0$, object A is dropped from the roof of a building. At the same instant, object B is dropped from a window 10 m below the roof. During their descent to the ground, the distance between the two objects (a) is proportional to t, (b) is proportional to t^2, (c) decreases, (d) remains 10 m throughout.

27 •• `SSM` Assume that the Porsche accelerates uniformly from 80.5 km/h (50 mi/h) at $t = 0$ to 113 km/h (70 mi/h) at $t = 9$ s. Which graph in Figure 2-28 best describes the motion of the car?

FIGURE 2-28 Problem 27

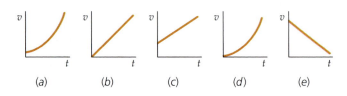

(a)   (b)   (c)   (d)   (e)

28 •• `SSM` An object is dropped from rest and falls a distance D in a given time. If the time during which it falls is doubled, the distance it falls will be (a) $4D$, (b) $2D$, (c) D, (d) $D/2$, (e) $D/4$.

29 •• A ball is thrown upward with an initial velocity v_0. Its velocity halfway to its highest point is (a) $0.25v_0$, (b) $0.5v_0$, (c) $0.707v_0$, (d) v_0, (e) cannot be determined from the information given.

30 • True or false:

(a) The equation $\Delta x = v_0 t + \frac{1}{2}at^2$ is valid for all particle motion in one dimension.

(b) If the velocity at a given instant is zero, the acceleration at that instant must also be zero.

(c) The equation $\Delta x = v_{av}\Delta t$ holds for all particle motion in one dimension.

31 • `SSM` If an object is moving at constant acceleration in a straight line, its instantaneous velocity halfway through any time interval is (a) greater than its average velocity, (b) less than its average velocity, (c) equal to its average velocity, (d) half its average velocity, (e) twice its average velocity.

32 • On a graph showing position on the vertical axis and time on the horizontal axis, a straight line with a negative slope represents motion with (a) constant positive acceleration, (b) constant negative acceleration, (c) zero velocity, (d) constant positive velocity, (e) constant negative velocity.

33 •• On a graph showing position on the vertical axis and time on the horizontal axis, a parabola that opens upward represents (a) a positive acceleration, (b) a negative acceleration, (c) no acceleration, (d) a positive followed by a negative acceleration, (e) a negative followed by a positive acceleration.

34 •• On a graph showing velocity on the vertical axis and time on the horizontal axis, a constant acceleration of zero is represented by (a) a straight line with positive slope, (b) a straight line with negative slope, (c) a straight line with zero slope, (d) either (a), (b), or (c), (e) none of the above.

35 •• On a graph showing velocity on the vertical axis and time on the horizontal axis, constant acceleration is represented by (a) a straight line with positive slope, (b) a straight line with negative slope, (c) a straight line with zero slope, (d) either (a), (b), or (c), (e) none of the above.

36 •• Which graph of v versus t in Figure 2-29 best describes the motion of a particle with positive velocity and negative acceleration?

FIGURE 2-29 Problem 36

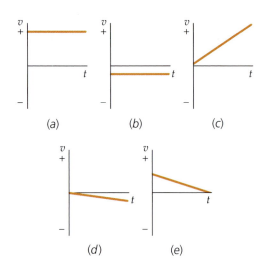

(a)   (b)   (c)

(d)   (e)

37 •• Which graph of v versus t in Figure 2-29 best describes the motion of a particle with negative velocity and negative acceleration?

38 •• A graph of the motion of an object is plotted with the velocity on the vertical axis and time on the horizontal axis. The graph is a straight line. Which of these quantities *can* be determined from this graph? (a) The displacement from time $t = 0$ to any other time shown. (b) The initial velocity at $t = 0$. (c) The acceleration as a function of time. (d) The average velocity for any specified time interval. (e) All of the above.

39 •• [SSM] Figure 2-30 shows the position of a car plotted as a function of time. At which times t_0 to t_7 is the velocity (a) negative? (b) positive? (c) zero? At which times is the acceleration (a) negative? (b) positive? (c) zero?

FIGURE 2-30 Problem 39

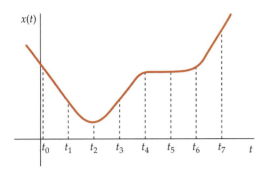

40 •• Sketch v-versus-t curves for each of the following conditions: (a) Acceleration is zero and constant while velocity is not zero. (b) Acceleration is constant but not zero. (c) Velocity and acceleration are both positive. (d) Velocity and acceleration are both negative. (e) Velocity is positive and acceleration is negative. (f) Velocity is negative and acceleration is positive. (g) Velocity is momentarily zero but the acceleration is not zero.

41 •• Figure 2-31 shows nine graphs of position, velocity, and acceleration for objects in motion along a straight line. Indicate the graphs that meet the following conditions: (a) velocity is constant, (b) velocity reverses its direction, (c) acceleration is constant, (d) acceleration is not constant. (e) Which graphs of position, velocity, and acceleration are mutually consistent?

FIGURE 2-31 Problem 41

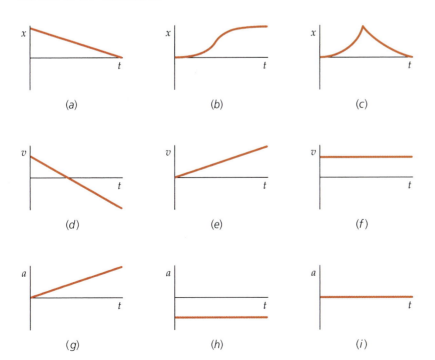

Estimation and Approximation

42 • Measure your own pulse rate (the number of heart beats per minute). Typical adult "resting rates" fall between 60 bpm (beats per min) and 80 bpm. (a) How many times will your heart beat during the time that it takes you to drive 1 mi at 60 mph? (b) How many times will your heart beat during your lifetime? (Assume a lifetime of 95 y.)

43 •• [SSM] [iSOLVE] Occasionally, people can survive after falling large distances if the surface they fall on is soft enough. During a traverse of the Eiger's infamous Nordvand, mountaineer Carlos Ragone's rock anchor pulled out and he plummeted 500 ft to land in snow. Amazingly, he suffered only a few bruises and a wrenched shoulder. (a) What final speed did he reach before impact? Ignore air resistance. (b) Assuming that his impact left a hole in the snow 4 ft deep, estimate his acceleration as he slowed to a stop. Assume that the acceleration was constant. Express this as a multiple of g (the magnitude of free-fall acceleration at the surface of the earth).

44 •• When we solve problems involving free-fall above the surface of the earth, it's important to remember that air resistance always exists; if we naively assume that objects always fall with constant acceleration, we may get answers that are wrong by orders of magnitude. How can we tell when it is valid to assume that a body is falling with (almost) constant acceleration? As a real body falls from rest through the air, as its speed increases, its acceleration downward decreases. The velocity will approach, but never quite reach, a *terminal velocity* that depends on the mass and cross-sectional area of the body; at the terminal velocity, the forces of gravity and air resistance exactly balance. For a "typical" skydiver falling through the air, a reasonable estimate for the terminal velocity is about 50 m/s (roughly 120 mph). At a speed of half the terminal velocity, the skydiver's acceleration will be $\frac{3}{4}g$. (a) Let's take half the terminal velocity as a reasonable "upper bound" beyond which we shouldn't use the constant acceleration formulas to calculate velocities and displacements. Roughly how far, and for how long, will the skydiver fall before we can't use these formulas anymore? (b) Repeat the analysis for a mouse, which has a terminal velocity of about 1 m/s.

45 •• On June 16, 1999 Maurice Greene of the United States set a new world's record for the 100-m dash with a time $t = 9.79$ s. Suppose that he started from rest at constant acceleration a and reached his maximum velocity in 3.00 s, which he then kept until the finish line. What was the acceleration a?

46 •• [SSM] The photograph in Figure 2-32 is a short-time exposure (1/30 s) of a juggler with two tennis balls in the air. The tennis ball near the top of its trajectory is less blurred than the lower one. Why is that? Can you estimate the speed of the lower ball from the picture?

FIGURE 2-32 Problem 46

47 •• Look up the speed at which a nerve impulse travels through the body. Estimate the time between stubbing your toe on a rock and feeling the pain due to this.

Speed, Displacement, and Velocity

48 • (a) An electron in a television tube travels the 16-cm distance from the grid to the screen at an average speed of 4×10^7 m/s. How long does the trip take? (b) An electron in a current-carrying wire travels at an average speed of 4×10^{-5} m/s. How long does it take to travel 16 cm?

49 • [SSM] A runner runs 2.5 km, in a straight line, in 9 min and then takes 30 min to walk back to the starting point. (a) What is the runner's average velocity for the first 9 min? (b) What is the average velocity for the time spent walking? (c) What is the average velocity for the whole trip? (d) What is the average speed for the whole trip?

50 • [i SOLVE] A car travels in a straight line with an average velocity of 80 km/h for 2.5 h and then with an average velocity of 40 km/h for 1.5 h. (a) What is the total displacement for the 4-h trip? (b) What is the average velocity for the total trip?

51 • One busy air route across the Atlantic Ocean is about 5500 km. (a) How long does it take for a supersonic jet flying at 2 times the speed of sound to make the trip? Use 340 m/s for the speed of sound. (b) How long does it take a subsonic jet flying at 0.9 times the speed of sound to make the same trip? (c) Allowing 2 h at each end of the trip for ground travel, check-in, and baggage handling, what is your average speed, door to door, when traveling on the supersonic jet? (d) What is your average speed taking the subsonic jet?

52 • [SSM] The speed of light, c, is 3×10^8 m/s. (a) How long does it take for light to travel from the sun to the earth, a distance of 1.5×10^{11} m? (b) How long does it take light to travel from the moon to the earth, a distance of 3.84×10^8 m? (c) A light-year is a unit of distance equal to that traveled by light in 1 year. Convert 1 light-year into kilometers and miles.

53 • Proxima Centauri, a dim companion to Alpha Centauri, is 4.1×10^{13} km away. From the vicinity of this star, Gregor places an order at Tony's Pizza in Hoboken, New Jersey, communicating via light signals. Tony's fastest delivery craft travels at $10^{-4}c$ (see Problem 52). (a) How long does it take Gregor's order to reach Tony's Pizza? (b) How long does Gregor wait between sending the signal and receiving the pizza? If Tony's has a 1000-years-or-it's-free delivery policy, does Gregor have to pay for the pizza?

54 • A car making a 100-km journey travels 40 km/h for the first 50 km. How fast must it go during the second 50 km to average 50 km/h?

55 •• [SSM] An archer fires an arrow, which produces a muffled "thwok" as it hits a target. If the archer hears the "thwok" exactly 1 s after firing the arrow and the average speed of the arrow was 40 m/s, what was the distance separating the archer and the target? Use 340 m/s for the speed of sound.

56 •• John can run 6 m/s. Marcia can run 15% faster than John. (a) By what distance does Marcia beat John in a 100-m race? (b) By what time does Marcia beat John in a 100-m race?

57 • [i SOLVE ✓] Figure 2-33 shows the position of a particle as a function of time. Find the average velocities for the time intervals a, b, c, and d indicated in the figure.

FIGURE 2-33 Problem 57

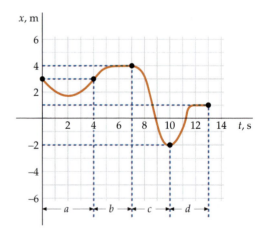

58 •• It has been found that galaxies are moving away from the earth at a speed that is proportional to their distance from the earth. This discovery is known as Hubble's law. The speed of a galaxy at a distance r from the earth is given by $v = Hr$, where H is the Hubble constant, equal to 1.58×10^{-18} s^{-1}. What is the speed of a galaxy (a) 5×10^{22} m from earth and (b) 2×10^{25} m from earth? (c) If each of these galaxies has traveled with constant speed, how long ago were they both located at the same place as the earth?

59 •• **SSM** **ISOLVE** The cheetah can run as fast as $v_1 = 113$ km/h, the falcon can fly as fast as $v_2 = 161$ km/h, and the sailfish can swim as fast as $v_3 = 105$ km/h. The three of them run a relay with each covering a distance L at maximum speed. What is the average speed v of this relay team? Compare this with the average of the three speeds.

60 •• Two cars are traveling along a straight road. Car A maintains a constant speed of 80 km/h; car B maintains a constant speed of 110 km/h. At $t = 0$, car B is 45 km behind car A. How much farther will car A travel before it is overtaken by car B?

61 •• **SSM** A car traveling at a constant speed of 20 m/s passes an intersection at time $t = 0$, and 5 s later another car traveling at a constant speed of 30 m/s passes the same intersection in the same direction. (a) Sketch the position functions $x_1(t)$ and $x_2(t)$ for the two cars. (b) Determine when the second car will overtake the first. (c) How far from the intersection will the two cars be when they pull even? (d) Where is the first car when the second car passes the intersection?

62 • Joe and Sally tend to argue when they travel. Just as they reached the moving sidewalk at the airport, their tempers flared to a point where neither was talking to the other. Though they stepped on the moving belt at the same time, Joe chose to stand and ride, while Sally opted to keep walking. Sally reached the end in 1 min, while Joe took 2 min. How long would it have taken Sally if she had walked twice as fast relative to the moving belt?

63 •• Margaret has just enough gas in her speedboat to get to the marina, an upstream journey that takes 4.0 h. Finding it closed for the season, she spends the next 8.0 h floating back downstream (out of gas) to her shack. The entire trip took 12.0 h. How long would it have taken if she had bought gas at the marina? Assume that the effect of the wind is negligible.

Acceleration

64 • **ISOLVE** A BMW-M3 sports car can accelerate in third gear from 48.3 km/h (30 mi/h) to 80.5 km/h (50 mi/h) in 3.7 s. (a) What is the average acceleration of this car in m/s²? (b) If the car continued at this acceleration for another second, how fast would it be moving?

65 • At $t = 5$ s, an object at $x = 3$ m has a velocity of $+5$ m/s. At $t = 8$ s, it is at $x = 9$ m and its velocity is -1 m/s. Find the average acceleration for this interval.

66 •• A particle moves with velocity $v = (8$ m/s²$)\, t - 7$ m/s. (a) Find the average acceleration for two 1-s intervals, one beginning at $t = 3$ s and the other beginning at $t = 4$ s. (b) Sketch v versus t. What is the instantaneous acceleration at any time?

67 •• **ISOLVE** ✔ The position of a certain particle depends on time according to the equation $x(t) = t^2 - 5t + 1$, where x is in meters if t is in seconds. (a) Find the displacement and average velocity for the interval 3 s $\leq t \leq 4$ s. (b) Find the general formula for the displacement for the time interval from t to $t + \Delta t$. (c) Use the limiting process to obtain the instantaneous velocity for any time t.

68 •• **SSM** **ISOLVE** The position of an object is related to time by $x = At^2 - Bt + C$, where $A = 8$ m/s², $B = 6$ m/s, and $C = 4$ m. Find the instantaneous velocity and acceleration as functions of time.

69 •• The one-dimensional motion of a particle is plotted in Figure 2-34. (a) What is the average acceleration in the intervals AB, BC, and CE? (b) How far is the particle from its starting point after 10 s? (c) Sketch the displacement of the particle as a function of time; label the instants A, B, C, D, and E on your figure. (d) At what time is the particle traveling most slowly?

FIGURE 2-34 Problem 69

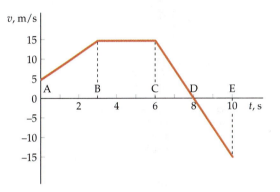

Constant Acceleration and Free-Fall

70 • **SSM** An object projected upward with initial velocity v_0 attains a height h. Another object projected up with initial velocity $2v_0$ will attain a height of (a) $4h$, (b) $3h$, (c) $2h$, (d) h.

71 • A car starting at $x = 50$ m accelerates from rest at a constant rate of 8 m/s². (a) How fast is it going after 10 s? (b) How far has it gone after 10 s? (c) What is its average velocity for the interval $0 \leq t \leq 10$ s?

72 • An object with an initial velocity of 5 m/s has a constant acceleration of 2 m/s². When its speed is 15 m/s, how far has it traveled?

73 • **SSM** An object with constant acceleration has a velocity of 10 m/s when it is at $x = 6$ m and of 15 m/s when it is at $x = 10$ m. What is its acceleration?

74 • The speed of an object increases at a constant rate of 4 m/s each second. At $t = 0$, its velocity is 1 m/s and its position is $x = 7$ m. How fast is it moving when it is at $x = 8$ m and what does t equal then?

75 •• **ISOLVE** ✔ A ball is thrown upward with an initial velocity of 20 m/s. (a) How long is the ball in the air? (Neglect the height of the release point.) (b) What is the greatest height reached by the ball? (c) How long after release is the ball 15 m above the release point?

76 •• **SOLVE** ✓ In the Blackhawk landslide in California, a mass of rock and mud fell 460 m down a mountain and then traveled 8 km across a level plain. It has been theorized that the rock and mud moved on a cushion of water vapor. Assume that the mass dropped with the free-fall acceleration and then slid horizontally, losing speed at a constant rate. (a) How long did the mud take to drop the 460 m? (b) How fast was it traveling when it reached the bottom? (c) How long did the mud take to slide the 8 km horizontally?

77 •• **SSM** A load of bricks is being lifted by a crane at a steady velocity of 5 m/s when one brick falls off 6 m above the ground. (a) Sketch $x(t)$ to show the motion of the free brick. (b) What is the greatest height the brick reaches above the ground? (c) How long does it take to reach the ground? (d) What is its speed just before it hits the ground?

78 •• A bolt comes loose from underneath an elevator that is moving upward at a speed of 6 m/s. The bolt reaches the bottom of the elevator shaft in 3 s. (a) How high up is the elevator when the bolt comes loose? (b) What is the speed of the bolt when it hits the bottom of the shaft?

79 •• **SSM** An object is dropped from rest at a height of 120 m. Find the distance it falls during its final second in the air.

80 •• An object is released from rest at a height h. During the final second of its fall, it traverses a distance of 38 m. What is h?

81 • **SSM** A stone is thrown vertically from a 200-m-tall cliff. During the last half second of its flight the stone travels a distance of 45 m. Find the initial speed of the stone.

82 •• An object is released from rest at a height h. It travels 0.4h during the first second of its descent. Determine the average velocity of the object during its entire descent.

83 •• A bus accelerates at 1.5 m/s² from rest for 12 s. It then travels at constant velocity for 25 s, after which it slows to a stop with an acceleration of −1.5 m/s². (a) How far does the bus travel? (b) What is its average velocity?

84 •• **SSM** It is relatively easy to use a spreadsheet program such as Microsoft Excel to solve certain types of physics problems. For example, you probably solved Problem 75 using algebra. Let's solve Problem 75 in a different way, this time using a spreadsheet program. While we can solve this problem using algebra, there are many places in physics where we can't get an alternative solution so easily, and have to rely on numerical methods like the one shown here. (a) Using Microsoft Excel or some other spreadsheet program, generate a graph of the height versus time for the ball in Problem 75 (thrown upward with an initial velocity of 20 m/s). Determine the maximum height, the time it was in the air, and the time(s) when the ball is 15 m above the ground by inspection (i.e., look at the graph and find them.) (b) Now change the initial velocity to 10 m/s and find the maximum height the ball reaches and the time the ball spends in the air.

85 •• **SSM** **SOLVE** Al and Bert are jogging side-by-side on a trail in the woods at a speed of 0.75 m/s. Suddenly Al

sees the end of the trail 35 m ahead and decides to speed up to reach it. He accelerates at a constant rate of 0.5 m/s², leaving Bert behind, who continues on at a constant speed. (a) How long does it take Al to reach the end of the trail? (b) Once he reaches the end of the trail, he immediately turns around and heads back along the trail with a constant speed of 0.85 m/s. How long does it take him to reach Bert? (c) How far are they from the end of the trail when they meet?

86 •• Solve Problem 85, parts (b) and (c), using a spreadsheet program.

87 •• A rocket is fired vertically with an upward acceleration of 20 m/s². After 25 s, the engine shuts off and the rocket then continues rising (while in free-fall). The rocket eventually stops rising and then falls back to the ground. Calculate (a) the highest point the rocket reaches, (b) the total time the rocket is in the air, (c) the speed of the rocket just before it hits the ground.

88 •• **SOLVE** A flowerpot falls from the ledge of an apartment building. A person in an apartment below, coincidentally holding a stopwatch, notices that it takes 0.2 s for the pot to fall past his window, which is 4 m high. How far above the top of the window is the ledge from which the pot fell?

89 •• **SSM** In a classroom demonstration, a glider moves along an inclined air track with constant acceleration. It is projected from the start of the track with an initial velocity. After 8 s have elapsed, it is 100 cm from its starting point and is moving along the track at a velocity of −15 cm/s. Find the initial velocity and the acceleration.

90 •• A rock dropped from a cliff falls one-third of its total distance to the ground in the last second of its fall. How high is the cliff?

91 •• A typical automobile under hard braking loses speed at a rate of about 7 m/s²; the typical reaction time to engage the brakes is 0.50 s. A school board sets the speed limit in a school zone to meet the condition that all cars should be able to stop in a distance of 4 m. (a) What maximum speed should be allowed for a typical automobile? (b) What fraction of the 4 m is due to the reaction time?

92 •• Two trains face each other on adjacent tracks. They are initially at rest, and their front ends are 40 m apart. The train on the left accelerates rightward at 1.4 m/s². The train on the right accelerates leftward at 2.2 m/s². How far does the train on the left travel before the front ends of the trains pass?

93 •• Two stones are dropped from the edge of a 60-m cliff, the second stone 1.6 s after the first. How far below the top of the cliff is the second stone when the separation between the two stones is 36 m?

94 •• **SSM** A motorcycle officer hidden at an intersection observes a car that ignores a stop sign, crosses the intersection, and continues on at constant speed. The police officer takes off in pursuit 2.0 s after the car has passed the stop sign, accelerates at 6.2 m/s² until her speed is 110 km/h, and then continues at this speed until she catches the car. At that instant, the car is 1.4 km from the intersection. How fast was the car traveling?

95 •• $\boxed{\text{SOLVE}}$ At $t = 0$, a stone is dropped from the top of a cliff above a lake. Another stone is thrown downward 1.6 s later from the same point with an initial speed of 32 m/s. Both stones hit the water at the same instant. Find the height of the cliff.

96 ••• $\boxed{\text{SOLVE}}$ A passenger train is traveling at 29 m/s when the engineer sees a freight train 360 m ahead of his train traveling in the same direction on the same track. The freight train is moving at a speed of 6 m/s. (a) If the reaction time of the engineer is 0.4 s, what is the minimum rate at which the passenger train must lose speed if a collision is to be avoided? (b) If the engineer's reaction time is 0.8 s and the train loses speed at the minimum rate described in part (a), what is the relative speed with which the two trains collide? How far will the passenger train have traveled in the time between the sighting of the freight train and the collision?

97 • Intent on studying the effects of gravity near the surface of the earth, a student launches a small projectile straight up with a speed of 300 m/s. How high will the projectile rise? (Ignore air resistance.)

98 • $\boxed{\text{SSM}}$ At the end of *Charlie and the Chocolate Factory*, Willie Wonka presses a button that shoots the great glass elevator through the roof of his chocolate factory. (a) If the elevator reaches a maximum height of 10 km above the roof, what was its speed immediately after crashing through the roof? Ignore air resistance, even though in this case it makes little sense to ignore it. (b) Assume that the elevator's speed just after it crashes through the roof was half of what it was just before its impact with the roof. Assuming that it started from rest on the ground floor of the chocolate factory, and that the height of the roof is 150 m above the ground floor, what uniform acceleration is needed for it to reach this high speed?

99 •• The click beetle can project itself vertically with an acceleration of about $a = 400g$ (an order of magnitude more than a human could stand). The beetle jumps by "unfolding" its legs, which are about $d = 0.6$ cm long. How high can the click beetle jump? How long is the beetle in the air? (Assume constant acceleration while in contact with the ground, and neglect air resistance.)

100 • $\boxed{\text{SOLVE}}$✓ A test of the prototype of a new automobile shows that the minimum distance for a controlled stop from 98 km/h to zero is 50 m. Find the acceleration, assuming it to be constant, and express your answer as a fraction of the free-fall acceleration. How much time does the car take to stop?

101 •• $\boxed{\text{SSM}}$ Consider the motion of a particle that experiences free-fall with a constant acceleration. Before the advent of computer-driven data-logging software, we used to do a free-fall experiment in which a coated tape was placed vertically next to the path of a dropped conducting puck. A high-voltage spark generator would cause an arc to jump between two vertical wires through the falling puck and through the tape, thereby marking the tape at fixed time intervals Δt. Show that the change in height in successive time intervals for an object falling from rest follows *Galileo's Rule of Odd Numbers*: $\Delta y_{21} = 3\Delta y_{10}$, $\Delta y_{32} = 5\Delta y_{10}$, . . . , where Δy_{10} is the change in y during the first interval of duration Δt, Δy_{21} is the change in y during the second interval of duration Δt, etc.

102 •• A particle moves with a constant acceleration of 3 m/s². At a time of 4 s, it is at a position of 100 m with respect to some coordinate system; at a time of 6 s, it has a velocity of 15 m/s. Find its position at a time of 6 s.

103 •• $\boxed{\text{SSM}}$ $\boxed{\text{SOLVE}}$✓ A plane landing on a small tropical island has just 70 m of runway on which to stop. If its initial speed is 60 m/s, (a) what is the minimum acceleration of the plane during landing, assuming it to be constant? (b) How long does it take for the plane to stop with this acceleration?

104 •• $\boxed{\text{SOLVE}}$ An automobile accelerates from rest at 2 m/s² for 20 s. The speed is then held constant for 20 s, after which there is an acceleration of -3 m/s² until the automobile stops. What is the total distance traveled?

105 •• $\boxed{\text{SSM}}$ $\boxed{\text{SOLVE}}$ If it were possible for a spacecraft to maintain a constant acceleration indefinitely, trips to the planets of the Solar System could be undertaken in days or weeks, while voyages to the nearer stars would only take a few years. (a) Show that g, the magnitude of free-fall acceleration on earth, is approximately 1 $c \cdot y/y^2$. (See Problem 52 for the definition of a light-year.) (b) Using data from the tables at the back of the book, find the time it would take for a one-way trip from Earth to Mars (at Mars' closest approach to Earth). Assume that the spacecraft starts from rest, travels along a straight line, accelerates halfway at $1g$, and then flips around and decelerates at $1g$ for the rest of the trip.

106 • $\boxed{\text{SOLVE}}$✓ The Stratosphere Tower in Las Vegas is 1137 ft high. It takes 1 min, 20 s to ascend from the ground floor to the top of the tower using the high-speed elevator. Assuming that the elevator starts and ends at rest, and its acceleration has a constant *magnitude* when moving, find the acceleration of the elevator. Express it in terms of a multiple of the acceleration of gravity.

107 •• $\boxed{\text{SOLVE}}$✓ A train pulls away from a station with a constant acceleration of 0.4 m/s². A passenger arrives at the track 6.0 s after the end of the train has passed the very same point. What is the slowest constant speed at which she can run and catch the train? Sketch curves for the motion of the passenger and the train as functions of time.

108 ••• Ball A is dropped from the top of a building at the same instant that ball B is thrown vertically upward from the ground. When the balls collide, they are moving in opposite directions, and the speed of A is twice the speed of B. At what fraction of the height of the building does the collision occur?

109 ••• Solve Problem 108 if the collision occurs when the balls are moving in the same direction and the speed of A is 4 times that of B.

110 •• $\boxed{\text{SSM}}$ Starting at one station, a subway train accelerates from rest at a constant rate of 1.0 m/s² for half the distance to the next station, then slows down at the same rate for the second half of the journey. The total distance between stations is 900 m. (a) Sketch a graph of the velocity v as a function of time over the full journey. (b) Sketch a graph of the distance covered as a function of time over the full journey. Place appropriate numerical values on both axes.

111 •• **SOLVE** A speeder traveling at a constant speed of 125 km/h races past a billboard. A patrol car pursues from rest with constant acceleration of (8 km/h)/s until it reaches its maximum speed of 190 km/h, which it maintains until it catches up with the speeder. (a) How long does it take the patrol car to catch the speeder if it starts moving just as the speeder passes? (b) How far does each car travel? (c) Sketch $x(t)$ for each car.

112 •• When the patrol car in Problem 111 (traveling at 190 km/h), pulls within 100 m behind the speeder (traveling at 125 km/h), the speeder sees the police car and slams on his brakes, locking the wheels. (a) Assuming that each car can brake at 6 m/s² and that the driver of the police car brakes instantly as she sees the brake lights of the speeder (reaction time = 0 s), show that the cars collide. (b) At what time after the speeder applies his brakes do the two cars collide? (c) Discuss how reaction time affects this problem.

113 •• Urgently needing the cash prize, Lou enters the Rest-to-Rest auto competition, in which each contestant's car begins and ends at rest, covering a distance L in as short a time as possible. The intention is to demonstrate mechanical and driving skills, and to consume the largest amount of fossil fuels in the shortest time possible. The course is designed so that maximum speeds of the cars are never reached. (a) If Lou's car has a maximum acceleration of a and a maximum deceleration of $2a$, then at what fraction of L should Lou move his foot from the gas pedal to the brake? (b) What fraction of the time for the trip has elapsed at that point?

114 •• **SOLVE** A physics professor demonstrates her new "anti-gravity parachute" by exiting from a helicopter at an altitude of 575 m with zero initial velocity. For 8 s, she falls freely. Then she switches on the "parachute" and her rate of descent slows at a constant rate of 15 m/s² until her downward speed reaches 5 m/s, whereupon she adjusts her controls to maintain that speed until she reaches the ground. (a) On a single graph, sketch her acceleration and velocity as functions of time. (Take upward to be positive.) (b) What is her speed at the end of the first 8 s? (c) For how long is she losing speed? (d) How far does she travel while losing speed? (e) How much time is required for the entire trip from the helicopter to the ground? (f) What is her average velocity for the entire trip?

Integration of the Equations of Motion

115 • **SSM** The velocity of a particle is given by $v(t) = (6 \text{ m/s}^2)t + (3 \text{ m/s})$. (a) Sketch v versus t and find the area under the curve for the interval $t = 0$ to $t = 5$ s. (b) Find the position function $x(t)$. Use it to calculate the displacement during the interval $t = 0$ to $t = 5$ s.

116 • Figure 2-35 shows the velocity of a particle versus time. (a) What is the magnitude in meters represented by the area of the shaded box? (b) Estimate the displacement of the particle for the two 1-s intervals beginning at $t = 1$ s and at $t = 2$ s. (c) Estimate the average velocity for the interval

$1 \text{ s} \leq t \leq 3 \text{ s}$. (d) The equation of the curve is $v = (0.5 \text{ m/s}^3)t^2$. Find the displacement of the particle for the interval $1 \text{ s} \leq t \leq 3$ s by integration and compare this answer with your answer for part (b). Is the average velocity equal to the mean of the initial and final velocities for this case?

FIGURE 2-35 Problem 116

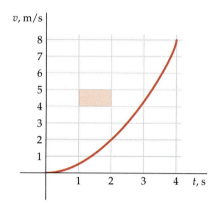

117 •• **SSM** The velocity of a particle is given by $v = (7 \text{ m/s}^3)t^2 - 5 \text{ m/s}$, where t is in seconds and v is in meters per second. If the particle starts from the origin, $x_0 = 0$, at $t_0 = 0$, find the general position function $x(t)$.

118 •• Consider the velocity graph in Figure 2-36. Assuming $x = 0$ at $t = 0$, write correct algebraic expressions for $x(t)$, $v(t)$, and $a(t)$ with appropriate numerical values inserted for all constants.

FIGURE 2-36 Problem 118

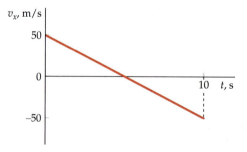

119 ••• Figure 2-37 shows the acceleration of a particle versus time. (a) What is the magnitude, in m/s, of the area of the shaded box? (b) The particle starts from rest at $t = 0$. Estimate the velocity at $t = 1$ s, 2 s, and 3 s by counting the boxes under the curve. (c) Sketch the curve v versus t from your results for part (b), and using it, estimate how far the particle travels in the interval $t = 0$ to $t = 3$ s.

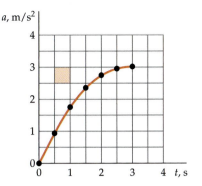

FIGURE 2-37 Problem 119

120 •• Figure 2-38 is a graph of v versus t for a particle moving along a straight line. The position of the particle at time $t_0 = 0$ is $x_0 = 5$ m. (a) Find x for various times t by counting boxes, and sketch x as a function of t. (b) Sketch a graph of the acceleration a as a function of the time t.

FIGURE 2-38 Problem 120

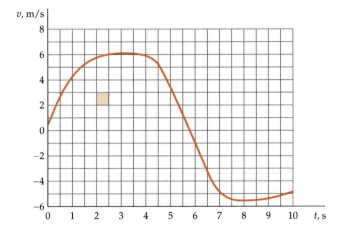

121 •• **SSM** Figure 2-39 shows a plot of x versus t for a body moving along a straight line. Sketch rough graphs of v as a function of t and a as a function of t for this motion.

FIGURE 2-39 Problem 121

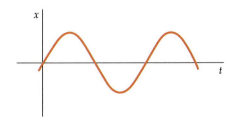

122 •• The acceleration of a certain rocket is given by $a = bt$, where b is a positive constant. (a) Find the general position function $x(t)$. (b) Find the position and velocity at $t = 5$ s if $x = 0$ and $v = 0$ at $t = 0$ and $b = 3$ m/s³.

123 •• **SOLVE** In the time interval from 0.0 s to 10.0 s, the acceleration of a particle is given by $a = (0.20 \text{ m/s}^3)t$ for one-dimensional motion. If a particle starts from rest at the origin, (a) calculate first its *instantaneous velocity* at any time during the interval, then (b) calculate its *average velocity* over the time interval from 2.0 s to 7.0 s.

124 • Consider the motion of a particle that experiences a nonconstant acceleration a given by $a = a_0 + bt$, where a_0 and b are constant. (a) Find the instantaneous velocity as a function of time. (b) Find the position as a function of time. (c) Find the average velocity over the time interval with an initial time of zero and arbitrary final time t.

General Problems

125 ••• **SOLVE** The following apparatus is used in a science class to determine the free-fall acceleration. Two photogates are set up, one at the edge of a table exactly 1.0 m high, the second one directly below it at a height of 0.5 m. A marble is dropped through the photogates from the height of 1 m above the floor (i.e., exactly the same as the height of the table), starting a timer when it enters the first photogate and stopping it when it enters the second one. The magnitude of free-fall acceleration is determined by $g_{exp} = (1 \text{ m})/(\Delta t)^2$, where Δt is the time measured by the timer. A careless student sets up the first photogate 0.5 cm below the edge of the table. (Assume that the second photogate is properly placed at a height of 0.5 m.) What value of g_{exp} will he determine? What percentage difference does this represent from the accepted value at sea level?

126 ••• **SSM** The position of a body oscillating on a spring is given by $x = A \sin \omega t$, where A and ω are constants with values $A = 5$ cm and $\omega = 0.175 \text{ s}^{-1}$. (a) Plot x as a function of t for $0 \leq t \leq 36$ s. (b) Measure the slope of your graph at $t = 0$ to find the velocity at this time. (c) Calculate the average velocity for a series of intervals beginning at $t = 0$ and ending at $t = 6, 3, 2, 1, 0.5,$ and 0.25 s. (d) Compute dx/dt and find the velocity at time $t = 0$. (e) Compare your results in parts (c) and (d).

127 ••• **SOLVE** Consider an object that is attached to a driving motor so that the object moves with a velocity given by $v = v_{max} \sin(\omega t)$, where ω is in radians/s. (a) What is the acceleration of the object and is it constant? (b) At $t = 0$, the position is x_0. What is the position as a function of time?

128 ••• Suppose the acceleration of a particle is a function of x, where $a(x) = (2 \text{ s}^{-2})\, x$. (a) If the velocity at $x = 1$ m is zero, what is the speed of the particle at $x = 3$ m? (b) How long does it take the particle to travel from $x = 1$ m to $x = 3$ m?

129 ••• Suppose that a particle moves in a straight line such that, at each instant of time, its position and velocity have the same numerical value if expressed in SI units. (*a*) Express the position x as a function of time t. (*b*) Show that at each instant of time the acceleration has the same numerical value as the position and velocity.

130 ••• A small rock sinking through water experiences an exponentially decreasing acceleration as a function of time given by $a(t) = ge^{-bt}$, where b is a positive constant that depends on the shape and size of the rock and the physical properties of the water. Based upon this result, derive an expression for the position of the rock as a function of time. Assume that its initial velocity is 0.

131 ••• **SSM** In Problem 130, a rock falls through water with a continuously decreasing acceleration of the form $a(t) = ge^{-bt}$, where b is a positive constant. In physics, we are not often given acceleration directly as a function of time, but usually either as a function of position or of velocity. Assume that the rock's acceleration as a function of *velocity* has the form $a = g - bv$ where g is the magnitude of free-fall acceleration and v is the rock's speed. Prove that if the rock has an initial velocity $v = 0$ at time $t = 0$, it will have the dependence on *time* given above.

132 ••• The acceleration of a skydiver jumping from an airplane is given by the formula $a = g - cv^2$, where c is a constant depending on the skydiver's cross-sectional area and the density of the surrounding atmosphere she is diving through. (*a*) If her initial speed is 0 when jumping from the plane, show that her speed as a function of time is given by the formula $v(t) = v_T \tanh(t/T)$, where v_T is the terminal velocity $(v_T = \sqrt{g/c})$ and $T = v_T/g$ is a time scale determining very roughly the time it takes for her speed to approach v_T. (*b*) Use a spreadsheet program such as Microsoft Excel to graph $v(t)$ as a function of time, using a terminal velocity of 56 m/s (use this to calculate c and T). Does the resulting curve make sense?

Motion in Two and Three Dimensions

SAILBOATS DON'T TRAVEL IN STRAIGHT LINES TO THEIR DESTINATIONS, BUT INSTEAD MUST "TACK" BACK AND FORTH ACROSS THE WIND. THIS BOAT MUST SAIL EAST, THEN SOUTH, AND THEN EAST AGAIN, IN ITS JOURNEY TO A SOUTHEASTERN PORT.

 How can we calculate its displacement and its average velocity? (See Example 3-3.)

Here we extend the ideas of Chapter 2 to two and three dimensions. To do this, we introduce vectors and show how they are used to analyze and describe motion.

➤ **Our major goal in this chapter is to develop the concept of the acceleration vector, a concept that is central to the development of Newton's laws in Chapters 4 and 5.**

3-1 The Displacement Vector

When motion occurs, the displacement of a particle has a direction in space as well as a magnitude. The quantity that gives the straight-line distance between two points in space as well as the direction is a line segment called the **displacement vector**. It is represented graphically by an arrow whose direction is the same as that of the displacement vector and whose length is proportional to the magnitude of the displacement vector. We denote vectors by italic boldface letters with an overhead arrow, $\vec{A}$. The magnitude of $\vec{A}$ is written $|\vec{A}|$, $\|\vec{A}\|$, or simply A, and has dimensions of length. The magnitude of a vector is never negative.

Addition of Displacement Vectors

Figure 3-1 shows the path of a particle that moves from point P_1 to a second point P_2 and then to a third point P_3. The vector $\vec{A}$ represents the displacement from P_1 to P_2, while $\vec{B}$ represents the displacement from P_2 to P_3. Note that the displacement vectors depend only on the end points and not on the actual path of the particle. The *resultant* displacement from P_1 to P_3, labeled $\vec{C}$, is called the sum of the two successive displacements $\vec{A}$ and $\vec{B}$:

$$\vec{C} = \vec{A} + \vec{B} \qquad\qquad 3\text{-}1$$

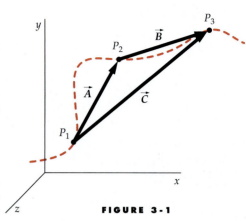

FIGURE 3-1

Two displacement vectors are added graphically by placing the tail of one at the head of the other (Figure 3-2). The resultant vector extends from the tail of the first to the head of the second. Note that C does not equal $A + B$ unless $\vec{A}$ and $\vec{B}$ are in the same direction. That is, $\vec{C} = \vec{A} + \vec{B}$ does not imply that $C = A + B$.

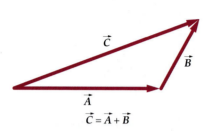

FIGURE 3-2
Head-to-tail method
of vector addition.

$$\vec{C} = \vec{A} + \vec{B}$$

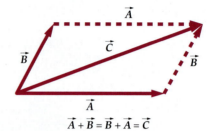

FIGURE 3-3
Parallelogram method
for adding vectors.

$$\vec{A} + \vec{B} = \vec{B} + \vec{A} = \vec{C}$$

An equivalent way of adding vectors, called the **parallelogram method,** is to move $\vec{B}$ so that it is tail-to-tail with $\vec{A}$. The diagonal of the parallelogram formed by $\vec{A}$ and $\vec{B}$ then equals $\vec{C}$. From Figure 3-3, we can see that it makes no difference in which order we add two vectors; that is, $\vec{A} + \vec{B} = \vec{B} + \vec{A}$.

YOUR DISPLACEMENT

EXAMPLE 3-1

You walk 3 km east and then 4 km north. What is your resultant displacement?

PICTURE THE PROBLEM Draw and label the two displacements $\vec{A}$ and $\vec{B}$ as well as the resultant displacement $\vec{C}$ (Figure 3-4). Include axes indicating the directions north and east. $\vec{A}$ and $\vec{B}$ are at right angles to each other, and $\vec{C} = \vec{A} + \vec{B}$ is the hypotenuse of the corresponding right triangle. The magnitude of $\vec{C}$ can be obtained using the Pythagorean theorem, and the direction of $\vec{C}$ is found using trigonometry.

FIGURE 3-4

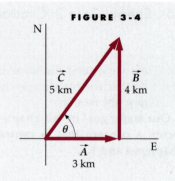

1. The magnitude of the resultant displacement is related to the magnitudes of the two sequential displacements by the Pythagorean theorem:

$$C^2 = A^2 + B^2$$
$$= (3\ \text{km})^2 + (4\ \text{km})^2$$
$$= 25\ \text{km}^2$$

so

$$C = \sqrt{25\ \text{km}^2} = \boxed{5\ \text{km}}$$

2. Let θ be the angle from the east axis to the resultant displacement $\vec{C}$. From the figure we find $\tan\theta$; using a calculator with trigonometric functions yields θ:

$$\tan\theta = \frac{B}{A}$$

so

$$\theta = \tan^{-1}\frac{B}{A} = \tan^{-1}\frac{4\ \text{km}}{3\ \text{km}} = \boxed{53.1°}$$

REMARKS A vector is described by its magnitude and its direction. Your resultant displacement is a vector of length 5 km in a direction 53.1° north of east.

3-2 General Properties of Vectors

Many quantities in physics have magnitude and direction, and add like displacements. Examples include velocity, acceleration, momentum, and force. Such quantities are called **vectors.** Quantities with magnitude but no associated direction—for example, distance and speed—are called **scalars.**

> Vectors are quantities with magnitude and direction that add like displacements.

DEFINITION—VECTORS

A vector is represented graphically by an arrow drawn in the same direction as that of the vector, and with a length that is proportional to the magnitude of the vector. When the magnitude of a vector is given, its units must also be given. The magnitude of a velocity vector, for example, requires units such as m/s. Two vectors are defined to be equal if they have the same magnitude and the same direction. Graphically, this means that they have the same length and are parallel to each other. A consequence of this definition is that moving a vector so that it remains parallel to itself does not change it. Thus, all the vectors in Figure 3-5 are equal. If we translate or rotate the coordinate system, all the vectors in Figure 3-5 remain equal. Vectors do not depend on the coordinate system used to represent them (except for position vectors, which are introduced in Section 3.3).

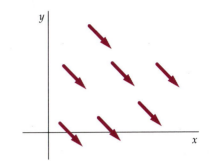

FIGURE 3-5 Vectors are equal if their magnitudes and directions are the same. All vectors in this figure are equal.

Multiplying a Vector by a Scalar

A vector $\vec{A}$ multiplied by a scalar s is the vector $\vec{B} = s\vec{A}$, which has magnitude $|s|A$ and is parallel to $\vec{A}$ if s is positive and antiparallel if s is negative. The vector $-\vec{A}$ is the vector you add to $\vec{A}$ to get zero. Thus, $-\vec{A}$ has the same magnitude as $\vec{A}$ but points in the opposite direction. The dimensions of $s\vec{A}$ are those of s multiplied by those of A.

Subtracting Vectors

We subtract vector $\vec{B}$ from vector $\vec{A}$ by adding $-\vec{B}$ to $\vec{A}$. The result is $\vec{C} = \vec{A} + (-\vec{B}) = \vec{A} - \vec{B}$ (Figure 3-6a). An equivalent way of subtracting $\vec{B}$ from $\vec{A}$ is to draw them tail-to-tail (Figure 3-6b) and noting that $\vec{C}$ is the vector that must be added to $\vec{B}$ to obtain the resultant vector $\vec{A}$. The rules for adding or subtracting any two vectors, such as two velocity vectors or two acceleration vectors, are the same as for adding or subtracting displacement vectors.

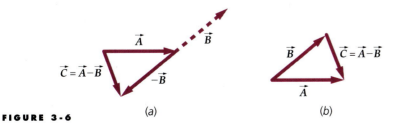

FIGURE 3-6
(a)   (b)

Components of Vectors

The component of a vector in a given direction is a number whose absolute value is the length of the projection of the vector onto an axis (directed line) in that direction. It is found by dropping a perpendicular from the head of the vector to the axis, as shown in Figure 3-7. The sign of the component is positive if the projection of the head of the vector is in the positive direction relative to the tail of the vector. Thus, A_S is positive and B_S is negative. The components of a vector along the x, y, and z directions, illustrated in Figure 3-8 for a vector in the xy plane, are called rectangular components. Note that the components of a vector *do* depend on the coordinate system used, although the vector itself does not.

Rectangular components are useful for the addition or subtraction of vectors. If θ is the angle between $\vec{A}$ and the x axis, then

$$A_x = A \cos \theta \qquad \text{3-2}$$

X COMPONENT OF A VECTOR

and

$$A_y = A \sin \theta \qquad \text{3-3}$$

Y COMPONENT OF A VECTOR

where A is the magnitude of $\vec{A}$.

If we know A_x and A_y, we can find the angle θ from

$$\tan \theta = \frac{A_y}{A_x}, \qquad \theta = \tan^{-1} \frac{A_y}{A_x} \qquad \text{3-4}$$

and the magnitude A from the Pythagorean theorem:

$$A = \sqrt{A_x^2 + A_y^2} \qquad \text{3-5}a$$

In three dimensions,

$$A = \sqrt{A_x^2 + A_y^2 + A_z^2} \qquad \text{3-5}b$$

Components can be positive or negative. For example, if $\vec{A}$ points in the negative x direction, then A_x is negative.

Consider two vectors $\vec{A}$ and $\vec{B}$ that lie in the xy plane. The rectangular components of each vector and those of the sum $\vec{C} = \vec{A} + \vec{B}$ are shown in Figure 3-9. We see that the vector equation $\vec{C} = \vec{A} + \vec{B}$ is equivalent to the two component equations

$$C_x = A_x + B_x \qquad \text{3-6}a$$

and

$$C_y = A_y + B_y \qquad \text{3-6}b$$

EXERCISE A car travels 20 km in a direction 30° north of west. Let the x axis point east and the y axis point north as in Figure 3-10. Find the x and y components of the displacement vector of the car. (*Answer* $A_x = -17.3$ km, $A_y = +10$ km)

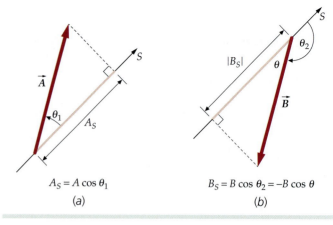

$$A_S = A \cos \theta_1 \qquad\qquad B_S = B \cos \theta_2 = -B \cos \theta$$
$$(a) \qquad\qquad\qquad (b)$$

FIGURE 3-7 Definition of the component of a vector. The component of the vector $\vec{A}$ in the positive S direction is A_S, and A_S is positive. The component of the vector $\vec{B}$ in the positive S direction is B_S, and B_S is negative.

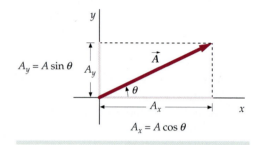

FIGURE 3-8 The rectangular components of a vector. $A_x = A \cos \theta$, $A_y = A \sin \theta$.

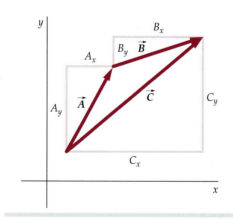

FIGURE 3-9

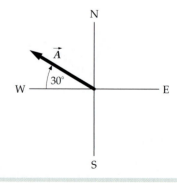

FIGURE 3-10

A TREASURE MAP   **EXAMPLE 3-2   Put It in Context**

You are working as a counselor at a tropical resort. You've been given a map and told to follow its directions in order to bury the "treasure" at a specific location. You don't want to waste time walking around the island, because you want to finish early and go surfing. The directions are to walk 3 km headed due west and then 4 km headed 60° north of east. Where should you head and how far must you walk to get the job done quickly? Find your answer (*a*) graphically and (*b*) using vector components.

PICTURE THE PROBLEM You need to find your resultant displacement, which is $\vec{C}$ in Figure 3-11. The triangle formed by the three vectors is not a right triangle, so the magnitudes of the vectors are not related by the Pythagorean theorem. We find the resultant graphically by drawing each of the displacements to scale and measuring the resultant displacement.

FIGURE 3-11

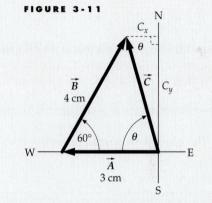

(*a*) If you draw the first displacement vector $\vec{A}$ 3 cm long and the second one $\vec{B}$ 4 cm long, you find the resultant vector $\vec{C}$ to be about 3.5 cm long. Thus, the magnitude of the resultant displacement is 3.5 km . The angle θ made between the resultant displacement and the west direction can then be measured with a protractor. It is about 75°. Therefore, you should walk 3.5 km headed 75° north of west .

(*b*) 1. To solve using vector components, let $\vec{A}$ be the first displacement and choose the direction of increasing *x* to be east and the direction of increasing *y* to be north. Compute A_x and A_y from Equations 3-2 and 3-3:

$A_x = -3$ km   and $A_y = 0$

2. Similarly, compute the components of the second displacement $\vec{B}$:

$B_x = (4 \text{ km}) \cos 60° = 2 \text{ km}$

$B_y = (4 \text{ km}) \sin 60° = 2\sqrt{3} \text{ km} = 3.46 \text{ km}$

3. The components of the resultant displacement $\vec{C} = \vec{A} + \vec{B}$ are found by addition:

$C_x = A_x + B_x = -3 \text{ km} + 2 \text{ km} = -1 \text{ km}$

$C_y = A_y + B_y = 0 + 2\sqrt{3} \text{ km} = 3.46 \text{ km}$

4. The Pythagorean theorem gives the magnitude of $\vec{C}$:

$C^2 = C_x^2 + C_y^2$

$\quad = (-1 \text{ km})^2 + (2\sqrt{3} \text{ km})^2 = 13 \text{ km}^2$

$C = \sqrt{13} \text{ km} = \boxed{3.61 \text{ km}}$

5. The ratio of C_y to $|C_x|$ equals the tangent of the angle θ between $\vec{C}$ and the negative *x* direction:

$\tan \theta = \dfrac{C_y}{|C_x|}$

so

$\theta = \tan^{-1} \dfrac{2\sqrt{3} \text{ km}}{1 \text{ km}} = \tan^{-1} (2\sqrt{3})$

$= \boxed{74° \text{ north of west}}$

REMARKS Since the displacement (which is a vector) was asked for, the answer must include either both the magnitude *and* direction, or both components. In (*b*) we could have stopped at step 3 because the *x* and *y* components completely define the displacement vector. We converted to the magnitude and direction to compare with the answer to Part (*a*). Note that step 5 of Part (*b*) gives the angle as 74°. This agrees with the results in (*a*) within the accuracy of our measurement.

Unit Vectors

A **unit vector** is a *dimensionless* vector with unit magnitude. The vector $A^{-1}\vec{A}$ is an example of a unit vector that points in the direction of $\vec{A}$. Unit vectors are often written with an overhead caret as in $\hat{A} = A^{-1}\vec{A}$. Unit vectors that point in the positive x, y, and z directions are convenient for expressing vectors in terms of their rectangular components. They are usually written $\hat{i}$, $\hat{j}$, and $\hat{k}$, respectively. For example, the vector $A_x\hat{i}$ has a magnitude $|A_x|$ and points in the positive x direction if A_x is positive (or the negative x direction if A_x is negative). A general vector $\vec{A}$ can be written as the sum of three vectors, each of which is parallel to a coordinate axis (Figure 3-12):

$$\vec{A} = A_x\hat{i} + A_y\hat{j} + A_z\hat{k} \qquad 3\text{-}7$$

The addition of two vectors $\vec{A}$ and $\vec{B}$ can be written in terms of unit vectors as

$$\vec{A} + \vec{B} = (A_x\hat{i} + A_y\hat{j} + A_z\hat{k}) + (B_x\hat{i} + B_y\hat{j} + B_z\hat{k})$$
$$= (A_x + B_x)\hat{i} + (A_y + B_y)\hat{j} + (A_z + B_z)\hat{k} \qquad 3\text{-}8$$

The general properties of vectors are summarized in Table 3-1.

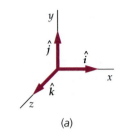

(a)

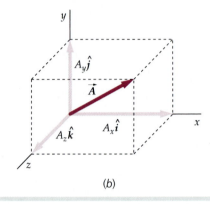

(b)

FIGURE 3-12 (*a*) The unit vectors $\hat{i}$, $\hat{j}$, and $\hat{k}$ in a rectangular coordinate system. (*b*) The vector $\vec{A}$ in terms of unit vectors: $\vec{A} = A_x\hat{i} + A_y\hat{j} + A_z\hat{k}$.

TABLE 3-1

Properties of Vectors

Property	Explanation	Figure	Component representation
Equality	$\vec{A} = \vec{B}$ if $|\vec{A}| = |\vec{B}|$ and their directions are the same		$A_x = B_x$ $A_y = B_y$ $A_z = B_z$
Addition	$\vec{C} = \vec{A} + \vec{B}$		$C_x = A_x + B_x$ $C_y = A_y + B_y$ $C_z = A_z + B_z$
Negative of a vector	$\vec{A} = -\vec{B}$ if $|\vec{B}| = |\vec{A}|$ and their directions are opposite		$A_x = -B_x$ $A_y = -B_y$ $A_z = -B_z$
Subtraction	$\vec{C} = \vec{A} - \vec{B}$		$C_x = A_x - B_x$ $C_y = A_y - B_y$ $C_z = A_z - B_z$
Multiplication by a scalar	$\vec{B} = s\vec{A}$ has magnitude $|\vec{B}| = |s||\vec{A}|$ and has the same direction as $\vec{A}$ if s is positive or as $-\vec{A}$ if s is negative		$B_x = sA_x$ $B_y = sA_y$ $B_z = sA_z$

EXERCISE Given two vectors

$$\vec{A} = (4\ \text{m})\hat{i} + (3\ \text{m})\hat{j} \qquad \text{and} \qquad \vec{B} = (2\ \text{m})\hat{i} - (3\ \text{m})\hat{j}$$

find (a) A, (b) B, (c) $\vec{A} + \vec{B}$, and (d) $\vec{A} - \vec{B}$. (Answers   (a) $A = 5$ m, (b) $B = 3.61$ m, (c) $\vec{A} + \vec{B} = (6\ \text{m})\hat{i}$, (d) $\vec{A} - \vec{B} = (2\ \text{m})\hat{i} + (6\ \text{m})\hat{j}$)

3-3 Position, Velocity, and Acceleration

Position and Velocity Vectors

The **position vector** of a particle is a vector drawn from the origin of a co-ordinate system to the position of the particle. For a particle at the point (x, y), the position vector $\vec{r}$ is

$$\vec{r} = x\hat{i} + y\hat{j} \tag{3-9}$$

DEFINITION—POSITION VECTOR

Figure 3-13 shows the actual path or trajectory of the particle. (Don't confuse the trajectory with the x-versus-t plots of Chapter 2.) At time t_1, the particle is at P_1, with position vector $\vec{r}_1$; by t_2, the particle has moved to P_2, with position vector $\vec{r}_2$. The particle's change in position is the displacement vector $\Delta\vec{r}$:

$$\Delta\vec{r} = \vec{r}_2 - \vec{r}_1 \tag{3-10}$$

DEFINITION—DISPLACEMENT VECTOR

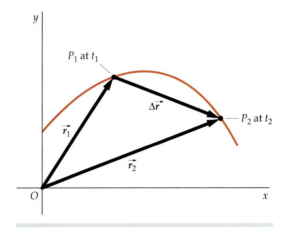

FIGURE 3-13 The displacement vector $\Delta\vec{r}$ is the difference in the position vectors, $\Delta\vec{r} = \vec{r}_2 - \vec{r}_1$. Equivalently, $\Delta\vec{r}$ is the vector that, when added to $\vec{r}_1$, yields the new position vector $\vec{r}_2$.

The ratio of the displacement vector to the time interval $\Delta t = t_2 - t_1$ is the **average-velocity vector**:

$$\vec{v}_{av} = \frac{\Delta\vec{r}}{\Delta t} \tag{3-11}$$

DEFINITION—AVERAGE-VELOCITY VECTOR

This vector points in the direction of the displacement.

The magnitude of the displacement vector is less than the distance traveled along the curve unless the particle moves in a straight line. However, if we consider smaller and smaller time intervals (Figure 3-14), the magnitude of the displacement approaches the distance along the curve, and the direction of $\Delta\vec{r}$ approaches the tangent to the curve at the beginning of the interval. We define the **instantaneous-velocity vector** as the limit of the average-velocity vector as Δt approaches zero:

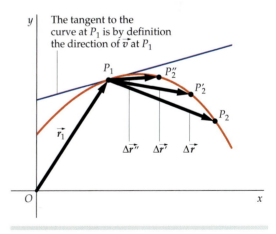

FIGURE 3-14 As the time interval is made smaller, the direction of the displacement approaches the tangent to the curve.

$$\vec{v} = \lim_{\Delta t \to 0} \frac{\Delta\vec{r}}{\Delta t} = \frac{d\vec{r}}{dt} \tag{3-12}$$

DEFINITION—INSTANTANEOUS-VELOCITY VECTOR

The instantaneous-velocity vector is the derivative of the position vector with respect to time. Its magnitude is the speed and its direction is the direction of motion of the particle along the line tangent to the curve.

To calculate the derivative in Equation 3-12, we write the position vectors in terms of their components:

$$\Delta \vec{r} = \vec{r}_2 - \vec{r}_1 = (x_2 - x_1)\hat{i} + (y_2 - y_1)\hat{j} = \Delta x \hat{i} + \Delta y \hat{j}$$

Then

$$\vec{v} = \lim_{\Delta t \to 0} \frac{\Delta \vec{r}}{\Delta t} = \lim_{\Delta t \to 0} \frac{\Delta x \hat{i} + \Delta y \hat{j}}{\Delta t} = \lim_{\Delta t \to 0}\left(\frac{\Delta x}{\Delta t}\right)\hat{i} + \lim_{\Delta t \to 0}\left(\frac{\Delta y}{\Delta t}\right)\hat{j}$$

or

$$\vec{v} = \frac{dx}{dt}\hat{i} + \frac{dy}{dt}\hat{j} = v_x \hat{i} + v_y \hat{j} \qquad\qquad 3\text{-}13$$

THE VELOCITY OF A SAILBOAT   **EXAMPLE 3-3**

A sailboat has coordinates $(x_1, y_1) = $ **(110 m, 218 m)** at $t_1 = $ **60 s.** Two minutes later, at time t_2, it has coordinates $(x_2, y_2) = $ **(130 m, 205 m).** (*a*) Find the average velocity for this 2-min interval. Express $\vec{v}_{av}$ in terms of its rectangular components. (*b*) Find the magnitude and direction of this average velocity. (*c*) For $t \geq 20$ s, the position of the sailboat as a function of time is $x(t) = b_1 + b_2 t$ and $y(t) = c_1 + c_2/t$, where $b_1 = 100$ m, $b_2 = \frac{1}{6}$ m/s, $c_1 = 200$ m, and $c_2 = 1080$ m·s. Find the instantaneous velocity at a general time $t \geq 20$ s.

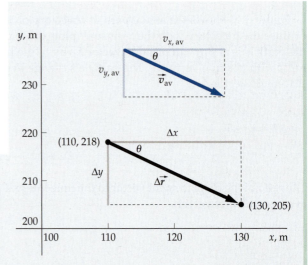

FIGURE 3-15

PICTURE THE PROBLEM The initial and final positions of the sailboat are given. (*a*) The average-velocity vector points from the initial toward the final position. (*b*) The instantaneous-velocity components are calculated from Equation 3-13: $v_x = dx/dt$ and $v_y = dy/dt$.

(*a*) 1. Draw a coordinate system (Figure 3-15) showing the displacement of the sailboat. The average-velocity vector and the displacement are in the same direction:

2. The x and y components of the average velocity $\vec{v}_{av}$ are calculated directly from their definitions:

$$\vec{v}_{av} = v_{x,av}\,\hat{i} + v_{y,av}\,\hat{j}$$

where

$$v_{x,av} = \frac{\Delta x}{\Delta t} = \frac{130\ \text{m} - 110\ \text{m}}{120\ \text{s}} = 0.167\ \text{m/s}$$

$$v_{y,av} = \frac{\Delta y}{\Delta t} = \frac{205\ \text{m} - 218\ \text{m}}{120\ \text{s}} = -0.108\ \text{m/s}$$

so

$$\vec{v}_{av} = \boxed{(0.167\ \text{m/s})\hat{i} - (0.108\ \text{m/s})\hat{j}}$$

(*b*) 1. The magnitude of $\vec{v}_{av}$ is found from the Pythagorean theorem:

$$v_{av} = \sqrt{(v_{x,av})^2 + (v_{y,av})^2} = \boxed{0.199\ \text{m/s}}$$

2. The ratio of $v_{y,\,av}$ to $v_{x,\,av}$ gives the tangent of the angle θ between $\vec{v}_{av}$ and the x axis:

$$\tan \theta = \frac{v_{y,av}}{v_{x,av}}$$

so

$$\theta = \tan^{-1} \frac{v_{y,av}}{v_{x,av}} = \tan^{-1} \frac{-0.108 \text{ m/s}}{0.167 \text{ m/s}} = \boxed{-33.0°}$$

(c) We find the instantaneous velocity $\vec{v}$ by calculating dx/dt and dy/dt:

$$\vec{v} = \frac{dx}{dt}\hat{i} + \frac{dy}{dt}\hat{j} = b_2\hat{i} - c_2 t^{-2}\hat{j} = \boxed{\left(\frac{1}{6}\text{ m/s}\right)\hat{i} - \frac{1080 \text{ m} \cdot \text{s}}{t^2}\hat{j}}$$

REMARKS The magnitude of $\vec{v}$ can be found from $v = \sqrt{v_x^2 + v_y^2}$ and its direction can be found from $\tan \theta = v_y/v_x$.

EXERCISE Find the x and y components and the magnitude and direction of the instantaneous velocity of the sailboat at time $t_1 = 60$ s. (*Answer* $\vec{v}_1 = (\frac{1}{6}\text{ m/s})\hat{i} - (0.30 \text{ m/s})\hat{j}, v_1 = 0.34 \text{ m/s}, \theta_1 = -60.9°$)

Relative Velocity

Relative velocities in two and three dimensions combine just as they do in one dimension, except that the velocity vectors are not necessarily along the same line. If a particle moves with velocity $\vec{v}_{pA}$ relative to reference frame A, which is in turn moving with velocity $\vec{v}_{AB}$ relative to reference frame B, the velocity of the particle relative to reference frame B is

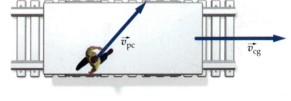

(a)

$$\vec{v}_{pB} = \vec{v}_{pA} + \vec{v}_{AB} \qquad\qquad 3\text{-}14$$

RELATIVE VELOCITY

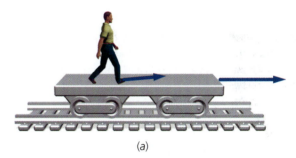

(b)

For example, if you are on a railroad flatcar that is moving with velocity $\vec{v}_{cg}$ relative to the ground (Figure 3-16a), and you are walking with velocity $\vec{v}_{pc}$ (Figure 3-16b) relative to the car, then your velocity relative to the ground is the vector sum of these two velocities: $\vec{v}_{pg} = \vec{v}_{pc} + \vec{v}_{cg}$ (Figure 3-16c).

The velocity of object A relative to object B is equal in magnitude and opposite in direction to the velocity of object B relative to object A. For example, $\vec{v}_{pc} = -\vec{v}_{cp}$ where $\vec{v}_{cp}$ is the velocity of the car relative to the person. The addition of relative velocities is done in the same way as the addition of displacements, either graphically, using either the head-to-tail or the parallelogram method, or analytically, using vector components.

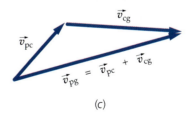

(c)

FIGURE 3-16

A FLYING PLANE   **EXAMPLE 3 - 4**

A pilot wants to fly a plane due north. The speed of the plane relative to the air is 200 km/h and the wind is blowing from west to east at 90 km/h. (*a*) In which direction should the plane head? (*b*) How fast does the plane travel relative to the ground?

PICTURE THE PROBLEM Because the wind is blowing toward the east, a plane headed due north will drift off course toward the east. To compensate for the crosswind, the plane must head west of due north. The velocity of the plane relative to the ground $\vec{v}_{pg}$ will be the sum of the velocity of the plane relative to the air $\vec{v}_{pa}$ and the velocity of the air relative to the ground $\vec{v}_{ag}$.

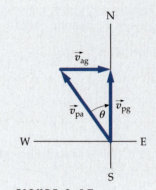

FIGURE 3-17

(a) 1. The velocity of the plane relative to the ground is given by Equation 3-14:

$$\vec{v}_{pg} = \vec{v}_{pa} + \vec{v}_{ag}$$

2. Make a velocity addition diagram showing the addition of the vectors in step 1. Include direction axes as shown in Figure 3-17.

3. The sine of the angle θ between the velocity of the plane and north equals the ratio of v_{ag} and v_{pa}:

$$\sin \theta = \frac{v_{ag}}{v_{pa}}$$

so

$$\theta = \sin^{-1} \frac{v_{ag}}{v_{pa}} = \sin^{-1} \frac{90 \text{ km/h}}{200 \text{ km/h}} = \boxed{26.7°}$$

(b) Because $\vec{v}_{ag}$ and $\vec{v}_{pg}$ are perpendicular, we can use the Pythagorean theorem to find the magnitude of $\vec{v}_{pg}$:

$$v_{pa}^2 = v_{pg}^2 + v_{ag}^2$$

so

$$v_{pg} = \sqrt{v_{pa}^2 - v_{ag}^2}$$

$$= \sqrt{(200 \text{ km/h})^2 - (90 \text{ km/h})^2} = \boxed{179 \text{ km/h}}$$

The Acceleration Vector

The **average-acceleration** is the ratio of the change in the instantaneous-velocity vector $\Delta \vec{v}$ to the time interval Δt:

$$\vec{a}_{av} = \frac{\Delta \vec{v}}{\Delta t} \qquad\qquad 3\text{-}15$$

DEFINITION—AVERAGE-ACCELERATION VECTOR

The **instantaneous-acceleration** is the limit of this ratio as Δt approaches zero; in other words, it is the derivative of the velocity vector with respect to time:

$$\vec{a} = \lim_{\Delta t \to 0} \frac{\Delta \vec{v}}{\Delta t} = \frac{d\vec{v}}{dt} \qquad\qquad 3\text{-}16$$

DEFINITION—INSTANTANEOUS-ACCELERATION VECTOR

To calculate the instantaneous acceleration, we express $\vec{v}$ in rectangular coordinates:

$$\vec{v} = v_x \hat{i} + v_y \hat{j} + v_z \hat{k} = \frac{dx}{dt} \hat{i} + \frac{dy}{dt} \hat{j} + \frac{dz}{dt} \hat{k}$$

Then

$$\vec{a} = \frac{dv_x}{dt} \hat{i} + \frac{dv_y}{dt} \hat{j} + \frac{dv_z}{dt} \hat{k} = \frac{d^2x}{dt^2} \hat{i} + \frac{d^2y}{dt^2} \hat{j} + \frac{d^2z}{dt^2} \hat{k}$$

$$= a_x \hat{i} + a_y \hat{j} + a_z \hat{k} \qquad\qquad 3\text{-}17$$

A THROWN BASEBALL   **EXAMPLE 3-5**

The position of a thrown baseball is given by $\vec{r} = 1.5\ \text{m}\ \hat{i} + (12\ \text{m/s}\ \hat{i} + 16\ \text{m/s}\ \hat{j})t - 4.9\ \text{m/s}^2\ \hat{j}\,t^2$. Find its velocity and acceleration.

PICTURE THE PROBLEM Because $\vec{r} = x\hat{i} + y\hat{j}$, we have $x = 1.5\ \text{m} + (12\ \text{m/s})t$ and $y = (16\ \text{m/s})t - (4.9\ \text{m/s}^2)t^2$. We can find the x and y components of the velocity and acceleration by differentiating x and y.

1. The x and y components of the velocity are found by differentiating x and y:

$$v_x = \frac{dx}{dt} = \frac{d}{dt}[1.5\ \text{m} + (12\ \text{m/s})t] = 12\ \text{m/s}$$

$$v_y = \frac{dy}{dt} = \frac{d}{dt}[(16\ \text{m/s})t - (4.9\ \text{m/s}^2)t^2]$$

$$= 16\ \text{m/s} - (9.8\ \text{m/s}^2)t$$

2. We differentiate again to obtain the components of the acceleration:

$$a_x = \frac{dv_x}{dt} = 0$$

$$a_y = \frac{dv_y}{dt} = -9.8\ \text{m/s}^2$$

3. In vector notation, the velocity and acceleration are:

$$\boxed{\vec{v} = (12\ \text{m/s})\hat{i} + [16\ \text{m/s} - (9.8\ \text{m/s}^2)t]\hat{j}}$$

$$\boxed{\vec{a} = (-9.8\ \text{m/s}^2)\hat{j}}$$

REMARKS This is an example of projectile motion, a topic we will study in Section 3-4.

For a vector to be constant, both its magnitude and direction must remain constant. If either changes, the vector changes. Thus, if a car rounds a curve in the road at constant speed, it is accelerating because the velocity is changing due to the change in direction of the velocity vector.

CORNERING A TURN   **EXAMPLE 3-6**

A car is traveling east at 60 km/h. It rounds a curve, and 5 s later it is traveling north at 60 km/h. Find the average acceleration of the car.

PICTURE THE PROBLEM The initial and final velocity vectors are shown in Figure 3-18. We choose the unit vector $\hat{i}$ to be east and $\hat{j}$ to be north, and calculate the average acceleration from its definition, $\vec{a}_{av} = \Delta\vec{v}/\Delta t$. Note that $\Delta\vec{v}$ is the vector that, when added to $\vec{v}_i$, results in $\vec{v}_f$.

1. The average acceleration is the ratio of the velocity change to the time interval:

$$\vec{a}_{av} = \frac{\Delta\vec{v}}{\Delta t}$$

2. The change in velocity is related to the initial and final velocities:

$$\Delta\vec{v} = \vec{v}_f - \vec{v}_i$$

3. Express the initial and final velocities as vectors:

$$\vec{v}_i = (60\ \text{km/h})\hat{i}, \qquad \vec{v}_f = (60\ \text{km/h})\hat{j}$$

4. Substitute these results to find the average acceleration:

$$\vec{a}_{av} = \frac{\vec{v}_f - \vec{v}_i}{\Delta t} = \frac{(60\ \text{km/h})\hat{j} - (60\ \text{km/h})\hat{i}}{5\ \text{s}}$$

$$= \boxed{-[(12\ \text{km/h})/\text{s}]\hat{i} + [(12\ \text{km/h})/\text{s}]\hat{j}}$$

FIGURE 3-18

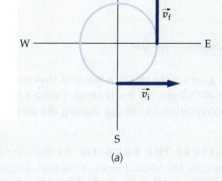

(a)

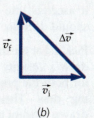

(b)

REMARKS Note that the car accelerates, even though its speed does not change.

EXERCISE Find the magnitude and direction of the average acceleration vector. (*Answer* $a_{av} = (17.0 \text{ km/h})/\text{s}$ at $45°$ west of north)

The motion of an object traveling in a circle is a common example of motion in which the velocity of an object changes even though its speed remains constant.

The Direction of the Acceleration Vector In the next few chapters we will want to determine the direction of the acceleration vector from a description of the motion. For example, consider a bungee jumper as she slows down prior to reversing direction at the lowest point of her jump. To find the direction of her acceleration as she loses speed we make a series of dots representing her position at successive ticks of a clock, as shown in Figure 3-19a. The faster she moves, the greater the distance she travels between ticks, and the greater the space between the dots in the diagram. Next we number the dots, starting with zero and increasing in the direction of her motion. At time t_0 she is at dot 0, at time t_1 she is at dot 1, and so forth. To determine the direction of the acceleration at time t_3 we draw vectors representing the jumper's velocities at times t_2 and t_4. The average acceleration during the interval from t_2 to t_4 equals $\Delta\vec{v}/\Delta t$, where $\Delta\vec{v} = \vec{v}_4 - \vec{v}_2$ and $\Delta t = t_4 - t_2$. We use this as an estimate of her acceleration at time t_3. That is, $\vec{a}_3 \approx \Delta\vec{v}/\Delta t$. Since $\vec{a}_3$ and $\Delta\vec{v}$ are in the same direction, by finding the direction of $\Delta\vec{v}$ we also find the direction of $\vec{a}_3$. The direction of $\Delta\vec{v}$ is obtained by using the relation $\vec{v}_2 + \Delta\vec{v} = \vec{v}_4$ and drawing the corresponding vector addition diagram (Figure 3-19b). Because the jumper is moving faster (the dots are farther apart) at t_2 than at t_4, we draw $\vec{v}_2$ longer than $\vec{v}_4$. From this figure we get the direction of $\Delta\vec{v}$, and thus the direction of $\vec{a}_3$.

EXERCISE Figure 3-20 is a motion diagram of the bungee jumper before, during, and after time t_6, when she momentarily comes to rest at the lowest point in her descent. During the part of her ascent shown she is moving upward with increasing speed. Use this diagram to determine the direction of the jumper's acceleration (a) at time t_6 and (b) at time t_9. (*Answer* (a) upward, (b) upward)

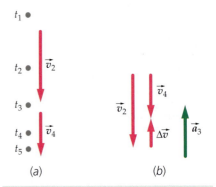

FIGURE 3-19 (a) A motion diagram of a bungee jumper losing speed as she descends. The dots are drawn at successive ticks of a clock. (b) We draw vectors $\vec{v}_2$ and $\vec{v}_4$ starting from the same point. Then we draw $\Delta\vec{v}$ from the head of $\vec{v}_2$ to the head of $\vec{v}_4$ to obtain the graphical expression of the relation $\vec{v}_2 + \Delta\vec{v} = \vec{v}_4$. The acceleration $\vec{a}_3$ is in the same direction as $\Delta\vec{v}$.

FIGURE 3-20 The dots for the bungee jumper's ascent are drawn to the right of those for her descent so that they do not overlap each other. Her motion, however, is straight down and then straight up.

ANOTHER FLYING CAP   **EXAMPLE 3-7**

A graduating physics student throws his cap into the air with an initial angle of 60° above the horizontal. Using a motion diagram, find the direction of the acceleration of the cap during the ascending portion of its flight.

PICTURE THE PROBLEM As the cap rises it both loses speed and changes direction. We should draw a motion diagram and then make a sketch of the relation $\vec{v}_i + \Delta\vec{v} = \vec{v}_f$ to find the direction of $\Delta\vec{v}$ and thus the direction of the acceleration.

1. Make a motion diagram (Figure 3-21a) of the cap's motion for the ascending portion of its flight. Because it slows as it ascends, the spacing between dots should decrease as it rises.

2. Pick a dot on the motion diagram and draw a velocity vector on the diagram for both the preceding and the following dots. These vectors should be drawn tangent to the path of the cap.

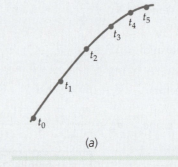

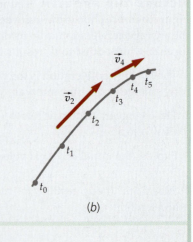

FIGURE 3-21

3. Draw the graphical expression of the relation $\vec{v}_i + \Delta\vec{v} = \Delta\vec{v}_f$. Begin by drawing the two velocity vectors from the same point. These vectors should have the same magnitude and direction as the vectors drawn for step 2. Then draw the $\Delta\vec{v}$ vector connecting their heads.

4. Draw the acceleration vector in the same direction as $\Delta\vec{v}$, but not necessarily the same length since $\vec{a} = \Delta\vec{v}/\Delta t$.

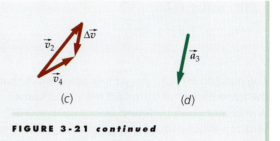

(c)   (d)

FIGURE 3-21 *continued*

REMARKS The process of finding the direction of the acceleration using a motion diagram is not precise. Therefore the result is an estimate of the direction of the acceleration, as opposed to a precise determination.

EXERCISE Use a motion diagram to find the direction of the acceleration of the cap in Example 3-7 during the descending portion of its flight. (*Answer*   directly downward)

3-4 Special Case 1: Projectile Motion

Figure 3-22 shows a particle launched with initial speed v_0 at angle θ_0 with the horizontal axis. Let the launch point be at (x_0, y_0); y is positive upward and x is positive to the right.

The initial velocity then has components

$$v_{0x} = v_0 \cos\theta_0 \qquad\qquad 3\text{-}18a$$

$$v_{0y} = v_0 \sin\theta_0 \qquad\qquad 3\text{-}18b$$

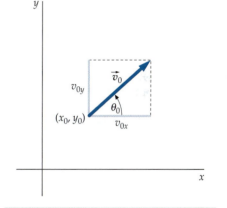

In the absence of air resistance, the acceleration is free-fall acceleration, vertically downward:

FIGURE 3-22

$$a_x = 0 \qquad\qquad 3\text{-}19a$$

and

$$a_y = -g \qquad\qquad 3\text{-}19b$$

Since the acceleration is constant, we can use the kinematics equations discussed in Chapter 2. The x component of the velocity is constant because there is no horizontal acceleration:

$$v_x = v_{0x} \qquad\qquad 3\text{-}20a$$

The y component of the velocity varies with time according to Equation 2-12, with $a = -g$:

$$v_y = v_{0y} - gt \qquad\qquad 3\text{-}20b$$

Notice that v_x does not depend on v_y and vice versa: *The horizontal and vertical components of projectile motion are independent.* (Dropping a ball from a desktop and projecting a second ball horizontally at the same time can demonstrate this. The two balls strike the floor simultaneously.) The displacements x and y are given by (see Equation 2-16)

$$x(t) = x_0 + v_{0x}t \qquad\qquad\qquad 3\text{-}21a$$

$$y(t) = y_0 + v_{0y}t - \tfrac{1}{2}gt^2 \qquad\qquad 3\text{-}21b$$

EQUATIONS OF MOTION FOR A PROJECTILE

The notation $x(t)$ and $y(t)$ simply emphasizes that x and y are functions of time. If the y component of the initial velocity is known, the time t for which the particle is at height y can be found from Equation 3-21b. The horizontal position at that time can then be found using Equation 3-21a. The total horizontal distance a projectile travels is called its **range.** (The vector forms of equations 3-19 to 3-21 are given on page 71.)

A CAP IN THE AIR

EXAMPLE 3-8

A delighted physics graduate throws her cap into the air with an initial velocity of 24.5 m/s at 36.9° above the horizontal. It is later caught by another student. Find (a) the total time the cap is in the air and (b) the total horizontal distance traveled.

PICTURE THE PROBLEM We choose the origin to be the initial position of the cap so that $x_0 = y_0 = 0$. We assume it is caught at the same height. The total time the cap is in the air is found by setting $y = 0$ in Equation 3-21b. We can then use this result in Equation 3-21a to find the total horizontal distance traveled.

(a) 1. Set $y = 0$ in Equation 3-21b and solve for t:

$$y = v_{0y}t - \tfrac{1}{2}gt^2 = t(v_{0y} - \tfrac{1}{2}gt) = 0$$

2. There are two solutions for t:

$$t = 0 \quad \text{(initial conditions)}$$

$$t = \frac{2v_{0y}}{g}$$

3. Compute the vertical component of the initial velocity vector:

$$v_{0y} = v_0 \sin\theta_0 = (24.5\ \text{m/s}) \sin 36.9° = 14.7\ \text{m/s}$$

4. Substitute for v_{0y} in the Step 2 result to find the total time t:

$$t = \frac{2v_{0y}}{g} = \frac{2v_0 \sin\theta_0}{g}$$

$$= \frac{2(24.5\ \text{m/s})\sin 36.9°}{9.81\ \text{m/s}^2} = \boxed{3.00\ \text{s}}$$

(b) Use this value for the time to calculate the total horizontal distance traveled:

$$x = v_{0x}t = (v_0 \cos\theta_0)t$$

$$= (24.5\ \text{m/s}) \cos 36.9°(3.00\ \text{s}) = \boxed{58.8\ \text{m}}$$

REMARKS The vertical component of the initial velocity of the cap is 14.7 m/s, the same as that of the cap in Example 2-9 (Chapter 2), where the cap was thrown straight up with $v_0 = 14.7$ m/s. The time the cap is in the air is also the same as in Example 2-9. Figure 3-23 shows the height y versus t for the cap. This curve is identical to Figure 2-11 (Example 2-9) because the caps each have the same vertical acceleration and vertical velocity. Figure 3-23 can be reinterpreted as a graph of y versus x if its time scale is converted to a distance scale, as shown in the figure. This can be done by multiplying the time values by 19.6 m/s, because the cap moves at $(24.5\ \text{m/s}) \cos 36.9° = 19.6$ m/s horizontally. The curve y versus x is a parabola.

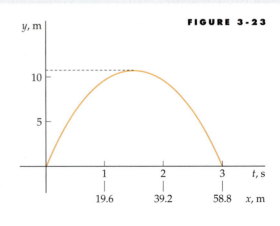

FIGURE 3-23

Figure 3-24 shows graphs of the vertical heights versus the horizontal distances for projectiles with an initial speed of 24.5 m/s and several different initial angles. The angles drawn are 45°, which has the maximum range, and pairs of angles at equal amounts above and below 45°. Notice that the paired angles have the same range. The green curve has an initial angle of 36.9° (0.64 rad), as in this example.

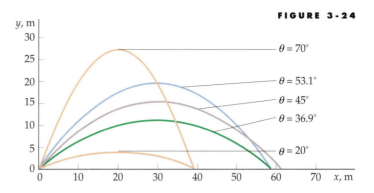

FIGURE 3-24

The general equation for the path $y(x)$ of a projectile can be obtained from Equations 3-21 by eliminating the variable t. Choosing $x_0 = 0$ and $y_0 = 0$, we obtain $t = x/v_{0x}$ from Equation 3-21a. Substituting this into Equation 3-21b gives

$$y(x) = v_{0y}\left(\frac{x}{v_{0x}}\right) - \frac{1}{2}g\left(\frac{x}{v_{0x}}\right)^2 = \left(\frac{v_{0y}}{v_{0x}}\right)x - \left(\frac{g}{2v_{0x}^2}\right)x^2$$

Writing out the velocity components yields

$$y(x) = (\tan \theta_0)x - \left(\frac{g}{2v_0^2 \cos^2 \theta_0}\right)x^2 \qquad\qquad 3\text{-}22$$

PATH OF A PROJECTILE

for the projectile's path. This is of the form $y = ax + bx^2$, the equation for a parabola passing through the origin. Figure 3-25 shows the path of a projectile with its velocity vector and components at several points. The path is for a projectile that impacts the ground at P. The horizontal distance between launch and impact is the range R.

If the initial and final elevations are equal, the range of a projectile can be written in terms of its initial speed and the angle of projection. As in the preceding examples, we find the range by multiplying the x component of the velocity by the total time that the projectile is in the air. The total flight time T is obtained by setting $y = 0$ in Equation 3-21b.

$$y(T) = v_{0y}T - \tfrac{1}{2}gT^2 = 0, \qquad T > 0$$

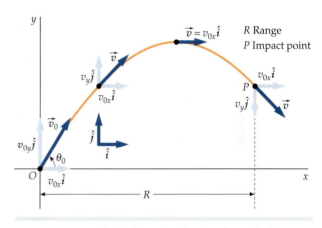

FIGURE 3-25 Path of a projectile showing velocity vectors.

Dividing through by T gives

$$v_{0y} - \tfrac{1}{2}gT = 0$$

The flight time of the projectile is thus

$$T = \frac{2v_{0y}}{g} = \frac{2v_0}{g}\sin \theta_0$$

and the range is

$$R = v_{0x}T = (v_0 \cos \theta_0)\left(\frac{2v_0}{g}\sin \theta_0\right) = \frac{2v_0^2}{g}\sin \theta_0 \cos \theta_0$$

This can be further simplified by using the following trigonometric identity:

$$\sin 2\theta = 2 \sin \theta \cos \theta$$

Thus,

$$R = \frac{v_0^2}{g} \sin 2\theta_0 \qquad\qquad 3\text{-}23$$

RANGE OF A PROJECTILE FOR EQUAL INITIAL AND FINAL ELEVATIONS

EXERCISE Use Equation 3-22 for the path to derive Equation 3-23. (*Answer* Set $y(x) = 0$ and solve for x)

Equation 3-23 is useful if you want to find the range for many projectiles with equal initial and final elevations. More importantly, this equation shows how the range depends on θ. Since the maximum value of $\sin 2\theta$ is 1 when $2\theta = 90°$ or $\theta = 45°$, the range is greatest when $\theta = 45°$. In many practical applications, the initial and final elevations may not be equal, and other considerations are important. For example, in the shot put, the ball ends its flight when it hits the ground, but it is projected from an initial height of about 2 m above the ground. This causes the range to be maximum at an angle somewhat lower than 45°, as shown in Figure 3-26. Studies of the best shot-putters show that maximum range occurs at an initial angle of about 42°.

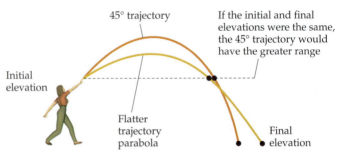

FIGURE 3-26 If a projectile lands at an elevation lower than the elevation of projection, the maximum range is achieved when the projection angle is somewhat lower than 45°.

A Supply Drop   **EXAMPLE 3 - 9**

A helicopter drops a supply package to flood victims on a raft on a swollen lake. When the package is released, the helicopter is 100 m above the raft and flying at 25 m/s at an angle $\theta_0 = 36.9°$ above the horizontal. (*a*) How long is the package in the air? (*b*) How far from the raft does the package land? (*c*) If the helicopter flies at constant velocity, where is the helicopter when the package lands?

FIGURE 3-27

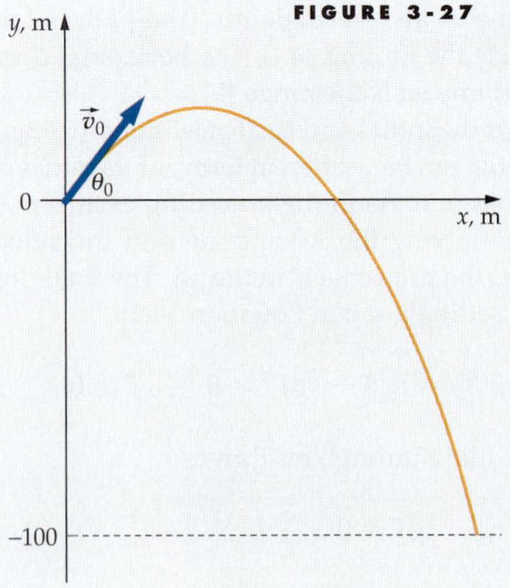

PICTURE THE PROBLEM The time in the air depends only on the vertical aspects of the motion. Using Equation 3-21b, solve for the time. Choose the origin to be at the location of the package when it is released. The initial velocity of the package is the initial velocity of the helicopter. The horizontal distance traveled by the package is given by Equation 3-21a, where t is the time the package is in the air.

(*a*) 1. Sketch the trajectory of the package during the time it is in the air. Include coordinate axes as shown in Figure 3-27.

2. To find the time of flight, write $y(t)$:

$$y(t) = y_0 + v_{0y}t - \tfrac{1}{2}gt^2$$

3. Set $y_0 = 0$ and apply the quadratic formula to solve for t:

$$y(t) = v_{0y}t - \tfrac{1}{2}gt^2 \quad \text{or} \quad 0 = \tfrac{1}{2}gt^2 - v_{0y}t + y(t)$$

so

$$t = \frac{v_{0y} \pm \sqrt{v_{0y}^2 - 2gy(t)}}{g}$$

4. Solve for the time when $y(t) = -100$ m. First solve for v_{0y}:

$$v_{0y} = v_0 \sin \theta_0 = (25 \text{ m/s})\sin 36.9° = 15 \text{ m/s}$$

so

$$t = \frac{15 \text{ m/s} \pm \sqrt{(15 \text{ m/s})^2 - 2(9.81 \text{ m/s}^2)(-100 \text{ m})}}{9.81 \text{ m/s}^2}$$

so

Because the package is released at $t = 0$, the time of impact cannot be negative. Hence:

$t = -3.24 \text{ s}$ or $t = 6.30 \text{ s}$

$t = \boxed{6.30 \text{ s}}$

(b) At impact the package has traveled a horizontal distance x, where x is the horizontal velocity times the time of flight. First solve for the horizontal velocity:

$v_{0x} = v_0 \cos \theta_0 = (25 \text{ m/s}) \cos 36.9° = 20 \text{ m/s}$

so

$x = v_{0x}t = (20 \text{ m/s})(6.3 \text{ s}) = \boxed{126 \text{ m}}$

(c) The coordinates of the helicopter at the time of impact are:

$x_h = v_{0x}t = (20 \text{ m/s})(6.3 \text{ s}) = \boxed{126 \text{ m}}$

$y_h = y_{h0} + v_{h0}t = 0 + (15 \text{ m/s})(6.30 \text{ s}) = \boxed{94.5 \text{ m}}$

The height of the helicopter is 194.5 m directly above the package.

REMARKS The positive time is appropriate because it corresponds to a time after the package is dropped (which occurs at $t = 0$). The negative time is when the package would have been at $y = -100$ m if its motion had started earlier as shown in Figure 3-28. Note that the helicopter is directly above the package when the package hits the water (and at all other times before then). Figure 3-29 shows a graph of y versus x for various initial angles and an initial speed of 25 m/s. The curve with an initial angle of 36.9° is the one given in this example. Note that the maximum range occurs at an angle less than 45°.

FIGURE 3-29

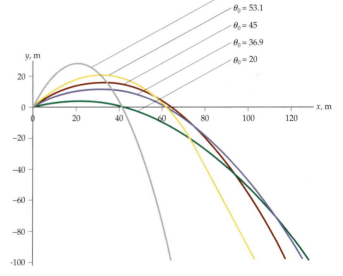

FIGURE 3-28

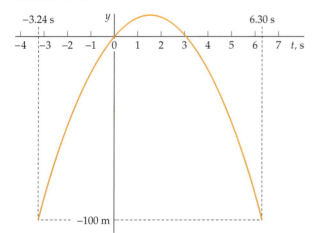

To Catch a Thief

EXAMPLE 3-10 **Try It Yourself**

FIGURE 3-30

A police officer chases a master jewel thief across city rooftops. They are both running when they come to a gap between buildings that is 4 m wide and has a drop of 3 m (Figure 3-30). The thief, having studied a little physics, leaps at 5 m/s and at 45° and clears the gap easily. The police officer did not study physics and thinks he should maximize his horizontal velocity, so he leaps at 5 m/s horizontally. (a) Does he clear the gap? (b) By how much does the thief clear the gap?

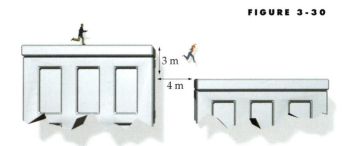

PICTURE THE PROBLEM Assuming they both clear the gap, the total time in the air depends only on the vertical aspects of the motion. Choose the origin at the launch point, with upward positive so that Equations 3-21 apply. Use Equation 3-21b for $y(t)$ and solve for the time when $y = -3$ m for $\theta_0 = 0°$ and again for $\theta_0 = 45°$. The horizontal distances traveled are the values of x at these times.

Cover the column to the right and try these on your own before looking at the answers.

Steps	Answers
(a) 1. Write $y(t)$ for the police officer and solve for t when $y = -3$ m.	$t = 0.782$ s
2. Substitute this into the equation for $x(t)$ and find the horizontal distance traveled during this time.	$x = \boxed{3.91 \text{ m}}$
	Since this is less than 4 m, it appears the police officer fails to make it across the gap between buildings.
(b) 1. Write $y(t)$ for the thief and solve for t when $y = -3$ m. $y(t)$ is a quadratic equation with two solutions, but only one of its solutions is acceptable.	$t = -0.5$ s or $t = 1.22$ s She must land after she leaps, so $t = 1.22$ s
2. Find the horizontal distance covered for the positive value of t.	$x = v_{0x}t = 4.31$ m
3. Subtract 4.0 m from this distance.	$4.31 \text{ m} - 4.0 \text{ m} = \boxed{0.31 \text{ m}}$

REMARKS The thief probably knew that she should jump at slightly less than 45°, but she didn't have time to solve the problem exactly. The police officer actually did make it across by tightening his abdominal muscles before impact. This raised his feet more than the 9 cm needed for him to complete the leap.

DROPPING SUPPLIES   **EXAMPLE 3-11**   **Try It Yourself**

In Example 3-9, find (a) the time t_1 for the package to reach its greatest height h above the water, (b) its greatest height h, and (c) the time t_2 for the package to fall to the water from its greatest height.

PICTURE THE PROBLEM The time t_1 is the time at which the vertical component of the velocity is zero. Using Equation 3-20b solve for t_1.

Cover the column to the right and try these on your own before looking at the answers.

Steps	Answers
(a) 1. Write $v_y(t)$ for the package.	$v_y(t) = v_{0y} - gt$
2. Set $v_y(t_1) = 0$ and solve for t_1.	$t_1 = \boxed{1.53 \text{ s}}$
(b) 1. Find $v_{y\,av}$ during the time the package is moving up.	$v_{y\,av} = 7.5$ m/s
2. Use $v_{y\,av}$ to find the distance traveled up. Then find h.	$\Delta y = 11.5$ m, $h = \boxed{111.5 \text{ m}}$
(c) Find the time for the package to fall a distance h.	$t_1 = \boxed{4.77 \text{ s}}$

REMARKS Note that $t_1 + t_2 = 6.3$ s, in agreement with Example 3-9.

EXERCISE Solve Part (b) of Example 3-11 using $y(t)$ (Equation 3-21b) instead of finding $v_{y\,av}$.

Equations 3-19a and b can be expressed in vector form. Multiplying through Equation 3-19a by $\hat{\imath}$, multiplying through Equation 3-19b by $\hat{\jmath}$, and then adding the two equations gives $a_x\hat{\imath} + a_y\hat{\jmath} = -g\hat{\jmath}$, or

$$\vec{a} = \vec{g} \qquad\qquad 3\text{-}19c$$

where $\vec{g}$ is the free-fall acceleration vector. At the surface of the earth the magnitude of $\vec{g}$ is $g = 9.81$ m/s^2. Equations 3-20a and b can also be expressed in vector form. Multiplying through Equation 3-20a by $\hat{\imath}$, multiplying through Equation 3-20b by $\hat{\jmath}$, and then adding the two equations gives $(v_x\hat{\imath} + v_y\hat{\jmath}) = (v_{0x}\hat{\imath} + v_{0y}\hat{\jmath}) - gt\hat{\jmath}$ or

$$\vec{v} = \vec{v}_0 + \vec{g}t \qquad \text{or} \qquad \Delta\vec{v} = \vec{g}t \qquad\qquad 3\text{-}20c$$

where $\vec{v} = v_x\hat{\imath} + v_y\hat{\jmath}$, $\vec{v}_0 = v_{0x}\hat{\imath} + v_{0y}\hat{\jmath}$, and $\vec{g} = -g\hat{\jmath}$. Repeating the process for Equations 3-21a and b we obtain

$$\vec{r} = \vec{r}_0 + \vec{v}_0t + \tfrac{1}{2}\vec{g}t^2 \qquad \text{or} \qquad \Delta\vec{r} = \vec{v}_0t + \tfrac{1}{2}\vec{g}t^2 \qquad\qquad 3\text{-}21c$$

where $\vec{r} = x\hat{\imath} + y\hat{\jmath}$ and $\vec{r}_0 = x_0\hat{\imath} + y_0\hat{\jmath}$. For a number of problems the vector forms of the kinematic equations (Equations 3-20c and 3-21c) are more suitable. This is the case in the following example.

THE RANGER AND THE MONKEY   **E X A M P L E   3 - 1 2**

A park ranger with a tranquilizer dart gun intends to shoot a monkey hanging from a branch. The ranger aims directly at the monkey, not realizing that the dart will follow a parabolic path that will pass below the present position of the creature. The monkey, seeing the gun discharge, lets go of the branch and drops out of the tree, expecting to avoid the dart. (a) Show that the monkey will be hit regardless of the initial speed of the dart so long as it is great enough for the dart to travel the horizontal distance to the tree. Assume the reaction time of the monkey is negligible. (b) Let $\vec{v}_{d0}$ be the velocity of the dart as it leaves the gun. Find the velocity of the dart relative to the monkey at an arbitrary time t during the dart's flight.

PICTURE THE PROBLEM We apply Equation 3-21c to both the monkey and the gun.

(a) 1. Apply Equation 3-21c to the monkey at time t:

$$\Delta\vec{r}_m = \tfrac{1}{2}\vec{g}t^2$$

where the initial velocity of the monkey is zero.

2. Apply Equation 3-21c to the dart at time t:

$$\Delta\vec{r}_d = \vec{v}_{d0}t + \tfrac{1}{2}\vec{g}t^2$$

where $\vec{v}_{d0}$ is the velocity of the dart as it leaves the gun.

3. Make a sketch of the monkey, the dart, and the gun, as shown in Figure 3-31. Show the dart and the monkey at their initial locations and at their locations a time t later. On the figure draw a vector representing each term in the step 1 and step 2 results. Note that at time t the dart and the monkey both are a distance $\tfrac{1}{2}gt^2$ below the line of sight of the gun. The dart will strike the monkey when it reaches the monkey's line of fall:

FIGURE 3-31

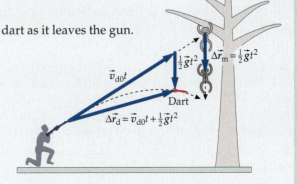

(b) 1. The velocity of the dart relative to the monkey equals the velocity of the dart relative to the gun plus the velocity of the gun relative to the monkey:

$$\vec{v}_{dm} = \vec{v}_{dg} + \vec{v}_{gm}$$

2. The velocity of the gun relative to the monkey is the negative of the velocity of the monkey relative to the gun:

$$\vec{v}_{dm} = \vec{v}_{dg} - \vec{v}_{mg}$$

3. Using Equation 3-20c, express the velocity of the dart relative to the gun and the velocity of the monkey relative to the gun:

$$\vec{v}_{dg} = \vec{v}_{d0} + \vec{g}t$$
$$\vec{v}_{mg} = \vec{g}t$$

4. Substitute these expressions into the Part (b) step 2 result:

$$\vec{v}_{dm} = (\vec{v}_{d0} + \vec{g}t) - (\vec{g}t) = \boxed{\vec{v}_{d0}}$$

REMARKS Relative to the monkey, the dart moves with constant speed v_{d0} in a straight line. The dart strikes the monkey at time $t = L/v_{d0}$, where L is the distance from the muzzle of the gun to the initial position of the monkey.

REMARKS In a familiar lecture demonstration, a target is suspended by an electromagnet. When the dart leaves the gun, the circuit to the magnet is broken and the target falls. The initial velocity of the dart is varied so that for large v_{d0} the target is hit very near its original height and for small v_{d0} it is hit just before it reaches the floor.

EXERCISE   A hockey puck at ice level is struck such that it misses the net and clears the top of the Plexiglass wall of height $h = 2.80$ m. The flight time at the moment the puck clears the wall is $t_1 = 0.650$ s, and the horizontal distance is $x_1 = 12.0$ m. (a) Find the initial speed and direction of the puck. (b) When does the puck reach its maximum height? (c) What is the maximum height of the puck? (*Answer*   (a) $\vec{v} = 20.0$ m/s, $\theta_0 = 22.0°$, (b) $t = 0.764$ s, (c) $v_{y\,av}\, t = 2.86$ m)

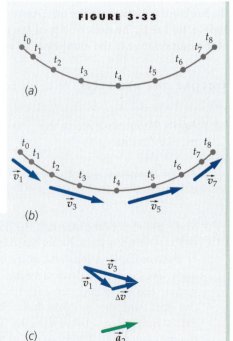

FIGURE 3-32

3-5 Special Case 2: Circular Motion

Figure 3-32 shows a pendulum bob swinging back and forth in a vertical plane. The path of the bob is a segment of a circular path. Motion along a circular path, or a segment of a circular path, is called **circular motion**.

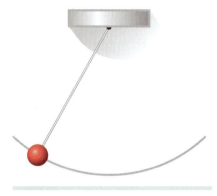

FIGURE 3-33

(a)

(b)

(c)

A SWINGING PENDULUM   **EXAMPLE 3-13**

Consider the motion of the pendulum bob shown in Figure 3-32. Using a motion diagram (Figure 3-33), find the direction of the acceleration vector when the bob is swinging from left to right and (a) on the descending portion of the path, (b) passing through the lowest point on the path, and (c) on the ascending portion of the path.

PICTURE THE PROBLEM As the bob descends it both gains speed and changes direction. The acceleration is related to the change in velocity by $\vec{a} \approx \Delta\vec{v}/\Delta t$. The direction of the acceleration at a point can be estimated by constructing a vector addition diagram for the relation $\vec{v}_i + \Delta\vec{v} = \vec{v}_f$ to find the direction of $\Delta\vec{v}$, and thus the direction of the acceleration vector.

(a) 1. Make a motion diagram for a full left-to-right swing of the bob. The spacing between dots is greatest at the lowest point where the speed is greatest.

2. Pick a dot on the descending portion of the motion diagram and draw a velocity vector on the diagram for both the preceding and the following dot. The velocity vectors should be drawn tangent to the path and with lengths proportional to the speed.

3. Draw the graphical expression of the relation $\vec{v}_i + \Delta\vec{v} = \vec{v}_f$. On this figure draw the acceleration vector. Since $\vec{a} \approx \Delta\vec{v}/\Delta t$, $\vec{a}$ is in the same direction as $\Delta\vec{v}$.

(b) Repeat steps 2 and 3 for the lowest point on the path.

(c) Repeat steps 2 and 3 for a point on the ascending portion of the path.

FIGURE 3-33 *continued*

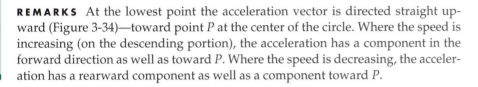

(d)   (e)

REMARKS At the lowest point the acceleration vector is directed straight upward (Figure 3-34)—toward point P at the center of the circle. Where the speed is increasing (on the descending portion), the acceleration has a component in the forward direction as well as toward P. Where the speed is decreasing, the acceleration has a rearward component as well as a component toward P.

If a particle moves along a circular arc, the direction from the particle toward the center of the circle is called the **centripetal direction.** In Example 3-13 the acceleration at the lowest point of the pendulum bob's path was found to be in the centripetal direction. At other points the acceleration was found to have a tangential component as well as a centripetal component.

FIGURE 3-34

Uniform Circular Motion

Motion in a circle at constant speed is called **uniform circular motion.** To find an expression for the acceleration of a particle moving in a circle at constant speed we will extend the method used in Example 3-13 to find the direction of the acceleration. The position and velocity vectors for a particle moving in a circle at constant speed are shown in Figure 3-35. The angle $\Delta\theta$ between $\vec{v}(t)$ and $\vec{v}(t + \Delta t)$ is the same as between $\vec{r}(t)$ and $\vec{r}(t + \Delta t)$ because the position and velocity vectors must both rotate through equal angles to remain mutually perpendicular. An isosceles triangle is formed by the two velocity vectors and $\Delta\vec{v}$, and a second isosceles triangle is formed by the two position vectors and $\Delta\vec{r}$. To find the direction of the acceleration vector we examine the triangle formed by the two velocity vectors and $\Delta\vec{v}$. The sum of the angles of any triangle is 180° and the base angles of any isosceles triangle are equal. In the limit that Δt approaches zero, $\Delta\theta$ also approaches zero, so in this limit the two base angles must each approach 90°. This means $\Delta\vec{v}$ is perpendicular to the velocity. If $\Delta\vec{v}$ is drawn from the position of the particle then it points in the centripetal direction.

The two triangles are similar, so $|\Delta\vec{v}|/v = |\Delta\vec{r}|/r$ (corresponding lengths of similar shapes are proportional). Dividing both sides by Δt and rearranging gives

$$\frac{|\Delta\vec{v}|}{\Delta t} = \frac{v|\Delta\vec{r}|}{r\ \Delta t}$$

In the limit as Δt approaches zero, the term $|\Delta\vec{v}|/\Delta t$ approaches a, the magnitude of the instantaneous acceleration, and the term $|\Delta\vec{r}|/\Delta t$ approaches v, the magnitude of the instantaneous velocity (the speed). With these substitutions we have

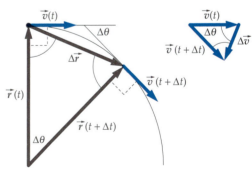

FIGURE 3-35 Position and velocity vectors for a particle moving in a circle at constant speed.

$$a = a_c = \frac{v^2}{r} \qquad\qquad 3\text{-}24$$

CENTRIPETAL ACCELERATION

The motion of a particle moving in a circle with constant speed is often described in terms of the time T required for one complete revolution, called the **period.** During one period, the particle travels a distance of $2\pi r$ (where r is the radius of the circle), so its speed is related to r and T by

$$v = \frac{2\pi r}{T} \qquad\qquad 3\text{-}25$$

A SATELLITE'S MOTION   **EXAMPLE 3-14**

A satellite moves at constant speed in a circular orbit about the center of the earth and near the surface of the earth. If its acceleration is $g = 9.81\ \text{m/s}^2$, find (a) its speed and (b) the time for one complete revolution.

PICTURE THE PROBLEM Since the satellite orbits near the surface of the earth, we take the radius of the orbit to be the radius of the earth, $r = 6370$ km.

(a) Set the centripetal acceleration v^2/r equal to g and solve for the speed v:

$$a = \frac{v^2}{r} = g \quad \text{or}$$

$$v = \sqrt{rg} = \sqrt{(6370\ \text{km})(9.81\ \text{m/s}^2)}$$

$$= \boxed{7.91\ \text{km/s} = 17{,}700\ \text{mi/h}}$$

(b) Use Equation 3-25 to get the period T:

$$T = \frac{2\pi r}{v} = \frac{2\pi(6370\ \text{km})}{7.91\ \text{km/s}} = \boxed{5060\ \text{s} = 84.3\ \text{min}}$$

REMARKS For actual satellites in orbit a few hundred kilometers above the earth's surface, the orbital radius r is slightly greater than 6370 km. As a result, the centripetal acceleration is slightly less than $9.81\ \text{m/s}^2$ because of the decrease in the gravitational force with distance from the center of the earth. Many satellites are launched into such orbits, and their periods are roughly 90 min.

EXERCISE A car rounds a curve of radius 40 m at 48 km/h. What is its centripetal acceleration? (*Answer* $4.44\ \text{m/s}^2$)

A particle moving in a circle with *varying speed* has a component of acceleration tangent to the circle, dv/dt, as well as the radially inward centripetal acceleration, v^2/r. For general motion along a curve, we can treat a portion of the curve as an arc of a circle (Figure 3-36). The particle then has acceleration v^2/r toward the center of curvature, and if the speed v is changing, it has tangential acceleration

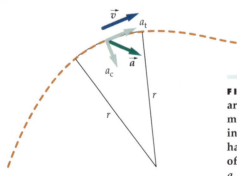

FIGURE 3-36 A particle moving along an arbitrary curve can be considered to be moving in a circular arc during a small time interval. Its instantaneous acceleration vector has a component $a_c = v^2/r$ toward the center of curvature of the arc and a component $a_t = dv/dt$ that is tangential to the curve.

$$a_t = \frac{dv}{dt} \qquad\qquad 3\text{-}26$$

TANGENTIAL ACCELERATION

Topic	Relevant Equations and Remarks
1. Vectors	
Definition	Vectors are quantities that have both magnitude and direction. Vectors add like consecutive displacements.
Components	The component of a vector in a direction in space is its projection on an axis in that direction. If $\vec{A}$ makes an angle θ with the positive x direction, its x and y components are
	$A_x = A \cos \theta$ **3-2**
	$A_y = A \sin \theta$ **3-3**
Magnitude	$A = \sqrt{A_x^2 + A_y^2}$ **3-5a**
Adding vectors graphically	Any two vectors whose magnitudes have the same units may be added graphically by placing the tail of one arrow at the head of the other.
Adding vectors using components	If $\vec{C} = \vec{A} + \vec{B}$, then
	$C_x = A_x + B_x$ **3-6a**
	and
	$C_y = A_y + B_y$ **3-6b**
Unit vectors	A vector $\vec{A}$ can be written in terms of unit vectors $\hat{i}$, $\hat{j}$, and $\hat{k}$, which have unit magnitude and lie along the x, y, and z axes, respectively
	$\vec{A} = A_x\hat{i} + A_y\hat{j} + A_z\hat{k}$ **3-7**
Position vector	The position vector $\vec{r}$ points from the origin of the coordinate system to the particle's position.
Instantaneous-velocity vector	The velocity vector $\vec{v}$ is the rate of change of the position vector. Its magnitude is the speed and it points in the direction of motion.
	$\vec{v} = \lim\limits_{\Delta t \to 0} \dfrac{\Delta \vec{r}}{\Delta t} = \dfrac{d\vec{r}}{dt}$ **3-12**
Instantaneous-acceleration vector	$\vec{a} = \lim\limits_{\Delta t \to 0} \dfrac{\Delta \vec{v}}{\Delta t} = \dfrac{d\vec{v}}{dt}$ **3-16**
2. Relative Velocity	If a particle moves with velocity $\vec{v}_{pA}$ relative to reference frame A, which is in turn moving with velocity $\vec{v}_{AB}$ relative to reference frame B, the velocity of the particle relative to B is
	$\vec{v}_{pB} = \vec{v}_{pA} + \vec{v}_{AB}$ **3-14**
3. Projectile Motion	The positive x direction is horizontal and the positive y direction is upward for the equations in this section.
Independence of motion	In projectile motion, the horizontal and vertical motions are independent. Thus,
	$a_x = 0$ and $a_y = -g$.

Dependence on time	$v_x(t) = v_{0x} + a_x t, \qquad v_y(t) = v_{0y} + a_y t$	2-14
	$x(t) = x_0 + v_{0x}t + \frac{1}{2}a_x t^2, \qquad y(t) = y_0 + v_{0y}t + \frac{1}{2}a_y t^2$	2-16
	where $a_x = 0$, $a_y = -g$, $v_{0x} = v_0 \cos\theta_0$, and $v_{0y} = v_0 \sin\theta_0$. Alternatively,	
	$\Delta\vec{v} = \vec{g}t, \qquad \Delta\vec{r} = \vec{v}_0 t + \frac{1}{2}\vec{g}\,t^2$	3-20c, 3-21c
	where $\vec{g} = -g\hat{\jmath}$	
Range	The range is found by multiplying v_x by the total time the projectile is in the air.	

3. Circular Motion

Centripetal acceleration	$a_c = \dfrac{v^2}{r}$	3-24
Tangential acceleration	$a_t = \dfrac{dv}{dt}$	
	where v is the speed.	3-26
Period	$v = \dfrac{2\pi r}{T}$	3-25

PROBLEMS

- • Single-concept, single-step, relatively easy
- •• Intermediate-level, may require synthesis of concepts
- ••• Challenging
- SSM Solution is in the *Student Solutions Manual*
- iSOLVE Problems available on iSOLVE online homework service
- iSOLVE ✔ These "Checkpoint" online homework service problems ask students additional questions about their confidence level, and how they arrived at their answer

In a few problems, you are given more data than you actually need; in a few other problems, you are required to supply data from your general knowledge, outside sources, or informed estimates.

Conceptual Problems

1 • SSM Can the magnitude of the displacement of a particle be less than the distance traveled by the particle along its path? Can its magnitude be more than the distance traveled? Explain.

2 • Give an example in which the distance traveled is a significant amount, yet the corresponding displacement is zero.

3 • What is the approximate average velocity of the race cars during the Indianapolis 500?

4 • iSOLVE True or false: The magnitude of the sum of two vectors *must be* greater than the magnitude of either vector.

5 • Can a component of a vector have a magnitude greater than the magnitude of the vector? Under what circumstances can a component of a vector have a magnitude equal to the magnitude of the vector?

6 • SSM Can a vector be equal to zero and still have one or more components not equal to zero?

7 • iSOLVE Are the components of $\vec{C}$, where $\vec{C} = \vec{A} + \vec{B}$, necessarily larger than the corresponding components of either $\vec{A}$ or $\vec{B}$?

8 • SSM True or false: The instantaneous-acceleration vector is *always* in the direction of motion.

9 • If an object is moving toward the west at some instant, in what direction is its acceleration? (a) North (b) East (c) West (d) South (e) May be any direction.

10 • iSOLVE A golfer drives the ball from the tee down the fairway in a high arcing shot. When the ball is at the highest point of its flight, (a) its velocity and acceleration are both zero, (b) its velocity is zero but its acceleration is nonzero, (c) its velocity is nonzero but its acceleration is zero, (d) its velocity and acceleration are both nonzero, (e) insufficient information is given to answer correctly.

11 • The velocity of a particle is in the eastward direction while the acceleration is directed toward the northwest as shown in Figure 3-37. The particle is (a) speeding up and turning toward the north, (b) speeding up and turning toward the south, (c) slowing down and turning toward the north, (d) slowing down and turning toward the south, (e) maintaining constant speed and turning toward the south.

FIGURE 3-37
Problem 11

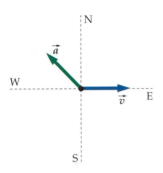

12 • **SSM** Assuming constant acceleration, if you know the position vectors of a particle at two points on its path and also know the time it took to move from one point to the other, you can then compute (a) the particle's average velocity, (b) the particle's average acceleration, (c) the particle's instantaneous velocity, (d) the particle's instantaneous acceleration, (e) insufficient information is given to describe the particle's motion.

13 •• Consider the path of a particle as it moves in space. (a) How is the velocity vector related geometrically to the path of the particle? (b) Sketch a curved path and draw the velocity vector for the particle for several positions along the path.

14 • The acceleration of a car is zero when it is (a) turning right at a constant speed, (b) driving up a long straight incline at constant speed, (c) topping the crest of a hill at constant speed, (d) bottoming out at the lowest point of a valley at constant speed, (e) speeding up as it descends a long straight decline.

15 • **SSM** Give examples of motion in which the directions of the velocity and acceleration vectors are (a) opposite, (b) the same, and (c) mutually perpendicular.

16 • **iSOLVE** How is it possible for a particle moving at constant speed to be accelerating? Can a particle with constant velocity be accelerating at the same time?

17 •• Imagine throwing a dart straight upward so that it sticks into the ceiling. After it leaves your hand, it steadily slows down as it rises before it sticks. (a) Draw the dart's velocity vector at times t_1 and t_2, where $\Delta t = t_2 - t_1$ is small. From your drawing find the direction of the change in velocity $\Delta \vec{v} = \vec{v}_2 - \vec{v}_1$, and thus the direction of the acceleration vector. (b) After it has stuck in the ceiling for a few seconds, the dart falls down to the floor. As it falls it speeds up, of course, until it hits the floor. Repeat part (a) to find the direction of its acceleration vector as it falls. (c) Now imagine tossing the dart horizontally. What is the direction of its acceleration vector after it leaves your hand, but before it strikes the floor?

18 •• **SSM** As a bungee jumper approaches the lowest point in her descent, the rubber band holding her stretches and she loses speed as she continues to move downward. Assuming that she is dropping straight down, make a motion diagram to find the direction of her acceleration vector as she slows down by drawing her velocity vectors at times t_1 and t_2, where $\Delta t = t_2 - t_1$ is small. From your drawing find the direction of the change in velocity $\Delta \vec{v} = \vec{v}_2 - \vec{v}_1$, and thus the direction of the acceleration vector.

19 •• After reaching the lowest point in her jump at time t_{low}, the bungee jumper in Problem 18 moves upward, gaining speed for a short time until gravity again dominates her motion. Draw her velocity vectors at times t_1 and t_2, where $\Delta t = t_2 - t_1$ is small and $t_1 < t_{low} < t_2$. From your drawing find the direction of the change in velocity $\Delta \vec{v} = \vec{v}_2 - \vec{v}_1$, and thus the direction of the acceleration vector.

20 • A river is 0.76 km wide. The banks are straight and parallel (Figure 3-38). The current is 4.0 km/h and is parallel to the banks. A boat has a maximum speed of 4 km/h in still water. The pilot of the boat wishes to go on a straight line from A to B, where AB is perpendicular to the banks. The pilot should (a) head directly across the river, (b) head 53° upstream from the line AB, (c) head 37° upstream from the line AB, (d) give up—the trip from A to B is not possible with a boat of this limited speed, (e) do none of the above.

FIGURE 3-38
Problem 20

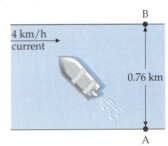

21 • **SSM** True or false: When a projectile is fired horizontally, it takes the same amount of time to reach the ground as an identical projectile dropped from rest from the same height. Ignore the effects of air resistance.

22 • **iSOLVE** A projectile is fired at 35° above the horizontal. At the highest point in its trajectory, its speed is 200 m/s. The initial velocity had a horizontal component of (a) 0; (b) (200 m/s) cos 35°; (c) (200 m/s) sin 35°; (d) (200 m/s)/cos 35°; (e) 200 m/s. Neglect the effects of air resistance.

23 • Figure 3-39 represents the parabolic trajectory of a ball going from A to E. What is the direction of the acceleration at point B? (a) Up and to the right (b) Down and to the left (c) Straight up (d) Straight down (e) The acceleration of the ball is zero.

FIGURE 3-39
Problems 23 and 24

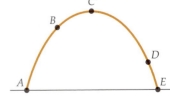

24 • Referring to the motion described in Problem 23, (a) at which point(s) is the speed the greatest? (b) At which point(s) is the speed the lowest? (c) At which two points is the speed the same? Is the velocity the same at those points?

25 • **SOLVE** True or false:

(*a*) If the speed is constant, the acceleration must be zero.
(*b*) If the acceleration is zero, the speed must be constant.

26 • The initial and final velocities of an object are as shown in Figure 3-40. Indicate the direction of the average acceleration.

FIGURE 3-40 Problem 26

27 • The velocities of objects A and B are shown in Figure 3-41. Draw a vector that represents the velocity of B relative to A.

FIGURE 3-41 Problem 27

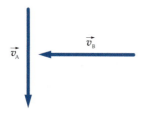

28 •• **SSM** A vector $\vec{A}(t)$ has a constant magnitude but is changing direction. (*a*) Find $d\vec{A}/dt$ in the following manner: Draw the vectors $\vec{A}(t + \Delta t)$ and $\vec{A}(t)$ for a small time interval Δt, and find the difference $\Delta \vec{A} = \vec{A}(t + \Delta t) - \vec{A}(t)$ graphically. How is the direction of $\Delta \vec{A}$ related to $\vec{A}$ for small time intervals? (*b*) Interpret this result for the special cases where $\vec{A}$ represents the position of a particle with respect to some coordinate system. (*c*) Could $\vec{A}$ represent a velocity vector? Explain.

29 •• The automobile path shown in Figure 3-42 is made up of straight lines and arcs of circles. The automobile starts from rest at point A. After it reaches point B, it travels at constant speed until it reaches point E. It comes to rest at point F. (*a*) At the middle of each segment (AB, BC, CD, DE, and EF), what is the direction of the velocity vector? (*b*) At which of these points does the automobile have an acceleration? In those cases, what is the direction of the acceleration? (*c*) How do the magnitudes of the acceleration compare for segments BC and DE?

FIGURE 3-42
Problem 29

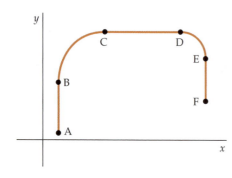

30 •• **SSM** Two cannons are pointed directly toward each other as shown in Figure 3-43. When fired, the cannonballs will follow the trajectories shown—P is the point where the trajectories cross each other. If we want the cannonballs to hit each other, should the gun crews fire cannon A first, cannon B first, or should they fire simultaneously? Ignore the effects of air resistance.

FIGURE 3-43 Problem 30

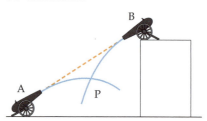

31 •• Galileo wrote the following in his *Dialogue concerning the two world systems:* "Shut yourself up . . . in the main cabin below decks on some large ship, and . . . hang up a bottle that empties drop by drop into a wide vessel beneath it. When you have observed [this] carefully . . . have the ship proceed with any speed you like, so long as the motion is uniform and not fluctuating this way and that. . . . The droplets will fall as before into the vessel beneath without dropping towards the stern, although while the drops are in the air the ship runs many spans." Explain this quotation.

32 • A man swings a stone attached to a rope in a horizontal circle at constant speed. Figure 3-44 represents the path of the rock looking down from above. (*a*) Which of the vectors $\vec{A}$ to $\vec{E}$ could represent the velocity of the stone? (*b*) Which could represent the acceleration?

FIGURE 3-44 Problem 32

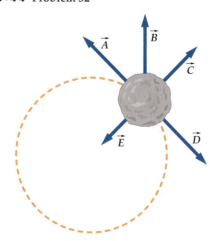

33 • True or false: An object cannot move in a circle unless it is accelerating.

34 •• Using a motion diagram, find the direction of the acceleration of the bob of a pendulum when the bob is at a point where it is just reversing its direction.

35 • The speed of a batted baseball immediately after being struck can reach 110 mph. Let's say that the baseball has a "launch angle" of 35°, which is fairly typical for the sport. Naively using the range equation (Equation 3-23) to calculate the distance the ball will travel, we find a range of 760 ft (232 m)! In reality, it will only travel about 400 ft. Can you give a reason why the range equation breaks down so badly here? Be specific: If you can, look up the *terminal speed* for a baseball.

Estimation and Approximation

36 •• **SSM** Estimate how far you can throw a ball if you throw it (*a*) horizontally while standing on level ground, (*b*) at θ = 45° while standing on level ground, (*c*) horizontally from the top of a building 12 m high, (*d*) at θ = 45° from the top of a building 12 m high.

37 •• **iSOLVE** In 1978, Geoff Capes of Great Britain threw a heavy brick a horizontal distance of 44.5 m. Find the approximate velocity of the brick at the highest point of its flight, neglecting the effects of air resistance.

Vectors, Vector Addition, and Coordinate Systems

38 • A wall clock has a minute hand that has a length of 0.5 m and an hour hand with a length of 0.25 m. Taking the center of the clock as the origin, and choosing an appropriate coordinate system, write the position of the hour and minute hands as vectors when the time reads (*a*) 12:00, (*b*) 3:30, (*c*) 6:30, (*d*) 7:15. (*e*) Call the position of the tip of the minute hand $\vec{A}$ and the position of the tip of the hour hand $\vec{B}$. Find $\vec{A} - \vec{B}$ for the times given in (*a*)–(*d*) above.

39 • **SSM** A bear walks northeast for 12 m and then east for 12 m. Show each displacement graphically and find the resultant displacement vector graphically, as in Example 3-2(*a*).

40 • **iSOLVE**✓ A circular arc is centered at x = 0, y = 0. (*a*) A student walks along the circular arc from the position x = 5 m, y = 0 to a final position x = 0, y = 5 m. What is her displacement? (*b*) A second student walks from the same initial position along the x axis to the origin and then along the y axis to y = 5 m and x = 0. What is his displacement?

41 • **SSM** **iSOLVE**✓ For the two vectors $\vec{A}$ and $\vec{B}$ of Figure 3-45, find the following graphically as in Example 3-2(*a*): (*a*) $\vec{A} + \vec{B}$, (*b*) $\vec{A} - \vec{B}$, (*c*) $2\vec{A} + \vec{B}$, (*d*) $\vec{B} - \vec{A}$, (*e*) $2\vec{B} - \vec{A}$.

FIGURE 3-45 Problem 41

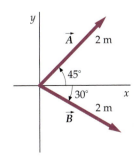

42 • A Scout walks 2.4 km due east from camp, then turns left and walks 2.4 km along the arc of a circle centered at the campsite, and finally walks 1.5 km directly toward the camp. (*a*) How far is the Scout from camp at the end of his walk? (*b*) In what direction is the Scout's position relative to the campsite? (*c*) What is the ratio of the final magnitude of the displacement to the total distance walked?

43 • A velocity vector has an x component of +5.5 m/s and a y component of −3.5 m/s. Which diagram in Figure 3-46 shows the direction of the vector correctly?

FIGURE 3-46 Problem 43

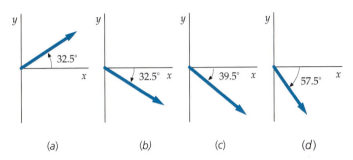

(*a*) (*b*) (*c*) (*d*)

(*e*) None of the above.

44 • **iSOLVE** Three vectors $\vec{A}$, $\vec{B}$, and $\vec{C}$ have the following x and y components: $A_x = 6$, $A_y = -3$; $B_x = -3$, $B_y = 4$; $C_x = 2$, $C_y = 5$. The magnitude of $\vec{A} + \vec{B} + \vec{C}$ is (*a*) 3.3, (*b*) 5.0, (*c*) 11, (*d*) 7.8, (*e*) 14.

45 • Find the rectangular components of the following vectors $\vec{A}$ which lie in the xy plane and make an angle θ with the x axis (Figure 3-47) if (*a*) A = 10 m, θ = 30°, (*b*) A = 5 m, θ = 45°, (*c*) A = 7 km, θ = 60°, (*d*) A = 5 km, θ = 90°, (*e*) A = 15 km/s, θ = 150°, (*f*) A = 10 m/s, θ = 240°, (*g*) A = 8 m/s², θ = 270°.

FIGURE 3-47
Problem 45

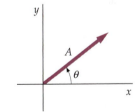

46 • **SSM** Vector $\vec{A}$ has a magnitude of 8 m at an angle of 37° with the positive x axis; vector $\vec{B} = (3 \text{ m})\hat{i} - (5 \text{ m})\hat{j}$; vector $\vec{C} = (-6 \text{ m})\hat{i} + (3 \text{ m})\hat{j}$. Find the following vectors: (*a*) $\vec{D} = \vec{A} + \vec{C}$, (*b*) $\vec{E} = \vec{B} - \vec{A}$, (*c*) $\vec{F} = \vec{A} - 2\vec{B} + 3\vec{C}$, (*d*) a vector $\vec{G}$ such that $\vec{G} - \vec{B} = \vec{A} + 2\vec{C} + 3\vec{G}$.

47 •• Find the magnitude and direction of the following vectors: (*a*) $\vec{A} = 5\hat{i} + 3\hat{j}$, (*b*) $\vec{B} = 10\hat{i} - 7\hat{j}$, (*c*) $\vec{C} = -2\hat{i} - 3\hat{j} + 4\hat{k}$.

48 • Find the magnitude and direction of $\vec{A}$, $\vec{B}$, and $\vec{C} = \vec{A} + \vec{B}$ for (*a*) $\vec{A} = -4\hat{i} - 7\hat{j}$, $\vec{B} = 3\hat{i} - 2\hat{j}$, and (*b*) $\vec{A} = 1\hat{i} - 4\hat{j}$, $\vec{B} = 2\hat{i} + 6\hat{j}$.

49 • Describe the following vectors using the unit vectors $\hat{i}$ and $\hat{j}$: (*a*) a velocity of 10 m/s at an angle of elevation of 60°, (*b*) a vector $\vec{A}$ of magnitude A = 5 m and θ = 225°, (*c*) a displacement from the origin to the point x = 14 m, y = −6 m.

50 • For the vector $\vec{A} = 3\hat{i} + 4\hat{j}$, find any three other vectors $\vec{B}$ that also lie in the xy plane and have the property that $A = B$ but $\vec{A} \neq \vec{B}$. Write these vectors in terms of their components and show them graphically.

51 •• **SSM** The faces of a cube with 3-m-long edges are parallel to the coordinate planes. The cube has one corner at the origin. A fly begins at the origin and walks along three edges until it is at the far corner. Write the displacement vector of the fly using the unit vectors $\hat{i}$, $\hat{j}$, and $\hat{k}$, and find the magnitude of this displacement.

52 • **SSM** A ship at sea receives radio signals from two transmitters A and B, which are 100 km apart, one due south of the other. The direction finder shows that transmitter A is $\theta = 30°$ south of east, while transmitter B is due east. Calculate the distance between the ship and transmitter B.

Velocity and Acceleration Vectors

53 • A stationary radar operator determines that a ship is 10 km south of him. An hour later the same ship is 20 km southeast. If the ship moved at constant speed and always in the same direction, what was its velocity during this time?

54 • A particle's position coordinates (x, y) are (2 m, 3 m) at $t = 0$; (6 m, 7 m) at $t = 2$ s; and (13 m, 14 m) at $t = 5$ s. (a) Find the average velocity v_{av} from $t = 0$ to $t = 2$ s. (b) Find v_{av} from $t = 0$ to $t = 5$ s.

55 • **SSM** **iSOLVE** ✓ A particle moving at a velocity of 4.0 m/s in the positive x direction is given an acceleration of 3.0 m/s^2 in the positive y direction for 2.0 s. The final speed of the particle is (a) −2.0 m/s, (b) 7.2 m/s, (c) 6.0 m/s, (d) 10 m/s, (e) none of the above.

56 • Initially, a particle is moving due west with a speed of 40 m/s; 5 s later it is moving due north with a speed of 30 m/s. (a) What was the change in the magnitude of the particle's velocity during this time? (b) What was the change in the direction of the velocity? (c) What are the magnitude and direction of $\Delta\vec{v}$ for this interval? (d) What are the magnitude and direction of $\vec{a}_{av}$ for this interval?

57 • At $t = 0$, a particle located at the origin has a velocity of 40 m/s at $\theta = 45°$. At $t = 3$ s, the particle is at $x = 100$ m and $y = 80$ m with a velocity of 30 m/s at $\theta = 50°$. Calculate (a) the average velocity and (b) the average acceleration of the particle during this interval.

58 •• **SSM** **iSOLVE** ✓ A particle moves in the xy plane with constant acceleration. At time zero, the particle is at $x = 4$ m, $y = 3$ m and has velocity $\vec{v} = (2 \text{ m/s})\hat{i} + (-9 \text{ m/s})\hat{j}$. The acceleration is given by $\vec{a} = (4 \text{ m/s}^2)\hat{i} + (3 \text{ m/s}^2)\hat{j}$. (a) Find the velocity at $t = 2$ s. (b) Find the position at $t = 4$ s. Give the magnitude and direction of the position vector.

59 •• **iSOLVE** A particle has a position vector given by by $\vec{r} = (30t)\hat{i} + (40t - 5t^2)\hat{j}$, where r is in meters and t is in seconds. Find the instantaneous-velocity and instantaneous-acceleration vectors as functions of time t.

60 •• A particle has a constant acceleration of $\vec{a} = (6 \text{ m/s}^2)\hat{i} + (4 \text{ m/s}^2)\hat{j}$. At time $t = 0$, the velocity is zero and the position vector is $\vec{r}_0 = (10 \text{ m})\hat{i}$. (a) Find the velocity and position vectors at any time t. (b) Find the equation of the particle's path in the xy plane and sketch the path.

61 ••• **iSOLVE** Starting from rest at a dock, a motor boat heads *north* while accelerating at a constant 3 m/s^2 for 20 s. The boat then turns *west* at the speed that it had at 20 s and travels west at this constant speed for 10 s. (a) What was the average velocity of the boat during the 30-s trip? (b) What was the average acceleration of the boat during the 30-s trip? (c) What is the displacement of the boat from the dock at the end of the 30-s trip?

62 ••• **SSM** Mary and Robert decide to rendezvous on Lake Michigan. Mary departs in her boat from Petoskey at 9:00 A.M. and travels due north at 8 mi/h. Robert leaves from his home on the shore of Beaver Island, 26 mi, 30° west of north of Petoskey, at 10:00 A.M. and travels at a constant speed of 6 mi/h. In what direction should Robert be heading to intercept Mary, and where and when will they meet?

Relative Velocity

63 •• A plane flies at an airspeed of 250 km/h. There is a wind blowing at 80 km/h in the northeast direction at exactly 45° to the east of north. (a) In what direction should the plane head in order to fly due north? (b) What is the speed of the plane relative to the ground?

64 •• **iSOLVE** ✓ A swimmer heads directly across a river, swimming at 1.6 m/s relative to the water. She arrives at a point 40 m downstream from the point directly across the river, which is 80 m wide. (a) What is the speed of the river current? (b) What is the swimmer's speed relative to the shore? (c) In what direction should the swimmer head to arrive at the point directly opposite her starting point?

65 •• **SSM** A small plane departs from point A heading for an airport 520 km due north at point B. The airspeed of the plane is 240 km/h and there is a steady wind of 50 km/h blowing northwest to southeast. Determine the proper heading for the plane and the time of flight.

66 •• **iSOLVE** ✓ Two boat landings are 2.0 km apart on the same bank of a stream that flows at 1.4 km/h. A motorboat makes the round trip between the two landings in 50 min. What is the speed of the boat relative to the water?

67 •• In radio-controlled model airplane competition each plane must fly from the center of a 1-km-radius circle to any point on the circle and back to the center. The winner is the plane with the shortest round-trip time. The contestants are free to fly their planes along any route so long as the plane begins at the center, travels to the circle, and then returns to the center. On the day of the race, a steady wind blows out of the north at 5 m/s. Your plane can maintain an airspeed of 15 m/s. Strategy is paramount. Should you fly your plane upwind on the first leg and downwind on the trip back, or across the wind flying east and then west? Optimize your chances by calculating the round-trip time for both routes.

68 • **iSOLVE** The pilot of a small plane maintains an air speed of 150 kts (knots, or nautical miles per hour) and wants to fly due north (000°) with respect to the earth. If a wind of 30 kts is blowing *from* the east (090°), calculate the heading (azimuth) the pilot must take.

69 •• **SSM** Car A is traveling east at 20 m/s toward an intersection. As car A crosses the intersection, car B starts from rest 40 m north of the intersection and moves south with a constant acceleration of 2 m/s². Six seconds after A crosses the intersection find (a) the position of B relative to A, (b) the velocity of B relative to A, (c) the acceleration of B relative to A.

70 ••• **SSM** A tennis racket is held horizontally and a tennis ball is held above the racket. When the ball is dropped from rest, and bounces off the strings of the racket, the ball always rebounds to 64% of its initial height. (a) Express the speed of the tennis ball just after it bounces as some fraction of the speed of the ball just before the bounce. (b) The tennis ball is now thrown up into the air and served using the same racket. Assuming the ball's pre-impact speed is zero, and that the racket's speed through impact is 25 m/s, with what speed does the tennis ball come off the racket strings? *Hint: Using the results of part (a), solve for the post-impact speed of the ball in the reference frame of the racket, and then calculate the ball's speed in the reference frame of the earth.* (c) From some well-established laws of physics, we never see a ball bounce higher than the point from which it was released. From this, can you give an upper bound on the speed of a served tennis ball in relation to the speed of the racket, no matter how well the racket is designed? (We will see later that these results can be explained in a different context: the idea of *conservation of momentum*.)

Circular Motion and Centripetal Acceleration

71 • What is the acceleration of the extreme tip of the minute hand of the clock in Problem 38? Express it as a fraction of the magnitude of free-fall acceleration g.

72 • A centrifuge spins at a rate of 15,000 rev/min. (a) Calculate the centripetal acceleration of a test-tube sample held in the centrifuge arm 15 cm from the rotation axis. (b) It takes 1 min, 15 s for the centrifuge to spin up to its maximum rate of revolution from rest. Calculate the *magnitude* of the tangential acceleration of the centrifuge while it is spinning up, assuming that the tangential acceleration is constant.

73 • An object resting on the equator has an acceleration toward the center of the earth due to the earth's rotational motion about its axis, and an acceleration toward the sun due to the earth's orbital motion. Calculate the magnitudes of both of these accelerations, and express them as a fraction of the magnitude of free-fall acceleration g. Use values from the physical-data table in the textbook.

74 •• **SSM** Determine the acceleration of the moon toward the earth, using values for its mean distance and orbital period from the physical-data table in the textbook. Assume a circular orbit. Express the acceleration as a fraction of the magnitude of free-fall acceleration g.

75 • A boy whirls a ball on a string in a horizontal circle of radius 0.8 m. How many revolutions per minute does the ball make if the magnitude of its centripetal acceleration is g (the magnitude of the free-fall acceleration)?

Projectile Motion and Projectile Range

76 • A pitcher throws a fastball at 140 km/h (about 87 mi/h) toward home plate, which is 18.4 m away. Neglecting air resistance (not a good idea if you are the batter), how far does the ball drop because of gravity by the time it reaches home plate?

77 • A projectile is launched with speed v_0 at an angle of θ_0 with the horizontal. Find an expression for the maximum height it reaches above its starting point in terms of v_0, θ_0, and g.

78 •• **SSM** A cannonball is fired with initial speed v_0 at an angle 30° above the horizontal from a height of 40 m above the ground. The projectile strikes the ground with a speed of $1.2v_0$. Find v_0.

79 •• In Figure 3-48, if x is 50 m and $h = 10$ m, what is the minimum initial speed of the dart if it is to hit the monkey before hitting the ground, which is 11.2 m below the initial position of the monkey?

FIGURE 3-48 Problem 79

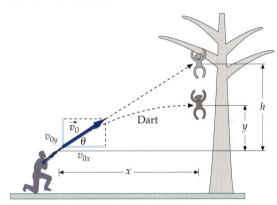

80 •• A projectile is fired with an initial speed of 53 m/s. Find the angle of projection such that the maximum height of the projectile is equal to its horizontal range.

81 •• A ball thrown into the air lands 40 m away 2.44 s later. Find the direction and magnitude of the initial velocity.

82 •• **SSM** **iSOLVE** Consider a ball that is thrown with initial speed v_0 at an angle θ above the horizontal. If we consider its speed v at some height h above the ground, show that $v(h)$ is independent of θ.

83 •• At $\frac{1}{2}$ of its maximum height, the speed of a projectile is $\frac{3}{4}$ of its initial speed. What was its launch angle?

84 • **iSOLVE** ✓ A cargo plane is flying horizontally at an altitude of 12 km with a speed of 900 km/h when a large crate falls out of the rear loading ramp. (a) How long does it take the crate to hit the ground? (b) How far horizontally is the crate from the point where it fell off when it hits the ground? (c) How far is the crate from the aircraft when the crate hits the ground, assuming that the plane continues to fly with constant velocity?

85 •• **SSM** **iSOLVE** Wiley Coyote (*Carnivorous hungribilous*) is chasing the Roadrunner (*Speedibus cantcatchmi*) yet again. While running down the road, they come to a deep gorge, 15 m straight across and 100 m deep. The Roadrunner launches itself across the gorge at a launch angle of 15° above the horizontal, and lands with 1.5 m to spare. (a) What was the Roadrunner's launch speed? Ignore air resistance. (b) Wiley Coyote launches himself across the gorge with the same initial speed, but at a different launch angle. To his horror, he is short the other lip by 0.5 m. What was his launch angle? (Assume that it was lower than 15°.)

86 • A cannon is elevated at an angle of 45°. It fires a ball with a speed of 300 m/s. (*a*) What height does the ball reach? (*b*) How long is the ball in the air? (*c*) What is the horizontal range of the cannon?

87 •• A stone thrown horizontally from the top of a 24-m tower hits the ground at a point 18 m from the base of the tower. (*a*) Find the speed with which the stone was thrown. (*b*) Find the speed of the stone just before it hits the ground.

88 •• **iSOLVE** A projectile is fired into the air from the top of a 200-m cliff above a valley (Figure 3-49). Its initial velocity is 60 m/s at 60° above the horizontal. Where does the projectile land?

FIGURE 3-49 Problem 88

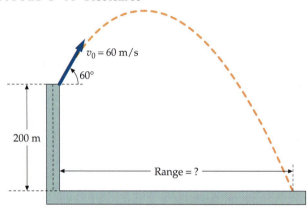

$v_0 = 60$ m/s

60°

200 m

Range = ?

89 •• The range of a cannonball fired horizontally from a cliff is equal to the height of the cliff. What is the direction of the velocity vector when the projectile strikes the ground?

90 • A projectile is fired at an angle of 60° above the horizontal with an initial speed of 300 m/s. Calculate (*a*) the horizontal distance traveled and (*b*) the vertical height attained in the first 6 s.

91 •• A cannonball is fired with an initial speed of 42.2 m/s at an angle of 30° above the horizontal from an initial height of 40 m. Find the range of the cannonball.

92 •• **SSM** **iSOLVE** The speed of an arrow fired from a compound bow is about 45 m/s. (*a*) A Tartar archer sits astride his horse and launches an arrow into the air, elevating the bow at an angle of 10° above the horizontal. If the bow is 2.25 m above the ground, what is the arrow's range? Assume that the ground is level, and ignore air resistance. (*b*) Now assume that his horse is at full gallop, moving in the same direction as he will fire the arrow, and that he elevates the bow in the same way as in part (*a*) and fires. If the horse's speed is 12 m/s, what is the arrow's range now?

93 • **iSOLVE** ✔ Using a potato cannon, Chuck launches a spud-plug horizontally with an initial velocity of 50.0 m/s (about 112 mi/h). (*a*) If Chuck holds the potato cannon so that it is 1.00 m above ground, how long is the spud-plug in the air? (*b*) How far does the plug travel before hitting the ground? (You can find more about potato cannons on the Internet.)

94 •• Compute $dR/d\theta$ from $R = (v_0^2/g)\sin(2\theta_0)$ and show that setting $dR/d\theta = 0$ gives $\theta = 45°$ for the maximum range.

95 • **SSM** In a science fiction short story written in the 1970s, Ben Bova described a conflict between two hypothetical colonies on the moon—one founded by the United States and the other by the USSR. In the story, colonists from each side started firing bullets at each other, only to find to their horror that their rifles had a high enough muzzle velocity that the bullets went into orbit. (*a*) If the magnitude of free-fall acceleration on the moon is 1.67 m/s², what is the maximum range of a rifle bullet with a muzzle velocity of 900 m/s? (Assume the curvature at the surface of the moon is negligible.) (*b*) What would the muzzle velocity have to be to send the bullet into a circular orbit just above the surface of the moon? (You will first need to look up the radius of the moon.)

96 ••• In the text, we calculated the range for a projectile that lands at the same elevation from which it is fired as $R = (v_0^2/g)\sin 2\theta_0$. Show that the change in the range for a small change in free-fall acceleration is given by $\Delta R/R = -\Delta g/g$.

97 ••• In the text, we calculated the range for a projectile that lands at the same elevation from which it is fired as $R = (v_0^2/g)\sin 2\theta_0$. Show that the change in the range for a small change in launch velocity is given by $\Delta R/R = 2\Delta v_0/v_0$.

98 ••• In the text, we calculated the range for a projectile that lands at the same elevation from which it is fired as $R = (v_0^2/g)\sin 2\theta_0$. Show that the range for the more general problem (Figure 3-50) where $\Delta y \neq 0$ is given by

FIGURE 3-50 Problem 98

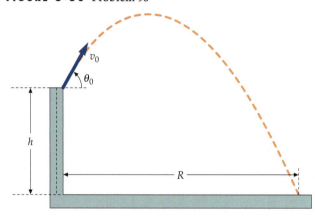

v_0

θ_0

h

R

$$R = \left(1 + \sqrt{1 + \frac{2gh}{v_0^2 \sin^2\theta_0}}\right)\frac{v_0^2}{2g}\sin 2\theta_0$$

99 •• **SSM** A projectile is launched over level ground at an elevation angle of θ. An observer standing at the launch site sights the projectile at the point of its highest elevation, and measures the angle ϕ shown in Figure 3-51. Show that $\tan \phi = \frac{1}{2}\tan \theta$.

FIGURE 3-51 Problem 99

y

h

θ ϕ

$R/2$ R x

100 • A projectile, fired with unknown initial velocity, lands 20 s later on the side of a hill, 3000 m away horizontally and 450 m vertically above its starting point. (*a*) What is the vertical component of its initial velocity? (*b*) What is the horizontal component of its initial velocity?

101 •• **SSM** A stone is thrown horizontally from the top of an incline that makes an angle θ with the horizontal. If the stone's initial speed is v_0, how far down the incline will it land?

102 ••• A toy cannon is placed on a ramp that has a slope of angle ϕ. (*a*) If the cannonball is projected up the hill at an angle of θ_0 above the horizontal (Figure 3-52) and has a muzzle speed of v_0, show that the range R of the cannonball (as measured along the ramp) is given by

$$R = \frac{2v_0^2 \cos^2\theta_0 \, (\tan\theta_0 - \tan\phi)}{g \cos\phi}$$

Ignore air resistance.

FIGURE 3-52 Problem 102

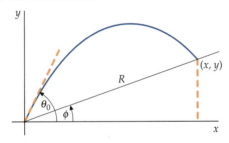

103 •• A rock is thrown from the top of a 20-m building at an angle of 53° above the horizontal. If the horizontal range of the throw is equal to the height of the building, with what speed was the rock thrown? What is the velocity of the rock just before it strikes the ground?

104 •• A girl throws a ball at a vertical wall 4 m away (Figure 3-53). The ball is 2 m above ground when it leaves the girl's hand with an initial velocity of $\vec{v}_0 = (10 \text{ m/s})(\hat{i} + \hat{j})$ or $10\sqrt{2}$ m/s at 45°. When the ball hits the wall, the horizontal component of its velocity is reversed; the vertical component remains unchanged. Where does the ball hit the ground? *Hint: The wall can be thought of as a mirror. Determine the range, neglecting the wall, and then consider the mirror-like reflection.*

FIGURE 3-53 Problem 104

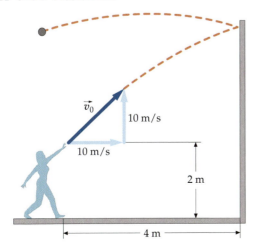

Hitting Targets and Related Problems

105 • A boy uses a slingshot to project a pebble at a shoulder-height target 40 m away. He finds that to hit the target he must aim 4.85 m above the target. Determine the velocity of the pebble on leaving the slingshot and the time of flight of the pebble.

106 •• **SSM** The distance from the pitcher's mound to home plate is 18.4 m. The mound is 0.2 m above the level of the field. A pitcher throws a fastball with an initial speed of 37.5 m/s. At the moment the ball leaves the pitcher's hand, it is 2.3 m above the mound. What should the angle between $\vec{v}_0$ and the horizontal be so that the ball crosses the plate 0.7 m above ground? (Neglect interaction with air.)

107 •• Suppose that a hockey puck is struck in such a way that, when it is at its highest point, it just clears a Plexiglass wall of height $h = 2.80$ m. Find v_{0y}, the time t to reach the wall, and v_{0x}, v_0, and θ_0 for this case. Assume that the horizontal distance is $x_1 = 12.0$ m.

108 •• Carlos is on his trail bike, approaching a creek bed that is 7 m wide. A ramp with an incline of 10° has been built for daring people who try to jump the creek. Carlos is traveling at his bike's maximum speed, 40 km/h. (*a*) Should Carlos attempt the jump or emphatically hit the brakes? (*b*) What is the minimum speed a bike must have to make this jump? Assume equal elevations on either side of the creek.

109 •• If a bullet that leaves the muzzle of a gun at 250 m/s is to hit a target 100 m away at the level of the muzzle, the gun must be aimed at a point above the target. How far above the target is that point?

General Problems

110 • The displacement vectors $\vec{A}$ and $\vec{B}$ in Figure 3-54 both have a magnitude of 1 m. (*a*) Find their x and y components. (*b*) Find the components, magnitude, and direction of the sum $\vec{S} = \vec{A} + \vec{B}$. (*c*) Find the components, magnitude, and direction of the difference $\vec{D} = \vec{A} - \vec{B}$.

FIGURE 3-54 Problem 110

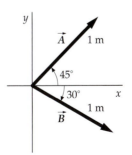

111 • **SSM** A plane is inclined at an angle of 30° from the horizontal. Choose the x axis pointing down the slope of the plane and the y axis perpendicular to the plane. Find the x and y components of the acceleration of gravity, which has the magnitude 9.81 m/s² and points vertically down.

112 • Two vectors $\vec{A}$ and $\vec{B}$ lie in the xy plane. Under what conditions does the ratio A/B equal A_x/B_x?

113 • The position vector of a particle is given by $\vec{r} = (5\text{ m/s})t\hat{i} + (10\text{ m/s})t\hat{j}$, where t is in seconds and $\vec{r}$ is in meters. (a) Draw the path of the particle in the xy plane. (b) Find $\vec{v}$ in component form and then find its magnitude.

114 •• **ISOLVE** A worker on the roof of a house drops her hammer, which slides down the roof at constant speed of 4 m/s. The roof makes an angle of 30° with the horizontal, and its lowest point is 10 m from the ground. What is the horizontal distance traveled by the hammer between the time it leaves the roof of the house and the time it hits the ground?

115 •• In 1940, Emanuel Zacchini flew about 53 m as a human cannonball, a record that remains unbroken. His initial velocity was 24.2 m/s at an angle θ. Find θ and the maximum height h Emanuel achieved during the record flight. Ignore the effects of air resistance.

116 •• A particle moves in the xy plane with constant acceleration. At $t = 0$ the particle is at $\vec{r}_1 = (4\text{ m})\hat{i} + (3\text{ m})\hat{j}$, with velocity $\vec{v}_1$. At $t = 2$ s the particle has moved to $\vec{r}_2 = (10\text{ m})\hat{i} - (2\text{ m})\hat{j}$ and its velocity has changed to $\vec{v}_2 = (5\text{ m/s})\hat{i} - (6\text{ m/s})\hat{j}$. (a) Find $\vec{v}_1$. (b) What is the acceleration of the particle? (c) What is the velocity of the particle as a function of time? (d) What is the position vector of the particle as a function of time?

117 •• **SSM** A small steel ball is projected horizontally off the top landing of a long, rectangular staircase. The initial speed of the ball is 3 m/s. Each step is 0.18 m high and 0.3 m wide. Which step does the ball strike first?

118 •• Suppose you can throw a ball a distance x_0 when standing on level ground. How far can you throw it from a building of height $h = x_0$ if you throw it at (a) 0°? (b) 30°? (c) 45°?

119 ••• Darlene is a stunt motorcyclist in a traveling circus. For the climax of her show, she takes off from the ramp at angle θ, clears a fiery ditch of width x, and lands on an elevated platform (height h) on the other side (Figure 3-55). (a) For a given height h, find the minimum necessary takeoff speed v_{min} needed to make the jump successfully. (b) What is v_{min} for a launch angle $\theta = 30°$ with a pit width of 8 m and a platform height $h = 4$ m? (c) Show that no matter what her takeoff speed is, the maximum height of the platform is $h_{max} < x \tan \theta$. Interpret this result physically. (Neglect the effects of air resistance and treat the bike as if it were a particle.)

FIGURE 3-55 Problem 119

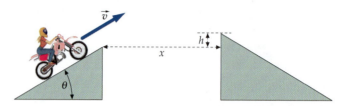

120 ••• A small boat is headed for a harbor 32 km northwest of its current position when it is suddenly engulfed in heavy fog. The captain maintains a compass bearing of northwest and a speed of 10 km/h relative to the water. The fog lifts 3 h later and the captain notes that he is now exactly 4.0 km south of the harbor. (a) What was the average velocity of the current during those 3 h? (b) In what direction should the boat have been heading to reach its destination along a straight course? (c) What would its travel time have been if it had followed a straight course?

121 •• **SSM** Galileo showed that, if air resistance is neglected, the ranges for projectiles whose angles of projection exceed or fall short of 45° by the same amount are equal. Prove Galileo's result.

122 •• Two balls are thrown with equal speeds from the top of a cliff of height h. One ball is thrown at an angle of α above the horizontal. The other ball is thrown at an angle of β below the horizontal. Show that each ball strikes the ground with the same speed, and find that speed in terms of h and the initial speed v_0.

Newton's Laws

THIS AIRPLANE IS ACCELERATING AS IT HEADS DOWN THE RUNWAY BEFORE TAKEOFF. NEWTON'S LAWS RELATE AN OBJECT'S ACCELERATION TO ITS MASS AND THE FORCES ACTING ON IT.

? **If you were a passenger on this plane, how might you use Newton's laws to determine the plane's acceleration? (See Example 4-9.)**

Now that we have studied how objects move in one, two, and three dimensions, we can ask the question, "why do objects start to move?" What causes a moving object to gain speed or change direction?

Classical mechanics relates the forces objects exert on each other, and relates changes in the motion of an object to the forces that act on it. It describes phenomena using Newton's three laws of motion. While we may already have an intuitive idea of a force as a push or a pull, like that exerted by our muscles or by stretched rubber bands and springs, Newton's laws allow us to refine our understanding of forces.

➤ **In this chapter, we describe Newton's three laws of motion and begin using them to solve problems involving objects in motion and at rest.**

A modern wording of Newton's laws is:

First law. An object at rest stays at rest *unless* acted on by an external force. An object in motion continues to travel with constant velocity *unless* acted on by an external force.

Second law. The direction of the acceleration of an object is in the direction of the net external force acting on it. The acceleration is proportional to the net external force $\vec{F}_{net}$, in accordance with $\vec{F}_{net} = m\vec{a}$, where m is the mass of the object. The net force acting on an object, also called the resultant force, is the vector sum of all the forces acting on it: $\vec{F}_{net} = \Sigma\vec{F}$. Thus,

$$\Sigma\vec{F} = m\vec{a}$$

4-1

NEWTON'S SECOND LAW

Third law. Forces always occur in equal and opposite pairs. If object A exerts a force $\vec{F}_{A,B}$ on object B, an equal but opposite force $\vec{F}_{B,A}$ is exerted by object B on object A. Thus,

$$\vec{F}_{B,A} = -\vec{F}_{A,B}$$

4-2

NEWTON'S THIRD LAW

4-1 Newton's First Law: The Law of Inertia

Push a piece of ice on a counter top: It slides, then stops. If the counter is wet, the ice will travel farther before stopping. A piece of dry ice (frozen carbon dioxide) riding on a cushion of carbon dioxide vapor slides quite far with little change in velocity. Before Galileo it was thought that a force, such as a push or pull, was always needed to keep an object moving with constant velocity. But Galileo, and later Newton, recognized that the slowing of objects in everyday experience is due to the force of friction. If friction is reduced, the rate of slowing is reduced. A water slick or a cushion of gas is especially effective at reducing friction, allowing the object to slide a great distance with little change in velocity. Galileo reasoned that, if we could remove from an object all external forces including friction, then the velocity of the object would never change—a property of matter he described as **inertia.** This conclusion, restated by Newton as his first law, is also called the **law of inertia.**

Friction is greatly reduced by a cushion of air that supports the hovercraft.

Inertial Reference Frames

Newton's first law makes no distinction between an object at rest and an object moving with constant (nonzero) velocity. Whether an object remains at rest or remains moving with constant velocity depends on the reference frame in which the object is observed. Suppose you are a passenger on an airplane that is flying along a straight path at constant altitude and you carefully place a small ball on your seat tray (which is horizontal). Relative to the plane, the ball will remain at rest as long as the plane continues to fly at constant velocity relative to the ground. Relative to the ground, the ball remains moving with the same velocity as the plane.

Now, suppose that the pilot opens the throttle and the plane suddenly accelerates forward (relative to the ground). You will then observe that the ball on your tray suddenly starts to roll backward, accelerating (relative to the plane) even though there is no horizontal force acting on it.

A reference frame accelerating relative to an inertial reference frame is not an inertial reference frame. *Newton's first law thus gives us the criterion for determining if a reference frame is an inertial frame.* In fact, it is useful to think of Newton's first law as a statement that defines inertial reference frames.

> If no forces act on an object, any reference frame with respect to which the acceleration of the object remains zero is an **inertial reference frame.**

DEFINITION—INERTIAL REFERENCE FRAME

Both the plane, when cruising at constant velocity, and the ground are, to a good approximation, inertial reference frames. Any reference frame moving with constant velocity relative to an inertial reference frame is also an inertial reference frame.

A reference frame attached to the surface of the earth is not quite an inertial reference frame because of the small acceleration of the surface of the earth due to the rotation of the earth and the small acceleration of the earth itself due to its revolution around the sun. However, these accelerations are of the order of 0.01 m/s^2 or less, so to a good approximation, a reference frame attached to the surface of the earth is an inertial reference frame.

The concept of inertial reference frame is of central importance because *Newton's first, second, and third law statements are valid only in inertial reference frames.*

4-2 Force, Mass, and Newton's Second Law

Newton's first and second laws allow us to define force. A **force** is an external influence on an object that causes it to accelerate relative to an inertial reference frame. (We assume there are no other forces acting.) The direction of the force is the direction of the acceleration it causes. The magnitude of the force is the product of the mass of the object and the magnitude of its acceleration. This definition is given in Equation 4-1.

Forces can be compared by stretching identical rubber bands. For example, if two identical rubber bands are stretched by the same amount, then they exert forces of equal magnitudes.

Objects intrinsically resist being accelerated. Imagine kicking a soccer ball or a bowling ball. The bowling ball resists being accelerated much more than does the soccer ball, as would be evidenced by your sore toes. This intrinsic property is called the object's **mass.** It is a measure of the object's inertia. The ratio of two masses is defined quantitatively by applying the same force to each and comparing their accelerations. If a force F produces acceleration a_1 when applied to an object of mass m_1, and an equal force produces acceleration a_2 when applied to an object of mass m_2, then the ratio of the two masses is defined by

$$\frac{m_2}{m_1} = \frac{a_1}{a_2}$$

4-3

DEFINITION—MASS

This definition agrees with our intuitive idea of mass. If a force is applied to an object and a force of equal magnitude is applied to a second object, then the object with more mass will accelerate less. The ratio a_1/a_2 produced by forces of equal magnitude acting on two objects is found experimentally to be independent of the magnitude, direction, or type of force used. Mass is an intrinsic property of an object that does not depend on its location—it remains the same whether the object is on the earth, on the moon, or in outer space.

If a direct comparison shows that $m_2/m_1 = 2$ and $m_3/m_1 = 4$, then m_3 will be twice m_2 when objects 2 and 3 are compared with each other. We can therefore establish a mass scale by choosing a standard object and assigning it a mass of 1 unit. As we noted in Chapter 1, the object chosen as the international standard for mass is a platinum-iridium alloy cylinder carefully preserved at the International Bureau of Weights and Measures at Sèvres, France. The mass of the standard object is 1 **kilogram** (kg), the SI unit of mass. The force required to produce an acceleration of 1 m/s² on the standard object is defined to be 1 **newton** (N). The force that produces an acceleration of 2 m/s² on the standard object is 2 N, and so on.

A SLIDING ICE CREAM CARTON   **EXAMPLE 4-1**

A given force produces an acceleration of 5 m/s² on the standard object of mass m_1. When an equal force is applied to a carton of ice cream of mass m_2, it produces an acceleration of 11 m/s². (a) What is the mass of the carton of ice cream? (b) What is the magnitude of the force?

PICTURE THE PROBLEM Apply $\Sigma \vec{F} = m\vec{a}$ to each object and solve for the mass of the ice-cream carton and the magnitude of the force.

(a) 1. Apply $\Sigma \vec{F} = m\vec{a}$ to each object. There is only one force and we only need to consider magnitudes of the vector quantities:

$$F_1 = m_1 a_1 \quad \text{and} \quad F_2 = m_2 a_2$$

2. The ratio of the masses varies inversely as the ratio of the accelerations under applied forces of equal magnitude:

$$F_1 = F_2 = F, \quad \text{so} \quad m_1 a_1 = m_2 a_2$$

and

$$\frac{m_2}{m_1} = \frac{a_1}{a_2} = \frac{5 \text{ m/s}^2}{11 \text{ m/s}^2}$$

3. Solve for m_2 in terms of m_1, which is 1 kg:

$$m_2 = \frac{5}{11} m_1 = \frac{5}{11}(1 \text{ kg}) = \boxed{0.45 \text{ kg}}$$

(b) The magnitude F is found by using the mass and acceleration of either object:

$$F = m_1 a_1 = (1 \text{ kg})(5 \text{ m/s}^2) = \boxed{5 \text{ N}}$$

EXERCISE A force of 3 N produces an acceleration of 2 m/s² on an object of unknown mass. (a) What is the mass of the object? (b) If the force is increased to 4 N, what is the acceleration? (*Answer* (a) 1.5 kg (b) 2.67 m/s²)

It is found experimentally that two or more forces acting on an object accelerate it as if the object were acted upon by a single force equal to the vector sum of the individual forces. That is, forces combine as vectors. Newton's second law is thus

$$\Sigma \vec{F} = \vec{F}_{\text{net}} = m\vec{a}$$

A WALK IN SPACE   **EXAMPLE 4-2**

You're stranded in space away from your spaceship. Fortunately, you have a propulsion unit that provides a constant force $\vec{F}$ for 3 s. After 3 s you have moved 2.25 m. If your mass is 68 kg, find $\vec{F}$.

The propulsion unit (not shown) is pushing the astronaut to the right.

PICTURE THE PROBLEM The force acting on you is constant, so your acceleration $\vec{a}$ is also constant. We can use the kinematic equations of Chapter 2 to find
$\vec{a}$, and then obtain the force from $\Sigma \vec{F} = m\vec{a}$. Choose $\vec{F}$ to be along the x axis, so
that $\vec{F} = F_x \hat{i}$ (Figure 4-1). The component of Newton's second law along the x axis
is then $F_x = ma_x$.

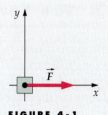

FIGURE 4-1

1. Apply $\Sigma \vec{F} = m\vec{a}$ to relate the net force to the mass and the acceleration:

$$F_x = ma_x$$

2. To find the acceleration, we use Equation 2-15 with $v_0 = 0$:

$$\Delta x = v_0 t + \tfrac{1}{2}a_x t^2 = \tfrac{1}{2}a_x t^2$$

$$a_x = \frac{2\Delta x}{t^2} = \frac{2(2.25 \text{ m})}{(3 \text{ s})^2} = 0.500 \text{ m/s}^2$$

3. Substitute $a_x = 0.500 \text{ m/s}^2$ and $m = 68$ kg to find the force:

$$F_x = ma_x = (68 \text{ kg})(0.500 \text{ m/s}^2) = \boxed{34.0 \text{ N}}$$

A PARTICLE SUBJECTED TO FORCES   **EXAMPLE 4-3**   **Try It Yourself**

A particle of mass 0.4 kg is subjected simultaneously to two forces $\vec{F}_1 = -2 \text{ N } \hat{i} - 4 \text{ N } \hat{j}$ and $\vec{F}_2 = -2.6 \text{ N } \hat{i} + 5 \text{ N } \hat{j}$. If the particle is at the origin and starts from rest at $t = 0$, find (*a*) its position vector $\vec{r}$ and (*b*) its velocity $\vec{v}$ at $t = 1.6$ s.

PICTURE THE PROBLEM Since $\vec{F}_1$ and $\vec{F}_2$ are constant, the acceleration of the
particle is constant. Hence, we use the kinematic equations of Chapter 2 to determine the particle's position and velocity as functions of time.

Cover the column to the right and try these on your own before looking at the answers.

Steps

(*a*) 1. Write the general equation for the position vector $\vec{r}$ as a function of time t for constant acceleration $\vec{a}$ in terms of $\vec{r}_0$, $\vec{v}_0$, and $\vec{a}$, and substitute $\vec{r}_0 = \vec{v}_0 = 0$.

2. Use $\Sigma \vec{F} = m\vec{a}$ to write the acceleration $\vec{a}$ in terms of the resultant force $\Sigma \vec{F}$ and the mass m.

3. Compute $\Sigma \vec{F}$ from the given forces.

4. Find the acceleration vector $\vec{a}$.

5. Find the position vector $\vec{r}$ for a general time t.

6. Find $\vec{r}$ at $t = 1.6$ s.

(*b*) Write the velocity vector $\vec{v}$ in terms of the acceleration and time and compute its components for the time
$t = 1.6$ s.

Answers

$$\vec{r} = \vec{r}_0 + \vec{v}_0 t + \tfrac{1}{2}\vec{a}t^2 = \tfrac{1}{2}\vec{a}t^2$$

$$\vec{a} = \frac{\Sigma \vec{F}}{m}$$

$$\Sigma \vec{F} = \vec{F}_1 + \vec{F}_2 = -4.6 \text{ N } \hat{i} + 1.0 \text{ N } \hat{j}$$

$$\vec{a} = \frac{\Sigma \vec{F}}{m} = -11.5 \text{ m/s}^2 \hat{i} + 2.5 \text{ m/s}^2 \hat{j}$$

$$\vec{r} = \tfrac{1}{2}\vec{a}t^2 = \tfrac{1}{2}a_x t^2 \hat{i} + \tfrac{1}{2}a_y t^2 \hat{j}$$

$$= -5.75 \text{ m/s}^2 \, t^2 \hat{i} + 1.25 \text{ m/s}^2 \, t^2 \hat{j}$$

$$\vec{r} = \boxed{-14.7 \text{ m } \hat{i} + 3.20 \text{ m } \hat{j}}$$

$$\vec{v} = \vec{a}t = (-11.5 \text{ m/s}^2 \hat{i} + 2.5 \text{ m/s}^2 \hat{j})t$$

$$= \boxed{-18.4 \text{ m/s } \hat{i} + 4.00 \text{ m/s } \hat{j}}$$

4-3 The Force Due to Gravity: Weight

If we drop an object near the earth's surface, it accelerates toward the earth. If we neglect air resistance, all objects have the same acceleration, called the free-fall acceleration $\vec{g}$, at any location. The force causing this acceleration is the gravitational force on the object, called its weight.[†] If its weight $\vec{w}$ is the *only* force acting on an object, the object is said to be in **free-fall.** If its mass is m, Newton's second law ($\Sigma \vec{F} = m\vec{a}$) defines the weight $\vec{w}$:

$$\vec{w} = m\vec{g} \qquad\qquad 4\text{-}4$$

WEIGHT

Since $\vec{g}$ is the same for all objects, it follows that the weight of an object is proportional to its mass. The vector $\vec{g}$ is the force per unit mass exerted by the earth on any object and is called the **gravitational field** of the earth. It is equal to the free-fall acceleration.[‡] Near the surface of the earth, g has the value

$$g = 9.81 \text{ N/kg} = 9.81 \text{ m/s}^2$$

Careful measurements show that $\vec{g}$ varies with location. In particular, at points above the surface of the earth, $\vec{g}$ points toward the center of the earth and varies inversely with the square of the distance to the center of the earth. Thus an object weighs slightly less at very high altitudes than it does at sea level. The gravitational field also varies slightly with latitude because the earth is not exactly spherical but is slightly flattened at the poles. **Thus weight, unlike mass, is not an intrinsic property of an object.** Although the weight of an object varies from place to place because of changes in g, this variation is too small to be noticed in most practical applications on or near the surface of the earth.

An example should help clarify the difference between mass and weight. Consider a bowling ball near the moon. Its weight is the gravitational force exerted on it by the moon, but that force is a mere sixth of the force exerted on the bowling ball when it is similarly positioned on earth. The ball weighs about one-sixth as much on the moon, and lifting the ball on the moon requires one-sixth the force. However, because the mass of the ball is the same on the moon as on the earth, throwing the ball with some horizontal acceleration requires the same force on the moon as on the earth or in free space.

Although the weight of an object may vary from one place to another, at any particular location the weight of the object is proportional to its mass. Thus we can conveniently compare the masses of two objects at a given location by comparing their weights.

Our sensation of our own weight comes from other forces that balance it. When you sit on a chair, you feel a force exerted by the chair that balances your weight and prevents you from falling to the floor. When you stand on a spring scale, your feet feel the force exerted by the scale. The scale is calibrated to read the force it must exert (by the compression of its springs) to balance your weight. This force is called your **apparent weight.** If there is no force to balance your weight, as in free-fall, your apparent weight is zero. This condition, called **weightlessness,** is experienced by astronauts in orbiting satellites. A satellite in a circular orbit near the surface of the earth is accelerating toward the earth. The only force acting on the satellite is that of gravity (its weight), so it is in free-fall.

[†] Referring to the gravitation force as "its weight" is unfortunate because it appears to imply that "its weight" is a property of the object rather than an external force acting on it. To avoid buying into the apparent implication of the wording, every time you read "its weight," mentally translate it to read "the gravitational force acting on it."

[‡] $\vec{g}$ refers to the free-fall acceleration, which is the acceleration (of an object in free-fall) relative to the ground. It is not entirely correct to attribute it to the gravitational attraction of the earth. This distinction is discussed further in Chapter 11.

Astronauts in the satellite are also in free-fall. The only force on them is their weight, which produces the acceleration g. Because there is no force balancing the force of gravity, the astronauts have zero apparent weight.

Units of Force and Mass

Like the second and the meter, the SI unit of mass, the kilogram, is a fundamental unit. The unit of force, the newton, and the units for other quantities that we will study such as momentum and energy are derived from the three fundamental units s, m, and kg.

As noted in Section 4-2, the newton is defined as the force that produces an acceleration of 1 m/s² when it acts on an object with a mass of 1 kg. Then Newton's second law gives

$$1 \text{ N} = (1 \text{ kg})(1 \text{ m/s}^2) = 1 \text{ kg·m/s}^2 \qquad\qquad 4\text{-}5$$

A convenient standard unit for mass in atomic and nuclear physics is the **unified mass unit** (u), which is defined as one-twelfth of the mass of the neutral carbon-12 (^{12}C) atom. The unified mass unit is related to the kilogram by

$$1 \text{ u} = 1.660\,540 \times 10^{-27} \text{ kg} \qquad\qquad 4\text{-}6$$

The mass of a hydrogen atom is approximately 1 u.

Although we will generally use SI units in this book, we need to know another scheme, the U.S. customary system, still used in the United States, which is based on the second, the foot, and the pound. The U.S. customary system differs from SI in that a unit of force, the pound, has been chosen as a fundamental unit rather than a unit of mass. The **pound** was originally defined as the weight of a particular standard object at a particular location. It is now defined as 4.448222 N. Rounding to three places, we have 1 lb ≈ 4.45 N. Since 1 kg weighs 9.81 N, its weight in pounds is

$$9.81 \text{ N} = 2.20 \text{ lb} \qquad\qquad 4\text{-}7$$

WEIGHT OF 1 KG

The unit of mass in the U.S. customary system is the rarely encountered slug, defined as the mass of an object that weighs 32.2 lb. When working problems in the U.S. customary system, we substitute w/g for mass m, where w is the weight in pounds and g is the acceleration due to gravity in feet per second per second:

$$g = 32.2 \text{ ft/s}^2 \qquad\qquad 4\text{-}8$$

AN ACCELERATING STUDENT   **E X A M P L E   4 - 4**

The net force acting on a 130-lb student is 25 lb. What is her acceleration?

PICTURE THE PROBLEM Apply $\Sigma F = ma$ and solve for the acceleration. The mass can be found from the student's weight.

According to Newton's second law, the student's acceleration is the force divided by her mass:

$$a = \frac{F}{m} = \frac{F}{w/g} = \frac{25 \text{ lb}}{(130 \text{ lb})/(32.2 \text{ ft/s}^2)}$$

$$= \boxed{6.19 \text{ ft/s}^2}$$

EXERCISE What force is needed to give an acceleration of 3 ft/s² to a 5-lb block? (*Answer* 0.466 lb)

4-4 Forces in Nature

The full power of Newton's second law emerges when it is combined with the force laws that describe the interactions of objects. For example, Newton's law for gravitation, which we study in Chapter 11, gives the gravitational force exerted by one object on another in terms of the distance between the objects and the masses of each. This, combined with Newton's second law, enables us to calculate the orbits of planets around the sun, the motion of the moon, and variations with altitude of g, the gravitational field strength.

(a)

The Fundamental Forces

All the different forces observed in nature can be explained in terms of four basic interactions that occur between elementary particles (see Figure 4.2):

1. The gravitational force—the force of mutual attraction between objects
2. The electromagnetic force—the force between electric charges
3. The strong nuclear force—the force between subatomic particles
4. The weak nuclear force—the force between subatomic particles during certain radioactive decay processes

The everyday forces that we observe between macroscopic objects are due to either the gravitational force or the electromagnetic force.

(b)

Action at a Distance

The fundamental forces of gravity and electromagnetism act between particles that are separated in space. This creates a philosophical problem referred to as **action at a distance.** Newton perceived action at

(c)

FIGURE 4-2 (*a*) The gravitational force between the earth and an object near the earth's surface is the weight of the object. The gravitational force exerted by the sun on the earth and the other planets is responsible for keeping the planets in their orbits around the sun. Similarly, the gravitational force exerted by the earth on the moon keeps the moon in its nearly circular orbit around the earth. The gravitational forces exerted by the moon and the sun on the oceans of the earth are responsible for the tides. Mont-Saint-Michel, France, shown in the photo, is an island when the tide is in. (*b*) The electromagnetic force includes both the electric and the magnetic forces. A familiar example of the electric force is the attraction between small bits of paper and a comb that is electrified after being run through hair. The lightning bolts above the Kitt Peak National Observatory, shown in the photo, are the result of the electromagnetic force. (*c*) The strong nuclear force occurs between elementary particles called hadrons, which include protons and neutrons, the constituents of atomic nuclei. This force results from the interaction of quarks, which are the building blocks of hadrons, and is responsible for holding nuclei together. The hydrogen bomb explosion shown here illustrates the strong nuclear force. (*d*) The weak nuclear force occurs between leptons (which include electrons and muons) and between hadrons (which include protons and neutrons). This false-color cloud chamber photograph illustrates the weak interaction between a cosmic ray muon (green) and an electron (red) knocked out of an atom.

(d)

a distance as a flaw in his theory of gravitation but avoided giving any other hypothesis. Today the problem is avoided by introducing the concept of a field, which acts as an intermediary agent. For example, we consider the attraction of the earth by the sun in two steps. The sun creates a condition in space that we call the gravitational field. This field then exerts a force on the earth. Similarly, the earth produces a gravitational field that exerts a force on the sun. Your weight is the force exerted by the gravitational field of the earth on you. When we study electricity and magnetism (Chapters 21–30) we will study electric fields, which are produced by electrical charges, and magnetic fields, which are produced by electrical charges in motion.

Contact Forces

Many forces we encounter are exerted by objects in direct contact—they touch. These forces are electromagnetic in origin. They are exerted between the surface molecules of the objects in contact.

Solids   If a surface is pushed against, it pushes back. Consider the ladder leaning against a wall shown in Figure 4-3. At the region of contact, the ladder pushes against the wall with a horizontal force, compressing the molecules in the surface of the wall. Like mattress springs, the compressed molecules in the wall push back on the ladder with a horizontal force. Such a force, *perpendicular* to the contacting surfaces, is called a **normal force** (the word *normal* means perpendicular). The wall bends slightly in response to a load, though this is rarely noticeable to the naked eye.

Normal forces can vary over a wide range of magnitudes. A table, for instance, will exert an upward normal force on any object resting on it. As long as the table doesn't break, this normal force will balance the downward weight force on the object. Furthermore, if you press down on the object, the upward normal force exerted by the table will increase, countering the extra force, thus preventing the object from accelerating downward.

Surfaces in contact can also exert forces on each other that are *parallel* to the contacting surfaces. Consider the large block on the floor shown in Figure 4-4. If the block is pushed sideways with a gentle enough force, it will not slide. The surface of the floor exerts a force back on the block, opposing its tendency to slide in the direction of the push. However, if the block is pushed sideways with a sufficiently hard force, it will start to slide. To keep it sliding it is necessary to continue to push it. If the sideways push is not sustained, the contact force will slow the motion of the box until it stops. A component of a contact force that opposes sliding, or the tendency to slide, is called a **frictional force;** it acts parallel to the contacting surfaces.

Although the frictional and normal forces are shown in diagrams as if they act at a single point, they are, in reality, distributed over the entire region of contact. Frictional forces are treated in more depth in Chapter 5.

Springs   When a spring is compressed or extended by a small amount Δx, the force it exerts is found experimentally to be

$$F_x = -k\,\Delta x \qquad\qquad 4\text{-}9$$

HOOKE'S LAW

where k is the force constant, a measure of the stiffness of the spring (Figure 4-5). The negative sign in Equation 4-9 signifies that when the spring is stretched or compressed, the force it exerts back is in the opposite direction. This relation, known as Hooke's law, turns out to be quite important. An object at rest under the influence of forces that balance is said to be in static equilibrium. If a small displacement results in a net restoring force toward

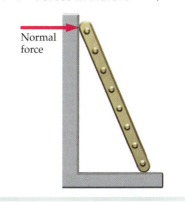

FIGURE 4-3 The wall supports the ladder by pushing on the ladder with a force normal to the wall.

FIGURE 4-4 The frictional force exerted by the floor on the block opposes its sliding motion or its tendency to slide.

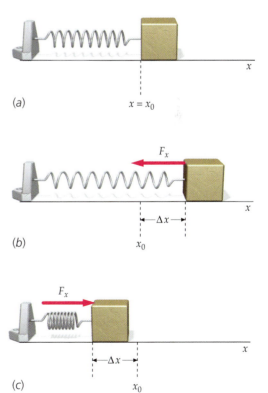

FIGURE 4-5 A horizontal spring. (*a*) When the spring is unstretched, it exerts no force on the block. (*b*) When the spring is stretched so that Δx is positive, it exerts a force of magnitude $k\,\Delta x$ in the negative x direction. (*c*) When the spring is compressed so that Δx is negative, the spring exerts a force of magnitude $k\,|\Delta x|$ in the positive direction.

the equilibrium position, the equilibrium is called stable equilibrium. For small displacements, nearly all restoring forces obey Hooke's law.

The molecular force of attraction between atoms in a molecule or solid varies approximately linearly with the change in separation (for small changes); the force varies much like that of a spring. We can therefore use two masses on a spring to model a diatomic molecule, or a set of masses connected by springs to model a solid as shown in Figure 4-6.

(a)

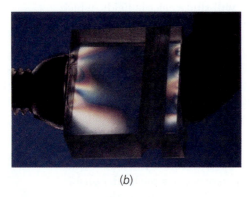

(b)

FIGURE 4-6 (*a*) Model of a solid consisting of atoms connected to each other by springs. The springs are very stiff (large force constant) so that when a weight is placed on the solid its deformation is not visible. However, compression such as that produced by the clamp on the plastic block in (*b*) leads to stress patterns that are visible when viewed with polarized light.

THE SLAM DUNK

EXAMPLE 4-5

A 110-kg basketball player hangs on the rim following a slam dunk (Figure 4-7). Prior to dropping to the floor, he hangs motionless with the front of the rim deflected down a distance of 15 cm. Assume the rim can be approximated by a spring and calculate the force constant k.

PICTURE THE PROBLEM Since the acceleration of the player is zero, the net force exerted on him must also be zero. The upward force exerted by the rim balances his weight. Let $y = 0$ be the original position of the rim and choose down to be the positive y direction. Then Δy is positive, the weight mg is positive, and the force exerted by the rim, $-k\,\Delta y$, is negative.

Apply $\Sigma \vec{F} = m\vec{a}$ to the player, and solve for k:

$$\Sigma F_y = w_y + F_y = ma_y$$
$$mg + (-k\Delta y) = 0$$
$$k = \frac{mg}{\Delta y} = \frac{(110 \text{ kg})(9.81 \text{ N/kg})}{0.15 \text{ m}}$$
$$= \boxed{7.19 \times 10^3 \text{ N/m}}$$

FIGURE 4-7

$\vec{F} = -k\Delta y\hat{j}$

$\vec{w} = m\vec{g}$

y

REMARKS Although a basketball rim doesn't look much like a spring, the rim is suspended by a hinge with a spring that is distorted when the front of the rim is pulled down. As a result, the upward force the rim exerts on the player's hands is proportional to the rim front's displacement and oppositely directed. Note that we used N/kg for the units of g so that kg cancels, giving N/m for the units of k. We can use either 9.81 N/kg or 9.81 m/s^2 for g, whichever is more convenient, because 1 N/kg = 1 m/s^2.

EXERCISE A 4-kg bunch of bananas is suspended motionless from a spring balance whose force constant is $k = 300$ N/m. By how much is the spring stretched? (*Answer* 13.1 cm)

EXERCISE A spring of force constant 400 N/m is attached to a 3-kg block that rests on a horizontal air track that renders friction negligible. What extension of the spring is needed to give the block an acceleration of 4 m/s^2 upon release? (*Answer* 3.0 cm)

EXERCISE IN DIMENSIONAL ANALYSIS An object of mass m oscillates at the end of an ideal spring of force constant k. The time for one complete oscillation is the period T. Assuming that T depends on m and k, use dimensional analysis to find the form of the relationship $T = f(m, k)$, ignoring numerical constants. This is most easily found by looking at the units. Note that the units of k are N/m = $(kg \cdot m/s^2)/m$ = kg/s^2, and the units of m are kg. (*Answer* $T = C\sqrt{m/k}$ where C is some dimensionless constant. The correct expression for the period, as we will see in Chapter 14, is $T = 2\pi\sqrt{m/k}$.

Strings   Strings (ropes) are used to pull things. We can think of a string as a spring with such a large force constant that the extension of the string is negligible. Strings are flexible, however, so unlike springs, they cannot push things. Instead they flex or bend. The magnitude of the force that one segment of a string exerts on an adjacent segment is called **tension.** It follows that if a string pulls on an object, the magnitude of the force equals the tension. The concept of tension in a string or rope is further developed in Section 4-7.

Constraints   A railroad car moves along a track. A wooden pony on a merry-go-round moves in a circle. A sled sliding along the surface of a frozen pond moves in a horizontal plane. Such conditions on the motion of objects are called **constraints.**

4-5 Problem Solving: Free-Body Diagrams

Imagine a dogsled being pulled across icy ground. The dog in front pulls on a rope attached to the sled (Figure 4-8a) with a horizontal force causing the sled to gain speed. We can think of the sled and rope together as a single object. What forces act on the sled-rope object? Both the dog and the ice touch the object, so we know that the dog and the ice exert contact forces on it. We also know that the earth exerts a gravitational force on the sled-rope (the object's weight). Thus, a total of three forces act on the object (assuming that friction is negligible):

1. The weight of the sled-rope $\vec{w}$.

2. The contact force $\vec{F}_n$ exerted by the ice (without friction, the contact force is directed normal to the ice.)

3. The contact force $\vec{F}$ exerted by the dog.

A diagram that shows schematically all the forces acting on a system, such as Figure 4-8b, is called a **free-body diagram.** It is called a free-body diagram because the body (object) is drawn free from (without) its surroundings. Drawing the force vectors on a free-body diagram to scale requires that we first determine the direction of the acceleration vector using kinematic methods. We know the object is moving to the right with increasing speed. It follows from kinematics that its acceleration vector is in the direction of its motion—to the right. Note that $\vec{F}_n$ and $\vec{w}$ in the diagram have equal magnitudes. The magnitudes must be equal because the vertical component of the acceleration is zero. As a check on the correctness of our free-body diagram, we draw a vector-addition diagram (Figure 4-9) verifying that the sum of the force vectors is in the direction of the acceleration vector.

The x component of Newton's second law gives

$$\Sigma F_x = F_{n,x} + w_x + F_x = ma_x$$
$$0 + 0 + F = ma_x$$

or

$$a_x = \frac{F}{m}$$

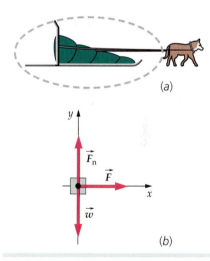

(a)

(b)

FIGURE 4-8 (*a*) A dog pulling a sled. The first step in problem solving is to isolate the system to be analyzed. In this case, the closed dashed curve represents the boundary between the sled-rope object and its surroundings. (*b*) The forces acting on the sled in Figure 4.8a.

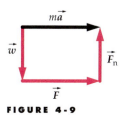

FIGURE 4-9

The y component of Newton's second law gives

$$\Sigma F_y = F_{n,y} + w_y + F_y = ma_y$$
$$F_n - w + 0 = 0$$

or

$$F_n = w$$

In this simple example, we found two things: the horizontal acceleration ($a_x = F/m$), and the vertical force $\vec{F}_n$ exerted by the ice ($F_n = w$).

A Dogsled Race   **EXAMPLE 4 - 6**

During your winter break, you enter a dogsled race in which students replace the dogs. Wearing cleats for traction, you begin the race by pulling on a rope attached to the sled with a force of 150 N at 25° with the horizontal. The mass of the sled-passenger-rope object is 80 kg and there is negligible friction between the sled runners and the ice. Find (*a*) the acceleration of the sled and (*b*) the normal force $\vec{F}_n$ exerted by the surface on the sled.

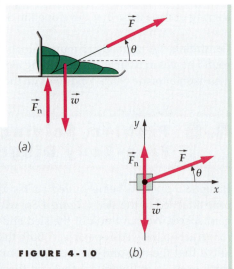

PICTURE THE PROBLEM Three forces act on the object: its weight $\vec{w}$, which acts downward; the normal force $\vec{F}_n$, which acts upward; and the force with which you pull the rope $\vec{F}$, directed 25° above the horizontal. Since the forces are not all parallel to a single line, we study the system by applying Newton's second law to the x and y directions separately.

(*a*) 1. Sketch a free-body diagram (Figure 4-10*b*) of the sled. Include a coordinate system with one of the coordinate axes in the direction of the sled's acceleration. The object moves to the right with increasing speed, so we know the acceleration is in that direction:

FIGURE 4-10   (*b*)

2. *Note:* Add the force vectors on the free-body diagram (Figure 4-11) to verify that their sum can be in the direction of the acceleration:

3. Apply Newton's second law to the object. Write out the equation in both vector and component form:

$$\vec{F}_n + \vec{w} + \vec{F} = m\vec{a}$$

or

$$F_{n,x} + w_x + F_x = ma_x$$
$$F_{n,y} + w_y + F_y = ma_y$$

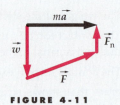

FIGURE 4-11

4. Express the x components of $\vec{F}_n$, $\vec{w}$, and $\vec{F}$:

$$F_{n,x} = 0, \quad w_x = 0, \quad \text{and} \quad F_x = F \cos \theta$$

5. Substitute the step 4 results into the x component equation in step 3. Then solve for the acceleration a_x:

$$0 + 0 + F \cos \theta = ma_x$$

$$a_x = \frac{F \cos \theta}{m} = \frac{(150 \text{ N}) \cos 25°}{80 \text{ kg}} = \boxed{1.70 \text{ m/s}^2}$$

(*b*) 1. Express the y component of $\vec{a}$:

$$a_y = 0$$

2. Express the y components of $\vec{F}_n$, $\vec{w}$, and $\vec{F}$:

$$F_{n,y} = F_n, \quad w_y = -mg, \quad \text{and} \quad F_y = F \sin \theta$$

3. Substitute the step *b*1 and *b*2 results into the y component equation in step *a*3. Then solve for F_n:

$$\Sigma F_y = F_n - mg + F \sin \theta = 0$$

$$F_n = mg - F \sin \theta$$

$$= (80 \text{ kg})(9.81 \text{ N/kg}) - (150 \text{ N}) \sin 25° = \boxed{721 \text{ N}}$$

REMARKS Note that only the x component of $\vec{F}$, which is $F\cos\theta$, causes the object to accelerate. Also note that the ice supports less than the full weight of the object since part of the weight, $F\sin\theta$, is supported by the rope.

⓿ **PLAUSIBILITY CHECK** If $\theta = 0$, the object is accelerated by a force F and the ice supports all the object's weight. Our results agree, giving $a_x = F/m$ and $F_n = mg$.

EXERCISE If $\theta = 25°$, what is the maximum force F that can be applied to the rope without lifting the sled off the surface? (*Answer* $F = 1.86$ kN)

Example 4-6 illustrates a general method for solving problems using Newton's laws:

1. Draw a neat diagram that includes the important features of the problem.

2. Isolate the object (particle) of interest, and draw a free-body diagram showing each external force that acts on it. If there is more than one object of interest in the problem, draw a separate free-body diagram for each. Choose a convenient coordinate system for each object and include it on that object's free-body diagram. If the direction of the acceleration is known, choose a coordinate axis that is parallel to it. For objects sliding along a surface, choose one coordinate axis parallel to the surface and the other perpendicular to it.

3. Apply Newton's second law, $\Sigma\vec{F} = m\vec{a}$, usually in component form.

4. For problems involving two or more objects, make use of Newton's third law, $\vec{F}_{A,B} = -\vec{F}_{B,A}$, and any constraints to simplify the equations obtained from applying $\Sigma\vec{F} = m\vec{a}$.

5. Solve the resulting equations for the unknowns.

6. Check to see whether your results have the correct units and seem plausible. Substituting extreme values into your solution is a good way to check your work for errors.

SOLVING PROBLEMS USING NEWTON'S LAWS

UNLOADING A TRUCK   **EXAMPLE 4-7**   **Put It in Context**

You are working for a big delivery company, and must unload a large, fragile package from your truck, using a delivery ramp (Figure 4-12). If the downward component of the velocity of the package when it reaches the bottom of the ramp is greater than 2.5 m/s (the speed an object would have if it were dropped from a height of about 1 ft), the package will break. What is the largest angle at which you can safely unload? The ramp is 1-m high, has rollers (i.e., the ramp is approximately frictionless), and is inclined at an angle θ to the horizontal.

PICTURE THE PROBLEM Two forces act on the box, the weight $\vec{w}$ and the normal force $\vec{F}_n$. Since these forces are not parallel to a single line, they cannot sum to zero, hence there is a net force on the box causing it to accelerate. The ramp constrains the box to move parallel to its surface, so we choose down the incline as the positive x direction. To determine the acceleration we apply Newton's second law to the box. Once the acceleration is known, we can use kinematics to determine the largest safe angle.

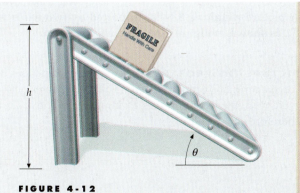

FIGURE 4-12

1. Relate the downward component of the velocity of the box to its speed v along the ramp:

$$v_d = v \sin \theta$$

2. The speed v is related to the displacement Δx along the ramp by the kinematic equation:

$$v^2 = v_0^2 + 2a_x \, \Delta x$$

3. To find a_x we apply Newton's second law ($\Sigma F_x = ma_x$) to the package. First we draw a free-body diagram (Figure 4-13). Two forces act on the package, the weight force and the normal force. We choose the direction of the acceleration, down the ramp, as the $+x$ direction.

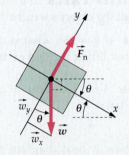

Note: The angle between $\vec{w}$ and the negative y axis is the same as the angle between the incline and the horizontal as we see from the free-body diagram. We can also see that $w_x = w \sin \theta$.

FIGURE 4-13

4. Applying Newton's second law gives:

$$F_{n,x} + w_x = ma_x$$

where

Note: F_n is perpendicular to the x axis and $w = mg$.

$$F_{n,x} = 0 \quad \text{and} \quad w_x = w \sin \theta = mg \sin \theta$$

5. Substituting and solving for the acceleration gives:

$$0 + mg \sin \theta = ma_x$$

so

$$a_x = g \sin \theta$$

6. Substituting for a_x in the kinematic equation (step 2) and setting v_0 to zero gives:

$$v^2 = 2g \sin \theta \, \Delta x$$

7. From Figure 4-12 we can see that when Δx equals the length of the ramp, $\Delta x \sin \theta = h$, where h is the height of the ramp:

$$v^2 = 2gh$$

8. Using $v_d = v \sin \theta$, solve for v_d:

$$v_d = \sqrt{2gh} \sin \theta$$

9. Solve for the maximum angle:

$$2.5 \text{ m/s} = \sqrt{2(9.81 \text{ m/s}^2)(1.0 \text{ m})} \sin \theta_{max}$$

$$\boxed{\theta_{max} = 34.4°}$$

REMARKS The acceleration down the incline is constant and equal to $g \sin \theta$. Also, the speed v at the bottom ($v = \sqrt{2gh}$) does not depend on the angle θ.

EXERCISE Apply $\Sigma F_y = ma_y$ to the package and show that $F_n = mg \cos \theta$.

PICTURE HANGING   **EXAMPLE 4-8**   **Try It Yourself**

A picture weighing 8 N is supported by two wires with tensions T_1 and T_2, as shown in Figure 4-14. Find each tension.

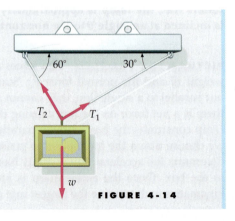

PICTURE THE PROBLEM Because the picture does not accelerate, the net force acting on it must be zero. The three forces acting on the picture (its weight $\vec{w}$, the tension forces $\vec{T}_1$ in one wire and $\vec{T}_2$ in the other wire) must therefore sum to zero.

FIGURE 4-14

Cover the column to the right and try these on your own before looking at the answers.

Steps:

Answers:

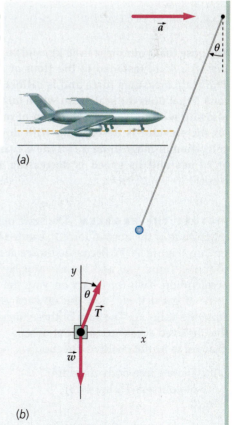

1. Draw a free-body diagram for the picture (Figure 4-15). On your diagram show the x and y components of the two tension forces.

2. Apply $\Sigma \vec{F} = m\vec{a}$ in vector form to the picture.

$$\vec{T}_1 + \vec{T}_2 + \vec{w} = m\vec{a}$$

3. Resolve each force into its x and y components. This gives you two equations for the two unknowns T_1 and T_2.

$$T_{1,x} + T_{2,x} + w_x = 0$$

$$T_{1,y} + T_{2,y} + w_y = 0$$

$$T_1 \cos 30° - T_2 \cos 60° + 0 = 0$$

$$T_1 \sin 30° + T_2 \sin 60° - w = 0$$

FIGURE 4-15

4. Solve the x component equation for T_2 in terms of T_1.

$$T_2 = T_1 \frac{\cos 30°}{\cos 60°} = T_1 \sqrt{3}$$

5. Substitute your result for T_2 (from step 4) into the y component equation and solve for T_1.

$$T_1 \sin 30° + (T_1 \sqrt{3})\sin 60° - w = 0$$

$$T_1 = \tfrac{1}{2}w = \boxed{4\text{N}}$$

6. Use your result for T_1 to find T_2.

$$T_2 = T_1 \sqrt{3} = \boxed{6.93 \text{ N}}$$

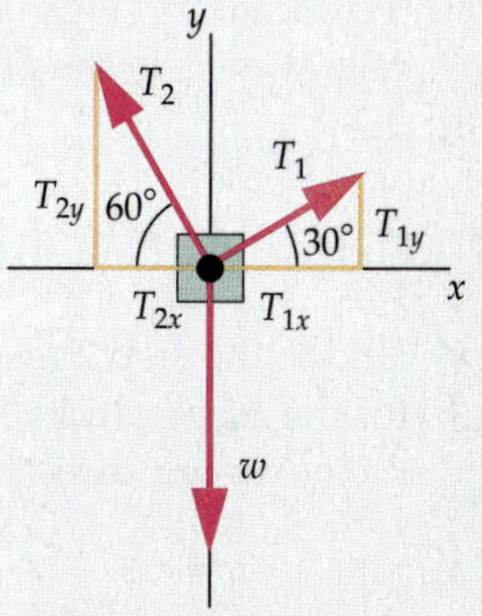

REMARKS Note that the more vertical of the two wires supports the greater share of the load, as you might expect. Also, we see that $T_1 + T_2 > 8$ N. The "extra" force is due to the wires pulling to the right and left.

AN ACCELERATING JET PLANE   **EXAMPLE 4-9**

As your jet plane speeds down the runway on takeoff, you decide to determine its acceleration, so you take out your yo-yo and note that when you suspend it, the string makes an angle of 22° with the vertical (Figure 4-16a). (a) What is the acceleration of the plane? (b) If the mass of the yo-yo is 40 g, what is the tension in the string?

(a)

PICTURE THE PROBLEM The yo-yo and plane have the same acceleration to the right. The net force on the yo-yo is in the direction of its acceleration. This force is supplied by the horizontal component of the tension force $\vec{T}$. The vertical component of $\vec{T}$ balances the weight of the yo-yo. We choose a coordinate system in which the x direction is parallel to the acceleration vector $\vec{a}$ and the y direction is vertical. Writing Newton's second law for both the x and y directions gives two equations to determine the two unknowns, a and T.

(a) 1. Draw a free-body diagram for the yo-yo (Figure 4-16b). Choose the postive x direction to be the direction of the acceleration.

2. Apply $\Sigma F_x = ma_x$ to the yo-yo. Then simplify using trigonometry:

$$T_x + w_x = ma_x$$

$$T \sin \theta + 0 = ma_x$$

or

$$T \sin \theta = ma_x$$

(b)

FIGURE 4-16

3. Apply $\Sigma F_y = ma_y$ to the yo-yo. Then simplify using trigonometry (Figure 4-16c) and $w = mg$. Since the acceleration is in the positive x direction, $a_y = 0$:

$$T_y + w_y = ma_y$$

$$T\cos\theta - mg = 0$$

or

$$T\cos\theta = mg$$

(c)

FIGURE 4-16

4. Divide the step 2 result by the step 3 result and solve for the acceleration. Since the acceleration vector is in the positive x direction, $a = a_x$:

$$\frac{T\sin\theta}{T\cos\theta} = \frac{ma_x}{mg}, \quad \text{so} \quad \tan\theta = \frac{a_x}{g}$$

and

$$a = g\tan\theta = (9.81 \text{ m/s}^2)\tan 22° = \boxed{3.96 \text{ m/s}^2}$$

(b) Using the step 3 result, solve for the tension:

$$T = \frac{mg}{\cos\theta} = \frac{(0.04 \text{ kg})(9.81 \text{ m/s}^2)}{\cos 22°} = \boxed{0.423 \text{ N}}$$

REMARKS Notice that T is greater than the weight of the yo-yo ($mg = 0.392$ N) because the cord not only keeps the yo-yo from falling but also accelerates it in the horizontal direction. Here we use the units m/s² for g because we are calculating acceleration.

PLAUSIBILITY CHECK At $\theta = 0$, we find that $T = mg$ and $a = 0$.

EXERCISE For what acceleration magnitude a would the tension in the string be equal to $3mg$? What is θ in this case? (*Answer* $a = 27.8$ m/s², $\theta = 70.5°$)

Our next example is the application of Newton's second law to objects that are at rest relative to a reference frame that is itself accelerating.

YOUR WEIGHT IN AN ELEVATOR   **EXAMPLE 4-10**

Suppose that your mass is 80 kg, and you are standing on a scale fastened to the floor of an elevator. The scale measures force and is calibrated in newtons. What does the scale read when (*a*) the elevator is rising with upward acceleration of magnitude *a*; (*b*) the elevator is descending with downward acceleration of magnitude *a'*; (*c*) the elevator is rising at 20 m/s and its speed is decreasing at a rate of 8 m/s²?

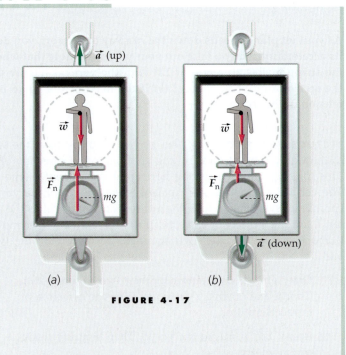

(a)   (b)

FIGURE 4-17

PICTURE THE PROBLEM The scale reading is the magnitude of the normal force F_n exerted by the scale on you (Figure 4-17). Because you are at rest relative to the elevator, you and the elevator have the same acceleration. Two forces act on you: the downward force of gravity, mg, and the upward normal force from the scale, F_n. The sum of these forces gives you the observed acceleration. We choose upward to be the positive direction.

(*a*) 1. Draw a free-body diagram of yourself (Figure 4-18):

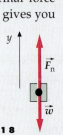

FIGURE 4-18

2. Apply $\Sigma\vec{F} = m\vec{a}$ in the y direction:

$$F_{n,y} + w_y = ma_y$$
$$F_n - mg = ma$$

3. Solve for F_n. This is the reading on the scale (your apparent weight):

$$F_n = mg + ma = \boxed{m(g + a)}$$

(b) 1. Apply $\Sigma\vec{F} = m\vec{a}$ in the y direction for the case in which the elevator accelerates downward with magnitude a':

$$F_{n,y} + w_y = ma_y$$
$$F_n - mg = m(-a')$$

2. Solve for F_n:

$$F_n = mg - ma' = \boxed{m(g - a')}$$

(c) 1. Apply $\Sigma\vec{F} = m\vec{a}$ in the y direction. Note that the acceleration is downward. (Why is that?) It follows that a_y is negative:

$$F_{n,y} + w_y = ma_y$$

2. Solve for F_n:

$$F_n - mg = ma_y$$
$$F_n = m(g + a_y) = (80 \text{ kg})(9.81 \text{ m/s}^2 - 8.00 \text{ m/s}^2) = \boxed{145 \text{ N}}$$

REMARKS Whether the elevator is ascending or descending, if it accelerates upward, then your apparent weight is greater than mg by ma. For you, it is as if gravity were increased from g to $g + a$. If it accelerates downward, then your apparent weight is less than mg by the amount ma'. You feel lighter, as if gravity were $g - a'$. If $a' = g$, the elevator is in free-fall, and you experience weightlessness.

EXERCISE An elevator descending comes to a stop with an acceleration of magnitude 4 m/s². If your mass is 70 kg and you are standing on a scale in the elevator, what does the scale read as the elevator is stopping? (*Answer* 967 N)

EXERCISE A man stands on a scale in an elevator that has an upward acceleration a. The scale reads 960 N. When he picks up a 20 kg box, the scale reads 1200 N. Find the mass of the man, his weight, and the acceleration a.

4-6 Newton's Third Law

When two objects interact, they exert forces on each other. Newton's third law states that these forces are equal in magnitude and opposite in direction. That is, if object A exerts a force on object B, then object B exerts a force on A that is equal in magnitude and opposite in direction. Thus forces always occur in pairs. It is common to refer to one force in the pair as an action and the other as a reaction. This terminology is unfortunate because it sounds like one force "reacts" to the other, which is not the case. The two forces occur simultaneously. **Either can be called the action and the other the reaction.** If we refer to an external force acting on a particular object as an action force, then the corresponding reaction force must act on a different object. Thus no two external forces acting on a single object can ever constitute an action–reaction pair.

In Figure 4-19, a block rests on a table. The force acting downward on the block is the weight $\vec{w}$ due to the attraction of the earth. An equal and opposite force $\vec{w}' = -\vec{w}$ is exerted by the block on the earth. These forces form an action–reaction pair. If they were the only forces present, the block would accelerate downward because it would have only a single force acting on it (and the earth would accelerate upward toward the block). However, the table exerts an

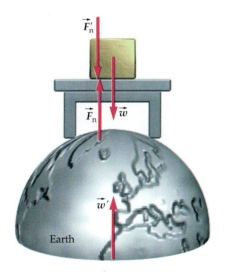

FIGURE 4-19

upward force $\vec{F}_n$ on the block that balances the block's weight. The block also exerts a force $\vec{F}'_n = -\vec{F}_n$ downward on the table. The forces $\vec{F}_n$ and $\vec{F}'_n$ also form an action–reaction pair.

EXERCISE Do the forces $\vec{w}$ and $\vec{F}_n$ in Figure 4-19 form an action–reaction pair? (*Answer* No, they do not. These forces are both external forces and they both act on the same object, the block. Thus, they cannot constitute an action–reaction pair.)

THE HORSE BEFORE THE CART **EXAMPLE 4-11**

A horse refuses to pull a cart (Figure 4-20*a*). The horse reasons, "according to Newton's third law, whatever force I exert on the cart, the cart will exert an equal and opposite force on me, so the net force will be zero and I will have no chance of accelerating the cart." What is wrong with this reasoning?

PICTURE THE PROBLEM Because we are interested in the motion of the cart, we draw a simple diagram for it (Figure 4-20*b*). The force exerted by the horse on the harness is labeled $\vec{F}$. (The harness is attached to the cart, and so we can consider it to be part of the cart.) Other forces acting on the cart are its weight $\vec{w}$, the vertical support force of the ground $\vec{F}_n$, and the horizontal force exerted by the pavement, labeled $\vec{f}$ (for friction).

1. Draw a free-body diagram for the cart (see Figure 4-20*c*). Because the cart does not accelerate vertically, the vertical forces must sum to zero. The horizontal forces are $\vec{F}$ to the right and $\vec{f}$ to the left. The cart will accelerate to the right if $\vec{F}$ is greater than $\vec{f}$.

2. Note that the reaction force to $\vec{F}$, which we call $\vec{F}'$, is exerted on the horse, not on the cart (Figure 4-20*d*). It has no effect on the motion of the cart, but it does affect the motion of the horse. If the horse is to accelerate to the right, there must be a force $\vec{F}_P$ (to the right) exerted by the pavement on the horse's hooves that is greater than $\vec{F}'$.

REMARKS This example illustrates the importance of drawing a simple diagram when solving mechanics problems. Had the horse done so, he would have seen that he need only push back hard against the pavement so that the pavement will push him forward.

EXERCISE As you stand facing a friend, place your palms against your friend's palms and push. Can your friend exert a force on you if you do not exert a force back? Try it.

EXERCISE True or false: The force exerted by the cart on the horse is equal and opposite to the force exerted by the horse on the cart, but only when the horse and cart are not accelerating. (*Answer* False! An action–reaction pair of forces describes the interaction of two objects. One force cannot exist without the other. They are *always* equal and opposite.)

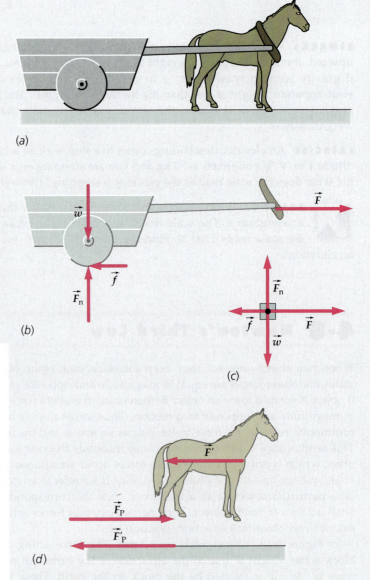

(a)

(b)

(c)

(d)

FIGURE 4-20

4-7 Problems With Two or More Objects

In some problems, two or more objects are in contact or are connected by a string or spring. Such problems are solved by drawing a separate free-body diagram for each object and then applying Newton's second law to each object. The resultant equations, together with any equations describing interactions and constraints, are solved simultaneously for the unknown quantities. If the objects are in direct contact, the forces they exert on each other must be equal and opposite, as stated in Newton's third law. For two objects, each moving in a straight line, that are connected by a taut nonstretching string, the acceleration components parallel to the string are the same for both objects. This is so because, for each object, its motion parallel to the string is identical with that of the other object. If the string passes over a pulley or peg, the phrase "parallel to the string" means parallel with that segment of the string attached to the object.

Consider the motion of Steve and Paul in Figure 4-21. The rate at which Paul descends equals the rate at which Steve slides along the glacier. That is, Paul's velocity component parallel with the segment of the rope attached to him equals Steve's velocity component parallel with the segment attached to him. These two velocity components must remain equal. If Steve and Paul are changing speed, they do so at the same rate. Their acceleration components parallel with the rope must be equal.

FIGURE 4-21

The tension in a string or rope is the magnitude of the force that one segment of the rope exerts on a neighboring segment. The tension can vary throughout the rope. For a rope dangling from a girder at the ceiling of a gymnasium, the tension is greatest near the top because the segment at the top has to support the weight of all the rope below it. However, for the problems in this book, the masses of strings and ropes are normally assumed to be so small that variations in tension due to the weight of a string or rope can be neglected. Conveniently, this also means that variations in the tension due to any acceleration of the rope can also be neglected. To see that this is so, consider the free-body diagram of a segment of the rope attached to Steve, where Δm_s is the segment's mass (Figure 4-22).

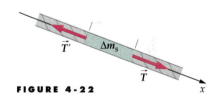

FIGURE 4-22

Applying Newton's second law to the segment gives $T - T' = \Delta m_s a_x$. If the mass of the segment is negligible, then $T = T'$. No net force is needed to give the segment an acceleration. (That is, only a negligible difference in tension is needed to give a rope segment of negligible mass any finite acceleration.)

Next we consider the entire rope connecting Steve and Paul. Neglecting gravity, there are three forces acting on the rope. Steve and Paul each exert a force, as does the ice at the edge of the glacier. Neglecting any friction between the ice and the rope means that the force exerted by the ice is always a normal force (Figure 4-23). A normal force has no component along the rope, so it cannot produce a change in the tension. Thus the tension is the same throughout the entire length of the rope. To summarize, if a taut rope of negligible mass changes direction by passing over a frictionless surface, the tension is the same throughout the rope.

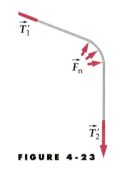

FIGURE 4-23

EXERCISE Suppose that instead of passing over the edge of a glacier, the rope passed around a pulley with frictionless bearings as shown in Figure 4-24. Would the tension then be the same throughout the length of the rope? (*Answer* No. Doing away with friction in the bearing is one thing, but the pulley still has inertia (mass). A difference in tension is needed to change the rate of rotation of the pulley.)

FIGURE 4-24

THE ROCK CLIMBERS   **EXAMPLE 4-12**

Paul (mass m_P) accidentally falls off the edge of a glacier as shown in Figure 4-21. Fortunately he is tied by a long rope to Steve (mass m_S), who has a climbing ax. Before Steve sets his ax to stop them, he slides without friction along the ice, attached by the rope to Paul. Assume there is no friction between the rope and the glacier. Find the acceleration of each person and the tension in the rope.

PICTURE THE PROBLEM The tension forces $\vec{T}_1$ and $\vec{T}_2$ have equal magnitudes because the rope is assumed to be massless and the glacier ice is assumed to be frictionless. The rope does not stretch or become slack, so Paul and Steve have equal speeds at all times. Their accelerations $\vec{a}_S$ and $\vec{a}_P$ must therefore be equal in magnitude, but not in direction. Steve accelerates down the face of the glacier whereas Paul accelerates vertically downward.

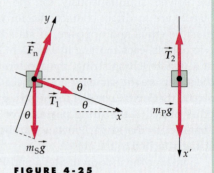

FIGURE 4-25

Newton's second law relates each person's acceleration to the forces acting on him. Apply $\Sigma\vec{F} = m\vec{a}$ to each, and solve for the accelerations and the tension.

1. Draw separate free-body diagrams for Steve and Paul (Figure 4-25). Put axes x and y on Steve's diagram, choosing the direction of Steve's acceleration as the positive x direction. Choose the direction of Paul's acceleration as the positive x' direction.

2. Apply $\Sigma\vec{F} = m\vec{a}$ in the x direction to Steve: $\qquad F_{n,x} + T_{1,x} + m_S g_x = m_S a_{S,x}$

3. Apply $\Sigma\vec{F} = m\vec{a}$ in the x' direction to Paul: $\qquad T_{2,x'} + m_P g_{x'} = m_P a_{P,x'}$

4. Because they are each moving in a straight line and are connected by a taut length of rope that does not stretch, the accelerations of Paul and Steve are related. Express this relation:

$$a_{P,x'} = a_{S,x} = a_x$$

5. Because the rope is of negligible mass and slides over the ice with negligible friction, the forces $\vec{T}_1$ and $\vec{T}_2$ are related. Express this relation:

$$T_2 = T_1 = T$$

6. Substitute the steps 4 and 5 results into the step 2 and step 3 equations:

$$T + m_S g \sin\theta = m_S a_x$$
$$-T + m_P g = m_P a_x$$

7. Solve the step 6 equations for the acceleration by eliminating T and solving for a_x:

$$a_x = \boxed{\frac{m_S \sin\theta + m_P}{m_S + m_P} g}$$

8. Substitute the step 7 result into either step 6 equation and solve for T:

$$T = \boxed{\frac{m_S m_P}{m_S + m_P}(1 - \sin\theta)g}$$

REMARKS In Step 3 we chose downward to be positive to keep the solution as simple as possible. With this choice, when Steve moves in the positive x direction (down the glacier), Paul moves in the positive x' direction (downward).

PLAUSIBILITY CHECK If m_P is very much greater than m_S, we expect the acceleration to be approximately g and the tension to be approximately zero. Substituting $m_S = 0$ does indeed give $a = g$ and $T = 0$ for this case. If m_P is much less than m_S, we expect the acceleration to be approximately $g \sin\theta$ (see Example 4-8) and the tension to be zero. Substituting $m_P = 0$ in steps 7 and 8, we indeed obtain $a_x = g \sin\theta$ and $T = 0$. At the extreme value of the inclination ($\theta = 90°$) we check our answers. Substituting $\theta = 90°$ in steps 7 and 8, we obtain $a_x = g$ and $T = 0$. This seems right since Steve and Paul would be in free-fall for $\theta = 90°$.

EXERCISE (*a*) Find the acceleration if $\theta = 15°$ and if the masses are $m_S = 78$ kg and $m_P = 92$ kg. (*b*) Find the acceleration if these two masses are interchanged. (*Answer* (*a*) $a_x = 0.660g$ (*b*) $a_x = 0.599g$)

BUILDING A SPACE STATION

EXAMPLE 4-13 **Try It Yourself**

You are an astronaut constructing a space station, and you push on a box of mass m_1 with a force of $\vec{F}_A$. The box is in direct contact with a second box of mass m_2 (Figure 4-26). (*a*) What is the acceleration of the boxes? (*b*) What is the magnitude of the force exerted by one box on the other?

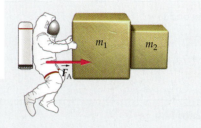

PICTURE THE PROBLEM Let $\vec{F}_{2,1}$ be the force exerted by box 2 on box 1, and $\vec{F}_{1,2}$ be the force exerted by box 1 on box 2. In accord with Newton's third law, these forces are equal and opposite ($\vec{F}_{2,1} = -\vec{F}_{1,2}$), so $F_{2,1} = F_{1,2}$. Apply Newton's second law to each box separately. The motions of the two boxes are identical, so the accelerations $\vec{a}_1$ and $\vec{a}_2$ are equal.

FIGURE 4-26

Cover the column to the right and try these on your own before looking at the answers.

Steps

Answers

(*a*) 1. Draw free-body diagrams for the two boxes (Figure 4-27).

2. Apply $\Sigma\vec{F} = m\vec{a}$ to box 1.

$$F_A - F_{2,1} = m_1 a_{1,x}$$

3. Apply $\Sigma\vec{F} = m\vec{a}$ to box 2.

$$F_{1,2} = m_2 a_{2,x}$$

4. Express both the relation between the two accelerations and the relation between the magnitudes of the forces the blocks exert on each other.

$$a_{2,x} = a_{1,x} = a_x$$
$$F_{2,1} = F_{1,2} = F$$

FIGURE 4-27

5. Substitute these back into the step 2 and step 3 results and solve for a_x.

$$a_x = \boxed{\frac{F_A}{m_1 + m_2}}$$

(*b*) Substitute your expression for a_x into either the step 2 or the step 3 result and solve for F.

$$F = \boxed{\frac{m_2}{m_1 + m_2} F_A}$$

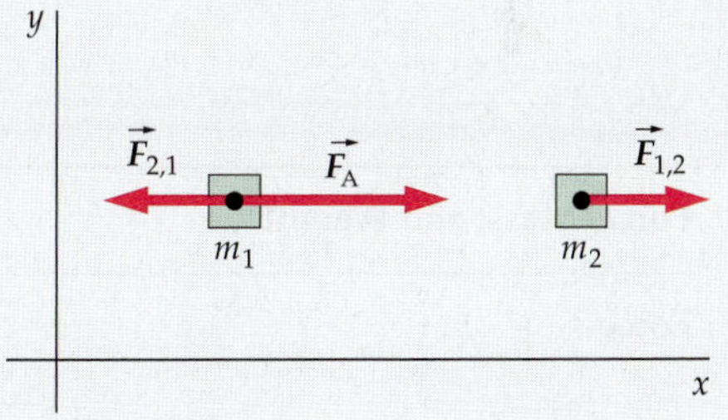

REMARKS Note that the result in step 5 is the same as if the force $\vec{F}_A$ had acted on a single mass equal to the sum of the masses of the two boxes. In fact, since the two boxes have the same acceleration, we can consider them to be a single object with mass $m_1 + m_2$.

EXERCISE (*a*) Find the acceleration and the contact force if $m_1 = 2$ kg, $m_2 = 3$ kg, and $F_A = 12$ N. (*b*) Find the contact force if the two boxes are interchanged so that the first block has a mass of 3 kg and the second block has a mass of 2 kg. (*Answer* (*a*) $a_x = 2.4$ m/s², $F = 7.2$ N (*b*) $F = 4.8$ N)

SUMMARY

1. Newton's laws of motion are fundamental laws of nature that serve as the basis for our understanding of mechanics.

2. Mass is an *intrinsic* property of an object.

3. Force is an important *derived* dynamic quantity.

Topic	Relevant Equations and Remarks
1. Newton's Laws	
First law	An object at rest stays at rest unless acted on by an external force. An object in motion continues to travel with constant velocity unless acted on by an external force. (Reference frames in which this occurs are called inertial reference frames.)
Second law	The magnitude of the acceleration is proportional to the magnitude of the net external force $\vec{F}_{net}$, in accordance with $\vec{F}_{net} = m\vec{a}$, where m is the mass of the object. The net force acting on an object, also called the resultant force, is the vector sum of all the forces acting on it: $\vec{F}_{net} = \Sigma\vec{F}$. Thus $$\Sigma\vec{F} = m\vec{a} \qquad \text{4-1}$$
Third law	Forces always occur in equal and opposite pairs. If object A exerts a force on object B, an equal but opposite force is exerted by object B on object A: $$\vec{F}_{A,B} = -\vec{F}_{B,A} \qquad \text{4-2}$$
2. Inertial Reference Frames	Our statements of Newton's laws are valid only in an inertial reference frame—a reference frame for which an object at rest remains at rest if no force acts on the object. Any reference frame that is moving with constant velocity relative to an inertial reference frame is itself an inertial reference frame, and any reference frame that is accelerating relative to an inertial frame is not an inertial reference frame. The earth's surface is, to a good approximation, an inertial reference frame.
3. Force, Mass, and Weight	
Force	Force is defined in terms of the acceleration it produces on a given object. A force of 1 newton (N) is that force which produces an acceleration of 1 m/s^2 on a mass of 1 kilogram (kg).
Mass	Mass is the intrinsic property of an object that measures its inertial resistance to acceleration. Mass does not depend on the location of the object. Applying identical forces to each of two objects and measuring their respective accelerations allows the masses of two objects to be compared. The ratio of the masses of the objects is equal to the inverse ratio of the accelerations produced: $$\frac{m_2}{m_1} = \frac{a_1}{a_2} \qquad \text{4-3}$$
Weight	The weight $\vec{w}$ of an object is the force of gravitational attraction exerted by the earth on the object. It is proportional to the mass m of the object and the gravitational field $\vec{g}$, which equals the free-fall acceleration: $$\vec{w} = m\vec{g} \qquad \text{4-4}$$ Weight is not an intrinsic property of an object; it depends on the location of the object.
4. Fundamental Forces	All the forces observed in nature can be explained in terms of four basic interactions: 1. The gravitational force 2. The electromagnetic force 3. The strong nuclear force (also called the hadronic force) 4. The weak nuclear force
5. Contact Forces	Contact forces of support and friction and those exerted by springs and strings are due to molecular forces that arise from the basic electromagnetic force.
Hooke's law	When a relaxed spring is compressed or extended by a small amount Δx, the force it exerts is proportional to Δx: $$F_x = -k\,\Delta x \qquad \text{4-9}$$

PROBLEMS

- Single-concept, single-step, relatively easy
- •• Intermediate-level, may require synthesis of concepts
- ••• Challenging
- **SSM** Solution is in the *Student Solutions Manual*
- **iSOLVE** Problems available on iSOLVE online homework service
- **iSOLVE ✔** These "Checkpoint" online homework service problems ask students additional questions about their confidence level, and how they arrived at their answer

In a few problems, you are given more data than you actually need; in a few other problems, you are required to supply data from your general knowledge, outside sources, or informed estimates.

For all problems, use $g = 9.81$ m/s² for the free-fall acceleration and neglect friction and air resistance unless instructed to do otherwise.

Conceptual Problems

1 •• **SSM** How can you tell if a particular reference frame is an inertial reference frame?

2 •• Suppose you observe an object from a reference frame and find that it has an acceleration $\vec{a}$ when there are no forces acting on it. How can you use this information to find an inertial frame?

3 • If an object has no acceleration when observed from an inertial reference frame, can you conclude that no forces are acting on it?

4 • **SSM** If only a single nonzero force acts on an object, must the object have an acceleration relative to any inertial reference frame? Can it ever have zero velocity?

5 • If an object is acted upon by a single known force, can you tell in which direction the object will move, using no other information?

6 • An object is observed to be moving at constant velocity in an inertial reference frame. It follows that (*a*) no forces act on the object, (*b*) a constant force acts on the object in the direction of motion, (*c*) the net force acting on the object is zero, (*d*) the net force acting on the object is equal and opposite to its weight.

7 • Suppose an object was sent far out in space, away from galaxies, stars, or other bodies. How would its mass change? Its weight?

8 • **SSM** How would an astronaut in apparent weightlessness be aware of her mass?

9 • **SSM** Under what circumstances would your apparent weight be greater than your true weight?

10 •• It is often said that Newton's first and second laws imply that it is impossible to use the laws of mechanics to tell if you are standing still or moving with a constant velocity. Explain.

11 • Suppose a block of mass m_1 rests on a block of mass m_2 and the combination rests on a table as shown in Figure 4-28. Find the force exerted (*a*) by m_1 on m_2, (*b*) by m_2 on m_1, (*c*) by m_2 on the table, (*d*) by the table on m_2.

FIGURE 4-28 Problem 11

12 • **SSM** True or false:

(*a*) If two external forces that are both equal in magnitude and opposite in direction act on the same object, the two forces can never be an action-reaction force pair.

(*b*) Action equals reaction only if the objects are not accelerating.

13 • An 80-kg man on ice skates pushes his 40-kg son, also on skates, with a force of 100 N. The force exerted by the boy on his father is (*a*) 200 N, (*b*) 100 N, (*c*) 50 N, (*d*) 40 N.

14 • A girl holds a bird in her hand. The reaction force to the weight of the bird is (*a*) the gravitational force of the earth on the bird, (*b*) the gravitational force of the bird on the earth, (*c*) the contact force of the hand on the bird, (*d*) the contact force of the bird on the hand, (*e*) the gravitational force of the earth on the hand.

15 • A baseball player hits a ball with a bat. If the force with which the bat hits the ball is considered the action force, what is the reaction force? (*a*) The force the bat exerts on the batter's hands. (*b*) The force on the ball exerted by the glove of the person who catches it. (*c*) The force the ball exerts on the bat. (*d*) The force the pitcher exerts on the ball while throwing it. (*e*) Friction, as the ball rolls to a stop.

16 • Consider any situation in which an external force, say a push, is applied to an object. If Newton's third law requires that *for every action there is an equal and opposite reaction*, why doesn't the reaction force always cancel out the applied force, leaving no acceleration at all?

17 • **SSM** A 2.5-kg object hangs at rest from a string attached to the ceiling. (*a*) Draw a diagram showing all the forces on the object and indicate the reaction force to each. (*b*) Do the same for each force acting on the string.

18 • Which of the free-body diagrams in Figure 4-29 represents a block sliding down a frictionless inclined surface?

FIGURE 4-29 Problem 18

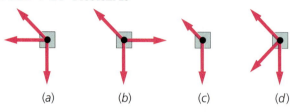

    (*a*)   (*b*)   (*c*)   (*d*)

19 • For an observer in an inertial reference frame, identify which (if any) of the following statements are true and which (if any) are false.

(*a*) If there are no forces acting on an object, it will not accelerate.
(*b*) If an object is not accelerating, there must be no forces acting on it.
(*c*) The motion of an object is always in the direction of the resultant force.
(*d*) The mass of an object depends on its location.

20 • A sky diver of weight *w* is descending near the surface of the earth. What is the magnitude of the force exerted by her body *on the earth*? (*a*) *w*. (*b*) Greater than *w*. (*c*) Less than *w*. (*d*) 9.8*w*. (*e*) 0. (*f*) It depends on the air resistance.

21 • **SSM** The net force on a moving object is suddenly reduced to zero and remains zero. As a consequence, the object (*a*) stops abruptly, (*b*) stops during a short time interval, (*c*) changes direction, (*d*) continues at constant velocity, (*e*) changes velocity in an unknown manner.

22 • A clothesline is stretched taut between two poles. Then a wet towel is hung at the center of the line. Can the line remain horizontal? Explain.

23 • What effect does the velocity of an elevator have on the apparent weight of a person in the elevator?

Estimation and Approximation

24 •• A car traveling 90 km/h crashes into the rear end of an unoccupied stalled vehicle. Fortunately, the driver is wearing a seat belt. Using reasonable values for the mass of the driver and the stopping distance, estimate the force (assuming it to be constant) exerted on the driver by the seat belt.

25 ••• **SSM** Making any necessary assumptions, find the normal force and the tangential force exerted by the road on the wheels of your bicycle (*a*) as you climb an 8% grade at constant speed, and (*b*) as you descend the 8% grade at constant speed. (An 8% grade means that the angle of inclination θ is given by $\tan\theta = 0.08$.)

Newton's First and Second Laws: Mass, Inertia, and Force

26 • A particle of mass *m* is traveling at an initial speed $v_0 = 25.0$ m/s. When a net force of 15.0 N acts on it, it comes to a stop in a distance of 62.5 m. What is *m*? (*a*) 37.5 kg. (*b*) 3.00 kg. (*c*) 1.50 kg. (*d*) 6.00 kg. (*e*) 3.75 kg.

27 • (*a*) An object has an acceleration of 3 m/s² when the only force acting on it is F_0. What is its acceleration when this force is doubled? (*b*) A second object has an acceleration of 9 m/s² under the influence of the force F_0. What is the ratio of the masses of the two objects? (*c*) If the two objects are glued together, what acceleration will the force F_0 produce?

28 • **ISOLVE ✓** A tugboat tows a ship with a constant force F_1. The increase in the ship's speed in a 10-s interval is 4 km/h. When a second tugboat applies an additional constant force F_2 in the same direction, the speed increases by 16 km/h in a 10-s interval. How do the magnitudes of the two forces compare? (Neglect water resistance.)

29 •• **SSM** **ISOLVE** A bullet of mass 1.8×10^{-3} kg moving at 500 m/s impacts a large fixed block of wood and travels 6 cm before coming to rest. Assuming that the acceleration of the bullet is constant, find the force exerted by the wood on the bullet.

30 •• **SSM** A cart on a horizontal, linear track has a fan attached to it. The cart is positioned at one end of the track, and the fan is turned on. Starting from rest, the cart takes 4.55 s to travel a distance of 1.5 m. The mass of the cart plus fan is 355 g. Assume that the cart travels with constant acceleration. (*a*) What is the net force exerted on the cart? (*b*) Weights are added to the cart until its mass is 722 g, and the experiment is repeated. How long does it take for the cart to travel 1.5 m now? Ignore the effects of friction.

31 • A horizontal force F_0 causes an acceleration of 3 m/s² when it acts on an object of mass *m* sliding on a frictionless surface. Find the acceleration of the same object in the circumstances shown in Figure 4-30*a* and *b*.

FIGURE 4-30 Problem 31

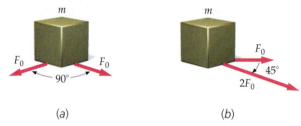

    (*a*)   (*b*)

32 • **ISOLVE** A force $\vec{F} = (6\text{ N})\hat{i} - (3\text{ N})\hat{j}$ acts on an object of mass 1.5 kg. Find the acceleration $\vec{a}$. What is the magnitude *a*?

33 • A single force of 12 N acts on a particle of mass *m*. The particle starts from rest and travels in a straight line a distance of 18 m in 6 s. Find *m*.

34 • **SSM** Al and Bert stand in the middle of a large frozen lake. Al pushes on Bert with a force of 20 N for a period of 1.5 s. Bert's mass is 100 kg. Assume that both are at rest before Al pushes Bert. (*a*) What is the speed that Bert reaches as he is pushed away from Al? Treat the ice as frictionless. (*b*) What speed does Al reach if his mass is 80 kg?

35 • If I push a block whose mass is m_1 across a frictionless floor with a force of a given magnitude, the block has acceleration 12 m/s². If I push on a different block whose mass is m_2 with a force of the same magnitude, its acceleration is 3 m/s². (Both forces are applied horizontally.) (*a*) What acceleration will this force give to a block with mass $m_2 - m_1$? The force is still applied horizontally. (*b*) What acceleration will this force give to a block with mass $m_2 + m_1$?

36 • To drag a 75-kg log along the ground at constant velocity, you have to pull on it with a horizontal force of 250 N. (*a*) What is the resistive force exerted by the ground? (*b*) What horizontal force must you exert if you want to give the log an acceleration of 2 m/s²?

37 • **iSOLVE✔** A 4-kg object is subjected to two forces, $\vec{F}_1 = (2\,\text{N})\hat{i} + (-3\,\text{N})\hat{j}$ and $\vec{F}_2 = (4\,\text{N})\hat{i} - (11\,\text{N})\hat{j}$. The object is at rest at the origin at time $t = 0$. (*a*) What is the object's acceleration? (*b*) What is its velocity at time $t = 3$ s? (*c*) Where is the object at time $t = 3$ s?

Mass and Weight

38 • **SSM** On the moon, the acceleration due to gravity is only about 1/6 of that on earth. An astronaut whose weight on earth is 600 N travels to the lunar surface. His mass as measured on the moon will be (*a*) 600 kg, (*b*) 100 kg, (*c*) 61.2 kg, (*d*) 9.81 kg, (*e*) 360 kg.

39 • **iSOLVE** Find the weight of a 54-kg student in (*a*) newtons and (*b*) pounds.

40 • Find the mass of a 165-lb engineer in kilograms.

Contact Forces

41 • **SSM** **iSOLVE✔** A vertical spring of force constant 600 N/m has one end attached to the ceiling and the other to a 12-kg block resting on a horizontal surface so that the spring exerts an upward force on the block. The spring stretches by 10 cm. (*a*) What force does the spring exert on the block? (*b*) What is the force that the surface exerts on the block?

42 • A 6-kg box on a frictionless horizontal surface is attached to a horizontal spring with a force constant of 800 N/m. If the spring is stretched 4 cm from its equilibrium length, what is the acceleration of the box?

Free-Body Diagrams: Static Equilibrium

43 • A traffic light is supported by two wires as in Figure 4-31. Is the tension in the wire that is more nearly vertical greater than or less than the tension in the other wire?

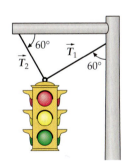

FIGURE 4-31 Problem 43

44 • **iSOLVE** A lamp with mass $m = 42.6$ kg is hanging from wires as shown in Figure 4-32. The ring has negligible mass. The tension T_1 in the vertical wire is (*a*) 209 N, (*b*) 418 N, (*c*) 570 N, (*d*) 360 N, (*e*) 730 N.

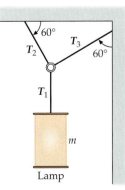

FIGURE 4-32 Problem 44

45 •• **SSM** **iSOLVE✔** In Figure 4-33*a*, a 0.500-kg block is suspended from a 1.25-m-long string. The ends of the string are attached to the ceiling at points separated by 1.00 m. (*a*) What angle does the string make with the ceiling? (*b*) What is the tension in the string? (*c*) The 0.500-kg block is removed and two 0.250-kg blocks are attached to the string such that the lengths of the three string segments are equal (Figure 4-33*b*). What is the tension in each segment of the string?

FIGURE 4-33 Problem 45

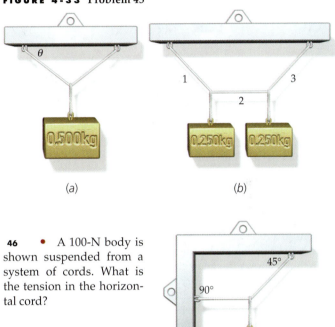

(*a*)   (*b*)

46 • A 100-N body is shown suspended from a system of cords. What is the tension in the horizontal cord?

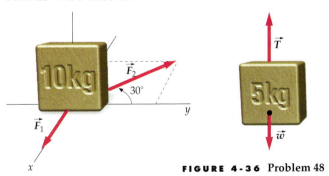

FIGURE 4-34 Problem 46

47 • **iSOLVE** A 10-kg object on a frictionless table is subjected to two horizontal forces, $\vec{F}_1$ and $\vec{F}_2$, with magnitudes $F_1 = 20$ N and $F_2 = 30$ N, as shown in Figure 4-35. (*a*) Find the acceleration $\vec{a}$ of the object. (*b*) A third force $\vec{F}_3$ is applied so that the object is in static equilibrium. Find $\vec{F}_3$.

FIGURE 4-35 Problem 47

FIGURE 4-36 Problem 48

48 • **SSM** **iSOLVE✔** A vertical force $\vec{T}$ is exerted on a 5-kg object near the surface of the earth, as shown in Figure 4-36. Find the acceleration of the object if (*a*) $T = 5$ N, (*b*) $T = 10$ N, and (*c*) $T = 100$ N.

49 •• A 2-kg picture is hung by two wires of equal length. Each makes an angle θ with the horizontal, as shown in Figure 4-37. (a) Find the general equation for the tension T, given θ and the weight w for the picture. For what angle θ is T the least? The greatest? (b) If $\theta = 30°$, what is the tension in the wires?

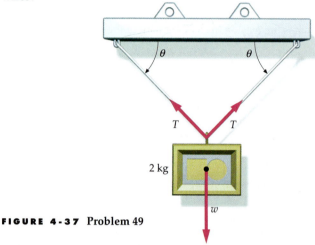

FIGURE 4-37 Problem 49

50 ••• SSM Balloon arches are often seen at festivals or celebrations; they are made by attaching helium-filled balloons to a rope that is fixed to the ground at each end. The lift from the balloons raises the structure into the arch shape. Figure 4-38a shows the geometry of such a structure: N balloons are attached at equally spaced intervals along a massless rope of length L, which is attached to two supports. Each balloon provides a lift force F. The horizontal and vertical coordinates of the point on the rope where the ith balloon is attached are x_i and y_i, and T_i is the tension in the ith segment (with segment 0 being the segment between the point of attachment and the first balloon, and segment N being the segment between the last balloon and the other point of attachment).

FIGURE 4-38 Problem 50

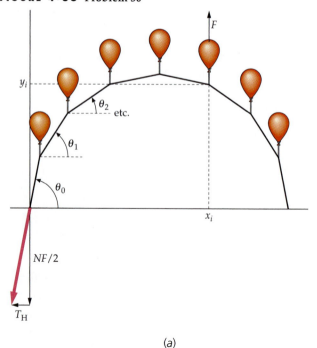

(a)

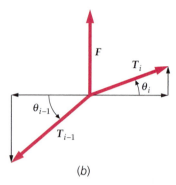

(b)

(a) Figure 4-38b shows a free-body diagram for the ith balloon. From this diagram, show that the horizontal component of the force T_i (call it T_H) is the same for all the balloons, and that by considering the vertical component of the force, one can derive the following equation relating the tension in the ith and $(i - 1)$th segments:

$$T_{i-1} \sin \theta_{i-1} - T_i \sin \theta_i = F$$

(b) Show that $\tan \theta_0 = -\tan \theta_{N+1} = NF/2T_H$.

(c) From the diagram and the two expressions above, show that

$$\tan \theta_i = (N - 2i)F/2T_H$$

and that

$$x_i = \frac{L}{N+1} \sum_{j=0}^{i-1} \cos \theta_j, \qquad y_i = \frac{L}{N+1} \sum_{j=0}^{i-1} \sin \theta_j$$

(d) Write a spreadsheet program to make a graph of the shape of a balloon arch with the following parameters: $N = 10$ balloons giving a lift force $F = 1$ N each attached to a rope length $L = 10$ m, with a horizontal component of tension $T_H = 10$ N. How far apart are the two points of attachment? How high is the arch at its highest point?

(e) Note that we haven't specified the spacing between the supports—it is determined by the other parameters. Vary T_H while keeping the other parameters the same until you create an arch that has a spacing of 8 m between the supports. What is T_H then? As you increase T_H, the arch should get flatter and more spread out. Does your spreadsheet model show this?

51 •• A 1000-kg load is being moved by a crane. Find the tension in the cable that supports the load as (a) it moves upward with a speed increasing by 2 m/s each second, (b) it is lifted at constant speed, and (c) it moves upward with speed decreasing by 2 m/s each second.

52 •• ISOLVE For the systems in equilibrium in Figures 4-39a, 4-39b, and 4-39c, find the unknown tensions and masses.

FIGURE 4-39 Problem 52

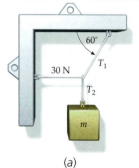

(a)

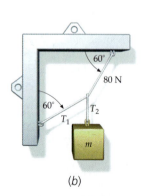

(b)

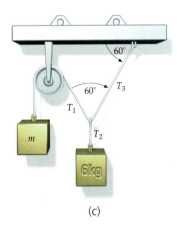

(c)

53 •• [SOLVE ✓] Your car is stuck in a mud hole. You are alone, but you have a long, strong rope. Having studied physics, you tie the rope tautly to a telephone pole and pull on it sideways, as shown in Figure 4-40. (a) Find the force exerted by the rope on the car when the angle θ is 3° and you are pulling with a force of 400 N but the car does not move. (b) How strong must the rope be if it takes a force of 600 N to move the car when θ is 4°?

FIGURE 4-40 Problem 53

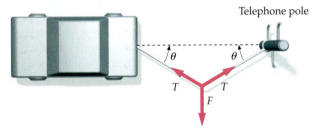

Free-Body Diagrams: Inclined Planes and the Normal Force

54 • [SSM] [SOLVE] A large box whose mass is 20 kg rests on a frictionless floor. A mover pushes on the box with a force of 250 N at an angle 35° below the horizontal. What is the acceleration of the box across the floor?

55 • [SOLVE] The box from Problem 54 now rests on a frictionless ramp with a 15° slope. The mover pulls up on a rope attached to the box to pull it up the incline (see Figure 4-41). If the rope makes an angle of 40° with the horizontal, what is the smallest force F the mover will have to exert to move the box up the ramp?

FIGURE 4-41 Problem 55

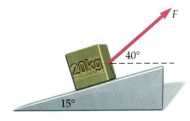

56 • A box slides down a frictionless inclined plane. Draw a diagram showing the forces acting on the box. For each force in your diagram, indicate the reaction force.

57 • The system shown in Figure 4-42 is in equilibrium and the incline is frictionless. It follows that the mass m is (a) 3.5 kg, (b) 3.5 sin 40° kg, (c) 3.5 tan 40° kg, (d) none of these answers.

FIGURE 4-42 Problem 57

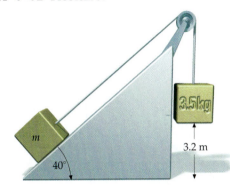

58 • [SSM] In Figure 4-43, the objects are attached to spring balances calibrated in newtons. Give the reading of the balance(s) in each case, assuming that the strings are massless.

FIGURE 4-43 Problem 58

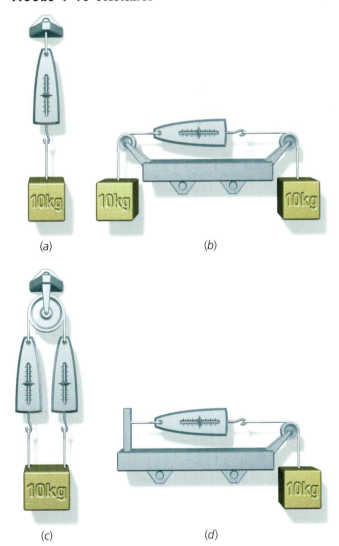

(a)   (b)

(c)   (d)

59 •• A box is held in position by a cable along a frictionless incline (Figure 4-44). (*a*) If $\theta = 60°$ and $m = 50$ kg, find the tension in the cable and the normal force exerted by the incline. (*b*) Find the tension as a function of θ and m, and check your result for $\theta = 0°$ and $\theta = 90°$.

FIGURE 4-44 Problem 59

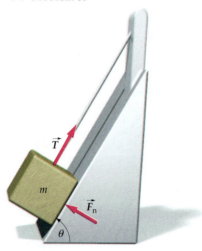

60 •• A horizontal force of 100 N pushes a 12-kg block up a frictionless incline that makes an angle of 25° with the horizontal. (*a*) What is the normal force that the incline exerts on the block? (*b*) What is the acceleration of the block?

61 •• **SSM** **iSOLVE** A 65-kg student weighs himself by standing on a scale mounted on a skateboard that is rolling down an incline, as shown in Figure 4-45. Assume there is no friction so that the force exerted by the incline on the skateboard is normal to the incline. What is the reading on the scale if $\theta = 30°$?

FIGURE 4-45 Problem 61

62 •• A block of mass m slides across a frictionless floor and then up a frictionless ramp (see Figure 4-46). The angle of the ramp is θ and the speed of the block before it starts up the ramp is v_0. The block will slide up to some maximum height h above the floor before starting to slide back down. Show that h is independent of θ.

FIGURE 4-46 Problem 62

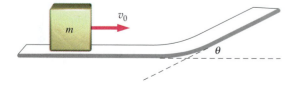

Free-Body Diagrams: Elevators

63 • An object is suspended from the ceiling of an elevator that is descending at a constant speed of 9.81 m/s. The tension in the string holding the object is (*a*) equal to the weight of the object, (*b*) less than the weight of the object, (*c*) greater than the weight of the object, (*d*) zero.

64 • Suppose you are standing on a force scale in a descending elevator as it comes to a stop on the ground floor. Will the scale's report of your weight be high, low, or correct as the elevator slows down?

65 • **SSM** A person of weight w is in an elevator going up when the cable suddenly breaks. What is the person's apparent weight immediately after the cable breaks? (*a*) w. (*b*) Greater than w. (*c*) Less than w. (*d*) $9.8w$. (*e*) Zero.

66 • **iSOLVE** A person in an elevator is holding a 10-kg block by a cord rated to withstand a tension of 150 N. When the elevator starts up, the cord breaks. What was the minimum acceleration of the elevator?

67 •• A 2-kg block hangs from a spring scale calibrated in newtons that is attached to the ceiling of an elevator (Figure 4-47). What does the scale read when (*a*) the elevator is moving up with a constant velocity of 30 m/s, (*b*) the elevator is moving down with a constant velocity of 30 m/s, (*c*) the elevator is ascending at 20 m/s and gaining speed at a rate of 3 m/s²? (*d*) From $t = 0$ to $t = 5$ s, the elevator moves up at 10 m/s. Its velocity is then reduced uniformly to zero in the next 4 s, so that it is at rest at $t = 9$ s. Describe the reading of the scale during the interval $0 < t < 9$ s.

FIGURE 4-47 Problem 67

Free-Body Diagrams: Ropes, Tension, and Newton's Third Law

68 • Two boxes of mass m_1 and m_2 connected by a massless string are accelerated uniformly on a frictionless surface, as shown in Figure 4-48. The ratio of the tensions T_1/T_2 is given by (a) m_1/m_2, (b) m_2/m_1, (c) $(m_1 + m_2)/m_2$, (d) $m_1/(m_1 + m_2)$, (e) $m_2/(m_1 + m_2)$.

FIGURE 4-48 Problem 68

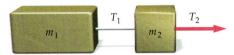

69 •• A box of mass $m_2 = 3.5$ kg rests on a frictionless horizontal shelf and is attached by strings to boxes of masses $m_1 = 1.5$ kg and $m_3 = 2.5$ kg, which hang freely, as shown in Figure 4-49. Both pulleys are frictionless and massless. The system is initially held at rest. After it is released, find (a) the acceleration of each of the boxes and (b) the tension in each string.

FIGURE 4-49 Problem 69

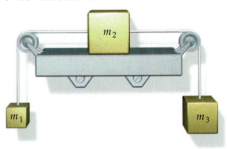

70 •• **SSM** Two blocks are in contact on a frictionless horizontal surface. The blocks are accelerated by a horizontal force $\vec{F}$ applied to one of them (Figure 4-50). Find the acceleration and the contact force for (a) general values of F, m_1, and m_2, and (b) for $F = 3.2$ N, $m_1 = 2$ kg, and $m_2 = 6$ kg.

FIGURE 4-50 Problems 70 and 71

71 •• Repeat Problem 70, but with the two blocks interchanged.

72 •• **iSOLVE✓** Two 100-kg boxes are dragged along a frictionless surface with a constant acceleration of 1.0 m/s², as shown in Figure 4-51. Each rope has a mass of 1 kg. Find the force F and the tension in the ropes at points A, B, and C.

FIGURE 4-51 Problem 72

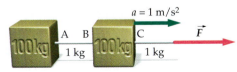

73 •• A block of mass m is being lifted vertically by a rope of mass M and length L. The rope is being held at its top end, and the rope and block are accelerating upward with acceleration a. The distribution of mass in the rope is uniform. Show that the tension in the rope at a distance x ($< L$) above the block is $(a + g)[m + (x/L)M]$.

74 •• **SSM** **iSOLVE** A chain consists of 5 links, each having a mass of 0.1 kg. The chain is lifted vertically with an upward acceleration of 2.5 m/s². The chain is held at the top link; no point of the chain touches the floor. Find (a) the force F exerted on the top of the chain, (b) the net force on each link, and (c) the force each link exerts on the link below it.

75 • A 40.0-kg object supported by a vertical rope is initially at rest. The object is then accelerated upward. The tension in the rope needed to give the object an upward speed of 3.50 m/s in 0.700 s is (a) 590 N, (b) 390 N, (c) 200 N, (d) 980 N, (e) 720 N.

76 • **iSOLVE✓** A hovering helicopter of mass m_h is lowering a truck of mass m_t. If the truck's downward speed is increasing at the rate of 0.1g, what is the tension in the supporting cable? (a) $1.1m_t g$. (b) $m_t g$. (c) $0.9m_t g$. (d) $1.1(m_h + m_t)g$. (e) $0.9(m_h + m_t)g$.

77 •• Two objects are connected by a massless string, as shown in Figure 4-52. The incline and pulley are frictionless. Find the acceleration of the objects and the tension in the string for (a) general values of θ, m_1, and m_2, and (b) $\theta = 30°$ and $m_1 = m_2 = 5$ kg.

FIGURE 4-52 Problem 77

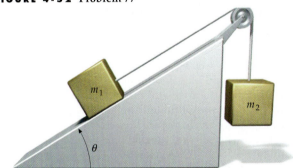

78 • **iSOLVE** In a stage production of Peter Pan, the 50-kg actress playing Peter has to fly in vertically, and to be in time with the music, she must be lowered, starting from rest, a distance of 3.2 m in 2.2 s. Backstage, a smooth surface sloped at 50° supports a counterweight of mass m, as shown in Figure 4-53. Show the calculations that the stage manager must perform to find (a) the mass of the counterweight that must be used and (b) the tension in the wire.

FIGURE 4-53 Problem 78

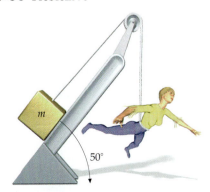

79 •• **ISOLVE** An 8-kg block and a 10-kg block connected by a rope that passes over a frictionless peg slide on frictionless inclines, as shown in Figure 4-54. (*a*) Find the acceleration of the blocks and the tension in the rope. (*b*) The two blocks are replaced by two others of masses m_1 and m_2 such that there is no acceleration. Find whatever information you can about the masses of these two new blocks.

FIGURE 4-54 Problem 79

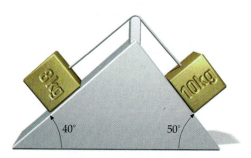

80 •• A heavy rope of length 5 m and mass 4 kg lies on a frictionless horizontal table. One end is attached to a 6-kg block. At the other end of the rope, a constant horizontal force of 100 N is applied. (*a*) What is the acceleration of the system? (*b*) Give the tension in the rope as a function of position along the rope.

81 •• **SSM** A 60-kg house-painter stands on a 15-kg aluminum platform. The platform is attached to a rope that passes through an overhead pulley, which allows the painter to raise herself and the platform (Figure 4-55). (*a*) To accelerate herself and the platform at a rate of 0.8 m/s², with what force *F* must she pull on the rope? (*b*) When her speed reaches 1 m/s, she pulls in such a way that she and the platform go up at a constant speed. What force is she exerting on the rope? (Ignore the mass of the rope.)

FIGURE 4-55 Problem 81

82 ••• Figure 4-56 shows a 20-kg block sliding on a 10-kg block. All surfaces are frictionless. Find the acceleration of each block and the tension in the string that connects the blocks.

FIGURE 4-56 Problem 82

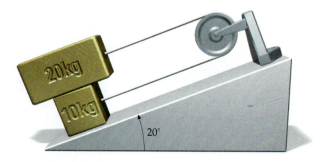

83 ••• **ISOLVE✓** A 20-kg block with a pulley attached slides along a frictionless ledge. It is connected by a massless string to a 5-kg block via the arrangement shown in Figure 4-57. (*a*) Find the horizontal distance the 20-kg block moves when the 5-kg block descends a distance of 10 cm. (*b*) Find the acceleration of each block and the tension in the connecting string.

FIGURE 4-57 Problem 83

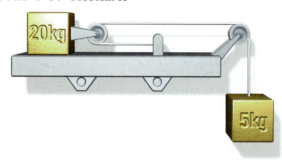

Free-Body Diagrams: The Atwood's Machine

84 •• **SSM** The apparatus in Figure 4-58 is called an *Atwood's machine* and is used to measure the free-fall acceleration *g* by measuring the acceleration of the two blocks. Assuming a massless, frictionless pulley and a massless string, show that the magnitude of the acceleration of either body and the tension in the string are

$$a = \frac{m_1 - m_2}{m_1 + m_2} g \quad \text{and} \quad T = \frac{2m_1 m_2 g}{m_1 + m_2}$$

FIGURE 4-58
Problems 84–87

85 •• **ISOLVE** If one of the masses of the Atwood's machine in Figure 4-58 is 1.2 kg, what should be the other mass so that the displacement of either mass during the first second following release is 0.3 m?

86 •• A very small pebble of mass *m* rests on the block of mass m_2 of the Atwood's machine in Figure 4-58. Find the force exerted by the pebble on m_2.

87 •• Find the force exerted by the Atwood's machine on the hanger to which the pulley is attached, as shown in Figure 4-58, while the blocks accelerate. Neglect the mass of the pulley. Check your answer by considering limiting values for m_1 and/or m_2 for which you can determine the answer by qualitative reasoning.

88 ••• The acceleration of gravity *g* can be determined by measuring the time *t* it takes for a mass m_2 in an Atwood's machine to fall a distance *L*, starting from rest. (*a*) Find an expression for *g* in terms of m_1, m_2, *L*, and *t*. (*b*) Show that if there is a small error in the time measurement *dt*, it will lead to an error in the determination of *g* by an amount *dg* given by $dg/g = -2dt/t$. (*c*) If *L* = 3 m and m_1 is 1 kg, find the value of m_2 such that *g* can be measured with an accuracy of ±5% with a time measurement that is accurate to 0.1 s. Assume that the only significant uncertainty in the measurement is the time of fall.

89 •• **SSM** You are given an Atwood's machine and a set of weights whose total mass is M. You are told to attach some of the weights to one side of the machine, and the rest to the other side. If m_1 represents the mass attached to the left side and m_2 is the mass attached to the right side, the tension in the rope is given by the expression

$$T = \frac{2m_1 m_2}{m_1 + m_2} g$$

as was shown in Problem 85. Show that the tension will be greatest when $m_1 = m_2 = M/2$.

90 ••• An Atwood's machine has a fixed mass m_1 attached on one side and variable mass m_2 ($> m_1$) on the other side. (a) Show that the largest possible magnitude of the tension in the rope is $2m_1 g$. (b) Interpret this result physically, without the use of calculus.

General Problems

91 • A redheaded woodpecker hits the bark of a tree extremely hard—the speed of its head reaches approximately $v = 3.5$ m/s before impact. If the mass of the bird's head is 0.060 kg, and the average force acting on the head during impact is $F = 6.0$ N, find (a) the acceleration of its head (assuming it is constant), (b) the depth of penetration into the bark, and (c) the time t it takes the woodpecker's head to stop.

92 •• **SSM** A simple accelerometer can be made by suspending a small object from a string attached to a fixed point on an accelerating object. Suppose such an accelerometer is attached to the ceiling of an automobile traveling on a large flat surface. When there is acceleration, the object will deflect and the string will make some angle with the vertical. (a) How is the direction in which the suspended object is deflected related to the direction of the acceleration? (b) Show that the acceleration a is related to the angle θ that the string makes by $a = g \tan \theta$. (c) Suppose the automobile brakes to rest from 50 km/h in a distance of 60 m. What angle will the accelerometer make? Will the object swing forward or backward?

93 •• **SOLVE** The mast of a sailboat is supported at bow and stern by stainless steel wires, the forestay and backstay, anchored 10 m apart (Figure 4-59). The 12-m long mast weighs 800 N and stands vertically on the deck of the boat. The mast is positioned 3.6 m behind where the forestay is attached. The tension in the forestay is 500 N. Find the tension in the backstay and the force that the mast exerts on the deck.

FIGURE 4-59 Problem 93

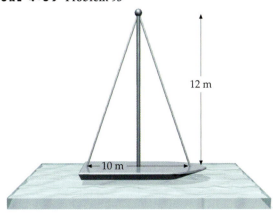

12 m

10 m

94 •• **SOLVE** A large uniform chain is hanging from the ceiling, supporting a block of mass 50 kg. The mass of the chain itself is 20 kg, and the length of the chain is 1.5 m. Determine the tension in the chain (a) at the point where the chain is supporting the block, (b) midway up the chain, and (c) at the top of the chain where it is attached to the ceiling.

95 ••• **SSM** **SOLVE** A man pushes a 24-kg box across a frictionless floor. The box begins moving from rest. He initially pushes on the box gently, but gradually increases his force so that the force he exerts on the box varies in time as $F = (8$ N/s$)t$. After 3 s, he stops pushing the box. The force is always exerted in the same direction. (a) What is the velocity of the box after 3 s? (b) How far has the man pushed the box in 3 s? (c) What is the average velocity of the box between 0 s and 3 s? (d) What is the average force that the man exerts on the box while he is pushing it?

96 •• Suppose that a frictionless surface is inclined at an angle of 30° to the horizontal. The 270-g block is attached to a 75-g hanging weight using a pulley, as shown in Figure 4-60. (a) Draw two free-body diagrams, one for the block and the other for the hanging weight. (b) Find the tension in the string and the acceleration of the block. (c) The block is released from rest. How long does it take for it to slide a distance of 1.00 m down the surface?

FIGURE 4-60 Problem 96

97 •• A box of mass m_1 is pulled along a smooth horizontal surface by a force F exerted at the end of a rope that has a much smaller mass m_2, as shown in Figure 4-61. (a) Find the acceleration of the rope and block, assuming them to be one object. (b) What is the net force acting on the rope? (c) Find the tension in the rope at the point where it is attached to the block. (d) The diagram, with the rope perfectly horizontal along its length, is not quite accurate. Correct the diagram and state how this correction affects your solution.

FIGURE 4-61 Problem 97

98 •• **SSM** **SOLVE** A 2-kg block rests on a frictionless wedge that has an inclination of 60° and an acceleration a to the right such that the mass remains stationary relative to the wedge (Figure 4-62). (a) Find a. (b) What would happen if the wedge were given a greater acceleration?

FIGURE 4-62 Problem 98

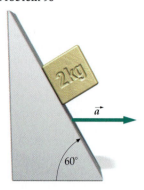

99 •• **iSOLVE** ✓ The masses attached to each side of an Atwood's machine consist of a stack of five washers, each of mass m, as shown in Figure 4-63. The tension in the string is T_0. When one of the washers is removed from the left side, the remaining washers accelerate and the tension decreases by 0.3 N. (a) Find m. (b) Find the new tension and the acceleration of each mass when a second washer is removed from the left side.

FIGURE 4-63 Problems 99 and 100

5m

5m

100 •• Consider the Atwood's machine in Figure 4-63. When N washers are transferred from the left side to the right side, the right side drops 47.1 cm in 0.40 s. Find N.

101 •• Blocks of mass m and $2m$ are connected by a string (Figure 4-64). (a) If the forces are constant, find the tension in the connecting string. (b) If the forces vary with time as $F_1 = Ct$ and $F_2 = 2Ct$, where C is a constant and t is time, find the time t_0 at which the tension in the string is T_0.

FIGURE 4-64 Problem 101

$\vec{F}_1$ m $2m$ $\vec{F}_2$

102 ••• **SSM** **iSOLVE** The pulley in an Atwood's machine is given an upward acceleration a, as shown in Figure 4-65. Find the acceleration of each mass and the tension in the string that connects them. *Hint: a constant upward acceleration has the same effect as an increase in the acceleration due to gravity.*

FIGURE 4-65 Problem 102

Applications of Newton's Laws

AS THIS CAR ROUNDS THE CURVE, IT IS PREVENTED FROM SLIDING RADIALLY BY THE STATIC FRICTIONAL FORCE EXERTED ON THE TIRES BY THE ROAD.

? **What factors determine how fast the car can go around the corner without skidding? (See Example 5-10.)**

In Chapter 4 we introduced Newton's laws and applied them to situations where action was restricted to straight-line motion, and the quantitative effects of friction were excluded.

➤ **In this chapter we will extend the application of Newton's laws to motion along curved paths, and we will include the quantitative effects of friction.**

5-1 Friction

Without friction our ground-based transportation system, from walking to automobiles, could not function. To start walking on a horizontal surface requires friction, and once you are already walking, friction is required if you are to change either speed or direction. Friction is required to hold a nut on a screw or a nail in wood. As important as friction is, it is often not desirable. Lubricants, such as motor oil in an automobile engine, or synovial fluid in our joints, can reduce friction.

Static Friction

When you apply a small horizontal force $\vec{F}$ (see Figure 5-1) to a large box resting on the floor, the box may not move because the force of **static friction** $\vec{f}_s$ exerted by the floor on the box, balances the force you are applying. The force of static friction, which opposes the applied force on the box, can vary from zero to some maximum force $f_{s,max}$, depending on how hard you push. Data show that $f_{s,max}$ is proportional to the normal force exerted by one surface on the other:

FIGURE 5-1

$$f_{s,max} = \mu_s F_n \qquad\qquad 5\text{-}1$$

DEFINITION—COEFFICIENT OF STATIC FRICTION

where the proportionality constant μ_s, called the **coefficient of static friction,** depends on the nature of the surfaces in contact. If you exert a horizontal force smaller than $f_{s,max}$ on the box, the frictional force will just balance this horizontal force. In general, we can write

$$f_s \leq \mu_s F_n \qquad\qquad 5\text{-}2$$

Kinetic Friction

If you push the box in Figure 5-1 hard enough, it will slide across the floor. As it slides, the floor exerts a force of **kinetic friction** $\vec{f}_k$ (also called sliding friction) that opposes the motion. To keep the box sliding with constant velocity, you must exert a force on the box that is equal in magnitude and opposite in direction to the force of kinetic friction exerted by the floor.

The **coefficient of kinetic friction** μ_k is the ratio of the magnitudes of the kinetic frictional force f_k and the normal force F_n:

$$f_k = \mu_k F_n \qquad\qquad 5\text{-}3$$

DEFINITION—COEFFICIENT OF KINETIC FRICTION

where μ_k depends on the nature of the surfaces in contact. Experimentally, it is found that μ_k is less than μ_s, and is approximately constant for speeds ranging from about 1 cm/s to several meters per second—the only situations we will consider.

Rolling Friction

When an ideal, rigid wheel rolls *at constant speed* along an ideal, rigid horizontal road without slipping, no frictional force slows its motion. However, because real tires and roads continually deform and because the tread and the road are continually peeled apart, the road exerts a force of **rolling friction** $\vec{f}_r$ that opposes the motion. To keep the wheel rolling with constant velocity, you must exert a force on the wheel that is equal in magnitude and opposite in direction to the force of rolling friction exerted by the road.

The **coefficient of rolling friction** μ_r is the ratio of the magnitudes of the rolling frictional force f_r and the normal force F_n:

$$f_r = \mu_r F_n \qquad\qquad 5\text{-}4$$

DEFINITION—COEFFICIENT OF ROLLING FRICTION

where μ_r depends on the nature of the surfaces in contact and the composition of the wheel and road. Typical values of μ_r are 0.01 to 0.02 for rubber tires on concrete and 0.001 to 0.002 for steel wheels on steel rails.

Friction Explained

Friction is a complex, incompletely understood phenomenon that arises from the attraction of molecules between two surfaces that are in close contact. The nature of this attraction is electromagnetic—the same as the molecular bonding that holds an object together. This attractive force is short ranged and becomes negligible at distances of only a few atomic diameters.

As shown in Figure 5-2, ordinary objects that look smooth and feel smooth are rough and bumpy at the microscopic (atomic) scale. This is the case even if the surfaces are highly polished. When surfaces come into contact, they touch only at widely spaced prominences, called asperities, shown in Figure 5-2. The normal force exerted by a surface is exerted at the tips of these asperities where the force per unit area is very large, large enough to flatten the tips of the asperities. As the normal force increases, so does this flattening, resulting in a larger microscopic contact area. Under a wide range of conditions the microscopic area of contact is proportional to the normal force. The frictional force is proportional to the microscopic contact area, so it is also proportional to the normal force.

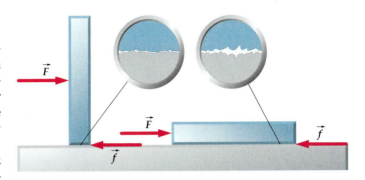

FIGURE 5-2 The microscopic area of contact between box and floor is only a small fraction of the macroscopic area of the box's bottom surface. This fraction is proportional to the normal force exerted between the surfaces. If the box rests on its side, the macroscopic area is increased, but the force per unit area is decreased so the microscopic area of contact is unchanged. Whether the box is upright or on its side, the same horizontal applied force F is required to keep it sliding at constant speed.

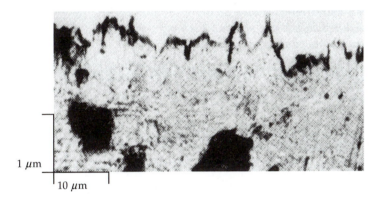

Magnified section of a polished steel surface showing surface irregularities. The irregularities are about 5×10^{-5} cm high, corresponding to several thousand atomic diameters.

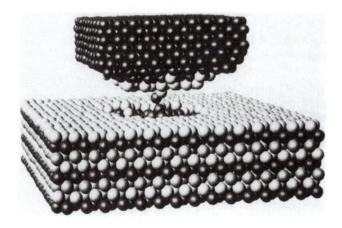

The computer graphic shows gold atoms (bottom) adhering to the fine point of a nickel probe (top) that has been in contact with the gold surface.

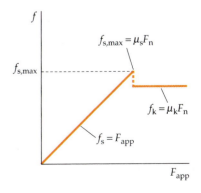

FIGURE 5-3

Figure 5-3 shows a plot of the frictional force exerted on the box by the floor as a function of the applied force. The force of friction balances the applied force until the applied force equals $\mu_s F_n$, at which point the box begins to slide. Then the frictional force is constant and equal to $\mu_k F_n$. Table 5-1 lists some approximate values of μ_s and μ_k for various surfaces.

TABLE 5-1

Approximate Values of Frictional Coefficients

Materials	μ_S	μ_k
Steel on steel	0.7	0.6
Brass on steel	0.5	0.4
Copper on cast iron	1.1	0.3
Glass on glass	0.9	0.4
Teflon on Teflon	0.04	0.04
Teflon on steel	0.04	0.04
Rubber on concrete (dry)	1.0	0.80
Rubber on concrete (wet)	0.30	0.25
Waxed ski on snow (0°C)	0.10	0.05

A GAME OF SHUFFLEBOARD   **EXAMPLE 5-1**

A cruise-ship passenger uses a shuffleboard cue to push a shuffleboard disk of mass 0.40 kg horizontally along the deck so that the disk leaves the cue with a speed of 5.5 m/s. The disk then slides a distance of 8 m before coming to rest. Find the coefficient of kinetic friction between the disk and the deck.

PICTURE THE PROBLEM The force of kinetic friction is the only horizontal force acting on the disk after it separates from the cue. Since the frictional force is constant, the acceleration is constant. We can find the acceleration using the constant-acceleration equations of Chapter 2 and relate it to μ_k using $\Sigma F_x = ma_x$.

FIGURE 5-4

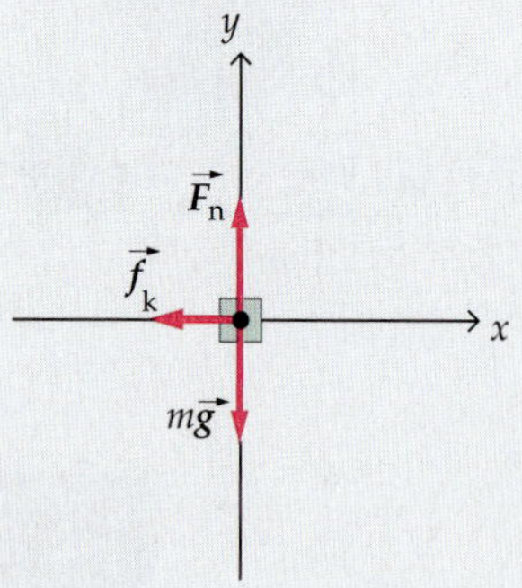

1. Draw a free-body diagram for the disk after it leaves the cue (Figure 5-4). The arrows on the x and y axes indicate the positive x and y directions.

2. The coefficient of friction relates the frictional and normal forces:

$$f_k = \mu_k F_n$$

3. Apply $\Sigma F_y = ma_y$ to the disk. Solve for the normal force. Then, using the step 2 result, solve for the frictional force:

$$\Sigma F_y = ma_y \Rightarrow F_n - mg = 0$$

so

$$F_n = mg \quad \text{and} \quad f_k = \mu_k mg$$

4. Apply $\Sigma F_x = ma_x$ to the disk. Using the step 3 result, solve for the acceleration:

$$\Sigma F_x = ma_x \Rightarrow -f_k = ma_x$$

so

$$-\mu_k mg = ma_x \quad \text{and} \quad a_x = -\mu_k g$$

5. The acceleration is constant. Relate it to the total distance traveled and the initial velocity using Equation 2-15. Using the step 4 result, solve for μ_k:

$$v^2 = v_0^2 + 2a_x\Delta x \Rightarrow 0 = v_0^2 - 2\mu_k g \, \Delta x$$

so

$$\mu_k = \frac{v_0^2}{2g \, \Delta x} = \frac{(5.5 \text{ m/s})^2}{2(9.81 \text{ m/s}^2)(8 \text{ m})} = \boxed{0.193}$$

REMARKS Note that the mass m of the disk cancels. The greater the mass, the harder it is to stop the disk, but the greater mass is also accompanied by greater friction. The net result is that mass has no effect.

Example 5-1 illustrates guidelines for solving problems involving friction. They are as follows:

1. Choose the y axis normal to the contacting surfaces. Choose the x axis parallel with the surface and either parallel or antiparallel with the frictional force.
2. Apply $\Sigma F_y = ma_y$ and solve for the normal force F_n.
   - If the friction is *kinetic*, solve for the frictional force using $f_k = \mu_k F_n$.
   - If the friction is *static*, relate the maximum frictional force to the normal force using $f_{s,max} = \mu_s F_n$.
   - If the friction is *rolling*, solve for the frictional forces using $f_r = \mu_r F_n$.
3. Apply $\Sigma F_x = ma_x$ to the object and solve for the desired quantity.

SOLVING PROBLEMS INVOLVING FRICTION

A SLIDING COIN   **EXAMPLE 5-2**

A hardcover book is resting on a table top with its front cover facing upward. You place a coin on this cover and very slowly open the book until the coin starts to slide. The angle θ_{max} is the angle the front cover makes with the horizontal just as the coin starts to slide. Find the coefficient of static friction μ_s between the book cover and the coin in terms of θ_{max}.

PICTURE THE PROBLEM The forces acting on the coin are its weight mg, the normal force F_n exerted by the plane, and the force of friction f. We should follow the guidelines for solving problems with static friction.

FIGURE 5-5

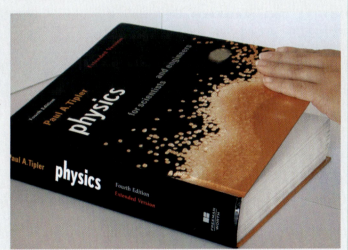

1. Draw a free-body diagram for the coin with the book cover inclined at angle θ, where $\theta \le \theta_{max}$ (Figure 5-6):

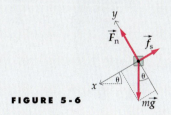

FIGURE 5-6

2. Apply $\Sigma F_y = ma_y$ to the coin and solve for the normal force. Then determine the maximum static frictional force:

$\Sigma F_y = ma_y \Rightarrow F_n - mg \cos \theta = 0$

so

$F_n = mg \cos \theta$

$f_{s,max} = \mu_s F_n$ so $f_{s,max} = \mu_s mg \cos \theta_{max}$

3. Apply $\Sigma F_x = ma_x$ to the coin and solve for the frictional force. Then substitute into the step 2 result:

$\Sigma F_x = ma_x \Rightarrow -f_{s,max} + mg \sin \theta_{max} = 0$

so

$f_{s,max} = mg \sin \theta_{max}$ and $mg \sin \theta_{max} = \mu_s mg \cos \theta_{max}$

4. Solve the step 3 result for μ_s:

$\mu_s = \dfrac{mg \sin \theta_{max}}{mg \cos \theta_{max}} = \boxed{\tan \theta_{max}}$

EXERCISE The coefficient of static friction between a car's tires and the road on a particular day is 0.7. What is the steepest angle of inclination of the road for which the car can be parked with its wheels locked and not slide down the hill? (*Answer* 35°)

From Example 5-2 we see that the coefficient of static friction is related to the angle θ_{max} at which an object begins to slip by

$$\mu_s = \tan \theta_{max} \qquad\qquad 5\text{-}5$$

ANGLE OF REPOSE

and θ_{max} is called the angle of repose.

Two children sitting on a sled at rest in the snow ask you to pull them. You oblige by pulling on the sled's rope, which makes an angle of 40° with the horizontal (Figure 5-7). The children have a combined mass of 45 kg and the sled has a mass of 5 kg. The coefficients of static and kinetic friction are $\mu_s = 0.2$ and $\mu_k = 0.15$. Find the frictional force exerted by the ground on the sled and the acceleration of the children and sled, starting from rest, if the tension in the rope is (*a*) 100 N and (*b*) 140 N.

PICTURE THE PROBLEM First we need to find out whether the frictional force is static or kinetic. To do this we solve for the maximum tension in the rope without the sled sliding.

FIGURE 5-7

(*a*) 1. Draw a free-body diagram for the sled (Figure 5-8):

FIGURE 5-8

2. Apply $\Sigma F_y = ma_y$ to the sled and solve for the normal force. Then solve for the maximum static frictional force:

$$\Sigma F_y = ma_y \Rightarrow F_n + T \sin \theta - mg = 0$$

so

$$F_n = mg - T \sin \theta$$

$$f_{s,max} = \mu_s F_n \quad \text{so} \quad f_{s,max} = \mu_s(mg - T_{max} \sin \theta)$$

3. Apply $\Sigma F_x = ma_x$ to the sled and solve for the frictional force. Then substitute into the step 2 result:

$$\Sigma F_x = ma_x \Rightarrow -f_{s,max} + T_{max} \cos \theta = 0$$

so

$$f_{s,max} = T_{max} \cos \theta \quad \text{and} \quad T_{max} \cos \theta = \mu_s(mg - T_{max} \sin \theta)$$

$$T_{max} \cos \theta = \mu_s(mg - T_{max} \sin \theta)$$

4. Solve the step 3 result for the maximum tension without slipping:

$$T_{max}(\mu_s \sin \theta + \cos \theta) = \mu_s mg$$

so

$$T_{max} = \frac{\mu_s mg}{\mu_s \sin \theta + \cos \theta}$$

$$= \frac{0.2(50\text{ kg})(9.81\text{ m/s}^2)}{0.2 \sin 40° + \cos 40°} = 110\text{ N}$$

5. The tension is 100 N, which is less than 110 N. The sled is *not* sliding. To find the frictional force, use the step 3 expression for f_s:

$$a_x = \boxed{0}$$

$$f_s = T \cos \theta = (100\text{ N})\cos 40° = \boxed{76.6\text{ N}}$$

(b) 1. The tension is 140 N, which is greater than $T_{max} = 110$ N, so the sled is sliding. Relate the kinetic frictional force f_k to the normal force:

$$f_k = \mu_k F_n$$

2. In step (a)2 we applied $\Sigma F_y = ma_y$ to the sled and found $F_n = mg - T \sin \theta$. Use this result along with the step (b)1 result to solve for the kinetic frictional force:

$$f_k = \mu_k(mg - T \sin \theta)$$

$$= 0.15\big[(50\text{ kg})(9.81\text{ N/kg}) - (140\text{ N}) \sin 40°\big]$$

$$= \boxed{60.1\text{ N}}$$

3. Apply $\Sigma F_x = ma_x$ to the sled and solve for the frictional force. Then substitute the step 2 result for f_k and solve for the acceleration:

$$\Sigma F_x = ma_x \Rightarrow -f_k + T \cos \theta = ma_x$$

so

$$a_x = \frac{-f_k + T \cos \theta}{m}$$

$$= \frac{(-60.1\text{ N}) + (140\text{ N})\cos 40°}{50\text{ kg}}$$

$$= \boxed{0.943\text{ m/s}^2}$$

REMARKS There are two important points to note about this example: (1) the normal force is not equal to the weight of the children and the sled. That is because the vertical component of the tension helps lift the sled off the ground. (2) In Part (a), the force of static friction is less than $\mu_s F_n$.

A SLIDING BLOCK　　　　　　　　　　　　　**EXAMPLE 5-4** **Try It Yourself**

The block of mass m_2 in Figure 5-9 has been adjusted so that the block of mass m_1 is on the verge of sliding. (a) If $m_1 = 7$ kg and $m_2 = 5$ kg, what is the coefficient of static friction between the table and the block? (b) With a slight nudge, the blocks move with acceleration a. Find a if the coefficient of kinetic friction between the table and the block is $\mu_k = 0.54$.

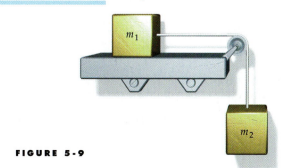

FIGURE 5-9

PICTURE THE PROBLEM Apply Newton's second law to each block. By neglecting the masses of both the rope and the pulley, and by neglecting friction in the pulley bearing, the tension has the same magnitude throughout the rope, so $T_1 = T_2 = T$, and, because the rope remains taut but does not stretch, the accelerations have the same magnitude, so $a_1 = a_2 = a$.

To find the coefficient of static friction μ_s, as required in Part (a), set the force of static friction on m_1 equal to its maximum value $f_{max} = \mu_s F_n$ and set the acceleration equal to zero.

Cover the column to the right and try these on your own before looking at the answers.

FIGURE 5-10

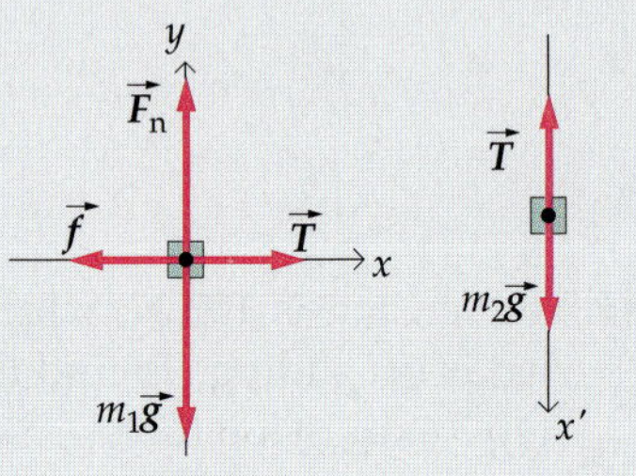

Block 1   Block 2

Steps	Answers
(a) 1. Draw a free-body diagram for each block (Figure 5-10). Choose the positive x and x' directions to be the same as the directions of the accelerations once the blocks are moving.	
2. Apply $\Sigma F_y = ma_y$ to block 1 and solve for the normal force. Then solve for the static frictional force.	$\Sigma F_y = m_1 a_{1y} \Rightarrow F_n - m_1 g = 0$ so $F_n = m_1 g$ $f_{s,max} = \mu_s F_n$ so $f_{s,max} = \mu_s m_1 g$
3. Apply $\Sigma F_x = ma_x$ to block 1 and solve for the frictional force. Then substitute into the step 2 result.	$\Sigma F_x = m_1 a_{1x} \Rightarrow T - f_{s,max} = 0$ so $f_{s,max} = T$ and $T = \mu_s m_1 g$
4. Apply $\Sigma F_x = ma_x$ to block 2 and solve for the tension. Then substitute into the step 3 result.	$\Sigma F_{x'} = m_2 a_{2x'} \Rightarrow m_2 g - T = 0$ so $T = m_2 g$ and $m_2 g = \mu_s m_1 g$
5. Solve the step 4 result for μ_s	$\mu_s = \dfrac{m_2}{m_1} = \dfrac{5\ \text{kg}}{7\ \text{kg}} = \boxed{0.714}$
(b) 1. During sliding the frictional force is kinetic. Relate the kinetic frictional force f_k to the normal force. The normal force was found in step 2 of Part (a).	$f_k = \mu_k F_n$ so $f_k = \mu_k m_1 g$
2. Apply $\Sigma F_x = ma_x$ to block 1. Then substitute for the frictional force using the result from step 1 of Part (b).	$\Sigma F_x = m_1 a_{1x} \Rightarrow T - f_k = m_1 a$ so $T - \mu_k m_1 g = m_1 a$
3. Apply $\Sigma F_{x'} = ma_{x'}$ to block 2.	$\Sigma F_{x'} = m_2 a_{2x'} \Rightarrow m_2 g - T = m_2 a$
4. Add the equations in steps 2 and 3 of Part (b) and solve for a.	$a = \dfrac{m_2 - \mu_k m_1}{m_1 + m_2} g = \boxed{0.997\ \text{m/s}^2}$

◑ PLAUSIBILITY CHECK Note that $\mu_k = 0$ gives the expression for the acceleration derived in Example 4-12 with $\theta = 0$.

EXERCISE What is the tension in the rope when the blocks are sliding? (*Answer* $T = m_2(g - a) = 44.1$ N)

THE RUNAWAY BUGGY   **EXAMPLE   5 - 5**

A runaway baby buggy is sliding without friction across a frozen pond toward
a hole in the ice (Figure 5-11). You race after the buggy on skates. As you grab
it, you and the buggy are moving toward the hole at speed v_0. The coefficient
of friction between your skates and the ice as you turn out the blades to brake
is μ_k. D is the distance to the hole when you reach the buggy, M is the total
mass of the buggy, and m is your mass. (*a*) What is the lowest value of D such
that you stop the buggy before it reaches the hole in the ice? (*b*) What force do
you exert on the buggy?

PICTURE THE PROBLEM Initially, you and the
buggy are moving toward the hole with speed v_0,
which we take to be in the positive x direction. If you
exert a force $\vec{F} = -F\hat{\imath}$ on the buggy, the buggy, in accord with Newton's third law, exerts a force $\vec{F}' = F\hat{\imath}$
on you. Apply Newton's second law to determine the
acceleration. After finding the acceleration, find the
distance D the buggy travels while slowing to a stop.
The lowest value of D is that for which your speed
reaches zero just as the buggy reaches the hole.

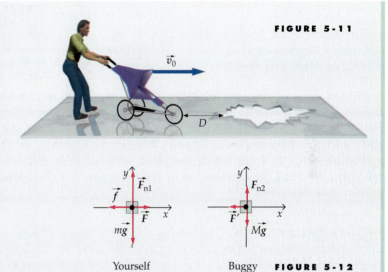

FIGURE 5-11

FIGURE 5-12

Yourself   Buggy

(*a*) 1. Draw separate free-body diagrams for yourself
       and the buggy.

2. Apply $\Sigma F_y = ma_y$ to yourself and solve first for the
   normal force and then for the frictional force:

$$\Sigma F_y = ma_y \Rightarrow F_n - mg = 0$$

and

$$f_k = \mu_k F_n \quad \text{so} \quad f_k = \mu_k mg$$

3. Apply $\Sigma F_x = ma_x$ to yourself. Then substitute in the
   step 2 result:

$$\Sigma F_x = ma_x \Rightarrow F - f_k = ma_x$$

so

$$F - \mu_k mg = ma_x$$

4. Apply $\Sigma F_x = ma_x$ to the buggy. Then substitute for F
   in the step 3 result:

$$\Sigma F_x = Ma_x \Rightarrow -F = Ma_x$$

so

$$-Ma_x - \mu_k mg = ma_x$$

5. Solve the step 4 result for a_x:

$$a_x = -\frac{\mu_k}{1 + M/m}g$$

(The acceleration is negative, as expected.)

6. Substitute the step 5 result into a kinematic equation
   and solve for D:

$$v_x^2 = v_{0x}^2 + 2a_x\Delta x \Rightarrow 0 = v_0^2 + 2a_x D$$

so

$$D = -\frac{v_0^2}{2a_x} = \boxed{\left(1 + \frac{M}{m}\right)\frac{v_0^2}{2\mu_k g}}$$

(*b*) F can be found from Newton's second law applied to
the buggy:

$$F = -Ma_x \Rightarrow F = \boxed{\frac{\mu_k M}{1 + M/m}g}$$

REMARKS The minimum value of D is proportional to v_0^2 and inversely proportional to μ_k. Figure 5-13 shows the stopping distance D versus initial velocity squared for values of M/m equal to 0.1, 0.3, and 1.0, with $\mu_k = 0.5$. Note that the larger the mass ratio M/m, the greater the distance D needed to stop for a given initial velocity. This is akin to stopping a car that is pulling a trailer that does not have its own brakes. The mass of the trailer increases the stopping distance for a given speed.

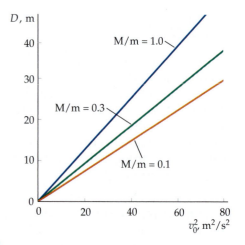

FIGURE 5-13

PULLING A CHILD ON A TOBOGGAN

EXAMPLE 5-6 **Try It Yourself**

A child of mass m_c sits on a toboggan of mass m_t, which in turn sits on a frozen pond assumed to be frictionless (Figure 5-14). The toboggan is pulled with a horizontal force as shown. The coefficients of static and sliding friction between the child and toboggan are μ_s and μ_k. (*a*) Find the maximum value of F for which the child will not slide relative to the toboggan. (*b*) Find the acceleration of the toboggan and child when F is greater than this value.

FIGURE 5-14

PICTURE THE PROBLEM The only force accelerating the child forward is the frictional force exerted by the toboggan on the child. In Part (*a*) the challenge is to find F when this frictional force is static and maximum. To do this, apply $\Sigma\vec{F} = m\vec{a}$ to the child and solve for the acceleration when static frictional force is maximum. Then apply $\Sigma\vec{F} = m\vec{a}$ to the toboggan and solve for F. In Part (*b*), we follow a parallel procedure. However, in this part F is given and we solve for the acceleration of the toboggan.

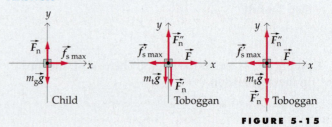

Cover the column to the right and try these on your own before looking at the answers.

Steps

(*a*) 1. Draw a free-body diagram for each object (Figure 5-15).

Two alternative styles for the free-body diagram are shown for the toboggan. In the first, the two downward forces are displaced slightly so they can be seen. In the second, they are drawn head to tail.

2. Equate the magnitudes of the forces in each action–reaction force pair appearing in the two free-body diagrams. Express the relation between the accelerations due to the nonslipping constraint.

3. Apply $\Sigma F_y = ma_y$ to the child. Solve first for the normal force, then the frictional force.

4. Apply $\Sigma F_x = ma_x$ to the child and solve for the acceleration.

Answers

FIGURE 5-15

$F_n' = F_n$ and $f_{s,\max}' = f_{s,\max}$

and

$a_{cx} = a_{tx} = a_x$

$\Sigma F_{cy} = m_c a_y \Rightarrow F_n - m_c g = 0$

$F_n = m_c g$ and

$f_{s,\max} = \mu_s F_n$ so $f_{s,\max} = \mu_s m_c g$

$\Sigma F_{cx} = m_c a_x \Rightarrow f_{s,\max} = m_c a_x$

and

$\mu_s m_c g = m_c a_x,$ so $a_x = \mu_s g$

5. Apply $\Sigma F_x = ma_x$ to the toboggan and, using the acceleration relations from step 2 and the result from step 3, solve for F.

$\Sigma F_{tx} = m_t a_x \Rightarrow F - f_{s,\text{max}} = m_t a_x$

and

$F - \mu_s m_c g = m_t \mu_s g$ so $F = \boxed{(m_c + m_t)\mu_s g}$

(b) 1. Equate the magnitudes of each action–reaction force pair and express the change in the relation between the accelerations due to the removal of the nonslipping constraint.

$F'_n = F_n$ and $f'_k = f_k$

but

$a_{cx} < a_{tx}$

2. Solve for the kinetic frictional force using the result from step 3 of Part (a) for the normal force.

$f_k = \mu_k F_n$ so $f_k = \mu_k m_c g$

3. Apply $\Sigma F_x = ma_x$ to the child and solve for *her acceleration.*

$\Sigma F_{cx} = m_c a_{cx} \Rightarrow f_k = m_c a_{cx}$

and

$\mu_k m_c g = m_c a_{cx}$ so $a_{cx} = \mu_k g$

4. Apply $\Sigma F_x = ma_x$ to the toboggan. Using the result from step 2 of Part (b), solve for *its acceleration.*

$\Sigma F_{tx} = m_t a_{tx} \Rightarrow F - f_k = m_t a_{tx}$

and

$F - \mu_k m_c g = m_t a_{tx}$ so $a_{tx} = \boxed{\dfrac{F - \mu_k m_c g}{m_t}}$

REMARKS Friction is a force between two surfaces in contact, and it is not correct that friction always opposes motion, or the tendency to motion. In this example, friction does not oppose the motion of the child, it causes it. It is correct that friction always opposes motion, or the tendency to motion, of each surface *relative to the other surface.* For example, even though the child moves forward *relative to the ice,* she moves, or tends to move, backward (leftward) *relative to the toboggan.* Friction opposes this relative motion or tendency to motion.

Figure 5-16 shows the forces acting on a front-wheel-drive car that is just starting to move from rest on a horizontal road. The weight of the car is balanced by the normal force F_n exerted on the tires. To start the car moving, the engine delivers power to the axle that makes the wheels rotate (we discuss power in Chapter 6). If the road were perfectly frictionless, the wheels would merely spin. When friction is present, the frictional force exerted by the road on the tires is in the forward direction, opposing the tendency of the tire surface to slip backward. This frictional force provides the acceleration needed for the car to start moving forward. If the power delivered by the engine is small enough so that the frictional force exerted by the tire surface on the road surface is sufficiently small, then the two surfaces do not slip. The wheels roll without slipping and the tire tread touching the road is at rest relative to the road. The friction between the road and the tire tread is then static friction. The largest frictional force that the tire can exert on the road (and that the road can exert on the tire) is $\mu_s F_n$.

For a car moving in a straight line with speed v relative to the road, the center of each of its wheels also moves with speed v, as shown in Figure 5-17. If a wheel is rolling without slipping, its top is moving faster than v whereas its bottom is moving slower than v. However, *relative to the car,* each point on the perimeter of the wheel moves in a circle with the same speed v. Moreover, the speed of the point on the tire momentarily in contact with the ground is zero *relative to the ground.* (Otherwise, the tire would be skidding along the road.)

If the engine power is great enough, the tire will slip and the wheels will spin. Then the force that accelerates the car is the force of kinetic friction, which is less than the force of static friction. If we are stuck on ice or snow, our chances of getting free are better if we use a light touch on the accelerator pedal. Similarly,

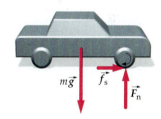

FIGURE 5-16 Forces acting on a car (front-wheel drive). The normal forces $\vec{F}_n$ are usually larger on the front tires because typically the engine of the car is mounted at the front of the car.

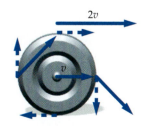

FIGURE 5-17 In this figure, dashed lines represent velocities relative to the body of the car, solid lines represent velocities relative to the ground.

when braking a car to a stop, the force exerted by the road on the tires may be either static friction or kinetic friction, depending on how the brakes are applied. If the brakes are applied so hard that the wheels lock, the tires will skid along the road and the stopping force will be that of kinetic friction. If the brakes are applied less forcefully, so that no slipping occurs between the tires and the road, the stopping force will be that of static friction. Antilock braking systems in cars have wheel-speed sensors. If the control unit senses that a wheel is about to lock, the module signals the brake pressure modulator to drop, hold, and then restore the pressure to that wheel up to 15 times per second. This varying of pressure is much like "pumping" the brake, but with the ABS system, the wheel that is locking is the only one being pumped. This is called threshold braking. With threshold braking, maximum friction for stopping is maintained.

When wheels do lock and tires skid, two undesirable things happen. The minimum stopping distance is increased *and* the ability to control the direction of the car's motion is greatly diminished. Obviously, this loss of directional control can have dire consequences.

THE EFFECT OF ANTILOCK BRAKES   **EXAMPLE 5 - 7**

A car is traveling at 30 m/s along a horizontal road. The coefficients of friction between the road and the tires are $\mu_s = 0.5$ and $\mu_k = 0.3$. How far does the car travel before stopping if (a) the car is braked with an antilock braking system so that threshold braking is sustained, and (b) the car is braked hard with no antilock braking system so that the wheels lock?

PICTURE THE PROBLEM The force that stops a car when it brakes without skidding is the force of static friction exerted by the road on the tires (Figure 5-18). We use Newton's second law to solve for the frictional force and the car's acceleration. Kinematics is then used to find the stopping distance.

FIGURE 5-18

(a) 1. Draw a free-body diagram for the car (Figure 5-19). Treat all four wheels as if they were a single point of contact with the ground. Assume further that the brakes are applied to all four wheels. Let $\vec{f} = \vec{f}' + \vec{f}''$.

FIGURE 5-19

2. Assuming that the acceleration is constant, we use Equation 2-15 to relate the stopping distance Δx to the initial speed v_0. The coefficients of friction change with temperature. Since skidding heats up the tires, these coefficients can be expected to change. Such changes are neglected here:

$$v^2 = v_0^2 + 2a_x \Delta x$$

When $v = 0$,

$$\Delta x = -\frac{v_0^2}{2a_x}$$

3. Apply $\Sigma F_y = ma_y$ to the car. Solve first for the normal force, then for the frictional force:

$$\Sigma F_y = ma_y \Rightarrow F_n - mg = 0, \quad \text{so}$$

$$F_n = mg \quad \text{and}$$

$$f_{s,max} = \mu_s F_n, \quad \text{so} \quad f_{s,max} = \mu_s mg$$

4. Apply $\Sigma F_x = ma_x$ to the car and solve for the acceleration:

$$\Sigma F_x = ma_x \Rightarrow -f_{s,max} = ma_x$$

and

$$-\mu_s mg = ma_x \quad \text{so} \quad a_x = -\mu_s g$$

5. Substituting these results in the equation for Δx in step 2 gives the stopping distance:

$$\Delta x = -\frac{v_0^2}{2a_x} = \frac{v_0^2}{2\mu_s g}$$

$$= \frac{(30 \text{ m/s})^2}{2(0.5)(9.81 \text{ m/s}^2)} = \boxed{91.8 \text{ m}}$$

(b) 1. When the wheels lock, the force exerted by the road on the car is that of kinetic friction. Using reasoning similar to that in Part (a), we obtain for the acceleration:

$$a_x = -\mu_k g$$

2. The stopping distance is then:

$$\Delta x = -\frac{v_0^2}{2a_x} = \frac{v_0^2}{2\mu_k g}$$

$$= \frac{(30 \text{ m/s})^2}{2(0.3)(9.81 \text{ m/s}^2)} = \boxed{153 \text{ m}}$$

REMARKS Notice that the stopping distance is more than 50% greater when the wheels are locked. Also note that the stopping distance is independent of the car's mass—the stopping distance is the same for a subcompact car as for a large truck—provided the coefficients of friction are the same.

EXERCISE What must be the coefficient of static friction between the road and the tires of a four-wheel-drive car if the car is to accelerate from rest to 25 m/s in 8 s? (*Answer* 0.319)

EXPLORING

Newton's laws are not valid in noninertial reference frames. Explore noninertial reference frames, pseudoforces, and cyclones at www.whfreeman.com/tipler5e.

5-2 Motion Along a Curved Path

In Chapter 3 we established that a particle moving with speed v along a curved path with a radius of curvature r has an acceleration component $a_c = v^2/r$ in the centripetal direction (toward the center of curvature), and an acceleration component $a_t = dv/dt$ in the tangential direction.

As with any acceleration, the net force is in the direction of the acceleration. The component of the net force in the centripetal direction is called the **centripetal force**. The centripetal force is not a new force, or a new kind of force. It is merely the name for the net-force component perpendicular to the direction of motion. It may be due to a string, spring, or other contact force such as a normal or frictional force; it may be an action-at-a-distance type of force such as a gravitational force; or it may be any combination of these. It is always directed inward—toward the center of curvature of the path.

SWINGING A PAIL   **EXAMPLE 5-8**

You swing a pail of water in a vertical circle of radius r (Figure 5-20). If its speed is v_t at the top of the circle, find (a) the force exerted on the water by the pail at the top of the circle, and (b) the minimum value of v_t for the water to remain in the pail. (c) What is the force exerted by the pail on the water at the bottom of the circle, where the pail's speed is v_b?

PICTURE THE PROBLEM We apply Newton's second law to find the force exerted by the pail on the water. Since the water moves in a circular path it has an acceleration component v^2/r toward the center of the circle.

(a) 1. Draw free-body diagrams for the water at the top and bottom of the circle (Figure 5-21). Choose the positive axis direction to be toward the center of the circle in each case.

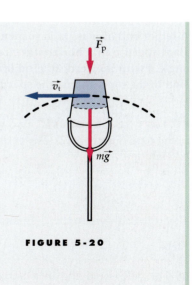

FIGURE 5-20

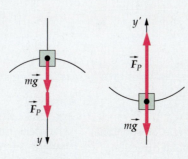

FIGURE 5-21

2. Apply $\Sigma F_y = ma_y$ to the water as it passes through the top of the circle with speed v_t. Solve for the force F_P exerted on the water by the pail:

$$\Sigma F_y = ma_y \Rightarrow F_P + mg = m\frac{v_t^2}{r}$$

so

$$F_P = \boxed{m\left(\frac{v_t^2}{r} - g\right)}$$

(b) The pail can push on the water, but not pull on it. The minimum force it can exert is zero. Set $F_P = 0$ and solve for the minimum speed:

$$0 = m\left(\frac{v_{t,min}^2}{r} - g\right) \Rightarrow v_{t,min} = \boxed{\sqrt{rg}}$$

(c) Apply $\Sigma F_{y'} = ma_{y'}$ to the water as it passes through the bottom of its path with speed v_b. Solve for F_P:

$$\Sigma F_{y'} = ma_{y'} \Rightarrow F_P - mg = \frac{mv_b^2}{r}$$

so

$$F_P = \boxed{m\left(\frac{v_b^2}{r} + g\right)}$$

REMARKS Note that there is no arrow for centripetal force in the free-body diagrams. Centripetal force is not a kind of force exerted by some agent; it is just the name for the component of the resultant force in the centripetal direction. When a whirling bucket is at the top of its circle, both gravity and the contact force of the pail contribute to the force in the centripetal direction acting on the water. When the water is moving at the minimum speed at the top of the circle it is in free-fall with acceleration $\vec{g}$. The only force acting on it at this point is its weight, $m\vec{g}$. At the bottom of the circle, F_P must be greater than the weight mg to provide a net force in the centripetal direction (upward).

PLAUSIBILITY CHECK When $v = 0$ at the bottom, $F_P = mg$.

EXERCISE Estimate (a) the minimum speed at the top of the circle and (b) the maximum period of revolution that will keep you from getting wet if you swing a pail of water in a vertical circle at constant speed. (*Answer* (a) Assuming $r \approx 1$ m, we find $v_{t,min} \approx 3$ m/s (b) $T = 2\pi r/v \approx 2$ s)

TETHER BALL　　　　　　　　　　**EXAMPLE 5-9**　**Try It Yourself**

A tether ball of mass m is suspended from a length of rope and travels at constant speed v in a horizontal circle of radius r as shown. The rope makes an angle θ with the vertical. Find (a) the direction of the acceleration, (b) the tension in the rope, and (c) the speed of the ball.

FIGURE 5-22

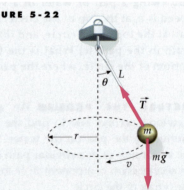

PICTURE THE PROBLEM Two forces act on the ball; its weight and the tension in the rope (Figure 5-22). The vector sum of these forces is in the direction of the acceleration.

Cover the column to the right and try these on your own before looking at the answers.

Steps

Answers

(a) The ball is moving in a horizontal circle at constant speed. The acceleration is in the centripetal direction.

The acceleration is horizontal and directed from the ball toward the center of the circle it is moving in.

(b) 1. Draw a free-body diagram for the ball. Choose the positive x direction as the direction of the ball's acceleration.

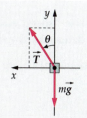

FIGURE 5-23

2. Apply $\Sigma F_y = ma_y$ to the ball and solve for the tension T.

$$\Sigma F_y = ma_y \Rightarrow T \cos \theta - mg = 0$$

so

$$\boxed{T = \frac{mg}{\cos \theta}}$$

(c) 1. Apply $\Sigma F_x = ma_x$ to the ball.

$$\Sigma F_x = ma_x \Rightarrow T \sin \theta = m\frac{v^2}{r}$$

2. Substitute $mg/\cos \theta$ for T and solve for v.

$$\frac{mg}{\cos \theta} \sin \theta = m\frac{v^2}{r} \Rightarrow \tan \theta = \frac{v^2}{rg}$$

so

$$v = \boxed{\sqrt{rg \tan \theta}}$$

REMARKS An object attached to a string and moving in a horizontal circle so that the string makes an angle θ with the vertical is called a *conical pendulum*.

When a car rounds a curve on a horizontal road, both the centripetal force and the forward forces are provided by the force of static friction exerted by the road on the tires of the car. If the car is traveling at constant speed, then the forward component of the frictional force is balanced by the rearward-directed forces of air drag and rolling friction. If air drag is negligible, then the forward component of the frictional force is zero.

A ROAD TEST   **EXAMPLE 5-10**   **Put It in Context**

You have a summer job as part of an automobile tire design team. You are testing new prototype tires to see whether or not the tires perform as well as predicted. In a skid test, a new BMW 530i was able to travel at constant speed in a circle of radius 45.7 m in 15.2 s without skidding. (a) What was its speed v? (b) What is the acceleration? (c) Assuming air drag and rolling friction to be negligible, what is the minimum value for the coefficient of static friction between the tires and the road?

PICTURE THE PROBLEM Figure 5-24 shows the forces acting on the car. The normal force $\vec{F}_n$ balances the downward force due to gravity $m\vec{g}$. The horizontal force is the force of static friction, which provides the centripetal acceleration. The faster the car travels, the greater the centripetal acceleration. The speed can be found from the circumference of the circle and the period T. This speed puts a lower limit on the maximum value of the coefficient of static friction.

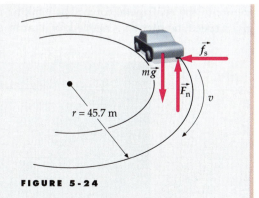

(a) 1. Draw a free-body diagram for the car (Figure 5-25). The positive r direction is away from the center of curvature.

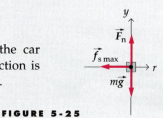

FIGURE 5-24

FIGURE 5-25

2. Use "speed equals distance divided by the time" to determine the speed v:

$$v = \frac{2\pi r}{T} = \frac{2\pi(45.7 \text{ m})}{15.2 \text{ s}} = \boxed{18.9 \text{ m/s}}$$

(b) Use v to calculate the acceleration:

$$a_c = \frac{v^2}{r} = \frac{(18.9 \text{ m/s})^2}{(45.7 \text{ m})} = \boxed{7.81 \text{ m/s}^2}$$

$$a_t = \frac{dv}{dt} = \boxed{0}$$

The acceleration is 7.81 m/s² in the centripetal direction.

(c) 1. Apply $\Sigma F_y = ma_y$ to the car. Solve for the normal and maximum frictional force:

$$\Sigma F_y = ma_y \Rightarrow F_n - mg = 0$$

so

$$F_n = mg \quad \text{and} \quad f_{s,\text{max}} = \mu_s mg$$

2. Apply $\Sigma F_r = ma_r$ to the car. Substituting from step (c), solve 1, for μ_s:

$$\Sigma F_r = ma_r \Rightarrow -f_{s,\text{max}} = m\left(-\frac{v^2}{r}\right)$$

so

$$\mu_s mg = m\frac{v^2}{r} \quad \text{and} \quad \mu_s = \frac{v^2}{rg}$$

$$\mu_s = \frac{(18.9 \text{ m/s})^2}{(45.7 \text{ m})(9.81 \text{ m/s}^2)} = \boxed{0.796}$$

REMARKS A good practice when calculating values to three figures is to calculate all intermediate values to at least four figures. For example, if you use the values shown in Part (b) you obtain $a_c = 7.816$ m/s². This should not be rounded to 7.82 m/s² because in step 2 of Part (a), substituting the exact values gives, to four figures, $v = 18.89$ m/s. Calculating a_c using $v = 18.89$ m/s (rather than 18.9 m/s) gives $a_c = 7.808$ m/s², which rounds up to $a_c = 7.81$ m/s². Storing intermediate values in the memory of your calculator and recalling them for later calculations facilitates this process.

❶ **PLAUSIBILITY CHECK** If μ_s were equal to 1, the inward force would be equal to mg and the centripetal acceleration would be g. Here μ_s is about 0.8 and the centripetal acceleration is about 0.8 g.

*Banked Curves

If a curved road is not horizontal but banked, the normal force of the road will have a component directed inward toward the center of the circle that will contribute to the centripetal force. The banking angle can be chosen so that, for a given speed, no friction is needed for a car to complete the curve.

As a car rounds the curve, the tire is distorted by the frictional force exerted by the road.

Large centrifuge used for research at
Sandia National Laboratories.

FIGURE 5-26

ROUNDING A BANKED CURVE   **EXAMPLE 5-11**

A curve of radius 30 m is banked at an angle θ. Find θ for which a car can round the curve at 40 km/h even if the road is covered with ice so that friction is negligible.

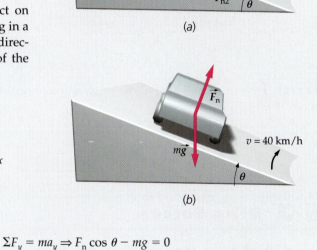

(a)

PICTURE THE PROBLEM In this case only two forces act on the car: gravity and the normal force. Since the car is traveling in a circle at constant speed, the acceleration is in the centripetal direction. The vector sum of the two forces is in the direction of the acceleration.

1. In Figure 5-26a the forces exerted by the road on the car are represented by $\vec{F}_{n1}$ and $\vec{F}_{n2}$. These forces are combined into $\vec{F}_n$ in Figure 5-26b. The angle between the normal force $\vec{F}_n$ and the vertical is θ, the same as the banking angle. Draw a free-body diagram for the car.

FIGURE 5-27

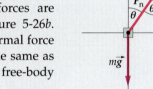
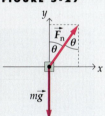

(b)

2. Apply $\Sigma F_y = ma_y$ to the car:

$$\Sigma F_y = ma_y \Rightarrow F_n \cos \theta - mg = 0$$

so

$$F_n = \frac{mg}{\cos \theta}$$

3. Apply $\Sigma F_x = ma_x$ to the car. Substitute for F_n using the step 2 result, then solve for θ:

$$\Sigma F_x = ma_x \Rightarrow F_n \sin \theta = m\frac{v^2}{r}$$

and

$$\frac{mg}{\cos \theta} \sin \theta = m\frac{v^2}{r} \quad so \quad \theta = \tan^{-1} \frac{v^2}{rg}$$

$$\theta = \tan^{-1} \frac{[(40{,}000 \text{ m})/(3600 \text{ s})]^2}{(30 \text{ m})(9.81 \text{ m/s}^2)} = \boxed{22.8°}$$

REMARKS The banking angle θ depends on v and r, but not the mass m; θ increases with increasing v, and decreases with increasing r. When the banking angle, speed, and radius satisfy $\tan \theta = v^2/rg$, the car rounds the curve smoothly, with no tendency to slide either inward or outward. If the car speed is greater than $\sqrt{rg \tan \theta}$, the road will exert a frictional force down the incline. This force has an inward horizontal component, which provides the additional centripetal force needed to keep the car from moving outward (sliding up the incline). If the car speed is less than this amount, the road must exert a frictional force up the incline.

ALTERNATIVE SOLUTION In the preceding solution we followed the guideline to choose one of the coordinate axis directions to be the direction of the acceleration vector, the centripetal direction. However, the solution is no more difficult if we choose one of the axis directions to be down the incline. This choice is taken in the following solution.

FIGURE 5-28

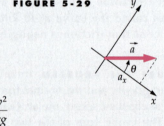

1. Draw a free-body diagram for the car (Figure 5-28). The x axis direction is down the incline and the y axis direction is the normal direction.

2. Apply $\Sigma F_x = ma_x$ to the car:

$$\Sigma F_x = ma_x \Rightarrow mg \sin \theta = ma_x$$

3. Draw a sketch and use trigonometry to obtain an expression for a_x in terms of a and θ (Figure 5-29):

$$a_x = a \cos \theta$$

FIGURE 5-29

4. Substitute the step 3 result into the step 2 result. Then substitute v^2/r for a and solve for θ:

$$mg \sin \theta = ma \cos \theta$$

$$g \sin \theta = \frac{v^2}{r} \cos \theta$$

$$\tan \theta = \frac{v^2}{rg} \Rightarrow \theta = \tan^{-1} \frac{v^2}{rg}$$

EXERCISE Find the component of the acceleration normal to the road surface. (*Answer* 1.60 m/s^2)

*5-3 Drag Forces

When an object moves through a fluid such as air or water, the fluid exerts a **drag force** or retarding force that opposes the motion of the object. The drag force depends on the shape of the object, the properties of the fluid, and the speed of the object relative to the fluid. Unlike ordinary friction, the drag force increases as the speed of the object increases. At low speeds, the drag force is approximately proportional to the speed of the object; at higher speeds, it is more nearly proportional to the square of the speed.

Consider an object dropped from rest and falling under the influence of the force of gravity, which we assume to be constant. Now add a drag force of magnitude bv^n, where b and n are constants. We then have a constant downward force mg and an upward force bv^n (Figure 5-30).

If we take the downward direction to be positive, we obtain from Newton's second law

$$\Sigma F_y = mg - bv^n = ma_y \qquad 5\text{-}6$$

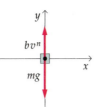

FIGURE 5-30 Free-body diagram showing forces on an object falling with air resistance.

At $t = 0$, the instant when the object is dropped, the speed is zero, so the drag force is zero and the acceleration is g downward. As the speed of the object increases, the drag force increases and the acceleration becomes less than g. Eventually, the speed is great enough for the magnitude of the drag force bv^n to approach the force of gravity mg. As this happens, the acceleration approaches zero and the speed approaches the **terminal speed** v_t. At terminal speed the drag force balances the weight force and the acceleration is zero. Setting the acceleration a_y in Equation 5-6 equal to zero, we obtain

$$bv_t^n = mg$$

Solving for the terminal speed, we get

$$v_t = \left(\frac{mg}{b}\right)^{1/n} \qquad 5\text{-}7$$

The larger the constant b, the smaller the terminal speed. A parachute is designed to maximize b so that the terminal speed will be small. On the other hand, cars are designed to minimize b to reduce the effect of wind resistance.

The terminal speed of a sky diver before release of the parachute is about 200 km/h, which is about 60 m/s. When the parachute is opened, the drag force rapidly increases, becoming greater than the force of gravity, and the sky diver experiences an upward acceleration while falling; that is, the speed of the descending sky diver decreases. As the speed of the sky diver decreases, the drag force decreases and the speed approaches a new terminal speed of about 20 km/h.

TERMINAL SPEED   **EXAMPLE 5-12**

A sky diver of mass 64 kg falls with a terminal speed of 180 km/h with her arms and legs outspread. (a) What is the magnitude of the upward drag force F_d on the sky diver? (b) If the drag force is equal to bv^2, what is the value of b?

(a) 1. Draw a free-body diagram.

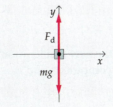

FIGURE 5-31

2. Apply $\Sigma F_y = ma_y$. Since the sky diver is moving with constant velocity, the acceleration is zero:

$$\Sigma F_y = ma_y \Rightarrow F_d - mg = 0$$

so

$$F_d = mg = (64 \text{ kg})(9.81 \text{ N/kg}) = \boxed{628 \text{ N}}$$

(b) 1. To find b we set $F_d = bv^2$:

$$F_d = mg = bv^2$$

so

$$b = \frac{mg}{v^2}$$

2. Find the speed in m/s, then calculate b:

$$180 \text{ km/h} = \frac{180 \text{ km}}{1 \text{ h}}\left(\frac{1000 \text{ m}}{1 \text{ km}}\right)\left(\frac{1 \text{ h}}{3600 \text{ s}}\right)$$

$$= \frac{180 \text{ km}}{1 \text{ h}}\left(\frac{1 \text{ m/s}}{3.6 \text{ km/h}}\right) = 50 \text{ m/s}$$

$$b = \frac{(64 \text{ kg})(9.81 \text{ N/kg})}{(50 \text{ m/s})^2} = \boxed{0.251 \text{ kg/m}}$$

*5-4 Numerical Integration: Euler's Method

If a particle moves under the influence of a *constant* force, its acceleration is constant and we can find its velocity and position from the constant-acceleration kinematic formulas in Chapter 2. But consider a particle moving through space where the force on it, and therefore its acceleration, depends on its position and its velocity. The position, velocity, and acceleration of the particle at one instant determine the position and velocity the next instant, which then determine its acceleration at that instant. The actual position, velocity, and acceleration of an object all change continuously with time. We can approximate this by replacing the continuous time variations with small time steps of duration Δt. The simplest approximation is to assume constant acceleration during each step. This approximation is called **Euler's method.** If the time interval is sufficiently short, the change in acceleration during the interval will be small and can be neglected.

Let x_0, v_0, and a_0 be the known position, velocity, and acceleration of a particle at some initial time t_0. If we assume constant acceleration during Δt, the velocity at time $t_1 = t_0 + \Delta t$ is given by

$$v_1 = v_0 + a_0 \Delta t$$

Similarly, if we neglect any change in velocity during the time interval, the new position is given by

$$x_1 = x_0 + v_0 \Delta t$$

We can use the values v_1 and x_1 to compute the new acceleration a_1 using Newton's second law, and then use x_1, v_1, and a_1 to compute x_2 and v_2.

$$x_2 = x_1 + v_1 \Delta t \qquad\qquad\qquad 5\text{-}8$$

$$v_2 = v_1 + a_1 \Delta t \qquad\qquad\qquad 5\text{-}9$$

The connection between the position and velocity at time t_n and time $t_{n+1} = t_n + \Delta t$ is given by

$$x_{n+1} = x_n + v_n \Delta t \qquad\qquad\qquad 5\text{-}10$$

and

$$v_{n+1} = v_n + a_n \Delta t \qquad\qquad\qquad 5\text{-}11$$

To find the velocity and position at some time t, we therefore divide the time interval $t - t_0$ into a large number of smaller intervals Δt and apply Equations 5-10 and 5-11, beginning at the initial time t_0. This involves a large number of simple, repetitive calculations that are most easily done on a computer. The technique of breaking the time interval into small steps and computing the acceleration, velocity, and position at each step using the values from the previous step is called numerical integration.

Drag Forces   To illustrate the use of numerical integration, let us consider a problem in which a sky diver is dropped from rest at some height under the influences of both gravity and a drag force that is proportional to the square of the speed. We will find the velocity v and the distance traveled x as functions of time.

The equation describing the motion of an object of mass m dropped from rest is Equation 5-6 with $n = 2$:

$$mg - bv^2 = ma$$

where down is the positive direction. The acceleration is thus

$$a = g - \frac{b}{m} v^2 \qquad\qquad\qquad 5\text{-}12$$

It is convenient to write the constant b/m in terms of the terminal speed v_t. Setting $a = 0$ in Equation 5-12 we obtain

$$0 = g - \frac{b}{m} v_t^2$$

$$\frac{b}{m} = \frac{g}{v_t^2}$$

Substituting g/v_t^2 for b/m in Equation 5-12 gives

$$a = g\left(1 - \frac{v^2}{v_t^2}\right) \qquad\qquad\qquad 5\text{-}13$$

The acceleration at time t_n is calculated using the values x_n and v_n.

To solve Equation 5-13 numerically, we need to use numerical values for g and v_t. A reasonable terminal speed for a sky diver is 60 m/s. If we choose $x_0 = 0$ for the initial position, the initial values are $x_0 = 0$, $v_0 = 0$, and $a_0 = g = 9.81$ m/s^2. To find the velocity v and position x after some time, say $t = 20$ s, we divide the time interval $0 < t < 20$ s into many small intervals Δt and apply Equations 5-10, 5-11, and 5-13. We do this by writing a computer program or by using a computer spreadsheet, shown in Figure 5-32. This spreadsheet has $\Delta t = 0.5$ s and at $t = 20$ s, the computed values are $x = 59.89$ m and $v = 939.9$ m/s.

FIGURE 5-32 (a) Spreadsheet to compute the position and speed of a sky diver with air drag proportional to v^2. (b) The same Excel spreadsheet displaying the formulas rather than the values.

(a)

	A	B	C	D
1	Δt =	0.5	s	
2	x0 =	0	m	
3	v0 =	0	m/s	
4	a0 =	9.81	m/s^2	
5	vt =	60	m/s	
6				
7	t	x	v	a
8	(s)	(m)	(m/s)	(m/s^2)
9	0.00	0.0	0.00	9.81
10	0.50	0.0	4.91	9.74
11	1.00	2.5	9.78	9.55
12	1.50	7.3	14.55	9.23
13	2.00	14.6	19.17	8.81
14	2.50	24.2	23.57	8.30
15	3.00	36.0	27.72	7.72
41	16.00	701.0	59.55	0.15
42	16.50	730.7	59.62	0.16
43	17.00	760.6	59.68	0.10
44	17.50	790.4	59.74	0.09
45	18.00	820.3	59.78	0.07
46	18.50	850.2	59.82	0.06
47	19.00	880.1	59.85	0.05
48	19.50	910.0	59.87	0.04
49	20.00	939.9	59.89	0.04
50				

(b)

	A	B	C	D
1	Δt =	0.5	s	
2	x0 =	0	m	
3	v0 =	0	m/s	
4	a0 =	9.81	m/s^2	
5	vt =	60	m/s	
6				
7	t	x	v	a
8	(s)	(m)	(m/s)	(m/s^2)
9	0	=B2	=B3	=$B$4*(1−C9^2/$B$5^2)
10	=A9+$B$1	=B9+C9*$B$1	=C9+D9*$B$1	=$B$4*(1−C10^2/$B$5^2)
11	=A10+$B$1	=B10+C10*$B$1	=C10+D10*$B$1	=$B$4*(1−C11^2/$B$5^2)
12	=A11+$B$1	=B11+C11*$B$1	=C11+D11*$B$1	=$B$4*(1−C12^2/$B$5^2)
13	=A12+$B$1	=B12+C12*$B$1	=C12+D12*$B$1	=$B$4*(1−C13^2/$B$5^2)
14	=A13+$B$1	=B13+C13*$B$1	=C13+D13*$B$1	=$B$4*(1−C14^2/$B$5^2)
15	=A14+$B$1	=B14+C14*$B$1	=C14+D14*$B$1	=$B$4*(1−C15^2/$B$5^2)
41	=A40+$B$1	=B40+C40*$B$1	=C40+D40*$B$1	=$B$4*(1−C41^2/$B$5^2)
42	=A41+$B$1	=B41+C41*$B$1	=C41+D41*$B$1	=$B$4*(1−C42^2/$B$5^2)
43	=A42+$B$1	=B42+C42*$B$1	=C42+D42*$B$1	=$B$4*(1−C43^2/$B$5^2)
44	=A43+$B$1	=B43+C43*$B$1	=C43+D43*$B$1	=$B$4*(1−C44^2/$B$5^2)
45	=A44+$B$1	=B44+C44*$B$1	=C44+D44*$B$1	=$B$4*(1−C45^2/$B$5^2)
46	=A45+$B$1	=B45+C45*$B$1	=C45+D45*$B$1	=$B$4*(1−C46^2/$B$5^2)
47	=A46+$B$1	=B46+C46*$B$1	=C46+D46*$B$1	=$B$4*(1−C47^2/$B$5^2)
48	=A47+$B$1	=B47+C47*$B$1	=C47+D47*$B$1	=$B$4*(1−C48^2/$B$5^2)
49	=A48+$B$1	=B48+C48*$B$1	=C48+D48*$B$1	=$B$4*(1+C49^2/$B$5^2)
50				

Figure 5-33 shows graphs of v versus t and x versus t plotted from this data.

But how accurate are our computations? We can estimate the accuracy by running the program again using a smaller time interval. If we use $\Delta t = 0.25$ s, one-half of the value we originally used, we obtain $v = 59.86$ m/s and $x = 943.1$ m at $t = 20$ s. The difference in v is about 0.05 percent and that in x is about 0.4 percent. These are our estimates of the accuracy of the original computations.

Since the difference between the value of a_{av} for some time interval Δt and the value of a at the beginning of the interval becomes smaller as the time interval becomes smaller, we might expect that it would be better to use very small time intervals, say $\Delta t = 0.000000001$ s. But there are two reasons for not using very small time intervals. First, the smaller the time interval, the larger the number of calculations that are required and the longer the program takes to run. Second, the computer keeps only a fixed number of digits at each step of the calculation, so that at each step there is a round-off error. These round-off errors add up. The larger the number of calculations, the more significant the total round-off errors become. When we first decrease the time interval, the accuracy improves because a_i more nearly approximates a_{av} for the interval. However, as the time interval is decreased further, the round-off errors build up and the accuracy of the computation decreases. A good rule of thumb to follow is to use no more than about 10^4 or 10^5 time intervals for the typical numerical integration.

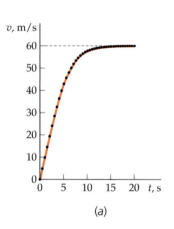

(a)

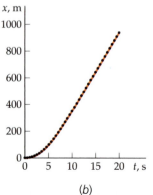

(b)

FIGURE 5-33 (a) Graph of v versus t for a sky diver, found by numerical integration using $\Delta t = 0.5$ s. The horizontal dashed line is the terminal speed $v_t = 60$ m/s. (b) Graph of x versus t using $\Delta t = 0.5$ s.

Friction and drag forces are complex phenomena empirically approximated by simple equations.

Topic	Relevant Equations and Remarks
1. Friction	Two objects in contact exert frictional forces on each other. These forces are parallel to the surfaces of the objects at the points of contact and directed opposite to the direction of sliding or tendency to slide.
Static friction	$f_s \leq \mu_s F_n$ **5-2** where F_n is the normal force of contact and μ_s is the coefficient of static friction.
Kinetic friction	$f_k = \mu_k F_n$ **5-3** where μ_k is the coefficient of kinetic friction. The coefficient of kinetic friction is slightly less than the coefficient of static friction.
2. Motion Along a Curved Path	A particle moving along an arbitrary curve can be considered to be moving along a circular arc during a small time interval. Its instantaneous acceleration vector has a component $a_c = v^2/r$ toward the center of curvature of the arc and a component $a_t = dv/dt$ that is tangential to the curve. If the particle is moving along a circular path of radius r at constant speed v, $a_t = 0$ and the speed, radius, and period T are related by $2\pi r = vT$.
3. *Drag Forces	When an object moves through a fluid it experiences a drag force that opposes its motion. The drag force increases with increasing speed. If the body is dropped from rest, its speed increases. As it does, the magnitude of the drag force comes closer and closer to the magnitude of the force of gravity, so the net force, and thus the acceleration, approaches zero. As the acceleration approaches zero, the speed approaches a constant value called its terminal speed. The terminal speed depends on the shape of the body and on the medium through which it falls.
4. *Numerical Integration: Euler's Method	To estimate the position x and velocity v at some time t, we first divide t into a large number of small intervals, each of length Δt. The initial acceleration a_0 is then calculated from the initial position x_0 and velocity v_0. The position x_1 and velocity v_1 a time Δt later are estimated using the relations $x_{n+1} = x_n + v_n \Delta t$ **5-10** and $v_{n+1} = v_n + a_n \Delta t$ **5-11** with $n = 0$. The acceleration a_{n+1} is calculated using the values for x_{n+1} and v_{n+1} and the process is repeated. This continues until estimations for the position and velocity at time t are calculated.

- Single-concept, single-step, relatively easy
- •• Intermediate-level, may require synthesis of concepts
- ••• Challenging
- SSM Solution is in the *Student Solutions Manual*
- iSOLVE Problems available on iSOLVE online homework service
- iSOLVE✓ These "Checkpoint" online homework service problems ask students additional questions about their confidence level, and how they arrived at their answer

In a few problems, you are given more data than you actually need; in a few other problems, you are required to supply data from your general knowledge, outside sources, or informed estimates.

Conceptual Problems

1 • Various objects lie on the floor of a truck moving along a straight horizontal road. If the truck gradually speeds up, what force acts on the objects to cause them to speed up too?

2 • SSM Any object resting on the floor of a truck will slide if the truck's acceleration is too great. How does the critical acceleration at which a light object slips compare with that at which a much heavier object slips?

3 • A block of mass m rests on a plane inclined at an angle θ with the horizontal. It follows that the coefficient of static friction between the block and plane is (a) $\mu_s \geq g$, (b) $\mu_s =$ tan θ, (c) $\mu_s \leq$ tan θ, (d) $\mu_s \geq$ tan θ.

4 • SSM A block of mass m is at rest on a plane inclined at an angle of 30° with the horizontal, as shown in Figure 5-34. Which of the following statements about the force of static friction is necessarily true? (a) $f_s > mg$. (b) $f_s > mg$ cos 30°. (c) $f_s = mg$ cos 30°. (d) $f_s = mg$ sin 30°. (e) None of these statements are true.

FIGURE 5-34 Problem 4

5 •• On an icy winter day, the coefficient of friction between the tires of a car and a roadway is reduced to one-half of its value on a dry day. As a result, the maximum speed at which a curve of radius R can be safely negotiated is (a) the same as on a dry day, (b) reduced to 71% of its value on a dry day, (c) reduced to 50% of its value on a dry day, (d) reduced to 25% of its value on a dry day, (e) reduced by an unknown amount depending on the car's mass.

6 •• SSM Show with a force diagram how a motorcycle can travel in a circle on the inside vertical wall of a hollow cylinder. Assume reasonable parameters (coefficient of friction, radius of the circle, mass of the motorcycle, or whatever is required), and calculate the minimum speed needed.

7 •• This is an interesting experiment that you can perform at home: take a wooden block and rest it on the floor or some other flat surface. Attach a spring (a slinky will work well) to the block and pull gently and steadily on the spring in the horizontal direction. At some point, the block will start moving, but it will not move smoothly. Instead, it will start moving, stop again, start moving again, stop again, etc. Explain why the motion of the block (called "stick–slip" motion) behaves this way.

8 • True or false: Viewed from an inertial reference frame, an object cannot move in a circle unless there is a net force acting on it.

9 •• A particle is traveling in a vertical circle at constant speed. One can conclude that the _____ is constant.

(a) velocity, (b) acceleration, (c) net force, (d) apparent weight, (e) none of the properties listed.

10 • SSM You place a lightweight piece of iron on a table and hold a small kitchen magnet above the iron at a distance of 1 cm. You find that the magnet cannot lift the iron, even though there is obviously a force between the iron and the magnet. Next, you again hold the piece of iron and the magnet 1 cm apart with the magnet above the iron, but this time you drop them from arm's length. As they fall, the magnet and the piece of iron are pulled together before hitting the floor. (a) Draw free-body diagrams illustrating all of the forces on the magnet and the iron for each demonstration. (b) Explain why the magnet and iron are pulled together when they are dropped, even though the magnet cannot pull up the piece of iron when it is sitting on the table.

11 ••• SSM The following question is an excellent "brain-twister" invented by Boris Korsunsky[†]: Two identical blocks are attached by a massless string running over a pulley as shown in Figure 5-35. The rope initially runs over the pulley at its (the rope's) midpoint, and the surface that block 1 rests on is frictionless. Blocks 1 and 2 are initially at rest when block 2 is released with the string taut and horizontal. Will block 1 hit the pulley before or after block 2 hits the wall? (Assume that the initial distance from block 1 to the pulley is the same as the initial distance from block 2 to the wall.) There is a very simple solution.

† Boris Korsunsky, "Braintwisters for Physics Students," *The Physics Teacher*, **33**, 550 (1995).

FIGURE 5-35
Problem 11

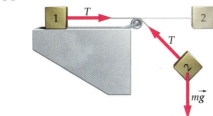

12 • True or false: The terminal speed of a falling object depends on its shape.

13 • **SSM** As a sky diver falls through the air, her terminal speed (a) depends on her mass, (b) depends on her orientation as she falls, (c) depends on the density of the air, (d) depends on all of the above.

14 •• You are sitting in the passenger seat in a car driving around a circular, horizontal, flat racetrack at a high speed. As you sit there, you "feel" a "force" pushing you toward the outside of the track. What is the true direction of the force acting on you, and where does it come from? (Assume that you don't slide across the seat.)

15 • **SSM** The mass of the moon is about 1% of that of the earth. The centripetal force that keeps the moon in its orbit around the earth (a) is much smaller than the gravitational force exerted on the moon by the earth, (b) depends on the phase of the moon, (c) is much greater than the gravitational force exerted on the moon by the earth, (d) is the gravitational force exerted on the moon by the earth, (e) cannot be answered; we haven't studied Newton's law of gravity yet.

16 • A block is sliding on a frictionless surface along a loop-the-loop as in Figure 5-36. The block is moving fast enough that it never loses contact with the track. Match the points along the track to the appropriate free-body diagrams in Figure 5-37.

Problem 16   FIGURE 5-36

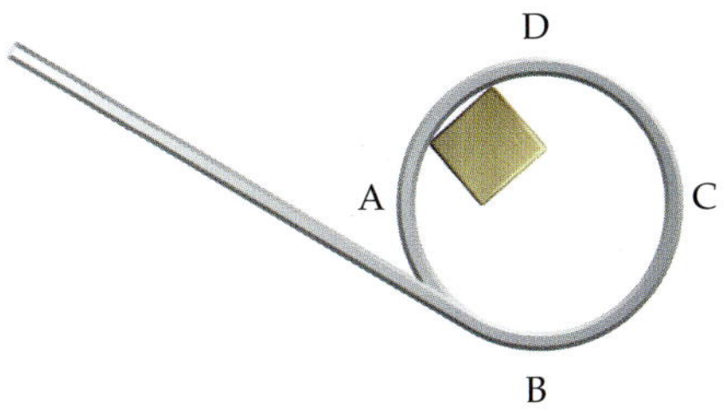

FIGURE 5-37

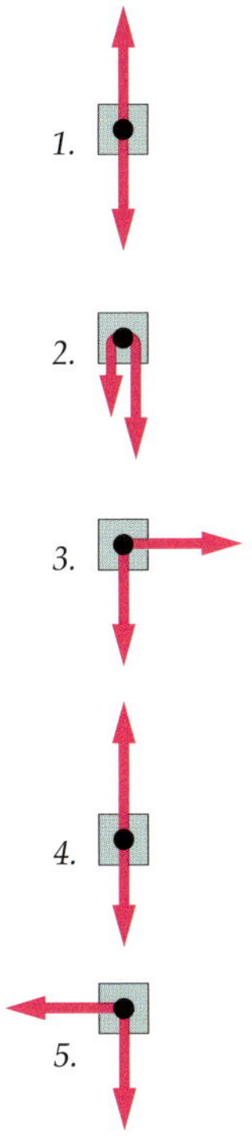

1.

2.

3.

4.

5.

17 •• If you hold a rock and a feather at the same height above the ground and drop them, the rock will hit the ground first. From this you might conclude that the average aerodynamic drag force on the feather as it falls is larger than that on the rock, but in fact the opposite is true. Explain in detail why the rock hits the ground first. Assume that the drag force is given by the Newtonian formula $F_D = (1/2) CA\rho v^2$ (see Problem 18(c) for an explanation of this formula).

Estimation and Approximation

18 • **SSM** To determine the aerodynamic drag on a car, the "coast-down" method is often used. The car is driven on a long, flat road at some convenient speed (60 mph is typical), shifted into neutral, and allowed to coast to a stop. The time that it takes for the speed to drop by successive 5-mph intervals is measured and used to compute the net force slowing the car down. (a) It was found that a Toyota Tercel with a mass of 1020 kg coasted down from 60 mph to 55 mph in 3.92 s. What is the average force slowing the car down? (b) If the coefficient of rolling friction for the car is 0.02, what is the force of rolling friction that is acting to slow the car down? If we assume that the only two forces acting on the car are rolling friction and aerodynamic drag, what is the average drag force acting on the car? (c) The drag force will have the form $\frac{1}{2}C\rho Av^2$, where A is the cross-sectional area of the car facing the wind, v is the car's speed, ρ is the density of air, and C is a dimensionless constant of order 1. If the cross-sectional area of the car is 1.91 m², determine C from the data given. (The density of air is 1.21 kg/m³; use 55 mph for the speed of the car in this computation.)

19 • Using dimensional analysis, determine the units and dimensions of the constant b in the retarding force bv^n if (a) $n = 1$ and (b) $n = 2$. (c) Newton showed that the air resistance of a falling object with a circular cross section should be approximately $\frac{1}{2}\rho\pi r^2 v^2$, where $\rho = 1.2$ kg/m³, the density of air. Show that this is consistent with your dimensional analysis for part (b). (d) Find the terminal speed for a 56-kg sky diver; approximate his or her cross-sectional area as a disk of radius 0.30 m. The density of air near the surface of the earth is 1.20 kg/m³. (e) The density of the atmosphere decreases with height above the surface of the earth; at a height of 8 km, the density is only 0.514 kg/m³. What is the terminal velocity at this height?

20 •• While hailstones the size of golf balls are a bit unusual, hailstones are generally substantially larger than raindrops. Estimate the terminal velocity of a raindrop and a golf-ball sized hailstone.

Friction

21 • **SSM** A block of mass m slides at constant speed down a plane inclined at an angle of θ with the horizontal. It follows that (a) $\mu_k = mg \sin \theta$, (b) $\mu_k = \tan \theta$, (c) $\mu_k = 1 - \cos \theta$, (d) $\mu_k = \cos \theta - \sin \theta$.

22 • A block of wood is pulled by a horizontal string across a horizontal surface at a constant velocity with a force of 20 N. The coefficient of kinetic friction between the surfaces is 0.3. The force of friction is (a) impossible to determine without knowing the mass of the block, (b) impossible to determine without knowing the speed of the block, (c) 0.3 N, (d) 6 N, (e) 20 N.

23 • SSM SOLVE✓ A 20-N block rests on a horizontal surface. The coefficients of static and kinetic friction between the surface and the block are $\mu_s = 0.8$ and $\mu_k = 0.6$. A horizontal string is attached to the block and a constant tension T is maintained in the string. What is the force of friction acting on the block if (a) $T = 15$ N or (b) $T = 20$ N.

24 • A block of mass m is pulled at a constant velocity across a horizontal surface by a string as shown in Figure 5-38. The magnitude of the frictional force is (a) $\mu_k mg$, (b) $T \cos \theta$, (c) $\mu_k (T - mg)$, (d) $\mu_k T \sin \theta$, (e) $\mu_k (mg - T \sin \theta)$.

FIGURE 5-38 Problem 24

25 • SOLVE A tired worker pushes with a horizontal force of 500 N on a 100-kg crate initially resting on a thick pile carpet. The coefficients of static and kinetic friction are 0.6 and 0.4, respectively. Find the frictional force exerted by the carpet on the crate.

26 • SOLVE A box weighing 600 N is pushed along a horizontal floor at constant velocity with a force of 250 N parallel to the floor. What is the coefficient of kinetic friction between the box and the floor?

27 • SOLVE✓ The coefficient of static friction between the tires of a car and a horizontal road is $\mu_s = 0.6$. If the net force on the car is the force of static friction exerted by the road, (a) what is the magnitude of the maximum acceleration of the car when it is braked? (b) What is the shortest distance in which the car can stop if it is initially traveling at 30 m/s?

28 • SSM The force that accelerates a car along a flat road is the frictional force exerted by the road on the car's tires. (a) Explain why the acceleration can be greater when the wheels do not slip. (b) If a car is to accelerate from 0 to 90 km/h in 12 s, what is the minimum coefficient of friction needed between the road and tires? Assume that half the weight of the car is supported by the drive wheels.

29 • A 5-kg block is held at rest against a vertical wall by a horizontal force of 100 N. (a) What is the frictional force exerted by the wall on the block? (b) What is the minimum horizontal force needed to prevent the block from falling if the coefficient of friction between the wall and the block is $\mu_s = 0.40$?

30 • A tired and overloaded student is attempting to hold a large physics textbook wedged under his arm as shown in Figure 5-39. The textbook has a mass of 10.2 kg, while the coefficient of static friction of the textbook against the student's underarm is 0.32 and the coefficient of static friction of the book against the student's shirt is 0.16. (a) What is the minimum horizontal force that the student must apply to the textbook to prevent it from falling? (b) If the student can only exert a force of 195 N, what is the acceleration of the textbook as it slides from under his arm? The coefficient of kinetic friction of arm against textbook is 0.20, while that of shirt against textbook is 0.09.

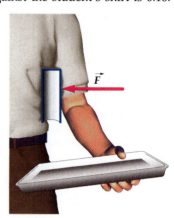

FIGURE 5-39 Problem 30

31 • On a snowy day with the temperature near the freezing point, the coefficient of static friction between a car's tires and an icy road is 0.08. What is the maximum incline that a vehicle with four-wheel drive can climb at constant speed?

32 • SSM A 50-kg box that is resting on a level floor must be moved. The coefficient of static friction between the box and the floor is 0.6. One way to move the box is to push down on the box at an angle θ with the horizontal. Another method is to pull up on the box at an angle θ with the horizontal. (a) Explain why one method is better than the other. (b) Calculate the force necessary to move the box by each method if $\theta = 30°$ and compare the answer with the results when $\theta = 0°$.

33 • SOLVE✓ A 3-kg block rests on a horizontal table, attached to a 2-kg block by a light string as shown in Figure 5-40. (a) What is the minimum coefficient of static friction such that the objects remain at rest? (b) If the coefficient of static friction is less than that found in part (a), and if the coefficient of kinetic friction between the block and the table is 0.3, find the time it takes for the 2-kg mass to fall 2 m to the floor if the system starts from rest.

FIGURE 5-40 Problem 33

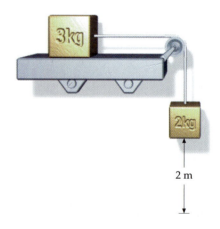

34 •• A block on a horizontal plane is given an initial speed v. Traveling in a straight line, it comes to rest after sliding a distance d. Show that the coefficient of kinetic friction is $\mu_k = v^2/2gd$.

35 •• **SSM** A block of mass $m_1 = 250$ g is at rest on a plane that makes an angle of $\theta = 30°$ above the horizontal. The coefficient of kinetic friction between the block and the plane is $\mu_k = 0.100$. The block is attached to a second block of mass $m_2 = 200$ g that hangs freely by a string that passes over a frictionless, massless pulley (Figure 5-41). When the second block has fallen 30.0 cm, its speed is (a) 83 cm/s, (b) 48 cm/s, (c) 160 cm/s, (d) 59 cm/s, (e) 72 cm/s.

FIGURE 5-41 Problems 35–37

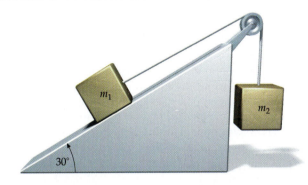

36 •• Returning to Figure 5-41, this time $m_1 = 4$ kg. The coefficient of static friction between the block and the incline is 0.4. (a) Find the range of possible values for m_2 for which the system will be in static equilibrium. (b) What is the frictional force on the 4-kg block if $m_1 = 1$ kg?

37 •• Returning once again to Figure 5-41, this time $m_1 = 4$ kg, $m_2 = 5$ kg, and the coefficient of kinetic friction between the inclined plane and the 4-kg block is $\mu_k = 0.24$. Find the acceleration of the masses and the tension in the cord.

38 •• **SSM** **SOLVE**✓ The coefficient of static friction between the bed of a truck and a box resting on it is 0.30. The truck is traveling at 80 km/h along a horizontal road. What is the shortest distance in which the truck can stop if the box is not to slide?

39 •• A 4.5-kg block is given an initial velocity of 14 m/s up an incline that makes an angle of 37° with the horizontal. When its displacement is 8.0 m, its velocity up the incline has diminished to 5.2 m/s. Find (a) the coefficient of kinetic friction between the block and plane, (b) the displacement of the block from its starting point at the time when it momentarily comes to rest, and (c) the speed of the block when it again reaches its starting point.

40 •• An automobile is going up a grade of 15° at a speed of 30 m/s. The coefficient of static friction between the tires and the road is 0.7. (a) What minimum distance does it take to stop the car? (b) What minimum distance would it take if the car were going down the grade?

41 •• A rear-wheel-drive car supports 40 percent of its weight on its two drive wheels and has a coefficient of static friction of 0.7 with a horizontal straight road. (a) Find the vehicle's maximum acceleration. (b) What is the shortest possible time in which this car can achieve a speed of 100 km/h? (Assume the engine has unlimited power.)

42 •• **SSM** Lou bets an innocent stranger that he can place a 2-kg box against the side of a cart, as in Figure 5-42, and that the box will not fall to the ground, even though Lou will use no hooks, ropes, fasteners, magnets, glue, or adhesives of any kind. When the stranger accepts the bet, Lou begins to push the cart in the direction shown. The coefficient of static friction between the box and the cart is 0.6. (a) Find the minimum acceleration for which Lou will win the bet. (b) What is the magnitude of the frictional force in this case? (c) Find the force of friction on the box if the acceleration is twice the minimum needed for the box not to fall. (d) Show that, for a box of any mass, the box will not fall if the acceleration is $a \geq g/\mu_s$, where μ_s is the coefficient of static friction.

FIGURE 5-42 Problem 42

43 •• **SOLVE** Two blocks attached by a string (Figure 5-43) slide down a 10° incline. The lower block has a mass of $m_1 = 0.25$ kg and a coefficient of kinetic friction $\mu_k = 0.2$. For the upper block, $m_2 = 0.8$ kg and $\mu_k = 0.3$. Find (a) the acceleration of the blocks and (b) the tension in the string.

44 •• **SSM** As in Problem 43, two blocks of masses m_1 and m_2 are sliding down an incline as shown in Figure 5-43. They are connected by a massless *rod* this time; the rod behaves in exactly the same way as a string, except that the force can be compressive as well as tensile. The coefficient of kinetic friction for block 1 is μ_1 and the coefficient of kinetic friction for block 2 is μ_2. (a) Determine the acceleration of the two blocks. (b) Determine the force that the rod exerts on the two blocks. Show that the force is 0 when $\mu_1 = \mu_2$ and give a simple, nonmathematical argument why this is true.

FIGURE 5-43 Problems 43 and 44

45 •• Two blocks attached by a string are at rest on an inclined surface. The lower block has a mass of $m_1 = 0.2$ kg and a coefficient of static friction $\mu_s = 0.4$. The upper block has a mass $m_2 = 0.1$ kg and $\mu_s = 0.6$. The angle θ is slowly increased. (a) At what angle θ_c do the blocks begin to slide? (b) What is the tension in the string just before sliding begins?

46 •• Two blocks connected by a massless, rigid rod slide on a surface inclined at an angle of 20°. The lower block has a mass $m_1 = 1.2$ kg, and the upper block's mass is $m_2 = 0.75$ kg. (a) If the coefficient of kinetic friction is $\mu_k = 0.3$ for the lower block and $\mu_k = 0.2$ for the upper block, what is the acceleration of the blocks? (b) Determine the force exerted by the rod on either block.

47 •• SSM A block of mass m rests on a horizontal table (Figure 5-44). The block is pulled by a massless rope with a force $\vec{F}$ at an angle θ. The coefficient of static friction is 0.6. The minimum value of the force needed to move the block depends on the angle θ. (a) Discuss qualitatively how you would expect this force to depend on θ. (b) Compute the force for the angles $\theta = 0°, 10°, 20°, 30°, 40°, 50°,$ and $60°$, and make a plot of F versus θ for $mg = 400$ N. From your plot, at what angle is it most efficient to apply the force to move the block?

FIGURE 5-44 Problem 47

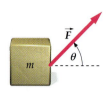

48 •• For the block in Problem 47, show that, in the general case for a block of mass m resting on a horizontal surface whose coefficient of static friction is μ_s: (a) If we want to apply the minimum possible force to move the block, it should be applied with the force pulling upward at an angle $\theta_{min} = \tan^{-1}\mu_s$, and (b) the minimum force necessary to start the block moving is $F_{min} = \left(\mu_s/\sqrt{1 + \mu_s^2}\right)mg$. (c) Once you start the block moving, if you want to apply the least possible force to *keep* it moving, should you keep the angle at which you are pulling the same, increase it, or decrease it?

49 •• Answer the questions in Problem 48, but for force $\vec{F}$ that pushes down on the block at an angle θ below the horizontal (Figure 5-45).

FIGURE 5-45 Problem 49

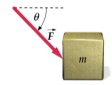

50 •• A 100-kg mass is pulled along a frictionless surface by a horizontal force $\vec{F}$ such that its acceleration is $a_1 = 6$ m/s² (Figure 5-46). A 20-kg mass slides along the top of the 100-kg mass and has an acceleration of $a_2 = 4$ m/s². (It thus slides backward relative to the 100-kg mass.) (a) What is the frictional force exerted by the 100-kg mass on the 20-kg mass? (b) What is the net force acting on the 100-kg mass? What is the force F? (c) After the 20-kg mass falls off the 100-kg mass, what is the acceleration of the 100-kg mass? (Assume that the force F does not change.)

FIGURE 5-46 Problem 50

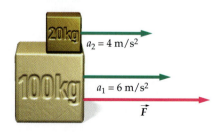

51 •• ISOLVE A 60-kg block slides along the top of a 100-kg block. The 60-kg block has an acceleration of 3 m/s² when a horizontal force $\vec{F}$ of 320 N is applied, as in Figure 5-47. There is no friction between the 100-kg block and a horizontal frictionless surface, but there is friction between the two blocks. (a) Find the coefficient of kinetic friction between the blocks. (b) Find the acceleration of the 100-kg block during the time that the 60-kg block remains in contact.

FIGURE 5-47 Problem 51

52 •• SSM ISOLVE✓ The coefficient of static friction between a rubber tire and the road surface is 0.85. What is the maximum acceleration of a 1000-kg four-wheel-drive truck if the road makes an angle of 12° with the horizontal and the truck is (a) climbing and (b) descending?

53 •• A 2-kg block sits on a 4-kg block that is on a frictionless table (Figure 5-48). The coefficients of friction between the blocks are $\mu_s = 0.3$ and $\mu_k = 0.2$. (a) What is the maximum horizontal force F that can be applied to the 4-kg block if the 2-kg block is not to slip? (b) If F has half this value, find the acceleration of each block and the force of friction acting on each block. (c) If F is twice the value found in (a), find the acceleration of each block.

FIGURE 5-48 Problem 53

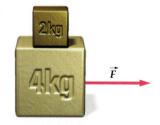

54 •• In Figure 5-49, the mass $m_2 = 10$ kg slides on a frictionless table. The coefficients of static and kinetic friction between m_2 and $m_1 = 5$ kg are $\mu_s = 0.6$ and $\mu_k = 0.4$. (a) What is the maximum acceleration of m_1? (b) What is the maximum value of m_3 if m_1 moves with m_2 without slipping? (c) If $m_3 = 30$ kg, find the acceleration of each body and the tension in the string.

FIGURE 5-49 Problem 54

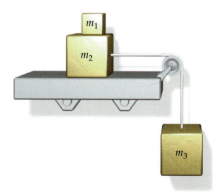

55 • A massive sandstone block is being dragged up a ramp by a counterweight suspended over a pulley, as shown in Figure 5-50. The mass of the block is 1600 kg, while the mass of the counterweight is 550 kg. The coefficient of kinetic friction of the block against the ramp is 0.15, and the ramp is sloped at an angle of 10°. (*a*) What is the acceleration of the block up the ramp? (*b*) Three seconds after the block begins moving up the ramp, the rope holding the counterweight breaks. How much farther will the block continue to slide up the ramp before stopping? (*c*) Unfortunately, the block begins to slide back down the ramp once it reaches its highest point. What is the acceleration of the block as it slides down the ramp?

FIGURE 5-50 Problem 55

56 ••• **iSOLVE** A 10-kg block rests on a 5-kg bracket shown in Figure 5-51. The 5-kg bracket sits on a frictionless surface. The coefficients of friction between the 10-kg block and the bracket on which it rests are $\mu_s = 0.40$ and $\mu_k = 0.30$. (*a*) What is the maximum force *F* that can be applied if the 10-kg block is not to slide on the bracket? (*b*) What is the corresponding acceleration of the 5-kg bracket?

FIGURE 5-51 Problem 56

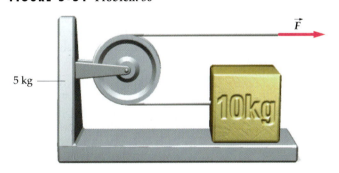

57 •• **SSM** On planet Vulcan, an introductory physics class performs several experiments involving friction. In one of these experiments the acceleration of a block is measured both when it is sliding up an incline and when it is sliding down the same incline. You copy the following data and diagram (Figure 5-52) out of one of the lab notebooks, but can't find any translations into metric units (Negative sign indicates that the acceleration is pointing down the incline.):

Acceleration of block
Going up inclined plane   −1.73 glapp/plip²
Going down plane   −1.42 glapp/plip²

FIGURE 5-52 Problem 57

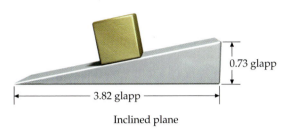

0.73 glapp

3.82 glapp

Inclined plane

From these data, determine the acceleration of gravity on Vulcan (in glapps/plip²) and the coefficient of kinetic friction between the block and the incline.

58 •• **SSM** A 100-kg block on an inclined plane is attached to another block of mass *m* via a string, as in Figure 5-53. The coefficients of static and kinetic friction of the block and the incline are $\mu_s = 0.4$ and $\mu_k = 0.2$. The angle of the incline is 18° above horizontal. (*a*) Determine the range of values for *m*, the mass of the hanging block, for which the block on the incline will not move unless disturbed, but if nudged, will slide *down* the incline. (*b*) Determine a range of values for *m* for which the block on the incline will not move unless nudged, but if nudged will slide *up* the incline.

FIGURE 5-53 Problem 58

59 ••• **iSOLVE**✓ A block of mass 0.5 kg rests on the inclined surface of a wedge of mass 2 kg, as in Figure 5-54. The wedge is acted on by a horizontal force $\vec{F}$ and slides on a frictionless surface. (*a*) If the coefficient of static friction between the wedge and the block is $\mu_s = 0.8$ and the angle of the incline is 35°, find the maximum and minimum values of *F* for which the block does not slip. (*b*) Repeat part (*a*) with $\mu_s = 0.4$.

FIGURE 5-54 Problem 59

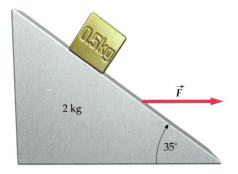

60 • The coefficient of kinetic friction between two surfaces is not truly independent of the relative velocity of the two objects in contact. In fact, it tends to decrease slightly with increasing velocity. For example, one series of experiments found that for wood sliding on wood, the coefficient of friction could be described by the equation

$$\mu_k = 0.11/(1 + 2.3\times10^{-4} v^2)^2$$

where v is measured in m/s. From this expression, find the force of kinetic friction acting on a 100-kg block of wood moving across a horizontal wooden surface at (a) 10 m/s and (b) 20 m/s.

61 •• In some experiments, it has been found that the coefficient of kinetic friction is not independent of the normal force acting on the object. Vaclav Konecny[†] found, in a set of experiments for wood sliding over wood, that the force of static friction between the two wood surfaces was given by the expression $f_k = 0.4 F_n^{0.91}$, where f_k is the force of friction and F_n is the normal force acting on the object (both forces measured in newtons). Because of this, the acceleration of a block down an inclined plane will not be independent of the mass of the block. Using this formula, calculate the acceleration and stopping distance for (a) a 10-kg block and (b) a 100-kg block sliding along a horizontal surface with an initial speed of 10 m/s.

∙62 ••• **SSM** A block of wood with a mass of 10 kg is pushed, starting from rest, with a constant horizontal force of 70 N across a wooden floor. Assuming that the coefficient of kinetic friction varies with particle speed as $\mu_k = 0.11/(1 + 2.3\times10^{-4} v^2)^2$ (see Problem 60), write a spreadsheet program using Euler's method to calculate and graph the speed of the block and its displacement as a function of time from 0 to 10 s. Compare this to the case where the coefficient of kinetic friction is equal to 0.11, independent of v.

63 •• To determine the coefficient of kinetic friction of a block of wood on a horizontal table surface, you are given the following assignment: Take the block of wood and shove it across the surface of the table. Using a stopwatch, measure the time it takes for the block to come to a stop (Δt) and the total distance that the block travels after the push (Δx). (a) Show that from these measurements, $\mu_k = 2\Delta x/[g(\Delta t)^2]$. If the block slides a distance of 1.37 m in 0.97 s, calculate μ_k. (b) What was the initial speed of the block?

64 •• **SSM** The following data show the acceleration of a block down an inclined plane as a function of the angle of incline θ[‡]:

θ (degrees)	Acceleration (m/s²)
25	1.6909
27	2.1043
29	2.4064
31	2.8883
33	3.1750
35	3.4886
37	3.7812
39	4.1486
41	4.3257
43	4.7178
45	5.1056

(a) Show that, for a block sliding down an incline, graphing $a/\cos\theta$ versus $\tan\theta$ should give a straight line with slope g and y-intercept $-\mu_k g$. (b) Using a spreadsheet program, graph these data and fit a straight line to them to determine μ_k and g. What is the percentage error in g from the commonly accepted value of 9.81 m/s²?

Motion Along a Curved Path

65 •• A stone with a mass $m = 95$ g is being whirled in a horizontal circle on the end of a string that is 85 cm long. The length of time required for the stone to make one complete revolution is 1.22 s. The angle that the string makes with the horizontal is (a) 52°, (b) 46°, (c) 26°, (d) 23°, (e) 3°.

66 •• A 0.20-kg stone attached to a 0.8-m string is rotated in the horizontal plane. The string makes an angle of 20° with the horizontal. Determine the speed of the stone.

67 •• A 0.75-kg stone attached to a string is whirled in a horizontal circle of radius 35 cm as in the conical pendulum of Example 5-9. The string makes an angle of 30° with the vertical. (a) Find the speed of the stone. (b) Find the tension in the string.

68 •• **SSM** **iSOLVE** A 50-kg pilot comes out of a vertical dive in a circular arc such that at the bottom of the arc her upward acceleration is 8.5g. (a) What is the magnitude of the force exerted by the airplane seat on the pilot at the bottom of the arc? (b) If the speed of the plane is 345 km/h, what is the radius of the circular arc?

69 •• **iSOLVE✓** A 65-kg airplane pilot pulls out of a dive by following, at constant speed, the arc of a circle whose radius is 300 m. At the bottom of the circle, where her speed is 180 km/h, (a) what are the direction and magnitude of her acceleration? (b) What is the net force acting on her at the bottom of the circle? (c) What is the force exerted on the pilot by the airplane seat?

70 •• **iSOLVE✓** Mass m_1 moves in a circular path of radius r on a frictionless horizontal table (Figure 5-55). It is attached to a string that passes through a frictionless hole in the center of the table. A second mass m_2 is attached to the other end of the string. Derive an expression for r in terms of m_1, m_2, and the time T for one revolution.

FIGURE 5-55 Problem 70

† Vaclav Konecny, "On the first law of friction," *American Journal of Physics*, **41**, 588 (1973).

‡ Data taken from Dennis W. Phillips, "Science Friction Adventure—Part II," *The Physics Teacher*, 553 (Nov. 1990).

71 •• [SSM] A block of mass m_1 is attached to a cord of length L_1, which is fixed at one end. The block moves in a horizontal circle on a frictionless table. A second block of mass m_2 is attached to the first by a cord of length L_2 and also moves in a circle, as shown in Figure 5-56. If the period of the motion is T, find the tension in each cord.

FIGURE 5-56 Problem 71

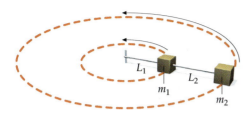

72 •• [SSM] A particle moves with constant speed in a circle of radius 4 cm. It takes 8 s to complete each revolution. Draw the path of the particle to scale, and indicate the particle's position at 1-s intervals. Draw displacement vectors for each interval. These vectors also indicate the directions for the average-velocity vectors for each interval. Find graphically the magnitude of the change in the average velocity $|\Delta\vec{v}|$ for two consecutive 1-s intervals. Compare $|\Delta\vec{v}|/\Delta t$, measured in this way, with the instantaneous acceleration computed from $a = v^2/r$.

73 •• A man swings his child in a circle of radius 0.75 m as shown in Figure 5-57. If the mass of the child is 25 kg and the child makes one revolution in 1.5 s, what is the magnitude and direction of the force that must be exerted by the man on the child? (Assume the child to be a point particle.)

FIGURE 5-57 Problem 73

74 •• The string of a conical pendulum is 50 cm long and the mass of the bob is 0.25 kg. Find the angle between the string and the horizontal when the tension in the string is six times the weight of the bob. Under those conditions, what is the period of the pendulum?

75 •• [iSOLVE✓] A 100-g coin sits on a horizontally rotating turntable. The turntable makes one revolution each second. The coin is located 10 cm from the axis of rotation of the turntable. (a) What is the frictional force acting on the coin?

(b) The coin will slide off the turntable if it is located more than 16 cm from the axis of rotation. What is the coefficient of static friction?

76 •• A 0.25-kg tether ball is attached to a vertical pole by a 1.2-m cord. Assume that the cord is attached to the center of the ball. If the ball moves in a horizontal circle with the cord making an angle of 20° with the vertical, (a) what is the tension in the cord? (b) What is the speed of the ball?

77 •• [SSM] An object on the equator has both an acceleration toward the center of the earth because of the earth's rotation, and an acceleration toward the sun because of the earth's motion along its orbit. Calculate the magnitudes of both accelerations, and express them as fractions of the free-fall acceleration g at the earth's surface.

78 • (a) Using data available from the textbook, calculate the net force acting on the earth that holds it in orbit around the sun. Assume that the orbit is circular. (b) Using data available from the textbook, calculate the force acting on the moon that holds it in orbit around the earth. Assume that the orbit is circular.

79 •• [iSOLVE✓] A small bead with a mass of 100 g slides without friction along a semicircular wire with a radius of 10 cm that rotates about a vertical axis at a rate of 2 revolutions per second, as in Figure 5-58. Find the value of θ for which the bead will remain stationary relative to the rotating wire.

FIGURE 5-58 Problem 79

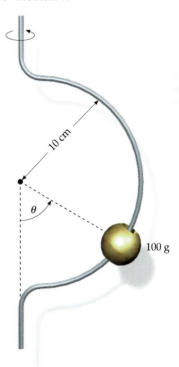

80 ••• Consider a bead of mass m that is free to move on a thin, circular wire of radius r. The bead is given an initial speed v_0, and there is a coefficient of kinetic friction μ_k. The experiment is performed in a spacecraft drifting in space. Find the speed of the bead at any subsequent time t.

81 ••• Revisiting Problem 80, (a) find the centripetal acceleration of the bead. (b) Find the tangential acceleration of the bead. (c) What is the magnitude of the resultant acceleration?

Concepts of Centripetal Force

82 • **SSM** A ride at an amusement park carries people in a vertical circle at constant speed such that the normal forces exerted by the seats are always inward—toward the center of the circle. At the top, the normal force exerted by a seat equals the person's weight, mg. At the bottom of the loop, the force exerted by the seat will be (a) 0, (b) mg, (c) $2mg$, (d) $3mg$, (e) greater than mg, but it cannot be calculated from the information given.

83 • The radius of curvature of a loop-the-loop roller coaster is 12.0 m. At the top of the loop, the force that the seat exerts on a passenger of mass m is $0.4mg$. Find the speed of the roller coaster at the top of the loop.

84 • A car speeds along the curved exit ramp of a freeway. The radius of the curve is 80 m. A 70-kg passenger holds the arm rest of the car door with a 220-N force in order to keep from sliding across the front seat of the car. (Assume the exit ramp is not banked and ignore friction with the car seat.) What is the car's speed? (a) 16 m/s. (b) 57 m/s. (c) 18 m/s. (d) 50 m/s. (e) 28 m/s.

85 ••• **SSM** Suppose you ride a bicycle on a horizontal surface in a circle with a radius of 20 m. The resultant force exerted by the road on the bicycle (normal force plus frictional force) makes an angle of 15° with the vertical. (a) What is your speed? (b) If the frictional force is half its maximum possible value, what is the coefficient of static friction?

86 •• An airplane is flying in a horizontal circle at a speed of 480 km/hr. Banked for this turn, the wings of the plane are tilted at an angle of 40° from the horizontal (Figure 5-59). Assume that a lift force acting perpendicular to the wings holds the aircraft in the sky. What is the radius of the circle in which the plane is flying?

FIGURE 5-59
Problem 86

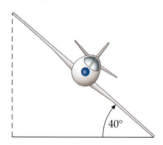

40°

87 • **iSOLVE** A 750-kg car travels around a curve of 160-m radius at 90 km/h. What should the banking angle of the curve be so that the force of the pavement on the tires of the car is in the normal direction?

88 •• **SSM** A curve of radius 150 m is banked at an angle of 10°. An 800-kg car negotiates the curve at 85 km/h without skidding. Find (a) the normal force exerted by the pavement on the tires, (b) the frictional force exerted by the pavement on the tires, (c) the minimum coefficient of static friction between the pavement and the tires.

89 •• On another occasion, the car in Problem 88 negotiates the curve at 38 km/h. Find (a) the normal force exerted on the tires by the pavement, and (b) the frictional force exerted on the tires by the pavement.

90 ••• **SSM** **iSOLVE** A civil engineer is asked to design a curved section of roadway that meets the following conditions: With ice on the road, when the coefficient of static friction between the road and rubber is 0.08, a car at rest must not slide into the ditch and a car traveling less than 60 km/h must not skid to the outside of the curve. What is the minimum radius of curvature of the curve and at what angle should the road be banked?

91 ••• **iSOLVE** A curve of radius 30 m is banked so that a 950-kg car traveling at 40 km/h can round it even if the road is so icy that the coefficient of static friction is approximately zero. Find the range of speeds at which a car can travel around this curve without skidding if the coefficient of static friction between the road and the tires is 0.3.

Drag Forces

92 • A small pollution particle settles toward the earth in still air with a terminal speed of 0.3 mm/s. The particle has a mass of 10^{-10} g and a retarding force of the form bv. What is the value of b?

93 • A Ping-Pong ball has a mass of 2.3 g and a terminal speed of 9 m/s. The retarding force is of the form bv^2. What is the value of b?

94 • **SSM** A sky diver of mass 60 kg can slow herself to a constant speed of 90 km/h by adjusting her form. (a) What is the magnitude of the upward drag force on the sky diver? (b) If the drag force is equal to bv^2, what is the value of b?

95 •• An 800-kg car rolls down a very long 6° grade. The drag force for motion of the car has the form $F_d = 100$ N $+ (1.2$ N·s²/m²$)v^2$. Neglect rolling friction. What is the terminal velocity for the car rolling down this grade?

96 ••• Small spherical particles experience a viscous drag force given by Stokes' law: $F_d = 6\pi\eta rv$, where r is the radius of the particle, v is its speed, and η is the viscosity of the fluid medium. (a) Estimate the terminal speed of a spherical pollution particle of radius 10^{-5} m and density of 2000 kg/m³. (b) Assuming that the air is still and that η is 1.8×10^{-5} N·s/m², estimate the time it takes for such a particle to fall from a height of 100 m.

97 ••• **SSM** A sample of air containing pollution particles of the size and density given in Problem 96 is captured in a test tube 8.0 cm long. The test tube is then placed in a centrifuge with the midpoint of the test tube 12 cm from the center of the centrifuge. The centrifuge spins at 800 revolutions per minute. Estimate the time required for nearly all of the pollution particles to settle at the end of the test tube and compare this to the time required for a pollution particle to fall 8.0 cm under the action of gravity and subject to the viscous drag of air.

Euler's Method

98 •• You are riding in a hot air balloon when you throw a baseball straight down with an initial speed of 35 km/h. A baseball falls with a terminal speed of 150 km/h. Assuming air drag is proportional to the speed squared, use Euler's method to estimate the speed of the ball after 10 s. What is the uncertainty in this estimate? You drop a second baseball, this one released from rest. How long does it take for it to reach 99% of its terminal speed? How far does it fall during this time?

99 •• **SSM** You throw a baseball straight up with an initial speed of 150 km/h. Its terminal speed when falling is also 150 km/h. Use Euler's method to estimate its height 3.5 s after release. What is the maximum height it reaches? How long after release does it reach its maximum height? How much later does it return to the ground? Is the time the ball spends on the way up less than, the same as, or greater than the time it spends on the way down?

100 •• A 0.80-kg block on a horizontal frictionless surface is held against a massless spring, compressing it 30 cm. The force constant of the spring is 50 N/m. The block is released and the spring pushes it 30 cm. Use Euler's method with $\Delta t = 0.005$ s to estimate the time it takes for the spring to push the block the 30 cm. How fast is the block moving at this time? What is the uncertainty in this speed?

General Problems

101 • **SOLVE** A 4.5-kg block slides down an inclined plane that makes an angle of 28° with the horizontal. Starting from rest, the block slides a distance of 2.4 m in 5.2 s. Find the coefficient of kinetic friction between the block and plane.

102 • **SOLVE** A model airplane of mass 0.4 kg is attached to a horizontal string and flies in a horizontal circle of radius 5.7 m. (The weight of the plane is balanced by the upward "lift" force of the air on the wings of the plane.) The plane makes 1.2 revolutions every 4 s. (a) Find the speed v of the plane. (b) Find the tension in the string.

103 •• **SSM** An 800-N box rests on an incline making a 30° angle with the horizontal. A physics student finds that she can prevent the box from sliding if she pushes on it with a force of at least 200 N parallel to the incline. (a) What is the coefficient of static friction between the box and the incline? (b) What is the greatest force that can be applied to the box parallel to the incline before the box slides up the incline?

104 •• The position of a particle is given by the vector $\vec{r} = -10$ m cos $\omega t\hat{i} + 10$ m sin $\omega t\hat{j}$, where $\omega = 2$ s^{-1}. (a) Show that the path of the particle is a circle. (b) What is the radius of the circle? (c) Does the particle move clockwise or counterclockwise around the circle? (d) What is the speed of the particle? (e) What is the time for one complete revolution?

105 •• **SOLVE** A crate of books is to be put on a truck with the help of some planks sloping up at 30°. The mass of the crate is 100 kg, and the coefficient of sliding friction between it and the planks is 0.5. You and your friends push *horizontally* with a force $\vec{F}$. Once the crate has started to move, how large must F be in order to keep the crate moving at constant speed?

106 •• An object with a mass of 5.5 kg is allowed to slide from rest down an inclined plane. The plane makes an angle of 30° with the horizontal and is 72 m long. The coefficient of kinetic friction between the plane and the object is 0.35. The speed of the object at the bottom of the plane is (a) 5.3 m/s, (b) 15 m/s, (c) 24 m/s, (d) 17 m/s, (e) 11 m/s.

107 •• **SSM** A brick slides down an inclined plank at constant speed when the plank is inclined at an angle θ_0. If the angle is increased to θ_1, the block accelerates down the plank with acceleration a. The coefficient of kinetic friction is the same in both cases. Given θ_0 and θ_1, calculate a.

108 •• Three forces act on an object, shown in Figure 5-60, that is in static equilibrium. (a) If F_1, F_2, and F_3 represent the magnitudes of the forces acting on the object, show that $F_1/\sin \theta_{23} = F_2/\sin \theta_{31} = F_3/\sin \theta_{12}$. (b) Show that $F_1^2 = F_2^2 + F_3^2 + 2F_2F_3 \cos \theta_{23}$.

FIGURE 5-60 Problem 108

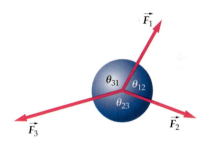

109 •• In a carnival ride, the passenger sits on a seat in a compartment that rotates with constant speed in a vertical circle of radius $r = 5$ m. The heads of the seated passengers always point toward the center of the circle. (a) If the carnival ride completes one full circle in 2 s, find the acceleration of the passenger. (b) Find the slowest rate of rotation (in other words, the longest time T_m to complete one full circle) if the seat belt is to exert no force on the passenger at the top of the ride.

110 •• **SSM** A flat-topped toy cart moves on frictionless wheels, pulled by a rope under tension T. The mass of the cart is m_1. A load of mass m_2 rests on top of the cart with a coefficient of static friction μ_s. The cart is pulled up a ramp that is inclined at angle θ above the horizontal. The rope is parallel to the ramp. What is the maximum tension T that can be applied without making the load slip?

111 •• A sled weighing 200 N rests on a 15° incline, held in place by static friction (Figure 5-61). The coefficient of static friction is 0.5. (*a*) What is the magnitude of the normal force on the sled? (*b*) What is the magnitude of the static frictional force on the sled? (*c*) The sled is now pulled up the incline at constant speed by a child. The child weighs 500 N and pulls on the rope with a constant force of 100 N. The rope makes an angle of 30° with the incline and has negligible mass. What is the magnitude of the kinetic frictional force on the sled? (*d*) What is the coefficient of kinetic friction between the sled and the incline? (*e*) What is the magnitude of the force exerted on the child by the incline?

FIGURE 5-61 Problem 111

112 • In 1976, Gerard O'Neill proposed the building of large space stations for human habitation in orbit around the earth and the moon. Because prolonged free-fall has adverse medical effects, he proposed making the stations in the form of long cylinders and spinning them around the cylinder axis to provide the inhabitants inside with the sensation of gravity. This idea has found its way into mainstream science fiction; for one example, the TV show *Babylon 5* was set on an O'Neill colony 5 mi long, with a diameter of 0.6 mi. Because of the rotation, someone on the inside of the colony would experience a sense of "gravity," because he or she would be in an accelerated frame of reference. (*a*) Show that the "acceleration of gravity" experienced by someone in an O'Neill colony is equal to his or her centripetal acceleration. *Hint: consider someone "looking in" from outside the colony.* (*b*) If we assume that the space station is composed of several decks which are at varying distances from the axis of rotation, show that the "acceleration of gravity" becomes weaker the closer one gets to the axis. (*c*) How many revolutions per minute would Babylon 5 have to make to give an "acceleration of gravity" of 9.8 m/s² at the outermost edge of the station?

113 •• A child slides down a slide inclined at 30° in time t_1. The coefficient of kinetic friction between her and the slide is μ_k. She finds that if she sits on a small cart with frictionless wheels, she slides down the same slide in time $\frac{1}{2}t_1$. Find μ_k.

114 •• **SSM** The position of a particle of mass $m = 0.8$ kg as a function of time is

$$\vec{r} = x\hat{i} + y\hat{j} = R \sin \omega t\, \hat{i} + R \cos \omega t\, \hat{j}$$

where $R = 4.0$ m and $\omega = 2\pi$ s⁻¹. (*a*) Show that the path of this particle is a circle of radius R with its center at the origin. (*b*) Compute the velocity vector. Show that $v_x/v_y = -y/x$. (*c*) Compute the acceleration vector and show that it is in the radial direction and has the magnitude v^2/r. (*d*) Find the magnitude and direction of the net force acting on the particle.

115 •• **SOLVE** In an amusement-park ride, riders stand with their backs against the wall of a spinning vertical cylinder. The floor falls away and the riders are held up by friction.

If the radius of the cylinder is 4 m, find the minimum number of revolutions per minute necessary when the coefficient of static friction between a rider and the wall is 0.4.

116 •• **SOLVE** A mass m_1 on a horizontal table is attached by a thin string that passes over a frictionless, massless pulley to a 2.5-kg mass m_2 that hangs over the side of the table 1.5 m above the ground (Figure 5-62). The system is released from rest at $t = 0$ and the 2.5-kg mass strikes the ground at $t = 0.82$ s. The system is now placed in its initial position and a 1.2-kg mass is placed on top of the block of mass m_1. Released from rest, the 2.5-kg mass now strikes the ground 1.3 s later. Determine the mass m_1 and the coefficient of kinetic friction between m_1 and the table.

FIGURE 5-62 Problem 116

117 ••• **SSM** (*a*) Show that a point on the surface of the earth at latitude θ, shown in Figure 5-63, has an acceleration of magnitude $(3.37 \text{ cm/s}^2)\cos \theta$ relative to a reference frame not rotating with the earth. What is the direction of this acceleration? (*b*) Discuss the effect of this acceleration on the apparent weight of an object near the surface of the earth. (*c*) The free-fall acceleration of an object at sea level measured *relative to the earth's surface* is 9.78 m/s² at the equator and 9.81 m/s² at latitude $\theta = 45°$. What are the values of the gravitational field g at these points?

FIGURE 5-63 Problem 117

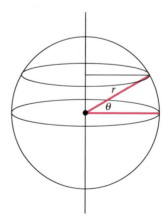

118 ••• **SSM** A small block of mass 0.01 kg is at rest atop a smooth (frictionless) sphere of radius 0.8 m. The block is given a tiny nudge and starts to slide down the sphere. The mass loses contact with the sphere when the angle between the vertical and the line from the center of the sphere to the position of the mass is θ. Find the angle θ.

Work and Energy

DOWNHILL RACING CAN LOOK LIKE A LOT OF FUN, OR A LOT OF WORK, DEPENDING UPON YOUR PERSPECTIVE.

? **Did you know that, in fact, work is being done on the skier? (See Example 6-12.)**

Work and energy are important concepts in physics as well as in our everyday lives. In physics, a force does **work** if its point of application moves through a distance and there is a component of the force in the direction of the velocity of the force's point of application. For a constant force in one dimension, the work done equals the force component in the direction of the displacement times the displacement. (This differs somewhat from the everyday use of the word work. When you study hard for an exam, the only work you do according to the use of the word in physics, is in pushing your pencil on the paper, or turning the pages of your book.)

Energy is closely associated with work. When work is done by one system on another, energy is transferred between the two systems. For example, when you do work pushing a swing, chemical energy of your body is transferred to the swing and appears as kinetic energy of motion or as gravitational potential energy of the earth–swing system. There are many forms of energy. Kinetic energy is associated with the motion of an object. Potential energy is associated with the configuration of a system, such as the separation distance between two objects that attract each other. Thermal energy is associated with the random motion of the molecules within a system and is closely connected with the temperature of the system.

➤ **In this chapter we study the concepts of work, kinetic energy, and potential energy. Because these concepts arise from Newton's laws, the concepts introduced in this chapter are a continuation of those introduced in previous chapters.**

These new concepts provide powerful methods of solving a wide class of problems. Many of the end-of-chapter problems for this chapter can be solved using the concepts and methods of the previous chapters. However, it is important that you resist any temptation to solve them in that way. The concepts and methods developed in this chapter are developed further in Chapter 7.

6-1 Work and Kinetic Energy

Motion in One Dimension With Constant Forces

The work W done by a constant force $\vec{F}$ moving through a displacement $\Delta x \hat{i}$ is given by

$$W = F_x \, \Delta x = F \cos \theta \, \Delta x \qquad\qquad 6\text{-}1$$

WORK BY A CONSTANT FORCE

where θ is the angle between the directions of $\vec{F}$ and $\hat{i}$, and $\Delta x \hat{i}$ is the displacement of the point of application of the force as shown in Figure 6-1.

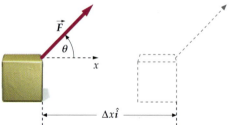

FIGURE 6-1

Work is a scalar quantity that is positive if Δx and F_x have the same signs and negative if they have opposite signs. The dimensions of work are those of force times distance. The SI unit of work and energy is the **joule** (J), which equals the product of a newton and a meter:

$$1\,\text{J} = 1\,\text{N·m} \qquad\qquad 6\text{-}2$$

(In the U.S. customary system, the unit of work is the foot-pound: 1 ft·lb = 1.356 J.) A convenient unit of work and energy in atomic and nuclear physics is the electron volt (eV):

$$1\,\text{eV} = 1.6 \times 10^{-19}\,\text{J} \qquad\qquad 6\text{-}3$$

Commonly used multiples are keV (10^3 eV) and MeV (10^6 eV). The work required to remove an electron from an atom is of the order of a few eV, whereas the work needed to remove a proton or a neutron from an atomic nucleus is of the order of several MeV.

EXERCISE A force of 12 N is exerted on a box at an angle of $\theta = 20°$, as in Figure 6-1. How much work is done by the force on the box as the box moves along the table a distance of 3 m? (*Answer* 33.8 J)

If there are several forces that do work, the total work is found by computing the work done by each force and summing.

$$W_{\text{total}} = F_{1x} \, \Delta x_1 + F_{2x} \, \Delta x_2 + F_{3x} \, \Delta x_3 + \ldots$$

A **particle** is any object that moves so that all of its parts undergo identical displacements during any interval of time. That is, an object can be modeled as a particle as long as it remains perfectly rigid and moves without rotating.

When several forces do work on a *particle,* the displacements of the points of application of each these forces are equal. Let the displacement of the point of application of any one of the forces be Δx. Then

$$W_{total} = F_{1x}\Delta x + F_{2x}\Delta x + \cdots = (F_{1x} + F_{2x} + \cdots)\Delta x = F_{net\,x}\Delta x \qquad 6\text{-}4$$

For a particle constrained to move along the x axis, the net force has only an x component. That is, $\vec{F}_{net} = F_{net\,x}\hat{i}$. Thus, for a particle the total work can be found by first finding the net force, and then multiplying the net force by the displacement.

The Work–Kinetic Energy Theorem

There is an important relation between the total work done on a particle and the initial and final speeds of the particle. If $F_{net\,x}$ is the net force acting on a particle, Newton's second law gives

$$F_{net\,x} = ma_x$$

For a constant force, the acceleration is constant, and we can relate the displacement to the initial speed v_i and final speed v_f by using the constant-acceleration formula $v_f^2 = v_i^2 + 2a_x\Delta x$ (Equation 2-16). Solving this for a_x gives

$$a_x = \frac{1}{2\Delta x}(v_f^2 - v_i^2)$$

Substituting for a_x in $F_{net\,x} = ma_x$ and then multiplying both sides by Δx gives

$$F_{net\,x}\Delta x = \tfrac{1}{2}mv_f^2 - \tfrac{1}{2}mv_i^2$$

We recognize the term on the left as the total work done on the particle. Thus

$$W_{total} = \tfrac{1}{2}mv_f^2 - \tfrac{1}{2}mv_i^2 \qquad 6\text{-}5$$

The quantity $\tfrac{1}{2}mv^2$ is a scalar quantity called the **kinetic energy** K of the particle:

$$K = \tfrac{1}{2}mv^2 \qquad 6\text{-}6$$

DEFINITION—KINETIC ENERGY

The quantity on the right side of Equation 6-5 is the change in the kinetic energy of the particle. Thus,

The total work done on a particle is equal to the change in its kinetic energy:

$$W_{total} = \Delta K \qquad 6\text{-}7$$

WORK–KINETIC ENERGY THEOREM

This result is known as the **work–kinetic energy theorem.** The derivation presented here is valid only if the net force remains constant. However, as we will see later in this chapter, this theorem is valid even when the net force varies and the motion is not along a straight line.

EXERCISE A woman of mass 50 kg is running at 3.5 m/s. What is her kinetic energy? (*Answer* 306 J)

EXAMPLE 6-1

A truck of mass 3000 kg is to be loaded onto a ship by a crane that exerts an upward force of 31 kN on the truck. This force, which is strong enough to overcome the gravitational force and get the truck started upward, is applied over a distance of 2 m. Find (*a*) the work done by the crane, (*b*) the work done by gravity, and (*c*) the upward speed of the truck after the 2 m.

PICTURE THE PROBLEM Sketch the truck at its initial and final positions and choose the positive *y* direction to be the direction of the displacement (Figure 6-2). Use the work–kinetic energy theorem to find the truck's final kinetic energy. The final speed of the truck can be obtained from the final kinetic energy. The total work is the sum of the results for (*a*) and (*b*).

FIGURE 6-2

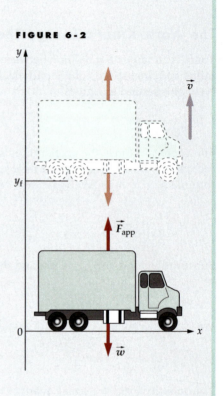

(*a*) Calculate the work done by the applied force:

$$W_{app} = F_{app\,y}\,\Delta y = F_{app}\cos 0°\,\Delta y$$
$$= (31\text{ kN})(1)(2\text{ m}) = \boxed{62.0\text{ kJ}}$$

(*b*) Calculate the work done by the force of gravity:

$$W_g = mg_y\,\Delta y = mg\cos 180°\,\Delta y$$
$$= (3000\text{ kg})(9.81\text{ N/kg})(-1)(2\text{ m})$$
$$= \boxed{-58.9\text{ kJ}}$$

(*c*) Apply the work–kinetic energy theorem and solve for v_f:

$$W_{total} = \Delta K$$
$$W_{app} + W_g = K_f - K_i$$
$$= \tfrac{1}{2}mv_f^2 - \tfrac{1}{2}mv_i^2$$
$$v_f^2 = v_i^2 + \frac{2(W_{app} + W_g)}{m}$$
$$= 0 + \frac{2(62{,}000\text{ J} - 58{,}900\text{ J})}{3000\text{ kg}}$$
$$= 2.09\text{ m}^2/\text{s}^2$$
$$v_f = \boxed{1.45\text{ m/s}}$$

REMARKS Notice that we treat each force separately when calculating the work done. We could also find the total work by first adding the forces to obtain the net force, then applying $W_{total} = F_{net\,y}\,\Delta y$. In either case, the work–kinetic energy theorem applies only to the total work.

EXERCISE Find the final speed of the truck if the same upward force were applied for 2 m after it was already moving upward at 1 m/s. (*Answer* 1.73 m/s. Note that the answer is *not* 1.45 m/s + 1.00 m/s. Why not?)

EXAMPLE 6-2

In a television tube, an electron is accelerated from rest to a kinetic energy of 2.5 keV over a distance of 80 cm. (The force that accelerates the electron is an electric force due to the electric field in the tube.) Find the force on the electron, assuming it to be constant and in the direction of motion.

PICTURE THE PROBLEM Because the electron starts from rest, the work done equals the final kinetic energy.

Set the work done to be equal to the change in kinetic energy and solve for the force. The initial and final kinetic energies are both given:

$$W_{\text{total}} = \Delta K$$

$$F_x \, \Delta x = K_f - K_i$$

$$F_x = \frac{K_f - K_i}{\Delta x}$$

$$= \frac{2500 \text{ eV} - 0}{0.8 \text{ m}} \times \frac{1.6 \times 10^{-19} \text{ J}}{1 \text{ eV}}$$

$$= \boxed{5.0 \times 10^{-16} \text{ N}}$$

REMARKS When we discuss electricity we will see that the work done per unit charge is called the potential difference and is measured in volts. Thus, 1 eV is the energy acquired or lost by a particle of charge e (an electron or proton, for example) when its potential difference changes by 1 V.

A DOGSLED RACE   **EXAMPLE 6-3**

During your winter break you enter a dogsled race across a frozen lake. To get started you pull the sled (total mass 80 kg) with a force of 180 N at 20° above the horizontal. Find (*a*) the work you do and (*b*) the final speed of the sled after it moves $\Delta x = 5$ m, assuming that it starts from rest and there is no friction.

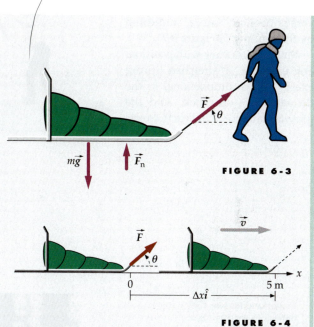

FIGURE 6-3

PICTURE THE PROBLEM The work done by you is $F_x \, \Delta x$, where we choose the direction of the displacement as the positive x direction. This is also the *total* work done on the sled because the other forces, mg and F_n, have no x components (Figure 6-3). The final speed of the sled is found by applying the work–kinetic energy theorem to the sled.

(*a*) 1. Sketch the sled both in its initial position and in its position after moving the 5 m. Draw the x axis in the direction of motion (Figure 6-4).

FIGURE 6-4

2. The work done by you on the sled is $F_x \, \Delta x$. (This is also the total work done on the sled since the other forces act perpendicular to the x direction):

$$W = F_x \, \Delta x = F \cos \theta \, \Delta x$$

$$= (180 \text{ N})(\cos 20°)(5 \text{ m}) = \boxed{846 \text{ J}}$$

(*b*) Apply the work–kinetic energy theorem to the sled and solve for the final speed:

$$W_{\text{total}} = \tfrac{1}{2} m v_f^2 - \tfrac{1}{2} m v_i^2$$

$$v_f^2 = v_i^2 + \frac{2W_{\text{total}}}{m}$$

$$= 0 + \frac{2(846 \text{ J})}{80 \text{ kg}}$$

$$= 21.1 \text{ m}^2/\text{s}^2$$

$$v_f = \boxed{4.60 \text{ m/s}}$$

REMARKS We do not need to work out the units. If we have a correct equation, and all quantities are in SI units, the result will be in the correct SI units. However, as a check on the equation, we can show that 1 J/kg = 1 m²/s². We have
1 J/kg = 1 N·m/kg = (1 kg·m/s²)·m/kg = 1 m²/s².

EXERCISE What is the magnitude of the force you exert if the sled starts with a speed of 2 m/s and its final speed is 4.5 m/s after you pull it through a distance of 5 m? (*Answer* 138 N)

What if you hold a massive object in a fixed position? You are expending energy, but are you doing work? According to the definition of work, you are not doing work *on the object* because the object does not move in the direction of the force you exert (Figure 6-5). But your muscles are continually contracting and relaxing as you hold the weight. In this process internal chemical energy in your body is converted to thermal energy (Figure 6-6).

Work Done by a Variable Force

In Figure 6-7 we plot a constant force F_x as a function of position x. The work done on a particle whose displacement is Δx is represented by the area under the force-versus-position curve, indicated by the shading in Figure 6-7.

Many forces vary with position. For example, a stretched spring exerts a force proportional to the distance it is stretched. And the gravitational force the earth exerts

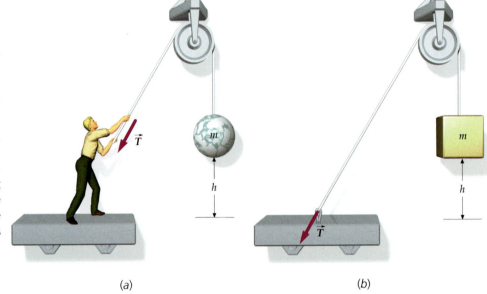

(a)

(b)

FIGURE 6-5 (*a*) The man standing on a ledge does not do work on the weight when holding it at a fixed position. (*b*) The same task could be accomplished by tying the rope to a fixed point.

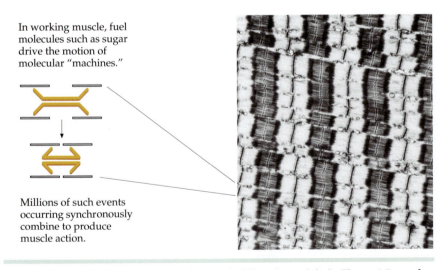

In working muscle, fuel molecules such as sugar drive the motion of molecular "machines."

Millions of such events occurring synchronously combine to produce muscle action.

FIGURE 6-6 Muscle work. While the man holding the weight in Figure 6-5 may be doing no work on the weight, his body is expending energy on the molecular level, as structures within the muscle slide over each other during muscular extension and contraction.

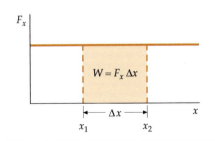

$$W = F_x \Delta x$$

FIGURE 6-7 The work done by a constant force is represented graphically as the area under the F_x-versus-x curve.

on a spaceship varies inversely with the square of the distance between the two bodies. We can approximate a variable force by a series of constant forces (Figure 6-8). The work done by a variable force is then

$$W = \lim_{\Delta x_i \to 0} \sum_i F_x \, \Delta x_i = \text{area under the } F_x\text{-versus-}x \text{ curve} \qquad 6\text{-}8$$

This limit is the integral of F_x over x. So the work done by a variable force F_x acting on a particle as it moves from x_1 to x_2 is

$$W = \int_{x_1}^{x_2} F_x \, dx = \text{area under the } F_x\text{-versus-}x \text{ curve} \qquad 6\text{-}9$$

WORK BY A VARIABLE FORCE

For each displacement interval Δx_i, the force is essentially constant. Therefore the work done equals the area of the rectangle of height $F_{x\,i}$ and width Δx_i. As was shown earlier in Section 6-1, this work equals the change in kinetic energy for this displacement interval (if the force is the net force). The total work done is the sum of the areas over all displacement intervals. It follows that the total work equals the change in kinetic energy for the entire displacement. Thus, $W_{\text{total}} = \Delta K$ holds for variable forces as well as for constant forces.

EXERCISE IN DIMENSIONAL ANALYSIS A spring is characterized by its force constant k, which has dimensions[†] N/m. How does the work required to stretch a spring by an amount x_0 depend on k and x_0? (*Answer* Because work has dimensions of N·m, the work must depend on k and x_0 in the combination kx_0^2. We will see in Example 6-5 that the actual expression is $W = \frac{1}{2}kx_0^2$. The factor $\frac{1}{2}$ arises because the force varies from 0 to a maximum value of kx_0, and has the average[‡] value $\frac{1}{2}kx_0$.)

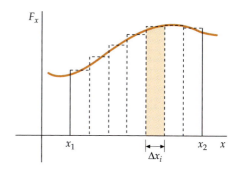

FIGURE 6-8 A variable force can be approximated by a series of constant forces over small intervals. The work done by the constant force in each interval is the area of the rectangle beneath the force curve. The sum of these rectangular areas is the sum of the work done by the set of constant forces that approximates the varying force. In the limit of infinitesimally small Δx_i, the sum of the areas of the rectangles equals the area under the complete force curve.

WORK DONE ON A PARTICLE **EXAMPLE 6-4**

A force F_x varies with x as shown in Figure 6-9. Find the work done by the force on a particle as the particle moves from $x = 0$ to $x = 6$ m.

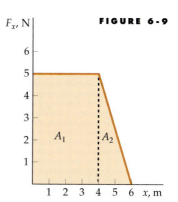

FIGURE 6-9

PICTURE THE PROBLEM The work is the area under the curve. Because the curve consists of straight-line segments, the easiest approach is to use the geometric formulas for area. (The alternative approach is to set up and execute an integration, which is done in Example 6-5.)

1. We find the work done by calculating the area under the F_x-versus-x curve: $\qquad W = A_{\text{total}}$

2. This area is the sum of the two areas shown. The area of a triangle is one half the altitude times the base:

$$W = A_{\text{total}} = A_1 + A_2$$
$$= (5\,\text{N})(4\,\text{m}) + \tfrac{1}{2}(5\,\text{N})(2\,\text{m})$$
$$= 20\,\text{J} + 5\,\text{J} = \boxed{25\,\text{J}}$$

EXERCISE The force shown is the only force that acts on a particle of mass 3 kg. If the particle starts from rest at $x = 0$, how fast is it moving when it reaches $x = 6$ m? (*Answer* 4.08 m/s)

† The SI units of k are N/m, but the dimensions of k are $[M][T]^{-2}$.
‡ Here average refers not to an average over time but to an average over distance.

WORK DONE ON A BLOCK BY A SPRING   EXAMPLE 6-5   **FIGURE 6-10**

A 4-kg block on a frictionless table is attached to a horizontal spring that obeys Hooke's law and exerts a force $\vec{F} = -kx\hat{i}$, where $k = 400$ N/m and x is measured from the equilibrium position of the block. The spring is originally compressed with the block at $x_1 = -5$ cm (Figure 6-10). Find (a) the work done by the spring on the block as the block moves from $x_1 = -5$ cm to its equilibrium position $x_2 = 0$ and (b) the speed of the block at $x_2 = 0$.

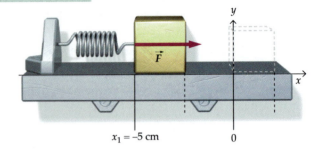

$x_1 = -5$ cm   0

PICTURE THE PROBLEM Make a graph of F_x versus x. The work done on the block as it moves from x_1 to $x_2 = 0$ equals the area under the F_x-versus-x curve between these limits, shaded in Figure 6-11, which can be calculated by integrating the force over the distance. The work done equals the change in kinetic energy, which is simply its final kinetic energy because the initial kinetic energy is zero. The speed of the block at $x = 0$ is found from the kinetic energy of the block.

FIGURE 6-11

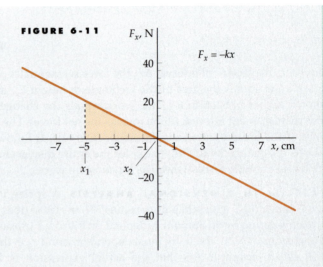

(a) The work W done by the spring on the block is the integral of $F_x\,dx$ from $x_1 = -5$ cm to $x_2 = 0$:

$$W = \int_{x_1}^{x_2} F_x\,dx = \int_{x_1}^{x_2} -kx\,dx = -k\int_{x_1}^{x_2} x\,dx$$

$$= -\tfrac{1}{2}kx^2\Big|_{x_1}^{x_2} = -\tfrac{1}{2}k(x_2^2 - x_1^2)$$

$$= -\tfrac{1}{2}(400\ \text{N/m})\left[0 - (0.05\ \text{m})^2\right] = \boxed{0.500\ \text{J}}$$

(b) Apply the work–kinetic energy theorem and solve for v_2:

$$W_{\text{total}} = \tfrac{1}{2}mv_2^2 - \tfrac{1}{2}mv_1^2$$

$$v_2^2 = v_1^2 + \frac{2W_{\text{total}}}{m} = 0 + \frac{2(0.500\ \text{J})}{4\ \text{kg}}$$

$$= 0.250\ \text{m}^2/\text{s}^2$$

$$v_2 = \boxed{0.500\ \text{m/s}}$$

REMARKS Besides the spring force, two other forces act on the block; the force of gravity, $m\vec{g}$, and the normal force of the table, $\vec{F}_n$. These forces do no work because they have no component in the direction of the displacement. Only the spring does work on the block. (The force it exerts does have a component in the direction of the displacement.)

EXERCISE Find the speed of the block when it reaches $x = 3$ cm if it starts from $x = 0$ with velocity $v_x = 0.5$ m/s. (*Answer* 0.4 m/s)

Note that we could *not* have solved Example 6-5 by first applying Newton's second law to find the acceleration, and then using the constant-acceleration kinematic equations. Because the force exerted by the spring on the block, $F_x = -kx$, varies with position, the acceleration also varies, and thus the constant-acceleration condition is not met.

6-2 The Dot Product

The component F_s in Figure 6-12 is related to the angle ϕ between the directions of $\vec{F}$ and $d\vec{s}$ by $F_s = F \cos \phi$, so the work done by $\vec{F}$ for the displacement $d\vec{s}$ is

$$W = F_s ds = F \cos \phi \, ds$$

This combination of two vectors and the cosine of the angle between their directions is called the **dot product** (or **scalar product**) of the vectors. The dot product of two general vectors $\vec{A}$ and $\vec{B}$ is written $\vec{A} \cdot \vec{B}$ and defined by

$$\vec{A} \cdot \vec{B} = AB \cos \phi \qquad \text{6-10}$$

DEFINITION—DOT PRODUCT

where ϕ is the angle between $\vec{A}$ and $\vec{B}$. (The angle between two vectors is defined as the angle between their directions in space.) The dot product $\vec{A} \cdot \vec{B}$ can be thought of either as A times the component of $\vec{B}$ in the direction of $\vec{A}$ ($A[B \cos \phi]$), or as B times the component of $\vec{A}$ in the direction of $\vec{B}$ ($B[A \cos \phi]$) (see Figure 6-13). Properties of the dot product are summarized in Table 6-1.

TABLE 6-1

Properties of Dot Products

If	then
$\vec{A}$ and $\vec{B}$ are perpendicular,	$\vec{A} \cdot \vec{B} = 0$ (since $\phi = 90°$, $\cos 90° = 0$)
$\vec{A}$ and $\vec{B}$ are parallel,	$\vec{A} \cdot \vec{B} = AB$ (since $\phi = 0°$, $\cos 0° = 1$)
$\vec{A} \cdot \vec{B} = 0$,	Either $\vec{A} = 0$ or $\vec{B} = 0$ or $\vec{A}$ and $\vec{B}$ are perpendicular
Furthermore,	
$\vec{A} \cdot \vec{A} = A^2$	Since $\vec{A}$ is parallel to itself
$\vec{A} \cdot \vec{B} = \vec{B} \cdot \vec{A}$	Commutative rule of multiplication
$(\vec{A} + \vec{B}) \cdot \vec{C} = \vec{A} \cdot \vec{C} + \vec{B} \cdot \vec{C}$	Distributive rule of multiplication

We can use unit vectors to write the dot product in terms of the rectangular components of the two vectors:

$$\vec{A} \cdot \vec{B} = (A_x \hat{i} + A_y \hat{j} + A_z \hat{k}) \cdot (B_x \hat{i} + B_y \hat{j} + B_z \hat{k})$$

The dot product of any unit vector with itself, like $\hat{i} \cdot \hat{i}$, is 1, so a term like $A_x \hat{i} \cdot B_x \hat{i}$ equals $A_x B_x$. Also, because the unit vectors $\hat{i}$, $\hat{j}$, and $\hat{k}$ are mutually perpendicular, the dot product of one of them with any other one, like $\hat{i} \cdot \hat{j}$, is zero. Thus, terms like $A_x \hat{i} \cdot B_y \hat{j}$ (called cross terms) each equal zero. The result is

$$\vec{A} \cdot \vec{B} = A_x B_x + A_y B_y + A_z B_z \qquad \text{6-11}$$

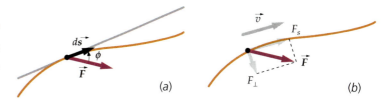

FIGURE 6-12 (*a*) A particle moving along an arbitrary curve in space. (*b*) The perpendicular component of the force $F_\perp$ changes the direction of the particle's motion but not its speed. The tangential component F_s changes the particle's speed but not its direction. F_s equals the mass m times the tangential acceleration dv/dt. Only this component does work.

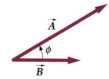

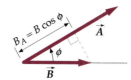

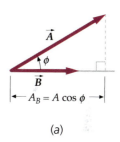

(*a*)

FIGURE 6-13 (*a*) The dot product $\vec{A} \cdot \vec{B}$ is the product of A and the projection of $\vec{B}$ on $\vec{A}$ or the product of B and the projection of $\vec{A}$ on $\vec{B}$. That is, $\vec{A} \cdot \vec{B} = AB \cos \phi = AB_A = BA_B$.

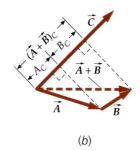

(*b*)

(*b*) $(\vec{A} + \vec{B}) \cdot \vec{C}$ equals $(\vec{A} + \vec{B})_C C$ (the projection of $\vec{A} + \vec{B}$ in the direction of $\vec{C}$ times C). However, $(\vec{A} + \vec{B})_C = A_C + B_C$, so $(\vec{A} + \vec{B}) \cdot \vec{C} = (A_C + B_C)C = A_C C + B_C C = \vec{A} \cdot \vec{C} + \vec{B} \cdot \vec{C}$. That is, for the dot product multiplication is distributive over addition.

The component of a vector in some direction can be written as the dot product of the vector and the unit vector in that direction. For example, the component A_x is found from

$$\vec{A} \cdot \hat{i} = (A_x\hat{i} + A_y\hat{j} + A_z\hat{k}) \cdot \hat{i} = A_x \qquad \text{6-12}$$

This suggests an algebraic procedure for obtaining a component equation, given a vector equation. That is, multiplying both sides of the vector equation $\vec{A} + \vec{B} = \vec{C}$ by $\hat{i}$ gives $\vec{A} \cdot \hat{i} + \vec{B} \cdot \hat{i} = \vec{C} \cdot \hat{i}$, which in turn gives $A_x + B_x = C_x$.

To establish the product rule for differentiating dot products we differentiate both sides of Equation 6-11. For brevity we do this for vectors with two dimensions.

$$\frac{d}{dt}(\vec{A} \cdot \vec{B}) = \frac{d}{dt}(A_xB_x + A_yB_y)$$

$$= \frac{dA_x}{dt}B_x + A_x\frac{dB_x}{dt} + \frac{dA_y}{dt}B_y + A_y\frac{dB_y}{dt}$$

Rearranging gives

$$\frac{d}{dt}(\vec{A} \cdot \vec{B}) = \frac{dA_x}{dt}B_x + \frac{dA_y}{dt}B_y + A_x\frac{dB_x}{dt} + A_y\frac{dB_y}{dt}$$

$$= \left(\frac{dA_x}{dt}\hat{i} + \frac{dA_y}{dt}\hat{j}\right) \cdot (B_x\hat{i} + B_y\hat{j}) + (A_x\hat{i} + A_y\hat{j}) \cdot \left(\frac{dB_x}{dt}\hat{i} + \frac{dB_y}{dt}\hat{j}\right)$$

so

$$\frac{d}{dt}(\vec{A} \cdot \vec{B}) = \frac{d\vec{A}}{dt} \cdot \vec{B} + \vec{A} \cdot \frac{d\vec{B}}{dt} \qquad \text{6-13}$$

USING THE DOT PRODUCT   **E X A M P L E   6 - 6**

(a) Find the angle between the vectors $\vec{A} = 3\text{ m}\hat{i} + 2\text{ m}\hat{j}$ and $\vec{B} = 4\text{ m}\hat{i} - 3\text{ m}\hat{j}$ (Figure 6-14). (b) Find the component of $\vec{A}$ in the direction of $\vec{B}$.

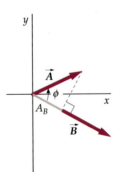

FIGURE 6-14

PICTURE THE PROBLEM We find the angle ϕ from the definition of the dot product. The component of $\vec{A}$ in the direction of $\vec{B}$ is found from the dot product $\vec{A} \cdot \hat{B}$, where $\hat{B} = \vec{B}/B$.

(a) 1. Write the dot product of $\vec{A}$ and $\vec{B}$ in terms of A, B, and $\cos \phi$ and solve for $\cos \phi$:

$$\vec{A} \cdot \vec{B} = AB\cos\phi, \quad \text{so}$$

$$\cos\phi = \frac{\vec{A} \cdot \vec{B}}{AB}$$

2. Find $\vec{A} \cdot \vec{B}$ from their components:

$$\vec{A} \cdot \vec{B} = A_xB_x + A_yB_y$$

$$= (3\text{ m})(4\text{ m}) + (2\text{ m})(-3\text{ m})$$

$$= 12\text{ m}^2 - 6\text{ m}^2 = 6\text{ m}^2$$

3. The magnitudes of the vectors are obtained from the dot product of the vector with itself:

$$\vec{A} \cdot \vec{A} = A^2 = A_x^2 + A_y^2$$
$$= (3\,\text{m})^2 + (2\,\text{m})^2 = 13\,\text{m}^2, \quad \text{so}$$
$$A = \sqrt{13}\,\text{m}$$

and

$$\vec{B} \cdot \vec{B} = B^2 = B_x^2 + B_y^2$$
$$= (4\,\text{m})^2 + (-3\,\text{m})^2 = 25\,\text{m}^2, \quad \text{so}$$
$$B = 5\,\text{m}$$

4. Substitute these values into the equation in step 1 for $\cos\phi$ to find ϕ:

$$\cos\phi = \frac{\vec{A} \cdot \vec{B}}{AB} = \frac{6\,\text{m}^2}{(\sqrt{13}\,\text{m})(5\,\text{m})} = 0.333$$
$$\phi = \boxed{70.6°}$$

(b) The component of $\vec{A}$ in the direction of $\vec{B}$ is the dot product of $\vec{A}$ with the unit vector $\hat{B} = \vec{B}/B$:

$$A_B = \vec{A} \cdot \hat{B} = \vec{A} \cdot \frac{\vec{B}}{B} = \frac{\vec{A} \cdot \vec{B}}{B}$$
$$= \frac{6\,\text{m}^2}{5\,\text{m}} = \boxed{1.2\,\text{m}}$$

PLAUSIBILITY CHECK The component of A along B is $A\cos\phi = (\sqrt{13}\,\text{m})\cos 70.6° = 1.2\,\text{m}$.

EXERCISE (a) Find $\vec{A} \cdot \vec{B}$ for $\vec{A} = 3\,\text{m}\hat{i} + 4\,\text{m}\hat{j}$ and $\vec{B} = 2\,\text{m}\hat{i} + 8\,\text{m}\hat{j}$. (b) Find A, B, and the angle between $\vec{A}$ and $\vec{B}$ for these vectors. (*Answer* (a) $38\,\text{m}^2$ (b) $A = 5\,\text{m}$, $B = 8.25\,\text{m}$, $\phi = 23°$)

In dot-product notation, the work dW done by a force $\vec{F}$ on a particle over a displacement $d\vec{s}$ is

$$dW = F\cos\phi\,ds = \vec{F} \cdot d\vec{s} \qquad 6\text{-}14$$

INCREMENTAL WORK

where $ds = |d\vec{s}|$ (the magnitude of $d\vec{s}$). The work done on the particle as it moves from point 1 to point 2 is

$$W = \int_{s_1}^{s_2} \vec{F} \cdot d\vec{s} \qquad 6\text{-}15$$

THE DEFINITION OF WORK

(If the force remains constant, the work can be expressed $W = \vec{F} \cdot \vec{s}$, where $\vec{s}$ is the net displacement.)

When several forces $\vec{F}_i$ act on a particle whose displacement is $d\vec{s}$, the total work done on it is

$$dW_{\text{total}} = \vec{F}_1 \cdot d\vec{s} + \vec{F}_2 \cdot d\vec{s} + \cdots = (\vec{F}_1 + \vec{F}_2 + \cdots) \cdot d\vec{s} = (\Sigma\vec{F}_i) \cdot d\vec{s} \qquad 6\text{-}16$$

EXAMPLE 6 - 7

You push a box up a ramp using a horizontal 100-N force $\vec{F}$. For each 5 m of distance along the ramp the box gains 3 m of height. Find the work done by $\vec{F}$ for each 5 m it moves along the ramp (a) by directly computing the dot product from the components of $\vec{F}$ and the displacement $\vec{s}$, (b) by multiplying the product of the magnitudes of $\vec{F}$ and $\vec{s}$ with the cosine of the angle between their directions, (c) by finding F_s (the component of the force in the direction of the displacement) and multiplying it by the magnitude of the displacement, and (d) by finding the component of the displacement in the direction of the force and multiplying it by the magnitude of the force.

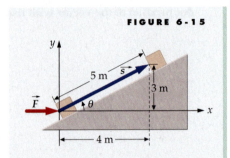

FIGURE 6-15

PICTURE THE PROBLEM Draw a sketch of the box in its initial and final positions. Place coordinate axes on the sketch with the x axis horizontal. Express the force and displacement vectors in component form and take the dot product. Then find the component of the force in the direction of the displacement, and vice versa.

(a) Express $\vec{F}$ and $\vec{s}$ in component form and take the dot product:

$$\vec{F} = 100\ N\hat{i} + 0\hat{j}$$

$$\vec{s} = 4\ m\hat{i} + 3\ m\hat{j}$$

$$W = \vec{F} \cdot \vec{s} = F_x\Delta x + F_y\Delta y$$

$$= (100\ N)(4\ m) + 0(3\ m) = \boxed{400\ J}$$

(b) Calculate $Fs \cos \phi$, where ϕ is the angle between the directions of the two vectors as shown. Equate this expression with the Part (a) result and solve for $\cos \phi$; then solve for the work:

$$\vec{F} \cdot \vec{s} = Fs \cos \phi \quad \text{and} \quad \vec{F} \cdot \vec{s} = F_x\Delta x + F_y\Delta y$$

so

$$\cos \phi = \frac{F_x\Delta x + F_y\Delta y}{Fs} = \frac{(100\ N)(4\ m) + 0}{(100\ N)(5\ m)} = 0.8$$

and

$$W = Fs \cos \phi$$

$$= (100\ N)(5\ m)0.8 = \boxed{400\ J}$$

(c) Find F_s and multiply it by s:

$$F_s = F \cos \phi = (100\ N)0.8 = 80\ N$$

$$W = F_s s = (80\ N)(5\ m) = \boxed{400\ J}$$

(d) Multiply F and s_F, where s_F is the component of $\vec{s}$ in the direction of $\vec{F}$. First calculate s_F:

$$s_F = s \cos \phi = (5\ m)0.8 = 4m$$

$$W = Fs_F = (100\ N)(4\ m) = \boxed{400\ J}$$

REMARKS For this problem, computing the work is easiest using the procedure in Part (d). For other problems, the procedure in Part (a), Part (b), or Part (c) may be the easiest. You should be comfortable with all four procedures. That way, for a given problem, you can choose the easiest procedure.

A DISPLACED PARTICLE   **E X A M P L E   6 - 8**   **Try It Yourself**

A particle is given a displacement $\vec{s} = 2\ m\hat{i} - 5\ m\hat{j}$ along a straight line. During the displacement, a constant force $\vec{F} = 3\ N\hat{i} + 4\ N\hat{j}$ acts on the particle. Find (a) the work done by the force and (b) the component of the force in the direction of the displacement.

FIGURE 6-16

$\vec{F} = 3\,N\hat{i} + 4\,N\hat{j}$

$|F_s|$

$\vec{s} = 2\,m\hat{i} - 5\,m\hat{j}$

PICTURE THE PROBLEM The work W is found by computing $W = \vec{F} \cdot \vec{s} = F_x \Delta x + F_y \Delta y$. Combining this with the relation $\vec{F} \cdot \vec{s} = F_s s$, we can find the component of $\vec{F}$ in the direction of the displacement. Make a sketch showing $\vec{F}, \vec{s}$, and $|F_s|$ (Figure 6-16).

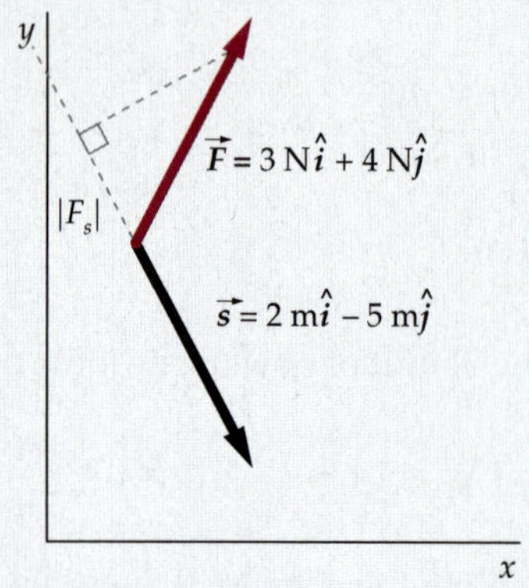

Cover the column to the right and try these on your own before looking at the answers.

Steps	Answers
(a) Compute the work done W.	$W = \vec{F} \cdot \vec{s} = \boxed{-14\,N\cdot m}$
(b) 1. Compute $\vec{s} \cdot \vec{s}$ and use your result to find the distance $|\vec{s}|$.	$|\vec{s}|^2 = \vec{s} \cdot \vec{s} = 29\,m^2$, so $|\vec{s}| = \sqrt{29}\,m$
2. Using $\vec{F} \cdot \vec{s} = F_s s$, solve for F_s.	$F_s = \boxed{-2.60\,N}$

REMARKS The component of the force in the direction of the displacement is negative, so the work done is negative.

EXERCISE Find the magnitude of $\vec{F}$ and the angle ϕ between $\vec{F}$ and $\vec{s}$. (*Answer* $F = 5\,N, \phi = 121°$)

DIFFERENTIATING A DOT PRODUCT   **EXAMPLE   6-9**

Show that $d(v^2)/dt = 2\vec{a} \cdot \vec{v}$, where v is the speed, $\vec{v}$ the velocity, and $\vec{a}$ the acceleration.

PICTURE THE PROBLEM Note that $v^2 = \vec{v} \cdot \vec{v}$, so the rule for differentiating dot products can be used here.

Apply the rule for differentiating dot products to the dot product $\vec{v} \cdot \vec{v}$:

$$\frac{d}{dt}(v^2) = \frac{d}{dt}(\vec{v} \cdot \vec{v})$$

$$= \frac{d\vec{v}}{dt} \cdot \vec{v} + \vec{v} \cdot \frac{d\vec{v}}{dt} = 2\frac{d\vec{v}}{dt} \cdot \vec{v}$$

so

$$\boxed{\frac{d}{dt}(v^2) = 2\vec{a} \cdot \vec{v}}$$

REMARKS This example involves only kinematic parameters, so the resulting relation is a strictly kinematic relation. Also, the result is unconditionally valid because the computation involved only the definition of acceleration and the rule for differentiating a dot product.

Power

The **power** P supplied by a force is the rate at which the force does work. Consider a particle moving with instantaneous velocity $\vec{v}$. In a short time interval dt, the particle has a displacement $d\vec{s} = \vec{v}\,dt$. The work done by a force $\vec{F}$ acting on the particle during this time interval is

$$dW = \vec{F} \cdot d\vec{s} = \vec{F} \cdot \vec{v}\,dt$$

The power delivered to the particle is then

$$P = \frac{dW}{dt} = \vec{F} \cdot \vec{v} \qquad\qquad 6\text{-}17$$

DEFINITION—POWER

The SI unit of power, one joule per second, is called a watt (W):

$1\,W = 1\,J/s$

Note the difference between power and work. Two motors that lift a given load a given distance expend the same amount of energy, but the one that does it in the least time is more powerful. Gas and electric companies charge for energy, not power, usually by the kilowatt-hour (kW·h). A kilowatt-hour of energy is

$1\,kW{\cdot}h = (10^3\,W)(3600\,s) = 3.6 \times 10^6\,W{\cdot}s = 3.6\,MJ$

In the U.S. customary system, the unit of energy is the foot-pound and the unit of power is the foot-pound per second. A commonly used multiple of this unit, called a horsepower (hp) is defined as

$1\,hp = 550\,ft{\cdot}lb/s = 746\,W$

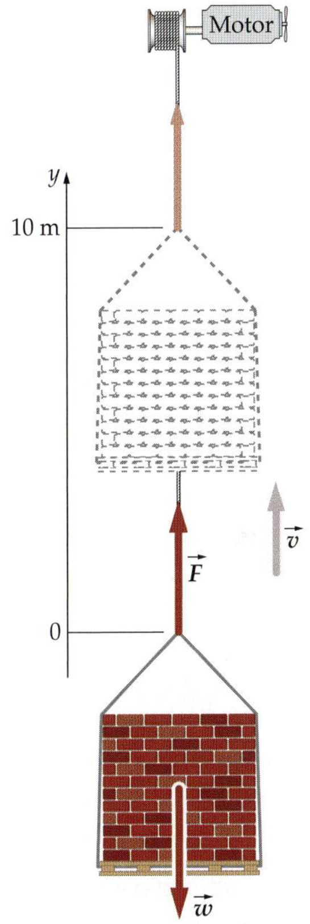

FIGURE 6-17

THE POWER OF A MOTOR   **EXAMPLE 6-10**

A small motor is used to operate a lift that raises a load of bricks weighing 800 N to a height of 10 m in 20 s (Figure 6-17). What is the minimum power the motor must deliver?

PICTURE THE PROBLEM To find the *minimum* power we assume that the bricks are lifted at constant speed. Because the acceleration is zero, the magnitude of the upward force $\vec{F}$ exerted by the motor is equal to the weight of the bricks, 800 N. The power delivered by the motor is the power supplied by $\vec{F}$.

The power is given by $\vec{F} \cdot \vec{v}$:

$$P = \vec{F} \cdot \vec{v} = Fv \cos\theta = Fv \cos 0$$

$$= (800\,N)\frac{10\,m}{20\,s}(1) = \boxed{400\,W}$$

REMARKS This minimum power output of 400 W is slightly more than $\frac{1}{2}$ hp.

EXERCISE (*a*) Find the total work done by the force. (*b*) Calculate the power by dividing the total work by the total time. (*Answer* (*a*) 8000 J (*b*) 400 W)

POWER AND KINETIC ENERGY   **EXAMPLE 6-11**

Show that the power delivered to a particle by the net force acting on it equals the rate at which the kinetic energy of the particle is changing.

PICTURE THE PROBLEM The power delivered by the net force equals $\vec{F}_{net} \cdot \vec{v}$. Show that $\vec{F}_{net} \cdot \vec{v} = dK/dt$, where $K = \frac{1}{2}mv^2$.

1. Apply the result of Example 6-9 together with $\vec{a} = \vec{F}_{net}/m$ and solve for $\vec{F}_{net} \cdot \vec{v}$:

$$2\vec{a} \cdot \vec{v} = \frac{d}{dt}(v^2)$$

$$2\frac{\vec{F}_{net}}{m} \cdot \vec{v} = \frac{d}{dt}(v^2)$$

$$\vec{F}_{net} \cdot \vec{v} = \frac{m}{2}\frac{d}{dt}(v^2)$$

2. The mass is constant so it can be moved inside the argument of the derivative:

$$\vec{F}_{net} \cdot \vec{v} = \frac{d}{dt}\left(\frac{1}{2}mv^2\right)$$

$$P_{net} = \frac{d}{dt}\left(\frac{1}{2}mv^2\right) = \boxed{\frac{dK}{dt}}$$

REMARKS In the next section we use the results of this example to obtain the work–kinetic energy theorem in three dimensions.

In Example 6-10, the power delivered to the bricks by the lower end of the rope was calculated. In that example the rate of change in kinetic energy of the rope is negligible, so the power delivered to the rope by the motor is equal to the power the rope delivers to the bricks.

6-3 Work and Energy in Three Dimensions

From Example 6-11 we have $\vec{F}_{net} \cdot \vec{v} = dK/dt$, where $K = \frac{1}{2}mv^2$. The work–kinetic energy theorem for three dimensions can be established by integrating both sides of this equation with respect to time. Integrating both sides gives

$$\int_{t_1}^{t_2} \vec{F}_{net} \cdot \vec{v}\, dt = \int_{t_1}^{t_2} \frac{dK}{dt}\, dt \qquad\qquad 6\text{-}18$$

Because $d\vec{s} = \vec{v}\, dt$, where $d\vec{s}$ is the displacement during time dt, and because $(dK/dt)\, dt = dK$, Equation 6-18 can be expressed

$$\int_1^2 \vec{F}_{net} \cdot d\vec{s} = \int_1^2 dK$$

where the integral on the left is the total work W_{total} done on the particle. (In Chapter 7 work–energy relations for objects that cannot be modeled as a particle are presented.) The integral on the right can be integrated giving

$$W_{total} = \int_1^2 \vec{F}_{net} \cdot d\vec{s} = K_2 - K_1 = \Delta K \qquad\qquad 6\text{-}19$$

WORK–KINETIC ENERGY EQUATION IN THREE DIMENSIONS

Equation 6-19 follows directly from Newton's second law of motion.

WORK DONE ON A SKIER   **EXAMPLE 6-12** **Put It in Context**

You and your friend are at ski resort with two ski runs, a beginner's run and an expert's run. Both runs begin at the top of the ski lift and end at a finish line at the bottom of the same lift. Let h be the vertical descent for both runs. The beginner's run is longer and less steep than the expert's run. You and your friend, who is a much better skier than you, are testing some experimental frictionless skis. To make things interesting, you offer a wager that if she takes the expert's run and you take the beginner's run, her speed at the finish line will not be greater than your speed at the finish line. She accepts the bet (momentarily forgetting that you are taking a physics course). The conditions are that you both start from rest at the top of the lift and both of you coast for the entire trip. Who wins the bet? (Assume airdrag is negligible.)

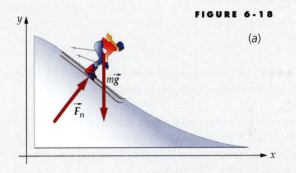

FIGURE 6-18

(a)

PICTURE THE PROBLEM Because you and your friend are coasting on the skis, you both can be modeled as particles. Two forces act on each of you, gravity $m\vec{g}$ and the normal force $\vec{F}_n$. Make a sketch of yourself and draw the two force vectors on the sketch (Figure 6-18a). Also include coordinate axes. The work–kinetic energy theorem, with $v_i = 0$, relates the final speed v_f to the total work. (The work–kinetic energy theorem works only for particles.)

1. The final speed is related to the final kinetic energy, which in turn is related to the total work by the work–kinetic energy theorem:

$$W_{total} = \tfrac{1}{2}mv_f^2 - \tfrac{1}{2}mv_i^2$$

2. For each of you, the total work is the work done by the normal force plus the work done by the gravitational force:

$$W_{total} = W_n + W_g$$

3. The force $m\vec{g}$ on you is constant, but the force $\vec{F}_n$ is not constant. First we calculate the work done by $\vec{F}_n$. Calculate the work dW_n done on you by $\vec{F}_n$ for an infinitesimal displacement $d\vec{s}$ (Figure 6-18b) at an arbitrary location along the run:

$$dW_n = \vec{F}_n \cdot d\vec{s} = F_n \cos\phi\, ds$$

(b)

4. Find the angle ϕ between the directions of $\vec{F}_n$ and $d\vec{s}$. The displacement $d\vec{s}$ is parallel to the slope:

$$\phi = 90°$$

5. Calculate the work done by $\vec{F}_n$ for the entire run:

$$dW_n = F_n \cos 90°\, ds = 0, \quad \text{so}$$

$$W_n = \int dW_n = 0$$

6. The force of gravity is constant, so the work done by gravity is $W_g = m\vec{g} \cdot \vec{s}$, where $\vec{s}$ (Figure 6-19) is the net displacement from the top to the bottom of the lift:

$$W_g = m\vec{g} \cdot \vec{s} = -mg\,\hat{j} \cdot (\Delta x\,\hat{i} + \Delta y\,\hat{j})$$
$$= -mg\,\Delta y$$

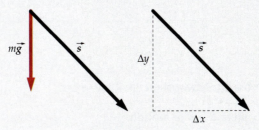

FIGURE 6-19

7. The skier is descending the hill, so Δy is negative. From Figure 6-18a, we see that $\Delta y = -h$:

$$\Delta y = -h$$

8. Substituting gives:

$$W_g = mgh$$

9. Apply the work–kinetic energy theorem to find v_f:

$$W_n + W_g = \Delta K$$

10. The final speed depends only on h, which is the same for both runs. Both of you will have the same final speeds.

$$0 + mgh = \tfrac{1}{2}mv_f^2 - 0$$

so

$$v_f = \sqrt{2gh}$$

YOU WIN! (The bet was that she would not be going faster than you.)

REMARKS Your friend on the steeper trail will cross the finish line in less time, but that was not the bet. What was shown here is that the work done by the gravitational force equals mgh. It does not depend upon the shape of the hill or upon the length of the path taken. It depends only upon the vertical drop h between the starting point and the finishing point.

In Example 6-12 we found that the work done by the gravitational force is independent of the path taken. This leads us to the concept of potential energy, which is the topic of the next section.

6-4 Potential Energy

The total work done on a particle equals the change in its kinetic energy. But we are often interested in the work done on a *system* consisting of two or more particles.[†] Often, the work done by external forces on a *system* does not increase the total kinetic energy *of the system*, but instead is stored as **potential energy**—energy associated with the configuration of the system.

Consider lifting a barbell of mass m to a height h. The work done by the gravitational force on the barbell is $-mgh$. The barbell starts at rest and ends at rest. Because the kinetic energy of the barbell does not change, we know the total work done on the barbell is zero. That means the work done by the force of your hands on the barbell is $+mgh$. Now consider the barbell and the earth to be a *system* of two particles. (You are not part of this system.) The external forces *on the earth–barbell system* are the gravitational attraction you exert *on the earth*, the force your feet exert *on the earth*, and the force mg exerted by your hands *on the barbell* (Figure 6-20). (We can neglect the gravitational force you exert on the barbell.) The barbell moves, but the motion of the earth is negligible, so only the force exerted on the barbell by your hands does work on the system. The total work done on the earth–barbell system by all forces *external* to the system is mgh. This work is stored as potential energy, which is energy associated with the position of the barbell relative to the earth. That is, it is energy associated with the configuration of the earth–barbell system. This kind of energy is called gravitational potential energy.

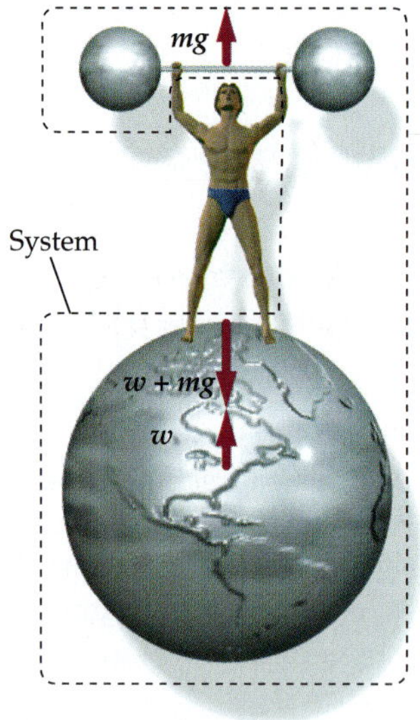

Another system that stores energy associated with its configuration is a spring. If you stretch or compress a spring, energy associated with the length of the spring is stored as potential energy. Consider the spring shown in Figure 6-21 as the system. You compress the spring, pushing it with equal and opposite forces $\vec{F}_1$ and $\vec{F}_2$. These forces sum to zero, so the net force on the spring remains

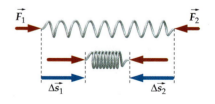

FIGURE 6-20 A system consisting of a barbell plus the earth, but not the person holding the barbell. When you lift the barbell, you do work on this system.

FIGURE 6-21 The spring is compressed by external forces $\vec{F}_1$ and $\vec{F}_2$. Both forces do positive work on the spring as they compress it, so the elastic potential energy of the spring increases as it is compressed.

[†] Systems of particles are discussed more thoroughly in Chapter 8.

zero. Thus, there is no change in the kinetic energy of the spring. The work you do on this system is stored not as kinetic energy, but as elastic potential energy. The configuration of this system has been changed, as evidenced by the change in the length of the spring. The total work done on the spring is positive because both $\vec{F}_1$ and $\vec{F}_2$ do positive work. (The work done by $\vec{F}_1$ is positive because both $\vec{F}_1$ and the displacement $\Delta\vec{s}_1$ of its point of application are in the same direction. The same can be said for $\vec{F}_2$ and $\Delta\vec{s}_2$.)

Conservative Forces

When you ride a ski lift to the top of a hill of height h, the work done by gravity on you is $-mgh$ and the work done by the lift on you is $+mgh$. When you ski down the hill to the bottom, the work done by gravity is $+mgh$ independent of the shape of the hill (as we saw in Example 6-12). The total work done by gravity on you during the round trip up and down the hill is zero, independent of the path you take. The force of gravity, exerted by the earth on you, is a **conservative force.**

> A force is conservative if the total work it does on a particle is zero when the particle moves around *any* closed path, returning to its initial position.

DEFINITION—CONSERVATIVE FORCE

From Figure 6-22 we see that this definition implies that:

> The work done by a conservative force on a particle is independent of the path taken as the particle moves from one point to another.

ALTERNATIVE DEFINITION—CONSERVATIVE FORCE

Now consider yourself and the earth to be a *two-particle system*. (The ski lift is not part of this system.) When a ski lift raises you to the top of the hill, it does work mgh on the you–earth system. This work is stored as the gravitational potential energy of the system. When you ski down the hill, this potential energy is converted to the kinetic energy of your motion.

Potential-Energy Functions

Because the work done by a conservative force on a particle does not depend on the path, it can depend only on the endpoints 1 and 2. We can use this property to define the **potential-energy function** U that is associated with a conservative force. Note that when you ski down the hill, the work done by gravity *decreases* the potential energy of the system. We define the potential-energy function such that the work done by a conservative force equals the decrease in the potential-energy function:

$$W = \int_1^2 \vec{F} \cdot d\vec{s} = -\Delta U$$

or

$$\Delta U = U_2 - U_1 = -\int_1^2 \vec{F} \cdot d\vec{s} \qquad 6\text{-}20a$$

DEFINITION—POTENTIAL-ENERGY FUNCTION

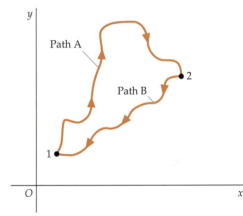

FIGURE 6-22 Two paths in space connecting the points 1 and 2. If the work done by a conservative force along path A from 1 to 2 is W, the work done on the return trip along path B must be $-W$ because the roundtrip work is zero. When traversing path B from 1 to 2, the force is the same at each point, but the displacement is opposite to that when going from 2 to 1. Then the work done along path B from 1 to 2 must also be W. It follows that the work done as a particle goes from point 1 to 2 is the same along any path connecting the two points.

For an infinitesimal displacement, we have

$$dU = -\vec{F} \cdot d\vec{s} \qquad\qquad 6\text{-}20b$$

We can calculate the potential-energy function associated with the gravitational force near the surface of the earth from Equation 6-20b. For the force $\vec{F} = -mg\hat{j}$, we have

$$dU = -\vec{F} \cdot d\vec{s} = -(-mg\hat{j}) \cdot (dx\hat{i} + dy\hat{j} + dz\hat{k}) = +mg\,dy$$

Integrating, we obtain

$$U = \int mg\,dy = mgy + U_0$$

$$U = U_0 + mgy \qquad\qquad 6\text{-}21$$

GRAVITATIONAL POTENTIAL ENERGY NEAR THE EARTH'S SURFACE

where U_0, the arbitrary constant of integration, is the value of the potential energy at $y = 0$. Because only a change in the potential energy is defined, the actual value of U is not important. We are free to choose U to be zero at any convenient reference point. For example, if the gravitational potential energy of the earth–skier system is chosen to be zero when the skier is at the bottom of the hill, its value when the skier is at a height h above that level is mgh. Or we could choose the potential energy to be zero when the skier is at point P half way down the ski slope, in which case its value at any other point would be mgy, where y is the height of the skier above point p. On the lower half of the slope the potential energy would then be negative.

EXERCISE A 55-kg window washer stands on a platform 8 m above the ground. What is the potential energy U of the window-washer–earth system if (a) U is chosen to be zero on the ground, (b) U is chosen to be zero 4 m above the ground, and (c) U is chosen to be zero 10 m above the ground? (*Answer*   (a) 4.32 kJ (b) 2.16 kJ (c) −1.08 kJ)

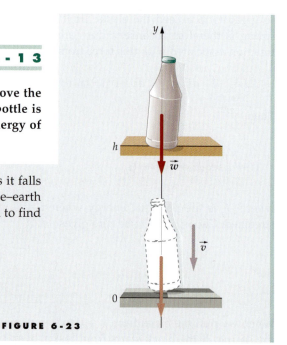

FIGURE 6-23

A FALLING BOTTLE	**EXAMPLE 6-13**

A bottle of mass 0.350 kg falls from rest from a shelf that is 1.75 m above the floor. Find the potential energy of the bottle–earth system when the bottle is on the shelf and just before impact with the floor. Find the kinetic energy of the bottle just before impact.

PICTURE THE PROBLEM The work done by the earth on the bottle as it falls equals the negative of the change in the potential energy of the bottle–earth system. Knowing the work, we can use the work–kinetic energy theorem to find the kinetic energy.

1. Make a sketch showing the bottle on the shelf and again when it is about to impact the floor (Figure 6-23). Choose the potential energy of the bottle–earth system to be zero when the bottle is on the floor, and place a y axis on the sketch with the origin at the floor:

2. The only force doing work on the falling bottle is the force of gravity, so $W_{total} = W_g$. Apply the work–kinetic energy theorem to the falling bottle:

$$W_{total} = W_g = \Delta K$$

3. The force exerted by the earth on the falling bottle is internal to the bottle–earth system. It is also a conservative force, so the work done by it equals the change in the potential energy of the system:

$$W_g = -\Delta U = -(U_f - U_i) = -(mgy_f - mgy_i)$$
$$= mg(y_i - y_f) = mg(h - 0) = mgh$$

4. Substitute the step 3 result into the step 2 result and solve for the final kinetic energy. The original kinetic energy is zero:

$$mgh = \Delta K$$
$$mgh = K_f - K_i$$
$$K_f = K_i + mgh$$
$$= 0 + (0.350\ kg)(9.81\ N/kg)(1.75\ m)$$
$$= \boxed{6.01\ J}$$

REMARKS In this example, the potential energy lost by the bottle–earth system is converted entirely to kinetic energy of the bottle as it falls. Note in step 4 that we have used the definition $1\ J = 1\ N \cdot m$.

Potential energy is associated with the configuration of a *system of particles,* but we sometimes have systems such as the bottle–earth system, in which only one particle moves (the earth's motion is negligible). For brevity, then, we sometimes refer to the potential energy of the bottle–earth system as simply the potential energy of the bottle.

Potential Energy of a Spring Another example of a conservative force is that of a stretched (or compressed) spring. Suppose we pull a block attached to a spring from a position $x = 0$ (equilibrium) to x_1 (Figure 6-24). The spring does negative work because the force it exerts on the block is opposite to the block's displacement. If we then release the block, the force of the spring does positive work on the block as it accelerates toward its initial position. The total work done by the spring on the block as it moves from its original position to $x = x_1$, and then back, is zero. This result is independent of the size of x_1 (as long as the stretching is not so great as to exceed the elastic limit of the spring). The force exerted by the spring is therefore a conservative force. We can calculate the potential-energy function associated with this force from Equation 6-20b:

$$dU = -\vec{F} \cdot d\vec{s} = -F_x\,dx = -(-kx)\,dx = +kx\,dx$$

Then

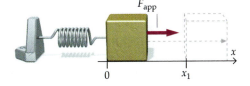

FIGURE 6-24 The applied force F_{app} pulls the block to the right, stretching the spring by x_1.

$$U = \int kx\,dx = \tfrac{1}{2}kx^2 + U_0$$

where U_0 is the potential energy when $x = 0$, that is, when the spring is unstretched. Choosing U_0 to be zero gives

$$U = \tfrac{1}{2}kx^2 \qquad\qquad\qquad 6\text{-}22$$

POTENTIAL ENERGY OF A SPRING

When we pull the block from $x = 0$ to $x = x_1$, we must exert an applied force to the spring. If the block starts from rest at $x = 0$ and ends at rest at $x = x_1$, the

change in its kinetic energy is zero. The work–energy theorem then implies that the total work done on the block is zero. That is, $W_{app} + W_{spring} = 0$, or

$$W_{app} = -W_{spring} = \Delta U_{spring} = \tfrac{1}{2}kx_1^2 - 0 = \tfrac{1}{2}kx_1^2$$

This work is stored as potential energy in the spring–block system.

POTENTIAL ENERGY OF A BASKETBALL PLAYER   **EXAMPLE 6-14**

A system consists of a basketball player, the rim of a basketball hoop, and the earth. Assume that the potential energy of this system is zero when the player is standing on the floor and the rim is horizontal. Find the total potential energy of this system when the player is hanging on the rim (as in Figure 6-25). Also assume that the player can be described as a point mass of 110 kg at 0.8 m above the floor when standing and at 1.3 m above the floor when hanging. The force constant of the rim is 7.2 kN/m and the front of the rim is displaced a distance $s = 15$ cm.

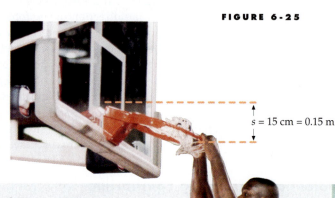

FIGURE 6-25

$s = 15\ \text{cm} = 0.15\ \text{m}$

PICTURE THE PROBLEM In the player's change in position from standing on the floor to hanging on the rim, the total change in potential energy consists of gravitational potential energy, $U_g = mgy$, and energy stored in the deformed rim, whose potential energy can be measured just as if it were a spring: $U_s = \tfrac{1}{2}ks^2$. Choose $y = 0$ at 0.8 m above the floor for the gravitational potential-energy reference point.

The total potential energy is the sum of gravitational potential energy and the elastic potential energy of the rim (see Figure 6-26):

$$U = U_g + U_s = mgy + \tfrac{1}{2}ks^2$$
$$= (110\ \text{kg})(9.81\ \text{N/kg})(0.5\ \text{m}) + \tfrac{1}{2}(7.2\ \text{kN/m})(0.15\ \text{m})^2$$
$$= 540\ \text{J} + 81\ \text{J} = \boxed{621\ \text{J}}$$

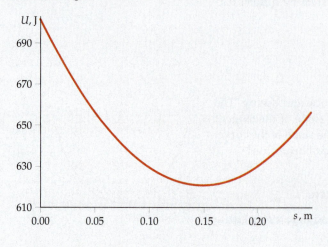

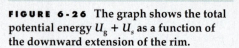

FIGURE 6-26 The graph shows the total potential energy $U_g + U_s$ as a function of the downward extension of the rim.

REMARKS Most of the potential energy is gravitational in this case, because of the choice of the potential-energy reference point.

EXERCISE A 3-kg block is hung vertically from a spring with a force constant of 600 N/m. (*a*) By how much is the spring stretched? (*b*) How much potential energy is stored in the spring? (*Answer* (*a*) 4.9 cm (*b*) 0.72 J)

Nonconservative Forces

Not all forces are conservative. Suppose that you push a box across a table along a straight line from point A to point B and back, so that the box ends up at its original position. Friction opposes the block's motion, so the force you push with is in the direction of motion and does positive work on both legs of the round trip. The total work done by the push does not equal zero. Thus, the push is an example of a **nonconservative force** and no potential-energy function can be defined for it.

Sometimes we can show that a given force is not conservative by computing the work done by the force around some chosen closed curve and showing that it is not zero. Consider the force $\vec{F} = F_0\hat{\phi}$, where $\hat{\phi}$ is a unit vector directed tangent to a circle of radius r. The work done by this force as we move around a circle of radius r is $+F_0 2\pi r$ if we move in the direction of the force (and $-F_0 2\pi r$ if we move opposite to the force). Since this work is not zero, we conclude that the force is not conservative. This method is of limited use in investigating whether a given force is conservative or not. If the work done around *any* particular closed path is not zero, we may conclude that the force is not conservative. However, if the force is conservative, the work must be zero around *all* possible closed paths. Since there are infinitely many possible closed paths, it is impossible to calculate the work done for each one. In more advanced physics courses, more sophisticated mathematical methods for testing forces are discussed.

Potential Energy and Equilibrium

For a general conservative force in one dimension, $\vec{F} = F_x\hat{i}$, Equation 6-20b is

$$dU = -\vec{F} \cdot d\vec{s} = -F_x\, dx$$

The force is therefore the negative derivative of the potential-energy function:

$$F_x = -\frac{dU}{dx} \qquad\qquad 6\text{-}23$$

We can illustrate this general relation for a block–spring system by differentiating the function $U = \frac{1}{2}kx^2$. We obtain

$$F_x = -\frac{dU}{dx} = -\frac{d}{dx}\left(\frac{1}{2}kx^2\right) = -kx$$

Figure 6-27 shows a plot of $U = \frac{1}{2}kx^2$ versus x for a block and spring. The derivative of this function is represented graphically as the slope of the tangent line to the curve. The force is thus equal to the negative of the slope of the curve. At $x = 0$, the force $F_x = -dU/dx$ is zero and the block is in equilibrium.

A particle is in equilibrium if the net force acting on it is zero.

CONDITION FOR EQUILIBRIUM

When x is positive in Figure 6-27, the slope is positive and the force F_x is negative. When x is negative, the slope is negative and the force F_x is positive. In either case, the force is in the direction that will accelerate the block toward lower potential energy. If the block is displaced slightly from $x = 0$, the force is directed back toward $x = 0$. The equilibrium at $x = 0$ is thus **stable equilibrium** because a small displacement results in a restoring force that accelerates the particle back toward its equilibrium position.

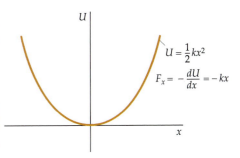

FIGURE 6-27 Plot of the potential-energy function U versus x for an object on a spring. A minimum in a potential-energy curve is a point of stable equilibrium. Displacement in either direction results in a force directed toward the equilibrium position.

In stable equilibrium, a small displacement in any direction results in a restoring force that accelerates the particle back toward its equilibrium position.

Figure 6-28 shows a potential-energy curve with a maximum rather than a minimum at the equilibrium point $x = 0$. Such a curve could represent the potential energy of a skier at the top of a hill. For this curve, when x is positive, the slope is negative and the force F_x is positive, and when x is negative, the slope is positive and the force F_x is negative. Again, the force is in the direction that will accelerate the particle toward lower potential energy, but this time the force is away from the equilibrium position. The maximum at $x = 0$ in Figure 6-28 is a point of **unstable equilibrium** because a small displacement results in a force that accelerates the particle away from its equilibrium position.

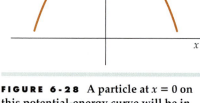

FIGURE 6-28 A particle at $x = 0$ on this potential-energy curve will be in unstable equilibrium because a displacement in either direction results in a force directed away from the equilibrium position.

In unstable equilibrium, a small displacement results in a force that accelerates the particle away from its equilibrium position.

Figure 6-29 shows a potential-energy curve that is flat in the region near $x = 0$. No force acts on a particle at $x = 0$, and hence the particle is at equilibrium; furthermore, there will be no resulting force if the particle is displaced slightly in either direction. This is an example of **neutral equilibrium.**

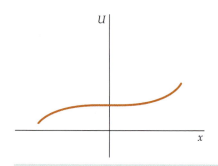

In neutral equilibrium, a small displacement results in zero force and the particle remains in equilibrium.

FIGURE 6-29 Neutral equilibrium. The force $F_x = -dU/dx$ is zero at $x = 0$ and at neighboring points, so displacement away from $x = 0$ results in no force, and the system remains in equilibrium.

FORCE AND THE POTENTIAL-ENERGY FUNCTION   **EXAMPLE 6-15**

In the region $-a < x < a$ the force on a particle is represented by the potential energy function

$$U = -b\left(\frac{1}{a + x} + \frac{1}{a - x}\right)$$

where a and b are positive constants. (*a*) Find the force F_x in the region $-a < x < a$. (*b*) At what value of x is the force zero? (*c*) At the location where the force equals zero, is the equilibrium stable or unstable?

Try It Yourself

PICTURE THE PROBLEM The force is the negative of the derivative of the potential-energy function. The equilibrium is stable where the potential-energy function is a minimum and it is unstable where the potential-energy function is a maximum.

Cover the column to the right and try these on your own before looking at the answers.

Steps	Answers

(*a*) Compute $F_x = -dU/dx$.

$$F_x = -\frac{d}{dx}\left[-b\left(\frac{1}{a + x} + \frac{1}{a - x}\right)\right]$$

$$= \boxed{-b\left(\frac{1}{(x + a)^2} - \frac{1}{(x - a)^2}\right)}$$

(b) Set F_x equal to zero and solve for x.

$F_x = 0$ at $\boxed{x = 0}$

(c) Compute d^2U/dx^2. If it is positive at the equilibrium position, then U is a minimum and the equilibrium is stable. If it is negative, then U is a maximum and the equilibrium is unstable.

At $x = 0$, $\dfrac{d^2U}{dx^2} = \dfrac{-4b}{a^3}$

Thus, $\boxed{\text{unstable}}$ equilibrium.

REMARKS This potential-energy function is that for a particle under the influence of the gravitational forces exerted by two identical fixed masses, one at $x = -a$ and the other at $x = +a$. The particle is located on the line joining the masses. Midway between the two masses the net force on the particle is zero. Otherwise, it is toward the closest mass.

SUMMARY

1. Work, kinetic energy, potential energy, and power are important derived dynamic quantities.
2. The work–kinetic energy theorem is an important relation derived from Newton's laws applied to a particle. (In this context, a particle is a perfectly rigid object that moves without rotating.)
3. The dot product of vectors is a mathematical definition that is useful throughout physics.

Topic	Relevant Equations and Remarks	
1. Work (definition)	$W = \displaystyle\int_1^2 \vec{F} \cdot d\vec{s}$	6-15
Constant force	$W = \vec{F} \cdot \vec{s}$	
In one dimension		
Constant force	$W = F_x \, \Delta x = F \cos\theta \, \Delta x$	6-1
Variable force	$W = \displaystyle\int_{x_1}^{x_2} F_x \, dx = $ area under the F_x-versus-x curve	6-9
2. Kinetic Energy (definition)	$K = \frac{1}{2}mv^2$	6-6
3. Work–Kinetic Energy Theorem	$W_{\text{total}} = \Delta K = \frac{1}{2}mv_f^2 - \frac{1}{2}mv_i^2$	6-7
4. Dot Product (definition)	$\vec{A} \cdot \vec{B} = AB \cos\phi$	6-10
In terms of components	$\vec{A} \cdot \vec{B} = A_x B_x + A_y B_y + A_z B_z$	6-11
Component of vector	$\vec{A} \cdot \hat{i} = A_x$	6-12
Derivative	$\dfrac{d}{dt}(\vec{A} \cdot \vec{B}) = \dfrac{d\vec{A}}{dt} \cdot \vec{B} + \vec{A} \cdot \dfrac{d\vec{B}}{dt}$	6-13
5. Power	$P = \dfrac{dW}{dt} = \vec{F} \cdot \vec{v}$	6-17

6. Conservative Force

A force is conservative if the total work it does on a particle is zero when the particle moves along any path that returns it to its initial position. Alternatively, the work done by a conservative force on a particle is independent of the path taken by the particle as it moves from one point to another.

7. Potential Energy

The potential energy of a system is the energy associated with the configuration of the system. The change in the potential energy of a system is defined as the negative of the work done by all internal conservative forces acting on the system.

Definition	$\Delta U = U_2 - U_1 = -\int_1^2 \vec{F} \cdot d\vec{s}$	**6-20**
	$dU = -\vec{F} \cdot d\vec{s}$	**6-20**
Gravitational	$U = U_0 + mgy$	**6-21**
Elastic (spring)	$U = \frac{1}{2}kx^2$	**6-22**
Conservative force	$F_x = -\dfrac{dU}{dx}$	**6-23**

Potential-energy curve	At a minimum on the curve of the potential-energy function versus the displacement, the force is zero and the system is in stable equilibrium. At a maximum, the force is zero and the system is in unstable equilibrium. A conservative force always tends to accelerate a particle toward a position of lower potential energy.

PROBLEMS

- • Single-concept, single-step, relatively easy
- •• Intermediate-level, may require synthesis of concepts
- ••• Challenging
- **SSM** Solution is in the *Student Solutions Manual*
- **iSOLVE** Problems available on iSOLVE online homework service
- **iSOLVE✓** These "Checkpoint" online homework service problems ask students additional questions about their confidence level, and how they arrived at their answer

In a few problems, you are given more data than you actually need; in a few other problems, you are required to supply data from your general knowledge, outside sources, or informed estimates.

Take $g = 9.81$ N/kg $= 9.81$ m/s^2 and neglect friction in all problems unless otherwise stated.

Conceptual Problems

1 • **SSM** True or false:
(a) Only the net force acting on an object can do work.
(b) No work is done on a particle that remains at rest.
(c) A force that is always perpendicular to the velocity of a particle never does work on the particle.

2 • You are to move a heavy box from the top of one table to the top of another table of the same height on the other side of the room. What is the minimum amount of work you must do on the box to accomplish the move? Explain.

3 • True or False: A person on a Ferris wheel is moving in a circle at constant speed. Thus, no force is doing work on the person.

4 • **SSM** By what factor does the kinetic energy of a car change when its speed is doubled?

5 • A particle moves in a circle at constant speed. Only one of the forces acting on the particle is in the centripetal direction. Does the net force on the particle do work on it? Explain.

6 • An object initially has kinetic energy K. The object then moves in the opposite direction with three times its initial speed. What is the kinetic energy now? (a) K. (b) $3K$. (c) $-3K$. (d) $9K$. (e) $-9K$.

7 • **SSM** How does the work required to stretch a spring 2 cm from its natural length compare with that required to stretch it 1 cm from its natural length?

8 • Suppose there is a net force acting on a particle but it does no work. Can the particle be moving in a straight line?

9 • The dimension of power is (a) $[M][L]^2[T]^2$, (b) $[M][L]^2/[T]$, (c) $[M][L]^2/[T]^2$, (d) $[M][L]^2/[T]^3$.

10 • Two knowledge seekers decide to ascend a mountain. Sal chooses a short, steep trail, while Joe, who weighs the same as Sal, goes up via a long, gently sloped trail. At the top, they get into an argument about who gained more potential energy. Which of the following is true:

(a) Sal gains more gravitational potential energy than Joe.
(b) Sal gains less gravitational potential energy than Joe.
(c) Sal gains the same gravitational potential energy as Joe.
(d) To compare energies, we must know the height of the mountain.
(e) To compare the energies, we must know the length of the two trails.

11 • True or false:

(a) Only conservative forces can do work.
(b) If only conservative forces act, the kinetic energy of a particle does not change.
(c) The work done by a conservative force equals the decrease in the potential energy associated with that force.

12 •• **SSM** Figure 6-30 shows the plot of a potential-energy function U versus x. (a) At each point indicated, state whether the force F_x is positive, negative, or zero. (b) At which point does the force have the greatest magnitude? (c) Identify any equilibrium points, and state whether the equilibrium is stable, unstable, or neutral.

FIGURE 6-30 Problem 12

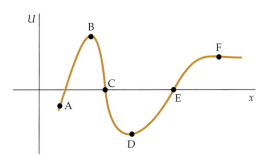

13 • True or false:

(a) The gravitational force cannot do work because it acts at a distance.
(b) Work is the area under the force-versus-time curve.

14 • Negative work means (a) the kinetic energy of the object increases, (b) the applied force is variable, (c) the angle between applied force is perpendicular to the displacement, (d) the applied force and the displacement is greater than 90°, (e) nothing; there is no such thing as negative work.

Estimation and Approximation

15 •• **SSM** A tightrope walker whose mass is 50 kg walks across a tightrope held between two supports 10 m apart; the tension in the rope is 5000 N. The height of the rope is 10 m above the ground. Estimate: (a) the sag in the tightrope when she stands in the exact center and (b) the change in her gravitational potential energy from just before stepping onto the tightrope to when she stands at its dead center.

16 • Estimate (a) the change in your potential energy on taking an elevator from the ground floor to the top of the Empire State building, (b) the average force acting on you by the elevator to bring you to the top, and (c) the average power due to that force. The building is 102 stories high.

17 • The nearest stars, apart from the sun, are at light-years away from the earth. (One light-year is the distance that light travels during a year: 9.47×10^{15} m.) If we ever want to send spacecraft to investigate the stars, they will have to have velocities that are an appreciable fraction of the speed of light. Calculate the kinetic energy of a 10,000-kg spacecraft traveling at a speed of 10 percent of the speed of light, and compare that to the amount of energy that the United States uses in a year (about 5×10^{20} J). NOTE At velocities approaching the speed of light, the theory of relativity tells us that the formula $\frac{1}{2}mv^2$ for kinetic energy is not correct. However, the correction is only about 1 percent for this particular speed ($0.1c$).

18 •• **SSM** The mass of the Space Shuttle orbiter is about 8×10^4 kg and the period of its orbit is 90 min. Estimate (a) the kinetic energy of the orbiter and (b) the change in its potential energy between resting on the surface of the earth and in its orbit, 200 mi above the surface of the earth. (c) Why is the change in potential energy much smaller than the shuttle's kinetic energy? Shouldn't they be equal?

19 • Ten inches of snow have fallen during the night, and you must shovel out your 50-ft-long driveway (Figure 6-31). Estimate how much work you do on the snow by completing this task. Make a plausible guess of any value(s) needed (the width of the driveway, for example), and state the basis for each guess.

FIGURE 6-31 Problem 19

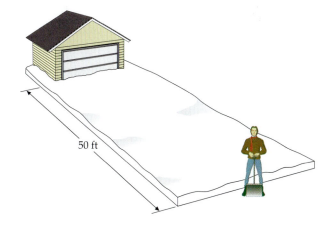

50 ft

Work and Kinetic Energy

20 • **SSM** A 15-g bullet has a speed of 1.2 km/s. (a) What is its kinetic energy in joules? (b) What is its kinetic energy if its speed is halved? (c) What is its kinetic energy if its speed is doubled?

21 • Find the kinetic energy in joules of (a) a 0.145-kg baseball moving with a speed of 45 m/s and (b) a 60-kg jogger running at a steady pace of 9 min/mi.

22 • A 6-kg box is raised a distance of 3 m from rest by a vertical applied force of 80 N. Find (*a*) the work done by the force, (*b*) the work done by gravity, and (*c*) the final kinetic energy of the box.

23 • A constant force of 80 N acts on a box of mass 5.0 kg that is moving in the direction of the force with a speed of 20 m/s. A few seconds later the box is moving with a speed of 68 m/s. Determine the work done by this force.

24 •• **SSM** You run a race with a friend. At first you each have the same kinetic energy, but you find that she is beating you. When you increase your speed by 25 percent, you are running at the same speed she is. If your mass is 85 kg, what is her mass?

Work Done by a Variable Force

25 •• A 3-kg particle is moving along the *x* axis with a velocity of 2 m/s as it passes through $x = 0$. It is subjected to a single force F_x that varies with position as shown in Figure 6-32. (*a*) What is the kinetic energy of the particle as it passes through $x = 0$? (*b*) How much work is done by the force as the particle moves from $x = 0$ to $x = 4$ m? (*c*) What is the speed of the particle when it is at $x = 4$ m?

FIGURE 6-32 Problem 25

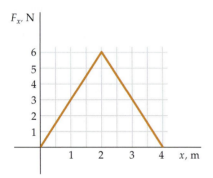

26 •• **SSM** **iSOLVE** ✓ A force F_x acts on a particle. The force is related to the position of the particle by the formula $F_x = Cx^3$, where *C* is a constant. Find the work done by this force on the particle as the particle moves from $x = 1.5$ m to $x = 3$ m.

27 •• **iSOLVE** Lou's latest invention, aimed at urban dog owners, is the X-R-Leash. It is made of a rubber-like material that exerts a force $F_x = -kx - ax^2$ when it is stretched a distance *x*, where *k* and *a* are constants. The ad claims, "You'll never go back to your old dog leash after you've had the thrill of an X-R-Leash experience. And you'll see a new look of respect in the eyes of your proud pooch." Find the work done on a dog by the leash if the person remains stationary, and the dog bounds off, stretching the X-R-Leash from $x = 0$ to $x = x_1$.

28 •• A 3-kg object moving along the *x* axis has a velocity of 2.40 m/s as it passes through the origin. It is acted on by a single force F_x that varies with *x* as shown in Figure 6-33. (*a*) Find the work done by the force from $x = 0$ to $x = 2$ m. (*b*) What is the kinetic energy of the object at $x = 2$ m? (*c*) What is the speed of the object at $x = 2$ m? (*d*) Find the work done on the object from $x = 0$ to $x = 4$ m. (*e*) What is the speed of the object at $x = 4$ m?

FIGURE 6-33 Problem 28

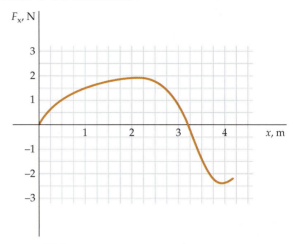

29 •• **SSM** Near Margaret's cabin is a 20-m water tower that attracts many birds during the summer months. During a hot spell, the tower went dry, and Margaret had to have her water hauled in. She was lonesome without the birds visiting, so she decided to carry some water up the tower to attract them back. Her bucket has a mass of 10 kg and holds 30 kg of water when it is full. However, the bucket has a hole, and as Margaret climbed at a constant speed, water leaked out at a constant rate. When she reached the top, only 10 kg of water remained for the birdbath. (*a*) Write an expression for the mass of the bucket plus water as a function of the height (*y*) climbed. (*b*) Find the work done by Margaret on the bucket.

Work, Energy, and Simple Machines

30 • **iSOLVE** ✓ A 6-kg block slides down a frictionless incline making an angle of 60° with the horizontal. (*a*) List all the forces acting on the block, and find the work done by each force when the block slides 2 m (measured along the incline). (*b*) What is the total work done on the block? (*c*) What is the speed of the block after it has slid 1.5 m if it starts from rest? (*d*) What is its speed after 1.5 m if it starts with an initial speed of 2 m/s?

31 • A 2-kg object attached to a horizontal string moves with a speed of 2.5 m/s in a circle of radius 3 m on a frictionless horizontal surface. (*a*) Find the tension in the string. (*b*) List the forces acting on the object, and find the work done by each force during one revolution.

32 • **SSM** *Simple machines* are used for reducing the amount of force that must be supplied to perform a task such as lifting a heavy weight. Such machines include the screw, block-and-tackle systems, and levers, but the simplest of the simple machines is the inclined plane. In Figure 6-34 you are raising the heavy box to the height of the truck bed by pushing it up an inclined plane (a ramp). (*a*) We define the *mechanical advantage M* of the inclined plane as the ratio of the force it would take to lift the block into the truck directly from the ground (at constant speed) to the force it takes to push it up the ramp (at constant speed). If the plane is frictionless, show that $M = 1/\sin\theta = L/H$, where *H* is the height of the truck bed and *L* is the length of the ramp. (*b*) Show that the work you do by moving the block into the truck is the same whether you lift it directly into the truck or push it up the frictionless ramp.

FIGURE 6-34 Problem 32

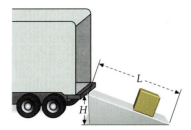

33 •• The screw is a form of the inclined plane. Figure 6-35 shows a type of car jack used for changing flat tires. The screw on the jack has a pitch (distance between threads) of p and a handle of length R. As the handle is turned through a full circle, the jack's height increases, raising the weight w by p. Assuming no friction, the work done turning the jack handle is the same as the increase in the potential energy of whatever the jack is lifting. Show that the jack has a mechanical advantage (defined in Problem 32) of $2\pi R/p$.

FIGURE 6-35 Problem 33

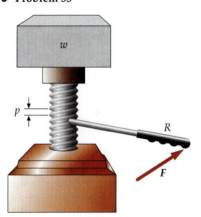

34 • Figure 6-36 shows two pulleys arranged to help lift a heavy load: a rope runs around two massless, friction-less pulleys and the weight $\vec{w}$ hangs from one pulley. You ex-ert a force of magnitude F on the free end of the cord. (a) If the weight is to move up a dis-tance h, through what distance must the force move? (b) How much work is done by the rope on the weight? (c) How much work do you do? (d) What is the mechanical advantage (defined in Problem 32) of this system?

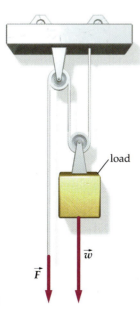

FIGURE 6-36 Problem 34

Dot Products

35 • **SSM** What is the angle between the vectors $\vec{A}$ and $\vec{B}$ if $\vec{A} \cdot \vec{B} = -AB$?

36 • Two vectors $\vec{A}$ and $\vec{B}$ each have magnitudes of 6 m and make an angle of 60° with each other. Find $\vec{A} \cdot \vec{B}$.

37 • Find $\vec{A} \cdot \vec{B}$ for the following vectors: (a) $\vec{A} = 3\hat{i} - 6\hat{j}$, $\vec{B} = -4\hat{i} + 2\hat{j}$; (b) $\vec{A} = 5\hat{i} + 5\hat{j}$, $\vec{B} = 2\hat{i} - 4\hat{j}$; and (c) $\vec{A} = 6\hat{i} + 4\hat{j}$, $\vec{B} = 4\hat{i} - 6\hat{j}$.

38 • Find the angles between the vectors $\vec{A}$ and $\vec{B}$ in Problem 37.

39 • **iSOLVE** ✓ A 2-kg object is given a displacement $\Delta\vec{s} = 3\,m\hat{i} + 3\,m\hat{j} - 2\,m\hat{k}$ along a straight line. During the displacement, a constant force $\vec{F} = 2\,N\hat{i} - 1\,N\hat{j} + 1\,N\hat{k}$ acts on the object. (a) Find the work done by $\vec{F}$ for this displace-ment. (b) Find the component of $\vec{F}$ in the direction of the dis-placement.

40 •• (a) Find the unit vector that is parallel to vector $\vec{A} = A_x\hat{i} + A_y\hat{j} + A_z\hat{k}$. (b) Find the component of the vec-tor $\vec{A} = 2\hat{i} - \hat{j} - \hat{k}$ in the direction of the vector $\vec{B} = 3\hat{i} + 4\hat{j}$.

41 •• **SSM** Given two vectors $\vec{A}$ and $\vec{B}$, show that if $|\vec{A} + \vec{B}| = |\vec{A} - \vec{B}|$, then $\vec{A} \perp \vec{B}$.

42 •• $\hat{A}$ and $\hat{B}$ are two unit vectors in the xy plane. They make angles of θ_1 and θ_2 with the positive x-axis, respectively. (a) Find the x and y components of the two vectors. (b) By considering the dot product of $\hat{A}$ and $\hat{B}$, show that $\cos(\theta_1 - \theta_2) = \cos\theta_1\cos\theta_2 + \sin\theta_1\sin\theta_2$.

43 • If $\vec{A} \cdot \vec{B} = \vec{A} \cdot \vec{C}$, must $\vec{B} = \vec{C}$? If no, give a coun-terexample, if yes, explain why.

44 •• (a) Let $\vec{A}$ be a constant vector in the xy plane with its tail at the origin. Let $\vec{r} = x\hat{i} + y\hat{j}$ be a vector in the xy plane that satisfies the relation $\vec{A} \cdot \vec{r} = 1$. Show that the points (x,y) lie on a straight line. (b) If $\vec{A} = 2\hat{i} - 3\hat{j}$, find the slope and y intercept of the line. (c) If we now let $\vec{A}$ and $\vec{r}$ be vectors in three-dimensional space, show that the relation $\vec{A} \cdot \vec{r} = 1$ specifies a plane.

45 •• **SSM** When a particle moves in a circle centered at the origin with constant speed, the magnitudes of its position vector and velocity vectors are constant. (a) Differentiate $\vec{r} \cdot \vec{r} = r^2 =$ constant with respect to time to show that $\vec{v} \cdot \vec{r} = 0$ and therefore $\vec{v} \perp \vec{r}$. (b) Differentiate $\vec{v} \cdot \vec{v} = v^2 =$ constant with re-spect to time to show that $\vec{a} \cdot \vec{v} = 0$ and therefore $\vec{a} \perp \vec{v}$. What do the results of (a) and (b) imply about the direction of $\vec{a}$? (c) Differentiate $\vec{v} \cdot \vec{r} = 0$ with respect to time and show that $\vec{a} \cdot \vec{r} + v^2 = 0$ and therefore $a_r = -v^2/r$.

Power

46 •• Force A does 5 J of work in 10 s. Force B does 3 J of work in 5 s. Which force delivers greater power?

47 • A 5-kg box is being lifted upward at a constant velocity of 2 m/s by a force equal to the weight of the box. (a) What is the power input of the force? (b) How much work is done by the force in 4 s?

48 • Fluffy has just caught a mouse and decides to bring it to the bedroom so that his human roommate can admire it when she wakes up. A constant horizontal force of 3 N is enough to drag the mouse across the rug at a constant speed v. If Fluffy's force does work at the rate of 6 W, (a) what is his speed v? (b) How much work does Fluffy do in 4 s?

49 • A single force of 5 N acts in the x direction on an 8-kg object. (a) If the object starts from rest at $x = 0$ at time $t = 0$, find its velocity v as a function of time. (b) Write an expression for the power input as a function of time. (c) What is the power input of the force at time $t = 3$ s?

50 • **ISOLVE** Find the power input of a force $\vec{F}$ acting on a particle that moves with a velocity $\vec{v}$ for (a) $\vec{F} = 4\,N\hat{i} + 3\,N\hat{k}$, $\vec{v} = 6\,m/s\hat{i}$; (b) $\vec{F} = 6\,N\hat{i} - 5\,N\hat{j}$, $\vec{v} = -5\,m/s\hat{i} + 4\,m/s\hat{j}$; and (c) $\vec{F} = 3\,N\hat{i} + 6\,N\hat{j}$, $\vec{v} = 2\,m/s\hat{i} + 3\,m/s\hat{j}$.

51 • **SSM** A small food service elevator (dumbwaiter) in a cafeteria is connected over a pulley system to a motor as shown in Figure 6-37; the motor raises and lowers the dumbwaiter. The mass of the dumbwaiter is 35 kg. In operation, it moves at a speed of 0.35 m/s upward, without accelerating (except for a brief initial period just after the motor is turned on). If the output power from the motor is 27 percent of its input power, what is the input power to the motor? Assume that the pulleys are frictionless.

FIGURE 6-37 Problem 51

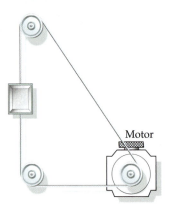

Motor

52 •• A sky diver falls though the air toward the ground at a constant speed of 120 mph, her terminal velocity, before opening her parachute. (a) If her mass is 55 kg, calculate the magnitude of the power due to the drag force. (b) After she opens her parachute, her speed slows to 15 mph. What is the magnitude of the power due to the drag force now?

53 •• **SSM** A cannon placed at the top of a cliff of height H fires a cannonball into the air with an initial speed v_0, shooting directly upward. The cannonball rises, falls back down, missing the cannon by a little bit, and lands at the foot of the cliff. Neglecting air resistance, calculate the velocity $\vec{v}(t)$ for all times while the cannonball is in the air, and show explicitly that the integral of $\vec{F} \cdot \vec{v}$ over the time that the cannonball spends in the air is equal to the change in the kinetic energy of the cannonball.

54 •• A particle of mass m moves from rest at $t = 0$ under the influence of a single constant force $\vec{F}$. Show that the power delivered by the force at time t is $P = F^2 t/m$.

Potential Energy

55 • **ISOLVE✓** An 80-kg man climbs up a 6-m high flight of stairs. What is the increase in the gravitational potential energy of the man–earth system?

56 • **ISOLVE** Water flows over Victoria Falls, which is 128 m high, at an average rate of 1.4×10^6 kg/s. If half the potential energy of this water were converted into electric energy, how much power would be produced by these falls?

57 • **ISOLVE** A 2-kg box slides down a long, frictionless incline of angle 30°. It starts from rest at time $t = 0$ at the top of the incline at a height of 20 m above the ground. (a) What is the original potential energy of the box relative to the ground? (b) From Newton's laws, find the distance the box travels during the interval $0 < t < 1$ s and its speed at $t = 1$ s. (c) Find the potential energy and the kinetic energy of the box at $t = 1$ s. (d) Find the kinetic energy and the speed of the box just as it reaches the bottom of the incline.

58 • A force $F_x = 6$ N is constant. (a) Find the potential-energy function $U(x)$ associated with this force for an arbitrary reference position x_0 at which $U = 0$. (b) Find $U(x)$ such that $U = 0$ at $x = 4$ m. (c) Find $U(x)$ such that $U = 14$ J at $x = 6$ m.

59 • **ISOLVE✓** A spring has a force constant $k = 10^4$ N/m. How far must it be stretched for its potential energy to be (a) 50 J and (b) 100 J?

60 •• **SSM** **ISOLVE** A simple Atwood's machine (Figure 6-38) uses two masses m_1 and m_2. Starting from rest, the speed of the two masses is 4.0 m/s at the end of 3.0 s. At that time, the kinetic energy of the system is 80 J and each mass has moved a distance of 6.0 m. Determine the values of m_1 and m_2.

FIGURE 6-38 Problem 60

m_1

m_2

61 •• A straight rod of negligible mass is mounted on a frictionless pivot as shown in Figure 6-39. Masses m_1 and m_2 are attached to the rod at distances ℓ_1 and ℓ_2. (a) Write an expression for the gravitational potential energy of the masses as a function of the angle θ made by the rod and the horizontal. (b) For what angle θ is the potential energy a minimum? Is the statement "systems tend to move toward a configuration of minimum potential energy" consistent with your result? (c) Show that if $m_1\ell_1 = m_2\ell_2$, the potential energy is the same for all values of θ. (When this holds, the system will balance at any angle θ. This result is known as *Archimedes' law of the lever*.)

FIGURE 6-39 Problem 61

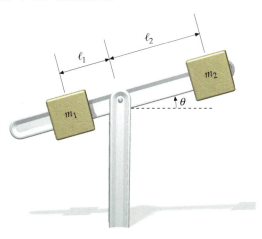

Force, Potential Energy, and Equilibrium

62 • (a) Find the force F_x associated with the potential-energy function $U = Ax^4$, where A is a constant. (b) At what point(s) is the force zero?

63 •• A potential-energy function is given by $U = C/x$, where C is a positive constant. (a) Find the force F_x as a function of x. (b) Is this force directed toward the origin or away from it? (c) Does the potential energy increase or decrease as x increases? (d) Answer parts (b) and (c) where C is a negative constant.

64 •• SSM On the potential-energy curve for U versus y shown in Figure 6-40, the segments AB and CD are straight lines. Sketch a plot of the force F_y versus y.

FIGURE 6-40 Problem 64

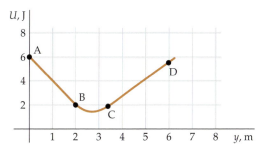

65 •• The force acting on an object is given by $F_x = a/x^2$. Determine the potential energy of the object as a function of x.

66 •• The potential energy of an object is given by $U(x) = 3x^2 - 2x^3$, where U is in joules and x is in meters. (a) Determine the force acting on this object. (b) At what positions is this object in equilibrium? (c) Which of these equilibrium positions are stable and which are unstable?

67 •• The potential energy of an object is given by $U(x) = 8x^2 - x^4$, where U is in joules and x is in meters. (a) Determine the force acting on this object. (b) At what positions is this object in equilibrium? (c) Which of these equilibrium positions are stable and which are unstable?

68 •• The force acting on an object is given by $F(x) = x^3 - 4x$. Locate the positions of unstable and stable equilibrium and show that at these points $U(x)$ is a local maximum or minimum, respectively.

69 •• The potential energy of a 4-kg object is given by $U = 3x^2 - x^3$ for $x \le 3$ m and $U = 0$ for $x \ge 3$ m, where U is in joules and x is in meters. (a) At what positions is this object in equilibrium? (b) Sketch a plot of U versus x. (c) Discuss the stability of the equilibrium for the values of x found in (a). (d) If the total energy of the particle is 12 J, what is its speed at $x = 2$ m?

70 •• A force is given by $F_x = Ax^{-3}$, where $A = 8$ N·m³. (a) For positive values of x, does the potential energy associated with this force increase or decrease with increasing x? (You can determine the answer to this question by imagining what happens to a particle that is placed at rest at some point x and is then released.) (b) Find the potential-energy function U associated with this force such that U approaches zero as x approaches infinity. (c) Sketch U versus x.

71 ••• SSM A novelty desk clock is shown in Figure 6-41: The clock (which has mass m) is supported by two light cables running over the two pulleys, which are attached to counterweights that each have mass M. (a) Find the potential energy of the system as a function of the distance y. (b) Find the value of y for which the potential energy of the system is a minimum. (c) If the potential energy is a minimum, then the system is in equilibrium. Apply Newton's second law to the clock and show that it is in equilibrium (the forces on it sum to zero) for the value of y obtained for part (b). Is this a point of stable or unstable equilibrium?

FIGURE 6-41 Problem 71

General Problems

72 • SSM In February 2002, a total of 60.7 billion kW·h of electrical energy was generated by nuclear power plants in the United States. At that time, the population of the United States was about 287 million people. If the average American has a mass of 60 kg, and if the entire energy output of all nuclear power plants was diverted to supplying energy for a single giant elevator, estimate the height h to which the entire population of the country could be lifted by the elevator. In your calculations, assume that 25 percent of the energy goes into lifting the people; assume also that g is constant over the entire height h.

73 • **ISOLVE** ✔ One of the most powerful cranes in the world, operating in Switzerland, can slowly raise a load of $M = 6000$ tonne to a height of $h = 12.0$ m (1 tonne $= 1000$ kg). (*a*) How much work is done by the crane? (*b*) If it takes 1.00 min to lift the load at constant velocity to this height, find the power developed by the crane.

74 • **ISOLVE** In Austria, there once was a ski lift of length 5.6 km. It took about 60 min for a gondola to travel all the way up. If there were 12 gondolas going up, each with a cargo of mass 550 kg, and if there were 12 empty gondolas going down, and the angle of ascent was 30°, estimate the power P of the engine needed to operate the ski lift.

75 • **ISOLVE** ✔ A 2.4-kg object attached to a horizontal string moves with constant speed in a circle of radius R on a frictionless horizontal surface. The kinetic energy of the object is 90 J and the tension in the string is 360 N. Find R.

76 • **SSM** The movie crew arrives in the Badlands ready to shoot a scene. The script calls for a car to crash into a vertical rock face at 100 km/h. Unfortunately, the car won't start, and there is no mechanic is sight. The crew are about to skulk back to the studio to face the producer's wrath when the cameraman gets an idea. They use a crane to lift the car by its rear end and then drop it vertically, filming at an angle that makes the car appear to be traveling horizontally. How high should the 800-kg car be lifted so that it reaches a speed of 100 km/h in the fall?

77 ••• The four strings pass over the bridge of a violin as shown in Figure 6-42. The angle that the strings make with the normal to the plane of the instrument is 72° on either side. The total normal force pressing the bridge into the violin is 103 N. The length of the strings from the bridge to the peg to which each is attached is 32.6 cm. (*a*) Determine the tension in the violin strings, assuming that the tension is the same for each string. (*b*) One of the strings is plucked out a distance of 4 mm, as shown in the figure. Make a free-body diagram showing all of the forces acting on the string at that point, and determine the force pulling the string back to its equilibrium position. Assume that the tension in the string remains constant. (*c*) Determine the work done on the string in plucking it out that distance. Remember that the net force pulling the string back to its equilibrium position is changing as the string is being pulled out, but assume that the magnitude of the tension in the string remains constant.

78 •• The force acting on a particle that is moving along the *x* axis is given by $F_x = -ax^2$, where *a* is a constant. Calculate the potential-energy function *U* relative to $U = 0$ at $x = 0$, and sketch a graph of *U* versus *x*.

79 •• **SSM** **ISOLVE** A horizontal force acts on a cart of mass *m* such that the speed *v* of the cart increases with distance *x* as $v = Cx$, where *C* is a constant. (*a*) Find the force acting on the cart as a function of position. (*b*) What is the work done by the force in moving the cart from $x = 0$ to $x = x_1$?

80 •• **ISOLVE** ✔ A force $\vec{F} = (2\ \text{N/m}^2)x^2\hat{i}$ is applied to a particle. Find the work done on the particle as it moves a total distance of 5 m (*a*) parallel to the *y* axis from point (2 m, 2 m) to point (2 m, 7 m) and (*b*) in a straight line from (2 m, 2 m) to (5 m, 6 m).

81 •• **ISOLVE** A particle of mass *m* moves along the *x* axis. Its position varies with time according to $x = 2t^3 - 4t^2$, where *x* is in meters and *t* is in seconds. Find (*a*) the velocity and acceleration of the particle at any time *t*, (*b*) the power delivered to the particle at any time *t*, and (*c*) the work done by the force from $t = 0$ to $t = t_1$.

82 •• **ISOLVE** A 3-kg particle starts from rest at $x = 0$ and moves under the influence of a single force $F_x = 6 + 4x - 3x^2$, where F_x is in newtons and *x* is in meters. (*a*) Find the work done by the force as the particle moves from $x = 0$ to $x = 3$ m. (*b*) Find the power delivered to the particle when it is at $x = 3$ m.

83 •• **SSM** **ISOLVE** ✔ The initial kinetic energy imparted to a 0.020-kg bullet is 1200 J. Neglecting air resistance, find the range of this projectile when it is fired at an angle such that the range equals the maximum height attained.

84 •• **ISOLVE** ✔ A force F_x acting on a particle is shown as a function of *x* in Figure 6-43. (*a*) From the graph, calculate the work done by the force when the particle moves from $x = 0$ to the following values of *x*: $-4, -3, -2, -1, 0, 1, 2, 3,$ and 4 m. (*b*) Plot the potential energy *U* versus *x* for the range of values of *x* from -4 m to $+4$ m, assuming that $U = 0$ at $x = 0$.

FIGURE 6-43 Problem 84

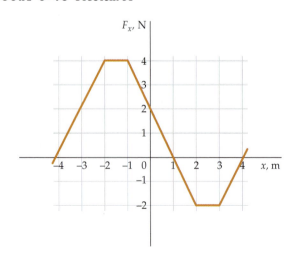

FIGURE 6-42 Problem 77

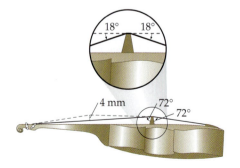

85 •• Repeat Problem 84 for the force F_x shown in Figure 6-44.

FIGURE 6-44 Problem 85

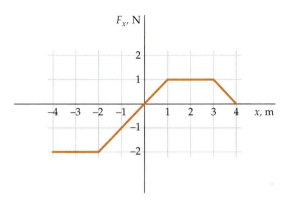

86 •• **ISOLVE** A box of mass M is at rest at the bottom of a frictionless inclined plane (Figure 6-45). The box is attached to a string that pulls with a constant tension T. (a) Find the work done by the tension T as the box moves through a distance x along the plane. (b) Find the speed of the box as a function of x and θ. (c) Determine the power produced by the tension in the string as a function of x and θ.

FIGURE 6-45 Problem 86

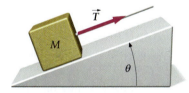

87 ••• A force in the xy plane is given by $\vec{F} = (F_0/r)(y\hat{i} - x\hat{j})$, where F_0 is a constant and $r = \sqrt{x^2 + y^2}$. (a) Show that the magnitude of this force is F_0 and that its direction is perpendicular to $\vec{r} = x\hat{i} + y\hat{j}$. (b) Find the work done by this force on a particle that moves once around a circle of radius 5 m centered at the origin. Is this force conservative?

88 ••• **SSM** A force in the xy plane is given by:

$F = -(b/r^3)(x\hat{i} + y\hat{j})$, where b is a positive constant and $r = \sqrt{x^2 + y^2}$. (a) Show that the magnitude of the force varies as the inverse of the square of the distance to the origin, and that its direction is antiparallel (opposite) to the radius vector $\vec{r} = x\hat{i} + y\hat{j}$. (b) If $b = 3$ N·m^2, find the work done by this force on a particle moving along a straight-line path between an initial position $x = 2$ m, $y = 0$ m and a final position $x = 5$ m, $y = 0$ m. (c) Find the work done by this force on a particle moving once around a circle of radius $r = 7$ m centered around the origin. (d) If this force is the only force acting on the particle, what is the particle's speed as it moves along this circular path? Assume that the particle's mass is $m = 2$ kg.

89 ••• In the words of Richard Feynman, one of the greatest physicists of the twentieth century, "If, in some cataclysm, all of scientific knowledge were to be destroyed, and only one sentence passed on to the next generation, . . . what statement would contain the most information in the fewest words? I believe it is . . . that all things are made of atoms—little particles . . . attracting each other when they are a little distance apart, but repelling upon being squeezed into each other."[†] When looking at the forces between atoms, modern physicists and chemists often model their interactions by the so-called "6–12" potential, where the potential energy function between two atoms is given by the functional form

$$U(r) = \frac{a}{r^{12}} - \frac{b}{r^6}$$

where r is the distance between atomic nuclei and a and b are constants that can be determined spectroscopically. Because they do not form atomic bonds, noble gas atoms have potential-energy functions that can be modeled well by a 6–12 potential and measured reasonably accurately. For argon, these parameters are $a = 1.09\times10^{-7}$ and $b = 6.84\times10^{-5}$, where r is measured in nm (1 nm = 10^{-9} m) and U is measured in eV (1 eV = 1.6×10^{-19} J.) (a) Using a spreadsheet program, make a graph of the interatomic potential energy function for two argon atoms as a function of separation, for values of r between 0.3 nm and 0.7 nm. Does the shape of the potential energy function support Feynman's claim? Explain. (b) What is the minimum value of the potential energy (compared to where the atoms are separated by a very large distance)? At what separation does this occur? Is this minimum a point of stable or unstable equilibrium? (c) From either the graph or the formula given above, estimate the force of attraction between two argon atoms separated by a distance of 5 Å and the force of repulsion for a separation of 3.5 Å. Make sure that you convert to MKS units!

90 ••• **SSM** A theoretical formula for the potential energy associated with the nuclear force between two protons, two neutrons, or a neutron and a proton is the Yukawa potential $U(r) = -U_0(a/r)e^{-r/a}$, where U_0 and a are constants, and r is the separation between the two nucleons. (a) Using a spreadsheet program such as Microsoft Excel™, make a graph of U versus r, using $U_0 = 4$ pJ (a picojoule, pJ, is 1×10^{-12} J) and $a = 2.5$ fm (a femtometer, fm, is 1×10^{-15} m). (b) Find the force $F(r)$ as a function of the separation of the two nucleons. (b) Find the force $F(r)$ as a function of the separation of the two nucleons. (c) Compare the magnitude of the force at the separation $r = 2a$ to that at $r = a$. (d) Compare the magnitude of the force at the separation $r = 5a$ to that at $r = a$.

† Richard P. Feynman, Matthew L. Sands, and Robert B. Leighton, *The Feynman Lectures on Physics*, Vol. 1, p. 1.1. Boston: Addison-Wesley (1970).

Conservation of Energy

7-1   The Conservation of Mechanical Energy

7-2   The Conservation of Energy

7-3   Mass and Energy

7-4   Quantization of Energy

ENERGY CAN BE CONVERTED FROM ONE FORM INTO ANOTHER. THIS ROLLER COASTER CONVERTS ELECTRICAL ENERGY PURCHASED FROM THE POWER COMPANY INTO GRAVITATIONAL POTENTIAL ENERGY OF THE PASSENGERS AND THEIR VEHICLES. THEN GRAVITATIONAL POTENTIAL ENERGY IS CONVERTED INTO KINETIC ENERGY, SOME OF WHICH IS CONVERTED BACK TO POTENTIAL ENERGY. EVENTUALLY FRICTION TRANSFORMS THE KINETIC AND POTENTIAL ENERGY INTO THERMAL ENERGY. CHEMICAL ENERGY STORED IN THE MUSCLES OF THE PASSENGERS IS CONVERTED INTO SOUND ENERGY EACH TIME SOMEONE SCREAMS.

? **How does conservation of mechanical energy allow us to determine how high up the cars must be when they start their descent, in order to complete the loop-the-loop? (See Example 7-5.)**

A system is a collection of particles. External forces are any forces exerted by particles not in the system on particles that are in the system, and internal forces are those exerted by particles in the system on other particles in the system. The many ways there are to change the total energy of the system can be divided into just two categories, work and heat. Total energy can change if external forces do work on the system or if energy is transferred because of a temperature difference between the system and its surroundings. (Energy transferred due to a temperature difference is called heat.) Because we do not analyze systems where heat plays a significant role until Chapter 18, here we will consider the change in energy of the system to be equal to the total work done on it by external forces. We define the potential energy of a system so that the change in the potential energy of the system equals the negative of the total work done by all internal conservative forces. If no external forces do work on the system, and if the internal conservative forces are the only internal forces that do work, then the work they do equals the change in the system's kinetic energy. Because the change in the system's potential energy equals the negative of the change in its kinetic energy, the sum of the potential and kinetic energies does not change. This is known as the conservation of mechanical energy. It follows from Newton's laws and is a useful alternative to them for solving many problems in mechanics.

The usefulness of mechanical energy conservation is limited by the presence of nonconservative forces such as friction. When kinetic friction is present in a system, the mechanical energy of the system does not stay the same but instead

decreases.[†] Because mechanical energy is often not conserved, the importance of energy conservation was not realized until the nineteenth century, when it was discovered that the disappearance of macroscopic mechanical energy is always accompanied by the appearance of some other kind of energy, often thermal energy, which is usually indicated by an increase in temperature or a change in phase (like the melting of ice). We now know that, on the microscopic scale, this thermal energy is associated with the kinetic and potential energies at the molecular level.

There are a number of different forms of energy, such as chemical energy, sound energy, electromagnetic energy, and nuclear energy. Whenever the energy of a system changes, we can account for the change by the appearance or disappearance of energy somewhere else. This experimental observation is the **law of conservation of energy,** one of the most fundamental and important laws in all of science. Although energy changes from one form to another, it is never created or destroyed.

➤ **In this chapter, we continue the study of energy begun in Chapter 6 by describing and applying the law of conservation of energy and examining different types of energy, including thermal energy. We also discuss Einstein's famous relation between mass and energy, and discover that energy changes for a system are not continuous, but occur in discrete "bundles" or "lumps" called quanta. Although in a macroscopic system the quantum of energy typically is so small that it goes unnoticed, its presence has profound consequences for microscopic systems such as atoms and molecules.**

7-1 The Conservation of Mechanical Energy

The total work done on each particle in a system equals the change in the kinetic energy ΔK_i of that particle, so the total work done by all the forces W_{total} equals the change in the total kinetic energy of the system ΔK_{sys}:

$$W_{total} = \sum \Delta K_i = \Delta K_{sys} \qquad \text{7-1}$$

Each internal force is either conservative or nonconservative. The negative of the total work done by all the conservative internal forces $-W_c$ equals the change in the potential energy of the system ΔU_{sys}:

$$-W_c = \Delta U_{sys} \qquad \text{7-2}$$

The total work done by all forces equals the work done by all external forces W_{ext}, plus the work done by all internal nonconservative forces W_{nc}, plus that done by all internal conservative forces W_c:

$$W_{total} = W_{ext} + W_{nc} + W_c$$

Rearranging gives $W_{ext} + W_{nc} = W_{total} - W_c$. Substituting from Equations 7-1 and 7-2 we have

$$W_{ext} + W_{nc} = \Delta K_{sys} + \Delta U_{sys} \qquad \text{7-3}$$

The right side of this equation can be simplified as

$$\Delta K_{sys} + \Delta U_{sys} = \Delta(K_{sys} + U_{sys}) \qquad \text{7-4}$$

The sum of the kinetic energy K_{sys} and the potential energy U_{sys} is called the **total mechanical energy** E_{mech}:

---

† Other factors in the system, such as a battery that runs a motor, can cause the mechanical energy to increase, so the presence of kinetic friction does not necessarily mean that the mechanical energy of the system will decrease.

$$E_{mech} = K_{sys} + U_{sys} \qquad\qquad 7\text{-}5$$

Combining Equations 7-4 and 7-5, and then substituting into Equation 7-3 gives

$$W_{ext} = \Delta E_{mech} - W_{nc}$$

The mechanical energy of a system is conserved (E_{mech} = constant) if the total work done by all external forces and by all internal nonconservative forces is zero.

$$E_{mech} = K_{sys} + U_{sys} = \text{constant} \qquad\qquad 7\text{-}6$$

This is **conservation of mechanical energy** and is the origin of the expression "conservative force."

If $E_{mech\,i} = K_i + U_i$ is the initial mechanical energy of the system and $E_{mech\,f} = K_f + U_f$ is the final mechanical energy of the system, conservation of mechanical energy implies that

$$E_{mech\,f} = E_{mech\,i} \qquad (\text{or } K_f + U_f = K_i + U_i) \qquad 7\text{-}7$$

Many mechanics problems can be solved by setting the final mechanical energy of a system equal to its initial mechanical energy.

Applications

Suppose that you are wearing skis and, starting at rest from a height h_0 above the bottom of a hill, you coast down the hill. Assuming that the skis are frictionless, what is your speed as you pass through a gate at height h above the bottom of the hill? The mechanical energy of the earth–skier system is conserved because the only force doing work is the internal, conservative force of gravity. If we choose $U = 0$ at the bottom of the hill, the initial potential energy is mgh_0. This is also the total mechanical energy because the initial kinetic energy is zero. Thus,

$$E_{mech\,i} = K_i + U_i = 0 + mgh_0$$

At height h, the potential energy is mgh and the speed is v. Hence,

$$E_{mech\,f} = K_f + U_f = \tfrac{1}{2}mv^2 + mgh$$

Setting $E_{mech\,f} = E_{mech\,i}$ we find

$$\tfrac{1}{2}mv^2 + mgh = mgh_0$$

or

$$v = \sqrt{2g(h_0 - h)}$$

Your speed is the same as if you had undergone free-fall through a distance $h_0 - h$. However, by skiing down the hill, you travel farther and take more time than you would if you were in free-fall and falling straight down.

Multiflash photograph of a simple pendulum. As the bob descends, gravitational potential energy is converted into kinetic energy, and the speed increases as indicated by the increased spacing of the recorded positions. The speed decreases as the bob moves up, and the kinetic energy is changed into potential energy.

KICKING A BALL   **EXAMPLE   7 - 1**

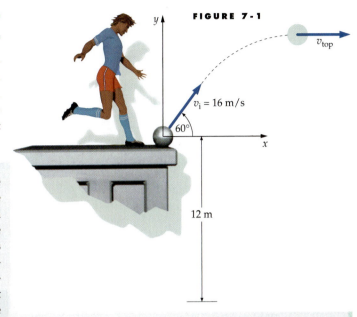

FIGURE 7-1

Standing near the edge of the roof of a 12-m-high building, you kick a ball with an initial speed of $v_i = 16$ m/s at an angle of 60° above the horizontal (Figure 7-1). Neglecting air resistance, find (a) how high above the height of the building the ball rises and (b) its speed just before it hits the ground.

PICTURE THE PROBLEM We choose the ball and the earth as the system. We consider this system during the interval from just after the kick to just before impact with the ground. There are no external forces doing work on the system, and there are no internal nonconservative forces doing work, so the mechanical energy of the system is conserved. At the top of its flight, the ball is moving horizontally with a speed v_{top}, equal to the horizontal component of its initial horizontal velocity v_{ix}. We choose $y = 0$ at the roof of the building.

(a) 1. Conservation of mechanical energy relates the height h above the top of the building to the initial speed v_i and the speed at the top of its flight v_{top}:

$$E_{mech\ top} = E_{mech\ i}$$
$$\tfrac{1}{2}mv_{top}^2 + mgy_{top} = \tfrac{1}{2}mv_i^2 + mgy_i$$
$$\tfrac{1}{2}mv_{top}^2 + mgh_{top} = \tfrac{1}{2}mv_i^2 + 0$$

2. Solve for h_{top}:

$$h_{top} = \frac{v_i^2 - v_{top}^2}{2g}$$

3. The velocity at the top of its flight equals its initial horizontal velocity:

$$v_{top} = v_{ix} = v_i \cos\theta$$

4. Substitute the step 3 result into the step 2 result and solve for h_{top}:

$$h_{top} = \frac{v_i^2 - v_{top}^2}{2g} = \frac{v_i^2 - v_i^2\cos^2\theta}{2g} = \frac{v_i^2(1 - \cos^2\theta)}{2g}$$

$$= \frac{(16\ \text{m/s})^2\,(1 - \cos^2 60°)}{2(9.81\ \text{m/s}^2)} = \boxed{9.79\ \text{m}}$$

(b) 1. If v_f is the speed of the ball just before it hits the ground, where $y = y_f = -12$ m, its energy is:

$$E_{mech\ f} = \tfrac{1}{2}mv_f^2 + mgy_f$$

2. Set the final mechanical energy equal to the initial mechanical energy:

$$\tfrac{1}{2}mv_f^2 + mgy_f = \tfrac{1}{2}mv_i^2 + 0$$

3. Solve for v_f, and set $y = -12$ m to find the final velocity:

$$v_f = \sqrt{v_i^2 - 2gy_f}$$

$$= \sqrt{(16\ \text{m/s})^2 - 2(9.81\ \text{m/s}^2)(-12\ \text{m})} = \boxed{22.2\ \text{m/s}}$$

A PENDULUM   **EXAMPLE   7 - 2**

A pendulum consists of a bob of mass m attached to a string of length L. The bob is pulled aside so that the string makes an angle θ_0 with the vertical, and is released from rest. As it passes through the bottom of the arc, find expressions for (a) the speed and (b) the tension. Air resistance is negligible.

PICTURE THE PROBLEM The system consists of the pendulum, the earth, and the pendulum mount and its supports. There are no external forces on this system. The two internal forces acting on the bob (air resistance being negligible) are the force of gravity $m\vec{g}$, which is conservative, and the tension force $\vec{T}$. The rate at which $\vec{T}$ does work is $\vec{T} \cdot \vec{v}$. Because $\vec{T}$ is perpendicular to $\vec{v}$, we know that $\vec{T} \cdot \vec{v} = 0$. Since only $m\vec{g}$ does work, we know that the mechanical energy of the bob–earth system is conserved. To find the speed of the bob, we equate the initial and final mechanical energies. The tension in the string is obtained using Newton's second law.

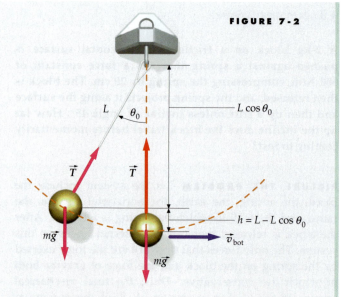

FIGURE 7-2

(a) 1. Make a sketch of the system in its initial and final configurations (Figure 7-2). We choose $y = 0$ at the bottom of the swing and $y = h$ at the initial position:

2. Apply conservation of mechanical energy. Initially the bob is at rest:

$$E_{\text{mech f}} = E_{\text{mech i}}$$

$$\tfrac{1}{2}mv_f^2 + mgy_f = \tfrac{1}{2}mv_i^2 + mgy_i$$

$$\tfrac{1}{2}mv_{\text{bot}}^2 + 0 = 0 + mgh$$

3. Conservation of mechanical energy thus relates the speed v_{bot} to the height h:

$$\tfrac{1}{2}mv_{\text{bot}}^2 = mgh$$

4. Solve for the speed v_{bot}:

$$v_{\text{bot}} = \sqrt{2gh}$$

5. To express speed in terms of the initial angle θ_0, we need to relate h to θ_0. This relation is illustrated in Figure 7-2:

$$L = L\cos\theta_0 + h$$

so

$$h = L - L\cos\theta_0 = L(1 - \cos\theta_0)$$

6. Substitute this value for h to express the speed at the bottom in terms of θ_0:

$$v_{\text{bot}} = \boxed{\sqrt{2gL(1 - \cos\theta_0)}}$$

(b) 1. When the bob is at the bottom of the circle, the forces on it are $m\vec{g}$ and $\vec{T}$. Apply $\Sigma F_y = ma_y$:

$$T - mg = ma_y$$

2. At the bottom, the bob has an acceleration v_{bot}^2/L toward the center of the circle, which is upward:

$$a_y = \frac{v_{\text{bot}}^2}{L} = \frac{2gL(1 - \cos\theta_0)}{L}$$

$$= 2g(1 - \cos\theta_0)$$

3. Substitute for a_y the Part (b) step 1 result and solve for T:

$$T = mg + ma_y = m(g + a_y)$$

$$= m[g + 2g(1 - \cos\theta_0)]$$

$$= \boxed{(3 - 2\cos\theta_0)mg}$$

REMARKS 1. The tension at the bottom is greater than the weight of the bob because the bob is accelerating upward. 2. Step 3 in Part (b) shows that for $\theta_0 = 0$, $T = mg$, the expected result for a stationary bob hanging from a string. 3. Step 4 in Part (a) shows that the speed at the bottom is the same as if the bob had dropped in free-fall from a height h. The speed of the bob at the bottom of the arc can also be found using Newton's laws directly (see Problem 7-92), but such a solution is more challenging because the acceleration component tangential to the circle varies with position, and therefore with time, so the constant-acceleration formulas do not apply.

EXAMPLE 7-3 Try It Yourself

A 2-kg block on a frictionless horizontal surface is pushed against a spring that has a force constant of 500 N/m, compressing the spring by 20 cm. The block is then released, and the spring projects it along the surface and then up a frictionless incline of angle 45°. How far up the incline does the block travel before momentarily coming to rest?

Answers

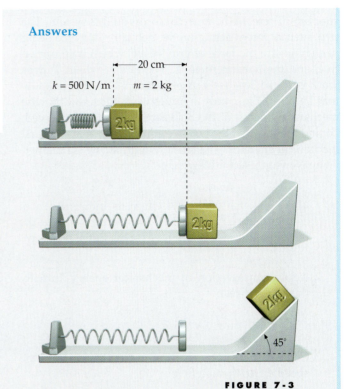

PICTURE THE PROBLEM Let the system include the block, the spring, the earth, the horizontal surface, the ramp, and the wall to which the spring is attached. After the block is released there are no external forces on this system. The only forces that do work are the force exerted by the spring on the block and the force of gravity, both of which are conservative. Thus, the total mechanical energy of the system is conserved. Find the maximum height h from the conservation of mechanical energy, and then the distance up the incline s from $\sin 45° = h/s$.

Cover the column to the right and try these on your own before looking at the answers.

FIGURE 7-3

Steps

1. Sketch the system with both the initial and final configurations (Figure 7-3).

2. Write the initial mechanical energy in terms of the compression distance x.

$$E_{\text{mech i}} = U_{\text{s i}} + U_{\text{g i}} + K_i = \tfrac{1}{2}kx^2 + 0 + 0$$

3. Write the final mechanical energy in terms of the height h.

$$E_{\text{mech f}} = U_{\text{s f}} + U_{\text{g f}} + K_f = 0 + mgh + 0$$

4. Apply conservation of mechanical energy and solve for h.

$$mgh = \tfrac{1}{2}kx^2$$

$$h = \frac{kx^2}{2mg} = 0.51 \text{ m}$$

5. Find the distance s from h and the angle of inclination.

$$h = s \sin \theta$$

$$s = \boxed{0.72 \text{ m}}$$

REMARKS In this problem, the initial mechanical energy of the system is the potential energy of the spring. This energy is converted first into kinetic energy and then into gravitational potential energy.

EXERCISE Find the speed of the block just after it leaves the spring. (*Answer* 3.16 m/s)

EXAMPLE 7-4

A spring with a force constant of k hangs vertically. A block of mass m is attached to the unstretched spring and allowed to fall from rest. Find an expression for the maximum distance the block falls before it begins moving upward.

PICTURE THE PROBLEM As the block drops, its speed first increases, then reaches some maximum value, and then decreases until it is again zero when the block is at its lowest point. There are only conservative forces present, so we apply the conservation of mechanical energy to the earth–spring–block system.

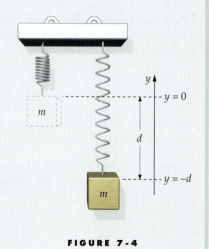

1. Sketch the system showing the initial and final positions of the block and spring (Figure 7-4). Include a y axis with up as the positive y direction. Choose the gravitational potential energy of the system to be zero at the original position $y = 0$ when the spring is unstretched. The potential energy of the spring is zero when the spring is unstretched. Let d be the distance the block falls.

2. Apply conservation of mechanical energy to relate the initial ($y = 0$) and final ($y = -d$) positions:

$$mgy_f + \tfrac{1}{2}ky_f^2 + \tfrac{1}{2}mv_f^2 = mgy_i + \tfrac{1}{2}ky_i^2 + \tfrac{1}{2}mv_i^2$$

$$mg(-d) + \tfrac{1}{2}k(-d)^2 + 0 = 0 + 0 + 0$$

FIGURE 7-4

3. Solve for d. There are two solutions. One gives $d = 0$ and the other is the solution we want—the maximum distance the block falls before it begins moving upward:

$$\tfrac{1}{2}kd^2 - mgd = 0$$

$$(\tfrac{1}{2}kd - mg)d = 0$$

Since $d \neq 0$, $\boxed{d = \dfrac{2mg}{k}}$

REMARKS Gravitational potential energy is converted into the kinetic energy of the block plus the potential energy of the spring. At the lowest point, where the block is momentarily at rest, the elastic potential energy of the spring equals the decrease in the gravitational potential energy of the system.

BACK TO THE FUTURE   **EXAMPLE 7-5**   **Put It in Context**

Imagine that you have time-traveled back to the late 1800s and are watching your great-great-great-grandparents on their honeymoon taking a ride on the **Flip Flap Railway,** a Coney Island roller coaster with a circular loop-the-loop. The car they are in is about to enter the loop-the-loop when a 100-lb sack of sand falls from a construction-site platform and lands in the back seat of the car. No one is hurt, but the impact causes the car to lose 25 percent of its speed. The car started from rest at a high point 2 times as high as the top of the loop-the-loop. Neglect any losses due to either friction or air drag. Will their car make it over the top of the loop-the-loop without falling off?

PICTURE THE PROBLEM The car has to have enough speed at the top of the loop-the-loop to maintain contact with the track. We can use conservation of mechanical energy to determine the speed just before the sandbag hits the car, and we can use it again to determine the speed it has at the top of the track. Then we can use Newton's second law to determine the normal force exerted on the car by the track.

FIGURE 7-5

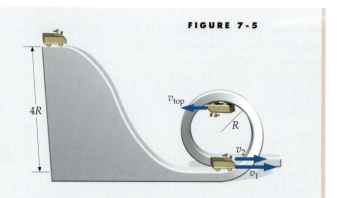

1. Draw a picture of the car and track, with the car at the starting point, at the bottom of the track, and at the top of the loop-the-loop (Figure 7-5):

2. Apply Newton's second law to relate the speed at the top of the loop-the-loop to the normal force:

$$F_n + mg = m\frac{v_{top}^2}{R}$$

3. Using conservation of mechanical energy, find the speed just prior to impact. The initial height is $4R$, where R is the radius of the loop-the-loop:

$$U_1 + K_1 = U_0 + K_0$$
$$0 + \tfrac{1}{2}mv_1^2 = mg4R + 0$$

so

$$v_1 = \sqrt{8Rg}$$

4. The impact with the sandbag results in a 25 percent decrease in speed. Find the speed after impact:

$$v_2 = 0.75v_1 = 0.75\sqrt{8Rg}$$

5. Using conservation of mechanical energy, find the speed just at the top of the loop-the-loop:

$$U_{\text{top}} + K_{\text{top}} = U_2 + K_2$$
$$mg\,2R + \tfrac{1}{2}mv_{\text{top}}^2 = 0 + \tfrac{1}{2}m(0.75^2\,8Rg)$$

so

$$v_{\text{top}}^2 = (0.75^2\,8 - 4)Rg = 0.5Rg$$

6. Substituting for v_{top}^2 in the step 2 result gives:

$$F_n + mg = 0.5mg$$

7. Solve for F_n:

$$F_n = -0.5mg$$

8. F_n is the magnitude of the normal force. It cannot be negative:

> Oops! The car has left the track.

REMARKS A loss of 25 percent of your speed means losing almost 44 percent of your kinetic energy. Fortunately, there were safety devices to prevent the cars from falling, so your ancestors likely would have survived. The biggest concern for riders on the Flip Flap Railway was a broken neck. In the loop-the-loop, the riders were subjected to up to $12gs$. This was the last of the circular loop-the-loop roller coasters. Loop-the-loops on more recent rides are oval shaped, higher than they are wide.

TWO BLOCKS ON A STRING **EXAMPLE 7-6** Try It Yourself

Two blocks are attached to a light string that passes over a massless, friction-less pulley. The two blocks have masses m_1 and m_2, where $m_2 > m_1$, and are initially at rest. Find the speed of either block when m_2 falls a distance h. Find the magnitude of the acceleration of the blocks as m_2 falls.

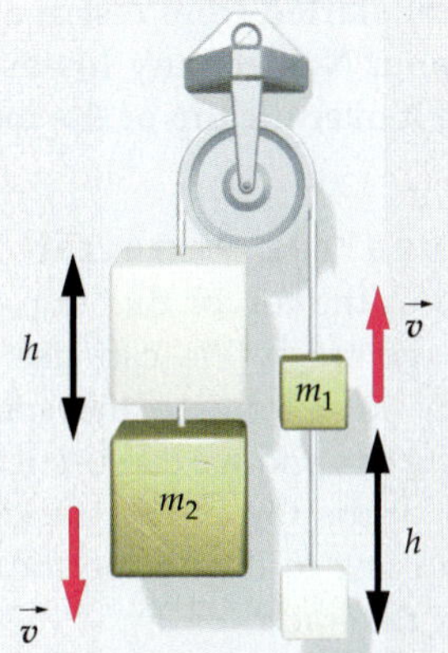

PICTURE THE PROBLEM Let the system be the two blocks, the string, the pulley, its supports, and the earth. There are no external forces on this system, hence no work is done on the system by external forces. Also, there is no kinetic friction. Thus the mechanical energy of the system is conserved.

Cover the column to the right and try these on your own before looking at the answers.

Steps

1. Make a sketch of the system showing it in both its initial and final configurations (Figure 7-6). Let h be the distance m_2 falls.

Answers

2. Initially, the total kinetic energy of the system is zero. Write the total kinetic energy K of the system when the blocks are moving with speed v.

$$K = \tfrac{1}{2}m_1v^2 + \tfrac{1}{2}m_2v^2 = \tfrac{1}{2}(m_1 + m_2)v^2$$

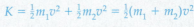

FIGURE 7-6

3. Initially, choose the total gravitational potential energy of the system U to be zero. Write an expression for U when m_1 has moved up a distance h (and m_2 has moved down the same distance).

$$U = m_1 gh - m_2 gh = (m_1 - m_2)gh$$

4. Add U and K to obtain the total mechanical energy E.

$$E = K + U$$
$$= \tfrac{1}{2}(m_1 + m_2)v^2 + (m_1 - m_2)gh$$

5. Apply conservation of mechanical energy.

$$E = E_i$$
$$\tfrac{1}{2}(m_1 + m_2)v^2 + (m_1 - m_2)gh = 0$$

6. Solve for v.

$$v = \sqrt{\frac{2(m_2 - m_1)}{(m_1 + m_2)}gh}$$

7. Find the acceleration a by dv/dt. Use the step 6 result and the chain rule for differentiation. Note that $v = dh/dt$.

$$a = \frac{dv}{dt} = \frac{dv}{dh}\frac{dh}{dt} = \frac{(m_2 - m_1)}{(m_1 + m_2)}g$$

❶ **PLAUSIBILITY CHECK** This device, called an *Atwood's machine,* is analyzed in terms of forces in Problems 84–90 in Chapter 4. There we see that the acceleration is given by the step 7 result.

EXERCISE What is the magnitude of the acceleration of either block if the masses are $m_1 = 3$ kg and $m_2 = 5$ kg? (*Answer* $a = 0.25g = 2.45$ m/s^2)

We've seen that mechanical energy conservation can be used as an alternative to Newton's laws for solving certain problems in mechanics. When we are not interested in the time t, the conservation of mechanical energy is often much easier to use than Newton's second law (Figure 7-7). Because mechanical energy conservation is derived from Newton's laws, any problem that can be solved using it can also be solved directly from Newton's laws, though often with much more difficulty.

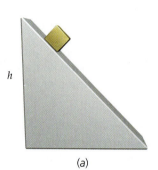

(a)

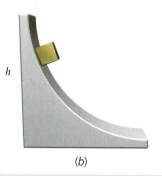

(b)

FIGURE 7-7 (*a*) One can easily find the speed of a block sliding down a frictionless incline of constant slope by applying Newton's second law or by using conservation of mechanical energy. However, if the incline is frictionless but not of constant slope, as in (*b*), the problem can still be solved easily using conservation of mechanical energy, whereas it can be solved using Newton's second law only if the slope of the incline is known at each point, and then the calculation is quite tedious.

7-2 The Conservation of Energy

In the macroscopic world, dissipative nonconservative forces, like kinetic friction, are always present to some extent. Such forces tend to decrease the mechanical energy of a system. However, any such decrease in mechanical energy is accompanied by a corresponding increase in thermal energy. (Consider how the tires of a car feel warm after a long trip.) Another type of nonconservative force is that involved in the deformations of objects. When you bend a metal coat hanger back and forth, you do work on the coat hanger but the work you do does not appear as mechanical energy. Instead the coat hanger becomes warm. The work done in deforming the hanger is dissipated as thermal energy. Similarly, when a falling ball of putty lands on the floor, it becomes warm as it deforms. The dissipated kinetic energy appears as thermal energy. For the ball–floor–earth system, the total energy is the sum of the thermal energy and the mechanical energy. The total energy of the system is conserved even though neither the total mechanical energy nor the total thermal energy are individually conserved.

A third type of nonconservative force is associated with chemical reactions. When we include systems in which chemical reactions take place, the sum of

mechanical energy plus thermal energy is not conserved. For example, suppose that you begin running from rest. Initially you have no kinetic energy. When you begin to run, internal chemical energy in your muscles is converted to kinetic energy of your body and thermal energy is produced. It is possible to identify and measure the chemical energy that is used. In this case, the sum of mechanical, thermal, and chemical energy is conserved.

Even when thermal energy and chemical energy are included, the total energy of the system does not always remain constant, because energy can be converted to radiation energy, such as sound waves or electromagnetic waves. But *the increase or decrease in the total energy of a system can always be accounted for by the appearance or disappearance of energy somewhere else.* This experimental result, known as the **law of conservation of energy,** is one of the most important laws in all of science. Let E_{sys} be the total energy of a given system, E_{in} be the energy that enters the system, and E_{out} be the energy that leaves the system. The law of conservation of energy then states

$$E_{in} - E_{out} = \Delta E_{sys} \qquad \qquad 7\text{-}8$$

LAW OF CONSERVATION OF ENERGY

Alternatively,

The total energy of the universe is constant. Energy can be converted from one form to another, or transmitted from one region to another, but energy can never be created or destroyed.

LAW OF CONSERVATION OF ENERGY

The total energy E of many systems familiar from everyday life can be accounted for completely by mechanical energy E_{mech}, thermal energy E_{therm}, and chemical energy E_{chem}. To be comprehensive and include other possible forms of energy, such as electromagnetic or nuclear energy, we include E_{other}, and write generally

$$E_{sys} = E_{mech} + E_{therm} + E_{chem} + E_{other} \qquad \qquad 7\text{-}9$$

The Work–Energy Theorem

A common way to transfer energy into or out of a system is to do work on the system from the outside. If this is the only method of energy transfer,[†] the law of conservation of energy becomes

$$W_{ext} = \Delta E_{sys} = \Delta E_{mech} + \Delta E_{therm} + \Delta E_{chem} + \Delta E_{other} \qquad 7\text{-}10$$

WORK–ENERGY THEOREM

where W_{ext} is the work done on the system by external forces and ΔE_{sys} is the change in the system's total energy. This work–energy theorem for systems, which we will call simply the work–energy theorem, is a powerful tool for studying a wide variety of systems. Note that if the system is just a single particle its energy can only be kinetic, so Equation 7-10 is equivalent to the work–kinetic energy theorem studied in Chapter 6.

EXPLORING

Devices that convert one form of energy to another are called transducers. To find out what your inner ear, a strain gauge, a microphone, and your sense of touch have in common, go to www.whfreeman.com/tipler5e.

---

† Energy can also be transferred when heat is exchanged between a system and its surroundings. Exchanges of heat energy, which occur when there is a temperature difference between a system and its surroundings, are discussed in Chapter 18.

FALLING CLAY   **E X A M P L E 7 - 7**

A ball of modeling clay with mass *m* is released from rest from a height *h* and falls to the perfectly rigid floor. Discuss the application of the law of conservation of energy to (*a*) the system consisting of the ball alone, and (*b*) the system consisting of the earth, the floor, and the ball.

PICTURE THE PROBLEM There are two forces that act on the ball: gravity and the contact force of the floor. Since the floor does not move, the contact force it exerts on the ball does no work. There are no chemical or other energy changes so we can neglect ΔE_{chem} and ΔE_{other}. We also neglect the sound energy radiated when the ball hits the floor. Thus, the only energy transferred to or from the ball is the work done by the force of gravity. Apply the work–energy theorem.

(*a*) 1. Write the work–energy theorem for the ball:

$$W_{ext} = \Delta E_{sys} = \Delta E_{mech} + \Delta E_{therm}$$

2. The two external forces on the system are gravity and the force exerted by the floor. The floor does not move and therefore does no work. The only work done on the ball is by gravity:

$$W_{ext} = mgh$$

3. Since the ball alone is our system, its mechanical energy is entirely kinetic, which is zero both initially and finally. Thus, the change in mechanical energy is zero:

$$\Delta E_{mech} = 0$$

4. Substitute *mgh* for W_{ext} and 0 for ΔE_{mech} in step 1:

$$W_{ext} = \Delta E_{mech} + \Delta E_{therm}$$

$$mgh = 0 + \Delta E_{therm}$$

so

$$\boxed{\Delta E_{therm} = mgh}$$

(*b*) 1. There are no external forces acting on the ball–earth–floor system (the force of gravity and the force of the floor are now internal to the system), so there is no external work done:

$$W_{ext} = 0$$

2. Write the work–energy theorem with $W_{ext} = 0$:

$$W_{ext} = \Delta E_{sys} = \Delta E_{mech} + \Delta E_{therm}$$

$$0 = \Delta E_{mech} + \Delta E_{therm}$$

3. The original mechanical energy of the ball–earth system is the original gravitational potential energy, and the final mechanical energy is zero:

$$E_{mech\,i} = mgh$$

$$E_{mech\,f} = 0$$

4. The change in mechanical energy of the ball–earth system is thus:

$$\Delta E_{mech} = 0 - mgh = -mgh$$

5. The work–energy theorem thus gives the same result as in Part (*a*):

$$\Delta E_{therm} = \boxed{-\Delta E_{mech} = mgh}$$

REMARKS In (*a*), energy is transferred to the ball by the work done on it by gravity. This energy appears as the kinetic energy of the ball before it hits the floor and as thermal energy after. The ball warms slightly and the energy is eventually transferred to the surroundings via heat. In (*b*), no energy is transferred to the ball–earth–floor system. The original potential energy of the system is converted to kinetic energy of the ball just before it hits, and then into thermal energy.

In this power plant in Kansas, energy stored in the fossil fuel coal (black area at the right) is released by burning the coal to produce steam; the steam is then used to drive turbines to produce electricity. The excess heat is dissipated by the cooling towers.

The potential energy of the water at the top of Niagara Falls is used to produce electrical energy.

Problems Involving Kinetic Friction

When surfaces slide across each other kinetic friction decreases the mechanical energy of the system and increases the thermal energy. Consider a block that begins with initial velocity v_i and slides along a table until it stops (Figure 7-8). We choose the block and table to be our system. Then $\Delta E_{chem} = \Delta E_{other} = 0$ and there is no external work done on this system. The work–energy theorem gives

$$0 = \Delta E_{mech} + \Delta E_{therm}$$

The mechanical energy lost is the initial kinetic energy of the block

$$\Delta E_{mech} = -\tfrac{1}{2}mv_i^2 \qquad\qquad 7\text{-}11$$

We can relate the loss in mechanical energy to frictional force. If f is the magnitude of the frictional force, Newton's second law gives

$$-f = ma$$

Multiplying both sides of this equation by Δs, we find

$$-f\Delta s = ma\,\Delta s = m(\tfrac{1}{2}v_f^2 - \tfrac{1}{2}v_i^2) = -\tfrac{1}{2}mv_i^2 \qquad\qquad 7\text{-}12$$

where Δs is the displacement of the block, and we have used the constant-acceleration formula $2a\,\Delta s = v_f^2 - v_i^2$ (with $v_f = 0$). Equating the left sides of Equations 7-11 and 7-12 we obtain

$$f\,\Delta s = -\Delta E_{mech} \qquad\qquad 7\text{-}13$$

Note that the quantity $-f\,\Delta s$ is *not* the work done by friction on the sliding block, because careful analysis shows the actual displacement of the kinetic frictional

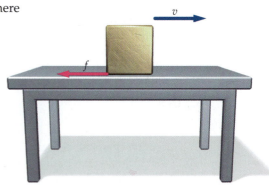

FIGURE 7-8 A block sliding on a table. The force of friction reduces the mechanical energy of the block–table system.

force on the block is not in general equal to the displacement of the block.[†] However, it can be shown that $f\Delta s$ does equal the increase in thermal energy of the block–table system due to the dissipation of the system's mechanical energy. This thermal energy is produced both on the bottom surface of the block and on the upper surface of the tabletop as the block slides across the tabletop. Thus,

$$f\,\Delta s = \Delta E_{\text{therm}} \qquad\qquad 7\text{-}14$$

ENERGY DISSIPATED BY FRICTION

Substituting this result into the work–energy theorem (with $E_{\text{chem}} = E_{\text{other}} = 0$), we obtain

$$W_{\text{ext}} = \Delta E_{\text{mech}} + \Delta E_{\text{therm}} = \Delta E_{\text{mech}} + f\,\Delta s \qquad\qquad 7\text{-}15$$

WORK–ENERGY THEOREM FOR PROBLEMS WITH KINETIC FRICTION

where Δs is the distance one contacting surface slides relative to the other. When there is no external work done on the system, the energy dissipated by friction equals the decrease in mechanical energy:

$$\Delta E_{\text{therm}} = f\,\Delta s = -\Delta E_{\text{mech}} \qquad (W_{\text{ext}} = 0) \qquad\qquad 7\text{-}16$$

Irish wind farm now under construction. The Arklow Banks project in the Irish Sea will have 200 turbines and produce 520 MW of electrical power—three times the combined capacity of all offshore wind farms currently in production in the world.

PUSHING A BOX

EXAMPLE 7-8

You push a 4-kg box, which is initially at rest on a horizontal table, a distance of 3 m with a horizontal force of 25 N. The coefficient of kinetic friction between the box and table is 0.35. Find (*a*) the external work done on the block–table system, (*b*) the energy dissipated by friction, (*c*) the final kinetic energy of the box, and (*d*) the speed of the box.

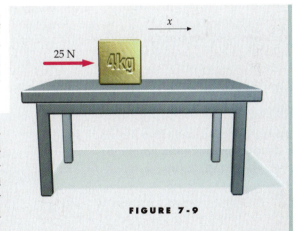

FIGURE 7-9

PICTURE THE PROBLEM The box plus table is the system (Figure 7-9). You are external to this system, so the force you push with is an external force. The final speed of the box is found from its final kinetic energy, which we find using the work–energy theorem with $\Delta E_{\text{chem}} = 0$ and $\Delta E_{\text{therm}} = f\Delta s$. The energy of the system is increased by the external work. Some of the energy increase is kinetic energy and some is thermal energy.

(*a*) There are four external forces acting on the system. However, only one of them does work. The total external work done is the product of the push force and the distance traveled:

$$\Sigma W_{\text{ext}} = W_{\text{by push on block}} + W_{\text{by gravity on block}}$$
$$+ W_{\text{by gravity on table}} + W_{\text{by floor on table}}$$
$$= F_{\text{push}}\,\Delta x + 0 + 0 + 0 = (25\text{ N})(3\text{ m})$$
$$= \boxed{75\text{ J}}$$

(*b*) The energy dissipated by friction is $f\Delta x$ (the magnitude of the normal force equals mg):

$$\Delta E_{\text{therm}} = f\,\Delta x = \mu_k F_n\,\Delta x = \mu_k mg\,\Delta x$$
$$= (0.35)(4\text{ kg})(9.81\text{ N/kg})\,(3\text{ m}) = \boxed{41.2\text{ J}}$$

[†] The work done by kinetic friction is examined in detail in B. A. Sherwood and W. H. Bernard, "Work and Heat Transfer in the Presence of Sliding Friction," *American Journal of Physics,* **52,** 1001 (1984).

(c) 1. Apply the work–energy theorem to find the final kinetic energy:

$$W_{ext} = \Delta E_{mech} + \Delta E_{therm}$$

2. There are no internal conservative forces doing work, so the change in potential energy is zero. The change in kinetic energy equals the change in mechanical energy:

$$\Delta E_{mech} = \Delta U + \Delta K$$
$$= 0 + (K_f - 0) = K_f$$

3. Substitute this result into the work–energy theorem:

$$W_{ext} = K_f + \Delta E_{therm}$$

so

$$K_f = W_{ext} - \Delta E_{therm}$$
$$= 75.0\,J - 41.2\,J = \boxed{33.8\,J}$$

(d) The final speed of the box is related to its kinetic energy. Solve for the final speed of the box:

$$K_f = \tfrac{1}{2}mv_f^2$$

so

$$v_f = \sqrt{\frac{2K_f}{m}} = \sqrt{\frac{2(33.8\,J)}{4\,kg}} = \boxed{4.11\,m/s}$$

A MOVING SLED   **EXAMPLE 7-9** **Try It Yourself**

A sled is coasting on a horizontal snow-covered surface with an initial speed of 4 m/s. If the coefficient of friction between the sled and the snow is 0.14, how far will the sled go before coming to rest?

PICTURE THE PROBLEM We choose the sled and snow as our system and then apply the work–energy theorem.

Cover the column to the right and try these on your own before looking at the answers.

Steps

Answers

1. Sketch the system in its initial and final configurations (Figure 7-10).

FIGURE 7-10

2. Apply the work–energy theorem.

$$W_{ext} = \Delta E_{mech} + \Delta E_{therm}$$
$$= (\Delta U + \Delta K) + f\,\Delta s$$

3. Solve for f. The normal force is equal to the weight.

$$f = \mu_k F_n = \mu_k mg$$

4. There are no conservative forces doing work and there are no external forces acting on the system. Use these observations to eliminate two terms from the step 2 result.

$$W_{ext} = \Delta U = 0$$

and

$$W_{ext} = \Delta U + \Delta K + f\,\Delta s$$
$$0 = 0 + \Delta K + \mu_k mg\,\Delta s$$

5. Express the change in kinetic energy in terms of the mass and the initial speed, and solve for Δs.

$$\Delta s = \frac{v_i^2}{2\mu_k g} = \boxed{5.82\,m}$$

A PLAYGROUND SLIDE   **EXAMPLE 7-10**

A child of mass 40 kg goes down an 8.0-m-long slide inclined at 30° with the horizontal. The coefficient of kinetic friction between the child and the slide is 0.35. If the child starts from rest at the top of the slide, how fast is she traveling when she reaches the bottom?

PICTURE THE PROBLEM As the child slides down, some of her potential energy is converted into kinetic energy and some into thermal energy because of friction. We choose the child–slide–earth as our system and apply the conservation of energy theorem.

1. Make a sketch of the system showing both its initial and final configurations (Figure 7-11).

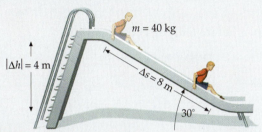

$m = 40$ kg

$|\Delta h| = 4$ m

$\Delta s = 8$ m

30°

FIGURE 7-11

2. Write out the conservation of energy equation:

$$W_{ext} = \Delta E_{mech} + \Delta E_{therm}$$
$$= (\Delta U + \Delta K) + f\,\Delta s$$

3. The initial kinetic energy is zero. The speed at the bottom is related to the final kinetic energy:

$$\Delta K = K_f - 0 = \tfrac{1}{2}mv_f^2$$

4. There are no external forces acting on the system:

$$W_{ext} = 0$$

5. The change in potential energy is related to the change in height Δh (which is negative).

$$\Delta U = mg\,\Delta h$$

6. To find f_k we apply Newton's second law to the child. First we draw a free-body diagram (Figure 7-12):

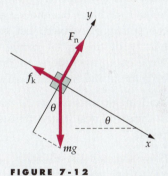

7. Next we apply Newton's second law. The normal component of the acceleration is zero. To find F_n we take components in the normal direction. Then we solve for f_k using $f_k = \mu_k F_n$:

$$F_n - mg\cos\theta = 0$$

so

FIGURE 7-12

$$f_k = \mu_k F_n = \mu_k mg\cos\theta$$

8. We use trigonometry to relate Δs to Δh:

$$|\Delta h| = \Delta s\sin\theta$$

9. Substituting into the step 2 result gives:

$$0 = mg\,\Delta h + \tfrac{1}{2}mv_f^2 + \mu_k mg\cos\theta\,\Delta s$$
$$= -mg\,\Delta s\sin\theta + \tfrac{1}{2}mv_f^2 + \mu_k mg\cos\theta\,\Delta s$$

10. Solving for v_f gives:

$$v_f^2 = 2g\,\Delta s(\sin\theta - \mu_k\cos\theta)$$
$$= 2(9.81\text{ m/s}^2)(8\text{ m})(\sin 30° - 0.35\cos 30°)$$
$$= 30.9\text{ m}^2/\text{s}^2$$

so

$$v_f = \boxed{5.60\text{ m/s}}$$

REMARKS Note that the result is independent of the mass of the child.

EXERCISE For the earth–child–slide system, calculate (*a*) the initial mechanical energy, (*b*) the final mechanical energy, and (*c*) the energy dissipated by friction. (*Answer* (*a*) 1570 J, (*b*) 618 J, (*c*) 952 J)

E X A M P L E   7 - 1 1   **Try It Yourself**

A 4-kg block hangs by a light string that passes over a massless, frictionless pulley and is connected to a 6-kg block that rests on a shelf. The coefficient of kinetic friction is 0.2. The 6-kg block is pushed against a spring, compressing it 30 cm. The spring has a force constant of 180 N/m. Find the speed of the blocks after the 6-kg block is released and the 4-kg block has fallen a distance of 40 cm.

PICTURE THE PROBLEM The speed of the blocks is obtained from their final kinetic energy. Consider the system to be everything shown in Figure 7-13 plus the earth. This system has both gravitational and elastic potential energy. Apply the work–energy theorem. Knowing the kinetic energy of the blocks means you can solve for their speed.

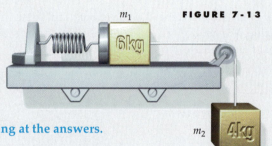

FIGURE 7-13

m_1

m_2

Cover the column to the right and try these on your own before looking at the answers.

Steps

Answers

1. Write out the equation for the conservation of energy of the system.

$$W_{ext} = \Delta E_{mech} + \Delta E_{therm}$$
$$= (\Delta U_s + \Delta U_g + \Delta K) + f\,\Delta s$$

2. There are no external forces on the system.

$$W_{ext} = 0$$

3. Make a table of the mechanical energy terms both initially, when the spring is compressed 30 cm, and finally, when each block has moved a distance $\Delta s = 40$ cm and the spring is relaxed. Let the gravitational potential energy of the initial configuration equal zero. Also, write down the difference (final minus initial) between each initial and final expression.

	U_s	U_g	K
Final	0	$-m_2 g\,\Delta s$	$\frac{1}{2}(m_1 + m_2)v_f^2$
Initial	$\frac{1}{2}kx_i^2$	0	0
Difference	$-\frac{1}{2}kx_i^2$	$-m_2 g\,\Delta s$	$\frac{1}{2}(m_1 + m_2)v_f^2$

4. Find an expression for f_k that includes μ_k. Substitute it, the step 2 result, and the step 3 results into the step 1 result.

$$0 = -\tfrac{1}{2}kx_i^2 - m_2 g\,\Delta s + \tfrac{1}{2}(m_1 + m_2)v_f^2 + \mu_k m_1 g\,\Delta s$$

5. Solve the step 4 result for v_f^2. Then substitute numerical values and solve for v_f:

$$v_f^2 = \frac{kx_i^2 + 2m_2 g\,\Delta s - 2\mu_k m_1 g\,\Delta s}{m_1 + m_2}$$

so

$$v_f = \boxed{1.95 \text{ m/s}}$$

REMARKS This solution assumes that the string remains taut at all times. This will be true if the acceleration of block 1 remains less than g, that is, if the net force on block 1 is less than $m_1 g = (6 \text{ kg})(9.81 \text{ N/kg}) = 58.9$ N. Initially the force exerted by the spring on block 1 has the magnitude $kx_1 = (180 \text{ N/m})(0.3 \text{ m}) = 54.0$ N and the frictional force has magnitude $f_k = \mu_k m_1 g = 0.2(58.9 \text{ N}) = 11.8$ N. These forces combine to produce a net force of 42.2 N directed to the right. Since the spring force decreases as block 1 moves following release, the acceleration of the 6-kg block will never exceed g, so the string will remain taut.

This pizza contains about 16 MJ of energy, about half the energy in 1 L (0.26 U.S. gal) of gasoline.

Systems With Chemical Energy

Sometimes a system's internal chemical energy is converted into mechanical energy and thermal energy with no work being done on the system by external forces. For example, at the beginning of this section we described the energy conversions that take place when you start running. In order to move forward, you push back on the floor and the floor pushes forward on you with a static frictional force. This force causes you to accelerate, but it does not do work. It does no work because the displacement of the point of application of the force is zero (assuming your shoes do not slip on the floor). Because no work is done, no energy is transferred from the floor to your body. The kinetic-energy increase of your body comes from the conversion of internal chemical energy derived from the food you eat. Consider the following example.

CLIMBING STAIRS   **E X A M P L E   7 - 1 2**

Suppose that you have mass m and you walk up a flight of stairs of height h. Discuss the application of energy conservation to a system consisting of you alone.

PICTURE THE PROBLEM There are two forces that act on you: gravity and the force of the stair treads on your feet. The force of gravity does negative work on you. To determine the work done by the force of the stair treads on your feet, consider the force of one stair tread on one of your feet. This force is a contact force. It does no work because the sole of your foot does not move while it is in contact with the tread. Thus, no work is done by the force of the stair treads on your feet.

1. Write the work–energy equation for the you-alone system: $W_{\text{ext}} = \Delta E_{\text{sys}} = \Delta E_{\text{mech}} + \Delta E_{\text{therm}} + \Delta E_{\text{chem}}$

2. There are two external forces acting on this system, the force of gravity and the force of the stair treads on your feet. The force of gravity does negative work because this force is directed opposite to your displacement. The force of the stair treads does no work because the point of application, the soles of your feet, do not move while this force is applied: $W_{\text{ext}} = -mgh$

3. As stated in the problem, the system consists of you alone. Because your configuration does not change, any change in your mechanical energy is entirely a change in your kinetic energy, which is the same initially and finally: $\Delta E_{\text{mech}} = 0$

4. Substitute these results into the work–energy equation: $-mgh = \boxed{\Delta E_{\text{therm}} + \Delta E_{\text{chem}}}$

REMARKS If there were no change in thermal energy, then your chemical energy would decrease by mgh. Because the human body is relatively inefficient, the increase in the amount of thermal energy will be considerably greater than mgh. The decrease in stored chemical energy equals mgh plus any thermal energy, which eventually is transferred from your body to your surroundings via heat.

EXERCISE Discuss the energy conservation for a system consisting of both you and the earth. (*Answer* For the you–earth system no external work is done, so the total energy, which now includes gravitational potential energy, is conserved. The change in mechanical energy is mgh, so the work–energy theorem gives $0 = mgh + \Delta E_{\text{therm}} + \Delta E_{\text{chem}}$.)

　　　　　　EXAMPLE 7-13

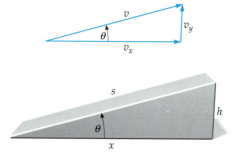

You are driving a 1000-kg gasoline-powered car at a constant speed of 100 km/h (= 27.8 m/s = 62.2 mi/h) up a 10-percent grade (Figure 7-14). (A 10-percent grade means that the road rises 1 m for each 10 m of horizontal distance—that is, the angle of inclination θ is given by $\tan \theta = 0.1$.) (a) If the efficiency is 15 percent, what is the rate at which the chemical energy of the car–earth–atmosphere system changes? (The efficiency is the fraction of the chemical energy consumed that appears as mechanical energy.) (b) What is the rate at which thermal energy is created?

FIGURE 7-14　　$\tan \theta = h/x \sim \sin \theta = h/s$

PICTURE THE PROBLEM Some of the chemical energy goes into increasing the potential energy of the car as it climbs the hill, and some goes into an increase in thermal energy, much of which is expelled by the car as exhaust. Apply the work–energy theorem to a system consisting of the car, the hill, the air, and the earth.

(a) 1. The rate of loss of chemical energy equals the negative of change in chemical energy per unit time:

$$\text{chemical energy loss rate} = \frac{-\Delta E_{\text{chem}}}{\Delta t}$$

2. The increase in mechanical energy equals 15% of the decrease in chemical energy:

$$\Delta E_{\text{mech}} = 0.15 |\Delta E_{\text{chem}}| = -0.15 \, \Delta E_{\text{chem}}$$

3. Solve for the loss rate of chemical energy:

$$\frac{-\Delta E_{\text{chem}}}{\Delta t} = \frac{1}{0.15} \frac{\Delta E_{\text{mech}}}{\Delta t}$$

4. The car moves at constant speed, so $\Delta K = 0$ and $\Delta E_{\text{mech}} = \Delta U$. Relate the change in mechanical energy to the change in height Δh and substitute it into the step 3 result:

$$\Delta E_{\text{mech}} = mg \, \Delta h$$

so

$$\frac{-\Delta E_{\text{chem}}}{\Delta t} = \frac{mg}{0.15} \frac{\Delta h}{\Delta t}$$

5. Convert the changes to time derivatives. That is, take the limit of both sides as Δt approaches zero:

$$-\frac{dE_{\text{chem}}}{dt} = \frac{mg}{0.15} \frac{dh}{dt}$$

6. The rate of change of h equals v_y, which is related to the speed v as shown in Figure 7-14:

$$\frac{dh}{dt} = v_y = v \sin \theta$$

7. We can approximate $\sin \theta$ by $\tan \theta$ because the angle is small:

$$\sin \theta \approx \tan \theta$$

8. Solve for the loss rate of chemical energy:

$$-\frac{dE_{\text{chem}}}{dt} = \frac{mg}{0.15} v \sin \theta \approx \frac{mg}{0.15} v \tan \theta$$

$$= \frac{(1000 \text{ kg})(9.81 \text{ N/kg})}{0.15}(27.8 \text{ m/s})0.1$$

$$= \boxed{182 \text{ kW}}$$

(b) 1. Write out the work–energy relation:

$$W_{\text{ext}} = \Delta E_{\text{mech}} + \Delta E_{\text{therm}} + \Delta E_{\text{chem}}$$

2. Convert to derivatives and solve for dE_{therm}/dt:

$$0 = \frac{dE_{\text{mech}}}{dt} + \frac{dE_{\text{therm}}}{dt} + \frac{dE_{\text{chem}}}{dt}$$

so

$$\frac{dE_{\text{therm}}}{dt} = -\frac{dE_{\text{mech}}}{dt} - \frac{dE_{\text{chem}}}{dt} = 0.15 \frac{dE_{\text{chem}}}{dt} - \frac{dE_{\text{chem}}}{dt}$$

$$= -0.85 \frac{dE_{\text{chem}}}{dt} = 0.85(182 \text{ kW}) = \boxed{155 \text{ kW}}$$

REMARKS Gasoline-powered cars are typically only about 15 percent efficient. About 85 percent of the chemical energy of the gasoline goes to thermal energy, most of which is expelled as heat exhaust. Additional thermal energy is created by rolling friction and wind resistance. The energy content of gasoline is about 31.8 MJ/L.

7-3 Mass and Energy

In 1905, Albert Einstein published his special theory of relativity, a result of which is the famous equation

$$E_0 = mc^2 \qquad\qquad 7\text{-}17$$

where $c = 3 \times 10^8$ m/s is the speed of light in vacuum. We will study this theory in some detail in later chapters.

According to Equation 7-17, a particle or system of mass m has "rest" energy mc^2. This energy is intrinsic to the particle. Consider the positron, a particle emitted in a nuclear process called beta decay. Positrons and electrons have identical masses, but equal and opposite electrical charge. When a positron encounters an electron in matter, electron–positron annihilation occurs, a process in which the electron and positron disappear and their energy appears as electromagnetic radiation. If the two particles are initially at rest, the energy of the electromagnetic radiation equals the rest energy of the electron plus that of the positron.

Energies in atomic and nuclear physics are usually expressed in units of electron volts (eV) or mega-electron-volts (1 MeV = 10^6 eV). A convenient unit for the masses of atomic particles is eV/c^2 or MeV/c^2. Table 7-1 lists rest energies (and therefore the masses) of some elementary particles and light nuclei. The total rest energy of a positron plus an electron is 2(0.511 MeV), which is the radiation energy emitted upon annihilation.

TABLE 7-1

Rest Energies of Some Elementary Particles and Light Nuclei

Particle	Symbol	Rest Energy (MeV)
Electron	e^-	0.5110
Positron	e^+	0.5110
Proton	p	938.272
Neutron	n	939.565
Deuteron	d	1875.613
Triton	t	2808.410
Helium-3	^{3}He	2808.39
Alpha particle	α	3727.379

The rest energy of a *system* can consist of the potential energy of the system or other internal energies of the system, in addition to the intrinsic rest energies of the particles in the system. If the system at rest absorbs energy ΔE and remains at rest, its rest energy increases by ΔE and its mass increases by

$$\Delta M = \frac{\Delta E}{c^2} \qquad\qquad 7\text{-}18$$

Consider two 1-kg blocks connected by a spring of force constant k. If we stretch the spring a distance x, the potential energy of the system increases by $\Delta U = \frac{1}{2}kx^2$. According to Equation 7-18, the mass of the system has also increased by $\Delta M = \Delta U/c^2$. Because c is such a large number, this increase in mass cannot be observed in macroscopic systems. For example, suppose $k = 800$ N/m and $x = 10$ cm $= 0.1$ m. The potential energy of the spring system is then $\frac{1}{2}kx^2 = \frac{1}{2}(800 \text{ N/m})(0.1 \text{ m})^2 = 4$ J. The increase in mass of the system is

$$\Delta M = \frac{\Delta U}{c^2} = \frac{4 \text{ J}}{(3 \times 10^8 \text{ m/s})^2} = 4.44 \times 10^{-17} \text{ kg}$$

The relative mass increase $\Delta M/M \approx 2 \times 10^{-17}$ is much too small to be observed.

Nuclear Energy

In nuclear reactions, the energy changes are often an appreciable fraction of the rest energy of the system. Consider the deuteron, which is the nucleus of deuterium, an isotope of hydrogen also called heavy hydrogen. The deuteron consists of a proton and neutron bound together. From Table 7-1 we see that the mass of the proton is 938.28 MeV/c^2 and the mass of the neutron is 939.57 MeV/c^2. The sum of these two masses is 1877.85 MeV/c^2. But the mass of the deuteron is 1875.63 MeV/c^2, which is less than the sum of the masses of the proton and neutron by 2.22 MeV/c^2. Note that this mass difference $\Delta M/M \approx 1.2 \times 10^{-3}$, much greater than any uncertainties in the measurement of these masses, and almost 14 orders of magnitude greater than the 2×10^{-17} discussed above for the spring–blocks system.

Heavy water (deuterium oxide) molecules are produced in the primary cooling water of a nuclear reactor when neutrons collide with the hydrogen nuclei (protons) of the water molecules. If a slow moving neutron is captured by a proton, 2.22 MeV of energy are released in the form of electromagnetic radiation. Thus, the mass of a deuterium atom is 2.22 MeV/c^2 less than the sum of the masses of an isolated hydrogen atom and an isolated neutron.

This process can be reversed by breaking a deuteron into its constituent parts if at least 2.22 MeV of energy is transferred to the deuteron with electromagnetic radiation or by collisions with other energetic particles. Any transferred energy in excess of 2.22 MeV appears as kinetic energy of the resulting proton and neutron.

The energy needed to completely separate a nucleus into individual neutrons and protons is called the **binding energy.** The binding energy of a deuteron is 2.22 MeV.

The deuteron is an example of a bound system. Its rest energy is less than the rest energy of its parts, so energy must be put into the system to break it apart. If the rest energy of a system is greater than the rest energy of its parts, the system is unbound. An example is uranium-236, which breaks apart or **fissions** into two smaller nuclei.[†] The sum of the masses of the resultant parts is less than the mass of the original nucleus. Thus the mass of the system decreases, and energy is released.

In nuclear fusion, two very light nuclei such as a deuteron and a triton (the nucleus of the hydrogen isotope tritium) fuse together. The rest mass of the resultant nucleus is less than that of the original parts and again energy is released. In a chemical reaction that produces energy, such as burning coal, the mass decrease is of the order of 1 eV/c^2 per atom. This is more than a million times smaller than the mass changes per nucleus in many nuclear reactions, and is not readily observable.

---

† Uranium-236, written ^{236}U, is made in a nuclear reactor when the stable isotope uranium-235 absorbs a neutron. This reaction will be discussed in Chapter 34.

BINDING ENERGY   **EXAMPLE 7-14**

A hydrogen atom consisting of a proton and an electron has a binding energy of 13.6 eV. By what percentage is the mass of a proton plus the mass of an electron greater than that of the hydrogen atom?

1. The fractional difference between the mass of the hydrogen atom and the masses of its parts is the ratio of the binding energy E_b divided by c^2 to $m_e + m_p$:

$$\text{Fractional difference} = \frac{E_b/c^2}{m_e + m_p} = \frac{13.6 \text{ eV}/c^2}{m_e + m_p}$$

2. Obtain the rest masses of the proton and electron from Table 7-1:

$$m_p = 938.28 \text{ MeV}/c^2; \quad m_e = 0.511 \text{ MeV}/c^2$$

3. Add to find the sum of these masses:

$$m_p + m_e = 938.79 \text{ MeV}/c^2$$

4. The rest mass of the hydrogen atom is less than this by 13.6 eV/c^2. The percentage difference is:

$$\text{Fractional difference} = \frac{13.6 \text{ eV}/c^2}{938.79 \times 10^6 \text{ eV}/c^2}$$

$$= 1.45 \times 10^{-8}$$

$$= \boxed{1.45 \times 10^{-6} \%}$$

REMARKS This mass difference is too small to be measured directly. However, binding energies can be accurately measured, so the mass difference can be found from $E_b = (\Delta m)c^2$.

NUCLEAR FUSION   **EXAMPLE 7-15** Try It Yourself

In a typical nuclear fusion reaction, a tritium nucleus (^{3}H) and a deuterium nucleus (^{2}H) fuse together to form a helium nucleus (^{4}He) plus a neutron. The reaction is written ^{2}H + ^{3}H → ^{4}He + n. If the initial kinetic energy of the particles is negligible, how much energy is released in this fusion reaction?

PICTURE THE PROBLEM Because energy is released, the total rest energy of the initial particles must be greater than that of the final particles. This difference equals the energy released.

Cover the column to the right and try these on your own before looking at the answers.

Steps

1. Write down the rest energies of ^{2}H and ^{3}H from Table 7-1 and add to find the total initial rest energy.

2. Do the same for ^{4}He and n to find the final rest energy.

3. Find the energy released from $E_{released} = E_0 (\text{initial}) - E_0 (\text{final})$.

Answers

$E_0 (\text{initial}) = 1875.628 \text{ MeV} + 2808.944 \text{ MeV}$
$= 4684.572 \text{ MeV}$

$E_0 (\text{final}) = 3727.409 \text{ MeV} + 939.573 \text{ MeV}$
$= 4666.982 \text{ MeV}$

$E_{released} = 4684.572 \text{ MeV} - 4666.982 \text{ MeV}$

$= \boxed{17.59 \text{ MeV} \approx 17.6 \text{ MeV}}$

REMARKS This and other fusion reactions occur in the sun. The energy that is released bathes the earth and is ultimately responsible for all life on the planet. The energy emitted by the sun is accompanied by a continuous decrease in the sun's rest mass.

Newtonian Mechanics and Relativity

When the speed of a particle approaches the speed of light, Newton's second law breaks down, and we must modify Newtonian mechanics according to Einstein's theory of relativity. The criterion for the validity of Newtonian mechanics can also be stated in terms of the energy of a particle. In Newtonian mechanics, the kinetic energy of a particle moving with speed v is

$$K = \tfrac{1}{2}mv^2 = \tfrac{1}{2}mc^2\frac{v^2}{c^2} = \tfrac{1}{2}E_0\frac{v^2}{c^2}$$

where $E_0 = mc^2$ is the rest energy of the particle. Then

$$\frac{v}{c} = \sqrt{\frac{2K}{E_0}}$$

Newtonian mechanics is valid if the speed of the particle is much less than the speed of light, or alternatively, the kinetic energy of a particle is much less than its rest energy.

7-4 Quantization of Energy

When energy is put into a system that remains at rest, the internal energy of the system increases. While it might seem that we could put any amount of energy into a system, this is found not to be true for microscopic systems such as atoms or molecules. The internal energy of a microscopic system can increase only by discrete increments.

If we have two blocks attached to a spring and we pull the blocks apart, we do work on the block–spring system, and its potential energy increases. If we then release the blocks, they oscillate back and forth. The energy of oscillation E—the kinetic energy of motion of the blocks plus the potential energy due to the stretching of the spring—equals the initial potential energy. In time, the energy of the system decreases because of various damping effects such as friction and air resistance. As closely as we can measure, the energy decreases continuously. Eventually all the energy is dissipated and the energy of oscillation is zero.

Now consider a diatomic molecule such as molecular oxygen, O_2. The force of attraction between the two oxygen atoms varies approximately linearly with the change in separation (for small changes), like that of two blocks connected by a spring. If a diatomic molecule is set oscillating with some energy E, the energy decreases with time as the molecule radiates or interacts with its surroundings, but the decrease is *not continuous*. The energy decreases in finite steps, and the lowest energy state, called the ground state, is not zero. The vibrational energy of a diatomic molecule is said to be **quantized;** that is, the molecule can absorb or release energies only in certain amounts, known as quanta.

When either blocks on a spring or diatomic molecules oscillate, the time for one oscillation is called the period T. The reciprocal of the period is the frequency of oscillation $f = 1/T$. We will see in Chapter 14 that the period and frequency of an oscillator do not depend on the energy of oscillation. As the energy decreases, the frequency remains the same. Figure 7-15 shows an **energy-level diagram** for an oscillator. The allowed energies are approximately equally spaced, and are given by[†]

$$E_n = (n + \tfrac{1}{2})hf, \qquad n = 0, 1, 2, 3, \dots \qquad \text{7-19}$$

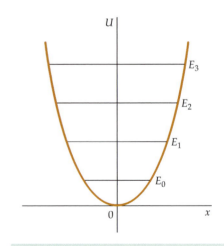

FIGURE 7-15 Energy-level diagram for an oscillator.

---

† A diatomic molecule can also have rotational energy. The rotational energy is also quantized, but the energy levels are not equally spaced, and the lowest possible energy is zero. We will study rotational energy in Chapters 9 and 10.

where f is the frequency of oscillation and h is a fundamental constant of nature called Planck's constant:[†]

$$h = 6.626 \times 10^{-34} \text{ J·s} \qquad 7\text{-}20$$

The integer n is called a **quantum number.** The lowest possible energy is the **ground state energy** $E_0 = \frac{1}{2}hf$.

Microscopic systems often gain or lose energy by absorbing or emitting electromagnetic radiation. By conservation of energy, if E_i and E_f are the initial and final energies of a system, the energy of the radiation emitted or absorbed is

$$E_{\text{rad}} = E_i - E_f$$

Since the system energies E_i and E_f are quantized, the radiated energy is also quantized.[‡] The quantum of radiation energy is called a **photon.** The energy of a photon is given by

$$E_{\text{photon}} = hf \qquad 7\text{-}21$$

where f is the frequency of the electromagnetic radiation.[§]

The quasar 3C 273 is shown imaged via X-ray energy. The X-ray energy emitted by this quasar is more than a million times that emitted by the entire Milky Way galaxy.

As far as we know, all bound systems exhibit energy quantization. For macroscopic bound systems, the steps between energy levels are so small that they are unobservable. For example, typical oscillation frequencies for two blocks on a spring are 1 to 10 times per second. If $f = 10$ oscillations per second, the spacing between allowed levels is $hf = (6.626 \times 10^{-34} \text{ J·s}) \times (10/s) \approx 6 \times 10^{-33}$ J. Since the energy of a macroscopic system is of the order of joules, a quantum step of 10^{-33} J is too small to be noticed. To put it another way, if the energy of a system is 1 J, the value of n is of the order of 10^{32} and changes of one or two quantum units will not be observable.

For a diatomic molecule, a typical frequency of vibration is 10^{14} vibrations per second, and a typical energy is 10^{-19} J. The spacing between allowed levels is then

$$E_{n+1} - E_n = hf \approx (6.63 \times 10^{-34} \text{ J·s})(10^{14} \text{ s}) \approx 6 \times 10^{-20} \text{ J}$$

Thus, changes in the energy of oscillation are of the same order of magnitude as the energy of the molecule, and quantization is definitely not negligible.

SUMMARY

1. The conservation of mechanical energy is an important relation derived from Newton's laws for conservative forces. It is useful in solving many problems.

2. The work–energy theorem and the conservation of energy are fundamental laws of nature that have applications in all areas of physics.

3. Einstein's equation $E_0 = mc^2$ is a fundamental relation between mass and energy.

4. Quantization is a fundamental property of the energy in bound systems.

---

† In 1900, the German physicist Max Planck introduced this constant as a calculational device to explain discrepancies between the theoretical curves and experimental data on the spectrum of blackbody radiation. The significance of Planck's constant was not appreciated by Planck or anyone else until Einstein postulated in 1905 that the energy of electromagnetic radiation is not continuous, but occurs in packets of size hf where f is the frequency of the radiation.

‡ Historically, the quantization of electromagnetic radiation, as proposed by Max Planck and Albert Einstein, was the first "discovery" of energy quantization.

§ Electromagnetic radiation includes light, microwaves, radio waves, television waves, X rays, and gamma rays. These differ from one another in their frequencies.

Topic	Relevant Equations and Remarks
1. Mechanical Energy	The sum of the kinetic and potential energy of a system is called the total mechanical energy $$E_{mech} = K_{sys} + U_{sys}$$ 7-5
Conservation of mechanical energy	If no external forces do work on a system, and if the internal forces that do work are all conservative, the total mechanical energy of the system remains constant: $$E_{mech} = K_{sys} + U_{sys} = \text{constant}$$ 7-6 $$K_f + U_f = K_i + U_i$$ 7-7
2. Total Energy of a System	The energy of a system consists of mechanical energy E_{mech}, thermal energy E_{therm}, chemical energy E_{chem}, and other types of energy E_{other}, such as sound radiation and electromagnetic radiation. $$E_{sys} = E_{mech} + E_{therm} + E_{chem} + E_{other}$$ 7-9
3. Conservation of Energy	
Universe	The total energy of the universe is constant. Energy can be converted from one form to another, or transmitted from one region to another, but energy can never be created or destroyed.
System	The energy of a system can be changed by work being done on the system and by energy transfer via heat. (These include the emission or absorption of radiation.) The increase or decrease in the energy of the system can always be accounted for by the appearance or disappearance of some kind of energy somewhere else: $$E_{in} - E_{out} = \Delta E_{sys}$$ 7-8
Work–energy theorem	$$W_{ext} = \Delta E_{sys} = \Delta E_{mech} + \Delta E_{therm} + \Delta E_{chem} + \Delta E_{other}$$ 7-10
4. Energy Dissipated by Friction	For a system that involves a pair of sliding surfaces, the total energy dissipated by friction on both surfaces equals the increase in thermal energy of the system and is given by $$f\,\Delta s = \Delta E_{therm}$$ 7-14 where Δs is the distance one surface slides relative to the other.
5. Problem Solving	The conservation of mechanical energy and the work–energy theorem can be used as an alternative to Newton's laws to solve mechanics problems that require the determination of the speed of a particle as a function of its position.
6. Mass and Energy	A particle with mass m has an intrinsic rest energy E_0 given by $$E_0 = mc^2$$ 7-17 where $c = 3 \times 10^8$ m/s is the speed of light in vacuum. A system with mass M also has a rest energy $E_0 = Mc^2$. If a system gains or loses internal energy ΔE, it simultaneously gains or loses mass $\Delta M = \Delta E/c^2$.
Binding energy	The energy required to separate a bound system into its constituent parts is called its binding energy. The binding energy is ΔMc^2, where ΔM is the sum of the masses of the constituent parts, less the mass of the bound system.
8. Newtonian Mechanics and Special Relativity	When the speed of a particle approaches the speed of light c (when the kinetic energy of the particle approaches its rest energy), Newtonian mechanics breaks down, and must be replaced by Einstein's special theory of relativity.

9. Energy Quantization

The internal energy of a system is found to have only a discrete set of possible values. For a system oscillating with frequency f, the allowed energy values are separated by an amount hf, where h is Planck's constant:

$$h = 6.626 \times 10^{-34} \text{ J·s}$$

7-20

Photons

Microscopic systems often exchange energy with their surroundings by emitting or absorbing electromagnetic radiation, which is also quantized. The quantum of energy of radiation is called the photon:

$$E_{\text{photon}} = hf$$

7-21

where f is the frequency of the electromagnetic radiation.

PROBLEMS

- Single-concept, single-step, relatively easy
- •• Intermediate-level, may require synthesis of concepts
- ••• Challenging
- **SSM** Solution is in the *Student Solutions Manual*
- **iSOLVE** Problems available on iSOLVE online homework service
- **iSOLVE ✓** These "Checkpoint" online homework service problems ask students additional questions about their confidence level, and how they arrived at their answer

In a few problems, you are given more data than you actually need; in a few other problems, you are required to supply data from your general knowledge, outside sources, or informed estimates.

Take $g = 9.81$ N/kg $= 9.81$ m/s^2 and neglect friction in all problems unless otherwise stated.

Conceptual Problems

1 • **SSM** Two objects of unequal mass are connected by a massless cord passing over a frictionless peg. After the objects are released from rest, which of the following statements are true? ($U =$ gravitational potential energy, $K =$ kinetic energy of the system.) (a) $\Delta U < 0$ and $\Delta K > 0$. (b) $\Delta U = 0$ and $\Delta K > 0$. (c) $\Delta U < 0$ and $\Delta K = 0$. (d) $\Delta U = 0$ and $\Delta K = 0$. (e) $\Delta U > 0$ and $\Delta K < 0$.

2 • Two stones are thrown with the same initial speed at the same instant from the roof of a building. One stone is thrown at an angle of 30° above the horizontal, the other is thrown horizontally. (Neglect air resistance.) Which statement below is true?

(a) The stones strike the ground at the same time and with equal speeds.
(b) The stones strike the ground at the same time with different speeds.
(c) The stones strike the ground at different times with equal speeds.
(d) The stones strike the ground at different times with different speeds.

3 • True or false:

(a) The total energy of a system cannot change.
(b) When you jump into the air, the floor does work on you increasing your mechanical energy.

4 • You stand on roller skates next to a rigid wall. To get started, you push off against the wall. Discuss the energy changes pertinent to this situation.

5 • **SSM** In *Surely You're Joking, Mr. Feynman,*[†] Richard Feynman described his annoyance at how the concept of energy was portrayed in a children's textbook in the following way: "There was a book which started out with four pictures: first, there was a wind-up toy; then there was an automobile; then there was a boy riding a bicycle; then there was something else. And underneath each picture it said 'What makes it go?' . . . I turned the page. The answer was . . . for everything, 'Energy makes it go' . . . It's also not even true that 'energy makes it go' because if it stops, you could say 'energy makes it stop' just as well. . . . Energy is neither increased or decreased in these examples; it's just changed from one form to another." Describe how energy changes from one form to another when a little girl pedals her bike up a hill, then freewheels down the hill, and brakes to a stop.

6 • **SSM** A body falling through the atmosphere (air resistance is present) gains 20 J of kinetic energy. The amount of gravitational potential energy that it lost is (a) 20 J, (b) more than 20 J, (c) less than 20 J, (d) impossible to tell without knowing the mass of the body, (e) impossible to tell without knowing how far the body falls.

---

† Richard P. Feynman and Ralph Leighton, *Surely You're Joking, Mr. Feynman,* New York: Bantam Books (1985).

7 •• Assume that, on applying the brakes, a constant frictional force acts on the wheels of a car. If that is so, it follows that (*a*) the distance the car travels before coming to rest is proportional to the speed of the car before the brakes are applied, (*b*) the car's kinetic energy diminishes at a constant rate, (*c*) the kinetic energy of the car is inversely proportional to the time that has elapsed since the application of the brakes, (*d*) none of the above are true.

8 • You are given two frictionless ramps and a block to slide down them (Figure 7-16). One ramp is in the form of an inclined plane with height *H* and length *L*. The other has a ramp cut in the form of a partial arc of a circle, but also has height *H* and length *L*. You slide the block down each ramp, releasing it from rest, and measure the time it takes to reach the bottom and the speed of the block upon getting there. You find that (*a*) the block takes the same time to slide down each ramp, (*b*) the block has the same speed on reaching the bottom of each ramp, (*c*) both (*a*) and (*b*) are correct, (*d*) neither (*a*) nor (*b*) is correct.

FIGURE 7-16 Problem 8

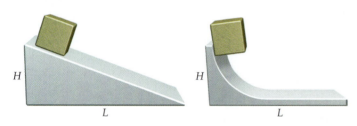

9 •• If a rock is attached to a massless, rigid rod and swung in a vertical circle at a constant speed, it will not have a constant total energy, as the kinetic energy of the rock will be constant, but the potential energy will be continually changing. Is any total work being done on the rock? Does the rod exert a tangential force on the rock?

Estimation and Approximation

10 •• **SSM** The *metabolic rate* is the rate at which the body uses chemical energy to sustain its life functions. Experimentally, the average metabolic rate is proportional to the total skin surface area of the body. The surface area for a 5-ft, 10-in. male weighing 175 lb is just about 2.0 m² and for a 5-ft, 4-in. female weighing 110 lb is approximately 1.5 m². There is about a 1 percent change in surface area for every 3 lb above or below the weights quoted here and a 1 percent change for every inch above or below the heights quoted. (*a*) Estimate your average metabolic rate over the course of a day using the following guide for physical activity: sleeping, metabolic rate = 40 W/m²; sitting, 60 W/m²; walking, 160 W/m²; moderate physical activity, 175 W/m²; and moderate aerobic exercise, 300 W/m². How does it compare to the power of a 100-W light bulb? (*b*) Express your average metabolic rate in terms of kcal/day (1 kcal = 4190 J). (A kcal is the "food calorie" used by nutritionists.) (*c*) An estimate used by nutritionists is that the "average person" must eat roughly 12–15 kcal/lb of body weight a day to maintain his or her weight. From the calculations in part (*b*), are these estimates reasonable?

11 • Assume that the maximum rate at which your body can expend energy is 250 W. Assuming a 20 percent efficiency for the conversion of chemical energy into mechanical energy, estimate how quickly you can run up four flights of stairs, with each flight 3.5 m high.

12 • How much rest mass is consumed in the core of a nuclear-fueled electric generating plant in producing (*a*) 1 J of thermal energy? (*b*) enough electrical energy to keep a 100-W light bulb burning for 10 y? (For each joule of electrical energy produced by the generator, the reactor core must produce 3 J of nuclear energy.)

13 • **SSM** The chemical energy released by burning a gallon of gasoline is approximately 2.6×10^5 kJ. Estimate the total energy used by all of the cars in the United States during the course of 1 y. What fraction does this represent of the total energy use by the United States in 1 y (about 5×10^{20} J)?

14 • The maximum efficiency of a solar energy panel in converting solar energy into useful electrical energy is about 12 percent. Using the known value of the solar intensity reaching the earth's surface (1.0 kW/m²), what area would have to be covered by solar panels in order to supply the energy requirements of the United States (approximately 5×10^{20} J/y)? Assume cloudless skies.

15 • Hydroelectric power plants convert gravitational potential energy into more useful forms by flowing water downhill through a turbine system to generate electrical energy. The Hoover Dam on the Colorado River is 211 m high and generates 4 billion kW·h/y (1 W · h = 3.6×10^3 J.) At what flow rate (in L/s) must water be flowing through the turbines to generate this power? The density of water is 1 kg/L. Assume a total efficiency of 20 percent in converting the water's potential energy into electrical energy.

The Conservation of Mechanical Energy

16 • A block of mass *m* is pushed against a spring, compressing it a distance *x*, and the block is then released. The spring projects the block along a frictionless horizontal surface, giving a speed *v*. The same spring projects a second block of mass 4*m*, giving it a speed 3*v*. What distance was the spring compressed in the second case?

17 • A bicyclist traveling at 10 m/s on a horizontal road stops pedaling as she starts up a hill inclined at 3.0° to the horizontal. Ignoring frictional forces, how far up the hill will she travel before stopping? (*a*) 5.1 m. (*b*) 30 m. (*c*) 97 m. (*d*) 10.2 m. (*e*) The answer depends on the mass of the person.

18 • **SSM** A pendulum of length *L* with a bob of mass *m* is pulled aside until the bob is at a height *L*/4 above its equilibrium position. The bob is then released. Find the speed of the bob as it passes the equilibrium position.

19 • **ISOLVE** A 3-kg block slides along a frictionless horizontal surface with a speed of 7 m/s (Figure 7-17). After sliding a distance of 2 m, the block makes a smooth transition to a frictionless ramp inclined at an angle of 40° to the horizontal. How far up the ramp does the block slide before coming momentarily to rest?

FIGURE 7-17 Problems 19, 46

20 • **i SOLVE ✓** The 3-kg object in Figure 7-18 is released from rest at a height of 5 m on a curved frictionless ramp. At the foot of the ramp is a spring of force constant $k = 400$ N/m. The object slides down the ramp and into the spring, compressing it a distance x before coming momentarily to rest. (*a*) Find x. (*b*) What happens to the object after it comes to rest?

FIGURE 7-18 Problem 20

$k = 400$ N/m

5 m

x

21 • **i SOLVE ✓** A 15-g ball is shot from a spring gun whose spring has a force constant of 600 N/m. The spring can be compressed 5 cm. How high will the ball go if the gun is aimed vertically?

22 • **i SOLVE** At a dock, a crane lifts a 4000-kg container 30 m, swings it out over the deck of a freighter, and lowers the container into the hold of the freighter, which is 8 m below the level of the dock. How much work is done by the crane on the container? (Neglect friction losses.)

23 • **i SOLVE** A 16-kg child on a playground swing moves with a speed of 3.4 m/s when the 6-m long swing is at its lowest point. What is the angle that the swing makes with the vertical when the swing is at its highest point?

24 •• **SSM** The system shown in Figure 7-19 is initially at rest when the lower string is cut. Find the speed of the objects when they are at the same height. The frictionless pulley has negligible mass.

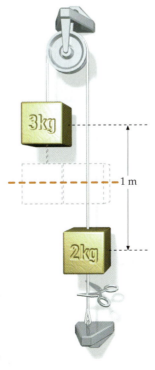

3kg

1 m

2kg

FIGURE 7-19 Problem 24

25 •• A block rests on an inclined plane as shown in Figure 7-20. A spring to which it is attached via a pulley is being pulled downward with gradually increasing force. The value of μ_s is known. Find the potential energy U of the spring at the moment when the block begins to move.

FIGURE 7-20 Problem 25

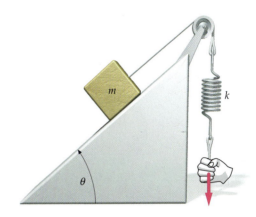

26 •• **i SOLVE** A 2.4-kg block is dropped onto a spring of spring constant 3955 N/m from a height of 5.0 m (Figure 7-21). When the block is momentarily at rest, the spring has been compressed by 25 cm. Find the speed of the block when the compression of the spring is 15 cm.

FIGURE 7-21 Problems 26, 91

27 •• **SSM** A ball at the end of a string moves in a vertical circle with constant mechanical energy E. What is the difference between the tension at the bottom of the circle and the tension at the top?

28 •• A girl of mass m is taking a picnic lunch to her grandmother. She ties a rope of length R to a tree branch over a creek and starts to swing from rest at a point that is a distance $R/2$ lower than the branch. What is the minimum breaking tension for the rope if it is not to break and drop the girl into the creek?

29 •• [SOLVE] A roller coaster car of mass 1500 kg starts at a distance $H = 23$ m above the bottom of a loop 15 m in diameter (Figure 7-22). If friction is negligible, the downward force of the rails on the car when it is upside down at the top of the loop is (a) 4.6×10^4 N, (b) 3.1×10^4 N, (c) 1.7×10^4 N, (d) 980 N, (e) 1.6×10^3 N.

FIGURE 7-22 Problem 29

30 • A single-car roller coaster pushes off, and on the first section of track, descends a 5-m-deep valley, then climbs to the top of a hill that is 9.5 m above the valley floor. (a) What is the minimum initial speed required to carry the coaster beyond the first hill? Assume that the track is frictionless. (b) Can we affect this speed by changing the depth of the valley to make the coaster pick up more speed at the bottom?

31 •• The Gravitron single-car roller coaster has a loop-the-loop that is constructed so that riders on the coaster will feel perfectly weightless when they reach the top of the circular arc. How heavy will they feel when they reach the bottom of the arc (that is, what is the normal force pressing up into their seats at the bottom of the loop)? Express the answer as a multiple of their normal weight. Assume that the arc is perfectly circular and no frictional forces act on the car.

32 •• [SSM] [SOLVE] A stone is thrown upward at an angle of 53° above the horizontal. Its maximum height during the trajectory is 24 m. What was the stone's initial speed?

33 •• [SOLVE✓] A baseball of mass 0.17 kg is thrown from the roof of a building 12 m above the ground. Its initial velocity is 30 m/s at an angle of 40° above the horizontal. (a) What is the maximum height the ball reaches? (b) What is the work done by gravity as the ball moves from the roof to its maximum height? (c) What is the speed of the ball as it strikes the ground?

34 •• An 80-cm-long pendulum with a 0.6-kg bob is released from rest at initial angle of θ_0 with the vertical. At the bottom of the swing, the speed of the bob is 2.8 m/s. (a) What was the initial angle of the pendulum? (b) What angle does the pendulum make with the vertical when the speed of the bob is 1.4 m/s?

35 •• [SSM] [SOLVE✓] The Royal Gorge bridge over the Arkansas River is $L = 310$ m above the river. A bungee jumper of mass 60 kg has an elastic cord of length $d = 50$ m attached to her feet. Assume the cord acts like a spring of force constant k. The jumper leaps, barely touches the water,

and after numerous ups and downs comes to rest at a height h above the water. (a) Find h. (b) Find the maximum speed of the jumper.

36 •• A pendulum consists of a 2-kg bob attached to a light string of length 3 m. The bob is struck horizontally so that it has an initial horizontal velocity of 4.5 m/s. For the point at which the string makes an angle of 30° with the vertical, what is (a) the speed, (b) the potential energy, and (c) the tension in the string? (d) What is the angle of the string with the vertical when the bob reaches its greatest height?

37 •• A pendulum consists of a string of length L and a bob of mass m. The string is brought to a horizontal position and the bob is given the minimum initial speed enabling the pendulum to make a full turn in the vertical plane. (a) What is the maximum kinetic energy K of the bob? (b) What is the tension in the string when the kinetic energy is maximum?

38 •• [SOLVE✓] A child whose weight is 360 N swings out over a pool of water using a rope attached to the branch of a tree at the edge of the pool. The branch is 12 m above ground level and the surface of the water is 1.8 m below ground level. The child holds on to the rope at a point 10.6 m from the branch and moves back until the angle between the rope and the vertical is 23°. When the rope is in the vertical position, the child lets go and drops into the pool. Find the speed of the child at the surface of the water.

39 •• [SSM] [SOLVE] Walking by a pond, you find a rope attached to a stout tree limb 5.2 m off the ground. You decide to use the rope to swing out over the pond. The rope is a bit frayed but supports your weight. You estimate that the rope might break if the tension is 80 N greater than your weight. You grab the rope at a point 4.6 m from the limb and move back to swing out over the pond. (a) What is the maximum safe initial angle between the rope and the vertical at which it will not break during the swing? (b) If you begin at this maximum angle, and the surface of the pond is 1.2 m below the level of the ground, with what speed will you enter the water if you let go of the rope when the rope is vertical?

40 •• A pendulum bob of mass m is attached to a light string of length L and is also attached to a spring of force constant k. With the pendulum in the position shown in Figure 7-23, the spring is at its unstretched length. If the bob is now pulled aside so that the string makes a *small* angle θ with the vertical and released, what is the speed of the bob as it passes through the equilibrium position? For θ in radians, if $|\theta| \ll 1$, then $\sin \theta \approx \tan \theta \approx \theta$ and $\cos \theta \approx 1$.

FIGURE 7-23 Problem 40

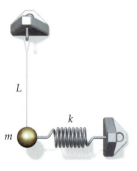

41 ••• A pendulum is suspended from the ceiling and attached to a spring fixed to the floor directly below the pendulum support (Figure 7-24). The mass of the pendulum bob is m, the length of the pendulum is L, and the spring constant is k. The unstretched length of the spring is $L/2$ and the distance between the floor and ceiling is 1.5L. The pendulum is pulled aside so that it makes an angle θ with the vertical and is then released from rest. Obtain an expression for the speed of the pendulum bob when $\theta = 0$.

FIGURE 7-24 Problem 41

The Conservation of Energy

42 • **ISOLVE** In a volcanic eruption, 4 km³ of mountain with a density of 1600 kg/m³ was lifted an average height of 500 m. (*a*) How much energy in joules was released in this eruption? (*b*) The energy released by thermonuclear bombs is measured in megatons of TNT, where 1 megaton of TNT = 4.2×10^{15} J. Convert your answer for (*a*) to megatons of TNT.

43 •• **ISOLVE** An 80-kg physics student climbs a 120-m-high hill. (*a*) What is the increase in the gravitational potential energy of the student? (*b*) Where does this energy come from? (*c*) The student's body is 20 percent efficient; that is, for every 20 J that are converted to mechanical energy, 100 J of chemical energy are expended, with 80 J going into thermal energy. How much chemical energy is expended by the student during the climb?

Kinetic Friction

44 • **ISOLVE** A 2000-kg car moving at an initial speed of 25 m/s along a horizontal road skids to a stop in 60 m. (*a*) Find the energy dissipated by friction. (*b*) Find the coefficient of kinetic friction between the tires and the road.

45 • An 8-kg sled is initially at rest on a horizontal road. The coefficient of kinetic friction between the sled and the road is 0.4. The sled is pulled a distance of 3 m by a force of 40 N applied to the sled at an angle of 30° above the horizontal. (*a*) Find the work done by the applied force. (*b*) Find the energy dissipated by friction. (*c*) Find the change in the kinetic energy of the sled. (*d*) Find the speed of the sled after it has traveled 3 m.

46 • **SSM** Returning to Problem 19 and Figure 7-17, suppose that the surfaces described are not frictionless and that the coefficient of kinetic friction between the block and the surfaces is 0.30. Find (*a*) the speed of the block when it reaches the ramp, and (*b*) the distance that the block slides up the ramp before coming momentarily to rest. (Neglect any energy dissipated along the transition curve.)

47 • The 2-kg block in Figure 7-25 slides down a frictionless curved ramp, starting from rest at a height of 3 m. The block then slides 9 m on a rough horizontal surface before coming to rest. (*a*) What is the speed of the block at the bottom of the ramp? (*b*) What is the energy dissipated by friction? (*c*) What is the coefficient of friction between the block and the horizontal surface?

FIGURE 7-25 Problem 47

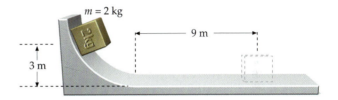

48 •• **ISOLVE** A 20-kg girl slides down a playground slide that is 3.2 m high. When she reaches the bottom of the slide, her speed is 1.3 m/s. (*a*) How much energy was dissipated by friction? (*b*) If the slide is inclined at 20°, what is the coefficient of friction between the girl and the slide?

49 •• In Figure 7-26, the coefficient of kinetic friction between the 4-kg block and the table is 0.35. (*a*) Find the energy dissipated by friction when the 2-kg block falls a distance y. (*b*) Find the total mechanical energy E of the two-block–earth system after the 2-kg block falls a distance y, assuming that $E = 0$ initially. (*c*) Use your result for (*b*) to find the speed of either block after the 2-kg block falls 2 m.

FIGURE 7-26 Problem 49

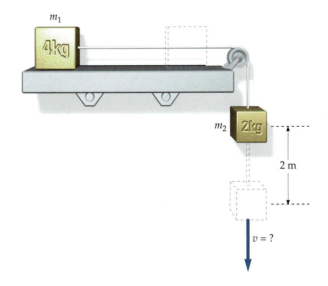

50 •• **SSM** A compact object of mass m moves in a horizontal circle of radius r on a rough table. It is attached to a horizontal string fixed at the center of the circle. The speed of the object is initially v_0. After completing one full trip around the circle, the speed of the object is $\frac{1}{2}v_0$. (*a*) Find the energy dissipated by friction during that one revolution in terms of m, v_0, and r. (*b*) What is the coefficient of kinetic friction? (*c*) How many more revolutions will the object make before coming to rest?

51 •• **iSOLVE** ✓ A 2.4-kg box has an initial velocity of 3.8 m/s upward along a plane inclined at 37° to the horizontal. The coefficient of kinetic friction between the box and the plane is 0.30. How far up the incline does the box travel? What is its speed when it passes its starting point on its way down the incline?

52 ••• A block of mass m rests on a plane inclined at θ with the horizontal. The block is attached to a spring of constant k as shown in Figure 7-27. The coefficients of static and kinetic friction between the block and plane are μ_s and μ_k, respectively. Very slowly, the spring is pulled upward along the plane until the block starts to move. (*a*) Obtain an expression for the extension d of the spring the instant the block moves. (*b*) Determine the value of μ_k such that the block comes to rest just as the spring is in its unstressed condition, that is, neither extended nor compressed.

FIGURE 7-27 Problem 52

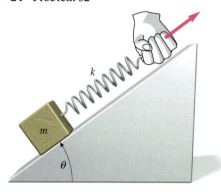

Mass and Energy

53 • (*a*) Calculate the rest energy in 1 g of dirt. (*b*) If you could convert this energy into electrical energy and sell it for $0.10/kW·h, how much money would you get? (*c*) If you could power a 100-W light bulb with this energy, for how long could you keep the bulb lit?

54 • One kiloton of TNT can be detonated, yielding an explosive energy of roughly 5×10^{12} J. What is the mass equivalent of the energy of the explosion?

55 • A muon has a rest energy of 105.7 MeV. Calculate its rest mass in kilograms.

56 • **SSM** If a black hole and a "normal" star orbit each other, gases from the star falling into the black hole can be heated millions of degrees by frictional heating in the black hole's accretion disk. When the gases are heated that much, they begin to radiate light in the X-ray region of the spectrum. Cygnus X-1, the second brightest source in the X-ray sky, is thought to be one such binary system; it radiates an estimated

power of 4×10^{31} W. If we assume that 1 percent of the infalling mass-energy escapes as X rays, at what rate is the black hole gaining mass?

57 • For the fusion reaction in Example 7-15, calculate the number of reactions per second that are necessary to generate 1 kW of power.

58 • Use Table 7-1 to calculate how much energy is needed to remove one neutron from ^{4}He, leaving ^{3}He plus a neutron.

59 • A free neutron at rest decays into a proton plus an electron:

$$n \rightarrow p + e$$

Use Table 7-1 to calculate the energy released in this reaction.

60 •• In one nuclear fusion reaction, two ^{2}H nuclei combine to produce ^{4}He. (*a*) How much energy is released in this reaction? (*b*) How many such reactions must take place per second to produce 1 kW of power?

61 •• A large nuclear power plant produces 3000 MW of electrical power by nuclear fission, which converts matter into energy. (*a*) How many kilograms of matter does the plant consume in one year? (Assume an efficiency of 33 percent for a nuclear power plant.) (*b*) In a coal-burning power plant, each kilogram of coal releases 31 MJ of thermal energy when burned. How many kilograms of coal are needed each year for a 3000-MW plant? (Assume an efficiency of 38 percent for a coal-burning power plant.)

General Problems

62 •• **SSM** A block of mass m, starting from rest, is pulled up a frictionless inclined plane that makes an angle θ with the horizontal by a string parallel to the plane. The tension in the string is T. After traveling a distance L, the speed of the block is v. The work done by the tension T is (*a*) $mgL \sin \theta$, (*b*) $mgL \cos \theta + \frac{1}{2}mv^2$, (*c*) $mgL \sin \theta + \frac{1}{2}mv^2$, (*d*) $mgL \cos \theta$, (*e*) $TL \cos \theta$.

63 •• A block of mass m slides with constant velocity v down a plane inclined at θ with the horizontal. During the time interval Δt, the energy dissipated by friction is (*a*) $mgv \Delta t \tan \theta$, (*b*) $mgv \Delta t \sin \theta$, (*c*) $\frac{1}{2}mv^3 \Delta t$, (*d*) cannot be determined without knowing the coefficient of kinetic friction.

64 • **iSOLVE** A 3.5-kg box rests on a horizontal frictionless surface in contact with a spring of spring constant 6800 N/m. The spring is fixed at its other end and is initially at its uncompressed length. A constant horizontal force of 70 N is applied to the box so that the spring compresses. Determine the distance the spring is compressed when the box is momentarily at rest.

65 • **SSM** **iSOLVE** The average energy per unit time per unit area that reaches the upper atmosphere of the earth from the sun, called the solar constant, is 1.35 kW/m². Because of absorption and reflection by the atmosphere, about 1 kW/m² reaches the surface of the earth on a clear day. How much energy is collected during 8 h of daylight by a window 1 m by 2 m? The window is on a mount that rotates, keeping the window facing the sun so the sun's rays remain perpendicular to the window.

66 •• (*a*) Using the value of the solar constant given in Problem 65 and the known distance from the earth to the sun, find the total energy that the sun radiates every second. (*b*) This energy is due to fusion reactions such as the one detailed in Example 7-15. There are several such reactions taking place in the "Phoenix cycle" in the sun, but the overall result is that 4 hydrogen nuclei fuse together to form 1 helium nucleus, liberating 26.7 MeV. If we assume that the sun is primarily made up of hydrogen, and that it will continue burning until roughly 10% of its hydrogen fuel is used up, use the known mass of the sun to determine approximately how long it can keep burning (that is, determine the solar lifetime).

67 • **ISOLVE**✓ After the 1250-kg jet-powered car *Spirit of America* went out of control during a test drive at Bonneville Salt Flats, Utah, it left skid marks about 9.5 km long. (*a*) If the car was moving initially at a speed of 708 km/h, estimate the coefficient of kinetic friction μ_k. (*b*) What was the kinetic energy K of the car 60 s after the brakes were applied?

68 •• **ISOLVE** A T-bar tow is required to pull 80 skiers up a 600-m slope inclined at 15° above the horizontal at a speed of 2.5 m/s. The coefficient of kinetic friction is 0.06. Find the motor power required if the mass of the average skier is 75 kg.

69 •• **ISOLVE** A 2-kg box is projected with an initial speed of 3 m/s up a rough plane inclined at 60° above the horizontal. The coefficient of kinetic friction is 0.3. (*a*) List all the forces acting on the box. (*b*) How far up the plane does the box slide before it stops momentarily? (*c*) What is the energy dissipated by friction as the box slides up the plane? (*d*) What is the speed of the box when it again reaches its initial position?

70 • **SSM** **ISOLVE** A 1200-kg elevator can safely carry a maximum load of 800 kg. What is the power provided by the electric motor powering the elevator when the elevator ascends with a full load at a speed of 2.3 m/s?

71 •• **ISOLVE** To reduce the power requirement of elevator motors, elevators are counterbalanced with weights connected to the elevator by a cable that runs over a pulley at the top of the elevator shaft. If the elevator in Problem 70 is counterbalanced with a mass of 1500 kg, what is the power provided by the motor when the elevator ascends fully loaded at a speed of 2.3 m/s? How much power is provided by the motor when the elevator ascends without a load at 2.3 m/s?

72 •• **ISOLVE**✓ The spring constant of a toy dart gun is 5000 N/m. To cock the gun the spring is compressed 3 cm. The 7-g dart, fired straight upward, reaches a maximum height of 24 m. Determine the energy dissipated by air friction during the dart's ascent. Estimate the speed of the projectile when it returns to its starting point.

73 •• **SSM** **ISOLVE**✓ In a volcanic eruption, a 2-kg piece of porous volcanic rock is thrown vertically upward with an initial speed of 40 m/s. It travels upward a distance of 50 m before it begins to fall back to the earth. (*a*) What is the initial kinetic energy of the rock? (*b*) What is the increase in thermal energy due to air friction during ascent? (*c*) If the increase in thermal energy due to air friction on the way down is 70% of that on the way up, what is the speed of the rock when it returns to its initial position?

74 •• A block of mass m starts from rest at a height h and slides down a frictionless plane inclined at θ with the horizontal as shown in Figure 7-28. The block strikes a spring of force constant k. Find the compression of the spring when the block is momentarily at rest.

FIGURE 7-28 Problem 74

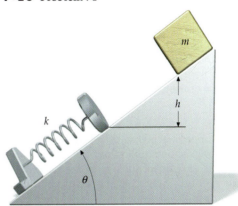

75 • **SSM** A 1.5×10^4-kg stone slab rests on a steel girder. On a very hot day, you find that the girder has expanded, lifting the slab by 0.1 cm. (*a*) What work does the girder do on the slab? (*b*) Where does the energy come from to lift the slab? Give a microscopic picture of what is happening in the steel.

76 •• **ISOLVE**✓ A 1500-kg car traveling at 24 m/s is at the foot of a hill that rises 120 m in 2.0 km. At the top of the hill, the speed of the car is 10 m/s. Assuming constant acceleration, find the average power delivered by the car's engine, neglecting any internal frictional losses.

77 •• **SSM** A block of mass m is suspended from the ceiling by a spring and is free to move vertically in the y direction as indicated (Figure 7-29). We are given that the potential energy as a function of position is $U = \frac{1}{2}ky^2 - mgy$. (*a*) Using a spreadsheet program or graphing calculator, make a graph of U as a function of y. For what value of y is the spring *unstretched*? (*b*) From the expression given for U, find the net force acting on m at any position y. (*c*) The block is released from rest at $y = 0$; if there is no friction, what is the maximum value of y, y_{max}, that will be reached by the mass? Indicate y_{max} on your graph. (*d*) Now consider the effect of friction. The mass ultimately settles into an equilibrium position y_{eq}. Find this point on your sketch. (*e*) Find the amount of thermal energy produced by friction from the start of the operation to the final equilibrium.

FIGURE 7-29 Problem 77

78 •• A spring-loaded gun is cocked by compressing a short, strong spring by a distance d. It fires a signal flare of mass m directly upward. The flare has speed v_0 as it leaves the spring and is observed to rise to a maximum height h above the point where it leaves the spring. After it leaves the spring, effects of drag force by the air on the flare are significant. (Express answers in terms of m, v_0, d, h, and g.) (a) How much work is done on the spring in the course of the compression? (b) What is the value of the spring constant k? (c) How much mechanical energy is converted to thermal energy because of the drag force of the air on the flare between the time of firing and the time at which maximum elevation is reached?

79 •• **iSOLVE** A roller-coaster car having a total mass (including passengers) of 500 kg travels freely along the winding frictionless track shown in Figure 7-30. Points A, E, and G are on horizontal straight sections, all at the same height of 10 m above the ground. Point C is at a height of 10 m above the ground on an inclined section of slope angle 30°. Point B is at the crest of a hill, while point D is at ground level at the bottom of a valley; the radius of curvature at both of these points is 20 m. Point F is at the middle of a banked horizontal curve with a radius of curvature of 30 m, and at the same height of 10 m above the ground as points A, E, and G. At point A the speed of the car is 12 m/s. (a) If the car is just barely able to make it over the hill at point B, what is the height of that point above the ground? (b) If the car is just barely able to make it over the hill at point B, what is the magnitude of the total force exerted on the car by the track at that point? (c) What is the acceleration of the car at point C? (d) What are the magnitude and direction of the total force exerted on the car by the track at point D? (e) What are the magnitude and direction of the total force exerted on the car by the track at point F? (f) At point G a constant braking force is applied to the car, bringing it to a halt in a distance of 25 m. What is the magnitude of the braking force?

FIGURE 7-30 Problem 79

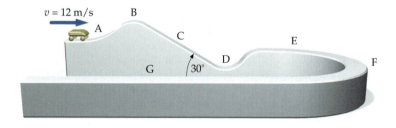

80 • **SSM** **iSOLVE** An elevator (mass $M = 2000$ kg) is moving downward at $v_0 = 1.5$ m/s. A braking system prevents the downward speed from increasing. (a) At what rate (in J/s) is the braking system converting mechanical energy to thermal energy? (b) While the elevator is moving downward at $v_0 = 1.5$ m/s, the braking system fails and the elevator is in free-fall for a distance $d = 5$ m before hitting the top of a large safety spring with force constant $k = 1.5 \times 10^4$ N/m. After the elevator hits the top of the spring, we want to know the distance Δy that the spring is compressed before the elevator is brought to rest. Write an algebraic expression for the value of Δy in terms of the known quantities M, v_0, g, k, and d, and substitute the given values to find Δy.

81 • To measure the force of friction (rolling friction plus air drag) on a moving car, engineers turn off the engine and allow the car to coast down hills of known steepness. The engineers collect the following data:

1. On a 2.87° hill, the car can coast at a steady 20 m/s.

2. On a 5.74° hill, the steady coasting speed is 30 m/s.

The total mass of the car is 1000 kg. (a) What is the force of friction at 20 m/s (F_{20}) and at 30 m/s (F_{30})? (b) How much useful power must the engine deliver to drive the car on a level road at steady speeds of 20 m/s (P_{20}) and 30 m/s (P_{30})? (c) At full throttle, the engine delivers 40 kW. What is the angle of the steepest incline up which the car can maintain a steady 20 m/s? (d) Assume that the engine delivers the same total useful work from each liter of gas, no matter what the speed. At 20 m/s on a level road, the car gets 12.7 km/L. How many kilometers per liter does it get if it goes 30 m/s instead?

82 •• A 2-kg block is released 4 m from a massless spring with a force constant $k = 100$ N/m that is along a plane inclined at 30°, as shown in Figure 7-31. (a) If the plane is frictionless, find the maximum compression of the spring. (b) If the coefficient of kinetic friction between the plane and the block is 0.2, find the maximum compression. (c) For the plane in Part (b), how far up the incline will the block travel after leaving the spring?

FIGURE 7-31 Problem 82

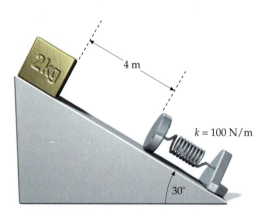

83 •• **iSOLVE** A train with a total mass of 2×10^6 kg rises 707 m while traveling a distance of 62 km at an average speed of 15.0 km/h. The frictional force is 0.8 percent of the weight. Find (a) the kinetic energy of the train, (b) the total change in its potential energy, (c) the energy dissipated by friction, and (d) the power output of the train's engines.

84 •• **SSM** While driving, one expects to expend more power when accelerating than when driving at a constant speed. (a) Neglecting friction, calculate the energy required to give a 1200-kg car a speed of 50 km/h. (b) If friction (rolling friction and air drag) results in a retarding force of 300 N at a speed of 50 km/h, what is the energy needed to move the car a distance of 300 m at a constant speed of 50 km/h? (c) Assuming the energy losses due to friction in Part (a) are 75 percent of those found in Part (b), estimate the ratio of the energy consumption for the two cases considered.

85 ••• **SSM** A pendulum consisting of a string of length L with a small bob on the end (mass M) is pulled horizontally, then released (Figure 7-32). At the lowest point of the swing, the string catches on a thin peg a distance R above the lowest point. Show that R must be smaller than $2L/5$ if the bob is to swing around the peg in a full circle.

FIGURE 7-32 Problem 85

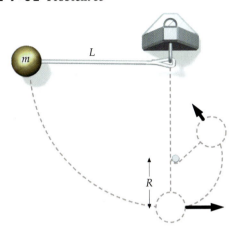

86 •• A rifle bullet is fired into a massive, stationary block of wood to determine its penetrating power. It is found that it penetrates the wood to a distance D. Another bullet that has the same mass, but double the velocity of the first bullet, is fired into an identical block. Assume that the average force acting on the bullet from the wood doesn't depend on the bullet's speed. The penetration depth is now found to be (a) D, (b) $2D$, (c) $4D$, (d) undeterminable from the information given.

87 •• A standard introductory physics lab to examine the conservation of energy and Newton's laws is shown in Figure 7-33. A glider is set up on a linear air track and is attached by a string over a massless pulley to a hanging weight. The mass of the glider is M, while the mass of the hanging weight is m. When the air supply to the air track is turned on, the track is essentially frictionless; you then release the hanging weight and measure the speed of the glider after the weight has fallen a given distance (Y). To show that the laws of physics are consistent, work out the speed of the glider in two different ways: (a) by conservation of mechanical energy; (b) by using Newton's second and third laws directly—make a free-body diagram for the two masses, find their acceleration, and use that to determine the speed of the glider.

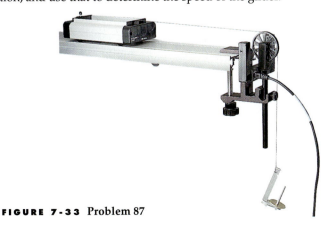

FIGURE 7-33 Problem 87

88 •• **SSM** In one model of jogging, the energy expended is assumed to go into accelerating and decelerating the legs. If the mass of the leg is m and the running speed is v, the energy needed to accelerate the leg from rest to v is $\frac{1}{2}mv^2$, and the same energy is needed to decelerate the leg back to rest for the next stride. Thus the energy required for each stride is mv^2. Assume that the mass of a man's leg is 10 kg and that he jogs at a speed of 3 m/s with 1 m between one footfall and the next. Therefore, the energy he must provide to his legs in each second is $3 \times mv^2$. Calculate the rate of the man's energy expenditure using this model and assuming that his muscles have an efficiency of 20 percent.

89 •• A high school chemistry teacher once suggested measuring the magnitude of free-fall acceleration by the following method: Hang a weight on a very fine thread (length L) to make a pendulum, which is then put on the edge of a table so that the weight is a height H above the floor when at its lowest point. Pull the pendulum back so that the thread makes an angle θ_0 with the vertical. At the bottom point of the pendulum, place a razor blade that is positioned to cut through the thread at its lowest point. Once the thread is cut, the weight is projected horizontally, and is thrown a distance D from the edge of the table. The idea was that the measurement of D as a function of θ_0 should somehow determine g. Apart from some obvious experimental difficulties, the experimental had one fatal flaw: D doesn't depend on g! Show that this is true, and that D depends only on the angle θ_0.

90 •• A 5-kg block is held against a spring of force constant 20 N/cm, compressing it 3 cm. The block is released and the spring extends, pushing the block along a horizontal surface. The coefficient of friction between the surface and the block is 0.2. (a) Find the work done on the block by the spring as it extends from its compressed position to its equilibrium position. (b) Find the energy dissipated by friction while the block moves the 3 cm to the equilibrium position of the spring. (c) What is the speed of the block when the spring is at its equilibrium position? (d) If the block is not attached to the spring, how far will it slide along the surface before coming to rest?

91 •• A block of mass m is dropped onto the top of a vertical spring whose force constant is k (Figure 7-21, Problem 26). If the block is released from a height h above the top of the spring, (a) what is the maximum kinetic energy of the block? (b) What is the maximum compression of the spring? (c) At what compression is the block's kinetic energy half its maximum value?

92 ••• The bob of a pendulum of length L is pulled aside so that the string makes an angle θ_0 with the vertical, and the bob is then released. In Example 7-2 the conservation of energy was used to obtain the speed of the bob at the bottom of its swing. In this problem, you are to obtain the same result using Newton's second law. (a) Show that the tangential component of Newton's second law gives $dv/dt = -g \sin \theta$, where v is the speed and θ is the angle between the string and the vertical. (b) Show that v can be written $v = L\,d\theta/dt$ (c) Use this result and the chain rule for derivatives to obtain

$$\frac{dv}{dt} = \frac{dv}{d\theta}\frac{d\theta}{dt} = \frac{dv}{d\theta}\frac{v}{L}$$

(d) Combine the results of (a) and (c) to obtain $v\,dv = -gL \sin \theta\,d\theta$. (e) Integrate the left side of the equation in Part (d) from

$v = 0$ to the final speed v and the right side from $\theta = \theta_0$ to $\theta = 0$, and show that the result is equivalent to $v = \sqrt{2gh}$, where h is the original height of the bob above the bottom.

93 ••• A rock climber is rappelling down the face of a cliff when his hold slips and he slides down over the rock face, supported only by the bungee cord he attached to the top of the cliff. The cliff face is in the form of a smooth quarter-cylinder with height (and radius) $H = 300$ m (see Figure 7-34). Treat the bungee cord as a spring with spring constant $k = 5$ N/m and unstretched length $L = 60$ m. The climber's mass is 85 kg. (*a*) Using a spreadsheet program, make a graph of the rock climber's potential energy as a function of s, his distance from the top of the cliff *measured along the curved surface*. Use values of s between 60 m and 200 m. (*b*) If his fall began when he was a distance $s_i = 60$ m from the top of the cliff, and ends when he is a distance $s_f = 110$ m from the top, determine how much energy is dissipated by friction between the time he initially slipped and finally came to a stop.

FIGURE 7-34 Problem 93

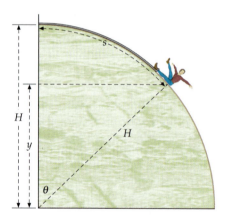

94 ••• [SSM] A block of wood (mass M) is connected to two massless springs as shown in Figure 7-35. Each spring has unstretched length L and spring constant k. (*a*) If the block is displaced a distance x, as shown, what is the change in the potential energy stored in the springs? (*b*) What is the magnitude of the force pulling the block back toward the equilibrium position? (*c*) Using a spreadsheet program or graphing calculator, make a graph of the potential energy U as a function of x for $0 \le x \le 0.2$ m. Assume $k = 1$ N/m, $L = 0.1$ m, and $M = 1$ kg. (*d*) If the block is displaced a distance $x = 0.1$ m and released, what is its speed as it passes the equilibrium point? Assume that the block is resting on a frictionless surface.

FIGURE 7-35 Problem 94

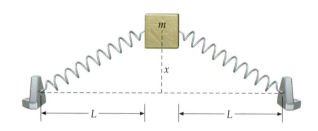

Systems of Particles and Conservation of Linear Momentum

THE MASS OF THIS GOLF CLUB TIMES ITS VELOCITY IS CALLED THE MOMENTUM OF THE CLUB. WHEN THE GOLFER STRIKES THE BALL WITH THE CLUB, SOME OF THE MOMENTUM OF THE CLUB IS TRANSFERRED TO THE BALL WITH DRAMATIC EFFECT.

? **If the golfer uses a more massive club, how will the distance traveled by the ball be affected? (See Example 8-12.)**

We saw in Chapters 6 and 7 that the work–kinetic energy theorem for particles and the work–energy theorem for systems of particles are useful tools for analyzing a variety of situations and solving a wide variety of problems. But in order to analyze the motion of extended objects, and to understand collisions between objects like the golf club and ball in the photo above, we need to add two more theorems to our problem-solving toolbox. These are **Newton's second law for systems** (Section 8-3) and **the impulse–momentum theorem** (Section 8-6). The impulse–momentum theorem for systems introduces the conservation of momentum, one of the universal laws of physics.

The mass of a particle times its velocity is called the **momentum** of the particle. The momentum of a system of particles is the vector sum of the momenta of the individual particles in the system. When the net external force acting on a system remains zero, the system's momentum remains constant. The momentum of an isolated system is a conserved quantity, just like the energy of an isolated system.

➤ In this chapter, we will use conservation of momentum to analyze collisions between billiard balls, cars, and subatomic particles, and the decay of radioactive nuclei. We will apply Newton's laws to the motion of extended objects, such as cars, rockets, and people, by realizing that there is one point of a system, the *center of mass*, that moves as if all the mass of the system were concentrated at that point and all the external forces acting on the system were acting exclusively on that point.

8-1 The Center of Mass

The motion of any object or system of particles can be described in terms of the motion of the center of mass (which may be thought of as the bulk motion of the system) plus the motion of individual particles in the system relative to the center of mass. Let's first consider a simple system of two particles in one dimension. If two point particles with masses m_1 and m_2 have coordinates x_1 and x_2 on the x axis, then the center-of-mass coordinate x_{cm} is defined by

$$Mx_{cm} = m_1x_1 + m_2x_2 \qquad \text{8-1}$$

where $M = m_1 + m_2$ is the total mass of the system. In the case of just two particles, the center of mass lies at some point on the line between the particles; if the particles have equal masses, then the center of mass is midway between them (Figure 8-1).

If the two particles are of unequal mass, then the center of mass is closer to the more massive particle (Figure 8-2).

If we choose the origin and direction of the x axis such that the position of m_1 is at the origin and that of m_2 is on the positive x axis, then $x_1 = 0$ and $x_2 = d$, where d is the distance between the particles (Figure 8-3) and the center of mass is given by

$$Mx_{cm} = m_1x_1 + m_2x_2 = m_1(0) + m_2d$$

$$x_{cm} = \frac{m_2}{M}d = \frac{m_2}{m_1 + m_2}d \qquad \text{8-2}$$

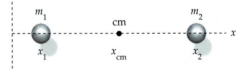

FIGURE 8-1

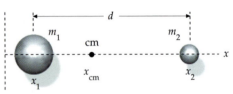

FIGURE 8-2

EXERCISE A 4-kg mass is at the origin and a 2-kg mass is at $x = 6$ cm. Find x_{cm}. (*Answer* $x_{cm} = 2$ cm)

We can generalize from two particles in one dimension to a system of many particles in three dimensions. For N particles,

$$Mx_{cm} = m_1x_1 + m_2x_2 + m_3x_3 + \cdots + m_Nx_N = \sum_i m_ix_i \qquad \text{8-3}$$

where again $M = \sum_i m_i$ is the total mass of the system. Similarly,

$$My_{cm} = \sum_i m_iy_i \quad \text{and} \quad Mz_{cm} = \sum_i m_iz_i \qquad \text{8-4}$$

In vector notation, $\vec{r}_i = x_i\hat{i} + y_i\hat{j} + z_i\hat{k}$ is the position vector of the ith particle. The position vector of the **center of mass**, $\vec{r}_{cm}$, is defined by

$$M\vec{r}_{cm} = m_1\vec{r}_1 + m_2\vec{r}_2 + \cdots = \sum_i m_i\vec{r}_i \qquad \text{8-5}$$

DEFINITION—CENTER OF MASS

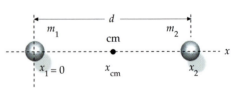

FIGURE 8-3

where $\vec{r}_{cm} = x_{cm}\hat{i} + y_{cm}\hat{j} + z_{cm}\hat{k}$.

To find the center of mass of a continuous object we replace the sum in Equation (8-5) with an integral:

$$M\vec{r}_{cm} = \int \vec{r}\, dm \qquad\qquad 8\text{-}6$$

CENTER OF MASS, CONTINUOUS OBJECT

where dm is an element of mass located at position $\vec{r}$, as shown in Figure 8-4. Examples involving Equation 8-6 are presented in Section 8-2. For highly symmetric objects, the center of mass is at the center of symmetry. For example, the center of mass of a uniform cylinder is located at its geometric center. Consider the following examples.

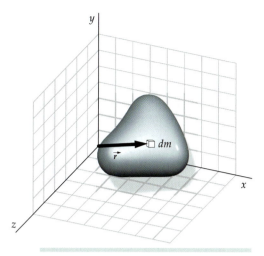

FIGURE 8-4 Mass element dm located at $\vec{r}$ for finding the center of mass by integration.

THE CENTER OF MASS OF A WATER MOLECULE　　　**EXAMPLE 8-1**

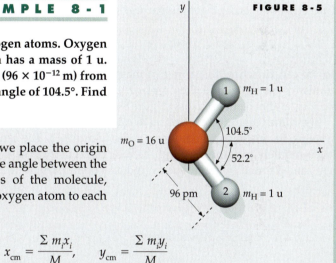

FIGURE 8-5

A water molecule consists of an oxygen atom and two hydrogen atoms. Oxygen has a mass of 16 unified mass units (u) and each hydrogen has a mass of 1 u. The hydrogen atoms are each at an average distance of 96 pm (96 × 10⁻¹² m) from the oxygen atom, and are separated from one another by an angle of 104.5°. Find the center of mass of a water molecule.

PICTURE THE PROBLEM The calculation is simplified if we place the origin at the location of the oxygen atom, with the x axis bisecting the angle between the hydrogen atoms (Figure 8-5). Then, given the symmetries of the molecule, the center of mass will be on the x axis, and the line from the oxygen atom to each hydrogen atom will make an angle of 52.2°.

1. The location of the center of mass is given by its coordinates, x_{cm} and y_{cm}:

$$x_{cm} = \frac{\Sigma\, m_i x_i}{M}, \qquad y_{cm} = \frac{\Sigma\, m_i y_i}{M}$$

2. Writing these out explicitly gives:

$$x_{cm} = \frac{m_{H1}x_{H1} + m_{H2}x_{H2} + m_O x_O}{m_{H1} + m_{H2} + m_O}$$

$$y_{cm} = \frac{m_{H1}y_{H1} + m_{H2}y_{H2} + m_O y_O}{m_{H1} + m_{H2} + m_O}$$

3. We have chosen the origin to be the location of the oxygen atom, so both the x and y coordinates of the oxygen atom are zero. The x and y coordinates of the hydrogen atoms are calculated from the 52.2° angle each hydrogen makes with the x axis:

$$x_O = y_O = 0$$

$$x_{H1} = x_{H2} = 96\ \text{pm}\ \cos 52.2° = 59\ \text{pm}$$

$$y_{H1} = 96\ \text{pm}\ \sin 52.2° = 76\ \text{pm}$$

$$y_{H2} = -96\ \text{pm}\ \sin 52.2° = -76\ \text{pm}$$

4. Substituting the coordinate and mass values into step 2 gives x_{cm}:

$$x_{cm} = \frac{(1\ \text{u})(59\ \text{pm}) + (1\ \text{u})(59\ \text{pm}) + (16\ \text{u})0}{1\ \text{u} + 1\ \text{u} + 16\ \text{u}}$$

$$= 6.6\ \text{pm}$$

$$y_{cm} = \frac{(1\ \text{u})(76\ \text{pm}) + (1\ \text{u})(-76\ \text{pm}) + (16\ \text{u})0}{1\ \text{u} + 1\ \text{u} + 16\ \text{u}}$$

$$= 0$$

5. The center of mass is on the x axis:

$$\vec{r}_{cm} = x_{cm}\,\hat{i} + y_{cm}\,\hat{j}$$
$$= 6.6\text{ pm }\hat{i} + 0\hat{j} = \boxed{6.6\text{ pm }\hat{i}}$$

REMARKS That $y_{cm} = 0$ can be seen from the symmetry of the mass distribution. Also, the center of mass is very close to the relatively massive oxygen atom, as expected.

We also can solve Example 8-1 by first finding the center of mass of just the two hydrogen atoms. For a system of three particles Equation 8-5 is

$$M\vec{r}_{cm} = m_1\vec{r}_1 + m_2\vec{r}_2 + m_3\vec{r}_3$$

The first two terms on the right side of this equation are related to the center of mass of the first two particles $\vec{r}'_{cm}$:

$$m_1\vec{r}_1 + m_2\vec{r}_2 = (m_1 + m_2)\vec{r}'_{cm}$$

The center of mass of the three-particle system can then be written

$$M\vec{r}_{cm} = (m_1 + m_2)\vec{r}'_{cm} + m_3\vec{r}_3$$

So we can first find the center of mass for two of the particles, the hydrogen atoms for example, and then replace them with a single particle of total mass $m_1 + m_2$ at that center of mass (Figure 8-6).

The same technique enables us to calculate centers of mass for more complex systems, for instance, two uniform rods (Figure 8-7). The center of mass of each rod separately is at the center of the rod. The center of mass of the two-rod system can be found by treating each rod as a point particle at its individual center of mass.

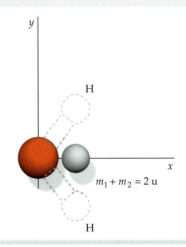

FIGURE 8-6 Example 8-1 with the two H atoms replaced by a single particle of mass $m_1 + m_2 = 2$ u on the x axis at the center of mass of the original atoms. The center of mass then falls between the oxygen atom at the origin and the calculated center of mass of the two hydrogen atoms.

FIGURE 8-7

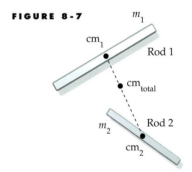

THE CENTER OF MASS OF A PLYWOOD SHEET

EXAMPLE 8-2

Find the center of mass of the uniform sheet of plywood shown in Figure 8-8*a*.

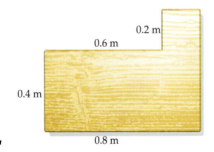

FIGURE 8-8*a*

Try It Yourself

PICTURE THE PROBLEM The sheet can be divided into two symmetrical parts (Figure 8-8*b*). The center of mass of each part is at its geometric center. Let m_1 be the mass of part 1 and m_2 be the mass of part 2. The total mass is $M = m_1 + m_2$. The masses are proportional to the areas.

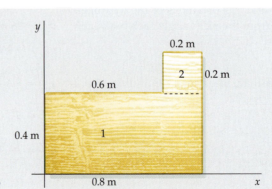

FIGURE 8-8*b*

Cover the column to the right and try these on your own before looking at the answers.

Steps	Answers
1. Write the x and y coordinates of the center of mass in terms of m_1 and m_2.	$Mx_{cm} = m_1 x_{1,cm} + m_2 x_{2,cm}$ $My_{cm} = m_1 y_{1,cm} + m_2 y_{2,cm}$
2. Divide through these equation by M and then substitute area ratios for the mass ratios.	$x_{cm} = \dfrac{A_1}{A} x_{1,cm} + \dfrac{A_2}{A} x_{2,cm}$ $y_{cm} = \dfrac{A_1}{A} y_{1,cm} + \dfrac{A_2}{A} y_{2,cm}$
3. Calculate the areas and the ratios of the areas.	$A_1 = 0.32 \text{ m}^2, \quad A_2 = 0.04 \text{ m}^2, \quad A = 0.36 \text{ m}^2$ $\dfrac{A_1}{A} = \dfrac{8}{9} \qquad \dfrac{A_2}{A} = \dfrac{1}{9}$
4. Write the x and y coordinates of the center-of-mass coordinates for each part by inspection of the figure.	$x_{1,cm} = 0.4 \text{ m}, \qquad y_{1,cm} = 0.2 \text{ m}$ $x_{2,cm} = 0.7 \text{ m}, \qquad y_{2,cm} = 0.5 \text{ m}$
5. Substitute these results to calculate x_{cm} and y_{cm}.	$x_{cm} = \boxed{0.433 \text{ m}}, \qquad y_{cm} = \boxed{0.233 \text{ m}}$

REMARKS 1. Note that the center of mass is very near the center of mass of part 1 because $m_1 = 8m_2$. 2. Placing the origin at the geometric center of part 1 of the sheet and drawing the x axis through the center of part 2 would make the calculation of the center of mass somewhat simpler.

Gravitational Potential Energy of a System

The gravitational potential energy of a system of particles in a uniform gravitational field is the same as if all the mass were concentrated at the center of mass. Let h_i be the height of the ith particle in a system above some reference level. The gravitational potential energy of the system is

$$U = \sum_i m_i g h_i = g \sum_i m_i h_i$$

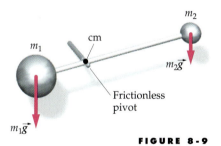

FIGURE 8-9

But, by definition of the center of mass, the height of the center of mass is given by

$$M h_{cm} = \sum_i m_i h_i$$

so

$$U = M g h_{cm} \qquad\qquad 8\text{-}7$$

We can use this result to locate the center of mass of an object experimentally. For example, two objects connected by a light rod will balance if the pivot is at the center of mass (Figure 8-9). If we pivot the system at any other point, the system will rotate until the potential energy is at a minimum, which occurs when the center of mass is at its lowest possible point directly below the pivot (Figure 8-10).

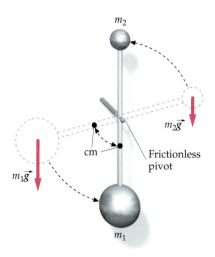

FIGURE 8-10

If we suspend any irregular object from a pivot, the object will hang so that its center of mass lies somewhere on the vertical line drawn directly downward from the pivot. Now suspend the object from another point and note where the vertical line now passes across the object. The center of mass will lie at the intersection of the two lines (Figure 8-11).

*8-2 Finding the Center of Mass by Integration

In this section we find the center of mass by integration (Equation 8-6):

$$M\vec{r}_{cm} = \int \vec{r} \, dm$$

We will use the simple problem of finding the center of mass of a uniform thin rod to illustrate the technique for setting up the integration. While we can find the answer to this problem by symmetry, here we will use integration.

Uniform Rod

We first choose a coordinate system. A good choice for a coordinate system is one with an x axis through the length of the rod, with the origin at one end of the rod (Figure 8-12). Shown on the figure is a mass element dm of length dx a distance x from the origin. Equation 8-6 thus gives

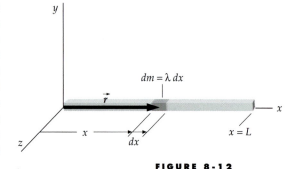

FIGURE 8-12

$$M\vec{r}_{cm} = \int \vec{r} \, dm = \int x\hat{i} \, dm$$

The mass is distributed on the x axis along the interval $0 \leq x \leq L$. To sweep dm along the mass distribution means the limits of the integral are 0 and L. (We sweep in the direction of increasing x.) The ratio dm/dx is the mass per unit length λ, so $dm = \lambda \, dx$:

$$M\vec{r}_{cm} = \hat{i} \int x \, dm = \hat{i} \int_0^L x\lambda \, dx$$

Because the rod is uniform, λ is constant and equal to M/L. Substituting for λ, we complete the calculation and obtain the expected result

$$\vec{r}_{cm} = \frac{1}{M} \hat{i} \, \lambda \, \frac{x^2}{2} \bigg|_0^L = \frac{1}{M} \hat{i} \, \frac{M}{L} \frac{L^2}{2} = \frac{1}{2} L\hat{i}$$

Semicircular Hoop

In calculating the center of mass of a uniform semicircular hoop, a good choice of coordinate axes is one with the origin at the center and with the x axis bisecting the semicircle (Figure 8-13). To find the center of mass we use Equation 8-6 $\left(M\vec{r}_{cm} = \int \vec{r} \, dm\right)$, where $\vec{r} = x\hat{i} + y\hat{j}$. The semicircular mass distribution suggests using polar coordinates, for which $x = r \cos \theta$ and $y = r \sin \theta$. With these substitutions we have

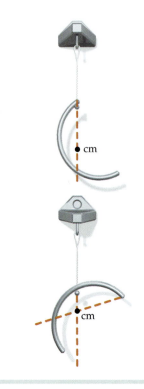

FIGURE 8-11 The center of mass of an irregular object can be found by suspending it from two points.

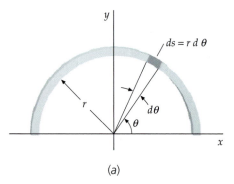

(a)

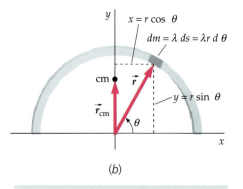

(b)

FIGURE 8-13 Geometry for calculating the center of mass of a semicircular hoop by integration.

$$M\vec{r}_{cm} = \int (x\hat{i} + y\hat{j})\,dm = \int r(\cos\theta\,\hat{i} + \sin\theta\,\hat{j})\,dm$$

Next we express dm in terms of $d\theta$. First, the mass element dm has length $ds = r\,d\theta$, so

$$dm = \lambda\,ds = \lambda r\,d\theta$$

where $\lambda = dm/ds$ is the mass per unit length. Thus, we have

$$M\vec{r}_{cm} = \int r(\cos\theta\,\hat{i} + \sin\theta\,\hat{j})\,\lambda r\,d\theta$$

Evaluating this integral involves sweeping dm along the semicircular mass distribution. This means $0 \le \theta \le \pi$. Sweeping in the direction of increasing θ, the integration limits go from 0 to π. That is,

$$\vec{r}_{cm} = \frac{r^2}{M}\int_0^\pi (\cos\theta\,\hat{i} + \sin\theta\,\hat{j})\lambda\,d\theta$$

Because the hoop is uniform, we know that $\lambda = M/\pi r$, where πr is the length of the semicircle. Substituting for λ and rearranging gives

$$\vec{r}_{cm} = \frac{r}{\pi}\hat{i}\int_0^\pi \cos\theta\,d\theta + \frac{r}{\pi}\hat{j}\int_0^\pi \sin\theta\,d\theta = \frac{r}{\pi}\hat{i}\,\sin\theta\,\Big|_0^\pi - \frac{r}{\pi}\hat{j}\,\cos\theta\,\Big|_0^\pi = \frac{2r}{\pi}\hat{j}$$

The center of mass is on the y axis a distance of $\dfrac{2}{\pi} r$ from the origin. Interestingly, it is outside of the material of the object.

8-3 Motion of the Center of Mass

Figure 8-14 is a multiflash photograph of a baton thrown into the air. Although the motion of the baton is complicated, the motion of the center of mass is simple. While the baton is in the air, the center of mass follows a parabolic path, the same path that would be followed by a point particle. We will show in general that the acceleration of the center of mass of a system of particles equals the net external force acting on the system divided by the total mass of the system. For the baton thrown into the air, the acceleration of the center of mass is g downward; to find it, we first find the velocity by differentiating Equation 8-5 with respect to time:

$$M\frac{d\vec{r}_{cm}}{dt} = m_1\frac{d\vec{r}_1}{dt} + m_2\frac{d\vec{r}_2}{dt} + \cdots = \sum_i m_i\frac{d\vec{r}_i}{dt}$$

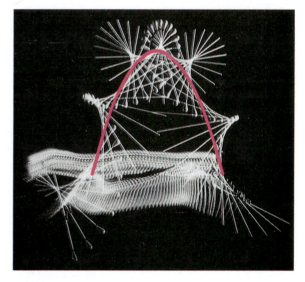

FIGURE 8-14

Because the time derivative of the position is the velocity, this gives

$$M\vec{v}_{cm} = m_1\vec{v}_1 + m_2\vec{v}_2 + \cdots = \sum_i m_i\vec{v}_i \qquad\qquad 8\text{-}8$$

Differentiating again, we obtain the accelerations:

$$M\vec{a}_{cm} = m_1\vec{a}_1 + m_2\vec{a}_2 + \cdots = \sum_i m_i\vec{a}_i \qquad\qquad 8\text{-}9$$

However, in accord with Newton's second law, $m_i\vec{a}_i$ equals the sum of the forces acting on the *i*th particle, so

$$\sum_i m_i\vec{a}_i = \sum_i \vec{F}_i$$

where the term on the right is the sum of all the forces acting on each and every particle in the system. Some of these forces are *internal* forces (exerted on a particle in the system by some other particle in the system) and others are *external* forces (exerted on a particle in the system by a particle not in the system). Thus,

$$M\vec{a}_{cm} = \sum_i \vec{F}_{i,int} + \sum_i \vec{F}_{i,ext} \qquad 8\text{-}10$$

According to Newton's third law, forces come in action–reaction pairs. Thus, for each internal force acting on a particle in the system there is an equal and opposite internal force acting on some other particle in the system. When we sum all the internal forces, each action–reaction force pair sums to zero, so $\sum \vec{F}_{i,int} = 0$. Equation 8-10 then becomes

$$\vec{F}_{net,ext} = \sum_i \vec{F}_{i,ext} = M\vec{a}_{cm} \qquad 8\text{-}11$$

NEWTON'S SECOND LAW FOR A SYSTEM

That is, the net external force acting on the system equals the product of the total mass M of the system and the acceleration of the center of mass $\vec{a}_{cm}$ of the system. Thus,

> The center of mass of a system moves like a particle of mass $M = \sum m_i$ under the influence of the net external force acting on the system.

This theorem is important because it describes the motion of the center of mass for *any* system of particles: The center of mass moves exactly like a single point particle of mass M acted on by only the external forces. The individual motion of any particle in the system is typically much more complex and is not described by Equation 8-11. The baton thrown into the air in Figure 8-14 is an example. The only external force acting is gravity, so the center of mass of the baton moves in a simple parabolic path, as would a point particle. (Equation 8-11 does not describe the rotational motion of one end of the baton about the center of mass.)

AN EXPLODING PROJECTILE   **E X A M P L E   8 - 3**

A projectile is fired into the air over level ground on a trajectory that would result in it landing 55 m away. However, at its highest point it explodes into two fragments of equal mass. Immediately following the explosion one fragment has a momentary speed of zero and then falls straight down to the ground. Where does the other fragment land? Neglect air resistance.

PICTURE THE PROBLEM Let the projectile be the system. Then the forces of the explosion are all internal forces. Since the only *external* force acting on the system is that due to gravity, the center of mass, which is midway between the two fragments, continues on its parabolic path as if there had been no explosion (Figure 8-15).

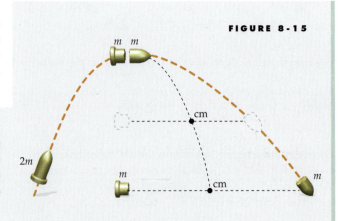

FIGURE 8-15

1. Let $x = 0$ be the initial position of the projectile. The landing positions x_1 and x_2 of the fragments are related to the final position of the center of mass by:

$$(2m)x_{cm} = mx_1 + mx_2$$

or

$$2x_{cm} = x_1 + x_2$$

2. At impact, $x_{cm} = R$ and $x_1 = 0.5R$, where $R = 55$ m is the range for the unexploded projectile. Solve for x_2:

$$x_2 = 2x_{cm} - x_1 = 2R - 0.5R = 1.5R$$

$$= 1.5(55 \text{ m}) = \boxed{82.5 \text{ m}}$$

REMARKS In Figure 8-16 height versus distance is plotted for exploding projectiles when fragment 1 has a horizontal velocity of half of the initial horizontal velocity. As in the original example in which fragment 1 falls straight down, the center of mass follows a normal parabolic trajectory.

EXERCISE If the fragment that falls straight down has twice the mass of the other fragment, how far from the launch position does the lighter fragment land? (*Answer*   2R)

REMARKS If both fragments have the same vertical component of velocity after the explosion, they land at the same time. If just after the explosion the vertical component of the velocity of one fragment is less than that of the other, the fragment with the smaller vertical velocity component will hit the ground first. As soon as it does, the ground exerts a force on it and the net external force on the system is no longer just the gravitational force. From that moment on, our analysis is invalid.

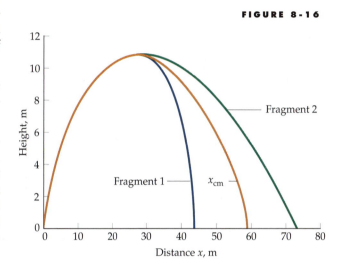

FIGURE 8-16

EXERCISE A cylinder rests on a sheet of paper on a table (Figure 8-17). You pull the paper to the right, causing the cylinder to roll leftward *relative to the paper*. How does the cylinder's center of mass move relative to the table? (*Answer*   It accelerates to the right, because the net external force acting on the cylinder is the frictional force to the right exerted on it by the paper. Try it. The cylinder may *appear* to move to the left, because relative to the paper it rolls leftward. However, relative to the table, which serves as an inertial reference frame, it moves to the right. If you mark the table with the original position of the cylinder, you will observe the center of mass move to the right *while the cylinder remains in contact with the paper*.)

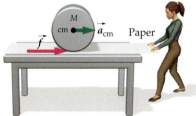

FIGURE 8-17

A special case of a system's center of mass in motion is the system with zero net external force acting on it. Here, $\vec{a}_{cm} = 0$, so the center of mass either remains at rest or moves with constant velocity. The internal forces and motion may be complex, but the behavior of the center of mass is simple. Further, if the net external force is not zero, but if a component of it in a given direction, say the x direction, remains zero, then $a_{cmx} = 0$ and v_{cmx} remains constant. An example of this is a projectile in the absence of air drag. The net external force on the projectile is the gravitational force. This force acts straight downward, so its component in any horizontal direction remains zero. It follows that the horizontal component of the velocity of the center of mass remains constant.

CHANGING PLACES IN A ROWBOAT **EXAMPLE 8-4**

Pete (mass 80 kg) and Dave (mass 120 kg) are in a rowboat (mass 60 kg) on a calm lake. Dave is near the bow of the boat, rowing, and Pete is at the stern, 2 m from the center. Dave gets tired and stops rowing. Pete offers to row, and after the boat comes to rest, they change places. How far does the boat move as Pete and Dave change places? (Neglect any horizontal force exerted by the water.)

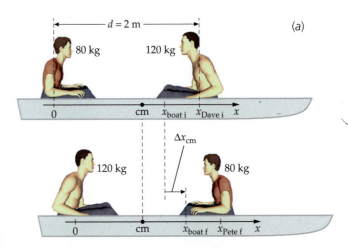

PICTURE THE PROBLEM Let the system be Dave, Pete, and the boat. There are no *external* forces in the horizontal direction, so the center of mass does not move horizontally relative to an inertial reference frame, such as the water. Relative to the boat, however, the center of mass does move. First find the displacement Δx_{cm} that the center of mass moves relative to the boat. Relative to the water, the boat must move the same distance Δx_{cm} in the opposite direction.

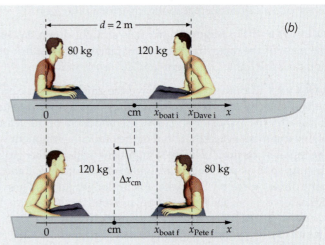

1. Make a sketch of the system in its initial and final configurations (Figure 8-18). Choose the origin at the back of the boat (Pete's initial position) and let $d = 2$ m so $x_{Pete\,i} = 0$ and $x_{Dave\,i} = d$. Let x_{boat} be the position of the center of mass of the boat. When Pete and Dave switch places they merely exchange positions.

FIGURE 8-18 (a) Pete and Dave changing places viewed from the reference frame of the water. (b) Pete and Dave changing places viewed from the reference frame of the boat.

2. Express the total mass M times the initial value of the center of mass. Repeat this for the final value of the center of mass.

$$Mx_{cm\,i} = m_{Pete}x_{Pete\,i} + m_{Dave}x_{Dave\,i} + m_{boat}x_{boat\,i}$$
$$= 0 + m_{Dave}d + m_{boat}x_{boat}$$
$$Mx_{cm\,f} = m_{Pete}x_{Pete\,f} + m_{Dave}x_{Dave\,f} + m_{boat}x_{boat\,f}$$
$$= m_{Pete}d + 0 + m_{boat}x_{boat}$$

3. Find the displacement of the center of mass relative to the boat by subtracting the first step 2 result from the second:

$$Mx_{cm\,f} - Mx_{cm\,i} = (m_{Pete} - m_{Dave})d + 0$$

so

$$x_{cm\,f} - x_{cm\,i} = \frac{(m_{Pete} - m_{Dave})d}{M}$$

4. Because the center of mass does not move relative to the water, the boat must move an equal distance in the opposite direction:

$$-(x_{cm\,f} - x_{cm\,i}) = \frac{(m_{Dave} - m_{Pete})d}{M}$$
$$= \frac{(120\ kg - 80\ kg)}{120\ kg + 80\ kg + 60\ kg}(2\ m)$$
$$= \boxed{0.308\ m}$$

REMARKS The location of the center of mass of the boat, relative to the boat, subtracted out in step 3. The displacement of the boat is independent of this location.

FIGURE 8-19

A SLIDING BLOCK **EXAMPLE 8-5**

A wedge of mass m_2 sits at rest on a scale as shown in Figure 8-19. A small block of mass m_1 slides down the frictionless incline of the wedge. Find the scale reading while the block slides.

PICTURE THE PROBLEM We choose the wedge plus block to be the system. Because the block accelerates down the wedge, the center of mass has acceleration components to the right and downward. The external forces on the system are the weights of the block and wedge, the force F_x exerted by the scale on the wedge, and the normal force F_n exerted upward by the scale. The scale reading is equal to the magnitude of F_n.

1. Draw a free-body diagram for the wedge–block system (Figure 8-20):

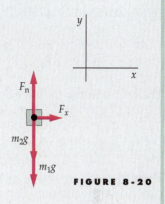

2. Write the vertical component of Newton's second law for the system and solve for F_n:

$$F_n - m_1g - m_2g = Ma_{cm,y} = (m_1 + m_2)a_{cm,y}$$

$$F_n = (m_1 + m_2)g + (m_1 + m_2)a_{cm,y}$$

3. Using Equation 8-9, express $a_{cm,y}$ in terms of the acceleration of the block a_{1y}:

$$Ma_{cm,y} = m_1a_{1y} + m_2a_{2y} = m_1a_{1y} + 0$$

$$a_{cm,y} = \frac{m_1}{m_1 + m_2}a_{1y}$$

4. From Example 4-7, a block sliding down a stationary incline has acceleration $g \sin \theta$ down the incline. Use trigonometry to find the y component of this acceleration and use it to find $a_{cm,y}$:

$$a_{1y} = -a_1 \sin \theta = -g \sin^2 \theta$$

so

$$a_{cm,y} = \frac{m_1}{m_1 + m_2}a_{1y}$$

$$= -\frac{m_1}{m_1 + m_2}g \sin^2 \theta$$

5. Substitute for $a_{cm,y}$ in the step 2 result and solve for F_n:

$$F_n = (m_1 + m_2)g + (m_1 + m_2)a_{cm,y}$$

$$= (m_1 + m_2)g - m_1g \sin^2 \theta = [(1 - \sin^2 \theta)m_1 + m_2]g$$

$$\boxed{= (m_1 \cos^2 \theta + m_2)g}$$

FIGURE 8-20

EXERCISE What values of F_n do you intuitively expect for $\theta = 0$ and for $\theta = 90°$? Show that $(m_1 \cos^2 \theta + m_2)g$ gives these expected values. (*Answer* For $\theta = 0$ we expect $F_n = (m_1 + m_2)g$; for $\theta = 90°$ we expect $F_n = m_2g$.)

EXERCISE Find the force component F_x exerted on the wedge by the scale. (*Answer* $F_x = m_1g \sin \theta \cos \theta$.)

8-4 Conservation of Linear Momentum

A particle's **linear momentum** $\vec{p}$ is defined as the product of its mass and velocity:

$$\vec{p} = m\vec{v}$$

8-12

DEFINITION—MOMENTUM OF A PARTICLE

Linear momentum is a vector quantity that may be thought of as a measurement of the effort needed to bring a particle to rest. For example, a heavy truck has more momentum than a light car traveling at the same speed. It takes a greater force to stop the truck in a given time than it does to stop the car in the same time. (The quantity $m\vec{v}$ is sometimes referred to as the *linear momentum* of a particle to distinguish it from the *angular momentum*, which is discussed in Chapter 10. When there is no need to clarify that the momentum being referred to is the linear momentum, the adjective *linear* will be suppressed and we will just refer to the *momentum*. Throughout the remainder of this chapter momentum is used when referring to linear momentum.)

Newton's second law can be written in terms of the momentum of a particle. Differentiating Equation 8-12 we obtain

$$\frac{d\vec{p}}{dt} = \frac{d(m\vec{v})}{dt} = m\frac{d\vec{v}}{dt} = m\vec{a}$$

Then substituting the force $\vec{F}_{net}$ for $m\vec{a}$,

$$\vec{F}_{net} = \frac{d\vec{p}}{dt} \qquad\qquad 8\text{-}13$$

Thus, the net force acting on a particle equals the time rate of change of the particle's linear momentum. (Newton's original statement of his second law was in fact in this form.)

The total momentum $\vec{P}_{sys}$ of a system of particles is the sum of the momenta of the individual particles:

$$\vec{P}_{sys} = \sum_i m_i\vec{v}_i = \sum_i \vec{p}_i$$

According to Equation 8-8, $\Sigma m_i\vec{v}_i$ equals the total mass M times the velocity of the center of mass:

$$\vec{P}_{sys} = \sum_i m_i\vec{v}_i = M\vec{v}_{cm} \qquad 8\text{-}14$$

TOTAL MOMENTUM OF A SYSTEM

Differentiating this equation, we obtain

$$\frac{d\vec{P}_{sys}}{dt} = M\frac{d\vec{v}_{cm}}{dt} = M\vec{a}_{cm}$$

But according to Newton's second law (Equation 8-11), $M\vec{a}_{cm}$ equals the net external force acting on the system. Thus,

$$\sum_i \vec{F}_{ext} = \vec{F}_{net,ext} = \frac{d\vec{P}_{sys}}{dt} \qquad 8\text{-}15$$

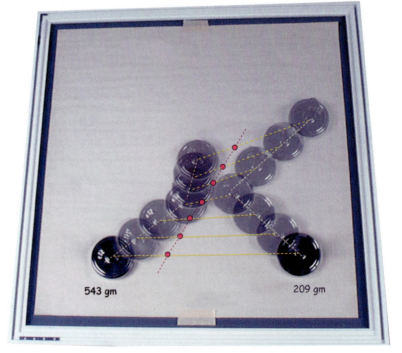

543 gm   209 gm

The two pucks are moving on an air cushion on a horizontal flat surface. (The hoses supplying the air are not shown.) The velocity of each puck changes in both magnitude and direction during the collision, but the velocity of the center of mass remains constant—unaffected by the internal forces of the collision.

When the net external force acting on a system of particles remains zero, the rate of change of the total momentum remains zero and the total momentum of the system remains constant:

$$\vec{P}_{sys} = \sum_i m_i\vec{v}_i = M\vec{v}_{cm} = \text{constant} \qquad (\vec{F}_{net,ext} = 0) \qquad 8\text{-}16$$

CONSERVATION OF MOMENTUM

This result is known as the **law of conservation of momentum:**

If the net external force on a system remains zero, the total momentum of the system remains constant.

This law is one of the most important in physics. It is more widely applicable than the law of conservation of mechanical energy because internal forces exerted by one particle in a system on another are often not conservative. Thus, these internal forces can change the total mechanical energy of the system, though they have no effect on the system's total momentum. If the total momentum of a system is constant, then the velocity of the center of mass of the system is constant. The law of conservation of momentum is a vector relation so it is valid component by component. For example, if the x component of the net external force on a system remains zero, then the x component of the total momentum of the system remains constant.

A SPACE REPAIR   **EXAMPLE 8-6**

During repair of the Hubble Space Telescope, an astronaut replaces a solar panel whose frame is bent. Pushing the detached panel away into space, she is propelled in the opposite direction. The astronaut's mass is 60 kg and the panel's mass is 80 kg. The astronaut is at rest relative to her spaceship when she shoves away the panel, and she shoves it at 0.3 m/s relative to the spaceship. What is her subsequent velocity relative to the space ship? (During this operation the astronaut is tethered to the ship; for our calculation, assume that the tether remains slack.)

PICTURE THE PROBLEM   The velocity of the astronaut can be found from the velocity of the panel using conservation of momentum. Choose the direction of motion of the panel to be positive.

1. Apply conservation of momentum to find the velocity of the astronaut. Since the total momentum is initially zero, it remains zero:

$$p_p + p_a = m_p v_p + m_a v_a = 0$$

2. Solve for the astronaut's velocity:

$$v_a = -\frac{m_p}{m_a} v_p$$

$$= -\frac{80 \text{ kg}}{60 \text{ kg}}(0.3 \text{ m/s}) = \boxed{-0.4 \text{ m/s}}$$

REMARKS   Although momentum is conserved, the mechanical energy of this system increased because chemical energy of the astronaut was converted to kinetic energy.

EXERCISE   Find the final kinetic energy of the astronaut–panel system. (*Answer*   8.4 J)

A RUNAWAY RAILROAD CAR   **EXAMPLE 8-7**

A runaway 14,000-kg railroad car is rolling horizontally at 4 m/s toward a switchyard. As it passes by a grain elevator, 2000 kg of grain are suddenly dropped into the car. How long does it take the car to cover the 500-m distance from the elevator to the switchyard? Assume that the grain falls straight down and that slowing due to rolling friction or air drag is negligible.

PICTURE THE PROBLEM We find the travel time that we seek from the distance traveled and the speed of the car. Consider the car and the grain as our system (Figure 8-21). There are no horizontal external forces acting on this system, so the horizontal component of the momentum of the system is conserved. The final speed of the grain-filled car is found from its final momentum, which equals the car's initial momentum. (The grain initially has no horizontal momentum.) Let m_c and m_g be the masses of the car and grain, respectively.

FIGURE 8-21

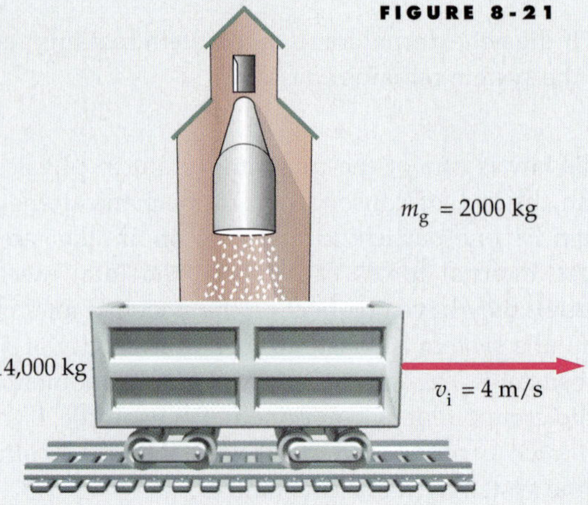

$m_g = 2000$ kg

$m_c = 14,000$ kg

$v_i = 4$ m/s

1. The time for the car to travel from the elevator to the yard is the distance to the yard d divided by the car's speed v_f following the grain dump:

$$\Delta t = \frac{d}{v_f}$$

2. Apply conservation of momentum to relate the final velocity v_f to the initial velocity v_i. Be careful, only the horizontal component of the system's momentum is conserved:

$$(m_c + m_g)v_f = m_c v_i + m_g(0)$$

3. Solve for v_f:

$$v_f = \frac{m_c v_i}{m_c + m_g}$$

4. Substitute the result for v_f into step 1 and solve for the time:

$$\Delta t = \frac{d}{v_f} = \frac{(m_c + m_g)d}{m_c v_i}$$

$$= \frac{(14000 \text{ kg} + 2000 \text{ kg})(500 \text{ m})}{(14000 \text{ kg})(4 \text{ m/s})}$$

$$= \boxed{143 \text{ s}}$$

REMARKS Mechanical energy of the system is converted to thermal energy. Let K_{gi} be the kinetic energy of the grain just as it hits the car. The initial mechanical energy is $K_{gi} + \frac{1}{2}m_c v_i^2 = K_{gi} + \frac{1}{2}(14,000 \text{ kg})(4 \text{ m/s})^2 = K_{gi} + 112$ kJ. The final kinetic energy is $\frac{1}{2}(m_c + m_g)v_f^2$, where $v_f = m_c v_i/(m_c + m_g) = 14(4 \text{ m/s})/16 = 3.5$ m/s, so the final kinetic energy is $\frac{1}{2}(16,000 \text{ kg})(3.5 \text{ m/s})^2 = 98$ kJ. The final kinetic energy is less than the initial kinetic energy by 14 kJ.

EXERCISE Suppose that there is a small vertical chute in the bottom of the car so that the grain leaks out at 10 kg/s. Now how long does it take the car to cover the 500 m? (*Answer* 143 s. The grain leaking out does not impart any momentum to the rest of the system. If the ground were frictionless and flat, all of the grain initially in the car would arrive at the switchyard along with the car.)

A Skateboard Workout **EXAMPLE 8-8**

A 40-kg skateboarder on a 3-kg board is training with two 5-kg weights. Beginning from rest, she throws the weights horizontally, one at a time from her board. The speed of each weight is 7 m/s relative to her and the board after it is thrown. Assume the board rolls without friction. (*a*) How fast is she propelled in the opposite direction after throwing the first weight? (*b*) the second weight?

PICTURE THE PROBLEM Because no external forces act on the skateboarder–weights–board system with horizontal components, the horizontal component of its momentum is conserved. We need to find the velocity of the skateboarder after throwing each weight (Figure 8-22).

(a) 1. The mass m of each weight is 5 kg and the mass M of the skateboard and skateboarder is 43 kg. Choose the direction of the skateboarder's motion to be the positive direction. Then $v_{ws} = -7 \text{ m/s}$ is the velocity of the thrown weight relative to the skateboarder. Let V_{sg1} and v_{wg1} be the respective velocities of the skateboarder and the thrown weight relative to the ground. Apply conservation of momentum for the first throw.

$$P_{sys1} = P_{sys0}$$

$$(M + m)V_{sg1} + mv_{wg1} = 0$$

(The numbers in the subscripts stand for the time. Time 0 is before the first throw, time 1 is between the two throws, and time 2 is following the second throw.)

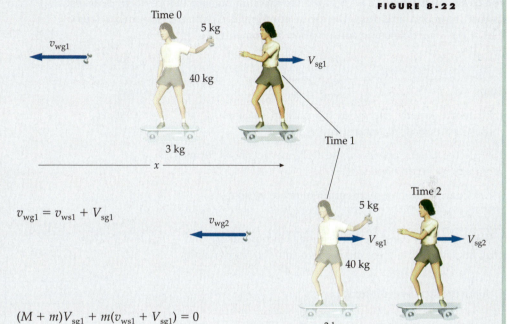

FIGURE 8-22

2. The velocity of the thrown weight relative to the ground equals the velocity of the weight relative to the skateboarder plus the velocity of the skateboarder relative to the ground:

$$v_{wg1} = v_{ws1} + V_{sg1}$$

3. Substitute for v_{wg1} in the step 1 result and solve for V_{sg1}:

$$(M + m)V_{sg1} + m(v_{ws1} + V_{sg1}) = 0$$

so

$$V_{sg1} = -\frac{m}{M + 2m}v_{ws1} = -\frac{5 \text{ kg}}{43 \text{ kg} + 10 \text{ kg}}(-7 \text{ m/s})$$

$$= \boxed{0.66 \text{ m/s}}$$

(b) 1. Repeat step 1 of Part (a) for the second throw. Let V_{sg2} and $v_{w'g2}$ be the respective velocities of the skateboarder and the second thrown weight relative to the ground:

$$P_{sys2} = P_{sys1}$$

$$MV_{sg2} + mv_{w'g2} = (M + m)V_{sg1}$$

2. Repeat step 2 of Part (a) for the second throw.

$$v_{w'g2} = v_{w's2} + V_{sg2}$$

3. Substitute for $v_{w'g2}$ in the Part (b) step 1 result and solve for V_{sg2}:

$$MV_{sg2} + m(v_{w's2} + V_{sg2}) = (M + m)V_{sg1}$$

so

$$V_{sg2} = \frac{(M + m)V_{sg1} - mv_{w's2}}{M + m} = V_{sg1} - \frac{m}{M + m}v_{w's2}$$

$$= 0.66 \text{ m/s} - \frac{5 \text{ kg}}{48 \text{ kg}}(-7 \text{ m/s}) = \boxed{1.39 \text{ m/s}}$$

REMARKS This example illustrates the principle of the rocket; a rocket moves forward by throwing its fuel out backward in the form of exhaust gases.

EXERCISE How fast is the skateboarder moving if, starting from rest, she throws both weights together, and the weights have speed 7 m/s relative to herself *after they are thrown*? (*Answer* 1.32 m/s)

EXAMPLE 8-9 Try It Yourself

A thorium-227 nucleus at rest decays into a radium-223 nucleus (mass 223 u) by emitting an α particle (mass 4 u) (Figure 8-23). The kinetic energy of the α particle is measured to be 6.00 MeV. What is the kinetic energy of the recoiling radium nucleus?

Thorium-227   Radium-223

 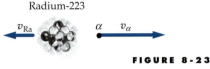

FIGURE 8-23

PICTURE THE PROBLEM Because the thorium nucleus before decay is at rest, its total momentum is zero. We can relate the velocity of the radium nucleus to that of the α particle using conservation of momentum.

Cover the column to the right and try these on your own before looking at the answers.

Steps	Answers
1. Write the kinetic energy of the radium nucleus K_{Ra} in terms of its mass m_{Ra} and speed v_{Ra}.	$K_{Ra} = \frac{1}{2}m_{Ra}v_{Ra}^2$
2. Write the kinetic energy of the alpha particle K_α in terms of its mass m_α and speed v_α.	$K_\alpha = \frac{1}{2}m_\alpha v_\alpha^2$
3. Use conservation of momentum to relate v_{Ra} to v_α.	$m_\alpha v_\alpha = m_{Ra} v_{Ra}$
4. Solve the step 1 and step 2 results for the speeds v_{Ra} and v_α, and substitute these expressions into the step 3 result.	$m_\alpha \left(\dfrac{2K_\alpha}{m_\alpha}\right)^{1/2} = m_{Ra} \left(\dfrac{2K_{Ra}}{m_{Ra}}\right)^{1/2}$
5. Solve the step 4 result for K_{Ra}.	$K_{Ra} = \dfrac{m_\alpha}{m_{Ra}} K_\alpha = \boxed{0.108 \text{ MeV}}$

REMARKS In this process, rest energy of the thorium nucleus is converted into kinetic energy of the alpha particle plus radium nucleus. The mass of the thorium nucleus is greater than that of the alpha particle plus radium nucleus by $6.1 \text{ MeV}/c^2$.

8-5 Kinetic Energy of a System

If the net external force on the system remains zero, then the total momentum of a system of particles must remain constant; however, the total mechanical energy of the system can change. As we saw in the examples of the previous section, internal forces that cannot change the total momentum may be nonconservative and thus change the total mechanical energy of the system. There is an important theorem concerning the kinetic energy of a system of particles that allows us to treat the energy of complex systems more easily and gives us insight into energy changes within a system:

The kinetic energy of a system of particles can be written as the sum of two terms: (1) the kinetic energy associated with the motion of the center of mass, $\frac{1}{2}Mv_{cm}^2$, where M is the total mass of the system; and (2) the kinetic energy associated with the motion of the particles of the system relative to the center of mass, $\Sigma \frac{1}{2} m_i u_i^2$, where $\vec{u}_i$ is the velocity of the ith particle relative to the center of mass.

THEOREM FOR THE KINETIC ENERGY OF A SYSTEM

The kinetic energy K of a system of particles is the sum of the kinetic energies of the individual particles:

$$K = \sum_i K_i = \sum_i \tfrac{1}{2}m_i v_i^2 = \sum_i \tfrac{1}{2}m_i (\vec{v}_i \cdot \vec{v}_i)$$

The velocity of each particle can be written as the sum of the velocity of the center of mass $\vec{v}_{cm}$ and the velocity of the particle relative to the center of mass $\vec{u}_i$:

$$\vec{v}_i = \vec{v}_{cm} + \vec{u}_i \qquad\qquad 8\text{-}17$$

Then

$$K = \sum_i \tfrac{1}{2}m_i(\vec{v}_i \cdot \vec{v}_i) = \sum_i \tfrac{1}{2}m_i(\vec{v}_{cm} + \vec{u}_i) \cdot (\vec{v}_{cm} + \vec{u}_i)$$

$$= \sum_i \tfrac{1}{2}m_i(v_{cm}^2 + 2\vec{v}_{cm} \cdot \vec{u}_i + u_i^2)$$

$$= \sum_i \tfrac{1}{2}m_i v_{cm}^2 + \vec{v}_{cm} \cdot \sum_i m_i\vec{u}_i + \sum_i \tfrac{1}{2}m_i u_i^2$$

where in the middle term we have factored $\vec{v}_{cm}$ from the sum because it is the same for each term in the sum. That is, $\vec{v}_{cm}$ does not change from particle to particle. The quantity $\Sigma m_i\vec{u}_i$ is the total momentum of the system *relative to the center of mass*. This quantity, which equals $M\vec{u}_{cm}$ is necessarily zero. (Relative to the center of mass, the velocity of the center of mass $\vec{u}_{cm}$ is zero, so the total momentum $M\vec{u}_{cm}$ is also zero.) Thus,

$$K = \sum_i \tfrac{1}{2}m_i v_{cm}^2 + \sum_i \tfrac{1}{2}m_i u_i^2 = \tfrac{1}{2} M v_{cm}^2 + K_{rel} \qquad\qquad 8\text{-}18$$

KINETIC ENERGY OF A SYSTEM OF PARTICLES

where M is the total mass and K_{rel} is the kinetic energy of the particles *relative to the center of mass*. If the net external force is zero, $\vec{v}_{cm}$ remains constant and the kinetic energy associated with bulk motion ($\tfrac{1}{2}Mv_{cm}^2$) does not change. Only the relative kinetic energy can change in an isolated system.

8-6 Collisions

In a **collision,** two objects approach and interact strongly for a very short time. During the brief time of collision, any external forces on the objects are assumed to be much weaker than the forces of interaction between the objects. Thus, during the collision the only important forces acting on the two-object system are the interaction forces, which are equal and opposite, so the total momentum of the system remains unchanged. Also, the collision time is so short that during the collision any displacements of the colliding objects can be neglected. Both prior to and following the collision, the interaction of the two objects is weak compared with their interaction during the collision. Examples of collisions are a cue ball hitting a billiard ball, a bat hitting a baseball, or a dart colliding with a dart board. **A comet swinging around the sun can be considered a collision, even though the two do not physically touch, and in the lab you may see a collision between two carts with magnetic bumpers that repel without touching.**

When the total kinetic energy of the two-object system is the same after the collision as before, the collision is called an **elastic collision.** Otherwise it is called an **inelastic collision.** An extreme case is the **perfectly inelastic collision,** in which all of the kinetic energy relative to the center of mass is converted to thermal or internal energy of the system, and the two objects stick together after the collision.

Impulse and Average Force

Figure 8-24 shows the time variation of the magnitude of a typical force exerted by one object on another during a collision. During the collision time $\Delta t = t_f - t_i$ the force is large. For other times the force is negligibly small. The **impulse $\vec{I}$** of the force is a vector defined as

$$\vec{I} = \int_{t_i}^{t_f} \vec{F}\, dt \qquad\qquad 8\text{-}19$$

DEFINITION — IMPULSE

The magnitude of the impulse of the force is the area under its F-versus-t curve. Units of impulse are N·s. The net force $\vec{F}_{net}$ acting on a particle is related to the rate of change of momentum of the particle by Newton's second law: $\vec{F}_{net} = d\vec{p}/dt$, so the impulse of the net force equals the total change in momentum $\Delta\vec{p}$ during the time interval:

$$\vec{I}_{net} = \int_{t_i}^{t_f} \vec{F}_{net}\, dt = \int_{t_i}^{t_f} \frac{d\vec{p}}{dt}\, dt = \vec{p}_f - \vec{p}_i = \Delta\vec{p} \qquad\qquad 8\text{-}20$$

IMPULSE–MOMENTUM THEOREM FOR A PARTICLE

Also, the net impulse on a system due to external forces equals the change in the total momentum of the system:

$$\vec{I}_{net,ext} = \int_{t_i}^{t_f} \vec{F}_{net,ext}\, dt = \Delta\vec{P}_{sys} \qquad\qquad 8\text{-}21$$

IMPULSE–MOMENTUM THEOREM FOR A SYSTEM

The **average force** for the interval $\Delta t = t_f - t_i$ is defined as

$$\vec{F}_{av} = \frac{1}{\Delta t} \int_{t_i}^{t_f} \vec{F}\, dt = \frac{\vec{I}}{\Delta t} \qquad\qquad 8\text{-}22$$

DEFINITION—AVERAGE FORCE

The average force is the constant force that gives the same impulse as the actual force in the time interval Δt, as shown by the rectangle in Figure 8-24. The average net force can be calculated from the change in momentum if a collision time is known. This time can often be estimated using the displacement of one of the bodies during the collision.

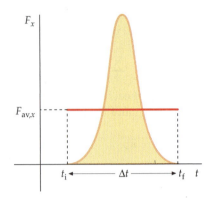

FIGURE 8-24 Typical time variation of force during a collision. The area under the F_x-versus-t curve is the x component of the impulse, I_x. $F_{av,x}$ is the average force for time interval Δt. The rectangular area $F_{av,x}\,\Delta t$ is the same as the area under the F_x-versus-t curve.

A KARATE COLLISION   **E X A M P L E   8 - 1 0**

With an expert karate blow, you shatter a concrete block. Consider your fist to have a mass 0.70 kg, to be moving 5.0 m/s as it strikes the block, and to stop within 6 mm of the point of contact. (a) What impulse does the block exert on your fist? (b) What is the approximate collision time and the average force the block exerts on your fist?

PICTURE THE PROBLEM The net impulse equals the change in momentum $\Delta\vec{p}$. We find $\Delta\vec{p}$ from the mass and velocity of the fist. The time of collision for Part (b) comes from the given displacement $\Delta y = -6$ mm and the average velocity v_{av} during the collision, which we can estimate by assuming constant acceleration. Make a sketch of the fist and block. Include a vertical coordinate axis on the sketch (Figure 8-25).

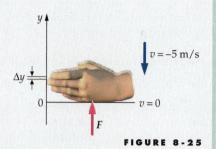

FIGURE 8-25

(a) 1. Set the impulse equal to the change in momentum:

$$\vec{I} = \Delta\vec{p} = \vec{p}_f - \vec{p}_i$$

2. The initial momentum is that of the fist just before it hits the block with velocity $\vec{v}$, and the final momentum is zero:

$$\vec{p}_i = m\vec{v} = (0.7 \text{ kg})(-5.0 \text{ m/s})\hat{j}$$
$$= -3.5 \text{ kg·m/s}\,\hat{j}$$
$$\vec{p}_f = 0$$

3. Find the impulse exerted by the block on the fist:

$$\vec{I} = \vec{p}_f - \vec{p}_i = 0 - (-3.5 \text{ kg} \cdot \text{m/s}\,\hat{j})$$
$$= \boxed{3.5 \text{ N·s}\,\hat{j}}$$

(b) 1. The collision time is the displacement divided by the average velocity:

$$\Delta t = \frac{\Delta y}{v_{av}}$$

2. We estimate the average speed by assuming constant acceleration, $v_{av} = \frac{1}{2}v$. Since we have chosen up to be positive, both Δy and v_{av} are negative. Calculate Δt:

$$\Delta t = \frac{\Delta y}{\frac{1}{2}v} = \frac{-0.006 \text{ m}}{-2.5 \text{ m/s}} = 0.0024 \text{ s}$$
$$= 2.4 \text{ ms}$$

3. The average force is the impulse divided by the collision time. It is upward, as expected:

$$\vec{F}_{av} = \frac{\vec{I}}{\Delta t} = \frac{3.5 \text{ N·s}\,\hat{j}}{0.0024 \text{ s}} = \boxed{1.46 \text{ kN}\,\hat{j}}$$

REMARKS Note that the average force is large—about 212 times the weight of the fist.

A CRASH TEST   **EXAMPLE 8-11**   **Try It Yourself**

A car equipped with an 80-kg crash-test dummy (Figure 8-26) drives into a wall at 25 m/s (about 56 mi/h). Estimate the force that the seatbelt exerts on the dummy upon impact.

PICTURE THE PROBLEM Assume that the car and dummy travel about 1 m as the front end of the car crumples, and that the acceleration is constant during the crash. To find the force, calculate the impulse I, then divide it by the collision time Δt. Choose forward as the positive x direction.

FIGURE 8-26

Cover the column on the right and try these on your own before looking at the answers.

Steps

1. Relate the average force to the impulse, and thus to the change in momentum.

2. Find the change in the dummy's momentum.

Answers

$$\vec{F}_{av}\Delta t = \vec{I} = \Delta\vec{p}, \quad \text{so} \quad \vec{F}_{av} = \frac{\Delta\vec{p}}{\Delta t}$$

$$\Delta\vec{p} = m\vec{v}_f - m\vec{v}_i = -2000 \text{ N·s}\,\hat{i}$$

3. Relate the time to the displacement, assuming constant acceleration.

$$\Delta t = \frac{\Delta x}{v_{av}}$$

4. Find the average velocity and use it and the step 3 result to find the time.

$$\vec{v}_{av} = \tfrac{1}{2}(\vec{v}_f + \vec{v}_i) = 12.5 \text{ m/s } \hat{i}, \quad \text{so}$$

$$\Delta t = 0.08 \text{ s}$$

5. Substitute the step 2 and step 4 results into the step 1 result and solve for the force.

$$\vec{F}_{av} = \boxed{-25 \text{ kN } \hat{i}}$$

REMARKS The magnitude of the average acceleration is $a_{av} = \Delta v/\Delta t = 313 \text{ m/s}^2$, or roughly $32g$. Such an acceleration means a net force about 32 times the weight of the dummy, clearly enough to cause serious injuries. An air bag increases the stopping distance somewhat, which helps to reduce the force and also allows the force to be distributed over a much larger area.

REMARKS In Figure 8-27, plot (a) shows the average force on the dummy as a function of the stopping distance. With no seat belt or air bag, you either fly though the windshield, or are stopped in a fraction of a meter by the dashboard or steering wheel. Plot (b) shows the force as a function of the initial velocity for three stopping distances: 2 m, 1.5 m, and 1 m.

FIGURE 8-27

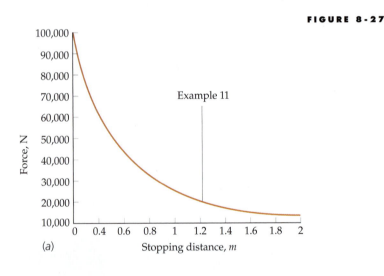

(a)

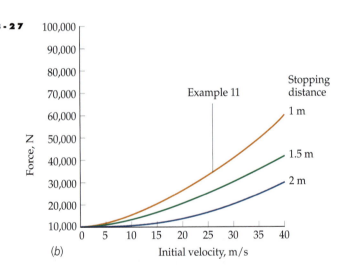

(b)

A *GOLF BALL*

EXAMPLE 8-12

You strike a golf ball with a driving iron. What are reasonable estimates for the (a) impulse I, (b) collision time Δt, and (c) average force F_{av}? A typical golf ball has a mass $m = 45$ g and a radius $r = 2$ cm. For a typical drive, the range R is roughly 192 m (210 yd) (Figure 8-28).

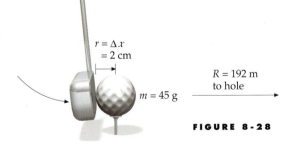

FIGURE 8-28

PICTURE THE PROBLEM Let v_0 denote the speed of the ball as it leaves the club face. The impulse on the ball equals its change in momentum, which is mv_0. We estimate v_0 from the range. We estimate the collision time from the distance traveled Δx and the average speed $\tfrac{1}{2}v_0$, assuming constant acceleration. Taking $\Delta x = 2$ cm, the average force is then obtained from the impulse I and collision time Δt.

(a) 1. Set the impulse equal to the change in momentum of the ball:

$$I = F_{av}\,\Delta t = \Delta p$$

2. The initial speed is related to the range R, which is given by Equation 2-23:

$$R = \frac{v_0^2}{g}\sin 2\theta_0$$

3. Take $\theta_0 = 13°$ and calculate the initial speed:

$$v_0 = \sqrt{\frac{Rg}{\sin 2\theta_0}}$$

$$= \sqrt{\frac{(192\text{ m})(9.81\text{ m/s}^2)}{\sin 26°}} = 65.5\text{ m/s}$$

4. Use this value of v_0 to calculate the magnitude of the impulse:

$$I = \Delta p = m(v_0 - 0) = (0.045\text{ kg})(65.5\text{ m/s})$$

$$= 2.95\text{ kg·m/s} = \boxed{2.95\text{ N·s}}$$

(b) Calculate the collision time Δt using $\Delta x = 2$ cm and $v_{av} = \frac{1}{2}(v_f - v_i)$:

$$\Delta t = \frac{\Delta x}{v_{av}} = \frac{\Delta x}{\frac{1}{2}v_0}$$

$$= \frac{0.02\text{ m}}{\frac{1}{2}(65.5\text{ m/s})} = \boxed{0.610 \times 10^{-3}\text{ s}}$$

(c) Use the calculated values of I and Δt to find the magnitude of the average force:

$$F_{av} = \frac{I}{\Delta t} = \frac{2.95\text{ N·s}}{6.10 \times 10^{-4}\text{ s}} = \boxed{4.83\text{ kN}}$$

REMARKS Again we see that forces exerted during a collision are very large. Here the force exerted on the golf ball by the club is roughly 10,000 times the weight of the ball, giving it an average acceleration of 10,000g for 0.61 ms. Also, the force of the air on the ball has been left out of our analysis. For an actual golf shot the effects of the air are not negligible.

Collisions in One Dimension (Head-on Collisions)

Consider an object of mass m_1 with initial velocity v_{1i} approaching a second object of mass m_2 that is moving in the same direction with initial velocity v_{2i}. If $v_{2i} < v_{1i}$, the objects collide. Let v_{1f} and v_{2f} be their final velocities after the collision. (These velocities can be positive or negative, depending on whether the objects are moving to the right or to the left.) Conservation of momentum gives one relation between the two unknown velocities v_{1f} and v_{2f}:

$$m_1 v_{1f} + m_2 v_{2f} = m_1 v_{1i} + m_2 v_{2i} \qquad\qquad 8\text{-}23$$

To determine v_{1f} and v_{2f}, we must have a second relation. That second relation, which we will develop here, depends on the type of collision.

Perfectly Inelastic Head-on Collisions In perfectly inelastic collisions, the particles stick together after the collision. For example, the collision between a catcher's mitt and a baseball is perfectly inelastic if the catcher does not drop the ball. The second kind of relation between the velocities is that the final velocities are equal to each other and to the velocity of the center of mass:

$$v_{1f} = v_{2f} = v_{cm}$$

Substituting this result combined with conservation of momentum gives

$$(m_1 + m_2)v_{cm} = m_1 v_{1i} + m_2 v_{2i} \qquad\qquad 8\text{-}24$$

Perfectly inelastic collision of two cars.

A CATCH IN SPACE

EXAMPLE 8-13

FIGURE 8-29

An astronaut of mass 60 kg is on a space walk to repair a communications satellite. Suddenly she needs to consult her physics book. You happen to have it with you, so you throw it to her with speed 4 m/s relative to your spacecraft. She is at rest relative to the spacecraft before catching the 3.0-kg book (Figure 8-29). Find (a) her velocity just after she catches the book, (b) the initial and final kinetic energies of the book–astronaut system, and (c) the impulse exerted by the book on the astronaut.

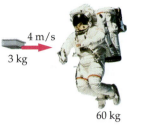
4 m/s
3 kg
60 kg

63 kg

PICTURE THE PROBLEM (a) The final velocity of the book and astronaut is the velocity of their center of mass. We find this using conservation of momentum, as expressed in Equation 8-24. The initial and final kinetic energies are calculated from the initial and final velocities. Because the book and astronaut move with the same final velocity, the collision is perfectly inelastic. (b) The kinetic energies of the book and astronaut are calculated directly from their masses and speeds. (c) The impulse exerted by the book on the astronaut equals the change in momentum of the astronaut.

(a) 1. Use conservation of momentum to relate the final velocity of the system v_{cm} to the initial velocities:

$$m_b v_b + m_a v_a = (m_b + m_a)v_{cm}$$

2. Solve for v_{cm}:

$$v_{cm} = \frac{m_b v_b + m_a v_a}{m_b + m_a}$$

$$= \frac{(3.0\ \text{kg})(4\ \text{m/s}) + 0}{60\ \text{kg} + 3\ \text{kg}} = \boxed{0.19\ \text{m/s}}$$

(b) 1. The initial mechanical energy of the book–astronaut system is the initial kinetic energy of the book:

$$K_i = K_b = \tfrac{1}{2}m_b v_b^2$$

$$= \tfrac{1}{2}(3.0\ \text{kg})(4\ \text{m/s})^2 = \boxed{24\ \text{J}}$$

2. The final kinetic energy is the kinetic energy of the book and astronaut moving together at v_{cm}:

$$K_f = \tfrac{1}{2}(m_b + m_a)v_{cm}^2$$

$$= \tfrac{1}{2}(63\ \text{kg})(0.19\ \text{m/s})^2 = \boxed{1.14\ \text{J}}$$

(c) Set the impulse exerted on the astronaut equal to the change in momentum of the astronaut:

$$I = \Delta p_a = m_a \Delta v_a$$

$$= (60\ \text{kg})(0.19\ \text{m/s} - 0)$$

$$= 11.4\ \text{kg·m/s} = \boxed{11.4\ \text{N·s}}$$

REMARKS Most of the initial kinetic energy in this collision is lost by conversion to thermal energy. The impulse exerted by the book on the astronaut is equal and opposite to that exerted by the astronaut on the book, so the total change in momentum of the book–astronaut system is zero.

It is useful to express the kinetic energy K of a particle in terms of its momentum p. For a mass m moving with speed v, we have

$$K = \tfrac{1}{2}mv^2 = \frac{(mv)^2}{2m}$$

Since $p = mv$,

$$K = \frac{p^2}{2m} \qquad\qquad 8\text{-}25$$

We can apply this to a perfectly inelastic collision where one object is initially at rest. The momentum of the system is that of the incoming object:

$$P_{sys} = p_{1i} = m_1 v_{1i}$$

The initial kinetic energy is

$$K_i = \frac{P_{sys}^2}{2m_1} \qquad\qquad 8\text{-}26$$

After colliding, the objects move together as a single mass $m_1 + m_2$ with v_{cm}. Momentum is conserved, so the final momentum equals P_{sys}. The final kinetic energy is then

$$K_f = \frac{P_{sys}^2}{2(m_1 + m_2)} \qquad\qquad 8\text{-}27$$

Comparing Equations 8-26 and 8-27, we see that the final kinetic energy is less than the initial kinetic energy.

A BALLISTIC PENDULUM   **E X A M P L E   8 - 1 4**

In a feat of public marksmanship, you fire a bullet into a hanging wood block. The block, with bullet embedded, swings upward. Noting the height reached at the top of the swing, you immediately inform the crowd of the bullet's speed. How fast was the bullet traveling?

PICTURE THE PROBLEM The initial speed of the bullet v_{1i} is related to the postcollision speed of the bullet–block system v_f by conservation of momentum. The speed v_f is related to the height h by conservation of mechanical energy (Figure 8-30). Let m_1 be the mass of the bullet and m_2 be the mass of the target.

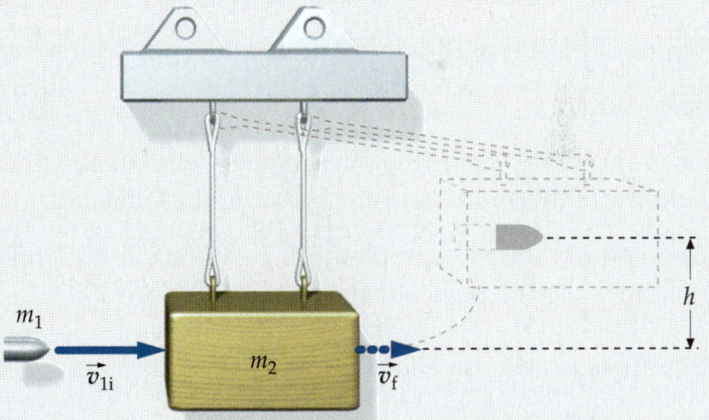

FIGURE 8-30

1. Using conservation of mechanical energy *after* the collision we relate the postcollision speed v_f to the maximum height h:

$$\tfrac{1}{2}(m_1 + m_2)v_f^2 = (m_1 + m_2)gh$$

2. Using conservation of momentum *during* the collision we relate v_{1i} to v_f:

$$m_1 v_{1i} = (m_1 + m_2)v_f$$

3. Solving the step 1 result for v_f and then substituting for v_f in the step 2 result, we can solve for v_{1i}:

$$v_f = \sqrt{2gh}$$

so

$$v_{1i} = \frac{m_1 + m_2}{m_1}v_f = \boxed{\frac{m_1 + m_2}{m_1}\sqrt{2gh}}$$

REMARKS We assumed that the time of the collision is so short that the displacement of the block during the collision is negligible. This meant the block had the postcollision speed while still at the lowest point in the arc.

Devices such as the one pictured are called *ballistic pendulums*.

EXERCISE Find the initial speed of the bullet if the mass of the bullet is 12 g, the mass of the block on the ballistic pendulum is 2 kg, and the final height is 10.4 cm. (*Answer* 240 m/s)

EXERCISE Can this example be solved by equating the initial kinetic energy of the bullet with the potential energy of the block–bullet composite at maximum height? That is, is mechanical energy conserved both during the inelastic collision and during the rise of the pendulum? (*Answer* No)

EXERCISE A 2000-kg car moving 25 m/s runs head-on into a 1500-kg car initially at rest. If the collision is perfectly inelastic, find (*a*) each car's speed after the collision, and (*b*) the ratio of the system's final kinetic energy to its initial kinetic energy. (*Answer* (*a*) 14.3 m/s, (*b*) 0.57)

COLLISION WITH AN EMPTY BOX   **EXAMPLE 8-15** Try It Yourself

You repeat your feat of Example 8-14, this time with an empty box as the target. The bullet strikes the box and passes through it completely. A laser ranging device indicates that the bullet emerged with half its initial velocity. Hearing this, you correctly report how high the target must have swung. How high did it swing?

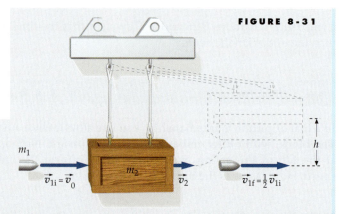

FIGURE 8-31

PICTURE THE PROBLEM The height h is related to the box's speed v_2 after colliding by conservation of mechanical energy (Figure 8-31). This speed can be determined using conservation of momentum.

Cover the column to the right and try these on your own before looking at the answers.

Steps

1. Use conservation of mechanical energy to relate the final height h to the speed v_2 of the box after the collision.

2. Use conservation of momentum to write an equation relating v_{1i} to the postcollision speed of the box v_2.

3. Eliminate v_2 and solve for h.

Answers

$$m_2gh = \tfrac{1}{2}m_2v_2^2$$

$$m_2v_2 + m_1(\tfrac{1}{2}v_{1i}) = m_1v_{1i}$$

$$h = \boxed{\dfrac{m_1^2v_{1i}^2}{8m_2^2g}}$$

REMARKS The collision of the bullet and the box is an inelastic collision, but not a perfectly inelastic collision because the two objects do not have the same velocity after the collision. Inelastic collisions also occur in microscopic systems. For example, when an electron collides with an atom, the atom is sometimes excited to a higher internal energy state. As a result, the total kinetic energy of the atom and the electron is less after the collision.

Elastic Head-on Collisions In some collisions, called **elastic collisions,** the initial and final kinetic energies are equal. Elastic collisions are an ideal that is sometimes approached but never realized in the macroscopic world. If a ball dropped onto a concrete platform bounces back to its original height, then the collision between the ball and the concrete would be elastic. That has never been observed. At the microscopic level, elastic collisions are common. For example, the collisions between air molecules are almost always elastic.

A bullet traveling 850 m/s collides inelastically with an apple, which moments later disintegrates completely. Exposure time is less than a millionth of a second.

For all elastic collisions we have:

$$\tfrac{1}{2}m_1v_{1f}^2 + \tfrac{1}{2}m_2v_{2f}^2 = \tfrac{1}{2}m_1v_{1i}^2 + \tfrac{1}{2}m_2v_{2i}^2 \qquad 8\text{-}28$$

This equation, together with the head-on conservation of momentum equation (Equation 8-23), is sufficient to determine the final velocities of the two objects. However, the quadratic nature of Equation 8-28 often complicates the solution of an elastic collision problem. Such problems can be treated more easily if we express the velocity of the two particles relative to each other after the collision in terms of their relative velocity before the collision. Rearranging Equation 8-28 gives

$$m_2(v_{2f}^2 - v_{2i}^2) = m_1(v_{1i}^2 - v_{1f}^2)$$

or

$$m_2(v_{2f} - v_{2i})(v_{2f} + v_{2i}) = m_1(v_{1i} - v_{1f})(v_{1i} + v_{1f}) \qquad 8\text{-}29$$

From conservation of momentum, we know that

$$m_1v_{1f} + m_2v_{2f} = m_1v_{1i} + m_2v_{2i}$$

or

$$m_2(v_{2f} - v_{2i}) = m_1(v_{1i} - v_{1f}) \qquad 8\text{-}30$$

Then dividing Equation 8-29 by Equation 8-30, we get

$$v_{2f} + v_{2i} = v_{1i} + v_{1f}$$

Rearranging, we obtain

$$v_{2f} - v_{1f} = -(v_{2i} - v_{1i}) \qquad 8\text{-}31$$

RELATIVE VELOCITIES IN AN ELASTIC COLLISION

If two objects are to collide, then $v_{2i} - v_{1i}$ must be negative (Figure 8-32), making their **speed of approach** $-(v_{2i} - v_{1i})$. After colliding, the objects' **speed of recession** is $v_{2f} - v_{1f}$. (Both of these terms give the speed of one object relative to the other.) Equation 8-31 states

In elastic collisions, the speed of recession equals the speed of approach.

Solving elastic-collision problems is usually easier using Equation 8-31 rather than Equation 8-28. But beware! Equation 8-31 is valid only if the initial and final kinetic energies are equal, so it applies *only* to elastic collisions.

FIGURE 8-32 Closing (approaching) and separating (receding) in a head-on elastic collision.

FIGURE 8-33

ELASTIC COLLISION OF TWO BLOCKS   **EXAMPLE 8-16**

A 4-kg block moving right at 6 m/s collides elastically with a 2-kg block moving right at 3 m/s (Figure 8-33). Find their final velocities.

PICTURE THE PROBLEM Conservation of momentum and the equality of the initial and final kinetic energies (expressed as a reversal of relative velocities) give two equations for the two unknown final velocities. Let subscript 1 denote the 4-kg block, subscript 2 the 2-kg block.

1. Apply conservation of momentum and simplify to obtain an equation relating the two final velocities:

$$m_1 v_{1f} + m_2 v_{2f} = m_1 v_{1i} + m_2 v_{2i}$$

$$(4 \text{ kg})v_{1f} + (2 \text{ kg})v_{2f} = (4 \text{ kg})(6 \text{ m/s}) + (2 \text{ kg})(3 \text{ m/s})$$

so

$$2v_{1f} + v_{2f} = 15 \text{ m/s}$$

2. The equality of the initial and final kinetic energies provides a second equation relating the two final velocities. This is implemented by equating the speeds of recession and approach:

$$v_{2f} - v_{1f} = -(v_{2i} - v_{1i})$$

$$= -(3 \text{ m/s} - 6 \text{ m/s}) = 3 \text{ m/s}$$

3. Subtract the step 2 result from the step 1 result and solve for v_{1f}:

$$2v_{1f} + v_{1f} = 12 \text{ m/s}$$

so

$$\boxed{v_{1f} = 4 \text{ m/s}}$$

4. Substitute into the step 2 result and solve for v_{2f}:

$$v_{2f} - 4 \text{ m/s} = 3 \text{ m/s}$$

so

$$\boxed{v_{2f} = 7 \text{ m/s}}$$

PLAUSIBILITY CHECK As a check, we calculate the initial and final kinetic energies.

$$K_i = \tfrac{1}{2}(4 \text{ kg})(6 \text{ m/s})^2 + \tfrac{1}{2}(2 \text{ kg})(3 \text{ m/s})^2 = 72 \text{ J} + 9 \text{ J} = 81 \text{ J}.$$

$$K_f = \tfrac{1}{2}(4 \text{ kg})(4 \text{ m/s})^2 + \tfrac{1}{2}(2 \text{ kg})(7 \text{ m/s})^2 = 32 \text{ J} + 49 \text{ J} = 81 \text{ J} = K_i.$$

ELASTIC COLLISION OF A NEUTRON AND A NUCLEUS   **EXAMPLE 8-17**

A neutron of mass m_n and speed v_{ni} collides elastically with a carbon nucleus of mass m_C initially at rest (Figure 8-34). (a) What are the final velocities of both particles? (b) What fraction f of its initial kinetic energy does the neutron lose?

FIGURE 8-34

PICTURE THE PROBLEM Conservation of momentum and conservation of energy allow us to find the final velocities. Since the initial kinetic energy of the carbon nucleus is zero, its final kinetic energy equals the energy lost by the neutron.

(a) 1. Use conservation of momentum to obtain one relation for the final velocities:

$$m_n v_{ni} = m_n v_{nf} + m_C v_{Cf}$$

2. The equality of the initial and final kinetic energies provides a second equation relating the two final velocities. This is implemented by equating the speeds of recession and approach:

$$v_{Cf} - v_{nf} = -(v_{Ci} - v_{ni}) = 0 + v_{ni}$$

so

$$v_{Cf} = v_{ni} + v_{nf}$$

3. To eliminate v_{Cf}, substitute the expression for v_{Cf} from step 2 into the Step 1 result:

$$m_n v_{ni} = m_n v_{nf} + m_C(v_{ni} + v_{nf})$$

4. Solve for v_{nf}: (Note that v_{nf} is negative. The neutron m_n bounces back from the more massive carbon nucleus m_C.)

$$v_{nf} = \boxed{-\frac{m_C - m_n}{m_n + m_C} v_{ni}}$$

5. Substitute the step 4 result into the step 2 result and solve for v_{Cf}:

$$v_{Cf} = v_{ni} - \frac{m_C - m_n}{m_n + m_C} v_{ni} = \boxed{\frac{2m_n}{m_n + m_C} v_{ni}}$$

(b) 1. The collision is elastic, so the kinetic energy lost by the neutron is the final kinetic energy of the carbon nucleus:

$$f = \frac{-\Delta K_n}{K_{ni}} = \frac{K_{Cf}}{K_{ni}}$$

$$= \frac{\frac{1}{2} m_C v_{Cf}^2}{\frac{1}{2} m_n v_{ni}^2} = \frac{m_C}{m_n} \left(\frac{v_{Cf}}{v_{ni}} \right)^2$$

2. Solve the Part (a) step 5 result for the ratio of the velocities, substitute into the Part (b) step 1 result, and solve for the fractional energy loss of the neutron:

$$f = \frac{m_C}{m_n} \left(\frac{2m_n}{m_n + m_C} \right)^2 = \boxed{\frac{4m_n m_C}{(m_n + m_C)^2}}$$

REMARKS An important application of energy transfer in elastic collisions is the slowing down of neutrons in a nuclear reactor. High-energy neutrons are emitted in the fission of a uranium nucleus. If these neutrons are to cause another uranium nucleus to fission, their energy must be reduced; that is, they must be slowed down or "moderated." One mechanism for slowing down neutrons is the elastic scattering of the neutrons with the nuclei in the reactor. The fractional energy loss $f = -\Delta K_n / K_n$ depends on the ratio of the mass of the moderator nucleus to that of the neutron, as shown in Figure 8-35. For uranium, $m_U \approx 235 m_n$ and $f \approx 0.017 = 1.7$ percent. For carbon, $m_C \approx 12 m_n$ and $f = 0.28 = 28$ percent; for deuterium, $m_D \approx 2 m_n$ and $f \approx 0.89 = 89$ percent; and for Hydrogen, $m_H \approx 1 m_n$ and $f \approx 1 = 100$ percent. A moderator such as graphite or water is added to a reactor to slow down the neutrons so that they can be captured by uranium nuclei.

EXERCISE Show that for an elastic head-on collision between a moving object and a stationary object of equal mass, the fractional loss in kinetic energy is one and the final speed of the initially stationary object equals the initial speed of the initially moving object.

EXERCISE A 2-kg box moving at 3 m/s makes an elastic collision with a stationary 4-kg box. (a) What is the original kinetic energy? (b) How much energy is transferred to the 4-kg box? (*Answer* (a) 9 J, (b) 8 J)

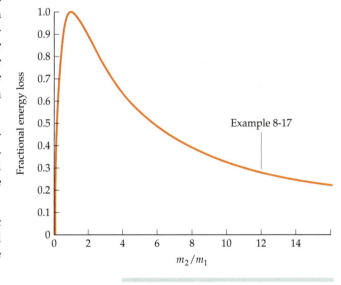

FIGURE 8-35 Fractional energy loss as a function of the ratio of the two masses. The maximum energy loss occurs when $m_1 = m_2$.

The results of Example 8-17 for the final velocities of an incoming particle colliding with a second particle initially at rest are worth noting. The final velocity of the incoming particle v_{1f} and that of the originally stationary particle v_{2f} are related to the initial velocity of the incoming particle by

$$v_{1f} = \frac{m_1 - m_2}{m_1 + m_2} v_{1i}$$

8-32a

and

$$v_{2f} = \frac{2m_1}{m_1 + m_2} v_{1i}$$

8-32b

When a very massive object (say a bowling ball) collides with a light stationery object (say a Ping-Pong ball), the massive object is essentially unaffected. Before the collision, the relative velocity of approach is v_{1i}. If the massive object continues with a velocity that is essentially v_{1i} after the collision, the velocity of the smaller object must be $2v_{1i}$ so that the speed of recession is equal to the speed of approach. This result also follows from Equations 8-32a and 8-32b if we take m_2 to be much smaller than m_1, in which case $v_{1f} \approx v_{1i}$ and $v_{2f} \approx 2v_{1i}$, as expected.

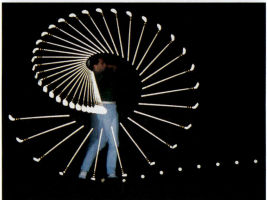

*The Coefficient of Restitution Most collisions lie somewhere between the extreme cases of elastic, in which the relative velocities are reversed, and perfectly inelastic, in which there is no relative velocity after the collision. The **coefficient of restitution** e, is a measure of the elasticity of a collision. It is defined as the ratio of the speed of recession to the speed of approach.

$$e = \frac{v_{rec}}{v_{app}} = -\frac{v_{2f} - v_{1f}}{v_{2i} - v_{1i}}$$

8-33

DEFINITION—COEFFICIENT OF RESTITUTION

For an elastic collision, $e = 1$. For a perfectly inelastic collision, $e = 0$.

Collisions in Three Dimensions

Perfectly Inelastic Collisions in Three Dimensions For collisions in three dimensions, the total initial momentum is the sum of the initial momentum vectors of each object involved in the collision. Because the objects stick together and because momentum is conserved we have $m_1\vec{v}_{1i} + m_2\vec{v}_{2i} = (m_1 + m_2)\vec{v}_f$. Because of this relation we know that the three velocity vectors, and thus the collision, are in the same plane. Also, from the definition of the center of mass we know that $\vec{v}_f = \vec{v}_{cm}$.

A CAR–TRUCK COLLISION   **EXAMPLE 8-18** **Put It in Context**

You are at the wheel of a 1200-kg car traveling east through an intersection when a 3000-kg truck traveling north through the intersection crashes into your car, as shown in Figure 8-36. Your car and the truck stick together after impact. The driver of the truck claims you were at fault because you were speeding. You look for evidence to disprove this claim. First, there are no skid marks, indicating that neither you nor the truck driver saw the accident coming and braked hard; second, there is a sign reading "Speed Limit 80 km/h" on the road you were driving on; third, the speedometer of the truck was smashed with the needle stuck at 50 km/h; and fourth, the wreck initially skidded from the impact zone at an angle of no less than 59° north of east. Does this evidence support or undermine the claim that you were speeding?

FIGURE 8-36

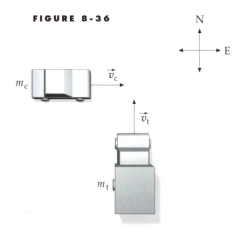

PICTURE THE PROBLEM Choose your coordinate system so that initially the car is traveling in the $+x$ direction and the truck is traveling in the $+y$ direction (Figure 8-37). Then write the momentum of each object in vector form, and use conservation of momentum.

1. Write out the conservation of momentum equation in vector form in terms of masses and velocities:

$$m_c \vec{v}_c + m_t \vec{v}_t = (m_c + m_t)\vec{v}_f$$

FIGURE 8-37

2. Equate the x component of the initial momentum to the x component of the final momentum:

$$m_c v_c + 0 = (m_c + m_t)v_f \cos \theta$$

3. Equate the y component of the initial momentum to the y component of the final momentum:

$$0 + m_t v_t = (m_c + m_t)v_f \sin \theta$$

4. Eliminate v_f by dividing the y component equation by the x component equation:

$$\frac{m_t v_t}{m_c v_c} = \frac{\sin \theta}{\cos \theta} = \tan \theta$$

so

$$v_c = \frac{m_t v_t}{m_c \tan \theta}$$

$$= \frac{(3000 \text{ kg})(50 \text{ km/h})}{(1200 \text{ kg}) \tan 59°}$$

$$= \boxed{75.1 \text{ km/h}}$$

5. Does this undermine the truck driver's claim that you were speeding?

| Because 75.1 km/h is less than the 80 km/h speed limit, the truck driver's claim is undermined by the careful application of physics. |

REMARKS A court would probably require that these arguments be presented by an expert witness.

*Elastic Collisions in Three Dimensions

Elastic collisions in three dimensions are more complicated than those we have already covered. Figure 8-38 shows an off-center collision between an object of mass m_1 moving with velocity $\vec{v}_{1i}$ parallel to the x axis toward an object of mass m_2 that is initially at rest at the origin. The distance b between the centers measured perpendicular to the direction of $\vec{v}_{1i}$ is called the **impact parameter.** After the collision, object 1 moves off with velocity $\vec{v}_{1f}$, making an angle θ_1 with its initial velocity, and object 2 moves with velocity $\vec{v}_{2f}$, making an angle θ_2 with $\vec{v}_{1f}$. Conservation of momentum gives

$$\vec{P}_{sys} = m_1 \vec{v}_{1i} = m_1 \vec{v}_{1f} + m_2 \vec{v}_{2f}$$

We can see from this equation that the vector $\vec{v}_{2f}$ must lie in the plane formed by $\vec{v}_{1i}$ and $\vec{v}_{1f}$, which we will take to be the xy plane. Assuming that we know the initial velocity $\vec{v}_{1i}$, we have four unknowns: the x and y components of both final velocities; or alternatively, the two final speeds and the two angles of deflection. The x and y components of the conservation-of-momentum equation give us two of the needed relations among these quantities. Conservation of mechanical energy gives a third relation. To find the four unknowns, we need another relation. The fourth relation depends on the impact parameter b and on the type of interacting force exerted by the bodies on each other. In practice, the fourth relation is often found experimentally, by measuring the angle of deflection or the angle of recoil. Such a measurement can then give us information about the type of interacting force between the bodies.

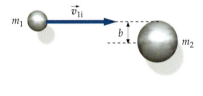

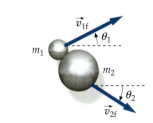

FIGURE 8-38 Off-center collision. The final velocities depend on the impact parameter b and on the type of force exerted by one object on the other.

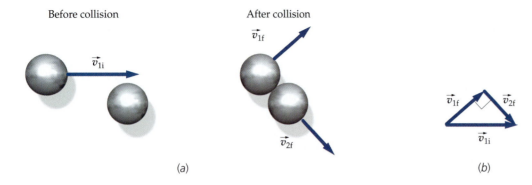

Before collision   After collision

$\vec{v}_{1f}$

$\vec{v}_{1i}$

$\vec{v}_{2f}$

$\vec{v}_{1f}$ $\vec{v}_{2f}$

$\vec{v}_{1i}$

(a)   (b)

FIGURE 8-39 (a) Off-center elastic collision of two spheres of equal mass when one sphere is initially at rest. After the collision, the spheres move off at right angles to each other. (b) The velocity vectors for this collision form a right triangle.

Consider the interesting special case of the off-center elastic collision of two objects *of equal mass* when one is initially at rest (Figure 8-39a). If $\vec{v}_{1i}$ and $\vec{v}_{1f}$ are the initial and final velocities of object 1, and $\vec{v}_{2f}$ is the final velocity of object 2, conservation of momentum gives

$$m\vec{v}_{1i} = m\vec{v}_{1f} + m\vec{v}_{2f}$$

or

$$\vec{v}_{1i} = \vec{v}_{1f} + \vec{v}_{2f}$$

These vectors form the triangle shown in Figure 8-39b. Because energy is conserved in the collision,

$$\tfrac{1}{2}mv_{1i}^2 = \tfrac{1}{2}mv_{1f}^2 + \tfrac{1}{2}mv_{2f}^2$$

or

$$v_{1i}^2 = v_{1f}^2 + v_{2f}^2 \qquad\qquad 8\text{-}34$$

Equation 8-34 is the Pythagorean theorem for a right triangle formed by the vectors $\vec{v}_{1f}$, $\vec{v}_{2f}$, and $\vec{v}_{1i}$, with the hypotenuse of the triangle being $\vec{v}_{1i}$. So for this special case, the final velocity vectors $\vec{v}_{1f}$ and $\vec{v}_{2f}$ are perpendicular to each other, as shown in Figure 8-39b.

Multiflash photograph of an off-center elastic collision of two balls of equal mass. The dotted ball, entering from the left, strikes the striped ball, which is initially at rest. The final velocities of the two balls are perpendicular to each other.

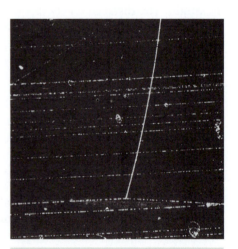

Proton–proton collision in a liquid-hydrogen bubble chamber. A proton entering from the left interacts with a stationary proton. The two then move off at right angles. The slight curvature of the tracks is due to a magnetic field.

*8-7 The Center-of-Mass Reference Frame

If the net external force on a system remains zero, the velocity of the center of mass remains constant. It is often convenient to choose a coordinate system with the origin at the center of mass. Then, relative to the original coordinate system, this coordinate system moves with a constant velocity $\vec{v}_{cm}$. The frame of reference that moves with the center of mass is called the **center-of-mass reference frame.** If a particle has velocity $\vec{v}$ in the original reference frame, then its velocity relative to the center of mass is $\vec{u} = \vec{v} - \vec{v}_{cm}$. In the center-of-mass frame, the velocity of the center of mass is zero. Since the total momentum of a system equals the total mass times the velocity of the center of mass, the total momentum is also zero in the center-of-mass frame. Thus, the center-of-mass reference frame is also a **zero-momentum reference frame.**

The mathematics of collisions is greatly simplified when considered within the center-of-mass reference frame. The momenta of the two incoming objects are equal and opposite. After a perfectly inelastic collision, the objects remain at rest. A perfectly elastic collision in one dimension between two objects reverses the direction of each velocity vector without changing the magnitudes of these vectors (you will derive this in *Problem 8-101*).

Consider a simple two-particle system in a reference frame in which one particle of mass m_1 is moving with a velocity $\vec{v}_1$ and a second particle of mass m_2 is moving with a velocity $\vec{v}_2$ (Figure 8-40). In this frame, the velocity of the center of mass is

$$\vec{v}_{cm} = \frac{m_1\vec{v}_1 + m_2\vec{v}_2}{m_1 + m_2}$$

We can transform the velocities of the two particles to their velocities in the center-of-mass reference frame by subtracting $\vec{v}_{cm}$. The velocities of the particles in the center-of-mass frame are $\vec{u}_1$ and $\vec{u}_2$, given by

$$\vec{u}_1 = \vec{v}_1 - \vec{v}_{cm} \qquad\qquad 8\text{-}35a$$

and

$$\vec{u}_2 = \vec{v}_2 - \vec{v}_{cm} \qquad\qquad 8\text{-}35b$$

Since the total momentum is zero in the center-of-mass frame, the particles have equal and opposite momenta in this frame.

Original reference frame

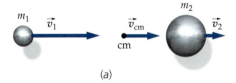

(a)

Center-of-mass reference frame

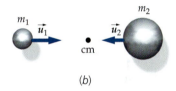

(b)

FIGURE 8-40 (a) Two particles viewed from a frame in which the center of mass has a velocity $\vec{v}_{cm}$. (b) The same two particles viewed from a reference frame for which the center of mass is at rest.

THE ELASTIC COLLISION OF TWO BLOCKS   **EXAMPLE 8-19**

Find the final velocities for the elastic head-on collision in Example 8-16 (in which a 4-kg block moving right at 6 m/s collides elastically with a 2-kg block moving right at 3 m/s) by transforming their velocities to the center-of-mass reference frame.

PICTURE THE PROBLEM We transform to the center-of-mass reference frame by first finding v_{cm} and subtracting it from each velocity. We then solve the collision by reversing the velocities and transforming back to the original frame.

1. Calculate the velocity of the center of mass v_{cm} (Figure 8-41):

$$v_{cm} = \frac{m_1 v_{1i} + m_2 v_{2i}}{m_1 + m_2}$$

$$= \frac{(4 \text{ kg})(6 \text{ m/s}) + (2 \text{ kg})(3 \text{ m/s})}{4 \text{ kg} + 2 \text{ kg}}$$

$$= 5 \text{ m/s}$$

FIGURE 8-41 Initial conditions

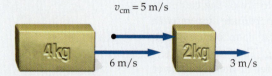

$v_{cm} = 5 \text{ m/s}$

2. Transform the initial velocities to the center-of-mass reference frame by subtracting v_{cm} from the initial velocities (Figure 8-42):

$$u_{1i} = v_{1i} - v_{cm}$$

$$= 6 \text{ m/s} - 5 \text{ m/s} = 1 \text{ m/s}$$

$$u_{2i} = v_{2i} - v_{cm}$$

$$= 3 \text{ m/s} - 5 \text{ m/s} = -2 \text{ m/s}$$

FIGURE 8-42 Transform to the center-of-mass frame by subtracting v_{cm}

$v_{cm} = 0$

3. Solve the collision in the center-of-mass reference frame by reversing the velocity of each object (Figure 8-43):

$$u_{1f} = -u_{1i} = -1 \text{ m/s}$$

$$u_{2f} = -u_{2i} = +2 \text{ m/s}$$

FIGURE 8-43 Solve collision

$v_{cm} = 0$

4. To find the final velocities in the original frame, add v_{cm} to each final velocity (Figure 8-44).

$$v_{1f} = u_{1f} + v_{cm}$$

$$= -1 \text{ m/s} + 5 \text{ m/s} = \boxed{4 \text{ m/s}}$$

$$v_{2f} = u_{2f} + v_{cm}$$

$$= 2 \text{ m/s} + 5 \text{ m/s} = \boxed{7 \text{ m/s}}$$

FIGURE 8-44 Transform back to the original frame by adding v_{cm}

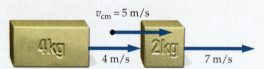

$v_{cm} = 5 \text{ m/s}$

REMARKS This is the same result found in Example 8-16.

EXERCISE Show that the total momentum of the system both before the collision and after the collision is zero in the center-of-mass reference frame. (*Answer* before: $P_{\text{sys i}} = (4 \text{ kg})(1 \text{ m/s}) + (2 \text{ kg})(-2 \text{ m/s}) = 0$; after: $P_{\text{sys f}} = (4 \text{ kg})(-1 \text{ m/s}) + (2 \text{ kg})(2 \text{ m/s}) = 0$)

*8-8 Systems With Continuously Varying Mass: Rocket Propulsion

A creative and important step in solving physics problems is specifying the system. In this section we explore situations in which the system has a continuously changing mass. One example of such a system is a rocket. For a rocket, we specify the system to be the rocket plus any unspent fuel in it. As the spent fuel (the exhaust) spews out the back, the mass of the system decreases.

Another example is a stream of effluence from a volcano on the Jovian moon Io impacting on an asteroid passing through the stream. We specify the system at time t to be the asteroid, plus the volcanic effluence accumulated on the asteroid, plus the effluence that will impact the asteroid between time t and time $t + \Delta t$. As the effluence continues to accumulate, the mass of this system increases. To solve such problems we will apply the impulse–momentum theorem to the system during the short time interval Δt, and then take the limit as $\Delta t \to 0$.

Consider the following derivation of Newton's second law for objects with continuously varying mass. A continuous stream of matter moving at velocity $\vec{u}$ is impacting an object of mass M that is moving with velocity $\vec{v}$ (Figure 8-45). The mass of particles that impact the object during time Δt is ΔM, and these impacting particles stick to the object, increasing its mass by ΔM during time Δt. Also, during time Δt the velocity $\vec{v}$ changes by $\Delta \vec{v}$, as shown. At time t we define the system to be the object (including all matter that impacted the object prior to time t) plus all matter that will impact the object between time t and time $t + \Delta t$. Applying the impulse–momentum theorem to this system gives

$$\vec{F}_{\text{net ext}} \Delta t = \Delta \vec{P} = [(M + \Delta M)(\vec{v} + \Delta \vec{v})] - [M\vec{v} + \Delta M \vec{u}]$$

where the first term (in square brackets) is the momentum at time $t + \Delta t$ and the second term in square brackets is the momentum at time t. Reorganizing terms gives

$$\vec{F}_{\text{net ext}} \Delta t = M \Delta \vec{v} + \Delta M(\vec{v} - \vec{u}) + \Delta M \Delta \vec{v} \qquad \text{8-36}$$

(where the net external force $\vec{F}_{\text{net ext}}$ would likely include a gravitational force plus one or more contact forces). Dividing through Equation 8-36 by Δt gives

$$\vec{F}_{\text{net ext}} = M\frac{\Delta \vec{v}}{\Delta t} + \frac{\Delta M}{\Delta t}(\vec{v} - \vec{u}) + \frac{\Delta M}{\Delta t} \Delta \vec{v}$$

and taking the limit as $\Delta t \to 0$ (which also means as $\Delta M \to 0$ and as $\Delta \vec{v} \to 0$) gives

$$\vec{F}_{\text{net ext}} = M\frac{d\vec{v}}{dt} - \frac{dM}{dt}(\vec{u} - \vec{v}) + 0$$

Rearranging, we have

$$\vec{F}_{\text{net ext}} + \frac{dM}{dt}\vec{v}_{\text{rel}} = M\frac{d\vec{v}}{dt} \qquad \text{8-37}$$

NEWTON'S SECOND LAW—CONTINUOUSLY VARIABLE MASS

where $\vec{v}_{\text{rel}} = \vec{u} - \vec{v}$, the velocity of dM relative to M. Note that except for the term $(dM/dt)\vec{v}_{\text{rel}}$, Equation 8-37 is just Newton's second law for a system with constant mass.

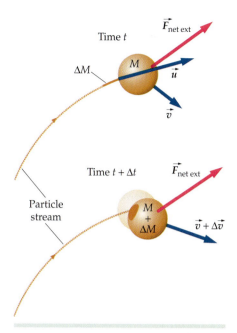

FIGURE 8-45 Particles in a continuous stream and moving at velocity $\vec{u}$ undergo perfectly inelastic collisions with an object of mass M moving at velocity $\vec{v}$.

A FALLING ROPE   **EXAMPLE 8-20**

A uniform rope of mass M and length L is held with its lower end just touching the surface of a scale. The rope is released and begins to fall. Find the force of the scale on the rope just as the midpoint of the rope first touches the scale.

PICTURE THE PROBLEM Apply Equation 8-37 to that portion of the rope on the scale. There are two external forces on that portion, the force of gravity and the normal force exerted by the scale. The impact velocities of the different points along the falling rope depend upon their initial heights above the scale.

1. Draw a sketch of the situation (Figure 8-46). Include the initial configuration and the configuration at an arbitrary time later. Include a coordinate axis.

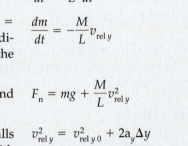

2. Express Equation 8-37 in component form. Let m denote the mass of the system (that part of the rope on the scale). The velocity of the system remains zero, so the dv_y/dt is zero:

$$F_{\text{net ext } y} + \frac{dm}{dt} v_{\text{rel } y} = m \frac{dv_y}{dt}$$

$$F_n - mg + \frac{dm}{dt} v_{\text{rel } y} = 0$$

3. Let dm denote the mass of a short rope segment of length $d\ell$ that falls on the scale during time dt. Since the rope is uniform, the relation between dm and $d\ell$ is:

$$\frac{dm}{d\ell} = \frac{M}{L}$$

FIGURE 8-46 A very flexible rope of length L and mass M is released from rest and falls on the surface of a scale.

4. Solve for dm/dt by multiplying both sides of the step 3 result by $d\ell/dt$:

$$\frac{dm}{dt} = \frac{M}{L} \frac{d\ell}{dt}$$

5. $d\ell/dt$ is the impact speed of the segment, so $v_{\text{rel } y} = -d\ell/dt$. ($v_{\text{rel } y}$ is negative because up is the positive y direction and the rope is falling.) Substituting this into the step 4 result gives:

$$\frac{dm}{dt} = -\frac{M}{L} v_{\text{rel } y}$$

6. Substituting the step 5 result into the step 2 result and solving for F_n gives:

$$F_n = mg + \frac{M}{L} v_{\text{rel } y}^2$$

7. Until it touches the scale, each point along the rope falls with constant acceleration g. Using Equation 2-23 with $\Delta y = -L/2$ gives:

$$v_{\text{rel } y}^2 = v_{\text{rel } y 0}^2 + 2a_y \Delta y$$
$$= 0 + 2(-g)(-L/2) = gL$$

8. Substituting the step 7 result into the step 6 result, with $m = M/2$, gives:

$$F_n = \frac{M}{2}g + \frac{M}{L}gL = \boxed{\frac{3}{2}Mg}$$

REMARKS As the midpoint of the rope strikes the surface, the weight of the system (the rope already on the scale) is $\frac{1}{2}Mg$, which is one third of the normal force. This means $F_{\text{net ext } y} = Mg$ (directed upward). This is the force required to stop the momentum of the falling rope striking the scale at that instant.

EXERCISE Find the normal force exerted by the scale on the rope (a) just before the upper end of the rope reaches the scale and (b) just after the upper end of the rope reaches the scale. (*Answer* (a) $3Mg$, (b) Mg)

Rocket propulsion is a striking example of the conservation of momentum in action. The mass of the rocket changes continuously as it burns fuel and expels exhaust gas. Consider a rocket moving straight up with velocity $\vec{v}$ relative to the earth, as shown in Figure 8-47. Assuming that the fuel is burned at a constant rate R, the rocket's mass at time t is

$$M = M_0 - Rt \qquad\qquad 8\text{-}38$$

where M_0 is the initial mass of the rocket. The exhaust gas leaves the rocket with velocity $\vec{u}_{\text{ex}}$ relative to the rocket, and the rate at which the fuel is burned is the rate at which the mass M decreases. We choose the rocket and the unspent fuel within it as the system. Neglecting air drag, the only external force on the system is that of gravity. With $\vec{F}_{\text{net ext}} = M\vec{g}$ and $dM/dt = -R$, Equation 8-37 becomes the **rocket equation:**

$$M\vec{g} - R\vec{u}_{ex} = M\frac{d\vec{v}}{dt}$$ 8-39

ROCKET EQUATION

The quantity $-R\vec{u}_{ex}$ is the force exerted on the rocket by the exhausting fuel. It is called the **thrust** $\vec{F}_{th}$:

$$\vec{F}_{th} = -R\vec{u}_{ex} = -\left|\frac{dM}{dt}\right|\vec{u}_{ex}$$ 8-40

DEFINITION—ROCKET THRUST

The rocket is moving straight up, so we choose upward as the positive y direction and express Equation 8-39 as

$$-Mg + Ru_{ex} = M\frac{dv_y}{dt}$$ 8-41

ROCKET EQUATION (VERTICAL COMPONENT)

Dividing through by M, rearranging terms, and substituting from Equation 8-38 gives

$$\frac{dv_y}{dt} = \frac{Ru_{ex}}{M} - g = \frac{Ru_{ex}}{M_0 - Rt} - g$$ 8-42

ACCELERATION OF ROCKET (VERTICAL COMPONENT)

where M_0 is the initial value of M. Equation 8-42 is solved by integrating both sides with respect to time. For a rocket starting at rest at $t = 0$, the result is

$$v_y = u_{ex}\ln\left(\frac{M_0}{M_0 - Rt}\right) - gt$$ 8-43

VELOCITY OF ROCKET (VERTICAL COMPONENT)

assuming the acceleration of gravity to be constant.

FIGURE 8-47

LIFTOFF

EXAMPLE 8-21   Try It Yourself

The Saturn V rocket used in the Apollo moon-landing program had an initial mass M_0 of 2.85×10^6 kg, a payload of 27%, a burn rate R of 13.84×10^3 kg/s, and a thrust F_{th} of 34×10^6 N. Find (a) the exhaust speed, (b) the burn time t_b, (c) the acceleration at liftoff, (d) the acceleration at burnout t_b, and (e) the final speed of the rocket.

PICTURE THE PROBLEM (a) The exhaust speed can be found from the thrust and burn rate. (b) To find the burn time, you need to find the total mass of fuel burned, which is the initial mass minus the payload. (c) The acceleration is found from Equation 8-42. (d) The final speed is given by Equation 8-43.

Cover the column to the right and try these on your own before looking at the answers.

Steps	Answers
(a) Calculate u_{ex} from the given thrust and burn rate.	$u_{ex} = \boxed{2.46 \text{ km/s}}$
(b) 1. Calculate the final mass M_f of the rocket.	$M_f = (0.27)M_0 = 7.70 \times 10^5 \text{ kg}$
2. Use your result to calculate the burn time t_b.	$t_b = \dfrac{M_0 - M_f}{R} = \boxed{150 \text{ s}}$
(c) Calculate dv_y/dt for $M = M_0$ and for $M = M_f$.	Initially, $\dfrac{dv_y}{dt} = 2.14 \text{ m/s}^2$;
	finally, $\dfrac{dv_y}{dt} = \boxed{34.3 \text{ m/s}^2}$
(d) Calculate the final speed from Equation 8-43.	$v_{yf} = \boxed{1.75 \text{ km/s}}$

REMARKS The initial acceleration is small—only $0.21g$. At burnout (also called flameout), the rocket's acceleration has increased to $3.5g$. Immediately following burnout the acceleration is $-g$. The speed of the rocket at burnout, after two and a half minutes of burning, is roughly 6,300 km/h (3900 mi/h).

SUMMARY

The conservation of momentum for an isolated system is a fundamental law of nature that has applications in all areas of physics.

Topic	Relevant Equations and Remarks	
1. Center of Mass		
System of particles	$M\vec{r}_{cm} = m_1\vec{r}_1 + m_2\vec{r}_2 + \cdots$	8-5
	$M\vec{v}_{cm} = m_1\vec{v}_1 + m_2\vec{v}_2 + \cdots$	8-8
	$M\vec{a}_{cm} = m_1\vec{a}_1 + m_2\vec{a}_2 + \cdots$	8-9
Continuous objects	$M\vec{r}_{cm} = \displaystyle\int \vec{r}\, dm$	8-6
Newton's second law for a system	$\vec{F}_{net\,ext} = \displaystyle\sum_i \vec{F}_{iext} = M\vec{a}_{cm}$	8-11
2. Momentum		
Definition for a particle	$\vec{p} = m\vec{v}$	8-12
Kinetic energy of a particle	$K = \dfrac{p^2}{2m}$	8-25
Momentum of a system	$\vec{P}_{sys} = \displaystyle\sum_i m_i\vec{v}_i = M\vec{v}_{cm}$	8-14

Newton's second law for a system	$\vec{F}_{\text{net ext}} = \dfrac{d\vec{P}_{\text{sys}}}{dt}$	8-15
Law of conservation of momentum	If the net external force acting on a system remains zero, the total momentum of the system is conserved.	

3. Energy of a System

Kinetic energy	The kinetic energy associated with the motion of the particles of a system relative to its center of mass is $K_{\text{rel}} = \Sigma \frac{1}{2} m_i u_i^2$, where u_i is the speed of the ith particle relative to the center of mass.	
	$K = \frac{1}{2} M v_{\text{cm}}^2 + K_{\text{rel}}$	8-18
Gravitational potential energy	$U = M g h_{\text{cm}}$	8-7

4. Collisions

Impulse	The impulse of a force is defined as the integral of the force over the time interval during which the force acts.	
	$\vec{I} = \displaystyle\int_{t_i}^{t_f} \vec{F}\, dt$	8-19
Impulse–momentum theorem	$\vec{I}_{\text{net}} = \displaystyle\int_{t_i}^{t_f} \vec{F}_{\text{net}}\, dt = \Delta\vec{p}$	8-20
Average force	$\vec{F}_{\text{av}} = \dfrac{1}{\Delta t}\displaystyle\int_{t_i}^{t_f} \vec{F}\, dt = \dfrac{\vec{I}}{\Delta t}$	8-22
Elastic collisions	An elastic collision is one in which the sum of the kinetic energies of the two objects is the same before and after the collision.	
Relative speeds of approach and recession	For an elastic collision, the speed of separation of the two objects following an elastic collision equals their speed of approach prior to the collision. For a head-on collision,	
	$v_{2\text{f}} - v_{1\text{f}} = -(v_{2\text{i}} - v_{1\text{i}})$	8-31
Perfectly inelastic collisions	Following a perfectly inelastic collision, the two objects stick together and move with the velocity of the center of mass.	
*Coefficient of restitution	The coefficient of restitution e is a measure of the elasticity. It is the ratio of the separation speed to the closing speed:	
	$e = -\dfrac{v_{2\text{f}} - v_{1\text{f}}}{(v_{2\text{i}} - v_{1\text{i}})}$	8-33
	For an elastic collision, $e = 1$; for a perfectly inelastic collision $e = 0$.	

*4. Continuously Variable Mass

Newton's second law	$\vec{F}_{\text{net ext}} + \dfrac{dM}{dt}\vec{v}_{\text{rel}} = M\dfrac{d\vec{v}}{dt}$	8-37
Rocket equation	$M\vec{g} - R\vec{u}_{\text{ex}} = M\dfrac{d\vec{v}}{dt}$	8-39
Thrust	$\vec{F}_{\text{th}} = -R\vec{u}_{\text{ex}} = -\left|\dfrac{dM}{dt}\right|\vec{u}_{\text{ex}}$	8-40

- Single-concept, single-step, relatively easy
- •• Intermediate-level, may require synthesis of concepts
- ••• Challenging
- **SSM** Solution is in the *Student Solutions Manual*
- **iSOLVE** Problems available on iSOLVE online homework service
- **iSOLVE** ✓ These "Checkpoint" online homework service problems ask students additional questions about their confidence level, and how they arrived at their answer

In a few problems, you are given more data than you actually need; in a few other problems, you are required to supply data from your general knowledge, outside sources, or informed estimates.

Take g = 9.81 N/kg = 9.81 m/s² and neglect friction in all problems unless otherwise stated.

Conceptual Problems

1 • Give an example of a three-dimensional object that has no matter at its center of mass.

2 • **SSM** A cannonball is dropped off a high tower while, simultaneously, an identical cannonball is launched directly upward into the air. The center of mass of the two cannonballs (a) stays in the same place, (b) initially rises, then falls, but begins to fall *before* the cannonball launched into the air starts falling, (c) initially rises, then falls, but begins to fall *at the same time* as the cannonball launched upward begins to fall, (d) initially rises, then falls, but begins to fall *after* the cannonball launched upward begins to fall.

3 • Two pucks of masses m_1 and m_2 lie unconnected on a frictionless table. A horizontal force F_1 is exerted on m_1 only. What is the magnitude of the acceleration of the center of mass of the two-puck system? (a) F_1/m_1. (b) $F_1/(m_1 + m_2)$. (c) F_1/m_2. (d) $(m_1 + m_2)F_1/m_1m_2$.

4 • The two pucks in Problem 3 are lying on a frictionless table and connected by a massless spring of force constant k. A horizontal force F_1 is again exerted only on m_1, along the spring and away from m_2. What is the magnitude of the acceleration of the center of mass? (a) F_1/m_1. (b) $F_1/(m_1 + m_2)$. (c) $(F_1 + kx)/(m_1 + m_2)$, where x is the amount the spring is stretched. (d) $(m_1 + m_2)F_1/m_1m_2$.

5 • **SSM** If two particles have equal kinetic energies, are the magnitudes of their momenta necessarily equal? Explain and give an example.

6 • True or false:

(a) The momentum of a heavy object is greater than that of a light object moving at the same speed.
(b) The momentum of a system may be conserved even when mechanical energy is not.
(c) The velocity of the center of mass of a system equals the total momentum of the system divided by its total mass.

7 • **iSOLVE** How is the recoil of a rifle related to momentum conservation?

8 • **SSM** A child jumps from a small boat to a dock. Why does she have to jump with more energy than she would

need if she were jumping the same distance from one dock to another?

9 •• **SSM** Much early research in rocket motion was done by Robert Goddard, physics professor at Clark College in Worcester, Massachusetts. A quotation from a 1920 editorial in the *New York Times* illustrates the public opinion of his work: "That Professor Goddard with his 'chair' at Clark College and the countenance of the Smithsonian Institution does not know the relation between action and reaction, and the need to have something better than a vacuum against which to react—to say that would be absurd. Of course, he only seems to lack the knowledge ladled out daily in high schools."[†] The belief that a rocket needs something to push against was a prevalent misconception before rockets in space were commonplace. Explain why that belief is wrong.

10 • Two bowling balls are moving with the same velocity, but one just slides down the alley, whereas the other rolls down the alley. Which ball has more kinetic energy? Explain.

11 • A philospher tells you, "All motion is impossible. Forces always come in equal but opposite action–reaction pairs. Therefore, all forces must cancel out." Answer his argument.

12 • If two objects collide and one is initially at rest, is it still possible for both to be at rest after the collision? (Any external forces acting on the two-object system are negligibly small.) Is it possible for one to be at rest? Explain.

13 • From the standpoint of physical laws, what is the problem with superhero comics where the hero flies around with no jets or hovers in midair while tossing huge objects like cars at the villains?

14 •• **SSM** If only external forces can cause the center of mass of a system of particles to accelerate, how can a car move? We normally think of the car's engine as supplying the force needed to accelerate the car, but is this true? Where does the external force that accelerates the car come from?

---

† On page 43 of the July 17, 1969, edition of the *New York Times* "A Correction" to their editorial of 1920 was printed. This commentary, which was published three days before man's first walk on the moon, stated that "it is now definitely established that a rocket can function in a vacuum as well as in an atmosphere. *The Times* regrets the error."

15 •• When we push on the brake pedal to slow down a car, a brake pad is pressed against the rotor so that the friction of the pad slows the wheel's rotation. However, the friction of the pad against the rotor can't be the force that slows the car down, because it is an internal force—both the rotor and the wheel are parts of the car, so any forces between them are purely internal to the system. What is the external force that slows down the car? Give a detailed explanation of how this force operates.

16 • Explain why a safety net can save the life of a circus performer.

17 • How might you estimate the collision time of a baseball and bat?

18 • **ISOLVE** Why does a wine glass survive a fall onto a carpet but not onto a concrete floor?

19 • **ISOLVE** True or false:

(a) In any perfectly inelastic collision, all the kinetic energy of the bodies is lost.

(b) In a head-on elastic collision, the relative speed of recession after the collision equals the relative speed of approach before the collision.

(c) Mechanical energy is conserved in an elastic collision.

20 •• **SSM** Under what conditions can all the initial kinetic energy of an isolated system consisting of two colliding bodies be lost in a collision?

21 •• Consider a perfectly inelastic collision of two objects of equal mass. (a) Is the loss of kinetic energy greater if the two objects have oppositely directed velocities of equal magnitude $v/2$ or if one of the two objects is initially at rest and the other has an initial velocity of v? (b) In which situation is the percentage loss in kinetic energy the greatest?

22 •• **SSM** A double-barrelled pea shooter is shown in Figure 8-48. Air is blown from the left end of the straw, and identical peas A and B are positioned inside the straw as shown. If the shooter is held horizontally while the peas are shot off, which pea, A or B, will travel farther after leaving the straw? Why? *Hint: The answer has to do with the impulse–momentum theorem.*

FIGURE 8-48 Problem 22

23 •• A particle of mass m_1 traveling with a speed v makes a head-on elastic collision with a stationary mass m_2. In which scenario will the energy imparted to the particle of mass m_2 be greatest? (a) $m_2 << m_1$. (b) $m_2 = m_1$. (c) $m_2 >> m_1$. (d) None of these.

Optional Section

24 • Describe a perfectly inelastic collision as viewed in the center-of-mass reference frame.

25 • Nozzles for a garden hose are often made with a right-angle shape as shown in Figure 8-49. If you open the nozzle and spray water out, you will find that the nozzle presses against your hand with a pretty strong force—much stronger than if you used a nozzle not bent into a right angle. Why is this?

FIGURE 8-49 Problem 25

26 • Why can friction and the force of gravity usually be neglected in collision problems?

27 • As a pendulum bob swings back and forth, is the momentum of the bob conserved? Explain.

28 •• **SSM** A railroad car is passing by a grain elevator, which is dumping grain into it at a constant rate. (a) Does momentum conservation imply that the railroad car should be slowing down as it passes the grain elevator? Assume that the track is frictionless and perfectly level. (b) If the car is slowing down, this implies that there is some external force acting on the car to slow it down. Where does this force come from? (c) After passing the elevator, the railroad car springs a leak, and grain starts leaking out of a vertical hole in its floor at a constant rate. Should the car speed up as it loses mass?

29 •• **SSM** To show that even really bright people can make mistakes, consider the following problem which was given to the freshman class at Caltech on an exam (paraphrased): *A sailboat is sitting in the water on a windless day. In order to make the boat move, a misguided sailor sets up a fan in the back of the boat to blow into the sails to make the boat move forward. Explain why the boat won't move.* The idea was that the net force of the wind pushing the sail forward would be counteracted by the force pushing the fan back (Newton's third law). However, as one of the students pointed out to his professor, the sailboat *could* in fact move forward. Why is that?

Estimation and Approximation

30 •• A 2000-kg car traveling at 90 km/h crashes into a concrete wall that does not give at all. (a) Estimate the time of collision, assuming that the center of the car travels halfway to the wall with constant deceleration. (Use any reasonable length for the car.) (b) Estimate the average force exerted by the wall on the car.

31 •• In hand-pumped railcar races, a speed of 32 km/h has been achieved by teams of four. A car of mass 350 kg is moving at that speed toward a river when Carlos, the chief pumper, notices that the bridge ahead is out. All four people (of mass 75 kg each) jump simultaneously backward off the car with a velocity that has a horizontal component of 4 m/s relative to the car after jumping. The car proceeds off the bank and falls into the water a horizontal distance of 25.0 m off the bank. (*a*) Estimate the time of the fall of the railcar. (*b*) What happens to the team of pumpers?

32 •• **SSM** **iSOLVE** A counterintuitive physics demonstration can be performed by firing a rifle bullet into a melon. (Don't try this at home!) When hit, nine times out of ten the melon will jump backward, toward the rifle, *opposite* to the direction in which the bullet was moving. (The tenth time, the melon simply explodes.) Doesn't this violate the laws of conservation of momentum? It doesn't, because we're not dealing simply with a two-body collision. Instead, a significant fraction of the energy of the bullet can be dumped into a jet of melon that is violently ejected out of the front of the melon. This jet can have a momentum greater than the momentum of the bullet, so that the rest of the melon must jump backward to conserve momentum. Let's make the following assumptions:

1. The mass of the melon is 2.50 kg.
2. The mass of the rifle bullet is 10.4 g and its velocity is 1800 ft/s.
3. 10 percent of the energy of the bullet is deposited as kinetic energy into a jet flying out of the front of the melon.
4. The mass of the matter in the jet is 0.14 kg.
5. All collisions occur in a straight line. What would be the speed of the melon's recoil? Compare this to a typical measured recoil speed of about 1.6 ft/s.

Finding the Center of Mass

33 • Three point masses of 2 kg each are located on the *x* axis at the origin, at *x* = 0.20 m, and at *x* = 0.50 m. Find the center of mass of the system.

34 • **SSM** Alley Oop's club-ax consists of a symmetrical 8-kg stone attached to the end of a uniform 2.5-kg stick that is 98 cm long. The dimensions of the club-ax are shown in Figure 8-50. How far is the center of mass from the handle end of the club-ax?

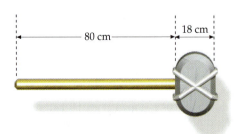

FIGURE 8-50 Problem 34

35 • **iSOLVE** ✓ Three balls A, B, and C, with masses of 3 kg, 1 kg, and 1 kg, respectively, are connected by massless rods. The balls are located as in Figure 8-51. What are the coordinates of the center of mass?

FIGURE 8-51 Problems 35, 45

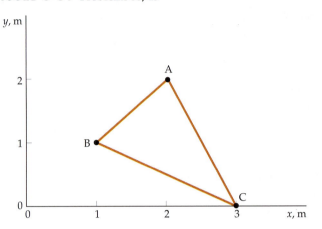

36 • **iSOLVE** By symmetry, locate the center of mass of an equilateral triangle of side length *a* with one vertex on the *y* axis and the others at $(-a/2, 0)$ and $(+a/2, 0)$.

37 •• **SSM** Find the center of mass of the uniform sheet of plywood in Figure 8-52. We shall consider this as two sheets, a square sheet of 3-m edge length and mass m_1 and a rectangular sheet 1 m × 2 m with a mass of $-m_2$. Let the coordinate origin be at the lower left-hand corner of the sheet.

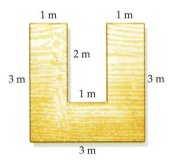

FIGURE 8-52 Problem 37

38 •• A can in the shape of a symmetrical cylinder with mass *M* and height *H* is filled with water. The initial mass of the water is *M*, the same mass as the can. A hole is punched in the bottom of the can, and the water drains out. (*a*) If the height of the water in the can is *x*, what is the height of the center of mass of the can + water? (*b*) What is the minimum height of the center of mass as the water drains out?

*Finding the Center of Mass by Integration

39 •• **SSM** Show that the center of mass of a uniform semicircular disk of radius *R* is at a point $4R/(3\pi)$ from the center of the circle.

40 ••• Find the center of mass of a homogeneous solid hemisphere of radius R and mass M.

41 ••• Find the center of mass of a thin homogeneous hemispherical shell.

42 ••• A sheet of metal is cut in the shape of a parabola (Figure 8-53). The curved edge of the sheet is specified by the equation $y = ax^2$, and y ranges from 0 to b. Find the center of mass in terms of a and b. (You will need to find the area first.)

FIGURE 8-53 Problem 42

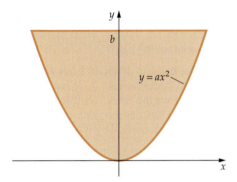

Motion of the Center of Mass

43 • [ISOLVE] ✓ Two 3-kg particles have velocities $\vec{v}_1 = 2 \text{ m/s } \hat{i} + 3 \text{ m/s } \hat{j}$ and $\vec{v}_2 = 4 \text{ m/s } \hat{i} - 6 \text{ m/s } \hat{j}$. Find the velocity of the center of mass for the system.

44 • [SSM] [ISOLVE] ✓ A 1500-kg car is moving westward with a speed of 20 m/s, and a 3000-kg truck is traveling east with a speed of 16 m/s. Find the velocity of the center of mass of the system.

45 • [ISOLVE] A force $\vec{F} = 12 \text{ N } \hat{i}$ is applied to the 3-kg ball in Figure 8-51 in Problem 35. What is the acceleration of the center of mass?

46 •• A block of mass m is attached to a string and suspended inside a hollow box of mass M. The box rests on a scale that measures the system's weight. (a) If the string breaks, does the reading on the scale change? Explain your reasoning. (b) Assume that the string breaks and the mass m falls with constant acceleration g. Find the acceleration of the center of mass of the box–block system, giving both direction and magnitude. (c) Using the result from (b), determine the reading on the scale while m is in free-fall.

47 •• [SSM] A massless, vertical spring of force constant k is attached at the bottom to a platform of mass m_p, and at the top to a massless cup, as in Figure 8-54. The platform rests on a scale. A ball of mass m_b is dropped into the cup from a negligible height. What is the reading on the scale (a) when the spring is compressed an amount $d = m_b g/k$; (b) when the ball comes to rest momentarily with the spring compressed; (c) when the ball again comes to rest in its original position?

FIGURE 8-54 Problems 47, 49, 136

48 •• [SSM] In the Atwood's machine in Figure 8-55, the string passes over a fixed cylinder of mass m_c. The cylinder does not rotate. Instead, the string slides on its frictionless surface. (a) Find the acceleration of the center of mass of the two-block-and-cylinder system. (b) Use Newton's second law for systems to find the force F exerted by the support. (c) Find the tension T in the string connecting the blocks and show that $F = m_c g + 2T$.

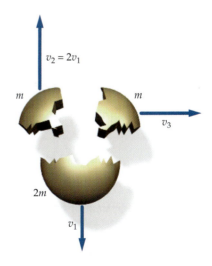

FIGURE 8-55 Problem 48

49 •• [ISOLVE] Repeat Problem 47(a) and 47(b) if the ball is dropped into the cup from a height h above the cup.

The Conservation of Momentum

50 • [ISOLVE] A woman of mass 55 kg jumps off the bow of a 75-kg canoe that is initially at rest. If her velocity is 2.5 m/s to the right, what is the velocity of the canoe after she jumps?

51 • [ISOLVE] ✓ A 5-kg object and a 10-kg object are connected by a massless compressed spring and rest on a frictionless table. After the spring is released, the object with the smaller mass has a velocity of 8 m/s to the left. What is the velocity of the object with the larger mass?

52 • [SSM] Figure 8-56 shows the behavior of a projectile just after it has broken up into three pieces. What was the speed of the projectile the instant before it broke up? (a) v_3. (b) $v_3/3$. (c) $v_3/4$. (d) $4v_3$. (e) $(v_1 + v_2 + v_3)/4$.

FIGURE 8-56
Problem 52

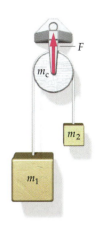

53 • A shell of mass m and speed v explodes into two identical fragments. If the shell was moving horizontally with respect to the earth, and one of the fragments is subsequently moving vertically with speed v, find the velocity $\vec{v}'$ of the other fragment.

54 •• **SSM** A block and a gun are firmly fixed to opposite ends of a long glider mounted on a frictionless air track (Figure 8-57). The block and gun are a distance L apart. The system is initially at rest. The gun is fired and the bullet leaves the muzzle with a velocity v_b and impacts the block, becoming imbedded in it. The mass of the bullet is m_b and the mass of the gun–glider–block system is m_p. (a) What is the velocity of the glider immediately after the bullet leaves the muzzle? (b) What is the velocity of the glider immediately after the bullet comes to rest in the block? (c) How far does the glider move while the bullet is in transit between the gun at rest and the block at rest?

FIGURE 8-57 Problem 54

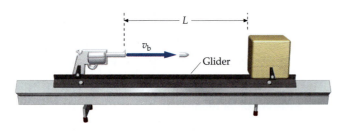

55 •• A small object of mass m slides down a wedge of mass $2m$ and exits smoothly onto a frictionless table. The wedge is initially at rest on the table. If the object is initially at rest at a height h above the table, find the velocity of the wedge when the object leaves it.

56 •• **SSM** Each glider on a frictionless air track (Figure 8-58) supports a strong magnet. These magnets attract each other. The mass of glider 1 is 0.10 kg and the mass of glider 2 is 0.20 kg. (Each mass includes the mass of the glider plus the supported magnet.) The origin is at the left end of the track, the center of glider 1 is at $x_1 = 0.10$ m and the center of glider 2 is at $x_2 = 1.60$ m. Glider 1 is 10 cm long, while glider 2 is 20 cm long. Each glider has its center of mass at its center. (a) When the two gliders are released from rest, they move toward each other and stick together. What is the position of the center of each glider when they first touch? (b) Will the two gliders continue to move after they stick together? Explain.

FIGURE 8-58 Problem 56

Kinetic Energy of a System of Particles

57 • **SSM** **iSOLVE** A 3-kg block is traveling to the right at 5 m/s, and a second 3-kg block is traveling to the left at 2 m/s. (a) Find the total kinetic energy of the two blocks. (b) Find the velocity of the center of mass of the two-block system. (c) Find the velocities of each block relative to the center of mass. (d) Find the kinetic energy of the motion of the blocks relative to the center of mass. (e) Show that your answer for part (a) is greater than your answer for part (d) by an amount equal to the kinetic energy associated with the motion of the center of mass.

58 • Repeat Problem 57 with the second 3-kg block replaced by a 5-kg block moving to the right at 3 m/s.

Impulse and Average Force

59 • You kick a soccer ball of mass 0.43 kg. The ball leaves your foot with an initial speed of 25 m/s. (a) What is the impulse imparted to the ball by you? (b) If your foot is in contact with the ball for 0.008 s, what is the average force exerted by your foot on the ball?

60 • A 0.3-kg brick is dropped from a height of 8 m. It hits the ground and comes to rest. (a) What is the impulse exerted by the ground on the brick? (b) If it takes 0.0013 s from the time the brick first touches the ground until it comes to rest, what is the average force exerted by the ground on the brick?

61 • **SSM** A meteorite of mass 30.8 tonne (1 tonne = 1000 kg) is exhibited in the American Museum of Natural History in New York City. Suppose that the kinetic energy of the meteorite as it hit the ground was 617 MJ. Find the impulse I experienced by the meteorite up to the time its kinetic energy was halved (which took about $t = 3.0$ s). Find also the average force F exerted on the meteorite during this time interval.

62 •• **iSOLVE** When a 0.15-kg baseball is hit, its velocity changes from +20 m/s to −20 m/s. (a) What is the magnitude of the impulse delivered by the bat to the ball? (b) If the baseball is in contact with the bat for 1.3 ms, what is the average force exerted by the bat on the ball?

63 •• **SSM** **iSOLVE**✔ A 60-g handball moving with a speed of 5.0 m/s strikes the wall at an angle of 40° and then bounces off with the same speed at the same angle. It is in contact with the wall for 2 ms. What is the average force exerted by the ball on the wall?

64 •• **iSOLVE** You throw a 150-g ball to a height of 40 m. (a) Use a reasonable value for the distance the ball moves while it is in your hand to calculate the average force exerted by your hand and the time the ball is in your hand while you throw it. (b) Is it reasonable to neglect the weight of the ball while it is being thrown?

65 •• A handball of mass 60 g is thrown straight against a wall with a speed of 10 m/s. It rebounds with a speed of 8 m/s. (a) What impulse is delivered to the wall? (b) If the ball is in contact with the wall for 0.003 s, what average force is exerted on the wall by the ball? (c) The ball is caught by a player who brings it to rest. In the process, her hand moves back 0.5 m. What is the impulse received by the player? (d) What average force was exerted on the player by the ball?

66 ••• **iSOLVE** The great limestone caverns were formed by dripping water. (a) If water droplets of 0.03 mL fall from a height of 5 m at a rate of 10 per minute, what is the average force exerted on the limestone floor by the droplets of water during a 1-min period? (b) Compare this force to the weight of a water droplet.

Collisions in One Dimension

67 • **SSM** A 2000-kg car traveling to the right at 30 m/s is chasing a second car of the same mass that is traveling to the right at 10 m/s. (*a*) If the two cars collide and stick together, what is their speed just after the collision? (*b*) What fraction of the initial kinetic energy of the cars is lost during this collision? Where does it go?

68 • An 85-kg running back moving at 7 m/s makes a perfectly inelastic collision with a 105-kg linebacker who is initially at rest. What is the speed of the players just after their collision?

69 • A 5.0-kg object with a speed of 4.0 m/s collides head-on with a 10-kg object moving toward it with a speed of 3.0 m/s. The 10-kg object stops dead after the collision. (*a*) What is the final speed of the 5-kg object? (*b*) Is the collision elastic?

70 • A ball of mass *m* moves with speed *v* to the right toward a much heavier bat that is moving to the left with speed *v*. Find the speed of the ball after it makes an elastic collision with the bat.

71 •• **SSM** **iSOLVE** A proton of mass *m* undergoes a head-on elastic collision with a stationary carbon nucleus of mass 12*m*. The speed of the proton is 300 m/s. (*a*) Find the velocity of the center of mass of the system. (*b*) Find the velocity of the proton after the collision.

72 •• **iSOLVE** A 3-kg block moving at 4 m/s makes a head-on elastic collision with a stationary block of mass 2 kg. Use conservation of momentum and the fact that the relative speed of recession equals the relative speed of approach to find the velocity of each block after the collision. Check your answer by calculating the initial and final kinetic energies of each block.

73 •• **iSOLVE** A block of mass $m_1 = 2$ kg slides along a frictionless table with a speed of 10 m/s. Directly in front of it, and moving in the same direction with a speed of 3 m/s, is a block of mass $m_2 = 5$ kg. A massless spring with spring constant $k = 1120$ N/m is attached to the second block as in Figure 8-59. (*a*) Before m_1 makes contact with the spring, what is the velocity of the center of mass of the system? (*b*) During the collision, the spring is compressed by a maximum amount Δx. What is the value of Δx? (*c*) The blocks will eventually separate again. What are the final velocities of the two blocks measured in the reference frame of the table?

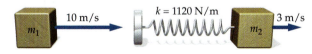

FIGURE 8-59 Problem 73

74 •• **SSM** **iSOLVE** A bullet of mass *m* is fired vertically from below into a thin sheet of plywood of mass *M* that is initially at rest, supported by a thin sheet of paper. The bullet blasts through the plywood, which rises to a height *H* above the paper before falling back down. The bullet continues rising to a height *h* above the paper. (*a*) Express the upward velocity of the bullet and the plywood immediately after the bullet exits the plywood in terms of *h* and *H*. (*b*) Use conservation of momentum to express the speed of the bullet before it enters the sheet of plywood in terms of *m*, *h*, *M*, and *H*. (*c*) Obtain expressions for the mechanical energy of the system before and after the inelastic collision. (*d*) Express the energy dissipated in terms of *m*, *h*, *M*, and *H*.

75 •• A proton of mass *m* is moving with initial speed v_0 directly toward the center of an α particle of mass 4*m*, which is initially at rest. Because both particles carry positive electrical charge, they repel each other. Find the speed v' of the α particle (*a*) when the distance between the two particles is least and (*b*) later when the two particles are far apart.

76 • An electron collides elastically with a hydrogen atom initially at rest. All motion occurs along a straight line. What fraction of the electron's initial kinetic energy is transferred to the atom? (Take the mass of the hydrogen atom to be 1840 times the mass of an electron.)

77 •• **iSOLVE** ✓ A 16-g bullet is fired into the bob of a ballistic pendulum of mass 1.5 kg. When the bob is at its maximum height, the strings make an angle of 60° with the vertical. The length of the pendulum is 2.3 m. Find the speed of the bullet.

78 •• **SSM** Show that in a one-dimensional elastic collision, if the mass and velocity of object 1 are m_1 and v_{1i}, and if the mass and velocity of object 2 are m_2 and v_{2i}, then the final velocities v_{1f} and v_{2f} are given by

$$v_{1f} = \frac{m_1 - m_2}{m_1 + m_2} v_{1i} + \frac{2m_2}{m_1 + m_2} v_{2i}$$

$$v_{2f} = \frac{2m_1}{m_1 + m_2} v_{1i} + \frac{m_2 - m_1}{m_1 + m_2} v_{2i}$$

79 •• Use the results of Problem 78 to investigate one-dimensional elastic collisions in the following limits: (*a*) When the two masses are equal, show that the particles "swap" velocities: $v_{1f} = v_{2i}$ and $v_{2f} = v_{1i}$. (*b*) If $m_2 \gg m_1$, show that $v_{1f} \approx -v_{1i} + 2v_{2i}$ and $v_{2f} \approx v_{2i}$.

Perfectly Inelastic Collisions and the Ballistic Pendulum

80 •• A bullet of mass m_1 is fired with a speed *v* into the bob of a ballistic pendulum of mass m_2. The bob is attached to a very light rod of length *L* that is pivoted at the other end. The bullet is stopped in the bob. Find the minimum *v* such that the bob will swing through a complete circle.

81 •• **SSM** A bullet of mass m_1 is fired with a speed *v* into the bob of a ballistic pendulum of mass m_2. Find the maximum height *h* attained by the bob if the bullet passes through the bob and emerges with a speed *v*/2.

82 • If we take a heavy wooden block normally used as a ballistic pendulum and put it on a flat table before firing a bullet into it, about how far will it slide before coming to a stop? Assume that the mass of the rifle bullet is 10.5 g, the mass of the wooden block is 10.5 kg, the bullet's velocity is 750 m/s, and the coefficient of kinetic friction of the block against the table is 0.22. Also assume that the bullet doesn't cause the block to rotate.

83 •• **SOLVE** A ball with mass $M = 0.425$ kg and speed $V = 1.3$ m/s rolls across a level table into an open box with mass $m = 0.327$ kg. The box with the ball inside it then slides across the table for a distance $x = 0.520$ m. What is the coefficient of kinetic friction of the table?

84 •• **SSM** Tarzan is in the path of a pack of stampeding elephants when Jane swings in to the rescue on a rope vine, hauling him off to safety. The length of the vine is 25 m, and Jane starts her swing with the rope horizontal. If Jane's mass is 54 kg, and Tarzan's is 82 kg, to what height above the ground will the pair swing after she grabs him?

Exploding Objects and Radioactive Decay

85 •• **SOLVE** The beryllium isotope ^{8}Be is unstable and decays into two α particles (helium nuclei of mass $m = 6.68 \times 10^{-27}$ kg) with the release of 1.5×10^{-14} J of energy. Determine the velocities of the two α particles that arise from the decay of a ^{8}Be nucleus at rest, assuming that all the energy appears as kinetic energy of the particles.

86 •• The light isotope of lithium, ^{5}Li, is unstable and breaks up spontaneously into a proton (hydrogen nucleus) and an α particle (helium nucleus). In this process, a total energy of 3.15×10^{-13} J is released, appearing as the kinetic energy of the two decay products. Determine the velocities of the proton and α particle that arise from the decay of a ^{5}Li nucleus at rest. (*Note*: The masses of the proton and alpha particle are $m_p = 1.67 \times 10^{-27}$ kg and $m_\alpha = 4m_p = 6.68 \times 10^{-27}$ kg.)

87 ••• **SOLVE** A projectile of mass $m = 3$ kg is fired with an initial speed of 120 m/s at an angle of 30° with the horizontal. At the top of its trajectory, the projectile explodes into two fragments of masses 1 kg and 2 kg. The 2-kg fragment lands on the ground directly below the point of explosion 3.6 s after the explosion. (a) Determine the velocity of the 1-kg fragment immediately after the explosion. (b) Find the distance between the point of firing and the point at which the 1-kg fragment strikes the ground. (c) Determine the energy released in the explosion.

88 ••• **SSM** The boron isotope ^{9}B is unstable and disintegrates into a proton and two α particles. The total energy released as kinetic energy of the decay products is 4.4×10^{-14} J. In one such event, with the ^{9}B nucleus at rest prior to decay, the velocity of the proton is measured as 6.0×10^6 m/s. If the two α particles have equal energies, find the magnitude and the direction of their velocities with respect to that of the proton.

*Coefficient of Restitution

89 • The coefficient of restitution for steel on steel is measured by dropping a steel ball onto a steel plate that is rigidly attached to the earth. If the ball is dropped from a height of 3 m and rebounds to a height of 2.5 m, what is the coefficient of restitution?

90 • **SSM** According to the official rules of racquetball, a ball acceptable for tournament play must bounce to a height of between 173 and 183 cm when dropped from a height of 254 cm at room temperature. What is the acceptable range of values for the coefficient of restitution for the racquetball–floor system?

91 • **SOLVE ✓** A ball bounces to 80 percent of its original height. (a) What fraction of its mechanical energy is lost each time it bounces? (b) What is the coefficient of restitution of the ball–floor system?

92 •• **SOLVE** A 2-kg object moving at 6 m/s collides with a 4-kg object that is initially at rest. After the collision, the 2-kg object moves backward at 1 m/s. (a) Find the velocity of the 4-kg object after the collision. (b) Find the energy lost in the collision. (c) What is the coefficient of restitution for this collision?

93 •• A 2-kg block moving to the right with speed 5 m/s collides with a 3-kg block that is moving in the same direction at 2 m/s, as in Figure 8-60. After the collision, the 3-kg block moves at 4.2 m/s. Find (a) the velocity of the 2-kg block after the collision and (b) the coefficient of restitution for the collision.

FIGURE 8-60 Problem 93

*Collisions in Three Dimensions

94 •• **SSM** In Section 8-6 it was proven by geometrical means that when a particle elastically collides with another particle of equal mass that is initially at rest, the two separate at right angles. In this problem we will examine another way of proving this, one that illustrates the power of abstract vector notation. (a) Given $\vec{A} = \vec{B} + \vec{C}$, show that $A^2 = B^2 + C^2 + 2\vec{B} \cdot \vec{C}$. (b) Let the momentum of the initially moving particle be $\vec{P}$ and the momenta of the particles after the collision be $\vec{p}_1$ and $\vec{p}_2$. By writing the vector equation for the conservation of momentum and obtaining the dot product of each side with itself, and then comparing it to the equation for the conservation of energy, show that $\vec{p}_1 \cdot \vec{p}_2 = 0$.

95 • **SOLVE** In a pool game, the cue ball, which has an initial speed of 5 m/s, makes an elastic collision with the eight ball, which is initially at rest. After the collision, the eight ball moves at an angle of 30° to the original direction of the cue ball. (a) Find the direction of motion of the cue ball after the collision. (b) Find the speed of each ball. Assume that the balls have equal mass.

96 •• Object A with mass m and velocity $v_0\hat{i}$ collides head-on with object B of mass $2m$ and velocity $\frac{1}{2}v_0\hat{j}$. Following the collision, object B has a velocity of $\frac{1}{4}v_0\hat{i}$. (a) Determine the velocity of object A after the collision. (b) Is the collision elastic? If not, express the change in the kinetic energy in terms of m and v_0.

97 •• **SSM** **iSOLVE** A puck of mass 5 kg moving at 2 m/s approaches an identical puck that is stationary on frictionless ice. After the collision, the first puck leaves with a speed v_1 at 30° to the original line of motion; the second puck leaves with speed v_2 at 60°, as in Figure 8-61. (a) Calculate v_1 and v_2. (b) Was the collision elastic?

FIGURE 8-61 Problem 97

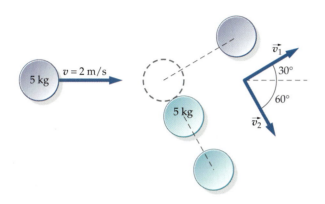

98 •• Figure 8-62 shows the result of a collision between two objects of unequal mass. (a) Find the speed v_2 of the larger mass after the collision and the angle θ_2. (b) Show that the collision is elastic.

FIGURE 8-62 Problem 98

99 •• **SSM** **iSOLVE** A ball moving at 10 m/s makes an off-center elastic collision with another ball of equal mass that is initially at rest. The incoming ball is deflected at an angle of 30° from its original direction of motion. Find the velocity of each ball after the collision.

100 •• A particle has initial speed v_0. It collides with a second particle *with the same mass* that is at rest, and is deflected through an angle ϕ. Its speed after the collision is v. The second particle recoils, and its velocity makes an angle θ with the initial direction of the first particle. (a) Show that

$$\tan \theta = \frac{v \sin \phi}{(v_0 - v \cos \phi)}$$

(b) Show that if the collision is elastic, $v = v_0 \cos \phi$.

Center-of-Mass Frame

101 •• In the center-of-mass reference frame a particle with mass m_1 and momentum p_1 makes an elastic head-on collision with a second particle of mass m_2 and momentum $p_2 = -p_1$. After the collision its momentum is p_1'. Write the total initial energy in terms of m_1, m_2, and p_1 and the total final energy in terms of m_1, m_2, and p_1', and show that $p_1' = \pm p_1$. If $p_1' = -p_1$, the particle is merely turned around by the collision and leaves with the speed it had initially. What is the significance of the plus sign in your solution?

102 •• **SSM** **iSOLVE** A 3-kg block is traveling in the negative x direction at 5 m/s, and a 1-kg block is traveling in the positive x direction at 3 m/s. (a) Find the velocity v_{cm} of the center of mass. (b) Subtract v_{cm} from the velocity of each block to find the velocity of each block in the center-of-mass reference frame. (c) After they make a head-on elastic collision, the velocity of each block is reversed (in this frame). Find the velocity of each block in the center-of-mass frame after the collision. (d) Transform back into the original frame by adding v_{cm} to the velocity of each block. (e) Check your result by finding the initial and final kinetic energies of the blocks in the original frame and comparing them.

103 •• Repeat Problem 102 with the second block having a mass of 5 kg and moving to the right at 3 m/s.

*Systems With Continuously Varying Mass: Rocket Propulsion

104 •• **iSOLVE** A rocket burns fuel at a rate of 200 kg/s and exhausts the gas at a relative speed of 6 km/s. Find the thrust of the rocket.

105 •• A rocket has an initial mass of 30,000 kg, of which 80 percent is the fuel. It burns fuel at a rate of 200 kg/s and exhausts its gas at a relative speed of 1.8 km/s. Find (a) the thrust of the rocket, (b) the time until burnout, and (c) its speed at burnout assuming it moves straight upward near the surface of the earth where the gravitational field g is constant.

106 •• **SSM** The *specific impulse* of a rocket propellant is defined as $I_{sp} = F_{th}/(Rg)$, where F_{th} is the thrust of the propellant, g the magnitude of free-fall acceleration at the surface of the earth, and R the rate at which the propellant is burned. The rate depends predominantly on the type and exact mixture of the propellant. (a) Show that the specific impulse has the dimension of time. (b) Show that $u_{ex} = gI_{sp}$, where u_{ex} is the exhaust velocity of the propellant. (c) What is the specific impulse (in seconds) of the propellant used in the Saturn V rocket of Example 8-21?

107 ••• **SSM** The initial *thrust-to-weight ratio* τ_0 of a rocket is $\tau_0 = F_{th}/(m_0 g)$, where F_{th} is the rocket's thrust and m_0 the initial mass of the rocket, including the propellant. (a) For a rocket launched straight up from the earth's surface, show that $\tau_0 = 1 + (a_0/g)$, where a_0 is the initial acceleration of the rocket. For manned rocket flight, τ_0 cannot be made much larger than 4 for the comfort and safety of the astronauts. (The astronauts will feel that their weight as the rocket lifts off is equal to τ_0 times their normal weight.) (b) Show that the final velocity of a rocket launched from the earth's surface, in terms of τ_0 and I_{sp} (see Problem 106) can be written as

$$v_f = gI_{sp}\left[\ln\left(\frac{m_0}{m_f}\right) - \frac{1}{\tau_0}\left(1 - \frac{m_f}{m_0}\right)\right]$$

where m_f is the mass of the rocket (not including the spent propellant). (c) Using a spreadsheet program or graphing calculator, graph v_f as a function of the mass ratio m_0/m_f for $I_{sp} = 250$ s and $\tau_0 = 2$ for values of the mass ratio from 2 to 10. (Note that the mass ratio cannot be less than 1.) (d) To lift a rocket into orbit, a final velocity after burnout of $v_f = 7$ km/s is needed. Calculate the mass ratio required of a single stage rocket to do this, using the values of specific impulse and thrust ratio given in Part (b). For engineering reasons, it is difficult to make a rocket with a mass ratio much greater than 10. Can you see why multistage rockets are usually used to put payloads into orbit around the earth?

108 •• The height that a model rocket launched from earth can reach can be estimated by assuming that the burn time is short compared to the total flight time, so for most of the flight the rocket is in free-fall. (This estimate neglects the burn time in calculations of both time and displacement.) For a model rocket with specific impulse $I_{sp} = 100$ s, mass ratio $m_0/m_f = 1.2$, and initial thrust-to-weight ratio $\tau_0 = 5$ (these parameters are defined in Problems 106 and 107), estimate (a) the height the rocket can reach, and (b) the total flight time. (c) Justify the assumption used in the estimates by comparing the flight time from Part (b) to the time it takes for the fuel to be spent.

General Problems

109 • A model-train car of mass 250 g traveling with a speed of 0.50 m/s links up with another car of mass 400 g that is initially at rest. What is the speed of the cars immediately after they have linked together? Find the initial and final kinetic energies.

110 • (a) Find the total kinetic energy of the two model-train cars of Problem 109 before they couple. (b) Find the initial velocities of the two cars relative to the center of mass of the system, and use them to calculate the initial center-of-mass kinetic energy of the system. (c) Find the kinetic energy of the center of mass. (d) Compare the sum of your answers for (b) and (c) with that for (a).

111 • **SSM** A 4-kg fish is swimming at 1.5 m/s to the right. He swallows a 1.2-kg fish swimming toward him at

3 m/s. Neglecting water resistance, what is the velocity of the larger fish immediately after his lunch?

112 • **iSOLVE✔** A 3-kg block moves at 6 m/s to the right while a 6-kg block moves at 3 m/s to the right. Find (a) the total kinetic energy of the two-block system, (b) the velocity of the center of mass, (c) the center-of-mass kinetic energy, and (d) the kinetic energy relative to the center of mass.

113 • A 1500-kg car traveling north at 70 km/h collides at an intersection with a 2000-kg car traveling west at 55 km/h. The two cars stick together. (a) What is the total momentum of the system before the collision? (b) Find the magnitude and direction of the velocity of the wreckage just after the collision.

114 •• **SSM** A 60-kg woman stands on the back of a 6-m-long, 120-kg raft that is floating at rest in still water with no friction. The raft is 0.5 m from a fixed pier, as shown in Figure 8-63. (a) The woman walks to the front of the raft and stops. How far is the raft from the pier now? (b) While the woman walks, she maintains a constant speed of 3 m/s relative to the raft. Find the total kinetic energy of the system (woman plus raft), and compare with the kinetic energy if the woman walked at 3 m/s on a raft tied to the pier. (c) Where does this energy come from, and where does it go when the woman stops at the front of the raft? (d) On land, the woman puts a lead shot 6 m. She stands at the back of the raft, aims forward, and puts the shot so that just after it leaves her hand, it has the same velocity relative to her as it did when she threw it from the ground. Approximately, where does her shot land?

FIGURE 8-63 Problem 114

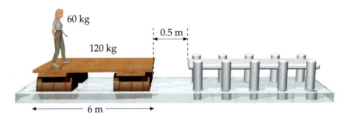

115 •• **iSOLVE** A 1-kg steel ball and a 2-m cord of negligible mass make up a simple pendulum that can pivot without friction about the point O, as in Figure 8-64. This pendulum is released from rest in a horizontal position and when the ball is at its lowest point it strikes a 1-kg block sitting at rest on a shelf. Assume that the collision is perfectly elastic and take the coefficient of friction between the block and shelf to be 0.1. (a) What is the velocity of the block just after impact? (b) How far does the block move before coming to rest?

FIGURE 8-64 Problem 115

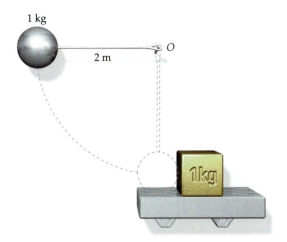

116 •• [SSM] Figure 8-65 shows a World War I cannon mounted on a railcar so that it will project a shell at an angle of 30°. With the car initially at rest, the cannon fires a 200-kg projectile at 125 m/s. (All values are for the frame of reference of the track.) Now consider a system composed of a cannon, shell, and railcar, all on the frictionless track. (*a*) Will the total vector momentum of that system be the same (that is, "conserved") before and after the shell is fired? Explain your answer. (*b*) If the mass of the railcar plus cannon is 5000 kg, what will be the recoil velocity of the car along the track after the firing? (*c*) The shell is observed to rise to a maximum height of 180 m as it moves through its trajectory. At this point, its speed is 80 m/s. On the basis of this information, calculate the amount of thermal energy produced by air friction on the shell on its way from firing to this maximum height.

FIGURE 8-65 Problem 116

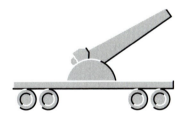

117 •• A 15-g bullet traveling at 500 m/s strikes an 0.8-kg block of wood that is balanced on a table edge 0.8 m above the ground (Figure 8-66). If the bullet buries itself in the block, find the distance *D* at which the block hits the floor.

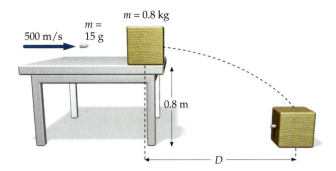

FIGURE 8-66 Problem 117

118 •• Two particles of masses *m* and 4*m* are moving in a vacuum at right angles as in Figure 8-67. A force $\vec{F}$ acts on the particle of mass *m* for a time *T*. As a result, the velocity of this particle becomes 4*v* in its original direction. A force equal to force $\vec{F}$ acts on the other particle for the same time *T*. Find the new velocity $\vec{v}'$ of the particle of mass 4*m*.

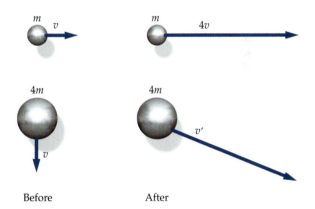

FIGURE 8-67 Problem 118

119 •• One popular, if dangerous, classroom demonstration involves holding a baseball an inch or so directly above a basketball, holding the basketball a few feet above a hard floor, and dropping the two balls simultaneously. The two balls will collide just after the basketball bounces from the floor; the baseball will then rocket off into the ceiling tiles with a hard "thud" while the basketball will stop in midair. (The author of this problem once broke a light doing this.) (*a*) Assuming that the collision of the basketball with the floor is elastic, what is the relation between the velocities of the balls just before they collide? (*b*) Assuming the collision between the two balls is elastic, use the result of Part (*a*) and the conservation of momentum and energy to show that, if the basketball is three times as heavy as the baseball, the final velocity of the basketball will be zero. (This is approximately the true mass ratio, which is why the demonstration is so dramatic.) (*c*) If the speed of the top ball is *v* just before the collision, what is its speed just after the collision?

120 ••• (*a*) In Problem 119, if we held a third ball above the other two, and wanted both the lowest and middle balls to stop in mid-air, what should be the ratio of the mass of the top ball to the mass of the middle ball? (*b*) If the speed of the top ball is *v* just before the collision, what is its speed just after the collision?

121 •• SSM iSOLVE✓ In the "slingshot effect," the transfer of energy in an elastic collision is used to boost the energy of a space probe so that it can escape from the solar system. All speeds are relative to an inertial frame in which the center of the sun remains at rest. Figure 8-68 shows a space probe moving at 10.4 km/s toward Saturn, which is moving at 9.6 km/s toward the probe. Because of the gravitational attraction between Saturn and the probe, the probe swings around Saturn and heads back in the opposite direction with speed v_f. (*a*) Assuming this collision to be a one-dimensional elastic collision with the mass of Saturn much greater than that of the probe, find v_f. (*b*) By what factor is the kinetic energy of the probe increased? Where does this energy come from?

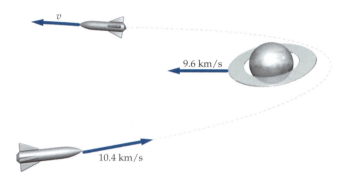

FIGURE 8-68 Problem 121

122 •• SSM Imagine that a flashlight is floating in intergalactic space, far from any planet or star. It has batteries in it that are charged with a total energy $E = 1.5$ kJ. When the flashlight is turned on, it loses energy in the form of light; because of this, it also loses mass slightly, from $E = mc^2$. Because the lost "mass" is traveling away at the speed of light, the flashlight should start moving in the opposite direction. (*a*) If the flashlight loses energy ΔE, argue that the change in momentum of the flashlight should be $P = \Delta E/c$. (*b*) If the mass of the flashlight is 1.5 kg, and it is turned on and left on until the batteries are discharged, what is its final velocity? Assume that the flashlight is 100 percent efficient in converting battery power into radiant power, and that it is initially at rest.

123 • A block has a mass $M = 1$ kg. If we could convert 1 g of the mass of the block into a well-collimated beam of light shining in one direction, calculate the speed attained by the rest of the block. (See Problem 122 for a discussion of the momentum carried by a light beam.)

124 •• You (mass 80 kg) and your friend (mass unknown) are in a rowboat (mass 60 kg) on a calm lake. You are at the center of the boat rowing and she is at the back, 2 m from the center. You get tired and stop rowing. She offers to row and after the boat comes to rest, you change places. You notice that after changing places the boat has moved 20 cm relative to a fixed log. What is your friend's mass?

125 •• iSOLVE✓ A small car of mass 800 kg is parked behind a small truck of mass 1550 kg on a level road (Figure 8-69). The brakes of both the car and the truck are off so that they are free to roll with negligible friction. A 50-kg woman sitting on the tailgate of the truck shoves the car away by exerting a constant force on the car with her feet. The car accelerates at 1.2 m/s². (*a*) What is the acceleration of the truck? (*b*) What is the magnitude of the force exerted on the car?

FIGURE 8-69 Problem 125

126 •• A 13-kg block is at rest on a level floor. A 400-g glob of putty is thrown at the block so that the putty travels horizontally, hits the block, and sticks to it. The block and putty slide 15 cm along the floor. If the coefficient of sliding friction is 0.4, what is the initial speed of the putty?

127 •• SSM A careless driver rear-ends a car that is halted at a stop sign. Just before impact, the driver slams on his brakes, locking the wheels. The driver of the struck car also has his foot solidly on the brake pedal, locking his brakes. The mass of the struck car is 900 kg, and that of the initially moving vehicle is 1200 kg. On collision, the bumpers of the two cars mesh. Police determine from the skid marks that after the collision the two cars moved 0.76 m together. Tests revealed that the coefficient of sliding friction between the tires and pavement was 0.92. The driver of the moving car claims that he was traveling at less than 15 km/h as he approached the intersection. Is he telling the truth?

128 •• A pendulum consists of a 0.4-kg bob attached to a string of length 1.6 m. A block of mass m rests on a horizontal frictionless surface (Figure 8-70). The pendulum is released from rest at an angle of 53° with the vertical and the bob collides elastically with the block. Following the collision, the maximum angle of the pendulum with the vertical is 5.73°. Determine the mass m.

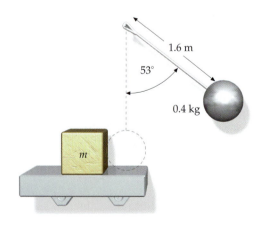

FIGURE 8-70 Problem 128

129 •• Initially, the block whose mass $m = 1.0$ kg and the block whose mass is M are both at rest on a frictionless inclined plane (Figure 8-71). The block of mass M rests against a spring that has a spring constant of 11,000 N/m. The distance along the plane between the two blocks is 4.0 m. The block of mass m is released, makes an elastic collision with the block of mass M, and rebounds a distance of 2.56 m back up the inclined plane. The block whose mass is M comes to rest momentarily 4.0 cm from its initial position. Find M.

FIGURE 8-71 Problem 129

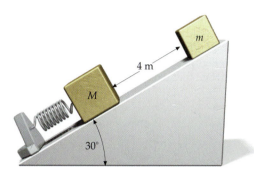

130 •• **SSM** A circular plate of radius r has a circular hole of radius $r/2$ cut out of it (Figure 8-72). Find the center of mass of the plate. *Hint: The hole can be represented by two disks superimposed, one of mass* m *and the other of mass* −m.

FIGURE 8-72
Problem 130

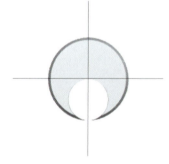

131 •• Using the hint from Problem 130, find the center of mass of a solid sphere of radius r that has a spherical cavity of radius $r/2$, as in Figure 8-73.

FIGURE 8-73
Problem 131

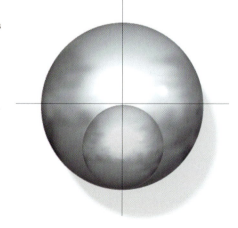

132 •• **SSM** A neutron of mass m makes an elastic head-on collision with a stationary nucleus of mass M. (a) Show that the energy of the nucleus after the collision is $K_{nucleus} = [4mM/(m + M)^2]K_n$, where K_n is the initial energy of the neutron. (b) Show that the fraction of energy lost by the neutron in this collision is

$$\frac{-\Delta K_n}{K_n} = \frac{4mM}{(m + M)^2} = \frac{4(m/M)}{(1 + m/M)^2}$$

133 •• The mass of a carbon nucleus is approximately 12 times the mass of a neutron. (a) Use the results of Problem 132 to show that after N head-on collisions of a neutron with carbon nuclei at rest, the energy of the neutron is approximately $0.716^N E_0$, where E_0 is its original energy. Neutrons emitted in the fission of a uranium nucleus have an energy of about 2 MeV. For such a neutron to cause the fission of another uranium nucleus in a reactor, its energy must be reduced to about 0.02 eV. (b) How many head-on collisions are needed to reduce the energy of a neutron from 2 MeV to 0.02 eV, assuming elastic head-on collisions with stationary carbon nuclei?

134 •• On average, a neutron loses 63 percent of its energy in an elastic collision with a hydrogen atom and 11 percent of its energy in an elastic collision with a carbon atom. (The numbers are lower than the ones we have been using in earlier problems because most collisions are not head-on.) Calculate the number of collisions, on average, needed to reduce the energy of a neutron from 2 MeV to 0.02 eV (a desirable outcome for reasons explained in Problem 133) if the neutron collides with (a) hydrogen atoms and (b) carbon atoms.

135 •• A rope of length L and mass M lies coiled on a table. Starting at $t = 0$, one end of the rope is lifted from the table with a force F such that it moves with a constant velocity v. (a) Find the height of the center of mass of the rope as a function of time. (b) Differentiate your result in (a) twice to find the acceleration of the center of mass. (c) Assuming that the force exerted by the table equals the weight of the rope still there, find the force F you exert on the top of the rope.

136 •• **iSOLVE** Repeat Problem 47 if the cup has a mass m_c and the ball collides with it inelastically.

137 •• Two astronauts at rest face each other in space. One, with mass m_1, throws a ball of mass m_b to the other, whose mass is m_2. She catches the ball and throws it back to the first astronaut. Following each throw the ball has a speed of v relative to the thrower. How fast are they moving after each has made one throw and one catch?

138 •• SSM The ratio of the mass of the earth to the mass of the moon is $M_e/m_m = 81.3$. The radius of the earth is about 6370 km, and the distance from the earth to the moon is about 384,000 km. (*a*) Locate the center of mass of the earth–moon system. Is it above or below the surface of the earth? (*b*) What external forces act on the earth–moon system? (*c*) In what direction is the acceleration of the center of mass of this system? (*d*) Assume that the center of mass of this system moves in a circular orbit around the sun. How far must the center of the earth move in the radial direction (toward or away from the sun) during the 14 days between the time the moon is farthest from the sun (full moon) and the time it is closest to the sun (new moon)?

139 •• ISOLVE✓ You want to enlarge a skating surface so you stand on the ice at one end and aim a hose horizontally to spray water on the schoolyard pavement. Water leaves the hose at 2.4 kg/s with a speed 30 m/s. If your mass is 75 kg, what is your recoil acceleration? (Neglect friction and the mass of the hose.)

140 ••• SSM A stream of glass beads, each with a mass of 0.5 g, comes out of a horizontal tube at a rate of 100 per second (Figure 8-74). The beads fall a distance of 0.5 m to a balance pan and bounce back to their original height. How much mass must be placed in the other pan of the balance to keep the pointer at zero?

FIGURE 8-74 Problem 140

141 ••• A dumbbell consisting of two balls of mass m connected by a massless rod of length L rests on a frictionless floor against a frictionless wall until it begins to slide down the wall as in Figure 8-75. Find the speed of the bottom ball at the moment when it equals the speed of the top ball.

FIGURE 8-75
Problem 141

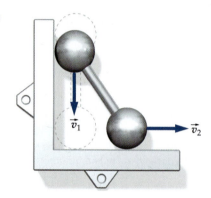

Rotation

THESE BOWLING BALLS ROLL ON A HORIZONTAL BALL RETURN WITHOUT SLIPPING.

? **How might you design a ball return to stop the balls' rotation? (See Example 9-13.)**

In Chapters 4 and 5 we explored Newton's laws. In Chapters 6 and 7 we examined the conservation of energy, and in Chapter 8 we studied the conservation of momentum. In those chapters we discovered tools (laws and theorems) that are useful in analyzing new situations and in solving new problems. We will use those tools now as we explore rotational motion.

Rotational motion is all around us. The earth rotates about its axis. Wheels, gears, propellers, motors, the drive shaft in a car, a CD in its player, a pirouetting ice skater, all rotate.

➤ **In this chapter, we consider rotation about an axis that is fixed in space, as in a spinning top or a merry-go-round, or an axis that is moving parallel to itself, as in a rolling ball. (More general examples of rotational motion will be discussed in Chapter 10.)**

9-1 Rotational Kinematics: Angular Velocity and Angular Acceleration

Every point in a body rotating about a fixed axis moves in a circle whose center is on the axis and whose radius is the distance of that point from the axis of

FIGURE 9-1

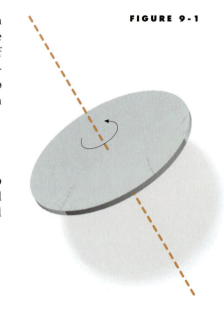

rotation. A line drawn from this axis to any point sweeps out the same angle in the same time. Imagine a disk spinning about a fixed axis perpendicular to the disk and through its center (Figure 9-1). Let r_i be the distance from the center of the disk to the ith particle (Figure 9-2), and let θ_i be the angle measured counterclockwise from a fixed reference line in space to a radial line from the center to the particle. As the disk rotates through an angle $d\theta$, the particle moves through a circular arc of length ds_i, such that

$$ds_i = r_i |d\theta| \qquad 9\text{-}1$$

where $d\theta$ is measured in radians. The distances ds_i and r_i vary from particle to particle, but their ratio, called the **angular displacement** $d\theta$, is the same for all particles of the disk. For one complete revolution, the arc length Δs_i is $2\pi r$ and the angular displacement $\Delta\theta$ is

$$\Delta\theta = \frac{2\pi r_i}{r_i} = 2\pi \text{ rad} = 360° = 1 \text{ rev}$$

The time rate of change of the angle is the same for all particles of the disk, and is called the **angular velocity** ω of the disk:

FIGURE 9-2

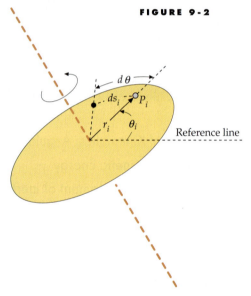

Reference line

$$\omega = \frac{d\theta}{dt} \qquad 9\text{-}2$$

DEFINITION—ANGULAR VELOCITY

where ω is a lowercase Greek omega. For counterclockwise rotation, θ increases, so ω is positive. For clockwise rotation, θ decreases, and ω is negative. The units of ω are radians per second. (In Chapter 10 we will see that, for rotation in general, the angular velocity is a vector quantity that points along the axis of rotation.) Since radians are dimensionless, the dimension of angular velocity is that of reciprocal time, $[T]^{-1}$. The magnitude of the angular velocity is called the **angular speed.** We often use revolutions per minute (rev/min or RPM) to describe rotation. To convert between revolutions, radians, and degrees, we use

$$1 \text{ rev} = 2\pi \text{ rad} = 360°$$

EXERCISE A compact disk is rotating at 3000 rev/min. What is its angular speed in radians per second? (*Answer*   314 rad/s)

The time rate of change of angular velocity is called the **angular acceleration** α:

$$\alpha = \frac{d\omega}{dt} = \frac{d^2\theta}{dt^2} \qquad 9\text{-}3$$

DEFINITION—ANGULAR ACCELERATION

The units of α are radians per second per second (rad/s²). If ω is increasing, α is positive; if ω is decreasing, α is negative.

The angular displacement θ, angular velocity ω, and angular acceleration α are analogous to the linear displacement x, linear velocity v, and linear

acceleration a in one-dimensional motion. If the angular acceleration α is constant, we can integrate Equation 9-3 to find ω:

$$\omega = \omega_0 + \alpha t \qquad 9\text{-}4$$

where the constant of integration ω_0 is the initial angular velocity. This is the rotational analog of $v = v_0 + at$. Integrating again, we obtain

$$\theta = \theta_0 + \omega_0 t + \tfrac{1}{2}\alpha t^2 \qquad 9\text{-}5$$

which is the rotational analog of $x = x_0 + v_0 t + \tfrac{1}{2}at^2$ with θ replacing x, ω replacing v, and α replacing a. Similarly, by eliminating t from Equations 9-4 and 9-5, we get

$$\omega^2 = \omega_0^2 + 2\alpha(\theta - \theta_0) \qquad 9\text{-}6$$

which is the rotational analog of $v^2 = v_0^2 + 2a(x - x_0)$. The equations for constant angular acceleration have the same form as those for constant linear acceleration.

Star tracks in a time exposure of the night sky.

A CD PLAYER

EXAMPLE 9-1

A compact disk rotates from rest to 500 rev/min in 5.5 s. (a) What is its angular acceleration, assuming it is constant? (b) How many revolutions does the disk make in 5.5 s? (c) How far does a point on the rim 6 cm from the center of the disk travel during the 5.5 s it takes to get to 500 rev/min?

PICTURE THE PROBLEM Part (*a*) is analogous to the linear problem of finding the acceleration given the time and the final velocity. To find α in rad/s² we first convert ω to rad/s. Part (*b*) is analogous to finding the distance traveled given the time and the final velocity.

(*a*) 1. The angular acceleration is related to the initial and final angular velocities:

$$\omega = \omega_0 + \alpha t = 0 + \alpha t$$

2. Solve for α:

$$\alpha = \frac{\omega}{t}$$

$$= \frac{500\ \text{rev/min}}{5.5\ \text{s}} \times \frac{2\pi\ \text{rad}}{1\ \text{rev}} \times \frac{1\ \text{min}}{60\ \text{s}}$$

$$= \boxed{9.52\ \text{rad/s}^2}$$

(*b*) 1. The angular displacement is related to the time by Equation 9-5:

$$\theta - \theta_0 = \omega_0 t + \tfrac{1}{2}\alpha t^2$$

$$= 0 + \tfrac{1}{2}(9.52\ \text{rad/s}^2)(5.5\ \text{s})^2$$

$$= 144\ \text{rad}$$

2. Convert radians to revolutions:

$$\theta - \theta_0 = 144 \text{ rad} \times \frac{1 \text{ rev}}{2\pi \text{ rad}} = \boxed{22.9 \text{ rev}}$$

(c) The distance traveled Δs is r times the angular displacement:

$$\Delta s = r \, \Delta\theta = (6 \text{ cm})(144 \text{ rad}) = \boxed{8.64 \text{ m}}$$

PLAUSIBILITY CHECK The average angular velocity in revolutions per minute is 250 rev/min. In 5.5 s, the compact disk rotates (250 rev/60 s)(5.5 s) = 22.9 rev.

REMARKS A compact disk is scanned by a laser that begins at the inner radius of about 2.4 cm and moves out to the edge at 6.0 cm. As the laser moves outward, the angular velocity of the disk decreases from 500 rev/min to 200 rev/min so that the linear (tangential) velocity of the disk at the point where the laser beam strikes remains constant.

EXERCISE (a) Convert 500 rev/min to rad/s. (b) Check the result of Part (b) in the example using $\omega^2 = \omega_0^2 + 2\alpha(\theta - \theta_0)$. (Answer (a) 500 rev/min = 52.4 rad/s)

The linear velocity v_t of a particle on the disk is tangent to the circular path of the particle and has magnitude ds_i/dt. We can relate this "tangential" velocity to the angular velocity of the disk using Equations 9-1 and 9-2:

$$v_t = \frac{r_i \, d\theta}{dt}$$

so

$$v_t = r_i \omega \qquad\qquad 9\text{-}7$$

Similarly, the tangential acceleration of a particle on the disk is

$$a_t = \frac{dv_t}{dt} = r_i \frac{d\omega}{dt}$$

so

$$a_t = r\alpha \qquad\qquad 9\text{-}8$$

Each particle of the disk also has a centripetal acceleration, which points inward along the radial line and has the magnitude

$$a_c = \frac{v_t^2}{r_i} = \frac{(r_i\omega)^2}{r_i}$$

so

$$a_c = r_i\omega^2 \qquad\qquad 9\text{-}9$$

EXERCISE A point on the rim of a compact disk is 6.0 cm from the axis of rotation. Find the tangential speed v_t, tangential acceleration a_t, and centripetal acceleration a_c of the point when the disk is rotating at a constant angular speed of 300 rev/min. (Answer $v_t = 188$ cm/s, $a_t = 0$, $a_c = 5.92 \times 10^3$ cm/s²)

EXERCISE Find the linear speed of a point on the CD in Example 9-1 at (a) $r =$ 2.4 cm when the disk rotates at 500 rev/min, and (b) $r = 6.0$ cm when the disk rotates at 200 rev/min. (*Answer* (a) 126 cm/s (b) 126 cm/s)

9-2 Rotational Kinetic Energy

The kinetic energy of a rigid object rotating about a fixed axis is the sum of the kinetic energies of the individual particles that collectively constitute the object. The kinetic energy of the *i*th particle, with mass m_i, is

$$K_i = \tfrac{1}{2}m_i v_i^2$$

Summing over all the particles and using $v_i = r_i\omega$ gives

$$K = \sum_i (\tfrac{1}{2}m_i v_i^2) = \tfrac{1}{2}\sum_i (m_i r_i^2 \omega^2) = \tfrac{1}{2}\left(\sum_i m_i r_i^2\right)\omega^2$$

The sum in the term on the right is the object's **moment of inertia** I for the axis of rotation.

$$I = \sum_i m_i r_i^2 \qquad\qquad 9\text{-}10$$

MOMENT OF INERTIA DEFINED

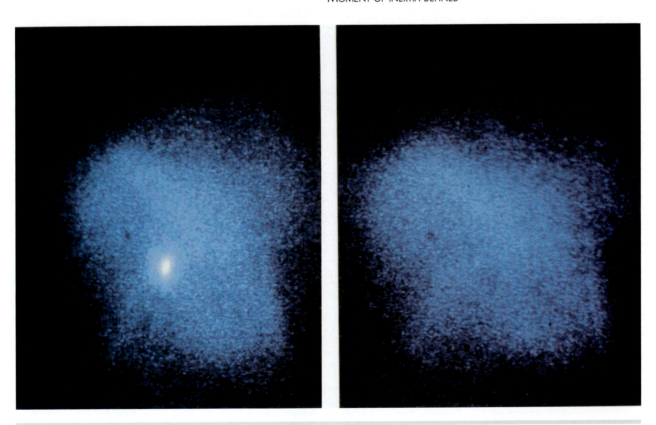

The Crab Pulsar is one of the fastest-rotating neutron stars known, but it is slowing down. It appears to blink on (left) and off (right) like the rotating lamp in a lighthouse, at the fast rate of about 30 times per second, but the period is increasing by about 10^{-5} s/y. The loss in rotational energy, which is equivalent to the power output of 100,000 suns, appears as light emitted by electrons accelerated in the magnetic field of the pulsar.

The kinetic energy is thus

$$K = \tfrac{1}{2}I\omega^2 \qquad\qquad 9\text{-}11$$

KINETIC ENERGY OF ROTATING OBJECT

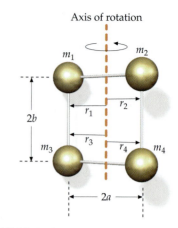

A Rotating System of Particles

EXAMPLE 9-2

An object consists of four point particles, each of mass m, that are connected by rigid massless rods to form a rectangle of sides $2a$ and $2b$ as shown in Figure 9-3. The system rotates with angular speed ω about an axis in the plane of the figure through the center as shown. (*a*) Find the kinetic energy of this object using Equations 9-10 and 9-11. (*b*) Check your result by individually calculating the kinetic energy of each particle and then taking their sum.

FIGURE 9-3

PICTURE THE PROBLEM Because we are given that the objects are particles, we use Equation 9-10 to calculate I and then use Equation 9-11. In Equation 9-10, r_i is the radial distance from the rotation axis to the particle of mass m_i.

(*a*) 1. Apply the definition of moment of inertia (Equation 9-10):

$$I = \sum_i m_i r_i^2$$

$$= m_1 r_1^2 + m_2 r_2^2 + m_3 r_3^2 + m_4 r_4^2$$

2. The masses m_i and the distances r_i are given:

$$m_1 = m_2 = m_3 = m_4 = m$$
$$r_1 = r_2 = r_3 = r_4 = a$$

3. Substitution gives the moment of inertia:

$$I = ma^2 + ma^2 + ma^2 + ma^2 = 4ma^2$$

4. Using Equation 9-11, solve for the kinetic energy:

$$K = \tfrac{1}{2}I\omega^2 = \tfrac{1}{2}4ma^2\omega^2 = \boxed{2ma^2\omega^2}$$

(*b*) 1. To find the kinetic energy of the *i*th particle we must first find its speed:

$$K_i = \tfrac{1}{2}m_i v_i^2$$

2. The particles are all moving in circles of radius a. Find the speed of each particle:

$$v_i = r_i \omega = a\omega \qquad (i = 1, \ldots, 4)$$

3. Substitute into the Part (*b*) step 1 result:

$$K_i = \tfrac{1}{2}m_i v_i^2 = \tfrac{1}{2}ma^2\omega^2$$

4. Each particle has the same kinetic energy. Sum the kinetic energies to get the total:

$$K = \sum_{i=1}^{4} K_i = \tfrac{1}{2}m_1 v_1^2 + \tfrac{1}{2}m_2 v_2^2 + \tfrac{1}{2}m_3 v_3^2 + \tfrac{1}{2}m_4 v_4^2$$

$$= 4(\tfrac{1}{2}ma^2\omega^2) = 2ma^2\omega^2$$

5. Compare with the Part (*a*) result:

The two calculations give the same result.

REMARKS Notice that I is independent of the length b, which has no effect on how far the masses are from the axis of rotation.

EXERCISE Find the moment of inertia of this system for rotation about an axis parallel to the first axis but passing through two of the particles, as shown in Figure 9-4. (*Answer* $I = 8ma^2$)

9-3 Calculating the Moment of Inertia

The moment of inertia about an axis is a measure of the inertial resistance of the object to changes in its rotational motion about the axis. It is the rotational analog of mass. The moment of inertia about an axis depends on the distribution of mass

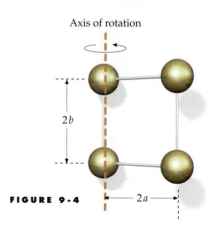

FIGURE 9-4

within the object relative to the axis. The farther an element of mass is from the axis, the greater its contribution to the moment of inertia about that axis. Thus, unlike the mass of an object, which is a property of the object itself, the moment of inertia depends on the location of the axis of rotation.

Systems of Discrete Particles

For systems consisting of discrete particles, we can compute the moment of inertia about a given axis directly from Equation 9-10.

Continuous Objects

To calculate the moment of inertia for continuous objects, we imagine the object to consist of a continuum of very small mass elements. Thus, the finite sum $\Sigma m_i r_i^2$ in Equation 9-10 becomes the integral

$$I = \int r^2 \, dm \qquad\qquad\qquad 9\text{-}12$$

where r is the radial distance from the axis to mass element dm.

MOMENT OF INERTIA OF A UNIFORM ROD   **EXAMPLE   9 - 3**

Find the moment of inertia of a uniform rod of length L and mass M about an axis perpendicular to the rod and through one end. Assume that the rod has negligible thickness.

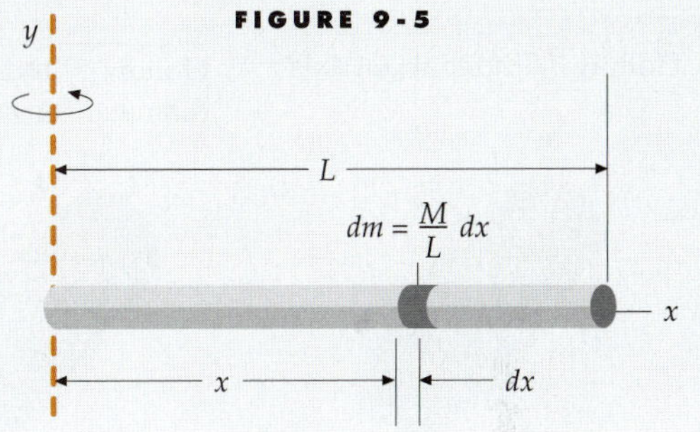

FIGURE 9-5

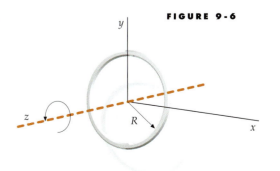

$dm = \dfrac{M}{L} \, dx$

PICTURE THE PROBLEM Let the rod lie along the x axis with its end at the origin. To calculate I about the y axis, we choose a mass element dm at a distance x from the axis (Figure 9-5). Since the total mass M is uniformly distributed along the length L, the mass per unit length (linear mass density) is $\lambda = M/L$.

1. The moment of inertia is given by the integral:

$$I = \int_0^L x^2 \, dm$$

2. To compute the integral, we first relate dm to dx. Express dm in terms of the mass density λ and dx:

$$dm = \lambda \, dx = \frac{M}{L} \, dx$$

3. Substitute and perform the integration. We choose integration limits so that we sweep dm through the mass distribution in the direction of increasing x:

$$I_y = \int x^2 \, dm = \int_0^L x^2 \frac{M}{L} \, dx = \frac{M}{L} \int_0^L x^2 \, dx$$

$$= \frac{M}{L} \frac{1}{3} x^3 \Big|_0^L = \frac{M}{L} \frac{L^3}{3} = \boxed{\frac{1}{3} ML^2}$$

REMARKS The moment of inertia about the z axis is also $\frac{1}{3} ML^2$, and that about the x axis is zero (assuming that all of the mass is right on the x axis).

We can calculate I for continuous objects of various shapes, again using Equation 9-12 (see Table 9-1).

FIGURE 9-6

*Hoop About a Perpendicular Axis Through Its Center Assume that a hoop has mass M and radius R (Figure 9-6). The axis of rotation is the axis of the hoop, which is perpendicular to the plane of the hoop. All the mass is at a distance $r = R$, and the moment of inertia is

$$I = \int r^2 \, dm = \int R^2 \, dm = R^2 \int dm = MR^2$$

TABLE 9-1

Moments of Inertia of Uniform Bodies of Various Shapes

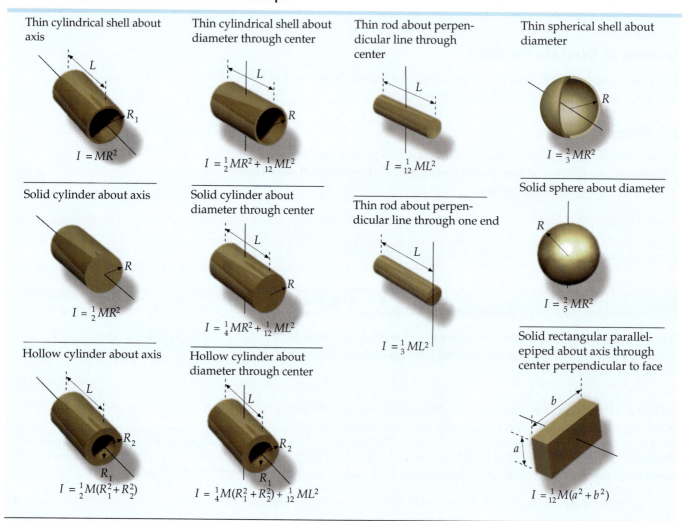

Thin cylindrical shell about axis

$$I = MR^2$$

Thin cylindrical shell about diameter through center

$$I = \tfrac{1}{2}MR^2 + \tfrac{1}{12}ML^2$$

Thin rod about perpendicular line through center

$$I = \tfrac{1}{12}ML^2$$

Thin spherical shell about diameter

$$I = \tfrac{2}{3}MR^2$$

Solid cylinder about axis

$$I = \tfrac{1}{2}MR^2$$

Solid cylinder about diameter through center

$$I = \tfrac{1}{4}MR^2 + \tfrac{1}{12}ML^2$$

Thin rod about perpendicular line through one end

$$I = \tfrac{1}{3}ML^2$$

Solid sphere about diameter

$$I = \tfrac{2}{5}MR^2$$

Hollow cylinder about axis

$$I = \tfrac{1}{2}M(R_1^2 + R_2^2)$$

Hollow cylinder about diameter through center

$$I = \tfrac{1}{4}M(R_1^2 + R_2^2) + \tfrac{1}{12}ML^2$$

Solid rectangular parallelepiped about axis through center perpendicular to face

$$I = \tfrac{1}{12}M(a^2 + b^2)$$

A disk is a cylinder whose length L is negligible. By setting $L = 0$, the above formulas for cylinders hold for disks.

*Uniform Disk About a Perpendicular Axis Through Its Center For the case of a uniform disk, we expect that I will be smaller than MR^2 since the mass is uniformly distributed from $r = 0$ to $r = R$ rather than being concentrated at $r = R$ as it is in a hoop. In Figure 9-7, each mass element is a hoop of radius r and thickness dr. The moment of inertia of any given mass element is $r^2\,dm$. Since the disk is uniform, mass per unit area σ is constant. $\sigma = M/A$, where $A = \pi R^2$ is the area of the disk. Since the area of each mass element is $dA = 2\pi r\,dr$, the mass of each element is

$$dm = \sigma\,dA = \frac{M}{A}\,2\pi r\,dr$$

We thus have

$$I = \int r^2\,dm = \int_0^R r^2\sigma\,2\pi r\,dr = 2\pi\sigma\int_0^R r^3\,dr$$

$$= \frac{2\pi M}{A}\frac{R^4}{4} = \frac{\pi M}{2\pi R^2}R^4 = \frac{1}{2}MR^2$$

FIGURE 9-7

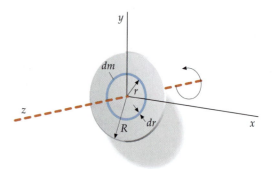

Uniform Solid Cylinder About Its Axis We consider a cylinder to be a set of disks, each with mass dm and moment of inertia $\frac{1}{2}dm\,R^2$ (Figure 9-8). The moment of inertia of the complete cylinder is then

$$I = \int \tfrac{1}{2}dm\,R^2 = \tfrac{1}{2}R^2 \int dm = \tfrac{1}{2}MR^2$$

where M is the total mass of the cylinder.

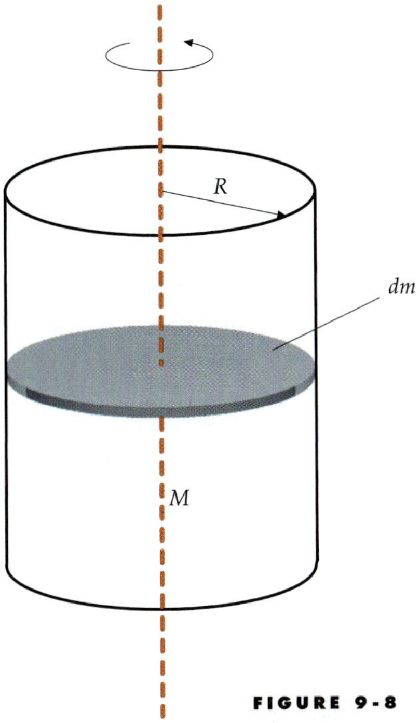

The Parallel-Axis Theorem

We can often simplify the calculation of moments of inertia for various bodies by using the **parallel-axis theorem**, which relates the moment of inertia about an axis through the center of mass of an object to the moment of inertia about a second, parallel axis (Figure 9-9). Let I_{cm} be the moment of inertia about an axis through the center of mass of an object of total mass M, and let I be that about a parallel axis a distance h away. The parallel-axis theorem states that

$$I = I_{cm} + Mh^2 \qquad\qquad 9\text{-}13$$

PARALLEL-AXIS THEOREM

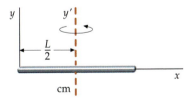

FIGURE 9-8

Example 9-2 and the exercise following it illustrate a special case of this theorem with $h = a$, $M = 4m$, and $I_{cm} = 4ma^2$. A proof of the parallel-axis theorem is given at the end of this section.

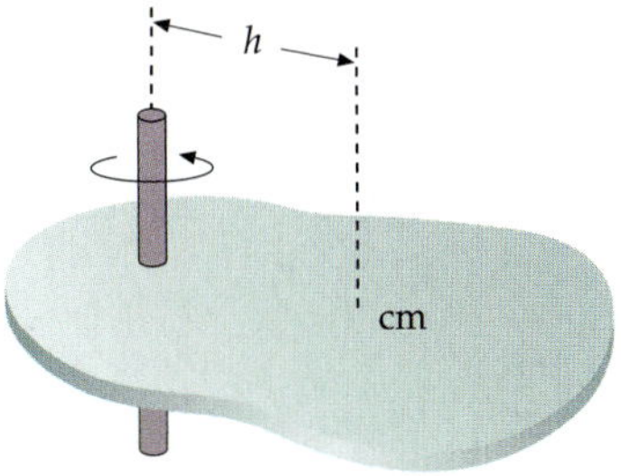

FIGURE 9-9 An object rotating about an axis parallel to an axis through the center of mass and a distance h from it.

MOMENT OF INERTIA ABOUT AN AXIS THROUGH THE CENTER OF MASS OF A UNIFORM ROD   **EXAMPLE 9-4**   **Try It Yourself**

Find the moment of inertia of a uniform rod about the y' axis through the center of mass (Figure 9-10).

FIGURE 9-10

PICTURE THE PROBLEM Here you know that $I = \frac{1}{3}ML^2$ about one end and want to find I_{cm}. Use the parallel-axis theorem with $h = \frac{1}{2}L$.

Cover the column to the right and try these on your own before looking at the answers.

Steps

1. Apply the parallel-axis theorem to write I about the end in terms of I_{cm}.

2. Substitute $I = \frac{1}{3}ML^2$ about the end and solve for I_{cm}.

Answers

$$I = I_{cm} + Mh^2$$

$$I_{cm} = I - Mh^2 = \frac{1}{3}ML^2 - M\left(\frac{L}{2}\right)^2$$

$$= \boxed{\frac{1}{12}ML^2}$$

REMARKS The moment of inertia is least when an object is rotated about its center of mass, as in this example. Compare this result to that of Example 9-3, where the uniform rod is rotated about an axis through one end.

EXERCISE Using the parallel-axis theorem, show that when comparing the moments of inertia of an object about two parallel axes, the moment of inertia is least about the axis that is nearest to the center of mass.

*Proof of the Parallel-Axis Theorem

To prove the parallel-axis theorem we start with an object and an axis A (Figure 9-11). We choose our origin to be at the center of mass with the z axis parallel with A. Then I is the moment of inertia about A and I_{cm} is the moment of inertia about the z axis. The mass m_i is located at (x_i, y_i, z_i), and the intersection of axis A with the xy plane is at coordinates x_A, y_A. The distances r_i and R_i are the perpendicular distances of m_i from A and the z axis, respectively, and h is the perpendicular distance between axes. It follows that r_i is the distance from $(x_A, y_A, 0)$ to $(x_i, y_i, 0)$, R_i is the distance from $(0, 0, 0)$ to $(x_i, y_i, 0)$, and h is the distance from $(0, 0, 0)$ to $(x_A, y_A, 0)$. Thus, $r_i^2 = (x_i - x_A)^2 + (y_i - y_A)^2$, $R_i^2 = x_i^2 + y_i^2$, and $h^2 = x_A^2 + y_A^2$. Consequently,

$$I_{cm} = \sum m_i R_i^2 = \sum m_i (x_i^2 + y_i^2)$$

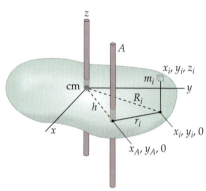

FIGURE 9-11

and

$$I = \sum m_i r_i^2 = \sum m_i \left[(x_i - x_A)^2 + (y_i - y_A)^2 \right]$$
$$= \sum m_i (x_i^2 + y_i^2) - \sum m_i 2 x_i x_A - \sum m_i 2 y_i y_A + \sum m_i (x_A^2 + y_A^2)$$

By factoring the common factors from these sums we have

$$I = \sum m_i (x_i^2 + y_i^2) - 2 x_A \sum m_i x_i - 2 y_A \sum m_i y_i + \left(\sum m_i \right)(x_A^2 + y_A^2)$$

The first term is I_{cm}. The second and third terms can be simplified using $\sum m_i x_i = M x_{cm}$ and $\sum m_i y_i = M y_{cm}$, and because $x_{cm} = y_{cm} = 0$, both the second and the third terms are equal to zero. The fourth term is Mh^2. Thus,

$$I = I_{cm} + Mh^2$$

which is the parallel-axis theorem.

A FLYWHEEL–POWERED CAR   **EXAMPLE 9-5**   **Put It in Context**

You are driving an experimental hybrid vehicle that is designed for use in stop-and-go traffic. In a conventional car, each time you brake to a stop, the kinetic energy is dissipated as heat. In this hybrid vehicle, the braking mechanism transforms the translational kinetic energy of the vehicle's motion to the rotational kinetic energy of a massive flywheel. When the car returns to cruising speed this energy is transferred back into the translational energy of the car. The 100-kg flywheel is a hollow cylinder with an inner diameter R_1 of 25 cm, an outer diameter R_2 of 40 cm, and a maximum angular speed of 30,000 rev/min. On a dark and dreary night, the car runs out of gas 15 mi from home with the flywheel spinning at maximum speed. Is there sufficient energy stored in the flywheel for you and your nervous grandmother to make it home? (When driving at the minimum highway speed of 40 mi/h, air drag and rolling friction dissipate energy at 10 kW.)

PICTURE THE PROBLEM The kinetic energy is calculated directly from $K = \frac{1}{2}I\omega^2$.

1. The kinetic energy of rotation is:

$$K = \frac{1}{2}I\omega^2$$

2. Calculate the moment of inertia of the hollow cylinder:

$$I = \frac{1}{2}m(R_1^2 + R_2^2) = 11.1 \text{ kg·m}^2$$

3. Convert ω to rad/s:

$$\omega = 30{,}000 \text{ rev/min} = 3140 \text{ rad/s}$$

4. Substitute these values to find the kinetic energy:

$$K = \frac{1}{2}I\omega^2 = 54.8 \text{ MJ}$$

5. Energy is dissipated at 10 kW at a speed of 40 mi/h. To find the energy dissipated during the 15-mi trip, we first need to find the time required for the trip:

$$\Delta x = v \, \Delta t, \quad \text{so} \quad \Delta t = 1350 \text{ s}$$

6. The energy is dissipated at 10 kW for 1350 s. The total energy dissipated is:

13.5 MJ

7. Is there enough energy stored in the flywheel?

54.8 MJ are available and 13.5 MJ are dissipated.

> Yes, there is more than enough energy stored.

REMARKS There are 130 MJ of energy in a gallon of gasoline. If the engine is 10% efficient, only 13 MJ/gal are available to move the car. We estimate the mileage for the car in this example to be about 15 mi/gal.

THE PIVOTED ROD **EXAMPLE 9-6**

A uniform thin rod of length L and mass M, pivoted at one end as shown in Figure 9-12, is held horizontal and then released from rest. Assuming the pivot to be frictionless, find (a) the angular velocity of the rod when it reaches its vertical position, and (b) the force exerted by the pivot at this time. (c) What initial angular velocity is needed for the rod to reach a vertical position at the top of its swing?

PICTURE THE PROBLEM (a) As the rod swings down, its potential energy decreases and its kinetic energy increases. Since the pivot is frictionless, we use conservation of mechanical energy. The angular velocity of the rod is then found from its rotational kinetic energy. (b) To find the force of the pivot use Newton's second law for a system. (c) As in Part (a), use conservation of mechanical energy.

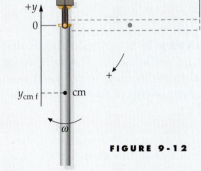

FIGURE 9-12

(a) 1. Make a diagram of the rod showing both its initial and its final configuration (Figure 9-12). Put a vertical coordinate axis on the figure with up as the positive direction and with its origin at the rotation axis.

2. Apply conservation of mechanical energy to relate the initial and final mechanical energies:

$$K_f + U_f = K_i + U_i$$

$$\frac{1}{2}I\omega_f^2 + Mgy_{cm\,f} = \frac{1}{2}I\omega_i^2 + Mgy_{cm\,i}$$

$$\frac{1}{2}I\omega_f^2 + Mg\left(-\frac{L}{2}\right) = 0 + 0$$

3. Solve for ω_f:

$$\omega_f = \sqrt{\frac{MgL}{I}}$$

4. Obtain I from Table 9-1 and substitute into the step 3 result:

$$\omega_f = \sqrt{\frac{MgL}{\frac{1}{3}ML^2}} = \boxed{\sqrt{\frac{3g}{L}}}$$

(b) 1. Make a free-body diagram of the rod as it passes through the vertical position at the bottom of its swing (Figure 9-13).

2. Apply Newton's second law for a system to the rod. At the bottom of the swing the acceleration of the center of mass is in the centripetal (upward) direction:

$$\Sigma F_{ext\,y} = Ma_{cm}$$

$$F_P - Mg = Ma_{cm}$$

3. Relate the acceleration of the center of mass to the angular speed using $a_c = r\omega^2$. Substitute the Part (a) step 4 result for ω and solve for a_{cm}:

$$a_c = r\omega^2$$

$$a_{cm} = \frac{L}{2}\frac{3g}{L} = \frac{3}{2}g$$

4. Substitute into the Part (b) step 2 result and calculate F_P:

$$F_P = Mg + Ma_{cm} = Mg + M\tfrac{3}{2}g$$

$$= \boxed{\tfrac{5}{2}Mg}$$

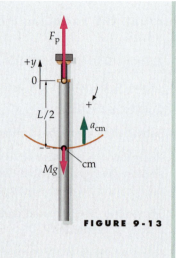

FIGURE 9-13

(c) 1. The initial angular velocity ω_i is related to the initial kinetic energy:

$$K_i = \tfrac{1}{2}I\omega_i^2$$

2. Make a diagram of the rod showing both its initial and its final configuration (Figure 9-14). Put a vertical coordinate axis on the figure with up as the positive direction and with its origin at the rotation axis.

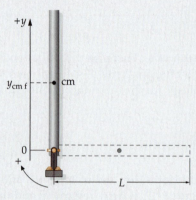

3. Apply conservation of mechanical energy with $K_f = 0$ and $U_i = 0$ to relate the initial kinetic energy to the final position:

$$K_f + U_f = K_i + U_i$$

$$\tfrac{1}{2}I\omega_f^2 + Mgy_{cm\,f} = \tfrac{1}{2}I\omega_i^2 + Mgy_{cm\,i}$$

$$0 + Mg\frac{L}{2} = \frac{1}{2}I\omega_i^2 + 0$$

FIGURE 9-14

4. Solve for the initial angular velocity:

$$\omega_i = \sqrt{\frac{MgL}{I}} = \sqrt{\frac{MgL}{\tfrac{1}{3}ML^2}} = \boxed{\sqrt{\frac{3g}{L}}}$$

REMARKS It is no coincidence that the answers to Part (a) and Part (c) are identical. The decrease in potential energy in Part (a) is equal to the increase in potential energy in Part (c). Thus the increase in kinetic energy in Part (a) is equal to the decrease in kinetic energy in Part (c).

Try It Yourself

A WINCH AND A BUCKET   **EXAMPLE 9-7**

A winch is at the top of a deep well. The drum of the winch has mass m_w and radius R. Virtually all its mass is concentrated a distance R from the axis. A cable wound around the drum suspends a bucket of water of mass m_b. The entire cable has mass m_c and length L. Just when you have the bucket at the highest point your hand slips and the bucket falls back down the well, unwinding the winch cable as it does. How fast is the bucket moving after it has fallen a distance d, where d is less than L?

PICTURE THE PROBLEM As the load falls, mechanical energy is conserved. Choose the initial potential energy to be zero. When the load has fallen a distance d, the center of mass of the hanging cable has dropped a distance $d/2$. Since the hanging part of the cable moves with speed v and the cable does not stretch or become slack, the entire cable must move at speed v. We find v from the conservation of mechanical energy.

Cover the column to the right and try these on your own before looking at the answers.

Steps

Answers

1. Make a diagram of the winch–cable–bucket system in both its initial and its final configuration, as shown in Figure 9-15. Include a y axis with the origin at the height of the center of the winch.

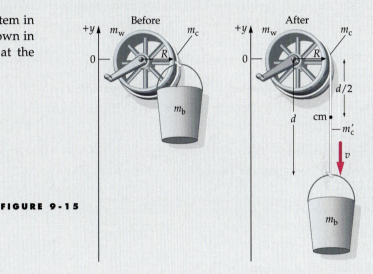

FIGURE 9-15

2. Apply conservation of mechanical energy. Choose the potential energy to be zero when the bucket of water is at the highest point.

$$U_f + K_f = U_i + K_i$$
$$= 0 + 0 = 0$$

3. Write an expression for the total potential energy when the bucket has fallen a distance d. Let m'_c denote the mass of the hanging part of the cable.

$$U_f = U_{bf} + U_{cf} + U_{wf} = m_b g(-d) + m'_c g\left(-\frac{d}{2}\right) + 0$$
$$= -(m_b + \tfrac{1}{2}m'_c)gd$$

4. Express the total kinetic energy when the bucket is falling with speed v. All the cable and the entire mass of the drum move with the same speed v as the bucket.

$$K_f = K_{fc} + K_{fb} + K_{fw} = \tfrac{1}{2}m_c v^2 + \tfrac{1}{2}m_b v^2 + \tfrac{1}{2}m_w v^2$$
$$= \tfrac{1}{2}(m_c + m_b + m_w)v^2$$

5. Substitute into the conservation of mechanical energy equation (step 2) and solve for v.

$$-(m_b + \tfrac{1}{2}m'_c)gd + \tfrac{1}{2}(m_c + m_b + m_w)v^2 = 0$$

so

$$v = \sqrt{\frac{(2m_b + m'_c)gd}{m_c + m_b + m_w}}$$

6. Assume the cable is uniform and express m'_c in terms of m_c, d, and L.

$$\frac{m'_c}{d} = \frac{m_c}{L} \implies m'_c = \frac{d}{L}m_c$$

7. Substitute the step 6 result into the step 5 result.

$$v = \sqrt{\frac{(2m_b L + m_c d)gd}{(m_c + m_b + m_w)L}}$$

REMARKS Because the entire mass of the drum is moving at the same speed v we can express its kinetic energy as $\tfrac{1}{2}m_w v^2$. However, we can express it as $\tfrac{1}{2}I_w \omega^2$, where $I_w = m_w R^2$ and $\omega = v/R$. With these substitutions $K_w = \tfrac{1}{2}I_w \omega^2 = \tfrac{1}{2}m_w R^2(v^2/R^2) = \tfrac{1}{2}m_w v^2$.

9-4 Newton's Second Law for Rotation

To set a top spinning, you twist it. In Figure 9-16, a disk is set spinning by the forces $\vec{F}_1$ and $\vec{F}_2$ exerted at the edges of the disk in the tangential direction. The directions of these forces are important. If the same forces are applied in the radial direction (Figure 9-17), the disk will not start to spin.

Figure 9-18 shows a particle of mass m attached to one end of a massless rigid rod of length r. There is an axis perpendicular to the rod and passing through its other end, and the rod is free to rotate about this axis. Consequently, the particle is constrained to move in a circle of radius r. A single force $\vec{F}$ is applied to the particle as shown. Applying Newton's second law to the particle and taking components in the tangential direction gives

$$F_t = ma_t$$

We wish to obtain an equation involving angular quantities. Substituting $r\alpha$ for a_t (Equation 9-8) and multiplying both sides by r gives

$$rF_t = mr^2\alpha \qquad \text{9-14}$$

The product rF_t is the **torque** τ associated with the force. That is,

$$\tau = F_t r \qquad \text{9-15}$$

TORQUE

Substituting into Equation 9-14 gives

$$\tau = mr^2\alpha \qquad \text{9-16}$$

A rigid object that rotates about a fixed axis is just a collection of individual particles, each of which is constrained to move in a circular path with the same angular velocity ω and acceleration α. Applying Equation 9-16 to the ith of these particles gives

$$\tau_{i\text{net}} = m_i r_i^2 \alpha$$

where $\tau_{i\text{net}}$ is the torque due to the net force on the ith particle. Summing both sides over all particles gives

$$\sum \tau_{i\text{net}} = \sum m_i r_i^2 \alpha = \left(\sum m_i r_i^2\right)\alpha = I\alpha \qquad \text{9-17}$$

In Chapter 8 we saw that the net force acting on a system of particles is equal to the net *external* force acting on the system because the internal forces (those exerted by the particles within the system on one another) cancel in pairs. The treatment of internal torques exerted by the particles within a system on one another leads to a similar result, that is, the net torque acting on a system equals the net *external* torque acting on the system. (We discuss this further in Chapter 10.) We can thus write Equation 9-17 as

$$\tau_{\text{net ext}} = \sum \tau_{\text{ext}} = I\alpha \qquad \text{9-18}$$

NEWTON'S SECOND LAW FOR ROTATION

This is the rotational analog of Newton's second law for linear motion, $\sum \vec{F} = m\vec{a}$.

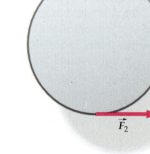

FIGURE 9-16

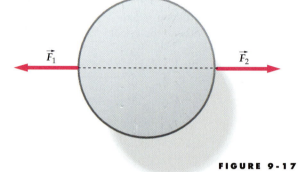

FIGURE 9-17

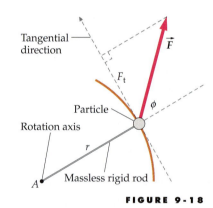

FIGURE 9-18

Calculating Torques

Figure 9-19 shows a force $\vec{F}$ acting on an object constrained to rotate about a fixed axis A, not shown, which passes through O and is perpendicular to the page. The positive tangential direction is shown at the point of application of the force, and r is the radial distance of this point of application from A. The torque τ due to this force about axis A is $\tau = F_t r$ (Equation 9-15). In principle, the expression $F_t r$ is all that is needed to calculate torques. However, in practice, calculations are often simpler if alternative expressions for torque are used. From the figure we can see that

$$F_t = F \sin \phi$$

where ϕ is the angle between the radial direction and the direction of the force. Thus, we can express the torque as $\tau = F_t r = (F \sin \phi)r$. The *line of action* of a force is the line parallel to the force and passing through the point of application of the force. From Figure 9-20 we can see that $r \sin \phi = \ell$, where the **lever arm** ℓ is the perpendicular distance between A and the line of action. Consequently, the torque is also given by $\tau = F\ell$. Putting all three equivalent expressions for the torque in one place, we have

$$\tau = F_t r = Fr \sin \phi = F\ell \qquad\qquad\text{9-19}$$

EQUIVALENT EXPRESSIONS FOR TORQUE

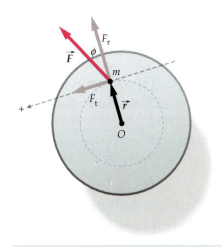

FIGURE 9-19 The force $\vec{F}$ produces a torque $F_t r$ about the center.

Torque Due to Gravity

We can model an extended object as an assembly of microscopic point particles, and there is a microscopic gravitational force on each particle. Each of these microscopic gravitational forces exerts a microscopic torque about a given axis, and the net gravitational torque on the object is the sum of these microscopic torques. The net gravitational torque can be calculated by considering the entire weight (the sum of the microscopic gravitational forces) to act at a single point—**the center of gravity.** Consider an object (Figure 9-21) constrained to rotate about a horizontal axis A coming out of the page. We choose the z axis of our coordinate system to coincide with axis A, and choose the x axis direction to be horizontal and the y axis vertical as shown. The torque on a particle of mass m_i due to gravity is $m_i g x_i$, where x_i is the lever arm of the force $m_i \vec{g}$. The net gravitational torque on the object is the sum of the gravitational torques on the particles that make it up. That is, $\tau_{grav} = \Sigma m_i g x_i$. If $\vec{g}$ is the same throughout the object, then g can be factored out of the sum. Factoring g out of the sum gives $\tau_{grav} = (\Sigma m_i x_i)g$. You should recognize the sum in the parentheses as $M x_{cm}$ (see Equation 8-1). Substituting this for the sum gives

$$\tau_{grav} = Mg x_{cm} \qquad\qquad\text{9-20}$$

TORQUE DUE TO GRAVITY

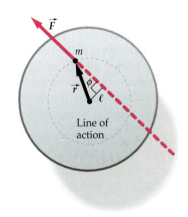

FIGURE 9-20 The force $\vec{F}$ produces a torque $F\ell$ about the center.

FIGURE 9-21

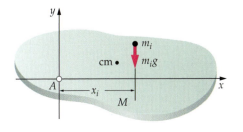

(a)

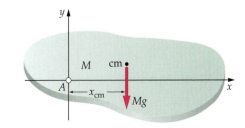

(b)

The torque due to a uniform gravitational field is calculated as if the entire gravitational force is applied at the center of mass. For an object in a uniform gravitational field, the center of gravity coincides with the center of mass.

9-5 Applications of Newton's Second Law for Rotation

In this section we give several applications of Newton's second law for rotation as expressed in Equation 9-18.

Problem-Solving Guidelines for Applying Newton's Second Law for Rotation

Free-Body Diagram Draw the free-body diagram with the object shown as a picture, not just a dot. Draw each force vector along the line of action of the force. On the diagram indicate the positive rotational direction (clockwise or counterclockwise).

A STATIONARY BIKE   **EXAMPLE 9-8**

To get some exercise without going anywhere, you set your bike on a stand so that the rear wheel is free to turn. As you pedal, the chain applies a force of 18 N to the sprocket at a distance of $r_s = 7$ cm from the axle of the wheel. Consider the wheel to be a hoop ($I = MR^2$) of radius $R = 35$ cm and mass 2.4 kg. What is the angular velocity of the wheel after 5 s?

FIGURE 9-22

$\vec{F} = 18$ N

35 cm

$M = 2.4$ kg

PICTURE THE PROBLEM The angular velocity is found from the angular acceleration, which is found from Newton's second law for rotation. Since the forces are constant, the torques are also constant and the constant angular acceleration equations apply. Note that $\vec{F}$ acts in the direction of the chain, so the line of force is tangent to the sprocket and the lever arm is the radius r_s of the sprocket (see Figure 9-22).

1. The angular velocity is related to the angular acceleration and the time:

$$\omega = \omega_0 + \alpha t = 0 + \alpha t$$

2. Apply Newton's second law for rotational motion to relate α to the net torque and the moment of inertia:

$$\Sigma \tau_{ext} = I\alpha$$

3. The only torque acting on the system is the applied force F with lever arm r_s:

$$\tau_{ext} = Fr_s$$

4. Substitute this value for the torque and $I = MR^2$ for the moment of inertia:

$$\alpha = \frac{\Sigma \tau_{ext}}{I} = \frac{Fr_s}{MR^2}$$

5. Substitute into the step 1 result and solve for the angular velocity after 5 s:

$$\omega = \alpha t = \frac{Fr_s}{MR^2} t = \frac{(18\,\text{N})(0.07\,\text{m})}{(2.4\,\text{kg})(0.35\,\text{m})^2} 5\,\text{s}$$

$$= \boxed{21.4\,\text{rad/s}}$$

A UNIFORM ROD, PIVOTED AT ONE END — **EXAMPLE 9-9**

A uniform thin rod of length L and mass M is pivoted at one end. It is held horizontal and released. The pivot is frictionless. Find (a) the angular acceleration of the rod immediately following its release and (b) the force F_A exerted on the rod by the pivot at this instant.

PICTURE THE PROBLEM The angular acceleration is found from Newton's second law for rotation (Equation 9-18). The force F_A is found from Newton's second law for a system (Equation 8-10). The tangential acceleration of the center of mass is related to the angular acceleration (Equation 9-5) and the centripetal acceleration of the center of mass is related to the angular speed (Equation 9-6).

FIGURE 9-23

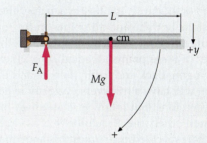

(a) 1. Sketch a free-body diagram of the rod (Figure 9-23).

2. Write Newton's second law for rotation:

$$\Sigma \tau_{\text{ext}} = I\alpha$$

3. Compute the torque due to gravity about the given axis. The rod is uniform so its center of mass is at its center, a distance $L/2$ from the axis:

$$\tau_{\text{grav}} = Mg\frac{L}{2}$$

4. Find the moment of inertia about the end of the rod from Table 9.1:

$$I = \tfrac{1}{3}ML^2$$

5. Substitute these values into the step 2 equation to compute α:

$$\alpha = \frac{\tau_{\text{grav}}}{I} = \frac{Mg(L/2)}{(1/3)ML^2} = \boxed{\frac{3g}{2L}}$$

(b) 1. Write Newton's second law for a system for the rod:

$$\Sigma F_{\text{exty}} = Ma_{\text{cmy}}$$

$$Mg - F_A = Ma_{\text{cmy}}$$

2. Use the relation $a_c = r\omega^2$ to find $a_{\text{cm c}}$. Just following release, $\omega = 0$:

$$a_c = r\omega^2$$

$$a_{\text{cm c}} = r_{\text{cm}}\omega^2 = \frac{L}{2}\omega^2 = 0$$

3. We now have two equations and three unknowns, α, a_{cmy}, and F_A. Use the relation $a_t = r\alpha$ to obtain an equation relating a_{cmy} to α:

$$a_t = r\alpha$$

$$a_{\text{cmy}} = r_{\text{cm}}\alpha = \frac{L}{2}\alpha$$

4. Substitute both the Part (a) step 5 result and the Part (b) step 1 result into the Part (b) step 3 result and solve for F_A:

$$\frac{Mg - F_A}{M} = \frac{L}{2}\frac{3g}{2L}$$

so

$$F_A = \boxed{\tfrac{1}{4}Mg}$$

REMARKS Just after the rod is released, the acceleration of the center of mass is directed straight down. Since the net external force and the acceleration must be in the same direction, it follows that $\vec{F}_A$ can not have a horizontal component at this moment.

EXERCISE A small pebble of mass $m \ll M$ is placed on top of the rod at its center. Find (a) the acceleration of the pebble and (b) the force it exerts on the rod just after the rod is released. (*Answer* (a) $a = 3g/4$ downward (b) $f = mg/4$ downward)

Nonslip Conditions

There are many situations in which a string is wrapped around a rotating wheel or cylinder. The string must move with a tangential velocity v_t that is equal to the tangential velocity of the rim of the wheel, provided the string remains taut and does not slip:

$$v_t = R\omega \qquad\qquad 9\text{-}21$$

NONSLIP CONDITION FOR v AND ω

For a point on a string, the velocity component tangent to the string is its *tangential* velocity, and the velocity perpendicular to the tangent is its *transverse* velocity. For example, if a bosun is being lifted straight upward by a rope and pulley system, both he and all points on the rope between him and the pulley have the same tangential velocity as the rim of the pulley wheel, but the transverse velocities of these points all equal zero. However, if, while being raised, the bosun is swinging back and forth, then he and the points along the swinging rope above him will have a transverse velocity in addition to their tangential velocity. In the discussions that follow it is assumed that there is no transverse motion (unless stated otherwise). Equation 9-21 is called a nonslip condition. Differentiating it with respect to time relates the tangential acceleration of the string to the angular acceleration of the rim.

$$a_t = R\alpha \qquad\qquad 9\text{-}22$$

a AND α UNDER NONSLIP CONDITIONS

Problem-Solving Guidelines for Applying Newton's Second Law for Rotation

Tension   Because of friction and inertia, if a string goes around a pulley wheel, the tension in it is greater on one side of the wheel than on the other. (This is necessary for the string to exert a torque on the wheel.) Use two different labels, T_1 and T_2, for these two tensions.

TENSION IN A STRING   **EXAMPLE 9-10**

An object of mass m is tied to a light string wound around a pulley wheel that has a moment of inertia I and radius R. The wheel bearing is frictionless and the string does not slip on the rim. Find the tension in the string and the acceleration of the object.

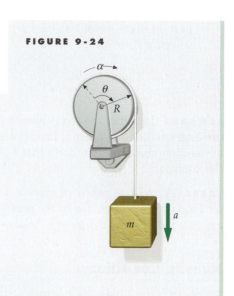

FIGURE 9-24

PICTURE THE PROBLEM   In this system, the object descends with a downward acceleration a, while the wheel turns with an angular acceleration α (Figure 9-24). We apply Newton's second law for rotation to the wheel to determine α, and Newton's second law to the object to obtain a. Relate a_t and α, using the nonslip condition.

1. Draw a free-body diagram of the pulley wheel, drawing each force vector with its tail at the point of application of the force. Put labels on the diagram and indicate the positive rotational direction as shown in Figure 9-25:

FIGURE 9-25

2. The only force that exerts a torque on the wheel is the tension T, which has lever arm R. Apply Newton's second law for rotational motion to relate T and the angular acceleration α:

$$\Sigma \tau_{\text{ext}} = I\alpha$$
$$TR = I\alpha$$

3. Draw a free-body diagram for the suspended object, and apply Newton's second law to relate T to the tangential acceleration a_t (Figure 9-26):

$$\Sigma F_{\text{exty}} = ma_y$$
$$mg - T = ma_t$$

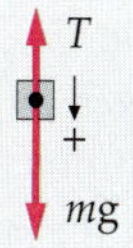

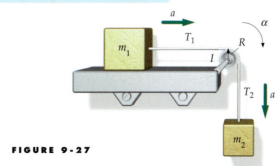

FIGURE 9-26

4. We have two equations for three unknowns, T, a_t, and α. A third equation is the relation between a_t and α for the nonslip condition:

$$a_t = R\alpha$$

5. We now have three equations enabling us to determine T, a_t, and α. To solve for T use the step 2 result to obtain an expression for α and use the step 3 result to obtain an expression for a_t. Substitute these expressions into the step 4 result and solve for T.

$$\frac{mg - T}{m} = R\frac{TR}{I}$$

so

$$\boxed{T = \frac{mg}{1 + (mR^2/I)}}$$

6. Substitute this result for T into the step 3 result and solve for a_t:

$$mg - \frac{mg}{1 + (mR^2/I)} = ma_t$$

so

$$\boxed{a_t = \frac{1}{1 + \dfrac{I}{mR^2}}g}$$

◑ PLAUSIBILITY CHECK Let's check a couple of extreme limits. If $I = 0$, the object should fall freely, and the string should be slack; our results give $T = 0$, $a_t = g$. What happens if I approaches ∞? For $I \gg mR^2$, our equations give $T \approx mg$ and $a_t \approx 0$.

A PULLEY

EXAMPLE 9-11 **Try It Yourself**

Two blocks are connected by a string that passes over a pulley of radius R and moment of inertia I. The block of mass m_1 slides on a frictionless, horizontal surface; the block of mass m_2 is suspended from the string (Figure 9-27). Find the acceleration a of the blocks and the tensions T_1 and T_2 assuming the string does not slip on the pulley.

FIGURE 9-27

PICTURE THE PROBLEM In this problem, the tensions T_1 and T_2 are not equal because the pulley has mass and because there is static friction between the string and the pulley (Figure 9-27). (Otherwise the pulley would not turn.) Note that T_2 exerts a clockwise torque and T_1 exerts a counterclockwise torque on the pulley. Use Newton's second law for each block and Newton's second law for rotational motion for the pulley. Relate α and a using the nonslip condition.

Cover the column to the right and try these on your own before looking at the answers.

Steps	Answers

FIGURE 9-28

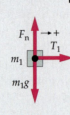

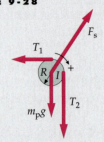

1. Draw a free-body diagram for each block and for the pulley, as shown in Figure 9-28. Note that the center of mass of the pulley does not accelerate, so the support must exert a force on the axle F_s that balances the resultant of the gravitational force and the forces exerted by the string.

2. Apply Newton's second law to each block.

$T_1 = m_1 a; \qquad m_2 g - T_2 = m_2 a$

3. Apply Newton's second law for rotation to the pulley wheel.

$T_2 R - T_1 R = I\alpha$

4. We have three equations and four unknowns. To get a fourth equation, use the nonslip condition to relate a and α.

$a = R\alpha$

5. Now we have four equations and four unknowns, so the rest is algebra. Do the algebra and obtain expressions for a, T_1, and T_2. (Hint: To find a, obtain expressions for T_1 and T_2 from the step 2 results. Substitute these into the step 3 result to obtain an equation with unknowns a and α. Use the step 4 result to eliminate α and solve for a.)

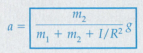

$$a = \boxed{\frac{m_2}{m_1 + m_2 + I/R^2}\, g}$$

$$T_1 = m_1 a = \boxed{\frac{m_2}{m_1 + m_2 + I/R^2}\, m_1 g}$$

$$T_2 = m_2(g - a) = \boxed{\frac{(m_1 + I/R^2)}{m_1 + m_2 + I/R^2}\, m_2 g}$$

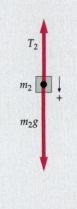

PLAUSIBILITY CHECK If $I = 0$, $T_1 = T_2$, and the acceleration is $a = m_2 g/(m_1 + m_2)$, as expected. If I is very large ($I/R^2 >> m_1 + m_2$), then $T_1 \approx 0$, $T_2 \approx m_2 g$, and $a \approx 0$.

Power

When you spin an object you do work on it, increasing its kinetic energy. Consider a force F_i acting on a rotating object. As the object rotates through an angle $d\theta$, the point of application of the force moves a distance $ds_i = r_i\, d\theta$, and the force does work

$$dW_i = F_{it}\, ds_i = F_{it} r_i\, d\theta = \tau_i\, d\theta$$

where τ_i is the torque exerted by the force F_i and F_{it} is the tangential component of F_i. In general, the work done by a torque τ when an object turns through a small angle $d\theta$ is

$$dW = \tau\, d\theta \qquad\qquad\qquad 9\text{-}23$$

The rate at which the torque does work is the power input of the torque:

$$P = \frac{dW}{dt} = \tau\frac{d\theta}{dt}$$

or

$$P = \tau\omega \qquad\qquad\qquad 9\text{-}24$$

POWER

Equations 9-23 and 9-24 are the rotational analogs of $dW = F_s\, ds$ and $P = F_s v_s$.

TORQUE EXERTED BY AN AUTOMOBILE ENGINE   **EXAMPLE 9 - 1 2**   **Try It Yourself**

The maximum torque produced by the 8.0-L V10 engine of a 2002 Dodge Viper is 675 N·m of torque at 3700 rev/min. Find the power output of the engine operating at these maximum torque conditions.

PICTURE THE PROBLEM The power equals the product of the torque and angular velocity, which are given. You must express ω in radians per second to obtain the power in watts.

Cover the column on the right and try these on your own before looking at the answers.

Steps	Answers
1. Write the power in terms of τ and ω.	$P = \tau\omega$
2. Convert rev/min to rad/s.	$\omega = 387 \text{ rad/s}$
3. Calculate the power.	$P = \boxed{262 \text{ kW}}$

REMARKS This power output is about 350 hp.

EXERCISE The maximum power produced by the Viper engine is 450 hp at 5200 rev/min. What is the torque when the engine is operating at maximum horsepower? (*Answer* 616 N·m)

There are many parallels between one-dimensional linear motion and rotational motion about a fixed axis. The similarities of the formulas can be seen in Table 9-2. The formulas are the same, but the symbols are different.

TABLE 9-2

Analogs in Rotational and Linear Motion

Rotational Motion		Linear Motion	
Angular displacement	$\Delta\theta$	Displacement	Δx
Angular velocity	$\omega = \dfrac{d\theta}{dt}$	Velocity	$v = \dfrac{dx}{dt}$
Angular acceleration	$\alpha = \dfrac{d\omega}{dt} = \dfrac{d^2\theta}{dt^2}$	Acceleration	$a = \dfrac{dv}{dt} = \dfrac{d^2x}{dt^2}$
Constant angular acceleration equations	$\omega = \omega_0 + \alpha t$	Constant acceleration equations	$v = v_0 + at$
	$\Delta\theta = \omega_{av}\,\Delta t$		$\Delta x = v_{av}\,\Delta t$
	$\omega_{av} = \frac{1}{2}(\omega_0 + \omega)$		$v_{av} = \frac{1}{2}(v_0 + v)$
	$\theta = \theta_0 + \omega_0 t + \frac{1}{2}\alpha t^2$		$x = x_0 + v_0 t + \frac{1}{2}at^2$
	$\omega^2 = \omega_0^2 + 2\alpha\,\Delta\theta$		$v^2 = v_0^2 + 2a\,\Delta x$
Torque	τ	Force	F
Moment of inertia	I	Mass	m
Work	$dW = \tau\,d\theta$	Work	$dW = F_s\,ds$
Kinetic energy	$K = \frac{1}{2}I\omega^2$	Kinetic energy	$K = \frac{1}{2}mv^2$
Power	$P = \tau\omega$	Power	$P = Fv$
Angular momentum†	$L = I\omega$	Momentum	$p = mv$
Newton's second law	$\tau_{net} = I\alpha = \dfrac{dL}{dt}$	Newton's second law	$F_{net} = ma = \dfrac{dp}{dt}$

† Angular momentum is introduced in Chapter 10.

9-6 Rolling Objects

Rolling Without Slipping

When a spool rolls without slipping down an incline (Figure 9-29), the points in contact with the incline are instantaneously at rest and the spool rotates about a rotation axis through the contact point. This can be observed because rapid motion causes blurring, so the part of the spool that is moving slowest is blurred the least. In Figure 9-30 a wheel of radius R is rolling without slipping along a flat surface. Point P on the wheel moves as shown with speed

$$v = r\omega \qquad 9\text{-}25$$

NONSLIP CONDITION FOR SPEED

FIGURE 9-29

where r is the perpendicular distance from P to the rotation axis. The center of mass of the wheel moves with speed

$$v_{cm} = R\omega \qquad 9\text{-}26$$

NONSLIP CONDITION FOR v_{CM}

Interestingly, for a point on the top of the wheel, $r = 2R$, so the top of the wheel is moving at twice the speed of the center of mass.

Differentiating each side of Equation (9-26) gives

$$a_{cm} = R\alpha \qquad 9\text{-}27$$

NONSLIP CONDITION FOR ACCELERATION

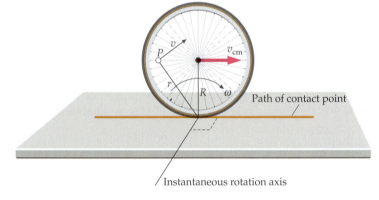

Instantaneous rotation axis

FIGURE 9-30

A falling yo-yo that is unwinding from a string, the top end of which is held fixed, follows the same nonslip conditions as the wheel.

A wheel of radius R is rolling without slipping along a straight path. As the wheel rotates through angle ϕ (Figure 9-31), the point of contact between the wheel and the plane moves a distance s that is related to ϕ by

$$s = R\phi \qquad 9\text{-}28$$

NONSLIP CONDITION FOR DISTANCE

FIGURE 9-31

If the wheel is rolling on a flat surface, the wheel's center of mass remains directly over the point of contact, so it also moves through s.

We saw in Chapter 8 that the kinetic energy of a system can be written as the sum of the kinetic energy of motion of the center of mass plus the kinetic energy relative to the center of mass. For a rotation object, the relative kinetic energy is $\frac{1}{2}I_{cm}\omega^2$. Thus, the kinetic energy of a rotating object is

$$K = \frac{1}{2}Mv_{cm}^2 + \frac{1}{2}I_{cm}\omega^2 \qquad 9\text{-}29$$

KINETIC ENERGY OF A ROTATING OBJECT

A BOWLING BALL   **EXAMPLE 9-13** **Try It Yourself**

A bowling ball of radius 11 cm and mass $M = 7.2$ kg is rolling without slipping on a horizontal ball return at 2 m/s. It then rolls without slipping up a hill to a height h before momentarily stopping before rolling back down the hill. Find h.

PICTURE THE PROBLEM Mechanical energy is conserved. The initial kinetic energy, which is the translational kinetic energy of the center of mass, $\frac{1}{2}mv_{cm}^2$, plus the kinetic energy of rotation about the center of mass, $\frac{1}{2}I_{cm}\omega^2$, is converted to potential energy mgh. Because the sphere rolls without slipping, the initial linear and angular speeds are related by $v_{cm} = R\omega$. Make a labeled sketch showing the ball in both its initial and final positions (Figure 9-32).

FIGURE 9-32

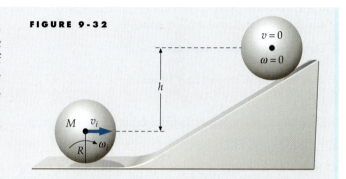

Cover the column to the right and try these on your own before looking at the answers.

Steps

1. Apply conservation of mechanical energy with $U_i = 0$ and $K_f = 0$. Write the total initial kinetic energy K_i in terms of the speed v_{cm} and the angular speed ω.

2. Substitute from $\omega = v_i/R$ and $I_{cm} = \frac{2}{5}MR^2$ and solve for h.

Answers

$$U_f + K_f = U_i + K_i$$

$$Mgh + 0 = 0 + \tfrac{1}{2}Mv_i^2 + \tfrac{1}{2}I_{cm}\omega_i^2$$

$$Mgh = \frac{1}{2}Mv_i^2 + \frac{1}{2}\left(\frac{2}{5}MR^2\right)\frac{v_i^2}{R^2}$$

$$= \tfrac{7}{10}Mv_i^2$$

so

$$h = \frac{7v_i^2}{10g} = 0.285 \text{ m} = \boxed{28.5 \text{ cm}}$$

REMARKS The height h is independent of the mass or radius of the ball.

EXERCISE Find the initial kinetic energy of the ball. (*Answer* 20.2 J)

PLAYING POOL   **EXAMPLE 9-14**

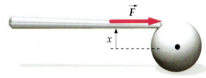

A cue stick hits a cue ball horizontally a distance x above the center of the ball (Figure 9-33). Find the value of x for which the cue ball will roll without slipping from the beginning. Express your answer in terms of the radius R of the ball.

FIGURE 9-33

PICTURE THE PROBLEM The lines of action of the weight and normal forces pass through the center of mass and thus exert no torque about it. The frictional force is much smaller than the collision force of the stick and can be neglected. If the stick hits the ball at the level of the ball's center, the ball initially translates with no rotation. If the stick hits below the center, the ball initially has backspin. At a certain value of x, the ball has just the right forward spin and forward acceleration to satisfy the nonslip condition. The value of x determines the torque to force ratio on the ball, and hence its angular acceleration α to linear acceleration ratio a. The linear acceleration a is F/m, independent of x. For the ball to roll without slipping from the start, we find α and a, then set $a = R\alpha$ (nonslip condition) to find x.

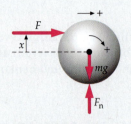

FIGURE 9-34

1. Sketch a free-body diagram of the ball (Figure 9-34). We are assuming friction between the ball and the table is negligible, so do not include a frictional force:

2. The torque about the axis through the center of the ball equals F times x:

$$\tau = Fx$$

3. Apply Newton's second law for a system and Newton's second law for rotational motion about the center of the ball:

$$F = ma_{cm} \quad \text{and} \quad \tau = I_{cm}\alpha$$

4. The nonslip condition relates a and α:

$$a_{cm} = R\alpha$$

5. Substitute from steps 2 and 3 into step 4:

$$\frac{F}{m} = R\frac{Fx}{I_{cm}}$$

6. Find the moment of inertia from Table 9.1 and solve for x:

$$x = \frac{I_{cm}}{mR} = \frac{(2/5)mR^2}{mR} = \boxed{(2/5)R}$$

REMARKS Striking the ball at a point either higher or lower than $2R/5$ from the center will result in the ball rolling *and* slipping (skidding). Skidding is often desirable in the game of pool. Rolling and slipping is discussed in the next subsection.

When an object rolls down an incline, its center of mass is accelerated. The analysis of such a problem is simplified by an important theorem concerning the center of mass:

> If the torques are computed from a reference frame moving with the center of mass, then Newton's second law for rotation holds even when the center of mass is accelerating, and the torques are computed in the center-of-mass reference frame. That is,
>
> $$\tau_{net,cm} = I_{cm}\alpha \qquad\qquad 9\text{-}30$$

This is the same as Equation 9-18 except that here the torques and the moment of inertia are computed from a reference frame moving with the center of mass. (A derivation may be found at www.whfreeman.com/tipler.) When the center of mass is accelerating (a ball rolling down an incline, for example), the center-of-mass reference frame is a noninertial one, where we would not necessarily expect our equations for Newton's second law for rotation to be valid. Nevertheless, they are.

ACCELERATION OF A BALL THAT IS ROLLING WITHOUT SLIPPING **EXAMPLE 9-15**

A uniform solid ball of mass m and radius R rolls without slipping down a plane inclined at an angle ϕ above the horizontal. Find the acceleration of the center of mass and the frictional force.

PICTURE THE PROBLEM From Newton's second law, the acceleration of the center of mass equals the net force divided by the mass. The forces acting are the weight $m\vec{g}$ downward, the normal force $\vec{F}_n$ that balances the normal component of the weight, and the force of friction $\vec{f}$ acting up the incline (Figure 9-35). As the object accelerates down the incline, its angular velocity must increase to maintain the nonslip condition. This angular acceleration requires a net external torque about the axis through the center of mass. We apply Newton's second law for rotation to find α. The nonslip condition relates α and a_{cm}.

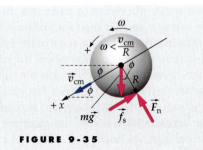

FIGURE 9-35

1. Apply Newton's second law for a system along the x axis:

$$\Sigma F_x = ma_{cmx}$$

$$mg \sin \phi - f_s = ma_{cm}$$

2. Apply Newton's second law for rotational motion, for rotations about an axis parallel to the instantaneous rotation axis and passing through the center of mass:

$$\Sigma \tau_i = I_{cm}\alpha$$

$$f_s R + 0 + 0 = I_{cm}\alpha$$

3. Relate a_{cm} and α using the nonslip condition:

$$a_{cm} = R\alpha$$

4. We now have three equations and three unknowns. Solve the step 1 result for f_s and the step 3 result for α, substitute for these quantities in the step 2 result and solve for a_{cm}:

$$f_s R = I_{cm}\alpha$$

$$(mg \sin \phi - ma_{cm})R = I_{cm}\frac{a_{cm}}{R}$$

so

$$a_{cm} = \frac{g \sin \phi}{1 + \dfrac{I_{cm}}{mR^2}}$$

5. Substitute the step 4 result into the step 1 result and solve for f_s:

$$f_s = mg \sin \phi - ma_{cm}$$

$$= mg \sin \phi - \frac{mg \sin \phi}{1 + \dfrac{I_{cm}}{mR^2}} = \frac{mg \sin \phi}{1 + \dfrac{mR^2}{I_{cm}}}$$

6. For a solid sphere, $I_{cm} = \frac{2}{5}mR^2$ (see Table 9.1):

$$a_{cm} = \frac{g \sin \phi}{1 + \frac{2}{5}} = \boxed{\frac{5}{7}g \sin \phi}$$

$$f_s = \frac{mg \sin \phi}{\frac{7}{2}} = \boxed{\frac{2}{7}mg \sin \phi}$$

REMARKS Because the ball rolls without slipping, the friction is static friction. Note that the result seems independent of the coefficient of static friction. However, we have assumed that the coefficient of static friction was large enough to prevent slipping.

The results of steps 5 and 6 in Example 9-15 apply to any round object with the center of mass at the geometric center that is rolling without slipping. For such objects, $I_{cm} = \beta mR^2$, where $\beta = \frac{2}{5}$ for a sphere, $\frac{1}{2}$ for a rolling solid cylinder, 1 for a thin cylindrical shell, and so forth. For such objects the step 5 and step 6 results can be expressed

$$f_s = \frac{mg \sin \phi}{1 + \beta^{-1}} \qquad\qquad\qquad 9\text{-}31$$

$$a_{cm} = \frac{g \sin \phi}{1 + \beta} \qquad\qquad\qquad 9\text{-}32$$

The linear acceleration of any object rolling down an incline is less than $g \sin \phi$ because of the frictional force directed up the incline. Note that these accelerations are independent of both the mass and the radius of the objects. That is, all solid spheres rolling without slipping down the same incline will have identical accelerations. However, if we release a sphere, a cylinder, and a hoop at the top of an incline, and if they all roll without slipping, the sphere will reach the bottom first because it has the greatest acceleration. The cylinder will be second and the hoop last (Figure 9-36). A block that slides without friction down the incline will arrive at the bottom ahead of all three rolling objects.

Because the friction is static, it does no work, and there is no dissipation of mechanical energy. We can therefore use the conservation of mechanical energy to find the speed of an object rolling without slipping down an incline. At the top of the incline, the total energy is the potential energy mgh. At the bottom, the total energy is kinetic energy. Conservation of mechanical energy therefore gives

$$\tfrac{1}{2} mv_{cm}^2 + \tfrac{1}{2} I_{cm}\omega^2 = mgh$$

We can use the nonslip condition to eliminate either v_{cm} or ω. Substituting $I_{cm} = \beta mR^2$ and $\omega = v_{cm}/R$, we obtain $\tfrac{1}{2} mv_{cm}^2 + \tfrac{1}{2} \beta mR^2 (v_{cm}^2/R^2) = mgh$. Solving for v_{cm}^2 gives

$$v_{cm}^2 = \frac{2gh}{1 + \beta} \qquad\qquad 9\text{-}33$$

For a cylinder, with $\beta = \tfrac{1}{2}$, we obtain $v_{cm} = \sqrt{\tfrac{4}{3} gh}$. Note that the speed is independent of the mass and radius of the cylinder, and is less than $\sqrt{2gh}$, the speed of an object sliding with no friction down the incline.

For an object rolling without slipping down an incline, the frictional force f_s is less than or equal to its maximum value; that is, $f_s \le \mu_s F_n$, where $F_n = mg \cos \phi$. Substituting the expression from Equation 9-31 for the frictional force, we have

$$\frac{mg \sin \phi}{1 + \beta^{-1}} \le \mu_s mg \cos \phi$$

or

$$\tan \phi \le (1 + \beta^{-1})\mu_s \qquad\qquad 9\text{-}34$$

(For a uniform cylinder, $I_{cm} = \tfrac{1}{2}MR^2$, so $\beta = \tfrac{1}{2}$ and $\tan \phi \le 3\mu_s$.) If the tangent of the angle of incline is greater than $(1 + \beta^{-1})\mu_s$, the object will slip as it moves down the incline.

EXERCISE A uniform cylinder rolls down a plane inclined at $\theta = 50°$. What is the minimum value of the coefficient of static friction for which the cylinder will roll without slipping? (*Answer* 0.40)

EXERCISE For a hoop rolling down an incline, (a) what is the force of friction and (b) what is the maximum value of $\tan \theta$ for which the hoop will roll without slipping? (*Answer* (a) $f = \tfrac{1}{2}mg \sin \theta$ (b) $\tan \theta \le 2\mu_s$)

*Rolling With Slipping

When an object slides as it rolls, the nonslip condition $v_{cm} = R\omega$ does not hold. Suppose a bowling ball is thrown with no initial rotation. As the ball slides along the bowling lane, $v_{cm} > R\omega$. However, the frictional force both reduces its linear speed v_{cm} (Figure 9-37) and increases its angular speed ω until the nonslip condition $v_{cm} = R\omega$ is reached, after which the ball rolls without slipping.

Another example of rolling with slipping is a ball with topspin, such as a cue ball struck at a point higher than $(2/5)R$ above the center (see Example 9-14) so that $v_{cm} < R\omega$. Then the frictional force both increases v and decreases ω until the nonslip condition $v_{cm} = R\omega$ is reached (Figure 9-38).

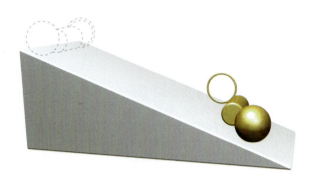

FIGURE 9-36 A sphere, a cylinder, and a hoop are released together from rest at the top of an incline. The sphere reaches the bottom first, followed by the cylinder and then the hoop.

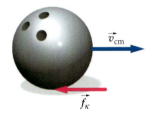

FIGURE 9-37 A bowling ball moving with no initial rotation. The frictional force $\vec{f}_\kappa$ exerted by the floor reduces the speed v_{cm} and increases the angular speed ω until $v_{cm} = R\omega$.

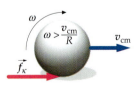

FIGURE 9-38 Ball with excess topspin. The frictional force accelerates the ball in the direction of motion.

EXAMPLE 9-16

A bowling ball of mass M and radius R is thrown so that the instant it touches the floor it is moving horizontally with speed $v_0 = 5$ m/s and is not rotating. The coefficient of kinetic friction between the ball and the floor is $\mu_k = 0.08$. Find (a) the time the ball slides and (b) the distance the ball slides before it rolls without slipping.

PICTURE THE PROBLEM We calculate v_{cm} and ω as functions of time, set $v_{cm} = R\omega$, and solve for t. The linear and angular accelerations are found from $\Sigma F = ma$ and $\tau = I\alpha$. Let the direction of motion be positive. There is slipping and kinetic friction, so mechanical energy is dissipated. Therefore, conservation of mechanical energy cannot be used to solve this problem.

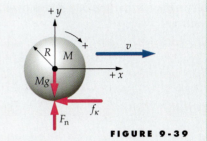

FIGURE 9-39

(a) 1. Sketch a free-body diagram of the ball (Figure 9-39).

2. The net force on the ball is the force of kinetic friction f_k, which acts in the negative x direction. Apply Newton's second law:

$$\Sigma F_x = Ma_{cmx}$$
$$-f_k = Ma_{cmx}$$

3. The acceleration is in the negative x direction and $a_{cmy} = 0$. Find f_k by first finding F_n:

$$\Sigma F_y = Ma_{cmy} = 0 \implies F_n = Mg$$

so

$$f_k = \mu_k F_n = \mu_k Mg$$

4. Find the acceleration using the step 2 and step 3 results:

$$-\mu_k Mg = Ma_{cmx} \implies a_{cmx} = -\mu_k g$$

5. Relate the linear velocity to the acceleration and the time using a kinematic equation:

$$v_{cmx} = v_0 + a_{cmx}t = v_0 - \mu_k gt$$

6. Find α by applying Newton's second law for rotational motion to the ball. Compute the torques about the axis through the center of mass. Note that the free-body diagram has clockwise as positive:

$$\Sigma\tau = I_{cm}\alpha$$
$$\mu_k MgR + 0 + 0 = \tfrac{2}{5} MR^2\alpha$$

so

$$\alpha = \frac{5}{2}\frac{\mu_k g}{R}$$

7. Relate the angular velocity to the angular acceleration and the time using a kinematic equation:

$$\omega = \omega_0 + \alpha t = 0 + \alpha t = \frac{5}{2}\frac{\mu_k g}{R}t$$

8. Solve for the time t at which $v_{cm} = R\omega$:

$$v_{cm} = R\omega$$
$$v_0 - \mu_k gt = \tfrac{5}{2}\mu_k gt$$

so

$$t = \frac{2v_0}{7\,\mu_k g} = \frac{2(5\,\text{m/s})}{7(0.08)(9.81\,\text{m/s}^2)} = \boxed{1.82\ \text{s}}$$

(b) The distance traveled in 1.82 s is:

$$\Delta x = v_0 t + \tfrac{1}{2}a_{cm}t^2$$

$$= v_0\frac{2v_0}{7\,\mu_k g} + \frac{1}{2}(-\mu_k g)\left(\frac{2v_0}{7\,\mu_k g}\right)^2 = \frac{12}{49}\frac{v_0^2}{\mu_k g}$$

$$= \frac{12}{49}\frac{(5\,\text{m/s})^2}{(0.08)(9.81\,\text{m/s}^2)} = \boxed{7.80\ \text{m}}$$

EXERCISE Find the speed of the bowling ball when it begins to roll without slipping. (*Answer* $v_{cm} = \frac{5}{7}v_0$). This result is independent of the coefficient of kinetic friction. The rolling speed is $\frac{5}{7}v_0$ whether μ_k is large or small. The total mechanical energy lost is thus independent of μ_k. (The time and the distance are sensitive to the value of μ_k, however.)

EXERCISE Find the total kinetic energy of the ball after it begins to roll without slipping. (*Answer* $K = \frac{5}{14}mv_0^2$)

REMARKS In a well-maintained bowling alley, the lanes are lightly oiled and very slick so that the ball slides over a great distance, giving the bowler added control.

SUMMARY

1. Angular displacement, angular velocity, and angular acceleration are fundamental defined quantities in rotational kinematics.

2. Torque and moment of inertia are important derived dynamic concepts. Torque is a measure of the effect of a force in causing an object to start or stop rotating. Moment of inertia is the measure of an object's inertial resistance to angular accelerations. The moment of inertia depends on the distribution of the mass relative to the rotation axis.

3. The parallel-axis theorem, which follows from the definition of the moment of inertia, often simplifies the calculation of I.

4. Newton's second law for rotation, $\Sigma\tau_{ext} = I\alpha$, is derived from Newton's second law and the definitions of τ, I, and α. It is an important relation for problems involving the rotation of a rigid object about a rotation axis of fixed direction.

Topic	Relevant Equations and Remarks	
1. Angular Velocity and Angular Acceleration		
Angular velocity	$\omega = \dfrac{d\theta}{dt}$ (Definition)	9-2
Angular acceleration	$\alpha = \dfrac{d\omega}{dt} = \dfrac{d^2\theta}{dt^2}$ (Definition)	9-3
Tangential velocity	$v_t = r\omega$	9-7
Tangential acceleration	$a_t = r\alpha$	9-8
Centripetal acceleration	$a_c = \dfrac{v^2}{r} = r\omega^2$	9-9

2. **Equations for Rotation With**

   Constant Angular Acceleration

$$\omega = \omega_0 + \alpha t \qquad \text{9-4}$$

$$\theta = \theta_0 + \omega_0 t + \tfrac{1}{2}\alpha t^2 \qquad \text{9-5}$$

$$\omega^2 = \omega_0^2 + 2\alpha(\theta - \theta_0) \qquad \text{9-6}$$

3. **Moment of Inertia**

   System of particles

$$I = \Sigma\, m_i r_i^2 \qquad \text{(Definition)} \qquad \text{9-10}$$

   Continuous object

$$I = \int r^2\, dm \qquad \text{9-12}$$

   Parallel-axis theorem

The moment of inertia about an axis a distance h from a parallel axis through the center of mass is

$$I = I_{cm} + Mh^2 \qquad \text{9-13}$$

where I_{cm} is the moment of inertia about the axis through the center of mass and M is the total mass of the object.

4. **Energy**

   Kinetic energy for rotation about a fixed axis

$$K = \tfrac{1}{2}I\omega^2 \qquad \text{9-11}$$

   Kinetic energy for rotating object

$$K = \tfrac{1}{2}Mv_{cm}^2 + \tfrac{1}{2}I_{cm}\omega^2 \qquad \text{9-29}$$

   Power

$$P = \tau\omega \qquad \text{9-24}$$

5. **Torque**

The torque due to a force equals the product of the tangential component of the force and the radial distance from the axis:

$$\tau = F_t r = Fr \sin \phi = F\ell \qquad \text{9-19}$$

6. **Newton's Second Law for Rotation**

$$\tau_{net,ext} = \sum_i \tau_{i,ext} = I\alpha \qquad \text{9-18}$$

If torques about an axis through the center of mass are computed from a reference frame moving with the center of mass, Newton's second law for rotation holds for rotation about an axis through the center of mass, even if the reference frame is noninertial.

7. **Nonslip Conditions**

When a string that is wrapped around a pulley or disk does not slip, the linear and angular quantities are related by

$$v_t = R\omega \qquad \text{9-21}$$

$$a_t = R\alpha \qquad \text{9-22}$$

8. **Rolling Objects**

   Rolling without slipping

$$v_{cm} = R\omega \qquad \text{9-26}$$

   *Rolling with slipping

When an object rolls and slips, $v_{cm} \neq R\omega$. Kinetic friction exerts a force that tends to change v_{cm}, and also exerts a torque that changes ω until $v_{cm} = R\omega$ and pure rolling sets in.

PROBLEMS

- Single-concept, single-step, relatively easy
- •• Intermediate-level, may require synthesis of concepts
- ••• Challenging
- **SSM** Solution is in the *Student Solutions Manual*
- Problems available on iSOLVE online homework service
- ✔ These "Checkpoint" online homework service problems ask students additional questions about their confidence level, and how they arrived at their answer

In a few problems, you are given more data than you actually need; in a few other problems, you are required to supply data from your general knowledge, outside sources, or informed estimates.

Take g = 9.81 N/kg = 9.81 m/s² and neglect friction in all problems unless otherwise stated.

Conceptual Problems

1 • **SSM** Two points are on a disk turning at constant angular velocity, one point on the rim and the other halfway between the rim and the axis. Which point moves the greater distance in a given time? Which turns through the greater angle? Which has the greater speed? The greater angular velocity? The greater tangential acceleration? The greater angular acceleration? The greater centripetal acceleration?

2 • True or false:

(a) Angular velocity and linear velocity have the same dimensions.
(b) All parts of a rotating wheel must have the same angular velocity.
(c) All parts of a rotating wheel must have the same angular acceleration.

3 •• Starting from rest, a disk takes 10 revolutions to reach an angular velocity ω at constant angular acceleration. How many additional revolutions are required to reach an angular velocity of 2ω? (a) 10 rev. (b) 20 rev. (c) 30 rev. (d) 40 rev. (e) 50 rev.

4 • **SSM** The dimension of torque is the same as that of (a) impulse, (b) energy, (c) momentum, (d) none of these.

5 • The moment of inertia of an object of mass M (a) is an intrinsic property of the object, (b) depends on the choice of axis of rotation, (c) is proportional to M regardless of the choice of axis, (d) both (b) and (c) are correct.

6 • **SSM** Can an object continue to rotate in the absence of torque?

7 • Does an applied net torque always increase the angular speed of an object?

8 • True or false:

(a) If the angular velocity of an object is zero at some instant, the net torque on the object must be zero at that instant.
(b) The moment of inertia of an object depends on the location of the axis of rotation.
(c) The moment of inertia of an object depends on the angular velocity of the object.

9 • A disk is free to rotate about an axis. A tangential force applied a distance d from the axis causes an angular acceleration α. What angular acceleration is produced if the same force is applied a distance $2d$ from the axis? (a) α. (b) 2α. (c) $\alpha/2$. (d) 4α. (e) $\alpha/4$.

10 • **SSM** The moment of inertia of an object about an axis that does not pass through its center of mass is _____ the moment of inertia about a parallel axis through its center of mass. (a) always less than, (b) sometimes less than, (c) sometimes equal to, (d) always greater than.

11 • A constant torque acts on a merry-go-round. The power input of the torque is (a) constant, (b) proportional to the angular speed of the merry-go-round, (c) zero, (d) none of these.

12 • True or false: When an object rolls without slipping, friction does no work on the object.

13 • Why do we put the knob on a door about as far away from the hinges as possible?

14 • **SSM** A wheel of radius R is rolling without slipping on a flat stationary surface. The velocity of the point on the rim that is in contact with the surface is (a) equal to $R\omega$ in the direction of motion of the center of mass, (b) equal to $R\omega$ opposite to the direction of motion of the center of mass, (c) zero, (d) equal to the velocity of the center of mass and in the same direction, (e) equal to the velocity of the center of mass but in the opposite direction.

15 •• A solid uniform cylinder and a solid uniform sphere have equal masses. Both roll without slipping on a horizontal surface. If their kinetic energies are the same, then (a) the translational speed of the cylinder is greater than that of the sphere, (b) the translational speed of the cylinder is less than that of the sphere, (c) the translational speeds of the two objects are the same, (d) (a), (b), or (c) could be correct depending on the radii of the objects.

16 • **SSM** Two identical-looking 1-m-long pipes enclose slugs of lead whose total mass is 10 kg (much larger than the mass of the pipe). In the first pipe the lead is concentrated at the center of the pipe, while in the second the lead is divided into two equal masses placed at opposite ends of the pipe. Without opening either pipe, how could you determine which is which?

17 •• Starting from rest at the same time, a coin and a ring roll down an incline without slipping. Which of the following is true? (*a*) The ring reaches the bottom first. (*b*) The coin reaches the bottom first. (*c*) The coin and ring arrive at the bottom simultaneously. (*d*) The race to the bottom depends on their relative masses. (*e*) The race to the bottom depends on their relative diameters.

18 •• For a hoop of mass M and radius R that is rolling without slipping, which is larger, its translational kinetic energy or its rotational kinetic energy? (*a*) Translational kinetic energy is larger. (*b*) Rotational kinetic energy is larger. (*c*) Both are the same size. (*d*) The answer depends on the radius. (*e*) The answer depends on the mass.

19 •• For a disk of mass M and radius R that is rolling without slipping, which is larger, its translational kinetic energy or its rotational kinetic energy? (*a*) Translational kinetic energy is larger. (*b*) Rotational kinetic energy is larger. (*c*) Both are the same size. (*d*) The answer depends on the radius. (*e*) The answer depends on the mass.

20 •• A perfectly rigid ball rolls without slipping along a perfectly rigid horizontal plane. Show that the frictional force acting on the ball must be zero. *Hint: Consider a possible direction for the action of the frictional force and what effects such a force would have on the velocity of the center of mass and on the angular velocity.*

21 • True or false: When a sphere rolls and slips on a rough surface, mechanical energy is dissipated.

22 • A cue ball is hit very near the top so that it starts to move with topspin. As it slides, the force of friction (*a*) increases v_{cm}, (*b*) decreases v_{cm}, (*c*) has no effect on v_{cm}.

Estimation and Approximation

23 •• A bicycle of mass 14 kg has 1.2-m-diameter wheels, each of mass 3 kg. The mass of the rider is 38 kg. Estimate the fraction of the total kinetic energy of bicycle and rider associated with rotation of the wheels.

24 •• Why does toast falling off a table always land jelly-side down? The question may sound silly, but it has been a subject of serious scientific enquiry. The analysis is too complicated to reproduce here, but R. D. Edge and Darryl Steinert showed that a piece of toast, pushed gently over the edge of a table until it tilts off, typically falls off the table when it makes an angle of about 30° with the horizontal (see Figure 9-40) and has an angular velocity of $\omega = 0.956\sqrt{g/\ell}$, where ℓ is the length of one side of the piece of toast (assumed square). Assuming that it starts jelly-side up, what side will it land on if it falls from a table of height 0.5 m? How about a height of 1 m? Assume that the toast has a side length $\ell = 0.1$ m, and that it will land jelly-side down if the angle is between 180° and 270°. (If the toast rotates through an angle greater than 360°, we will need to reduce it to the range 0°–360°.) Ignore any forces due to air resistance. For readers interested in this problem and a host of others, we highly recommend Robert Erlich's wonderful book, *Why Toast Lands Jelly-Side Down: Zen and the Art of Physics Demonstrations.*†

---

† Robert Erlich, *Why Toast Lands Jelly-Side Down;* Princeton, NJ: Princeton University Press (1997).

FIGURE 9-40 Problem 24

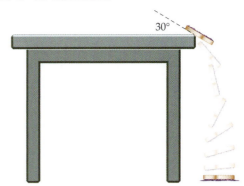

30°

25 •• **SSM** Consider the moment of inertia of an average adult man about an axis running vertically through the center of his body when he is standing straight up with arms flat at his sides and again when he is standing straight up holding his arms straight out. Estimate the ratio of his moment of inertia with his arms straight out to the moment of inertia with his arms flat against his sides.

Angular Velocity and Angular Acceleration

26 • A particle moves in a circle of radius 90 m with a constant speed of 25 m/s. (*a*) What is its angular velocity in radians per second about the center of the circle? (*b*) How many revolutions does it make in 30 s?

27 • **ISOLVE ✔** A wheel starts from rest with constant angular acceleration of 2.6 rad/s². After 6 s: (*a*) What is its angular velocity? (*b*) Through what angle has the wheel turned? (*c*) How many revolutions has it made? (*d*) What is the speed and acceleration of a point 0.3 m from the axis of rotation?

28 • **SSM** When a turntable rotating at $33\frac{1}{3}$ rev/min is shut off, it comes to rest in 26 s. Assuming constant angular acceleration, find (*a*) the angular acceleration, (*b*) the average angular velocity of the turntable, and (*c*) the number of revolutions it makes before stopping.

29 • A disk of radius 12 cm, initially at rest, begins rotating about its axis with a constant angular acceleration of 8 rad/s². At $t = 5$ s, what are (*a*) the angular velocity of the disk and (*b*) the tangential acceleration a_t and the centripetal acceleration a_c of a point on the edge of the disk?

30 • **ISOLVE** A Ferris wheel of radius 12 m rotates once in 27 s. (*a*) What is its angular velocity in radians per second? (*b*) What is the linear speed of a passenger? What is the centripetal acceleration of a passenger?

31 • A cyclist accelerates from rest. After 8 s, the wheels have made 3 revolutions. (*a*) What is the angular acceleration of the wheels? (*b*) What is the angular velocity of the wheels after 8 s?

32 • What is the angular velocity of the earth in radians per second as it rotates about its axis?

33 • A wheel rotates through 5.0 rad in 2.8 s as it is brought to rest with constant angular acceleration. The initial angular velocity of the wheel before braking began was (*a*) 0.6 rad/s, (*b*) 0.9 rad/s, (*c*) 1.8 rad/s, (*d*) 3.6 rad/s, (*e*) 7.2 rad/s.

34 • A bicycle has wheels of 1.2-m diameter. The bicyclist accelerates from rest with constant acceleration to 24 km/h in 14.0 s. What is the angular acceleration of the wheels?

35 •• **SSM** **iSOLVE**✓ The tape in a standard VHS video-tape cassette has a length $L = 246$ m; the tape plays for 2.0 h (Figure 9-41). As the tape starts, the full reel has an outer radius of about $R = 45$ mm and an inner radius of about $r = 12$ mm. At some point during the play, both reels have the same angular speed. Calculate this angular speed in radians per second and revolutions per minute.

FIGURE 9-41 Problem 35

45 mm
12 mm

Torque, Moment of Inertia, and Newton's Second Law for Rotation

36 • **iSOLVE** A disk-shaped grindstone of mass 1.7 kg and radius 8 cm is spinning at 730 rev/min. After the power is shut off, a woman continues to sharpen her ax by holding it against the grindstone for 9 s until the grindstone stops rotating. (a) What is the angular acceleration of the grindstone? (b) What is the torque exerted by the ax on the grindstone? (Assume constant angular acceleration and a lack of other frictional torques.)

37 • **SSM** A 2.5-kg cylinder of radius 11 cm, initially at rest, is free to rotate about the axis of the cylinder. A rope of negligible mass is wrapped around it and pulled with a force of 17 N. Find (a) the torque exerted by the rope, (b) the angular acceleration of the cylinder, and (c) the angular velocity of the cylinder at $t = 5$ s.

38 •• **iSOLVE**✓ A wheel free to rotate about its axis that is not frictionless is initially at rest. A constant external torque of 50 N·m is applied to the wheel for 20 s, giving the wheel an angular velocity of 600 rev/min. The external torque is then removed, and the wheel comes to rest 120 s later. Find (a) the moment of inertia of the wheel and (b) the frictional torque, which is assumed to be constant.

39 •• A pendulum consisting of a string of length L attached to a bob of mass m swings in a vertical plane. When the string is at an angle θ to the vertical, (a) what is the tangential component of acceleration of the bob? (b) What is the torque exerted about the pivot point? (c) Show that $\tau = I\alpha$ with $a_t = L\alpha$ gives the same tangential acceleration as found in Part (a).

40 ••• **SSM** A uniform rod of mass M and length L is pivoted at one end and hangs as in Figure 9-42 so that it is free to rotate without friction about its pivot. It is struck by a horizontal force F_0 for a short time Δt at a distance x below the pivot as shown. (a) Show that the speed of the center of mass of the rod just after being struck is given by $v_0 = 3F_0 x \, \Delta t / 2ML$. (b) Find the horizontal component of the force delivered by the pivot, and show that this force component is zero if $x = \frac{2}{3}L$. (Note: The point $x = \frac{2}{3}L$ is called the center of percussion of the rod.)

FIGURE 9-42
Problem 40

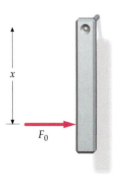

x

F_0

41 ••• A uniform horizontal disk of mass M and radius R is rotating about the vertical axis through its center with an angular velocity ω. When it is placed on a horizontal surface, the coefficient of kinetic friction between the disk and the surface is μ_k. (a) Find the torque $d\tau$ exerted by the force of friction on a circular element of radius r and width dr. (b) Find the total torque exerted by friction on the disk. (c) Find the time required for the disk to stop rotating.

Calculating the Moment of Inertia

42 • **iSOLVE** A tennis ball has a mass of 57 g and a diameter of 7 cm. Find the moment of inertia about its diameter. Assume that the ball is a thin spherical shell.

43 • **SSM** **iSOLVE** Four particles at the corners of a square with side length $L = 2$ m are connected by massless rods (Figure 9-43). The masses of the particles are $m_1 = m_3 = 3$ kg and $m_2 = m_4 = 4$ kg. Find the moment of inertia of the system about the z axis.

FIGURE 9-43 Problems 43–45

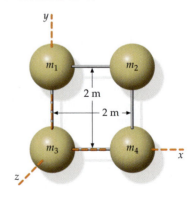

y

m_1 m_2

2 m

2 m

m_3 m_4 x

z

44 • Use the parallel-axis theorem and your results for Problem 43 to find the moment of inertia of the four-particle system in Figure 9-43 about an axis that is perpendicular to the plane of the configuration and passes through the center of mass of the system. Check your result by direct computation.

45 • For the four-particle system of Figure 9-43, (a) find the moment of inertia I_x about the x axis, which passes through m_3 and m_4, and (b) find I_y about the y axis, which passes through m_1 and m_4.

46 • **iSOLVE** Use the parallel-axis theorem to find the moment of inertia of a solid sphere of mass M and radius R about an axis that is tangent to the sphere (Figure 9-44).

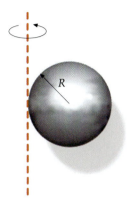

FIGURE 9-44 Problem 46

47 •• **iSOLVE** ✓ A 1.0-m-diameter wagon wheel consists of a thin rim having a mass of 8 kg and six spokes, each having a mass of 1.2 kg. Determine the moment of inertia of the wagon wheel for rotation about its axis.

48 •• **SSM** **iSOLVE** Two point masses m_1 and m_2 are separated by a massless rod of length L. (a) Write an expression for the moment of inertia about an axis perpendicular to the rod and passing through it at a distance x from mass m_1. (b) Calculate dI/dx and show that I is at a minimum when the axis passes through the center of mass of the system.

49 •• A uniform rectangular plate has mass m and sides of lengths a and b. (a) Show by integration that the moment of inertia of the plate about an axis that is perpendicular to the plate and passes through one corner is $m(a^2 + b^2)/3$. (b) What is the moment of inertia about an axis that is perpendicular to the plate and passes through its center of mass?

50 •• **SSM** Tracey and Corey are doing intensive research on baton-twirling. Each is using "The Beast" as a model baton: two uniform spheres, each of mass 500 g and radius 5 cm, mounted at the ends of a 30-cm uniform rod of mass 60 g (Figure 9-45). They want to calculate the moment of inertia of The Beast about an axis perpendicular to the rod and passing through its center. Corey uses the approximation that the two spheres can be treated as point particles that are 20 cm from the axis of rotation, and that the mass of the rod is negligible. Tracey, however, makes her calculations without approximations. (a) Compare the two results. (b) If the spheres retained the same mass but were hollow, would the rotational inertia increase or decrease? Justify your choice with a sentence or two. It is not necessary to calculate the new value of I.

FIGURE 9-45 Problem 50

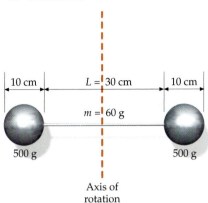

10 cm | $L = 30$ cm | 10 cm

$m = 60$ g

500 g · · · · · · · · · · · · 500 g

Axis of rotation

51 •• The methane molecule (CH_4) has four hydrogen atoms located at the vertices of a regular tetrahedron of side length 0.18 nm, with the carbon atom at the center of the tetrahedron (Figure 9-46). Find the moment of inertia of this molecule for rotation about an axis that passes through the centers of the carbon atom and one of the hydrogen atoms.

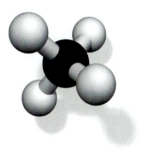

FIGURE 9-46 Problem 51

52 •• A hollow cylinder has mass m, an outside radius R_2, and an inside radius R_1. Show that its moment of inertia about its symmetry axis is given by $I = \frac{1}{2}m (R_2^2 + R_1^2)$. In Section 9-3 the moment of inertia was calculated for a solid cylinder by direct integration, first finding the moment of inertia of a disk. This calculation is the same as that one, except for the value of one of the integration limits and the expression for the area of a cross section.

53 ••• Show that the moment of inertia of a spherical shell of radius R and mass m is $2mR^2/3$. This can be done by direct integration or, more easily, by finding the increase in the moment of inertia of a solid sphere when its radius changes. To do this, first show that the moment of inertia of a solid sphere of density ρ is $I = (8/15)\pi\rho R^5$. Then compute the change dI in I for a change dR, and use the fact that the mass of this shell is $dm = 4\pi R^2\rho\, dR$.

54 ••• **SSM** The density of the earth is not quite uniform. It varies with the distance r from the center of the earth as $\rho = C [1.22 - (r/R)]$, where R is the radius of the earth and C is a constant. (a) Find C in terms of the total mass M and the radius R. (b) Find the moment of inertia of the earth. (See Problem 53.)

55 ••• Use integration to determine the moment of inertia of a right circular homogeneous solid cone of height H, base radius R, and mass density ρ about its symmetry axis.

56 ••• Use integration to determine the moment of inertia of a thin uniform disk of mass M and radius R for rotation about a diameter. Check your answer by referring to Table 9-1.

57 ••• A roadside ice-cream stand uses rotating solid cones to catch the eye of travelers. Each cone rotates about an axis perpendicular to its axis of symmetry and passing through its apex. The sizes of the cones vary, and the owner wonders if it would be more energy-efficient to use several smaller cones or a few big ones. To answer this, he must calculate the moment of inertia of a homogeneous right circular cone of height H, base radius R, and mass density ρ. What is the result?

Rotational Kinetic Energy

58 • The particles in Figure 9-47 are connected by a very light rod whose moment of inertia can be neglected. They rotate about the y axis with angular velocity $\omega = 2$ rad/s. (a) Find the speed of each particle, and use it to calculate the kinetic energy of this system directly from $\Sigma \frac{1}{2} m_i v_i^2$. (b) Find the moment of inertia about the y axis, and calculate the kinetic energy from $K = \frac{1}{2} I \omega^2$.

FIGURE 9-47 Problem 58

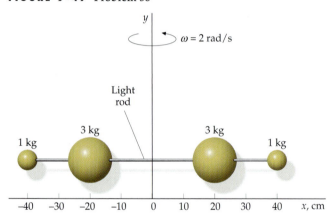

59 • **SSM** **iSOLVE** A solid ball of mass 1.4 kg and diameter 15 cm is rotating about its diameter at 70 rev/min. (a) What is its kinetic energy? (b) If an additional 2 J of energy are supplied to the rotational energy, what is the new angular speed of the ball?

60 • An engine develops 400 N·m of torque at 3700 rev/min. Find the power developed by the engine.

61 •• Two point particles with masses m_1 and m_2 are connected by a massless rod of length L to form a dumbbell that rotates about its center of mass with angular velocity ω. Show that the ratio of kinetic energies of the particles is $K_1/K_2 = m_2/m_1$.

62 •• **iSOLVE** ✓ Calculate the kinetic energy of rotation of the earth about its axis, and compare it with the kinetic energy of the orbital motion of the earth's center of mass about the sun. Assume the earth to be a homogeneous sphere of mass 6.0×10^{24} kg and radius 6.4×10^6 m. The radius of the earth's orbit is 1.5×10^{11} m.

63 •• **SSM** **iSOLVE** A 2000-kg block is lifted at a constant speed of 8 cm/s by a steel cable that passes over a massless pulley to a motor-driven winch (Figure 9-48). The radius of the winch drum is 30 cm. (a) What force must be exerted by the cable? (b) What torque does the cable exert on the winch drum? (c) What is the angular velocity of the winch drum? (d) What power must be developed by the motor to drive the winch drum?

FIGURE 9-48 Problem 63

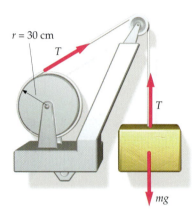

64 •• A uniform disk of mass M and radius R can rotate freely about a horizontal axis through its center and perpendicular to the plane of the disk. A small particle of mass m is attached to the rim of the disk at the top, directly above the pivot. The system is given a gentle nudge, and the disk begins to rotate. (a) What is the angular velocity of the disk when the particle is at its lowest point? (b) At this point, what force must be exerted by the disk on the particle to keep it stuck to the disk?

65 •• A uniform ring 1.5 m in diameter is pivoted at one point on its perimeter so that it is free to rotate about a horizontal axis. Initially, the line joining the support point and the center is horizontal (Figure 9-49). (a) If the ring is released from rest, what is its maximum angular velocity? (b) What minimum initial angular velocity must it be given if it is to rotate a full 360°?

FIGURE 9-49
Problem 65

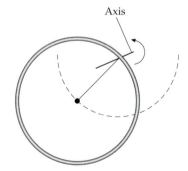

66 •• **iSOLVE** You set out to design a car that uses the energy stored in a flywheel consisting of a uniform 100-kg cylinder of radius R. The flywheel must deliver an average of 2 MJ/km of mechanical energy, with a maximum angular velocity of 400 rev/s. Find the smallest value of R for which the car can travel 300 km without the flywheel having to be recharged.

67 •• A ladder that is 8.6 m long and has a mass of 60 kg is placed in a nearly vertical position against the wall of a building. You stand on a rung with your center of mass at the top of the ladder. Assume that your mass is 80 kg. As you lean back slightly, the ladder begins to rotate about its base away from the wall. Is it better to quickly step off the ladder and drop to the ground or to hold onto the ladder and step off just before the top end hits the ground?

Pulleys, Yo-Yos, and Hanging Things

68 •• **SSM** **iSOLVE**✓ A 4-kg block resting on a friction-less horizontal ledge is attached to a string that passes over a pulley and is attached to a hanging 2-kg block (Figure 9-50). The pulley is a uniform disk of radius 8 cm and mass 0.6 kg. (*a*) Find the speed of the 2-kg block after it falls from rest a distance of 2.5 m. (*b*) What is the angular velocity of the pulley at this time?

FIGURE 9-50
Problems 68–70

69 •• For the system in Problem 68, find the linear acceleration of each block and the tension in the string.

70 •• Solve Problem 68 for the case in which the coefficient of friction between the ledge and the 4-kg block is 0.25.

71 •• A 1200-kg car is being unloaded by a winch. At the moment shown in Figure 9-51, the gearbox shaft of the winch breaks and the car falls from rest. During the car's fall, there is no slipping between the (massless) rope, the pulley, and the winch drum. The moment of inertia of the winch drum is 320 kg·m² and that of the pulley is 4 kg·m². The radius of the winch drum is 0.80 m and that of the pulley is 0.30 m. Find the speed of the car as it hits the water.

FIGURE 9-51
Problem 71

Pulley

Winch drum

5.0 m

72 •• **SSM** The system in Figure 9-52 is released from rest. The 30-kg block is 2 m above the ledge. The pulley is a uniform disk with a radius of 10 cm and a mass of 5 kg. Find (*a*) the speed of the 30-kg block just before it hits the ledge, (*b*) the angular speed of the pulley at that time, (*c*) the tensions in the strings, and (*d*) the time it takes for the 30-kg block to reach the ledge. Assume that the string does not slip on the pulley.

FIGURE 9-52
Problem 72

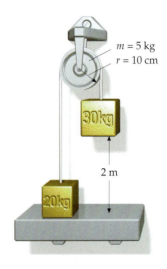

$m = 5$ kg
$r = 10$ cm

30kg

2 m

20kg

73 •• A uniform sphere of mass *M* and radius *R* is free to rotate about a horizontal axis through its center. A string is wrapped around the sphere and is attached to an object of mass *m* as shown in Figure 9-53. Find (*a*) the acceleration of the object and (*b*) the tension in the string.

FIGURE 9-53 **Problem 73**

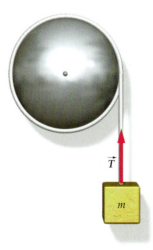

$\vec{T}$

m

74 •• **iSOLVE** An Atwood's machine has two objects of masses $m_1 = 500$ g and $m_2 = 510$ g, connected by a string of negligible mass that passes over a pulley (Figure 9-54) with frictionless bearings. The pulley is a uniform disk with a mass of 50 g and a radius of 4 cm. The string does not slip on the pulley. (*a*) Find the acceleration of the objects. (*b*) What is the tension in the string supporting m_1? In the string supporting m_2? By how much do they differ? (*c*) What would your answers have been if you had neglected the mass of the pulley?

FIGURE 9-54 Problem 74

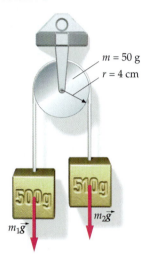

$m = 50$ g
$r = 4$ cm

500g 510g

$m_2\vec{g}$

$m_1\vec{g}$

75 •• **SSM** Two objects are attached to ropes that are attached to wheels on a common axle as shown in Figure 9-55. The two wheels are glued together so that they form a single object. The total moment of inertia of the object is 40 kg·m². The radii of the wheels are R_1 = 1.2 m and R_2 = 0.4 m. (*a*) If m_1 = 24 kg, find m_2 such that there is no angular acceleration of the wheels. (*b*) If 12 kg is gently added to the top of m_1, find the angular acceleration of the wheels and the tensions in the ropes.

R_1
R_2
m_1
m_2

FIGURE 9-55 Problem 75

76 •• The string wrapped around the cylinder in Figure 9-56 is held by a hand that is accelerated upward so that the center of mass of the cylinder does not move. Find (*a*) the tension in the string, (*b*) the angular acceleration of the cylinder, and (*c*) the acceleration of the hand.

FIGURE 9-56
Problems 76, 80

$\vec{T}$

77 •• A uniform cylinder of mass m_1 and radius R is pivoted on frictionless bearings. A massless string wrapped around the cylinder is connected to a block of mass m_2 that is on a frictionless incline of angle θ as shown in Figure 9-57.

The system is released from rest with m_2 at a height h above the bottom of the incline. (*a*) What is the acceleration of the block? (*b*) What is the tension in the string? (*c*) What is the total energy of the cylinder–block–earth system when the block is at height h? (*d*) What is the total energy of the system when the block is at the bottom of the incline and has a speed v? (*e*) What is the speed v? (*f*) Evaluate your answers for the extreme cases of $\theta = 0°$, $\theta = 90°$, and $m_1 = 0$.

FIGURE 9-57
Problem 77

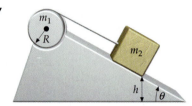

m_1
R
m_2
h
θ

78 •• **SSM** A device for measuring the moment of inertia of an object is shown in Figure 9-58. A circular platform has a concentric drum of radius 10 cm about which a string is wound. The string passes over a frictionless pulley to a weight of mass M. The weight is released from rest, and the time required for it to drop a distance D is measured. The system is then rewound, the object placed on the platform, and the system is again released from rest. The time required for the weight to drop the same distance D then provides the data needed to calculate I. With M = 2.5 kg and D = 1.8 m, the time is 4.2 s. (*a*) Find the combined moment of inertia of the platform, drum, shaft, and pulley. (*b*) With the object placed on the platform, the time is 6.8 s for D = 1.8 m. Find I of that object about the axis of the platform.

FIGURE 9-58
Problem 78

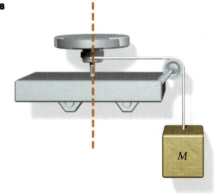

M

Objects Rolling Without Slipping

79 •• **iSOLVE** ✔ In 1993, a giant yo-yo of mass 400 kg measuring about 1.5 m in radius was dropped from a crane 57 m high. One end of the string was tied to the top of the crane, so the yo-yo unwound as it descended. Assuming that the axle of the yo-yo had a radius of r = 0.1 m, find the velocity of descent v at the end of the fall.

80 •• A uniform cylinder of mass M and radius R has a string wrapped around it. The string is held fixed, and the cylinder falls vertically as shown in Figure 9-56. (*a*) Show that the acceleration of the cylinder is downward with a magnitude $a = 2g/3$. (*b*) Find the tension in the string.

81 •• A 0.1-kg yo-yo consists of two solid disks of radius 10 cm joined together by a massless rod of radius 1 cm and a string wrapped around the rod. One end of the string is held fixed and is under constant tension T as the yo-yo is released. Find the acceleration of the yo-yo and the tension T.

82 • **SSM** A homogeneous solid cylinder rolls without slipping on a horizontal surface. The total kinetic energy is K. The kinetic energy due to rotation about its center of mass is $(a)\frac{1}{2}K$, $(b)\frac{1}{3}K$, $(c)\frac{4}{7}K$, (d) none of these.

83 • **iSOLVE** A homogeneous cylinder of radius 18 cm and mass 60 kg is rolling without slipping along a horizontal floor at 5 m/s. How much work was required to give it this motion?

84 • Find the percentages of the total kinetic energy associated with rotation and translation for an object that is rolling without slipping if the object is (a) a uniform sphere, (b) a uniform cylinder, or (c) a hoop.

85 • A hoop of radius 0.40 m and mass 0.6 kg is rolling without slipping at a speed of 15 m/s toward an incline of slope 30°. How far up the incline will the hoop roll, assuming that it rolls without slipping?

86 •• **SSM** A uniform sphere rolls without slipping down an incline. What must be the angle of the incline if the linear acceleration of the center of mass of the sphere is $0.2g$?

87 •• Repeat Problem 86 for a thin spherical shell.

88 •• A basketball rolls without slipping down an incline of angle θ. The coefficient of static friction is μ_s. Find (a) the acceleration of the center of mass of the ball, (b) the frictional force acting on the ball, and (c) the maximum angle of the incline for which the ball will roll without slipping.

89 •• Repeat Problem 88 for a solid cylinder of wood.

90 •• **SSM** A hollow sphere and uniform sphere of the same mass m and radius R roll down an inclined plane from the same height H without slipping (Figure 9-59). Each is moving horizontally as it leaves the ramp. When the spheres hit the ground, the range of the hollow sphere is L. Find the range L' of the uniform sphere.

FIGURE 9-59 Problem 90

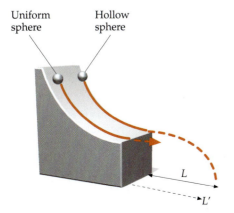

Uniform sphere   Hollow sphere

L

L'

91 •• A uniform thin-walled cylinder and a uniform solid cylinder are rolling horizontally without slipping. The speed of the thin-walled cylinder is v. The cylinders encounter an incline that they climb without slipping. If the maximum height they reach is the same, find the initial speed v' of the solid cylinder.

92 •• A hollow, thin-walled cylinder and a solid sphere start from rest and roll without slipping down an inclined plane of length 3 m. The cylinder arrives at the bottom of the plane 2.4 s after the sphere. Determine the angle between the inclined plane and the horizontal.

93 ••• A wheel has a thin 3.0-kg rim and four spokes, each of mass 1.2 kg. Find the kinetic energy of the wheel when it rolls at 6 m/s on a horizontal surface.

94 •• **iSOLVE** ✔ Two uniform 20-kg disks of radius 30 cm are connected by a short rod of radius 2 cm and mass 1 kg. When the rod is placed on a plane inclined at 30°, so that the disks hang over the sides, the assembly rolls without slipping. Find (a) the linear acceleration of the system and (b) the angular acceleration of the system. (c) Find the kinetic energy of translation of the system after it has rolled 2 m down the incline, starting from rest. (d) Find the kinetic energy of rotation of the system at the same point.

95 ••• A wheel of radius R rolls without slipping at a speed V. The coordinates of the center of the wheel are X, Y. (a) Show that the x and y coordinates of point P in Figure 9-60 are $X + r_0 \cos \theta$ and $R + r_0 \sin \theta$, respectively. (b) Show that the total velocity $\vec{v}$ of point P has the components $v_x = V + (r_0 V \sin \theta)/R$ and $v_y = -(r_0 V \cos \theta)/R$. (c) Show that at the instant that $X = 0$, $\vec{v}$ and $\vec{r}$ are perpendicular to each other by calculating $\vec{v} \cdot \vec{r}$. (d) Show that $v = r\omega$, where $\omega = V/R$ is the angular velocity of the wheel. These results demonstrate that, in the case of rolling without slipping, the motion is the same as if the rolling object were instantaneously rotating about the point of contact with an angular speed $\omega = V/R$.

FIGURE 9-60
Problem 95

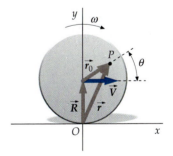

96 ••• **SSM** A uniform cylinder of mass M and radius R is at rest on a block of mass m, which in turn rests on a horizontal, frictionless table (Figure 9-61). If a horizontal force $\vec{F}$ is applied to the block, it accelerates and the cylinder rolls without slipping. Find the acceleration of the block.

FIGURE 9-61
Problems 96–98

97 ••• (a) Find the angular acceleration of the cylinder in Problem 96. Is the cylinder rotating clockwise or counterclockwise? (b) What is the cylinder's linear acceleration relative to the table? Let the direction of $\vec{F}$ be the positive direction. (c) What is the linear acceleration of the cylinder relative to the block?

98 ••• If the force in Problem 96 acts over a distance d, find (a) the kinetic energy of the block and (b) the kinetic energy of the cylinder. (c) Show that the total kinetic energy is equal to the work done on the system.

99 •• Two large gears are shown in Figure 9-62; each is free to rotate about a fixed axis running through its center. The radius of the first gear is $R_1 = 0.5$ m and the radius of the second gear is $R_2 = 1$ m. The moment of inertia of gear 1 is $I_1 = 1$ kg·m² and the moment of inertia of gear 2 is $I_2 = 16$ kg·m². The lever attached to gear 1 is 1 m long. (a) If a force of 2 N is applied to the end of the lever, what will the angular accelerations of gears 1 and 2 be? (b) What force must be applied tangentially to the edge of gear 2 to keep the gear system from rotating?

FIGURE 9-62
Problem 99

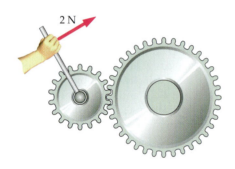

2 N

100 •• SSM A marble of radius 1 cm rolls from rest from the top of a large sphere of radius 80 cm, which is held fixed. (a) Assuming that the marble rolls without slipping while it is in contact with the sphere (which is unrealistic), find the angle from the top of the sphere to the point where the marble breaks contact with the sphere. (b) Why is it unrealistic to assume that the marble rolls without slipping all the way down to the point where it breaks contact?

Rolling With Slipping

101 • A bowling ball of mass M and radius R is released so that at the instant it touches the floor it is moving horizontally with a speed v_0 and is not rotating. It slides for a time t_1 a distance s_1 before it begins to roll without slipping. (a) If μ_k is the coefficient of kinetic friction between the ball and the floor, find s_1, t_1, and the final speed v_1 of the ball. (b) Find the ratio of the final mechanical energy to the initial mechanical energy of the ball. (c) Evaluate these quantities for $v_0 = 8$ m/s and $\mu_k = 0.06$.

102 •• SSM A cue ball of radius r is initially at rest on a horizontal pool table (Figure 9-63). It is struck by a horizontal cue stick that delivers a force of magnitude F_0 for a very short time Δt. The stick strikes the ball at a point h above the ball's point of contact with the table. Show that the ball's initial angular velocity ω_0 is related to the initial linear velocity of its center of mass v_0 by $\omega_0 = (5/2)v_0(h - r)/r^2$.

FIGURE 9-63
Problem 102

h

r

103 •• A uniform solid sphere is set rotating about a horizontal axis with an angular speed ω_0 and is placed on the floor. If the coefficient of sliding friction between the sphere and the floor is μ_k, find the speed of the center of mass of the sphere when it begins to roll without slipping.

104 •• A uniform solid ball resting on a horizontal surface has a mass of 20 g and a radius of 5 cm. A sharp force is applied to the ball in a horizontal direction 9 cm above the horizontal surface. The force increases linearly from 0 to a peak value of 40,000 N in 10^{-4} s and then decreases linearly to 0 in 10^{-4} s. (a) What is the velocity of the ball after impact? (b) What is the angular velocity of the ball after impact? (c) What is the velocity of the ball when it begins to roll without sliding? (d) For how long does the ball slide on the surface? Assume that $\mu_k = 0.5$.

105 •• iSOLVE A 0.16-kg billiard ball of radius 3 cm is given a sharp blow by a cue stick. The applied force is horizontal and passes through the center of the ball. The initial velocity of the ball is 4 m/s. The coefficient of kinetic friction is 0.6. (a) For how many seconds does the ball slide before it begins to roll without slipping? (b) How far does it slide? (c) What is its velocity once it begins rolling without slipping?

106 •• A billiard ball initially at rest is given a sharp blow by a cue stick. The force is horizontal and is applied at a distance $2R/3$ below the centerline, as shown in Figure 9-64. The initial speed of the ball is v_0 and the coefficient of kinetic friction is μ_k. (a) What is the initial angular speed ω_0? (b) What is the speed of the ball once it begins to roll without slipping? (c) What is the initial kinetic energy of the ball? (d) What is the frictional work done as it slides on the table?

FIGURE 9-64
Problem 106

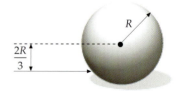

$\dfrac{2R}{3}$

R

107 •• A bowling ball of radius R is given an initial velocity v_0 down the lane and a forward spin $\omega_0 = 3v_0/R$. The coefficient of kinetic friction is μ_k. (a) What is the speed of the ball when it begins to roll without slipping? (b) For how long does the ball slide before it begins to roll without slipping? (c) What distance does the ball slide down the lane before it begins rolling without slipping?

108 •• **SSM** A solid cylinder of mass M resting on its side on a horizontal surface is given a sharp blow by a cue stick. The applied force is horizontal and passes through the center of the cylinder so that the cylinder begins translating with initial velocity v_0. The coefficient of sliding friction between the cylinder and surface is μ_k. (a) What is the translational velocity of the cylinder when it is rolling without slipping? (b) How far does the cylinder travel before it rolls without slipping? (c) What fraction of its initial mechanical energy is dissipated in friction?

109 •• Consider a ball of radius r and total mass m, with a nonuniform but radially symmetric mass distribution inside it, so that it can have an almost arbitrary moment of inertia I.

(a) Show that if this ball is projected across a floor with an initial velocity v, and is at first purely sliding across the floor (that is, its initial rotational velocity $\omega = 0$), its final velocity when it is rolling without slipping will be

$$v_f = \frac{1}{1 + I/mr^2} v,$$

independent of the coefficient of kinetic friction of the floor.

(b) Show that the total kinetic energy of the ball will be

$$K = \frac{1}{2} \frac{m}{(1 + I/mr^2)} v^2$$

(Note that I/mr^2 is independent of the mass and radius of the ball, but depends only on the distribution of mass inside the ball.)

General Problems

110 • **SSM** The moon rotates as it revolves around the earth so that we always see the same side. Use this fact to find the angular velocity in radians per second of the moon about its axis. (The period of revolution of the moon about the earth is 27.3 d.)

111 • Find the moment of inertia of a hoop about an axis perpendicular to the plane of the hoop and through its edge.

112 •• The radius of a park merry-go-round is 2.2 m. To start it rotating, you wrap a rope around it and pull with a force of 260 N for 12 s. During this time, the merry-go-round makes one complete rotation. (a) Find the angular acceleration of the merry-go-round. (b) What torque is exerted by the rope on the merry-go-round? (c) What is the moment of inertia of the merry-go-round?

113 • A uniform stick of length 2 m is raised at an angle of 30° to the horizontal above a sheet of ice. The bottom end of the stick rests on the ice. When dropped from rest, the bottom of the stick remains in contact with the ice at all times. How far will the bottom of the stick have moved when the stick falls to the ice? Assume that the ice is frictionless.

114 •• **ISOLVE** ✓ A uniform disk of radius 0.12 m and mass 5 kg is pivoted so that it rotates freely about its central axis (Figure 9-65). A string wrapped around the disk is pulled with a force of 20 N. (a) What is the torque exerted on the disk? (b) What is the angular acceleration of the disk? (c) If the

disk starts from rest, what is its angular velocity after 5 s? (d) What is its kinetic energy after 5 s? (e) What is the total angle θ that the disk turns through in 5 s? (f) Show that the work done by the torque, $\tau \Delta \theta$, equals the kinetic energy.

FIGURE 9-65 Problem 114

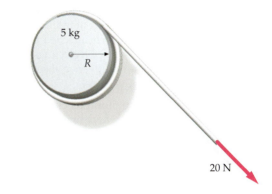

5 kg

R

20 N

115 •• A 0.25-kg rod of length 80 cm is suspended by a frictionless pivot at one end. It is held horizontal and released. Immediately after it is released, what is (a) the acceleration of the center of the rod and (b) the initial acceleration of a point on the end of the rod? (c) Find the linear velocity of the center of mass of the rod when it is vertical.

116 •• A marble of mass M and radius R rolls without slipping down the track on the left from a height h_1 as shown in Figure 9-66. The marble then goes up the *frictionless* track on the right to a height h_2. Find h_2.

FIGURE 9-66
Problem 116

h_1

h_2

117 •• **SSM** **ISOLVE** A uniform disk with a mass of 120 kg and a radius of 1.4 m rotates initially with an angular speed of 1100 rev/min. (a) A constant tangential force is applied at a radial distance of 0.6 m. How much work must this force do to stop the wheel? (b) If the wheel is brought to rest in 2.5 min, what torque does the force produce? What is the magnitude of the force? (c) How many revolutions does the wheel make in these 2.5 min?

118 •• **ISOLVE** A park merry-go-round consists of a 240-kg circular wooden platform 4.00 m in diameter. Four children running alongside push tangentially along the platform's circumference until, starting from rest, the merry-go-round reaches a steady speed of one complete revolution every 2.8 s. (a) If each child exerts a force of 26 N, how far does each child run? (b) What is the angular acceleration of the merry-go-round? (c) How much work does each child do? (d) What is the kinetic energy of the merry-go-round?

119 •• [iSOLVE] A hoop of mass 1.5 kg and radius 65 cm has a string wrapped around its circumference and lies flat on a horizontal frictionless table. The string is pulled with a constant force of 5 N. (*a*) How far does the center of the hoop travel in 3 s? (*b*) What is the angular velocity of the hoop about its center of mass after 3 s?

120 •• A vertical grinding wheel is a uniform disk of mass 60 kg and radius 45 cm. It has a handle of radius 65 cm of negligible mass. A compact 25-kg load is attached to the handle when it is in the horizontal position. Neglecting friction, find (*a*) the initial angular acceleration of the wheel and (*b*) the maximum angular velocity of the wheel.

121 •• [SSM] Consider two uniform blocks of wood, identical in shape and composition, where one is larger than the other by a factor *S* in all dimensions. (*a*) What is the ratio of the surface areas of the two blocks? (*b*) What is the ratio of the masses of the two blocks? (*c*) What is the ratio of the moments of inertia about some axis running through the block (in the same relative position and orientation in each)? These are examples of *scaling laws:* How do surface area, mass, and moment of inertia vary with the size of an object?

122 •• In this problem, you are to derive the perpendicular-axis theorem for planar objects, which relates the moments of inertia about two perpendicular axes in the plane of Figure 9-67 to the moment of inertia about a third axis that is perpendicular to the plane of the figure. Consider the mass element *dm* for the figure shown in the *xy* plane. (*a*) Write an expression for the moment of inertia of the figure about the *z* axis in terms of *dm* and *r*. (*b*) Relate the distance *r* of *dm* to the distances *x* and *y* and show that $I_z = I_y + I_x$. (*c*) Apply your result to find the moment of inertia of a uniform disk of radius *R* about a diameter of the disk.

FIGURE 9-67
Problem 122

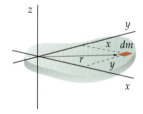

123 •• [iSOLVE] A uniform disk of radius *R* and mass *M* is pivoted about a horizontal axis parallel to its symmetry axis and passing through its edge so that it can swing freely in a vertical plane (Figure 9-68). It is released from rest with its center of mass at the same height as the pivot. (*a*) What is the angular velocity of the disk when its center of mass is directly below the pivot? (*b*) What force is exerted by the pivot at this time?

FIGURE 9-68
Problem 123

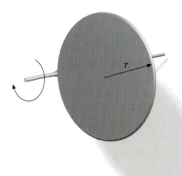

124 •• The roof of the student dining hall at St. Mary's College is supported by high cross-braced wooden beams attached together in the shape of an upside-down L (see Figure 9-69). Each vertical beam is 12 ft high and 2 ft wide and the horizontal cross-member is 6 ft long. The mass of the vertical beam is 350 kg and the mass of the horizontal beam is 175 kg. When workers were building the hall, one of the structures started to fall over before it was anchored into place. (Luckily, the workers stopped it before it fell.) (*a*) If it started falling from an upright position, what was the initial angular acceleration of the structure? Assume that the bottom didn't slide across the floor and that it didn't fall "out of plane," that is, that the fall occurred in the vertical plane defined by the vertical and horizontal beams. (*b*) If a sparrow were sitting on the right end of the horizontal beam, what would the magnitude of its initial linear acceleration be when the structure started falling? (*c*) What would the horizontal component of the sparrow's initial linear acceleration be?

FIGURE 9-69 Problem 124

125 •• A spool of mass *M* rests on an inclined plane at a distance *D* from the bottom. The ends of the spool have radius *R*, the center has radius *r*, and the moment of inertia of the spool about its axis is *I*. A long string of negligible mass is wound many times around the center of the spool. The other end of the string is fastened to a hook at the top of the inclined plane so that the string always pulls parallel to the slope as shown in Figure 9-70. (*a*) Suppose that initially the slope is so icy that there is *no* friction. How does the spool move as it slips down the slope? Use energy considerations to determine the speed of the center of mass of the spool when it reaches the bottom of the slope. Give your answer in terms of *M*, *I*, *r*, *R*, *g*, *D*, and *θ*. (*b*) Now suppose that the ice is gone and that when the spool is set up in the same way, there is enough friction to keep it from slipping on the slope. What is the direction and magnitude of the friction force in this case?

FIGURE 9-70 Problem 125

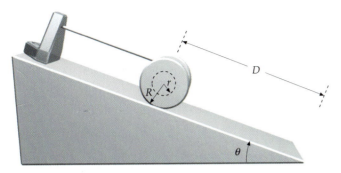

126 •• ▮**SOLVE**✓ A solid metal rod 1.5 m long is free to rotate without friction about a fixed horizontal axis perpendicular to the rod and passing through one end. The other end is held in a horizontal position. Small coins of mass m are placed on the rod 25 cm, 50 cm, 75 cm, 1 m, 1.25 m, and 1.5 m from the bearing. If the free end is now released, calculate the initial force exerted on each coin by the rod. Assume that the mass of the coins may be neglected in comparison to the mass of the rod.

127 •• ▮SSM▮ A popular classroom demonstration involves taking a meterstick and holding it horizontally at one end with a number of pennies spaced evenly along the meterstick. If the hand is relaxed so that the meterstick pivots about the hand under the influence of gravity, an interesting thing is seen: Pennies near the pivot point stay on the meterstick, while those farther away than a certain distance from the pivot are left behind by the falling meterstick. (This is often called the "faster than gravity" demonstration.) (a) What is the acceleration of the far end of the meterstick? (b) How far should a penny be from the end of the meterstick in order for it to be "left behind"?

128 •• ▮**SOLVE**▮ Figure 9-71 shows a hollow cylinder of length 1.8 m, mass 0.8 kg, and radius 0.2 m. The cylinder is free to rotate about a vertical axis that passes through its center and is perpendicular to the cylinder's axis. Inside the cylinder are two masses of 0.2 kg each, attached to springs of spring constant k and unstretched lengths 0.4 m. The inside walls of the cylinder are frictionless. (a) Determine the value of the spring constant if the masses are located 0.8 m from the center of the cylinder when the cylinder rotates at 24 rad/s. (b) How much work was needed to bring the system from $\omega = 0$ to $\omega = 24$ rad/s?

FIGURE 9-71 Problem 128

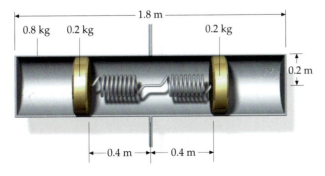

129 •• Suppose that for the system described in Problem 128, the spring constants are each $k = 60$ N/m. The system starts from rest and slowly accelerates until the masses are 0.8 m from the center of the cylinder. How much work was done in the process?

130 •• A string is wrapped around a uniform cylinder of radius R and mass M that rests on a horizontal frictionless surface. The string is pulled horizontally from the top with force F. (a) Show that the angular acceleration of the cylinder is twice that needed for rolling without slipping, so that the bottom point on the cylinder slides backward against the table. (b) Find the magnitude and direction of the frictional force between the table and cylinder needed for the cylinder to roll without slipping. What is the acceleration of the cylinder in this case?

131 •• ▮SSM▮ Let's calculate the position y of the falling load attached to the winch in Example 9-7 as a function of time by numerical integration. To do this, note that $v(y) = dy/dt$, or

$$t = \int_0^y \frac{1}{v(y')}\, dy' \approx \sum_{i=1}^N \frac{1}{v(y_i')}\, \Delta y'$$

where $\Delta y'$ is some (small) increment, $y_i' = i(\Delta y)$, and $y = y_N' = N(\Delta y)$. Hence, we can calculate t as a function of y by numerical summation. Make a graph of y' versus t for t between 0 s and 2 s. Assume $M = 10$ kg, $R = 0.5$ m, $m = 5$ kg, $L = 10$ m, and $m_c = 3.5$ kg. Use $\Delta y' = 0.1$ m. Compare this to the position of the falling load if it were in free-fall.

132 •• Figure 9-72 shows a solid cylinder of mass M and radius R to which a hollow cylinder of radius r is attached. A string is wound about the hollow cylinder. The solid cylinder rests on a horizontal surface. The coefficient of static friction between the cylinder and surface is μ_s. If a light tension is applied to the string in the vertical direction, the cylinder will roll to the left; if the tension is applied with the string horizontally to the right, the cylinder rolls to the right. Find the angle of the string with the horizontal that will allow the cylinder to remain stationary when a small tension is applied to the string.

FIGURE 9-72
Problem 132

133 •• **SSM** In problems dealing with a pulley with a nonzero moment of inertia, the magnitude of the tensions in the ropes hanging on either side of the pulley are not equal. The difference in the tension is due to the static frictional force between the rope and the pulley; however, the static frictional force cannot be made arbitrarily large. If you consider a massless rope wrapped partly around a cylinder through an angle $\Delta\theta$ (measured in radians), then you can show that if the tension on one side of the pulley is T, while the tension on the other side is T' ($T' > T$), the maximum value of T' in relation to T that can be maintained without the rope slipping is $T'_{max} = Te^{\mu_s \Delta\theta}$, where μ_s is the coefficient of static friction. Consider the Atwood's machine in Figure 9-73: the pulley has a radius $r = 0.15$ m, the moment of inertia $I = 0.35$ kg·m², and the coefficient of static friction $\mu_s = 0.30$. (a) If the tension on one side of the pulley is 10 N, what is the maximum tension on the other side that will prevent the rope from slipping on the pulley? (b) If the mass of one of the hanging blocks is 1 kg, what is the maximum mass of the other block if, after the blocks are released, the pulley is to rotate without slipping? (c) What is the acceleration of the blocks in this case?

FIGURE 9-73 Problem 133

134 ••• A heavy, uniform cylinder has a mass m and a radius R (Figure 9-74). It is accelerated by a force $\vec{T}$ that is applied through a rope wound around a light drum of radius r that is attached to the cylinder. The coefficient of static friction is sufficient for the cylinder to roll without slipping. (a) Find the frictional force. (b) Find the acceleration a of the center of the cylinder. (c) Show that it is possible to choose r so that a is greater than T/m. (d) What is the direction of the frictional force in the circumstances of Part (c)?

FIGURE 9-74
Problem 134

135 ••• A uniform rod of length L and mass M is free to rotate about a horizontal axis through one end as shown in Figure 9-75. The rod is released from rest at $\theta = \theta_0$. Show that the force exerted by the axis on the rod is given by $F_\parallel = \frac{1}{2}Mg(5\cos\theta - 3\cos\theta_0)$ and $F_\perp = \frac{1}{4}Mg\sin\theta$, where $F_\parallel$ is the component of the force parallel with the rod and $F_\perp$ is the component of the force perpendicular to the rod.

FIGURE 9-75
Problem 135

Conservation of Angular Momentum

ANGULAR MOMENTUM IS THE
ROTATIONAL ANALOG OF LINEAR
MOMENTUM.

? **When the riders move closer
to the center, the merry-go-round
gains speed. Why does this
happen? (See Example 10-4.)**

> In this chapter, we extend our study of rotational motion to situations in which the direction of the axis of rotation may change.

We begin with an examination of the vector properties of angular velocity and torque and then introduce the concept of angular momentum, which is the rotational analog of linear momentum. We then show that the net torque acting on a system equals the rate of change of its angular momentum, a result that is equivalent to Newton's second law for rotational motion. Angular momentum is therefore conserved in systems with zero net external torque. Like conservation of linear momentum, conservation of angular momentum is a fundamental law of nature, applying even in the atomic domain where Newtonian mechanics fails.

10-1 The Vector Nature of Rotation

In Chapter 9, we indicated the direction of rotation about an axis with a fixed direction by assigning plus and minus signs to indicate the direction of the angular velocity, just as we used them to indicate the direction of the velocity in one-dimensional motion. But when the direction of the axis of rotation is *not* fixed in space, plus and minus signs are not adequate to describe the direction of the angular velocity. This inadequacy is overcome by treating the angular velocity as a vector $\vec{\omega}$ directed along the rotation axis. Consider, for example, the rotating disk in Figure 10-1. We determine the direction of $\vec{\omega}$ by a convention known as the

FIGURE 10-1

309

right-hand rule, which is illustrated in Figure 10-2. Thus, if the rotation is directed as shown in Figure 10-1, $\vec{\omega}$ is directed as shown; if the direction of the rotation is reversed, so is the direction of $\vec{\omega}$.

We apply similar considerations to the torque. Figure 10-3 shows a force $\vec{F}$ acting on a particle at some position $\vec{r}$ relative to the origin O. The torque $\vec{\tau}$ exerted by this force relative to the origin O is defined as a vector that is perpendicular to the plane formed by $\vec{F}$ and $\vec{r}$ and has magnitude $Fr \sin \phi$, where ϕ is the angle between $\vec{F}$ and $\vec{r}$. (The angle between two vectors refers to the angle *between the directions* of the two vectors.) If $\vec{F}$ and $\vec{r}$ are in the xy plane, as in Figure 10-3, the torque vector is along the z axis. If $\vec{F}$ is applied to the rim of a disk of radius r, as shown in Figure 10-4, the torque vector has the magnitude Fr, and is along the axis of rotation as shown.

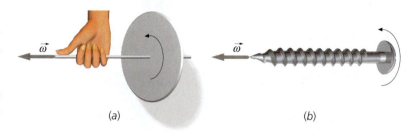

FIGURE 10-2 (*a*) When the fingers of the right hand curl in the direction of rotation, the thumb points in the direction of $\vec{\omega}$. (*b*) Looked at another way, the direction of $\vec{\omega}$ is that of the advance of a rotating right-hand screw.

The Cross Product

Torque is expressed mathematically as the **cross product** (or **vector product**) of $\vec{r}$ and $\vec{F}$:

$$\vec{\tau} = \vec{r} \times \vec{F} \qquad \text{10-1}$$

The cross product of two vectors $\vec{A}$ and $\vec{B}$ is defined to be a vector $\vec{C} = \vec{A} \times \vec{B}$ whose magnitude equals the area of the parallelogram formed by the two vectors (Figure 10-5). The vector $\vec{C}$ is perpendicular to the plane containing $\vec{A}$ and $\vec{B}$ in the direction given by the right-hand rule, that is, as the fingers curl from the direction of $\vec{A}$ toward the direction of $\vec{B}$ (Figure 10-6). If ϕ is the angle between the two vectors and $\hat{n}$ is a unit vector that is perpendicular to each vector in the direction of $\vec{C}$, the cross product of $\vec{A}$ and $\vec{B}$ is

$$\vec{A} \times \vec{B} = AB \sin \phi \, \hat{n} \qquad \text{10-2}$$

DEFINITION—CROSS PRODUCT

If $\vec{A}$ and $\vec{B}$ are parallel, $\vec{A} \times \vec{B}$ is zero. It follows from the definition of the cross product that

$$\vec{A} \times \vec{A} = 0 \qquad \text{10-3}$$

and

$$\vec{A} \times \vec{B} = -\vec{B} \times \vec{A} \qquad \text{10-4}$$

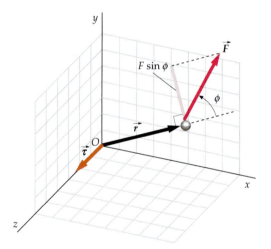

FIGURE 10-3

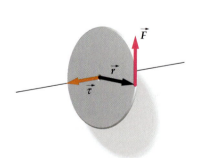

FIGURE 10-4

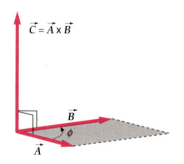

FIGURE 10-5 The cross product $\vec{A} \times \vec{B}$ is a vector $\vec{C}$ that is perpendicular to both $\vec{A}$ and $\vec{B}$ and has a magnitude $AB \sin \phi$, which equals the area of the parallelogram shown.

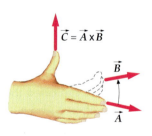

FIGURE 10-6 The direction of $\vec{A} \times \vec{B}$ is given by the right-hand rule when the fingers are rotated from the direction of $\vec{A}$ toward $\vec{B}$ through the angle ϕ.

Note that the order in which two vectors are multiplied in a cross product makes a difference. Below are some properties of the cross product of two vectors:

1. The cross product obeys a distributive law under addition:

$$\vec{A} \times (\vec{B} + \vec{C}) = (\vec{A} \times \vec{B}) + (\vec{A} \times \vec{C})$$
10-5

2. If $\vec{A}$ and $\vec{B}$ are functions of some variable such as t, the derivative of $\vec{A} \times \vec{B}$ follows the usual product rule for derivatives:

$$\frac{d}{dt}(\vec{A} \times \vec{B}) = \left(\vec{A} \times \frac{d\vec{B}}{dt}\right) + \left(\frac{d\vec{A}}{dt} \times \vec{B}\right)$$
10-6

3. The unit vectors $\hat{i}$, $\hat{j}$, and $\hat{k}$ (Figure 10-7), which are mutually perpendicular, have cross products given by

$$\hat{i} \times \hat{j} = \hat{k}, \qquad \hat{j} \times \hat{k} = \hat{i}, \qquad \text{and} \qquad \hat{k} \times \hat{i} = \hat{j}$$
10-7a

Furthermore,

$$\hat{i} \times \hat{i} = \hat{j} \times \hat{j} = \hat{k} \times \hat{k} = 0$$
10-7b

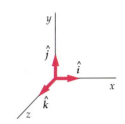

FIGURE 10-7

10-2 Torque and Angular Momentum

Figure 10-8 shows a particle of mass m moving with a velocity $\vec{v}$ at a position $\vec{r}$ relative to the origin O. The linear momentum of the particle is $\vec{p} = m\vec{v}$. The **angular momentum** $\vec{L}$ of the particle relative to the origin O is defined to be the cross product of $\vec{r}$ and $\vec{p}$:

$$\vec{L} = \vec{r} \times \vec{p}$$
10-8

ANGULAR MOMENTUM OF A PARTICLE DEFINED

If $\vec{r}$ and $\vec{p}$ are in the xy plane, as in Figure 10-8, $\vec{L}$ is parallel with the z axis and is given by $\vec{L} = \vec{r} \times \vec{p} = mvr \sin \phi \, \hat{k}$. Like torque, angular momentum is defined *relative to a point in space.*

Figure 10-9 shows a particle of mass m attached to a circular disk of negligible mass in the xy plane with its center at the origin. The disk is spinning about its axis with angular speed ω. The speed v of the particle and its angular speed are related by $v = r\omega$. The angular momentum of the particle relative to the center of the disk is

$$\vec{L} = \vec{r} \times \vec{p} = \vec{r} \times m\vec{v} = rmv \sin 90° \, \hat{k} = rmv\hat{k} = mr^2\omega\hat{k} = mr^2\vec{\omega}$$

FIGURE 10-8

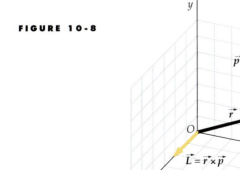

FIGURE 10-9

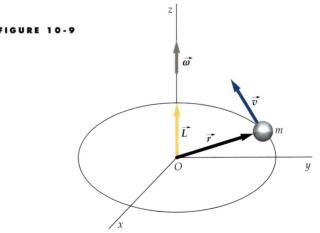

The angular momentum vector is in the same direction as the angular velocity vector.

Since mr^2 is the moment of inertia for a single particle about the z axis, we have

$$\vec{L} = mr^2\vec{\omega} = I\vec{\omega}$$

This result does not hold for the angular momentum about a general point on the z axis. Figure 10-10 shows the angular momentum vector $\vec{L}'$ for the same particle attached to the same disk but with $\vec{L}'$ computed about a point on the z axis that is not at the center of the circle. In this case, the angular momentum is not parallel to the angular velocity vector $\vec{\omega}$, which is parallel with the z axis.

In Figure 10-11, we attach a second particle of equal mass to the spinning disk. The angular momentum vectors $\vec{L}'_1$ and $\vec{L}'_2$ are shown relative to the same point O'. The total angular momentum $\vec{L}'_1 + \vec{L}'_2$ of the two-particle system is again parallel to the angular velocity vector $\vec{\omega}$. In this case, the axis of rotation, the z axis, passes through the center of mass of the two-particle system, and the mass distribution is symmetric about this axis. Such an axis is called a **symmetry axis.** For any system of particles that rotates about a symmetry axis, the total angular momentum (which is the sum of the angular momenta of the individual particles) is parallel to the angular velocity and is given by

$$\vec{L} = I\vec{\omega} \qquad\qquad 10\text{-}9$$

ANGULAR MOMENTUM OF A SYSTEM ROTATING ABOUT A SYMMETRY AXIS

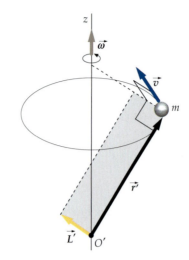

FIGURE 10-10

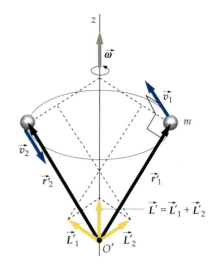

FIGURE 10-11

ANGULAR MOMENTUM ABOUT THE ORIGIN   **EXAMPLE 10-1**

Find the angular momentum about the origin for the following situations. (a) A car of mass 1200 kg moves in a circle of radius 20 m with a speed of 15 m/s. The circle is in the xy plane, centered at the origin. When viewed from a point on the positive z axis, the car moves counterclockwise. (b) The same car moves in the xy plane with velocity $\vec{v} = -(15\ \text{m/s})\hat{i}$ along the line $y = y_0 = 20$ m parallel to the x axis. (c) A disk in the xy plane of radius 20 m and mass 1200 kg rotates at 0.75 rad/s about its axis, which is also the z axis. When viewed from a point on the positive z axis, the disk rotates counterclockwise.

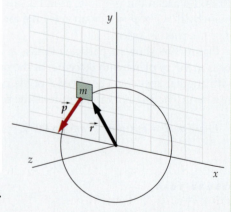

PICTURE THE PROBLEM For (a) and (b) use $\vec{L} = \vec{r} \times \vec{p}$. For (c) use $\vec{L} = I\vec{\omega}$. Draw a figure and apply the right-hand rule to find the direction of $\vec{L}$.

FIGURE 10-12

(a) $\vec{r}$ and $\vec{p}$ are perpendicular and $\vec{r} \times \vec{p}$ is along the z axis (Figure 10-12):

$$\vec{L} = \vec{r} \times \vec{p} = rmv \sin 90°\, \hat{k}$$

$$= (20\ \text{m})(1200\ \text{kg})(15\ \text{m/s})\hat{k}$$

$$= \boxed{3.6 \times 10^5\ \text{kg}\cdot\text{m}^2/\text{s}\,\hat{k}}$$

(b) 1. For the same car moving in the direction of decreasing x along the line $y = 20$ m, we express $\vec{r}$ and $\vec{p}$ in terms of unit vectors:

$$\vec{r} = x\hat{i} + y\hat{j} = x\hat{i} + y_0\hat{j}$$

$$\vec{p} = m\vec{v} = -mv\hat{i}$$

2. Now compute $\vec{r} \times \vec{p}$ (Figure 10-13):

$$\vec{L} = \vec{r} \times \vec{p} = (x\hat{i} + y_0\hat{j}) \times (-mv\hat{i})$$
$$= -xmv(\hat{i} \times \hat{i}) - y_0\,mv(\hat{j} \times \hat{i})$$
$$= 0 - y_0mv(-\hat{k}) = y_0mv\hat{k}$$
$$= (20\text{ m})(1200\text{ kg})(15\text{ m/s})\hat{k}$$
$$= \boxed{3.6 \times 10^5\text{ kg·m}^2\text{/s}\,\hat{k}}$$

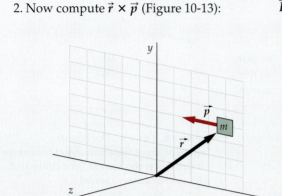

FIGURE 10-13

(c) Use $\vec{L} = I\vec{\omega}$ (Figure 10-14):

$$\vec{L} = I\vec{\omega} = I\omega\hat{k} = \tfrac{1}{2}mR^2\omega\,\hat{k}$$
$$= \tfrac{1}{2}(1200\text{ kg})(20\text{ m})^2\,(0.75\text{ rad/s})\hat{k}$$
$$= \boxed{1.8 \times 10^5\text{ kg·m}^2\text{/s}\,\hat{k}}$$

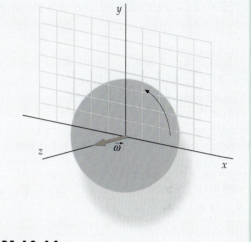

FIGURE 10-14

REMARKS The angular momentum of the car moving in a circle in (a) is the same as that of the car moving along a straight line in (b). In (c), the velocity of a point on the rim is $v = R\omega = (20\text{ m})(0.75\text{ rad/s}) = 15\text{ m/s}$, the same as the velocity of the car in Parts (a) and (b). The moment of inertia of a 1200-kg disk of radius 20 m is less than that of a 1200-kg car at 20 m from the axis because much of the mass of the disk is closer to the axis of rotation.

There are several additional results concerning torque and angular momentum for a system of particles. The first of these is

$$\vec{\tau}_{\text{net ext}} = \frac{d\vec{L}_{\text{sys}}}{dt} \qquad\qquad 10\text{-}10$$

The net external torque acting on a system equals the rate of change of the angular momentum of the system.

NEWTON'S SECOND LAW FOR ANGULAR MOTION

In Equation 10-10 the net external torque is the vector sum of the external torques acting on the system. Integrating both sides of this equation with respect to time gives

$$\Delta\vec{L}_{\text{sys}} = \int_{t_i}^{t_f} \vec{\tau}_{\text{net ext}}\,dt \qquad\qquad 10\text{-}11$$

ANGULAR IMPULSE–ANGULAR MOMENTUM EQUATION

It is often useful to split the total angular momentum of a system about an arbitrary point O into orbital angular momentum and spin angular momentum:

$$\vec{L}_{\text{sys}} = \vec{L}_{\text{orbit}} + \vec{L}_{\text{spin}} \qquad \text{10-12}$$

The earth has spin angular momentum due to its spinning motion about its rotational axis and it has orbital angular momentum about the center of the sun due to its orbital motion around the sun (Figure 10-15). The total angular momentum of the earth relative to the sun is the vector sum of the spin and orbital angular momenta. $\vec{L}_{\text{spin}}$ is the angular momentum of a system about its center of mass and $\vec{L}_{\text{orbit}}$ is the angular momentum that a point particle of mass M located at the center of mass and moving at the velocity of the center of mass would have. That is,

$$\vec{L}_{\text{orbit}} = \vec{r}_{\text{cm}} \times M\vec{v}_{\text{cm}} \qquad \text{10-13}$$

In Chapter 9 torques are computed about axes instead of about points. The relation between the torque about an axis and the torque about a point is straightforward. If point O is the origin and if force $\vec{F}$ exerts torque $\vec{\tau}$ about O, then the corresponding torque exerted about the z axis is τ_z (the z component of $\vec{\tau}$.) Taking components of cross products requires some care. If $\vec{\tau} = \vec{r} \times \vec{F}$, then

$$\vec{\tau}_z = \vec{r}_{\text{rad}} \times \vec{F}_{xy} \qquad \text{10-14}$$

where $\vec{\tau}_z$, $\vec{r}_{\text{rad}}$, and $\vec{F}_{xy}$ (see Figure 10-16) are vector components of $\vec{\tau}$, $\vec{r}$, and $\vec{F}$. (The vector component in a given direction is the scalar component in that direction times the unit vector in that direction. For example, $\vec{\tau}_z = \tau_z \hat{k}$.) $\vec{r}_{\text{rad}}$ is the component of $\vec{r}$ directed radially away from the z axis and $\vec{F}_{xy}$ is the component of $\vec{F}$ in the xy plane (the plane perpendicular to the z axis). The relation between angular momentum about an axis and angular momentum about a point is also straightforward. The angular momentum about an axis is

$$\vec{L}_z = \vec{r}_{\text{rad}} \times \vec{p}_{xy} \qquad \text{10-15}$$

ANGULAR MOMENTUM ABOUT z AXIS

where $\vec{p}_{xy}$ is the projection of the linear momentum $\vec{p}$ in the xy plane. Taking the z vector components of both sides of Equation 10-10 gives

$$\vec{\tau}_{\text{net ext},z} = \frac{d\vec{L}_{\text{sys},z}}{dt} \qquad \text{10-16}$$

For a rigid object rotating about the z axis, $\vec{L}_{\text{sys},z} = I_z\vec{\omega}$, where I_z is the moment of inertia about the z axis. Substituting this into Equation 10-16 gives

$$\vec{\tau}_{\text{net ext},z} = \frac{d\vec{L}_{\text{sys},z}}{dt} = \frac{d}{dt}(I_z\vec{\omega}) = I_z\vec{\alpha} \qquad \text{10-17}$$

where the angular acceleration vector $\vec{\alpha}$ is defined by $\vec{\alpha} = d\vec{\omega}/dt$. (Equation 10-17 is the vector form of Equation 9-18.)

For a system of particles, the total angular momentum about the z axis equals the sum of the angular momenta about the z axis of the individual particles. Also, the total torque about the z axis is the sum of the torques about the z axis acting on the system.

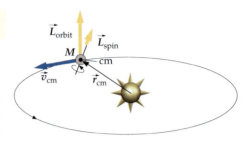

FIGURE 10-15 The angular momentum of the earth about the center of the sun is the sum of the orbital and the spin angular momenta.

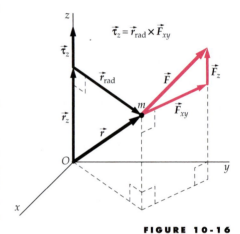

FIGURE 10-16

THE ATWOOD'S MACHINE REVISITED

EXAMPLE 10-2

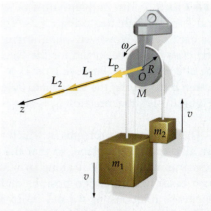

An Atwood's machine has two blocks with masses m_1 and m_2 ($m_1 > m_2$) connected by a string of negligible mass that passes over a pulley with frictionless bearings. The pulley is a uniform disk of mass M and radius R. The string does not slip on the pulley. Apply Equation 10-16 to the system consisting of the two blocks, the string, and the pulley, to find the angular acceleration of the pulley and the linear acceleration of the blocks.

FIGURE 10-17

PICTURE THE PROBLEM Let the pulley and blocks be in the xy plane with the z axis out of the page through the center of the pulley at point O as shown in Figure 10-17. We compute the torques and angular momenta about the z axis and apply Newton's second law for angular motion (Equation 10-10). Since m_1 is greater than m_2, the disk will rotate counterclockwise corresponding to $\vec{\omega}$ out of the page in the positive z direction. All the forces are in the xy plane so all torques are along the z axis. Also, all the velocities are in the xy plane so all the angular momentum vectors are along the z axis. Since the torque, angular velocity, and angular momentum vectors are all along the z axis, we can treat this as a one-dimensional problem with positive assigned to counterclockwise motion and negative to clockwise motion. The acceleration a of the blocks is related to the angular acceleration α of the pulley by the nonslip condition $a = R\alpha$.

1. Draw a free-body diagram of the system (Figure 10-18). The only thing touching the system is the pulley bearings. The external forces on the system are the normal force of the pulley bearings on the pulley and the gravity forces on the two blocks and the pulley:

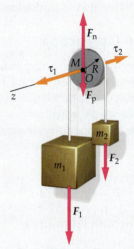

2. Express Newton's second law for rotation, z components only (Equation 10-16):

$$\sum \tau_{ext,z} = \frac{dL_z}{dt}$$

3. The total external torque about the z axis is the sum of the torques exerted by the external forces. The lever arms F_1 and F_2 each equal R. (The lever arms of the other two forces are each zero.) $F_1 = m_1 g$ and $F_2 = m_2 g$:

$$\sum \tau_{ext,z} = \tau_n + \tau_p + \tau_1 + \tau_2$$

$$= 0 + 0 + m_1 g R - m_2 g R$$

FIGURE 10-18

4. The total angular momentum about the z axis equals the angular momentum of the pulley $\vec{L}_p$ plus the angular momenta of block 1, $\vec{L}_1$, and block 2, $\vec{L}_2$, each in the positive z direction. The pulley has spin angular momentum but no orbital angular momentum, whereas each block has orbital angular momentum but no spin angular momentum.

$$L_z = L_p + L_1 + L_2$$

$$= I\omega + m_1 v R + m_2 v R$$

5. Substitute these results into Newton's second law for rotation in step 2:

$$\sum \tau_{ext,z} = \frac{dL_z}{dt}$$

$$m_1 g R - m_2 g R = \frac{d}{dt}(I\omega + m_1 v R + m_2 v R)$$

$$m_1 g R - m_2 g R = I\alpha + (m_1 + m_2)Ra$$

6. Relate I to M and R, and use the nonslip condition to relate α to a and solve for both a and α:

$$m_1 g R - m_2 g R = \frac{1}{2}MR^2 \frac{a}{R} + (m_1 + m_2)Ra$$

so

$$a = \boxed{(m_1 - m_2)g/(\tfrac{1}{2}M + m_1 + m_2)}, \text{ and}$$

$$\alpha = a/R = \boxed{(m_1 - m_2)g/[(\tfrac{1}{2}M + m_1 + m_2)R]}$$

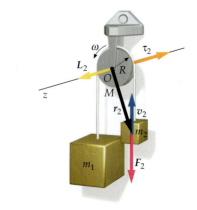

REMARKS 1. This problem could be solved by writing the tensions T_1 on the left and T_2 on the right and using $\tau = I\alpha$ (Equation 10-17) for the pulley and $\Sigma F_y = ma_y$ for each block. However, using angular momentum (Equation 10-16) is easier, and once you have solved for the acceleration, it is straightforward to solve for the two tensions. 2. Because $\vec{L}_2 = \vec{r}_2 \times m_2\vec{v}_2$ (Figure 10-19), the direction of $\vec{L}_2$ is gotten by applying the right-hand rule (Figure 10-6). And because $\vec{\tau}_2 = \vec{r}_2 \times \vec{F}_2$ (Figure 10-19), the direction of $\vec{\tau}_2$ also is gotten by applying the right-hand rule.

There are many problems in which the forces, position vectors, and velocities are all parallel to a plane, so that the torques, angular velocities, and angular momentum vectors are all along an axis of rotation that remains fixed in space. In such cases we can assign positive and negative values to counterclockwise or clockwise rotations, as we did in Example 10-2, and treat the case like a one-dimensional problem. However, there are other situations, such as the motion of a gyroscope, where torque, angular velocity, and angular momentum must be considered as multidimensional vectors.

FIGURE 10-19

The Gyroscope

A gyroscope is a common example of motion in which the axis of rotation changes direction. Figure 10-20 shows a gyroscope consisting of a bicycle wheel that is free to turn on its axle. The axle is pivoted at a point a distance D from the center of the wheel, and is free to rotate about the pivot in any direction. We can give a qualitative understanding of the complex motion of such a system by using Newton's second law for rotation,

$$\vec{\tau}_{net} = \frac{d\vec{L}}{dt} \quad (\text{or} \quad d\vec{L} = \vec{\tau}_{net}dt)$$

along with the relations

$$\vec{\tau}_{net} = \vec{r}_{cm} \times M\vec{g}$$

and

$$\vec{L} = I_s\vec{\omega}_s$$

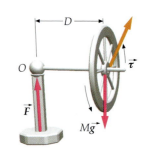

FIGURE 10-20

where I_s and ω_s are the moment of inertia and angular velocity of the wheel about its spin axis. All we need to remember in order to describe the motion of a gyroscope is that the *change* in angular momentum of the wheel must be in the direction of the net torque acting on it.

Suppose the axle is held horizontally and then released. If the wheel isn't spinning, it simply falls, rotating about a horizontal axis through O and perpendicular to $\vec{r}$. The torque vector is horizontal, into the page. For this case the initial angular momentum is zero, so the *change* in angular momentum equals the angular momentum itself, which, in this case, is also horizontal and into the page. However, if the wheel *is* spinning and has a large angular momentum along its axle, it does not fall when the axle is released. If it were to fall, the axle would point downward, resulting in a large component of angular momentum in the downward direction. But there is no torque in the downward direction. The torque is horizontal. What actually happens is that the axle moves horizontally (into the paper in the figure). The wheel must move this way so that the *change* in angular momentum is in the direction of the net torque. This is illustrated in Figure 10-21a, where we see a large angular momentum $\vec{L}$ along the axis of the wheel and a change in angular momentum $d\vec{L}$ in the direction of the torque $\vec{\tau}$. This motion, which is always surprising when first encountered, is called **precession**. We can calculate the angular velocity ω_p of precession. In a small time interval dt, the change in the angular momentum has a magnitude dL:

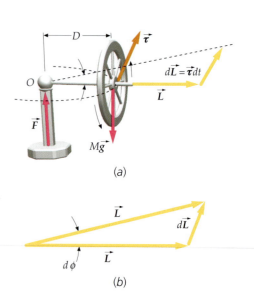

(a)

(b)

FIGURE 10-21

$$dL = \tau \, dt = MgD \, dt$$

where MgD is the magnitude of the torque about the pivot point. From Figure 10-21b, the angle $d\phi$ through which the axle moves is

$$d\phi = \frac{dL}{L} = \frac{\tau \, dt}{L} = \frac{MgD \, dt}{L}$$

The angular velocity of the precession is thus

$$\omega_p = \frac{d\phi}{dt} = \frac{MgD}{L} = \frac{MgD}{I_s \omega_s} \qquad \text{10-18}$$

If the angular momentum due to the spin of the wheel is large, the precession can be very slow.

In the preceding analysis it is assumed that the spin angular momentum of the wheel is very large compared to the orbital angular momentum associated with the precessional motion.

If you prevent a spinning gyroscope from precessing using your hand, upon release it will initiate precessional motion with an up and down bouncing motion called nutation. This initial bouncing motion can be avoided by giving the gyroscope axis an initial angular velocity equal to that associated with its precessional motion.

10-3 Conservation of Angular Momentum

When the net external torque acting on a system remains zero, we have

$$\frac{d\vec{L}_{sys}}{dt} = 0$$

or

$$\vec{L}_{sys} = \text{constant} \qquad \text{10-19}$$

Equation 10-19 is a statement of the **law of conservation of angular momentum.**

> If the net external torque acting on a system is zero, the total angular momentum of the system is constant.

CONSERVATION OF ANGULAR MOMENTUM

This is the rotational analog of the law of conservation of linear momentum. If a system is isolated from its surroundings, so that there are no external forces or torques acting on it, three quantities are conserved: energy, linear momentum, and angular momentum. The law of conservation of angular momentum is a fundamental law of nature. Even on the microscopic scale of atomic and nuclear physics, where Newtonian mechanics does not hold, the angular momentum of an isolated system is found to be constant over time.

Although conservation of angular momentum is an experimental law, independent of Newton's laws of motion, the fact that the internal torques of a system cancel is suggested by Newton's third law. Consider the two particles shown in Figure 10-22. Let $\vec{F}_{1,2}$ be the force exerted by particle 1 on particle 2 and $\vec{F}_{2,1}$ be

The Segway™ HT, a Dean Kamen invention, has been described as "the world's first self-balancing human transporter." When the person leans forward, the rotational motion of the device is detected by silicon "gyroscopes," which detect the rotational motion and then supply the internal control system with a signal. The control system interprets the signal and then activates motors that drive the wheels at just the right speed to maintain balance.

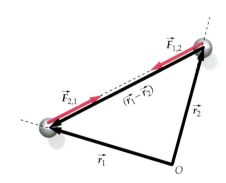

FIGURE 10-22

that exerted by particle 2 on particle 1. By Newton's third law, $\vec{F}_{2,1} = -\vec{F}_{1,2}$. The sum of the torques exerted by these forces about the origin O is

$$\vec{\tau}_1 + \vec{\tau}_2 = \vec{r}_1 \times \vec{F}_{2,1} + \vec{r}_2 \times \vec{F}_{1,2} = \vec{r}_1 \times \vec{F}_{2,1} + \vec{r}_2 \times (-\vec{F}_{2,1}) = (\vec{r}_1 - \vec{r}_2) \times \vec{F}_{2,1}$$

The vector $\vec{r}_1 - \vec{r}_2$ is along the line joining the two particles. If $\vec{F}_{2,1}$ acts parallel to the line joining m_1 and m_2, $\vec{F}_{2,1}$ and $\vec{r}_1 - \vec{r}_2$ are either parallel or antiparallel and

$$(\vec{r}_1 - \vec{r}_2) \times \vec{F}_{2,1} = 0$$

If this is true for all the internal forces, the internal torques cancel in pairs.

There are many examples of the conservation of angular momentum in everyday life. Figures 10-23 and 10-24 illustrate angular momentum conservation in diving and ice skating.

FIGURE 10-23 Multiflash photograph of a diver. The diver's center of mass moves along a parabolic path after he leaves the board. The angular momentum is provided by the initial external torque due to the force of the board, which does not pass through the diver's center of mass if he leans forward as he jumps. If the diver wanted to undergo one or more somersaults in the air, he would draw in his arms and legs, decreasing his moment of inertia to increase his angular velocity.

FIGURE 10-24 A spinning skater. Because the torque exerted by the ice is small, the angular momentum of the skater is approximately constant. When she reduces her moment of inertia by drawing in her arms, her angular velocity increases.

A ROTATING DISK   **EXAMPLE 10-3**

A disk is rotating with an initial angular velocity ω_i about a frictionless shaft through its symmetry axis as shown in Figure 10-25. Its moment of inertia about this axis is I_1. It drops onto another disk of moment of inertia I_2 that is initially at rest on the same shaft. Because of surface friction, the two disks eventually attain a common angular velocity ω_f. Find ω_f.

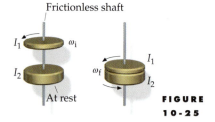

FIGURE 10-25

PICTURE THE PROBLEM We find the final angular velocity from the final angular momentum, which is equal to the initial angular momentum because there are no external torques acting on the two-disk system. Note that we do *not* use conservation of mechanical energy. The angular speed of the upper disk is reduced while that of the lower disk is increased by the forces of kinetic friction. Because kinetic friction dissipates mechanical energy, we expect that the total mechanical energy is decreased.

1. The final angular velocity is related to the initial angular velocity by conservation of angular momentum:

$$L_f = L_i$$
$$(I_1 + I_2)\omega_f = I_1 \omega_i$$

2. Solve for the final angular velocity:

$$\boxed{\omega_f = \frac{I_1}{I_1 + I_2} \omega_i}$$

PLAUSIBILITY CHECK If $I_2 << I_1$, the collision should have little effect on disk 1. Our results agree, and give $\omega_f \to \omega_i$. If $I_2 >> I_1$, then disk 1 should slow to a stop without causing disk 2 to rotate appreciably. Our results give $\omega_f \to 0$, as expected.

In the collision of the two disks in Example 10-3, mechanical energy is not conserved. We can see this by writing the energy in terms of the angular momentum. An object rotating with an angular velocity ω has kinetic energy

$$K = \frac{1}{2} I\omega^2 = \frac{(I\omega)^2}{2I}$$

Using $L = I\omega$ gives

$$K = \frac{L^2}{2I} \qquad\qquad\qquad 10\text{-}20$$

(Compare this result with that for linear motion $K = p^2/2m$, Equation 8-25.) The initial kinetic energy in Example 10-3 is

$$K_i = \frac{L_i^2}{2I_1}$$

and the final kinetic energy is

$$K_f = \frac{L_f^2}{2(I_1 + I_2)}$$

Since $L_f = L_i$, the ratio of the final to the initial kinetic energy is

$$\frac{K_f}{K_i} = \frac{I_1}{I_1 + I_2}$$

which is less than one. This interaction of the disks is analogous to a one-dimensional perfectly inelastic collision of two objects.

The rotating plates in the transmission of a truck make inelastic collisions when engaged.

MUD IN YOUR EYE   **EXAMPLE   10 - 4**   **Put It in Context**

Benny, a high school physics student, has been the schoolyard bully for many years, so four of his fellow students decide to teach him a lesson using conservation of angular momentum. Here is their plan. The schoolyard has a small merry-go-round (Figure 10-26), a 3-m diameter turntable with a 130-kg·m² moment of inertia, and Benny loves to ride it. Initially, they need to get all five students to stand on the merry-go-round next to the rim while the merry-go-round is rotating at a modest 20 rev/min. When the signal is given, the four students will quickly move to the center of the merry-go-round leaving Benny near the rim. The merry-go-round will speed up, throwing Benny off and into the mud. (They plan to do this after a heavy rain.) Benny is very quick and very strong, so throwing him off will require that the centripetal acceleration of the rim be at least 4g. Will this plan work? (Assume that each student has a mass of 60 kg.)

FIGURE 10-26

PICTURE THE PROBLEM By moving to the center of the merry-go-round, the four students are decreasing the moment of inertia of the students–merry-go-round system. No external torques act on the system, so the angular momentum about the axis remains constant. The angular momentum is the moment of inertia times the angular velocity, so a decrease in the moment of inertia means an increase in the angular velocity. The angular velocity can be used to find the centripetal acceleration at the rim.

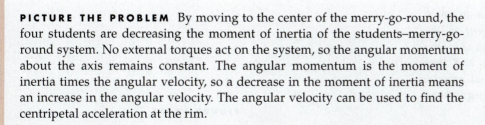

1. The centripetal acceleration depends on the angular speed ω and the radius R:

$$a_c = \omega^2 R$$

2. Angular momentum is conserved. For rotations about a fixed axis, $L = I\omega$:

$$L_f = L_i$$
$$I_f\omega_f = I_i\omega_i$$

3. The moment of inertia of the system is the sum of the moments of inertia of the students plus that of the merry-go-round. Each student has mass m:

$$I_i = 5 \times mR^2 + I_{mgr}$$
$$= 5(60 \text{ kg})(1.5 \text{ m})^2 + 130 \text{ kg} \cdot \text{m}^2$$
$$= 805 \text{ kg} \cdot \text{m}^2$$

4. To find the final moment of inertia, assume that the four students are 30 cm ($\approx$1 ft) from the center:

$$I_f = mR^2 + 4 \times mr^2 + I_{mgr}$$
$$= (60 \text{ kg})(1.5 \text{ m})^2 + 4(60 \text{ kg})(0.3 \text{ m})^2 + 130 \text{ kg} \cdot \text{m}^2$$
$$= 287 \text{ kg} \cdot \text{m}^2$$

5. Using conservation of angular momentum, solve for the final angular velocity:

$$\omega_f = \frac{I_i}{I_f}\omega_i = \frac{805 \text{ kg} \cdot \text{m}^2}{287 \text{ kg} \cdot \text{m}^2} 20 \text{ rev/min}$$
$$= 56.2 \text{ rev/min} = 5.88 \text{ rad/s}$$

6. Solve for the centripetal acceleration of the rim:

$$a_c = \omega^2 R = (5.88 \text{ rad/s})^2(1.5 \text{ m}) = 51.9 \text{ m/s}^2$$

7. Convert to gs:

$$a_c = 51.9 \text{ m/s}^2 \times \frac{1 g}{9.81 \text{ m/s}^2} = 5.29 g$$

8. Does Benny end up in the mud?

> Success! The acceleration is much greater than $4g$, so Benny flies off and lands in the mud.

REMARKS The linear speed of the rotating merry-go-round is greatest at the rim and decreases to zero at the center. At the rim the students are moving in a circle. As they walk toward the center they are stepping onto a part of the merry-go-round that is moving more slowly than they are, so they drag the merry-go-round in the direction of its motion, thus speeding it up. Also, the merry-go-round drags the students back, slowing their motion. The static frictional forces exerted by their feet exerts a net torque on the merry-go-round, increasing its angular momentum about the rotation axis. The merry-go-round exerts equal and opposite static frictional forces on the students' feet, so the torques associated with these forces decreases their angular momentum. The two torques are equal and opposite, as are the associated angular momentum changes. Thus the angular momentum of the students–merry-go-round system remains constant.

   The moment of inertia of the students–merry-go-round system decreases as the students walk toward the center. Thus, the system's moment of inertia decreases while its angular momentum remains constant. As a result, we can see from Equation 10-20 that the kinetic energy of the students–merry-go-round system increases. The energy for this kinetic energy increase comes from the internal energy of the students.

ANOTHER RIDE ON THE MERRY-GO-ROUND   **EXAMPLE 10-5** Try It Yourself

A 25-kg child in a playground runs with an initial speed of 2.5 m/s along a path *tangent* to the rim of a merry-go-round with a moment of inertia of 500 kg·m² that is initially at rest, and then jumps on (Figure 10-27). Find the final angular velocity of the child and the merry-go-round together.

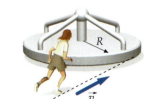

FIGURE 10-27

PICTURE THE PROBLEM Once the child's feet leave the ground, no external torques act on the child–merry-go-round system, hence the total angular momentum of the system about the rotation axis of the merry-go-round is conserved. The mass of the child is $m = 25$ kg, her initial speed is $v = 2.5$ m/s, the moment of inertia of the merry-go-round is $I = 500$ kg $\cdot$ m^2, and the radius of the merry-go-round is $R = 2.0$ m. The initial angular speed of the merry-go-round is $\omega_i = 0$.

Cover the column at the right and solve it yourself before looking at the answers.

Steps	Answers
1. Write an expression for the initial angular momentum of the child running about the axis of the merry-go-round.	$L_i = mvR$
2. Write an expression for the total final angular momentum of the child–merry-go-round system in terms of the final angular velocity ω_f.	$L_f = (mR^2 + I_m)\omega_f$
3. Set your expressions in steps 1 and 2 equal and solve for ω_f.	$\omega_f = \dfrac{mvR}{mR^2 + I_m} = \boxed{0.208 \text{ rad/s}}$

■ **EXERCISE** Calculate the initial and final kinetic energies of the child–merry-go-round system. (*Answer* $K_i = 78.2$ J, $K_f = 13.0$ J)

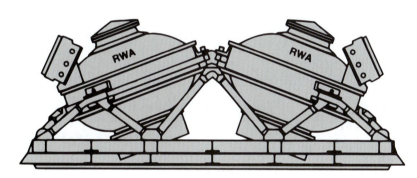

The Hubble Space Telescope is aimed by regulating the spin rates of 45-kg flywheels arranged off-axis from each other and spinning at up to 3000 rpm. Software-controlled changes in the spin rates create angular momentum that causes the satellite to slew into new positions. This aiming mechanism can achieve and hold a target to within 0.005 arcsec—equivalent to holding a flashlight beam in Los Angeles on a dime in San Francisco.

PULLING THROUGH A HOLE   **EXAMPLE 10-6**

A particle of mass m moves with speed v_0 in a circle of radius r_0 on a frictionless tabletop. The particle is attached to a string that passes through a hole in the table, as shown in Figure 10-28. The string is slowly pulled downward so that the particle moves in a smaller circle of radius r_f. (a) Find the final velocity in terms of r_0, v_0, and r_f. (b) Find the tension when the particle is moving in a circle of radius r in terms of m, r, and the angular momentum $\vec{L}$. (c) Calculate the work done on the particle by the tension $\vec{T}$ by integrating $\vec{T} \cdot d\vec{\ell}$. Express your answer in terms of r and L_0.

FIGURE 10-28

PICTURE THE PROBLEM The speed of the particle is related to its angular momentum. The net torque is equal to the rate of change of the angular momentum. Since the net force acting on the particle is the tension force $\vec{T}$ exerted by the string, which is always directed toward the hole, the torque about the axis through the hole is zero. Thus, the angular momentum about this axis remains constant.

An astronaut examines the flywheel of the Hubble Space Telescope.

(a) Conservation of angular momentum relates the final speed to the initial speed and the initial and final radii:

$$L_f = L_0$$

$$mv_f r_f = mv_0 r_0$$

so

$$v_f = \boxed{\dfrac{r_0}{r_f} v_0}$$

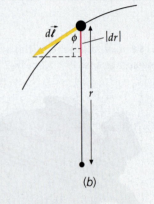

(a)

(b) 1. Apply Newton's second law to relate T to v and r:

$$T = m\dfrac{v^2}{r}$$

2. Obtain a relation between L, r, and v using the definition of angular momentum:

$$\vec{L} = \vec{r} \times \vec{p}$$

$$L = rmv \sin 90° = mvr$$

3. Eliminate v by solving the Part (b) step 2 result for v and then substituting into the Part (a) result:

$$T = m\dfrac{v^2}{r} = \dfrac{m}{r}\left(\dfrac{L}{mr}\right)^2 = \boxed{\dfrac{L^2}{mr^3}}$$

(c) 1. Make a drawing of the particle as it moves closer to the hole (Figure 10-29). When the particle undergoes displacement $d\vec{\ell}$, its distance r from the axis changes by dr. Since r is decreasing, dr is negative. Thus:

$$dr = -|dr|$$

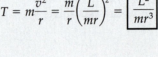

(b)

FIGURE 10-29

2. Using step 1, write $dW = \vec{T} \cdot d\vec{\ell}$ in terms of T and dr, with $T = L^2/mr^3$ from Part (b):

$$dW = \vec{T} \cdot d\vec{\ell} = T\, d\ell \cos \phi$$

Since

$$|dr| = d\ell \cos \phi$$

$$dW = T|dr| = -T\, dr$$

3. Integrate from r_0 to r_f after substituting for T from the Part (b) step 3 result:

$$W = -\int_{r_0}^{r_f} T\, dr = -\int_{r_0}^{r_f} \dfrac{L^2}{mr^3}\, dr$$

$$= -\dfrac{L^2}{m}\int_{r_0}^{r_f} r^{-3}\, dr = -\dfrac{L^2}{m}\dfrac{r^{-2}}{-2}\Big|_{r_0}^{r_f}$$

$$= \boxed{\dfrac{L^2}{2m}\left(\dfrac{1}{r_f^2} - \dfrac{1}{r_0^2}\right)}$$

PLAUSIBILITY CHECK Note that work must be done to pull the string downward. Since r_f is less than r_0, the work is positive. This work is converted into increased kinetic energy. We can calculate the change in kinetic energy of the particle directly. Using $K = L^2/2I$, with $L_0 = L_f = L$, and $I = mr^2$, the change in kinetic energy is $K_f - K_i = (L^2/2mr_f^2) - (L^2/2mr_0^2) = (L^2/2m)(r_f^{-2} - r_0^{-2})$, which is the same as the result found by direct integration.

REMARKS The increment of work dW can also be obtained by expressing the increment of displacement $d\vec{\ell}$ as $d\vec{r}$, the change in the position vector $\vec{r}$. The dot product $\vec{T} \cdot d\vec{r}$ is then expanded via components giving $dW = \vec{T} \cdot d\vec{r} = T_r\, dr = -T\, dr$. In this expansion $T_r = -T$ is the radial component of $\vec{T}$ and dr is the radial component of $d\vec{r}$.

EXERCISE At what radius r_N is the tension N times the tension at radius r_0? (*Answer* $r_N = r_0/\sqrt[3]{N}$)

In Figure 10-30, a puck on a frictionless plane is given an initial speed v_0. The puck is attached to a string that wraps around a vertical post. This situation looks similar to Example 10-6, but it is not the same. **There is no agent that can do work on the puck, nor is there any mechanism for energy dissipation.** Thus, mechanical energy must be conserved. Since $K = L^2/2I$ is constant and I decreases as r_0 decreases, L must also decrease. Note that the tension force does not act toward the axis of the post. The tension force on the puck produces a torque vector about the axis of the post in the downward direction, which reduces the angular momentum vector of the puck, which is in the upward direction.

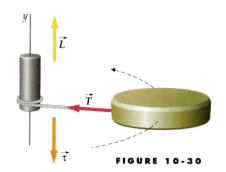

FIGURE 10-30

THE BALLISTIC PENDULUM REVISITED   **EXAMPLE 10-7   Try It Yourself**

A thin rod of mass M and length d is attached to a pivot at the top. A piece of clay of mass m and speed v hits the rod a distance x from the pivot and sticks to it (Figure 10-31). Find the ratio of the system's kinetic energy just after the collision to the kinetic energy just before the collision.

PICTURE THE PROBLEM The collision is inelastic, so we do not expect mechanical energy to be conserved. During the collision, the pivot exerts a large force on the rod, so linear momentum is also not conserved. However, there are no external torques about the pivot point of the clay–rod system, so angular momentum about the pivot point is conserved. The kinetic energy after the inelastic collision can be written in terms of the angular momentum L_{sys} and the moment of inertia I' of the combined clay–rod system. Conservation of angular momentum allows you to relate L_{sys} to the mass m and velocity v of the clay.

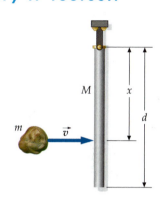

FIGURE 10-31

Cover the column on the right and try these on your own before looking at the answers.

Steps	Answers
1. Before the collision the kinetic energy of the system is that of the moving clay ball.	$K_i = \frac{1}{2}mv^2$
2. After the collision it is that of the swinging clay–rod object. Write the kinetic energy after the collision in terms of the angular momentum L_{sys} and the moment of inertia I' of the clay–rod object.	$K_f = \dfrac{L_{sys}^2}{2I'}$
3. During the collision angular momentum is conserved. Write the angular momentum L_{sys} in terms of m, v, and x.	$L_{sys} = mvx$
4. Write I' in terms of m, x, M, and d.	$I' = mx^2 + \frac{1}{3}Md^2$
5. Substitute these expressions for L_f and I' into your equation for K_f.	$K_f = \dfrac{L_{sys}^2}{2I'} = \dfrac{(mvx)^2}{2(mx^2 + \frac{1}{3}Md^2)}$ $= \dfrac{3}{2}\dfrac{m^2x^2v^2}{(3mx^2 + Md^2)}$
6. Divide the kinetic energy after the collision by the initial kinetic energy of the clay.	$\dfrac{K_f}{K_i} = \dfrac{\frac{3}{2}\dfrac{m^2x^2v^2}{(3mx^2 + Md^2)}}{\frac{1}{2}mv^2} = \boxed{\dfrac{1}{1 + \dfrac{Md^2}{3mx^2}}}$

REMARKS This example is the rotational analog of the ballistic pendulum discussed in Example 8-14. In that example we used conservation of linear momentum to find the energy of the pendulum after the collision.

Figure 10-32 shows a student demonstrating conservation of angular momentum about an axis. She is sitting on a turntable, holding a bicycle wheel. We choose the z axis to coincide with the axis of the turntable. Because the turntable bearings are frictionless, the vertical component of the torque exerted on the turntable–student–wheel system remains zero. Consequently, the vertical component of the total angular momentum of the system remains constant. Initially, the turntable is stationary and the bicycle wheel is spinning rapidly about its axis, which is initially horizontal, as shown in Figure 10-32a. The system's initial angular momentum $\vec{L}_{\text{sys i}}$ is just the initial spin angular momentum of the wheel $\vec{L}_{\text{w,s i}}$, which is horizontal. This means that the vertical component of $\vec{L}_{\text{sys i}}$ is zero. If the student now (Figure 10-32b) tips the axis of the spinning wheel upward, the wheel's spin angular momentum vector $\vec{L}_{\text{w,s}}$ tips upward with it. As she tips the wheel upward the system starts rotating clockwise (viewed from above) about the turntable axis. The angular momentum vector associated with this clockwise rotation of the system about the turntable axis has a z component $L_{\text{t}z}$ directed downward. Since the vertical component of $\vec{L}_{\text{sys}}$ remains zero, we can conclude that $L_{\text{t}z}\hat{k}$ is equal and opposite to $L_{\text{w,s}z}\hat{k}$.

Unless you have seen this demonstration before, the rotation of the turntable is unexpected. Let's examine what causes it. The student must exert an upward torque on the spinning wheel in order to tip its angular momentum vector upward. (Due to the cross product, an upward torque requires horizontal forces.) The wheel exerts a downward torque (also horizontal forces) of equal magnitude back on the student. It is this downward torque on the student that causes the system to start rotating clockwise about the turntable axis.

EXERCISE The bicycle wheel is spinning with its axis vertical and with its angular momentum vector upward when it is handed to the student on the turntable. In what direction will the turntable rotate when the student tips and then rotates the axis of the wheel toward the horizontal? (*Answer* Counterclockwise, if viewed from above)

Proofs of Equations 10-10, 10-12, 10-13, 10-14, and 10-15

Proof of Equation 10-10 We will now show that Newton's second law implies that the rate of change of the angular momentum of a particle equals the net torque acting on the particle. If more than one force acts on a particle, then the net torque relative to the origin O is the sum of the torques due to each force:

$$\vec{\tau}_{\text{net}} = \vec{r} \times \vec{F}_1 + \vec{r} \times \vec{F}_2 + \cdots = \vec{r} \times \sum_i \vec{F}_i = \vec{r} \times \vec{F}_{\text{net}}$$

According to Newton's second law, the net force equals the rate of change of the linear momentum $d\vec{p}/dt$. Thus

$$\vec{\tau}_{\text{net}} = \vec{r} \times F_{\text{net}} = \vec{r} \times \frac{d\vec{p}}{dt} \qquad\qquad 10\text{-}21$$

We now compare this with the expression for the time rate of change of the angular momentum. The definition of the angular momentum of a particle (Equation 10-8) is

$$\vec{L} = \vec{r} \times \vec{p}$$

We can compute $d\vec{L}/dt$ using the product rule for derivatives:

$$\frac{d\vec{L}}{dt} = \frac{d}{dt}(\vec{r} \times \vec{p}) = \left(\frac{d\vec{r}}{dt} \times \vec{p}\right) + \left(\vec{r} \times \frac{d\vec{p}}{dt}\right)$$

The second term from the right of this equation is zero because $\vec{p} = m\vec{v}$ and $\vec{v} = d\vec{r}/dt$, so

$$\frac{d\vec{r}}{dt} \times \vec{p} = \vec{v} \times m\vec{v} = 0$$

$$\vec{L}_{\text{sys i}} = \vec{L}_{\text{w,s i}}$$

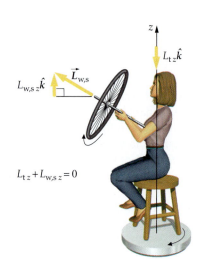

$$L_{\text{t}z} + L_{\text{w,s}z} = 0$$

FIGURE 10-32

Thus

$$\frac{d\vec{L}}{dt} = \vec{r} \times \frac{d\vec{p}}{dt}$$

Comparing with Equation 10-21 gives

$$\vec{\tau}_{net} = \frac{d\vec{L}}{dt} \qquad\qquad 10\text{-}22$$

The net torque acting on a system of particles is the sum of the individual torques. The generalization of Equation 10-22 to a system of particles is then

$$\sum_i \vec{\tau} = \sum_i \frac{d\vec{L}_i}{dt} = \frac{d}{dt} \sum_i \vec{L}_i = \frac{d\vec{L}_{sys}}{dt}$$

In this equation, the sum of the torques may include internal as well as external torques. The sum of the internal torques equals zero, so

$$\vec{\tau}_{net\ ext} = \frac{d\vec{L}_{sys}}{dt} \qquad\qquad 10\text{-}10$$

NEWTON'S SECOND LAW FOR ANGULAR MOTION

*Proofs of Equations 10-12 and 10-13   We will now show that the angular momentum of a system of particles can be written as the sum of the orbital angular momentum and the spin angular momentum.

Figure 10-33 shows a system of particles. The angular momentum $\vec{L}_i$ of the ith particle about arbitrary point O is given by

$$\vec{L}_i = \vec{r}_i \times \vec{p}_i = \vec{r}_i \times m_i\vec{v}_i \qquad\qquad 10\text{-}23$$

and the angular moment of the system about O is

$$\vec{L} = \Sigma\vec{L}_i = \Sigma(\vec{r}_i \times m_i\vec{v}_i)$$

The angular momentum about the center of mass is given by

$$\vec{L}_{cm} = \Sigma(\vec{r}_i' \times m_i\vec{u}_i)$$

where $\vec{r}_i'$ and $\vec{u}_i$ are the position and velocity of the ith particle relative to the center of mass. It can be seen from the figure that

$$\vec{r}_i = \vec{r}_{cm} + \vec{r}_i'$$

Differentiating both sides gives

$$\vec{v}_i = \vec{v}_{cm} + \vec{u}_i$$

Substituting these into Equation 10-23, we have

$$\vec{L}_i = \vec{r}_i \times m_i\vec{v}_i = (\vec{r}_{cm} + \vec{r}_i') \times m_i(\vec{v}_{cm} + \vec{u}_i)$$

Expanding the right side, we obtain

$$\vec{L}_i = (\vec{r}_{cm} \times m_i\vec{v}_{cm}) + (\vec{r}_{cm} \times m_i\vec{u}_i) + (m_i\vec{r}_i' \times \vec{v}_{cm}) + (\vec{r}_i' \times m_i\vec{u}_i)$$

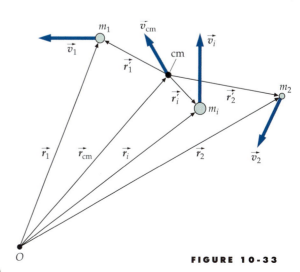

FIGURE 10-33

Summing both sides and factoring common terms out of the sums gives

$$\vec{L}_{sys} = \Sigma\vec{L}_i = \left(\vec{r}_{cm} \times \Sigma(m_i)\,\vec{v}_{cm}\right) + \left(\vec{r}_{cm} \times \Sigma(m_i\vec{u}_i)\right) + \left(\Sigma(m_i\vec{r}_i') \times \vec{v}_{cm}\right) + \left(\Sigma(\vec{r}_i' \times m_i\vec{u}_i)\right)$$

Because $\Sigma(m_i\vec{r}_i')$ and $\Sigma(m_i\vec{u}_i)$ are both zero, and because $\Sigma m_i = M$ and $\Sigma(\vec{r}_i' \times m_i\vec{u}_i) = \vec{L}_{cm}$, we have $\vec{L}_{sys} = \vec{r}_{cm} \times M\vec{v}_{cm} + \vec{L}_{cm}$, or

$$\vec{L}_{sys} = \vec{L}_{orbit} + \vec{L}_{spin} \qquad\qquad 10\text{-}12$$

where $\vec{L}_{spin} = \vec{L}_{cm}$ and $\vec{L}_{orbit} = \vec{v}_{cm} \times M\vec{v}_{cm}$. $\qquad$ 10-13

*Proofs of Equations 10-14 and 10-15 We will now take the z components of the vectors for the torque and the angular momentum about a point to obtain the formulas for the torque and angular momentum about a fixed axis. The angular momentum of a particle about the origin is $\vec{L} = \vec{r} \times \vec{p}$, so finding the z component of the angular momentum means finding the z component of the product $\vec{r} \times \vec{p}$. To do this we express $\vec{r}$ and $\vec{p}$ as

$$\vec{r} = \vec{r}_{rad} + \vec{r}_z \qquad \text{and} \qquad \vec{p} = \vec{p}_{xy} + \vec{p}_z$$

where $\vec{r}_{rad}, \vec{r}_z, \vec{p}_{xy}$, and $\vec{p}_z$ are vector components (see Figure 10-34) of $\vec{r}$ and $\vec{p}$. Substituting $\vec{r}$ and $\vec{p}$ gives

$$\vec{L} = \vec{r} \times \vec{p} = (\vec{r}_{rad} + \vec{r}_z) \times (\vec{p}_{xy} + \vec{p}_z)$$

and expanding the right side, we have

$$\vec{L} = (\vec{r}_{rad} \times \vec{p}_{xy}) + (\vec{r}_{rad} \times \vec{p}_z) + (\vec{r}_z \times \vec{p}_{xy}) + (\vec{r}_z \times \vec{p}_z)$$

The cross product of any two vectors is perpendicular to both vectors, so the product $\vec{r}_{rad} \times \vec{p}_{xy}$ is parallel to the z axis. In each of the other three products at least one of the two vectors is parallel to the z axis, so the z component of each of these cross products is zero. Therefore,

$$\vec{L}_z = \vec{r}_{rad} \times \vec{p}_{xy} \qquad\qquad 10\text{-}14$$

ANGULAR MOMENTUM ABOUT Z AXIS

The torque about the origin associated with a force acting on the particle is given by $\vec{\tau} = \vec{r} \times \vec{F}$ (Equation 10-1). Following the same procedure with the torque that we followed with the angular momentum gives

$$\vec{\tau}_z = \vec{r}_{rad} \times \vec{F}_{xy} \qquad\qquad 10\text{-}15$$

TORQUE ABOUT Z AXIS

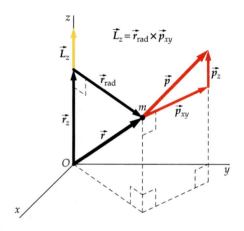

FIGURE 10-34 The vector components $\vec{r}_{rad}, \vec{r}_z, \vec{p}_{xy}$, and $\vec{p}_z$ of $\vec{r}$ and $\vec{p}$ that are used for calculating the angular momentum about the z axis $\vec{L}_z$.

*10-4 Quantization of Angular Momentum

Angular momentum plays an important role in the description of atoms, molecules, nuclei, and elementary particles. Like energy, angular momentum is **quantized,** that is, changes in angular momentum occur only in discrete amounts.

The angular momentum of a particle due to its motion is its orbital angular momentum. The magnitude of the orbital angular momentum L of a particle can have only the values

$$L = \sqrt{\ell(\ell + 1)}\hbar, \qquad \ell = 0, 1, 2, \ldots \qquad\qquad 10\text{-}24$$

where $\hbar$ (read "h-bar") is the **fundamental unit of angular momentum,** which is related to Planck's constant h:

$$\hbar = \frac{h}{2\pi} = 1.05 \times 10^{-34}\,\text{J}\cdot\text{s} \qquad\qquad 10\text{-}25$$

The component of orbital angular momentum along any line in space is also quantized and can have only the values $\pm m\hbar$, where m is a nonnegative integer that is less than or equal to ℓ. For example, if $\ell = 2$, m can equal 2, 1, or 0.

Because the quantum of angular momentum $\hbar$ is so small, the quantization of angular momentum is not noticed in the macroscopic world. Consider a particle of mass $1\,\text{g} = 10^{-3}\,\text{kg}$ moving in a circle of radius 1 cm with a period of 1 s. Its orbital angular momentum is

$$L = mvr = mr^2\omega = mr^2\frac{2\pi}{T} = (10^{-3}\,\text{kg})(10^{-2}\,\text{m})^2\frac{2\pi}{1\,\text{s}} = 6.28 \times 10^{-7}\,\text{J}\cdot\text{s}$$

If we divide by $\hbar$, we obtain

$$\frac{L}{\hbar} = \frac{6.28 \times 10^{-7}\,\text{J}\cdot\text{s}}{1.05 \times 10^{-34}\,\text{J}\cdot\text{s}} = 6 \times 10^{27}$$

Thus this typical macroscopic angular momentum contains 6×10^{27} units of the fundamental unit of angular momentum. Even if we could measure L to one part in a billion, we would never notice the quantization of this macroscopic angular momentum.

The quantization of orbital angular momentum leads to the quantization of rotational kinetic energy. Consider a molecule rotating about its center of mass with angular momentum L (Figure 10-35). Let I be its moment of inertia. Its kinetic energy is

$$K = \frac{L^2}{2I} \qquad\qquad 10\text{-}26$$

But L^2 is quantized to the values $L^2 = \ell(\ell + 1)\hbar^2$ with $\ell = 0, 1, 2, \ldots$. Thus, the kinetic energy is quantized to the values K_ℓ given by

$$K_\ell = \frac{L^2}{2I} = \frac{\ell(\ell + 1)\hbar^2}{2I} = \ell(\ell + 1)E_{0r} \qquad\qquad 10\text{-}27a$$

where

$$E_{0r} = \frac{\hbar^2}{2I} \qquad\qquad 10\text{-}27b$$

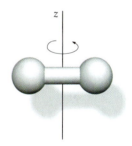

FIGURE 10-35 Model of a rigid diatomic molecule rotating about the z axis.

Figure 10-36 shows an energy-level diagram for a rotating molecule with constant moment of inertia I. Note that, unlike the energy levels for a vibrating system (Section 7-4), the rotational energy levels are not equally spaced, and the lowest level is zero.

Stable matter contains just three kinds of particles: electrons, protons, and neutrons. In addition to its orbital angular momentum each of these particles also has an intrinsic angular momentum called its **spin.** The spin angular momentum of a particle, like its mass and electric charge, is a fundamental property of the particle that cannot be changed. The magnitude of the spin angular momentum vector for electrons and other fermions is $s = \sqrt{\frac{1}{2}(\frac{1}{2} + 1)}\,\hbar$ and the component along any line in space can have just two values: $+\frac{1}{2}\hbar$ and $-\frac{1}{2}\hbar$. Such particles are called "spin one-half" particles. Electrons, protons, neutrons, and other spin-one-half particles are called **fermions.** Other particles such as photons and α particles, called

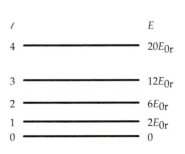

FIGURE 10-36 Energy-level diagram for a rotating molecule.

bosons, have zero spin or integral spin. Spin is a quantum property of the particle that has nothing to do with the motion of the particle.

The picture of an electron as a spinning ball that orbits the nucleus in an atom (like the spinning earth orbiting the sun) is often a useful visualization. However, the angular momentum of a spinning ball can be increased or decreased, whereas the spin of the electron is a fixed property like its charge and mass that does not change. Furthermore, as far as we know, electrons are point particles that have no size.

SUMMARY

1. Angular momentum is an important derived dynamic quantity in macroscopic physics. In microscopic physics, angular momentum is an intrinsic, fundamental property of elementary particles.
2. Conservation of angular momentum is a fundamental law of nature.
3. Quantization of angular momentum is a fundamental law of nature.

Topic	Relevant Equations and Remarks	
1. Vector Nature of Rotation	When the direction of the axis of rotation is not fixed in space, plus and minus signs are inadequate to describe the direction of the angular velocity direction.	
Angular velocity $\vec{\omega}$	The direction of the angular velocity $\vec{\omega}$ is along the axis of rotation in the sense given by the right-hand rule.	
Torque $\vec{\tau}$	$\vec{\tau} = \vec{r} \times \vec{F}$	10-1
2. Vector Product	$\vec{A} \times \vec{B} = AB \sin \phi \, \hat{n}$	10-2
	where ϕ is the angle between the vectors and $\hat{n}$ is a unit vector perpendicular to the plane of $\vec{A}$ and $\vec{B}$ in the sense given by the right-hand rule as $\vec{A}$ is rotated into $\vec{B}$.	
Properties	$\vec{A} \times \vec{A} = 0$	10-3
	$\vec{A} \times \vec{B} = -\vec{B} \times \vec{A}$	10-4
	$\dfrac{d}{dt}(\vec{A} \times \vec{B}) = \left(\vec{A} \times \dfrac{d\vec{B}}{dt}\right) + \left(\dfrac{d\vec{A}}{dt} \times \vec{B}\right)$	10-6
	$\hat{i} \times \hat{j} = \hat{k} \quad \hat{j} \times \hat{k} = \hat{i} \quad \hat{k} \times \hat{i} = \hat{j}$	10-7a
3. Angular Momentum		
For a particle	$\vec{L} = \vec{r} \times \vec{p}$	10-8
For a system rotating about a symmetry axis	$\vec{L} = I\vec{\omega}$	10-9
For any system	The angular momentum about any point O is the angular momentum about the center of mass (spin angular momentum) plus the angular momentum associated with center-of-mass motion about O (orbital angular momentum).	
	$\vec{L} = \vec{L}_{\text{orbit}} + \vec{L}_{\text{spin}} = \left(\vec{r}_{\text{cm}} \times M\vec{v}_{\text{cm}}\right) + \left(\displaystyle\sum_i \vec{r}_i \times m_i\vec{u}_i\right)$	10-12
Newton's second law for angular motion	$\vec{\tau}_{\text{net ext}} = \dfrac{d\vec{L}}{dt}$	10-10

Conservation of angular momentum	If the net external torque is zero, the angular momentum of the system is conserved. (If the component of the net external torque in a given direction is zero, the component of the angular momentum of the system in that direction is conserved.)	
Kinetic energy of a rotating object	$K = \dfrac{L^2}{2I}$	**10-20**
Quantization of angular momentum	The magnitude of the orbital angular momentum of a particle can have only the values $$L = \sqrt{\ell(\ell + 1)}\hbar, \qquad \ell = 0, 1, 2, \ldots$$ where	
Fundamental unit of angular momentum	$\hbar = \dfrac{h}{2\pi} = 1.05 \times 10^{-34}\,\text{J·s}$ is the fundamental unit of angular momentum and h is Planck's constant.	**10-25**
*Quantization of any component of orbital angular momentum	The component of orbital angular momentum along any line in space is also quantized and can have only the values $\pm m\,\hbar$, where m is a nonnegative integer that is less than or equal to ℓ.	
Spin	Electrons, protons, and neutrons have an intrinsic angular momentum called spin. The magnitude of the spin angular momentum vector for these particles is $$s = \sqrt{\tfrac{1}{2}(\tfrac{1}{2} + 1)}\,\hbar$$ and the component along any line in space can have just two values, $+\tfrac{1}{2}\hbar$ and $-\tfrac{1}{2}\hbar$.	

PROBLEMS

- • Single-concept, single-step, relatively easy
- •• Intermediate-level, may require synthesis of concepts
- ••• Challenging, for advanced students
- **SSM** Solution is in the *Student Solutions Manual*
- **iSOLVE** Problems available on iSOLVE online homework service
- **iSOLVE ✓** These "Checkpoint" online homework service problems ask students additional questions about their confidence level, and how they arrived at their answer

In a few problems, you are given more data than you actually need; in a few other problems, you are required to supply data from your general knowledge, outside sources, or informed estimates.

Take g = 9.81 N/kg = 9.81 m/s² and neglect friction in all problems unless otherwise stated.

Conceptual Problems

1 • **SSM** True or false:

(a) If two vectors are parallel, their cross product must be zero.

(b) When a disk rotates about its symmetry axis, $\vec{\omega}$ is along the axis.

(c) The torque exerted by a force is always perpendicular to the force.

2 • Consider two nonzero vectors $\vec{A}$ and $\vec{B}$. Their cross product has the greatest magnitude if $\vec{A}$ and $\vec{B}$ are (a) parallel, (b) equal in magnitude, (c) perpendicular, (d) antiparallel, (e) at an angle of 45° to each other.

3 • What is the angle between a particle's linear momentum $\vec{p}$ and its angular momentum $\vec{L}$?

4 • A particle of mass m is moving with speed v along a line that passes through point P. What is the angular momentum of the particle about point P? (a) mv. (b) Zero. (c) It changes sign as the particle passes through point P. (d) It depends on the distance of point P from the origin of the coordinates.

5 •• [SSM] A particle travels in a circular path and point P is at the center of the circle. (*a*) If its linear momentum $\vec{p}$ is doubled, how is its angular momentum about P affected? (*b*) If the radius of the circle is doubled but the speed is unchanged, how is the angular momentum of the particle about P affected?

6 •• A particle moves along a straight line at constant speed. How does its angular momentum about any fixed point vary over time?

7 • True or false: If the net torque on a rotating system is zero, the angular velocity of the system cannot change.

8 •• [SSM] Standing on a turntable that is initially not rotating, can you rotate yourself through 180°? Assume that no external torques act on the you–turntable system. *Hint: While you cannot change your mass easily, there are ways to change your moment of inertia.*

9 • If the angular momentum of a system is constant, which of the following statements must be true?

(*a*) No torque acts on any part of the system.
(*b*) A constant torque acts on each part of the system.
(*c*) Zero net torque acts on each part of the system.
(*d*) A constant external torque acts on the system.
(*e*) Zero net external torque acts on the system.

10 •• In Example 10-4, does the force exerted by the merry-go-round on Benny do work?

11 •• Is it easier to crawl radially outward or radially inward on a rotating merry-go-round? Why?

12 •• [SSM] A block sliding on a frictionless table is attached to a string that passes through a hole in the table. Initially, the block is sliding with speed v_0 in a circle of radius r_0. A student under the table pulls slowly on the string. What happens as the block spirals inward? Give supporting arguments for your choice. (*a*) Its energy and angular momentum are conserved. (*b*) Its angular momentum is conserved and its energy increases. (*c*) Its angular momentum is conserved and its energy decreases. (*d*) Its energy is conserved and its angular momentum increases. (*e*) Its energy is conserved and its angular momentum decreases.

13 •• [SSM] How can you tell a hardboiled egg from an uncooked one without breaking it? One way is to lay the egg flat on a hard surface and try to spin it. A hardboiled egg will spin easily, while it takes a lot of effort to make an uncooked egg spin. However, once spinning, the uncooked egg will do something unusual: If you stop it with your finger, it may start spinning again. Explain the difference in the behavior of the two types of eggs.

14 • True or false:

When a gyroscope is not spinning, $\vec{\tau} = \dfrac{d\vec{L}}{dt}$ does not hold.

15 • [iSOLVE] An object of mass M is rotating about a fixed axis with angular momentum L. Its moment of inertia about this axis is I. What is its kinetic energy? (*a*) $IL^2/2$. (*b*) $L^2/2I$. (*c*) $ML^2/2$. (*d*) $IL^2/2M$.

16 • Explain why a helicopter with just one main rotor has a second smaller rotor mounted on a horizontal axis at the rear as in Figure 10-37. Describe the resultant motion of the helicopter if this rear rotor fails during flight.

FIGURE 10-37 Problem 16

17 •• The angular momentum vector for a spinning wheel lies along its axle and is pointed east. To make this vector point south, it is necessary to exert a force on the east end of the axle in which direction? (*a*) Up. (*b*) Down. (*c*) North. (*d*) South. (*e*) East.

18 •• [SSM] You are walking north and with your left hand you are carrying a suitcase that contains a spinning gyroscope mounted on an axle attached to the front and back of the case. The angular velocity of the gyroscope points north. You now begin to turn to walk east. As a result, the front end of the suitcase will (*a*) resist your attempt to turn and will try to remain pointed north, (*b*) fight your attempt to turn and will pull to the west, (*c*) rise upward, (*d*) dip downward, (*e*) show no effect whatsoever.

19 •• The angular momentum of the propeller of a small single-engine airplane points forward. The propeller rotates clockwise if viewed from behind. (*a*) As the plane takes off, the nose lifts and the airplane tends to veer to one side. To which side does it veer and why? (*b*) If the plane is flying horizontally and suddenly turns to the right, does the nose of the plane tend to move up or down? Why?

20 •• A car is powered by the energy stored in a single flywheel with an angular momentum $\vec{L}$. Discuss the problems that would arise for various orientations of $\vec{L}$ and various maneuvers of the car. For example, what would happen if $\vec{L}$ points vertically upward and the car travels over a hilltop or through a valley? What would happen if $\vec{L}$ points forward or to one side and the car attempts to turn to the left or right? In each case that you examine, consider the direction of the torque exerted on the car by the road.

21 •• [iSOLVE] You sit on a spinning piano stool with your arms folded. When you extend your arms out to the side, your kinetic energy (*a*) increases, (*b*) decreases, (*c*) remains the same.

22 •• [SSM] In tetherball, a ball is attached to a string that is attached to a pole. When the ball is hit, the string wraps around the pole and the ball spirals inward. Neglecting air resistance, what happens as the ball swings around the pole? Give supporting arguments for your choice. (*a*) The mechanical energy and angular momentum of the ball are conserved. (*b*) The angular momentum of the ball is conserved, but the mechanical energy of the ball increases. (*c*) The angular momentum of the ball is conserved, and the mechanical energy of the ball decreases. (*d*) The mechanical energy of the ball is conserved and the angular momentum of the ball increases. (*e*) The mechanical energy of the ball is conserved and the angular momentum of the ball decreases.

23 •• A uniform rod of mass M and length L lies on a horizontal frictionless table. A piece of putty of mass $m = M/4$ moves along a line perpendicular to the rod, strikes the rod near its end, and sticks to the rod. Describe qualitatively the subsequent motion of the rod and putty.

24 • **SSM** True or false:

(a) The rate of change of a system's angular momentum is always parallel to the net external torque.
(b) If the net torque on a body is zero, the angular momentum must be zero.

Estimation and Approximation

25 •• **SSM** **iSOLVE** ✓ An ice skater starts her pirouette with arms outstretched, rotating at 1.5 rev/s. Estimate her rotational speed (in revolutions per second) when she brings her arms flat against her body.

26 •• **iSOLVE** The polar ice caps contain about 2.3×10^{19} kg of ice. This mass contributes negligibly to the moment of inertia of the earth because it is located at the poles, close to the axis of rotation. Estimate the change in the length of the day that would be expected if the polar ice caps were to melt and the water were distributed uniformly over the surface of the earth. (The moment of inertia of a spherical shell of mass m and radius r is $2mr^2/3$.)

27 • A 2-g particle moves at a constant speed of 3 mm/s around a circle of radius 4 mm. (a) Find the magnitude of the angular momentum of the particle. (b) If $L = \sqrt{\ell(\ell + 1)}\hbar$, where ℓ is an integer, find the value of $\ell(\ell + 1)$ and the approximate value of ℓ. (c) Explain why the quantization of angular momentum is not noticed in macroscopic physics.

28 •• **SSM** One problem in astrophysics in the 1960s was explaining pulsars—extremely regular astronomical sources of radio pulses whose periods ranged from seconds to milliseconds. At one point, these radio sources were given the acronym LGM, standing for "Little Green Men," a reference to the idea that they might be signals of extraterrestrial civilizations. The explanation given today is no less interesting: The sun, which is a fairly typical star, has a mass of 1.99×10^{30} kg and a radius of 6.96×10^8 m. While it doesn't rotate uniformly, because it isn't a solid body, its average rate of rotation can be taken as about 1 rev/25 d. Stars somewhat larger than the sun can end their life in spectacular explosions—supernovae—leaving behind a collapsed remnant of the star called a neutron star. These neutron-star remnants have masses comparable to the original mass of the star, but radii of only a few kilometers! The high rotation rate is due to the conservation of angular momentum during the collapse. These stars emit beams of radio waves. Because of the rapid angular speed of the stars, the beam sweeps past the earth at regular intervals. To produce the observed radio-wave pulses, the star has to rotate at rates from about 1 rev/s to 1 rev/ms. (a) Using data from the textbook, estimate the rotation rate of the sun if it were to collapse into a neutron star of radius 10 km. Because the sun is not a uniform sphere of gas, its moment of inertia is given by the formula $I = 0.059MR^2$. Assume that the neutron star is spherical and has a uniform mass distribution. (b) Is the rotational kinetic energy of the sun greater or smaller after the collapse? By what factor does it change, and where does the energy go to or come from?

29 •• The moment of inertia of the earth about its spin axis is approximately 8.03×10^{37} kg·m². (a) Since the earth is nearly spherical, assume that the moment of inertia can be written as $I = CMR^2$, where C is a dimensionless constant, $M = 5.98 \times 10^{24}$ kg is the mass of the earth, and $R = 6370$ km is its radius. Determine C. (b) If the earth's mass were distributed uniformly, C would equal $2/5$. From the value of C calculated in Part (a), is the earth's mass density greater near the core or near the crust?

30 •• **SSM** Estimate the angular velocity and angular momentum of the diver in Figure 10-23 (page 318) about his center of mass. Make any approximations that you think reasonable.

31 •• Estimate the angular velocity of the diver in Figure 10-23 (page 318) if he curled himself into a ball in middive.

32 •• **SSM** Estimate Timothy Goebel's initial takeoff speed, rotational velocity, and angular momentum when he performs a quadruple Lutz (Figure 10-38). Make any assumptions you think reasonable, but be prepared to justify them. Goebel's mass is about 60 kg and the height of the jump is about 0.6 m. Note that the angular velocity will change quite a bit during the jump, as he begins with arms outstretched and pulls them in. Your answer should be accurate to within a factor of 2 if you're careful.

FIGURE 10-38
Problem 32

Vector Nature of Rotation

33 • A force of magnitude F is applied horizontally in the negative x direction to the rim of a disk of radius R as shown in Figure 10-39. Write $\vec{F}$ and $\vec{r}$ in terms of the unit vectors $\hat{i}$, $\hat{j}$, and $\hat{k}$, and compute the torque produced by the force about the origin at the center of the disk.

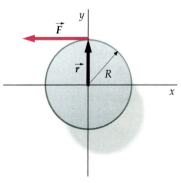

FIGURE 10-39 **Problem 33**

34 • Compute the torque about the origin for the force $\vec{F} = -mg\hat{j}$ acting on a particle at $\vec{r} = x\hat{i} + y\hat{j}$ and show that this torque is independent of the y coordinate.

35 • **ISOLVE** ✓ Find $\vec{A} \times \vec{B}$ for (a) $\vec{A} = 4\hat{i}$ and $\vec{B} = 6\hat{i} + 6\hat{j}$, (b) $\vec{A} = 4\hat{i}$ and $\vec{B} = 6\hat{i} + 6\hat{k}$, and (c) $\vec{A} = 2\hat{i} + 3\hat{j}$ and $\vec{B} = 3\hat{i} + 2\hat{j}$.

36 • **SSM** Under what conditions is the magnitude of $\vec{A} \times \vec{B}$ equal to $\vec{A} \cdot \vec{B}$?

37 •• A particle moves in a circle that is centered at the origin. The particle has position $\vec{r}$ and angular velocity $\vec{\omega}$. (a) Show that its velocity is $\vec{v} = \vec{\omega} \times \vec{r}$. (b) Show that its centripetal acceleration is $\vec{a}_c = \vec{\omega} \times \vec{v} = \vec{\omega} \times (\vec{\omega} \times \vec{r})$.

38 •• **ISOLVE** ✓ If $\vec{A} = 4\hat{i}$, $B_z = 0$, $|\vec{B}| = 5$, and $\vec{A} \times \vec{B} = 12\hat{k}$, determine $\vec{B}$.

39 • If $\vec{A} = 3\hat{j}$, $\vec{A} \times \vec{B} = 9\hat{i}$, and $\vec{A} \cdot \vec{B} = 12$, find $\vec{B}$.

40 •• Use the rules for evaluating a determinant to show that if

$$\vec{A} = a_x\hat{i} + a_y\hat{j} + a_z\hat{k}$$

$$\vec{B} = b_x\hat{i} + b_y\hat{j} + b_z\hat{k}$$

$$\vec{C} = c_x\hat{i} + c_y\hat{j} + c_z\hat{k}$$

then

$$\vec{A} \cdot (\vec{B} \times \vec{C}) = \begin{vmatrix} a_x & a_y & a_z \\ b_x & b_y & b_z \\ c_x & c_y & c_z \end{vmatrix} = \vec{C} \cdot (\vec{A} \times \vec{B}) = \vec{B} \cdot (\vec{C} \times \vec{A})$$

41 •• Given three noncoplanar vectors $\vec{A}$, $\vec{B}$, and $\vec{C}$, show that $\vec{A} \cdot (\vec{B} \times \vec{C})$ is the volume of the parallelopiped formed by the three vectors.

42 •• **SSM** Using the cross product, prove the *law of sines* for the triangle shown in Figure 10-40: if A, B, and C are the lengths of each side of the triangle, show that $A/\sin a = B/\sin b = C/\sin c$.

FIGURE 10-40
Problem 42

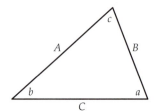

Angular Momentum

43 • A particle moving at constant velocity has zero angular momentum about a particular point. Show that the particle is moving either directly toward the point, directly away from the point, or through the point.

44 • **ISOLVE** A 2-kg particle moves at a constant speed of 3.5 m/s around a circle of radius 4 m. (a) What is its angular momentum about the center of the circle? (b) What is its moment of inertia about an axis through the center of the circle

and perpendicular to the plane of the motion? (c) What is the angular speed of the particle?

45 • **ISOLVE** ✓ A 2-kg particle moves at constant speed of 4.5 m/s along a straight line. (a) What is the magnitude of its angular momentum about a point 6 m from the line? (b) Describe qualitatively how its angular speed about that point varies with time.

46 •• **SSM** A particle is traveling with a constant velocity $\vec{v}$ along a line that is a distance b from the origin O (Figure 10-41). Let dA be the area swept out by the position vector from O to the particle in time dt. Show that dA/dt is constant and is equal to $\frac{1}{2}L/m$, where L is the angular momentum of the particle about the origin.

FIGURE 10-41 Problem 46

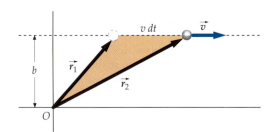

47 •• **ISOLVE** A 15-g coin of diameter 1.5 cm is spinning at 10 rev/s about a vertical diameter at a fixed point on a tabletop. (a) What is the angular momentum of the coin about its center of mass? (A coin is a cylinder of length L and radius R, where L is negligible compared to R. You can obtain its moment of inertia about an axis through a diameter in Table 9-1.) (b) What is its angular momentum about a point on the table 10 cm from the coin? If the coin spins about a vertical diameter at 10 rev/s while its center of mass travels in a straight line across the tabletop at 5 cm/s, (c) what is the angular momentum of the coin about a point on the line of motion of the center of mass? (d) What is the angular momentum of the coin about a point 10 cm to either side of the line of motion of the center of mass? (There are two answers to this question. Explain why and give both.)

48 •• Two particles of masses m_1 and m_2 are located at $\vec{r}_1$ and $\vec{r}_2$ relative to some origin O as in Figure 10-42. They exert equal and opposite forces on each other. Calculate the resultant torque exerted by these internal forces about the origin O and show that it is zero if the forces $\vec{F}_1$ and $\vec{F}_2$ lie along the line joining the particles.

FIGURE 10-42
Problem 48

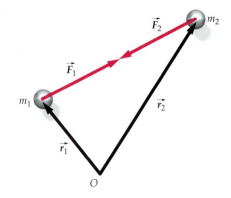

Special Relativity

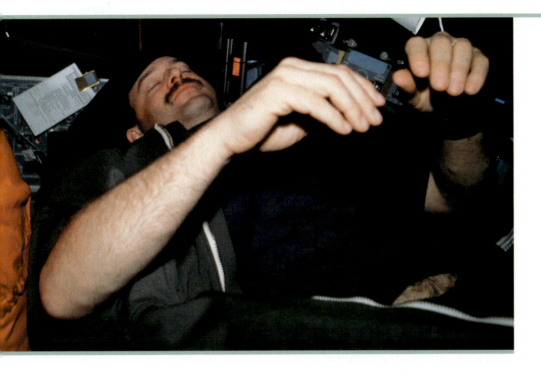

STS109-335-020 (1-12 MARCH 2002)—
ASTRONAUT SCOTT D. ALTMAN,
STS-109 MISSION COMMANDER,
SLEEPS ON THE FLIGHT DECK OF THE
SPACE SHUTTLE COLUMBIA. THE
ORBITAL SPEED OF THE SHUTTLE IS
ABOUT 7.7 KM/S (5 MI/S). THIS IS A
SMALL FRACTION OF THE SPEED OF
LIGHT, WHICH IS 3×10^5 KM/S.

? **If an astronaut on a spaceship that is traveling at 0.6c relative to the earth takes a one-hour-long nap, does that nap take one hour according to observers on earth? (See Example R-1.)**

The theory of relativity consists of two rather different theories, the special theory and the general theory. The special theory, developed by Albert Einstein and others in 1905, concerns the comparison of measurements made in different inertial reference frames moving with constant velocity relative to one another. Its consequences, which can be derived with a minimum of mathematics, are applicable in a wide variety of situations encountered in physics and engineering. On the other hand, the general theory, also developed by Einstein and others around 1916, is concerned with accelerated reference frames and gravity. A thorough understanding of the general theory requires sophisticated mathematics, and the applications of this theory are chiefly in the area of gravitation. ➤ **In this chapter we concentrate on the special theory of relativity (often referred to as *special relativity*). In the early 1900s the special theory of relativity was accepted with enthusiasm by some, but by many it was either reluctantly accepted or dismissed as folly. Today it is not only widely accepted, but embraced as a window into the workings of nature. We will see how this theory challenges our everyday experience of time and distance, as we describe the slowing down of moving clocks, the shortening of moving sticks, the relativity of** simultaneity for events that occur in different locations, and the relativity of momentum and energy relation.

R-1 The Principle of Relativity and the Constancy of the Speed of Light

The principle of relativity can be stated as follows:

> It is impossible to devise an experiment that determines whether you are at rest or moving uniformly.

POSTULATE I, THE PRINCIPLE OF RELATIVITY

Moving uniformly means moving at constant velocity relative to an inertial reference frame. For example, suppose that you are in your seat on board a high-speed airplane moving uniformly relative to the surface of the earth. If you drop your fork, it will fall to the floor in exactly the same way that it would if the plane were parked on a runway. When the airplane is in flight, you can consider yourself and the airplane to be at rest and the surface of the earth below you to be moving. There is nothing to distinguish whether you and the plane are moving and the surface of the earth is at rest, or vice versa.

Any reference frame in which a particle with no forces acting on it moves with constant velocity is, by definition, an inertial reference frame.[†] The surface of the earth is, to a good approximation, an inertial reference frame. The airplane is also an inertial reference frame as long as it moves with constant velocity relative to the surface of the earth. As long as you remain seated or standing still on the airplane you can consider yourself and the airplane to be at rest and the surface of the earth to be moving, or you can consider the surface of the earth to be at rest and yourself and the airplane to be moving.

In the nineteenth century the existence of a preferred frame of reference that could be considered to be at rest was widely accepted. This was thought to be the reference frame of the *ether,* the medium filling all of space through which light was thought to propagate. (It was then accepted that light waves needed a medium to propagate through, just as it is now accepted that sound waves need air or some other material medium through which to propagate.) The ether was considered to be the preferred "at rest" reference frame.

A carefully devised series of measurements to measure the orbital speed of the earth relative to the ether were carried out in 1887 by Albert Michelson and Edward Morley. These measurements were considered challenging because the orbital speed of the earth is less than 1/10,000 the speed of light in vacuum. Much to the surprise of nearly everyone, the observations always found the speed of the earth relative to the ether to be zero. It was Albert Einstein who came up with a theory that was consistent with these observations. His explanation was that light is capable of traveling through empty space and that the ether was an unnecessary construct that did not exist. Einstein also postulated:

> The speed of light is independent of the speed of the light source.

POSTULATE II

The *speed of light* refers to the speed at which light travels through the vacuum of empty space.

A consequence of Postulate II and the principle of relativity is that all inertial observers measure the same value for the speed of light. (An inertial observer

---

† Further discussion on reference frames can be found in Chapters 2 and 4.

is one that remains at rest in an inertial reference frame.) To establish that all inertial observers measure the same value for the speed of light, we consider inertial observers A and B, where observer A is moving relative to observer B. The principle of relativity states that it is impossible to devise an experiment that determines whether an inertial observer is at rest or moving uniformly. If observer A measures a different value for the speed of light than observer B, then observers A and B could not both consider themselves to be at rest—a result in direct contradiction with the principle of relativity. Thus, a consequence of both the principle of relativity and Postulate II (that the speed of light is independent of the speed of the source) leads to the **constancy of the speed of light:**

The speed of light c is the same in any inertial reference frame.

THE CONSTANCY OF THE SPEED OF LIGHT

That is, anything (not just light) that travels at speed c relative to one inertial reference frame travels at the same speed c relative to any inertial reference frame.

Suppose you are in your backyard here on earth and Bob is on a spaceship moving away from you at half the speed of light ($\frac{1}{2}c$). You point a flashlight in Bob's direction and turn it on. The light leaves the flashlight, traveling at speed c (relative to the flashlight), and passes by your neighbor Keisha who is standing on the roof of her house next door. Keisha measures the speed of the light going by and finds it to be traveling at speed c. A few minutes later the light travels past Bob and his spaceship. Like Keisha, Bob measures the speed of the light going by him and also finds it to be traveling at speed c. This surprises Bob because he expected the light to be traveling past him at speed $\frac{1}{2}c$ rather than at speed c—after all, Bob is moving at speed $\frac{1}{2}c$ relative to the source of the light (the flashlight in your backyard). Like many people, Bob finds the constancy of the speed of light to be counterintuitive. This leaves him with a dilemma. Should he trust his measuring instruments or trust his intuition? It turns out that it is Bob's intuition that needs adjusting, not his instruments. Bob must change his concepts of both space and time.

Suppose that instead of pointing a flashlight you point a high-speed particle beam in his direction, where by "high speed" we mean a speed very close to the speed of light c. (A particle such as an electron or proton cannot travel at the speed of light, but it can travel at speeds extremely close to the speed of light.) If Keisha measures the particles going by her to be traveling at $0.9999c$ (relative to her), then how fast will Bob measure the particles going by him? Bob's intuition tells him that, because he is moving away from the source of the particles, they will be traveling past him at the slower speed of $0.4999c$, but that is not the case. When Bob measures the speed of the particles (relative to him) he finds it to be extremely close to $0.9999c$. (The actual value is $0.9997c$.)[†]

We tend to think of distances between cities as fixed. However, this too is not the case. According to a certain road map the distance between Baltimore and Philadelphia is 160 km. However, if you travel from Baltimore to Philadelphia at a significant fraction of the speed of light, the distance between the two cities will be much shorter than it is if you travel at 100 km/h (62 mi/h). For someone driving at 100 km/h, the distance between Baltimore and Philadelphia is very close to 160 km. However, for someone traveling at a speed of $0.866c$ (relative to the earth's surface), the distance is only 80 km, and for someone traveling at $0.9999c$, the distance is only 2.2 km.

The fastest speed that a human being has ever traveled relative to the earth (which occurred during the Apollo missions to the moon) is only about 10 km/s = $3.3 \times 10^{-5}c$. This speed is so slow compared to the speed of light, that for someone traveling from Baltimore to Philadelphia at that speed, the distance between those cities would be shorter by less than the diameter of a human hair. The logic explaining how this is determined is presented in the next three sections.

[†] Relative velocity in special relativity is covered in Chapter 37.

R-2 Moving Sticks

We wish to show that if a stick moves perpendicular to its length, its length does not change. We do this by showing that any increase or decrease in length contradicts the principle of relativity. Showing that a stick does not change its length may seem mundane. However, we show it because an immediate consequence is that moving clocks run slow.

Suppose that we have two identical metersticks, stick A and stick B. We verify that the sticks are identical by placing them side by side in a reference frame in which they are both at rest and examining them visually. We then give stick B to Bob before he takes off on another trip in his spaceship. On this trip Bob makes sure to hold the stick at right angles to the velocity of the spaceship relative to the earth. Is stick B now shorter than stick A, which remains back on the earth with us?

To answer this question we conduct a thought experiment. We attach felt-tipped marking pens to stick A, one at the 20-cm mark, the other at the 80-cm mark. Then Bob and his spaceship execute a flyby during which Bob holds stick B out a porthole, keeping it at right angles to the ship's velocity. During the flyby we hold up our stick (stick A), keeping it parallel with stick B. As the sticks pass by each other, two marks are drawn on stick B (Figure R-1). Bob then returns to earth with stick B and the two sticks are again placed side by side (Figure R-2), and the distance between the two marks on stick B and the distance between the two marking pens on stick A are compared. Let us assume that a stick moving perpendicular to its length is shorter than is an identical stationary stick. Then the distance between the two pens will be less than the distance between the two marks (Figure R-2)—clear evidence that during the flyby the moving stick (stick B) was shorter than the stationary stick. However, according to the principle of relativity it is equally valid to think of stick B as stationary and stick A as moving during the flyby. From this perspective, the same evidence (Figure R-2) demonstrates that the moving stick—now stick A—is longer than the stationary stick. Thus our assumption—that a stick moving perpendicular to its length is shorter than an identical stationary stick—leads to a contradiction and must be rejected. The assumption that a stick moving perpendicular to its length is longer than is an identical stationary stick also leads to a contradiction, as can be shown using an analogous argument. Thus we conclude:

> A stick moving perpendicular to its length has the same length as an identical stationary stick.

This rule is established without any consideration of the material from which the two sticks are made. Thus, the rule does not reflect a property of sticks. Instead it reflects a property of space.

The frame of reference in which the stick is at rest is called its **proper frame** or **rest frame,** and the length of a stick in its proper reference frame is called its **proper length** or **rest length.**

R-3 Moving Clocks

Clocks are used to measure time. In this section we will show that clocks moving at high speeds run slow, so if a high-speed spaceship travels by us, we would observe that all the clocks on the ship run slower than our clocks. However, the people on the ship are free to consider themselves to be at rest and us to be moving, and they would observe our clocks to run slow compared to their clocks. Let's examine how these observations are consistent with the constancy of the speed of light and the principle of relativity.

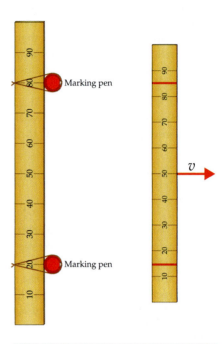

FIGURE R-1 During the flyby marks like this would be made on stick B by marking pens attached to stick A if the moving stick was shortened.

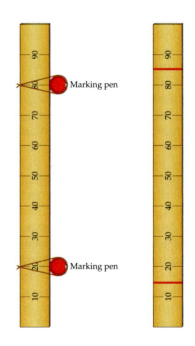

FIGURE R-2 If the distance between marks is greater than the distance between marker pens, this would demonstrate that stick B was shorter than stick A when the marks were made.

We construct a clock, called a *light clock*, using a stick of proper length L_0 and two mirrors (Figure R-3). The two mirrors face each other, and a light pulse bounces back and forth between them. Each time the light pulse strikes one of the mirrors, say the lower mirror, the clock is said to tick. Between successive ticks the light pulse travels a distance $2L_0$ in the proper reference frame of the clock. Thus the time between ticks T_0 is related to L_0 by

$$2L_0 = cT_0 \qquad\qquad \text{R-1}$$

Next we consider the time between ticks T of the same light clock, but this time we observe it from a reference frame in which the clock is moving with speed v perpendicular to the stick (Figure R-4). In this reference frame the clock moves a distance vT between ticks and the light pulse moves a distance cT between ticks. The distance the pulse moves in traveling from the bottom mirror to the top mirror is $\sqrt{L_0^2 + (\tfrac{1}{2}vT)^2}$. The light pulse travels the same distance in traveling from the top mirror to the bottom mirror. Thus,

$$2\sqrt{L_0^2 + (\tfrac{1}{2}vT)^2} = cT \qquad\qquad \text{R-2}$$

(Note that we have used the same symbol c for the speed of light in Equations R-1 and R-2.) Solving Equation R-1 for L_0 and substituting into Equation R-2 gives

$$\sqrt{(\tfrac{1}{2}cT_0)^2 + (\tfrac{1}{2}vT)^2} = \tfrac{1}{2}cT$$

Solving for T gives

$$T = \frac{T_0}{\sqrt{1 - (v^2/c^2)}} \qquad\qquad \text{R-3}$$

TIME DILATION

According to Equation R-3, the time between ticks in the reference frame in which the clock moves at speed v is greater than the time between ticks in the proper reference frame of the clock.

This raises the question, do other clocks run slow when they move with speed v according to Equation R-3 or is Equation R-3 valid only for light clocks? To answer this question we attach a conventional clock (with a conventional clock mechanism) to the lower mirror of the light clock (Figure R-5). The conventional clock has no minute or hour hands. Instead of a second hand it has an opaque disk with a narrow slot to indicate the time. The clock's face contains 60 equal-spaced marks (called tick marks) around its perimeter—one for each second. The clock ticks each time the slot passes over one of the tick marks. We adjust the length L_0 of the light-clock stick so that the time between ticks of both clocks is the same in the proper reference frame of the clocks. Next, we synchronize the clocks so each tick of the light clock occurs simultaneously with a tick of the conventional clock.

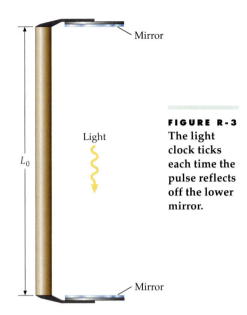

FIGURE R-3
The light clock ticks each time the pulse reflects off the lower mirror.

Mirror

Light

L_0

Mirror

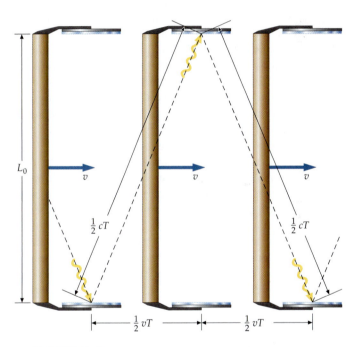

FIGURE R-4

L_0

v v v

$\tfrac{1}{2}cT$ $\tfrac{1}{2}cT$

$\tfrac{1}{2}vT$ $\tfrac{1}{2}vT$

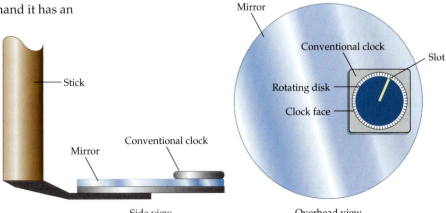

Stick

Mirror

Conventional clock

FIGURE R-5

Side view

Mirror

Conventional clock

Slot

Rotating disk

Clock face

Overhead view

We then ask, "If the ticks of the two clocks occur simultaneously in the proper reference frame of the clocks, do they also occur simultaneously in a reference frame in which the clocks are moving at speed v?"

The answer is yes. To understand why this is so, consider the following experiment. In the proper reference frame of the clocks, the time between ticks of both clocks is exactly one second. A light-sensitive film is placed on the face of the conventional clock, behind the rotating disk. Each time the light pulse reflects off the lower mirror, the narrow region of the light-sensitive paper directly behind the slot gets exposed. These exposed regions will be aligned with the tick marks as shown in Figure R-6, and all observers must agree with this permanent record.

In reference frame A in which the clocks are both moving, the light pulse exposes the film behind the slot on the clock face each time the pulse reflects off the lower mirror. Because the light clock is moving, the time between these reflections is greater than 1 s, in accordance with Equation R-3. When an observer of reference frame A sees that the lines appearing where the film was exposed are aligned with the tick marks, she realizes that in her reference frame the conventional clock runs slow in exactly the same manner that the light clock runs slow—in accord with Equation R-3. Thus we conclude that all moving clocks run slow in exactly the same manner that a moving light clock runs slow. Because this is the case, we conclude that it is time itself that runs slow, a phenomenon known as **time dilation.**

Something that occurs at a specific instant in time and at a specific location in space is called a **spacetime event,** or just an **event.** Each reflection of the light pulse off the lower mirror of the light clock is a spacetime event. If we call one of these reflections event 1, and the next reflection event 2, then the time between events 1 and 2 in a frame of reference in which the two events occur at the same location is called the **proper time interval** T_0 between the two events. Let T be the time between the same two events in a reference frame in which they occur at different locations. Equation R-3 relates the time T between two events to the proper time T_0 between the same two events.

Each time the light pulse reflects off the lower mirror, the slot (second hand) of the conventional clock passes directly over a tick mark. In the proper frame of the two clocks, these two events—the arrival of the light pulse and the passing of the slot over a tick mark—occur at the same time *and* at the same place. Any two events that occur both at the same time and at the same place in one reference frame will occur both at the same time and at the same place in all reference frames. This is because such events can have lasting consequences—like producing lines on the light-sensitive film aligned with the tick marks on the clock face. We cannot have the marks aligned with the ticks marks in one reference frame and not aligned with the tick marks in another reference frame. After all, there is only one clock face and one set of marks. This conclusion can be generalized into a principle, called the **principle of invariance of coincidences:**

> If two events occur at the same time and at the same place in one reference frame, then they occur at the same time and at the same place in all reference frames.

PRINCIPLE OF INVARIANCE OF COINCIDENCES

We can better visualize this principle by considering two automobiles passing through an intersection at the same time. The two events are: (1) automobile A passes through the intersection and (2) automobile B passes through the intersection. If these two events occur at the same time in one reference frame, then they must occur at the same time in all reference frames. Either a fender becomes dented or it doesn't. That is, if the automobiles collide, then there is no question that the two cars were in the intersection at the same time. The lasting evidence dictates that observers in all reference frames must agree on this fact. Any pair of events that occur at the same time *and* at the same location are referred to as a **spacetime coincidence.**

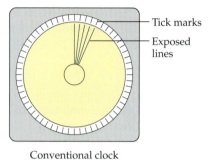

Tick marks

Exposed lines

Conventional clock

FIGURE R-6

THE NAPPING ASTRONAUTS   **EXAMPLE R-1**   **Put It in Context**

Astronauts in a spaceship traveling at $v = 0.6c$ relative to the earth sign off from space control, saying that they are going to nap for 1 h and then will call back. How long does their nap last according to observers on the earth?

PICTURE THE PROBLEM Clock S on the ship reads t_0 when the nap begins (a spacetime coincidence) and reads $t_0 + 1$ h when the nap ends (also a spacetime co-incidence). Observers on the ship agree that, because clock S is stationary it does not run slow, so the nap lasted 1 h. In the reference frame of the ship the two events (the beginning of the nap and the end of the nap) occur at the same location, so the time interval between the events is the proper time interval between them. Observers on the earth agree that clock S reads t_0 when the nap begins and it reads $t_0 + 1$ h when the nap ends. However, the observers on the earth agree that because clock S is moving at speed v, it is running slow so the nap lasted more than 1 h. In the reference frame of the earth the ship is moving so the nap begins and ends at different locations. Therefore, in the reference frame of the earth the time interval between the events is not the proper time interval between the events.

1. Event 1 is the beginning of the nap and event 2 is the end of the nap. Clock S on the ship advances 1 h between these events. Determine the proper time interval T_0 between these events:

$T_0 = 1$ h

2. Find the time interval T between events 1 and 2 for observers on earth:

$$T = \frac{T_0}{\sqrt{1 - (v^2/c^2)}} = 1\,\text{h}/\sqrt{1 - \frac{(0.6c)^2}{c^2}}$$

$$= \frac{1\,\text{h}}{\sqrt{1 - 0.36}} = \frac{1\,\text{h}}{\sqrt{0.64}} = \frac{1\,\text{h}}{0.8} = \boxed{1.25\,\text{h}}$$

REMARKS Clock S is an unnecessary construct as the astronauts themselves serve as clocks. What it is necessary to realize is that the proper time between the beginning and the end of the nap is 1 h, so the time T between the same events in a reference frame where the clocks (astronauts) are moving with speed v is given by Equation R-3.

EXERCISE A pion[†] has a median proper lifetime of 26 ns (1 ns = 1×10^{-9} s) (measured when the pion is at rest). What is the median lifetime if measured when the pion is moving at $0.995c$? (*Answer*   260 ns)

EXERCISE A beam of pions (see previous exercise) moving at $0.995c$ passes point P. How far from P do the pions travel before only half of the pions in the beam remain? (*Answer*   78 m)

R-4 Moving Sticks Again

In Section R-2 the length of a stick moving perpendicular to its length and the length of an identical stationary stick are compared and found to be equal. However, the technique used for this comparison works only if the velocity of the moving stick is perpendicular to its length. Here we apply a different technique to compare the length of a stick at rest to its length when it is moving parallel to its velocity.

A light clock is shown in its proper frame in Figure R-7. This clock ticks each time the light pulse reflects off the mirror on the left. In its proper reference frame the length of the clock is L_0 and the time between ticks is $T_0 = 2L_0/c$ (Equation R-1). To find the

FIGURE R-7

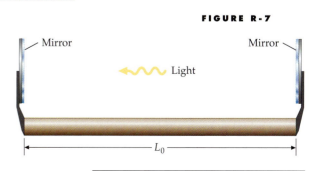

---

† A pion (short for pi meson) is a subatomic particle.

length of the clock in a reference frame in which it is moving to the right at speed v we consider three sequential events:

Event 0   Light pulse reflects off the mirror at the left end.

Event 1   Light pulse reflects off the mirror at the right end.

Event 2   Light pulse reflects off at the mirror at the left end.

In Figure R-8 the clock is shown at the time of each of these events in a reference frame in which the clocks move to the right with speed v. (The clocks are drawn lower down the page at later times to avoid visual overlap.) The times of occurrence for events 0, 1, and 2 in this reference frame are t'_0, t'_1, and t'_2, respectively. In the time between events 0 and 1 the clock moves a distance $v(t'_1 - t'_0)$ and the light pulse travels a distance $c(t'_1 - t'_0)$. Thus,

$$c(t'_1 - t'_0) = L + v(t'_1 - t'_0) \qquad \text{R-4}$$

In the time between events 1 and 2 the clock moves a distance $v(t'_2 - t'_1)$ and the light pulse travels a distance $c(t'_2 - t'_1)$, so

$$c(t'_2 - t'_1) = L - v(t'_2 - t'_1) \qquad \text{R-5}$$

Eliminating t'_1 by solving Equation R-4 for t'_1, substituting the result into Equation R-5, and then solving for $t'_2 - t'_0$ gives

$$t'_2 - t'_0 = \frac{2L/c}{1 - (v^2/c^2)} \qquad \text{R-6}$$

The time interval $t'_2 - t'_0$ is related to the proper time interval $t_2 - t_0$ between events 0 and 2 (Equation R-3) by

$$t'_2 - t'_0 = \frac{t_2 - t_0}{\sqrt{1 - (v^2/c^2)}} \qquad \text{R-7}$$

where $t_2 - t_0 = 2L_0/c$ (Equation R-1). Substituting $2L_0/c$ for $t_2 - t_0$ gives

$$t'_2 - t'_0 = \frac{2L_0/c}{\sqrt{1 - (v^2/c^2)}} \qquad \text{R-8}$$

Equating the right sides of Equations R-6 and R-8, and then solving for L gives

$$L = L_0\sqrt{1 - (v^2/c^2)} \qquad \text{R-9}$$

LENGTH CONTRACTION

Establishing this result did not involve any properties of the stick. Thus Equation R-9 reflects the nature of space and time, and not the nature of the sticks.

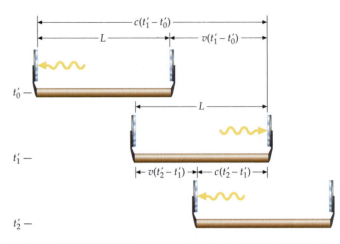

FIGURE R-8 A light clock moving to the right at speed v is shown at 3 instants of time.

R-5 Distant Clocks and Simultaneity

We have established three useful relations: (1) that the length of a stick moving perpendicular to its length is the same as its rest length; (2) that the time T between two ticks of a moving clock is greater than the time between the two ticks of the same clock for an observer moving with the clock according to $T = T_0/\sqrt{1 - (v^2/c^2)}$; and (3) that the length L of a stick moving parallel to its length is less than its rest length L_0 according to $L = L_0\sqrt{1 - (v^2/c^2)}$. But in order to analyze events from the perspective of observers in reference frames moving at different velocities, we need one more relation, one that concerns the readings on clocks at different locations.

Clocks A and B (Figure R-9a) are at rest relative to each other, and in their rest frame the clocks are separated by a distance L_0. To synchronize these clocks there is a flash lamp on clock A and a light sensitive film on the face of clock B. The alarm on clock A is set to energize the flash lamp when the second hand on clock A passes zero. Like the conventional clock in Section R-3, the second hand on clock B is a rotating opaque disk with a slot to indicate the time. Behind the disk is a light-sensitive film. When the light from the flash reaches clock B, the film is illuminated on the narrow region behind the slot. This provides a lasting record of the reading on clock B when the light from the flash lamp reaches it. Let this reading be t_1. In the rest frame of the clocks, the time for the light to travel at speed c from clock A to clock B is L_0/c, so when the light arrives at clock B, clock A reads L_0/c and clock B reads t_1. To synchronize the two clocks we turn clock B back by $\Delta t = t_1 - L_0/c$.

With the two clocks synchronized in their rest frame (frame 1), we then determine whether they are also synchronized in a reference frame (frame 2) in which they are moving at speed v parallel to the line joining them, shown in Figure R-9b. We reset the alarm to energize the flash lamp when clock A next reads zero. These two events—clock A reads zero and the lamp flashes—are a spacetime coincidence, so we know they occur simultaneously in all reference frames. Also, the light reaching clock B and clock B reading L_0/c, are a spacetime coincidence, so we know they occur simultaneously in all reference frames.

In frame 2, the distance L between the clocks is given by

$$L = L_0\sqrt{1 - (v^2/c^2)}$$

and clock B is moving toward the flash lamp. In this frame the light traveling from clock A to clock B travels a distance $L - vt$, where t is the time required for the light to travel the distance. The time t, the distance L, and the speed v are related by

$$ct = L - vt$$

Solving for the time gives $t = L/(c + v)$.

Moving clocks run slow, so during time t the readings on both clocks advance not by $L/(c + v)$ but by

$$\frac{L}{c + v}\sqrt{1 - (v^2/c^2)} = \frac{L_0}{c + v}\left(1 - \frac{v^2}{c^2}\right) = \frac{L_0}{(c + v)}\frac{(c + v)(c - v)}{c^2} = \frac{L_0}{c} - \frac{vL_0}{c^2}$$

Thus, when the light arrives at clock B, clock B reads L_0/c and clock A reads $L_0/c - vL_0/c^2$. Therefore, in frame 2 clock B is ahead of clock A by vL_0/c^2:

> If two clocks are synchronized in their rest frame, then in a frame where they move with speed v parallel to the line joining them the clock in the rear is ahead of the clock in the front by vL_0/c^2.

THE RELATIVITY OF SIMULTANEITY

In this case L_0 is the distance between the clocks in their rest frame. It is also true that if two clocks are synchronized in their rest frame, they are also synchronized in a frame in which they move with speed v perpendicular to the line joining them. This follows from the symmetry of the situation. (For one thing, there is no way to state a rule specifying which of the two clocks is ahead.)

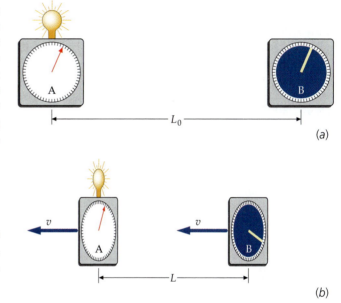

FIGURE R-9 (a) The clocks are synchronized in the reference frame in which they are at rest. (b) Are the clocks synchronized in the reference frame in which they are moving with speed v parallel to the line joining them?

R-6 Applying the Rules

A Train Through a Tunnel

EXAMPLE R-2

A high-speed train is about to enter a tunnel through a mountain. The tunnel has a proper length of 1.2 km. The length of the train in the reference frame of the mountain is also 1.2 km, and the proper length of the train is 2.0 km. Clock A is attached to the mountain at the entrance to the tunnel, and clock B is attached to the mountain at the exit to the tunnel. In the reference frame of the mountain, when the front of the train enters the tunnel both clocks read zero. (*a*) In the reference frame of the mountain, what is the speed of the train and what is the reading of both clocks at the instant the front of the train exits the tunnel (Figure R-10*a*)? (*b*) In the reference frame of the train, what is the length of the tunnel, what is the reading of both clocks at the instant the front of the train enters the tunnel (Figure R-10*b*), and what is the reading of both clocks at the instant the front of the train exits the tunnel? (*c*) For a passenger on the train, how long does it take for the front of the train to pass through the tunnel?

FIGURE R-10

(a)

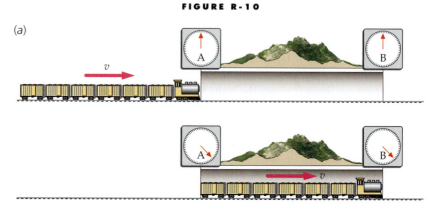

(b)

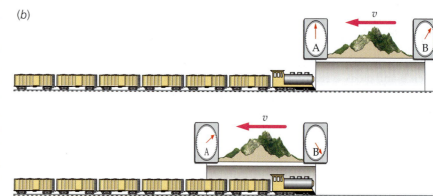

PICTURE THE PROBLEM The speed of the train and the length of the train are related by the length-contraction formula. Some of the clock readings in the two reference frames can be equated because they are event pairs that form spacetime coincidences. Other clock readings can be related by the relativity–of–simultaneity relation.

(*a*) 1. Using the length-contraction formula, solve for the speed of the train:

$$L = L_0\sqrt{1 - (v^2/c^2)}$$

$$1.2 \text{ km} = 2.0 \text{ km}\sqrt{1 - (v^2/c^2)}$$

so

$$v = 0.8c = 0.8(3.00 \times 10^8 \text{ m/s}) = \boxed{2.4 \times 10^8 \text{ m/s}}$$

2. The length of the tunnel equals its proper length and because the clocks are not moving they do not run slow. The reading on both clocks is the time t that it takes for the front of the train to travel the length of the tunnel:

$$L_{\text{tunnel},0} = vt$$

so

$$t = \frac{L_{\text{tunnel},0}}{v} = \frac{1.2 \times 10^3 \text{ m}}{2.4 \times 10^8 \text{ m/s}} = 5 \times 10^{-6} \text{ s} = 5\ \mu\text{s}$$

3. The clocks are synchronized, so when the front of the train exits the tunnel both clocks read 5 μs:

$$\text{Clock A reading} = \text{Clock B reading} = \boxed{5\ \mu\text{s}}$$

(*b*) 1. In this frame the mountain is moving at 0.8*c*. Using the length-contraction formula, solve for the length of the tunnel:

$$L_{\text{tunnel}} = L_{\text{tunnel},0}\sqrt{1 - (v^2/c^2)} = 1.2 \text{ km}\sqrt{1 - \frac{(0.8c)^2}{c^2}}$$

$$= 1.2 \text{ km}\sqrt{1 - 0.8^2} = \boxed{0.72 \text{ km} = 720 \text{ m}}$$

2. The front of the train entering the tunnel and a zero reading on clock A are a spacetime coincidence:

> Clock A reads zero

3. The two clocks are moving toward the train with clock B in the rear, so clock B is ahead of clock A by vL_0/c^2. When the train enters the tunnel clock A reads zero, so clock B reads vL_0/c^2:

$$\frac{vL_{\text{tunnel},0}}{c^2} = \frac{0.8cL_{\text{tunnel},0}}{c^2} = \frac{0.8L_{\text{tunnel},0}}{c}$$

$$= \frac{0.8(1.2 \times 10^3\text{ m})}{3.0 \times 10^8\text{ m/s}} = 3.2\ \mu s$$

> Clock B reading = $3.2\ \mu s$

4. The front of the train exiting the tunnel and a 5-μs reading of clock B are a spacetime coincidence:

> Clock B reading = $5\ \mu s$

5. Clock B is in the rear, so clock A lags behind clock B by vL_0/c^2.

$$\text{Clock A reading} = \text{Clock B reading} - \frac{vL_{\text{tunnel},0}}{c^2}$$

$$= 5\ \mu s - 3.2\ \mu s = \boxed{2.6\ \mu s}$$

(c) For an observer in the reference frame of the train, the mountain is traveling at 0.8c and the tunnel is 720-m long:

$$L_{\text{tunnel}} = vt$$

so

$$t = \frac{L_{\text{tunnel}}}{v} = \frac{L_{\text{tunnel}}}{0.8c} = \frac{720\text{ m}}{2.4 \times 10^8\text{ m/s}} = \boxed{3\ \mu s}$$

REMARKS In the frame of reference of the mountain, the length of the train and the length of the tunnel are both 1.2 km. In the frame of reference of the train, the length of the train is 2.0 km and the length of the tunnel is 720 m. In the frame of reference of the mountain, when both clocks read 5 μs the entire train is in the tunnel. In the reference frame of the train, the train is longer than the tunnel so at no time is the entire train within the tunnel.

EXERCISE Event 1 is the front of the train entering the tunnel, and event 2 is the front of the train exiting the tunnel. (a) In which reference frame do these two events occur at the same location? (b) What is the proper time interval between events 1 and 2? (*Answer* (a) The reference frame of the train, because both events occur at the front end of the train. (b) 3 μs)

It is often convenient to measure large distances in light-years, where a light-year is the distance traveled when traveling at the speed of light for a time of one year. That is,

$$1 \text{ light-year} = 1\ c\cdot y$$

where $1\ c\cdot y = c\ (1\ y)$. This notation is particularly convenient when distances are divided by speeds. For example, the time T for a particle traveling at $v = 0.1c$ to travel a distance of $L = 25$ light-years is

$$T = \frac{L}{v} = \frac{25\ c\cdot y}{0.1c} = 250\ y$$

where the c's cancel.

EXERCISE In the reference frame of the earth, it takes 8 minutes for light to travel from the sun to the earth, so the distance between the sun and the earth is 8 $c\cdot$min. How many minutes does it take a particle from the sun to reach the earth if the particle travels at 0.1c? (*Answer* (8 $c\cdot$min)/0.1c = (8 min)/0.1 = 80 min)

EXPLORING

PDF files of two worked-out examples on the twin paradox are available at
www.whfreeman.com/tipler5e.

R-7 Relativistic Momentum, Mass, and Energy

Momentum and Mass

In special relativity both momentum and energy are conserved, just as they are in classical physics. However, there are differences that have to be accounted for. The momentum of a particle moving with velocity v is given by

$$p = \frac{mv}{\sqrt{1 - (v^2/c^2)}} \qquad \text{R-10}$$

RELATIVISTIC MOMENTUM

where m is the mass of the particle. Equation R-10 is sometimes written $p = m_r v$, where m_r is called the relativistic mass: $m_r = m/\sqrt{1 - (v^2/c^2)}$. In the rest frame of the particle, $v = 0$ and the relativistic mass equals m. (The mass m is sometimes called the rest mass to differentiate it from the relativistic mass.) Relativistic momentum and mass are discussed further in Chapter 39.

Energy

In relativistic mechanics, as in classical mechanics, the net force on a particle is equal to the time rate of change of the momentum of the particle. Considering one-dimensional motion only, we have

$$F_{net} = \frac{dp}{dt} \qquad \text{R-11}$$

We wish to find an expression for the kinetic energy. To do this we will multiply both sides of Equation R-11 by the displacement ds. This gives

$$F_{net}\, ds = \frac{dp}{dt} ds \qquad \text{R-12}$$

where we identify the term on the left as the work and the term on the right as the change in kinetic energy dK. Substituting $v\, dt$ for ds in the term on the right we obtain

$$dK = \frac{dp}{dt} v\, dt = v\, dp$$

Integrating both sides gives

$$K = \int v\, dp \qquad \text{R-13}$$

This integral is evaluated in Chapter 37. The result is

$$K = \frac{mc^2}{\sqrt{1 - (v^2/c^2)}} - mc^2 \qquad \text{R-14}$$

Defining $\dfrac{mc^2}{\sqrt{1 - (v^2/c^2)}}$ as the **total relativistic energy** E gives

$$E = K + mc^2 = \frac{mc^2}{\sqrt{1 - (v^2/c^2)}} \qquad \text{R-15}$$

where $m_0 c^2$, called the **rest energy** E_0, is energy the particle has when it is at rest.

By multiplying both sides of Equation R-10 by c and then dividing the resulting equation by Equation R-15 we obtain

$$\frac{v}{c} = \frac{pc}{E} \qquad \text{R-16}$$

which can be useful when trying to solve for the speed v. Eliminating v from Equations R-10 and R-16, and solving for E^2 (see Problem R-41) gives

$$E^2 = p^2c^2 + m^2c^4 \qquad \text{R-17}$$

The relation between mass and energy is discussed in Section 3 of Chapter 7.

SUMMARY

Topic	Relevant Equations and Remarks
1. Postulates of Special Relativity	
Postulate I: Principle of relativity	It is impossible to devise an experiment that determines whether you are at rest or moving uniformly, where moving uniformly means moving at constant velocity relative to an inertial reference frame.
Postulate II	The speed of light is independent of the speed of the source.
Constancy of the speed of light	It follows that the speed of light is the same in any inertial reference frame.
2. Moving Sticks	The length of a stick moving perpendicular to its length is equal to its proper length.
	The length of a stick moving with speed v parallel to its length is shorter than its proper length L_0 according to $$L = L_0\sqrt{1 - (v^2/c^2)} \qquad \text{R-9}$$
3. Moving Clocks	
Time dilation	The time between ticks of a clock moving with speed v is longer than the proper time between ticks of the same clock by $$T = \frac{T_0}{\sqrt{1 - (v^2/c^2)}} \qquad \text{R-3}$$
Relativity of simultaneity	If two clocks are synchronized in their rest frame, in a frame where they move with speed v parallel to the line joining them the clock in the rear is ahead of the clock in front by vL_0/c^2, where L_0 is the distance between them in their rest frame.
	If two clocks are synchronized in their rest frame they are also synchronized in a frame where they move with speed v perpendicular to the line joining them.
4. Spacetime Coincidence	If two events occur both at the same time and at the same place in one reference frame, they occur both at the same time and at the same place in any reference frame.
5. Momentum, Mass, and Energy	
Momentum	The momentum of a particle is given by $$p = \frac{mv}{\sqrt{1 - (v^2/c^2)}} \qquad \text{R-10}$$

Mass and energy	The total relativistic energy of a particle equals its rest energy plus its kinetic energy.	
	$$E = K + mc^2 = \frac{mc^2}{\sqrt{1 - (v^2/c^2)}}$$	**R-15**
	where mc^2 is the rest energy E_0.	

Momentum and Energy	$\dfrac{v}{c} = \dfrac{pc}{E}$ and $E^2 = p^2c^2 + m^2c^4$	**R-16, R-17**

PROBLEMS

- Single-concept, single-step, relatively easy
- ● ● Intermediate-level, may require synthesis of concepts
- ● ● ● Challenging
- **SSM** Solution is in the *Student Solutions Manual*
- **iSOLVE** Problems available on iSOLVE online homework service
- **iSOLVE** ✔ These "Checkpoint" online homework service problems ask students additional questions about their confidence level, and how they arrived at their answer

In a few problems, you are given more data than you actually need; in a few other problems, you are required to supply data from your general knowledge, outside sources, or informed estimates.

Conceptual Problems

1 ● You are standing on a corner and a friend is driving past in an automobile. Each of you is wearing a wrist watch. Both of you note the times when the car passes two different intersections and determine from your watch readings the time that elapses between the two events. Which of you has determined the proper time interval?

2 ● **SSM** If event A occurs before event B in some frame, might it be possible for there to be a reference frame in which event B occurs before event A?

3 ● Two events are simultaneous in a frame in which they also occur at the same point in space. Are they simultaneous in all other reference frames?

4 ● ● Two inertial observers are in relative motion. In what circumstances can they agree on the simultaneity of two different events?

5 ● The approximate total energy of a particle of mass m moving at speed $v \ll c$ is (a) $mc^2 + \frac{1}{2}mv^2$, (b) mv^2, (c) cmv, (d) $\frac{1}{2}mc^2$, (e) cmv.

6 ● **SSM** True or false:

(a) The speed of light is the same in all reference frames.
(b) Proper time is the shortest time interval between two events.
(c) Absolute motion can be determined by means of length contraction.
(d) The light-year is a unit of distance.
(e) For two events to form a spacetime coincidence they must occur at the same place.
(f) If two events are not simultaneous in one frame, they cannot be simultaneous in any other frame.

Estimation and Approximation

7 ● ● The satellites used for the Global Positioning System (GPS) have extremely accurate clocks on board. In fact, the positioning system is so accurate that the relativistic time dilation of the clocks with respect to earth-bound observers must be taken into account. If the GPS satellite orbits have an average radius of about 26,000 km, estimate the time-dilation factor of the satellite clocks with respect to an earth-bound observer.

8 ● ● It is said that exercising by riding a bike will extend your life. From the standpoint of relativity, this is certainly true. Estimate by what factor your life will be extended through relativistic time dilation with respect to a stationary observer if you ride a bike as part of your regular exercise program. Make any assumptions that you feel are reasonable, but be certain to state them in your answer.

9 ● ● **SSM** **iSOLVE** ✔ In 1975, an airplane carrying an atomic clock flew back and forth for 15 h at an average speed of 140 m/s as part of a time-dilation experiment. The time on the clock was compared to the time on an atomic clock kept on the ground. How much time did the airborne clock "lose" with respect to the clock on the ground?

Length Contraction and Time Dilation

10 ● **iSOLVE** ✔ The proper mean lifetime of subnuclear particles called pions is 2.6×10^{-8} s. A beam of pions has a speed of $0.85c$ relative to a laboratory. (a) What would be their mean lifetime as measured in the laboratory? (b) How far would they travel, on average, before they decay? (c) What would be your answer to Part (b) if you neglect time dilation?

11 • [SOLVE] (a) In the reference frame of the pion in Problem 10, how far does the laboratory travel in a typical lifetime of 2.6×10^{-8} s? (b) What is this distance in the laboratory's frame?

12 • [SSM] [SOLVE]✓ The proper mean lifetime of a subnuclear particle called a muon is 2 μs. Muons in a beam are traveling at 0.999c relative to a laboratory. (a) What is their mean lifetime as measured in the laboratory? (b) How far do they travel, on average, before they decay?

13 • [SOLVE] (a) In the reference frame of the muon in Problem 12, how far does the laboratory travel in a typical lifetime of 2 μs? (b) What is this distance in the laboratory's frame?

14 • You have been posted to a remote region of space to monitor traffic. Toward the end of a quiet shift, a spacecraft goes by and you measure its length using a laser device, which reports a length of 85 m. You flip open your handy reference catalogue and identify the craft as a CCCNX-22, which has a proper length of 100 m. When you phone in your report, what speed should you give for this spacecraft?

15 • [SSM] A spaceship travels from earth to a star 95 light-years away at a speed of 2.2×10^8 m/s. How long does it take to get there (a) as measured on the earth and (b) as measured by a passenger on the spaceship?

16 • The average lifetime of a beam of subnuclear particles called pions traveling at high speed is measured to be 7.5×10^{-8} s. Their average lifetime when measured at rest is 2.6×10^{-8} s. How fast is the pion beam traveling?

17 • A meterstick moves with speed 0.8c relative to you in the direction parallel to the stick. (a) Find the length of the stick as measured by you. (b) How long does it take for the stick to pass you?

18 • The mean lifetime of charged subnuclear particles called pions in their rest frame is 1.8×10^{-8} s (that is, in the rest frame of the pions, if there are N pions at time $t = 0$, there will be only $N/2$ pions at time $t = 1.8 \times 10^{-8}$ s). Pions are produced in an accelerator and emerge with a speed of 0.998c. How far do these particles travel in the accelerator laboratory before half of them have decayed?

19 •• [SOLVE] Your friend, who is the same age as you, travels to the star Alpha Centauri, which is 4 light-years away, and returns immediately. He claims that the entire trip took just 6 y. How fast did he travel?

20 •• [SSM] Two spaceships pass each other traveling in opposite directions. A passenger in ship A, who happens to know that her ship is 100 m long, notes that ship B is moving with a speed of 0.92c relative to A and that the length of B is 36 m. What are the lengths of the two spaceships as measured by a passenger in ship B?

21 •• Supersonic jets achieve maximum speeds of about $3 \times 10^{-6}c$. (a) By what percentage would a jet traveling at this speed contract in length? (b) During a time of 1 y $= 3.15 \times 10^7$ s on your clock, how much time would elapse on the pilot's clock? How many minutes are lost by the pilot's clock in 1 y of your time?

The Relativity of Simultaneity

Problems 22 through 26 refer to the following situation: Mary is a worker on a large space platform. She places a clock A at point A and clock B at point B, which is 100 light-minutes from point A (Figure R-12). She also places a flashbulb at a point midway between points A and B. Jamal, a worker on a different platform, is standing next to clock C. Each clock immediately starts at zero when the flash reaches it. Mary's platform moves at speed of 0.6c relative to Jamal's. As Mary's platform passes by, clock B, then the flashbulb, and then clock A pass directly over clock C—just missing it as they go by. As the flashbulb passes next to clock C, it flashes and clock C immediately starts at zero.

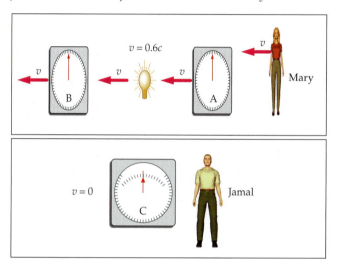

FIGURE R-12

22 • [SSM] According to Jamal, (a) what is the distance between the flashbulb and clock A, (b) how far does the flash travel to reach clock A, and (c) how far does clock A travel while the flash is traveling from the flashbulb to it?

23 •• According to Jamal, how long does it take the flash to travel to clock A, and what does clock C read as the flash reaches clock A.

24 •• Show that clock C reads 100 min as the light flash reaches clock B, which is traveling away from clock C with speed 0.6c.

25 •• According to Jamal, the reading on clock C advances from 25 min to 100 min between the reception of the flashes by clocks A and B in Problems 23 and 24. Again according to Jamal, how much will the reading on clock A advance during this 75-min interval?

26 •• [SSM] The advance of clock A calculated in Problem 29 is the amount that clock A leads clock B according to Jamal. Compare this result with vL_0/c^2, where $v = 0.6c$.

27 •• In inertial reference frame S, event B occurs 2 μs after event A, which occurs 1.5 km from event A. How fast must an observer be moving along the line joining the two events so that the two events occur simultaneously? For an observer traveling fast enough is it possible for event B to precede event A?

28 •• A large flat space platform has an x axis painted on it. A firecracker explodes on the x axis at $x_1 = 480$ m, and a second firecracker explodes on the x axis 5 μs later at $x_2 = 1200$ m. In the reference frame of a train traveling alongside the x axis at speed v relative to the platform, these two explosions occur at the same place in space. What is the separation in time between the two explosions in the reference frame of the train?

29 • **SOLVE** Herb and Randy are twin jazz musicians who perform as a trombone–saxophone duo. At the age of twenty, however, Randy got an irresistible offer to join a road trip to perform on a star 15 light-years away. To celebrate his good fortune, he bought a new vehicle for the trip—a deluxe space-coupe that travels at $0.999c$. Each of the twins promises to practice diligently, so they can reunite afterward. However, Randy's gig goes so well that he stays for a full 10 y before returning to Herb. After their reunion, (*a*) how many years of practice will Randy have had; (*b*) how many years of practice will Herb have had?

30 •• **SSM** Al and Bert are twins. Al travels at $0.6c$ to Alpha Centauri (which is 4 $c\cdot$y from the earth as measured in the reference frame of the earth) and returns immediately. Each twin sends the other a light signal every 0.01 y as measured in his own reference frame. (*a*) At what rate does Bert receive signals as Al is moving away from him? (*b*) How many signals does Bert receive at this rate? (*c*) How many total signals are received by Bert before Al has returned? (*d*) At what rate does Al receive signals as Bert is receding from him? (*e*) How many signals does Al receive at this rate? (*f*) How many total signals are received by Al? (*g*) Which twin is younger at the end of the trip and by how many years?

Relativistic Energy and Momentum

31 • **SOLVE** Find the ratio of the total energy to the rest energy of a particle of rest mass m_0 moving with speed (*a*) $0.1c$, (*b*) $0.5c$, (*c*) $0.8c$, and (*d*) $0.99c$.

32 • **SOLVE**✓ A proton (rest energy 938 MeV) has a total energy of 1400 MeV. (*a*) What is its speed? (*b*) What is its momentum?

33 • **SSM** How much energy would be required to accelerate a particle of mass m from rest to (*a*) $0.5c$, (*b*) $0.9c$, and (*c*) $0.99c$? Express your answers as multiples of the rest energy.

34 • If the kinetic energy of a particle equals its rest energy, what error is made by using $p = mv$ for its momentum?

35 • What is the total energy of a proton whose momentum is $3mc$?

36 •• **SSM** Using a spreadsheet program or graphing calculator, make a graph of the kinetic energy of a particle with mass $m = 100$ MeV/c^2 for speeds between 0 and c. On the same graph, plot $\frac{1}{2}mv^2$ by way of comparison. Using the graph, estimate at about what velocity this formula is no longer a good approximation to the kinetic energy. As a suggestion, plot the energy in units of MeV and the velocity in the dimensionless form v/c.

37 •• Derive the equation $E^2 = p^2c^2 + m^2c^4$ (Equation R-17) by eliminating v from Equations R-10 and R-16.

38 •• Use the binomial expansion and Equation R-17 to show that when $pc \ll mc^2$, the total energy is given approximately by $E \approx mc^2 + p^2/(2m)$

39 •• (*a*) Show that the speed v of a particle of mass m and total energy E is given by

$$\frac{v}{c} = \left[1 - \frac{(mc^2)^2}{E^2}\right]^{1/2}$$

and that when E is much greater than mc^2, this can be approximated by

$$\frac{v}{c} \approx 1 - \frac{(mc^2)^2}{2E^2}.$$

Find the speed of an electron with kinetic energy of (*b*) 0.51 MeV and (*c*) 10 MeV.

40 •• **SSM** The rest energy of a proton is about 938 MeV. If its kinetic energy is also 938 MeV, find (*a*) its momentum and (*b*) its speed.

41 •• What percentage error is made in using $\frac{1}{2}m_0v^2$ for the kinetic energy of a particle if its speed is (*a*) $0.1c$ and (*b*) $0.9c$?

General Problems

42 • A spaceship departs from Earth for the star Alpha Centauri, which is 4 light-years away in the reference frame of Earth. The spaceship travels at $0.75c$. How long does it take to get there (*a*) as measured on Earth and (*b*) as measured by a passenger on the spaceship?

43 • The total energy of a particle is three times its rest energy. (*a*) Find v/c for the particle. (*b*) Show that its momentum is given by $p = \sqrt{8}mc$.

44 • **SOLVE** A subnuclear particle called a muon has a mean lifetime of 2 μs when stationary. If you measure the mean lifetime of the muons coming out of a nuclear reactor port to be 46 μs, how fast are they moving?

45 •• **SSM** The rest mass of the neutrino (the "ghost particle" of physics) is known to have a small but as yet unmeasured value. Both high-energy and light neutrinos are produced by a supernova explosion, so it is possible to estimate the neutrino's mass by timing the relative arrival of light versus neutrinos from a supernova. If a supernova explodes 100,000 light-years from the earth, calculate the rest mass necessary for a neutrino with total energy of 100 MeV to trail the light by (*a*) 1 min, (*b*) 1 s, and (*c*) 0.01 s.

46 • **SOLVE**✓ Relative to you, how fast must a meter-stick travel in the direction parallel to itself so that its length as measured by you is 50 cm?

47 ••• **SSM** Keisha and Ernie are trying to fit a 15-ft-long ladder into a 10-ft-long shed with doors at each end. Recalling her physics lessons, Keisha suggests to Ernie that they open the front door to the shed and have Ernie run toward it with the ladder at a speed such that the length contraction of the ladder shortens it enough so that it fits in the shed. As soon as the back end of the ladder passes through the door, Keisha will slam it shut. (*a*) What is the minimum speed at which Ernie must run to fit the ladder into the shed? Express it as a fraction of the speed of light. (*b*) As Ernie runs toward the shed at a speed of $0.866c$, he realizes that in the reference frame of the ladder and himself, it is the *shed* which is shorter, not the ladder. How long is the shed in the rest frame of the ladder? (*c*) In the reference frame of the ladder is there any instant that both ends of the ladder are simultaneously inside the shed? Examine this from the point of view of relativistic simultaneity.

Gravity

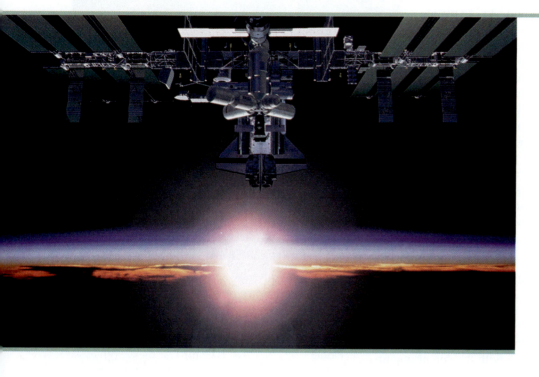

ON A CLEAR AUTUMN NIGHT YOU ARE ON THE ROOF OF YOUR APARTMENT BUILDING IN NEW YORK CITY, TALKING ON YOUR CELL PHONE TO YOUR FRIEND IN KANSAS CITY, WHEN YOU SEE THE INTERNATIONAL SPACE STATION (ISS) PASS DIRECTLY OVERHEAD.

? **How might you use your understanding of gravity to determine when the ISS will next pass directly over Kansas City? (See Example 11-3.)**

Gravity is the weakest of the four basic forces. It is negligible in the interactions of elementary particles and thus plays no role in the behavior of molecules, atoms, and nuclei. The gravitational attraction between objects of the size we ordinarily encounter, for example, the gravitational force of attraction exerted by a building on a car, is too small to be readily noticed. Yet when we consider objects of astronomical size, such as moons, planets, and stars, gravity is of primary importance. The gravitational force exerted by the earth on us and on the objects around us is a fundamental part of our experience. It is gravity that binds us to the earth and keeps the earth and the other planets on course within the solar system. The gravitational force plays an important role in the life history of stars and in the behavior of galaxies. On the largest of all scales, it is gravity that controls the evolution of the universe.

At the time of Newton many believed that nature followed different rules in the larger universe than here on earth. Newton's law of universal gravity, along with his three laws of motion, revealed that nature follows the same rules everywhere, and this revelation has had a profound effect on our view of the universe. ➤ **In this chapter we use the tools of conservation of angular momentum, conservation of energy, Newton's laws of motion, and Newton's law of gravity to predict the motion of the planets and other celestial bodies, including those we have put there.**

11-1 Kepler's Laws

The nighttime sky with its myriad stars and shining planets has always fascinated us on earth. Toward the end of the sixteenth century, the astronomer Tycho Brahe studied the motions of the planets and made observations that were considerably more accurate than those previously available. Using Brahe's data, Johannes Kepler discovered that the paths of the planets about the sun are ellipses (Figure 11-1). He also showed that each planet moves faster when its orbit brings it closer to the sun and slower when its orbit takes it farther away. Finally, Kepler developed a precise mathematical relation between the period of a planet and its average distance from the sun (see Table 11.1). He stated these results in three empirical laws of planetary motion. Ultimately, these laws provided the basis for Newton's discovery of the law of gravity. Kepler's three laws follow.

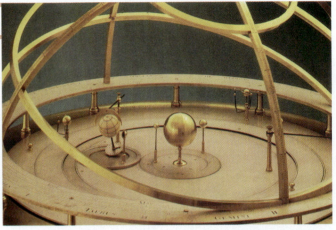

A mechanical model of the solar system, called an orrery, in the collection of Historical Scientific Instruments at Harvard University.

> **Law 1.** All planets move in elliptical orbits with the sun at one focus.

An ellipse is the locus of points for which the sum of the distances from two foci F is constant, as shown in Figure 11-2. Figure 11-3 shows a planet following an elliptical path with the sun at one focus. The earth's orbit is nearly circular, with the distance to the sun at perihelion (closest point) being 1.48×10^{11} m and at aphelion (farthest point) being 1.52×10^{11} m. The semimajor axis equals the average of these two distances, which is 1.50×10^{11} m (93 million miles) for the earth's orbit. The mean earth–sun distance defines the astronomical unit (AU):

$$1 \text{ AU} = 1.50 \times 10^{11} \text{ m} = 93.0 \times 10^6 \text{ mi} \qquad 11\text{-}1$$

The AU is used frequently in problems dealing with the solar system.

TABLE 11-1

Mean Orbital Radii and Orbital Periods for the Planets

Planet	Mean Radius r ($\times 10^{10}$ m)	Period T (y)
Mercury	5.79	0.241
Venus	10.8	0.615
Earth	15.0	1.00
Mars	22.8	1.88
Jupiter	77.8	11.9
Saturn	143	29.5
Uranus	287	84
Neptune	450	165
Pluto	590	248

Mercury
Venus
Earth
Mars
Jupiter
Saturn
Uranus
Neptune
Pluto

FIGURE 11-1 Orbits of the planets around the sun.

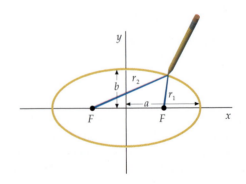

FIGURE 11-2 An ellipse is the locus of points for which $r_1 + r_2 =$ constant. The distance a is called the *semimajor* axis, and b is the *semiminor* axis. You can draw an ellipse with a piece of string by fixing each end at a focus F and using it to guide the pencil. Circles are special cases in which the two foci coincide.

Law 2. A line joining any planet to the sun
sweeps out equal areas in equal times.

Figure 11-4 illustrates Kepler's second law, the
law of equal areas. A planet moves faster when it is
closer to the sun than when it is farther away, so
that the area swept out by the radius vector in a
given time interval is the same throughout the orbit. The law of equal areas is a consequence of the
conservation of angular momentum, as we will see
in the next section.

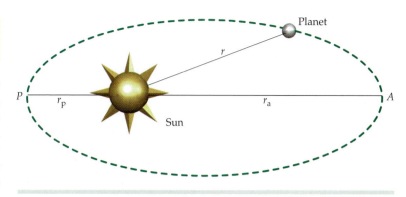

FIGURE 11-3 The elliptical path of a planet with the sun at one focus.
Point P, where the planet is closest to the sun, is called the perihelion, and
point A, where it is farthest, is called the aphelion. The average distance
between the planet and the sun, defined as $(r_p + r_a)/2$, is equal to the
semimajor axis.

Law 3. The square of the period of any
planet is proportional to the cube of
the semimajor axis of its orbit.

Kepler's third law relates the period of any
planet to its mean distance from the sun, which
equals the semimajor axis of its elliptical path. In
algebraic form, if r is the mean distance between a
planet and the sun and T is the planet's period of
revolution, Kepler's third law states that

$$T^2 = Cr^3 \qquad\qquad 11\text{-}2$$

where the constant C has the same value for all the
planets. This law is a consequence of the fact that
the force exerted by the sun on a planet varies inversely with the square of the distance from the
sun to the planet. We will demonstrate this in Section 11-2 for the special case of a circular orbit.

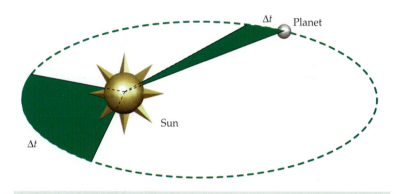

FIGURE 11-4 When a planet is close to the sun, it moves faster than
when it is farther away. The areas swept out by the radius vector in a given
time interval are equal.

JUPITER'S ORBIT   **EXAMPLE 11-1**

**The mean distance from the sun to Jupiter is 5.20 AU. What is the period of
Jupiter's orbit around the sun?**

PICTURE THE PROBLEM We use Kepler's third law to relate the period of
Jupiter to its mean orbital radius. The constant C can be obtained from the known
mean distance and period of the earth. Let $T_E = 1$ y and $r_E = 1$ AU be the period
and mean distance for the earth, and let T_J and $r_J = 5.20$ AU be the period and
mean distance for Jupiter.

1. Kepler's third law relates Jupiter's period T_J and mean distance r_J: $T_J^2 = Cr_J^3$

2. Apply Kepler's third law to the earth to obtain a second equation $T_E^2 = Cr_E^3$
   relating the same constant C to T_E and r_E:

3. Divide the two equations, eliminating C, and solve for T_J: $\dfrac{T_J^2}{T_E^2} = \dfrac{r_J^3}{r_E^3}$

   so

$$T_J = T_E\left(\frac{r_J}{r_E}\right)^{3/2} = (1\text{ y})\left(\frac{5.20\text{ AU}}{1\text{ AU}}\right)^{3/2}$$

$$= \boxed{11.9\text{ y}}$$

REMARKS The periods of the planets Earth, Jupiter, Saturn, Uranus, and Neptune are plotted in Figure 11-5 as functions of their mean distances from the sun. In (a), periods are plotted versus mean distances from the sun. In (b), the squares of the periods are plotted versus the cubes of the mean distances from the sun. Here the points fall on a straight line.

FIGURE 11-5

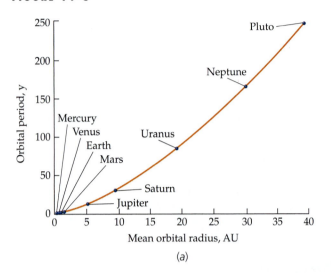

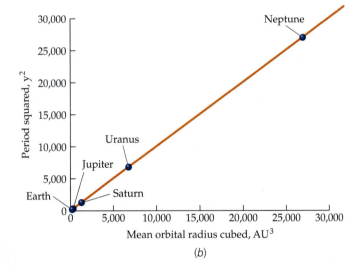

(a)

(b)

EXERCISE The period of Neptune is 164.8 y. What is its mean distance from the sun? (*Answer* 30.1 AU)

EXERCISE If the logs of the periods of the planets Earth, Jupiter, Saturn, Uranus, and Neptune are plotted versus the logs of their mean distances from the sun, the points fall on a curve. What is the shape of this curve? (*Answer* A straight line)

11-2 Newton's Law of Gravity

Although Kepler's laws were an important first step in understanding the motion of planets, they were nothing more than empirical rules obtained from the astronomical observations of Brahe. It remained for Newton to take the next giant step by attributing the acceleration of a planet in its orbit to a specific force exerted on it by the sun. Newton proved that a force that varies inversely with the square of the distance between the sun and a planet results in an elliptical orbit, as observed by Kepler. He then made the bold assumption that this force acts between any two objects in the universe. Before Newton, it was not even generally accepted that the laws of physics observed on earth were applicable to the heavenly bodies. **Newton's law of gravity** postulates that there is a force of attraction between each pair of point particles that is proportional to the product of the masses of the particles and inversely proportional to the square of the distance separating them. Let m_1 and m_2 be the masses of point particles 1 and 2 (at positions $\vec{r}_1$ and $\vec{r}_2$, respectively) and $\vec{r}_{1,2}$ be the vector pointing from particle 1 to particle 2 (Figure 11-6a).

The force $\vec{F}_{1,2}$ exerted by particle 1 on particle 2 is then

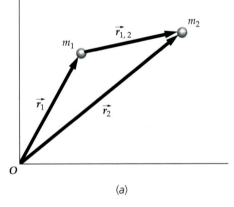

(a)

FIGURE 11-6 (a) Particles at $\vec{r}_1$ and $\vec{r}_2$. (b) The particles exert equal and opposite forces on each other.

$$\vec{F}_{1,2} = -\frac{Gm_1m_2}{r_{1,2}^2}\hat{r}_{1,2}$$

11-3

NEWTON'S LAW OF GRAVITY

where $\hat{r}_{1,2} = \vec{r}_{1,2}/r_{1,2}$ is a unit vector pointing from 1 to 2 and G is the **universal gravitational constant,** which has the value

$$G = 6.67 \times 10^{-11} \, \text{N·m}^2/\text{kg}^2 \qquad \text{11-4}$$

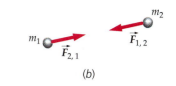

(b)

The force $\vec{F}_{2,1}$ exerted by 2 on 1 is the negative of $\vec{F}_{1,2}$, according to Newton's third law (Figure 11-6b). The magnitude of the gravitational force exerted by a point particle of mass m_1 on another point particle of mass m_2 a distance r away is thus given by

$$F = \frac{Gm_1m_2}{r^2} \qquad \text{11-5}$$

Newton published his theory of gravitation in 1686, but it was not until a century later that an accurate experimental determination of G was made by Cavendish, as will be discussed in Section 11-4.

We can use the known value of G to compute the gravitational attraction between two ordinary objects.

EXERCISE Find the gravitational force that attracts a 65-kg man to a 50-kg woman when they are 0.5 m apart. Model them as point particles. (*Answer* 8.67×10^{-7} N)

This exercise demonstrates that the gravitational force exerted by an object of ordinary size on another such object is so small as to be unnoticeable. For comparison, a mosquito weighs about 1×10^{-7} N. The weight of a 50-kg person is 491 N, about half a billion times the force of attraction calculated in the exercise! Gravitational attraction is easily noticed only when at least one of the objects is astronomically massive. The gravitational attraction between the girl and the earth for example, is readily apparent.

To check the validity of the inverse-square nature of the gravitational force, Newton compared the acceleration of the moon in its orbit with the free-fall acceleration of objects near the surface of the earth (such as the legendary apple). He assumed that the gravitational attraction due to the earth causes both accelerations. He first assumed that the earth and moon could be treated as point particles with their total masses concentrated at their centers. The force on a particle of mass m a distance r from the center of the earth is

$$F = \frac{GM_Em}{r^2} \qquad \text{11-6}$$

If this is the only force acting on the particle, then its acceleration is

$$a = \frac{F}{m} = \frac{GM_E}{r^2} \qquad \text{11-7}$$

For objects near the surface of the earth, $r = R_E$ and the free-fall acceleration is g:

$$g = \frac{GM_E}{R_E^2} \qquad \text{11-8}$$

The distance to the moon is about 60 times the radius of the earth ($r = 60R_E$). Substituting this into Equation 11-7 gives $a = g/60^2$, so the acceleration of the moon in its near-circular orbit is the free-fall acceleration g near the surface of the earth divided by 60^2. That is, the acceleration of the moon a_m should be $(9.81 \, \text{m/s}^2)/60^2$. The moon's acceleration can be calculated from its known distance from the center of the earth, $r = 3.84 \times 10^8$ m, and its period $T = 27.3 \, \text{d} = 2.36 \times 10^6$ s:

$$a_m = \frac{v^2}{r} = \frac{(2\pi r/T)^2}{r} = \frac{4\pi^2 r}{T^2} = \frac{4\pi^2(3.84 \times 10^8 \text{ m})}{(2.36 \times 10^6 \text{ s})^2} = 2.72 \times 10^{-3} \text{ m/s}^2$$

Then

$$\frac{g}{a_m} = \frac{9.81 \text{ m/s}^2}{2.72 \times 10^{-3} \text{ m/s}^2} = 3604 \approx 60^2$$

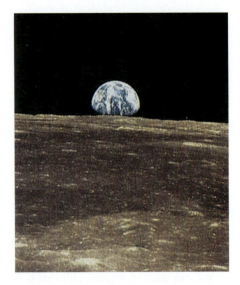

Earth as seen from Apollo 11 orbiting the moon on July 16, 1969.

In Newton's words, "I thereby compared the force requisite to keep the Moon in her orb with the force of gravity at the surface of the Earth, and found them answer pretty nearly."

The assumption that the earth and moon can be treated as point particles in the calculation of the force on the moon is reasonable because the earth-to-moon distance is large compared with the radius of either the earth or the moon, but such an assumption is certainly questionable when applied to an object near the earth's surface. After considerable effort, Newton was able to prove that the force exerted by any object with a spherically symmetric mass distribution on a point mass either on or outside its surface is the same as if all the mass of the object were concentrated at its center. The proof involves integral calculus, which Newton developed to solve this problem.

Because $g = 9.81 \text{ m/s}^2$ is readily measured and the radius of the earth is known, Equation 11-8 can be used to determine the value of the product GM_E. Newton estimated the value of G from an approximation of the mass of the earth. When Cavendish determined G some 100 years later by measuring the force between small spheres of known mass and separation, he called his experiment "weighing the earth." Knowing the value of G meant that the mass of the sun and the mass of any planet with a satellite could be determined. The method for doing this is described in Section 11-4.

FALLING TO EARTH   **EXAMPLE 1 1 - 2**

What is the free-fall acceleration of an object at the altitude of the space shuttle's orbit, about 400 km above the earth's surface?

PICTURE THE PROBLEM The force is given by Equation 11-6 with $r = R_E + 400$ km.

1. The acceleration is given by $a = F/m$, where F is given by Newton's law of gravity:

$$a = \frac{F}{m} = \frac{GmM_E/r^2}{m} = \frac{GM_E}{r^2}$$

2. The distance r is related to the radius of the earth R_E and the altitude h:

$$r = R_E + h = 6370 \text{ km} + 400 \text{ km}$$
$$= 6770 \text{ km}$$

3. The acceleration is then:

$$a = \frac{GM_E}{r^2}$$

$$= \frac{(6.67 \times 10^{-11} \text{ N·m}^2/\text{kg}^2)(5.98 \times 10^{24} \text{ kg})}{(6.77 \times 10^6 \text{ m})^2}$$

$$= \boxed{8.70 \text{ m/s}^2}$$

REMARKS This is also the acceleration of the "weightless" shuttle astronauts as they accelerate in their circular orbit.

The calculation in Example 11-2 can be simplified by using Equation 11-8 to eliminate GM_E from Equation 11-7. Then the acceleration at a distance r is

$$a = \frac{F}{m} = \frac{GM_E}{r^2} = g\frac{R_E^2}{r^2} \qquad 11\text{-}9$$

EXERCISE At what distance h above the surface of the earth is the acceleration of gravity half its value at sea level? (*Answer* 2640 km)

Measurement of G

The universal gravitational constant G was first measured in 1798 by Henry Cavendish, who used the apparatus shown in Figure 11-7. Cavendish's measurement of G has been repeated by other experimenters with various improvements and refinements. All measurements of G are difficult because of the extreme weakness of the gravitational attraction. Consequently, the value of G is known today only to about 1 part in 10,000. Although G was one of the first physical constants ever measured, it remains one of the least accurately known.

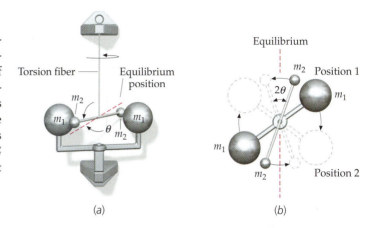

(a)   *(b)*

Gravitational and Inertial Mass

The property of an object responsible for the gravitational force it exerts on another object is its *gravitational* mass. On the other hand, the property of an object that measures its resistance to acceleration is its *inertial* mass. We have used the same symbol m for these two properties because, experimentally, they are proportional. For convenience, units are judiciously defined to make the proportionality constant one. The fact that the gravitational force exerted on an object is proportional to its inertial mass is a characteristic unique to the force of gravity. One consequence is that all objects near the surface of the earth fall with the same acceleration if air resistance is neglected. The well-known story of Galileo dropping objects from the Leaning Tower of Pisa to demonstrate that the free-fall acceleration is the same for objects with different inertial masses is just one example of the excitement this discovery aroused in the sixteenth century.

We could easily imagine that the gravitational and inertial masses of an object were not the same. Suppose we write m_G for the gravitational mass and m for the inertial mass. The force exerted by the earth on an object near its surface would then be

$$F = \frac{GM_E m_G}{R_E^2} \qquad 11\text{-}10$$

where M_E is the gravitational mass of the earth. The free-fall acceleration of the object near the earth's surface would then be

$$a = \frac{F}{m} = \left(\frac{GM_E}{R_E^2}\right)\frac{m_G}{m} \qquad 11\text{-}11$$

If gravity were just another property of matter, like color or hardness, it might be

FIGURE 11-7 (*a*) Two small spheres, each of mass m_2, are at the ends of a light rod that is suspended by a fine fiber. Careful measurements determine the torque required to turn the fiber through a given angle. Two large spheres, each of mass m_1, are then placed near the small spheres. Because of the gravitational attraction of the large spheres of mass m_1 for the small spheres, the fiber is turned through a very small angle θ from its equilibrium position. (*b*) The apparatus as seen from above. After the apparatus comes to rest, the positions of the large spheres are reversed, as shown by the dashed lines, so that they are at the same distance from the equilibrium position of the balance but on the other side. If the apparatus is again allowed to come to rest, the fiber will turn through angle 2θ in response to the reversal of the torque. Once the torsion constant has been determined, the forces between the masses m_1 and m_2 can be determined from the measurement of this angle. Since the masses and their separations are known, G can be calculated. Cavendish obtained a value for G within about 1 percent of the presently accepted value given by Equation 11-4.

A gravitational torsion balance used in student labs for the measurement of G. A tiny angular deflection of the balance results in a large angular deflection of the laser beam that reflects from a mirror on the balance.

reasonable to expect that the ratio m_G/m would depend on such things as the chemical composition of the object or its temperature. The free-fall acceleration would then be different for different objects. **The experimental evidence, however, is that a is the same for all objects.** Thus we need not maintain the distinction between m_G and m and can set $m_G = m$. We must keep in mind, however, that the equivalence of gravitational and inertial mass is an empirical law that is limited by the accuracy of experiment. Experiments testing this equivalence were carried out by Simon Stevin in the 1580s. Galileo publicized this law widely, and his contemporaries made considerable improvements in the experimental accuracy with which the law was established.

The most precise early comparisons of gravitational and inertial mass were made by Newton. Through experiments using simple pendulums rather than falling bodies, Newton was able to establish the equivalence between gravitational and inertial mass to an accuracy of about 1 part in 1000. Experiments comparing gravitational and inertial mass have improved steadily over the years. Their equivalence is now established to about 1 part in 10^{12}. Thus, the equivalence of gravitational and inertial mass is one of the best established of all physical laws. It is the basis for the principle of equivalence, which is the foundation of Einstein's general theory of relativity.

Derivation of Kepler's Laws

Newton showed that when an object such as a planet or comet moves around a $1/r^2$ force center such as the sun, the object's path is a conic section (an ellipse, a parabola, or a hyperbola). The parabolic and hyperbolic paths apply to objects that make one pass by the sun and never return. Such orbits are not closed. The only closed orbits in an inverse-square force field are ellipses. Thus, Kepler's first law is a direct consequence of Newton's law of gravity. Kepler's second law, the law of equal areas, follows from the fact that the force exerted by the sun on a planet is directed toward the sun. Such a force is called a **central force**. Figure 11-8a shows a planet moving in an elliptical orbit around the sun. In time dt, the planet moves a distance $v\, dt$ and the radius vector $\vec{r}$ sweeps out the area shaded in the figure. This is half the area of the parallelogram formed by the vectors $\vec{r}$ and $\vec{v}\, dt$, which is $|\vec{r} \times \vec{v}\, dt|$. Thus the area dA swept out by the radius vector $\vec{r}$ in time dt is

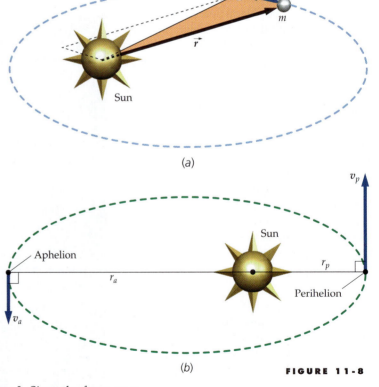

(a)

$$dA = \frac{1}{2}|\vec{r} \times \vec{v}\, dt| = \frac{1}{2m}|\vec{r} \times m\vec{v}|\, dt$$

or

$$\frac{dA}{dt} = \frac{1}{2m}L \qquad \text{11-12}$$

where $L = |\vec{r} \times m\vec{v}|$ is the magnitude of the orbital angular momentum of the planet about the sun. The area dA swept out in a given time interval dt is therefore proportional to the magnitude of the angular momentum L. Since the force on a planet is along the line from the planet to the sun, it has no torque about the sun. Thus the magnitude of the angular momentum of the planet is conserved; that is, L is constant. Therefore, the rate at which area is swept out is the same for all parts of the orbit, which is Kepler's second law. Also, the fact that L is constant

(b)

FIGURE 11-8

means that $rv \sin \phi$ is constant. At aphelion and perihelion $\phi = 90°$ (Figure 11-8*b*), so $r_a v_a = r_p v_p$.

We will now show that Newton's law of gravity implies Kepler's third law for the special case of a circular orbit. Consider a planet moving with speed v in a circular orbit of radius r about the sun. The gravitational force of attraction between the sun and the planet provides the centripetal acceleration v^2/r. Newton's second law gives

$$F = M_p a$$

$$\frac{GM_s M_p}{r^2} = M_p \frac{v^2}{r} \qquad\qquad 11\text{-}13$$

where M_s is the mass of the sun and M_p is that of the planet. Solving for v^2,

$$v^2 = \frac{GM_s}{r} \qquad\qquad 11\text{-}14$$

Because the planet moves a distance $2\pi r$ in time T, its speed is related to the period by

$$v = \frac{2\pi r}{T} \qquad\qquad 11\text{-}15$$

Substituting this expression for v in Equation 11-14, we obtain

$$v^2 = \frac{4\pi^2 r^2}{T^2} = \frac{GM_s}{r}$$

or

$$T^2 = \frac{4\pi^2}{GM_s} r^3 \qquad\qquad 11\text{-}16$$

KEPLER'S THIRD LAW

Equation 11-16 is Kepler's third law, which is the same as Equation 11-2 with $C = 4\pi^2/GM_s$. Equation 11-16 also applies to the orbits of the satellites of any planet if we replace the mass of the sun M_s with the mass of the planet.

THE ORBITING SPACE STATION **EXAMPLE 11-3** Put It in Context

The International Space Station travels in a roughly circular orbit around the earth. If its altitude is 385 km above the earth's surface, how long do you have to wait between sightings? (Assume that air resistance can be neglected.)

PICTURE THE PROBLEM The sightings occur only at night and then only if the space station is above the horizon at your location. Thus, the minimum time between sightings is approximately equal to the orbital period. To find the orbital period we use Kepler's third law with M_s in Equation 11-16 replaced by the mass of the earth M_E. The numerical calculation is simplified somewhat by using $GM_E = R_E^2 g$ from Equation 11-8.

1. Apply Kepler's third law to the space station:

$$T^2 = \frac{4\pi^2}{GM_E} r^3$$

2. At an altitude $h = 385$ km, $r = R_E + h = 6760$ km. Substitute $r = R_E + h$ and solve for the period:

$$T^2 = \frac{4\pi^2}{GM_E}(R_E + h)^3$$

3. Use $GM_E = R_E^2 g$ to write T in terms of g:

$$T^2 = \frac{4\pi^2}{R_E^2 g}(R_E + h)^3$$

so

$$T = \frac{2\pi}{R_E\sqrt{g}}(R_E + h)^{3/2}$$

$$= \frac{2\pi(6.755 \times 10^6 \text{ m})^{3/2}}{(6.37 \times 10^6 \text{ m})(9.81 \text{ m/s}^2)^{1/2}}$$

$$= 5529 \text{ s} = \boxed{92.1 \text{ min}}$$

EXERCISE How many degrees does the earth rotate in a single orbital period of the ISS? (*Answer* The earth rotates $360°$ in 24 h = 1440 min. In 92.1 min it rotates $23.0°$.)

REMARKS The near-circular orbit of the ISS, which is inclined $\approx52°$ to the equatorial plane, does not rotate with the earth. If the ISS is directly over your home at time t, 92.1 min later it will be directly over a location $23.0°$ due west of your home. For example, Kansas City is $23°$ due west of New York City. If the ISS passes over your home in New York City at midnight Eastern Time, you could tell your friend in Kansas City that it will pass over Kansas City at 12:32 A.M. Central Time (1:32 A.M. Eastern Time).

EXERCISE Find the radius of a circular orbit of a satellite that orbits the earth with a period of 1 d. (*Answer* $r = 6.63R_E = 4.22 \times 10^7$ m = 26,200 mi. If such a satellite is in orbit over the equator and moves in the same direction as the rotation of the earth, it appears stationary relative to the earth. Many satellites are "parked" in such an orbit, called a geosynchronous orbit.)

Because G is known, by using Equation 11-16 we can determine the mass of an astronomical object by measuring the period T and the mean orbital radius r of a satellite orbiting it. In establishing Equation 11-16 the mass of the satellite was assumed negligible compared to the mass of the central object. This means that the central object remains stationary as the satellite revolves around it. In fact, the central object and satellite both revolve around a common point, their center of mass. If the mass of the satellite is not assumed negligible, the result is

$$T^2 = \frac{4\pi^2}{G(M_1 + M_2)}r^3 \qquad\qquad 11\text{-}17$$

where r is the center-to-center separation of the objects. (For the more general elliptical orbits, the math is more challenging but the result is the same.) If the mass of the satellite is not negligible, as is the case with most binary star systems, then only the sum of the masses is determined, as revealed by Equation 11-17. The moon, along with planets Mercury and Venus, have no natural satellites, so their masses were not well known until the 1960s when artificial satellites were first placed in orbit around them.

EXERCISE The Martian moon Phobos has a period of 460 min and a mean orbital radius of 9400 km. What is the mass of Mars? (*Answer* 6.45×10^{23} kg = $0.108 M_E$)

EXPLORING

What is the difference between gravitational mass and inertial mass? Find out this, and more, at www.whfreeman.com/tipler5e.

11-3 Gravitational Potential Energy

Near the surface of the earth, the gravitational force exerted by the earth on an object is essentially uniform because the distance to the center of the earth, $r = R_E + h$, is always approximately R_E for $h \ll R_E$. The potential energy of an object near the earth's surface is $mgh = mg(r - R_E)$, where we have chosen $U = 0$ at the earth's surface, $r = R_E$. When we are far from the surface of the earth, we must take into account the fact that the gravitational force exerted by the earth is not uniform but decreases as $1/r^2$. The general definition of potential energy (Equation 6-19b) is

$$dU = -\vec{F} \cdot d\vec{s}$$

where $\vec{F}$ is the conservative force on a particle and $d\vec{s}$ is a general displacement of the particle. For the radial gravitational force $\vec{F}$ given by Equation 11-6 we have

$$dU = -\vec{F} \cdot d\vec{s} = -F\hat{r} \cdot d\vec{s} = -F_r\, dr = -\left(-\frac{GM_E m}{r^2}\right) dr = \frac{GM_E m}{r^2}\, dr \qquad 11\text{-}18$$

Integrating both sides of this equation we obtain

$$U = -\frac{GM_E m}{r} + U_0 \qquad 11\text{-}19$$

where U_0 is a constant of integration. Since only changes in potential energy are important, we can choose the potential energy to be zero at any position. The earth's surface is a good choice for many everyday problems, but it is not always a convenient choice. For example, when considering the potential energy associated with a planet and the sun, there is no reason to want the potential energy to be zero at the surface of the sun. In fact, it is nearly always more convenient to choose the gravitational potential energy of a two-object system to be zero when the separation of the objects is infinite. Thus, $U_0 = 0$ is often a convenient choice. Then

$$U(r) = -\frac{GMm}{r} \qquad 11\text{-}20$$

GRAVITATIONAL POTENTIAL ENERGY WITH $U = 0$ AT INFINITE SEPARATION

where $U = 0$ if $r = \infty$. Figure 11-9 is a plot of $U(r)$ versus r for this choice of $U = 0$ at $r = \infty$ for an object of mass m and the earth of mass M_E. This function begins at the negative value $U = -GM_E m/R_E = -mgR_E$ at the earth's surface and increases as r increases, approaching zero at infinite r. The slope of this curve at $r = R_E$ is

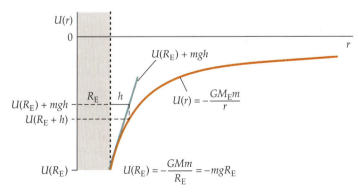

FIGURE 11-9

$GM_Em/R_E^2 = mg$, so the equation of the tangent line, drawn in blue, is $f(h) = U(R_E) + mgh$, where $h = r - R_E$ is the distance above the earth's surface. From the figure you can see that for small h, $U(R_E) + mgh \approx U(r)$.

Escape Speed

During the last half-century, the idea of escaping from the earth's gravity has changed from fantasy to reality. Space probes have been sent out to the far reaches of the solar system. Some of these probes orbit the sun, while others leave the solar system and drift on into outer space. We will see that there is a minimum initial speed, called the **escape speed,** that is required for an object in free-fall to escape from the earth.

If we project an object upward from the earth with some initial kinetic energy, the kinetic energy decreases and the potential energy increases as the object rises. But the maximum increase in potential energy is GM_Em/R_E. Therefore, this amount is the most that the kinetic energy can decrease. If the initial kinetic energy is greater than GM_Em/R_E, then the total energy E will be greater than zero (E_2 in Figure 11-10), and the object will still have some kinetic energy when r is very large (or even when r approaches infinity). Thus, if the initial kinetic energy is greater than GM_Em/R_E, the object will escape from the earth's gravity. Since the potential energy at the earth's surface is $-GM_Em/R_E$, the total energy $E = K + U$ must be greater than or equal to zero in order for the object to escape. The speed near the earth's surface corresponding to zero total energy is called the escape speed v_e. It is found from

$$K_f + U_f = K_i + U_i$$

$$0 + 0 = \frac{1}{2}mv_e^2 - \frac{GM_Em}{R_E}$$

so

$$v_e = \sqrt{\frac{2GM_E}{R_E}} = \sqrt{2gR_E} \qquad 11\text{-}21$$

ESCAPE SPEED

Using $g = 9.81 \text{ m/s}^2$ and $R_E = 6.37 \times 10^6$ m, we obtain

$$v_e = \sqrt{2(9.81 \text{ m/s}^2)(6.37 \times 10^6 \text{ m})} = 11.2 \text{ km/s}$$

This is about 7 mi/s or 25,000 mi/h. An object with this speed will escape the earth's gravity but it will not escape the solar system because we have neglected the gravitational attraction of the sun and other planets (see Problems 46 and 48).

The escape speed for a planet or moon relative to the thermal speeds of gas molecules determines the kind of atmosphere a planet or moon can have. The average kinetic energy of gas molecules, $(\frac{1}{2}mv^2)_{av}$, is proportional to the absolute temperature T (Chapter 18). Near the surface of the earth, the speeds of nearly all of the oxygen and nitrogen molecules are much lower than the escape speed, so these gases are retained in our atmosphere. For the lighter molecules hydrogen and

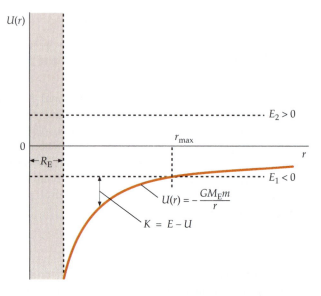

FIGURE 11-10 The kinetic energy of an object at a distance r from the center of the earth is $E - U(r)$. When the total energy is less than zero (E_1 in the figure), the kinetic energy is zero at $r = r_{max}$ and the object is bound to the earth. When the total energy is greater than zero (E_2 in the figure), the object can escape the earth.

helium, however, a significant fraction have speeds greater than the escape speed. Hydrogen and helium gases are therefore not found in our atmosphere. The escape speed at the surface of the moon is 2.3 km/s, which can be calculated from Equation 11-21 with the mass and radius of the moon replacing M_E and R_E. This is considerably smaller than the escape speed for earth and, in fact, is too small for any atmosphere to exist.

EXERCISE Find the escape speed at the surface of Mercury, which has a mass $M = 3.31 \times 10^{23}$ kg and a radius $R = 2440$ km. (*Answer* $v_e = \sqrt{2GM/R} = 4.25$ km/s)

Classification of Orbits by Energy

In Figure 11-10, two possible values for the total energy E are indicated on a graph of U versus r: E_1, which is negative, and E_2, which is positive. A negative total energy simply means that the kinetic energy at the earth's surface is less than $+GM_Em/R_E$, so that $K + U$ is never greater than zero. From this figure we see that, if the total energy is negative, the total-energy line intersects the potential-energy curve at some maximum separation r_{max} and the system is bound. On the other hand, if the total energy is zero or positive, there is no such intersection and the system is unbound. The criteria for a bound or unbound system are simply stated:

> If $E < 0$, the system is bound.
>
> If $E \geq 0$, the system is unbound.

When E is negative, its absolute value $|E|$ is called the binding energy. The binding energy is the energy that must be added to the system to bring the total energy up to zero.

The potential energy of an object such as a planet or comet of mass m at a distance r from the sun is

$$U(r) = -\frac{GM_s m}{r} \qquad\qquad 11\text{-}22$$

where M_s is the mass of the sun. The kinetic energy of the object is $\frac{1}{2}mv^2$. If the total energy, kinetic plus potential, is less than zero, then the orbit will be an ellipse (possibly a circle), and the object will be bound to the sun. If, instead, the total energy is positive, then the orbit will be a hyperbola, and the object will make one swing around the sun and leave the solar system, never to return. If the total energy is exactly zero, the orbit will be a parabola, and again the object will make one pass and then escape. To summarize, when the total energy is zero or positive the object is not bound to the sun, but will escape. Curiously, there have not been any measurements of the energy E of a comet or an asteroid that are definitely nonnegative. Thus, all observed objects of this type are bound to the solar system.

HEIGHT OF A PROJECTILE **EXAMPLE 11-4**

A projectile is fired straight up from the surface of the earth with an initial speed $v_i = 8$ km/s. Find the maximum height it reaches, neglecting air resistance.

PICTURE THE PROBLEM The maximum height is found using energy conservation. We take the surface of the earth as the initial point, with $U_i = -GM_Em/R_E$ and $K_i = \frac{1}{2}mv_i^2$. At the greatest height, $K_f = 0$.

1. Apply conservation of mechanical energy:

$$K_f + U_f = K_i + U_i$$

$$0 - \frac{GM_E m}{r_f} = \frac{1}{2}mv_i^2 - \frac{GM_E m}{R_E}$$

2. Multiply through by $-1/(GM_E m)$, use $GM_E = gR_E^2$, and solve for r_f:

$$\frac{1}{r_f} = -\frac{v_i^2}{2GM_E} + \frac{1}{R_E} = -\frac{v_i^2}{2gR_E^2} + \frac{1}{R_E}$$

so

$$r_f = 1/\left(\frac{1}{R_E} - \frac{v_i^2}{2gR_E^2}\right) = R_E/\left(1 - \frac{v_i^2}{2gR_E}\right)$$

3. Substitute numerical values to find r and $h = r - R_E$:

$$r_f = R_E/\left(1 - \frac{(8000 \text{ m/s})^2}{2(9.81 \text{ m/s}^2)(6.37 \times 10^6 \text{ m})}\right)$$

$$= 2.05R_E$$

$$h = r - R_E = \boxed{1.05R_E = 6.69 \times 10^6 \text{ m}}$$

SPEED OF A PROJECTILE   **EXAMPLE 11-5**   **Try It Yourself**

A projectile is fired straight up from the surface of the earth with an initial speed $v_i = 15$ km/s. Find the speed of the projectile when it is very far from the earth, neglecting air resistance.

PICTURE THE PROBLEM Very far from earth means $r \gg R_E$. The initial speed is greater than the escape speed of 11 km/s, so the total energy of the projectile is positive and the projectile will escape the earth with some final kinetic energy. Use conservation of mechanical energy to find this kinetic energy and then solve for the final speed.

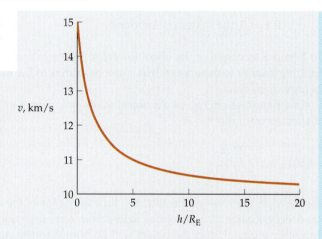

FIGURE 11-11

Cover the column to the right and try these on your own before looking at the answers.

Steps

1. Apply conservation of mechanical energy, noting that $r_f = \infty$, so $U_f = 0$.

2. Solve for v_f^2 using $GM_E/R_E^2 = g$ to simplify.

3. Substitute known values for g and R_E to calculate v_f.

Answers

$$K_f + U_f = K_i + U_i$$

$$\frac{1}{2}mv_f^2 + 0 = \frac{1}{2}mv_i^2 - \frac{GM_E m}{R_E}$$

$$v_f^2 = v_i^2 - \frac{2GM_E}{R_E} = v_i^2 - \frac{2gR_E^2}{R_E}$$

$$= v_i^2 - 2gR_E$$

$$v_f = \sqrt{(15 \times 10^3 \text{ m/s})^2 - 2(9.81 \text{ m/s}^2)(6.37 \times 10^6 \text{ m})}$$

$$= 10^4 \text{ m/s} = \boxed{10 \text{ km/s}}$$

REMARKS In Figure 11-11, the speed of the projectile in kilometers per second is plotted versus h/R_E, where h is the height above the earth's surface. At very large values of h/R_E, the speed approaches the horizontal line $v = 10$ km/s.

| TOTAL ENERGY OF A SATELLITE | EXAMPLE 11-6 Try It Yourself |

Show that the total energy of a satellite in a circular orbit is half its potential energy.

PICTURE THE PROBLEM The total energy of a satellite is the sum of its potential and kinetic energies, $E = U + K$. Newton's second law will allow us to find the speed v of the satellite in terms of its orbital radius r. The kinetic energy depends on the speed, so we can find the kinetic energy in terms of r. Assume the mass of the earth is much greater than that of the satellite.

Cover the column to the right and try these on your own before looking at the answers.

Steps	Answers
1. Write the total energy equal to the sum of the potential energy and the kinetic energy.	$E = K + U$ $$= \frac{1}{2}mv^2 - \frac{GM_E m}{r}$$
2. Apply Newton's second law to the satellite and solve for the square of the speed.	$F = ma$ $$\frac{GM_E m}{r^2} = m\frac{v^2}{r}$$ so $$v^2 = \frac{GM_E}{r}$$
3. Substitute into the step 1 result and simplify.	$$E = \frac{1}{2}m\frac{GM_E}{r} - \frac{GM_E m}{r}$$ $$= \frac{GM_E m - 2GM_E m}{2r} = -\frac{GM_E m}{2r}$$
4. Compare the step 3 result with U in step 1.	$$E = -\frac{GM_E m}{2r} = \frac{1}{2}\left(-\frac{GM_E m}{r}\right) = \boxed{\frac{1}{2}U}$$

EXERCISE A satellite of mass 450 kg orbits the earth in a circular orbit at 6830 km above the earth's surface. Find (a) the potential energy, (b) the kinetic energy, and (c) the total energy of the satellite. (*Answer* Note that $r = R_E + h = $ 13,200 km (a) $U = -13.6 \times 10^9$ J (b) $K = 6.80 \times 10^9$ J (c) $E = -6.80 \times 10^9$ J)

11-4 The Gravitational Field $\vec{g}$

The gravitational force exerted by a point particle of mass m_1 on a second point particle of mass m_2 a distance $r_{1,2}$ away is given by

$$\vec{F}_{1,2} = -\frac{Gm_1 m_2}{r_{1,2}^2}\hat{r}_{1,2}$$

where $\hat{r}_{1,2} = \vec{r}_{1,2}/r_{1,2}$ is a unit vector pointing from particle 1 to particle 2. The gravitational field at point P is determined by placing a point particle of mass m at P and calculating the gravitational force $\vec{F}$ on it due to all other particles. The gravitational force $\vec{F}$ divided by the mass m is the **gravitational field** $\vec{g}$ at P:

$$\vec{g} = \frac{\vec{F}}{m}$$

11-23

DEFINITION—GRAVITATIONAL FIELD

The point P is called a **field point**. The gravitational field at a field point due to the masses of a collection of point particles is the vector sum of the fields due to the individual masses at that point:

$$\vec{g} = \sum_i \vec{g}_i$$

11-24a

The points where these point particles are located are called **source points**. To find the gravitational field at a field point due to a continuous object, we find the field $d\vec{g}$ due to a small element of volume with mass dm and integrate over the entire mass distribution of the object (the entire set of source points).

$$\vec{g} = \int d\vec{g}$$

11-24b

The gravitational field of the earth at a distance $r \geq R_E$ points toward the earth and has the magnitude $g(r)$ given by

$$g(r) = \frac{F}{m} = \frac{GM_E}{r^2}$$

11-25

GRAVITATIONAL FIELD OF THE EARTH

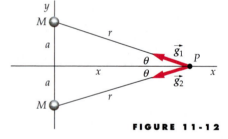

FIGURE 11-12

GRAVITATIONAL FIELD OF TWO POINT PARTICLES   **EXAMPLE 11-7**

Two point particles, each of mass M, are fixed on the y axis at $y = +a$ and $y = -a$ (see Figure 11-12). Find the gravitational field at all points on the x axis as a function of x.

PICTURE THE PROBLEM Two particles of mass M each produce a gravitational field at point P located at $x\hat{i}$. The distance between P and either particle is $r = \sqrt{x^2 + a^2}$. The resultant field $\vec{g}$ is the vector sum of the fields $\vec{g}_1$ and $\vec{g}_2$ due to each particle.

1. Calculate the magnitude of either $\vec{g}_1$ or $\vec{g}_2$:
$$g_1 = \frac{GM}{r^2}$$

2. The y component of the resultant field is zero. The x component is the sum of g_{1x} and g_{2x}:
$$g_x = g_{1x} + g_{2x} = 2g_{1x} = 2g_1 \cos\theta$$

3. Express $\cos\theta$ in terms of x and r from the figure:
$$\cos\theta = \frac{x}{r}$$

4. Combining the last two results yields $\vec{g}$. To express $\vec{g}$ as a function of x, substitute $(x^2 + a^2)^{1/2}$ for r:
$$\vec{g} = g_x\hat{i} = -2\frac{GM}{r^2}\frac{x}{r}\hat{i} = -\frac{2GMx}{r^3}\hat{i}$$

$$= -\frac{2GMx}{(x^2 + a^2)^{3/2}}\hat{i}$$

☻ PLAUSIBILITY CHECK For $x < 0$, $\vec{g}$ is in the positive x direction and for $x > 0$, $\vec{g}$ is in the negative direction, as expected. If $x = 0$, we find that $\vec{g} = 0$; the fields due to m_1 and m_2 are equal and opposite at $x = 0$, and hence they cancel. For $x \gg a$, $\vec{g} \approx -(2GM/x^2)\hat{i}$. The field is the same as if a single mass of $2M$ were at the origin.

GRAVITATIONAL FIELD OF A UNIFORM ROD   **EXAMPLE 11-8**

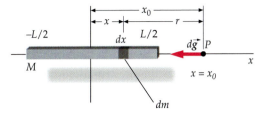

A uniform rod of mass M and length L is centered on the origin and lies along the x axis (Figure 11-13). Find the gravitational field due to the rod at all points on the x axis for $x > L/2$.

FIGURE 11-13

PICTURE THE PROBLEM Choose a mass element dm of length dx at $x\hat{\imath}$, and choose a field point P on the x axis at $x = x_0$ in the region $x > L/2$. Each mass element produces a gravitational field at P that points in the negative x direction. We can calculate the total field by integrating the x component of the field produced by dm from $x = -L/2$ to $x = +L/2$.

1. Find the x component of the field at P due to the element dm:

$$dg_x = -\frac{G\,dm}{r^2}$$

2. The mass dm is proportional to the size of the element dx:

$$dm = \frac{M}{L}\,dx$$

3. Write the distance r between dm and point P in terms of x and x_0:

$$r = x_0 - x$$

4. Substitute these results to express dg in terms of x:

$$dg_x = -\frac{G\,dm}{r^2} = -\frac{G(M/L)dx}{(x_0 - x)^2}$$

5. Integrate to find the x component of the total field:

$$g_x = \int dg_x = -\frac{GM}{L}\int_{-L/2}^{L/2}\frac{dx}{(x_0 - x)^2} = -\frac{GM}{L}\left[\frac{1}{x_0 - x}\right]_{-L/2}^{L/2}$$

$$= -\frac{GM}{L}\left(\frac{1}{x_0 - L/2} - \frac{1}{x_0 + L/2}\right) = -\frac{GM}{x_0^2 - (L/2)^2}$$

6. Express the resultant field as a vector:

$$\vec{g} = g_x\hat{\imath} = -\frac{GM}{x_0^2 - (L/2)^2}\,\hat{\imath}$$

7. x_0 is an arbitrary point on the x axis in the region $x > L/2$, so we can replace it with x:

$$\boxed{\vec{g} = -\frac{GM}{x^2 - (L/2)^2}\,\hat{\imath}}$$

PLAUSIBILITY CHECK For $x \gg L/2$, the field approaches that of a point particle of mass M, $\vec{g} = -(GM/x^2)\hat{\imath}$.

$\vec{g}$ of a Spherical Shell and of a Solid Sphere

One of Newton's motivations for developing calculus was to prove that the gravitational field outside a solid sphere is the same as if all the mass of the sphere were concentrated at its center. We will prove this in the next section. Here we merely discuss the results of this proof. We first consider a uniform spherical shell of mass M and radius R (Figure 11-14). We will show that the gravitational field due to the shell at a distance r from the center of the shell is given by

FIGURE 11-14 A uniform spherical shell of mass M and radius R.

$$\vec{g} = -\frac{GM}{r^2}\hat{r} \quad \text{for} \quad r > R \qquad \qquad 11\text{-}26a$$

$$\vec{g} = 0 \quad \text{for} \quad r < R \qquad \qquad 11\text{-}26b$$

GRAVITATIONAL FIELD OF A SPHERICAL SHELL

We can understand the result that $\vec{g} = 0$ inside the shell from Figure 11-15, which shows a point mass m_0 inside a spherical shell. In this figure, the masses of the shell segments with masses m_1 and m_2 are related by

$$\frac{m_1}{m_2} = \frac{r_1^2}{r_2^2} \quad \text{or} \quad \frac{m_1}{r_1^2} = \frac{m_2}{r_2^2}$$

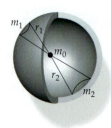

Since the gravitational force falls off inversely as the square of the distance, the force due to the smaller mass on the left is exactly balanced by that due to the more distant, larger mass on the right.

The gravitational field outside a solid sphere is a simple extension of Equation 11-26a. We merely consider the solid sphere to consist of a continuous set of spherical shells. Since the field due to each shell is the same as if its mass were concentrated at the center of the shell, the field due to the entire sphere is the same as if the entire mass of the sphere were concentrated at its center:

FIGURE 11-15 A point mass m_0 inside a uniform spherical shell feels no net force.

$$g_r = -\frac{GM}{r^2} \quad \text{for} \quad r > R \qquad\qquad 11\text{-}27$$

This result holds whether or not the sphere has a constant density, as long as the density depends only on r so that spherical symmetry is maintained.

M = total mass

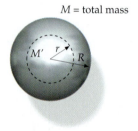

$\vec{g}$ Inside a Solid Sphere

We now use Equations 11-26a and 11-26b to find the gravitational field inside a solid sphere of constant density at a point a distance r from the center, where r is less than the radius R of the sphere. This would apply, for example, to finding the weight of an object at the bottom of a deep mine shaft. As we have seen, the field inside a spherical shell is zero. Thus, in Figure 11-16 the mass of that part of the sphere outside r exerts no force at or inside r. Therefore, only the mass M' within the radius r contributes to the gravitational field at r. This mass produces a field equal to that of a point mass M' at the center of the sphere. The fraction of the total mass of the sphere within r is equal to the ratio of the volume of a sphere of radius r to that of a sphere of radius R. Thus, for a uniform mass distribution, if M is the total mass of the sphere, M' is given by

FIGURE 11-16 A uniform solid sphere of radius R and mass M. Only the mass M', which is inside the sphere of radius r, contributes to the gravitational field at the distance r.

$$M' = M\frac{\frac{4}{3}\pi r^3}{\frac{4}{3}\pi R^3} = M\frac{r^3}{R^3} \qquad\qquad 11\text{-}28$$

The gravitational field at the distance r is thus

$$g_r = -\frac{GM'}{r^2} = -\frac{GM}{r^2}\frac{r^3}{R^3}$$

or

$$g_r = -\frac{GM}{R^3}r \quad \text{for} \quad r < R \qquad\qquad 11\text{-}29$$

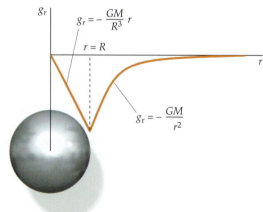

The magnitude of the field is zero at the center and increases with distance r inside the sphere. Figure 11-17 shows a plot of the field g_r as a function of r for a solid sphere of uniform mass density.

FIGURE 11-17 A plot of g_r versus r for a uniform solid sphere of mass M. The magnitude of the field increases linearly with r inside the sphere and decreases as $1/r^2$ outside the sphere.

A HOLLOW PLANET   **EXAMPLE   11-9**

A planet with a hollow core consists of a uniform, thick spherical shell with mass M, outer radius R, and inner radius $R/2$. (*a*) What amount of mass is closer than $\frac{3}{4}R$ to the center of the planet? (*b*) What is the gravitational field a distance $\frac{3}{4}R$ from the center?

PICTURE THE PROBLEM The mass of that part of the thick shell that is closer to the center than $\frac{3}{4}R$ is the density times the volume of the thick shell with outer radius $\frac{3}{4}R$ and inner radius $\frac{1}{2}R$. First find the density and the volume, then find the mass. The gravitational field at $r = \frac{3}{4}R$ is due only to the mass closer than this to the center.

(*a*) 1. The mass M' (the mass of the thick shell with outer radius $\frac{3}{4}R$ and inner radius $\frac{1}{2}R$) is the density ρ times the volume V':

$$M' = \rho V'$$

2. The density is the mass M divided by the volume V:

$$\rho = \frac{M}{V} = M / \left[\frac{4}{3}\pi R^3 - \frac{4}{3}\pi\left(\frac{R}{2}\right)^3 \right] = \frac{M}{7\pi R^3/6}$$

3. Find the volume V' of the thick shell with outer radius $\frac{3}{4}R$ and inner radius $\frac{1}{2}R$:

$$V' = \frac{4}{3}\pi\left(\frac{3R}{4}\right)^3 - \frac{4}{3}\pi\left(\frac{R}{2}\right)^3 = \frac{19}{48}\pi R^3$$

4. Find the mass M':

$$M' = \rho V' = \frac{6}{7}\frac{M}{\pi R^3}\frac{19}{48}\pi R^3 = \boxed{\frac{19}{56}M}$$

(*b*) The gravitational field at $r = \frac{3}{4}R$ is due only to the mass M':

$$\vec{g} = -\frac{GM'}{r^2}\hat{r} = -\frac{G\frac{19}{56}M}{(\frac{3}{4}R)^2}\hat{r}$$

$$= \boxed{-\frac{38}{63}\frac{GM}{R^2}\hat{r}}$$

PLAUSIBILITY CHECK We expect less than half the mass M to be in the region $\frac{1}{2}R < r < \frac{3}{4}R$, and that is the case here, $19/56$ is less than half.

REMARKS Note that the volume V in the denominator of Part (*a*) step 2 is $7\pi R^3 16$. That is more than half the volume of a sphere of radius R. We expect the volume of the region $\frac{1}{2}R < r < R$ to be more than half the volume of a sphere of radius R. Also, note that the volume V' (Part (*a*) step 3) and mass M' (Part (*a*) step 4) are the same fraction of V and M. This is as expected because the shell is uniform.

RADIALLY DEPENDENT DENSITY   **EXAMPLE   11-10**   Try It Yourself

A solid sphere of radius R and mass M is spherically symmetric but not uniform. Its density ρ, defined as its mass per unit volume, is proportional to the distance from the center r for $r \leq R$. That is, $\rho = Cr$ for $r \leq R$, where C is a constant. (*a*) Find C. (*b*) Find g_r for $r \geq R$. (*c*) Find g_r at $r = R/2$.

PICTURE THE PROBLEM (*a*) You can find C by integrating the density over the volume of the sphere and setting the result equal to M. For a volume element, take a spherical shell of radius r and thickness dr (Figure 11-18). Its volume is $4\pi r^2\,dr$ and its mass is $dm = \rho\,dV = Cr(4\pi r^2\,dr)$. (*b*) The field at $r \geq R$ is the same as if the total mass M were at the center of the sphere. (*c*) The field at $r = R/2$ is

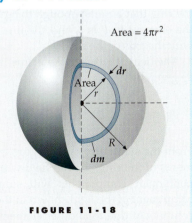

Area $= 4\pi r^2$

FIGURE 11-18

the same as if mass M' were at the center of the sphere, where M' is the amount of mass within the sphere of radius $R/2$. The mass between $r = R/2$ and $r = R$ produces zero field at $r = R/2$.

Cover the column to the right and try these on your own before looking at the answers.

Steps

Answers

(a) 1. Integrate dm from $r = 0$ to $r = R$.

$$M = \int dM = \int \rho \, dV = \int_0^R Cr(4\pi r^2 \, dr) = C\pi R^4$$

2. Solve for C in terms of the given quantities M and R.

$$C = \boxed{\dfrac{M}{\pi R^4}}$$

(b) Write an expression for the field outside the sphere in terms of the mass M, the distance r from the center, and the unit vector in the $\hat{r}$ direction of increasing r.

$$\vec{g} = \boxed{-\dfrac{GM}{r^2}\hat{r}} \quad (r > R)$$

(c) 1. Compute the mass M' that is within the radius $R/2$ by integrating $dm = \rho \, dV$ from $r = 0$ to $r = R/2$ and use the value of C found in Part (a) step 2.

$$M' = \dfrac{M}{16}$$

2. Write an expression for the field at $r = R/2$ in terms of M and R.

$$\vec{g} = -\dfrac{GM'}{r^2}\hat{r} = \boxed{-\dfrac{GM}{4R^2}\hat{r}} \quad \text{at } r = \dfrac{R}{2}$$

❶ PLAUSIBILITY CHECK For a uniform sphere, Equation 11-29 gives the field at $r = R/2$ as $g_r = -GM/(2R^2)$, twice as large as our Part (c) result. This is as expected since a uniform sphere has a larger fraction of its total mass in the region $0 < r < R/2$ than does the sphere in Equation 11-29.

REMARKS Note that the units for C are kg/m⁴, so the units for ρ are kg/m³, which is mass per volume.

*11-5 Finding the Gravitational Field of a Spherical Shell by Integration

We will derive the equation for the gravitational field of a spherical shell in two steps. First, we find the gravitational field on the axis of a ring of uniform mass. We then apply our result to a spherical shell, which we can consider to be a set of coaxial rings.

Figure 11-19 shows a ring of total mass m and radius a and a field point P on the axis of the ring a distance x from its center. We choose a mass element dm on the ring that is small enough to be considered a point particle. The distance from the element to P is s, and the line joining the element and P makes an angle α with the axis of the ring.

The field at P due to the element dm is toward the element and has magnitude dg given by

$$dg = \frac{G \, dm}{s^2}$$

From the symmetry of the figure, we can see that when we sum over all the elements of the ring, the net field will be along the axis of the ring; that is, the perpendicular components will sum to zero. For example, the perpendicular component of the field shown in the figure will be canceled by the perpendicular component due to another element of the ring directly opposite the one shown.

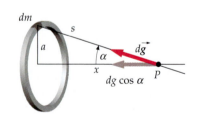

FIGURE 11-19 The gravitational field at a point P a distance x from a uniform ring. The field due to the element dm points toward the element. The total field due to the ring is along the axis of the ring.

The net field will therefore be in the negative x direction. The x component of the field due to the element dm is

$$dg_x = -dg \cos \alpha = -\frac{G\,dm}{s^2} \cos \alpha$$

We obtain the total field by summing over all the elements of the ring:

$$g_x = -\int \frac{G \cos \alpha}{s^2}\,dm$$

Since s and α are the same for all points on the ring, they are constants as far as the integration is concerned. Thus,

$$g_x = -\frac{G \cos \alpha}{s^2} \int dm = -\frac{Gm}{s^2} \cos \alpha \qquad \text{11-30}$$

where $m = \int dm$ is the total mass of the ring.

We now use this result to calculate the gravitational field of a spherical shell of mass M and radius R at a point a distance r from the center of the shell. We first consider the case in which the field point P is outside the shell, as in Figure 11-20. By symmetry, the field must be radial. We choose for our element of mass the strip shown, which can be considered to be a ring of mass dM. The field due to this strip is given by Equation 11-30 with m replaced by dM.

$$dg_r = -\frac{G\,dM}{s^2} \cos \alpha \qquad \text{11-31}$$

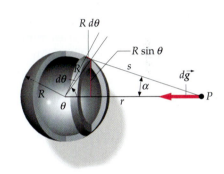

The mass dM is proportional to the area of the strip dA, which equals the circumference times the width. The radius of the strip is $R \sin \theta$, so the circumference is $2\pi R \sin \theta$. The width is $R\,d\theta$. If M is the total mass of the shell and $A = 4\pi R^2$ is its total area, the mass of the strip of area dA is

FIGURE 11-20 A uniform thin spherical shell of radius R and total mass M. The strip shown can be considered to be a ring of width $R\,d\theta$ and circumference $2\pi R \sin \theta$.

$$dM = \frac{M}{A}\,dA = \frac{M}{4\pi R^2}\,2\pi R^2 \sin \theta\,d\theta = \frac{1}{2} M \sin \theta\,d\theta \qquad \text{11-32}$$

Substituting this result into Equation 11-31 gives

$$dg_r = -\frac{G\,dM}{s^2} \cos \alpha = -\frac{GM \sin \theta\,d\theta}{2s^2} \cos \alpha \qquad \text{11-33}$$

Before integrating over the entire shell, we must eliminate two of the three related variables s, θ, and α. It turns out to be easiest to write everything in terms of s. By the law of cosines, we have

$$s^2 = r^2 + R^2 - 2rR \cos \theta$$

Differentiating gives

$$2s\,ds = +2rR \sin \theta\,d\theta$$

or

$$\sin \theta\,d\theta = \frac{s\,ds}{rR}$$

An expression for $\cos \alpha$ can be obtained by again applying the law of cosines to the same triangle. We have

$$R^2 = s^2 + r^2 - 2sr \cos \alpha$$

or

$$\cos \alpha = \frac{s^2 + r^2 - R^2}{2sr}$$

Substituting these results into Equation 11-33 gives

$$dg_r = -\frac{GM \sin \theta \, d\theta}{2s^2} \cos \alpha = -\frac{GM}{2s^2} \left(\frac{s \, ds}{rR} \right) \frac{s^2 + r^2 - R^2}{2sr}$$

$$= -\frac{GM \, ds}{4s^2r^2R} (s^2 + r^2 - R^2) = -\frac{GM}{4r^2R} \left(1 + \frac{r^2 - R^2}{s^2} \right) ds \qquad \text{11-34}$$

To find the field at P we integrate this over the entire shell. The integration limits for this step depend on whether the field point P lies outside the shell or inside it. For P outside the shell s varies from $r - R$ ($\theta = 0$) to $s = r + R$ ($\theta = 180°$), so the field due to the entire shell is found by integrating from $s = r - R$ to $s = r + R$.

$$g_r = -\frac{GM}{4r^2R} \int_{r-R}^{r+R} \left(1 + \frac{(r-R)(r+R)}{s^2} \right) ds = -\frac{GM}{4r^2R} \left[s - \frac{(r-R)(r+R)}{s} \right]_{r-R}^{r+R}$$

Substitution of the upper and lower limits yields $4R$ for the quantity in brackets. Thus,

$$g_r = -\frac{GM}{r^2} \qquad \text{for} \quad r > R$$

which is the same result as Equation 11-26a.

If the field point P is inside the shell (Figure 11-21), the calculation is identical except that s now varies from $R - r$ to $R + r$. Thus,

$$g_r = -\frac{GM}{4r^2R} \left[s - \frac{(r-R)(r+R)}{s} \right]_{R-r}^{R+r}$$

Substitution of these upper and lower limits yields 0. Therefore,

$$g_r = 0 \qquad \text{for} \quad r < R$$

which is the same as Equation 11-26b.

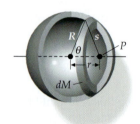

FIGURE 11-21

EXPLORING

Are tidal forces just as strong as gravitational forces? Find out this, and more, at www. whfreeman.com/tipler5e.

SUMMARY

1. Kepler's laws are *empirical* observations. They can also be derived from Newton's laws.

2. Newton's law of gravity is a *fundamental law* of physics.

3. The gravitational potential energy of a two-particle system relative to $U = 0$ at infinite separation is given by $U = -Gm_1m_2/r$. If the system is bound, its total energy is negative.

4. The gravitational field is a *fundamental physical concept* that describes the condition in space set up by a mass distribution.

Topic	Relevant Equations and Remarks
1. Kepler's Three Laws	Law 1. All planets move in elliptical orbits with the sun at one focus. Law 2. A line joining any planet to the sun sweeps out equal areas in equal times. Law 3. The square of the period of any planet is proportional to the cube of the planet's mean distance from the sun:

$$T^2 = Cr^3 \qquad \text{11-2}$$

where C has almost the same value for all planets; from Newton's law of gravity, C can be shown to be $4\pi^2/[G(M_s + M_p)]$. If $M_s \gg M_p$, this can be expressed as

$$T^2 = \frac{4\pi^2}{GM_s} r^3 \qquad \text{11-16}$$

Kepler's laws can be derived from Newton's law of gravity. The first and third laws follow from the fact that the force exerted by the sun on the planets varies inversely as the square of the separation distance. The second law follows from the fact that the force exerted by the sun on a planet is along the line joining them, so the orbital angular momentum of the planet is conserved. Kepler's laws also hold for any object orbiting another in an inverse-square field, such as a satellite orbiting a planet.

| **2. Newton's Law of Gravity** | Every point particle exerts on every other point particle an attractive force that is proportional to the masses of the two particles and inversely proportional to the square of the distance separating them. |

$$\vec{F}_{1,2} = -\frac{Gm_1m_2}{r_{1,2}^2} \hat{r}_{1,2} \qquad \text{11-3}$$

| Universal gravitational constant | $G = 6.67 \times 10^{-11} \ \text{N·m}^2/\text{kg}^2$ **11-4** |

| **3. Gravitational Potential Energy** | The gravitational potential energy U for a system consisting of a particle of mass m outside a spherically symmetric object of mass M and at a distance r from its center is |

$$U(r) = -\frac{GMm}{r} \qquad \text{11-20}$$

This potential-energy function approaches zero as r approaches infinity.

| **4. Mechanical Energy** | The mechanical energy E for a system consisting of a particle of mass m outside a spherically symmetric object of mass M and at a distance r from its center is |

$$E = \frac{1}{2}mv^2 - \frac{GMm}{r}$$

Escape speed	For a given value of r, the speed of the particle for which $E = 0$ is called the escape speed v_e. That is, if $v = v_e$, then $E = 0$.	
5. Classification of Orbits	If $E < 0$, the system is bound and the orbit is an ellipse (or circle, which is a type of ellipse). If $E \geq 0$, the system is unbound and the orbit is a hyperbola (or a parabola for $E = 0$).	
6. Gravitational Field		
Definition	$$\vec{g} = \frac{\vec{F}}{m}$$	**11-23**
Due to the earth	$$\vec{g}(r) = \frac{\vec{F}}{m} = -\frac{GM_E}{r^2}\hat{r} \quad (r \geq R_E)$$	**11-29**
Due to thin spherical shell	Outside the shell the gravitational field is the same as if all the mass of the shell were concentrated at the center. The field inside the shell is zero.	
	$$\vec{g} = -\frac{GM}{r^2}\hat{r} \quad \text{for} \quad r > R$$	**11-26a**
	$$\vec{g} = 0 \quad \text{for} \quad r < R$$	**11-26b**

PROBLEMS

- Single-concept, single-step, relatively easy
- • Intermediate-level, may require synthesis of concepts
- • • Challenging

SSM Solution is in the *Student Solutions Manual*

iSOLVE Problems available on iSOLVE online homework service

iSOLVE✓ These "Checkpoint" online homework service problems ask students additional questions about their confidence level, and how they arrived at their answer

In a few problems, you are given more data than you actually need; in a few other problems, you are required to supply data from your general knowledge, outside sources, or informed estimates.

Take g = 9.81 N/kg = 9.81 m/s² **and neglect friction in all problems unless otherwise stated.**

Conceptual Problems

1 • **SSM** True or false:

(a) Kepler's law of equal areas implies that gravity varies inversely with the square of the distance.
(b) The planet closest to the sun, on the average, has the shortest orbital period.

2 • If the mass of a satellite is doubled, the radius of its orbit can remain constant if the speed of the satellite (a) increases by a factor of 8, (b) increases by a factor of 2, (c) does not change, (d) is reduced by a factor of 8, (e) is reduced by a factor of 2.

3 • • **SSM** At the surface of the moon, the acceleration due to the gravity of the moon is a. At a distance from the center of the moon equal to four times the radius of the moon, the acceleration due to the gravity of the moon is (a) $16a$, (b) $a/4$, (c) $a/3$, (d) $a/16$, (e) none of the answers given.

4 • Why is G so difficult to measure?

5 • If the mass of a planet is doubled with no increase in its radius, the escape speed for that planet will be (a) increased by a factor of 1.4, (b) increased by a factor of 2, (c) unchanged, (d) reduced by a factor of 1.4, (e) reduced by a factor of 2.

6 • • A newly discovered comet-like object is observed to make a pass around the sun. How can we tell if the object will return many years later, or if it will never return?

7 • • Explain why the gravitational field is directly proportional to r rather than inversely proportional to r inside a solid sphere of uniform mass.

8 • **SSM** If K is the kinetic energy of Mercury in its orbit around the sun, and U is the potential energy of the Mercury–sun system, what is the relationship between K and U?

9 • • A student whose weight on the surface of the earth is w is lifted to a height that is two earth radii above the surface. Her weight there is (a) $w/2$, (b) $w/4$, (c) $w/3$, (d) $w/9$.

10 •• In the novel "First Men in the Moon" by H.G. Wells, the trip was accomplished in a slightly different manner than by Jules Verne's heroes. The novel's hero, Professor Cavour, invented a material, Cavourite, that shielded objects from the gravitational force. Simply by surrounding an object with Cavourite, the gravitational pull on it by its surroundings was nullified. The adventurers in this book built a ship whose bottom was covered by Cavourite, but whose top was unshielded. Then they pointed the top of the ship at the moon to let its attraction pull them up. As far as real physics goes, Cavourite is an impossibility, because it would allow the violation of the law of conservation of energy. To show this, invent a perpetual motion machine using Cavourite.

Estimation and Approximation

11 • Estimate the mass of the galaxy if the sun orbits the center of the galaxy with a period of 250 million years at a mean distance of 30,000 light-years. Express the mass in terms of multiples of the solar mass M_s. (Neglect the mass farther from the center than the sun, and assume that the mass closer to the center than the sun exerts the same force on the sun as would a point particle of the same mass.)

12 ••• SSM One of the great discoveries in astronomy in recent years is the detection of planets outside the solar system. Since 1996, 100 planets have been detected orbiting stars other than the sun. While the planets themselves cannot be seen directly, telescopes can detect the small periodic motion of the star as the star and planet orbit around their common center of mass. (This is measured using the *Doppler effect,* which will be discussed in Chapter 15.) Both the period of this motion and the variation in the velocity of the star over the course of time can be determined observationally. The mass of the star is found from its observed luminance and from the theory of stellar structure. Iota Draconis is the 8th brightest star in the constellation Draco. Observations show that there is a planet orbiting this star with an orbital period of 1.50 y. The mass of Iota Draconis is $1.05M_{sun}$. (a) What is the size (in AU) of the semimajor axis of this planet's orbit? (b) The radial velocity of the star is observed to vary by 592 m/s. Use conservation of momentum to find the mass of the planet. Assume the orbit is circular, we are observing the orbit edge-on, and no other planets orbit Iota Draconis. Express the mass as a multiple of the mass of Jupiter.

13 ••• One of the biggest unresolved problems in the theory of the formation of the solar system is that, while the mass of the sun is 99.9 percent of the total mass of the solar system, it carries only about 2 percent of the total angular momentum. The most widely accepted theory of solar system formation has as its central hypothesis the collapse of a cloud of dust and gas under the force of gravity, with most of the mass forming the sun. However, because the net angular momentum of this cloud is conserved, a simple theory would indicate that the sun should be rotating much more rapidly than it currently is. In this problem, you will show why it is important that most of the angular momentum was somehow transferred to the planets. (a) The sun is a cloud of gas held together by the force of gravity. If the sun were rotating too rapidly, gravity couldn't hold it together. Using the known mass of the sun (1.99×10^{30} kg) and its radius (6.96×10^8 m),

estimate the maximum angular velocity that the sun can have if it is to stay intact. What is the period of rotation corresponding to this rotation rate? (b) Calculate the orbital angular momentum of Jupiter and of Saturn from their masses (318 and 95.1 Earth masses, respectively), mean distances from the sun (778 and 1430 million km, respectively), and orbital periods (11.9 and 29.5 y, respectively). Compare them to the experimentally measured value of the sun's angular momentum of 1.91×10^{41} kg m²/s. (c) If we were to somehow transfer all of Jupiter's and Saturn's angular momentum to the sun, what would be the sun's new rotational period? Because the sun is not a uniform sphere of gas, its moment of inertia is given by the formula $I = 0.059MR^2$. Compare this to the maximum rotation rate of Part(a).

Kepler's Laws

14 • SSM The new comet Alex-Casey is discovered on a very elliptical orbit with a period of 127.4 y. If the closest approach of Alex-Casey to the sun is 0.1 AU, what is its greatest distance from the sun?

15 • The radius of the earth's orbit is 1.496×10^{11} m and that of Uranus is 2.87×10^{12} m. What is the period of Uranus?

16 • ISOLVE✔ The asteroid Hektor, discovered in 1907, is in a nearly circular orbit of radius 5.16 AU about the sun. Determine the period of this asteroid.

17 •• The asteroid Icarus, discovered in 1949, was so named because its highly eccentric elliptical orbit brings it close to the sun at perihelion. The eccentricity e of an ellipse is defined by the relation $r_p = a(1 - e)$, where r_p is the perihelion distance and a is the semimajor axis. Icarus has an eccentricity of 0.83 and a period of 1.1 y. (a) Determine the semimajor axis of the orbit of Icarus. (b) Find the perihelion and aphelion distances of the orbit of Icarus.

18 •• There has been much discussion of a manned mission to Mars and the problems that it would entail due to the extremely long time spent in space by the astronauts. To examine this issue in a simple way, consider one possible trajectory for the spacecraft: the "Hohmann transfer orbit." This orbit consists of an elliptical orbit tangent to the orbit of Earth at its point nearest the sun and tangent to the orbit of Mars at the point farthest from the sun. Given that Mars has a mean distance from the sun of 1.52 times the mean sun–Earth distance, calculate the total time spent by the astronauts in transit.

19 •• SSM Kepler determined distances in the solar system from his data. For example, he found the relative distance from the sun to Venus (as compared to the distance from the sun to Earth) as follows. Because Venus's orbit is closer to the sun than is Earth's, Venus is a morning or evening star— its position in the sky is never very far from the sun (see Figure 11-22). If we consider the orbit of Venus as a perfect circle, then consider the relative orientation of Venus, Earth, and the sun at maximum extension—when Venus is farthest from the sun in the sky. (a) Under this condition, show that angle b in Figure 11-22 is 90°. (b) If the maximum elongation angle between Venus and the sun is 47°, what is the distance between Venus and the sun in AU?

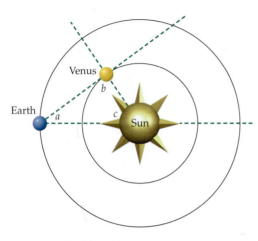

FIGURE 11-22 Problem 19

20 •• At apogee the moon is 406,395 km from the earth and at perigee it is 357,643 km. What is the velocity of the moon at perigee and apogee? Its orbital period is 27.3 d.

Newton's Law of Gravity

21 •• **SSM** Jupiter's satellite Europa orbits Jupiter with a period of 3.55 d at an average distance of 6.71×10^8 m. (a) Assuming that the orbit is circular, determine the mass of Jupiter from the data given. (b) Another satellite of Jupiter, Callisto, orbits at an average distance of 18.8×10^8 m with an orbital period of 16.7 d. Determine the acceleration of Callisto and Europa from the data given, and show that this data is consistent with an inverse square law for gravity [Note: Do NOT use the value of G anywhere in Part (b)].

22 • **SSM** Some people think that shuttle astronauts are "weightless" because they are "beyond the pull of Earth's gravity." In fact, this is completely untrue. (a) What is the magnitude of the acceleration of gravity for shuttle astronauts? A shuttle orbit is about 400 km above the ground. (b) Given the answer in Part (a), why are shuttle astronauts "weightless"?

23 • **SOLVE** The mass of Saturn is 5.69×10^{26} kg. (a) Find the period of its moon Mimas, whose mean orbital radius is 1.86×10^8 m. (b) Find the mean orbital radius of its moon Titan, whose period is 1.38×10^6 s.

24 • **SOLVE** Calculate the mass of the earth from the period of the moon, $T = 27.3$ d; its mean orbital radius, $r_m = 3.84 \times 10^8$ m; and the known value of G.

25 • **SOLVE** Use the period of the earth (1 y), its mean orbital radius (1.496×10^{11} m), and the value of G to calculate the mass of the sun.

26 • **SSM** **SOLVE** An object is dropped from a height of 6.37×10^6 m above the surface of the earth. What is its initial acceleration?

27 • Suppose you leave the solar system and arrive at a planet that has the same mass-to-volume ratio as the earth but has 10 times the earth's radius. What would you weigh on this planet compared with what you weigh on the earth?

28 • Suppose that the earth retained its present mass but was somehow compressed to half its present radius. What would be the value of g at the surface of this new, compact planet?

29 • A planet orbits a massive sun. When the planet is at perihelion, it has a speed of 5×10^4 m/s and is 1.0×10^{15} m from the sun. The orbital radius increases to 2.2×10^{15} m at aphelion. What is the planet's speed at aphelion?

30 • What is the acceleration of gravity at the surface of a neutron star whose mass is 1.60 times the mass of the sun and whose radius is 10.5 km?

31 •• **SSM** The speed of an asteroid is 20 km/s at perihelion and 14 km/s at aphelion. Determine the ratio of the aphelion to perihelion distances.

32 •• A satellite with a mass of 300 kg moves in a circular orbit 5×10^7 m above the earth's surface. (a) What is the gravitational force on the satellite? (b) What is the speed of the satellite? (c) What is the period of the satellite?

33 •• **SSM** **SOLVE** A superconducting gravity meter can measure changes in gravity of the order $\Delta g/g = 10^{-11}$. (a) You are hiding behind a tree holding the meter, and your 80-kg friend approaches the tree from the other side. How close to you can your friend get before the meter detects a change in g due to his presence? (b) You are in a hot air balloon and are using the meter to determine the rate of ascent (presumed constant). What is the smallest change in altitude that results in a detectable change in the gravitational field of the earth?

34 •• Suppose that the attractive interaction between a star of mass M and a planet of mass $m \ll M$ were of the form $F = KMm/r$, where K is the gravitational constant. What would be the relation between the radius of the planet's circular orbit and its period?

35 •• **SSM** The mass of the earth is 5.97×10^{24} kg and its radius is 6370 km. The radius of the moon is 1738 km. The acceleration of gravity at the surface of the moon is 1.62 m/s². What is the ratio of the average density of the moon to that of the earth?

Measurement of G

36 • **SOLVE** The large and small spheres in a Cavendish balance (Figure 11-7) have masses $m_1 = 10$ kg and $m_2 = 0.010$ kg, respectively, and the center-to-center separation of the two small spheres is 20 cm. The center-to-center separation of each large sphere with the small sphere adjacent to it is 6 cm. (a) What is the force by which each large sphere attracts the small sphere adjacent to it? (b) What torque must be exerted by the suspension to balance the torque due to these forces?

Gravitational and Inertial Mass

37 • **SOLVE** A standard object defined as having a mass of exactly 1 kg is given an acceleration of 2.6587 m/s² when a certain force is applied to it. A second object of unknown mass acquires an acceleration of 1.1705 m/s² when the same force is applied to it. (a) What is the mass of the second object? (b) Is the mass that you determined in Part (a) gravitational or inertial mass?

38 • **iSOLVE** ✓ The weight of a standard object defined as having a mass of exactly 1 kg is measured to be 9.81 N. In the same laboratory, a second object weighs 56.6 N. (a) What is the mass of the second object? (b) Is the mass you determined in Part (a) gravitational or inertial mass?

39 • **SSM** The principle of equivalence states that the free-fall acceleration of any object in a gravitational field is independent of the mass of the object. This can be seen from the form of the law of universal gravitation—but how well does it hold experimentally? The Roll-Krotkov-Dicke experiment performed in the 1960s indicates that the free-fall acceleration is independent of mass to at least 1 part in 10^{12}. Suppose two objects are simultaneously released from rest in a uniform gravitational field. Also, suppose one of the objects falls with a constant acceleration of exactly 9.8m/s^2 while the other falls with a constant acceleration that is greater than 9.8 m/s^2 by one part in 10^{12}. How far will the first object have fallen when the second object has fallen 1 mm further than it has? Note that this estimate is an upper bound on the difference in the accelerations; most physicists believe that there is no difference whatsoever.

Gravitational Potential Energy

40 • (a) Taking the potential energy to be zero at infinite separation, find the potential energy of a 100-kg object at the surface of the earth. (Use 6.37×10^6 m for the earth's radius.) (b) Find the potential energy of the same object at a height above the earth's surface equal to the earth's radius. (c) Find the escape speed for a body projected from this height.

41 • A point particle of mass m_0 is initially at the surface of a large sphere of mass M and radius R. How much work is needed to remove it to a very large distance away from the large sphere?

42 • Suppose that in space there is a duplicate earth, except that it has no atmosphere, is not rotating, and is not in motion around any sun. What initial velocity must a spacecraft on this planet's surface have to travel vertically upward a distance equal to one earth radius above the surface of the planet?

43 •• **SSM** **iSOLVE** ✓ An object is dropped from rest from a height of 4×10^6 m above the surface of the earth. If there is no air resistance, what is its speed when it strikes the earth?

pg 351

44 •• **iSOLVE** An object is projected upward from the surface of the earth with an initial speed of 4 km/s. Find the maximum height it reaches.

45 •• **iSOLVE** A uniform thin spherical shell has a radius R and a mass M. (a) Write expressions for the force exerted by the shell on a point mass m_0 when m_0 is outside the shell and when it is inside the shell. (b) What is the potential-energy function $U(r)$ for this system when the mass m_0 is at a distance r ($r \geq R$) if $U = 0$ at $r = \infty$? Evaluate this function at $r = R$. (c) Using the general relation for $dU = -\vec{F} \cdot d\vec{r} = -F_r\, dr$, show that U is constant everywhere inside the shell. (d) Using the fact that U is continuous everywhere, including at $r = R$, find the value of the constant U inside the shell. (e) Sketch $U(r)$ versus r for all possible values of r.

46 • **iSOLVE** The planet Saturn has a mass 95.2 times that of the earth and a radius 9.47 times that of the earth. Find the escape speed for objects on the surface of Saturn.

47 • **iSOLVE** ✓ Find the escape speed for a rocket leaving the moon. The acceleration of gravity on the moon is 0.166 times that on earth and the moon's radius is $0.273R_E$.

48 • **SSM** The science fiction writer Robert Heinlein once said, "If you can get into orbit, then you're halfway to anywhere." Justify this statement by comparing the kinetic energy needed to place a satellite into low earth orbit ($h = 400$ km) to that needed to set it completely free from the bonds of earth's gravity.

49 •• A particle is projected from the surface of the earth with a speed twice the escape speed. When it is very far from the earth, what is its speed?

50 •• What initial speed should a particle be given if it is to have a final speed when it is very far from the earth that is equal to its escape speed?

51 •• **iSOLVE** (a) Calculate the energy in joules necessary to launch an object whose mass is 1 kg from the earth at escape speed. (b) Convert this energy to kilowatt-hours. (c) If energy can be obtained at $0.10/kW·h, what is the minimum cost of giving an 80-kg astronaut enough energy to escape the earth's gravitational field?

52 •• An object is projected vertically from the surface of the earth at less than the escape speed. Show that the maximum height reached by the object is $H = R_E H'/(R_E - H')$, where H' is the height that it would reach if the gravitational field were constant.

Orbits

53 •• **iSOLVE** ✓ A spacecraft of 100-kg mass is in a circular orbit about the earth at a height $h = 2R_E$. (a) What is the period of the spacecraft's orbit about the earth? (b) What is the spacecraft's kinetic energy? (c) Express the angular momentum L of the spacecraft about the center of the earth in terms of its kinetic energy K and find its numerical value.

54 • **SSM** We normally say that the moon orbits the earth, but this is not quite true: instead, both the moon and the earth orbit their common center of mass, which is not at the center of the earth. (a) The mass of the earth is 5.98×10^{24} kg and the mass of the moon is 7.36×10^{22} kg. The mean distance from the center of the earth to the center of the moon is 3.82×10^8 m. How far above the surface of the earth is the center of mass of the earth–moon system? (b) Estimate the mean "orbital speed" of the earth as it orbits the center of mass of the earth–moon system, using the moon's mass, period, and average distance from the earth. Ignore any external forces acting on the system. The orbital period of the moon is 27.3 d.

55 •• **iSOLVE** ✓ Many satellites orbit the earth about 1000 km above the earth's surface. Geosynchronous satellites orbit at a distance of 4.22×10^7 m from the center of the earth. How much more energy is required to launch a 500-kg satellite into a geosynchronous orbit than into an orbit 1000 km above the surface of the earth?

56 •• Using the period of the moon's orbit around the earth (27.3 d), its average distance from the center of the earth (3.82×10^8 m), the length of a year (365.25 d), and the average distance from the earth to the sun (1.50×10^{11} m), establish the ratio of the mass of the sun to the mass of the earth. Compare this to the measured ratio of 3.33×10^5.

The Gravitational Field

57 • **SOLVE** A 3-kg mass experiences a gravitational force of 12 N $\hat{\imath}$ at some point P. What is the gravitational field at that point?

58 • **SSM** **SOLVE** The gravitational field at some point is given by $\vec{g} = 2.5 \times 10^{-6}$ N/kg $\hat{\jmath}$. What is the gravitational force on a mass of 4 g at that point?

59 •• A point mass m is on the x axis at $x = L$ and an identical point mass is on the y axis at $y = L$. (a) Find the gravitational field at the origin. (b) What is the magnitude of this field?

60 •• Five objects of mass M are equally spaced on the arc of a semicircle of radius R as in Figure 11-23. An object of mass m is located at the center of curvature of the arc. (a) If M is 3 kg, m is 2 kg, and R is 10 cm, what is the force on m due to the five objects? (b) If the object whose mass is m is removed, what is the gravitational field at the center of curvature of the arc?

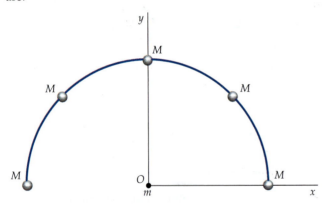

FIGURE 11-23 Problem 60

61 •• A point particle of mass $m_1 = 2$ kg is at the origin and a second point particle of mass $m_2 = 4$ kg is on the x axis at $x = 6$ m. Find the gravitational field at (a) $x = 2$ m and (b) $x = 12$ m. (c) Find the point on the x axis for which $g = 0$.

62 •• Show that the maximum value of $|g_x|$ for the field of Example 11-7 occurs at the points $x = \pm a/\sqrt{2}$.

63 •• A nonuniform rod of length L lies on the x axis with one end at the origin and the other at $x = L$. Its mass per unit length λ varies as $\lambda = Cx$, where C is a constant. (Thus an element of the rod has mass $dm = \lambda\, dx$.) (a) What is the total mass of the rod? (b) Find the gravitational field due to the rod on the x axis at a point x_0, where $x_0 > L$.

64 ••• A uniform rod of mass M and length L lies along the x axis with its center at the origin. Consider an element of length dx at point x, where $-\frac{1}{2}L < x < \frac{1}{2}L$. (a) Show that this element produces a gravitational field at a point x_0 on the x axis ($x_0 > \frac{1}{2}L$) given by

$$dg_x = -\frac{GM}{L(x_0 - x)^2}\, dx$$

(b) Integrate this result over the length of the rod to find the total gravitational field at the point x_0 due to the rod. (c) What is the force on an object of mass m_0 at x_0? (d) Show that for $x_0 \gg L$, the field is approximately equal to that of a point particle of mass M at $x = 0$.

$\vec{g}$ due to Spherical Objects

65 • **SOLVE** A thin spherical shell has a radius of 2 m and a mass of 300 kg. What is the gravitational field at the following distances from the center of the shell: (a) 0.5 m, (b) 1.9 m, (c) 2.5 m?

66 • **SOLVE** A thin spherical shell has a radius of 2 m and a mass of 300 kg, and its center is located at the origin of a coordinate system. Another thin spherical shell with a radius of 1 m and mass 150 kg is inside the larger shell with its center at 0.6 m on the x axis. What is the gravitational force of attraction between the two shells?

67 • **SSM** Two solid spheres, S_1 and S_2, have equal radii R and equal masses M. The density of sphere S_1 is constant, whereas that of sphere S_2 depends on the radial distance according to $\rho(r) = C/r$. If the acceleration of gravity at the surface of sphere S_1 is g_1, what is the acceleration of gravity at the surface of sphere S_2?

68 •• Two homogeneous solid spheres, S_1 and S_2, have equal masses but different radii, R_1 and R_2. If the acceleration of gravity on the surface of sphere S_1 is g_1, what is the acceleration of gravity on the surface of sphere S_2?

69 •• Two concentric uniform thin spherical shells have masses M_1 and M_2 and radii a and $2a$ as in Figure 11-24. What is the magnitude of the gravitational force on a point particle of mass m located (a) a distance $3a$ from the center of the shells? (b) a distance $1.9a$ from the center of the shells? (c) a distance $0.9a$ from the center of the shells?

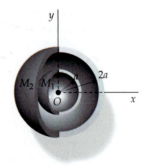

FIGURE 11-24 Problems 69, 70

70 •• The inner spherical shell in Problem 69 is shifted so that its center is now on the x axis at $x = 0.8a$. What is the magnitude of the gravitational force on a particle of point mass m located on the x axis at (a) $x = 3a$, (b) $x = 1.9a$, (c) $x = 0.9a$?

$\vec{g}$ Inside Solid Spheres

71 •• **SSM** Is your "weight" (as measured on a spring scale) at the bottom of a deep mine shaft greater than or less than your weight at the surface of the earth? Model the earth as a homogeneous sphere. Consider the effects in (a) and (b).
(a) Show that the force of gravity on you from a uniform-density, perfectly spherical planet is proportional to your distance from the center of the planet.
(b) Show that your effective "weight" increases linearly due to the effects of rotation as you approach the center. (Consider a mine shaft located on the equator.)
(c) Which effect is more important for the earth? Use $M = 5.98 \times 10^{24}$ kg, $R = 6370$ km, and $T = 24$ h.

72 •• Suppose the earth were a nonrotating uniform sphere. If there were an elevator shaft going 15,000 m into the earth, what would be the decrease in the weight of a student who weighs 800 N at the surface of the earth as he is lowered from the surface to the bottom of the shaft?

73 •• A solid sphere of radius R has its center at the origin. It has a uniform mass density ρ_0, except that there is a spherical cavity in it of radius $r = \frac{1}{2}R$ centered at $x = \frac{1}{2}R$ as in Figure 11-25. Find the gravitational field at points on the x axis for $|x| > R$. (*Hint: The cavity may be thought of as a sphere of mass $m = (4/3)\pi r^3 \rho_0$ plus a sphere of mass $-m$.*)

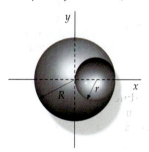

FIGURE 11-25 Problems 73, 74

74 ••• For the sphere with the cavity in Problem 73, show that the gravitational field inside the cavity is uniform and find its magnitude and direction.

75 ••• A straight, smooth tunnel is dug through a spherical planet whose mass density ρ_0 is constant. The tunnel passes through the center of the planet and is perpendicular to the planet's axis of rotation, which is fixed in space. The planet rotates with a constant angular velocity ω so that objects in the tunnel have no apparent weight. Find ω.

76 ••• **ISOLVE** The density of a sphere is given by $\rho(r) = C/r$. The sphere has a radius of 5 m and a mass of 1011 kg. (a) Determine the constant C. (b) Obtain expressions for the gravitational field for (1) $r > 5$ m and (2) $r < 5$ m.

77 ••• **SSM** **ISOLVE** A small diameter hole is drilled into the sphere of Problem 76 toward the center of the sphere to a depth of 2 m below the sphere's surface. A small mass is dropped from the surface into the hole. Determine the speed of the small mass as it strikes the bottom of the hole.

78 ••• **ISOLVE✔** The crust of the earth is about 40 km thick and has a density of about 3000 kg/m³. A spherical de-

posit of heavy metals with a density of 8000 kg/m³ and radius of 1000 m is centered 2000 m below the surface. Find $\Delta g/g$ at the surface directly above this deposit, where Δg is the increase in the gravitational field due to the deposit.

79 ••• **SSM** Two identical spherical cavities are made in a lead sphere of radius R. The cavities have a radius $R/2$. They touch the outside surface of the sphere and its center as in Figure 11-26. The mass of a solid uniform lead sphere is M.
(a) Find the force of attraction of a point particle of mass m at the position shown in the figure to the cavitated lead sphere.
(b) What is the attractive force on the point particle if it is located on the x axis at $x = R$?

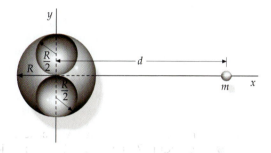

FIGURE 11-26 Problem 79

80 •• A globular cluster is a roughly spherical collection of up to millions of stars bound together by the force of gravity. Astronomers can measure the velocities of stars in the cluster to study its composition and to get an idea of the mass distribution within the cluster. Assuming that all of the stars have about the same mass and are distributed uniformly within the cluster, show that the mean speed of a star in a circular orbit around the center of the cluster should increase linearly with its distance from the center.

General Problems

81 • **SSM** The mean distance of Pluto from the sun is 39.5 AU. Find the period of Pluto.

82 • Calculate the mass of the earth using the known values of G, g, and R_E.

83 •• The force exerted by the earth on a particle of mass m at a distance r ($r > R_E$) from the center of the earth has the magnitude mgR_E^2/r^2, where $g = GM_E/R_E^2$. (a) Calculate the work you must do to move the particle from distance r_1 to distance r_2. (b) Show that when $r_1 = R_E$ and $r_2 = R_E + h$, the result can be written $W = mgR_E^2[(1/R_E) - 1/(R_E + h)]$. (c) Show that when $h \ll R_E$, the work is given approximately by $W = mgh$.

84 •• A uniform sphere of radius 100 m and density 2000 kg/m³ is in free space far from other massive objects. (a) Find the gravitational field outside of the sphere as a function of r. (b) Find the gravitational field inside the sphere as a function of r.

85 •• Jupiter has a mass 320 times that of Earth and a volume 1320 times that of Earth. A "day" on Jupiter is 9 h 50 min long. Find the height h above Jupiter at which a satellite must be orbiting to have a period equal to one Jovian day.

86 •• The average density of the moon is $\rho = 3340\ \text{kg/m}^3$. Find the minimum possible period T of a spacecraft orbiting the moon.

87 •• A satellite is circling around the moon (radius 1700 km) close to the surface at a speed v. A projectile is launched from the moon vertically up at the same initial speed v. How high will it rise?

88 •• **SSM** In the novel "A Voyage to the Moon" by Jules Verne, astronauts were launched by a giant cannon from the earth to the moon. (a) If we estimate the velocity needed to reach the moon as the earth's escape velocity, and the length of the cannon as 900 ft (as stated in the book), what is the probability that the astronauts would have survived "liftoff"? (b) In the book, the astronauts in their ship feel the effects of the earth's gravity until they reach the balance point where the moon's gravitational pull on the ship equals that of the earth. At this point, everything flips around and what was the ceiling in the ship becomes the floor. How far away from the center of the earth is this balance point? (c) Is this description of what the astronauts experienced reasonable? Is this what actually happened to the Apollo astronauts on their visit to the moon?

89 •• In a binary star system, two stars follow circular orbits about their common center of mass. If the stars have masses m_1 and m_2 and are separated by a distance r, show that the period of rotation is related to r by $T^2 = 4\pi r^3/[G(m_1 + m_2)]$.

90 •• Two particles of masses m_1 and m_2 are released from rest at a large separation distance. Find their speeds v_1 and v_2 when their separation distance is r. The initial separation distance is given as large, but large is a relative term. Relative to what distance is it large?

91 •• **SSM** Four identical planets are arranged in a square as shown in Figure 11-27. If the mass of each planet is M and the edge length of the square is a, what must be their speed if they are to orbit their common center under the influence of their mutual attraction?

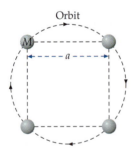

FIGURE 11-27 Problem 91

92 •• A hole is drilled from the surface of the earth to its center as in Figure 11-28. Ignore the earth's rotation and air resistance, and model the earth as a uniform sphere. (a) How much work is required to lift a particle of mass m from the center of the earth to the earth's surface? (b) If the particle is dropped from rest at the surface of the earth, what is its speed when it reaches the center of the earth? (c) What is the escape speed for a particle projected from the center of the earth? Express your answers in terms of m, g, and R_E.

FIGURE 11-28 Problem 92

93 •• A thick spherical shell of mass M and uniform density has an inner radius R_1 and an outer radius R_2. Find the gravitational field g_r as a function of r for $0 < r < \infty$. Sketch a graph of g_r versus r.

94 •• (a) Sketch a plot of the gravitational field g_x versus x due to a uniform ring of mass M and radius R whose axis is the x axis. (b) At what points is the magnitude of g_x a maximum?

95 ••• Find the magnitude of the gravitational field a distance r from an infinitely long wire whose mass per unit length is λ.

96 ••• One big question in planetary science is whether each of the rings of Saturn is solid or composed of many smaller satellites. There is a simple observation that can now be made to resolve this issue. Measure the velocity of the inner and outer portion of the ring: if the inner portion of the ring moves more slowly than the outer portion, then the ring is solid; if the opposite is true, then it is composed of many separate chunks. (a) If the thickness of the ring is r, the average distance of the ring from the center of Saturn is R, and the average velocity of the ring is v, show that $v_{out} - v_{in} \approx rv/R$ if the ring is solid. Here, v_{out} is the speed of the outermost portion of the ring, v_{in} is the speed of the innermost portion, and v is the average velocity of the ring. (b) If, however, the ring is composed of many small chunks, show that $v_{out} - v_{in} \approx -\frac{1}{2}rv/R$. (Assume that $r \ll R$.)

97 ••• In this problem you are to find the gravitational potential energy of the rod in Example 11-8 and a point mass m_0 that is on the x axis at x_0. (a) Show that the potential energy shared by an element of the rod of mass dm (shown in Figure 11-13) and a point particle of mass m_0 located on the x axis at $x_0 \geq \frac{1}{2}L$ is given by

$$dU = -\frac{Gm_0\,dm}{x_0 - x} = \frac{GMm_0}{L(x_0 - x)}\,dx$$

where $U = 0$ at $x_0 = \infty$. (b) Integrate your result for Part (a) over the length of the rod to find the total potential energy for the system. Write your result as a general function $U(x)$ by setting x_0 equal to a general point x. (c) Compute the force on m_0 at a general point x from $F_x = -dU/dx$ and compare your result with $m_0 g$, where g is the field at x_0 calculated in Example 11-8.

98 ••• **SSM** A uniform sphere of mass M is located near a thin, uniform rod of mass m and length L as in Figure 11-29. Find the gravitational force of attraction exerted by the sphere on the rod.

FIGURE 11-29 Problem 98

99 ••• A uniform rod of mass $M = 20$ kg and length $L = 5$ m is bent into a semicircle. What is the gravitational force exerted by the rod on a point mass $m = 0.1$ kg located at the center of curvature of the circular arc?

100 ••• |SSM| Both the sun and the moon exert gravitational forces on the oceans of the earth, causing tides. (*a*) Show that the ratio of the force exerted on a point particle on earth by the sun to that exerted by the moon is $M_s r_m^2 / M_m r_s^2$, where M_s and M_m are the masses of the sun and moon and r_s and r_m are the distances from the earth to the sun and to the moon. Evaluate this ratio. (*b*) Even though the sun exerts a much greater force on the oceans than does the moon, the moon has a greater ef-

fect on the tides because it is the difference in the force from one side of the earth to the other that is important. Differentiate the expression $F = Gm_1 m_2 / r^2$ to calculate the change in F due to a small change in r. Show that $dF/F = (-2\ dr)/r$. (*c*) During one full day, the rotation of the earth can cause the distance from the sun or moon to an ocean to change by, at most, the diameter of the earth. Show that for a small change in distance, the change in the force exerted by the sun is related to the change in the force exerted by the moon by $\Delta F_s / \Delta F_m \approx (M_s r_m^3)/(M_m r_s^3)$. Calculate this ratio.

101 •• United Federation Spaceship *Excelsior* is dropping two robot probes to the surface of a neutron star for exploration. The mass of the star is the same as that of the sun, but the star's diameter is only 10 km. The robot probes are linked together by a 1-m-long steel cord, and are dropped vertically (that is, one always above the other). (*a*) Explain why there seems to be a "force" trying to pull the robots apart. (See Problem 99 for a hint.) (*b*) How close will the robots be to the surface of the star before the cord breaks? Assume that the cord has a breaking tension of 25,000 N and that the robots each have a mass of 1 kg.

Static Equilibrium and Elasticity

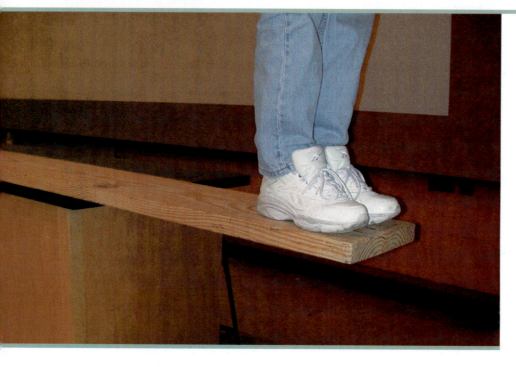

THE PLANK IN THIS PHOTO IS IN STATIC EQUILIBRIUM. THE OVERHANG IS ONE-SIXTH THE LENGTH OF THE PLANK, AND THE PLANK REMAINS BALANCED WHEN A STUDENT STANDS WITH HIS CENTER OF MASS OVER THE END OF THE PLANK.

? **What is the maximum ratio of the mass of the student to the mass of the plank if the plank remains balanced when the student stands with his center of mass over the very end of the plank? (See Example 12-1.)**

I f an object is stationary and remains stationary, it is said to be in static equilibrium. Being able to determine the forces acting on an object in static equilibrium has many important applications. For example, the forces exerted by the cables of a suspension bridge must be known so that the cables can be designed to be strong enough to support the bridge. Similarly, cranes must be designed so that they do not topple over when lifting a weight.

The forces exerted by the cables and beams in a structure are called elastic forces. They are the result of slight deformations—the stretching or compression of solid objects under stress from bearing loads.

➤ **In this chapter, we study the equilibrium of rigid bodies and then briefly consider the deformations and elastic forces that arise when real solids are under stress.**

12-1 Conditions for Equilibrium

A necessary condition for a particle at rest to remain at rest is that the net force acting on it remain zero. Similarly, a necessary condition for the center of mass of a rigid object to remain at rest is that the net force acting on the object remain zero. A rigid object can rotate, even when its center of mass is at rest, but then the object is not in static equilibrium. Therefore, a second necessary condition for a rigid object to remain in static equilibrium is that the net torque acting on it about *any* point must remain zero. This gives us the option to choose the point P when calculating torques, which greatly simplifies the solution of most static problems.

The two necessary conditions for a rigid body to be in static equilibrium are as follows:

1. The net external force acting on the body must remain zero:

$$\sum \vec{F} = 0 \qquad\qquad 12\text{-}1$$

2. The net external torque about *any* point must remain zero:

$$\sum \vec{\tau} = 0 \qquad\qquad 12\text{-}2$$

CONDITIONS FOR EQUILIBRIUM

12-2 The Center of Gravity

Figure 12-1*a* shows a rigid object in static equilibrium and a point O. We consider the object to be composed of many small particles. The weight of the ith small particle is $\vec{w}_i$, and the total weight of the object is $\vec{W} = \Sigma \vec{w}_i$. If $\vec{r}_i$ is the position vector of the ith particle relative to O, then $\vec{\tau}_i = \vec{r}_i \times \vec{w}_i$, where $\vec{\tau}_i$ the torque due to $\vec{w}_i$ about O. The net gravitational torque about O is then $\vec{\tau}_{net} = \Sigma(\vec{r}_i \times \vec{w}_i)$. Conveniently, the net torque due to gravity about any point can be calculated as if the entire weight $\vec{W}$ were applied at a single point, the **center of gravity** (see Figure 12-1*b*). That is,

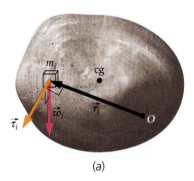

$$\vec{\tau}_{net} = \vec{r}_{cg} \times \vec{W} \qquad\qquad 12\text{-}3$$

CENTER OF GRAVITY DEFINED

(a)

where $\vec{r}_{cg}$ is the position vector of the center of gravity relative to O.

If the gravitational field $\vec{g}$ is uniform over the object (as is nearly always the case for objects of less than astronomical size), we can write $\vec{w}_i = m_i\vec{g}$. Summing both sides of this gives $\vec{W} = M\vec{g}$, where $M = \Sigma m_i$ is the mass of the object. The net torque is the sum of the individual torques. That is,

$$\vec{\tau}_{net} = \sum_i (\vec{r}_i \times \vec{w}_i) = \sum_i (\vec{r}_i \times m_i\vec{g}) = \sum_i (m_i\vec{r}_i \times \vec{g})$$

Factoring $\vec{g}$ from the term on the right gives

$$\vec{\tau}_{net} = \left(\sum_i m_i\vec{r}_i \right) \times \vec{g}$$

and substituting $M\vec{r}_{cm}$ for $\Sigma m_i\vec{r}_i$ using the definition of center of mass ($M\vec{r}_{cm} = \Sigma m_i\vec{r}_i$), we obtain

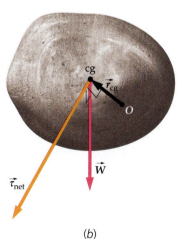

(b)

FIGURE 12-1

$$\vec{\tau}_{net} = M\vec{r}_{cm} \times \vec{g} = \vec{r}_{cm} \times M\vec{g} = \vec{r}_{cm} \times \vec{W} \qquad\qquad 12\text{-}4$$

Equations 12-3 and 12-4 are valid for any choice of the point O only if $\vec{r}_{cg} = \vec{r}_{cm}$. That is, the center of gravity and the center of mass coincide if the object is in a uniform gravitational field.

If O is directly above the center of gravity, then $\vec{r}_{cg}$ and $\overrightarrow{W}$ are both in the same direction (downward), so $\vec{\tau}_{net} = \vec{r}_{cg} \times \overrightarrow{W} = 0$. For example, when a mobile is suspended with its center of gravity directly below its suspension point, the net torque on the mobile about the suspension point is zero, so it is in static equilibrium.

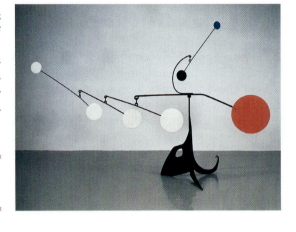

12-3 Some Examples of Static Equilibrium

For most examples and problems in this chapter, all the forces are perpendicular to the z axis. For such problems it is best to calculate torques about an axis parallel to the z axis. Out of the page is frequently chosen as the positive z direction. This is equivalent to choosing counterclockwise as positive and clockwise as negative.

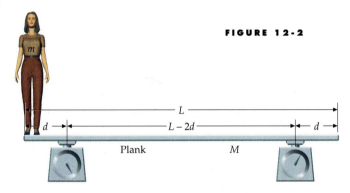

FIGURE 12-2

WALKING THE PLANK   **EXAMPLE 12-1**

A uniform plank of length $L = 3$ m and mass $M = 35$ kg is supported by scales a distance $d = 0.5$ m from the ends of the board, as shown in Figure 12-2. (*a*) Find the reading on the scales when Mary, whose mass $m = 45$ kg, stands on the left end of the plank. (*b*) Sergio climbs onto the plank and walks toward Mary, who jumps to the floor when the plank starts to tip. Sergio keeps walking all the way to the left end of the plank, and when he gets there the scale supporting the right end of the plank reads zero. What is Sergio's mass?

PICTURE THE PROBLEM The readings on the scales are the magnitudes of the forces they exert on the boards. To find these magnitudes we apply the two conditions for equilibrium.

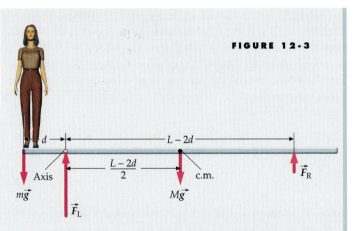

FIGURE 12-3

(*a*) 1. Draw a free-body diagram of the system consisting of Mary and the plank (Figure 12-3). Forces $\overrightarrow{F}_L$ and $\overrightarrow{F}_R$ are the forces exerted by the left and right scales:

2. Set the net force equal to zero, taking upward as positive:

$$\sum F_y = 0$$

$$F_L + F_R - Mg - mg = 0$$

3. Calculate the net torque about the axis directed out of the page (making counterclockwise positive) and through the point of application of $\overrightarrow{F}_L$:

$$\sum \tau = F_R(L - 2d) - Mg\frac{L - 2d}{2} + mgd$$

4. Set the net torque equal to zero and solve for F_R:

$$0 = F_R(L - 2d) - Mg\frac{L - 2d}{2} + mgd$$

so

$$F_R = \left(\frac{1}{2}M - \frac{d}{L - 2d}m\right)g$$

5. Substitute this result for F_R into step 2 and solve for F_L:

$$F_L = Mg + mg - F_R = \left(\frac{1}{2}M + \frac{L-d}{L-2d}m\right)g$$

6. Substitute numerical values to obtain numerical values for the forces:

$$F_R = \left(\frac{1}{2}(35 \text{ kg}) - \frac{0.5 \text{ m}}{2 \text{ m}}45 \text{ kg}\right)(9.81 \text{ N/kg})$$

$$= \boxed{61.3 \text{ N}}$$

$$F_L = \left(\frac{1}{2}(35 \text{ kg}) + \frac{2.5 \text{ m}}{2 \text{ m}}45 \text{ kg}\right)(9.81 \text{ N/kg})$$

$$= \boxed{723.5 \text{ N}}$$

(b) Using the Part (a) step 4 result, set $F_R = 0$ and solve for m:

$$0 = \left(\frac{1}{2}M - \frac{d}{L-2d}m\right)g$$

so

$$m = \frac{L-2d}{2d}M = \frac{2 \text{ m}}{1 \text{ m}}(35 \text{ kg}) = \boxed{70 \text{ kg}}$$

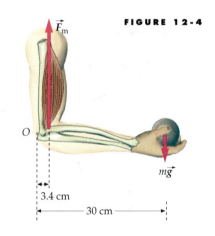

FIGURE 12-4

PLAUSIBILITY CHECK The sum of the two forces in the Part (a) step 6 results should equal Mary's weight plus the weight of the plank. The total weight is $(M + m)g = (35 \text{ kg} + 45 \text{ kg})(9.81 \text{ N/kg}) = 785 \text{ N}$. Also, $F_L + F_R = 723.5 \text{ N} + 61.3 \text{ N} = 785 \text{ N}$. Also, $F_R < F_L$ as one would expect.

REMARKS Sergio is 0.5 m from the axis and the center of mass of the plank is 1 m from the axis when the system is balanced with $F_R = 0$. Thus, Sergio's mass is twice the mass of the plank.

Example 12-1 can be solved using an axis through the center of the plank, but in this case both F_L and F_R appear in the torque equation, hence the algebra is a bit more complex. In general, a statics problem can be simplified by computing the torques about an axis through the line of action of one of the unknown forces, as when we chose the axis through the point of application of force F_L in the example.

> Obtain a simple solution by choosing an axis through the point of application of the force you have the least information about to calculate the torques.
>
> PROBLEM SOLVING GUIDELINE

FORCE ON AN ELBOW   **EXAMPLE 12-2**

You hold a 6-kg weight in your hand with your forearm making a 90° angle with your upper arm, as shown in Figure 12-4. Your biceps muscle exerts an upward force $\vec{F}_m$ that acts 3.4 cm from the pivot point O at the elbow joint. Model the forearm and hand as a 30-cm-long uniform rod with a mass of 1 kg. (a) Find the magnitude of $\vec{F}_m$ if the distance from the weight to the pivot point (elbow joint) is 30 cm and (b) find the magnitude and direction of the force exerted on the elbow joint by the upper arm.

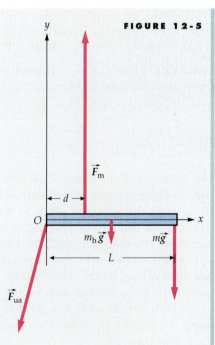

FIGURE 12-5

PICTURE THE PROBLEM To find the two forces, apply the two conditions for static equilibrium ($\Sigma F = 0$ and $\Sigma\tau = 0$) to the forearm.

(a) 1. Draw a free-body diagram of the forearm (Figure 12-5). Model the forearm as a horizontal rod.

2. The force we know least about is the force of the upper arm on the elbow joint $\vec{F}_{ua}$ (we know neither its magnitude nor its direction). Apply $\Sigma\tau = 0$ about an axis directed out of the page and through the point of application of $\vec{F}_{ua}$:

$$F_{ua}(0) - m_h g \frac{L}{2} + F_m d - mgL = 0$$

so

$$F_m = \left(\frac{1}{2}m_h + m\right)g\frac{L}{d}$$

$$= \left(\frac{1}{2}(1\text{ kg}) + 6\text{ kg}\right)(9.81\text{ N/kg})\frac{30\text{ cm}}{3.4\text{ cm}}$$

$$= \boxed{563\text{ N}}$$

(b) Apply $\Sigma F_x = 0$ and $\Sigma F_y = 0$ to obtain $\vec{F}_{ua}$:

$$F_{ua,x} + 0 + 0 + 0 = 0$$

and

$$F_{ua,y} + F_m - m_h g - mg = 0$$

so

$$F_{ua,x} = 0$$

and

$$F_{ua,y} = (m + m_h)g - F_m$$
$$= (7\text{ kg})(9.81\text{ N/kg}) - 563\text{ N}$$
$$= -494\text{ N}$$

Therefore

$$\vec{F}_{ua} = \boxed{494\text{ N, down}}$$

REMARKS The force that must be exerted by the muscle is 9.6 times the weight of the object! In addition, as the muscle pulls upward, the upper arm must push downward to keep the forearm in equilibrium. The force exerted by the upper arm is 8.4 times greater than the object's weight.

EXERCISE Show that F_{ua} can be found in one step by choosing the pivot point to be where the biceps attaches to the forearm. (*Answer* Setting net torque equal to zero gives $F_{ua}(3.4\text{ cm}) + F_m(0) - (6\text{ kg})(9.81\text{ N/kg})(30\text{ cm} - 3.4\text{ cm}) - (1\text{ kg})(9.81\text{ N/kg})(15\text{ cm} - 3.4\text{ cm}) = 0$. This yields $F_{ua} = 494\text{ N}$.)

REMARKS This example and exercise show that we can choose the pivot point wherever it is convenient for our calculation.

HANGING A SIGN   **EXAMPLE 12-3** **Try It Yourself**

The manager of the campus bookstore has ordered a new sign to hang in front of the store. The sign has a mass of 20 kg, and the manager plans to hang the sign from the end of a rod that will be attached to the wall by a wire (Figure 12-6). The manager needs to know how strong a wire is needed, so she asks you, a physics student, to calculate the tension in the wire. She is also concerned about how much force the rod puts on the wall, so she asks you to calculate that as well. The rod has a length of 2 m and a mass of 4 kg, and the wire is attached to a point 1 m above the rod on the wall.

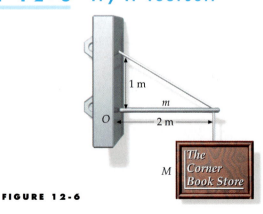

FIGURE 12-6

PICTURE THE PROBLEM There are three conditions for the rod to be in equilibrium: $\Sigma F_x = 0$, $\Sigma F_y = 0$, and $\Sigma \tau = 0$, and we have three unknowns: T and the components F_x and F_y of the force exerted by the wall on the rod. The force exerted by the rod on the wall is equal but opposite to the force exerted by the wall on the rod.

Cover the column to the right and try these on your own before looking at the answers.

Steps

Answers

1. Draw a free-body diagram for the rod (Figure 12-7).

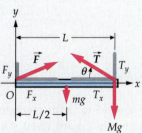

2. Set $\Sigma \tau = 0$ about point O.

$$TL \sin \theta - MgL - mg\frac{L}{2} = 0$$

3. Use trigonometry to solve for θ.

$$\theta = \tan^{-1}\tfrac{1}{2} = 26.6°$$

4. Solve the step 2 result for T.

$$T = \boxed{483 \text{ N}}$$

FIGURE 12-7

5. Set $\Sigma F_x = 0$ and $\Sigma F_y = 0$ and, using your values for T and θ, solve for F_x and F_y.

$$F_x + T_x = 0$$
$$F_y + T_y - Mg - mg = 0$$

so

$$F_x = 432 \text{ N}, \qquad F_y = 19.2 \text{ N}$$

6. Solve for the force $\vec{F}'$ exerted by the rod on the wall. The force exerted by the rod on the wall and that by the wall on the rod constitute an action–reaction pair.

$$\vec{F}' = -\vec{F} = \boxed{-432 \text{ N}\hat{i} - 19.2 \text{ N}\hat{j}}$$

RAISING A WHEEL   **EXAMPLE 12-4 Try It Yourself**

A wheel of mass M and radius R (Figure 12-8) rests on a horizontal surface against a step of height h ($h < R$). The wheel is to be raised over the step by a horizontal force $\vec{F}$ applied to the axle of the wheel as shown. Find the minimum force F_{min} necessary to raise the wheel over the step.

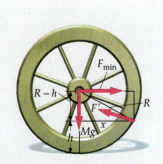

PICTURE THE PROBLEM If the magnitude of F is less than F_{min}, the surface at the bottom of the wheel exerts an upward normal force on the wheel. If F is increased, this normal force decreases. Apply the conditions for static equilibrium to find the value of F that will hold the wheel in place when the normal force is zero.

FIGURE 12-8

Cover the column to the right and try these on your own before looking at the answers.

Steps

Answers

1. Draw a free-body diagram of the wheel (Figure 12-9).

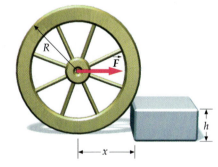

2. Apply $\Sigma \tau = 0$ to the wheel. Both the direction and the magnitude of $\vec{F}'$ are unknown, so follow the guidelines and calculate torques about an axis through its point of application. Obtain expressions for the lever arms from the free-body diagram and solve for F_{min}.

$$F_{min}(R - h) - Mgx = 0$$

so

$$F_{min} = \frac{Mgx}{R - h}$$

FIGURE 12-9

3. Use the Pythagorean theorem to express x in terms of h and R. $x = \sqrt{h(2R - h)}$

4. Set the magnitudes of the torques equal to each other and solve for F. $F_{min} = \dfrac{Mg\sqrt{h(2R - h)}}{R - h}$

REMARKS Applying $\Sigma\tau = 0$ about the axis through the center of the wheel shows that $\vec{F}'$ is directed toward the wheel's center; otherwise there would be a nonzero net torque.

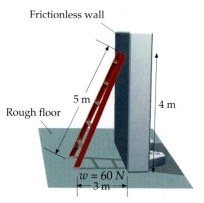

Frictionless wall

Rough floor

5 m   4 m

$w = 60 \text{ N}$
3 m

FIGURE 12-10

A LEANING LADDER   **EXAMPLE 12-5**

A uniform 5-m ladder weighing 60 N leans against a frictionless vertical wall, as shown in Figure 12-10. The foot of the ladder is 3 m from the wall. What is the minimum coefficient of static friction necessary between the ladder and the floor if the ladder is not to slip?

PICTURE THE PROBLEM There are three conditions for the ladder to be in equilibrium: $\Sigma F_x = 0$, $\Sigma F_y = 0$, and $\Sigma\tau = 0$. Apply these along with $f_s \leq \mu_s F_n$ to solve for μ_s.

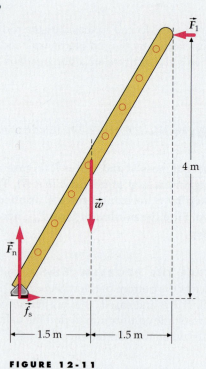

$\vec{F}_1$

4 m

$\vec{w}$

$\vec{F}_n$

$\vec{f}_s$

1.5 m   1.5 m

FIGURE 12-11

1. Draw a free-body diagram of the ladder as shown in Figure 12-11. The forces acting on the ladder are the force due to gravity $\vec{w}$, the force $\vec{F}_1$ exerted by the wall (since the wall is frictionless, it exerts only a normal force), and the force exerted by the floor, which consists of a normal component F_n and a static frictional component f_s. The three conditions for static equilibrium determine F_1, f_s, and F_n.

2. The minimum coefficient of static friction relates the frictional force f_s and normal force F_n: $\mu_s \geq \dfrac{f_s}{F_n}$ so $\mu_{s,min} = \dfrac{f_s}{F_n}$

3. Set $\Sigma F_x = 0$ and $\Sigma F_y = 0$: $f_s - F_1 = 0$ and $F_n - w = 0$

4. Solve for f_s and F_n: $f_s = F_1$ and $F_n = w = 60 \text{ N}$

5. Following the guideline, we set $\Sigma\tau = 0$ about an axis directed out of the page and through the foot of the ladder, the point of application of the force we know the least about: $F_1(4 \text{ m}) - w(1.5 \text{ m}) = 0$

6. Solve for the force F_1: $F_1 = \dfrac{w(15 \text{ m})}{4 \text{ m}} = \dfrac{(60 \text{ N})(1.5 \text{ m})}{4 \text{ m}} = 22.5 \text{ N}$

7. Use this result for F_1, and $f_s = F_1$ from step 4, to find f_s: $f_s = F_1 = 22.5 \text{ N}$

8. Use the results for f_s and F_n to obtain the minimum value of μ_s from step 2: $\mu_{s,min} = \dfrac{f_s}{F_n} = \dfrac{22.5 \text{ N}}{60 \text{ N}} = \boxed{0.375}$

REMARKS There is another way to look at this problem. In the free-body diagram for the ladder shown in Figure 12-12, the lines of action of the weight $\vec{w}$ and the force $\vec{F}_1$ exerted by the wall intersect at point P. The line of action of the resultant force exerted by the ground, $\vec{f}_s + \vec{F}_n$, must also go through point P or there would be an unbalanced torque about this point. The tangent of θ' equals 4 m/1.5 m = 2.67 = F_n/f_s. Whenever an object is in static equilibrium under the influence of three nonparallel forces, the lines of action of the forces must intersect at one point.

12-4 Couples

The forces $\vec{F}_n$ and $\vec{w}$ in Figure 12-11 of Example 12-5 are equal and opposite. Such a pair of forces, called a couple, tends to produce an angular acceleration, but its net force is zero. The forces $\vec{f}_s$ and $\vec{F}_1$ in those figures also constitute a couple. Figure 12-13 shows a couple consisting of forces $\vec{F}_1$ and $\vec{F}_2$ a distance D apart. The torque produced by this couple about an arbitrary point O is

$$\vec{\tau} = \vec{r}_1 \times \vec{F}_1 + \vec{r}_2 \times \vec{F}_2 = \vec{r}_1 \times \vec{F}_1 + \vec{r}_2 \times (-\vec{F}_1) = (\vec{r}_1 - \vec{r}_2) \times \vec{F}_1 \qquad 12\text{-}5$$

This result does not depend on the choice of the point O. The magnitude of this torque is

$$\tau = FD \qquad 12\text{-}6$$

where F is the magnitude of either force and D is the distance between their lines of action.

> The torque produced by a couple is the same about all points in space.

EXERCISE Show that the magnitude of the torque exerted by the couple in Figure 12-13 is given by FD, where D is the distance between the lines of action of the two forces and F is the magnitude of either force. (*Answer* The angle β between $\vec{r}_1 - \vec{r}_2$ and $\vec{F}_1$ is $90° - \phi$, so $|\vec{\tau}| = |(\vec{r}_1 - \vec{r}_2) \times \vec{F}_1| = |\vec{r}_1 - \vec{r}_2| F \sin(90° - \phi) = F|\vec{r}_1 - \vec{r}_2| \cos \phi = FD$.)

12-5 Static Equilibrium in an Accelerated Frame

By an accelerated frame we mean a reference frame that is accelerating relative to an inertial frame of reference. The net force on an object that remains at rest relative to an accelerated reference frame is not equal to zero. An object at rest relative to the accelerated frame has the same acceleration as the frame. The two conditions for an object to be in static equilibrium in an accelerated reference frame are

1. $\Sigma\vec{F} = m\vec{a}_{cm}$

    where $\vec{a}_{cm}$ is the acceleration of the center of mass, which is the acceleration of the reference frame.

2. $\Sigma\vec{\tau}_{cm} = 0$

    The sum of the torques about the center of mass must be zero.

The second condition follows from the fact that Newton's second law for rotation, $\Sigma\vec{\tau}_{cm} = I_{cm}\vec{\alpha}$, holds for torques about the center of mass whether or not the center of mass is accelerating.[†]

---

† See the discussion surrounding Equation 9-30.

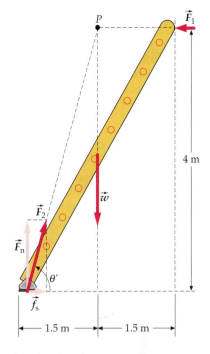

FIGURE 12-12

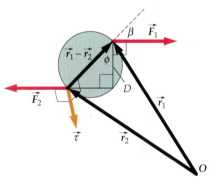

FIGURE 12-13

EXAMPLE 12-6

A truck (Figure 12-14*a*) carries a uniform box of mass *m*, height *h*, and square cross section of edge-length *L*. What is the greatest acceleration the truck can have without the box tipping over? Assume that the box tips before it slides.

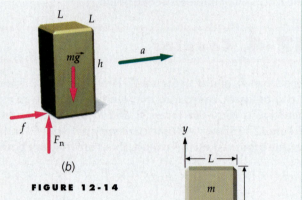

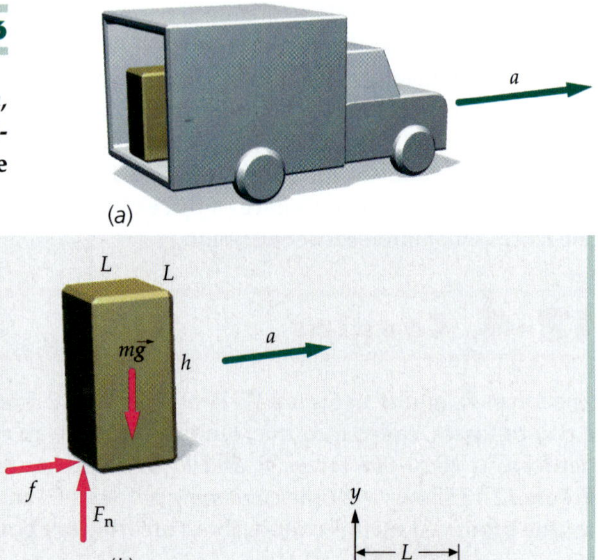

(*a*)

PICTURE THE PROBLEM The acceleration of the box is due to the frictional force, as shown in Figure 12-14*b*. This force exerts a counterclockwise torque about the center of mass of the box. The only other force that exerts a torque about the center of mass of the box is the normal force. If the box is not accelerating, the normal force is distributed uniformly across the bottom of the box. If the acceleration is small, this distribution shifts and the effective normal force moves to the left to provide a balancing torque about the center of mass. The greatest balancing torque this force can exert is when the effective normal force is at the edge of the box, as shown.

(*b*)

FIGURE 12-14

1. Draw a free-body diagram of the box (Figure 12-15).

2. Apply $\Sigma F_y = ma_{cmy}$ to the box and then solve for the normal force:

$$F_n - mg = 0 \quad \text{so} \quad F_n = mg$$

3. Apply $\Sigma F_x = ma_{cmx}$ to the box:

$$f_s = ma$$

4. Apply $\Sigma\tau_{cm} = 0$:

$$f_s\frac{h}{2} - F_n d = 0$$

5. Substitute for d, f_s, and F_n and solve for a:

$$ma\frac{h}{2} - mg\frac{L}{2} = 0 \quad \text{so} \quad \boxed{a = \frac{L}{h}g}$$

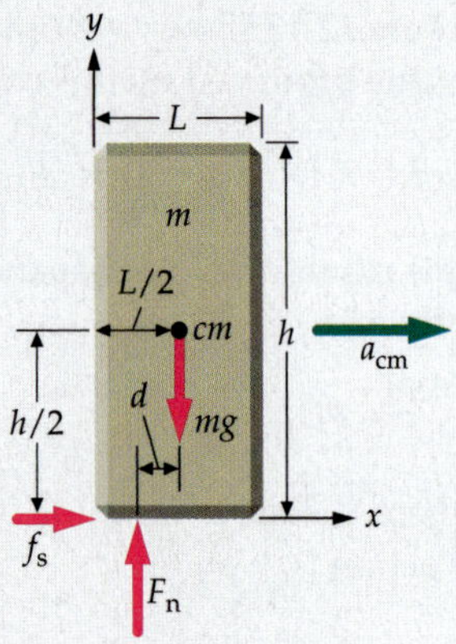

FIGURE 12-15

REMARKS The maximum acceleration is proportional to L/h. This maximum acceleration is small for a tall, narrow box (L/h small) and large for a short, wide box (L/h large). Thus, a short, wide box is more stable.

12-6 Stability of Rotational Equilibrium

There are three categories of rotational equilibrium for an object: stable, unstable, or neutral. **Stable rotational equilibrium** occurs when the torques that arise from a small angular displacement of the object urge the object back toward its equilibrium position. Stable equilibrium is illustrated in Figure 12-16*a*. When the

FIGURE 12-16 If a slight rotation raises the center of gravity, as in (*a*), the equilibrium is stable. If a slight rotation lowers the center of gravity, as in (*b*), the equilibrium is unstable. If a slight rotation neither raises nor lowers the center of gravity, as in (*c*), the equilibrium is neutral.

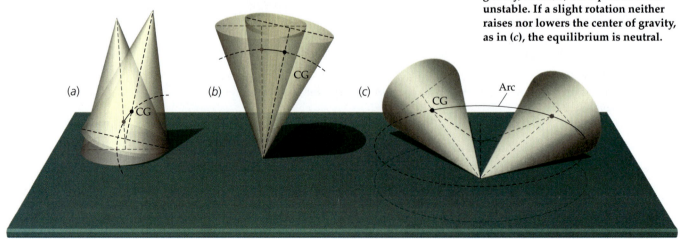

cone is tipped slightly as shown, the resulting torque about the pivot point tends to restore the cone to its original position. Note that this slight tipping lifts the center of gravity, increasing the potential energy of the cone.

Unstable rotational equilibrium, illustrated in Figure 12-16b, occurs when the torques that arise from a small angular displacement of the object urge the object away from its equilibrium position. A slight tipping of the cone causes it to fall over because the torque due to its weight tends to rotate it away from its original position. Here the rotation lowers the center of gravity and decreases the potential energy of the cone.

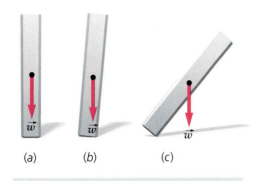

(a)   (b)   (c)

FIGURE 12-17 Stability of equilibrium is relative. If the rod in (a) is rotated slightly, as in (b), it returns to its original equilibrium position as long as the center of gravity lies over the base of support. (c) If the rotation is too great, the center of gravity is no longer over the base of support, and the rod falls over.

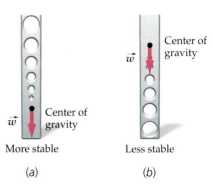

More stable   Less stable

(a)   (b)

FIGURE 12-18 When a nonuniform rod rests on its heavy end with its center of gravity low, as in (a), the equilibrium is more stable than when its center of gravity is high, as in (b).

The cone resting on a horizontal surface in Figure 12-16c illustrates **neutral rotational equilibrium.** If the cone is rolled slightly, there is no torque or force that urges it either back toward or away from its original position. As the cone rotates, the height of the center of gravity remains unchanged, so the potential energy does not change.

In summary, if a system is disturbed slightly from its equilibrium position, the equilibrium is stable if the system returns to its original position, unstable if it moves farther away, and neutral if there are no torques or forces tending to rotate it in either direction.

Because "disturbed slightly" is a relative term, stability is also relative. One example of equilibrium may be more or less stable than another. A rod is balanced on one end, as in Figure 12-17a. Here, if the disturbance is very small (Figure 12-17b), the rod will move back toward its original position, but if the disturbance is great enough so that the center of gravity no longer lies over the base of support (Figure 12-17c), the rod will fall.

We can improve the stability of a system by either lowering the center of gravity or widening the base of support. Figure 12-18 shows a nonuniform rod that is loaded so that its center of gravity is near one end. If it stands on its heavy end so that the center of gravity is low (Figure 12-18a), it is much more stable than if it stands on the other end so that the center of gravity is high (Figure 12-18b).

In Figure 12-19 the system is stable for any angular displacement because the resulting torque always rotates the system back toward its equilibrium position.

Standing or walking upright is difficult for a human because the center of gravity is high, and must be kept over a relatively small base of support, the feet. Human infants take about a year to learn to walk. A four-footed creature has a much easier time because its base of support is larger and its center of gravity is lower. Newborn kittens can walk almost immediately.

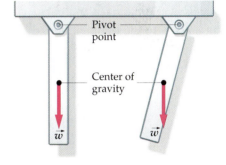

Pivot point

Center of gravity

FIGURE 12-19

12-7 Indeterminate Problems

When objects are not rigid, but deformable, we need more information to determine the forces required for equilibrium. Consider a car resting on a horizontal surface. Suppose there is a very heavy object on one side of the trunk.

We wish to find the vertical support force exerted by the road on each tire. Let the road be in the *xy* plane. If we choose one of the tires as our origin, the torque exerted by all the forces about that point has *x* and *y* components, but no *z* component because there are no horizontal forces. We thus obtain two equations by setting the net torque equal to zero, and a third equation by setting the net vertical force equal to zero. We need another equation to find the force exerted by the road on each of the four tires. If we let air out of one of the tires and pump up another tire to a greater pressure, the car remains in equilibrium, but the force exerted on each tire changes. Clearly, the forces on the tires in this problem are not determined by the information given. The tires are not rigid bodies. To some extent, every object is deformable.

12-8 Stress and Strain

If a solid object is subjected to forces that tend to stretch, shear, or compress the object, its shape changes. If the object returns to its original shape when the forces are removed, it is said to be **elastic**. Most objects are elastic for forces up to a certain maximum, called the **elastic limit**. If the forces exceed the elastic limit, the object does not return to its original shape but is permanently deformed.

Figure 12-20 shows a solid bar subjected to a stretching or **tensile force *F*** acting equally to the right and to the left. The bar is in equilibrium, but the forces acting on it tend to increase its length. The fractional change in the length $\Delta L/L$ of a segment of the bar is called the **strain:**

$$\text{Strain} = \frac{\Delta L}{L} \qquad \qquad 12\text{-}7$$

The ratio of the force *F* to the cross-sectional area *A* is called the **tensile stress:**

$$\text{Stress} = \frac{F}{A} \qquad \qquad 12\text{-}8$$

Figure 12-21 shows a graph of stress versus strain for a typical solid bar. The graph is linear until point A. Up to this point, known as the proportional limit, the strain is proportional to the stress. The result that strain varies linearly with stress is known as Hooke's law. Point B in Figure 12-21 is the elastic limit of the material. If the bar is stretched beyond this point, it is permanently deformed. If an even greater stress is applied, the material eventually breaks, shown happening at point C. The ratio of stress to strain in the linear region of the graph is a constant called **Young's modulus *Y*:**

$$Y = \frac{\text{stress}}{\text{strain}} = \frac{F/A}{\Delta L/L} \qquad \qquad 12\text{-}9$$

YOUNG'S MODULUS DEFINED

The units of Young's modulus are newtons per square meter (or pounds per square inch). Approximate values of Young's modulus for various materials are listed in Table 12-1.

EXERCISE Suppose that the biceps muscle of your right arm has a maximum cross-sectional area of $12 \text{ cm}^2 = 1.2 \times 10^{-3} \text{ m}^2$. What is the stress in the muscle if it exerts a force of 300 N? (*Answer* Stress $= F/A = 2.5 \times 10^5 \text{ N/m}^2$. The maximum stress that can be exerted is approximately the same for all human muscles. Greater forces can be exerted by muscles with greater cross-sectional areas.)

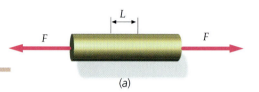

(a)

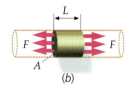

(b)

FIGURE 12-20 (*a*) A solid bar subjected to stretching forces of magnitude *F* acting on each end. (*b*) A small section of the bar of length *L*. The elements of the bar to the left and right of this section exert forces on the section. If the section is not too close to the end, these forces are distributed equally over the cross-sectional area. The force per unit area is the stress.

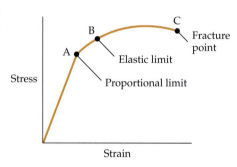

FIGURE 12-21 A graph of stress versus strain. Up to point A, the strain is proportional to the stress. Beyond the elastic limit at point B, the bar will not return to its original length when the stress is removed. At point C, the bar fractures.

TABLE 12-1

Young's Modulus Y and Strengths of Various Materials[†]

Material	Y, GN/m²[‡]	Tensile strength, MN/m²	Compressive strength, MN/m²
Aluminum	70	90	
Bone			
Tensile	16	200	
Compressive	9		270
Brass	90	370	
Concrete	23	2	17
Copper	110	230	
Iron (wrought)	190	390	
Lead	16	12	
Steel	200	520	520

[†] These values are representative. Actual values for particular samples may differ.
[‡] 1 GN = 10^3 MN = 1×10^9 N.

If a bar is subjected to forces that tend to compress it rather than stretch it, the stress is called **compressive stress.** For many materials, Young's modulus for compressive stress is the same as that for tensile stress. Note that ΔL in Equation 12-7 is then taken to be the *decrease* in the length of the bar. If the tensile or compressive stress is too great, the bar breaks. The stress at which breakage occurs is called the **tensile strength,** or in the case of compression, the **compressive strength.** Approximate values of the tensile and compressive strengths of various materials are listed in Table 12-1. Note from the table that the compressive strength of bone is greater than the tensile strength. Also note that, for bone, Young's modulus is significantly larger for tensile stress than for compressive stress. These differences have biological significance, because the major job of bone is to resist the compressive load exerted by contracting muscles.

ELEVATOR SAFETY   **EXAMPLE 12-7**   **Put It in Context**

While working with an engineering company during the summer, you are assigned to check the safety of an elevator system in a new office building. The elevator has a maximum load of 1000 kg including its own mass, and is supported by a steel cable 3.0 cm in diameter and 300 m long at full extension. There will be safety concerns if the steel stretches more than 3.0 cm. Your job is to determine whether or not the elevator is safe as planned, given a maximum acceleration of the system of 1.5 m/s².

PICTURE THE PROBLEM L is the unstretched length of cable, F is the force acting on it, and A is its cross-sectional area. The stretch in the cable ΔL is related to Young's modulus by $Y = (F/A)/(\Delta L/L)$. From Table 12-1 we find the numerical value of Young's modulus for steel, $Y = 2.0 \times 10^{11}$ N/m².

1. The amount the cable is stretched, ΔL, is found from Young's modulus:

$$Y = \frac{F/A}{\Delta L/L} \quad \text{so} \quad \Delta L = \frac{FL}{AY}$$

2. To find the force acting on the cable we apply Newton's second law to the elevator. There are two forces on the elevator, the force F of the cable and the weight:

$$F - mg = ma_y$$

so

$$
\begin{aligned}
F_{\max} &= m(g + a_{y,\max}) \\
&= (1000 \text{ kg})(9.81 \text{ N/kg} + 1.5 \text{ N/kg}) \\
&= 1.13 \times 10^4 \text{ N}
\end{aligned}
$$

3. Substitute into the step 1 result and obtain the maximum amount of stretch:

$$
\begin{aligned}
\Delta L &= \frac{F_{\max}L}{AY} = \frac{F_{\max}L}{\pi r^2 Y} \\
&= \frac{(1.13 \times 10^4 \text{ N})(300 \text{ m})}{\pi (0.015 \text{ m})^2 (2.0 \times 10^{11} \text{ N/m}^2)} \\
&= 2.40 \text{ cm}
\end{aligned}
$$

4. Report your results to your boss:

> According to my calculations, the most the cable will stretch is 2.4 cm, only 20% less than the 3.0-cm limit. However, in reading the footnote to the table, I note that the values given for Young's modulus are representative values, and that actual values vary from sample to sample. I recommend that you consult a engineer and get a professional evaluation.

EXERCISE A wire 1.5 m long has a cross-sectional area of 2.4 mm². It is hung vertically and stretches 0.32 mm when a 10-kg block is attached to it. Find (a) the stress, (b) the strain, and (c) Young's modulus for the wire. (*Answer* (a) $4.09 \times 10^7 \text{ N/m}^2$ (b) 2.13×10^{-4} (c) 192 GN/m^2)

In Figure 12-22, a force F_s is applied tangentially to the top of a book. Such a force is called a **shear force**. The ratio of the shear force F_s to the area A is called the **shear stress**:

$$\text{Shear stress} = \frac{F_s}{A} \qquad \text{12-10}$$

A shear stress tends to deform an object, as shown in Figure 12-22. The ratio $\Delta X/L$ is called the **shear strain**:

$$\text{Shear strain} = \frac{\Delta X}{L} = \tan\theta \qquad \text{12-11}$$

where θ is the shear angle shown in the figure. The ratio of the shear stress to the shear strain is called the **shear modulus** M_s:

$$M_s = \frac{\text{shear stress}}{\text{shear strain}} = \frac{F_s/A}{\Delta X/L} = \frac{F_s/A}{\tan\theta} \qquad \text{12-12}$$

DEFINITION — SHEAR MODULUS

The shear modulus is also known as the **torsion modulus**. The torsion modulus is approximately constant for small stresses, which implies that the shear strain varies linearly with the shear stress. This observation is known as Hooke's law for torsional stress. In a torsion balance, such as that used in Cavendish's apparatus for measuring the universal gravitational constant G, the torque (which is related to the stress) is proportional to the angle of twist (which equals the strain for small angles). Approximate values of the shear modulus for various materials are listed in Table 12-2.

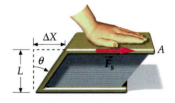

FIGURE 12-22 The application of the horizontal force $\vec{F}_s$ to the book causes a shear stress defined as the force per unit area. The ratio $\Delta X/L = \tan\theta$ is the shear strain.

TABLE 12-2

Approximate Values of the Shear Modulus M_s of Various Materials

Material	M_s, GN/m²
Aluminum	30
Brass	36
Copper	42
Iron	70
Lead	5.6
Steel	84
Tungsten	150

Topic	Relevant Equations and Remarks
1. Equilibrium of a Rigid Object	
Conditions	1. The net external force acting on the object must be zero:
	$$\Sigma \vec{F} = 0 \qquad \text{12-1}$$
	2. The net external torque about any point must be zero:
	$$\Sigma \vec{\tau} = 0 \qquad \text{12-2}$$
	If all the forces acting on a rigid object are perpendicular to an axis, then the sum of the torques about that axis equals zero.
Stability	The equilibrium of an object can be classified as stable, unstable, or neutral. An object resting on some surface will be in equilibrium if its center of gravity lies over its base of support. Stability can be improved by lowering the center of gravity or by increasing the size of the base.
2. Center of Gravity	The force of gravity exerted on the various parts of an object can be replaced by a single force, the total weight of the object $\vec{W}$, acting at the center of gravity.
	$$\vec{\tau}_{\text{net}} = \sum_{i} (\vec{r}_i \times \vec{w}_i) = \vec{r}_{\text{cg}} \times \vec{W} \qquad \text{12-3}$$
	For an object in a uniform gravitational field, the center of gravity coincides with the center of mass.
3. Couples	A pair of equal and opposite forces constitutes a couple. The torque produced by a couple is the same about any point in space.
	$$\vec{\tau} = (\vec{r}_1 - \vec{r}_2) \times \vec{F}_1, \quad \text{so} \quad \tau = FD \qquad \text{12-5, 12-6}$$
	where D is the distance between the lines of action of the forces.
5. Accelerated Reference Frame	The conditions for static equilibrium in an accelerated reference frame are
	1. $\Sigma \vec{F} = m\vec{a}_{\text{cm}}$
	where $\vec{a}_{\text{cm}}$ is the acceleration of the center of mass, which is the acceleration of the reference frame.
	2. $\Sigma \vec{\tau}_{\text{cm}} = 0$
	The sum of the torques about the center of mass must be zero.
6. Stress and Strain	
Young's modulus	$$Y = \frac{\text{stress}}{\text{strain}} = \frac{F/A}{\Delta L/L} \qquad \text{12-9}$$
Shear modulus	$$M_s = \frac{\text{shear stress}}{\text{shear strain}} = \frac{F_s/A}{\Delta X/L} = \frac{F_s/A}{\tan \theta} \qquad \text{12-12}$$

PROBLEMS

- Single-concept, single-step, relatively easy
- •• Intermediate-level, may require synthesis of concepts
- ••• Challenging
- **SSM** Solution is in the *Student Solutions Manual*
- **iSOLVE** Problems available on iSOLVE online homework service
- **iSOLVE✓** These "Checkpoint" online homework service problems ask students additional questions about their confidence level, and how they arrived at their answer

In a few problems, you are given more data than you actually need; in a few other problems, you are required to supply data from your general knowledge, outside sources, or informed estimates.

Take $g = 9.81$ N/kg $= 9.81$ m/s² and neglect friction in all problems unless otherwise stated.

Conceptual Problems

1 • **SSM** True or false:

(a) $\sum_i \vec{F}_i = 0$ is sufficient for static equilibrium to exist.

(b) $\sum_i \vec{F}_i = 0$ is necessary for static equilibrium to exist.

(c) In static equilibrium, the net torque about any point is zero.

(d) An object is in equilibrium only when there are no forces acting on it.

2 • True or false: The center of gravity is always at the geometric center of a body.

3 • Must there be any material at the center of gravity of an object?

4 • **SSM** If the gravitational field $\vec{g}$ is not constant over an object, is it the center of mass or the center of gravity that is the pivot point when the object is balanced?

5 •• To find the center of mass of an irregular plane figure, the following method can be used: hang the figure from one corner, by a string and extend the line that the string makes downward across the figure. Repeat, hanging it from another corner. The intersection of the two lines will be at the center of mass. Explain why this technique works.

6 • **SSM** Is it possible to climb a ladder placed against a wall if the ground is frictionless but the wall is not? Explain.

7 • An aluminum wire and a steel wire of the same length L and diameter D are joined to form a wire of length $2L$. The wire is fastened to the ceiling and a weight W is attached to the other end. Neglecting the mass of the wires, which of the following statements is true? (a) The aluminum portion will stretch by the same amount as the steel portion. (b) The tensions in the aluminum portion and the steel portion are equal. (c) The tension in the aluminum portion is greater than that in the steel portion. (d) None of these statements is true.

8 • If the net torque about some point is zero, must it be zero about any other point? Explain.

9 • **SSM** The horizontal bar in Figure 12-23 will remain horizontal if (a) $L_1 = L_2$ and $R_1 = R_2$, (b) $L_1 = L_2$ and $M_1 = M_2$, (c) $R_1 = R_2$ and $M_1 = M_2$, (d) $L_1M_1 = L_2M_2$, (e) $R_1L_1 = R_2L_2$.

FIGURE 12-23 Problem 9

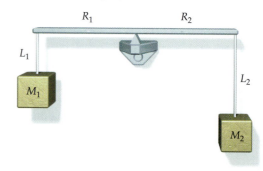

10 •• Sit in a chair with your back straight. Now try to stand up without leaning forward. Explain why you cannot do it.

11 •• **SSM** The great engineering feats of ancient times (Roman arch bridges, the great cathedrals, and the pyramids, to name a few) all have two things in common: they are made of stone, and they are compressive structures—that is, they are built so that all strains in the structure are compressive rather than tensile in nature. Look up the tensile and compressive strengths of stone and cement to explain why this is true.

Estimation and Approximation

12 •• A large crate weighing 4500 N rests on four 12-cm-high blocks on a horizontal surface (Figure 12-24). The crate is 2 m long, 1.2 m high, and 1.2 m deep. You are asked to lift one end of the crate using a long steel pry bar. The fulcrum on the pry bar is 10 cm from the end that lifts the crate. Estimate the length of the bar you will need to lift the end of the crate.

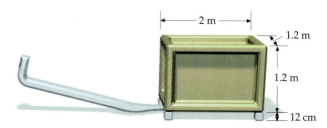

FIGURE 12-24 Problem 12

13 •• **SSM** Consider an atomic model for Young's modulus: assume that we have a large number of atoms arranged in a cubic array separated by distance a. Imagine that each atom is attached to its 6 nearest neighbors by little springs with spring constant k. (Atoms are not really attached by springs, but the forces between them act enough like springs to make this a good model.) (a) Show that this material, if stretched, will have a Young's modulus $Y = k/a$. (b) From Table 12-1, and assuming that $a \approx 1$ nm, estimate a typical value for the "atomic spring constant" k in a metal.

Conditions for Equilibrium

14 • A seesaw consists of a 4-m board pivoted at the center. A 28-kg child sits on one end of the board. Where should a 40-kg child sit to balance the seesaw?

15 • **iSOLVE** ✓ In Figure 12-25, Misako is about to do a push-up. Her center of gravity lies directly above point P on the floor, which is 0.9 m from her feet and 0.6 m from her hands. If her mass is 54 kg, what is the force exerted by the floor on her hands?

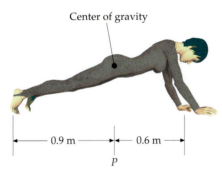

Center of gravity

← 0.9 m → ← 0.6 m →

P

FIGURE 12-25 Problem 15

16 • **SSM** **iSOLVE** ✓ Misako wants to measure the strength of her biceps muscle by exerting a force on a test strap as shown in Figure 12-26. The strap is 28 cm from the pivot point at the elbow, and her biceps muscle is attached at a point 5 cm from the pivot point. If the scale reads 18 N when she exerts her maximum force, what force is exerted by the biceps muscle?

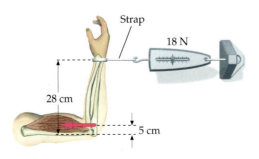

Strap

18 N

28 cm

5 cm

FIGURE 12-26 Problem 16

17 • A crutch is pressed against the sidewalk with a force $\vec{F}_c$ along its own direction as in Figure 12-27. This force is balanced by the normal force $\vec{F}_n$ and a frictional force $\vec{f}_s$. (a) Show that when the force of friction is at its maximum value, the coefficient of friction is related to the angle θ by $\mu_s = \tan \theta$. (b) Explain how this result applies to the forces on your foot when you are not using a crutch. (c) Why is it advantageous to take short steps when walking on ice?

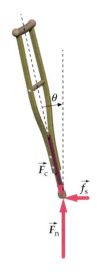

θ

$\vec{F}_c$

$\vec{f}_s$

$\vec{F}_n$

FIGURE 12-27 Problem 17

The Center of Gravity

18 • **iSOLVE** An automobile has 58 percent of its weight on the front wheels. The front and back wheels are separated by 2 m. Where is the center of gravity located with respect to the front wheels?

19 • **SSM** Each of the objects shown in Figure 12-28 is suspended from the ceiling by a thread attached to the point marked × on the object. Describe the orientation of each suspended object with a diagram.

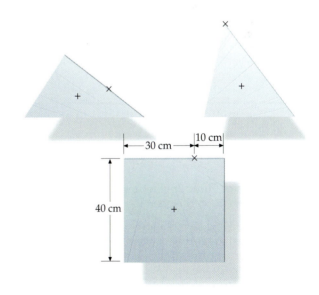

×

+ × +

← 30 cm → |10 cm|

×

40 cm

+

FIGURE 12-28 Problem 19

20 •• ![SOLVE✓] A square plate is produced by welding together four smaller square plates, each of side a as shown in Figure 12-29. Plate 1 weighs 40 N; plate 2, 60 N; plate 3, 30 N; and plate 4, 50 N. Find the center of gravity (x_{cg}, y_{cg}).

FIGURE 12-29 Problem 20

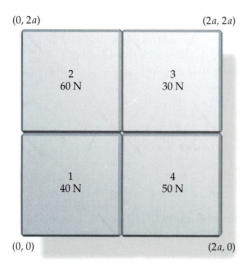

(0, 2a)   (2a, 2a)

2 60 N	3 30 N
1 40 N	4 50 N

(0, 0)   (2a, 0)

21 •• ![SOLVE] A uniform rectangular plate has a circular section of radius R cut out as shown in Figure 12-30. Find the center of gravity of the system. *Hint: Do not integrate. Use superposition of a rectangular plate minus a circular plate.*

FIGURE 12-30 Problem 21

Some Examples of Static Equilibrium

22 • Figure 12-31 shows a lever with a force f being applied to lift a load F. (*a*) If we define the mechanical advantage of the lever as $M = F/f_{min}$, where f_{min} is the smallest force necessary to lift the load F, show that $M = x/X$, where x is the moment arm (distance to the pivot) for the applied force and X is the moment arm for the load. (*b*) To lift a heavy load, the moment arm for the applied force is usually larger than the load, so that the lifting force is smaller than the applied force. However, sometimes the moment arm for the applied force is much smaller than that for the load, so that the applied force must be much larger than the load (the arrangements of muscles in the forearm, as shown in Figure 12-4, is a good example). Why is this arrangement useful under some circumstances?

FIGURE 12-31 Problem 22

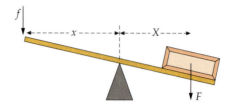

23 • ![SOLVE] Figure 12-32 shows a 25-foot sailboat. The mast is a uniform pole of 120 kg and is supported on the deck and held fore and aft by wires as shown. The tension in the forestay (wire leading to the bow) is 1000 N. Determine the tension in the backstay and the force that the deck exerts on the mast. Is there a tendency for the mast to slide forward or aft? If so, where should a block be placed to prevent the mast from moving?

FIGURE 12-32 Problem 23

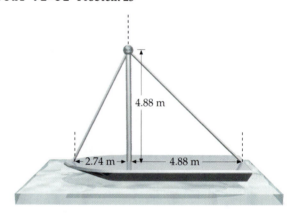

4.88 m

←2.74 m→   ←———4.88 m———→

24 •• ![SOLVE] A 10-m beam of mass 300 kg extends over a ledge as in Figure 12-33. The beam is not attached, but simply rests on the surface. A 60-kg student intends to position the beam so that he can walk to the end of it. How far from the edge of the ledge can the beam extend?

FIGURE 12-33
Problem 24

x

25 •• SSM A gravity board for locating the center of gravity of a person consists of a horizontal board supported by a fulcrum at one end and by a scale at the other end. A physics student lies horizontally on the board with the top of his head above the fulcrum point as shown in Figure 12-34. The scale is 2 m from the fulcrum. The student has a mass of 70 kg, and when he is on the gravity board, the scale advances 250 N. Where is the center of gravity of the student?

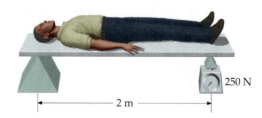

250 N

2 m

FIGURE 12-34 Problem 25

26 •• [SOLVE]✓ A 3-m board of mass 5 kg is hinged at one end. A force $\vec{F}$ is applied vertically at the other end to lift a 60-kg block, which rests on the board 80 cm from the hinge, as shown in Figure 12-35. (a) Find the magnitude of the force needed to hold the board stationary at $\theta = 30°$. (b) Find the force exerted by the hinge at this angle. (c) Find the magnitude of the force $\vec{F}$ and the force exerted by the hinge if $\vec{F}$ is exerted perpendicular to the board when $\theta = 30°$.

FIGURE 12-35
Problem 26

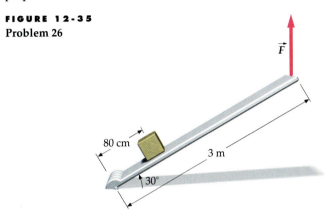

27 •• [SSM] [SOLVE] A cylinder of weight W is supported by a frictionless trough formed by a plane inclined at 30° to the horizontal on the left and one inclined at 60° on the right as shown in Figure 12-36. Find the force exerted by each plane on the cylinder.

FIGURE 12-36
Problem 27

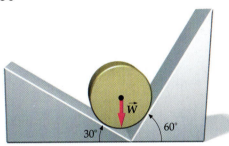

28 •• [SOLVE] An 80-N weight is supported by a cable attached to a strut hinged at point A as in Figure 12-37. The strut is supported by a second cable under tension T_2. The mass of the strut is negligible. (a) What are the three nonnegligible forces acting on the strut? (b) Show that the vertical component of the tension T_2 must equal 80 N. (c) Find the force exerted on the strut by the hinge.

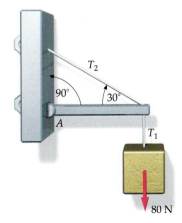

FIGURE 12-37 **Problem 28**

29 •• A horizontal board 8.0 m long is used by pirates to make their victims walk the plank. A pirate of mass 105 kg stands on the shipboard end of the plank to prevent it from tipping. Find the maximum distance the plank can overhang for a 63-kg victim to be able to walk to the end if (a) the mass of the plank is negligible and (b) the mass of the plank is 25 kg.

30 •• A uniform 18-kg door that is 2.0 m high by 0.8 m wide is hung from two hinges that are 20 cm from the top and 20 cm from the bottom. If each hinge supports half the weight of the door, find the magnitude and direction of the horizontal components of the forces exerted by the two hinges on the door.

31 • Find the force exerted by the edge of the block on the wheel in Example 12-4, just as the wheel lifts off the surface.

32 •• [SSM] [SOLVE] The diving board shown in Figure 12-38 has a mass of 30 kg. Find the force on the supports when a 70-kg diver stands at the end of the diving board. Give the direction of each support force as a tension or a compression.

FIGURE 12-38 **Problem 32**

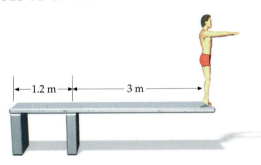

33 •• Find the force exerted on the strut by the hinge at A for the arrangement in Figure 12-39 if (a) the strut is weightless and (b) the strut weighs 20 N.

FIGURE 12-39
Problem 33

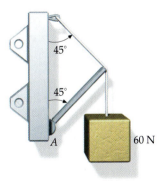

34 •• Julie has been hired to help paint the trim of a building, but she is not convinced of the safety of the apparatus. A 5.0-m plank is suspended horizontally from the top of the building by ropes attached at each end. Julie knows from previous experience that the ropes being used will break if the tension exceeds 1 kN. Her 80-kg boss dismisses Julie's worries and begins painting while standing 1 m from the end of the plank. If Julie's mass is 60 kg and the plank has a mass of 20 kg, then over what range of positions can Julie stand to join her boss without causing the ropes to break?

35 •• A cylinder of mass M and radius R rolls against a step of height h as shown in Figure 12-40. When a horizontal force $\vec{F}$ is applied to the top of the cylinder, the cylinder remains at rest. (a) What is the normal force exerted by the floor on the cylinder? (b) What is the horizontal force exerted by the edge of the step on the cylinder? (c) What is the vertical component of the force exerted by the edge of the step on the cylinder?

FIGURE 12-40
Problems 35, 36

$\vec{F}$

R

h

36 •• For the cylinder in Problem 35, find the minimum horizontal force $\vec{F}$ that will roll the cylinder over the step if the cylinder does not slide on the edge.

37 •• SSM Figure 12-41 shows a hand holding an epee, a weapon used in the sport of fencing. The center of mass of the epee, indicated in Figure 12-41, is 24 cm from the pommel; its total mass is 0.700 kg and its length is 110 cm. (a) Apply one of the conditions for static equilibrium to find the (total) force exerted by the hand on the epee. (b) Apply the other condition for static equilibrium to find the torque exerted by the hand on the epee. (c) Model the forces exerted by the hand as two oppositely directed forces whose lines of action are separated by the width of the fencer's hand (≈ 10 cm). What are the magnitudes and directions of these two forces?

FIGURE 12-41 Problem 37

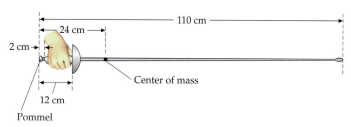

110 cm

24 cm

2 cm

Center of mass

12 cm

Pommel

38 •• A large gate weighing 200 N is supported by hinges at the top and bottom and is further supported by a wire as shown in Figure 12-42. (a) What must be the tension in the wire for the force on the upper hinge to have no horizontal component? (b) What is the horizontal force on the lower hinge? (c) What are the vertical forces on the hinges?

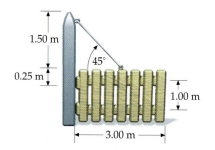

1.50 m

45°

0.25 m

1.00 m

3.00 m

FIGURE 12-42 Problem 38

39 ••• SSM A uniform log with a mass of 100 kg, a length of 4 m, and a radius of 12 cm is held in an inclined position, as shown in Figure 12-43. The coefficient of static friction between the log and the horizontal surface is 0.6. The log is on the verge of slipping to the right. Find the tension in the support wire and the angle the wire makes with the vertical wall.

FIGURE 12-43 Problem 39

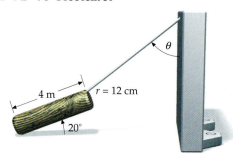

θ

4 m

$r = 12$ cm

20°

40 ••• A tall, uniform, rectangular block sits on an inclined plane as shown in Figure 12-44. A cord is attached to the top of the block to prevent it from falling down the incline. What is the maximum angle θ for which the block will not slide on the incline? Let b/a be 4 and $\mu_s = 0.8$.

FIGURE 12-44 Problem 40

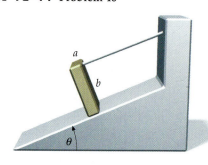

a

b

θ

41 •• SSM A boat is moored at the end of a dock in a rapidly flowing river by a chain 5 m long, as shown in Figure 12-45. To give the chain some flexibility, a 100-N weight is attached in the center in the chain, to allow for variations in the force pulling the boat away from the dock. (a) If the drag force on the boat is 50 N, what is the tension in the chain? (b) How far will the chain sag? Ignore the weight of the chain itself. (c) How far is the boat from the dock? (d) If the maximum tension that the chain can support is 500 N, what is the maximum value of the force that the river can exert on the boat?

2.5 m   2.5 m

100 N

50 N

FIGURE 12-45 Problem 41

42 •• **iSOLVE** A thin rod of length 10 m and mass 20 kg is supported at a 30° incline. One support is 2 m and the other is 6 m from the lower end of the rod. Friction prevents the rod from sliding off the supports. Find the normal force exerted on the rod by each support.

43 • Two 80-N forces are applied to opposite corners of a rectangular plate as shown in Figure 12-46. Find the torque produced by this couple.

FIGURE 12-46 Problems 43, 45

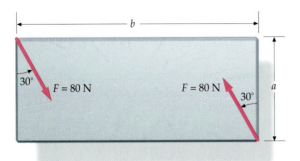

44 •• **SSM** A uniform cube of side a and mass M rests on a horizontal surface. A horizontal force $\vec{F}$ is applied to the top of the cube as in Figure 12-47. This force is not sufficient to move or tip the cube. (a) Show that the force of static friction exerted by the surface and the applied force constitute a couple, and find the torque exerted by the couple. (b) This couple is balanced by the couple consisting of the normal force exerted by the surface and the weight of the cube. Use this fact to find the effective point of application of the normal force when $F = Mg/3$. (c) What is the greatest magnitude of $\vec{F}$ for which the cube will not tip?

FIGURE 12-47
Problem 44

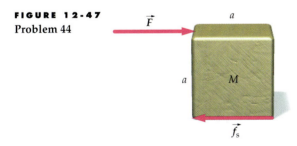

45 •• Resolve each force in Problem 43 into its horizontal and vertical components, producing two couples. The algebraic sum of the two component couples equals the resultant couple. Use this result to find the perpendicular distance between the lines of action of the two forces.

46 •• **SSM** A section of a cathedral wall is shown in Figure 12-48. The arch attached to the wall exerts a force of 2×10^5 N directed at an angle of 30° below the horizontal at a point 10 m above the ground and the mass of the wall itself is 30,000 kg. The coefficient of static friction between the wall and the ground is $\mu_k = 0.8$ and the base of the wall is 1.25 m long. (a) Calculate the effective normal force, the frictional force, and the point at which the effective normal force acts on the wall. (This is called the *thrust point* by architects and civil

engineers.) (b) If the thrust point ever moves outside the base of the wall, the wall will overturn. Apart from aesthetics, placing a heavy statue on top of the wall has good engineering practicality: it will move the thrust point toward the center of the wall. Explain why.

FIGURE 12-48
Problems 46, 47

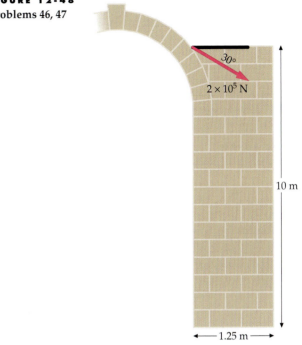

47 •• In civil engineering, the *thrust line* can be found by determining the thrust point at *any* height inside a wall or similar structure. Determine the thrust line for the wall in Problem 46 and graph it using a spreadsheet program or graphing calculator. (Despite its name, the thrust line is really a curve.)

Ladder Problems

48 •• **SSM** **iSOLVE** ✔ Romeo takes a uniform 10-m ladder and leans it against the smooth (frictionless) wall of the Capulet residence. The ladder's mass is 22.0 kg and the bottom rests on the ground 2.8 m from the wall. When Romeo, whose mass is 70 kg, gets 90 percent of the way to the top, the ladder begins to slip. What is the coefficient of static friction between the ground and the ladder?

49 •• **SSM** A massless ladder of length L leans against a smooth wall making an angle of θ with the horizontal floor. The coefficient of friction between the ladder and the floor is μ_s. A man of mass M climbs the ladder. What height h can he reach before the ladder slips?

50 •• A uniform ladder of length L and mass m leans against a frictionless vertical wall with its lower end on the ground. It makes an angle of 60° with the horizontal ground. The coefficient of static friction between the ladder and the ground is 0.45. If your mass is four times that of the ladder, how far up the ladder can you climb before it begins to slip?

51 •• A ladder making an angle θ with the horizontal, of mass m and length L, leans against a frictionless vertical wall. The center of mass is at a height h from the floor. A force F pulls horizontally against the ladder at the midpoint. Find the minimum coefficient of static friction μ_s for which the top end of the ladder will separate from the wall while the lower end does not slip.

52 •• **ISOLVE** ✓ A 900-N man sits on top of a stepladder of negligible weight that rests on a frictionless floor as in Figure 12-49. There is a cross brace halfway up the ladder. The angle at the apex is $\theta = 30°$. (a) What is the force exerted by the floor on each leg of the ladder? (b) Find the tension in the cross brace. (c) If the cross brace is moved down toward the bottom of the ladder (maintaining the same angle θ), will its tension be greater or less?

FIGURE 12-49
Problem 52

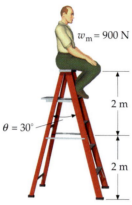

$w_m = 900$ N

2 m

$\theta = 30°$

2 m

53 •• **SSM** A uniform ladder rests against a frictionless vertical wall. The coefficient of static friction between the ladder and the floor is 0.3. What is the smallest angle at which the ladder will remain stationary?

Stress and Strain

54 • **SSM** A 50-kg ball is suspended from a steel wire of length 5 m and radius 2 mm. By how much does the wire stretch?

55 • **ISOLVE** ✓ Copper has a breaking stress of about 3×10^8 N/m². (a) What is the maximum load that can be hung from a copper wire of diameter 0.42 mm? (b) If half this maximum load is hung from the copper wire, by what percentage of its length will it stretch?

56 • **ISOLVE** A 4-kg mass is supported by a steel wire of diameter 0.6 mm and length 1.2 m. How much will the wire stretch under this load?

57 • **SSM** **ISOLVE** As a runner's foot pushes off on the ground, the shearing force acting on an 8-mm-thick sole is shown in Figure 12-50. If the force of 25 N is distributed over an area of 15 cm², find the angle of shear θ, given that the shear modulus of the sole is 1.9×10^5 N/m².

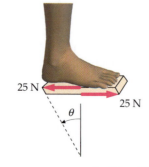

25 N

25 N

θ

FIGURE 12-50 Problem 57

58 •• **ISOLVE** A steel wire of length 1.5 m and diameter 1 mm is joined to an aluminum wire of identical dimensions to make a composite wire of length 3.0 m. What is the length of the composite wire if it is used to support a mass of 5 kg?

59 •• A force F is applied to a long wire of length L and cross-sectional area A. Show that if the wire is considered to be a spring, the force constant k is given by $k = AY/L$ and the energy stored in the wire is $U = \frac{1}{2}F\Delta L$, where Y is Young's modulus and ΔL is the amount the wire has stretched.

60 •• **ISOLVE** The steel E string of a violin is under a tension of 53 N. The diameter of the string is 0.20 mm and its length under tension is 35.0 cm. Find (a) the unstretched length of this string and (b) the work needed to stretch the string.

61 •• **SSM** When a rubber strip with a cross section of 3 mm × 1.5 mm is suspended vertically and various masses are attached to it, a student obtains the following data for length versus load:

Load, kg	0	0.1	0.2	0.3	0.4	0.5
Length, cm	5.0	5.6	6.2	6.9	7.8	10.0

(a) Find Young's modulus for the rubber strip for small loads.
(b) Find the energy stored in the strip when the load is 0.15 kg. (See Problem 59.)

62 •• A large mirror is hung from a nail as shown in Figure 12-51. The supporting steel wire has a diameter of 0.2 mm and an unstretched length of 1.7 m. The distance between the points of support at the top of the mirror's frame is 1.5 m. The mass of the mirror is 2.4 kg. What is the distance between the nail and the top of the frame when the mirror is hung?

FIGURE 12-51
Problems 62, 86

63 •• **ISOLVE** ✓ Two masses, M_1 and M_2, are supported by wires that have equal lengths when unstretched. The wire supporting M_1 is an aluminum wire 0.7 mm in diameter, and the one supporting M_2 is a steel wire 0.5 mm in diameter. What is the ratio M_1/M_2 if the two wires stretch by the same amount?

64 •• A 0.5-kg ball is attached to an aluminum wire having a diameter of 1.6 mm and an unstretched length of 0.7 m. The other end of the wire is fixed to a post. The ball rotates about the post in a horizontal plane at a rotational speed such that the angle between the wire and the horizontal is 5.0°. Find the tension in the wire and its length.

65 •• **SSM** An elevator cable is to be made of a new type of composite developed by Acme Laboratories. In the lab, a sample of the cable that is 2 m long and has a cross-sectional area of 0.2 mm² fails under a load of 1000 N. The cable in the elevator will be 20 m long and have a cross-sectional area of 1.2 mm². It will need to support a load of 20,000 N safely. Will it?

66 ••• **SSM** When a material is stretched in one direction, if its density remains constant, then (because its total volume remains constant), its length must decrease in one or both of the other directions. Take a rectangular block of length x, width y, and depth z, and pull on it so that its new length $x'' = x + \Delta x$. If $\Delta x \ll x$ and $\Delta y/y = \Delta z/z$, show that $\Delta y/y = -\frac{1}{2} \Delta x/x$.

67 •• If you are given a wire with a circular cross-section of radius r and a length L, show that $\Delta r/r = -\frac{1}{2} \Delta L/L$, assuming that $\Delta L \ll L$. (See Problem 66.)

68 ••• **SSM** For most materials listed in Table 12-1, the tensile strength is two to three orders of magnitude lower than Young's modulus. Consequently, most of these materials will break before their strain exceeds 1 percent. Of man-made materials, nylon has about the greatest extensibility—it can take strains of about 0.2 before breaking. But spider silk beats anything man-made. Certain forms of spider silk can take strains on the order of 10 before breaking! (a) If such a thread has a circular cross-section of radius r_0 and unstretched length L_0, find its new radius r when stretched to a length $L = 10L_0$. (b) If the Young's modulus of the spider thread is Y, calculate the tension needed to break the thread in terms of Y and r_0.

General Problems

69 • A 90-N board 12 m long rests on two supports, each 1 m from the end of the board. A 360-N block is placed on the board 3 m from one end as shown in Figure 12-52. Find the force exerted by each support on the board.

FIGURE 12-52
Problem 69

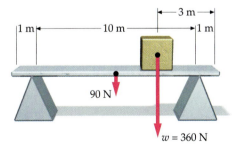

70 • The height of the center of gravity of a man standing erect is determined by weighing the man as he lies on a board of negligible weight supported by two scales as shown in Figure 12-53. If the man's height is 188 cm and the left scale reads 445 N while the right scale reads 400 N, where is his center of gravity relative to his feet? Would the reading on the scales be different if he were holding his head up—keeping it slightly above the board?

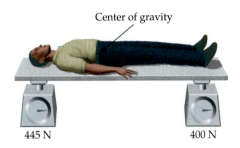

Center of gravity

445 N   400 N

FIGURE 12-53 Problem 70

71 • **SSM** **iSOLVE** ✓ Figure 12-54 shows a mobile consisting of four weights hanging on three rods of negligible mass. Find the value of each of the unknown weights if the mobile is to balance. *Hint: Find the weight w_1 first.*

FIGURE 12-54 Problem 71

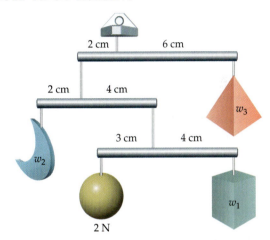

72 • A rope and pulley system, called a block and tackle, is used to support an object of mass m, as shown in Figure 12-55. The object is being raised at constant speed. When a segment of rope of length L passes over the uppermost pulley wheel, the height of the lower pulley is increased by h. (a) What is the ratio L/h? (b) Assume that the block and tackle are massless and frictionless. Show that $FL = mgh$ by applying the work–energy principle to the block–tackle object.

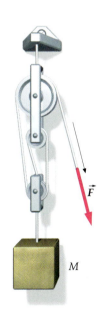

FIGURE 12-55 Problem 72

73 •• A plate of mass M in the shape of an equilateral triangle is suspended from one corner and a mass m is suspended from another of its corners. If the base of the triangle makes an angle of 6.0° with the horizontal, what is the ratio m/M?

74 •• A standard six-sided pencil is placed on a pad of paper (Figure 12-56). Find the minimum coefficient of static friction μ_s such that, if the pad is inclined, the pencil rolls down rather than slides.

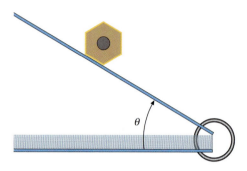

75 •• SSM A uniform box of mass 8 kg that is twice as tall as it is wide rests on the floor of a truck. What is the maximum coefficient of static friction between the box and floor so that the box will slide toward the rear of the truck rather than tip when the truck accelerates on a level road?

76 •• A balance scale has unequal arms. A 1.5-kg block appears to have a mass of 1.95 kg on the left pan of the scale (Figure 12-57). Find its apparent mass if the 1.5-kg block is placed on the right pan.

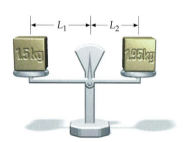

FIGURE 12-57 Problem 76

77 •• SSM A cube of mass M leans against a frictionless wall making an angle of θ with the floor as shown in Figure 12-58. Find the minimum coefficient of static friction μ_s between the cube and the floor that allows the cube to stay at rest.

FIGURE 12-58 Problem 77

78 •• Figure 12-59 shows a steel meter stick hinged to a vertical wall and supported by a thin wire. The wire and meter stick make angles of 45° with the vertical. The mass of the meter stick is 5.0 kg. When a 10.0-kg block is suspended from the midpoint of the meter stick, the tension T in the supporting wire is 52 N. If the wire will break should the tension exceed 75 N, what is the maximum distance along the meter stick at which the block can be suspended?

FIGURE 12-59 Problem 78

79 •• **iSOLVE** Figure 12-60 shows a 20-kg ladder leaning against a frictionless wall and resting on a frictionless horizontal surface. To keep the ladder from slipping, the bottom of the ladder is tied to the wall with a thin wire; the tension in the wire is 29.4 N. The wire will break if the tension exceeds 200 N. (a) If an 80-kg person climbs halfway up the ladder, what force will be exerted by the ladder against the wall? (b) How far up can an 80-kg person climb this ladder?

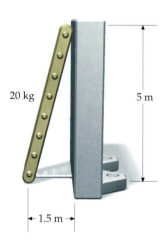

FIGURE 12-60 Problem 79

80 •• SSM A uniform cube can be moved along a horizontal plane either by pushing the cube so that it slips or by turning it over ("rolling"). What coefficient of kinetic friction μ_k between the cube and the floor makes both ways equal in terms of the work needed?

81 •• A tall, uniform, rectangular block sits on an inclined plane as shown in Figure 12-61. If $\mu_s = 0.4$, does the block slide or fall over as the angle θ is slowly increased?

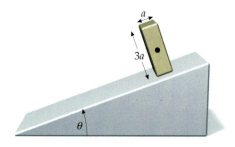

FIGURE 12-61 Problem 81

82 •• A 360-kg object is supported on a wire attached to a 15-m-long steel bar that is pivoted at a vertical wall and supported by a cable as shown in Figure 12-62. The mass of the bar is 85 kg. (*a*) With the cable attached to the bar 5.0 m from the lower end as shown, find the tension in the cable and the force exerted by the wall on the steel bar. (*b*) Repeat if a somewhat longer cable is attached to the steel bar 5.0 m from its upper end, maintaining the same angle between the bar and the wall.

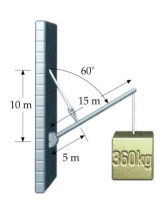

FIGURE 12-62 Problem 82

83 •• Repeat Problem 75 if the truck accelerates up a hill that makes an angle of 9.0° with the horizontal.

84 •• **SSM** A thin rod 60 cm long is balanced 20 cm from one end when an object whose mass is (2*m* + 2 grams) is at the end nearest the pivot and an object of mass *m* is at the opposite end (Figure 12-63*a*). Balance is again achieved if the object whose mass is (2*m* + 2 grams) is replaced by the object of mass *m* and no object is placed at the other end (Figure 12-63*b*). Determine the mass of the rod.

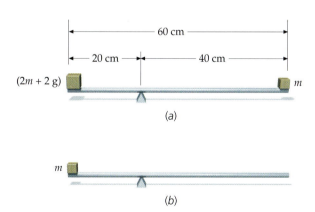

FIGURE 12-63 Problem 84

85 •• **SSM** If you balance a meter stick across two fingers (one from each hand) and slowly bring your hands together, the fingers will always meet in the middle of the stick, no matter where they are initially placed. (*a*) Explain why this is true. (*b*) As you move your fingers together, first one will move, and then the other; both will not move at the same time. Explain quantitatively why this is true; assume that the coefficient of static friction between the meter stick and your fingers is μ_s and the coefficient of kinetic friction is μ_k.

86 •• When a picture is hung on a smooth vertical wall using a wire and a nail, as in Figure 12-51, the picture almost always tips slightly forward, that is, the plane of the picture makes a small angle with the vertical. (*a*) Explain why pictures supported in this manner generally do not hang flush against the wall. (*b*) A framed picture 1.5 m wide and 1.2 m high with a mass of 8.0 kg is hung as in Figure 12-51 using a wire 1.7 m long. The ends of the wire are fastened to the sides of the frame at the rear, 0.4 m below the top. When the picture is hung, the angle between the plane of the frame and the wall is 5.0°. Determine the force that the wall exerts on the bottom of the frame.

87 •• **SSM** If a train travels around an unbanked bend in the rail bed too fast, the freight cars will tip over. Assume that the cargo portion of each freight car is a regular parallelepiped of uniform density and 1.5×10^4 kg mass, 10 m long, 3.0 m high, and 2.20 m wide, and that its base is 0.65 m above the rails. The car axles are 7.6 m apart, 1.2 m from each end of the boxcar. The separation between the rails is 1.55 m. Find the maximum safe speed of the train if the radius of curvature of the bend is (*a*) 150 m and (*b*) 240 m. (Neglect the mass of the flatcar.)

88 •• For balance, a tightrope walker uses a thin rod 8 m long and bowed in a circular arc shape. At each end of the rod is a lead mass of 20 kg. The tightrope walker, whose mass is 58 kg and whose center of gravity is 0.90 m above the rope, holds the rod tightly at its center 0.65 m above the rope. What should be the radius of curvature of the arc of the rod so that the tightrope walker will be in neutral equilibrium as he slowly makes his way across the rope? Neglect the mass of the rod.

89 ••• **SSM** We have a large number of identical uniform bricks, each of length *L*. If we stack one on top of another lengthwise (see Figure 12-64), the maximum offset that will allow the top brick to rest on the bottom brick is *L*/2. (*a*) Show that if we place this two-brick stack on top of a third brick, the maximum offset of the second brick on the third brick is *L*/3. (*b*) Show that, in general, if we have a stack of *N* bricks, the maximum offset of the (*n* − 1)th brick (counting down from the top) on the *n*th brick is *L*/*n*, where *n* ≤ *N*. (*c*) Write a spreadsheet program to calculate total offset (the sum of the individual offsets) for a stack of *N* bricks, and calculate this for *L* = 1 m and *N* = 5, 10, and 100. (*d*) Does the sum of the individual offsets approach a finite limit as *N* → ∞? If so, what is that limit?

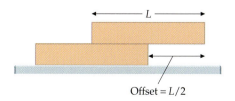

FIGURE 12-64 Problem 89

90 ••• A uniform sphere of radius R and mass M is held at rest on an inclined plane of angle θ by a horizontal string, as shown in Figure 12-65. Let $R = 20$ cm, $M = 3$ kg, and $\theta = 30°$. (*a*) Find the tension in the string. (*b*) What is the normal force exerted on the sphere by the inclined plane? (*c*) What is the frictional force acting on the sphere?

FIGURE 12-65
Problem 90

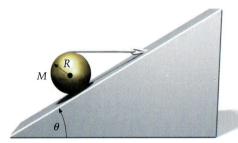

91 ••• The legs of a tripod make equal angles of 90° with each other at the apex, where they join together. A 100-kg block hangs from the apex. What are the compressional forces in the three legs?

92 •• Figure 12-66 shows a 20-cm-long uniform beam resting on a cylinder of 4-cm radius. The mass of the beam is 5.0 kg and that of the cylinder is 8.0 kg. The coefficient of static friction between beam and cylinder is zero, whereas the coefficient of static friction between the cylinder and the floor, and between the beam and the floor, are each greater than zero. Are there any values for these coefficients of static friction such that the system is in static equilibrium? If so, what are these values? If not, explain.

FIGURE 12-66
Problem 92

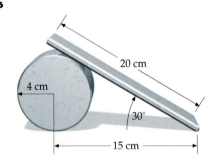

93 ••• Two solid smooth (frictionless) spheres of radius r are placed inside a cylinder of radius R as in Figure 12-67. The mass of each sphere is m. Find the force exerted by the bottom of the cylinder on the bottom sphere, the force exerted by the wall of the cylinder on each sphere, and the force exerted by one sphere on the other.

FIGURE 12-67
Problem 93

94 ••• **SSM** A solid cube of side length a balanced atop a cylinder of diameter d is in unstable equilibrium if $d \ll a$ and is in stable equilibrium if $d \gg a$ (Figure 12-68). Determine the minimum value of d/a for which the cube is in stable equilibrium.

FIGURE 12-68 Problem 94

Fluids

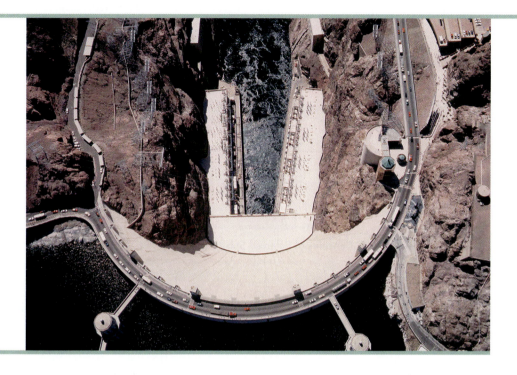

THE HOOVER DAM IS 660 FEET THICK AT THE BOTTOM, BUT ONLY 45 FEET THICK AT THE TOP OF THE SPAN.

? **Dams are designed to be thicker at the bottom than at the top. Why is this so? (See Example 13.2.)**

13-1 Density

13-2 Pressure in a Fluid

13-3 Buoyancy and Archimedes' Principle

13-4 Fluids in Motion

luids include both liquids and gases. Liquids flow under gravity until they occupy the lowest possible regions of their containers. Gases expand to fill their containers. Fluids permeate our environment as well as our bodies. To understand fluid behavior is to understand much about ourselves and our interactions with the world around us.

➤ **We begin this chapter by studying fluids at rest, followed by the study of steady-state flow with an emphasis on laminar flow.**

In a gas, the average distance between two molecules is large compared with the size of a molecule. The molecules have little influence on one another except during their brief collisions. In a liquid or solid, the molecules are close together and exert forces on one another that are comparable to the forces that bind atoms into molecules. Molecules in a liquid form temporary short-range bonds that are continually broken and reformed due to the thermal kinetic energy of the molecules. These bonds hold the liquid together; if the bonds were not present, the liquid would immediately evaporate and the molecules would escape as a vapor. The strength of the bonds in a liquid depends on the type of molecule that makes up the liquid. For example, the bonds between helium molecules are very weak and, for this reason, helium does not liquefy at atmospheric pressure unless the temperature is 4.2 K or lower.

13-1 Density

An important property of a substance is the ratio of its mass to its volume, which is called its **density:**

$$\text{Density} = \frac{\text{mass}}{\text{volume}}$$

The Greek letter ρ (rho) is usually used to denote density:

$$\rho = \frac{m}{V} \qquad\qquad 13\text{-}1$$

DEFINITION — DENSITY

Because the gram was originally defined as the mass of one cubic centimeter of water, the density of water in cgs units is 1 g/cm^3. Converting to SI units, we obtain for the density of water

$$\rho_w = \frac{1\text{ g}}{\text{cm}^3} \times \frac{1\text{ kg}}{10^3\text{ g}} \times \left(\frac{100\text{ cm}}{1\text{ m}}\right)^3 = 10^3 \text{ kg/m}^3 \qquad 13\text{-}2$$

Precise measurements of density must take temperature into account, since the densities of most materials, including water, vary with temperature. Equation 13-2 gives the maximum value for the density of water, which occurs at 4°C. Table 13-1 lists the densities of some common materials.

A convenient unit of volume for fluids is the **liter** (L):

$$1\text{ L} = 10^3 \text{ cm}^3 = 10^{-3} \text{ m}^3$$

In terms of this unit, the density of water at 4°C is 1.00 kg/L = 1.00 g/mL. When an object's density is greater than that of water, it sinks in water. When its density is less, it floats. The ratio of the density of a substance to that of water is called its **specific gravity.** For example, the specific gravity of aluminum is 2.7, meaning that a volume of aluminum has 2.7 times the mass of an equal volume of water. The specific gravities of objects that sink in water range from 1 to about 22.5 (for the densest element, osmium).

Most solids and liquids expand only slightly when heated, and contract slightly when subjected to an increase in external pressure. Since these changes in volume are relatively small, we often treat the densities of solids and liquids as approximately independent of temperature and pressure. The density of a gas, on the other hand, depends strongly on the pressure and temperature, so these variables must be specified when reporting the densities of gases. By convention, the standard conditions for the measurement of physical properties are atmospheric pressure at sea level and a temperature of 0°C. The densities for the substances listed in Table 13-1 are for these conditions. Note that the densities of liquids and solids are considerably greater than those of gases. For example, the density of water is about 800 times that of air under standard conditions.

TABLE 13-1

Densities of Selected Substances

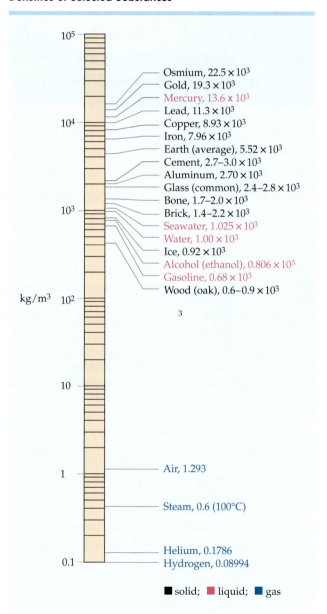

kg/m³

- Osmium, 22.5×10^3
- Gold, 19.3×10^3
- Mercury, 13.6×10^3
- Lead, 11.3×10^3
- Copper, 8.93×10^3
- Iron, 7.96×10^3
- Earth (average), 5.52×10^3
- Cement, $2.7–3.0 \times 10^3$
- Aluminum, 2.70×10^3
- Glass (common), $2.4–2.8 \times 10^3$
- Bone, $1.7–2.0 \times 10^3$
- Brick, $1.4–2.2 \times 10^3$
- Seawater, 1.025×10^3
- Water, 1.00×10^3
- Ice, 0.92×10^3
- Alcohol (ethanol), 0.806×10^3
- Gasoline, 0.68×10^3
- Wood (oak), $0.6–0.9 \times 10^3$

- Air, 1.293
- Steam, 0.6 (100°C)
- Helium, 0.1786
- Hydrogen, 0.08994

■ solid; ■ liquid; ■ gas

CALCULATING DENSITY   **E X A M P L E   1 3 - 1**

A 200-mL flask is filled with water at 4°C. When the flask is heated to 80°C, 6 g of water spill out. What is the density of water at 80°C? (Assume that the expansion of the flask is negligible.)

PICTURE THE PROBLEM The density of water at 80°C is $\rho' = m'/V$, where $V = 200$ mL $= 200$ cm^3 is the volume of the flask, and m' is the mass remaining in the flask after 6 g spill out. We find m' by first finding the mass of water originally in the flask.

1. Calculate the original mass m of water in the flask at 4°C using $\rho = 1.00$ g/cm:

$$m = \rho V = (1.00 \text{ g/cm}^3)(200 \text{ cm}^3) = 200 \text{ g}$$

2. Calculate the mass of water remaining m' after 6 g spill out:

$$m' = m - 6 \text{ g} = 200 \text{ g} - 6 \text{ g} = 194 \text{ g}$$

3. Use this value of m' to find the density of water at 80°:

$$\rho' = \frac{m'}{V} = \frac{194 \text{ g}}{200 \text{ cm}^3} = \boxed{0.97 \text{ g/cm}^3}$$

EXERCISE A solid metal cube 8 cm on a side has a mass of 4.08 kg. (*a*) What is the density of the cube? (*b*) If the cube is made from a single element listed in Table 13-1, what is the element? (*Answer* (*a*) 7.97 kg/L (*b*) iron)

EXERCISE A gold brick is 5 cm × 10 cm × 20 cm. What is its mass? (*Answer* 19.3 kg)

13-2 Pressure in a Fluid

When a body is submerged in a fluid such as water, the fluid exerts a force perpendicular to the surface of the body at each point on the surface. This force per unit area is called the **pressure** P of the fluid:

$$P = \frac{F}{A}$$

13-3

DEFINITION — PRESSURE

The SI unit of pressure is the newton per square meter (N/m^2), which is called the **pascal** (Pa):

$$1 \text{ Pa} = 1 \text{ N/m}^2$$

13-4

In the U.S. customary system, pressure is usually given in pounds per square inch (lb/in.^{2}). Another common unit of pressure is the atmosphere (atm), which equals approximately the air pressure at sea level. One atmosphere is now defined to be 101.325 kilopascals, which is approximately 14.70 lb/in.^{2}:

$$1 \text{ atm} = 101.325 \text{ kPa} \approx 14.70 \text{ lb/in.}^2$$

13-5

Other units of pressure in common use will be discussed later in this chapter.

The pressure due to a fluid pressing in on an object tends to compress the object. The ratio of the increase in pressure ΔP to the fractional decrease in volume $(-\Delta V/V)$ is called the **bulk modulus**:

$$B = -\frac{\Delta P}{\Delta V/V}$$

13-6

DEFINITION — BULK MODULUS

(The minus sign in Equation 13-6 is introduced to make B positive, since all materials decrease in volume when subjected to an increase in external pressure.)

The more difficult it is to compress a material, the smaller is the fractional volume decrease $-\Delta V/V$ for a given pressure increase ΔP, and hence the greater the bulk modulus. The **compressibility** is the reciprocal of the bulk modulus. (The easier it is to compress a material, the larger the compressibility.) Liquids, gases, and solids all have a bulk modulus. Since liquids and solids are relatively incompressible, they have large values of B, and these values are relatively independent of temperature and pressure. Gases, on the other hand, are easily compressed, and their values for B depend strongly on pressure and temperature. Table 13-2 charts values for the bulk modulus of various materials.

As any scuba diver knows, the pressure in a lake or ocean increases with depth. Similarly, the pressure of the atmosphere decreases with altitude. For a liquid such as water, whose density is approximately constant throughout, the pressure increases linearly with depth. We can see this by considering a column of liquid of cross-sectional area A, as shown in Figure 13-1. To support the weight of the liquid in the column of height Δh, the pressure at the bottom of the column must be greater than the pressure at the top by the weight of the column. The weight of the liquid in the column is

$$w = mg = (\rho V)g = \rho A\,\Delta hg$$

If P_0 is the pressure at the top and P is the pressure at the bottom, the net upward force exerted by this pressure difference is $PA - P_0A$. Setting this net upward force equal to the weight of the column, we obtain

$$PA - P_0A = \rho A \Delta hg$$

or

$$P = P_0 + \rho g\Delta h \qquad (\rho\text{ constant}) \qquad\qquad 13\text{-}7$$

TABLE 13-2

Approximate Values for the Bulk Modulus B of Various Materials

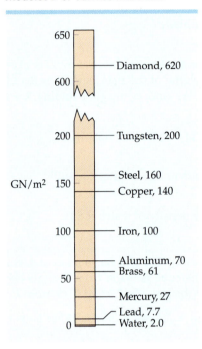

GN/m²	
650	
600	Diamond, 620
200	Tungsten, 200
150	Steel, 160
	Copper, 140
100	Iron, 100
	Aluminum, 70
50	Brass, 61
	Mercury, 27
	Lead, 7.7
0	Water, 2.0

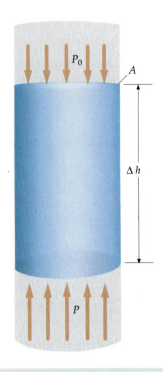

FIGURE 13-1 Column of water of height Δh and cross-sectional area A. The pressure P at the bottom must be greater than the pressure P_0 at the top to balance the weight of the water.

EXERCISE How far below the surface of a lake is the pressure equal to 2 atm? (The pressure at the surface is 1 atm.) (*Answer* With $P_0 = 1$ atm $= 101$ kPa, $P = 2$ atm, $\rho = 1000$ kg/m³, and $g = 9.81$ N/kg, we have $\Delta h = \Delta P/\rho g = 10.3$ m. The pressure at a depth of 10.3 m is twice that at the surface.)

FORCE ON A DAM　　　　　　　　**EXAMPLE 13-2**

A rectangular dam 30 m wide supports a body of water to a depth of 25 m. Find the total horizontal force on the dam.

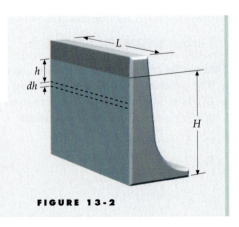

FIGURE 13-2

PICTURE THE PROBLEM Because the pressure varies with depth, we cannot merely multiply the pressure times the area of the dam to find the force exerted by the water. Instead we can consider the force exerted on a strip of width $L = 30$ m, height dh, and area $dA = L\,dh$ at a depth h (Figure 13-2), and then integrate from $h = 0$ to $h = H = 25$ m. The water pressure at depth h is $P_{atm} + \rho gh$. We can omit the atmospheric pressure because it is exerted on each side of the wall.

1. Express the force dF on the element of width L and height dh in terms of the net pressure ρhg

$$dF = P\, dA = \rho g h L\, dh$$

2. Integrate from $h = 0$ to $h = H$:

$$F = \int_{h=0}^{h=H} dF = \int_0^H \rho g h L\, dh = \rho g L \left. \frac{h^2}{2}\right|_0^H = \frac{1}{2}\rho g L H^2$$

3. Substitute the given values to find the numerical result:

$$F = \tfrac{1}{2}\rho g L H^2 = \tfrac{1}{2}(1000\ \text{kg/m}^3)(9.81\ \text{N/kg})(30\ \text{m})(25\ \text{m})^2$$

$$= \boxed{9.20 \times 10^7\ \text{N}}$$

REMARKS Dams typically are thicker at the bottom than at the top because the pressure on the dam increases with the depth of the water. Because the pressure on the dam is *proportional* to the depth of the water, the average pressure is $P_{\text{atm}} + \rho g H/2$ (the pressure at depth $h = H/2$).

The result that the pressure increases linearly with depth holds for a liquid in any container, independent of the shape of the container. Furthermore, the pressure is the same at all points at the same depth. We can see this by comparing the pressure at point 1 in Figure 13-3a with the pressure at point 2, which is inside an underwater cave. First we compare the pressure at points 1 and 3, a point directly below 1 at the same depth as 2, as shown in Figure 13-3b. Consider the vertical forces on the vertical column of water of height Δh and cross-sectional area A joining points 1 and 3. The upward force on the column, $P_3 A$, balances the two downward forces $P_1 A$ and mg, where $m = \rho A \Delta h$ is the mass of the water in the column ($A\Delta h$ is the volume of the column). That is, $P_3 A = P_1 A + \rho A \Delta h g$. Dividing through by A gives

$$P_3 = P_1 + \rho g \Delta h$$

Next consider the forces on the horizontal cylinder of water, also of cross-sectional area A, connecting points 2 and 3 (Figure 13-3c). There are two forces with components along the

FIGURE 13-3

(a)

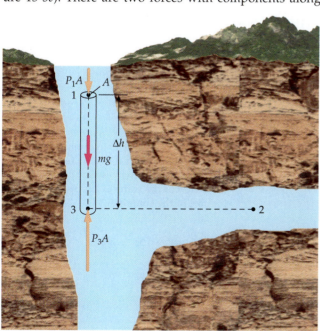

(b)

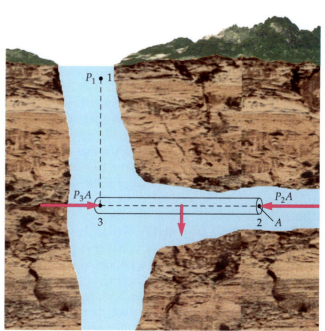

(c)

cylinder's axis, P_3A and P_2A. The fact that these forces balance each other means that $P_3 = P_2$. It follows that

$$P_2 = P_1 + \rho g \Delta h$$

If we increase the pressure in a container of water by pressing down on the top surface with a piston, the increase in pressure is the same throughout the liquid. This is known as **Pascal's principle,** named after Blaise Pascal (1623–1662):

> A pressure change applied to an enclosed liquid is transmitted undiminished to every point in the liquid and to the walls of the container.

PASCAL'S PRINCIPLE

A common application of Pascal's principle is the hydraulic lift shown in Figure 13-4.

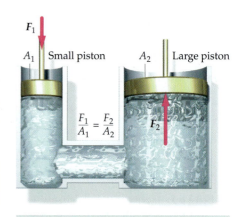

FIGURE 13-4 Hydraulic lift. A small force F_1 on the small piston produces a change in pressure F_1/A_1 that is transmitted by the liquid to the large piston. Since the pressures at the small and large pistons are the same, the forces are related by $F_2/A_2 = F_1/A_1$. Since the area of the large piston is much greater than that of the small piston, the force on the large piston $F_2 = (A_2/A_1)F_1$ is much greater than F_1.

A Hydraulic Lift   **EXAMPLE 1 3 - 3**

The large piston in a hydraulic lift has a radius of 20 cm. What force must be applied to the small piston of radius 2 cm to raise a car of mass 1500 kg?

PICTURE THE PROBLEM The pressure P times the area A_2 of the large piston must equal the weight mg of the car. The force F_1 that must be exerted on the small piston is this pressure times the area A_1. (See Figure 13-4.)

1. The force F_1 is the pressure P times the area A_1:

$$F_1 = PA_1$$

2. The pressure P times the area A_2 equals the weight of the car:

$$PA_2 = mg \quad \text{so} \quad P = \frac{mg}{A_2}$$

3. Substitute this result for P into the step 1 result and calculate F_1:

$$F_1 = PA_1$$

$$= \frac{mg}{A_2} A_1 = mg\frac{A_1}{A_2} = mg\frac{\pi r_1^2}{\pi r_2^2}$$

$$= (1500 \text{ kg})(9.81 \text{ N/kg})\left(\frac{2 \text{ cm}}{20 \text{ cm}}\right)^2$$

$$= \boxed{147 \text{ N}}$$

Figure 13-5 shows water in a container with sections of different shapes. At first glance, it might seem that the pressure at the bottom of section 3 of the container would be greatest and that water would therefore be forced to a greater height in section 2, which is smaller. But that does not happen, a result known as the **hydrostatic paradox. The pressure depends only on the depth of the water, not on the shape of the container, so at the same depth the pressure is the same in all parts of the container, a finding that can be shown experimentally.** Although the water in section 4 of the container weighs more than that in section 2, the portion of the water in section 4 that is not above the opening at the bottom is supported by the horizontal shelf of the section. In fact, the water above the opening at the bottom of section 5 weighs less than the water above an opening of the same size at the bottom of section 1. However, the

FIGURE 13-5 The hydrostatic paradox. The water level is the same regardless of the shape of the vessel. The shaded portion of the water is supported by the sides of the container.

1   2   3   4   5

horizontal shelf of section 5 exerts a downward force on the water—exactly compensating for the shortfall of weight.

We can use the fact that the pressure difference is proportional to the depth of a liquid to measure unknown pressures. Figure 13-6 shows a simple pressure gauge, the open-tube manometer. The top of the tube is open to the atmosphere at pressure P_{at}. The other end of the tube is at pressure P, which is to be measured. The difference $P - P_{at}$, called the **gauge pressure** P_{gauge}, is equal to $\rho g h$, where ρ is the density of the liquid in the tube. The pressure you measure in your automobile tire is gauge pressure. When the tire is entirely flat, the gauge pressure is zero, and the absolute pressure in the tire is atmospheric pressure. The absolute pressure P is obtained from the gauge pressure by adding atmospheric pressure to it:

$$P = P_{gauge} + P_{at} \tag{13-8}$$

Figure 13-7 shows a mercury barometer, which is used to measure atmospheric pressure. The top end of the tube has been closed off and evacuated so that the pressure there is zero. The other end is submerged in a pool of mercury that is open to the atmosphere at pressure P_{at}. The pressure P_{at} is $\rho g h$, where ρ is the density of mercury.

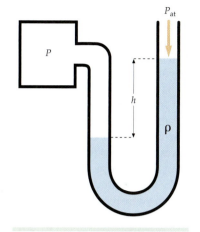

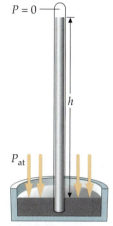

FIGURE 13-6 Open-tube manometer for measuring an unknown pressure P. The difference $P - P_{at}$ equals $\rho g h$.

FIGURE 13-7

EXERCISE At 0°C, the density of mercury is 13.595×10^3 kg/m³. What is the height of the mercury column in a barometer if the pressure is 1 atm = 101.325 kPa? (*Answer* $h = P/\rho g = 0.760$ m = 760 mm)

In practice, pressure is often measured in millimeters of mercury, a unit called the **torr,** after the Italian physicist Evangelista Torricelli, or in inches of mercury (written inHg). The various units of pressure are related as follows:

$$1 \text{ atm} = 760 \text{ mmHg} = 760 \text{ torr} = 29.9 \text{ in.Hg}$$

$$= 101.325 \text{ kPa} = 14.7 \text{ lb/in.}^2 \tag{13-9}$$

Other units commonly used on weather maps are the **bar** and the **millibar,** which are defined as

$$1 \text{ bar} = 10^3 \text{ millibars} = 100 \text{ kPa} \tag{13-10}$$

A pressure of 1 atm is about 1.3% greater than a pressure of 1 bar.

Tire-pressure gauge. The piston pushes the rod to the right until the force of the spring plus the force due to atmospheric pressure balances the force due to the air pressure in the tire.

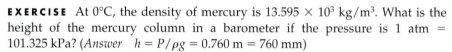

Blood Pressure in the Aorta   **EXAMPLE   13-4**

The average gauge pressure in the human aorta is about 100 mmHg. Convert this average blood pressure to pascals.

We use the conversion factors that can be obtained from Equation 13-9: $P = 100 \text{ mmHg} \times \dfrac{101.325 \text{ kPa}}{760 \text{ mmHg}}$

$$= \boxed{13.3 \text{ kPa}}$$

EXERCISE Convert a pressure of 45 kPa to (*a*) millimeters of mercury and (*b*) atmospheres. (*Answer* (*a*) 338 mmHg (*b*) 0.444 atm)

The relation between pressure and altitude (or depth) is more complicated for a gas than for a liquid because the density of a gas is not constant like that of a liquid, but is approximately proportional to the pressure. As you go up from the surface of the earth, pressure in a column of air decreases, just as the pressure would decrease as you go up from the bottom in a water column. But the decrease in air pressure is not linear with distance. Instead, the air pressure decreases by a constant fraction for a given increase in height, as shown in Figure 13-8. At a height of about 5.5 km (18,000 ft), the air pressure is half its value at sea level. If we go up another 5.5 km to an altitude of 11 km (a typical altitude for airliners), the pressure is again halved so that it is one-fourth its value at sea level, and so on. This example of an *exponential decrease* is called the law of atmospheres. At the high altitudes at which commercial jets fly, the cabins must be pressurized. The density of air is approximately proportional to the pressure, so the density of air decreases with altitude. There is less oxygen available on a mountain top than at normal elevations. This makes exercising in the Rockies difficult, and climbing in the Himalayas dangerous.

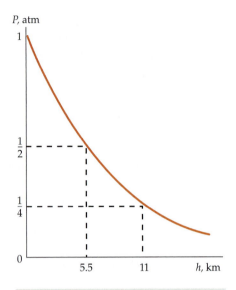

FIGURE 13-8 Variation in pressure with height above the earth's surface. For each 5.5-km increase in height, the pressure decreases by half.

13-3 Buoyancy and Archimedes' Principle

If a dense object submerged in water is weighed by suspending it from a spring scale (Figure 13-9a), the apparent weight of the object when submerged (the reading on the scale) is less than the weight of the object. This is so because the water exerts an upward force that partially balances the force of gravity. This upward force is even more evident when we submerge a piece of cork. When completely submerged, the cork experiences an upward force from the water pressure that is greater than the force of gravity, so when released it accelerates up toward the surface. The force exerted by a fluid on a body wholly or partially submerged in it is called the **buoyant force.**[†] It is equal to the weight of the fluid displaced by the body.

> A body wholly or partially submerged in a fluid is buoyed up by a force equal to the weight of the displaced fluid.
>
> ARCHIMEDES' PRINCIPLE

This result is known as **Archimedes' principle.**

We can derive Archimedes' principle from Newton's laws by considering the forces acting on a portion of a fluid and noting that in static equilibrium the net force must be zero. Figure 13-9b shows the vertical forces acting on an object being weighed while submerged. These are the force of gravity $\vec{w}$ acting down, the force of the spring scale $\vec{F}_s$ acting up, a force $\vec{F}_1$ acting down because of the fluid pressure on the top surface of the object, and a force $\vec{F}_2$ acting up because of the fluid pressing on the bottom surface of the object. Since the spring scale reads a force less than the weight, the force $\vec{F}_2$ must be greater in magnitude than the force $\vec{F}_1$. The difference in magnitude of these two forces is the buoyant force $\vec{B} = \vec{F}_2 - \vec{F}_1$ (Figure 13-9c). The buoyant force occurs because the pressure of the fluid at the bottom of the object is greater than that at the top.

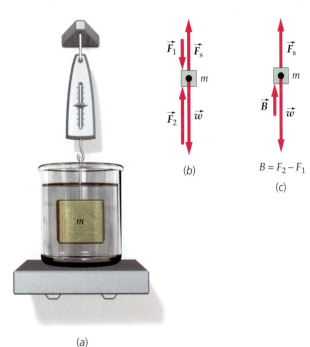

(a)

(b)

(c)

$B = F_2 - F_1$

FIGURE 13-9 (a) Weighing an object submerged in a fluid. (b) Free-body diagram showing the weight $\vec{w}$, the force $\vec{F}_s$ of the spring, and the forces $\vec{F}_1$ and $\vec{F}_2$ exerted by the surrounding field. (c) The buoyant force $B = F_2 - F_1$ is the net force exerted on the object by the fluid.

[†] The definition of buoyant force is further refined later in this section.

In Figure 13-10, the spring scale has been eliminated and the submerged object has been replaced by an equal volume of fluid (outlined by the dashed lines). The buoyant force $\vec{B} = \vec{F}_2 - \vec{F}_1$ acting on this volume of fluid is the same as the buoyant force that acted on our original object since the fluid surrounding the space is the same. Because this volume of fluid is in equilibrium, the net force acting on it must be zero. The upward buoyant force thus equals the downward weight of this volume of the fluid:

$$B = w_f \qquad \text{13-11}$$

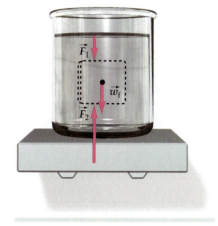

FIGURE 13-10 Figure 13-9 with the submerged body replaced by an equal volume of fluid. The forces $\vec{F}_1$ and $\vec{F}_2$, due to the pressure of the fluid are the same as in Figure 13-9. The magnitude of the buoyant force is thus equal to the weight w_f of the displaced fluid.

Note that this result does not depend on the shape of the submerged object. If we consider any irregularly shaped portion of a static fluid, there must be a buoyant force acting on it by the surrounding fluid that exactly supports its weight. Thus, we have derived Archimedes' principle.

Archimedes (287–212 B.C.) had been given the task of determining whether a crown made for King Hieron II was of pure gold or had been adulterated with some cheaper metal such as silver, and to do this without destroying the crown. For Archimedes, the problem was to determine if the density of the irregularly shaped crown was the same as the density of gold. As the story goes, he came upon the solution while sinking himself into a bathtub and immediately rushed naked through the streets of Syracuse shouting "Eureka!" ("I have found it!"). This flash of insight preceded Newton's laws, which we used to derive Archimedes' principle, by some 1900 years. What Archimedes found was a simple and accurate way to compare the density of the crown with the density of gold, using a pan balance. He placed the balance in a large basin, and placed the crown on one pan and an equal mass of pure gold on the other pan. He then added water to the basin, submerging the crown and the pure gold. The balance tilted, with the crown rising—indicating that the buoyant force on the crown was greater than that on the pure gold because the volume of water displaced by the crown was greater than that displaced by the pure gold. The crown was less dense than the pure gold.

The specific gravity of an object equals the weight of the object divided by the weight of an equal volume of water. According to Archimedes' principle, the weight of an equal volume of water equals the buoyant force on the object when it is submerged in water. Therefore, the specific gravity is equal to the weight of the object divided by the buoyant force on it when it is submerged in water:

$$\text{Specific gravity} = \frac{\text{weight}}{\text{buoyant force when submerged in water}} = \frac{w}{B_{\text{water}}} \qquad \text{13-12}$$

The apparent weight w_{app} of an object submerged in a fluid is the difference between its weight w and the buoyant force B:

$$w_{\text{app}} = w - B \qquad \text{13-13}$$

Hot-air balloons rising in the night sky over Albuquerque during a balloon festival.

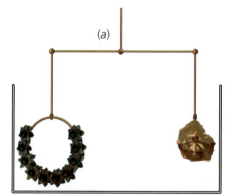

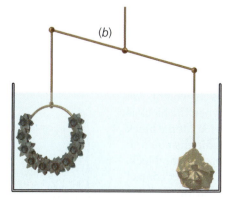

(a) The crown and the gold nugget have equal weight. (b) The wreath displaces more water than the gold nugget.

Your friend is concerned about a gold ring she bought on a recent trip. The ring was expensive, and she would like to know whether it is really made of gold or of something else. You decide to help her, using your knowledge of physics. You weigh the ring and find that it has a weight of 0.158 N. Using a string, you suspend the ring from the scale and, with the ring submerged in water, weigh it again to find a new reading of 0.150 N. Is the ring pure gold?

PICTURE THE PROBLEM If the ring is pure gold, its specific gravity (its density relative to that of water) is 19.3 (see Table 13-1). Using Equation 13-12 as a guide, determine the specific gravity of the ring.

1. Equation 13-12 relates the specific gravity of the ring to the ratio of its weight w to the buoyant force B when submerged in water:

$$\text{Specific gravity} = \frac{w}{B_{\text{water}}} = \frac{w}{B}$$

2. B equals the weight minus the apparent weight w_{app} when submerged:

$$B = w - w_{\text{app}}$$

3. Combine steps 1 and 2 and solve for the specific gravity:

$$\text{Specific gravity} = \frac{w}{B} = \frac{w}{w - w_{\text{app}}}$$

$$= \frac{0.158 \text{ N}}{0.158 \text{ N} - 0.150 \text{ N}} = 19.3$$

4. Compare the specific gravity of the ring with the specific gravity of gold, which is 19.3:

$$19.3 \sim 19.3$$

The ring is pure gold.

EXERCISE A block of an unknown material weighs 3 N and has an apparent weight of 1.89 N when submerged in water. What is the material? (*Answer* The specific gravity of the material is 2.70, which is the specific gravity of aluminum. The material is aluminum.)

EXERCISE An aluminum block has an apparent weight of 3 N when surrounded by air. What is the weight of the block? (*Answer* $w_{\text{app}} = w - B$ where $B = \rho_{\text{air}} Vg$ and $w = \rho_{\text{alum}} Vg$. From Table 13-1 we obtain $\rho_{\text{air}} = 1.293 \text{ kg/m}^3$ and $\rho_{\text{alum}} = 2.70 \times 10^3 \text{ kg/m}^3$. We have three equations and three unknowns: w, B, and V. Thus, $w = 3.0014 \text{ N}$, which is only 0.048 percent greater than the apparent weight in air. Clearly, buoyant forces on solids and liquids due to air can usually be ignored.)

EXERCISE A piece of lead (specific gravity = 11.3) weighs 80 N in air. What does it weigh when submerged in water? (*Answer* 72.9 N)

The density of the block shown in Figure 13-11 is greater than the density of the surrounding fluid, and both the block and the scale pan are completely submerged in the fluid. The gravitation force on the block is its weight, $\vec{w}$, and the scale is adjusted, so it reads zero when the block is not being supported by the pan (Figure 11-13b). If the block is on the pan (Figure 13-11a), the scale reading is equal to the magnitude of the apparent weight w_{app} of the block. When the block is on the pan, the fluid is in direct contact with the entire surface of the block—except for those regions of the bottom surface of the block that are in direct contact with the pan. We assume the pan is not perfectly flat, but instead has some high and some low regions, and that the pan is in direct contact with the bottom surface of

FIGURE 13-11

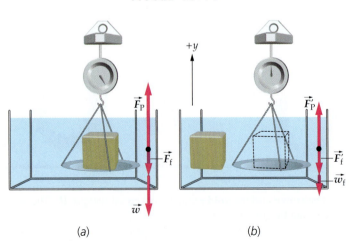

(a)   (b)

the block only at the high regions of the pan. (At the low regions of the pan, there is fluid in direct contact with bottom surface of the block.) We now analyze this situation to determine if the scale reading depends on the area of contact of the bottom surface of the block with the pan.

While resting on the pan, the net force exerted by the fluid on the block $\vec{F}_f$ is a combination of the downward force of the fluid on the top surface of the block and the upward force of the fluid on those regions of the bottom surface of the block that are in direct contact with the fluid. (We have drawn $\vec{F}_f$ downward. However, if the fluid were in direct contact with a large enough area of the bottom surface of the block, this force would be upward.) The two other vertical forces acting on the block are the gravitational force $\vec{w}$ and the upward force $\vec{F}_P$ exerted on the block by the pan at the regions of direct contact. In Figure 13-11b, the block has been moved off the pan, and in its place is a sample of fluid of identical size and shape (outlined by the dashed lines). The same regions of the surface of the pan are in direct contact with the bottom surface of this sample of fluid as if it were in direct contact with the block before it was moved. The forces acting on the fluid sample are the forces acting on it by the surrounding fluid $\vec{F}_f'$, the upward force on it by the pan $\vec{F}_P'$, and the gravitational force $\vec{w}_f$. The forces $\vec{F}_f$ and $\vec{F}_f'$ are equal because, at every point where the sample and the surrounding fluid are in direct contact, the pressure of the surrounding fluid is the same as it was at the same point before the block was moved off the pan.

When the submerged block rests on the pan, the block is in equilibrium, so $F_{f,y} + F_{P,y} + w_y = 0$, or

$$F_{f,y} + F_P - w = 0$$

and when the block is moved off the pan, the fluid sample in its place is in equilibrium, so

$$F_{f,y}' + F_P' - w_f = 0$$

Subtracting these equations, exploiting that $F_{f,y}' = F_{f,y}$, and rearranging the result gives

$$F_P - F_P' = w - w_f$$

where $F_P - F_P'$ is the change in the scale reading when the block is moved off the pan, so $F_P - F_P'$ is the magnitude of the apparent weight of the submerged block. Thus,

$$w_{app} = w - w_f$$

It is common parlance to refer to $w - w_{app}$ as the buoyant force B. Rearranging gives

$$B = w - w_{app} = w_f$$

This is the same expression for the buoyant force as is in Equation 13-13, which was established with the fluid in direct contact with 100 percent of surface of the submerged object.

MEASURING THE FAT   **EXAMPLE 13-6**

The percentage of body fat can be estimated by measuring the density of the body. Fat is less dense than muscle or bone. The density of fat $\approx 0.9 \times 10^3$ kg/m^3 and the density of lean tissue (everything except fat) $\approx 1.1 \times 10^3$ kg/m^3. Measuring the density of the body involves measuring the apparent weight while the body is submerged in water with the air completely exhaled from the lungs. (In practice, the amount of air remaining in the lungs is estimated and corrected for.) Suppose that your apparent weight when submerged in water is 5 percent of your weight. What percentage of your body mass is fat?

To determine the percentage of fat in this man's body, his density is measured by weighing him while he is submerged under water.

PICTURE THE PROBLEM For the person, the total volume equals the volume of the fat plus the volume of the lean, and the total mass equals the mass of the fat plus the mass of the lean. The volume and density are related to the mass by $m = \rho V$. The fraction of fat equals the mass of the fat divided by the total mass and the fraction of lean equals the mass of the lean divided by the total mass. Also, the fraction of fat plus the fraction of lean equals 1.

1. Using Equations 13-2 and 13-3, find the ratio of your body's density to the density of water:

$$\frac{\rho}{\rho_{water}} = \frac{w}{w - w_{app}} = \frac{w}{w - 0.05w} = 1.05$$

2. Your total body volume equals the volume of fat plus the volume of lean tissue:

$$V_{tot} = V_{fat} + V_{lean}$$

3. Because density equals mass times volume, volume equals mass divided by density. Substitute the corresponding mass-to-density ratio for each volume in the step 2 result:

$$\frac{m_{tot}}{\rho} = \frac{m_{fat}}{\rho_{fat}} + \frac{m_{lean}}{\rho_{lean}}$$

4. The ratio m_{fat}/m_{tot} is the fraction of fat f_{fat}, and m_{lean}/m_{tot} is the fraction of lean f_{lean}. Substitute for m_{fat} and m_{lean} in the step 3 result:

$$\frac{m_{tot}}{\rho} = \frac{f_{fat}m_{tot}}{\rho_{fat}} + \frac{f_{lean}m_{tot}}{\rho_{lean}}$$

5. The fraction of fat plus the fraction of lean tissue equals 1:

$$f_{fat} + f_{lean} = 1$$

6. Divide both sides of the step 4 result by m_{tot} and substitute $1 - f_{fat}$ for f_{lean}:

$$\frac{1}{\rho} = \frac{f_{fat}}{\rho_{fat}} + \frac{(1 - f_{fat})}{\rho_{lean}}$$

7. Solve the step 6 result for f_{fat}:

$$f_{fat} = \frac{1 - (\rho_{lean}/\rho)}{1 - (\rho_{lean}/\rho_{fat})}$$

8. Using the step 1 result, substitute for ρ in the step 7 result and solve for f_{fat}:

$$f_{fat} = \frac{1 - (\rho_{lean}/1.05\rho_{water})}{1 - (\rho_{lean}/\rho_{fat})} = \frac{1 - (1.1/1.05)}{1 - (1.1/0.9)} = 0.21$$

9. Convert to a percentage:

$$100\% \times f_{fat} = \boxed{21\%}$$

EXERCISE If Ed's apparent weight when submerged is zero, what is his body-fat percentage? (*Answer* 45 percent)

FIGURE 13-12

FLOATING ON A RAFT **EXAMPLE 13-7**

A raft of area A, thickness h, and mass $M = 600$ kg floats in calm water with 7 cm submerged (Figure 13-12). When Bob stands on the raft, 8.4 cm are submerged. What is Bob's mass m?

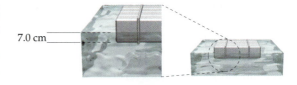

7.0 cm

PICTURE THE PROBLEM Let A be the area of the raft. The weight of the displaced fluid is then $\rho_w A d_1 g$ with just the raft and $\rho_w A d_2 g$ with Bob on the raft, where ρ_w is the density of the water, $d_1 = 7$ cm, and $d_2 = 8.4$ cm. If we set the weight of displaced fluid equal to the weight of the floating objects in each case, we can eliminate A and ρ_w and solve for m.

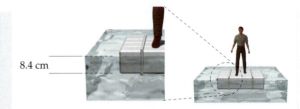

8.4 cm

1. Set the buoyant force with $d_1 = 7$ cm submerged equal to the weight of the raft, and with $d_2 = 8.4$ cm submerged equal to the weight of the raft plus Bob:

$$\rho_w A d_1 g = Mg$$
$$\rho_w A d_2 g = (M + m)g$$

2. Divide these two equations to eliminate the unknowns, A and ρ_w:

$$\frac{d_2}{d_1} = \frac{M + m}{M}$$

3. Solve for m:

$$m = \left(\frac{d_2}{d_1} - 1\right)M = \left(\frac{8.4 \text{ cm}}{7.0 \text{ cm}} - 1\right)(600 \text{ kg}) = \boxed{120 \text{ kg}}$$

A BOBBING CORK   **EXAMPLE 13-8**

A cork has a density of 200 kg/m³. Find the fraction of the volume of the cork that is submerged when the cork floats in water.

PICTURE THE PROBLEM Let V be the volume of the cork and V_{sub} be the volume that is submerged when the cork floats. The weight of the cork is $\rho_c V g$ and the buoyant force due to the water is $\rho_w V_{sub} g$.

1. Since the cork is in equilibrium, the buoyant force equals its weight:

$$w = B$$

$$\rho_c V g = \rho_w V_{sub} g$$

2. Solve for V_{sub}/V:

$$\frac{V_{sub}}{V} = \frac{\rho_c}{\rho_w} = \frac{200 \text{ kg/m}^3}{1000 \text{ kg/m}^3} = \boxed{\frac{1}{5}}$$

REMARKS We see that only one-fifth of the cork is submerged. This result is independent of the shape of the cork.

If we replace ρ_w in the calculation above with ρ_f, the density of the fluid, we can determine the submerged fraction of an object floating in any fluid. From Example 13-8, the fraction of a floating object that is submerged equals the ratio of its density to the density of the fluid.

$$\frac{V_{sub}}{V} = \frac{\rho}{\rho_f}$$

13-14

Because the density of ice is 920 kg/m³ and that of sea water is 1025 kg/m³, the fraction of an iceberg that is submerged in sea water is

$$\frac{V_{sub}}{V} = \frac{\rho}{\rho_f} = \frac{920 \text{kg/m}^3}{1025 \text{kg/m}^3} = 0.898$$

The great danger that icebergs pose to ships springs directly from the fact that only about 10 percent of an iceberg is visible above the water, and the visible portion gives little hint of where the submerged portion may extend, posing extreme danger to passing vessels.

Smoke from a burning cigarette. At first the smoke rises in a regular stream, but the simple streamlined flow quickly becomes turbulent and the smoke begins to swirl irregularly.

13-4 Fluids in Motion

The behavior of a fluid in motion can be complex. Consider, for example, the rise of smoke from a burning cigarette. At first the smoke rises in a regular stream of warm gas, but the simple streamlined flow quickly becomes turbulent and the smoke begins to swirl irregularly. Turbulent flow is very difficult to describe, even qualitatively. We will therefore consider only nonturbulent, steady-state flow of an "ideal" fluid, one that is nonviscous; that is, one that flows with no dissipation of mechanical energy. We also assume that the fluid is incompressible, which is an excellent approximation for liquids in most situations. In an incompressible fluid, the density is constant throughout the fluid.

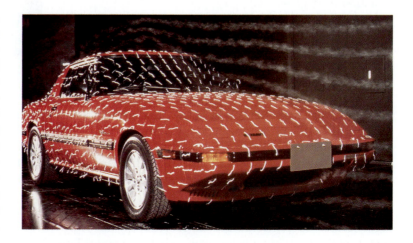

Figure 13-13 shows a tube full of liquid. The tube contains a tapered section with decreasing cross-sectional area. The fluid is flowing from left to right, and the shaded portion on the left depicts the volume ΔV of fluid that flows past point 1 during time Δt. If the speed of the fluid at this point is v_1 and the cross-sectional area of the tube is A_1, then the volume flowing past point 1 in time Δt is

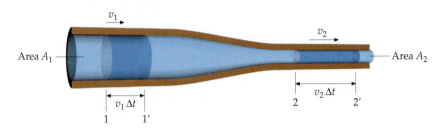

FIGURE 13-13

$$\Delta V = A_1 v_1 \, \Delta t$$

Since we assume that the fluid is incompressible, an equal volume of fluid must flow past any point along the tube during the same time Δt. The volume that flows past point 2 is depicted by the shaded portion on the right. If the speed of the fluid at this point is v_2 and the cross-sectional area is A_2, then the volume is $\Delta V = A_2 v_2 \, \Delta t$. Since these volumes must be equal, we have

$$A_1 v_1 \, \Delta t = A_2 v_2 \, \Delta t \qquad \text{so} \qquad A_1 v_1 = A_2 v_2 \qquad\qquad \text{13-15}$$

The quantity Av is called the **volume flow rate** I_v. The dimensions of I_v are volume divided by time. In the flow of an incompressible fluid, the volume flow rate is the same at any point in the fluid:

$$I_v = Av = \text{constant} \qquad\qquad \text{13-16}$$

CONTINUITY EQUATION FOR INCOMPRESSIBLE FLUID

Equation 13-16 is called the **continuity equation** for an incompressible fluid.

EXERCISE Blood flows in an aorta of radius 1.0 cm at 30 cm/s. What is the volume flow rate? (*Answer* $I_v = vA = 9.42 \times 10^{-5} \text{ m}^3/\text{s}$. It is customary to give the pumping rate of the heart in liters per minute. Using 1 m^3 = 1000 L, we have $I_v = 5.65$ L/min.)

EXERCISE Blood flows from a large artery of radius 0.3 cm, where its speed is 10 cm/s, into a region where the radius has been reduced to 0.2 cm because of thickening of the arterial walls (arteriosclerosis). What is the speed of the blood in the narrower region? (*Answer* If v_1 and v_2 are the initial and final speeds and A_1 and A_2 are the initial and final areas, Equation 13-16 gives

$$v_2 = \frac{A_1}{A_2} v_1 = \frac{\pi (0.3 \text{ cm})^2}{\pi (0.2 \text{ cm})^2} (10 \text{ cm/s}) = 22.5 \text{ cm/s} \Big)$$

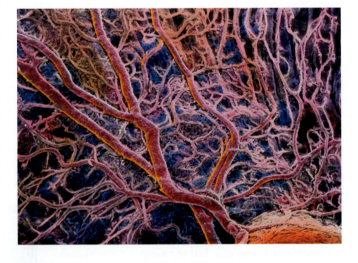

Bernoulli's Equation

When a small sample of a fluid flows into a narrow region of a tube, it gains speed because the pressure behind it that pushes it forward is greater than the pressure in front of it opposing its motion. Bernoulli's equation relates the pressure, elevation, and speed of an *incompressible* fluid in *steady* flow. It follows from Newton's laws and is most easily derived by applying the work–energy theorem to a segment of the fluid.

Consider a fluid flowing in a tube that varies in elevation as well as in cross-sectional area, as shown in Figure 13-14. We apply the work–energy theorem to a sample of fluid that initially is contained between points 1 and 2 in Figure 13-14*a*.

During time Δt this sample moves along the tube to the region between points 1′ and 2′ as shown in Figure 13-14b. Let ΔV be the volume of fluid passing point 1′ during time Δt. The same volume passes point 2 during the same time. Also, let $\Delta m = \rho \Delta V$ be the mass of the fluid with volume ΔV. The net effect on the sample during time Δt is that a mass Δm initially at height h_1 moving with speed v_1 is "transferred" to height h_2 with speed v_2. The change in the potential energy of the sample is thus

$$\Delta U = (\Delta m)gh_2 - (\Delta m)gh_1 = \rho \, \Delta V \, g (h_2 - h_1)$$

and the change in its kinetic energy is

$$\Delta K = \tfrac{1}{2}(\Delta m)v_2^2 - \tfrac{1}{2}(\Delta m)v_1^2 = \tfrac{1}{2}\rho \, \Delta V(v_2^2 - v_1^2)$$

The fluid behind the sample (to the sample's left in Figure 13-15) pushes on the sample with a force of magnitude $F_1 = P_1 A_1$, where P_1 is the pressure at point 1. This force does work:

$$W_1 = F_1 \, \Delta x_1 = P_1 A_1 \, \Delta x_1 = P_1 \, \Delta V$$

At the same time, the fluid in front of the sample (to the sample's right) exerts a force $F_2 = P_2 A_2$ opposing the sample's motion (pushing it to the left). This force does negative work:

$$W_2 = -F_2 \, \Delta x_2 = -P_2 A_2 \, \Delta x_2 = -P_2 \, \Delta V$$

The total work done by these forces is

$$W_{\text{total}} = P_1 \, \Delta V - P_2 \, \Delta V = (P_1 - P_2)\Delta V$$

The work–energy theorem gives

$$W_{\text{total}} = \Delta U + \Delta K$$

so

$$(P_1 - P_2)\Delta V = \rho \, \Delta V \, g(h_2 - h_1) + \tfrac{1}{2}\rho \, \Delta V(v_2^2 - v_1^2)$$

If we divide both sides by ΔV, we obtain

$$P_1 - P_2 = \rho gh_2 - \rho gh_1 + \tfrac{1}{2}\rho v_2^2 - \tfrac{1}{2}\rho v_1^2$$

When we collect all the quantities having a subscript 1 on one side and those having a subscript 2 on the other, this equation becomes

$$P_1 + \rho gh_1 + \tfrac{1}{2}\rho v_1^2 = P_2 + \rho gh_2 + \tfrac{1}{2}\rho v_2^2 \qquad \text{13-17}a$$

This result can be restated as

$$P + \rho gh + \tfrac{1}{2}\rho v^2 = \text{constant} \qquad \text{13-17}b$$

BERNOULLI'S EQUATION

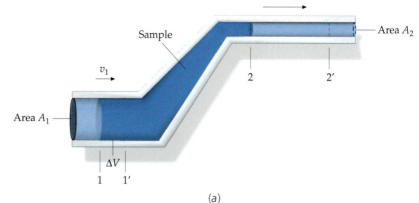

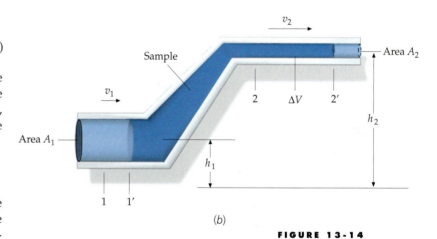

FIGURE 13-14

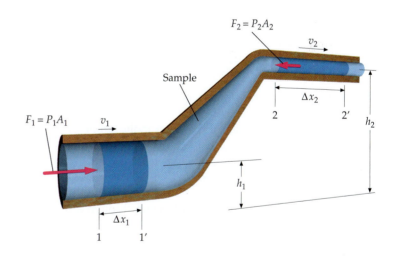

FIGURE 13-15 Fluid moving in a pipe that varies in both height and cross-sectional area. The total work done by the forces $F_1 = P_1 A_1$ and $F_2 = P_2 A_2$ has the effect of raising the shaded portion of the fluid from height h_1 to h_2 and changing its speed from v_1 to v_2.

which means that this combination of quantities has the same value at any point along the tube. Equations 13-17a and 13-17b are known as **Bernoulli's equation** for the steady, nonviscous flow of an incompressible fluid.

A special application of Bernoulli's equation is for a fluid at rest. Then $v_1 = v_2 = 0$, and we obtain

$$P_1 - P_2 = \rho gh_2 - \rho gh_1 = \rho g\Delta h$$

This is the same as Equation 13-7.

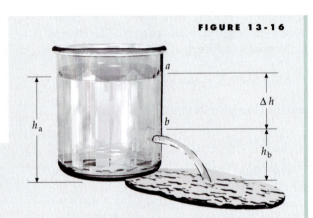

FIGURE 13-16

A LEAKING TANK   **EXAMPLE 13-9**

A large tank of water, open at the top, has a small hole through its side a distance h below the surface of the water. Find the speed of the water as it flows out the hole.

PICTURE THE PROBLEM We apply Bernoulli's equation to points a and b in Figure 13-16. Because the diameter of the hole is much smaller than the diameter of the tank, we can neglect the velocity of the water at the top (point a).

1. Bernoulli's equation with $v_a = 0$ gives:

$$P_a + \rho gh_a + 0 = P_b + \rho gh_b + \tfrac{1}{2}\rho v_b^2$$

2. The pressure at point a and at point b is the same, P_{at}, since both points are open to the atmosphere:

$$P_a = P_{at} \quad \text{and} \quad P_b = P_{at}$$

so

$$P_{at} + \rho gh_a + 0 = P_{at} + \rho gh_b + \tfrac{1}{2}\rho v_b^2$$

3. Solve the step 2 result for the speed v_b of the water flowing from the hole:

$$v_b^2 = 2g(h_a - h_b) = 2g\Delta h$$

so

$$\boxed{v_b = \sqrt{2g\Delta h}}$$

EXERCISE If the water flowing out of the hole is directed vertically upward, how high does it rise? (*Answer*   The water shoots upward a distance h; that is, to the same level as the surface of the water in the tank.)

In Example 13-9, the water emerges from the hole with a speed equal to the speed it would have if it dropped in free-fall a distance h. This finding is known as *Torricelli's law*.

In Figure 13-17, water is shown flowing through a horizontal pipe that has a constricted section. Because both sections of the pipe are at the same elevation, $h_1 = h_2$ in Equation 13-17a. Then Bernoulli's equation becomes

$$P + \tfrac{1}{2}\rho v^2 = \text{constant} \qquad\qquad 13\text{-}18$$

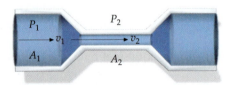

When the fluid moves into the constriction, the area A gets smaller, so the speed v must get larger since Av remains constant. But because $P + \tfrac{1}{2}\rho v^2$ is constant, when the speed gets larger the pressure must get smaller. Thus, the pressure in the constriction is reduced.

FIGURE 13-17 Constriction in a pipe carrying a moving fluid. The pressure is lower in the narrow section of the pipe where the fluid is moving faster.

When the speed of a fluid increases, the pressure drops.

VENTURI EFFECT

This result is often referred to as the **Venturi effect.** Equation 13-18 is an important result that applies to many situations in which we can ignore changes in height.

In Figure 13-18, lines, called **streamlines,** are drawn to pictorially represent the flow of the fluid. The direction of the lines denotes the direction of flow and the spacing between lines represents the speed of the flow. The smaller the spacing, the greater the speed. For horizontal flow, where the speed increases, the pressure decreases, so a decrease in streamline spacing is accompanied by a decrease in pressure.

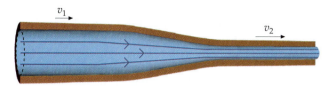

FIGURE 13-18

A VENTURI METER   **EXAMPLE   13-10**   **Try It Yourself**

A *Venturi meter,* which is used to measure the flow rate of a fluid, is shown in Figure 13-19. The fluid of density ρ_F passes through a pipe of cross-sectional area A_1 that has a constriction of cross-sectional area A_2. The two parts of the pipe are connected with a U-tube manometer partially filled with a liquid of density ρ_L. Since the velocity of flow is greater in the constricted region, the pressure in that section is less than in the other portion of the pipe. The pressure difference is measured by the difference in the levels of the liquid in the U-tube, Δh. Express the velocity v_1 in terms of the measured height Δh and the known quantities ρ_F, ρ_L, and $r = A_1/A_2$.

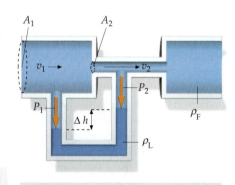

FIGURE 13-19 A Venturi meter.

PICTURE THE PROBLEM The pressures P_1 and P_2 in the two regions are related to the speeds v_1 and v_2 by Bernoulli's equation. The pressure difference is related to the height h by $P_1 - P_2 = \rho_L gh$. You can express v_2 in terms of v_1 and the areas A_1 and A_2 by the continuity equation.

Cover the column to the right and try these on your own before looking at the answers.

Steps   **Answers**

1. Write Bernoulli's equation for constant elevation for the two regions.

$$P_1 + \tfrac{1}{2}\rho_F v_1^2 = P_2 + \tfrac{1}{2}\rho_F v_2^2$$

2. Write the continuity equation for the two regions, and solve for v_2 in terms of v_1 and the areas A_1 and A_2.

$$v_2 A_2 = v_1 A_1$$

so

$$v_2 = \frac{A_1}{A_2}v_1 = rv_1$$

3. Substitute your result for v_2 into the equation in step 1 and obtain an equation for $P_1 - P_2$.

$$P_1 - P_2 = \tfrac{1}{2}\rho_F(v_2^2 - v_1^2)$$
$$= \tfrac{1}{2}\rho_F(r^2 - 1)v_1^2$$

$$P_1 - P_2 = \rho_L g\Delta h - \rho_F g\Delta h$$

4. Write $P_1 - P_2$ in terms of the difference in height Δh of the liquid in the arms of the U-tube. This pressure difference equals the pressure drop in a column of the liquid of height Δh and that in a column of the fluid of the same height.

$$= (\rho_L - \rho_F)g\Delta h$$

$$\tfrac{1}{2}\rho_F(r^2 - 1)v_1^2 = (\rho_L - \rho_F)g\Delta h$$

5. Equate the two expressions for $P_1 - P_2$ and solve for v_1 in terms of Δh.

so

$$v_1 = \boxed{\sqrt{\frac{2(\rho_L - \rho_F)g\Delta h}{\rho_F(r^2 - 1)}}}$$

EXERCISE Find v_1 if $\Delta h = 3$ cm, $r = 4$, the fluid is air ($\rho_F = 1.29$ kg/m³), and the liquid in the U-tube portion of the Venturi meter is water ($\rho_w = 10^3$ kg/m³). (*Answer* $v_1 = 5.51$ m/s)

REMARKS Air is not an incompressible fluid, so the calculation in the Exercise is not as accurate as the calculation in the Example 13-9. Strictly speaking, Bernoulli's equation and the continuity equation hold only for incompressible fluids.

The Venturi effect can be used to give a qualitative understanding of the lift of an airplane wing and the path of a pitcher's curveball. An airplane wing is designed so that air moves faster over the top of the wing than it does under the wing, thus making the air pressure less on top than underneath. This difference in pressure results in a net force upward on the wing. Figure 13-20a shows a top view of the motion of a curveball. As the ball spins, it tends to drag air around with it. Figure 13-20b is drawn from the point of view of a stationary (but spinning) ball with the air rushing past it. The air movement caused by the spinning ball adds to the velocity of the air rushing by on the left side of the ball (the top in the figure) and subtracts from it on the right (the bottom in the figure). Thus, the air speed is greater on the left side of the ball than on the right, causing the pressure on the left to be less than that on the right. The ball therefore curves to the left. The atomizer shown in Figure 13-21 also works on the principle of the Venturi effect.

Although Bernoulli's equation is very useful for qualitative descriptions of many features of fluid flow, such descriptions are often grossly inaccurate when compared with the quantitative results of experiments. Prominent reasons for the discrepancies are that gases like air are hardly incompressible, and liquids like water have viscosity, which invalidates the assumption of the conservation of mechanical energy. In addition, it is often difficult to maintain steady-state, streamlined flow without turbulence, and the introduction of turbulence can greatly affect the results.

Airflow above and below the wing of this Indy race car creates greater pressure above the wing, increasing the effective weight of the car for better control at high speeds. An airplane wing is designed so that the flow creates greater pressure below the wing to lift the plane.

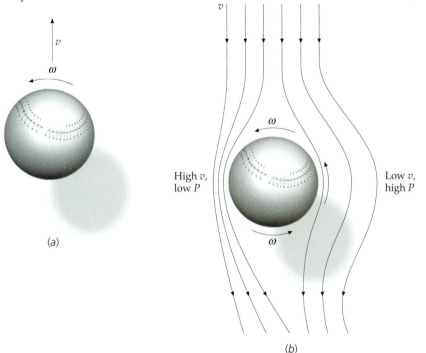

(a)

High v, low P Low v, high P

(b)

FIGURE 13-21 When the bulb of an atomizer is squeezed, the air is forced through the constriction in the horizontal tube, which reduces the pressure there below atmospheric pressure. Because of the resulting pressure difference, the liquid in the jar is pumped up through the vertical tube, enters the air stream, and emerges from the nozzle. A similar effect occurs in the carburetor of a gasoline lawnmower engine.

FIGURE 13-20 (a) Top view of a baseball thrown with a counterclockwise spin ω, like an overhead view of a curveball thrown by a right-handed pitcher. (b) In the frame moving with the ball, the ball is stationary (but spinning) and the air rushes past it. Because of its rough cover, the spinning ball drags the air around with it, making the air speed higher on the left side and lower on the right. The pressure is therefore lower on the left side, so the ball curves to the left.

*Viscous Flow

According to Bernoulli's equation, when a fluid flows steadily through a long, narrow, horizontal pipe of constant cross section, the pressure along the pipe will be constant. In practice, however, we observe a pressure drop as we move along the direction of the flow. Put another way, a pressure difference is required to push a fluid through a horizontal pipe. This pressure difference is needed because of the drag force that is exerted by the pipe on the layer of fluid in contact with the pipe, and because of the drag force exerted by each layer of the fluid on an adjacent layer that is moving with a slightly different velocity. These drag forces are called **viscous forces**. As a result of viscous forces, the velocity of the fluid is not constant across the diameter of the pipe. Instead, it is greatest near the center of the pipe and approaches zero where the fluid is in contact with the walls of the pipe (Figure 13-22). Let P_1 be the pressure at point 1 and P_2 be that at point 2, a distance L downstream from point 1. The pressure drop $\Delta P = P_1 - P_2$ is proportional to the volume flow rate:

$$\Delta P = P_1 - P_2 = I_V R \qquad\qquad 13\text{-}19$$

where $I_V = vA$ is the volume flow rate and the proportionality constant R is the resistance to flow, which depends on the length of the pipe L, the radius r, and the viscosity of the fluid.

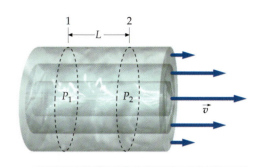

FIGURE 13-22 When a viscous fluid flows through a pipe, the speed is greatest at the center of the pipe. At the walls of the pipe, the fluid flow rate approaches zero.

RESISTANCE TO BLOOD FLOW   **EXAMPLE 13-11**

Blood flows from the aorta through the major arteries, the small arteries, the capillaries, and the veins until it reaches the right atrium. In the course of that flow, the (gauge) pressure drops from about 100 torr to zero. If the volume flow rate is 0.8 L/s, find the total resistance of the circulatory system.

PICTURE THE PROBLEM The resistance is related to the pressure drop and volume flow rate by Equation 13-19. We can use Equation 13-9 to convert from torr to kPa.

Write the resistance in terms of the pressure drop and volume flow rate, and convert all terms to SI units:

$$R = \frac{\Delta P}{I_v}$$

$$= \frac{100 \text{ torr}}{0.8 \text{ L/s}} \times \frac{101 \text{ kPa}}{760 \text{ torr}} \times \frac{1 \text{ L}}{10^3 \text{ cm}^3} \times \frac{1 \text{ cm}^3}{10^{-6} \text{ m}^3}$$

$$= \boxed{16.6 \text{ kPa·s/m}^3}$$

REMARKS We could have used 1 Pa = 1 N/m² to write the result as 16.6 kN·s/m⁵.

To define the coefficient of viscosity of a fluid, we consider a fluid that is confined between two parallel plates, each of area A, separated by a distance z as shown in Figure 13-23. The upper plate is pulled at a constant speed v by a force $\vec{F}$ while the bottom plate is held at rest. A force is needed to pull the upper plate because the fluid next to the plate exerts a viscous drag force opposing its motion. The speed of the fluid between the plates approaches v near the upper plate and zero near the lower plate, and it varies linearly with separation between the plates. The force $\vec{F}$ is found to be directly

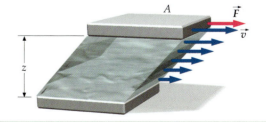

FIGURE 13-23 Two plates of equal area with a viscous fluid between them. When the upper plate is moved relative to the lower one, each layer of fluid exerts a drag force on adjacent layers. The force needed to pull the upper plate is directly proportional to v and the area A and inversely proportional to z, the separation between the plates.

proportional to v and A and inversely proportional to the plate separation z. The proportionality constant is the **coefficient of viscosity** η:

$$F = \eta \frac{vA}{z}$$

13-20

The SI unit of viscosity is the N·s/m² = Pa·s. An older cgs unit still in common use is the **poise,** named after the French physicist Jean Poiseuille. These units are related by

$$1 \text{ Pa·s} = 10 \text{ poise}$$

13-21

Table 13-3 lists the coefficients of viscosity for several fluids at various temperatures. Typically, the viscosity of a liquid increases as the temperature decreases. Thus, in cold climates, a less viscous grade of oil is used to lubricate automobile engines in the winter than in summer.

Poiseuille's Law   The resistance to flow R in Equation 13-19 for steady flow through a circular tube of radius r can be shown to be

$$R = \frac{8\eta L}{\pi r^4}$$

13-22

Equations 13-19 and 13-22 can be combined to give the pressure drop over a length L of a circular tube of radius r:

$$\Delta P = \frac{8\eta L}{\pi r^4} I_v$$

13-23

POISEUILLE'S LAW

Equation 13-23 is known as **Poiseuille's law.** Note the inverse r^4 dependence of the pressure drop. If the radius of the tube is halved, the pressure drop for a given volume flow rate is increased by a factor of 16; or a pressure 16 times as

TABLE 13-3

Coefficients of Viscosity for Various Fluids

Fluid	t, °C	η, mPa·s
Water	0	1.8
	20	1.00
	60	0.65
Blood (whole)	37	4.0
Engine oil (SAE 10W)	30	200
Glycerin	0	10,000
	20	1,410
	60	81
Air	20	0.018

great is needed to pump the fluid through the tube at the original volume flow rate. Thus, for example, if the diameter of a person's blood vessels or arteries is reduced for some reason, either the volume flow rate of the blood is greatly reduced, or the blood pressure must escalate to maintain the volume flow rate. For water flowing through a long garden hose, the pressure drop is pretty much fixed. It equals the difference in pressure between that at the water source and atmospheric pressure at the open end. The volume flow rate is then proportional to the fourth power of the radius. Thus, if the radius is halved, the volume flow rate drops by a factor of 16.

Poiseuille's law applies only to the laminar (nonturbulent) flow of a fluid of constant viscosity. In some fluids, viscosity changes with velocity, violating Poiseuille's law. Blood, for example, is a complex fluid consisting of solid particles of various shapes suspended in a liquid. Red blood cells are disk-shaped objects that are randomly oriented at low velocities but at high velocities tend to become oriented to facilitate the flow. Thus, the viscosity of blood decreases as the flow velocity increases, so Poiseuille's law cannot be strictly applied. Nevertheless, Poiseuille's law is a good approximation that is very useful for obtaining a qualitative understanding of blood flow.

In Chapter 25 the flow of electrical current I through metal wires is studied. One of the basic relations in that chapter is Ohm's law, $\Delta V = IR$, where ΔV is the potential difference and R is the electrical resistance of the wire. As we shall see, Ohm's law is analogous to Poiseuille's law $\Delta P = I_V R$.

Turbulence—Reynolds Number   When the flow velocity of a fluid becomes sufficiently great, laminar flow breaks down and turbulence sets in. The critical velocity above which the flow through a tube is turbulent depends on the density and viscosity of the fluid and on the radius of the tube. The flow of a fluid can be characterized by a dimensionless number called the **Reynolds number** N_R, which is defined by

$$N_R = \frac{2r\rho v}{\eta} \qquad\qquad 13\text{-}24$$

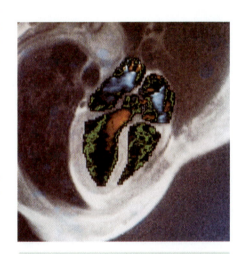

where v is the average velocity of the fluid. Experiments have shown that the flow will be laminar if the Reynolds number is less than about 2000 and turbulent if it is greater than 3000. Between these values, the flow is unstable and may change from one type to the other.

BLOOD FLOW IN THE AORTA   **E X A M P L E   1 3 - 1 2**

False-color view of turbulence of blood flowing into and out of the heart as seen by magnetic resonance imaging (MRI). Systolic ejection from the left ventricle into the aorta is seen in red, and diastolic filling of the ventricles in blue.

Calculate the Reynolds number for blood flowing at 30 cm/s through an aorta of radius 1.0 cm. Assume that blood has a viscosity of 4 mPa·s and a density of 1060 kg/m³.

PICTURE THE PROBLEM   Because N_R is dimensionless, we can use any set of units as long as we are consistent.

Write Equation 13-24 for the Reynolds number, expressing each quantity in SI units:

$$N_R = \frac{2r\rho v}{\eta}$$

$$= \frac{2(0.01\ \text{m})(1060\ \text{kg/m}^3)(0.3\ \text{m/s})}{4 \times 10^{-3}\ \text{Pa·s}} = \boxed{1590}$$

REMARKS   Since the Reynolds number is less than 2000, this flow will be laminar rather than turbulent.

1. Density, specific gravity, and pressure are defined quantities that are important in fluid statics and dynamics.
2. Pascal's principle and Archimedes' principle are derived from Newton's laws.
3. Bernoulli's equation is derived from the conservation of mechanical energy.
*4. The Venturi effect is a special case of Bernoulli's equation.
*5. Poiseuille's law accounts for pressure drops due to viscosity, Reynolds number is used to predict whether flow is laminar or turbulent.

Topic	Relevant Equations and Remarks
1. Density	The density of a substance is the ratio of its mass to its volume:
	$$\rho = \frac{m}{V} \qquad \text{13-1}$$
	The densities of most solids and liquids are approximately independent of temperature and pressure, whereas those of gases depend strongly on these quantities.
2. Specific Gravity	The specific gravity of a substance is the ratio of its density to that of water. An object sinks or floats in a given fluid depending on whether its density is greater than or less than that of the fluid.
3. Pressure	$$P = \frac{F}{A} \qquad \text{13-3}$$
Units	$1\,\text{Pa} = 1\,\text{N/m}^2 \qquad \text{13-4}$
	$1\,\text{atm} = 760\,\text{mmHg} = 760\,\text{torr} = 29.9\,\text{in.Hg}$
	$\qquad = 101.325\,\text{kPa} = 14.7\,\text{lb/in.}^2 \qquad \text{13-9}$
	$1\,\text{bar} = 10^3\,\text{millibars} = 100\,\text{kPa} \qquad \text{13-10}$
Gauge pressure	Gauge pressure is the difference between the absolute pressure and atmospheric pressure:
	$$P = P_{\text{gauge}} + P_{\text{at}} \qquad \text{13-8}$$
In a liquid	$P = P_0 + \rho g \Delta h \quad (\rho \text{ constant}) \qquad \text{13-7}$
In a gas	In a gas such as air, pressure decreases exponentially with altitude.
Bulk modulus	$$B = -\frac{\Delta P}{\Delta V/V} \qquad \text{13-6}$$
4. Pascal's Principle	Pressure changes applied to an enclosed liquid are transmitted undiminished to every point in the fluid and to the walls of the container.
5. Archimedes' Principle	A body wholly or partially submerged in a fluid is buoyed up by a force equal to the weight of the displaced fluid.
*6. Fluid Flow	
Volume flow rate	$I_v = Av$

Continuity equation for an incompressible fluid	$I_v = Av = $ constant	13-16
Bernoulli's equation	$P + \rho gh + \frac{1}{2}\rho v^2 = $ constant	13-17b
Venturi effect	When the speed of a fluid increases, the pressure drops.	
Resistance to fluid flow	$\Delta P_2 = I_v R$	13-19
Coefficient of viscosity	$F = \eta \dfrac{vA}{z}$	13-20
Poiseuille's law for viscous flow	$\Delta P = RI_v = \dfrac{8\eta L}{\pi r^4} I_v$	13-23
Laminar flow, turbulent flow, and the Reynolds number	The flow will be laminar if the Reynolds number N_R is less than about 2000 and turbulent if it is greater than 3000, where N_R is given by $$N_R = \dfrac{2r\rho v}{\eta}$$	13-24

PROBLEMS

- • Single-concept, single-step, relatively easy
- •• Intermediate-level, may require synthesis of concepts
- ••• Challenging
- **SSM** Solution is in the *Student Solutions Manual*
- Problems available on iSOLVE online homework service
- ✓ These "Checkpoint" online homework service problems ask students additional questions about their confidence level, and how they arrived at their answer

In a few problems, you are given more data than you actually need; in a few other problems, you are required to supply data from your general knowledge, outside sources, or informed estimates.

Conceptual Problems

1 • If the gauge pressure is doubled, the absolute pressure will be (*a*) halved, (*b*) doubled, (*c*) unchanged, (*d*) squared, (*e*) not enough information is given to determine the effect.

2 • **SSM** Does Archimedes' principle hold in a satellite orbiting the earth in a circular orbit? Explain.

3 •• A rock of mass M with a density three times that of water is suspended by a thread. Holding the free end of the thread in your hand, you lower the rock into an aquarium tank with water almost up to the rim. The tank is resting on the platform of a scale. When the rock is just above the bottom of the tank the thread breaks. From the moment the thread breaks to the time the rock is resting on the bottom of the tank the reading on the scale increases by (*a*) $2Mg$, (*b*) Mg, (*c*) $\frac{2}{3}Mg$, (*d*) $\frac{1}{3}Mg$, (*e*) zero.

4 •• A rock is thrown into a swimming pool filled with water of uniform temperature. Which of the following statements is true?

(*a*) The buoyant force on the rock is zero as it sinks.

(*b*) The buoyant force on the rock increases as it sinks.

(*c*) The buoyant force on the rock decreases as it sinks.

(*d*) The buoyant force on the rock is constant as it sinks.

(*e*) The buoyant force on the rock as it sinks is nonzero at first but becomes zero once the terminal velocity is reached.

5 •• A fishbowl rests on a scale. The fish suddenly swims upward to get food. What happens to the scale reading?

6 •• **SSM** Two objects are balanced as in Figure 13-24. The objects have identical volumes but different masses. Will the equilibrium be disturbed if the entire system is completely immersed in water? Explain.

FIGURE 13-24 Problem 6

7 •• A 200-g block of lead and a 200-g block of copper are completely under water. Each is suspended by a thread just above the bottom of an aquarium filled with water. Which of the following is true?

(a) The buoyant force is greater on the lead than on the copper.
(b) The buoyant force is greater on the copper than on the lead.
(c) The buoyant force is the same on both blocks.
(d) More information is needed to choose the correct answer.

8 •• A 20-cm^3 block of lead and a 20-cm^3 block of copper are completely under water. Each is suspended by a thread just above the bottom of an aquarium filled with water. Which of the following is true?

(a) The buoyant force is greater on the lead than on the copper.
(b) The buoyant force is greater on the copper than on the lead.
(c) The buoyant force is the same on both blocks.
(d) More information is needed to choose the correct answer.

9 • In a department store, a beach ball is supported by the airstream from a hose connected to the exhaust of a vacuum cleaner. Does the air blow under or over the ball to support it? Why?

10 • A horizontal pipe narrows from a diameter of 10 cm to 5 cm. For a nonviscous liquid flowing without turbulance from the larger diameter to the smaller, (a) the velocity and pressure both increase, (b) the velocity increases and the pressure decreases, (c) the velocity decreases and the pressure increases, (d) the velocity and pressure both decrease, (e) either the velocity or pressure changes but not both.

11 • **SSM** True or false: The buoyant force on a submerged object depends on the shape of the object.

12 • Figure 13-25 shows a Cartesian diver. The diver consists of a small tube, open at the bottom, with an air bubble at the top, inside a closed plastic soda bottle that is partly filled with water. Normally, the diver floats, but sinks when the bottle is squeezed hard. Explain why this happens.

FIGURE 13-25
Problem 12

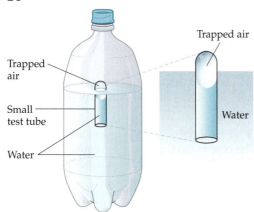

Trapped air
Trapped air
Small test tube
Water
Water

13 • A glass of water has ice cubes floating in it. What happens to the water level when the ice melts?

14 • Why do you float higher out of the water in salt water than in fresh water?

15 •• A certain object has a density just slightly less than that of water so that it floats almost completely submerged. However, the object is more compressible than water. What happens if the floating object is given a slight push to submerge it?

16 •• In Example 13-10, the fluid is accelerated to a greater speed as it enters the narrow part of the pipe. Identify the forces that act on the fluid to produce this acceleration.

17 •• A glass of water is accelerating to the right along a horizontal surface. What is the origin of the force that produces the acceleration on a small element of water in the middle of the glass? Explain with a picture.

18 •• **SSM** You are sitting in a boat floating on a very small pond. You take the anchor out of the boat and drop it into the water. What happens to the water level in the pond?

19 •• Figure 13-26 is a diagram of a prairie dog tunnel. The geometry of the two holes and their positioning ensures that the wind blowing above hole number 2 will always have a lower speed than that above hole number 1. Explain how Bernoulli's principle keeps the tunnel ventilated, and indicate in which direction air will flow through the tunnel.

FIGURE 13-26
Problem 19

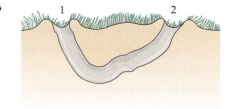

20 • **SSM** Three bottles of different shapes are filled to the same level with water, as shown in Figure 13-27. The area of the bottom of each bottle is the same. The hydrostatic pressure is the same on the bottom of each of the bottles, but the total force must be different, as each bottle has a different amount of liquid in it. Explain this apparent paradox.

FIGURE 13-27
Problem 20

Density

21 • Find the mass of a copper cylinder that is 6 cm long with a radius of 2 cm.

22 • Find the mass of a lead sphere of radius 2 cm.

23 • **iSOLVE** Find the mass of air in a room 4 m × 5 m × 4 m.

24 •• **SSM** A 60-mL flask is filled with mercury at 0°C (Figure 13-28). When the temperature rises to 80°C, 1.47 g of mercury spill out of the flask. Assuming that the volume of the flask is constant, find the density of mercury at 80°C if its density at 0°C is 13,645 kg/m^3.

0°C   80°C

FIGURE 13-28 Problem 24

Pressure

25 • **SOLVE** Barometer readings are commonly given in inches of mercury. Find the pressure in inches of mercury equal to 101 kPa.

26 • The pressure on the surface of a lake is atmospheric pressure $P_{at} = 101$ kPa. (a) At what depth is the pressure twice atmospheric pressure? (b) If the pressure at the top of a deep pool of mercury is P_{at}, at what depth is the pressure $2P_{at}$?

27 • **SSM** **SOLVE** A hydraulic lift is used to raise an automobile of mass 1500 kg. The radius of the shaft of the lift is 8 cm and that of the piston is 1 cm. How much force must be applied to the piston to raise the automobile?

28 •• When a woman in high-heeled shoes takes a step, she momentarily places her entire weight on one heel of her shoe. If her mass is 56 kg and if the area of the heel is 1 cm^2, what is the pressure exerted on the floor by her heel?

29 • **SSM** What pressure increase is required to compress the volume of 1 kg of water from 1.00 L to 0.99 L?

30 • **SOLVE** A 1500-kg car rests on four tires, each of which is inflated to a gauge pressure of 200 kPa. What is the area of contact of each tire with the road, if the four tires support the weight equally?

31 •• In the seventeenth century, Blaise Pascal performed the experiment shown in Figure 13-29. A wine barrel filled with water was coupled to a long tube. Water was added to the tube until the barrel burst. (a) If the radius of the lid was 20 cm and the height of the water in the tube was 12 m, calculate the force exerted on the lid. (b) If the tube had an inner radius of 3 mm, what mass of water in the tube caused the pressure that burst the barrel?

FIGURE 13-29 Problem 31

32 •• **SOLVE** Blood plasma flows from a bag through a tube into a patient's vein, where the blood pressure is 12 mmHg. The specific gravity of blood plasma at 37°C is 1.03. What is the minimum elevation the bag must have so the plasma flows into the vein?

33 •• Many people have imagined that if they were to float the top of a flexible snorkel tube out of the water, they would be able to breathe through it while walking underwater (Figure 13-30). However, they generally do not take into account just how much water pressure opposes the expansion of the chest and the inflation of the lungs. Suppose you can just breathe while lying on the floor with a 400-N (90-lb) weight on your chest. How far below the surface of the water could your chest be for you still to be able to breathe, assuming your chest has a frontal area of 0.09 m^2?

FIGURE 13-30 Problem 33

34 •• In Example 13-3, a force of 147 N is applied to a small piston to lift a car that weighs 14,700 N. Demonstrate that this does not violate the law of conservation of mechanical energy by showing that, when the car is lifted some distance h, the work done by the 147-N force acting on the small piston equals the work done by the large piston on the car.

35 • **SOLVE** A hollow cube with edge length a is half-filled with water of density ρ. Find the force exerted on a side of the cube by the water. (The edges of the cube are either horizontal or vertical.)

36 ••• **SSM** The volume of a cone of height h and base radius r is $V = \pi r^2 h/3$. A conical vessel of height 25 cm resting on its base of radius 15 cm is filled with water. (a) Find the volume and weight of the water in the vessel. (b) Find the force exerted by the water on the base of the vessel. Explain how this force can be greater than the weight of the water.

Buoyancy

37 • **SSM** **SOLVE** A 500-g piece of copper (specific gravity 9.0) is suspended from a spring scale and is submerged in water (Figure 13-31). What force does the spring scale read?

0.5kg

FIGURE 13-31 Problem 37

38 • [SOLVE] When a 60-N stone is attached to a spring scale and is submerged in water, the spring scale reads 40 N. What is the density of the stone?

39 • [SOLVE] A block of an unknown material weighs 5 N in air and 4.55 N when submerged in water. (*a*) What is the density of the material? (*b*) Of what material is the block made?

40 • A solid piece of metal weighs 90 N in air and 56.6 N when submerged in water. Determine the density of this metal.

41 •• [SOLVE]✓ A homogeneous solid object floats on water with 80 percent of its volume below the surface. The same object when placed in a second liquid floats on that liquid with 72 percent of its volume below the surface. Determine the density of the object and the specific gravity of the liquid.

42 •• [SSM] A 5-kg iron block is suspended from a spring scale and is submerged in a fluid of unknown density. The spring scale reads 6.16 N. What is the density of the fluid?

43 •• A large piece of cork weighs 0.285 N in air. When held submerged underwater by a spring scale as shown in Figure 13-32, the spring scale reads 0.855 N. Find the density of the cork.

FIGURE 13-32 Problem 43

44 •• [SOLVE]✓ A helium balloon lifts a basket and cargo of total weight 2000 N under standard conditions, at which the density of air is 1.29 kg/m³ and the density of helium is 0.178 kg/m³. What is the minimum volume of the balloon?

45 •• [SSM] An object has neutral buoyancy when its density equals that of the liquid in which it is submerged, which means that it neither floats nor sinks. If the average density of an 85-kg diver is 0.96 kg/L, what mass of lead should be added to give him neutral buoyancy?

46 •• [SOLVE]✓ A beaker of mass 1 kg containing 2 kg of water rests on a scale. A 2-kg block of aluminum (density 2.70 × 10³ kg/m³) suspended from a spring scale is submerged in the water as in Figure 13-33. Find the readings of both scales.

FIGURE 13-33 Problem 46

Aluminum

47 ••• A ship sails from seawater (specific gravity 1.025) into freshwater and therefore sinks slightly. When its load of 600,000 kg is removed, it returns to its original level. Assuming that the sides of the ship are vertical at the water line, find the mass of the ship before it was unloaded.

48 ••• [SSM] The hydrometer shown in Figure 13-34 is a device for measuring the specific gravity of liquids. The bulb contains lead shot, and the specific gravity can be read directly from the liquid level on the stem after the hydrometer has been calibrated. The volume of the bulb is 20 mL, the stem is 15 cm long and has a diameter of 5.00 mm, and the mass of the glass is 6.0 g. (*a*) What mass of lead shot must be added so that the lowest specific gravity of liquid that can be measured is 0.9? (*b*) What is the maximum specific gravity of liquid that can be measured?

FIGURE 13-34 Problems 48, 98

49 • When cracks form at the base of a dam, the water seeping in the cracks exerts a buoyant force that tends to lift the dam. This can topple the dam. Estimate the buoyant force exerted on a 2-m thick by 5-m long dam wall by water seeping into cracks at its base. The water level is 5 m above the cracks.

50 •• A large helium weather balloon is spherical in shape, with a radius of 2.5 m and a total mass of 15 kg (balloon plus helium plus equipment). (*a*) What is the initial upward acceleration of the balloon when it is released from sea level? (*b*) If the drag force on the balloon is given by $F_D = \frac{1}{2}\pi r^2 \rho v^2$, where r is the balloon radius, ρ is the density of air, and v the balloon's ascension speed, calculate the terminal velocity of the ascending balloon. (*c*) Roughly how long will it take for the balloon to ascend to a height of 10 km?

Continuity and Bernoulli's Equation

51 •• **SSM** Water exits a circular tap moving straight down with a flow rate of 10.5 cm³/s. (*a*) If the diameter of the tap is 1.2 cm, what is the speed of the water? (*b*) As the fluid falls from the tap, the stream of water narrows. Find the new diameter of the stream at a point 7.5 cm below the tap. Assume that the stream still has a circular cross section and neglect any drag forces acting on the water. (*c*) If turbulent flows are characterized by Reynolds numbers above 2300 or so, how far does the water have to fall before it becomes turbulent? Does this match everyday experience?

52 • **iSOLVE✓** Water flows at 0.65 m/s through a 3-cm-diameter hose. At the end of the hose is a 0.30-cm-diameter nozzle. (*a*) At what speed does the water pass through the nozzle? (*b*) If the pump at one end of the hose and the nozzle at the other end are at the same height, and if the pressure at the nozzle is 1 atm, what is the pressure at the pump? Assume laminar nonviscous flow.

53 • **iSOLVE✓** Water is flowing at 3 m/s in a horizontal pipe under a pressure of 200 kPa. The pipe narrows to half its original diameter. (*a*) What is the speed of flow in the narrow section? (*b*) What is the pressure in the narrow section? (*c*) How do the volume flow rates in the two sections compare? Assume laminar nonviscous flow.

54 • **iSOLVE** The pressure in a section of horizontal pipe with a diameter of 2 cm is 142 kPa. Water flows through the pipe at 2.80 L/s. If the pressure at a certain point is to be reduced to 101 kPa by constricting a section of the pipe, what should the diameter of the constricted section be? Assume laminar nonviscous flow.

55 •• **SSM** Blood flows in an aorta of radius 9 mm at 30 cm/s. (*a*) Calculate the volume flow rate in liters per minute. (*b*) Although the cross-sectional area of a capillary is much smaller than that of the aorta, there are many capillaries, so their total cross-sectional area is much larger. If all the blood from the aorta flows into the capillaries and the speed of flow through the capillaries is 1.0 mm/s, calculate the total cross-sectional area of the capillaries.

56 •• **iSOLVE✓** A large tank of water is tapped a distance h below the water surface by a small pipe as in Figure 13-35. Find the distance x reached by the water flowing out of the pipe. Assume laminar nonviscous flow.

FIGURE 13-35 Problems 56, 61

57 •• **iSOLVE** The $8-billion, 800-mile long Alaskan Pipeline has a capacity of 240,000 m³ of oil per day. Along most of the pipeline the radius is 60 cm. Find the pressure P' at a point where the pipe has a 30-cm radius. Take the pressure in the 60-cm-diameter sections to be $P = 180$ kPa and the density of oil to be 800 kg/m³. Assume laminar nonviscous flow.

58 •• **SSM** Water flows through a Venturi meter like that in Example 13-10 with a pipe diameter of 9.5 cm and a constriction diameter of 5.6 cm. The U-tube manometer is partially filled with mercury. Find the volume flow rate of the water if the difference in the mercury level in the U-tube is 2.40 cm.

59 •• A firefighter holds a hose with a bend in it as in Figure 13-36. Water is expelled from the hose in a stream of radius 1.5 cm at a speed of 30 m/s. (*a*) What mass of water emerges from the hose in 1 s? (*b*) What is the horizontal momentum of this water? (*c*) Before reaching the bend, the water has momentum upward, whereas afterward, its momentum is horizontal. Draw a vector diagram of the initial and final momentum vectors, and find the change in the momentum of the water at the bend during 1 s. From this, find the force exerted on the water by the hose.

FIGURE 13-36 Problem 59

60 •• **iSOLVE✓** A fountain designed to spray a column of water 12 m into the air has a 1-cm-diameter nozzle at ground level. The water pump is 3 m below the ground. The pipe to the nozzle has a diameter of 2 cm. Find the necessary pump pressure. Assume laminar nonviscous flow.

61 ••• In Figure 13-35, H is the depth of the liquid and h is the distance of the opening below the surface of the liquid. (*a*) Find the distance x at which the water strikes the ground as a function of h and H. (*b*) Show that, for a given value of H, there are two values of h (whose average value is $\frac{1}{2}H$) that give the same distance x. (*c*) Show that x is a maximum when $h = \frac{1}{2}H$. What is the value of this maximum distance x? Assume laminar nonviscous flow.

62 •• **SSM** Figure 13-37 shows a Pitot-static tube, a device used for measuring the velocity of a gas. The inner pipe faces the incoming fluid, while the ring of holes in the outer tube are parallel to the gas flow. Show that the speed of the gas is given by $v^2 = 2gh(\rho - \rho_g)/\rho_g$, where ρ is the density of the liquid used in the manometer and ρ_g is the density of the gas.

FIGURE 13-37 Problem 62

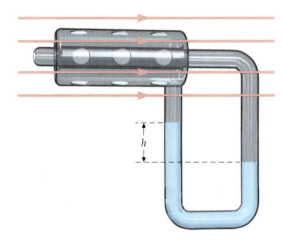

63 •• A siphon is a device for transferring a liquid from container to container. The tube shown in Figure 13-38 must be filled to start the siphon, but once this has been done, fluid will flow through the tube until the liquid surfaces in the containers are at the same level. (a) Using Bernoulli's equation, show that the velocity of water in the tube is $v = \sqrt{2gd}$ (b) What is the pressure at the highest part of the tube? Assume laminar nonviscous flow.

FIGURE 13-38 Problem 63

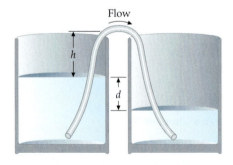

Flow

*Viscous Flow

64 • **SOLVE** A horizontal tube with an inside diameter of 1.2 mm and a length of 25 cm has water flowing through it at 0.30 mL/s. Find the pressure difference required to drive this flow if the viscosity of water is 1.00 mPa·s. Assume laminar flow.

65 • Find the diameter of a tube that would give double the flow rate for the pressure difference in Problem 64.

66 • **SSM** Blood takes about 1.0 s to pass through a 1-mm-long capillary of the human circulatory system. If the diameter of the capillary is 7 μm and the pressure drop is 2.60 kPa, find the viscosity of blood. Assume laminar flow.

67 • **SSM** At Reynolds numbers of about 3×10^5 there is an abrupt transition, where the drag on a sphere abruptly decreases. Estimate the velocity at which this drag crisis occurs for a baseball, and comment on whether or not it should play a role in the physics of the game.

68 ••• Stokes' law states that the drag force on a sphere at very low Reynolds number is given by $F_D = 6\pi\eta av$, where η is the viscosity of the surrounding fluid and a is the radius of the sphere. Using this, find the terminal velocity of ascent for a spherical carbon dioxide bubble of 1-mm diameter rising in a glass of soda (density 1.1×10^3 kg/m³). How long should it take for the bubble to rise the height of a "typical" soda glass? Is this consistent with your experience?

General Problems

69 •• **SSM** Very roughly speaking, the mass of a person should increase as the cube of his or her height—that is, $M = C\rho h^3$, where M is the mass, h is the height, ρ is body density, and C is a person's "coefficient of roundness." Estimate C for an adult male and female, using "typical" values for height and weight. Assume $\rho = 1000$ kg/m³.

70 • Using that weight is proportional to height cubed (see problem 69) what should be the difference in weight for two men, one of whom is 5′ 9″ tall and the other 6′ 0″?

71 • The top of a card table is 80 cm × 80 cm. What is the force exerted on it by the atmosphere? Why doesn't the table collapse?

72 • **SOLVE** ✔ A 4.0-g Ping-Pong ball is attached by a thread to the bottom of a beaker. When the beaker is filled with water so that the ball is totally submerged, the tension in the thread is 2.8×10^{-2} N. Determine the diameter of the ball.

73 • **SOLVE** Seawater has a bulk modulus of 2.3×10^9 N/m². Find the density of seawater at a depth where the pressure is 800 atm if the density at the surface is 1025 kg/m³.

74 • A solid cubical block of edge length 0.6 m is suspended from a spring balance. When the block is submerged in water, the spring balance reads 80 percent of the reading when the block is in air. Determine the density of the block.

75 • **SSM** When submerged in water, a block of copper has an apparent weight of 56 N. What fraction of this copper block will be submerged if it is floated on a pool of mercury?

76 • **SOLVE** A 4.5-kg block of material floats on ethanol with 10 percent of its volume above the liquid surface. What fraction of the volume of this block will be submerged if it is floated on water?

77 • What is the buoyant force on your body when you are floating (a) in a freshwater lake (specific gravity = 1.00) and (b) in the ocean (specific gravity = 1.03)?

78 • Suppose that when you are floating in fresh water, 96 percent of your body volume is submerged. What is the volume of water your body displaces when it is fully submerged?

79 •• A block of wood of 1.5-kg mass floats on water with 68 percent of its volume submerged. A lead block is placed on the wood and the wood is then fully submerged with the lead entirely out of the water. Find the mass of the lead block.

80 •• **SSM** **SOLVE** ✓ A Styrofoam cube, 25 cm on an edge, is placed on one pan of a balance. The balance is in equilibrium when a 20-g mass of brass is placed on the opposite pan of the balance. Find the mass of the Styrofoam cube.

81 •• **SOLVE** A spherical shell of copper with an outer diameter of 12 cm floats on water with half its volume above the water's surface. Determine the inner diameter of the shell.

82 •• A beaker filled with water is balanced on the left pan of a balance. A cube 4 cm on an edge is attached to a string and lowered into the water so that it is completely submerged. The cube is not touching the bottom of the beaker. A weight of mass m is added to the right pan to restore equilibrium. What is m?

83 •• **SSM** Crude oil has a viscosity of about 0.8 Pa·s at normal temperature. A 50-km pipeline is to be constructed from an oil field to a tanker terminal. The pipeline is to deliver oil at the terminal at a rate of 500 L/s and the flow through the pipeline is to be laminar to minimize the pressure needed to push the fluid through the pipeline. Assuming that the density of crude oil is 700 kg/m³, estimate the diameter of the pipeline that should be used.

84 •• **SOLVE** Water flows through the pipe in Figure 13-39 and exits to the atmosphere at C. The diameter of the pipe is 2.0 cm at A, 1.0 cm at B, and 0.8 cm at C. The gauge pressure in the pipe at A is 1.22 atm and the flow rate is 0.8 L/s. The vertical pipes are open to the air. Find the level of the liquid–air interfaces in the two vertical pipes. Assume laminar nonviscous flow.

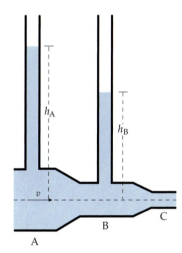

FIGURE 13-39 Problems 84, 85

85 •• Repeat Problem 84 with the flow rate reduced to 0.6 L/s and the size of the opening at C reduced so that the pressure in the pipe at A remains unchanged.

86 •• **SSM** **SOLVE** Figure 13-40 is a sketch of an *aspirator*, a simple device that can be used to achieve a partial vacuum in a reservoir connected to the vertical tube at B. An aspirator attached to the end of a garden hose may be used to deliver soap or fertilizer from the reservoir. Suppose that the diameter at A is 2.0 cm and at C, where the water exits to the atmosphere, it is 1.0 cm. If the flow rate is 0.5 L/s and

the gauge pressure at A is 0.187 atm, what diameter of the constriction at B will achieve a pressure of 0.1 atm in the container? Assume laminar nonviscous flow.

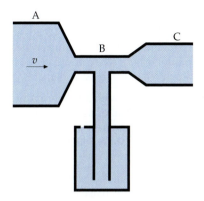

FIGURE 13-40 Problem 86

87 •• A cylindrical buoy at the entrance of a harbor has a diameter of 0.9 m and a height of 2.6 m. The mass of the buoy is 600 kg. It is attached to the bottom of the sea with a nylon cable of negligible mass. The specific gravity of the seawater is 1.025. (*a*) How much of the buoy is visible when the cable is slack? (*b*) If a tsunami completely submerges the buoy, what is the tension in the taut cable? (*c*) If the cable breaks, what is the initial upward acceleration of the buoy?

88 •• **SOLVE** Two connected vessels contain a liquid of density ρ_0 (Figure 13-41). The cross-sectional areas of the vessels are A and $3A$. Find the change in elevation of the liquid level if an object of mass m and density $\rho' = 0.8\rho_0$ is placed into one of the vessels.

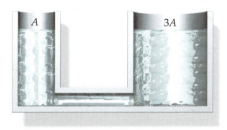

FIGURE 13-41 Problem 88

89 •• If an oil-filled manometer ($\rho = 900$ kg/m³) can be read to ±0.05 mm, what is the smallest pressure change that can be detected?

90 •• **SOLVE** ✓ A U-tube is filled with water until the liquid level is 28 cm above the bottom of the tube. An oil of specific gravity 0.78 is now poured into one arm of the U-tube until the level of the water in the other arm of the tube is 34 cm above the bottom of the tube. Find the levels of the oil–water and oil–air interfaces in the other arm of the tube.

91 •• A U-tube contains liquid of unknown specific gravity. An oil of density 800 kg/m³ is poured into one arm of the tube until the oil column is 12 cm high. The oil–air interface is then 5.0 cm above the liquid level in the other arm of the U-tube. Find the specific gravity of the liquid.

92 •• A lead block is suspended from the underside of a 0.5-kg block of wood of specific gravity 0.7. If the upper surface of the wood is just level with the surface of the water, what is the mass of the lead block?

93 •• [SSM] A helium balloon can just lift a load of 750 N. The skin of the balloon has a mass of 1.5 kg. (a) What is the volume of the balloon? (b) If the volume of the balloon were twice that found in Part (a), what would be the initial acceleration of the balloon when it carried a load of 900 N?

94 •• A hollow sphere with an inner radius R and outer radius $2R$ is made of material of density ρ_0 and is floating in a liquid of density $2\rho_0$. The interior is now filled with material of density ρ' so that the sphere just floats completely submerged. Find ρ'.

95 •• [SSM] As mentioned in the discussion of *the law of atmospheres*, the fractional decrease in atmospheric pressure is proportional to the change in altitude. Expressed as a differential equation we have $dP/P = -C/dh$, where C is a constant. (a) Show that $P(h) = P_0 e^{-Ch}$ is a solution of the differential equation. (b) Show that if $\Delta h \ll h_0$ where $h_0 = 1/C$, then $P(h + \Delta h) \approx P(h)(1 - \Delta h/h_0)$. (c) Given that the pressure at $h = 5.5$ km is half that at sea level, find the constant C.

96 •• A submarine has a total mass of 2.4×10^6 kg, including crew and equipment. The vessel consists of two parts, the pressure hull, which has a volume of 2×10^3 m³, and the ballast tanks, which have a volume of 4×10^2 m³. When the sub cruises on the surface, the ballast tanks are filled with air; when it is cruising below the surface, seawater is admitted into the tanks. (a) What fraction of the submarine's volume is above the water surface when the tanks are filled with air? (b) How much water must be admitted into the tanks to give the submarine neutral buoyancy? Neglect the mass of air in the tanks and use 1.025 as the specific gravity of seawater.

97 •• A marine salvage crew raises a crate that measures 1.4 m × 0.75 m × 0.5 m. The average density of the empty crate is the same as seawater, 1.025×10^3 kg/m³, and its mass when empty is 32 kg. The crate contains gold bullion that fills 36 percent of its volume; the remaining volume is filled with seawater. (a) What is the tension in the cable that raises the crate and bullion while the crate is below the surface of the sea? (b) What is the tension in the cable while the crate is lifted to the deck of the ship if (1) none of the seawater leaks out of the crate, and (2) the crate is lifted so slowly that all of the seawater leaks out of the crate?

98 ••• When the hydrometer of Problem 48 (Figure 13-34) is placed in a liquid whose specific gravity is greater than some minimum value, the device floats with part of the glass tube above the liquid level. Consider a hydrometer that has a spherical bulb 2.4 cm in diameter. The glass tube attached to the bulb is 20 cm long and has a diameter of 7.5 mm. The mass of the hydrometer before lead pellets are dropped into the bulb and the tube is sealed is 7.28 g. (a) What mass of lead should be placed in the bulb so that the hydrometer just floats in a liquid of specific gravity 0.78? (b) If the hydrometer is now placed in water, what is the length of the tube that shows above the surface of the water? (c) The hydrometer is placed in a liquid of unknown specific gravity; the length of the tube above the surface of the liquid is 12.2 cm. Determine the specific gravity of the liquid.

99 ••• A large root beer keg of height H and cross-sectional area A_1 is filled with root beer. The top is open to atmospheric pressure. At the bottom is a spigot opening of area A_2, which is much smaller than A_1. (a) Show that when the height of the root beer is h, the speed of the root beer leaving the spigot is approximately $\sqrt{2gh}$. (b) Show that for the approximation $A_2 \ll A_1$, the rate of change of the height h of the root beer is given by

$$\frac{dh}{dt} = -\frac{A_2}{A_1}(2gh)^{1/2}$$

(c) Find h as a function of time if $h = H$ at $t = 0$. (d) Find the total time needed to drain the keg if $H = 2$ m, $A_1 = 0.8$ m², and $A_2 = 10^{-4}A_1$. Assume laminar nonviscous flow.

Oscillations

THIS BOAT IS RISING AND FALLING ON THE SWELLS OF THE SEA. ITS MOTION IS AN EXAMPLE OF OSCILLATORY MOTION. THE MAXIMUM CHANGE IN THE VERTICAL POSITION OF THE BOAT CAN BE READILY MEASURED, AS CAN THE TIME FOR THE BOAT TO COMPLETE ONE CYCLE OF THIS UP AND DOWN MOTION.

? **How can the vertical position of the boat be expressed as a function of time? (See Example 14-1.)**

Oscillation occurs when a system is disturbed from a position of stable equilibrium. There are many familiar examples: boats bob up and down, clock pendulums swing back and forth, and the strings and reeds of musical instruments vibrate. Other, less familiar examples are the oscillations of air molecules in a sound wave and the oscillations of electric currents in radios and television sets.

➤ **In this chapter, we deal mostly with simple harmonic motion—the most basic type of oscillatory motion. Applying the kinematics and dynamics of simple harmonic motion provides the analysis of the oscillatory motion of a variety of interesting systems. In some situations dissipative forces dampen the oscillatory motion, but in other situations driving forces sustain the motion by compensating for the damping.**

14-1 Simple Harmonic Motion

A common, very important, and very basic kind of oscillatory motion is **simple harmonic motion** such as the motion of an object attached to a spring (Figure 14-1). In equilibrium, the spring exerts no force on the object. When the object is displaced an amount x from its equilibrium position, the spring exerts a force $-kx$, as given by Hooke's law[†]:

$$F_x = -kx \qquad\qquad 14\text{-}1$$

where k is the force constant of the spring, a measure of the spring's stiffness. The minus sign indicates that the force is a restoring force; that is, it is opposite to the direction of the displacement from the equilibrium position. Combining Equation 14-1 with Newton's second law ($F_x = ma_x$), we have

$$-kx = ma_x$$

or

$$a_x = -\frac{k}{m}x \quad \left(\text{or}\quad \frac{d^2x}{dt^2} = -\frac{k}{m}x\right) \qquad\qquad 14\text{-}2$$

The acceleration is proportional to the displacement and is oppositely directed. This is the defining characteristic of simple harmonic motion and can be used to identify systems that will exhibit it:

> Whenever the acceleration of an object is proportional to its displacement and is oppositely directed, the object will move with simple harmonic motion.

CONDITIONS FOR SIMPLE HARMONIC MOTION IN TERMS OF ACCELERATION

Because the acceleration is proportional to the net force, whenever the net force on an object is proportional to its displacement and is oppositely directed, the object will move with simple harmonic motion.

The time it takes for a displaced object to execute a complete cycle of oscillatory motion—from one extreme to the other extreme and back—is called the **period** T. The reciprocal of the period is the **frequency** f, which is the number of cycles per second:

$$f = \frac{1}{T} \qquad\qquad 14\text{-}3$$

The unit of frequency is the cycle per second (cy/s), which is called a **hertz** (Hz). For example, if the time for one complete cycle of oscillation is 0.25 s, the frequency is 4 Hz.

Figure 14-2 shows how we can experimentally obtain x versus t for a mass on a spring. The general equation for such a curve is

$$x = A\cos(\omega t + \delta) \qquad\qquad 14\text{-}4$$

POSITION IN SIMPLE HARMONIC MOTION

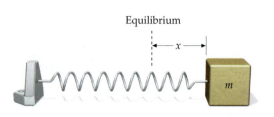

FIGURE 14-1 An object and spring on a frictionless surface. The displacement x, measured from the equilibrium position, is positive if the spring is stretched and negative if the spring is compressed.

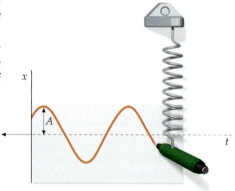

FIGURE 14-2 A marking pen is attached to a mass on a spring, and the paper is pulled to the left. As the paper moves with constant speed, the pen traces out the displacement x as a function of time t. (Here we have chosen x to be positive when the spring is compressed.)

[†] Hooke's law is introduced in Chapter 4, Section 4.

where A, ω, and δ are constants. The maximum displacement x_{max} from equilibrium is called the **amplitude** A. The argument of the cosine function, $\omega t + \delta$, is called the **phase** of the motion, and the constant δ is called the **phase constant,** which is the phase at $t = 0$. (Note that $\cos(\omega t + \delta) = \sin(\omega t + \delta + \pi/2)$; thus, whether the equation is expressed as a cosine function or a sine function simply depends on the phase of the oscillation at the moment we designate to be $t = 0$.) If we have just one oscillating system, we can always choose $t = 0$ at which $\delta = 0$. If we have two systems oscillating with the same amplitude and frequency but different phase, we can choose $\delta = 0$ for one of them. The equations for the two systems are then

$$x_1 = A \cos(\omega t)$$

and

$$x_2 = A \cos(\omega t + \delta)$$

If the phase difference δ is 0 or an integer times 2π, then $x_2 = x_1$ and the systems are said to be in phase. If the phase difference δ is π or an odd integer times π, then $x_2 = -x_1$, and the systems are said to be 180° out of phase.

The swaying of the Citicorp Building in New York City during high winds is reduced by this tuned-mass damper mounted on an upper floor. It consists of a 400-ton sliding block connected to the building by a spring. The spring constant is chosen so that the natural frequency of the spring–block system is the same as the natural sway frequency of the building. Set into motion by winds, the building and damper oscillate 180° out of phase with each other, thereby significantly reducing the swaying.

We can show that Equation 14-4 is a solution of Equation 14-2 by differentiating x twice with respect to time. The first derivative of x gives the velocity v:

$$v = \frac{dx}{dt} = -\omega A \sin(\omega t + \delta) \qquad 14\text{-}5$$

VELOCITY IN SIMPLE HARMONIC MOTION

Differentiating velocity with respect to time gives the acceleration:

$$a = \frac{dv}{dt} = \frac{d^2x}{dt^2} = -\omega^2 A \cos(\omega t + \delta) \qquad 14\text{-}6$$

Substituting x for $A \cos(\omega t + \delta)$ (see Equation 14-4) gives

$$a = -\omega^2 x \qquad\qquad \text{14-7}$$

Comparing $a = -\omega^2 x$ (Equation 14-7) with $a = -(k/m)x$ (Equation 14-2), we see that $x = A \cos(\omega t + \delta)$ is a solution of Equation 14-2 (which can be expressed $d^2x/dt^2 = -(k/m)x$) if

$$\omega = \sqrt{\frac{k}{m}} \qquad\qquad \text{14-8}$$

The amplitude A and the phase constant δ can be determined from the initial position x_0 and the initial velocity v_0 of the system. Setting $t = 0$ in $x = A \cos(\omega t + \delta)$ gives

$$x_0 = A \cos \delta \qquad\qquad \text{14-9}$$

Similarly, setting $t = 0$ in $v = dx/dt = -A\omega \sin(\omega t + \delta)$ gives

$$v_0 = -A\omega \sin \delta \qquad\qquad \text{14-10}$$

These equations can be solved for A and δ in terms of x_0 and v_0.

The period T is the shortest time satisfying the relation

$$x(t) = x(t + T)$$

for all t. Substituting into this relation using Equation 14-4 gives

$$A \cos(\omega t + \delta) = A \cos[\omega(t + T) + \delta]$$
$$= A \cos(\omega t + \delta + \omega T)$$

The cosine (and sine) function repeats in value when the phase increases by 2π, so

$$\omega T = 2\pi \qquad \left(\text{or} \quad \omega = \frac{2\pi}{T} \right)$$

The constant ω is called the **angular frequency.** It has units of radians per second and dimensions of inverse time, the same as angular speed, which is also designated by ω. Substituting $2\pi/T$ for ω in Equation 14-4 gives

$$x = A \cos\left(2\pi \frac{t}{T} + \delta \right)$$

We can see by inspection that each time t increases by T, the phase increases by 2π and one cycle of the motion is completed.

The frequency is the reciprocal of the period:

$$f = \frac{1}{T} = \frac{\omega}{2\pi} \qquad\qquad \text{14-11}$$

Because $\omega = \sqrt{k/m}$, the frequency and period of an object on a spring are related to the force constant k and the mass m by

$$f = \frac{1}{T} = \frac{1}{2\pi}\sqrt{\frac{k}{m}}$$ 14-12

FREQUENCY AND PERIOD FOR AN OBJECT ON A SPRING

The frequency increases with increasing k (spring stiffness) and decreases with increasing mass.

Astronaut Alan L. Bean measures his body mass during the second Skylab mission by sitting in a seat attached to a spring and oscillating back and forth. The total mass of the astronaut plus the seat is related to his frequency of vibration by Equation 14-12.

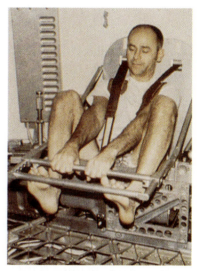

RIDING THE WAVES   **E X A M P L E   1 4 - 1**

You are on a boat, which is bobbing up and down. The boat's vertical displacement y is given by

$$y = (1.2\ \text{m})\cos\left(\frac{1}{2\,\text{s}}t + \frac{\pi}{6}\right)$$

(a) Find the amplitude, angular frequency, phase constant, frequency, and period of the motion. (b) Where is the boat at $t = 1$ s? (c) Find the velocity and acceleration as functions of time t. (d) Find the initial values of the position, velocity, and acceleration of the boat.

PICTURE THE PROBLEM We find the quantities asked for in (a) by comparing the equation of motion

$$y = (1.2\ \text{m})\cos\left(\frac{1}{2\,\text{s}}t + \frac{\pi}{6}\right)$$

with the standard equation for simple harmonic motion, Equation 14-4. The velocity and acceleration are found by differentiating $y(t)$.

(a) 1. Compare this equation with Equation 14-4, $y = A\cos(\omega t + \delta)$, to get A, ω, and δ:

$$y = (1.2\ \text{m})\cos\left(\frac{1}{2\,\text{s}}t + \frac{\pi}{6}\right)$$

$$A = \boxed{1.2\ \text{m}} \qquad \omega = \boxed{1/2\ \text{rad/s}} \qquad \delta = \boxed{\pi/6\ \text{rad}}$$

2. The frequency and period are found from ω:

$$f = \frac{\omega}{2\pi} = \boxed{0.0796\ \text{Hz}} \qquad T = \frac{1}{f} = \boxed{12.6\ \text{s}}$$

(b) Set $t = 1$ s to find the boat's position:

$$y = (1.2\ \text{m})\cos\left[\frac{1}{2\,\text{s}}(1\ \text{s}) + \frac{\pi}{6}\right] = \boxed{0.624\ \text{m}}$$

(c) The velocity and acceleration are obtained from the position by differentiation with respect to time:

$$v_y = \frac{dy}{dt} = \frac{d}{dt}[A\cos(\omega t + \delta)]$$

$$= -\omega A \sin(\omega t + \delta)$$

$$= -\frac{1}{2\,\text{s}}(1.2\ \text{m})\sin\left(\frac{1}{2\,\text{s}}t + \frac{\pi}{6}\right)$$

$$= \boxed{-(0.6\ \text{m/s})\sin\left(\frac{1}{2\,\text{s}}t + \frac{\pi}{6}\right)}$$

$$a_y = \frac{dv_y}{dt} = \frac{d}{dt}[-\omega A \sin(\omega t + \delta)]$$

$$= -\omega^2 A \cos(\omega t + \delta)$$

$$= -\left(\frac{1}{2\,\text{s}}\right)^2 (1.2\,\text{m}) \cos\left(\frac{1}{2\,\text{s}}t + \frac{\pi}{6}\right)$$

$$= \boxed{-(0.3\,\text{m/s}^2)\cos\left(\frac{1}{2\,\text{s}}t + \frac{\pi}{6}\right)}$$

(d) Set $t = 0$ to find y_0, v_{y0}, and a_{y0}:

$$y_0 = (1.2\,\text{m})\cos\frac{\pi}{6} = \boxed{1.04\,\text{m}}$$

$$v_{y0} = -(0.6\,\text{m/s})\sin\frac{\pi}{6} = \boxed{-0.300\,\text{m/s}}$$

$$a_{y0} = -(0.3\,\text{m/s}^2)\cos\frac{\pi}{6} = \boxed{-0.260\,\text{m/s}^2}$$

EXERCISE A 0.8-kg object is attached to a spring of force constant $k = 400\,\text{N/m}$. Find the frequency and period of motion of the object when it is displaced from equilibrium. (*Answer* $f = 3.56\,\text{Hz}$, $T = 0.281\,\text{s}$)

Figure 14-3 shows two identical masses attached to identical springs and resting on a frictionless surface. One spring is stretched 10 cm and the other 5 cm. If they are released at the same time, which object reaches the equilibrium position first?

According to Equation 14-12, the period depends only on k and m and not on the amplitude. Since k and m are the same for both systems, the periods are the same. Thus, the objects reach the equilibrium position at the same time. The second object has twice as far to go to reach equilibrium, but it will also have twice the speed at any given instant. Figure 14-4 shows a sketch of the position functions for the two objects. This illustrates an important general property of simple harmonic motion:

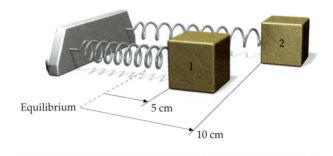

FIGURE 14-3 Two identical mass–spring systems.

The frequency and period of simple harmonic motion are independent of the amplitude.

The fact that the frequency in simple harmonic motion is independent of the amplitude has important consequences in many fields. In music, for example, it means that when a note is struck on the piano, the pitch (which corresponds to the frequency) does not depend on how loudly the note is played (which corresponds to the amplitude).[†] If changes in amplitude had a large effect on the frequency, then musical instruments would be unplayable.

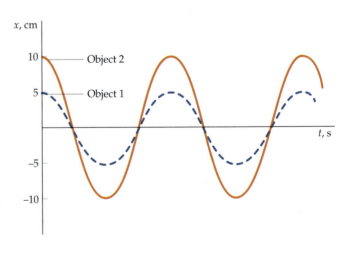

FIGURE 14-4 Plots of x versus t for the systems in Figure 14-3. Both reach their equilibrium positions at the same time.

---

† For many musical instruments, there is a slight dependence of frequency on amplitude. The vibration of an oboe reed, for example, is not exactly simple harmonic, thus its pitch depends slightly on how hard it is blown. This effect can be corrected for by a skilled musician.

AN OSCILLATING OBJECT   **EXAMPLE   14-2**

An object oscillates with angular frequency $\omega = 8.0$ rad/s. At $t = 0$, the object is at $x = 4$ cm with an initial velocity $v = -25$ cm/s. (*a*) Find the amplitude and phase constant for the motion. (*b*) Write x as a function of time.

PICTURE THE PROBLEM The initial position and velocity give us two equations from which to determine the amplitude A and the phase constant δ.

(*a*) 1. The initial position and velocity are related to the amplitude and phase constant. The position is given by Equation 14-4. The velocity is found by taking the derivative with respect to time:

$$x = A \cos(\omega t + \delta)$$

and

$$v = \frac{dx}{dt} = -\omega A \sin(\omega t + \delta)$$

2. At $t = 0$ the position and velocity are:

$$x_0 = A \cos \delta \quad \text{and} \quad v_0 = -\omega A \sin \delta$$

3. Divide these equations to eliminate A:

$$\frac{v_0}{x_0} = \frac{-\omega A \sin \delta}{A \cos \delta} = -\omega \tan \delta$$

4. Substituting numerical values yields δ:

$$\tan \delta = -\frac{v_0}{\omega x_0}$$

so

$$\delta = \tan^{-1}\left(-\frac{v_0}{\omega x_0}\right)$$

$$= \tan^{-1}\left[-\frac{-25 \text{ cm/s}}{(8.0 \text{ rad/s})(4 \text{ cm})}\right]$$

$$= \boxed{0.663 \text{ rad}}$$

5. The amplitude can be found using either the x_0 or v_0 equation. Here we use x_0:

$$A = \frac{x_0}{\cos \delta} = \frac{4 \text{ cm}}{\cos 0.663} = \boxed{5.08 \text{ cm}}$$

(*b*) Comparing with Equation 14-4 yields x:

$$x = \boxed{(5.08 \text{ cm}) \cos[(8.0 \text{ s}^{-1})t + 0.663]}$$

When the phase constant is $\delta = 0$, Equations 14-4, 14-5, and 14-6 then become

$$x = A \cos \omega t \qquad\qquad\qquad 14\text{-}13a$$
$$v = -\omega A \sin \omega t \qquad\qquad\quad 14\text{-}13b$$

and

$$a = -\omega^2 A \cos \omega t \qquad\qquad 14\text{-}13c$$

These functions are plotted in Figure 14-5.

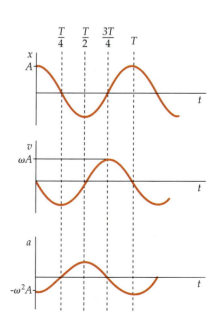

FIGURE 14-5 Plots of x, v, and a as functions of time t for $\delta = 0$. At $t = 0$, the displacement is maximum, the velocity is zero, and the acceleration is negative and equal to $-\omega^2 A$. The velocity becomes negative as the object moves back toward its equilibrium position. After one quarter-period ($t = T/4$), the object is at equilibrium, $x = 0$, $a = 0$, and the speed has its maximum value of ωA. At $t = T/2$, the displacement is $-A$, the velocity is again zero, and the acceleration is $+\omega^2 A$. At $t = 3T/4$, $x = 0$, $a = 0$, and $v = +\omega A$.

A BLOCK ON A SPRING   **E X A M P L E   1 4 - 3**   **Try It Yourself**

A 2-kg block is attached to a spring as in Figure 14-1. The force constant of the spring is $k = 196$ N/m. The block is held a distance 5 cm from the equilibrium position and is released at $t = 0$. (*a*) Find the angular frequency ω, the frequency f, and the period T. (*b*) Write x as a function of time.

Cover the column to the right and try these on your own before looking at the answers.

Steps   **Answers**

(*a*) 1. Calculate ω from $\omega = \sqrt{k/m}$. $\omega = \boxed{9.90 \text{ rad/s}}$

    2. Use your result to find f and T. $f = \boxed{1.58 \text{ Hz}}$ $T = \boxed{0.635 \text{ s}}$

    3. Find A and δ from the initial conditions. $A = 5$ cm $\delta = 0$

(*b*) Write $x(t)$ using your results for A, ω, and δ. $x = \boxed{(5 \text{ cm}) \cos[(9.90 \text{ s}^{-1})t]}$

SPEED AND ACCELERATION OF AN OBJECT ON A SPRING   **E X A M P L E   1 4 - 4**

Consider an object on a spring whose position is given by $x = (5$ cm$) \cos$ $(9.90 \text{ s}^{-1} t)$. (*a*) What is the maximum speed of the object? (*b*) When does this maximum speed first occur? (*c*) What is the maximum of the acceleration of the object? (*d*) When does maximum acceleration first occur after $t = 0$?

PICTURE THE PROBLEM Because the object is released from rest, $\delta = 0$, and the position, velocity, and acceleration are given by Equations 14-13*a*, *b*, and *c*.

(*a*) 1. Equation 14-13*a*, with $\delta = 0$, gives the position. We get the velocity by taking the derivative with respect to time:

$$x = A \cos \omega t$$

so

$$v = \frac{dx}{dt} = -\omega A \sin \omega t$$

    2. Maximum speed occurs when $|\sin \omega t| = 1$:

$$|v| = \omega A |\sin \omega t|$$

so

$$|v|_{max} = \omega A = (9.90 \text{ rad/s})(5 \text{ cm})$$

$$= \boxed{49.5 \text{ cm/s}}$$

(*b*) 1. $|\sin \omega t| = 1$ first occurs when $\omega t = \pi/2$:

$$|\sin \omega t| = 1 \Rightarrow \omega t = \frac{\pi}{2}, \frac{3\pi}{2}, \frac{5\pi}{2}, \cdots$$

    2. Solve for t when $\omega t = \pi/2$:

$$t = \frac{\pi}{2\omega} = \frac{\pi}{2(9.90 \text{ s}^{-1})} = \boxed{0.159 \text{ s}}$$

(*c*) 1. We find the acceleration by taking the derivative of the velocity, obtained in step 1 of Part (*a*):

$$a = \frac{dv}{dt} = -\omega^2 A \cos \omega t$$

2. Maximum acceleration corresponds to cos $\omega t = -1$. $a_{max} = \omega^2 A = (9.90 \text{ rad/s})^2 (5 \text{ cm})$

$$= \boxed{490 \text{ cm/s}^2 \approx \tfrac{1}{2} g}$$

(d) The maximum acceleration occurs when $|\cos \omega t| = 1$, $\omega t = \pi$
which is when $\omega t = 0, \pi, 2\pi, \ldots$:

so

$$t = \frac{\pi}{\omega} = \frac{\pi}{9.90 \text{ s}^{-1}} = \boxed{0.317 \text{ s}}$$

REMARKS The maximum speed first occurs after one quarter-period

$$t = \frac{\pi}{2\omega} = \frac{\pi}{2(2\pi/T)} = \frac{1}{4}T$$

The maxima of the magnitude of the acceleration occur when $\omega t = 0, \pi, 2\pi, \ldots$.
These correspond to $t = 0, \frac{1}{2}T, \frac{2}{2}T, \frac{3}{2}T, \ldots$.

Simple Harmonic Motion and Circular Motion

There is a relation between simple harmonic motion and circular motion with constant speed. Imagine a particle moving with constant speed v in a circle of radius A (Figure 14-6a). Its angular displacement relative to the x axis is

$$\theta = \omega t + \delta \qquad\qquad 14\text{-}14$$

where δ is the angular displacement at time $t = 0$ and $\omega = v/A$ is the angular speed of the particle. The x component of the particle's position (Figure 14-6b) is

$$x = A \cos \theta = A \cos(\omega t + \delta)$$

which is the same as Equation 14-4 for simple harmonic motion.

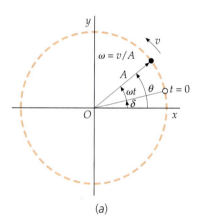

When a particle moves with constant speed in a circle, its projection onto a diameter of the circle moves with simple harmonic motion (Figure 14-6).

The speed of a particle moving in a circle is $r\omega$, where r is the radius. For the particle in Figure 14-6b, $r = A$, so its speed is $A\omega$. The projection of the velocity vector onto the x axis gives $v_x = -v \sin \theta$. Substituting for v and θ gives

$$v_x = -\omega A \sin \theta = -\omega A \sin(\omega t + \delta)$$

which is the same as Equation 14-5 for simple harmonic motion.

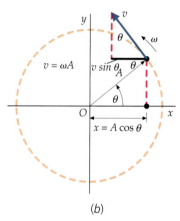

FIGURE 14-6 A particle moves in a circular path with constant speed. (*a*) Its x component of position describes simple harmonic motion, and (*b*) its x component of velocity describes the velocity of the simple harmonic motion.

Bubbles foaming off the edge of a rotating propeller that is moving through water produce a sinusoidal pattern.

14-2 Energy in Simple Harmonic Motion

When an object on a spring undergoes simple harmonic motion, the system's potential energy and kinetic energy vary with time. Their sum, the total mechanical energy $E = K + U$, is constant. Consider an object a distance x from equilibrium, acted on by a restoring force $-kx$. The system's potential energy is

$$U = \tfrac{1}{2}kx^2$$

This is Equation 6-21. For simple harmonic motion, $x = A\cos(\omega t + \delta)$. Substituting gives

$$U = \tfrac{1}{2}kA^2\cos^2(\omega t + \delta) \qquad\qquad 14\text{-}15$$

POTENTIAL ENERGY IN SIMPLE HARMONIC MOTION

The kinetic energy of the system is

$$K = \tfrac{1}{2}mv^2$$

where m is the object's mass and v is its speed. For simple harmonic motion, $v_x = -\omega A\sin(\omega t + \delta)$. Substituting gives

$$K = \tfrac{1}{2}m\omega^2 A^2\sin^2(\omega t + \delta)$$

Then using $\omega^2 = k/m$,

$$K = \tfrac{1}{2}kA^2\sin^2(\omega t + \delta) \qquad\qquad 14\text{-}16$$

KINETIC ENERGY IN SIMPLE HARMONIC MOTION

The total mechanical energy is the sum of the potential and kinetic energies:

$$E_{\text{total}} = U + K = \tfrac{1}{2}kA^2\cos^2(\omega t + \delta) + \tfrac{1}{2}kA\sin^2(\omega t + \delta)$$
$$= \tfrac{1}{2}kA^2\left[\cos^2(\omega t + \delta) + \sin^2(\omega t + \delta)\right]$$

Since $\sin^2(\omega t + \delta) + \cos^2(\omega t + \delta) = 1$,

$$E_{\text{total}} = \tfrac{1}{2}kA^2 \qquad\qquad 14\text{-}17$$

TOTAL MECHANICAL ENERGY IN SIMPLE HARMONIC MOTION

This equation reveals an important general property of simple harmonic motion:

The total mechanical energy in simple harmonic motion is proportional to the square of the amplitude.

For an object at its maximum displacement, the total energy is all potential energy. As the object moves toward its equilibrium position, the kinetic energy of the system increases and its potential energy decreases. As it moves through its equilibrium position, the kinetic energy of the object is maximum, the potential energy of the system is zero, and the total energy is kinetic.

As the object moves past the equilibrium point, its kinetic energy begins to decrease, and the potential energy of the system increases until the object again stops momentarily at its maximum displacement (now in the other direction). At all times, the sum of the potential and kinetic energies is constant. Figures 14-7b and c show plots of U and K versus time. These curves have the same shape except that one is zero when the other is maximum. Their average values over one or more cycles are equal, and because U + K = E, their average values are given by

$$U_{av} = K_{av} = \tfrac{1}{2}E \qquad\qquad 14\text{-}18$$

In Figure 14-8, the potential energy U is graphed as a function of x. The total energy E_{total} is constant and is therefore plotted as a horizontal line. This line intersects the potential-energy curve at $x = A$ and $x = -A$. These are the points at which oscillating objects reverse direction and head back toward the equilibrium position, and are called the **turning points**. Because $U \le E_{total}$, the motion is restricted to $A \le x \le +A$.

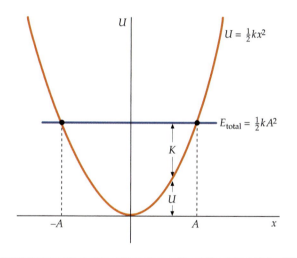

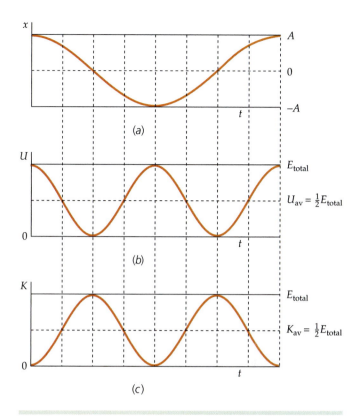

FIGURE 14-7 Plots of x, U, and K versus t.

FIGURE 14-8 The potential-energy function $U = \tfrac{1}{2}kx^2$ for an object of mechanical mass m on a (massless) spring of force constant k. The horizontal blue line represents the total mechanical energy E_{total} for an amplitude of A. The kinetic energy K is represented by the vertical distance $K = E_{total} - U$. Since $E_{total} \ge U$, the motion is restricted to $-A \le x \le +A$.

ENERGY AND SPEED OF AN OSCILLATING OBJECT   **EXAMPLE   14-5**

A 3-kg object attached to a spring oscillates with an amplitude of 4 cm and a period of 2 s. (*a*) What is the total energy? (*b*) What is the maximum speed of the object? (*c*) At what position x_1 is the speed equal to half its maximum value?

PICTURE THE PROBLEM (*a*) The total energy can be found from the amplitude and force constant, and the force constant can be found from the mass and period. (*b*) The maximum speed occurs when the kinetic energy equals the total energy. (*c*) We can relate the position to the speed by using conservation of energy.

(*a*) 1. Write the total energy E in terms of the force constant k and amplitude A:

$$E = \tfrac{1}{2}kA^2$$

2. The force constant is related to the period and mass:

$$k = m\omega^2 = m\left(\frac{2\pi}{T}\right)^2$$

3. Substitute the given values to find E:

$$E = \frac{1}{2}kA^2 = \frac{1}{2}m\left(\frac{2\pi}{T}\right)^2 A^2$$

$$= \frac{1}{2}(3\text{ kg})\left(\frac{2\pi}{2\text{ s}}\right)^2 (0.04\text{ m})^2 = \boxed{2.37 \times 10^{-2}\text{ J}}$$

(b) To find v_{max}, set the kinetic energy equal to the total energy and solve for v:

$$\tfrac{1}{2} mv_{max}^2 = E$$

so

$$v_{max} = \sqrt{\frac{2E}{m}} = \sqrt{\frac{2(2.37 \times 10^{-2}\,\text{J})}{3\,\text{kg}}}$$

$$= \boxed{0.126\ \text{m/s}}$$

(c) 1. Conservation of energy relates the position x to the speed v:

$$E = \tfrac{1}{2} mv^2 + \tfrac{1}{2} kx^2$$

2. Substitute $v = \tfrac{1}{2} v_{max}$ and solve for x_1. It is convenient to find x in terms of E and then write $E = \tfrac{1}{2} kA^2$ to obtain an expression for x in terms of A:

$$E = \tfrac{1}{2} m \left(\tfrac{1}{2} v_{max}\right)^2 + \tfrac{1}{2} kx_1^2$$

$$= \tfrac{1}{4}\left(\tfrac{1}{2} mv_{max}^2\right) + \tfrac{1}{2} kx_1^2 = \tfrac{1}{4} E + \tfrac{1}{2} kx_1^2$$

so

$$\tfrac{1}{2} kx_1^2 = E - \tfrac{1}{4} E = \tfrac{3}{4} E$$

and

$$x_1 = \sqrt{\frac{3E}{2k}} = \sqrt{\frac{3}{2k}\left(\tfrac{1}{2} kA^2\right)} = \frac{\sqrt{3}}{2} A$$

$$= \frac{\sqrt{3}}{2}(4\ \text{cm}) = \boxed{3.46\ \text{cm}}$$

EXERCISE Calculate ω for this example and find v_{max} from $v_{max} = \omega A$. (*Answer* $\omega = 3.14\ \text{rad/s}$, $v_{max} = 0.126\ \text{m/s}$)

EXERCISE An object of mass 2 kg is attached to a spring of force constant 40 N/m. The object is moving at 25 cm/s when it is at its equilibrium position. (*a*) What is the total energy of the object? (*b*) What is the amplitude of the motion? (*Answer* (*a*) $E_{total} = \tfrac{1}{2} mv_{max}^2 = 0.0625\ \text{J}$ (*b*) $A = \sqrt{2E_{total}/k} = 5.59\ \text{cm}$)

*General Motion Near Equilibrium

Simple harmonic motion typically occurs when a particle is displaced slightly from a position of stable equilibrium. Figure 14-9 is a graph of the potential energy U versus x for a force that has a position of stable equilibrium and a position of unstable equilibrium. As discussed in Chapter 6, the maximum at x_2 on Figure 14-9 corresponds to unstable equilibrium, whereas the minimum at x_1 corresponds to stable equilibrium. Many smooth curves with a minimum as in Figure 14-9 can be approximated near the minimum by a parabola. The dashed curve in this figure is a parabolic curve that approximately fits U near the stable equilibrium point. The general equation for a parabola that has a minimum at point x_1 can be written

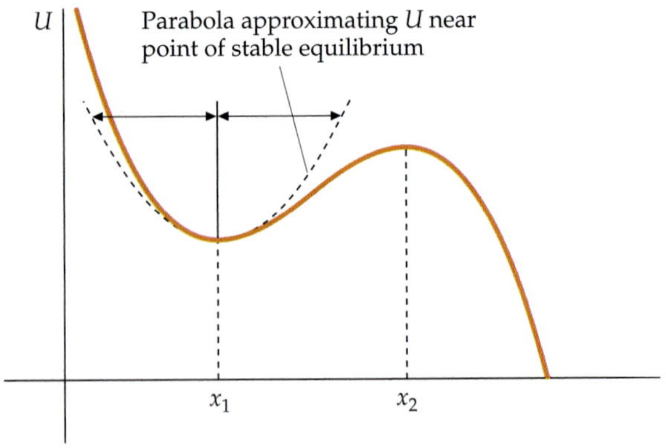

$$U = A + B(x - x_1)^2 \qquad\qquad 14\text{-}19$$

where A and B are constants. The constant A is the value of U at the equilibrium position $x = x_1$. The force is related to the potential energy curve by $F_x = -dU/dx$. Then

$$F_x = -\frac{dU}{dx} = -2B(x - x_1)$$

FIGURE 14-9 Plot of U versus x for a force that has a position of stable equilibrium (x_1) and a position of unstable equilibrium (x_2).

If we set $2B = k$, this equation reduces to

$$F_x = -\frac{dU}{dx} = -k(x - x_1)$$ 14-20

According to Equation 14-20, the force is proportional to the displacement and oppositely directed, so the motion will be simple harmonic. Figure 14-9 shows a graph of this system's potential energy function $U(x)$, which has a position of stable equilibrium at $x = x_1$. Figure 14-10 shows a potential energy function that has a position of stable equilibrium at $x = 0$. The system for this function is a small particle of mass m oscillating back and forth at the bottom of a frictionless spherical bowl.

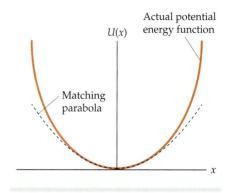

FIGURE 14-10 Plot of U versus x for a small particle oscillating back and forth at the bottom of a spherical bowl.

14-3 Some Oscillating Systems

Object on a Vertical Spring

When an object hangs from a vertical spring, there is a downward force mg in addition to the force of the spring (Figure 14-11). If we choose downward as the positive y direction, then the spring's force on the object is $-ky$, where y is the extension of the spring. The net force on the object is then

$$\sum F_y = -ky + mg$$ 14-21

We can simplify this equation by changing to a new variable $y' = y - y_0$, where $y_0 = mg/k$ is the amount the spring is stretched when the object is in equilibrium. Substituting $y = y' + y_0$ gives

$$\sum F_y = -k(y' + y_0) + mg$$

But $ky_0 = mg$, so

$$\sum F_y = -ky'$$ 14-22

Newton's second law ($\sum F_y = ma_y$) gives

$$-ky' = m\frac{d^2y}{dt^2}$$

However, $y = y' + y_0$, where $y_0 = mg/k$ is a constant. Thus $d^2y/dt^2 = d^2y'/dt^2$, so

$$-ky' = m\frac{d^2y'}{dt^2}$$

Rearranging gives

$$\frac{d^2y'}{dt^2} = -\frac{k}{m}y'$$

which is the same as Equation 14-2 with y' replacing x. It has the now familiar solution

$$y' = A\cos(\omega t + \delta)$$

where $\omega = \sqrt{k/m}$.

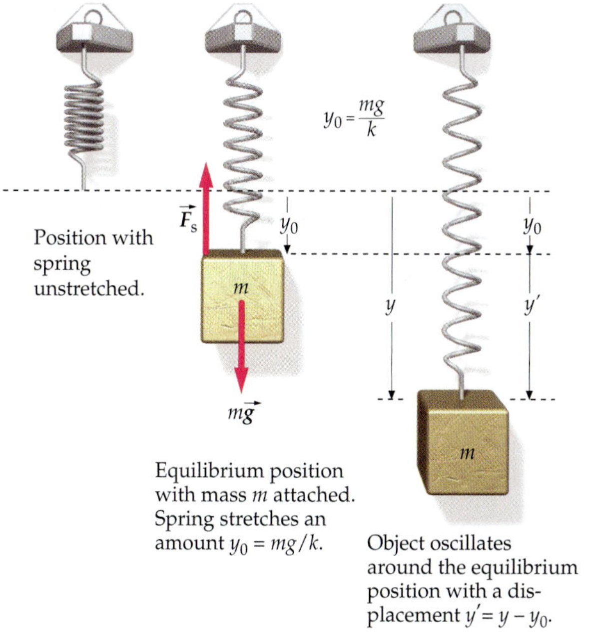

Position with spring unstretched.

$$y_0 = \frac{mg}{k}$$

Equilibrium position with mass m attached. Spring stretches an amount $y_0 = mg/k$.

Object oscillates around the equilibrium position with a displacement $y' = y - y_0$.

FIGURE 14-11 The problem of a mass on a vertical spring is simplified if the displacement (y') is measured from the equilibrium position of the spring with the mass attached.

Thus the effect of the gravitational force mg is merely to shift the equilibrium position from $y = 0$ to $y' = 0$. When the object is displaced from this equilibrium position by the amount y', the net force is $-ky'$. The object oscillates about this equilibrium position with an angular frequency $\omega = \sqrt{k/m}$, the same angular frequency as that for an object on a horizontal spring.

A force is conservative if the work done by it is independent of path. Both the force of the spring and the force of gravity are conservative, and the sum of these forces (Equations 14-21 and 14-22) also is conservative. The potential energy function U associated with the sum of these forces is the negative of the work done plus an arbitrary integration constant. That is,

$$U = -\int -ky' \, dy' = \tfrac{1}{2}ky'^2 + U_0$$

where the integration constant U_0 is the value of U at the equilibrium position ($y' = 0$). Thus,

$$U = \tfrac{1}{2}ky'^2 + U_0 \qquad\qquad 14\text{-}23$$

PAPER SPRINGS   **EXAMPLE 14-6**   **Put It in Context**

You are showing your nieces how to make paper party decorations using paper springs. One niece makes a paper spring and, with a single sheet of colored paper suspended from it, the spring is stretched 8 cm. You want the decorations to bounce at approximately 1 cy/s. How many sheets of colored paper should be used for the decoration on that spring if it is to bounce at 1 cy/s?

PICTURE THE PROBLEM The frequency depends on the ratio of the spring constant to the suspended mass (Equation 14-12), and you do not know either the spring constant or the mass. However, Hooke's law (Equation 14-1) can be used to find the required ratio from the information given.

1. Write the frequency in terms of the force constant k and the mass M (Equation 14-12), where M is the mass of N sheets. We need to find N:

$$f = \frac{\omega}{2\pi} = \frac{1}{2\pi}\sqrt{\frac{k}{M}}$$

2. The spring stretches a distance of $y_0 = 8$ cm when a single sheet of mass m is suspended:

$$ky_0 = mg \quad \text{so} \quad \frac{k}{m} = \frac{g}{y_0}$$

3. The mass of N sheets equals N times the mass of a single sheet:

$$M = Nm$$

4. Using the steps 2 and 3 results, solve for k/M:

$$\frac{k}{M} = \frac{k}{Nm} = \frac{1}{N}\frac{g}{y_0}$$

5. Substitute the step 4 result into the step 1 result and solve for N:

$$f = \frac{1}{2\pi}\sqrt{\frac{k}{M}} = \frac{1}{2\pi}\sqrt{\frac{1}{N}\frac{g}{y_0}}$$

so

$$N = \frac{g}{(2\pi f)^2 y_0} = \frac{9.81 \text{ m/s}^2}{4\pi^2\,(1 \text{ Hz})^2\,(0.08 \text{ m})} = 3.11$$

$$\boxed{\text{Three sheets are needed.}}$$

REMARKS Note that in this example we didn't need to use the value of m or k because the frequency depends on the ratio k/m, which equals g/y_0. Also, the units work out since 1 Hz = 1 cy/s, and a cycle is a dimensionless unit.

How much is the paper spring stretched when a decoration made from three sheets of paper is suspended from it and the paper is in equilibrium? (*Answer*   24 cm)

A BEAD ON A BLOCK   **EXAMPLE   14 - 7**

A block attached to a spring oscillates vertically with a frequency of 4 Hz and an amplitude of 7 cm. A tiny bead is placed on top of the oscillating block just as it reaches its lowest point. Assume that the bead's mass is so small that its effect on the motion of the block is negligible. At what distance from the block's equilibrium position does the bead lose contact with the block?

PICTURE THE PROBLEM The forces on the bead are its weight *mg* downward and the upward normal force exerted by the block. The magnitude of this normal force changes as the acceleration changes. As the block moves upward *from equilibrium*, its acceleration and that of the bead is *downward* and increasing in magnitude. When the acceleration reaches *g* downward, the normal force will be zero. If the block's downward acceleration becomes even slightly larger, the bead will leave the block.

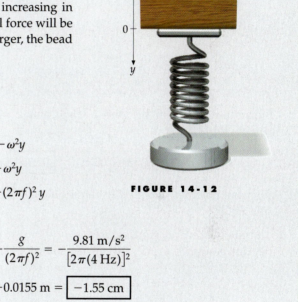

1. Draw a sketch of the system (Figure 14-12). Include a *y* coordinate axis with its origin at the equilibrium position and with down as the positive direction:

2. We are looking for the value of *y* when the acceleration is *g* downward. Use Equation 14-7:

$$a_y = -\omega^2 y$$
$$g = -\omega^2 y$$

3. Substitute $2\pi f$ for ω and solve for *y*:

$$g = -(2\pi f)^2 y$$

so

FIGURE 14-12

$$y = -\frac{g}{(2\pi f)^2} = -\frac{9.81 \text{ m/s}^2}{[2\pi(4 \text{ Hz})]^2}$$

$$= -0.0155 \text{ m} = \boxed{-1.55 \text{ cm}}$$

PLAUSIBILITY CHECK The bead leaves the block when *y* is negative, which is when the bead is above the equilibrium position, as expected.

POTENTIAL ENERGY OF THE SPRING–EARTH SYSTEM   **EXAMPLE   14 - 8**

A 3-kg object stretches a spring 16 cm when it hangs vertically in equilibrium. The spring is then stretched an additional 5 cm and the object is released. Let *U* be the total potential energy of the spring-object-planet system. When the mass is at its maximum displacement from equilibrium find *U* (*a*) with *U* = 0 at the equilibrium position and (*b*) with *U* = 0 when the spring is unstretched.

PICTURE THE PROBLEM (*a*) With *U* = 0 at the equilibrium position the total potential energy *U* is $\frac{1}{2}ky'^2$, where *y'* is the displacement from the equilibrium position. (*b*) With *U* = 0 when the spring is unstretched the total potential energy is the potential energy of the spring plus the gravitational potential energy.

FIGURE 14-13

(*a*) 1. Make three sketches of the system, one with the spring unstretched, one with it stretched 16 cm, and a third with it stretched 21 cm (Figure 14-13).

2. Let the positive y' direction be downward and let $y' = 0$ at the equilibrium position. The total potential energy function (Equation 14-23) is:

$$U = \tfrac{1}{2}ky'^2$$

3. To determine U we first need to find the spring constant k. At equilibrium the upward force of the spring equals the downward force of gravity. Use this to calculate the value of k:

$$ky_0 = mg$$

so

$$k = \frac{mg}{y_0} = \frac{(3\ \text{kg})(9.81\ \text{m/s}^2)}{0.16\ \text{m}} = 184\ \text{N/m}$$

4. Substituting for k in the step 1 result and solving for U gives:

$$U = \tfrac{1}{2}ky'^2 = \tfrac{1}{2}(184\ \text{N/m})(0.05\ \text{m})^2 = \boxed{0.230\ \text{J}}$$

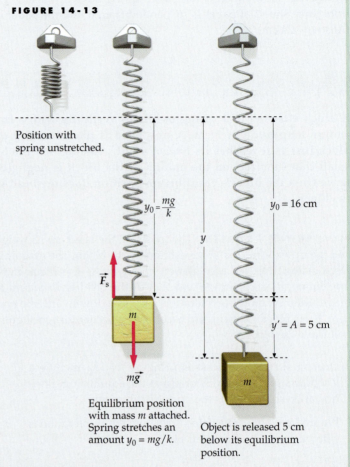

Position with spring unstretched.

$y_0 = \dfrac{mg}{k}$

$\vec{F}_s$

m

$m\vec{g}$

Equilibrium position with mass m attached. Spring stretches an amount $y_0 = mg/k$.

$y_0 = 16$ cm

y

$y' = A = 5$ cm

m

Object is released 5 cm below its equilibrium position.

(*b*) 1. The total potential energy function is given by Equation 14-23:

$$U = \tfrac{1}{2}ky'^2 + U_0$$

2. The potential energy equals zero at $y' = y'_{\text{ref}} = -16$ cm. Set $U = 0$, set $y' = y'_{\text{ref}} = -16$ cm, and solve for U_0:

$$0 = \tfrac{1}{2}ky'^2_{\text{ref}} + U_0$$

so

$$U_0 = -\tfrac{1}{2}ky'^2_{\text{ref}} = -\tfrac{1}{2}(184\ \text{N/m})(-0.16\ \text{m})^2$$

$$= -2.35\ \text{J}$$

3. Substituting for U_0 in the Part (*b*) step 1 result gives:

$$U = \tfrac{1}{2}ky'^2 + U_0$$

$$= \tfrac{1}{2}(184\ \text{N/m})(0.05\ \text{m})^2 - 2.35\ \text{J} = \boxed{-2.12\ \text{J}}$$

MASTER the CONCEPT / WEB

PLAUSIBILITY CHECK The potential energy calculated in Part (*b*) should equal the sum of the spring's potential energy U_s at $y = 21$ cm plus the gravitational potential energy U_g at $y = 21$ cm, where each of these potential energies is zero if the spring is unstretched, the positive y direction is downward, and $y = 0$ if the spring is unstretched. $U_s = \tfrac{1}{2}ky^2 = \tfrac{1}{2}(184\ \text{N/m})(0.21\ \text{m})^2 = 4.06\ \text{J}$ and $U_g = mg(-y) = (3\ \text{kg})(9.81\ \text{N/kg})(-0.21\ \text{m}) = -6.18\ \text{J}$. Adding these gives $4.06\ \text{J} - 6.18\ \text{J} = -2.12\ \text{J}$, which agrees with the Part (*b*) result.

The Simple Pendulum

A simple pendulum consists of a string of length L and a bob of mass m. When the bob is released from an initial angle ϕ_0 with the vertical, it swings back and forth with some period T.

EXERCISE IN DIMENSIONAL ANALYSIS We might expect the period of a simple pendulum to depend on the mass m of a pendulum bob, the length L of the pendulum, the acceleration due to gravity g, and the initial angle ϕ_0. Find a simple combination of some or all of these quantities that gives the correct dimensions for the period. (*Answer* $\sqrt{L/g}$)

REMARKS The units of length, mass, and g, are m, kg, and m/s², respectively. If we divide L by g, the meters cancel and we are left with seconds squared, suggesting the form $\sqrt{L/g}$. If the formula for the period contains the mass, then the unit kg must be canceled by some other quantity. But there is no combination of L and g that can cancel mass units. So the period cannot depend on the mass of the bob. Since the initial angle ϕ_0 is dimensionless, we cannot tell whether or not it is a factor in the period. We will see below that for small ϕ_0, the period is given by $T = 2\pi\sqrt{L/g}$.

The forces on the bob are its weight $m\vec{g}$ and the string tension $\vec{T}$ (Figure 14-14). At an angle ϕ with the vertical, the weight has components $mg \cos \phi$ along the string and $mg \sin \phi$ tangential to the circular arc in the direction of decreasing ϕ. Using tangential components, Newton's second law ($\Sigma F_t = m\,a_t$) gives

$$-mg \sin \phi = m\frac{d^2s}{dt^2} \qquad \text{14-24}$$

where the arc length s is related to the angle ϕ by $s = L\phi$. Repeatedly differentiating both sides of $s = L\phi$ gives

$$\frac{d^2s}{dt^2} = L\frac{d^2\phi}{dt^2}$$

Substituting $L\,d^2\phi/dt^2$ into Equation 14-24 for d^2s/dt^2 and rearranging gives

$$\frac{d^2\phi}{dt^2} = -\frac{g}{L}\sin \phi \qquad \text{14-25}$$

Note that the mass m does not appear in Equation 14-25—the motion of a pendulum does not depend on its mass. For small ϕ, $\sin \phi \approx \phi$, and

$$\frac{d^2\phi}{dt^2} \approx -\frac{g}{L}\phi \qquad \text{14-26}$$

Equation 14-26 is of the same form as Equation 14-2 for an object on a spring. Thus, the motion of a pendulum approximates simple harmonic motion for small angular displacements.

Equation 14-26 can be written

$$\frac{d^2\phi}{dt^2} = -\omega^2\phi, \qquad \text{where} \quad \omega^2 = \frac{g}{L} \qquad \text{14-27}$$

and ω is the angular frequency—not the angular speed—of the motion of the pendulum. The period of the motion is thus

$$T = \frac{2\pi}{\omega} = 2\pi\sqrt{\frac{L}{g}} \qquad \text{14-28}$$

PERIOD OF A SIMPLE PENDULUM

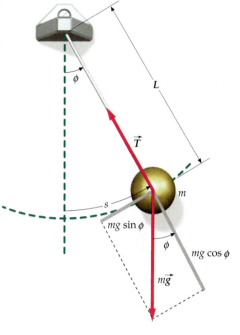

FIGURE 14-14 Forces on a pendulum bob.

The solution of Equation 14-27 is

$$\phi = \phi_0 \cos(\omega t + \delta)$$

where ϕ_0 is the maximum angular displacement.

According to Equation 14-28, the greater the length of a pendulum, the greater the period, which is consistent with experimental observation. The period, and therefore the frequency, are independent of the amplitude of oscillation (as long as the amplitude is small), a general feature of simple harmonic motion.

EXERCISE Find the period of a simple pendulum of length 1 m. (*Answer* 2.01 s)

The acceleration due to gravity can be measured using a simple pendulum. We need only measure the length L and period T of the pendulum, and using Equation 14-28, solve for g. (When finding T, we usually measure the time for n oscillations and then divide by n, which minimizes measurement error.)

Pendulum in an Accelerated Reference Frame Figure 14-15a shows a simple pendulum suspended from the ceiling of a boxcar that has acceleration $\vec{a}_0$ relative to the ground, to the right, and $\vec{a}$ is the acceleration of the bob relative to the ground. Applying Newton's second law to the bob gives

$$\sum \vec{F} = \vec{T} + m\vec{g} = m\vec{a} \qquad \text{14-29}$$

If the bob remains at rest relative to the boxcar, then $\vec{a} = \vec{a}_0$ and

$$\sum F_x = T \sin \theta_0 = ma_0$$

$$\sum F_y = T \cos \theta_0 - mg = 0$$

where θ_0 is the equilibrium angle. θ_0 is thus given by $\tan \theta_0 = a_0/g$. If the bob is moving relative to the boxcar, then $\vec{a}' = \vec{a} - \vec{a}_0$, where $\vec{a}'$ is the acceleration of the bob relative to the boxcar. Substituting for $\vec{a}$ in Equation 14-29 gives

$$\sum \vec{F} = \vec{T} + m\vec{g} = m(\vec{a}' + \vec{a}_0)$$

Subtracting $m\vec{a}_0$ from both sides of this equation and rearranging terms gives

$$\vec{T} + m\vec{g}' = m\vec{a}'$$

where $\vec{g}' = \vec{g} - \vec{a}_0$. Thus by replacing $\vec{g}$ by $\vec{g}'$ and $\vec{a}$ by $\vec{a}'$ in Equation 14-29 we can solve for the motion of the bob relative to the boxcar. The vectors $\vec{T}$ and $m\vec{g}'$ are shown in Figure 14-15b. If the string breaks so that $\vec{T} = 0$, then our equation gives $\vec{a}' = \vec{g}'$, which means that $\vec{g}'$ is the free-fall acceleration in the reference frame of the boxcar. If the bob is displaced slightly from equilibrium, it will oscillate with a period T given by Equation 14-28 with g replaced by g'.

EXERCISE A simple pendulum of length 1 m is in a boxcar that is accelerating horizontally with acceleration $a_0 = 3$ m/s^2. Find g' and the period T. (*Answer* $g' = 10.26$ m/s^2, $T = 1.96$ s)

Large-Amplitude Oscillations When the amplitude of a pendulum's oscillation becomes large, its motion continues to be periodic, but it is no longer simple harmonic. A slight dependence on the amplitude

All mechanical clocks keep time because the period of the oscillating part of the mechanism remains constant. The period of any pendulum changes with changes in amplitude. However, the driving mechanism of a pendulum clock maintains the amplitude at a constant value.

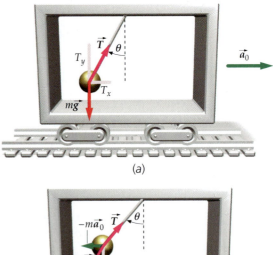

(a)

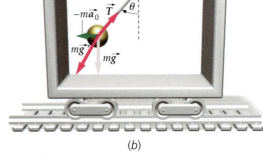

(b)

FIGURE 14-15 (a) Simple pendulum in apparent equilibrium in an accelerating boxcar. Forces are those as seen from a separate stationary frame. (b) Forces on the bob as seen in the accelerated frame. Adding the pseudoforce $-m\vec{a}_0$ is equivalent to replacing $\vec{g}$ by $\vec{g}'$.

FIGURE 14-16 Note that the ordinate values range from 1 to 1.06. Over a range of ϕ from 0 to 0.8 rad (46°), the period varies by about 5 percent.

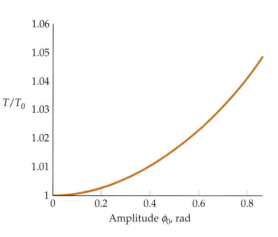

must be accounted for when determining the period. For a general angular amplitude ϕ_0, the period can be shown to be

$$T = T_0\left[1 + \frac{1}{2^2}\sin^2\frac{1}{2}\phi_0 + \frac{1}{2^2}\left(\frac{3}{4}\right)^2\sin^4\frac{1}{2}\phi_0 + \cdots\right]$$ 14-30

PERIOD FOR LARGE-AMPLITUDE OSCILLATIONS

where $T_0 = 2\pi\sqrt{L/g}$ is the period for very small amplitudes. Figure 14-16 shows T/T_0 as a function of amplitude ϕ_0.

A PENDULUM CLOCK   **EXAMPLE 14-9** Try It Yourself

A simple pendulum clock is calibrated to keep accurate time at an angular amplitude of $\phi_0 = 10°$. When the amplitude has decreased to the point where it is very small, does the clock gain or lose time? How much time will the clock gain or lose in one day if the amplitude remains very small.

Cover the column to the right and try these on your own before looking at the answers.

Steps

1. Answer the first question by finding out if the period increases or decreases.

2. Use Equation 14-30 to find the percentage change $[(T - T_0)/T] \times 100\%$ for $\phi = 10°$. Use only the first correction term.

3. Find the number of minutes in a day.

4. Combine the steps 2 and 3 to find the change in the number of minutes in a day.

Answers

T decreases as ϕ decreases, so the clock gains time.

0.190%

There are 1440 minutes in a day.

The gain is 2.73 min/d

REMARKS To avoid this gain, pendulum-clock mechanisms are designed to keep the amplitude fairly constant.

*The Physical Pendulum

A rigid object free to rotate about a horizontal axis that is not through its center of mass will oscillate when displaced from equilibrium. Such a system is called a **physical pendulum.** Consider a plane figure with a rotation axis a distance D from the figure's center of mass and displaced from equilibrium by the angle ϕ (Figure 14-17). The torque about the axis has a magnitude $MgD \sin \phi$ and tends to decrease $|\phi|$. Newton's second law applied to rotation is

$$\tau = I\alpha$$

where α is the angular acceleration and I is the moment of inertia about the axis. Substituting $-MgD \sin \phi$ for the net torque and $d^2\phi/dt^2$ for α, we have

$$-MgD \sin \phi = I\frac{d^2\phi}{dt^2}$$

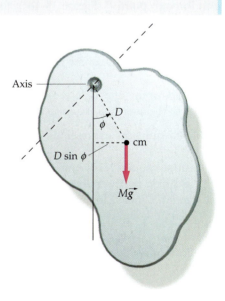

FIGURE 14-17 A physical pendulum.

or

$$\frac{d^2\phi}{dt^2} = -\frac{MgD}{I}\sin\phi \qquad\qquad \text{14-31}$$

As with the simple pendulum, the motion is approximately simple harmonic if the angular displacements are small, so $\sin\phi \approx \phi$. In this case, we have

$$\frac{d^2\phi}{dt^2} \approx -\frac{MgD}{I}\phi = -\omega^2\phi \qquad\qquad \text{14-32}$$

where $\omega = \sqrt{MgD/I}$ is the angular frequency—not the angular speed—of the motion. The period is therefore

$$T = \frac{2\pi}{\omega} = 2\pi\sqrt{\frac{I}{MgD}} \qquad\qquad \text{14-33}$$

PERIOD OF A PHYSICAL PENDULUM

For large amplitudes, the period is given by Equation 14-30, with T_0 given by Equation 14-33. For a simple pendulum of length L, the moment of inertia is $I = ML^2$ and $D = L$. Then Equation 14-33 gives $T = 2\pi\sqrt{ML^2/(MgL)} = 2\pi\sqrt{L/g}$, the same as Equation 14-28.

A Rotating Rod   **EXAMPLE 14-10**

A uniform rod of mass M and length L is free to rotate about a horizontal axis perpendicular to the rod and through one end. (*a*) Find the period of oscillation for small angular displacements. (*b*) Find the period of oscillation if the rotation axis is a distance x from the center of mass.

PICTURE THE PROBLEM (*a*) The period is given by Equation 14-33. The center of mass is at the center of the rod, so the distance from the center of mass to the rotation axis is half the length of the rod (Figure 14-18*a*). The moment of inertia of a uniform rod can be found in Table 9-1. (*b*) For rotations around an axis through point P (Figure 14-18*b*), the moment of inertia can be found from the parallel-axis theorem $I = I_{cm} + MD^2$ (Equation 9-44), where I_{cm} can be found in Table 9-1.

FIGURE 14-18 Plot of the period versus the distance from the pivot to the center of mass. For $x > 0.5$ m the pivot is beyond the end of the rod.

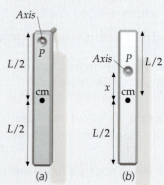

(*a*) 1. The period is given by Equation 14-33:

$$T = 2\pi\sqrt{\frac{I}{MgD}}$$

2. I about the end is found in Table 9-1 and D is half the length of the rod:

$$I = \tfrac{1}{3}ML^2; \qquad D = \tfrac{1}{2}L$$

3. Substitute the expressions for I and D to find T:

$$T = 2\pi\sqrt{\frac{\tfrac{1}{3}ML^2}{Mg(\tfrac{1}{2}L)}} = \boxed{2\pi\sqrt{\frac{2L}{3g}}}$$

(*b*) 1. About point P, $D = x$, and the moment of inertia is given by the parallel-axis theorem. The moment of inertia about a parallel axis through the center of mass is found in Table 9-1:

$$D = x$$

$$I = I_{cm} + MD^2 = \tfrac{1}{12}ML^2 + Mx^2$$

2. Substitute these values to find T:

$$T = 2\pi\sqrt{\frac{I}{MgD}} = 2\pi\sqrt{\frac{(\tfrac{1}{12}ML^2 + Mx^2)}{Mgx}}$$

$$= \boxed{2\pi\sqrt{\frac{(\tfrac{1}{12}L^2 + x^2)}{gx}}}$$

① PLAUSIBILITY CHECK As $x \to 0$, $T \to \infty$ as expected. (If the rotation axis of the rod passes through its center of mass, we do not expect gravity to exert a restoring torque.) Also, if $x = L/2$, we get the same result as found in Part (a), and if $x \gg L$, the expression for the period approaches $T = 2\pi\sqrt{x/g}$, which is the expression for the period of a simple pendulum of length x (Equation 14-28).

EXERCISE What is the period of oscillation for small angular displacements of a 1-m-long uniform rod about an axis through one end? (*Answer* $T = 1.64$ s) Note that this is a smaller period than for a simple pendulum of length $L = 1$ m. The period of the simple pendulum is greater because the ratio of its moment of inertia to the restoring torque is greater.

EXERCISE Show that when $x = L/6$, the period is the same as when $x = L/2$.

REMARKS The period T versus distance x from the center of mass for a rod of length 1 m is shown in Figure 14-19.

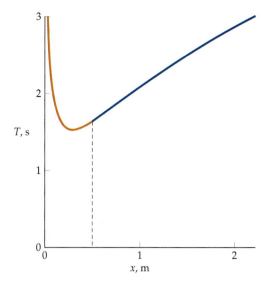

FIGURE 14-19

THE ROTATING ROD REVISITED   **EXAMPLE 14-11 Try It Yourself**

Find the value of x in Example 14-10 for which the period is a minimum.

PICTURE THE PROBLEM At the value of x for which T is a minimum, $dT/dx = 0$.

Cover the column to the right and try these on your own before looking at the answers.

Steps

1. The period, given by the Example 14-10 Part (b) result, equals $T = 2\pi\sqrt{Z/g}$, where $Z = (\frac{1}{12}L^2 + x^2)/x$. Find the period both as x approaches zero and as x approaches infinity.

2. The period goes to infinity as x approaches zero and as x approaches infinity. Somewhere in the range $0 < x < \infty$ the period is a minimum. To find the minimum, evaluate dT/dx, set it equal to zero, and solve for x.

Answers

As $x \to 0$, $Z \to \infty$, and $T \to \infty$.

As $x \to \infty$, $Z \to \infty$, and $T \to \infty$.

$$\frac{dT}{dx} = \frac{dT}{dZ}\frac{dZ}{dx} = \frac{\pi}{\sqrt{g}}Z^{-1/2}\frac{dZ}{dx}$$

$Z > 0$ throughout the range $0 < x < \infty$, so $\dfrac{dT}{dx} = 0 \Rightarrow \dfrac{dZ}{dx} = 0.$

$$\frac{dZ}{dx} = 0 \Rightarrow x = \boxed{\frac{L}{\sqrt{12}} = 0.289L}$$

14-4 Damped Oscillations

Left to itself, a spring or a pendulum eventually stops oscillating because the mechanical energy is dissipated by frictional forces. Such motion is said to be **damped.** If the damping is large enough, as, for example, a pendulum submerged in molasses, the oscillator fails to complete even one cycle of oscillation. Instead it just moves toward the equilibrium position with a speed that approaches zero as the object approaches the equilibrium position. This type of motion is referred to as **overdamped.** If the damping is small enough that the system oscillates with an amplitude that decreases slowly with time—like a child on a playground swing when Mom stops providing a push each cycle—the motion is said to be **underdamped.** Motion with the minimum damping for nonoscillatory motion is said to be **critically damped.** (With any less damping, the motion would be underdamped.)

Underdamped Motion The damping force exerted on an oscillator such as the one shown in Figure 14-20a can be represented by the empirical expression

$$\vec{F}_d = -b\vec{v}$$

(a)

(b)

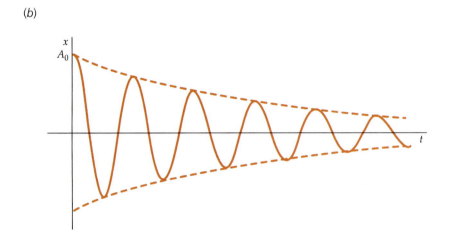

FIGURE 14-20 (*a*) A damped oscillator. The motion is damped by the plunger immersed in the liquid. (*b*) Damped oscillation curve.

where b is a constant. Such a system is said to be linearly damped. The discussion here is for linearly damped motion. Because the damping force is opposite to the direction of motion, it does negative work and causes the mechanical energy of the system to decrease. This energy is proportional to the square of the amplitude (Equation 14-17), and the square of the amplitude decreases exponentially with increasing time. That is,

$$A^2 = A_0^2 e^{-t/\tau} \tag{14-34}$$

DEFINITION — TIME CONSTANT

where A is the amplitude, A_0 is the amplitude at $t = 0$, and τ is the **decay time** or **time constant.** The time constant is the time for the energy to decrease by a factor of e.

The motion of a damped system can be obtained from Newton's second law. For an object of mass m on a spring of force constant k, the net force is $-kx - b(dx/dt)$. Setting the net force equal to the mass times the acceleration d^2x/dt^2, we obtain

$$-kx - b\frac{dx}{dt} = m\frac{d^2x}{dt^2} \tag{14-35}$$

DIFFERENTIAL EQUATION FOR A DAMPED OSCILLATOR

The exact solution of this equation can be found using standard methods for solving differential equations. The solution for the underdamped case is

$$x = A_0 e^{-(b/2m)t} \cos(\omega' t + \delta)$$ 14-36

where A_0 is the initial amplitude. The frequency ω' is given by

$$\omega' = \omega_0 \sqrt{1 - \left(\frac{b}{2m\omega_0}\right)^2}$$ 14-37

where ω_0 is the frequency with no damping ($\omega_0 = \sqrt{k/m}$ for a mass on a spring). For weak damping, $b/(2m\omega_0) << 1$ and ω' is nearly equal to ω_0. The dashed curves in Figure 14-20b correspond to $x = A$ and $x = -A$, where A is given by

$$A = A_0 e^{-(b/2m)t}$$ 14-38

By squaring both sides of this equation and comparing the results with Equation 14-34 we have

$$\tau = \frac{m}{b}$$ 14-39

If the damping constant b is gradually increased, the angular frequency ω' decreases until it becomes zero at the critical value

$$b_c = 2m\omega_0$$ 14-40

When b is greater than or equal to b_c, the system does not oscillate. If $b > b_c$, the system is overdamped. The smaller b is, the more rapidly the object returns to equilibrium. If $b = b_c$, the system is said to be critically damped and the object returns to equilibrium (without oscillation) most rapidly. Figure 14-21 shows plots of the displacement versus time for a critically damped and an overdamped oscillator. We often use critical damping when we want a system to avoid oscillations and yet return to equilibrium quickly. For example, shock absorbers are used to damp the oscillations of an automobile on its springs. You can test the damping of a car's shock absorbers by pushing down on one fender of the car and then releasing it. If the car returns to equilibrium with no oscillation, then the system is critically damped or overdamped. (You will usually observe one or two oscillations for an unoccupied vehicle, indicating that the damping constant is just under the critical value.)

Because the energy of an oscillator is proportional to the square of its amplitude, the energy of an underdamped oscillator (averaged over a cycle) also decreases exponentially with time:

$$E = \tfrac{1}{2}m\omega^2 A^2 = \tfrac{1}{2}m\omega^2(A_0 e^{-(b/2m)t})^2 = \tfrac{1}{2}m\omega^2 A_0^2 e^{-(b/m)t} = E_0 e^{-t/\tau}$$ 14-41

where $E_0 = \tfrac{1}{2}m\omega^2 A_0^2$ and

$$\tau = \frac{m}{b}$$ 14-42

A damped oscillator is often described by its Q factor (for quality factor),

$$Q = \omega_0 \tau$$ 14-43

DEFINITION—Q FACTOR

Shock absorbers (yellow cylinders) are used to damp the oscillations of this truck.

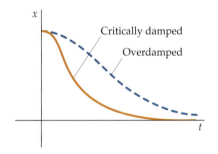

FIGURE 14-21 Plots of displacement versus time for a critically damped and an overdamped oscillator.

Weights are placed in automobile wheels when the wheels are "balanced." The purpose of balancing the wheels is to prevent vibrations that will drive oscillations of the wheel assembly.

The Q factor is dimensionless. (Since ω_0 has dimensions of reciprocal time, $\omega_0\tau$ is without dimension.) We can relate Q to the fractional energy loss per cycle. Differentiating Equation 14-41 gives

$$dE = -(1/\tau)E_0 e^{-t/\tau} dt = -(1/\tau)E\, dt$$

If the damping is weak so that the energy loss per cycle is small, we can replace dE by ΔE and dt by the period T. Then $|\Delta E|/E$ in one cycle (one period) is given by

$$\left(\frac{|\Delta E|}{E}\right)_{\text{cycle}} = \frac{T}{\tau} = \frac{2\pi}{\omega_0\tau} = \frac{2\pi}{Q} \qquad\qquad 14\text{-}44$$

by

$$Q = \frac{2\pi}{(|\Delta E|/E)_{\text{cycle}}}, \qquad \frac{|\Delta E|}{E} \ll 1 \qquad\qquad 14\text{-}45$$

PHYSICAL INTERPRETATION OF Q FOR WEAK DAMPING

Q is thus inversely proportional to the fractional energy loss per cycle.

MAKING MUSIC **EXAMPLE 1 4 - 1 2**

When middle C on a piano (frequency 262 Hz) is struck, it loses half its energy after 4 s. (a) What is the decay time τ? (b) What is the Q factor for this piano wire? (c) What is the fractional energy loss per cycle?

PICTURE THE PROBLEM (a) We use $E = E_0 e^{-t/\tau}$ and set E equal to $\frac{1}{2}E_0$. (b) The Q value can then be found from the decay time and the frequency.

(a) 1. Set the energy at time $t = 4$ s equal to half the original energy:
$$E = E_0 e^{-t/\tau} \quad\text{so}\quad \tfrac{1}{2}E_0 = E_0 e^{-4\,\text{s}/\tau}$$
$$\tfrac{1}{2} = e^{-4\,\text{s}/\tau}$$

2. Solve for the time τ by taking the natural log of both sides:
$$\ln\frac{1}{2} = -\frac{4\,\text{s}}{\tau}$$
so
$$\tau = \frac{4\,\text{s}}{\ln 2} = \boxed{5.77\ \text{s}}$$

(b) Calculate Q from τ and ω_0:
$$Q = \omega_0\tau = 2\pi f\tau$$
$$= 2\pi(262\ \text{Hz})(5.77\ \text{s}) = \boxed{9.50 \times 10^3}$$

(c) The fractional energy loss in a cycle is given by Equation 14-44 and the frequency $f = 1/T$:
$$\left(\frac{|\Delta E|}{E}\right)_{\text{cycle}} = \frac{T}{\tau} = \frac{1}{f\tau} = \frac{1}{(262\ \text{Hz})(5.77\ \text{s})} = \boxed{6.61 \times 10^{-4}}$$

PLAUSIBILITY CHECK Q can also be calculated from $Q = 2\pi/(\Delta E/E)_{\text{cycle}} = 2\pi/(6.61 \times 10^{-4}) = 9.50 \times 10^3$. Note that the fractional energy loss after 4 s is not just the number of cycles (4×262) times the fractional energy loss per cycle, because the energy decrease is exponential, not constant.

REMARKS Figure 14-22 shows the relative amplitude A/A_0 versus time and the relative energy E/E_0 versus time for the oscillation of a piano string after middle C is struck. After 4 s, the amplitude has decreased to about 0.7 times its initial value, and the energy, which is proportional to the amplitude squared, drops to about half its initial value.

Note that Q is quite large. You can estimate τ and Q of various oscillating systems. Tap a crystal wine glass and see how long it rings. The longer it rings, the greater the value of τ and Q and the lower the damping. Glass beakers from the laboratory may also have a high Q. Try tapping a plastic cup. How does the damping compare to that of the beaker?

In terms of Q, the exact frequency of an underdamped oscillator is

$$\omega' = \omega_0\sqrt{1 - \left(\frac{b}{2m\omega_0}\right)^2} = \omega_0\sqrt{1 - \frac{1}{4Q^2}} \qquad 14\text{-}46$$

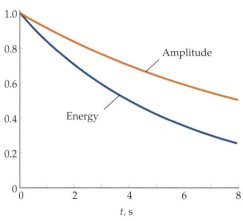

FIGURE 14-22 Plots of A/A_0 and E/E_0 for a struck piano string.

Because b is quite small (and Q is quite large) for a weakly damped oscillator (Example 14-12), we see that ω' is nearly equal to ω_0.

We can understand much of the behavior of a weakly damped oscillator by considering its energy. The power dissipated by the damping force equals the instantaneous rate of change of the total mechanical energy

$$P = \frac{dE}{dt} = \vec{F}_d \cdot \vec{v} = -b\vec{v} \cdot \vec{v} = -bv^2 \qquad 14\text{-}47$$

For a weakly damped oscillator, the total mechanical energy decreases slowly with time. The average kinetic energy per cycle equals half the total energy

$$\left(\frac{1}{2}mv^2\right)_{av} = \frac{1}{2}E \qquad \text{or} \qquad (v^2)_{av} = \frac{E}{m}$$

If we substitute $(v^2)_{av} = E/m$ for v^2 in Equation 14-47, we have

$$\frac{dE}{dt} = -bv^2 \approx -b(v^2)_{av} = -\frac{b}{m}E \qquad 14\text{-}48$$

Rearranging Equation 14-48 gives

$$\frac{dE}{E} = -\frac{b}{m}dt$$

which upon integration gives

$$E = E_0 e^{-(b/m)t} = E_0 e^{-t/\tau}$$

which is Equation 14-41.

14-5 Driven Oscillations and Resonance

To keep a damped system going, mechanical energy must be put into the system. When this is done, the oscillator is said to be driven or forced. When you keep a swing going by "pumping," that is, by moving your body and legs, you are driving an oscillator. If you put mechanical energy into the system faster than it is dissipated, the mechanical energy increases with time, and the amplitude increases. If you put mechanical energy in at the same rate it is being dissipated,

By pumping the swing, she is transferring her internal energy into the mechanical energy of the oscillator.

the amplitude remains constant over time. The motion of the oscillator is then said to be in steady state.

Figure 14-23 shows a system consisting of an object on a spring that is being driven by moving the point of support up and down with simple harmonic motion of frequency ω. At first the motion is complicated, but eventually steady-state motion is reached in which the system oscillates with the same frequency as that of the driver and with a constant amplitude and, therefore, at constant energy. In the steady state, the energy put into the system per cycle by the driving force equals the energy dissipated per cycle due to the damping.

The amplitude, and therefore the energy, of a system in the steady state depends not only on the amplitude of the driving force, but also on its frequency. The **natural frequency** of an oscillator, ω_0, is its frequency when no driving or damping forces are present. (In the case of a spring, for example, $\omega_0 = \sqrt{k/m}$.) If the driving frequency is approximately equal to the natural frequency of the system, the system will oscillate with a relatively large amplitude. For example, if the support in Figure 14-23 oscillates at a frequency close to the natural frequency of the mass–spring system, the mass will oscillate with a much greater amplitude than it would if the support oscillates at higher or lower frequencies. This phenomenon is called **resonance**. When the driving frequency equals the natural frequency of the oscillator, the energy per cycle transferred to the oscillator is maximum. The natural frequency of the system is thus called the **resonance frequency**. (Mathematically, the angular frequency ω is more convenient to use than the frequency $f = \omega/(2\pi)$. Because ω and f are proportional, most statements concerning angular frequency also hold for frequency. In verbal descriptions, we usually omit the word angular when the omission will not cause confusion.) Figure 14-24 shows plots of the average power delivered to an oscillator as a function of the driving frequency for two different values of damping. These curves are called **resonance curves**. When the damping is weak (large Q), the width of the peak of the resonance curve is correspondingly narrow, and we speak of the resonance as being sharp. For strong damping, the resonance curve is broad. The width of each resonance curve $\Delta\omega$, indicated in the figure, is the width at half the maximum height. For weak damping, the ratio of the width of the resonance to the resonant frequency can be shown to equal the reciprocal of the Q factor (see Problem 126):

$$\frac{\Delta\omega}{\omega_0} = \frac{1}{Q} \qquad\qquad 14\text{-}49$$

RESONANCE WIDTH FOR WEAK DAMPING

FIGURE 14-23 An object on a vertical spring can be driven by moving the support up and down.

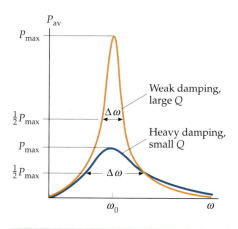

FIGURE 14-24 Resonance for an oscillator. The width $\Delta\omega$ of the resonance peak for a high-Q oscillator is small compared to the natural frequency of ω_0.

Thus, the Q factor is a direct measure of the sharpness of resonance.

You can do a simple experiment to demonstrate resonance. Hold a meterstick at one end between two fingers so that it acts like a pendulum. (If a meterstick is not available, use whatever is convenient. A golf club works fine.) Release the stick from some initial angular displacement and observe the natural frequency of its motion. Then move your hand back and forth horizontally, driving it at its natural frequency. Even if the amplitude of the motion of your hand is small, the stick will oscillate with a substantial amplitude. Now move your hand back and forth at a frequency two or three times the natural frequency and note the decrease in amplitude of the oscillating stick.

There are many familiar examples of resonance. When you sit on a swing, you learn intuitively to pump with the same frequency as the natural frequency of the swing. Many machines vibrate because they have rotating parts that are not in perfect balance. (Observe a washing machine in the spin cycle for an example.) If such a machine is attached to a structure that can vibrate, the structure becomes a driven oscillatory system that is set in motion by the machine. Engineers pay great attention to balancing the rotary parts of such machines, damping their vibrations, and isolating them from building supports.

A glass with low damping can be broken by an intense sound wave at a frequency equal to or very nearly equal to the natural frequency of vibration of the glass. This is often done in physics demonstrations using an audio oscillator and an amplifier.

*Mathematical Treatment of Resonance

We can treat a driven oscillator mathematically by assuming that, in addition to the restoring force and a damping force, the oscillator is subject to an external driving force that varies harmonically with time:

$$F_{ext} = F_0 \cos \omega t \qquad 14\text{-}50$$

where F_0 and ω are the amplitude and angular frequency of the driving force. This frequency is generally not related to the natural angular frequency of the system ω_0.

Newton's second law applied to an object of mass m attached to a spring of force constant k and subject to a damping force $-bv_x$ and an external force $F_0 \cos \omega t$ gives

$$\sum F_x = ma_x$$

$$-kx - bv_x + F_0 \cos \omega t = m\frac{d^2x}{dt^2}$$

where we have used $a_x = d^2x/dt^2$. Substituting $m\omega_0^2$ for k (Equation 14-8) and rearranging gives

$$m\frac{d^2x}{dt^2} + b\frac{dx}{dt} + m\omega_0^2 x = F_0 \cos \omega t \qquad 14\text{-}51$$

DIFFERENTIAL EQUATION FOR A DRIVEN OSCILLATOR

We will discuss the general solution of Equation 14-51 qualitatively. It consists of two parts, the **transient solution** and the **steady-state solution.** The transient part of the solution is identical to that for a damped oscillator given in Equation 14-36. The constants in this part of the solution depend on the initial conditions. Over time, this part of the solution becomes negligible because of the exponential decrease of the amplitude. We are then left with the steady-state solution, which can be written as

$$x = A \cos(\omega t - \delta) \qquad 14\text{-}52$$

POSITION FOR A DRIVEN OSCILLATOR

where the angular frequency ω is the same as that of the driving force. The amplitude A is given by

268 Hz ($Q = 52$)

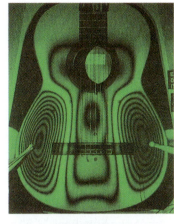

553 Hz ($Q = 66$)

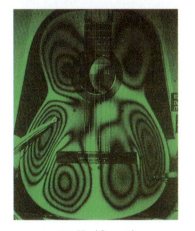

672 Hz ($Q = 61$)

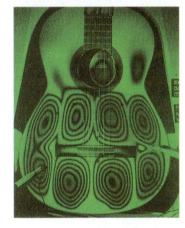

1010 Hz ($Q = 80$)

Extended objects have more than one resonance frequency. When plucked, a guitar string transmits its energy to the body of the guitar. The body's oscillations, coupled to those of the air mass it encloses, produce the resonance patterns shown.

$$A = \frac{F_0}{\sqrt{m^2(\omega_0^2 - \omega^2)^2 + b^2\omega^2}}$$ 14-53

AMPLITUDE FOR A DRIVEN OSCILLATOR

and the phase constant δ is given by

$$\tan \delta = \frac{b\omega}{m(\omega_0^2 - \omega^2)}$$ 14-54

PHASE CONSTANT FOR A DRIVEN OSCILLATOR

Comparing Equations 14-50 and 14-52, we can see that the displacement and the driving force oscillate with the same frequency, but they differ in phase by δ. When the driving frequency ω is much less than the natural frequency ω_0, $\delta \approx 0$, as can be seen from Equation 14-54. At resonance, $\omega = \omega_0$, $\delta = \pi/2$, and when ω is much greater than ω_0, $\delta \approx \pi$. At the beginning of this chapter the displacement of a particle undergoing simple harmonic motion is written $x = A \cos(\omega t + \delta)$ (Equation 14-4). This equation is identical with Equation 14-52 except for the sign preceding the phase constant δ. The phase of a driven oscillator always lags behind the phase of the driving force. The negative sign in Equation 14-52 ensures that δ is always positive (rather than always negative).

In your simple experiment to drive a meterstick by moving your hand back and forth, you should note that at resonance the oscillation of your hand is neither in phase nor 180° out of phase with the oscillation of the stick. If you move your hand back and forth at a frequency several times the natural frequency of the stick, the stick's steady state motion will be almost 180° out of phase with your hand.

The velocity of the object in the steady state is obtained by differentiating x with respect to t:

$$v = \frac{dx}{dt} = -\omega A \sin(\omega t - \delta)$$

At resonance, $\delta = \pi/2$, and the velocity is in phase with the driving force:

$$v = -\omega A \sin\left(\omega t - \frac{\pi}{2}\right) = +\omega A \cos \omega t$$

Thus, at resonance the object is always moving in the direction of the driving force, as would be expected for maximum power input. The velocity amplitude ωA is maximum at $\omega = \omega_0$.

EXPLORING

Is there a numerical solution to the equations for damped and driven oscillators? Find out this, and more, at www.whfreeman.com/tipler5e.

AN OBJECT ON A SPRING   **EXAMPLE 14-13**   Try It Yourself

An object of mass 1.5 kg on a spring of force constant 600 N/m loses 3 percent of its energy in each cycle. The same system is driven by a sinusoidal force with a maximum value of $F_0 = 0.5$ N. (*a*) What is Q for this system? (*b*) What is the resonance (angular) frequency? (*c*) If the driving frequency is slowly varied through resonance, what is the width $\Delta\omega$ of the resonance? (*d*) What is the amplitude at resonance? (*e*) What is the amplitude if the driving frequency is $\omega = 19$ rad/s?

PICTURE THE PROBLEM The energy loss per cycle is only 3 percent, so the damping is weak. We can find Q from $Q = 2\pi/ (\Delta E/E)_{cycle}$ (Equation 14-45) and then use this result and $\Delta\omega/\omega_0 = 1/Q$ (Equation 14-49) to find the width $\Delta\omega$ of the resonance. The resonance frequency is the natural frequency. The amplitude both at resonance and off resonance can be found from Equation 14-53, with the damping constant calculated from Q using the definition of Q (Equation 14-43) $Q = \omega_0\tau = \omega_0 m/b$.

Cover the column to the right and try these on your own before looking at the answers.

Steps	Answers
(a) The damping is weak. Relate Q to the fractional energy loss using Equation 14-45 and solve for Q.	$Q \approx \dfrac{2\pi}{(\|\Delta E\|/E)_{cycle}} = \dfrac{2\pi}{0.03} = \boxed{209}$
(b) Relate the resonance frequency to the natural frequency of the system.	$\omega_0 = \sqrt{\dfrac{k}{m}} = \boxed{20 \text{ rad/s}}$
(c) Relate the width of the resonance $\Delta\omega$ to Q.	$\Delta\omega = \dfrac{\omega_0}{Q} = \boxed{0.0957 \text{ rad/s}}$
(d) 1. Write an expression for the amplitude A for any driving frequency ω.	$A(\omega) = \dfrac{F_0}{\sqrt{m^2(\omega_0^2 - \omega^2)^2 + b^2\omega^2}}$
2. Substitute $\omega = \omega_0$ to calculate A at resonance.	$A(\omega_0) = \dfrac{F_0}{b\omega_0}$
3. Use Equation 14-43 to relate the damping constant b to Q.	$b = \dfrac{m\omega_0}{Q} = 0.144 \text{ kg/s}$
4. Use the results of the previous two steps to calculate the amplitude at resonance.	$A(\omega_0) = \dfrac{F_0}{b\omega} = \boxed{17.4 \text{ cm}}$
(e) Calculate the amplitude for $\omega = 19$ rad/s. (We can omit the units to simplify the equation. Since all quantities are in SI units, A will be in meters.)	$A(19 \text{ s}^{-1}) = \dfrac{0.5}{\sqrt{1.5^2(20^2 - 19^2)^2 + 0.144^2(19)^2}}$ $= \boxed{0.854 \text{ cm}}$

REMARKS At just 1 rad/s off resonance, the amplitude drops by a factor of 20. This is not surprising, because the width $\Delta\omega$ of the resonance is only 0.0957 rad/s. Note that off resonance the term $b^2\omega^2$ is negligible compared with the other term in the denominator of the expression for A. When $\omega - \omega_0$ is more than several times the half width $\Delta\omega$, as it was in this example, we can neglect the $b^2\omega^2$ term and calculate A from $A \approx F_0/[m(\omega_0^2 - \omega^2)]$. Figure 14-25 shows the amplitude versus driving frequency ω. Note that the horizontal scale is over a small range of ω.

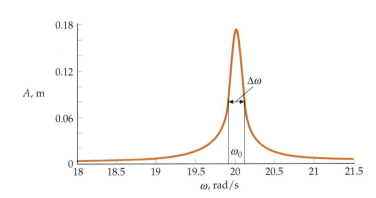

FIGURE 14-25

1. Simple harmonic motion occurs whenever the restoring force is proportional to the displacement from equilibrium. It has wide application in the study of oscillations, waves, electrical circuits, and molecular dynamics.

2. Resonance is an important phenomenon in many areas of physics. It occurs when the frequency of the driving force is close to the natural frequency of the oscillating system.

Topic	Relevant Equations and Remarks	
1. Simple Harmonic Motion	In simple harmonic motion, the net force and acceleration are both proportional to the displacement and oppositely directed.	
Position function	$x = A \cos(\omega t + \delta)$	**14-4**
Velocity	$v = -\omega A \sin(\omega t + \delta)$	**14-5**
Acceleration	$a_x = -\omega^2 A \cos(\omega t + \delta)$	**14-6**
	$a_x = -\omega^2 x$	**14-7**
Angular frequency	$\omega = 2\pi f = \dfrac{2\pi}{T}$	**14-11**
Total energy	$E_{\text{total}} = K + U = \frac{1}{2} k A^2$	**14-17**
Average kinetic or potential energy	$K_{\text{av}} = U_{\text{av}} = \frac{1}{2} E_{\text{total}}$	**14-18**
Circular motion	If a particle moves in a circle with constant speed, the projection of the particle onto a diameter of the circle moves in simple harmonic motion.	
General motion near equilibrium	If an object is given a small displacement from a position of stable equilibrium, it typically oscillates about this position with simple harmonic motion.	
2. Angular Frequencies for Various Systems		
Mass on a spring	$\omega = \sqrt{\dfrac{k}{m}}$	**14-8**
Simple pendulum	$\omega = \sqrt{\dfrac{g}{L}}$	**14-27**
*Physical pendulum	$\omega = \sqrt{\dfrac{MgD}{I}}$	**14-32**
	where D is the distance from the center of mass to the rotation axis and I is the moment of inertia about the rotation axis.	
3. Damped Oscillations	In the oscillations of real systems, the motion is damped because of dissipative forces. If the damping is greater than some critical value, the system does not oscillate when disturbed but merely returns to its equilibrium position. The motion of a weakly damped system is nearly simple harmonic with an amplitude that decreases exponentially with time.	

Frequency	$\omega' = \omega_0\sqrt{1 - \dfrac{1}{4Q^2}}$	14-46
Amplitude	$A = A_0 e^{-(b/2m)t}$	14-38
Energy	$E = E_0 e^{-t/\tau}$	14-41
Decay time	$\tau = \dfrac{m}{b}$	14-42
Q factor defined	$Q = \omega_0\tau$	14-43
Q factor for weak damping	$Q = \dfrac{2\pi}{(|\Delta E|/E)_{\text{cycle}}}, \qquad \dfrac{|\Delta E|}{E} \ll 1$	14-45

4. Driven Oscillations   When an underdamped ($b < b_c$) system is driven by an external sinusoidal force $F_{\text{ext}} = F_0 \cos \omega t$, the system oscillates with a frequency ω equal to the driving frequency and an amplitude A that depends on the driving frequency.

Resonance frequency	$\omega = \omega_0$	
Resonance width for weak damping	$\dfrac{\Delta\omega}{\omega_0} = \dfrac{1}{Q}$	14-49
*Position function	$x = A\cos(\omega t - \delta)$	14-52
*Amplitude	$A = \dfrac{F_0}{\sqrt{m^2(\omega_0^2 - \omega^2)^2 + b^2\omega^2}}$	14-53
*Phase constant	$\tan\delta = \dfrac{b\omega}{m(\omega_0^2 - \omega^2)}$	14-54

PROBLEMS

- Single-concept, single-step, relatively easy
- •• Intermediate-level, may require synthesis of concepts
- ••• Challenging
- **SSM** Solution is in the *Student Solutions Manual*
- **iSOLVE** Problems available on iSOLVE online homework service
- **iSOLVE✓** These "Checkpoint" online homework service problems ask students additional questions about their confidence level, and how they arrived at their answer

In a few problems, you are given more data than you actually need; in a few other problems, you are required to supply data from your general knowledge, outside sources, or informed estimates.

Conceptual Problems

1   •   What is the magnitude of the acceleration of an oscillator of amplitude A and frequency f when its speed is maximum? When its displacement from equilibrium is maximum?

2   •   Are the acceleration and the displacement (from equilibrium) of a simple harmonic oscillator ever in the same direction? The acceleration and the velocity? The velocity and the displacement? Explain.

3   •   True or false:

(a) For a simple harmonic oscillator, the period is proportional to the square of the amplitude.
(b) For a simple harmonic oscillator, the frequency does not depend on the amplitude.
(c) If the acceleration of a particle undergoing 1-dimensional motion is proportional to the displacement from equilibrium and oppositely directed, the motion is simple harmonic.

4 • $\boxed{\text{SSM}}$ If the amplitude of a simple harmonic oscillator is tripled, by what factor is the energy changed?

5 •• An object attached to a spring has simple harmonic motion with an amplitude of 4.0 cm. When the object is 2.0 cm from the equilibrium position, what fraction of its total energy is potential energy? (a) One-quarter. (b) One-third. (c) One-half. (d) Two-thirds. (e) Three-quarters.

6 • True or false:

(a) For a given object on a given spring, the period is the same whether the spring is vertical or horizontal.
(b) For a given object oscillating with amplitude A on a given spring, the maximum speed is the same whether the spring is vertical or horizontal.

7 • True or false: The motion of a simple pendulum is simple harmonic for any initial angular displacement.

8 • True or false: The motion of a simple pendulum is periodic for any initial angular displacement.

9 •• $\boxed{\text{SSM}}$ Two identical carts on a frictionless air track are attached by a spring. One is suddenly struck a blow that sends it moving away from the other cart. The motion of the carts is seen to be very jerky—first one cart moves, then stops as the other cart moves and stops in its turn. Explain the motion in a qualitative way.

10 •• $\boxed{\text{SSM}}$ The length of the string or wire supporting a pendulum bob increases slightly when its temperature is raised. How would this affect a clock operated by a simple pendulum?

11 • True or false: The mechanical energy of a damped, undriven oscillator decreases exponentially with time.

12 • True or false:

(a) Resonance occurs when the driving frequency equals the natural frequency.
(b) If the Q value is high, the resonance is sharp.

13 • Give some examples of common systems that can be considered to be driven oscillators.

14 • A crystal wineglass shattered by an intense sound is an example of (a) resonance, (b) critical damping, (c) an exponential decrease in energy, (d) overdamping.

15 • $\boxed{\text{SSM}}$ The effect of the mass of a spring on the motion of an object attached to it is usually neglected. Describe qualitatively its effect when it is not neglected.

16 •• A lamp hanging from the ceiling of the club car in a train oscillates with period T_0 when the train is at rest. The period will be (match left and right columns)

1. greater than T_0 when	A. the train moves horizontally with constant velocity.
2. less than T_0 when	B. the train rounds a curve of radius R with speed v.
3. equal to T_0 when	C. the train climbs a hill of inclination θ at constant speed.
	D. the train goes over the crest of a hill of radius of curvature R with constant speed.

17 •• Two mass–spring systems oscillate at frequencies f_A and f_B. If $f_A = 2f_B$ and the spring constants of the two springs are equal, it follows that the masses are related by (a) $M_A = 4M_B$, (b) $M_A = M_B/\sqrt{2}$, (c) $M_A = M_B/2$, (d) $M_A = M_B/4$.

18 •• Two mass–spring systems A and B oscillate so that their energies are equal. If $M_A = 2M_B$, which formula relates the amplitudes of oscillation? (a) $A_A = A_B/4$. (b) $A_A = A_B/\sqrt{2}$. (c) $A_A = A_B$. (d) Not enough information is given to determine the ratio of the amplitudes.

19 •• Two mass–spring systems A and B oscillate so that their energies are equal. If $k_A = 2k_B$, then which formula relates the amplitudes of oscillation? (a) $A_A = A_B/4$. (b) $A_A = A_B/\sqrt{2}$. (c) $A_A = A_B$. (d) Not enough information is given to determine the ratio of the amplitudes.

20 •• Pendulum A has a bob of mass M_A and a length L_A; pendulum B has a bob of mass M_B and a length L_B. If the period of A is twice that of B, then (a) $L_A = 2L_B$ and $M_A = 2M_B$, (b) $L_A = 4L_B$ and $M_A = M_B$, (c) $L_A = 4L_B$ whatever the ratio M_A/M_B, (d) $L_A = \sqrt{2}\,L_B$ whatever the ratio M_A/M_B.

Estimation and Approximation

21 •• For a child on a swing, the amplitude drops by a factor of $1/e$ in about eight periods if no mechanical energy is fed in. Estimate the Q factor for this system.

22 •• $\boxed{\text{SSM}}$ (a) Estimate the natural period of oscillation for swinging your arms as you walk, with your hands empty. (b) Now estimate it when carrying a heavy briefcase. Look around at other people as they walk by—do these two estimates seem on target?

Simple Harmonic Motion

23 • $\boxed{\text{iSOLVE}}$ The position of a particle is given by $x = (7\ \text{cm})\cos 6\pi t$, where t is in seconds. What are (a) the frequency, (b) the period, and (c) the amplitude of the particle's motion? (d) What is the first time after $t = 0$ that the particle is at its equilibrium position? In what direction is it moving at that time?

24 • What is the phase constant δ in Equation 14-4 if the position of the oscillating particle at time $t = 0$ is (a) 0, (b) $-A$, (c) A, (d) A/2?

25 • $\boxed{\text{SSM}}$ A particle of mass m begins at rest from $x = +25$ cm and oscillates about its equilibrium position at $x = 0$ with a period of 1.5 s. Write equations for (a) the position x as a function of t, (b) the velocity v as a function of t, and (c) the acceleration a as a function of t.

26 • Find (a) the maximum speed and (b) the maximum acceleration of the particle in Problem 23. (c) What is the first time that the particle is at $x = 0$ and moving to the right?

27 •• Work Problem 25 with the particle initially at $x = 25$ cm and moving with velocity $v_0 = +50$ cm/s.

28 •• The period of an oscillating particle is 8 s and its amplitude is 12 cm. At $t = 0$ it is at its equilibrium position. Find the distance traveled during the intervals (a) $t = 0$ to

$t = 2$ s, (b) $t = 2$ s to $t = 4$ s, (c) $t = 0$ to $t = 1$ s, and (d) $t = 1$ s to $t = 2$ s.

29 •• The period of an oscillating particle is 8 s. At $t = 0$, the particle is at rest at $x = A = 10$ cm. (a) Sketch x as a function of t. (b) Find the distance traveled in the first, second, third, and fourth second after $t = 0$.

30 •• **SSM** **iSOLVE** Military specifications often call for electronic devices to be able to withstand accelerations of $10g = 98.1$ m/s^2. To make sure that their products meet this specification, manufacturers test them using a shaking table that can vibrate a device at various specified frequencies and amplitudes. If a device is given a vibration of amplitude 1.5 cm, what should its frequency be in order to test for compliance with the $10g$ military specification?

31 •• **iSOLVE✓** The position of a particle is given by $x = 2.5 \cos \pi t$, where x is in meters and t is in seconds. (a) Find the maximum speed and maximum acceleration of the particle. (b) Find the speed and acceleration of the particle when $x = 1.5$ m.

32 •• **SSM** (a) Show that $A_0 \cos(\omega t + \delta)$ can be written as $A_s \sin(\omega t) + A_c \cos(\omega t)$, and determine A_s and A_c in terms of A_0 and δ. (b) Relate A_c and A_s to the initial position and velocity of a particle undergoing simple harmonic motion.

Simple Harmonic Motion and Circular Motion

33 • **iSOLVE** A particle moves in a circle of radius 40 cm with a constant speed of 80 cm/s. Find (a) the frequency of the motion and (b) the period of the motion. (c) Write an equation for the x component of the position of the particle as a function of time t, assuming that the particle is on the positive x axis at time $t = 0$.

34 • **SSM** A particle moves in a circle of radius 15 cm, making 1 revolution every 3 s. (a) What is the speed of the particle? (b) What is its angular velocity ω? (c) Write an equation for the x component of the position of the particle as a function of time t, assuming that the particle is on the positive x axis at time $t = 0$.

Energy in Simple Harmonic Motion

35 • A 2.4-kg object is attached to a horizontal spring of force constant $k = 4.5$ kN/m. The spring is stretched 10 cm from equilibrium and released. Find its total energy.

36 • Find the total energy of a 3-kg object oscillating on a horizontal spring with an amplitude of 10 cm and a frequency of 2.4 Hz.

37 • A 1.5-kg object oscillates with simple harmonic motion on a spring of force constant $k = 500$ N/m. Its maximum speed is 70 cm/s. (a) What is the total mechanical energy? (b) What is the amplitude of the oscillation?

38 • A 3-kg object oscillating on a spring of force constant 2 kN/m has a total energy of 0.9 J. (a) What is the amplitude of the motion? (b) What is the maximum speed?

39 • An object oscillates on a spring with an amplitude of 4.5 cm. Its total energy is 1.4 J. What is the force constant of the spring?

40 •• **SSM** **iSOLVE✓** A 3-kg object oscillates on a spring with an amplitude of 8 cm. Its maximum acceleration is 3.50 m/s^2. Find the total energy.

Springs

41 • A 2.4-kg object is attached to a horizontal spring of force constant $k = 4.5$ kN/m. The spring is stretched 10 cm from equilibrium and released. What are (a) the frequency of the motion, (b) the period, (c) the amplitude, (d) the maximum speed, and (e) the maximum acceleration? (f) When does the object first reach its equilibrium position? What is its acceleration at this time?

42 • Answer the questions in Problem 41 for a 5-kg object attached to a spring of force constant $k = 700$ N/m when the spring is initially stretched 8 cm from equilibrium.

43 • A 3-kg object attached to a horizontal spring oscillates with an amplitude $A = 10$ cm and a frequency $f = 2.4$ Hz. (a) What is the force constant of the spring? (b) What is the period of the motion? (c) What is the maximum speed of the object? (d) What is the maximum acceleration of the object?

44 • **SSM** **iSOLVE✓** An 85-kg person steps into a car of mass 2400 kg, causing it to sink 2.35 cm on its springs. Assuming no damping, with what frequency will the car and passenger vibrate on the springs?

45 • A 4.5-kg object oscillates on a horizontal spring with an amplitude of 3.8 cm. Its maximum acceleration is 26 m/s^2. Find (a) the force constant k, (b) the frequency, and (c) the period of the motion.

46 • An object oscillates with an amplitude of 5.8 cm on a horizontal spring of force constant 1.8 kN/m. Its maximum speed is 2.20 m/s. Find (a) the mass of the object, (b) the frequency of the motion, and (c) the period of the motion.

47 •• **iSOLVE✓** A 0.4-kg block attached to a spring of force constant 12 N/m oscillates with an amplitude of 8 cm. Find (a) the maximum speed of the block, (b) the speed and acceleration of the block when it is at $x = 4$ cm from the equilibrium position, and (c) the time it takes the block to move from $x = 0$ to $x = 4$ cm.

48 •• **SSM** An object of mass m is supported by a vertical spring of force constant 1800 N/m. When pulled down 2.5 cm from equilibrium and released from rest, the object oscillates at 5.5 Hz. (a) Find m. (b) Find the amount the spring is stretched from its natural length when the object is in equilibrium. (c) Write expressions for the displacement x, the velocity v, and the acceleration a as functions of time t.

49 •• **iSOLVE✓** An object of unknown mass is hung on the end of an unstretched spring and is released from rest. If the object falls 3.42 cm before first coming to rest, find the period of the motion.

50 •• A spring of force constant $k = 250$ N/m is suspended from a rigid support. An object of mass 1 kg is attached to the unstretched spring and the object is released from rest. (a) How far below the starting point is the equilibrium position for the object? (b) How far down does the object move before it starts up again? (c) What is the period of oscillation? (d) What is the speed of the object when it first reaches its equilibrium position? (*e*) When does it first reach its equilibrium position?

51 •• [I SOLVE] The St. Louis Arch has a height of 192 m. Suppose that a stunt woman of mass 60 kg jumps off the top of the arch with an elastic band attached to her feet. She reaches the ground at zero speed. Find her kinetic energy *K* after 2.00 s of the flight. (Assume that the elastic band obeys Hooke's law, and neglect its length when relaxed.)

52 •• [SSM] A suitcase of mass 20 kg is hung from two bungie cords, as shown in Figure 14-26. Each cord is stretched 5 cm when the suitcase is in equilibrium. If the suitcase is pulled down a little and released, what will be its oscillation frequency?

FIGURE 14-26 Problem 52

53 •• A 0.12-kg block is suspended from a spring. When a small stone of mass 30 g is placed on the block, the spring stretches an additional 5 cm. With the stone on the block, the spring oscillates with an amplitude of 12 cm. (*a*) What is the frequency of the motion? (*b*) How long does the block take to travel from its lowest point to its highest point? (*c*) What is the net force on the stone when it is at the point of maximum upward displacement?

54 •• In Problem 53, find the maximum amplitude of oscillation at which the stone will remain in contact with the block.

55 •• An object of mass 2.0 kg is attached to the top of a vertical spring that is anchored to the floor. The uncompressed length of the spring is 8.0 cm and the length of the spring when the object is in equilibrium is 5.0 cm. When the object is resting at its equilibrium position, it is given a downward impulse with a hammer so that its initial speed is 0.3 m/s. (*a*) To what maximum height above the floor does the object eventually rise? (*b*) How long does it take for the object to reach its maximum height for the first time? (*c*) Does the spring ever become uncompressed? What minimum initial velocity must be given to the object for the spring to be uncompressed at some time?

56 •• [SSM] A winch cable has a cross-sectional area of 1.5 cm² and a length of 2.5 m. Young's modulus for the cable is 150 GN/m². A 950-kg engine block is hung from the end of the cable. (*a*) By what length does the cable stretch? (*b*) Treating the cable as a simple spring, what is the oscillation frequency of the engine block at the end of the cable?

Energy of an Object on a Vertical Spring

57 •• A 2.5-kg object hanging from a vertical spring of force constant 600 N/m oscillates with an amplitude of 3 cm. When the object is at its maximum downward displacement, find (*a*) the total energy of the system, (*b*) the gravitational potential energy, and (*c*) the potential energy in the spring. (*d*) What is the maximum kinetic energy of the object? (Choose $U = 0$ when the object is in equilibrium.)

58 •• [I SOLVE] A 1.5-kg object that stretches a spring 2.8 cm from its natural length when hanging at rest oscillates with an amplitude of 2.2 cm. (*a*) Find the total energy of the system. (*b*) Find the gravitational potential energy at maximum downward displacement. (*c*) Find the potential energy in the spring at maximum downward displacement. (*d*) What is the maximum kinetic energy of the object?

59 •• [SSM] A 1.2-kg object hanging from a spring of force constant 300 N/m oscillates with a maximum speed of 30 cm/s. (*a*) What is its maximum displacement? When the object is at its maximum displacement, find (*b*) the total energy of the system, (*c*) the gravitational potential energy, and (*d*) the potential energy in the spring.

Simple Pendulums

60 • Find the length of a simple pendulum if the period is 5 s at a location where $g = 9.81$ m/s².

61 • What would be the period of the pendulum in Problem 60 if the pendulum were on the moon, where the acceleration due to gravity is one-sixth that on earth?

62 • [I SOLVE ✓] If the period of a pendulum 70 cm long is 1.68 s, what is the value of g at the location of the pendulum?

63 • [SSM] [I SOLVE] A pendulum set up in the stairwell of a 10-story building consists of a heavy weight suspended on a 34.0-m wire. If $g = 9.81$ m/s², what is the period of oscillation?

64 •• Show that the total energy of a simple pendulum undergoing oscillations of small amplitude ϕ_0 is approximately $E \approx \frac{1}{2}mgL\phi_0^2$. *Hint: Use the approximation* $\cos \phi \approx 1 - \frac{1}{2}\phi^2$ *for small* ϕ.

65 •• A simple pendulum of length L is attached to a massive cart that slides without friction down a plane inclined at angle θ with the horizontal as shown in Figure 14-27. Find the period of oscillation of the pendulum on the sliding cart.

FIGURE 14-27 Problem 65

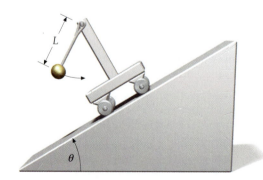

66 •• A simple pendulum of length L is released from rest from an angle ϕ_0. (a) Assuming that the pendulum undergoes simple harmonic motion, find its speed as it passes through $\phi = 0$. (b) Using the conservation of energy, find this speed exactly. (c) Show that your results for (a) and (b) are the same when ϕ_0 is small. (d) Find the difference in your results for $\phi_0 = 0.20$ rad and $L = 1$ m.

*Physical Pendulums

67 • [SOLVE]✔ A thin disk of mass 5 kg and radius 20 cm is suspended by a horizontal axis perpendicular to the disk through its rim. The disk is displaced slightly from equilibrium and released. Find the period of the subsequent simple harmonic motion.

68 • A circular hoop of radius 50 cm is hung on a narrow horizontal rod and allowed to swing in the plane of the hoop. What is the period of its oscillation, assuming that the amplitude is small?

69 • A 3-kg plane figure is suspended at a point 10 cm from its center of mass. When it is oscillating with small amplitude, the period of oscillation is 2.6 s. Find the moment of inertia I about an axis perpendicular to the plane of the figure through the pivot point.

70 •• The pendulum bob of a large town-hall clock has a length of 4 m. (a) What is its period of oscillation? Treat it as a simple pendulum with small amplitude oscillations. (b) To regulate the period of the pendulum there is a tray attached to its shaft, halfway up. The tray holds a stack of coins. To change the period by a little bit, coins are added or removed from the tray. Explain in detail why this works. Will adding coins increase or decrease the period of the pendulum?

71 •• Figure 14-28 shows a dumbbell with two equal masses, to be considered as point masses attached to a thin massless rod of length L. (a) Show that the period of this pendulum is a minimum when the pivot point P is at one of the masses. (b) Find the period of this physical pendulum if the distance between P and the upper mass is $L/4$.

FIGURE 14-28 Problem 71

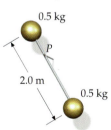

2.0 m

0.5 kg

0.5 kg

72 •• Suppose the rod in Problem 71 has a mass of $2m$ (Figure 14-29). Determine the distance between the upper mass and the pivot point P when the period of this physical pendulum is a minimum.

FIGURE 14-29 Problem 72

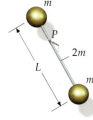

m

2m

L

m

73 •• [SSM] You are given a meterstick and asked to drill a narrow hole in it so that, when the stick is pivoted about the hole, the period of the pendulum will be a minimum. Where should you drill the hole?

74 •• [SOLVE] Figure 14-30 shows a uniform disk of radius $R = 0.8$ m and 6-kg mass with a small hole a distance d from the disk's center that can serve as a pivot point. (a) What should be the distance d so that the period of this physical pendulum is 2.5 s? (b) What should be the distance d so that this physical pendulum will have the shortest possible period? What is this period?

FIGURE 14-30 Problem 74

d
0.8 m
6 kg

75 ••• A plane object has a moment of inertia I about its center of mass. When pivoted at point P_1, as shown in Figure 14-31, it oscillates about the pivot with a period T. There is a second point P_2 on the opposite side of the center of mass about which the object can be pivoted so that the period of oscillation is also T. Show that $h_1 + h_2 = gT^2/(4\pi^2)$.

FIGURE 14-31 Problem 75

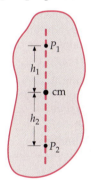

P_1
h_1
cm
h_2
P_2

76 ••• A physical pendulum consists of a spherical bob of radius r and mass m suspended from a string (Figure 14-32). The distance from the center of the sphere to the point of support is L. When r is much less than L, such a pendulum is often treated as a simple pendulum of length L. (a) Show that the period for small oscillations is given by

FIGURE 14-32 Problem 76

L

r m

$$T = T_0\sqrt{1 + \frac{2r^2}{5L^2}}$$

where $T_0 = 2\pi\sqrt{L/g}$ is the period of a simple pendulum of length L. (b) Show that when r is much smaller than L, the period is approximately $T \approx T_0(1 + r^2/5L^2)$. (c) If $L = 1$ m and

$r = 2$ cm, find the error when the approximation $T = T_0$ is used for this pendulum. How large must be the radius of the bob for the error to be 1 percent?

77 ••• Figure 14-33 shows the pendulum of a clock. The uniform rod of length $L = 2.0$ m has a mass $m = 0.8$ kg. Attached to the rod is a uniform disk of mass $M = 1.2$ kg and radius 0.15 m. The clock is constructed to keep perfect time if the period of the pendulum is exactly 3.50 s. (a) What should be the distance d so that the period of this pendulum is 2.50 s? (b) Suppose that the pendulum clock loses 5.0 min/d. How far and in what direction should the disk be moved to ensure that the clock will keep perfect time?

FIGURE 14-33 Problem 77

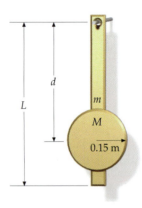

78 •• **SSM** A pendulum clock loses 48 s/d when the amplitude of the pendulum is 8.4°. What should be the amplitude of the pendulum so that the clock keeps perfect time?

79 •• **iSOLVE✓** A pendulum clock that has run down to a very small amplitude gains 5 min each day. What angular amplitude should the pendulum have to keep the correct time?

Damped Oscillations

80 • Show that the damping constant b has units of kg/s.

81 • An oscillator has a Q factor of 200. By what percentage does its energy decrease during one period?

82 • **iSOLVE** A 2-kg object oscillates with an initial amplitude of 3 cm on a spring of force constant $k = 400$ N/m. Find (a) the period and (b) the total initial energy. (c) If the energy decreases by 1 percent per period, find the damping constant b and the Q factor.

83 •• Show that the ratio of the amplitudes for two successive oscillations is constant for a damped oscillator.

84 •• An oscillator has a period of 3 s. Its amplitude decreases by 5 percent during each cycle. (a) By how much does its energy decrease during each cycle? (b) What is the time constant τ? (c) What is the Q factor?

85 •• An oscillator has a Q factor of 20. (a) By what fraction does the energy decrease during each cycle? (b) Use Equation 14-37 to find the percentage difference between ω' and ω_0. Hint: Use the approximation $(1 + x)^{\frac{1}{2}} \approx 1 + \frac{1}{2}x$ for small x.

86 •• **iSOLVE** A damped mass–spring system oscillates at 200 Hz. The time constant of the system is 2.0 s. At $t = 0$ the

amplitude of oscillation is 6.0 cm and the energy of the oscillating system is 60 J. (a) What are the amplitudes of oscillation at $t = 2.0$ s and $t = 4.0$ s? (b) How much energy is dissipated in the first 2-s interval and in the second 2-s interval?

87 •• **SSM** **iSOLVE** It has been stated that the vibrating earth has a resonance period of 54 min and a Q factor of about 400 and that after a large earthquake, the earth "rings" (continues to vibrate) for about 2 months. (a) Find the percentage of the energy of vibration lost to damping forces during each cycle. (b) Show that after n periods the energy is $E_n = (0.984)^n E_0$, where E_0 is the original energy. (c) If the original energy of vibration of an earthquake is E_0, what is the energy after 2 d?

88 •• A compact pendulum used in a physics experiment has a mass of 15 g; the length of the pendulum is 75 cm. To start the bob oscillating, a physics student puts a fan next to it that blows a horizontal stream of air on the bob. With the fan on, the bob is in equilibrium when the pendulum is displaced by an angle of 5° from the vertical. The speed of the air from the fan is 7 m/s. The fan is then turned off, and the pendulum is allowed to oscillate. (a) If we assume that the drag force due to the air is of the form $-bv$, what is the decay time constant τ for the oscillations of the pendulum? (b) How long will it take for the amplitude of oscillation of the pendulum to reach 1°?

Driven Oscillations and Resonance

89 • Find the resonance frequency for each of the three systems shown in Figure 14-34.

FIGURE 14-34 Problem 89

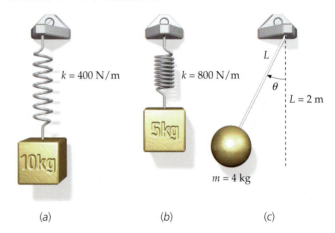

(a)   (b)   (c)

90 • A damped oscillator loses 2 percent of its energy during each cycle. (a) What is its Q factor? (b) If its resonance frequency is 300 Hz, what is the width of the resonance curve $\Delta\omega$ when the oscillator is driven?

91 •• A 2-kg object oscillates on a spring of force constant $k = 400$ N/m. The damping constant has a value of $b = 2.00$ kg/s. The system is driven by a sinusoidal force of maximum value 10 N and angular frequency $\omega = 10$ rad/s. (a) What is the amplitude of the oscillations? (b) If the driving frequency is varied, at what frequency will resonance occur? (c) What is the amplitude of oscillation at resonance? (d) What is the width of the resonance curve $\Delta\omega$?

92 •• A damped oscillator loses 3.5 percent of its energy during each cycle. (*a*) How many cycles elapse before half of its original energy is dissipated? (*b*) What is its *Q* factor? (*c*) If the natural frequency is 100 Hz, what is the width of the resonance curve when the oscillator is driven?

Collisions

93 ••• Figure 14-35 shows a vibrating mass–spring system supported on a frictionless surface and a second, equal mass that is moving toward the vibrating mass with velocity *v*. The motion of the vibrating mass is given by $x(t) = (0.1 \text{ m}) \cos(40 \text{ s}^{-1} t)$, where *x* is the displacement of the mass from its equilibrium position. The two masses collide elastically just as the vibrating mass passes through its equilibrium position traveling to the right. (*a*) What should be the velocity *v* of the second mass so that the mass–spring system is at rest following the elastic collision? (*b*) What is the velocity of the second mass after the elastic collision?

FIGURE 14-35 Problem 93

94 ••• Following the elastic collision in Problem 93, the kinetic energy of the recoiling mass is 8.0 J. Find the masses *m* and the spring constant *k*.

95 ••• An object of mass 2 kg resting on a frictionless horizontal surface is attached to a spring of force constant 600 N/m. A second object of mass 1 kg slides along the surface toward the first object at 6 m/s. (*a*) Find the amplitude of oscillation if the objects make a perfectly inelastic collision and remain together on the spring. What is the period of oscillation? (*b*) Find the amplitude and period of oscillation if the collision is elastic. (*c*) For each type of collision, write an expression for the position *x* as a function of time *t* for the object attached to the spring, assuming that the collision occurs at time $t = 0$.

General Problems

96 • A particle has a displacement $x = 0.4 \cos(3t + \pi/4)$, where *x* is in meters and *t* is in seconds. (*a*) Find the frequency *f* and period *T* of the motion. (*b*) Where is the particle at $t = 0$? (*c*) Where is the particle at $t = 0.5$ s?

97 • (*a*) Find an expression for the velocity of the particle whose position is given in Problem 96. (*b*) What is the velocity at time $t = 0$? (*c*) What is the maximum velocity? (*d*) At what time after $t = 0$ does this maximum velocity first occur?

98 • An object on a horizontal spring oscillates with a period of 4.5 s. If the object is suspended from the spring vertically, by how much is the spring stretched from its natural length when the object is in equilibrium?

99 •• **SSM** A small particle of mass *m* slides without friction in a spherical bowl of radius *r*. (*a*) Show that the motion

of the particle is the same as if it were attached to a string of length *r*. (*b*) Figure 14-36 shows a particle of mass m_1 that is displaced a small distance s_1 from the bottom of the bowl, where s_1 is much smaller than *r*. A second particle of mass m_2 is displaced in the opposite direction a distance $s_2 = 3s_1$, where s_2 is also much smaller than *r*. If the particles are released at the same time, where do they meet? Explain.

FIGURE 14-36 Problems 99, 100

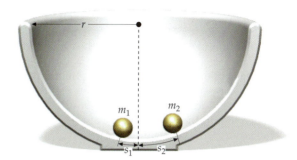

100 •• Now consider a very small uniform ball of mass *m* and radius *R* rolling without slipping near the bottom of the bowl in Figure 14-36. (*a*) Write an expression for the total energy of the ball in terms of its velocity and the distance (assumed small) from the center of the bowl. (*b*) By comparing this expression with that for the total energy of a frictionless ball of mass *m* sliding down the side of the bowl, determine the oscillation frequency of the ball about the center of the bowl.

101 •• **iSOLVE** ✓ As your jet plane speeds down the runway on takeoff, you measure its acceleration by suspending your yo-yo as a simple pendulum and noting that when the bob (mass 40 g) is at rest relative to you, the string (length 70 cm) makes an angle of 22° with the vertical. Find the period *T* for small oscillations of this pendulum.

102 •• If a wire is twisted, there will be a restoring torque $\tau = -\kappa\theta$, where κ is a torsional spring constant and θ is the total twist angle. A torsion balance consists of an object with moment of inertia *I* hung at the end of a wire. If the object is given a twist, show that the frequency of small torsional oscillations is $\omega = \sqrt{\kappa/I}$.

103 •• A simple torsion balance (see Problem 102) used in a variety of physics experiments is shown in Figure 14-37. There is a cross-arm of negligible mass at the end of the wire with identical particles attached at each end. If each particle has a mass of 50 g, the length of the cross-arm is 5.0 cm, and the oscillation period of the balance is 80 s, what is the wire's torsion constant κ?

FIGURE 14-37 Problem 103

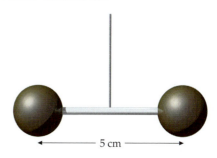

5 cm

104 •• **SSM** **iSOLVE** A wooden cube with edge length a and mass m floats in water with one of its faces parallel to the water surface. The density of the water is ρ. Find the period of oscillation in the vertical direction if the cube is pushed down slightly.

105 •• A clock with a pendulum keeps perfect time on the earth's surface. In which case will the error be greater: if the clock is placed in a mine of depth h or if the clock is elevated to a height h? Assume that $h \ll R_E$.

106 •• **iSOLVE** Figure 14-38 shows a pendulum of length L with a bob of mass M. The bob is attached to a spring of spring constant k as shown. When the bob is directly below the pendulum support, the spring is at its equilibrium length. (a) Derive an expression for the period of this oscillating system for small amplitude vibrations. (b) Suppose that $M = 1$ kg and L is such that in the absence of the spring the period is 2.0 s. What is the spring constant k if the period of the oscillating system is 1.0 s?

FIGURE 14-38
Problem 106

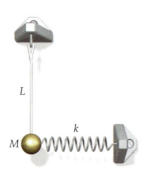

107 •• An object of mass m_1 sliding on a frictionless horizontal surface is attached to a spring of force constant k and oscillates with an amplitude A. When the spring is at its greatest extension and the mass is instantaneously at rest, a second object of mass m_2 is placed on top of it. (a) What is the smallest value for the coefficient of static friction μ_s such that the second object does not slip on the first? (b) Explain how the total energy E, the amplitude A, the angular frequency ω, and the period T of the system are changed by placing m_2 on m_1, assuming that the friction is great enough so that there is no slippage.

108 •• A box with a mass of 100 kg hangs from the ceiling of a room by a spring with a spring constant of 500 N/m. The uncompressed length of the spring is 0.5 m. (a) Find the equilibrium position of the box. (b) An identical spring is stretched and attached to the ceiling and box in parallel with the first spring. Find the frequency of the oscillations when the box is released. (c) What is the new equilibrium position of the box once it comes to rest?

109 •• The acceleration due to gravity g varies with geographical location because of the earth's rotation and because the earth is not exactly spherical. This was first discovered in the seventeenth century, when it was noted that a pendulum clock carefully adjusted to keep correct time in Paris lost about 90 s/d near the equator. (a) Show that a small change in the acceleration of gravity Δg produces a small change in the period ΔT of a pendulum given by $\Delta T/T \approx -\frac{1}{2}\Delta g/g$. (Use differentials to approximate ΔT and Δg.) (b) How great a change in g is needed to account for a change in the period of 90 s/d?

110 •• Figure 14-39 shows two 0.6-kg blocks glued to each other and connected to a spring of spring constant $k = 240$ N/m. The blocks, which rest on a frictionless horizontal surface, are displaced 0.6 m from their equilibrium position and released. Before they are released, a few drops of solvent are deposited on the glue. (a) Find the frequency of vibration and total energy of the vibrating system before the glue has dissolved. (b) Find the frequency, amplitude of vibration, and energy of the vibrating system if the glue dissolves when the spring is (1) at maximum compression and (2) at maximum extension.

FIGURE 14-39 Problem 110

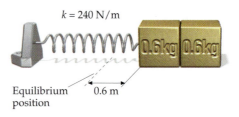

$k = 240$ N/m

Equilibrium position   0.6 m

111 •• Show that for the situations in Figure 14-40(a) and (b), the object oscillates with a frequency $f = (1/2\pi)\sqrt{k_{eff}/m}$, where k_{eff} is given by (a) $k_{eff} = k_1 + k_2$ and (b) $1/k_{eff} = 1/k_1 + 1/k_2$. *Hint: Find the net force* F *on the object for a small displacement* x *and write* $F = -k_{eff}x$. *Note that in (b) the springs stretch by different amounts, the sum of which is x.*

FIGURE 14-40 Problem 111

(a)

k_1 k_2 m

(b)

k_1 k_2 m

112 •• **SSM** A small block of mass m_1 rests on a piston that is vibrating vertically with simple harmonic motion given by $y = A \sin \omega t$. (a) Show that the block will leave the piston if $\omega^2 A > g$. (b) If $\omega^2 A = 3g$ and $A = 15$ cm, at what time will the block leave the piston?

113 •• **iSOLVE** The plunger of a pinball machine has mass m_p and is attached to a spring of force constant k (Figure 14-41). The spring is compressed a distance x_0 from its equilibrium position $x = 0$ and released. A ball of mass m_b is next to the plunger. (a) Where does the ball leave the plunger? (b) What is the speed v_s of the ball when it separates? (c) At what distance x_f does the plunger come to rest momentarily? (Assume that the surface is horizontal and frictionless so that the ball slides rather than rolls.)

FIGURE 14-41 Problem 113

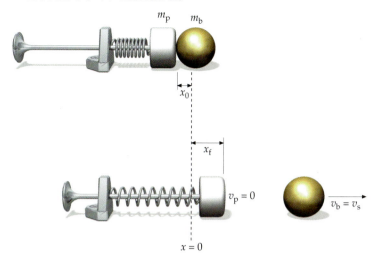

x_0

x_f

$v_p = 0$

$v_b = v_s$

$x = 0$

114 •• A level platform vibrates horizontally with simple harmonic motion with a period of 0.8 s. (a) A box on the platform starts to slide when the amplitude of vibration reaches 40 cm; what is the coefficient of static friction between the box and the platform? (b) If the coefficient of friction between the box and platform were 0.40, what would be the maximum amplitude of vibration before the box would slip?

115 ••• The potential energy of a particle of mass m as a function of position is given by $U(x) = U_0(\alpha + 1/\alpha)$, where $\alpha = x/a$ and a is a constant. (a) Plot $U(x)$ versus x for $0.1a < x < 3a$. (b) Find the value of $x = x_0$ at stable equilibrium. (c) Write the potential energy $U(x)$ for $x = x_0 + \varepsilon$, where ε is a small displacement from the equilibrium position x_0. (d) Approximate the $1/x$ term using the binomial expansion

$$(1 + r)^n = 1 + nr + \frac{n(n-1)}{(2)(1)}r^2 + \frac{n(n-1)(n-2)}{(3)(2)(1)}r^3 + \cdots,$$

with $r = \varepsilon/x_0 \ll 1$ and discarding all terms of power greater than r^2. (e) Compare your result with the potential for a simple harmonic oscillator. Show that the mass will undergo simple harmonic motion for small displacements from equilibrium and determine the frequency of this motion.

116 ••• A solid cylindrical drum of mass 6.0 kg and diameter 0.06 m rolls without slipping on a horizontal surface (Figure 14-42). The axle of the drum is attached to a spring of spring constant $k = 4000$ N/m as shown. (a) Determine the frequency of oscillation of this system for small displacements from equilibrium. (b) What is the minimum value of the coefficient of static friction so that the drum will not slip when the vibrational energy is 5.0 J?

FIGURE 14-42 Problem 116

117 ••• SSM If we attach two blocks of masses m_1 and m_2, to either end of a spring of spring constant k and set them into oscillation, show that the oscillation frequency $\omega = (k/\mu)^{1/2}$, where $\mu = m_1 m_2/(m_1 + m_2)$ is the reduced mass of the system.

118 •• One of the vibrational modes of the HCl molecule has a frequency of 8.969×10^{13} s^{-1}. Using the relation stated in Problem 117, find the "spring constant" for the HCl molecule.

119 •• In Problem 118, if we were to replace the hydrogen atom in HCl by a deuterium atom, what would be the new vibration frequency of the molecule? Deuterium consists of 1 proton and 1 neutron.

120 ••• A block of mass m on a horizontal table is attached to a spring of force constant k as shown in Figure 14-43. The coefficient of kinetic friction between the block and the table is μ_k. The spring is unstretched if the block is at the origin ($x = 0$). The block is released with the spring stretched a distance A, where $kA > \mu_k mg$. (a) Apply Newton's second law to the block to obtain an equation for its acceleration d^2x/dt^2 for the first half-cycle, during which the block is moving to the left. Show that the resulting equation can be written $d^2x'/dt^2 = -\omega^2 x'$, where $\omega = \sqrt{k/m}$ and $x' = x - x_0$, with $x_0 = \mu_k mg/k = \mu_k g/\omega^2$. (b) Repeat Part (a) for the second half-cycle as the block moves to the right, and show that $d^2x''/dt^2 = -\omega^2 x''$, where $x'' = x + x_0$ and x_0 has the same value. (c) Use a spreadsheet program to graph the first 5 half-cycles for $A = 10x_0$. Describe the motion, if any, after the fifth half-cycle.

FIGURE 14-43
Problem 120

121 ••• Figure 14-44 shows a uniform solid half-cylinder of mass M and radius R resting on a horizontal surface. If one side of this cylinder is pushed down slightly and then released, the object will oscillate about its equilibrium position. Determine the period of this oscillation.

FIGURE 14-44
Problem 121

122 ••• SSM A straight tunnel is dug through the earth as shown in Figure 14-45. Assume that the walls of the tunnel are frictionless. (a) The gravitational force exerted by the earth on a particle of mass m at a distance r from the center of the earth when $r < R_E$ is $F_r = -(GmM_E/R_E^3)r$, where M_E is the mass of the earth and R_E is its radius. Show that the net force on a particle of mass m at a distance x from the middle of the tunnel is given by $F_x = -(GmM_E/R_E^3)x$, and that the motion of the particle is therefore simple harmonic motion. (b) Show that the period of the motion is given by $T = 2\pi\sqrt{R_E/g}$ and find its value in minutes. (This is the same period as that of a satellite orbiting near the surface of the earth and is independent of the length of the tunnel.)

Tunnel

x m

$\vec{F}$

r

M_E

R_E

FIGURE 14-45
Problem 122

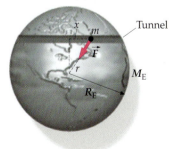

123 ••• A damped oscillator has a frequency ω' that is 10 percent less than its undamped frequency. (*a*) By what factor is the amplitude of the oscillator decreased during each oscillation? (*b*) By what factor is its energy reduced during each oscillation?

124 ••• Show by direct substitution that Equation 14-52 is a solution of Equation 14-51.

125 ••• **SSM** In this problem, you will derive the expression for the average power delivered by a driving force to a driven oscillator (Figure 14-24, page 450).

(*a*) Show that the instantaneous power input of the driving force is given by

$$P = Fv = -A\omega F_0 \cos \omega t \sin(\omega t - \delta).$$

(*b*) Use the trigonometric identity $\sin(\theta_1 - \theta_2) = \sin \theta_1 \cos \theta_2 - \cos \theta_1 \sin \theta_2$ to show that the equation in (*a*) can be written

$$P = A\omega F_0 \sin \delta \cos^2 \omega t - A\omega F_0 \cos \delta \cos \omega t \sin \omega t.$$

(*c*) Show that the average value of the second term in your result for (*b*) over one or more periods is zero and that therefore

$$P_{av} = \tfrac{1}{2} A\omega F_0 \sin \delta.$$

(*d*) From Equation 14-54 for $\tan \delta$, construct a right triangle in which the side opposite the angle δ is $b\omega$ and the side adjacent is $m(\omega_0^2 - \omega^2)$, and use this triangle to show that

$$\sin \delta = \frac{b\omega}{\sqrt{m^2(\omega_0^2 - \omega^2)^2 + b^2\omega^2}} = \frac{b\omega A}{F_0}$$

(*e*) Use your result for (*d*) to eliminate ωA from your result for (*c*) so that the average power input can be written

$$P_{av} = \frac{1}{2}\frac{F_0^2}{b} \sin^2 \delta = \frac{1}{2}\left[\frac{b\omega^2 F_0^2}{m^2(\omega_0^2 - \omega^2)^2 + b^2\omega^2} \right] \qquad 14\text{-}55$$

126 ••• In this problem, you are to use the result of Problem 125 to derive Equation 14-49, which relates the width of the resonance curve to the Q value when the resonance is sharp. At resonance, the denominator of the fraction in brackets in Equation 14-55 is $b^2\omega_0^2$ and P_{av} has its maximum value. (Equation 14-55 can be found in Problem 125.) For a sharp resonance, the variation in ω in the numerator in Equation 14-55 can be neglected. Then the power input will be half its maximum value at the values of ω, for which the denominator is $2b^2\omega_0^2$.

(*a*) Show that ω then satisfies $m^2(\omega - \omega_0)^2(\omega + \omega_0)^2 \approx b^2\omega_0^2$.

(*b*) Using the approximation $\omega + \omega_0 \approx 2\omega_0$, show that $\omega - \omega_0 \approx \pm b/2m$.

(*c*) Express b in terms of Q.

(*d*) Combine the results of (*b*) and (*c*) to show that there are two values of ω for which the power input is half that at resonance and that they are given by

$$\omega_1 = \omega_0 - \frac{\omega_0}{2Q} \quad \text{and} \quad \omega_2 = \omega_0 + \frac{\omega_0}{2Q}$$

Therefore, $\omega_2 - \omega_1 = \Delta\omega = \omega_0/Q$, which is equivalent to Equation 14-49.

127 ••• The Morse potential, which is often used to model interatomic forces, can be written in the form $U(r) = D(1 - e^{-\beta(r-r_0)})^2$, where r is the distance between the two atomic nuclei. (*a*) Using a spreadsheet program or graphing calculator, make a graph of the Morse potential using $D = 5$ eV, $\beta = 0.2$ nm^{-1}, and $r_0 = 0.75$ nm. (*b*) Determine the equilibrium separation and "spring constant" for small displacements from equilibrium for the Morse potential. (*c*) Determine a formula for the oscillation frequency for a homonuclear diatomic molecule (that is, two of the same atoms), where the atoms have mass m.

Traveling Waves

THIS POLICE RADAR UNIT SENDS OUT ELECTROMAGNETIC WAVES THAT TRAVEL AT THE SPEED OF LIGHT AND REFLECT FROM THE MOVING CAR.

? **How does the police officer determine the speed of the car? (See Example 15-12.)**

Waves transport energy and momentum through space without transporting matter. As a water wave moves across a pond, for example, the molecules of water oscillate up and down, but do not cross the pond with the wave. Energy and momentum are transported by the wave, but matter is not. A rowboat will bob up and down on the waves but will not be moved by them across the pond. Water waves, waves on a stretched guitar string, and sound waves all involve oscillation.

➤ **In this chapter we continue the study of oscillatory motion that we began in Chapter 14 by examining periodic waves, particularly harmonic waves. We will see that mechanical waves occur when there is a disturbance in a medium, such as air or water, while electromagnetic waves exist without a material medium.**

15-1 Simple Wave Motion

Transverse and Longitudinal Waves

A mechanical wave is caused by a disturbance in a medium. For example, when a taut string is plucked, the pulses produced travel down the string as waves. The disturbance in this case is the change in shape of the string from its equilibrium shape. Its propagation arises from the interaction of each string segment with the adjacent segments. The segments of the string (the medium) move in the direction perpendicular to the string as the pulses propagate up and down the string. Waves such as these, in which the disturbance is perpendicular to the direction of propagation, are called **transverse** (Figure 15-1). Waves in which the disturbance is parallel to the direction of propagation are called **longitudinal** (Figure 15-2). Sound waves are examples of longitudinal waves—the molecules of a gas, liquid, or solid, through which sound travels oscillate (move back and forth) along the line of propagation, alternately compressing and rarefying (expanding) the medium.

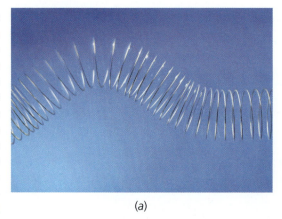

(a)

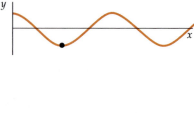

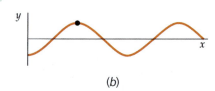

(b)

FIGURE 15-1 (*a*) Transverse wave pulse on a spring. The disturbance is perpendicular to the direction of the motion of the wave. (*b*) Three successive drawings of a transverse wave on a string traveling to the right. An element of the string moves up and down.

Wave Pulses

Figure 15-3*a* shows a pulse on a string at time $t = 0$. The shape of the string at this time can be represented by some function $y = f(x)$. At some later time (Figure 15-3*b*), the pulse is farther down the string. In a new coordinate system with origin O' that moves with the speed of the pulse, the pulse is stationary. The string is described in this frame by $f(x')$ for all times. The x coordinates of the two reference frames are related by

$$x' = x - vt$$

so $f(x') = f(x - vt)$.

Thus, the shape of the string in the original frame is

$$y = f(x - vt), \quad \text{wave moving in the positive } x \text{ direction} \qquad 15\text{-}1$$

The same line of reasoning for a pulse moving to the left leads to

$$y = f(x + vt), \quad \text{wave moving in the negative } x \text{ direction} \qquad 15\text{-}2$$

In both expressions, v is the speed of propagation of the wave. The function $y = f(x - vt)$ is called a **wave function.** For waves on a string, the wave function represents the transverse displacement of the string. For sound waves in air, the wave function can be the longitudinal displacement of the air molecules, or the pressure of the air. These wave functions are solutions of a differential equation called the wave equation, which can be derived from Newton's laws.

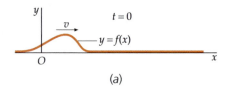

FIGURE 15-2 Longitudinal wave pulse on a spring. The disturbance is in the direction of the motion of the wave.

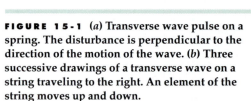

(a)

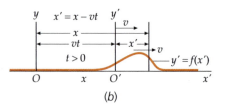

(b)

FIGURE 15-3

Speed of Waves

A general property of waves is that their speed relative to the medium depends on the properties of the medium but is independent of the motion of the source of the waves. For example, the speed of a sound from a car horn depends only on the properties of air and not on the motion of the car.

For wave pulses on a rope, we can easily demonstrate that the greater the tension, the faster the propagation of the waves. Furthermore, waves propagate faster in a light rope than in a heavy rope under the same tension. If F_T is the tension (we use F_T for tension because we use T for the period) and μ is the linear mass density (mass per unit length), then the wave speed is

$$v = \sqrt{\frac{F_T}{\mu}}$$

15-3

SPEED OF WAVES ON A STRING

INCHY RUNS FOR HIS LIFE　　　　　　**EXAMPLE　15-1　Put It in Context**

Inchy, an inchworm, is inching along a cotton clothesline. The 25-m-long clothesline has a mass of 0.25 kg and is kept taut by a hanging object of mass 10 kg as shown in Figure 15-4. Vivian is hanging up her swimsuit 5 m from one end when she sees Inchy 2.5 cm from the opposite end. She plucks the line sending a terrifying 3-cm-high pulse toward Inchy. If Inchy crawls at 1 in./s, will he get to the end of the clothesline before the pulse reaches him?

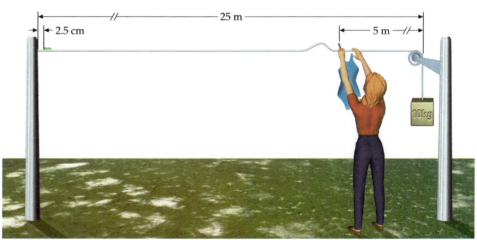

FIGURE 15-4

PICTURE THE PROBLEM We need to know how fast the wave travels. To find the wave speed we use the formula $v = \sqrt{F_T/\mu}$.

1. The speed of the pulse is related to the tension F_T and mass density μ:

$$v = \sqrt{\frac{F_T}{\mu}}$$

2. Express the mass density and tension in terms of the given parameters:

$$\mu = \frac{m_s}{L} \quad \text{and} \quad F_T = mg$$

3. Substitute these values to calculate the speed:

$$v = \sqrt{\frac{F_T}{\mu}} = \sqrt{\frac{mgL}{m_s}} = \sqrt{\frac{(10\text{ kg})(9.81\text{ m/s}^2)(25\text{ m})}{0.25\text{ kg}}}$$

$$= 99.0\text{ m/s}$$

4. Use this speed to find the time for the pulse to travel the 20 m to the far end.

$$\Delta t = \frac{\Delta x}{v} = \frac{20\text{ m}}{99.0\text{ s}} = 0.202\text{ s}$$

5. Find the time it takes Inchy to travel the 2.5 cm to the end traveling at 1 in./s:

$$\Delta t' = \frac{\Delta x'}{v'} = \frac{2.5\text{ cm}}{1\text{ in./s}} \times \frac{1\text{ in.}}{2.54\text{ cm}} = 0.984\text{ s}$$

Inchy does not beat the pulse.

EXERCISE If the 10-kg mass is replaced with a 20-kg mass, what is the speed of waves on the clothesline? (*Answer* 140 m/s)

EXERCISE Show that the units of $\sqrt{F_T/\mu}$ are m/s when F_T is in newtons and μ is in kg/m.

For sound waves in a fluid such as air or water, the speed v is given by

$$v = \sqrt{\frac{B}{\rho}}$$ 15-4

where ρ is the equilibrium density of the medium and B is the bulk modulus[†] (Equation 13-6). Comparing Equations 15-3 and 15-4, we can see that, in general, the speed of waves depends on an elastic property of the medium (the tension for string waves and the bulk modulus for sound waves) and on an inertial property of the medium (the linear mass density or the volume mass density).

For sound waves in a gas such as air, the bulk modulus[‡] is proportional to the pressure, which in turn is proportional to the density ρ and to the absolute temperature T of the gas. The ratio B/ρ is thus independent of density and is merely proportional to the absolute temperature T. In Chapter 17 we will show that, in this case, Equation 15-4 is equivalent to

$$v = \sqrt{\frac{\gamma RT}{M}}$$ 15-5

SPEED OF SOUND IN A GAS

In this equation T is the absolute temperature measured in kelvins (K), which is related to the Celsius temperature t_C by

$$T = t_C + 273$$ 15-6

The constant γ depends on the kind of gas. For diatomic molecules, such as O_2 and N_2, γ has the value 1.4, and since O_2 and N_2 comprise 98 percent of the atmosphere, that is the value for air. (For monatomic molecules such as He, γ has the value 1.67.) The constant R is the universal gas constant

$$R = 8.314 \text{ J}/(\text{mol·K})$$ 15-7

and M is the molar mass of the gas (that is, the mass of 1 mol of the gas), which for air is

$$M = 29 \times 10^{-3} \text{ kg/mol}$$

---

† The bulk modulus is the negative ratio of the pressure change to the fractional change in volume (Chapter 13):

$$B = -\frac{\Delta P}{\Delta V/V}$$

‡ The **isothermal bulk modulus**, which describes changes that occur at constant temperature, differs from the **adiabatic bulk modulus**, which describes changes with no heat transfer. For sound waves at audible frequencies the changes occur too rapidly for appreciable heat flow, so the appropriate bulk modulus is the adiabatic bulk modulus.

E X A M P L E   1 5 - 2   Try It Yourself

Calculate the speed of sound in air at (*a*) 0°C and (*b*) 20°C.

Cover the column to the right and try these on your own before looking at the answers.

Steps	Answers
(*a*) 1. Write Equation 15-5.	$v_a = \sqrt{\dfrac{\gamma R T_a}{M}}$
2. Enter the given values into the equation and solve for the speed. (Be sure to convert the temperature to kelvins.)	$v_a = \boxed{331 \text{ m/s}}$
(*b*) 1. From Equation 15-5 we can see that *v* is proportional to $\sqrt{T}$. Use this to express the ratio of the speed at 293 K to the speed at 273 K.	$\dfrac{v_b}{v_a} = \sqrt{\dfrac{T_b}{T_a}}$
2. Calculate *v* at 293 K.	$v_b = \boxed{343 \text{ m/s}}$

REMARKS We see from this example that the speed of sound in air is about 340 m/s at normal temperatures.

EXERCISE For helium, $M = 4 \times 10^{-3}$ kg/mol and $\gamma = 1.67$. What is the speed of sound waves in helium at 20°C? (*Answer* 1.01 km/s)

Derivation of *v* for Waves on a String Equation 15-3 can be obtained from Newton's laws. Consider a pulse traveling along a string with a speed *v* to the right (Figure 15-5*a*). If the amplitude of the pulse is small compared to the length of the string, then the tension F_T will be approximately constant along the string. In a reference frame moving with speed *v* to the right, the pulse is stationary and the string moves with a speed *v* to the left. Figure 15-5*b* shows a small segment of the string of length Δs at the top of the pulse. The segment forms part of a circular arc of radius *R*. Instantaneously it is moving with speed *v* in a circular path, so it has an acceleration v^2/R in the centripetal direction. The magnitudes of the forces acting on the segment are the tension F_T at each end. The horizontal components of these forces are equal and opposite and thus cancel. The vertical components of these forces point radially inward toward the center of the circular arc for sufficiently small Δs. These radial forces provide the centripetal acceleration.

Let the angle subtended by the string be θ. The centripetal component of the net force on the segment is

$$\sum F_c = 2F_T \sin\tfrac{1}{2}\theta \approx 2F_T(\tfrac{1}{2}\theta) = F_T\theta$$

where we have used the small angle approximation $\sin\tfrac{1}{2}\theta \approx \tfrac{1}{2}\theta$. If μ is the mass per unit length of the string, the mass of a segment of length Δs is $m = \mu \, \Delta s$. The angle θ is related to Δs by

$$\theta = \frac{\Delta s}{R}$$

The mass of the element is thus

$$m = \mu \, \Delta s = \mu R \theta$$

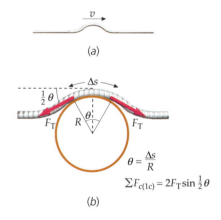

FIGURE 15-5 (*a*) Wave pulse moving with a speed *v* along a string. (*b*) In a frame in which the wave pulse of (*a*) is at rest, the string is moving with a speed *v* to the left. A small segment of the string of length Δs is moving in a circular arc of radius *R*. The centripetal acceleration of the segment is provided by the radial components of the tension.

Newton's second law ($\Sigma F_c = ma_c$) gives

$$F_T\theta = \mu R\theta \frac{v^2}{R}$$

Solving for v, we obtain $v = \sqrt{F_T/\mu}$.

In the original frame, the string is fixed, and the pulse moves with speed $v = \sqrt{F_T/\mu}$, which is Equation 15-3. Since v is independent of R and θ, this result holds for the peak of any pulse. In the following discussion we will show that this result is true not only for the peak but for other parts of the pulse.

*The Wave Equation

We can apply Newton's laws to a segment of the string to derive a differential equation known as the wave equation, which relates the spatial derivatives of $y(x,t)$ to its time derivatives. Figure 15-6 shows one segment of a string. We consider only small angles θ_1 and θ_2. Then the length of the segment is approximately Δx and its mass is $m = \mu \Delta x$, where μ is the string's mass per unit length. First we will show that, for small vertical displacements, the net horizontal force on a segment is zero and the tension is uniform and constant. The net force in the horizontal direction is zero. That is,

$$\Sigma F_x = F_{T2} \cos \theta_2 - F_{T1} \cos \theta_1 = 0$$

where θ_2 and θ_1 are the angles shown and F_T is the tension in the string. Since the angles are assumed to be small, we may approximate $\cos \theta$ by 1 for each angle. Then the net horizontal force on the segment can be written

$$\Sigma F_x = F_{T2} - F_{T1} = 0$$

Thus

$$F_{T2} = F_{T1} = F_T$$

The segment moves vertically, and the net force in this direction is

$$\Sigma F_y = F_T \sin \theta_2 - F_T \sin \theta_1$$

Since the angles are assumed to be small, we may approximate $\sin \theta$ by $\tan \theta$ for each angle. Then the net vertical force on the string segment can be written

$$\Sigma F_y = F_T(\sin \theta_2 - \sin \theta_1) \approx F_T(\tan \theta_2 - \tan \theta_1)$$

The tangent of the angle made by the string with the horizontal is the slope of the curve formed by the string. The slope S is the first derivative of $y(x,t)$ with respect to x for constant t. A derivative of a function of two variables with respect to one of the variables with the other held constant is called a **partial derivative.** The partial derivative of y with respect to x is written $\partial y/\partial x$. Thus we have

$$S = \tan \theta = \frac{\partial y}{\partial x}$$

Then

$$\Sigma F_y = F_T(S_2 - S_1) = F_T \Delta S$$

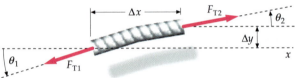

FIGURE 15-6 Segment of a stretched string used for the derivation of the wave equation. The net vertical force on the segment is $F_{T2} \sin \theta_2 - F_{T1} \sin \theta_1$, where F is the tension in the string. The wave equation is derived by applying Newton's second law to the segment.

where S_1 and S_2 are the slopes of either end of the string segment and ΔS is the change in the slope. Setting this net force equal to the mass $\mu \, \Delta x$ times the acceleration $\partial^2 y / \partial t^2$ gives

$$F_T \, \Delta S = \mu \, \Delta x \frac{\partial^2 y}{\partial t^2}$$

or

$$F_T \frac{\Delta S}{\Delta x} = \mu \frac{\partial^2 y}{\partial t^2} \qquad \text{15-8}$$

In the limit $\Delta x \rightarrow 0$, we have

$$\lim_{\Delta x \to 0} \frac{\Delta S}{\Delta x} = \frac{\partial S}{\partial x} = \frac{\partial}{\partial x} \frac{\partial y}{\partial x} = \frac{\partial^2 y}{\partial x^2}$$

Thus Equation 15-8 becomes

$$\frac{\partial^2 y}{\partial x^2} = \frac{\mu}{F_T} \frac{\partial^2 y}{\partial t^2} \qquad \text{15-9}a$$

Equation 15-9a is the **wave equation** for a stretched string.

We now show that the wave equation is satisfied by any function $x - vt$. Let $\alpha = x - vt$ and consider any wave function

$$y = y(x - vt) = y(\alpha)$$

We will use y' for the derivative of y with respect to α. Then, by the chain rule for derivatives,

$$\frac{\partial y}{\partial x} = \frac{\partial y}{\partial \alpha} \frac{\partial \alpha}{\partial x} = y' \frac{\partial \alpha}{\partial x}$$

and

$$\frac{\partial y}{\partial t} = \frac{\partial y}{\partial \alpha} \frac{\partial \alpha}{\partial t} = y' \frac{\partial \alpha}{\partial t}$$

Since

$$\frac{\partial \alpha}{\partial x} = \frac{\partial (x - vt)}{\partial x} = 1 \qquad \text{and} \qquad \frac{\partial \alpha}{\partial t} = \frac{\partial (x - vt)}{\partial t} = -v$$

we have

$$\frac{\partial y}{\partial x} = y' \qquad \text{and} \qquad \frac{\partial y}{\partial t} = -vy'$$

Taking the second derivatives, we obtain

$$\frac{\partial^2 y}{\partial x^2} = y''$$

and

$$\frac{\partial^2 y}{\partial t^2} = -v\frac{\partial y'}{\partial t} = -v\frac{\partial y'}{\partial \alpha}\frac{\partial \alpha}{\partial t} = +v^2 y''$$

Thus

$$\frac{\partial^2 y}{\partial x^2} = \frac{1}{v^2}\frac{\partial^2 y}{\partial t^2}$$ 15-9b

WAVE EQUATION

The same result can be obtained for any function of $x + vt$. Comparing Equations 15-9a and 15-9b, we see that the speed of propagation of the wave is $v = \sqrt{F_T/\mu}$, which is Equation 15-3.

HARMONIC WAVE FUNCTION   **EXAMPLE   15-3**

In the following section harmonic waves are defined by the wave function $y(x,t) = A \sin(kx - \omega t)$, where $v = \omega/k$. Show by explicitly calculating the derivatives that this wave function satisfies Equation 15-9b.

1. Calculate the second derivative of y with respect to x:

$$\frac{\partial y}{\partial x} = \frac{\partial}{\partial x}[A\sin(kx - \omega t)] = A\cos(kx - \omega t)\frac{\partial(kx - \omega t)}{\partial x}$$

$$= kA\cos(kx - \omega t)$$

$$\frac{\partial^2 y}{\partial x^2} = \frac{\partial}{\partial x}\frac{\partial y}{\partial x} = \frac{\partial}{\partial x}kA\cos(kx - \omega t)$$

$$= -kA\sin(kx - \omega t)\frac{\partial(kx - \omega t)}{\partial x}$$

$$= -k^2 A\sin(kx - \omega t)$$

2. Similarly, the two partial derivatives with respect to t are:

$$\frac{\partial y}{\partial t} = \frac{\partial}{\partial t}[A\sin(kx - \omega t)] = A\cos(kx - \omega t)\frac{\partial(kx - \omega t)}{\partial t}$$

$$= -\omega A\cos(kx - \omega t)$$

$$\frac{\partial^2 y}{\partial t^2} = \omega A\sin(kx - \omega t)\frac{\partial(kx - \omega t)}{\partial t} = -\omega^2 A\sin(kx - \omega t)$$

3. Substituting these results in Equation 15-9b gives:

$$-k^2 A\sin(kx - \omega t) = \frac{1}{v^2}[-\omega^2 A\sin(kx - \omega t)]$$

or

$$A\sin(kx - \omega t) = \frac{\omega^2/k^2}{v^2}A\sin(kx - \omega t)$$

4. Substituting for k using $k = \omega/v$ gives:

$$A\sin(kx - \omega t) = \frac{v^2}{v^2}A\sin(kx - \omega t) = \boxed{A\sin(kx - \omega t)}$$

REMARKS We have shown that the function $y = A\sin(kx - \omega t)$ is a solution to the wave equation provided $v = \omega/k$.

EXERCISE Show that any function $y(x + vt)$ satisfies Equation 15-9b.

A wave equation for sound waves can also be derived using Newton's laws. In one dimension, this equation is

$$\frac{\partial^2 s}{\partial x^2} = \frac{1}{v_s^2}\frac{\partial^2 s}{\partial t^2}$$

where s is the displacement of the medium in the x direction and v_s is the speed of sound.

15-2 Periodic Waves

If one end of a long taut string is shaken back and forth in periodic motion, then a **periodic wave** is generated. If a periodic wave is traveling along a taut string or any other medium, each point along the medium oscillates with the same period.

Harmonic Waves

Harmonic waves are the most basic type of periodic waves. All waves, whether they are periodic or not, can be modeled as a superposition of harmonic waves. Consequently, an understanding of harmonic wave motion can be generalized to form an understanding of any type of wave motion. If a **harmonic wave** is traveling through a medium, each point of the medium oscillates in simple harmonic motion.

If one end of a string is attached to a vibrating tuning fork that is moving up and down with simple harmonic motion, a sinusoidal wave train propagates along the string. This wave train is a harmonic wave. As shown in Figure 15-7, the shape of the string is that of a sine function. The minimum distance after which the wave repeats (the distance between crests, for example) in this figure is called the **wavelength** λ.

As the wave propagates along the string, each point on the string moves up and down—perpendicular to the direction of propagation—in simple harmonic motion with the frequency f of the tuning fork. During one period $T = 1/f$, the wave moves a distance of one wavelength, so its speed is given by

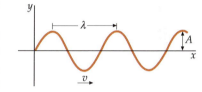

FIGURE 15-7 Harmonic wave at some instant in time. A is amplitude and λ is the wavelength. For a wave on a string, this figure can be obtained by taking a photographic snapshot of the string.

$$v = \frac{\lambda}{T} = f\lambda \qquad\qquad 15\text{-}10$$

Since this relation arises only from the definitions of wavelength and frequency, it applies to all harmonic waves.

The sine function that describes the displacements in Figure 15-7 is

$$y(x) = A\sin\left(2\pi\frac{x}{\lambda} + \delta\right)$$

where A is the amplitude, λ is the wavelength, and δ is a phase constant that depends on the choice of the origin $x = 0$. This equation is expressed more simply as

$$y(x) = A\sin(kx + \delta) \qquad\qquad 15\text{-}11$$

where k, called the **wave number,** is given by

$$k = \frac{2\pi}{\lambda} \qquad\qquad 15\text{-}12$$

Note that k has dimensions of m^{-1}. (Because the angle must be in radians, we sometimes write the units of k as rad/m.) When dealing with a single harmonic wave we usually choose the location of the origin so that $\delta = 0$.

For a wave traveling in the direction of increasing x with speed v, replace x in Equation 15-11 with $x - vt$ (see "Wave Pulses" in Section 15-1). With δ chosen to be zero, this gives

$$y(x,t) = A \sin k(x - vt) = A \sin(kx - kvt)$$

or

$$y(x,t) = A \sin(kx - \omega t) \qquad \text{15-13}$$

HARMONIC WAVE FUNCTION

where

$$\omega = kv \qquad \text{15-14}$$

is the angular frequency, and the argument of the sine function, $(kx - wt)$, is called the **phase.** The angular frequency is related to the frequency f and period T by

$$\omega = 2\pi f = \frac{2\pi}{T} \qquad \text{15-15}$$

Substituting $\omega = 2\pi f$ into Equation 15-14 and using $k = 2\pi/\lambda$, we obtain

$$2\pi f = kv = \frac{2\pi}{\lambda} v$$

or $v = f\lambda$, which is Equation 15-10.

If a harmonic wave traveling along a string is described by $y(x,t) = A \sin(kx - \omega t)$, the velocity of a point on the string at a fixed value of x is

$$v_y = \frac{\partial y}{\partial t} = \frac{\partial}{\partial t}\left[A \sin(kx - \omega t)\right] = -\omega A \cos(kx - \omega t) \qquad \text{15-16}$$

TRANSVERSE VELOCITY

The acceleration of this point is given by $\partial^2 y / \partial t^2$.

A Harmonic Wave on a String   **EXAMPLE 15-4**

The wave function for a harmonic wave on a string is $y(x,t) = (0.03 \text{ m}) \times \sin[(2.2 \text{ m}^{-1})x - (3.5 \text{ s}^{-1})t]$. (*a*) In what direction does this wave travel and what is its speed? (*b*) Find the wavelength, frequency, and period of this wave. (*c*) What is the maximum displacement of any string segment? (*d*) What is the maximum speed of any short string segment?

PICTURE THE PROBLEM (*a*) To find the direction of travel, express $y(x,t)$ as either a function of $(x - vt)$ or as a function of $(x + vt)$ and use Equations 15-1 and 15-2. To find the wave speed use $\omega = kv$ (Equation 15-14). (*b*) The wavelength, frequency, and period can be found from the wave number k and the angular frequency ω. (*c*) The maximum displacement of a point on the string is the amplitude A. (*d*) The velocity of any short string segment is $\partial y / \partial t$.

(a) 1. The given wave function is of the form $y(x,t) = A \sin(kx - \omega t)$. Using $\omega = kv$ (Equation 15-14), write the wave function as a function of $x - vt$. Then use Equations 15-1 and 15-2 to find the direction of travel:

$y(x,t) = A \sin(kx - \omega t)$ and $\omega = kv$

so

$y(x,t) = A \sin(kx - kvt) = A \sin[k(x - vt)]$

The wave travels in the $\boxed{+x \text{ direction}}$

2. Since the form is $y = A \sin(kx - \omega t)$, we know A as well as both ω and k. Use these to calculate the speed:

$v = \dfrac{\lambda}{T} = \dfrac{\lambda}{2\pi}\dfrac{2\pi}{T} = \dfrac{\omega}{k} = \dfrac{3.5 \text{ s}^{-1}}{2.2 \text{ m}^{-1}} = \boxed{1.59 \text{ m/s}}$

(b) The wavelength λ is related to the wave number k, and the period T and frequency f are related to ω:

$\lambda = \dfrac{2\pi}{k} = \dfrac{2\pi}{2.2 \text{ m}^{-1}} = \boxed{2.86 \text{ m}}$

$T = \dfrac{2\pi}{\omega} = \dfrac{2\pi}{3.5 \text{ s}^{-1}} = \boxed{1.80 \text{ s}}$

$f = \dfrac{1}{T} = \dfrac{1}{1.80 \text{ s}} = \boxed{0.557 \text{ Hz}}$

(c) The maximum displacement of a string segment is the amplitude A:

$A = \boxed{0.03 \text{ m}}$

(d) 1. Compute $\partial y / \partial t$ to find the velocity of a point on the string:

$v_y = \dfrac{\partial y}{\partial t} = (0.03 \text{ m})\dfrac{\partial[\sin(2.2 \text{ m}^{-1} x - 3.5 \text{ s}^{-1} t)]}{\partial t}$

$= (0.03 \text{ m})(-3.5 \text{ s}^{-1}) \cos(2.2 \text{ m}^{-1} x - 3.5 \text{ s}^{-1} t)$

$= -(0.105 \text{ m/s}) \cos(2.2 \text{ m}^{-1} x - 3.5 \text{ s}^{-1} t)$

2. The maximum transverse speed occurs when the cosine function has the value of ± 1:

$v_{y,max} = \boxed{0.105 \text{ m/s}}$

REMARKS We have included the units explicitly to show how they work out. Often we will omit the units for simplicity.

Energy Transfer via Waves on a String Consider again a string attached to a tuning fork. As the fork vibrates, it transfers energy to the segment of the string attached to it. For example, as the fork moves through its equilibrium position, it stretches the adjacent string segment slightly, increasing its potential energy, and the fork imparts a transverse speed to it, increasing its kinetic energy. As a wave moves along the string, energy is transferred to the other segments of the string.

   Power is the rate of energy transfer. We can calculate the power by considering work done by the force that one segment of the string exerts on a neighboring segment. The rate of work done by this force is the power. Figure 15-8 shows a harmonic wave moving to the right along a string segment. The tension force $\vec{F}_T$ on the left end of the segment is directed tangent to the string as shown. To calculate the power transferred by this force we use the formula $P = \vec{F}_T \cdot \vec{v}_t$ (Equation 6-16), where F_T is the tension and $\vec{v}_t$, the transverse velocity, is the velocity of the end of the segment. To obtain an expression for the power we first express the vectors in component form. That is, $\vec{F}_T = F_{Tx}\hat{i} + F_{Ty}\hat{j}$ and $\vec{v}_t = v_y\hat{j}$, so $P = F_{Ty}v_y$. We obtain v_y from Equation 15-16. From the figure we see that $F_{Ty} = -F_T \sin \theta \approx -F_T \tan \theta$, where we have used the small angle approximation $\sin \theta \approx \tan \theta$. Since $\tan \theta$ is the slope of the string, we have $\tan \theta = \partial y / \partial x$. Thus

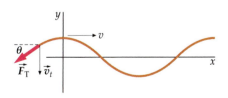

FIGURE 15-8

$P = F_{Ty}v_y \approx -F_T \tan \theta v_y = -F_T \dfrac{\partial y}{\partial x}\dfrac{\partial y}{\partial t}$

$= -F_T[kA \cos(kx - \omega t)][-A\omega \cos(kx - \omega t)]$

Using Equations 15-3 and 15-14, we substitute for F_T and the leading k to obtain

$$P = \mu v \omega^2 A^2 \cos^2 (kx - \omega t) \qquad \qquad 15\text{-}17$$

where v is the wave speed. The average power is

$$P_{av} = \tfrac{1}{2} \mu v \omega^2 A^2 \qquad \qquad 15\text{-}18$$

FIGURE 15-9

since the average value of $\cos^2(kx - \omega t)$ is $\tfrac{1}{2}$ if the average is taken over an entire period of the motion and x remains constant.

The energy travels along a string at the wave speed v, so the average energy $(\Delta E)_{av}$ flowing past point P_1 during time Δt (Figures 15-9a and 15-9b) is

$$(\Delta E)_{av} = P_{av} \, \Delta t = \tfrac{1}{2} \mu v \omega^2 A^2 \, \Delta t$$

This energy is distributed over a length $\Delta x = v \, \Delta t$ so the average energy in length Δx is

$$(\Delta E)_{av} = \tfrac{1}{2} \mu \omega^2 A^2 \, \Delta x \qquad \qquad 15\text{-}19$$

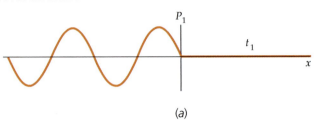

(a)

Note that both the average energy and the power transmitted are proportional to the square of the amplitude of the wave.

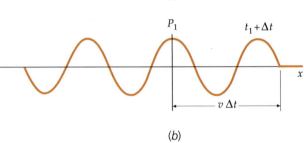

(b)

AVERAGE TOTAL ENERGY OF A WAVE ON A STRING   **EXAMPLE 15-5**

A harmonic wave of wavelength 25 cm and amplitude 1.2 cm moves along a 15-m-long segment of a 60-m-long string that has a mass of 320 g and a tension of 12 N. (a) What is the speed and angular frequency of the wave? (b) What is the average total energy of the wave?

PICTURE THE PROBLEM The speed of the waves is $v = \sqrt{F_T/\mu}$, where F_T is given and $\mu = m/L$. We find ω from $\omega = 2\pi f$, where $f = v/\lambda$. The energy is found using Equation 15-19.

(a) 1. The speed is related to the tension and mass density:

$$v = \sqrt{\frac{F_T}{\mu}}$$

2. Calculate the linear mass density:

$$\mu = \frac{m}{L}$$

so

$$v = \sqrt{\frac{F_T}{\mu}} = \sqrt{\frac{F_T L}{m}} = \sqrt{\frac{(12\text{ N})(60\text{ m})}{(0.32\text{ kg})}} = \boxed{47.4\text{ m/s}}$$

3. The angular frequency is found from the frequency, which is found from the speed and wavelength:

$$\omega = 2\pi f = 2\pi \frac{v}{\lambda} = 2\pi \frac{47.4\text{ m/s}}{0.25\text{ m}} = \boxed{1190\text{ rad/s}}$$

(b) The average total energy of waves on the string is given by Equation 15-19 with $\mu \, \Delta x = m = 80$ g:

$$(\Delta E)_{av} = \frac{1}{2}\mu\omega^2 A^2 \Delta x = \frac{1}{2}\frac{m}{L}\omega^2 A^2 \Delta x$$

$$= \frac{1}{2}\frac{0.32\text{ kg}}{60\text{m}}(1190\text{ s}^{-1})^2 (0.012\text{ m})^2 (15\text{ m})$$

$$= \boxed{8.19\text{ J}}$$

EXERCISE   Calculate the rate at which energy is transmitted along the string.
(*Answer*   25.9 W)

Harmonic Sound Waves

Harmonic sound waves can be generated by a tuning fork or loudspeaker that is vibrating with simple harmonic motion. The vibrating source causes the air molecules next to it to oscillate with simple harmonic motion about their equilibrium positions. These molecules collide with neighboring molecules, causing them to oscillate, thereby propagating the sound wave. Equation 15-13 describes a harmonic sound wave if the wave function $y(x,t)$ is replaced by $s(x,t)$, the displacement of the molecules from equilibrium:

$$s(x, t) = s_0 \sin(kx - \omega t) \qquad \text{15-20}$$

These displacements are along the direction of the motion of the wave, and lead to variations in the density and pressure of the air. Figure 15-10 shows the displacement of air molecules and the density changes caused by a sound wave at some fixed time. Because the pressure in a gas is proportional to its density, the change in pressure is maximum where the change in density is maximum. We see from this figure that the pressure or density wave is 90° out of phase with the displacement wave. (In the arguments of sine or cosine functions we will always express phase angles in radians. However, in verbal descriptions we usually say that "two waves are 90° out of phase" rather than "two waves are out of phase by $\pi/2$ rad.) When the displacement is zero, the pressure and density changes are either maximum or minimum. When the displacement is a maximum or minimum, the pressure and density changes are zero. A displacement wave given by Equation 15-20 thus implies a pressure wave given by

$$p = p_0 \sin\left(kx - \omega t - \frac{\pi}{2}\right) \qquad \text{15-21}$$

where p stands for the *change* in pressure from the equilibrium pressure and p_0, the maximum value of this change, is called the pressure amplitude. It can be shown that the pressure amplitude p_0 is related to the displacement amplitude s_0 by

$$p_0 = \rho \omega v s_0 \qquad \text{15-22}$$

where v is the speed of propagation and ρ is the equilibrium density of the gas. Thus, as a sound wave moves in time, the displacement of air molecules, the pressure, and the density all vary sinusoidally with the frequency of the vibrating source.

EXERCISE   We can hear sound of frequencies from about 20 Hz to about 20,000 Hz (although many people have rather limited hearing above 15,000 Hz). If the speed of sound in air is 340 m/s, what are the wavelengths that correspond to these extreme frequencies? (*Answer* $\lambda = 17$ m at 20 Hz, 1.7 cm at 20,000 Hz)

Energy of Sound Waves   The average energy of a harmonic sound wave in a volume element ΔV is given by Equation 15-19 with A replaced by s_0 and $\mu \, \Delta x$ replaced by $\rho \, \Delta V$, where ρ is the average density of the medium.

$$(\Delta E)_{av} = \tfrac{1}{2} \rho \omega^2 s_0^2 \, \Delta V \qquad \text{15-23}$$

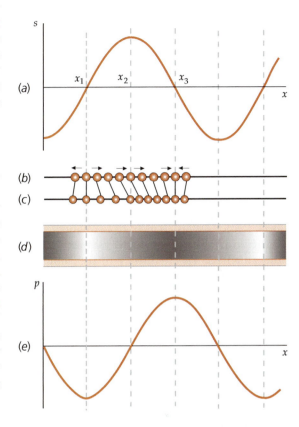

FIGURE 15-10 (*a*) Displacement from equilibrium of air molecules in a harmonic sound wave versus position at some instant. Points x_1 and x_3 are points of zero displacement. (*b*) Some representative molecules equally spaced at their equilibrium positions 1/4 cycle earlier. The arrows indicate the directions of their velocities at that instant. (*c*) Molecules near points x_1, x_2, and x_3 after the sound wave arrives. Just to the left of x_1, the displacement is negative, indicating that the gas molecules are displaced to the left, away from point x_1, at this time. Just to the right of x_1, the displacement is positive, indicating that the molecules are displaced to the right, which is again away from point x_1. So at point x_1, the density is a minimum because the gas molecules on both sides are displaced away from that point. At point x_3, the density is a maximum because the molecules on both sides of that point are displaced toward point x_3. At point x_2, the density does not change because the gas molecules on both sides of that point have equal displacements in the same direction. (*d*) Density of the air at this instant. The density is maximum at x_3 and minimum at x_1, which are both points of zero displacement. It is equal to the equilibrium value at point x_2, which is a maximum in displacement. (*e*) Pressure change, which is proportional to the density change, versus position. The pressure change and displacement (position change) are 90° out of phase.

The energy per unit volume is the average **energy density** η_{av}:

$$\eta_{av} = \frac{\Delta E_{av}}{\Delta V} = \frac{1}{2}\rho\omega^2 s_0^2 \qquad\qquad 15\text{-}24$$

Electromagnetic Waves

Electromagnetic waves include light, radio waves, X rays, gamma rays, and microwaves, among others. The various types of electromagnetic waves differ only in wavelength and frequency. Unlike mechanical waves, electromagnetic waves do not require a medium for propagation. They travel through a vacuum with speed c, which is a universal constant, $c \approx 3 \times 10^8$ m/s. The wave function for electromagnetic waves is an electric field associated with the wave, $\vec{E}(x,t)$. (Electric fields are discussed in Chapter 21. A wave equation, similar to those for string waves and sound waves, is derived from the laws of electricity and magnetism in Chapter 30.) The electric field is perpendicular to the direction of propagation, so electromagnetic waves are transverse waves.

Electromagnetic waves are produced when free electric charges accelerate or when electrons bound to atoms and molecules make transitions to lower energy states. Radio waves, which have frequencies of about 1 MHz for AM and 100 MHz for FM, are produced by macroscopic electric currents oscillating in radio antennas. The frequency of the emitted waves equals the frequency of oscillation of the charges. Light waves, which have frequencies of the order of 10^{14} Hz, are generally produced by atomic or molecular transitions involving bound electrons. The spectrum of electromagnetic waves is discussed in Chapter 31.

15-3 Waves in Three Dimensions

Figure 15-11 shows two-dimensional circular waves on the surface of water in a ripple tank. These waves are generated by a point source moving up and down with simple harmonic motion. The wavelength is the distance between successive wave crests, which in this case are concentric circles. These circles are called **wavefronts.** For a point source of sound, the waves move out in three dimensions, and the wavefronts are concentric spherical surfaces.

The motion of any set of wavefronts can be indicated by **rays,** which are directed lines perpendicular to the wavefronts (Figure 15-12). For circular or spherical waves, the rays are radial lines.

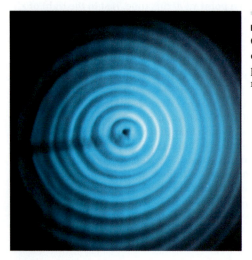

FIGURE 15-11 Circular wavefronts diverging from a point source in a ripple tank.

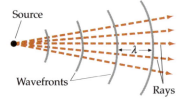

FIGURE 15-12 The motion of wavefronts can be represented by rays drawn perpendicular to the wavefronts. For a point source, the rays are radial lines diverging from the source.

In a homogeneous medium, such as air at constant density, a wave travels in straight lines in the direction of the rays, much like a beam of particles. At a great distance from a point source, a small part of the wavefront can be approximated by a plane, and the rays are approximately parallel lines; such a wave is called a **plane wave** (Figure 15-13). The two-dimensional analog of a plane wave is a line wave, which is a small part of a circular wavefront at a great distance from the source. Such waves can also be produced in a ripple tank by a line source, as in Figure 15-14.

Wave Intensity

If a point source emits waves uniformly in all directions, then the energy at a distance r from the source is distributed uniformly on a spherical surface of radius r and area $A = 4\pi r^2$. If P is the power emitted by the source, then the power per unit area at a distance r from the source is $P/(4\pi r^2)$. The average power per unit area that is incident perpendicular to the direction of propagation is called the **intensity**:

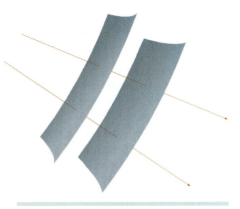

FIGURE 15-13 Plane waves. At great distances from a point source, the wavefronts are approximately parallel planes, and the rays are approximately parallel lines perpendicular to the wavefronts.

$$I = \frac{P_{av}}{A} \qquad\qquad 15\text{-}25$$

INTENSITY DEFINED

The units of intensity are watts per square meter. At a distance r from a point source, the intensity is

$$I = \frac{P_{av}}{4\pi r^2} \qquad\qquad 15\text{-}26$$

INTENSITY DUE TO A POINT SOURCE

FIGURE 15-14 A two-dimensional analog of a plane wave can be generated in a ripple tank by a flat board that oscillates up and down in the water to produce the wavefronts, which are straight lines.

The intensity of a three-dimensional wave varies inversely with the square of the distance from a point source.

There is a simple relation between the intensity of a wave and the energy density in the medium through which it propagates. Figure 15-15 shows a spherical wave that has just reached the radius r_1. The volume inside the radius r_1 contains energy because the particles in that region are oscillating with simple harmonic motion. The region outside r_1 contains no energy because the wave has not yet reached it. After a short time Δt, the wave moves out a short distance $\Delta r = v\,\Delta t$ past r_1. The average energy in the spherical shell of surface area A, thickness $v\,\Delta t$, and volume $\Delta V = A\,\Delta r = Av\,\Delta t$ is

$$(\Delta E)_{av} = \eta_{av}\,\Delta V = \eta_{av}\,Av\,\Delta t$$

The rate of transfer of energy is the power passing into the shell. The average incident power is

$$P_{av} = \frac{(\Delta E)_{av}}{\Delta t} = \eta_{av}Av$$

and the intensity of the wave is

$$I = \frac{P_{av}}{A} = \eta_{av}v \qquad\qquad 15\text{-}27$$

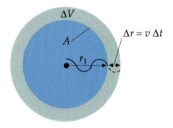

Volume of shell $= \Delta V = A\,\Delta r = Av\,\Delta t$

FIGURE 15-15

Thus the intensity equals the product of the wave speed v and the average energy density η_{av}. Substituting $\eta_{av} = \frac{1}{2}\rho\omega^2 s_0^2$ from Equation 15-24 for the energy density in a sound wave, we obtain

$$I = \eta_{av}v = \frac{1}{2}\rho\omega^2 s_0^2 v = \frac{1}{2}\frac{p_0^2}{\rho v} \qquad\qquad 15\text{-}28$$

where we have used $s_0 = p_0/(\rho\omega v)$ from Equation 15-22. This result—that the intensity of a sound wave is proportional to the square of the amplitude—is a general property of harmonic waves.

The human ear can accommodate a large range of sound-wave intensities, from about 10^{-12} W/m^2 (which is usually taken to be the threshold of hearing) to about 1 W/m^2 (an intensity great enough to stimulate pain in most people). The pressure variations that correspond to these extreme intensities are about 3×10^{-5} Pa for the hearing threshold and 30 Pa for the pain threshold. (Recall that a pascal is a newton per square meter.) These very small pressure variations add to or subtract from the normal atmospheric pressure of about 101 kPa.

Sound waves from a telephone handset spreading out in the air. The waves have been made visible by sweeping out the space in front of the handset with a light source whose brightness is controlled by a microphone.

A LOUDSPEAKER

EXAMPLE 15-6

A loudspeaker diaphragm 30 cm in diameter is vibrating at 1 kHz with an amplitude of 0.020 mm. Assuming that the air molecules in the vicinity have the same amplitude of vibration, find (*a*) the pressure amplitude immediately in front of the diaphragm, (*b*) the sound intensity in front of the diaphragm, and (*c*) the acoustic power being radiated. (*d*) If the sound is radiated uniformly into the forward hemisphere, find the intensity at 5 m from the loudspeaker.

PICTURE THE PROBLEM (*a*) and (*b*) The pressure amplitude is calculated directly from $p_0 = \rho\omega v s_0$ (Equation 15-22), and the intensity from $I = \frac{1}{2}\rho\omega^2 s_0^2 v$ (Equation 15-28). (*c*) The power radiated is the intensity times the area of the diaphragm. (*d*) The area of a hemisphere of radius r is $2\pi r^2$. We can use Equation 15-25 with $A = 2\pi r^2$.

(*a*) Equation 15-22 relates the pressure amplitude to the displacement amplitude, frequency, wave velocity, and air density:

$$p_0 = \rho\omega v s_0$$
$$= (1.29\ \text{kg/m}^3)2\pi(10^3\ \text{Hz})(340\ \text{m/s})(2 \times 10^{-5}\ \text{m})$$
$$= \boxed{55.1\ \text{N/m}^2}$$

(*b*) Equation 15-28 relates the intensity to these same known quantities:

$$I = \frac{1}{2}\rho\omega^2 s_0^2 v$$
$$= \frac{1}{2}(1.29\ \text{kg/m}^3)[2\pi(10^3\ \text{Hz})]^2\,(2 \times 10^{-5}\ \text{m})^2\,(340\ \text{m/s})$$
$$= \boxed{3.46\ \text{W/m}^2}$$

(*c*) The power is the intensity times the area of the diaphragm:

$$P = IA = (3.46\ \text{W/m}^2)\pi(0.15\ \text{m})^2 = \boxed{0.245\ \text{W}}$$

(*d*) Calculate the intensity at $r = 5$ m, assuming uniform radiation into the forward hemisphere:

$$I = \frac{P_{av}}{A} = \frac{0.245\ \text{W}}{2\pi(5\ \text{m})^2} = \boxed{1.56 \times 10^{-3}\ \text{W/m}^2}$$

REMARKS The assumption of uniform radiation in the forward hemisphere is not a very good one because the wavelength in this case [$\lambda = v/f = (340\ \text{m/s})/(1000\text{s}^{-1}) = 34$ cm] is not large compared with the speaker diameter. There is also some radiation in the backward direction, as can be observed if you stand behind a loudspeaker.

Loudspeakers at a rock concert may put out more than 100 times as much power as the speaker in this example.

Intensity Level and Loudness Our perception of loudness is not proportional to the intensity but varies logarithmically. We therefore use a logarithmic scale to describe the **intensity level** β of a sound wave, which is measured in **decibels** (dB) and defined by

$$\beta = 10 \log \frac{I}{I_0}$$

15-29

DEFINITION—INTENSITY LEVEL IN DB

Here I is the intensity of the sound and I_0 is a reference level, which usually is taken to be the threshold of hearing:

$$I_0 = 10^{-12} \text{ W/m}^2$$

15-30

On this scale, the threshold of hearing is $\beta = 10 \log(I/I_0) = 0$ dB and the pain threshold ($I = 1$ W/m²) is $\beta = 10 \log(1/10^{-12}) = 10 \log 10^{12} = 120$ dB. Thus, the range of sound intensities from 10^{-12} W/m² to 1 W/m² corresponds to intensity levels from 0 dB to 120 dB. Table 15-1 lists the intensity levels of some common sounds.

SOUNDPROOFING

E X A M P L E   1 5 - 7

A sound absorber attenuates the sound level by 30 dB. By what factor is the intensity decreased?

From Table 15-1, we can see that for every 10-dB drop in the intensity level, the intensity decreases by a factor of 10. Thus, if the sound level drops 30 db then the intensity drops by a factor of $10^3 = \boxed{1000}$.

TABLE 15-1

Intensity and Intensity Level of Some Common Sounds ($I_0 = 10^{-12}$ W/m²)

Source	I/I_0	dB	Description
	10^0	0	Hearing threshold
Normal breathing	10^1	10	Barely audible
Rustling leaves	10^2	20	
Soft whisper (at 5 m)	10^3	30	Very quiet
Library	10^4	40	
Quiet office	10^5	50	Quiet
Normal conversation (at 1 m)	10^6	60	
Busy traffic	10^7	70	
Noisy office with machines; average factory	10^8	80	
Heavy truck (at 15 m); Niagara Falls	10^9	90	Constant exposure endangers hearing
Old subway train	10^{10}	100	
Construction noise (at 3 m)	10^{11}	110	
Rock concert with amplifiers (at 2 m); jet takeoff (at 60 m)	10^{12}	120	Pain threshold
Pneumatic riveter; machine gun	10^{13}	130	
Jet takeoff (nearby)	10^{15}	150	
Large rocket engine (nearby)	10^{18}	180	

The sensation of loudness depends on the frequency as well as the intensity of a sound. Figure 15-16 is a plot of intensity level versus frequency for sounds of equal loudness to the human ear. (In this figure, the frequency is plotted on a logarithmic scale to display the wide range of frequencies from 20 Hz to 10 kHz.) We note from this figure that the human ear is most sensitive at about 4 kHz for all intensity levels.

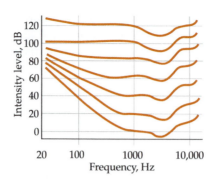

FIGURE 15-16 Intensity level versus frequency for sounds perceived to be of equal loudness. The lowest curve is below the threshold for hearing of all but about 1 percent of the population. The second lowest curve is approximately the hearing threshold for about 50 percent of the population.

BARKING DOGS

EXAMPLE 15-8

A barking dog delivers about 1 mW of power. (*a*) If this power is uniformly distributed in all directions, what is the sound intensity level at a distance of 5 m? (*b*) What would be the intensity level of two dogs barking at the same time if each delivered 1 mW of power?

PICTURE THE PROBLEM The intensity level is found from the intensity, which is found from $I = P/(4\pi r^2)$. For two dogs, the intensities are added.

(*a*) 1. The intensity level is related to the intensity:

$$\beta = 10 \log \frac{I}{I_0}$$

2. Calculate the intensity at $r = 5$ m:

$$I_1 = \frac{P_1}{4\pi r^2} = \frac{10^{-3}\,\text{W}}{4\pi(5\,\text{m})^2} = 3.18 \times 10^{-6}\,\text{W/m}^2$$

3. Use your result to find the intensity level at 5 m:

$$\beta_1 = 10 \log \frac{I_1}{I_0} = 10 \log \frac{3.18 \times 10^{-6}}{10^{-12}} = \boxed{65.0\,\text{dB}}$$

(*b*) If I_1 is the intensity for one dog, the intensity for two dogs is $I_2 = 2I_1$:

$$\beta_2 = 10 \log \frac{I_2}{I_0} = 10 \log \frac{2I_1}{I_0} = 10 \log 2 + 10 \log \frac{I_1}{I_0}$$

$$= 10 \log 2 + \beta_1 = 3.01 + 65.0 = \boxed{68.0\,\text{dB}}$$

REMARKS We can see from this example that whenever the intensity is doubled, the intensity level increases by 3 dB.

15-4 Waves Encountering Barriers

Reflection and Refraction

When a wave is incident on a boundary that separates two regions of differing wave speed, part of the wave is reflected and part is transmitted. Figure 15-17*a* shows a pulse on a light string that is attached to a heavier string. In this case, the pulse reflected at the boundary is inverted. If the second string is lighter than the first (Figure 15-17*b*), then the reflected pulse is not inverted. In either case,

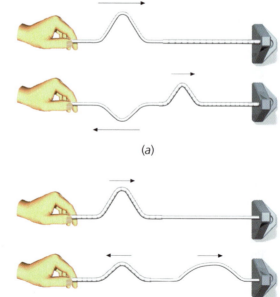

(a)

(b)

FIGURE 15-17 (*a*) A wave pulse traveling on a light string attached to a heavier string in which the wave speed is smaller. The reflected pulse is inverted, whereas the transmitted pulse is not. (*b*) A wave pulse traveling on a heavy string attached to a light string in which the wave speed is greater. In this case, the reflected pulse is not inverted.

the transmitted pulse is not inverted. If the string is tied to a fixed point, then the pulse is reflected and inverted. If it is tied to a string of negligible mass, then the pulse is reflected, but not inverted.

TWO SOLDERED WIRES   **EXAMPLE 15-9**

Two wires of different linear mass densities are soldered together end to end and then stretched under a tension F_T (the tension is the same in both wires). The wave speed in the first wire is twice that in the second. If a harmonic wave traveling in the first wire is incident on the junction of the wires, the amplitude of the reflected wave is half the amplitude of the transmitted wave. (*a*) If the amplitude of the incident wave is A, what are the amplitudes of the reflected and transmitted waves? (*b*) What fraction of the incident power is reflected at the junction and what fraction is transmitted?

PICTURE THE PROBLEM By conservation of energy, the power incident on the junction equals the power reflected plus the power transmitted. Each power is expressed in Equation 15-18 as a function of the density μ, amplitude A, frequency ω, and wave speed v (Figure 15-18). The angular frequencies of all the waves are equal. Since the reflected wave and incident wave are in the same medium, they have the same wave speed v_1. We are given that the speed in the second wire is $v_2 = \frac{1}{2} v_1$.

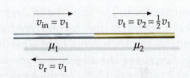

FIGURE 15-18

(*a*) 1. By conservation of energy, the incident power equals the transmitted power plus the reflected power:

$$P_{in} = P_t + P_r$$

2. Write Equation 15-18:

$$P_{av} = \tfrac{1}{2} \mu v \omega^2 A^2$$

3. Substitute into the step 1 result and simplify: The angular frequency is the same for all three waves.

$$\tfrac{1}{2} \mu_1 \omega^2 A_{in}^2 v_1 = \tfrac{1}{2} \mu_2 \omega^2 A_t^2 v_2 + \tfrac{1}{2} \mu_1 \omega^2 A_r^2 v_1$$

$$\mu_1 A_{in}^2 v_1 = \mu_2 A_t^2 v_2 + \mu_1 A_r^2 v_1$$

4. Using the relation $v = \sqrt{F_T/\mu}$ (Equation 15-3), substitute for μ_1 and μ_2 and simplify. F_T is the same on either side of the junction:

$$\frac{F_T}{v_1^2} A_{in}^2 v_1 = \frac{F_T}{v_2^2} A_t^2 v_2 + \frac{F_T}{v_1^2} A_r^2 v_1$$

$$\frac{A_{in}^2}{v_1} = \frac{A_t^2}{v_2} + \frac{A_r^2}{v_1}$$

5. Using the given relations $v_2 = \frac{1}{2} v_1$ and $A_r = \frac{1}{2} A_t$, substitute and solve for the amplitudes:

$$\frac{A_{in}^2}{v_1} = \frac{A_t^2}{\frac{1}{2} v_1} + \frac{(\frac{1}{2} A_t)^2}{v_1} = \frac{9}{4} \frac{A_t^2}{v_1}$$

so

$$A_t = \boxed{\tfrac{2}{3} A_{in}} \quad \text{and} \quad A_r = \boxed{\tfrac{1}{3} A_{in}}$$

(*b*) 1. In Part (*a*) steps 1–4 it was shown that the power is proportional to A^2/v. Express each of the three powers, using b as the proportionality constant:

$$P_{in} = b \frac{A_{in}^2}{v_1} \qquad P_t = b \frac{A_t^2}{v_2} \qquad P_r = b \frac{A_r^2}{v_1}$$

2. Using the Part (*a*) step 5 results, eliminate v_2, A_t, and A_r from the expressions for P_t and P_r:

$$P_t = b \frac{(\frac{2}{3} A_{in})^2}{\frac{1}{2} v_1} = \frac{8}{9} b \frac{A_{in}^2}{v_1} = \boxed{\frac{8}{9} P_{in}}$$

$$P_r = b \frac{(\frac{1}{3} A_{in})^2}{v_1} = \frac{1}{9} b \frac{A_{in}^2}{v_1} = \boxed{\frac{1}{9} P_{in}}$$

REMARKS The reflected wave is inverted relative to the incident wave, so it is 180° out of phase with it. When the displacement of the wire just to the left of the junction would be y_1 due only to the incident wave, it would be $-(y_1/3)$ due only to the reflected wave. These add (according to the principle of superposition to be studied in the next chapter), giving a resultant displacement of $2y_1/3$, which is also the displacement that occurs just to the right of the junction due to the transmitted wave. It can be shown that, given the ratio of the wave speeds, the amplitudes of the transmitted and reflected waves can be determined by requiring that the displacement and the slope be continuous at the junction.

In three dimensions, a boundary between two regions of differing wave speed is a surface. Figure 15-19 shows a ray incident on such a boundary surface. This example could be a sound wave in air striking a solid or liquid surface. The reflected ray makes an angle with the normal to the surface equal to that of the incident ray, as shown.

The transmitted ray is bent toward or away from the normal—depending on whether the wave speed in the second medium is less or greater than that in the incident medium. The bending of the transmitted ray is called **refraction.** When the wave speed in the second medium is greater than that in the incident medium (as occurs when a light wave in glass or water is refracted into the air), the ray describing the direction of propagation is bent away from the normal, as shown in Figure 15-20. As the angle of incidence is increased, the angle of refraction increases, until a critical angle of incidence is reached for which the angle of refraction is 90°. For incident angles greater than the critical angle, there is no refracted ray, a phenomenon known as **total internal reflection.**

The amount of energy reflected from a surface depends on the surface. Flat walls, floors, and ceilings make good reflectors for sound waves, whereas porous and less rigid materials, such as cloth in draperies and furniture coverings, absorb much of the incident sound. The reflection of sound waves plays an important role in the design of a lecture hall, a library, or a music auditorium. If a lecture hall has many flat reflecting surfaces, speech is difficult to understand because of the many echoes that arrive at different times at the listener's ear. Absorbent material is often placed on the walls and ceiling to reduce such reflections. In a concert hall, a reflecting shell is placed behind the orchestra, and reflecting panels are hung from the ceiling to reflect and direct the sound back toward the listeners.

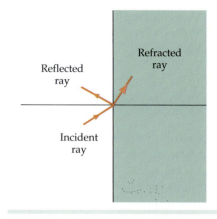

FIGURE 15-19 A wave striking a boundary surface between two media in which the wave speed differs. Part of the wave is reflected and part is transmitted. The change in direction of the transmitted (refracted) ray is called refraction.

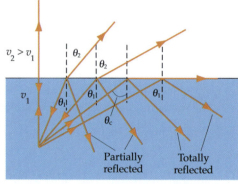

FIGURE 15-20 Light from a source in the water is bent away from the normal when it enters the air. For angles of incidence above a critical angle, there is no transmitted ray, a condition known as total internal reflection.

Diffraction

If a wavefront is partially blocked by an obstacle, the unblocked part of the wavefront bends behind the obstacle. This bending of the wavefronts is called **diffraction.** Almost all of the diffraction occurs for that part of the wavefront that passes within a few wavelengths of the edge of the obstacle. For the parts of the wavefront that pass farther than a few wavelengths from the edge, diffraction is negligible and the wave propagates in straight lines in the direction of the incident rays. When wavefronts encounter a barrier with an aperture (hole) only a few wavelengths across, the part of the wavefronts passing through the aperture all pass within a few wavelengths of an edge. Thus, flat

wavefronts bend and spread out and become spherical or circular (Figure 15-21). In contrast, for a beam of *particles* falling upon a barrier with an aperture, the part of the beam passing through the aperture does so with no change in the direction of the particles (Figure 15-22). Diffraction is one of the key characteristics that distinguishes waves from particles. We will discuss how diffraction arises when we study the interference and diffraction of light in Chapter 35.

Though waves passing through an aperture always bend, or diffract, to some extent, the amount of diffraction depends on whether the wavelength is small or large relative to the size of the aperture. If the wavelength is large relative to the aperture, as in Figure 15-23, the diffraction effects are large, and the waves spread out as they pass through the aperture—as if the waves were originating from a point source. On the other hand, if the wavelength is small relative to the aperture, the effect of diffraction is small, as shown in Figure 15-21. Near the edges of the aperture the wavefronts are distorted and the waves appear to bend slightly. For the most part, however, the wavefronts are not affected and the waves propagate in straight lines, much like a beam of particles. The approximation that waves propagate in straight lines in the direction of the rays with no diffraction is known as the **ray approximation.** Wavefronts are distorted *near* the edges of any obstacle blocking part of the wavefronts. By *near* we mean within a few wavelengths of the edges.

Because the wavelengths of audible sound (which range from a few centimeters to several meters) are generally large compared with apertures and obstacles (doors or windows, and people, for example), diffraction of sound waves is a phenomenon that is often observed. On the other hand, the wavelengths of visible light (4×10^{-7} to 7×10^{-7} m) are so small compared with the size of ordinary objects and apertures that the diffraction of light is not easily noticed; light appears to travel in straight lines. Nevertheless, the diffraction of light is an important phenomena, one we will study in detail in Chapter 35.

Diffraction places a limitation on how accurately small objects can be located by reflecting waves off them and on how well details of the objects can be resolved. Waves are not reflected appreciably from objects smaller than the wavelength, so detail cannot be observed on a scale smaller than the wavelength used. If waves of wavelength λ are used to locate an object, then its position can be known only to within an uncertainty of one wavelength.

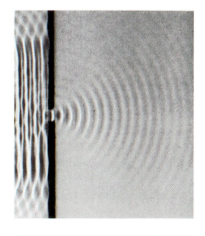

FIGURE 15-21 Plane waves in a ripple tank meeting a barrier with an opening that is only a few wavelengths wide. Beyond the barrier are circular waves that are concentric about the opening, just as if there were a point source at the opening.

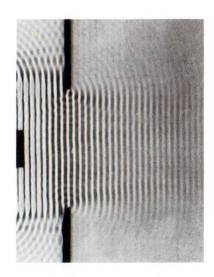

FIGURE 15-23 Plane waves in a ripple tank meeting a barrier with an opening width that is large compared to **λ**. The wave continues in the forward direction, with only a small amount of spreading into the regions to either side of the opening.

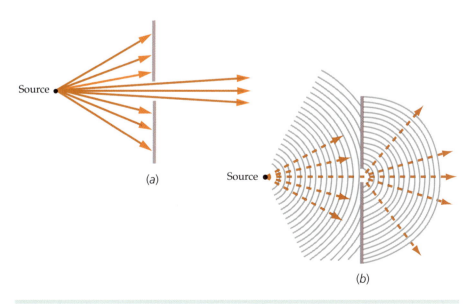

(a)

Source

(b)

Source

FIGURE 15-22 Comparison of particles and waves passing through a narrow opening in a barrier. (*a*) Transmitted particles are confined to a narrow-angle beam. (*b*) Transmitted waves spread out (radiate widely) from the aperture, which acts like a point source of circular waves.

Sound waves with frequencies above 20,000 Hz are called **ultrasonic waves.** Because of their very small wavelengths, narrow beams of ultrasonic waves can be sent out and reflected from small objects. Bats can emit and detect frequencies up to about 120,000 Hz, corresponding to a wavelength of 2.8 mm, which they use to locate small prey such as moths. Echolocation systems, called sonar (from *sound* and *na*vigation *r*anging), are used to detect the outlines of submerged objects with sound waves. The frequency used by commercially available fish finders ranges from about 25 to 200 kHz, and porpoises produce echolocation clicks in the same frequency range. In medicine, ultrasonic waves are used for diagnostic purposes. Ultrasonic waves are passed through the human body and information about the frequency and intensity of the transmitted and reflected waves is processed to construct a three-dimensional picture of the body's interior, called a sonogram.

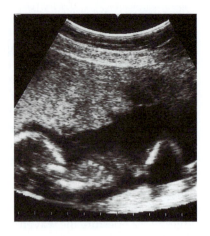

15-5 The Doppler Effect

If a wave source and a receiver are moving relative to each other, the received frequency is not the same as the frequency of the source. If they are moving closer together, the received frequency is greater than the source frequency; and if they are moving farther apart, the received frequency is less than the source frequency. This is called the **Doppler effect.** A familiar example is the drop in pitch of the sound of the horn of an approaching car as the car passes by—and then recedes.

In the following discussion, all motions are relative to the medium. Consider the source moving with speed u_s, shown in Figure 15-24a and b, and a stationary receiver. The source has frequency f_s (and period $T_s = 1/f_s$). The received frequency f_r, the number of wave crests passing the receiver per unit time, is

$$f_r = \frac{v}{\lambda} \quad \text{(stationary receiver)} \qquad \qquad 15\text{-}31$$

where v is the wave speed and λ is the wavelength (the distance between successive crests). To find f_r we first need to find λ. Consider event 1—a wave crest leaves the source—and event two—the next wave crest leaves the source—as shown in Figure 15-24c. The time between these two events is T_s, and between these events the crest leaving the source first travels a distance vT_s while the source itself travels a distance u_sT_s. Consequently, at the time of the second event, the distance between the source and the crest leaving first equals the wavelength λ.

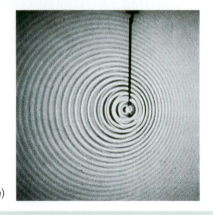

(a)

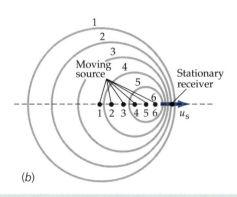

(b)

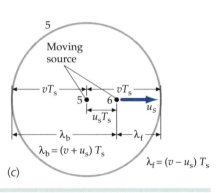

(c)

FIGURE 15-24 (a) Waves in a ripple tank produced by a point source moving to the right. The wavefronts are closer together in front of the source and farther apart behind the source. (b) Successive wavefronts emitted by a point source moving with speed u_s to the right. The numbers of the wavefronts correspond to the positions of the source when the wave was emitted. (c) The source vibrates one cycle in time T_s. During time T_s the source moves a distance u_sT_s and the 5th wavefront travels a distance vT_s. In front of the source the wavelength $\lambda_f = (v - u_s)T_s$, while behind the source $\lambda_b = (v + u_s)T_s$.

Behind the source $\lambda = \lambda_b = (v + u_s)T_s$, and in front of the source $\lambda = \lambda_f = (v - u_s)T_s$, provided $u_s < v$. (If $u_s \geq v$, no wavefronts reach the region ahead of the source.) We can express both these wavelengths as

$$\lambda = (v \pm u_s)T_s = \frac{v \pm u_s}{f_s} \qquad\qquad 15\text{-}32$$

where we have substituted $1/f_s$ for T_s. In front of the source the wavelength is shortest, so the minus sign applies. Behind the source the plus sign applies. Substituting for λ in Equation 15-31 gives

$$f_r = \frac{v}{\lambda} = \frac{v}{v \pm u_s}f \qquad \text{(stationary receiver)} \qquad 15\text{-}33$$

When the receiver moves relative to the medium, the received frequency is different simply because the receiver moves past more or fewer wave crests in a given time. For a receiver moving with speed u_r, let T_r denote the time between arrivals of successive crests. Then, during the time between the arrivals of two successive crests, each crest will have traveled a distance vT_r, and during the same time the receiver will have traveled a distance u_rT_r. If the receiver moves in the direction opposite to that of the wave (Figure 15-25), then during time T_r the distance each crest moves plus the distance the receiver moves equals the wavelength. That is, $vT_r + u_rT_r = \lambda$, or $T_r = \lambda/(v + u_r)$. [If the receiver moves in the same direction as the wave, then $vT_r - \lambda = u_rT_r$, so $T_r = \lambda/(v - u_r)$]. Since $f_r = 1/T_r$ we have

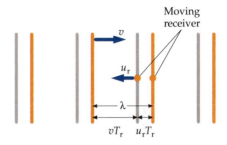

FIGURE 15-25

$$f_r = \frac{1}{T_r} = \frac{v \pm u_r}{\lambda} \qquad\qquad 15\text{-}34$$

where, if the receiver moves in the same direction as the wave, the received frequency is lower, so we choose the negative sign. If the receiver moves in the direction opposite to that of the wave, the frequency is higher, so we choose the positive sign. Substituting for λ from Equation 15-32 we obtain

$$f_r = \frac{v \pm u_r}{v \pm u_s}f_s \qquad\qquad 15\text{-}35a$$

The correct choices for the plus or minus signs are most easily determined by remembering that the frequency tends to increase both when the source moves toward the receiver and when the receiver moves toward the source. For example, if the receiver is moving toward the source the plus sign is selected in the numerator, which tends to increase the received frequency, and if the source is moving away from the receiver the plus sign is selected in the denominator, which tends to decrease the received frequency. Equation 15-35a appears more symmetric, and thus is easier to remember, if expressed in the form

$$\frac{f_r}{v \pm u_r} = \frac{f_s}{v \pm u_s} \qquad\qquad 15\text{-}35b$$

It can be shown (see Problem 89) that if both u_s and u_r are much smaller than the wave speed v, then the shift in frequency $\Delta f = f_r - f_s$ is given approximately by

$$\frac{\Delta f}{f_s} \approx \pm\frac{u}{v} \qquad (u \ll v) \qquad\qquad 15\text{-}36$$

where $u = u_s \pm u_r$ is the speed of the source relative to the receiver.

Equations 15-31 through 15-36 are valid *only* in the reference frame of the medium. In a reference frame in which the medium is moving (for example, the reference frame of the ground if air is the medium and there is a wind blowing), the wave speed v is replaced by $v' = v \pm u_w$, where u_w is the speed of the wind relative to the ground.

SOUNDING THE HORN   **EXAMPLE 15-10**

The frequency of a car horn is 400 Hz. If the horn is honked as the car moves with a speed $u_s = 34$ m/s (about 122 km/h) through still air toward a stationary receiver, find (*a*) the wavelength of the sound passing the receiver and (*b*) the frequency received. Take the speed of sound in air to be 340 m/s. (*c*) Find the wavelength of the sound passing the receiver and find the frequency received if the car is stationary as the horn is honked and a receiver moves with a speed $u_r = 34$ m/s toward the car.

PICTURE THE PROBLEM (*a*) The waves in front of the source are compressed, so we use the minus sign in Equation 15-32. (*b*) We calculate the received frequency from Equation 15-35*a*. (*c*) For a moving receiver we use the same equations as in Parts (*a*) and (*b*).

(*a*) Using Equation 15-32, calculate the wavelength in front of the car. In front of the source the wavelength is shorter, so choose the sign accordingly:

$$\lambda = \frac{v - u_s}{f_s} = \frac{340 \text{ m/s} - 34 \text{ m/s}}{400 \text{ Hz}} = \boxed{0.765 \text{ m}}$$

(*b*) Using Equation 15-35*a*, solve for the received frequency:

$$f_r = \frac{v \pm u_r}{v \pm u_s} f_s = \frac{v + 0}{v - u_s} f_s$$

$$= \left(\frac{340}{340 - 34} \right)(400 \text{ Hz}) = \boxed{444 \text{ Hz}}$$

(*c*) 1. Using Equation 15-32, calculate the wavelength in the vicinity of the receiver:

$$\lambda = \frac{v \pm u_s}{f_s} = \frac{340 \text{ m/s} \pm 0}{400 \text{ Hz}} = \boxed{0.850 \text{ m}}$$

2. The received frequency is given by Equation 15-35*a*. The source is approaching the receiver so the frequency is shifted upward. Choose the sign accordingly:

$$f_r = \frac{v \pm u_r}{v \pm u_s} f_s = \frac{v + u_r}{v \pm 0} f_s = \left(1 + \frac{u_r}{v} \right) f_s$$

$$= \left(1 + \frac{34}{340} \right)(400 \text{ Hz}) = \boxed{440 \text{ Hz}}$$

REMARKS The frequency f_r can also be obtained using Equation 15-34.

EXERCISE As a train moving at 90 km/h is approaching a stationary listener, it blows its horn, which has a frequency of 630 Hz. There is no wind. (*a*) What is the wavelength of the sound waves in front of the train? (*b*) What frequency is heard by the listener? (Use 340 m/s for the speed of sound.) (*Answer* (*a*) $\lambda = 0.5$ m (*b*) $f_r = 680$ Hz)

ANOTHER CAR HORN   **EXAMPLE 15-11**   **Try It Yourself**

The ratio of the frequency of a note to the frequency of the semitone above it on the diatonic scale is about 15:16. How fast is a car going if its horn drops a semitone as it passes you? There is no wind and you are standing next to the road. Take the speed of sound in air to be 340 m/s.

PICTURE THE PROBLEM Let u_s be the speed of the car and f_s be the frequency of the horn. The frequency received as the car approaches, f_r, is greater than f_s and the frequency received as the car recedes, f_r', is less than f_s. Set the ratio $f_r'/f_r = 15/16$ and solve for u_s.

Cover the column to the right and try these on your own before looking at the answers.

Steps   **Answers**

1. Write the frequency received as the car approaches in terms of f_s.

$$f_r = \frac{v \pm u_r}{v \pm u_s} f_s = \frac{v}{v - u_s} f_s$$

2. Write the frequency received as the car recedes in terms of f_s.

$$f_r' = \frac{v \pm u_r}{v \pm u_s} f_s = \frac{v}{v + u_s} f_s$$

3. Set the ratio f_r'/f_r equal to 15/16.

$$\frac{f_r'}{f_r} = \frac{v - u_s}{v + u_s} = \frac{15}{16}$$

4. Solve for u_s.

$$u_s = 0.0323v = \boxed{39.5 \text{ km/h} = 24.5 \text{ mi/h}}$$

REMARKS   The wavelength of the sound behind the car is longer than the wavelength of the sound in front of the car. Also, 1 m/s = 3.6 km/h.

Another familiar example of the Doppler effect is the radar used by police to measure the speed of a car. Electromagnetic waves emitted by the radar transmitter strike the moving car. The car acts as both a moving receiver and a moving source as the waves reflect off it back to the radar receiver. Since electromagnetic waves travel at the speed of light, $v = c = 3 \times 10^8$ m/s, the condition $u \ll v$ is certainly met and Equation 15-36 can be used to calculate the Doppler shift.

POLICE RADAR   **EXAMPLE 15-12**   **Try It Yourself**

The radar unit in a police car sends out electromagnetic waves that travel at the speed of light c. The electric current in the antenna of the radar unit oscillates at frequency f_s. The waves reflect from a speeding car moving away from the police car at speed u relative to the police car. There is a frequency difference of Δf between f_s and f_r', the frequency received at the police car. Find u in terms of f_s and Δf.

PICTURE THE PROBLEM   The radar wave strikes the speeding car at frequency f_r. This frequency is less than f_s because the car is moving away from the source. The frequency shift is given by Equation 15-36. The car then acts as a moving source emitting waves of frequency f_r. The police unit detects waves of frequency $f_r' < f_r$ because the source (the speeding car) is moving away from the police car. The frequency difference is $f_r' - f_s$.

Cover the column to the right and try these on your own before looking at the answers.

Steps   **Answers**

1. The radar unit must be able to determine the speed based only on what it transmits and what it detects.

The radar unit must determine u in terms of f_s and f_r'. Because of the way Equation 15-36 is written, we will solve for u in terms of f_s and $\Delta f = f_r' - f_s$.

2. The frequency difference Δf is the frequency difference $\Delta f_1 = f_r - f_s$ plus the frequency difference $\Delta f_2 = f_r' - f_r$.

$$\Delta f = \Delta f_1 + \Delta f_2$$

3. Using Equation 15-36, substitute for the frequency differences in step 2.

$$\Delta f = -\frac{u}{c} f_s - \frac{u}{c} f_r = -\frac{u}{c}(f_s + f_r)$$

4. Again using Equation 15-36, solve for f_r in terms of f_s.

$$\frac{\Delta f_1}{f_s} = -\frac{u}{c} \quad \text{so} \quad f_r = \left(1 - \frac{u}{c}\right)f_s$$

5. Substitute your step 4 result into your step 3 result and simplify.

$$\Delta f = -\frac{u}{c}\left(2 - \frac{u}{c}\right)f_s$$

6. u/c is negligible compared to 2. Use this to simplify the step 5 result and solve for u in terms of Δf and f_s.

$$\Delta f \approx -2f_s\frac{u}{c} \quad \text{so} \quad u = -\frac{\Delta f}{2f_s}c = \boxed{\frac{|\Delta f|}{2f_s}c}$$

REMARKS The difference in frequency between two waves of nearly equal frequency is easy to detect because the two waves interfere to produce a wave whose amplitude oscillates with frequency $|\Delta f|$, which is called the beat frequency. Interference and beats are discussed in Chapter 16.

EXERCISE Calculate Δf if $f_s = 1.5 \times 10^9$ Hz, $c = 3 \times 10^8$ m/s, and $u = 50$ m/s (*Answer* $\Delta f = 500$ Hz)

The Doppler Shift and Relativity We see from Example 15-10 (and Equations 15-33, 15-34, and 15-35) that the magnitude of the Doppler shift in frequency depends on whether it is the source or the receiver that is moving relative to the medium. For sound, these two situations are physically different. For example, if you move relative to still air, you feel air rushing past you. In your reference frame, there is a wind. For sound waves in air, therefore, we can tell whether the source or receiver is moving by noting if there is a wind in the reference frame of the source or the receiver. However, light and other electromagnetic waves propagate through empty space in which there is no medium. There is no "wind" to tell us whether the source or receiver is moving. According to Einstein's theory of relativity, absolute motion cannot be detected, and all observers measure the same speed c for light, independent of their motion relative to the source. Thus Equation 15-35 cannot be correct for the Doppler shift for light. Two modifications must be made in calculating the relativistic Doppler effect for light. First, the speed of waves passing a receiver is c independent of the motion of the receiver. Second, the time interval between the emission of successive wave crests, which is $T_s = 1/f_s$ in the reference frame of the source, is different in the reference frame of the receiver when the two reference frames are in relative motion because of relativistic time dilation and length contraction (Equations R-9 and R-3). (We will discuss the relativistic Doppler effect in Chapter 39.) The result is that the frequency received depends only on the relative speed of approach (or recession) u, and is related to the frequency emitted by

$$f_r = \sqrt{\frac{c \pm u}{c \mp u}}f_s \tag{15-37}$$

Choose the signs that give an up-shift in frequency when the source and receiver are approaching, and vice versa. (The upper signs are used if the source and receiver are approaching, and the lower signs are used if they are separating.) Again, when $u \ll c$, $\Delta f/f_s \approx \pm u/c$, as given by Equation 15-36.

Shock Waves

In our derivations of the Doppler-shift expressions, we assumed that the speed u of the source was less than the wave speed v. If a source moves with speed greater than the wave speed, then there will be no waves in front of the source. Instead, the waves pile up behind the source to form a shock wave. In the case of sound waves, this shock wave is heard as a sonic boom when it arrives at the receiver.

(a)

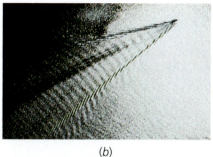

(b)

(c)

Figure 15-26 shows a source originally at point P_1 moving to the right with speed u. After some time t, the wave emitted from point P_1 has traveled a distance vt. The source has traveled a distance ut and will be at point P_2. The line from this new position of the source to the wavefront emitted when the source was at P_1 makes an angle θ with the path of the source, given by

$$\sin \theta = \frac{vt}{ut} = \frac{v}{u} \qquad \text{15-38}$$

Thus the shock wave is confined to a cone that narrows as u increases. The ratio of the source speed u to the wave speed v is called the Mach number:

$$\text{Mach number} = \frac{u}{v} \qquad \text{15-39}$$

(a) Shock waves from a supersonic airplane. (b) Bow waves from a boat. (c) Shock waves produced by a bullet traversing a helium balloon.

Equation 15-38 also applies to the electromagnetic radiation called Cerenkov radiation, which is given off when a charged particle moves in a medium with speed u that is greater than the speed of light v in that medium. (According to the special theory of relativity, it is impossible for a particle to move faster than c, the speed of light in vacuum. In a medium such as glass however, electrons and other particles can move faster than the speed of light in that medium.) The blue glow surrounding the fuel elements of a nuclear reactor is an example of Cerenkov radiation.

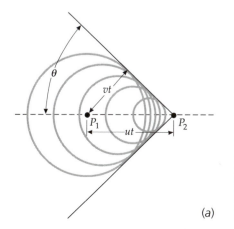

(a)

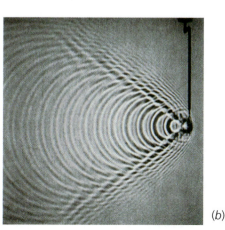

(b)

FIGURE 15-26 (a) Source moving with a speed u that is greater than the wave speed v. The envelope of the wavefronts forms a cone with the source at the apex. (b) Waves in a ripple tank produced by a source moving with a speed $u > v$.

A SONIC BOOM   **EXAMPLE 15-13**   Try It Yourself

A supersonic plane flying due east at an altitude of 15 km passes directly over point P. The sonic boom is heard at point P when the plane is 22 km east of point P. What is the speed of the supersonic plane?

PICTURE THE PROBLEM The speed of the plane is related the sine of the Mach angle (Equation 15-38). Draw a picture so the sine of the Mach angle can be calculated.

Cover the column to the right and try these on your own before looking at the answers.

Steps

Answers

FIGURE 15-27

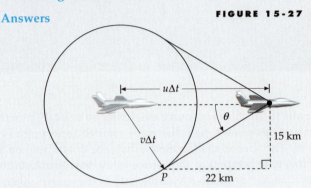

1. Sketch the position of the plane (Figure 15-27) both at the instant the sonic boom is heard at point P and at the instant that sound was produced. Label the distance the sound travels $v\Delta t$, and the distance the plane travels $u\Delta t$.

2. From your sketch and Equation 15-38, calculate u.

$$\tan \theta = \frac{15 \text{ km}}{22 \text{ km}} \quad \text{so} \quad \theta = 34.3°$$

$$\sin \theta = \frac{v\Delta t}{u\Delta t} = \frac{v}{u} \quad \text{so}$$

$$u = \frac{v}{\sin \theta} = \boxed{604 \text{ m/s}}$$

SUMMARY

1. In wave motion, energy and momentum are transported from one point in space to another without the transport of matter.

2. The relation $v = f\lambda$ holds for all harmonic waves.

Topic	Relevant Equations and Remarks	
1. **Transverse and Longitudinal Waves**	In transverse waves, such as waves on a string, the disturbance is perpendicular to the direction of propagation. In longitudinal waves, such as sound waves, the disturbance is along the direction of propagation.	
2. **Speed of Waves**	The wave speed v is independent of the motion of the wave source. The speed of a wave relative to the medium depends on the density and elastic properties of the medium.	
Waves on a string	$v = \sqrt{F_T/\mu}$	15-3
Sound waves	$v = \sqrt{B/\rho}$	15-4

Sound waves in a gas	$v = \sqrt{\gamma RT/M}$	15-5
	where T is the absolute temperature,	
	$T = t_C + 273$	15-6
	R is the universal gas constant,	
	$R = 8.314\,\text{J/mol·K}$	15-7
	M is the molar mass of the gas, which for air is $29 \times 10^{-3}\,\text{kg/mol}$, and γ is a constant that depends on the kind of gas. For a diatomic gas such as air, $\gamma = 1.4$. For a monatomic gas such as helium, $\gamma = 1.67$.	
Electromagnetic waves	The speed of electromagnetic waves in vacuum is a universal constant	
	$c = 3 \times 10^8\,\text{m/s}$	
*3. Wave Equation	$\dfrac{\partial^2 y}{\partial x^2} = \dfrac{1}{v^2}\dfrac{\partial^2 y}{\partial t^2}$	15-9b
4. Harmonic Waves		
Wave function	$y(x,t) = A\sin(kx \pm \omega t)$	15-13
	where A is the amplitude, k is the wave number, and ω is the angular frequency. Use $-$ for a wave traveling in the positive x direction, and $+$ for a wave traveling in the negative x direction.	
Wave number	$k = \dfrac{2\pi}{\lambda}$	15-12
Angular frequency	$\omega = 2\pi f = \dfrac{2\pi}{T}$	15-15
Speed	$v = f\lambda = \omega/k$	15-10, 15-14
Energy	The energy in a harmonic wave is proportional to the square of the amplitude.	
Power for harmonic waves on a string	$p_{av} = \frac{1}{2}\mu v \omega^2 A^2$	15-18
5. Harmonic Sound Waves	Sound waves can be considered to be either displacement waves or pressure waves. The human ear is sensitive to sound waves of frequencies from about 20 Hz to 20 kHz. In a harmonic sound wave, the pressure and displacement are 90° out of phase.	
Amplitudes	The pressure and displacement amplitudes are related by	
	$p_0 = \rho \omega v s_0$	15-22
	where ρ is the density of the medium.	
Energy density	$\eta_{av} = \dfrac{(\Delta E)_{av}}{\Delta V} = \dfrac{1}{2}\rho\omega^2 s_0^2$	15-24
6. Intensity	The intensity of a wave is the average power per unit area.	
	$I = \dfrac{P_{av}}{A}$	15-25

Average energy density η_{av} of a sound wave	$I = \eta_{av} v = \dfrac{1}{2} \rho \omega^2 s_0^2 v = \dfrac{1}{2} \dfrac{p_0^2}{\rho v}$	15-28

---

*Intensity level β in dB

Sound intensity levels are measured on a logarithmic scale.

$$\beta = 10 \log \frac{I}{I_0} \qquad \text{15-29}$$

where $I_0 = 10^{-12}\,\text{W/m}^2$ is approximately the threshold of hearing.

---

7. Reflection and Refraction

When a wave is incident on a boundary surface that separates two regions of differing wave speed, part of the wave is reflected and part is transmitted.

---

8. Diffraction

If a wavefront is partially blocked by an obstacle, the unblocked part of the wavefront diffracts (bends) into the region behind the obstacle.

Ray approximation

If a wavefront is partially blocked by an obstacle, almost all of the diffraction occurs for that part of the wavefront that passes within a few wavelengths of the edge. For those parts of the wavefront that pass farther from the edge than a few wavelengths, diffraction is negligible and the wave propagates in straight lines in the direction of the incident rays.

---

9. Doppler Effect

When a sound source and receiver are in relative motion, the received frequency f_r is higher than the frequency of the source f_s if their separation is decreasing, and lower if their separation is increasing.

Moving source	$\lambda = \dfrac{v \pm u_s}{f_s}$	15-32†

Moving receiver	$f_r = \dfrac{v \pm u_r}{\lambda}$	15-34†

Either source or receiver moving

$$f_r = \frac{v \pm u_r}{v \pm u_s} f_s \qquad \text{or} \qquad \frac{f_r}{v \pm u_r} = \frac{f_s}{v \pm u_s} \qquad \text{15-35†}$$

Choose the signs that give an up-shift in frequency for an approaching source or receiver, and vice versa.

Small speeds of source or receiver	$\dfrac{\Delta f}{f_s} \approx \pm \dfrac{u}{v} \qquad (u << v)$	15-36†

Relativistic Doppler shift

$$f_r = \sqrt{\frac{c \pm u}{c \mp u}}\, f_s \qquad \text{15-37}$$

Choose the signs that give an up-shift in frequency for an approaching source or receiver, and vice versa.

---

10. Shock Waves

When the source speed is greater than the wave speed, the waves behind the source are confined to a cone of angle θ given by

---

Mach angle	$\sin \theta = \dfrac{v}{u}$	15-38

Mach number	$\text{Mach number} = \dfrac{u}{v}$	15-39

---

† Equations 15-32 through 15-36 are valid *only* in the reference frame of the medium. If the medium is moving, the wave speed v is replaced by $v' = v \pm u_m$, where u_m is the speed of the medium.

- Single-concept, single-step, relatively easy
- •• Intermediate-level, may require synthesis of concepts
- ••• Challenging
- **SSM** Solution is in the *Student Solutions Manual*
- **iSOLVE** Problems available on iSOLVE online homework service
- **iSOLVE✔** These "Checkpoint" online homework service problems ask students additional questions about their confidence level, and how they arrived at their answer

In a few problems, you are given more data than you actually need; in a few other problems, you are required to supply data from your general knowledge, outside sources, or informed estimates.

Use $v = 340$ m/s for the speed of sound in air unless otherwise indicated.

Conceptual Problems

1 • **SSM** A rope hangs vertically from the ceiling. Do waves on the rope move faster, slower, or at the same speed as they move from bottom to top? Explain.

2 • A traveling wave passes a point of observation. At this point, the time between successive crests is 0.2 s. Which of the following is true? (*a*) The wavelength is 5 m. (*b*) The frequency is 5 Hz. (*c*) The velocity of propagation is 5 m/s. (*d*) The wavelength is 0.2 m. (*e*) There is not enough information to justify any of these statements.

3 • True or false: The energy in a wave is proportional to the square of the amplitude of the wave.

4 • A rope hangs vertically. You shake the bottom back and forth, creating a sinusoidal wave train. Is the wavelength near the top the same as, less than, or greater than the wavelength near the bottom?

5 • **SSM** The crack of a bullwhip is caused by the speed of the tip breaking the sound barrier. Explain how the tapered shape of the whip helps the tip move much faster than the hand holding the whip.

6 • True or false: A 60-dB sound has twice the intensity of a 30-dB sound.

7 • If the source and receiver are at rest relative to each other but the wave medium is moving relative to them, will there be any Doppler shift in frequency?

8 • The frequency of a car horn is f_0. What frequency is observed if both the car and the observer are at rest, but a wind blows toward the observer? (*a*) f_0. (*b*) Greater than f_0. (*c*) Less than f_0. (*d*) It could be either greater or less than f_0. (*e*) It could be f_0 or greater than f_0, depending on how wind speed compares to speed of sound.

9 •• **SSM** Stars often occur in pairs revolving around their common center of mass. If one of the stars is a black hole, it is invisible. Explain how the existence of such a black hole might be inferred from the light observed from the other, visible star.

10 • When a guitar string is plucked, is the wavelength of the wave it produces in air the same as the wavelength of the wave on the string?

11 • True or false:
(*a*) Wave pulses on strings are transverse waves.
(*b*) Sound waves in air are transverse waves of compression and rarefaction.
(*c*) The speed of sound in air at 20°C is twice that at 5°C.

12 • Sound travels at 340 m/s in air and 1500 m/s in water. A sound of 256 Hz is made under water. In the air, the frequency will be (*a*) the same, but the wavelength will be shorter, (*b*) higher, but the wavelength will stay the same, (*c*) lower, but the wavelength will be longer, (*d*) lower, and the wavelength will be shorter, (*e*) the same, and the wavelength too will stay the same.

13 • **SSM** While out on patrol, the battleship *Rodger Young* hits a mine and begins to burn, ultimately exploding. Sailor Abel jumps into the water and begins swimming away from the doomed ship, while Sailor Baker gets into a life raft. Comparing their experiences later, Abel tells Baker, "I was swimming underwater, and heard a big explosion. When I surfaced, I heard a second explosion. What do you think it could be?" Baker says, "I think it was your imagination—I only heard one explosion." Explain why Baker only heard one explosion, while Abel heard two.

14 •• Figure 15-28 shows a wave pulse at time $t = 0$ moving to the right. At this particular time, which segments of the string are moving up? Which are moving down? Is there any segment of the string at the pulse that is instantaneously at rest? Answer these questions by sketching the pulse at a slightly later time and a slightly earlier time to see how the segments of the string are moving.

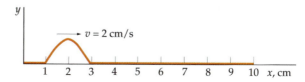

FIGURE 15-28 Problems 14, 15

15 •• Make a sketch of the velocity of each string segment versus position for the pulse shown in Figure 15-28.

16 •• In a classic physics experiment, a bell is placed in a sealed jar and rung while the air is slowly removed. After a while, the bell becomes inaudible. This is commonly cited as proof that sound waves can't travel through a vacuum, but in fact the sound becomes inaudible well before the jar is completely evacuated. Can you give another reason why the sound from the bell can't be heard?

17 •• **SSM** The explosion of a depth charge beneath the surface of the water is recorded by a helicopter hovering above its surface, as shown in Figure 15-29. Along which path, A, B, or C, will the sound wave take the least time to reach the helicopter?

Helicopter

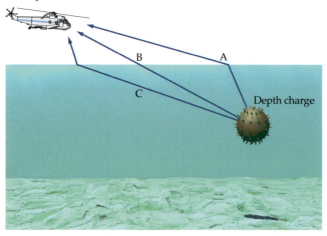

FIGURE 15-29 Problem 17

Estimation and Approximation

18 •• Normal human speech has a sound intensity level of about 65 dB at a distance of 1 m. Estimate the power of human speech.

19 •• A man drops a stone from a high bridge and hears it strike the water below exactly 4 s later. (a) Estimate the distance to the water based on the assumption that the travel time for the sound to reach the man is negligible. (b) Improve your estimate by using your result from Part (a) for the distance to the water to estimate the time it takes for sound to travel this distance and then calculate the distance the rock falls in 4 s minus this time. (c) Calculate the exact distance and compare your result with your previous estimates.

20 •• **SSM**
Estimate the speed of the bullet as it passes through the helium balloon in Figure 15-30 from the angle of its shock cone.

FIGURE 15-30
Problem 20

21 •• The new student townhouses at a local college are in the form of a semicircle half-enclosing the track field. To estimate the speed of sound in air, an ambitious physics student stood at the center of the semicircle and clapped his hands rhythmically at a frequency at which he couldn't hear the echo of the clap, because it reached him at the same time as his next clap. This frequency was about 2.5 claps/s. Once he established this frequency, he paced off the distance to the townhouses, which was 30 double paces. Assuming that the distance of a double-pace stride is the same as his height (5 ft 11 in), estimate the speed of sound in air using this data. How far off is this from the commonly accepted value?

Speed of Waves

22 • **SOLVE** ✔ (a) The bulk modulus for water is 2.0×10^9 N/m². Use it to find the speed of sound in water. (b) The speed of sound in mercury is 1410 m/s. What is the bulk modulus for mercury ($\rho = 13.6 \times 10^3$ kg/m³)?

23 • **SSM** **SOLVE** ✔ Calculate the speed of sound waves in hydrogen gas at $T = 300$ K. (Take $M = 2$ g/mol and $\gamma = 1.4$.)

24 • A steel wire 7 m long has a mass of 100 g. It is under a tension of 900 N. What is the speed of a transverse wave pulse on this wire?

25 • Transverse waves travel at 150 m/s on a wire of length 80 cm that is under a tension of 550 N. What is the mass of the wire?

26 • **SSM** A wave pulse propagates along a wire in the positive x direction at 20 m/s. What will be the pulse velocity if we (a) double the length of the wire but keep the tension and mass per unit length constant? (b) double the tension while holding the length and mass per unit length constant? (c) double the mass per unit length while holding the other variables constant?

27 • **SOLVE** A steel piano wire is 0.7 m long and has a mass of 5 g. It is stretched with a tension of 500 N. (a) What is the speed of transverse waves on the wire? (b) To reduce the wave speed by a factor of 2 without changing the tension, what mass of copper wire would have to be wrapped around the steel wire?

28 •• A common method for estimating the distance to a lightning bolt is to begin counting when the flash is observed and continue until the thunder clap is heard. The number of seconds counted is then divided by 3 to get the distance in kilometers. (a) What is the velocity of sound in kilometers per second? (b) How accurate is this procedure? (c) Is a correction for the time it takes for the light to reach you important? (The speed of light is 3×10^8 m/s.)

29 •• **SSM** (a) Compute the derivative of the speed of a wave on a string with respect to the tension dv/dF, and show that the differentials dv and dF obey $dv/v = \frac{1}{2}dF/F$. (b) A wave moves with a speed of 300 m/s on a wire that is under a tension of 500 N. Using dF to approximate a change in tension, determine how much the tension must be changed to increase the speed to 312 m/s.

30 •• (a) Compute the derivative of the velocity of sound in air with respect to the absolute temperature, and show that the differentials dv and dT obey $dv/v = \frac{1}{2}dT/T$. (b) Use this result to compute the percentage change in the velocity of sound when the temperature changes from 0 to 27°C. (c) If the speed of sound is 331 m/s at 0°C, what is it (approximately) at 27°C? How does this approximation compare with the result of an exact calculation?

31 ••• In this problem, you will derive a convenient formula for the speed of sound in air at temperature t in Celsius degrees. Begin by writing the temperature as $T = T_0 + \Delta T$, where $T_0 = 273$ K corresponds to 0°C and $\Delta T = t$, the Celsius temperature. The speed of sound is a function of T, $v(T)$. To a first-order approximation, you can write $v(T) \approx v(T_0) + (dv/dT)_{T_0} \Delta T$, where $(dv/dT)_{T_0}$ is the derivative evaluated at $T = T_0$. Compute this derivative, and show that the result leads to $v = (331 \text{ m/s})\left(1 + \dfrac{t}{2T_0}\right) = (331 + 0.606t)$ m/s.

32 •• Various stories of psychic phenomena can often be explained by considering physical phenomena. For example, there is the often repeated story of a man being woken from a deep sleep for no reason, getting out of bed and walking to the window just in time to hear the sound of the explosion of a munitions plant across town. The story is often cited to give credence to the idea of clairvoyance, but can be explained instead by assuming that the man was woken by the tremor of the sound wave traveling through the earth, and then walked to the window in time to hear the sound wave traveling through the air. If he took 3 s to move from his bed to the window, and the average speed of sound through solid rock is 3000 m/s, how far was his house from the munitions plant?

33 ••• While studying physics in her dorm room, a student is listening to a live radio broadcast of a baseball game. She is 1.6 km due south of the baseball field. Over her radio, the student hears a noise generated by the electromagnetic pulse of a lightning bolt. Two seconds later, she hears over the radio the thunder picked up by the microphone at the baseball field. Four seconds after she hears the noise of the electromagnetic pulse over the radio, thunder rattles her windows. Where, relative to the ballpark, did the lightning bolt occur?

34 ••• SSM Weather station Beta is located 0.75 mi due east of weather station Alpha. Observers at the two stations see a lightning strike to the north of the stations; observers at station Alpha hear the thunder 3.4 s after seeing the strike, while observers at Beta hear it 2.5 s after seeing the strike. Locate the coordinates of the lightning strike relative to the position of station Alpha.

35 ••• A coiled spring, such as a Slinky, is stretched to a length L. It has a force constant k and a mass m. (a) Show that the velocity of longitudinal compression waves along the spring is given by $v = L\sqrt{k/m}$. (b) Show that this is also the velocity of transverse waves along the spring if the natural length of the spring is much less than L.

The Wave Equation

36 • Show explicitly that the following functions satisfy the wave equation: (a) $y(x,t) = k(x + vt)^3$; (b) $y(x, t) = Ae^{ik(x - vt)}$

where A and k are constants and $i = \sqrt{-1}$; and (c) $y(x,t) = \ln k(x - vt)$.

37 • SSM Show that the function $y = A \sin kx \cos \omega t$ satisfies the wave equation.

Harmonic Waves on a String

38 • SOLVE One end of a string 6 m long is moved up and down with simple harmonic motion at a frequency of 60 Hz. The waves reach the other end of the string in 0.5 s. Find the wavelength of the waves on the string.

39 • Equation 15-13 expresses the displacement of a harmonic wave as a function of x and t in terms of the wave parameters k and ω. Write the equivalent expressions that contain the following pairs of parameters instead of k and ω: (a) k and v, (b) λ and f, (c) λ and T, (d) λ and v, and (e) f and v.

40 • SSM Equation 15-10 applies to all types of periodic waves, including electromagnetic waves such as light waves and microwaves, which travel at 3×10^8 m/s in a vacuum. (a) The range of wavelengths of light to which the eye is sensitive is about 4×10^{-7} to 7×10^{-7} m. What are the frequencies that correspond to these wavelengths? (b) Find the frequency of a microwave that has a wavelength of 3 cm.

41 • SOLVE A harmonic wave on a string with a mass per unit length of 0.05 kg/m and a tension of 80 N has an amplitude of 5 cm. Each section of the string moves with simple harmonic motion at a frequency of 10 Hz. Find the power propagated along the string.

42 • SOLVE ✓ A rope 2 m long has a mass of 0.1 kg. The tension is 60 N. A power source at one end sends a harmonic wave with an amplitude of 1 cm down the rope. The wave is extracted at the other end without any reflection. What is the frequency of the power source if the power transmitted is 100 W?

43 •• The wave function for a harmonic wave on a string is $y(x, t) = (0.001 \text{ m}) \sin(62.8 \text{ m}^{-1} x + 314 \text{ s}^{-1} t)$. (a) In what direction does this wave travel, and what is its speed? (b) Find the wavelength, frequency, and period of this wave. (c) What is the maximum speed of any string segment?

44 •• A harmonic wave with a frequency of 80 Hz and an amplitude of 0.025 m travels along a string to the right with a speed of 12 m/s. (a) Write a suitable wave function for this wave. (b) Find the maximum speed of a point on the string. (c) Find the maximum acceleration of a point on the string.

45 •• SOLVE Waves of frequency 200 Hz and amplitude 1.2 cm move along a 20-m string that has a mass of 0.06 kg and a tension of 50 N. (a) What is the average total energy of the waves on the string? (b) Find the power transmitted past a given point on the string.

46 •• SSM In a real string, a wave loses some energy as it travels down the string. Such a situation can be described by a wave function whose amplitude $A(x)$ depends on x: $y = A(x) \sin (kx - \omega t) = (A_0 e^{-bx}) \sin (kx - \omega t)$ (a) What is the power transported by the wave at the origin? (b) What is the power transported by the wave at point x, where $x > 0$?

47 •• Power is to be transmitted along a stretched wire by means of transverse harmonic waves. The wave speed is 10 m/s and the linear mass density of the wire is 0.01 kg/m. The power source oscillates with an amplitude of 0.50 mm. (a) What average power is transmitted along the wire if the frequency is 400 Hz? (b) The power transmitted can be increased by increasing the tension in the wire, the frequency of the source, or the amplitude of the waves. By how much would each of these quantities have to increase to cause an increase in power by a factor of 100 if it is the only quantity changed? (c) Which of the quantities would probably be the easiest to vary?

48 ••• [SSM] Two very long strings are tied together at the point $x = 0$. In the region $x < 0$, the wave speed is v_1, while in the region $x > 0$, the speed is v_2. A sinusoidal wave is incident from the left ($x < 0$); part of the wave is reflected and part is transmitted. For $x < 0$, the displacement of the wave is describable by $y(x,t) = A \sin(k_1 x - \omega t) + B \sin(k_1 x + \omega t)$, while for $x > 0$, $y(x,t) = C \sin(k_2 x - \omega t)$, where $\omega/k_1 = v_1$ and $\omega/k_2 = v_2$. (a) If we assume that both the wave function y and its first spatial derivative $\partial y/\partial x$ must be continuous at $x = 0$, show that $C/A = 2/(1 + v_1/v_2)$, and that $B/A = (1 - v_1/v_2)/(1 + v_1/v_2)$. (b) Show that $B^2 + (v_1/v_2)C^2 = A^2$.

Harmonic Sound Waves

49 • [SSM] A sound wave in air produces a pressure variation given by

$$p(x,t) = 0.75 \cos \frac{\pi}{2}(x - 340t)$$

where p is in pascals, x is in meters, and t is in seconds. Find (a) the pressure amplitude of the sound wave, (b) the wavelength, (c) the frequency, and (d) the speed.

50 • [SOLVE] (a) Middle C on the musical scale has a frequency of 262 Hz. What is the wavelength of this note in air? (b) The frequency of the C an octave above middle C is twice that of middle C. What is the wavelength of this note in air?

51 • (a) What is the displacement amplitude for a sound wave having a frequency of 100 Hz and a pressure amplitude of 10^{-4} atm? (b) The displacement amplitude of a sound wave of frequency 300 Hz is 10^{-7} m. What is the pressure amplitude of this wave?

52 • [SOLVE✓] (a) Find the displacement amplitude of a sound wave of frequency 500 Hz at the pain-threshold pressure amplitude of 29 Pa. (b) Find the displacement amplitude of a sound wave with the same pressure amplitude but a frequency of 1 kHz.

53 • A typical loud sound wave with a frequency of 1 kHz has a pressure amplitude of about 10^{-4} atm. (a) At $t = 0$, the pressure is a maximum at some point x_1. What is the displacement at that point at $t = 0$? (b) What is the maximum value of the displacement at any time and place? (Take the density of air to be 1.29 kg/m³.)

54 • [SSM] An octave represents a change in frequency by a factor of two. Over how many octaves can a typical person hear?

Waves in Three Dimensions: Intensity

55 • A piston at one end of a long tube filled with air at room temperature and normal pressure oscillates with a frequency of 500 Hz and an amplitude of 0.1 mm. The area of the piston is 100 cm². (a) What is the pressure amplitude of the sound waves generated in the tube? (b) What is the intensity of the waves? (c) What average power is required to keep the piston oscillating (neglecting friction)?

56 • A spherical source radiates sound uniformly in all directions. At a distance of 10 m, the sound intensity level is 10^{-4} W/m². (a) At what distance from the source is the intensity 10^{-6} W/m²? (b) What power is radiated by this source?

57 • [SSM] [SOLVE] A loudspeaker at a rock concert generates 10^{-2} W/m² at 20 m at a frequency of 1 kHz. Assume that the speaker spreads its energy uniformly in three dimensions. (a) What is the total acoustic power output of the speaker? (b) At what distance will the intensity be at the pain threshold of 1 W/m²? (c) What is the intensity at 30 m?

58 •• When a pin of mass 0.1 g is dropped from a height of 1 m, 0.05 percent of its energy is converted into a sound pulse with a duration of 0.1 s. (a) Estimate the range at which the dropped pin can be heard if the minimum audible intensity is 10^{-11} W/m². (b) Your result in (a) is much too large in practice because of background noise. If you assume that the intensity must be at least 10^{-8} W/m² for the sound to be heard, estimate the range at which the dropped pin can be heard. (In both parts, assume that the intensity is $P/4\pi r^2$.)

*Intensity Level

59 • What is the intensity level in decibels of a sound wave of intensity (a) 10^{-10} W/m² and (b) 10^{-2} W/m²?

60 • [SOLVE] Find the intensity of a sound wave if (a) $\beta = 10$ dB and (b) $\beta = 3$ dB. (c) Find the pressure amplitudes of sound waves in air for each of these intensities.

61 • [SSM] The sound level of a dog's bark is 50 dB. The intensity of a rock concert is 10,000 times that of the dog's bark. What is the sound level of the rock concert?

62 • [SOLVE] Two sounds differ by 30 dB. The intensity of the louder sound is I_L and that of the softer sound is I_s. The value of the ratio I_L/I_s is (a) 1000, (b) 30, (c) 9, (d) 100, (e) 300.

63 • Show that if the intensity is doubled, the intensity level increases by 3.0 dB.

64 • [SSM] What fraction of the acoustic power of a noise would have to be eliminated to lower its sound intensity level from 90 to 70 dB?

65 •• [SOLVE] A spherical source radiates sound uniformly in all directions. At a distance of 10 m, the sound intensity level is 80 dB. (a) At what distance from the source is the intensity level 60 dB? (b) What power is radiated by this source?

66 •• A spherical source of intensity I_0 radiates sound uniformly in all directions. Its intensity level is β_1 at a distance r_1 and β_2 at a distance r_2. Find β_2/β_1.

67 •• ISOLVE✓ A loudspeaker at a rock concert generates 10^{-2} W/m² at 20 m at a frequency of 1 kHz. Assume that the speaker spreads its energy uniformly in all directions. (a) What is the intensity level at 20 m? (b) What is the total acoustic power output of the speaker? (c) At what distance will the intensity level be at the pain threshold of 120 dB? (d) What is the intensity level at 30 m?

68 •• An article on noise pollution claims that sound intensity levels in large cities have been increasing by about 1 dB annually. (a) To what percentage increase in intensity does this correspond? Does this increase seem plausible? (b) In about how many years will the intensity of sound double if it increases at 1 dB annually?

69 •• Three noise sources produce intensity levels of 70, 73, and 80 dB when acting separately. When the sources act together, their intensities add. (a) Find the sound intensity level in decibels when the three sources act at the same time. (b) Discuss the effectiveness of eliminating the two least intense sources in reducing the intensity level of the noise.

70 •• SSM If you double the distance between a source of sound and a receiver, the intensity at the receiver drops by approximately (a) 2 dB, (b) 3 dB, (c) 6 dB, (d) Amount cannot be determined from the information given.

71 ••• ISOLVE Everyone at a party is talking equally loudly. If only one person were talking, the sound level would be 72 dB. Find the sound level when all 38 people are talking.

72 ••• SSM When a violinist pulls the bow across a string, the force with which the bow is pulled is fairly small, about 0.6 N. Suppose the bow travels across the A string, which vibrates at 440 Hz, at 0.5 m/s. A listener 35 m from the performer hears a sound of 60 dB intensity. With what efficiency is the mechanical energy of bowing converted to sound energy? (Assume that the sound radiates uniformly in all directions.)

73 ••• The noise level in an empty examination hall is 40 dB. When 100 students are writing an exam, the sounds of heavy breathing and pens traveling rapidly over paper cause the noise level to rise to 60 dB (not counting the occasional groans). Assuming that each student contributes an equal amount of noise power, find the noise level to the nearest decibel when 50 students have left.

The Doppler Effect

In Problems 74 through 79, a source emits sounds of frequency 200 Hz that travel through still air at 340 m/s.

74 • The sound source described moves with a speed of 80 m/s relative to still air toward a stationary listener. (a) Find the wavelength of the sound in the region between the source and the listener. (b) Find the frequency heard by the listener.

75 • Consider the situation in Problem 74 from the reference frame in which the source is at rest. In this frame, the listener moves toward the source with a speed of 80 m/s, and there is a wind blowing at 80 m/s from the listener to the source. (a) What is the speed at which the sound travels from the source to the listener in this frame? (b) Find the wavelength of the sound in the region between the source and the listener. (c) Find the frequency heard by the listener.

76 • The source moves away from the stationary listener at 80 m/s. (a) Find the wavelength of the sound waves in the region between the source and the listener. (b) Find the frequency heard by the listener.

77 • ISOLVE✓ The listener moves at 80 m/s relative to still air toward the stationary source. (a) What is the wavelength of the sound in the region between the source and the listener? (b) What is the frequency heard by the listener?

78 • Consider the situation in Problem 77 in a reference frame in which the listener is at rest. (a) What is the wind velocity in this frame? (b) What is the speed of the sound as it travels from the source to the listener in this frame, that is, relative to the listener? (c) Find the wavelength of the sound in the region between the source and the listener in this frame. (d) Find the frequency heard by the listener.

79 • The listener moves at 80 m/s relative to the still air away from the stationary source. Find the frequency heard by the listener.

80 • A jet is traveling at Mach 2.5 at an altitude of 5000 m. (a) What is the angle that the shock wave makes with the track of the jet? (Assume that the speed of sound at this altitude is still 340 m/s.) (b) Where is the jet when a person on the ground hears the shock wave?

81 • If you are running at top speed toward a source of sound at 1000 Hz, estimate the frequency of the sound that you hear. Suppose that you can recognize a change in frequency of 3 percent. Can you use your sense of pitch to estimate your running speed?

82 •• ISOLVE A radar device emits microwaves with a frequency of 2.00 GHz. When the waves are reflected from a car moving directly away from the emitter, a frequency difference of 293 Hz is detected. Find the speed of the car.

83 •• SSM The Doppler effect is routinely used to measure the speed of winds in storm systems. A weather station uses a Doppler radar system of frequency $f = 625$ MHz to bounce a radar pulse off of the raindrops in a swirling thunderstorm system 50 km away; the reflected radar pulse is found to be up-shifted in frequency by 325 Hz. Assuming the wind is headed directly toward the radar antenna, how fast are the winds in the storm system moving? (The radar system can only measure the radial component of the velocity.)

84 •• ISOLVE A stationary destroyer is equipped with sonar that sends out pulses of sound at 40 MHz. Reflected pulses are received from a submarine directly below with a time delay of 80 ms at a frequency of 39.958 MHz. If the speed of sound in seawater is 1.54 km/s, find (a) the depth of the submarine and (b) its vertical speed.

85 •• A police radar unit transmits microwaves of frequency 3×10^{10} Hz. The speed of these waves in air is 3.0×10^8 m/s. Suppose a car is receding from the stationary police car at a speed of 140 km/h. What is the frequency difference between the transmitted signal and the signal received from the receding car?

86 •• Suppose the police car of Problem 85 is moving in the same direction as the other vehicle at a speed of 60 km/h. What then is the difference in frequency between the emitted and the reflected signals?

87 •• **ISOLVE** At time $t = 0$, a supersonic plane is directly over point P, flying due west at an altitude of 12 km and a speed of Mach 1.6. Where is the plane when the sonic boom is heard?

88 •• A small radio of mass 0.10 kg is attached to one end of an air track by a spring. The radio emits a sound of 800 Hz. A listener at the other end of the air track hears a sound whose frequency varies between 797 and 803 Hz. (*a*) Determine the energy of the vibrating mass–spring system. (*b*) If the spring constant is 200 N/m, what is the amplitude of vibration of the mass and what is the period of the oscillating system?

89 •• A sound source of frequency f_0 moves with speed u_s relative to still air toward a receiver who is moving with speed u_r relative to still air away from the source. (*a*) Write an expression for the received frequency f'. (*b*) Use the result that $(1 - x)^{-1} \approx 1 + x$ to show that if both u_s and u_r are small compared to v, then the received frequency is approximately

$$f' \approx \left(1 + \frac{u_s - u_r}{v}\right)f_0 = \left(1 + \frac{u_{rel}}{v}\right)f_0$$

where $u_{rel} = u_s - u_r$ is the relative velocity of approach of the source and receiver.

90 •• Two students with vibrating 440-Hz tuning forks walk away from each other with equal speeds. How fast must they walk so that they each hear a frequency of 438 Hz from the other fork?

91 •• A physics student walks down a long hall carrying a vibrating 512-Hz tuning fork. The end of the hall is closed so that sound reflects from it. The student hears a sound of 516 Hz from the wall. How fast is the student walking?

92 •• **SSM** A small speaker radiating sound at 1000 Hz is tied to one end of an 0.8-m-long rod that is free to rotate about its other end. The rod rotates in the horizontal plane at 4.0 rad/s. Derive an expression for the frequency heard by a stationary observer far from the rotating speaker.

93 •• A balloon carried along by a 36-km/h wind emits a sound of 800 Hz as it approaches a tall building. (*a*) What is the frequency of the sound heard by an observer at the window of this building? (*b*) What is the frequency of the reflected sound heard by a person riding in the balloon?

94 •• **ISOLVE**✓ A car is approaching a reflecting wall. A stationary observer behind the car hears a sound of frequency 745 Hz from the car horn and a sound of frequency 863 Hz from the wall. (*a*) How fast is the car traveling? (*b*) What is the frequency of the car horn? (*c*) What frequency does the car driver hear reflected from the wall?

95 •• The driver of a car traveling at 100 km/h toward a vertical cliff briefly sounds the horn. Exactly 1 s later she hears the echo and notes that its frequency is 840 Hz. How far from the cliff was the car when the driver sounded the horn and what is the frequency of the horn?

96 •• You are on a transatlantic flight traveling due west at 800 km/h. A Concorde flying at Mach 1.6 and 3 km to the north of your plane is also on an east-to-west course. What is the distance between the two planes when you hear the sonic boom from the Concorde?

97 •• **SSM** The Hubble space telescope has been used to determine the existence of planets orbiting around distant stars. The planet orbiting the star will cause the star to "wobble" with the same period as the planet's orbit; because of this, light from the star will be Doppler-shifted up and down periodically. Estimate the maximum and minimum wavelengths of light of nominal wavelength 500nm emitted by the sun that is Doppler-shifted by the motion of the sun due to the planet Jupiter.

98 ••• **ISOLVE**✓ A physics student drops a vibrating 440-Hz tuning fork down the elevator shaft of a tall building. When the student hears a frequency of 400 Hz, how far has the tuning fork fallen?

99 •• The SuperKamiokande neutrino detector in Japan is a water tank the size of a 14-story building. It detects neutrinos, the "ghost particles" of physics, by the shock wave produced when a neutrino imparts most of its energy to an electron, which then goes flying off at near light-speed through the water. If the maximum angle of the Cerenkov shock-wave cone is 48.75°, what is the speed of light in water?

General Problems

100 • At time $t = 0$, the shape of a wave pulse on a string is given by the function

$$y(x,0) = \frac{0.12 \text{ m}^3}{(2.00 \text{ m})^2 + x^2}$$

where x is in meters. (*a*) Sketch $y(x, 0)$ versus x. Give the wave function $y(x,t)$ at a general time t if (*b*) the pulse is moving in the positive x direction with a speed of 10 m/s and (*c*) the pulse is moving in the negative x direction with a speed of 10 m/s.

101 • **ISOLVE** A wave with a frequency of 1200 Hz propagates along a wire that is under a tension of 800 N. Its wavelength is 24 cm. What will be the wavelength if the tension is decreased to 600 N and the frequency is kept constant?

102 • **ISOLVE**✓ In a common lecture demonstration of wave pulses, a length of rubber tubing is tied at one end to a fixed post and is passed over a pulley to a weight hanging at the other end. Suppose that the distance from the fixed support to the pulley is 10 m, the mass of this length of tubing is 0.7 kg, and the suspended weight is 110 N. If the tubing is given a transverse blow at one end, how long will it take the resulting pulse to reach the other end?

103 • A boat traveling at 10 m/s on a still lake makes a bow wave at an angle of 20° with its direction of motion. What is the speed of the bow wave?

104 • If a wavelength is much larger than the diameter of a loudspeaker, the speaker radiates in all directions, much like a point source. On the other hand, if the wavelength is much smaller than the loudspeaker diameter, the sound travels in a beam in the forward direction—and does not spread out. Find the frequency of a sound wave that has a wavelength (*a*) 10 times the diameter of a 30-cm speaker and (*b*) one-tenth the diameter of a 30-cm speaker. (*c*) Repeat this problem for a 6-cm speaker.

105 • A whistle of frequency 500 Hz moves in a circle of radius 1 m at 3 rev/s. What are the maximum and minimum frequencies heard by a stationary listener in the plane of the circle and 5 m away from its center?

106 • Ocean waves move toward the beach with a speed of 8.9 m/s and a crest-to-crest separation of 15.0 m. You are in a small boat anchored off shore. (*a*) What is the frequency of the ocean waves? (*b*) You now lift anchor and head out to sea at a speed of 15 m/s. What frequency of the waves do you observe?

107 •• A 12.0-m wire of mass 85 g is stretched under a tension of 180 N. A pulse is generated at the left end of the wire, and 25 ms later a second pulse is generated at the right end of the wire. Where do the pulses first meet?

108 •• **SSM** **iSOLVE** ✓ Find the speed of a car the tone of whose horn will drop by 10 percent as it passes you.

109 •• A loudspeaker driver 20 cm in diameter is vibrating at 800 Hz with an amplitude of 0.025 mm. Assuming that the air molecules in the vicinity have the same amplitude of vibration, find (*a*) the pressure amplitude immediately in front of the driver, (*b*) the sound intensity, and (*c*) the acoustic power being radiated.

110 •• **iSOLVE** A plane, harmonic, acoustical wave that oscillates in air with an amplitude of 1 μm has an intensity of 10 mW/m². What is the frequency of the sound wave?

111 •• Water flows at 7 m/s in a pipe of radius 5 cm. A plate having an area equal to the cross-sectional area of the pipe is suddenly inserted to stop the flow. Find the force exerted on the plate. Take the speed of sound in water to be 1.4 km/s. *Hint: When the plate is inserted, a pressure wave propagates through the water at the speed of sound v_s. The mass of water brought to a stop in time Δt is the water in a length of tube equal to $v_s \Delta t$.*

112 •• A high-speed flash photography setup to capture a picture of a bullet exploding a soap bubble is shown in Figure 15-31. The shock wave from the bullet is to be detected by a microphone that will trigger the flash. The microphone is placed on a track that is parallel to the path of the bullet, at the same height as the bullet. The track is used to adjust the position of the microphone. If the bullet is traveling at 1.25 times the speed of sound, and the distance between the lab bench and the track is 0.35 m, how far back from the soap bubble must the microphone be set to trigger the flash? (Assume that the flash itself is instantaneous once the microphone is triggered.)

113 •• A column of precision marchers keeps in step by listening to the band positioned at the head of the column. The beat of the music is for 100 paces/min. A television camera shows that only the marchers at the front and the rear of the column are actually in step. The marchers in the middle section are striding forward with the left foot when

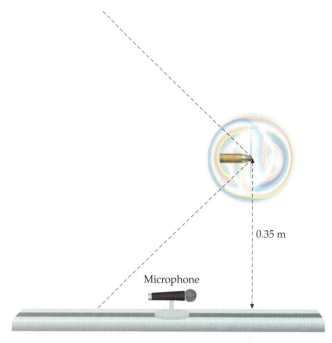

FIGURE 15-31 Problem 112

those at the front and rear are striding forward with the right foot. The marchers are so well trained, however, that they are all certain that they are in proper step with the music. Explain the source of the problem, and calculate the length of the column.

114 •• A bat flying toward an obstacle at 12 m/s emits brief, high-frequency sound pulses at a repetition frequency of 80 Hz. What is the interval between the arrival times of the echo pulses heard by the bat?

115 •• **SSM** Laser ranging to the moon is done routinely to accurately determine the earth–moon distance. However, to determine the distance accurately, corrections must be made for the speed of light in the earth's atmosphere, which is 99.997 percent of the speed of light in vacuum. Assuming that the earth's atmosphere is 8 km high, estimate the length of the correction.

116 •• A tuning fork attached to a stretched wire generates transverse waves. The vibration of the fork is perpendicular to the wire. Its frequency is 400 Hz and the amplitude of its oscillation is 0.50 mm. The wire has a linear mass density of 0.01 kg/m and is under a tension of 1 kN. Assume that there are no reflected waves. (*a*) Find the period and frequency of waves on the wire. (*b*) What is the speed of the waves? (*c*) What are the wavelength and wave number? (*d*) Write a suitable wave function for the waves on the wire. (*e*) Calculate the maximum speed and acceleration of a point on the wire. (*f*) At what average rate must energy be supplied to the fork to keep it oscillating at a steady amplitude?

117 ••• If a loop of chain is spun at high speed, it will roll like a hoop without collapsing. Consider a chain of linear mass density μ that is rolling without slipping at a high speed v_0. (a) Show that the tension in the chain is $F = \mu v_0^2$. (b) If the chain rolls over a small bump, a transverse wave pulse will be generated in the chain. At what speed will it travel along the chain? (c) How far around the loop (in degrees) will a transverse wave pulse travel in the time the hoop rolls through one complete revolution?

118 ••• A long rope with a mass per unit length of 0.1 kg/m is under a constant tension of 10 N. A motor at the point $x = 0$ drives one end of the rope with harmonic motion at 5 oscillations per second and an amplitude of 4 cm. (a) What is the wave speed? (b) What is the wavelength? (c) What is the maximum transverse linear momentum of a 1-mm segment of the rope? (d) What is the maximum net force on a 1-mm segment of the rope?

119 ••• **SSM** A heavy rope 3 m long is attached to the ceiling and is allowed to hang freely. (a) Show that the speed of transverse waves on the rope is independent of its mass and length but does depend on the distance y from the bottom according to the formula $v = \sqrt{gy}$. (b) If the bottom end of the rope is given a sudden sideways displacement, how long does it take the resulting wave pulse to go to the ceiling, reflect, and return to the bottom of the rope?

120 ••• In this problem you will derive an expression for the potential energy of a segment of a string carrying a traveling wave (Figure 15-32). The potential energy of a segment equals the work done by the tension in stretching the string, which is $\Delta U = F(\Delta\ell - \Delta x)$, where F is the tension, $\Delta\ell$ is the length of the stretched segment, and Δx is its original length. From the figure we see that

$$\Delta\ell = \sqrt{(\Delta x)^2 + (\Delta y)^2} = \Delta x[1 + (\Delta y/\Delta x)^2]^{1/2}$$

(a) Use the binomial expansion to show that $\Delta\ell - \Delta x \approx \frac{1}{2}(\Delta y/\Delta x)^2\,\Delta x$, and therefore $\Delta U \approx \frac{1}{2}F(\Delta y/\Delta x)^2\,\Delta x$. (b) Compute dy/dx from the wave function in Equation 15-13 and show that $\Delta U \approx \frac{1}{2}Fk^2\cos^2(kx - \omega t)\,\Delta x$.

FIGURE 15-32
Problem 120

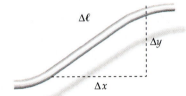

Superposition and Standing Waves

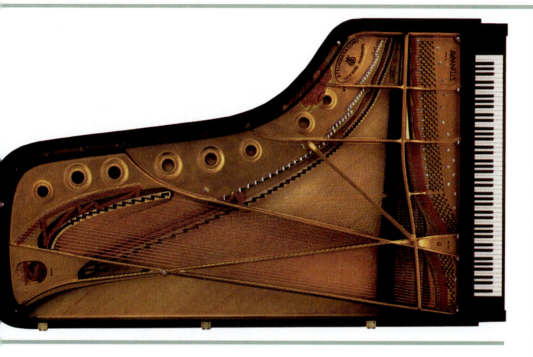

THE STRINGS IN THIS STEINWAY GRAND PIANO VIBRATE WHEN STRUCK BY THE HAMMERS, WHICH ARE CONTROLLED BY THE KEYS. THE LONGER STRINGS VIBRATE AT LOWER FREQUENCIES THAN THE SHORTER STRINGS.

? **What other factors come into play when tuning a piano? (See Example 16-6.)**

When two or more waves overlap in space, their individual disturbances (represented mathematically by their wave functions) superimpose and add algebraically, creating a resultant wave. This property of waves is called the principle of superposition. Under certain circumstances the superposition of harmonic waves of the same frequency produces sustained wave patterns in space. This phenomenon is called interference. Interference and diffraction are what distinguish wave motion from particle motion. Thomas Young's observation in 1801 of interference of light led to the understanding that light propagates via wave motion, not via particle motion as had been proposed by Newton. The observation of interference of electron waves by C. J. Davisson and L. H. Germer in 1927 led to our understanding of the wave nature of electrons and other material objects. These ideas are integral to understanding quantum physics, which is presented in Chapter 34.

➤ **In this chapter, we begin with the superposition of wave pulses on a string and then consider the superposition and interference of harmonic waves. We examine the phenomenon of beats, which result from the interference of two waves of slightly different frequencies, and study standing waves, which occur when harmonic waves are confined in space. We also consider the analysis of complex musical tones in terms of their component harmonic waves, and** the inverse problem of the synthesis of harmonic waves to produce complex tones. We conclude with a qualitative discussion of the extension of harmonic analysis to aperiodic waves such as wave pulses.

16-1 Superposition of Waves

Figure 16-1a shows small wave pulses moving in opposite directions on a string. The shape of the string when they meet can be found by adding the displacements produced by each pulse separately. The **principle of superposition** is a property of wave motion which states:

> When two or more waves overlap, the resultant wave is the algebraic sum of the individual waves.

PRINCIPLE OF SUPERPOSITION

Mathematically, when there are two pulses on the string, the total wave function is the algebraic sum of the individual wave functions.

In the special case of two pulses that are identical except that one is inverted relative to the other, as in Figure 16-1b, there is an instant when the pulses exactly overlap and add to zero. At this instant the string is horizontal, but it is not stationary. At the right edge of the overlap region the string is moving upward and at the left edge it is moving downward. A short time later the individual pulses emerge, each continuing in its original direction.

Superposition is a characteristic and unique property of wave motion. There is no analogous situation in Newtonian particle motion; that is, two Newtonian particles never overlap or add together in this way.

*Superposition and the Wave Equation

The principle of superposition follows from the fact that the wave equation (Equation 15-9) is linear for small transverse displacements. That is, the function $y(x,t)$ and its derivatives occur only to the first power. The defining property of a linear equation is that if y_1 and y_2 are two solutions of the equation, then the linear combination

$$y_3 = C_1 y_1 + C_2 y_2 \qquad \text{16-1}$$

is also a solution, where C_1 and C_2 are any constants. The linearity of the wave equation can be shown by the direct substitution of y_3 into the equation. The result is the mathematical statement of the principle of superposition. If any two waves satisfy a wave equation, then their algebraic sum also satisfies the same wave equation.

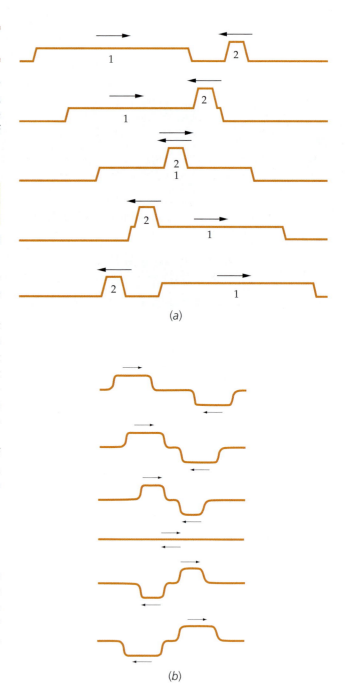

(a)

(b)

FIGURE 16-1 Wave pulses moving in opposite directions on a string. The shape of the string when the pulses meet is found by adding the displacements of each separate pulse. (a) Superposition of pulses having displacements in the same direction. (b) Superposition of pulses having equal but opposite displacements. Here the algebraic addition of the displacement amounts to the subtraction of the magnitudes.

SUPERPOSITION AND THE WAVE EQUATION   **EXAMPLE 16-1**

Show that if functions y_1 and y_2 both satisfy wave equation

$$\frac{\partial^2 y}{\partial x^2} = \frac{1}{v^2}\frac{\partial^2 y}{\partial t^2}   \text{(Equation 15-9}b\text{)}$$

then the function y_3 given by Equation 16-1 also satisfies it.

PICTURE THE PROBLEM Substitute y_3 into the wave equation, assume that y_1 and y_2 each satisfy the wave equation, and show that, as a consequence, the linear combination $C_1 y_1 + C_2 y_2$ satisfies the wave equation.

1. Substitute the expression for y_3 in Equation 16-1 into the left side of the wave equation, then break it into separate terms for y_1 and y_2:

$$\frac{\partial^2 y_3}{\partial x^2} = \frac{\partial^2}{\partial x^2}(C_1 y_1 + C_2 y_2) = C_1\frac{\partial^2 y_1}{\partial x^2} + C_2\frac{\partial^2 y_2}{\partial x^2}$$

2. Both y_1 and y_2 satisfy the wave function. Write the wave equation for both y_1 and y_2:

$$\frac{\partial^2 y_1}{\partial x^2} = \frac{1}{v^2}\frac{\partial^2 y_1}{\partial t^2} \quad \text{and} \quad \frac{\partial^2 y_2}{\partial x^2} = \frac{1}{v^2}\frac{\partial^2 y_2}{\partial t^2}$$

3. Substitute the step 2 results into the step 1 result and factor out any common terms:

$$\frac{\partial^2 y_3}{\partial x^2} = C_1\frac{1}{v^2}\frac{\partial^2 y_1}{\partial t^2} + C_2\frac{1}{v^2}\frac{\partial^2 y_2}{\partial t^2} = \frac{1}{v^2}\left(C_1\frac{\partial^2 y_1}{\partial t^2} + C_2\frac{\partial^2 y_2}{\partial t^2}\right)$$

4. Move the constants inside the arguments of the derivatives and express the sum of the derivatives as the derivative of the sum:

$$\frac{\partial^2 y_3}{\partial x^2} = \frac{1}{v^2}\left(\frac{\partial^2 C_1 y_1}{\partial t^2} + \frac{\partial^2 C_2 y_2}{\partial t^2}\right) = \frac{1}{v^2}\frac{\partial^2}{\partial t^2}(C_1 y_1 + C_2 y_2)$$

5. The argument of the derivative in step 4 is y_3:

$$\therefore \quad \boxed{\frac{\partial^2 y_3}{\partial x^2} = \frac{1}{v^2}\frac{\partial^2 y_3}{\partial t^2}}$$

Interference of Harmonic Waves

The result of the superposition of two harmonic waves of the same frequency depends on the phase difference δ between the waves. Let $y_1(x,t)$ be the wave function for a harmonic wave traveling to the right with amplitude A, angular frequency ω, and wave number k:

$$y_1 = A\sin(kx - \omega t) \tag{16-2}$$

For this wave function, we have chosen the phase constant to be zero.[†] If we have another harmonic wave also traveling to the right with the same amplitude, frequency, and wave number, then the general equation for its wave function can be written

$$y_2 = A\sin(kx - \omega t + \delta) \tag{16-3}$$

where δ is the phase constant. The two waves described by Equations 16-2 and 16-3 differ in phase by δ. Figure 16-2 shows a plot of the two wave functions versus position for a fixed time. The resultant wave is the sum

$$y_1 + y_2 = A\sin(kx - \omega t) + A\sin(kx - \omega t + \delta) \tag{16-4}$$

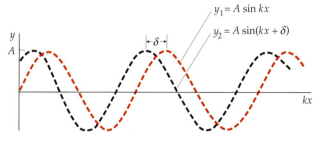

FIGURE 16-2 Displacement versus position for two harmonic waves having the same amplitude, frequency, and wavelength, but differing in phase by δ.

[†] This choice is convenient but not mandatory. If, for example, we chose $t = 0$ when the displacement was maximum at $x = 0$, we would write $y_1 = A\cos(kx - \omega t) = A\sin(kx - \omega t - \pi/2)$.

We can simplify Equation 16-4 by using the trigonometric identity

$$\sin \theta_1 + \sin \theta_2 = 2 \cos \tfrac{1}{2}(\theta_1 - \theta_2) \sin \tfrac{1}{2}(\theta_1 + \theta_2)$$ 16-5

For this case, $\theta_1 = kx - \omega t$ and $\theta_2 = kx - \omega t + \delta$, so that

$$\tfrac{1}{2}(\theta_1 - \theta_2) = -\tfrac{1}{2}\delta$$

and

$$\tfrac{1}{2}(\theta_1 + \theta_2) = kx - \omega t + \tfrac{1}{2}\delta$$

Thus, Equation 16-4 becomes

$$y_1 + y_2 = \left[2A \cos \tfrac{1}{2}\delta\right] \sin(kx - \omega t + \tfrac{1}{2}\delta)$$ 16-6

SUPERPOSITION OF TWO WAVES OF THE SAME AMPLITUDE AND FREQUENCY

where we have used $\cos(-\tfrac{1}{2}\delta) = \cos \tfrac{1}{2}\delta$. We see that the result of the superposition of two harmonic waves of equal wave number and frequency is a harmonic wave having the same wave number and frequency. The resultant wave has amplitude $2A \cos \tfrac{1}{2}\delta$ and a phase equal to half the difference between the phases of the original waves. The phenomenon of two or more waves of the same, or almost the same, frequency superposing to produce an observable pattern in the intensity is called **interference.** In this example, the intensity, which is proportional to the square of the amplitude, is uniform. If the two waves are in phase, then $\delta = 0$, $\cos 0 = 1$, and the amplitude of the resultant wave is $2A$. The interference of two waves in phase is called **constructive interference** (Figure 16-3). If the two waves are 180° out of phase, then $\delta = \pi$, $\cos(\pi/2) = 0$, and the amplitude of the resultant wave is zero. The interference of two waves 180° out of phase is called **destructive interference** (Figure 16-4).

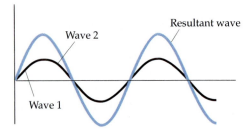

FIGURE 16-3 Constructive interference. When two waves are in phase, the amplitude of the resultant wave is the sum of the amplitudes of the individual waves.

EXERCISE Two waves with the same frequency, wavelength, and amplitude are traveling in the same direction. (*a*) If they differ in phase by $\pi/2$ and each has an amplitude of 4.0 cm, what is the amplitude of the resultant wave? (*b*) For what phase difference δ will the resultant amplitude be equal to 4.0 cm? (*Answer* (*a*) 5.66 cm (*b*) 120° or 240°)

Beats The interference of two sound waves with slightly different frequencies produces the interesting phenomenon known as **beats.** Consider two sound waves that have angular frequencies of ω_1 and ω_2 and the same pressure amplitude p_0. What do we hear? At a fixed point, the spatial dependence of the wave merely contributes a phase constant, so we can neglect it. The pressure at the ear due to either wave acting alone will be a simple harmonic function of the type

$$p_1 = p_0 \sin \omega_1 t$$

and

$$p_2 = p_0 \sin \omega_2 t$$

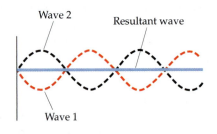

FIGURE 16-4 Destructive interference. When two waves have a phase difference of π, the amplitude of the resultant wave is the difference between the amplitudes of the individual waves. If the original waves have equal amplitudes, they cancel completely.

where we have chosen sine functions for convenience and have assumed that the waves are in phase at time $t = 0$. Using the trigonometry identity

$$\sin \theta_1 + \sin \theta_2 = 2 \cos \tfrac{1}{2}(\theta_1 - \theta_2) \sin \tfrac{1}{2}(\theta_1 + \theta_2)$$

for the sum of two sine functions, we obtain for the resultant wave

$$p = p_0 \sin \omega_1 t + p_0 \sin \omega_2 t = 2p_0 \cos \tfrac{1}{2}(\omega_1 - \omega_2)t \sin \tfrac{1}{2}(\omega_1 + \omega_2)t$$

If we write $\omega_{av} = \frac{1}{2}(\omega_1 + \omega_2)$ for the average angular frequency and $\Delta\omega = \omega_1 - \omega_2$ for the difference in angular frequencies, the resultant wave function is

$$p = 2p_0 \cos(\tfrac{1}{2}\Delta\omega\, t)\,\sin \omega_{av}t =$$
$$2p_0 \cos(2\pi \tfrac{1}{2}\Delta f\, t)\sin 2\pi f_{av}t \qquad \text{16-7}$$

where $\Delta f = \Delta\omega/(2\pi)$ and $f_{av} = \omega_{av}/(2\pi)$.

Figure 16-5 shows a plot of pressure variations as a function of time. The waves are originally in phase and add constructively at time $t = 0$. Because their frequencies differ, the waves gradually become out of phase, and at time t_1 they are 180° out of phase and interfere destructively.[†] An equal time interval later (time t_2 in the figure), the two waves are again in phase and interfere constructively. The greater the difference in the frequencies of the two waves, the more rapidly they oscillate in and out of phase.

The tone we hear has a frequency of $f_{av} = \frac{1}{2}(f_1 + f_2)$ and amplitude $2p_0 \cos(2\pi\frac{1}{2}\Delta f\, t)$. (For some values of t the amplitude is negative. Since $-\cos\theta = \cos(\theta + \pi)$, a change in the sign of the amplitude is equivalent to a 180° phase change.) The amplitude oscillates with the frequency $\frac{1}{2}\Delta f$. Since the sound intensity is proportional to the square of the amplitude, the sound is loud whenever the amplitude function is either a maximum or a minimum. Thus the frequency of this variation in intensity, called the **beat frequency,** is twice $\frac{1}{2}\Delta f$:

$$f_{beat} = \Delta f \qquad \text{16-8}$$

BEAT FREQUENCY

The beat frequency equals the difference in the individual frequencies of the two waves. If we simultaneously strike two tuning forks having the frequencies 241 Hz and 243 Hz, we will hear a pulsating tone at the average frequency of 242 Hz that has a maximum intensity 2 times per second; that is, the beat frequency is 2 Hz. The ear can detect up to about 15 to 20 beats per second. Above this frequency, the fluctuations in loudness are too rapid to be distinguished.

The phenomenon of beats is often used to compare an unknown frequency with a known frequency, as when a tuning fork is used to tune a piano string. Pianos are tuned by simultaneously ringing the tuning fork and striking a key, while at the same time adjusting the tension of the piano string until the beats are far apart, indicating that the difference in frequency of the two sound generators is very small.

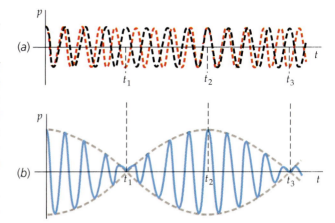

FIGURE 16-5 Beats. (a) Two waves of different but nearly equal frequencies that are in phase at $t_0 = 0$ are 180° out of phase at some time later t_1. At a still later time, t_2, they are back in phase. (b) The resultant of the two waves shown in (a). The frequency of the resultant wave is about the same as those of the original waves, but the amplitude is modulated as indicated by the dashed envelope. The amplitude is maximum at times t_0 and t_2 and zero at times t_1 and t_3.

TUNING A GUITAR **EXAMPLE 16-2**

When a 440-Hz (concert A) tuning fork is struck simultaneously with the playing of the A string of a slightly out-of-tune guitar, 3 beats per second are heard. The guitar string is tightened a little to increase its frequency. As the guitar string is slowly tightened, you hear the beat frequency slowly increase. What is the frequency of the guitar string after it is tightened?

PICTURE THE PROBLEM Because 3 beats per second were heard initially, the original frequency of the guitar string was either 443 Hz or 437 Hz. Had it been 437 Hz, slowly increasing the string's frequency by tightening it would decrease the beat frequency.

Because the beat frequency increases as the tension increases, from 3 to 6 beats per second, the original frequency must have been 443 Hz. $\quad f = f_A + f_{beat} = 440\text{ Hz} + 3\text{ Hz} = \boxed{443\text{ Hz}}$

[†] Complete cancellation occurs only when the pressure amplitudes of the two waves are equal.

Phase Difference due to Path Difference   A common cause of a phase difference between two waves is different path lengths between the sources of the waves and the point of interference. Suppose that two sources oscillate in phase (for example, positive crests leave the sources at the same time) and emit harmonic waves of the same frequency and wavelength. Now consider a point in space for which the path lengths to the two sources differ. If the path difference is one wavelength, as is the case in Figure 16-6a, or an integral number of wavelengths, the interference is constructive. If the path difference is one half of a wavelength or an odd number of half wavelengths, as in Figure 16-6b, the maximum of one wave falls at the minimum of the other and the interference is destructive.

The wave functions for waves from two sources oscillating in phase can be written

$$p_1 = p_0 \sin(kx_1 - \omega t)$$

and

$$p_2 = p_0 \sin(kx_2 - \omega t)$$

The phase difference for these two wave functions is

$$\delta = (kx_2 - \omega t) - (kx_1 - \omega t) = k(x_2 - x_1) = k\Delta x$$

Using $k = 2\pi/\lambda$, we have

$$\delta = k\Delta x = 2\pi\frac{\Delta x}{\lambda} \qquad\qquad 16\text{-}9$$

PHASE DIFFERENCE DUE TO PATH DIFFERENCE

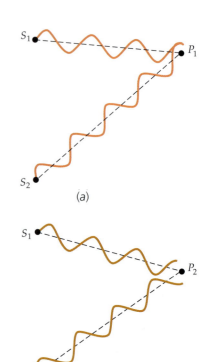

FIGURE 16-6 Waves from two sources S_1 and S_2 that are in phase when they meet at a point P_1. (a) When the path difference is one wavelength λ, the waves are in phase at P_1 and interfere constructively. (b) When the path difference is $\frac{1}{2}\lambda$, the waves at P_2 are out of phase by 180° and therefore interfere destructively. If the waves are of equal amplitude at P_2, they will cancel completely at this point.

A Resultant Sound Wave   **EXAMPLE   16-3**

Two sound sources oscillate in phase. At a point 5.00 m from one source and 5.17 m from the other, the amplitude of the sound from each source separately is p_0. Find the amplitude of the resultant wave if the frequency of the sound waves is (a) 1000 Hz, (b) 2000 Hz, and (c) 500 Hz. (Use 340 m/s for the speed of sound.)

PICTURE THE PROBLEM The amplitude of the resultant wave due to superposition of two waves differing in phase by δ is given by $A = 2p_0 \cos\frac{1}{2}\delta$ (Equation 16-6), where p_0 is the amplitude of either wave and $\delta = 2\pi\,\Delta x/\lambda$ is the phase difference. We are given the path difference, $\Delta x = 5.17\text{ m} - 5\text{ m} = 0.17\text{ m}$, so all that is needed is the wavelength λ.

(a) 1. The wavelength equals the speed divided by the frequency. Calculate λ for $f = 1000$ Hz:
$\lambda = \dfrac{v}{f} = \dfrac{340 \text{ m/s}}{1000 \text{ Hz}} = 0.34 \text{ m}$

2. For $\lambda = 0.34$ m, the given path difference ($\Delta x = 0.17$ m) is $\frac{1}{2}\lambda$, so we expect destructive interference. Use this value of λ to calculate the phase difference δ and use δ to calculate the amplitude A:
$\delta = 2\pi\dfrac{\Delta x}{\lambda} = 2\pi\dfrac{0.17 \text{ m}}{0.34 \text{ m}} = \pi$

so

$A = 2p_0 \cos\dfrac{1}{2}\delta = 2p_0 \cos\dfrac{\pi}{2} = \boxed{0}$

(b) 1. Calculate λ for $f = 2000$ Hz:
$\lambda = \dfrac{v}{f} = \dfrac{340 \text{ m/s}}{2000 \text{ Hz}} = 0.17 \text{ m}$

2. For $\lambda = 0.17$ m, the path difference equals λ, so we expect constructive interference. Calculate the phase difference and amplitude:

$$\delta = 2\pi\frac{\Delta x}{\lambda} = 2\pi\frac{0.17\text{ m}}{0.17\text{ m}} = 2\pi$$

so

$$A = 2p_0\cos\tfrac{1}{2}\delta = 2p_0\cos\pi = \boxed{-2p_0}$$

(c) 1. Calculate λ for $f = 500$ Hz:

$$\lambda = \frac{v}{f} = \frac{340\text{ m/s}}{500\text{ Hz}} = 0.68\text{ m}$$

2. Calculate the phase difference and amplitude:

$$\delta = 2\pi\frac{\Delta x}{\lambda} = 2\pi\frac{0.17\text{ m}}{0.68\text{ m}} = \frac{\pi}{2}$$

so

$$A = 2p_0\cos\frac{1}{2}\delta = 2p_0\cos\frac{\pi}{4} = \boxed{\sqrt{2}p_0}$$

REMARKS In part (b), A is found to be negative. Equation 16-6 can be written $y_1 + y_2 = A'\sin(kx - \omega t + \frac{\delta}{2})$, which can also be written $y_1 + y_2 = -A'\sin(kx - \omega t + \frac{\delta}{2} + \pi)$. A phase shift of $\pi = 180°$ is equivalent to multiplying by -1.

SOUND INTENSITY OF TWO LOUDSPEAKERS   **EXAMPLE 16-4**

Two identical loudspeakers face each other at a distance of 180 cm and are driven by a common audio oscillator at 680 Hz. Locate the points between the speakers along a line joining them for which the sound intensity is (a) maximum and (b) minimum. (Neglect the variation in intensity from either speaker with distance, and use 340 m/s for the speed of sound.)

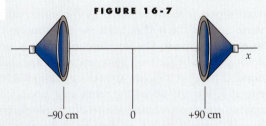
FIGURE 16-7

PICTURE THE PROBLEM We choose the origin to be at the midpoint between the speakers (Figure 16-7). Since this point is equidistant from the speakers, it is a point of maximum intensity. When we move a distance x toward one of the speakers, the path difference is $2x$. The intensity will be maximum when $2x = 0, \lambda, 2\lambda, 3\lambda, \ldots$, and minimum when $2x = (1/2)\lambda, (3/2)\lambda, (5/2)\lambda, \ldots$.

(a) 1. The intensity will be maximum when $2x$ equals an integral number of wavelengths:

$$2x = 0, \pm\lambda, \pm 2\lambda, \pm 3\lambda, \ldots$$

2. Calculate the wavelength:

$$\lambda = \frac{v}{f} = \frac{340\text{ m/s}}{680\text{ Hz}} = 0.5\text{ m} = 50\text{ cm}$$

3. Solve for x using the calculated wavelength:

$$x = 0, \pm\tfrac{1}{2}\lambda, \pm\lambda, \pm\tfrac{3}{2}\lambda, \ldots$$

$$= \boxed{0, \pm 25\text{ cm}, \pm 50\text{ cm}, \pm 75\text{ cm}}$$

(b) 1. The intensity will be minimum when $2x$ equals an odd number of half wavelengths:

$$2x = \pm\tfrac{1}{2}\lambda, \pm\tfrac{3}{2}\lambda, \pm\tfrac{5}{2}\lambda, \ldots$$

2. Solve for x using the calculated wavelength:

$$x = \pm\tfrac{1}{4}\lambda, \pm\tfrac{3}{4}\lambda, \pm\tfrac{5}{4}\lambda, \ldots$$

$$= \boxed{\pm 12.5\text{ cm}, \pm 37.5\text{ cm}, \pm 62.5\text{ cm}, \pm 87.5\text{ cm}}$$

REMARKS The maxima and minima will be relative maxima and relative minima because the amplitude from the near speaker will be slightly greater than that from the far speaker. Only seven terms were used for the maxima and only eight terms for the minima because any additional terms would be at a distance beyond one speaker.

Figure 16-8*a* shows the wave pattern produced by two point sources in a ripple tank that are oscillating in phase. Each source produces waves with circular wavefronts. The circular wavefronts shown all have the same phase and are separated by one wavelength. We can construct a similar pattern with a compass by drawing circular arcs representing the wave crests from each source at some particular time (Figure 16-8*b*). Where the crests from each source overlap, the waves interfere constructively. At these points, the path lengths from the two sources are either equal or they differ by an integral number of wavelengths. The dashed lines indicate the points that are equidistant from the sources or whose path differences are one wavelength, two wavelengths, or three wavelengths. At each point along any of these lines the interference is constructive, so these are lines of interference maxima. Between the lines of interference maxima are lines of interference minima. On a line of interference minima, the path length from any point on the line to each of the two sources differs by an odd number of half wavelengths. Throughout the region where the two waves are superposed, the amplitude of the resultant wave is given by $A = 2p_0 \cos \frac{1}{2}\delta$, where p_0 is the amplitude of each wave separately and δ is related to the path difference Δr by $\delta = 2\pi \Delta r/\lambda$ (Equation 16-9).

Figure 16-9 shows the intensity I of the resultant wave from two sources as a function of path difference Δx. At points where the interference is constructive, the amplitude of the resultant wave is twice that of either wave alone, and since the intensity is proportional to the square of the amplitude, the intensity is $4I_0$, where I_0 is the intensity due to either source alone. At points of destructive interference, the intensity is zero. The average intensity, shown by the dashed line in the figure, is twice the intensity due to either source alone, a result required by the conservation of energy. The interference of the waves from the two sources thus redistributes the energy in space. The interference of two sound sources can be demonstrated by driving two separated speakers with the same amplifier (so that they are always in phase) fed by an audio-signal generator. Moving about the room, one can detect by ear the positions of constructive and destructive interference.[†] This is best done in a room called an anechoic chamber, where reflections (echoes) off the walls of the room are minimized.

Coherence

Two sources need not be in phase to produce an interference pattern. Consider two sources that are 180° out of phase. (Two speakers that are in phase can be made to be out of phase by 180° merely by switching the leads to one of the speakers.) The intensity pattern is the same as that in Figure 16-9 except that the maxima and minima are interchanged. At points for which the distance differs by an integral number of wavelengths, the interference is destructive because the waves are 180° out of phase. At points where the path difference is an odd number of half wavelengths, the waves are now in phase because the 180° phase difference of the sources is offset by the 180° phase difference due to the path difference.

Similar interference patterns will be produced by any two sources whose phase difference remains constant. Two

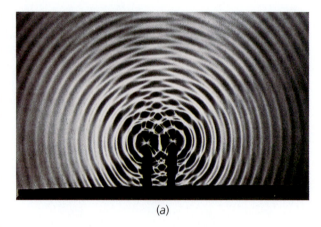

(a)

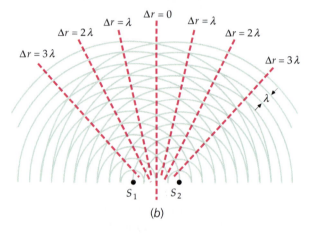

(b)

FIGURE 16-8 (*a*) Water waves in a ripple tank produced by two nearby sources oscillating in phase. (*b*) Drawing of wave crests for the sources in (*a*). The dashed lines indicate points for which the path difference is an integral number of wavelengths.

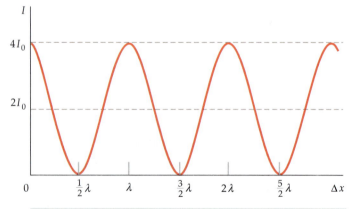

FIGURE 16-9 Intensity versus path difference for two sources that are in phase. I_0 is the intensity due to each source individually.

---

† In this demonstration, the sound intensity will be not quite zero at the points of destructive interference because of sound reflections from the walls or objects in the room.

sources that remain in phase or maintain a constant phase difference are said to be **coherent.** Coherent sources of water waves in a ripple tank are easy to produce by driving both sources with the same motor. Coherent sound sources are obtained by driving two speakers with the same signal source and amplifier.

Wave sources whose difference in phase is not constant but varies randomly are said to be **incoherent sources.** There are many examples of incoherent sources, such as two speakers driven by different amplifiers or two violins played by different violinists. For incoherent sources, the interference at a particular point varies rapidly back and forth from constructive to destructive, and no interference pattern is maintained long enough to be observed. The resultant intensity of waves from two or more incoherent sources is simply the sum of the intensities due to the individual sources.

16-2   Standing Waves

If waves are confined in space, like the waves on a piano string, sound waves in an organ pipe, or light waves in a laser, reflections at both ends cause the waves to travel in both directions. These superposing waves interfere in accordance with the principle of superposition. For a given string or pipe, there are certain frequencies for which superposition results in a stationary vibration pattern called a **standing wave.** Standing waves have important applications in musical instruments and in quantum theory.

Standing Waves on Strings

String Fixed at Both Ends   If we fix both ends of a string and move a portion of the string up and down with simple harmonic motion of small amplitude, we find that at certain frequencies, standing-wave patterns such as those shown in Figure 16-10 are produced. The frequencies that produce these patterns are called the **resonance frequencies** of the string system. Each such frequency, with its accompanying wave function, is called a **mode of vibration.** The lowest resonance frequency is called the **fundamental** frequency f_1. It produces the standing-wave pattern shown in Figure 16-10a, which is called the **fundamental mode** of vibration or the **first harmonic.** The second lowest frequency f_2 produces the pattern shown in Figure 16-10b. This mode of vibration has a frequency twice that of the fundamental frequency and is called the second harmonic. The third lowest frequency f_3 is three times the fundamental frequency, and it produces the third harmonic pattern shown in Figure 16-10c. The set of all resonant frequencies is called the **resonant frequency spectrum** of the string.

Not all resonant frequencies are called harmonics. Only if each frequency of a resonant frequency spectrum is an integral multiple of the fundamental (lowest) frequency are the frequencies referred to as harmonics. Many systems that support standing waves have resonant frequency spectra in which the resonant frequencies are not integral multiples of the lowest frequency. In all resonant frequency spectra the lowest resonant frequency is called the fundamental frequency (or just the fundamental), the next lowest resonant frequency is called the first **overtone,** the next lowest the second overtone, and so forth. This terminology has its

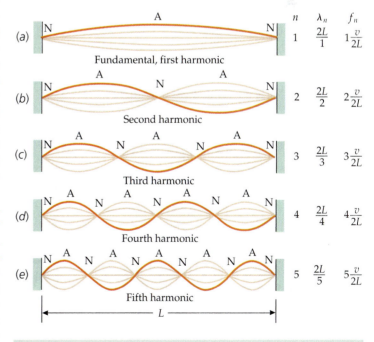

FIGURE 16-10 Standing waves on a string that is fixed at both ends. Points labeled A are antinodes and those labeled N are nodes. In general, the nth harmonic has n antinodes.

roots in music. Only if each resonant frequency is an integral multiple of the fundamental frequency are they referred to as harmonics.

We note from Figure 16-10 that for each harmonic there are certain points on the string (the midpoint in Figure 16-10b, for example) that do not move. Such points are called **nodes.** Midway between each pair of nodes is a point of maximum amplitude of vibration called an **antinode.** Both fixed ends of the string are, of course, nodes. (If one end is attached to a tuning fork or other vibrator rather than being fixed, it will still be approximately a node because the amplitude of the vibration at that end is so much smaller than the amplitude at the antinodes.) We note that the first harmonic has one antinode, the second harmonic has two antinodes, and so on.

We can relate the resonance frequencies to the wave speed in the string and the length of the string. The distance from a node to the nearest antinode is one-fourth of the wavelength. Therefore, the length of the string L equals one-half the wavelength in the fundamental mode of vibration (Figure 16-11) and, as Figure 16-10 reveals, L equals two half-wavelengths for the second harmonic, three half-wavelengths for the third harmonic, and so forth. In general, if λ_n is the wavelength of the nth harmonic, we have

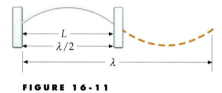

FIGURE 16-11

$$L = n\frac{\lambda_n}{2}, \qquad n = 1, 2, 3, \ldots \qquad \text{16-10}$$

STANDING-WAVE CONDITION, BOTH ENDS FIXED

This result is known as the **standing-wave condition.** We can find the frequency of the nth harmonic from the fact that the wave speed v equals the frequency f_n times the wavelength. Thus,

$$f_n = \frac{v}{\lambda_n} = \frac{v}{2L/n} \qquad n = 1, 2, 3, \ldots$$

or

$$f_n = n\frac{v}{2L} = nf_1 \qquad n = 1, 2, 3, \ldots \qquad \text{16-11}$$

RESONANCE FREQUENCIES, BOTH ENDS FIXED

where $f_1 = v/(2L)$ is the fundamental frequency.

> You shouldn't bother to memorize Equation 16-11. Just sketch Figure 16-10 to remind yourself of the standing-wave condition, $\lambda_n = 2L/n$, and then use $v = f_n\lambda_n$.
>
> PROBLEM-SOLVING GUIDELINE

We can understand standing waves in terms of resonance. Consider a string of length L that is attached at one end to a vibrator (Figure 16-12) and is fixed at the other end. The first wave crest sent out by the vibrator travels down the string a distance L to the fixed end, where it is reflected and inverted. It then travels back a distance L and is again reflected and inverted at the vibrator. The total time for the round trip is $2L/v$. If this time equals the period

FIGURE 16-12

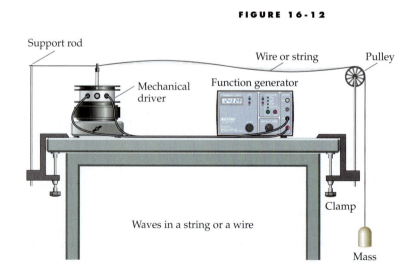

Waves in a string or a wire

of the vibrator, the twice-reflected wave crest exactly overlaps the second wave crest produced by the vibrator, and the two crests interfere constructively, producing a crest with twice the original amplitude. The combined wave crest travels down the string and back and is added to by the third crest produced by the vibrator, increasing the amplitude three-fold, and so on. Thus, the vibrator is in resonance with the string. The wavelength is equal to $2L$ and the frequency is equal to $v/(2L)$.

Resonance also occurs at other vibrator frequencies. The vibrator is in resonance with the string if the time it takes for the first wave crest to travel the distance $2L$ is any integer n times the period T_n of the vibrator. That is, if $2L/v = nT_n$, where $2L/v$ is the round trip time for a wave crest. Thus,

$$f_n = \frac{1}{T_n} = n\frac{v}{2L}, \qquad n = 1, 2, 3, \ldots$$

is the condition for resonance. This is the same result we found by fitting an integral number of half-wavelengths into the distance L. Various damping effects, such as the loss of energy during reflection and the imperfect flexibility of the string, put a limit on the maximum amplitude that can be reached.

The resonance frequencies given by Equation 16-11 are also called the **natural frequencies** of the string. When the frequency of the vibrator is not one of the natural frequencies of the vibrating string, standing waves are not produced. After the first wave travels the distance $2L$ and is reflected from the fork, it differs in phase from the wave being generated at the vibrator (Figure 16-13). When this resultant wave has traveled the distance $2L$ and is again reflected at the vibrator, it will differ in phase from the next wave generated. In some cases, the new resultant wave will have an amplitude greater than that of the previous wave, in other cases the new amplitude will be less. On the average, the amplitude will not increase but will remain on the order of the amplitude of the first wave generated, which is the amplitude of the vibrator. This amplitude is very small compared with the amplitudes attained at resonance frequencies.

The resonance of standing waves is analogous to the resonance of a simple harmonic oscillator with a harmonic driving force. However, a vibrating string has not just one natural frequency but a sequence of natural frequencies that are integral multiples of the fundamental frequency. This sequence is called a **harmonic series.**

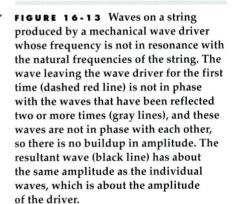

FIGURE 16-13 Waves on a string produced by a mechanical wave driver whose frequency is not in resonance with the natural frequencies of the string. The wave leaving the wave driver for the first time (dashed red line) is not in phase with the waves that have been reflected two or more times (gray lines), and these waves are not in phase with each other, so there is no buildup in amplitude. The resultant wave (black line) has about the same amplitude as the individual waves, which is about the amplitude of the driver.

Turbulent winds set up standing waves in the Tacoma Narrows suspension bridge, leading to its collapse on November 7, 1940, just four months after it had been opened for traffic.

GIVE ME AN A

EXAMPLE 16-5

A string is stretched between two fixed supports 0.7 m apart and the tension is adjusted until the fundamental frequency of the string is concert A, 440 Hz. What is the speed of transverse waves on the string?

PICTURE THE PROBLEM The wave speed equals the frequency times the wavelength. For a string fixed at both ends, in the fundamental mode there is a single antinode in the middle of the string. Thus the length of the string equals one half-wavelength.

1. The wave speed is related to the frequency and wave-length: $\quad v = f_1 \lambda_1$

2. For the fundamental, the length of the string is one half-wavelength: $\quad L = \lambda_1 / 2$

3. Use this wavelength and the given frequency to find the speed: $\quad v = f_1 \lambda_1 = f_1 2L = (440 \text{ Hz}) \times 2(0.7 \text{ m})$

$$= \boxed{616 \text{ m/s}}$$

 EXERCISE The speed of transverse waves on a stretched string is 200 m/s. If the string is 5 m long, find the frequencies of the fundamental and the second and third harmonics. (*Answer* $f_1 = 20$ Hz, $f_2 = 40$ Hz, $f_3 = 60$ Hz)

TESTING PIANO WIRE

EXAMPLE 16-6

You have a summer job at a music shop, helping the owner build instruments. He asks you to test a new wire for possible use in pianos. He tells you that the 3-m-long wire has a linear mass density of 0.0025 kg/m, and he has found two adjacent resonant frequencies at 252 Hz and at 336 Hz. He wants you to determine the fundamental frequency of the wire and determine whether or not the wire is a good choice for piano strings. You know that safety issues start to arise if the tension in the wire gets above 700 N.

PICTURE THE PROBLEM The tension F_T is found from $v = \sqrt{F_T/\mu}$, where the speed v can be found from $v = f\lambda$ using any harmonic. The wavelength for the fundamental is twice the length of the wire. To find the fundamental frequency let 252 Hz be the nth harmonic. Then $f_n = nf_1$ and $f_{n+1} = (n+1)f_1$, where $f_{n+1} = 336$ Hz. We can solve these two equations for f_1.

1. The tension is related to the wave speed: $\quad v = \sqrt{F_T/\mu} \quad$ or $\quad F_T = \mu v^2$

2. Use the fundamental f_1, with $\lambda_1 = 2L$, to obtain the speed: $\quad v = f_1 \lambda_1 = f_1(2L)$

3. Combine the two previous results to find the tension: $\quad F_T = \mu v^2 = \mu f_1^2 (2L)^2$

4. The consecutive harmonics f_n and f_{n+1} are related to the fundamental frequency f_1: $\quad nf_1 = 252 \text{ Hz}$

$$(n+1)f_1 = 336 \text{ Hz}$$

5. Dividing these equations eliminates f_1 and allows us to determine n: $\quad \dfrac{n}{n+1} = \dfrac{252 \text{ Hz}}{336 \text{ Hz}} = 0.75 = \dfrac{3}{4}$

$$4n = 3n + 3, \quad \text{so} \quad n = 3$$

6. Solve for f_1:

$$f_n = n f_1 \quad \text{so} \quad f_1 = \frac{f_n}{n} = \frac{f_3}{3} = \frac{252 \text{ Hz}}{3} = 84 \text{ Hz}$$

7. Using the step 3 result, solve for F_T:

$$F_T = \mu f_1^2 (2L)^2 = (0.0025 \text{ kg/m})(84 \text{ Hz})^2 (6 \text{ m})^2$$
$$= 635 \text{ N}$$

8. Is the tension safe?

The tension is less than 700 N. It seems it is safe to use as long as the tension is not increased significantly

String Fixed at One End, Free at the Other   Figure 16-14 shows a string that has one end fixed and one end attached to a massless ring that is free to slide up and down on a friction-free pole. The vertical motion of the end of the string attached to the ring is unconstrained, so it is said to be a free end. The ring is massless, that is, a finite vertical force on it by the string would give the ring an infinite vertical acceleration. This acceleration will remain finite if the slope of the string at its free end remains horizontal. This means that the free end of the string is an antinode. In the fundamental mode of vibration for a string fixed at one end only, there is a node at the fixed end and an antinode at the free end, so $L = \lambda_1/4$ (Figure 16-15). (The distance from a node to an adjacent antinode is equal to one-quarter wavelength.)

In each mode of vibration shown in Figure 16-16 there are an odd number of quarter-wavelengths in the length L. That is, $L = n\lambda_n/4$, where $n = 1, 3, 5, \ldots$. The standing-wave condition can thus be written

FIGURE 16-14 An approximation of a string fixed at one end and free at the other end can be produced by connecting the "free" end of the string to a ring that is free to move on a post. The end attached to the mechanical wave driver is approximately fixed because the amplitude of the driver is very small.

$$L = n\frac{\lambda_n}{4}, \qquad n = 1, 3, 5, \ldots \qquad \text{16-12}$$

STANDING-WAVE CONDITION, ONE END FREE

so $\lambda_n = 4L/n$. The resonance frequencies are therefore given by

$$f_n = \frac{v}{\lambda_n} = n\frac{v}{4L} = nf_1, \qquad n = 1, 3, 5, \ldots \qquad \text{16-13}$$

RESONANCE FREQUENCIES, ONE END FREE

where

$$f_1 = \frac{v}{4L} \qquad \text{16-14}$$

is the fundamental frequency. The natural frequencies of this system occur in the ratios 1:3:5:7:..., which means that the even harmonics are missing.

Again, an easy way to remember the resonance frequencies given by Equation 16-13 is to sketch Figure 16-16 to remind yourself of the standing-wave condition and use $v = f_n\lambda_n$

PROBLEM-SOLVING GUIDELINE

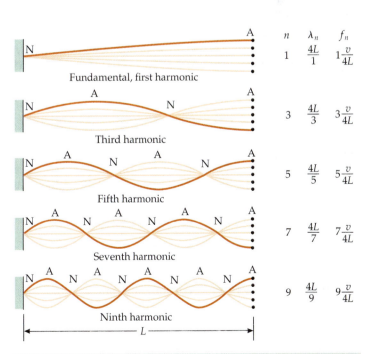

FIGURE 16-15

Fundamental, first harmonic

Third harmonic

Fifth harmonic

Seventh harmonic

Ninth harmonic

n	λ_n	f_n
1	$\frac{4L}{1}$	$1\frac{v}{4L}$
3	$\frac{4L}{3}$	$3\frac{v}{4L}$
5	$\frac{4L}{5}$	$5\frac{v}{4L}$
7	$\frac{4L}{7}$	$7\frac{v}{4L}$
9	$\frac{4L}{9}$	$9\frac{v}{4L}$

FIGURE 16-16 Standing waves on a string fixed at only one end. The free end is an antinode.

Wave Functions for Standing Waves If a string vibrates in its nth mode, each point on the string moves with simple harmonic motion. Its displacement $y_n(x,t)$ is given by

$$y_n(x,t) = A_n(x) \cos(\omega_n t + \delta_n)$$

where ω_n is the angular frequency, δ_n is the phase constant, which depends on the initial conditions, and $A_n(x)$ is the amplitude, which depends on the position x of the segment. The function $A_n(x)$ is the shape of the string when $\cos(\omega_n t + \delta_n) = 1$ (the instant that the vibration has its maximum displacement). The amplitude of a string vibrating in its nth mode is described by

$$A_n(x) = A_n \sin k_n x \qquad\qquad 16\text{-}15$$

where $k_n = 2\pi/\lambda_n$ is the wave number. The wave function for a standing wave in the nth harmonic can thus be written

$$y_n(x,t) = A_n \sin(k_n x)\cos(\omega_n t + \delta_n) \qquad\qquad 16\text{-}16$$

It is useful to remember the two conditions necessary for standing-wave motion, which are as follows:

1. Each point on the string either remains at rest or oscillates in simple harmonic motion. (Those points remaining at rest are at nodes.)

2. The motions of any two points on the string not at nodes oscillate either in phase or 180° out of phase.

NECESSARY CONDITIONS FOR A STANDING WAVE MOTION ON A LENGTH OF STRING

STANDING WAVES   **EXAMPLE 16-7**   **Try It Yourself**

(a) The wave functions for two waves that have equal amplitude, frequency, and wavelength, but that travel in opposite directions, are given by $y_1 = y_0 \sin(kx - \omega t)$ and $y_2 = y_0 \sin(kx + \omega t)$. Show that the superposition of these two waves is a standing wave. (b) A standing wave on a string that is fixed at both ends is given by $y(x,t) = (0.024\ \text{m}) \sin(52.3\ \text{m}^{-1}\ x) \cos(480\ \text{s}^{-1}\ t)$. Find the speed of waves on the string and find the distance between adjacent nodes for the standing waves.

PICTURE THE PROBLEM To show that the superposition of the two given waves is a standing wave is to show that the algebraic sum of y_1 and y_2 can be written in the form of Equation 16-16. To find the wave speed and the wavelength we compare the given wave function with Equation 16-16 and identify the wave number and angular frequency. Knowing these, we can determine the wavelength and wave speed.

Cover the column to the right and try these on your own before looking at the answers.

Steps	Answers
(a) 1. Write Equation 16-16. If the sum of y_1 and y_2 can be written in this form, then the superposition of the two traveling waves is a standing wave.	$y(x,t) = A \sin kx \cos \omega t$
2. Add the two wave functions and use the trigonometric identity $\sin\theta_1 + \sin\theta_2 = 2 \sin\frac{1}{2}(\theta_1 + \theta_2) \cos\frac{1}{2}(\theta_1 - \theta_2)$.	$y = y_0 \sin(kx - \omega t) + y_0 \sin(kx + \omega t)$ $= 2y_0 \sin kx \cos \omega t$

Note: This is of the form given by Equation 16-16 with $A = 2y_0$, so the superposition is a standing wave.

(b) 1. Identify the wave number and the angular frequency.

$k = \boxed{52.3 \text{ m}^{-1}}$, $\omega = \boxed{480 \text{ s}^{-1}}$

2. Calculate the speed from $v = \omega/k$.

$v = \boxed{9.18 \text{ m/s}}$

3. Find the wavelength $\lambda = 2\pi/k$, and use it to find the distance between nodes.

$\dfrac{\lambda}{2} = \boxed{6.01 \text{ cm}}$

Standing Sound Waves

An organ pipe is a familiar example of the use of standing waves in air columns. In the flue-type organ pipe, a stream of air is directed against the sharp edge of an opening (point A in Figure 16-17). The complicated swirling motion of the air near the edge sets up vibrations in the air column. The resonance frequencies of the pipe depend on the length of the pipe and on whether the top is stopped (closed) or open.

FIGURE 16-17 Flue-type organ pipe. Air is blown against the edge, causing a swirling motion of the air near point A that excites standing waves in the pipe. There is a pressure node near point A, which is open to the atmosphere.

In an open organ pipe, the pressure remains at one atmosphere near each open end. Since the pressure just beyond the ends does not vary, we say that there is a pressure node near each end. If the sound wave in the tube is a one-dimensional wave, which is largely correct if the tube diameter is much smaller than the wavelength, then the pressure node is at the opening of the tube. In practice, however, the pressure nodes lie slightly beyond the ends of the tube. The effective length of the pipe is $L_{\text{eff}} = L + \Delta L$ where ΔL is the end correction, which is of the order of the tube diameter. The standing-wave condition for this system is the same as that for a string fixed at both ends, where L is replaced by L_{eff} (the effective length of the tube), and all the same equations apply.

In a stopped organ pipe (open at one end, closed at the other), there is a pressure node near the opening (point A in Figure 16-17 and a pressure antinode at the closed end. The standing-wave condition for this system is the same as that for a string with one end fixed and one end free. The effective length of the tube is equal to an odd integer times $\lambda/4$. That is, the wavelength of the fundamental mode is 4 times the effective length of the tube, and only the odd harmonics are present.

As we saw in Chapter 15, a sound wave can be thought of as either a pressure wave or a displacement wave. The pressure and displacement variations in a sound wave are 90° out of phase. Thus, in a standing sound wave, the pressure nodes are displacement antinodes and vice versa. Near the open end of an organ pipe there is a pressure node and a displacement antinode, whereas at a stopped end there is a pressure antinode and a displacement node.

STANDING SOUND WAVES IN AN AIR COLUMN I   **EXAMPLE 16-8** **Try It Yourself**

If the speed of sound is 340 m/s, what are the allowed frequencies and wavelengths for standing sound waves in an unstopped (open at both ends) organ pipe whose effective length is 1 m?

PICTURE THE PROBLEM There is a displacement antinode (and a pressure node) at each end. Therefore, the effective length of the pipe is equal to an integral number of half-wavelengths.

Cover the column to the right and try these on your own before looking at the answers.

Steps	**Answers**
1. Calculate the fundamental wavelength from $\lambda_1 = 2L_{eff}$.	$\lambda_1 = 2L_{eff} = 2$ m
2. Use your value of λ_1 to calculate the fundamental frequency f_1.	$f_1 = \dfrac{v}{\lambda_1} = 170$ Hz
3. Write expressions for the frequencies f_n and wavelengths λ_n of the other harmonics in terms of n.	$\boxed{f_n = nf_1 = n(170 \text{ Hz}), \qquad n = 1, 2, 3, \dots}$
	$\boxed{\lambda_n = \dfrac{2L}{n} = \dfrac{2 \text{ m}}{n}, \qquad n = 1, 2, 3, \dots}$

STANDING SOUND WAVES IN AN AIR COLUMN II   **EXAMPLE   16 - 9**

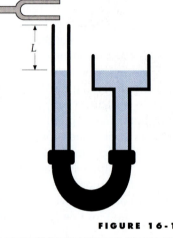

FIGURE 16-18

When a tuning fork of frequency 500 Hz is held above a tube that is partly filled with water as in Figure 16-18, resonances are found when the water level is at distances $L = 16.0, 50.5, 85.0,$ and 119.5 cm from the top of the tube. (a) What is the speed of sound in air? (b) How far from the open end of the tube is the displacement antinode?

PICTURE THE PROBLEM Sound waves of frequency 500 Hz are excited in the air column whose length L can be adjusted (by adjusting the water level). The air column is stopped at one end, open at the other. Thus at resonance, the number of quarter-wavelengths in the effective length L_{eff} of the tube is equal to an odd integer (Figure 16-19). There is a displacement node at the surface of the water and a displacement antinode a short distance above the open end of the tube. Since the frequency is fixed, so is the wavelength. The speed is then found from $v = f\lambda$, where f is 500 Hz.

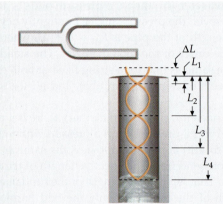

FIGURE 16-19

(a) 1. The speed of sound in air is related to the frequency and wavelength:

$$v = f\lambda$$

2. The wavelength is twice the distance between successive water levels at which resonance occurs:

$$\lambda = 2\,(L_{n+1} - L_n), \, n = 1, 2, 3, 4$$

3. The distance between successive levels is found from the data given in the problem:

$$L_{n+1} - L_n = L_4 - L_3 = 119.5 \text{ cm} - 85 \text{ cm} = 34.5 \text{ cm}$$

so

$$\lambda = 2(34.5 \text{ cm}) = 69 \text{ cm} = 0.69 \text{ m}$$

4. Substitute the values of f and λ to determine v:

$$v = f\lambda = (500 \text{ Hz})(0.69 \text{ m}) = \boxed{345 \text{ m/s}}$$

(b) There will be a displacement antinode one quarter-wavelength above the displacement node at the surface of the water. Thus, the distance from the highest water level supporting resonance and the displacement antinode above the opening of the tube is one-quarter wavelength. $\frac{1}{4}\lambda = L_1 + \Delta L$:

$$\Delta L = \tfrac{1}{4}\lambda - L_1 = \tfrac{1}{4}(69.0 \text{ cm}) - (16.0 \text{ cm})$$

$$= \boxed{1.25 \text{ cm}}$$

Most musical wind instruments are much more complicated than simple cylindrical tubes. The conical tube, which is the basis for the oboe, bassoon, English horn, and saxophone, has a complete harmonic series with its fundamental wavelength equal to twice the length of the cone. Brass instruments are combinations of cones and cylinders. The analysis of these instruments is extremely complex. The fact that they have nearly harmonic series is a triumph of educated trial and error rather than mathematical calculation.

Holographic interferograms showing standing waves in a handbell. The "bull's eyes" locate the antinodes.

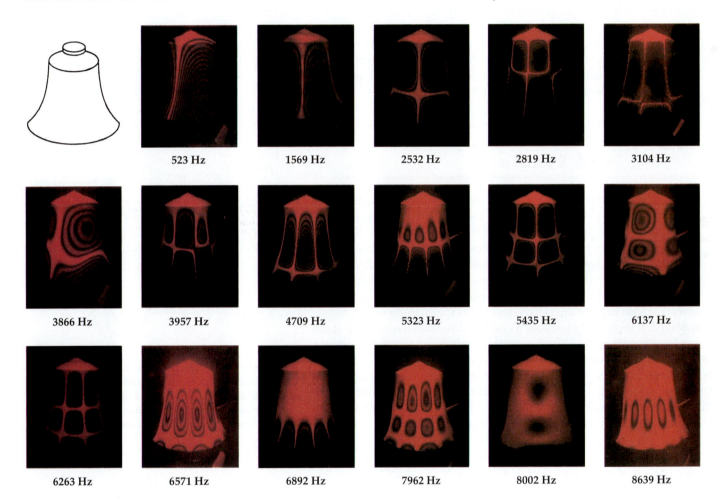

523 Hz	1569 Hz	2532 Hz	2819 Hz	3104 Hz
3866 Hz	3957 Hz	4709 Hz	5323 Hz	5435 Hz	6137 Hz
6263 Hz	6571 Hz	6892 Hz	7962 Hz	8002 Hz	8639 Hz

*16-3 The Superposition of Standing Waves

As we saw in the preceding section, there is a set of natural resonance frequencies that produce standing waves for sound waves in air columns or vibrating strings that are fixed at one or both ends. For example, for a string fixed at both ends, the frequency of the fundamental mode of vibration is $f_1 = v/(2L)$, where L is the length of the string and v is the wave speed and the wave function is Equation 16-16:

$$y_1(x,t) = A_1 \sin k_1 x \cos(\omega_1 t + \delta_1)$$

In general, a vibrating system does not vibrate in a single harmonic mode. Instead, the motion consists of a mixture of the allowed harmonics. The wave function is a linear combination of the harmonic wave functions:

$$y(x,t) = \sum_n A_n \sin(k_n x) \cos(\omega_n t + \delta_n) \qquad 16\text{-}17$$

where $k_n = 2\pi/\lambda_n$, $\omega_n = 2\pi f_n$, and A_n and δ_n are constants. The constants A_n and δ_n depend on the initial position and velocity of the string. If a harp string, for example, is plucked at the center and released, as in Figure 16-20, the initial shape of the string is symmetric about the point $x = \frac{1}{2}L$ and the initial velocity is zero throughout the length of the string. The motion of the string after it has been released will remain symmetric about $x = \frac{1}{2}L$. Only the odd harmonics, which are also symmetric about $x = \frac{1}{2}L$, will be excited. The even harmonics, which are antisymmetric about $x = \frac{1}{2}L$, are not excited; that is, the constant A_n is zero for all even n. The shapes of the first four harmonics are shown in Figure 16-21. Most of the energy of the plucked string is associated with the fundamental, but small amounts of energy are associated with the third, fifth, and other odd harmonic modes. Figure 16-22 shows an approximation to the initial shape of the string using the superposition of only the first three odd harmonics.

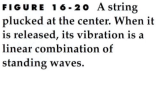

FIGURE 16-20 A string plucked at the center. When it is released, its vibration is a linear combination of standing waves.

*16-4 Harmonic Analysis and Synthesis

When a clarinet and an oboe play the same note, say, concert A, they sound quite different. Both notes have the same **pitch,** a physiological sensation of the highness or lowness of the note that is strongly correlated with frequency. However, the notes differ in what is called **tone quality.** The principal reason for the difference in tone quality is that, although both the clarinet and oboe are producing vibrations at the same fundamental frequency, each instrument is also producing harmonics whose relative intensities depend on the instrument and how it is played. If the sound produced by each instrument were entirely at the fundamental frequency of the instrument, they would sound identical.

Figure 16-23 shows plots of the pressure variations versus time for the sound from a tuning fork, a clarinet, and an oboe, each playing the same note. These patterns are called **waveforms.** The waveform for the sound from the tuning fork is nearly a pure sine wave, but those from the clarinet and the oboe are clearly more complex.

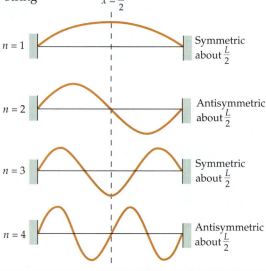

FIGURE 16-21 The first four harmonics for a string fixed at both ends. The odd harmonics are symmetrical about the center of the string, whereas the even harmonics are not. When a string is plucked at the center, it vibrates only in its odd harmonics.

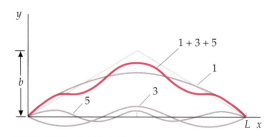

FIGURE 16-22 Approximating the shape of a string plucked at the center, as in Figure 16-20, using harmonics. The red line is an approximation of the original shape of the string based on the first three odd harmonics. The height of the string is exaggerated in this drawing to show the relative amplitudes of the harmonics. Most of the energy is associated with the fundamental, but there is some energy in the third, fifth, and other odd harmonics.

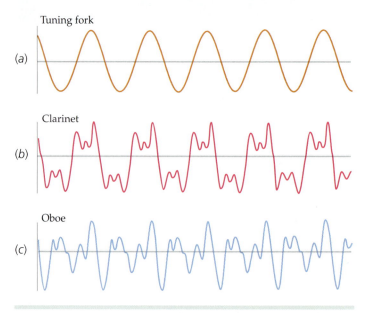

FIGURE 16-23 Waveforms of (a) a tuning fork, (b) a clarinet, and (c) an oboe, each at a fundamental frequency of 440 Hz and at approximately the same intensity.

Waveforms can be analyzed in terms of the harmonics that comprise them by means of **harmonic analysis.** (Harmonic analysis is also called **Fourier analysis** after the French mathematician J.B.J. Fourier, who developed the techniques for analyzing periodic functions.) Figure 16-24 shows a plot of the relative intensities of the harmonics of the waveforms in Figure 16-23. The waveform of the sound from the tuning fork contains only the fundamental frequency. That for the sound from the clarinet contains the fundamental, large amounts of the third, fifth, and seventh harmonics, and lesser amounts of the second, fourth, and sixth harmonics. For the sound from the oboe, there is more intensity in the second and third harmonics than in the fundamental.

The inverse of harmonic analysis is **harmonic synthesis,** which is the construction of a periodic wave from harmonic components. Figure 16-25a shows the first three odd harmonics used to synthesize a square wave and Figure 16-25b shows the square wave that results from the sum of the three harmonics. The more harmonics used in a synthesis, the closer the approximation will be to the actual waveform (the gray line in the Figure). The relative amplitudes of the harmonics needed to synthesize the square wave are shown in Figure 16-26.

*16-5 Wave Packets and Dispersion

The waveforms discussed in Section 16.4 are periodic in time. Pulses, which are aperiodic, can also be represented by a group of harmonic waves of different frequencies. However, the synthesis of a pulse requires a continuous

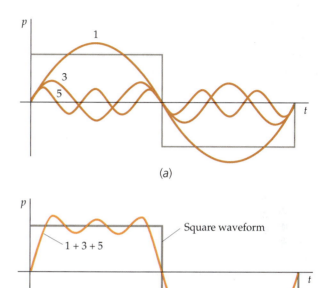

(a)

(b)

FIGURE 16-25 (a) The first three odd harmonics of a single sine wave, used to synthesize a square wave. (b) The approximation of a square wave that results from summing the first three odd harmonics in (a).

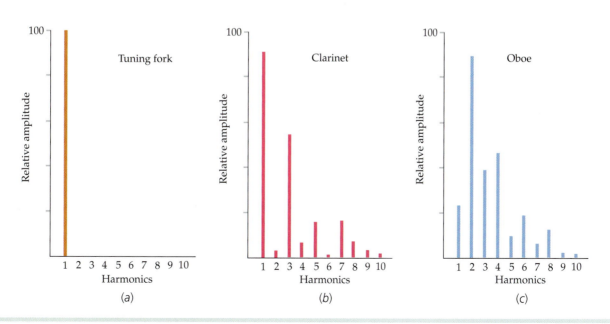

FIGURE 16-24 Relative intensities of the harmonics in the waveforms shown in Figure 16-23 for (a) the tuning fork, (b) the clarinet, and (c) the oboe.

distribution of frequencies rather than a discrete set of harmonics as in Figure 16-26. Such a group is called a **wave packet.** The characteristic feature of a wave pulse is that it has a beginning and an end, whereas a harmonic wave repeats over and over. If the duration Δt of the pulse is very short, the range of frequencies $\Delta \omega$ needed to describe the pulse is very large. The general relation between Δt and $\Delta \omega$ is

$$\Delta \omega \, \Delta t \sim 1 \qquad\qquad 16\text{-}18$$

where the tilde $\sim$ means "of the order of."

The exact value of this product depends on just how the quantities $\Delta \omega$ and Δt are defined. For any reasonable definitions, $\Delta \omega$ and $1/\Delta t$ have the same order of magnitude. A wave pulse produced by a source of short duration Δt, like the crack of a bat on a ball, has a narrow width in space $\Delta x = v \, \Delta t$, where v is the wave speed. Each harmonic wave of frequency ω has a wave number $k = \omega/v$. A range of frequencies $\Delta \omega$ implies a range of wave numbers $\Delta k = \Delta \omega/v$. Substituting $v \, \Delta k$ for $\Delta \omega$ in Equation 16-18 gives $v \, \Delta k \, \Delta t \sim 1$ or

$$\Delta k \, \Delta x \sim 1 \qquad\qquad 16\text{-}19$$

If a wave packet is to maintain its shape as it travels, all of the component harmonic waves that make up the packet must travel with the same speed. This occurs if the speed of the component waves in a given medium is independent of frequency or wavelength. Such a medium is called a **nondispersive medium.** Air is, to an excellent approximation, a nondispersive medium for sound waves, but solids and liquids generally are not. (Probably the most familiar example of dispersion is the formation of a rainbow, due to the fact that the velocity of light waves in water depends slightly on the frequency of the light, so the different colors, corresponding to different frequencies, have slightly different angles of refraction.)

When the wave speed in a dispersive medium depends only slightly on the frequency (or wavelength), a wave packet changes shape very slowly as it travels, and it covers a considerable distance as a recognizable entity. But the speed of the packet, called the **group velocity,** is not the same as the (average) speed of the individual component harmonic waves, called the **phase velocity.** (By the speed of an individual harmonic wave we mean the speed of its wavefronts. Because wavefronts are lines or surfaces of constant phase, their speed is called the phase velocity of the wave.)

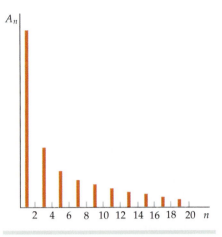

FIGURE 16-26 Relative amplitudes A_n of the first ten harmonics needed to synthesize a square wave. The more harmonics that are used, the closer the approximation is to the square wave.

SUMMARY

1. The principle of superposition, which holds for all electromagnetic waves in empty space, for waves on a flexible taut string in the small angle approximation, and for sound waves of small amplitude, follows from the linearity of the corresponding wave equations.

2. Interference is an important wave phenomenon that applies to all coherent superposing waves. It follows from the principle of superposition. Diffraction and interference distinguish wave motion from particle motion.

3. The standing-wave conditions can be recalled by sketching a string or tube and drawing waves that have nodes at a fixed or stopped end, and antinodes at a free or open end.

Topic	Relevant Equations and Remarks
1. Superposition and Interference	The superposition of two harmonic waves of equal amplitude, wave number, and frequency but phase difference δ results in a harmonic wave of the same wave number and frequency, but differing in phase and amplitude from each of the two waves

$$y = y_1 + y_2 = y_0 \sin(kx - \omega t) + y_0 \sin(kx - \omega t + \delta)$$
$$= \left[2y_0 \cos \tfrac{1}{2}\delta\right] \sin(kx - \omega t + \tfrac{1}{2}\delta) \qquad \text{16-6}$$

Constructive interference	If waves are in phase or differ in phase by an integer times 2π, then the amplitudes of the waves add and the interference is constructive.
Destructive interference	If waves differ in phase by π or by an odd integer times π, then the amplitudes subtract and the interference is destructive.
Beats	Beats are the result of the interference of two waves of slightly different frequencies. The beat frequency equals the difference in the frequencies of the two waves:

$$f_{\text{beat}} = \Delta f \qquad \text{16-8}$$

| Phase difference δ due to path difference Δx | $$\delta = k\,\Delta x = 2\pi \frac{\Delta x}{\lambda} \qquad \text{16-9}$$ |

2. Standing Waves	Standing waves occur for certain frequencies and wavelengths when waves are confined in space. They occur only if each point of the system oscillates in simple harmonic motion and any two moving points move either in phase or 180° out of phase.
Wavelength	The distance between a node and an adjacent antinode is a quarter-wavelength.
String fixed at both ends	For a string fixed at both ends, there is a node at each end so that an integral number of half-wavelengths must fit into the length of the string. The standing-wave condition in this case is

$$L = n\frac{\lambda_n}{2}, \qquad n = 1, 2, 3, \ldots \qquad \text{16-10}$$

| Standing wave function for a string fixed at both ends | The allowed waves form a harmonic series, with the frequencies given by |

$$f_n = \frac{v}{\lambda_n} = n\frac{v}{2L} = nf_1, \qquad n = 1, 2, 3, \ldots \qquad \text{16-11}$$

where $f_1 = v/2L$ is the lowest frequency, called the fundamental.

| Organ pipe open at both ends | Standing sound waves in the air in a pipe that is open at both ends have a pressure node (and a displacement antinode) near each end so that the standing wave condition is the same as for a string fixed at both ends. |
| String fixed at one end and free at the other | For a string with one end fixed and one end free, there is a node at the fixed end and an antinode at the free end, so that an integral number of quarter-wavelengths must fit into the length of the string. The standing-wave condition in this case is |

$$L = n\frac{\lambda_n}{4}, \qquad n = 1, 3, 5, \ldots \qquad \text{16-12}$$

Only the odd harmonics are present. Their frequencies are given by

$$f_n = \frac{v}{\lambda_n} = n\frac{v}{4L} = nf_1, \qquad n = 1, 3, 5, \ldots \qquad \text{16-13}$$

where $f_1 = v/4L$.

| Organ pipe open at one end and stopped at the other | Standing sound waves in a pipe that is open at one end and stopped at the other end have a displacement antinode at the open end and a displacement node at the stopped end. The standing wave condition is the same as for a string fixed at one end. |

Wave Functions for Standing Waves	$y_n(x,t) = A_n \sin(k_n x) \cos(\omega_n t + \delta_n)$	**16-16**
	where $k_n = 2\pi/\lambda_n$ and $\omega_n = 2\pi f_n$.	

The necessary conditions for standing waves on a string are

1. Each point on the string either remains at rest or oscillates with simple harmonic motion. (Those points remaining at rest are nodes.)
2. The motions of any two points on the string that are not nodes oscillate either in phase or 180° out of phase.

*3. Superposition of Standing Waves	A vibrating system typically does not vibrate in a single harmonic mode but in a superposition of the allowed harmonic modes.	
*4. Harmonic Analysis and Synthesis	Sounds of different tone quality contain different mixtures of harmonics. The analysis of a particular tone in terms of its harmonic content is called harmonic analysis. Harmonic synthesis is the construction of a tone by the addition of harmonics.	
*5. Wave Packets	A wave pulse can be represented by a continuous distribution of harmonic waves. The range of frequencies $\Delta\omega$ is related to the width in time Δt, and the range of wave numbers Δk is related to the width in space Δx.	
Frequency and time ranges	$\Delta\omega \, \Delta t \sim 1$	**16-18**
Wave number and space ranges	$\Delta k \, \Delta x \sim 1$	**16-19**
*6. Dispersion	In a nondispersive medium, the phase velocity is independent of frequency, and a pulse (wave packet) travels without change in shape. In a dispersive medium, the phase velocity varies with frequency, and the pulse changes shape as it moves. The pulse moves with a velocity called the group velocity of the packet.	

PROBLEMS

- Single-concept, single-step, relatively easy
- •• Intermediate-level, may require synthesis of concepts
- ••• Challenging
- **SSM** Solution is in the *Student Solutions Manual*
- **iSOLVE** Problems available on iSOLVE online homework service
- **iSOLVE ✓** These "Checkpoint" online homework service problems ask students additional questions about their confidence level, and how they arrived at their answer

Conceptual Problems

1 •• **SSM** Two rectangular wave pulses are traveling in opposite directions along a string. At $t = 0$, the two pulses are as shown in Figure 16-27. Sketch the wave functions for $t = 1$, 2, and 3 s.

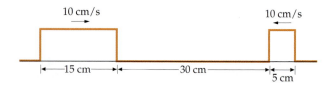

10 cm/s   10 cm/s

|←—15 cm—→|←————30 cm————→|←—→|
                                    5 cm

FIGURE 16-27 Problems 1, 2

2 •• Repeat Problem 1 for the case in which the pulse on the right is inverted.

3 • Beats are produced by the superposition of two harmonic waves only if (a) their amplitudes and frequencies are equal, (b) their amplitudes are the same but their frequencies differ slightly, (c) their frequencies differ slightly even if their amplitudes are not equal, (d) their frequencies are equal but their amplitudes differ slightly.

4 • True or false:

(a) The frequency of the third harmonic is three times that of the first harmonic.
(b) The frequency of the fifth harmonic is five times that of the fundamental.
(c) In a pipe that is open at one end and closed at the other, the even harmonics are not excited.

5 •• Standing waves result from the superposition of two waves of (a) the same amplitude, frequency, and direction of propagation, (b) the same amplitude and frequency and opposite directions of propagation, (c) the same amplitude, slightly different frequency, and the same direction of propagation, (d) the same amplitude, slightly different frequency, and opposite directions of propagation.

6 • SSM The resonant frequencies of a violin string are all integer multiples of the fundamental frequency, while the resonant frequencies of a circular drumhead are irregularly spaced. Given this information, explain the difference in the sounds of a violin and a drum.

7 • An organ pipe open at both ends has a fundamental frequency of 400 Hz. If one end of this pipe is now closed, the fundamental frequency will be (a) 200 Hz, (b) 400 Hz, (c) 546 Hz, (d) 800 Hz.

8 •• A string fixed at both ends resonates at a fundamental frequency of 180 Hz. Which of the following will reduce the fundamental frequency to 90 Hz? (a) Double the tension and double the length. (b) Halve the tension and keep the length fixed. (c) Keep the tension fixed and double the length. (d) Keep the tension fixed and halve the length.

9 •• How do the resonance frequencies of an organ pipe change when the air temperature increases?

10 • SSM When two waves moving in opposite directions superimpose as in Figure 16-1, does either impede the progress of the other?

11 • When a guitar string is plucked, is the wavelength of the wave it produces in air the same as the wavelength of the wave on the string?

12 • When two waves interfere constructively or destructively, is there any gain or loss in energy? Explain.

13 • A musical instrument consists of drinking glasses partially filled with water that are struck with a small mallet. Explain how this works.

14 •• During an organ recital, the air compressor that drives the organ pipes suddenly fails. An enterprising physics student in the audience comes to the rescue by connecting a tank of pure nitrogen gas under high pressure to the output of the compressor. What effect, if any, will this change have on the operation of the organ? What if the tank contained helium?

15 •• SSM When the tension on a piano wire is increased, which of the following occurs? (a) Its wavelength decreases. (b) Its wavelength remains the same while its frequency increases. (c) Its wavelength and frequency increase. (d) None of the above occur.

16 •• The following instructions are given for connecting stereo speakers to an amplifier so that they are in phase: "After both speakers are connected, play a monophonic record or program with the bass control turned up and the treble control turned down. While listening to the speakers, turn the balance control so that first one speaker is heard separately, then the two together, and then the other separately. If the bass is stronger when both speakers play together, they are connected properly. If the bass is weaker when both play together than when each plays separately, reverse the connections on one speaker." Explain why this method works. In particular, explain why a stereo source is not used and why only the bass is compared.

17 •• The constant γ for helium (and all monatomic gases) is 1.67. If a man inhales helium and then speaks, he sounds like Alvin of the Chipmunks. Why?

18 •• SSM Figure 16-28 is a photograph of two pieces of very finely woven silk placed one on top of the other. Where the pieces overlap, a series of light and dark lines are seen. This moiré pattern can also be seen when a scanner is used to copy photos from a book or newspaper. What causes the moiré pattern, and how is it similar to the phenomenon of interference?

FIGURE 16-28 Problem 18

Estimation and Approximation

19 •• About how accurately can you tune a piano string to a tuning-fork frequency?

20 • SSM The shortest pipes used in organs are about 7.5 cm long. (a) What is the fundamental frequency of a pipe this long that is open at both ends? (b) For such a pipe, what is the highest harmonic that is within the audible range? (The normal range of hearing is about 20 to 20,000 Hz.)

21 •• On a windy day, a drain pipe will sometimes resonate. Estimate the resonance frequency of a drain pipe on a single-story house. How much might this frequency change from winter to summer in your region?

Superposition and Interference

22 • Two waves traveling on a string in the same direction both have a frequency of 100 Hz, a wavelength of 2 cm, and an amplitude of 0.02 m. What is the amplitude of the resultant wave if the original waves differ in phase by (a) $\pi/6$ and (b) $\pi/3$?

23 • Two waves having the same frequency, wavelength, and amplitude are traveling in the same direction. If they differ in phase by $\pi/2$ and each has an amplitude of 0.05 m, what is the amplitude of the resultant wave?

24 • SSM ISOLVE Two sound sources oscillate in phase with the same amplitude A. They are separated in space by $\lambda/3$. What is the amplitude of the resultant wave formed from the two sources at a point that is on the line that passes through the sources but is not between the sources?

25 • Two sound sources oscillate in phase with a frequency of 100 Hz. At a point 5.00 m from one source and 5.85 m from the other, the amplitude of the sound from each source separately is A. (a) What is the phase difference in the sound waves from the two sources at that point? (b) What is the amplitude of the resultant wave at that point?

26 • SSM With a compass, draw circular arcs representing wave crests originating from each of two point sources a distance $d = 6$ cm apart for $\lambda = 1$ cm. Connect the intersections corresponding to points of constant path difference and label the path difference for each line. (See Figure 16-8.)

27 • Two speakers separated by some distance emit sound waves of the same frequency. At some point P, the intensity due to each speaker separately is I_0. The path distance from P to one of the speakers is $\frac{1}{2}\lambda$ greater than that from P to the other speaker. What is the intensity at P if (a) the speakers are coherent and in phase, (b) the speakers are incoherent, and (c) the speakers are coherent but have a phase difference of π rad?

28 • Answer the questions of Problem 27 for a point P' for which the distance to the far speaker is 1λ greater than the distance to the near speaker. Assume that the intensity at point P' due to each speaker separately is again I_0.

29 • Two speakers separated by some distance emit sound waves of the same frequency, but the speakers are out of phase by 90°. Let r_1 be the distance from some point to speaker 1 and r_2 be the distance from that point to speaker 2. Find the smallest value of $r_2 - r_1$ at which the sound at that point will be (a) maximum and (b) minimum. (Express your answers in terms of the wavelength.)

30 •• SSM Show that, if the separation between two sound sources radiating coherently in phase is less than half a wavelength, complete destructive interference will not be observed in any direction.

31 •• ISOLVE A transverse wave of frequency 40 Hz propagates down a string. Two points 5 cm apart are out of phase by $\pi/6$. (a) What is the wavelength of the wave? (b) At a given point, what is the phase difference between two displacements for times 5 ms apart? (c) What is the wave velocity?

32 •• It is thought that the brain determines the direction of the source of a sound by sensing the phase difference between the sound waves striking the eardrums. A distant source emits sound of frequency 680 Hz. When you are directly facing a sound source there should be no phase difference. Estimate the phase difference between the sounds received by each ear as you turn from facing directly toward the source through 90°.

33 •• ISOLVE✔ Sound source A is located at $x = 0$, $y = 0$, and sound source B is placed at $x = 0$, $y = 2.4$ m. The two sources radiate coherently in phase. An observer at $x = 40$ m, $y = 0$ notes that as she takes a few steps in either the positive

or negative y direction away from $y = 0$, the sound intensity diminishes. What are the lowest and the next higher frequencies of the sources that can account for that observation?

34 •• Suppose that the observer in Problem 33 finds herself at a point of minimum intensity at $x = 40$ m, $y = 0$. What are then the lowest and next higher frequencies of the sources consistent with this observation?

35 •• SSM Two harmonic water waves of equal amplitudes but different frequencies, wave vectors, *and velocities* are superposed on each other. The total displacement of the wave can be written as $y(x,t) = A[\cos(k_1 x - \omega_1 t) + \cos(k_2 x - \omega_2 t)]$, where $\omega_1 k_1 = v_1$ (the speed of the first wave) and $\omega_2/k_2 = v_2$ (the speed of the second wave). (a) Show that $y(x,t)$ can be written in the form $y(x,t) = 2A\,\cos[(\Delta k/2)x - (\Delta\omega/2)t]\,\cos(k_{av}x - \omega_{av}t)$ where $\omega_{av} = (\omega_1 + \omega_2)/2$. $k_{av} = (k_1 + k_2)/2$, $\Delta\omega = \omega_1 - \omega_2$, and $\Delta k = k_1 - k_2$. The factor $2A\,\cos[(\Delta k/2)x - (\Delta\omega/2)t]$ is called the *envelope* of the wave. (b) Using a spreadsheet program or graphing calculator, make a graph of $y(x,t)$ for $A = 1$, $\omega_1 = 1$ rad/s, $k_1 = 1$ m^{-1}, $\omega_2 = 0.9$ rad/s, and $k_2 = 0.8$ m^{-1} at $t = 0$ s, 0.5 s, and 1 s for x between 0 m and 50 m. (c) What is the speed at which the envelope moves?

36 •• Two point sources that are in phase are separated by a distance d. An interference pattern is detected along a line parallel to the line through the sources and a large distance D from the sources, as shown in Figure 16-29. (a) Show that the path difference from the two sources to some point on the line at a small angle θ is given approximately by $\Delta s = d \sin \theta$. Hint: Assume that $D \gg d$, so the lines from the sources to P are approximately parallel. (b) Show that the distance y_m from the central maximum point to the mth interference maximum is given approximately by $y_m = m(D\lambda/d)$.

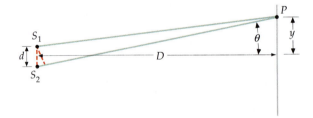

FIGURE 16-29 Problems 36–40

37 •• ISOLVE✔ Two sound sources radiating in phase at a frequency of 480 Hz interfere such that maxima are heard at angles of 0° and 23° from a line perpendicular to that joining the two sources. Find the separation between the two sources and any other angles at which a maximum intensity will be heard. (Use the result of Problem 36.)

38 ••• SSM Two loudspeakers are driven in phase by an audio amplifier at a frequency of 600 Hz. The speakers are on the y axis, one at $y = +1.00$ m and the other at $y = -1.00$ m. A listener begins at $y = 0$ a very large distance D away and walks along a line parallel to the y axis. (See Problem 36.) (a) At what angle θ will she first hear a minimum in the sound intensity? (b) At what angle will she first hear a maximum (after $\theta = 0$)? (c) How many maxima can she possibly hear if she keeps walking in the same direction?

39 ••• [i·SOLVE] Two sound sources driven in phase by the same amplifier are 2 m apart on the y axis. At a point a very large distance from the y axis, constructive interference is first heard at an angle $\theta_1 = 0.140$ rad with the x axis and is next heard at $\theta_2 = 0.283$ rad (see Figure 16-29). (a) What is the wavelength of the sound waves from the sources? (b) What is the frequency of the sources? (c) At what other angles is constructive interference heard? (d) What is the smallest angle for which the sound waves cancel?

40 ••• The two sound sources from Problem 39 are now driven 90° out-of-phase, but at the same frequency as in Problem 39. At what angles are constructive and destructive interference heard?

41 ••• A radio telescope consists of two antennas separated by a distance of 200 m. Both antennas are tuned to a particular frequency, such as 20 MHz. The signals from each antenna are fed into a common amplifier, but one signal first passes through a phase adjuster that delays its phase by a chosen amount so that the telescope can "look" in different directions. When the phase delay is zero, plane radio waves that are incident vertically on the antennas produce signals that add constructively at the amplifier. What should be the phase delay so that signals coming from an angle $\theta = 10°$ with the vertical (in the plane formed by the vertical and the line joining the antennas) will add constructively at the amplifier?

Beats

42 • [i·SOLVE] When two tuning forks are struck simultaneously, 4 beats per second are heard. The frequency of one fork is 500 Hz. (a) What are the possible values for the frequency of the other fork? (b) A piece of wax is placed on the 500-Hz fork to lower its frequency slightly. Explain how the measurement of the new beat frequency can be used to determine which of your answers to Part (a) is the correct frequency of the second fork.

43 •• [SSM] Two ambulances move toward each other on a straight road. Each travels at a speed of 50 mph. The siren on each ambulance produces a sound wave of frequency 500 Hz. (a) The driver of each ambulance hears the other's siren and hears a beat note between the frequency of his own siren and the siren of the other ambulance. What is the frequency of the beat note? (b) A passerby stands midway between the two ambulances. What is the frequency of the beat note between the two sirens that he hears?

Standing Waves

44 • [SSM] [i·SOLVE]✓ A string fixed at both ends is 3 m long. It resonates in its second harmonic at a frequency of 60 Hz. What is the speed of transverse waves on the string?

45 • [i·SOLVE] A string 3 m long and fixed at both ends is vibrating in its third harmonic. The maximum displacement of any point on the string is 4 mm. The speed of transverse waves on this string is 50 m/s. (a) What are the wavelength and frequency of this wave? (b) Write the wave function for this wave.

46 • [i·SOLVE]✓ Calculate the fundamental frequency for a 10-m organ pipe that is (a) open at both ends and (b) closed at one end.

47 • A steel wire having a mass of 5 g and a length of 1.4 m is fixed at both ends and has a tension of 968 N. (a) Find the speed of transverse waves on the wire. (b) Find the wavelength and frequency of the fundamental. (c) Find the frequencies of the second and third harmonics.

48 • A rope 4 m long is fixed at one end; the other end is attached to a light string so that it is free to move. The speed of waves on the rope is 20 m/s. Find the frequency of (a) the fundamental, (b) the second harmonic, and (c) the third harmonic.

49 • A piano wire without windings has a fundamental frequency of 200 Hz. When it is wound with wire, its linear mass density is doubled. What is its new fundamental frequency, assuming that the tension is unchanged?

50 • [SSM] What is the greatest length that an organ pipe can have in order to have its fundamental note in the audible range (20 to 20,000 Hz) if (a) the pipe is closed at one end and (b) it is open at both ends?

51 •• [i·SOLVE]✓ The wave function $y(x,t)$ for a certain standing wave on a string fixed at both ends is given by $y(x,t) = 4.2 \sin 0.20x \cos 300t$, where y and x are in centimeters and t is in seconds. (a) What are the wavelength and frequency of this wave? (b) What is the speed of transverse waves on this string? (c) If the string is vibrating in its fourth harmonic, how long is it?

52 •• The wave function $y(x,t)$ for a certain standing wave on a string fixed at both ends is given by $y(x,t) = (0.05 \text{ m}) \sin 2.5 \text{ m}^{-1} x \cos 500 \text{ s}^{-1} t$. (a) What are the speed and amplitude of the two traveling waves that result in this standing wave? (b) What is the distance between successive nodes on the string? (c) What is the shortest possible length of the string?

53 •• A 2.51-m-long string has the wave function given in Problem 52. (a) Sketch the position of the string at the times $t = 0$, $t = T/4$, $t = T/2$, and $t = 3T/4$, where $T = 1/f$ is the period of the vibration. (b) Find T in seconds. (c) At a time t when the string is horizontal, that is, $y(x) = 0$ for all x, what has become of the energy in the wave?

54 •• [SSM] [i·SOLVE] Three successive resonance frequencies for a certain string are 75, 125, and 175 Hz. (a) Find the ratios of each pair of successive resonance frequencies. (b) How can you tell that these frequencies are for a string fixed at one end only rather than for a string fixed at both ends? (c) What is the fundamental frequency? (d) Which harmonics are these resonance frequencies? (e) If the speed of transverse waves on this string is 400 m/s, find the length of the string.

55 •• The space above the water in a tube like that shown in Example 16-9 is 120 cm long. Near the open end there is a loudspeaker that is driven by an audio oscillator whose frequency can be varied from 10 to 5000 Hz. (a) What is the lowest frequency of the oscillator that will produce resonance within the tube? (b) What is the highest frequency that will produce resonance? (c) How many different frequencies of the oscillator will produce resonance? (Neglect the end correction.)

56 •• A 460-Hz tuning fork causes resonance in the tube in Example 16-9 when the top of the tube is 18.3 and 55.8 cm above the water surface. (a) Find the speed of sound in air. (b) What is the end correction to adjust for the fact that the antinode does not occur exactly at the end of the open tube?

57 •• [SSM] [ISOLVE]✓ At 16°C, the fundamental frequency of an organ pipe is 440.0 Hz. What will be the fundamental frequency of the pipe if the temperature increases to 32°C? Would it be better to construct the pipe with a material that expands substantially as the temperature increases or should the pipe be made of material that maintains the same length at all normal temperatures?

58 •• The end correction for a circular pipe is approximately $\Delta L = 0.3186D$, where D is the pipe diameter. Find the length of a pipe open at both ends that will produce a middle C (256 Hz) as its fundamental mode for pipes of diameter $D = 1$ cm, 10 cm, and 30 cm.

59 •• [ISOLVE] A violin string of length 40 cm and mass 1.2 g has a frequency of 500 Hz when it is vibrating in its fundamental mode. (a) What is the wavelength of the standing wave on the string? (b) What is the tension in the string? (c) Where should you place your finger to increase the frequency to 650 Hz?

60 •• The G string on a violin is 30 cm long. When played without fingering, it vibrates at a frequency of 196 Hz. The next higher notes on the C-major scale are A (220 Hz), B (247 Hz), C (262 Hz), and D (294 Hz). How far from the end of the string must a finger be placed to play each of these notes?

61 •• A string with a mass density of 4×10^{-3} kg/m is under a tension of 360 N and is fixed at both ends. One of its resonance frequencies is 375 Hz. The next higher resonance frequency is 450 Hz. (a) What is the fundamental frequency of this string? (b) Which harmonics are the ones given? (c) What is the length of the string?

62 •• [ISOLVE]✓ A string fastened at both ends has successive resonances with wavelengths of 0.54 m for the nth harmonic and 0.48 m for the (n + 1)th harmonic. (a) Which harmonics are these? (b) What is the length of the string?

63 •• The strings of a violin are tuned to the tones G, D, A, and E, which are separated by a fifth from one another. That is, $f(D) = 1.5f(G)$, $f(A) = 1.5f(D) = 440$ Hz, and $f(E) = 1.5f(A)$. The distance between the two fixed points, the bridges at the scroll and over the body of the instrument, is 30 cm. The tension on the E string is 90 N. (a) What is the mass per meter of the E string? (b) To prevent distortion of the instrument over time, it is important that the tension on all strings be the same. Find the masses per meter of the other strings.

64 •• An ambulance is driving at 50 mph towards the brick wall of the hospital, which reflects the sound of the siren back toward the ambulance. When the ambulance is stationary, the siren's frequency is 500 Hz. (a) What is the spatial period of the standing wave caused by the sound of the siren and its reflection? (b) A doctor standing between the ambulance and the wall will hear the siren grow alternately louder and softer as the ambulance drives toward her. Why is this?

65 •• To tune a violin, the violinist first tunes the A string to the correct pitch of 440 Hz and then bows two adjoining strings simultaneously and listens for a beat pattern. While bowing the A and E strings, the violinist hears a beat frequency of 3 Hz and notes that the beat frequency increases as the tension on the E string is increased. (The E string is to be tuned to 660 Hz.) (a) Why is a beat produced by these two strings bowed simultaneously? (b) What is the frequency of the E string vibration when the beat frequency is 3 Hz? (c) If the tension on the E string is 80.0 N when the beat frequency is 3 Hz, what tension corresponds to perfect tuning of that string?

66 •• Suppose that you carry a small oscillator and speaker as you walk very slowly down a long hall. The speaker emits a sound of frequency 680 Hz, which is reflected from the walls at each end of the hall. As you walk along, you note that the sound intensity that you hear passes through successive maxima and minima. What distance must you walk to pass from one maximum to the next?

67 •• [SSM] Show that the standing wave function $A' \sin kx \cos(\omega t + \delta)$ can be written as the sum of two harmonic wave functions—one for a wave traveling in the positive x direction and the other for a wave of the same amplitude traveling in the negative x direction. The traveling waves each have the same wave number and angular frequency as does the standing wave.

68 •• A 2-m string is fixed at one end and is vibrating in its third harmonic with amplitude 3 cm and frequency 100 Hz. (a) Write the wave function for this vibration. (b) Write an expression for the kinetic energy of a segment of the string of length dx at a point x at some time t. At what time is this kinetic energy maximum? What is the shape of the string at this time? (c) Find the maximum kinetic energy of the string by integrating your expression for Part (b) over the total length of the string.

69 •• [SSM] A commonly used physics experiment that examines resonances of transverse waves on a string is shown in Figure 16-30. A weight is attached to the end of a string draped over a pulley; the other end of the string is attached to a mechanical oscillator that moves the string up and down at a set frequency f. The length L between the oscillator and the pulley is fixed. For certain values of the weight the string resonates. If $L = 1$ m, $f = 80$ Hz, and the mass density of the string is $\mu = 0.75$ g/m, what weights are needed for each of the first three modes (standing waves) of the string?

FIGURE 16-30 Problem 69

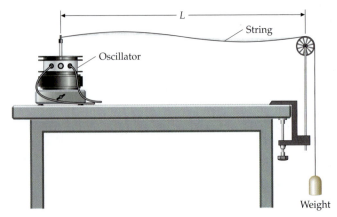

*Wave Packets

70 • $\boxed{\text{i SOLVE}}$ Information used by computers is transmitted along a cable in the form of short electric pulses at the rate of 10^7 pulses per second. (*a*) What is the maximum duration of each pulse if no two pulses overlap? (*b*) What is the range of frequencies to which the receiving equipment must respond?

71 • $\boxed{\text{SSM}}$ A tuning fork of frequency f_0 begins vibrating at time $t = 0$ and is stopped after a time interval Δt. The waveform of the sound at some later time is shown as a function of x. Let N be the (approximate) number of cycles in this waveform. (*a*) How are N, f_0, and Δt related? (*b*) If Δx is the length in space of this wave packet, what is the wavelength in terms of Δx and N? (*c*) What is the wave number k in terms of N and Δx? (*d*) The number N is uncertain by about ± 1 cycle. Use Figure 16-31 to explain why. (*e*) Show that the uncertainty in the wave number due to the uncertainty in N is $2\pi/\Delta x$.

FIGURE 16-31 Problem 71

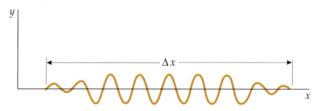

General Problems

72 • Middle C on the equal-temperament scale used by modern instrument makers has a frequency of 261.63 Hz. If a 7-g piano wire that is 80 cm long is to be tuned so that 261.63 is its fundamental frequency, what should be the tension in the wire?

73 • $\boxed{\text{i SOLVE}}$ The ear canal, which is about 2.5 cm long, roughly approximates a pipe that is open at one end and closed at the other. (*a*) What are the resonance frequencies of the ear canal? (*b*) Describe the possible effect of the resonance modes of the ear canal on the threshold of hearing.

74 • A 4-m-long, 160-g rope is fixed at one end and is tied to a light string at the other end. Its tension is 400 N. (*a*) What are the wavelengths of the fundamental and the next two harmonics? (*b*) What are the frequencies of these standing waves?

75 •• Two waves from two coherent sources have the same wavelength λ, frequency ω, and amplitude A. What is the path difference if the resultant wave at some point has amplitude A?

76 •• $\boxed{\text{i SOLVE}}$ A 35-m string has a linear mass density of 0.0085 kg/m and is under a tension of 18 N. Find the frequencies of the lowest four harmonics if (*a*) the string is fixed at both ends and (*b*) the string is fixed at one end and attached to a long, thin, massless thread at the other end.

77 •• $\boxed{\text{i SOLVE}}✓$ You find an abandoned mine shaft and decide to measure its depth. Using an audio oscillator of

variable frequency, you note that you can produce successive resonances at frequencies of 63.58 and 89.25 Hz. What is the depth of the shaft?

78 •• A string 5 m long that is fixed at one end only is vibrating in its fifth harmonic with a frequency of 400 Hz. The maximum displacement of any segment of the string is 3 cm. (*a*) What is the wavelength of this wave? (*b*) What is the wave number k? (*c*) What is the angular frequency? (*d*) Write the wave function for this standing wave.

79 •• The wave function for a standing wave on a string is described by $y(x,t) = 0.02 \sin 4\pi x \cos 60\pi t$, where y and x are in meters and t is in seconds. Determine the maximum displacement and maximum speed of a point on the string at (*a*) $x = 0.10$ m, (*b*) $x = 0.25$ m, (*c*) $x = 0.30$ m, and (*d*) $x = 0.50$ m.

80 •• A 2.5-m-long wire with a mass of 0.10 kg is fixed at both ends and is under tension of 30 N. When the nth harmonic is excited, there is a node 0.50 m from one end. (*a*) What is n? (*b*) What are the frequencies of the first three allowed modes of vibration?

81 •• $\boxed{\text{SSM}}$ In an early method used to determine the speed of sound in gases, powder was spread along the bottom of a horizontal, cylinderical glass tube. One end of the tube was closed by a piston that oscillated at a known frequency f. The other end was closed by a movable piston whose position was adjusted until resonance occurred. At resonance, the powder collected in equally spaced piles along the bottom of the tube. (*a*) Explain why the powder collects in this way. (*b*) Derive a formula that gives the speed of sound in the gas in terms of f and the distance between the piles of powder. (*c*) Give suitable values for the frequency f and the distance between the piles of powder. (*d*) Give suitable values for the frequency f and the length L of the tube for which the speed of sound could be measured in either air or helium.

82 •• In a lecture demonstration of standing waves, a string is attached to a tuning fork that vibrates at 60 Hz and sets up transverse waves of that frequency on the string. The other end of the string passes over a pulley, and the tension is varied by attaching weights to that end. The string has approximate nodes at the tuning fork and at the pulley. (*a*) If the string has a linear mass density of 8 g/m and is 2.5 m long (from the tuning fork to the pulley), what must be the tension for the string to vibrate in its fundamental mode? (*b*) Find the tension necessary for the string to vibrate in its second, third, and fourth harmonic.

83 •• $\boxed{\text{i SOLVE}✓}$ Three successive resonance frequencies in an organ pipe are 1310, 1834, and 2358 Hz. (*a*) Is the pipe closed at one end or open at both ends? (*b*) What is the fundamental frequency? (*c*) What is the length of the pipe?

84 •• $\boxed{\text{i SOLVE}✓}$ A wire of mass 1 g and length 50 cm is stretched with a tension of 440 N. It is then placed near the open end of the tube in Example 16-9 and stroked with a violin bow so that it oscillates at its fundamental frequency. The water level in the tube is then lowered until a resonance is obtained, which occurs at 18 cm below the top of the tube. Use the data given to determine the speed of sound in air. Why is this method not very accurate?

85 •• A standing wave on a rope is represented by the wave function $y(x,t) = 0.02 \sin \frac{1}{2}\pi x \cos 40\pi t$, where x and y are in meters and t is in seconds. (a) Write wave functions for two traveling waves that, when superimposed, will produce the resultant standing-wave pattern. (b) What is the distance between the nodes of the standing wave? (c) What is the velocity of a segment of the rope at $x = 1$ m? (d) What is the acceleration of a segment of the rope at $x = 1$ m?

86 •• **SOLVE** Two identical speakers emit sound waves of frequency 680 Hz uniformly in all directions. The total audio output of each speaker is 1 mW. A point P is 2.00 m from one speaker and 3.00 m from the other. (a) Find the intensities I_1 and I_2 from each speaker separately at point P. (b) If the speakers are driven coherently and are in phase, what is the intensity at point P? (c) If they are driven coherently but are 180° out of phase, what is the intensity at point P? (d) If the speakers are incoherent, what is the intensity at point P?

87 •• Three waves with the same frequency, wavelength, and amplitude are traveling in the same direction. The three waves are given by $y_1(x,t) = 0.05 \sin\left(kx - \omega t - \frac{\pi}{3}\right)$, $y_2(x,t) = 0.05 \sin(kx - \omega t)$, and $y_3(x,t) = 0.05 \sin\left(kx - \omega t + \frac{\pi}{3}\right)$. Find the resultant wave.

88 •• A plane wave has the form $f(x, y, t) = A \cos(k_x x + k_y y - \omega t)$. Show that the direction in which the wave is traveling makes an angle $\theta = \tan^{-1}(k_y/k_x)$ with the positive x direction and that the wave speed is $v = \omega/\sqrt{k_x^2 + k_y^2}$.

89 •• **SSM** The speed of sound is proportional to the square root of the absolute temperature T (Equation 15-5). (a) Show that if the temperature changes by a small amount ΔT, the fundamental frequency of an organ pipe changes by approximately Δf, where $\Delta f/f = \frac{1}{2}\Delta T/T$. (b) Suppose that an organ pipe that is closed at one end has a fundamental frequency of 200 Hz when the temperature is 20°C. What will be its fundamental frequency when the temperature is 30°C? (Ignore any change in the length of the pipe due to thermal expansion.)

90 •• Two traveling wave pulses on a string are represented by the wave functions

$$y_1(x,t) = \frac{0.02}{2 + (x - 2t)^2}$$

and

$$y_2(x,t) = \frac{-0.02}{2 + (x + 2t)^2}$$

where x is in meters and t is in seconds. (a) Using a spreadsheet program or graphing calculator, make a graph of each wave function separately as a function of x at $t = 0$ and describe the behavior of each as time increases. (b) Find the resultant wave function at $t = 0$. (c) Find the resultant wave function at $t = 1$ s. (d) Graph the resultant wave function at $t = 1$ s.

91 •• If you put your ear and your hand near the end of a long, open-ended tube and snap your fingers, you will hear a sound similar to that of a guitar string being plucked. (Tubes of about 1-m length are best.) (a) Explain what causes this sound. (b) What effective tube length do you need to make a sound like that of a guitar string with a pitch of A above middle C (440 Hz)?

92 •• The kinetic energy of a segment of length Δx and mass Δm of a vibrating string is given by $\Delta K = \frac{1}{2}\Delta m(\partial y/\partial t)^2 = \frac{1}{2}\mu(\partial y/\partial t)^2 \Delta x$, where $\mu = \Delta m/\Delta x$. (a) Find the total kinetic energy of the nth mode of vibration of a string of length L fixed at both ends. (b) Give the maximum kinetic energy of the string. (c) What is the wave function when the kinetic energy has its maximum value? (d) Show that the maximum kinetic energy in the nth mode is proportional to $n^2 A_n^2$.

93 •• (a) Show that when the tension in a string fixed at both ends is changed by a small amount dF, the frequency of the fundamental is changed by approximately df, where $df/f = \frac{1}{2}dF/F$. Does this result apply to all harmonics? (b) Use this result to find the percentage change in the tension needed to increase the frequency of the fundamental of a piano wire from 260 to 262 Hz.

94 •• **SSM** Two sources of harmonic waves on the x axis have a phase difference that is proportional to time: $\delta_s = Ct$, where C is a constant. The amplitude of the wave from each source at some point P on the x axis is A_0. (a) Write the wave functions for each of the two waves at point P, assuming this point to be a distance x_1 from one source and $x_1 + \Delta x$ from the other. (b) Find the resultant wave function and show that its amplitude is $2A_0 \cos[\frac{1}{2}(\delta + \delta_0)]$, where δ is the phase difference at P due to the path difference. (c) Using a spreadsheet program or graphing calculator, graph the intensity at point P versus time for a zero path difference. (Let I_0 be the intensity due to each wave separately.) What is the time average of the intensity? (d) Make the same graph for the intensity at a point for which the path difference is $\lambda/2$.

95 ••• The wave functions of two standing waves on a string of length L are $y_1(x,t) = A_1 \cos \omega_1 t \sin k_1 x$ and $y_2(x,t) = A_2 \cos \omega_2 t \sin k_2 x$, where $k_n = n\pi/L$ and $\omega_n = n\omega_1$. The wave function of the resultant wave is $y_r(x,t) = y_1(x,t) + y_2(x,t)$. (a) Find the velocity of a segment dx of the string. (b) Find the kinetic energy of this segment. (c) By integration, find the total kinetic energy of the resultant wave. Notice the disappearance of the cross terms so that the total kinetic energy is proportional to $(n_1 A_1)^2 + (n_2 A_2)^2$.

96 ••• A 2-m wire fixed at both ends is vibrating in its fundamental mode. The tension in the wire is 40 N and the mass of the wire is 0.1 kg. At the midpoint of the wire, the amplitude is 2 cm. (a) Find the maximum kinetic energy of the wire. (b) At the instant that the transverse displacement is given by $(0.02 \text{ m}) \sin(\pi x/2)$, what is the kinetic energy of the wire? (c) At what position on the wire does the kinetic energy per unit length have its largest value? (d) Where does the potential energy per unit length have its maximum value?

97 ••• In principle, a wave with almost any arbitrary shape can be expressed as a sum of harmonic waves of different frequencies. (a) Consider the function defined by

$$f(x) = \frac{4}{\pi}\left(\frac{\cos x}{1} + \frac{\cos 3x}{3} + \frac{\cos 5x}{5} + \cdots\right)$$

$$= \frac{4}{\pi}\sum_{n=0}^{\infty}(-1)^n \frac{\cos[(2n + 1)x]}{2n + 1}$$

Write a spreadsheet program to calculate this series using a finite number of terms, and make three graphs of the function in the range $x = 0$ to $x = 4\pi$. For the first graph approximate the sum from $n = 0$ to $n = \infty$ with the first term of the sum. For the second and third graphs use only the first five term and the first ten terms, respectively. This function is sometimes called the *square wave* (or *θ function*). (*b*) What is the relation between this function and Liebnitz' series for π,

$$\frac{\pi}{4} = 1 - \frac{1}{3} + \frac{1}{5} - \frac{1}{7} + \cdots ?$$

98 ••• Write a spreadsheet program to calculate and graph the function

$$f(x) = \frac{4}{\pi}\left(\sin x - \frac{\sin 3x}{9} + \frac{\sin 5x}{25} - \cdots\right)$$

$$= \frac{4}{\pi}\sum_n \frac{(-1)^n \sin(2n + 1)x}{(2n + 1)^2}$$

What kind of wave is this?

99 ••• If you clap your hands at the end of a long, cylindrical tube, the echo you hear back will not sound like the handclap; instead, you will hear what sounds like a whistle, initially at a very high frequency, but descending rapidly down to almost nothing. This "culvert whistler" can be explained by thinking of the sound from the clap as a single compression radiating outward from the hands. The echoes of the handclap arriving at your ear have traveled along different paths through the tube, as shown in Figure 16-32. The first echo to arrive travels straight down and straight back along the tube, while the second echo reflects once off of the center of the tube going out, and again going back, the third echo reflects twice at points 1/4 and 3/4 of the distance, etc. The tone of the sound you hear reflects the frequency at which these echoes reach your ears. (*a*) Show that the time delay between the n_{th} echo and the $n+1_{th}$ is

$$\Delta t_n = \frac{2}{v}\left(\sqrt{(2n)^2 r^2 + L^2} - \sqrt{[2(n-1)]^2 r^2 + L^2}\right),$$

where v is the speed of sound, L is the length of the tube and r is its radius. (*b*) Using a spreadsheet program or graphing calculator, graph Δt_n versus n for $L = 90$ m, $r = 1$ m. (These are the approximate length and diameter of the long tube in the San Francisco Exploratorium.) Go to at least $n = 100$. (*c*) From your graph, explain why the frequency decreases over time. What are the highest and lowest frequencies you will hear in the whistler?

FIGURE 16-32 Problem 99

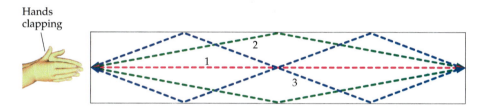

Hands clapping

CHAPTER

17

Temperature and the Kinetic Theory of Gases

THESE HELIUM BALLOONS ARE WELL INFLATED ON A WARM SUMMER DAY.

? **What might happen to them if they are taken indoors to an air-conditioned room? (See Example 17-6.)**

Temperature is familiar to us as the measure of the hotness or coldness of objects or of our surroundings.

➤ **In this chapter, we will show that a consistent temperature scale can be defined in terms of the properties of gases at low densities, and that temperature is a measure of the average internal molecular kinetic energy of an object.**

17-1 Thermal Equilibrium and Temperature

Our sense of touch can usually tell us if an object is hot or cold. Early in childhood we learn that to make a cold object warmer, we place it in contact with a hot object. To make a hot object cooler, we place it in contact with a cold object.

When an object is heated or cooled, some of its physical properties change. Most solids and liquids expand when they are heated. A gas, if its pressure is kept constant, will also expand when it is heated, or, if its volume is kept constant, its pressure will rise. If an electrical conductor is heated, its electrical resistance changes. (This is discussed in Chapter 25.) A physical property that changes with temperature is called a **thermometric property.** A change in a thermometric property indicates a change in the temperature of the object.

Suppose that we place a warm copper bar in close contact with a cold iron bar so that the copper bar cools and the iron bar warms. We say that the two bars are in **thermal contact.** The copper bar contracts slightly as it cools, and the iron bar expands slightly as it warms. Eventually this process stops and the lengths of the bars remain constant. The two bars are then in **thermal equilibrium** with each other.

Suppose instead that we place the warm copper bar in a cool running stream. The bar cools until it stops contracting, at the point at which the bar and the water are in thermal equilibrium. Next we place a cold iron bar in the stream on the side opposite the copper bar. The iron bar will warm until it and the water are also in thermal equilibrium. If we remove the bars and place them in thermal contact with each other, we find that their lengths do not change. They are in thermal equilibrium with each other. Though it is common sense, there is no logical way to deduce this fact, which is called the **zeroth law of thermodynamics** (Figure 17-1):

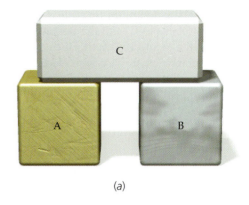

(a)

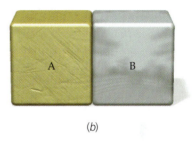

(b)

> If two objects are in thermal equilibrium with a third, then they are in thermal equilibrium with each other.

ZEROTH LAW OF THERMODYNAMICS

FIGURE 17-1 The zeroth law of thermodynamics. (*a*) Systems A and B are in thermal contact with system C but not with each other. When A and B are each in thermal equilibrium with C, they are in thermal equilibrium with each other, which can be checked by placing them in contact with each other as in (*b*).

Two objects are defined to have the same *temperature* if they are in thermal equilibrium with each other. The zeroth law, as we will see, enables us to define a temperature scale.

17-2 The Celsius and Fahrenheit Temperature Scales

Any thermometric property can be used to establish a temperature scale. The common mercury thermometer consists of a glass bulb and tube containing a fixed amount of mercury.[†] When this thermometer is put in contact with a warmer object, the mercury expands, increasing the length of the mercury column (the glass expands too, but by a negligible amount). We can create a scale along the glass tube as follows. First the thermometer is placed in ice and water in equilibrium[‡] at a pressure of 1 atm. When the thermometer is in thermal equilibrium with the ice water, the position of the mercury column is marked on the glass tube. This is the **ice-point temperature** (also called the **normal freezing point** of water). Next, the thermometer is placed in boiling water at a pressure of 1 atm. When the thermometer is in thermal equilibrium with the boiling water, the new position of the mercury column is marked. This is the **steam-point temperature** (also called the **normal boiling point** of water).

---

† Because mercury is highly toxic, mercury thermometers are no longer sold in the United States. Today, alcohol is commonly used in thermometers.

‡ Water and ice in equilibrium provide a constant-temperature bath. When ice is placed in warm water, the water cools as some of the ice melts. Eventually, thermal equilibrium is reached and no more ice melts. If the system is heated slightly, some more of the ice melts, but the temperature does not change as long as some ice remains.

The **Celsius temperature scale** defines the ice-point temperature as zero degrees Celsius (0°C) and the steam-point temperature as 100°C. The space between the 0° and 100° marks is divided into 100 equal intervals (degrees). Degree markings are also extended below and above these points. If L_t is the length of the mercury column, the Celsius temperature t_C is given by

$$t_C = \frac{L_t - L_0}{L_{100} - L_0} \times 100° \qquad \qquad \text{17-1}$$

where L_0 is the length of the mercury column when the thermometer is in an ice bath and L_{100} is its length when the thermometer is in a steam bath. The normal temperature of the human body measured on the Celsius scale is about 37°C.

The **Fahrenheit temperature scale** (which is used in the United States) defines the ice-point temperature as 32°F and the steam-point temperature as 212°F.[†] To convert temperatures between Fahrenheit and Celsius scales, we note there are 100 Celsius degrees and 180 Fahrenheit degrees between the ice and steam points. A temperature change of one Celsius degree therefore equals a change of 1.8 = 9/5 Fahrenheit degrees. To convert a temperature from one scale to the other, we must also take into account the fact that the zero temperatures of the two scales are not the same. The general relation between a Fahrenheit temperature t_F and Celsius temperature t_C is

$$t_C = \tfrac{5}{9}(t_F - 32°) \qquad \qquad \text{17-2}$$

FAHRENHEIT–CELSIUS CONVERSION

FIGURE 17-2 A bimetallic strip. When heated or cooled, the two metals expand or contract by different amounts, causing the strip to bend.

CONVERTING FAHRENHEIT AND CELSIUS TEMPERATURES   **E X A M P L E   1 7 - 1**

(a) **Find the temperature on the Celsius scale equivalent to 41°F.** (b) **Find the temperature on the Fahrenheit scale equivalent to 37.0°C.**

(a) Apply Equation 17-2 with $t_F = 41°F$: $t_C = \tfrac{5}{9}(t_F - 32°) = \tfrac{5}{9}(41° - 32°) = \tfrac{5}{9}(9°) = \boxed{5°C}$

(b) 1. Solve Equation 17-2 for t_F in terms of t_C: $t_F = \tfrac{9}{5}t_C + 32°$

2. Substitute $t_C = 37°C$: $t_F = \tfrac{9}{5}(37.0°) + 32° = 66.6° + 32° = \boxed{98.6°F}$

EXERCISE (a) Find the Celsius temperature equivalent to 68°F. (b) Find the Fahrenheit temperature equivalent to −40°C. (*Answer*   (a) 20°C (b) −40°F)

Other thermometric properties can be used to set up thermometers and construct temperature scales. Figure 17-2 shows a bimetallic strip consisting of two different metals bonded together. When the strip is heated or cooled, it bends to accommodate the difference in the thermal expansion of the two metals. Figure 17-3 shows a thermometer consisting of a bimetallic coil with a pointer attached to indicate the temperature. When the thermometer is heated, the coil bends and the pointer moves. Like mercury thermometers, it is calibrated by dividing the interval between the ice point and the steam point into 100 Celsius degrees (or 180 Fahrenheit degrees).

---

† When the German physicist Daniel Fahrenheit devised his temperature scale, he wanted all measurable temperatures to be positive. Originally, he chose 0°F for the coldest temperature he could obtain with a mixture of ice and salt water and 96°F (a convenient number with many factors for subdivision) for the temperature of the human body. He then modified his scale slightly to make the ice-point and steam-point temperatures whole numbers. This resulted in the average temperature of the human body being between 98° and 99°F.

FIGURE 17-3 (*a*) A thermometer using a bimetallic strip in the form of a coil. (The red pointer is attached to one end of the coil.) When the temperature of the coil increases, the needle rotates clockwise because the outer metal expands more than the inner metal. (*b*) A home thermostat controls the central air conditioner. When the air gets warmer, the coil expands, the glass bulb mounted on it tilts, and mercury in the tube slides to close an electrical switch, turning on the air conditioning. A slide lever (at the lower right), used to rotate the coil mount, is used to set the desired temperature. The circuit will be broken when the cooler air causes the bimetallic coil to contract.

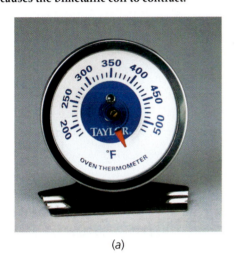

(a)

Glass Bulb
Mercury switch

Bimetalic
strip coil

Slide
lever

(b)

17-3 Gas Thermometers and the Absolute Temperature Scale

When different types of thermometers are calibrated in ice water and steam, they agree (by definition) at 0°C and 100°C, but they give slightly different readings at points in between. Discrepancies increase markedly above the steam point and below the ice point. However, in one group of thermometers, gas thermometers, measured temperatures agree closely with each other even far from the calibration points. In a **constant-volume gas thermometer,** the gas volume is kept constant, and change in gas pressure is used to indicate a change in temperature (Figure 17-4). An ice-point pressure P_0 and steam-point pressure P_{100} are determined by placing the thermometer in ice–water and water–steam baths, and the interval between is divided into 100 equal degrees (for the Celsius scale). If the pressure is P_t in a bath whose temperature is to be determined, that temperature in degrees Celsius is defined to be

$$t_C = \frac{P_t - P_0}{P_{100} - P_0} \times 100°$$ 17-3

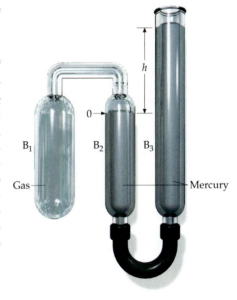

FIGURE 17-4 A constant-volume gas thermometer. The volume is kept constant by raising or lowering tube B_3 so that the mercury in tube B_2 remains at the zero mark. The temperature is chosen to be proportional to the pressure of the gas in tube B_1, which is indicated by the height h of the mercury column in tube B_3.

Suppose we measure a specific temperature, say the boiling point of sulfur at 1 atm pressure, using four constant-volume gas thermometers, each containing one of four gasses—air, hydrogen, nitrogen, and oxygen. The thermometers are calibrated, meaning values for P_{100} and P_0 are determined for each. Each thermometer is then immersed in boiling sulfur, and when it is in thermal equilibrium with the sulfur, the pressure in the thermometer is measured. Next, the temperature is calculated using Equation 17-3. Will this process give the same result for each of the four thermometers? Surprisingly perhaps, the answer is yes. All four thermometers measure the same temperature so long as the density of the gas in each is sufficiently low.

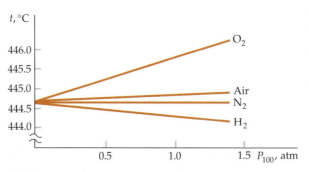

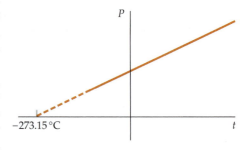

FIGURE 17-5 Temperature of the boiling point of sulfur measured with constant-volume gas thermometers filled with various gases. Increasing or decreasing the amount of gas in the thermometer varies the pressure P_{100} at the steam point of water. As the amount of gas is reduced, the temperatures measured by all the thermometers approach the value 444.60°C.

One measure of the density of the gas in the thermometer is its pressure at the steam point, P_{100}. If we vary the amount of gas in a constant-volume gas thermometer, by either adding or removing gas, we change both P_{100} and P_0. As a result, each time the amount of gas is varied, the thermometer must be recalibrated. Figure 17-5 shows the results of measurements of the boiling point of sulfur using four constant-volume gas thermometers, each filled with air, hydrogen, nitrogen, or oxygen. For each thermometer the measured temperature is plotted as a function of the steam-point pressure P_{100} of the thermometer. As the amount of a gas is reduced, its density and the steam-point pressure both decrease. We see that agreement among the thermometers is very close at low gas densities (low P_{100}). In the limit as gas density goes to zero, all gas thermometers give the same value for the temperature of boiling sulfur. This low-density temperature measurement is independent of the properties of any particular gas. Of course, there is nothing special about the boiling point of sulfur. Constant-volume gas thermometers at low densities are in agreement at any temperature. Thus, low-density gas thermometers can be used to define temperature.

Now consider a series of temperature measurements with a constant-volume gas thermometer that has a very small but fixed amount of gas. According to Equation 17-3, the pressure in the thermometer P_t varies linearly with the measured temperature t_C. Figure 17-6 shows a plot of pressure versus measured temperature in a constant-volume gas thermometer. When we extrapolate this straight line to zero pressure, the temperature approaches −273.15°C. This limit is the same no matter what kind of gas is used.

A reference state that is much more precisely reproducible than either the ice or steam points is the **triple point of water**—the unique temperature and pressure at which water, water vapor, and ice coexist in equilibrium (see Figure 17-7). This equilibrium state occurs at 4.58 mmHg and 0.01°C. The **ideal-gas temperature scale** is defined so that the temperature of the triple point is 273.16 kelvins (K). (The kelvin is a degree unit that is the same size as the Celsius degree.) The temperature T of any other state is defined to be proportional to the pressure in a constant-volume gas thermometer:

$$T = \frac{273.16 \text{ K}}{P_3} P \qquad 17\text{-}4$$

IDEAL-GAS TEMPERATURE SCALE

where P is the observed pressure of the gas in the thermometer and P_3 is the pressure when the thermometer is immersed in a water–ice–vapor bath at its triple point. The value of P_3 depends

FIGURE 17-6 Plot of pressure versus temperature as measured by a constant-volume gas thermometer. When extrapolated to zero pressure, the plot intersects the temperature axis at the value −273.15°C.

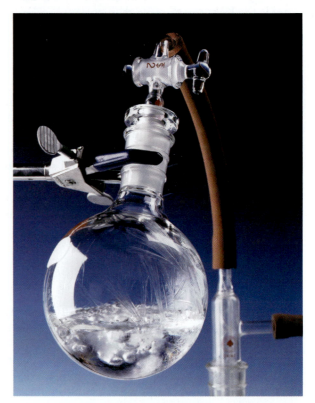

FIGURE 17-7 H_2O at its triple point. The spherical flask contains water, ice, and water vapor in equilibrium.

on the amount of gas in the thermometer. The ideal-gas temperature scale, defined by Equation 17-4, has the advantage that any measured temperature does not depend on the properties of the particular gas that is used, but depends only on the general properties of gases.

The lowest temperature that can be measured with a gas thermometer is about 1 K, and requires helium for the gas. Below this temperature helium liquefies; all other gases liquefy at higher temperatures (see Table 17-1). In Chapter 19 we will see that the second law of thermodynamics can be used to define the **absolute temperature scale** independent of the properties of any substance, and with no limitations on the range of temperatures that can be measured. Temperatures as low as a millionth of a kelvin have been measured. The absolute scale so defined is identical to that defined by Equation 17-4 for the range of temperatures for which gas thermometers can be used. The symbol T is used when referring to absolute temperature.

Because the Celsius degree and the kelvin are the same size, temperature *differences* are the same on both the Celsius and the absolute temperature scales (also called the **Kelvin scale**). That is, a temperature *change* of 1 K is identical to a temperature *change* of 1C°.[†] The two scales differ only in the choice of zero temperature. To convert from degrees Celsius to kelvins, we merely add 273.15:[‡]

$$T = t_C + 273.15 \text{ K} \qquad\qquad 17\text{-}5$$

CELSIUS–ABSOLUTE CONVERSION

Although the Celsius and Fahrenheit scales are convenient for everyday use, the absolute scale is much more convenient for scientific purposes, partly because many formulas are more simply expressed in it, and partly because the absolute temperature can be given a more fundamental interpretation.

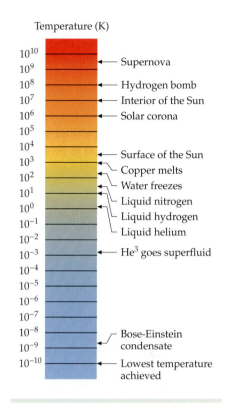

Temperature (K)

- 10^{10} — Supernova
- 10^9
- 10^8 — Hydrogen bomb
- 10^7 — Interior of the Sun
- 10^6 — Solar corona
- 10^5
- 10^4
- 10^3 — Surface of the Sun
- 10^2 — Copper melts
- 10^1 — Water freezes
- 10^0 — Liquid nitrogen
- 10^{-1} — Liquid hydrogen
- 10^{-2} — Liquid helium
- 10^{-3} — He^3 goes superfluid
- 10^{-4}
- 10^{-5}
- 10^{-6}
- 10^{-7}
- 10^{-8} — Bose-Einstein condensate
- 10^{-9}
- 10^{-10} — Lowest temperature achieved

TABLE 17-1 The temperatures of various places and phenomena.

CONVERTING FROM KELVIN TO FAHRENHEIT   **EXAMPLE 17-2**

What is the Kelvin temperature corresponding to 70°F?

PICTURE THE PROBLEM First convert to degrees Celsius, then to kelvins.

1. Convert to degrees Celsius:

$$t_C = \tfrac{5}{9}(70° - 32°) = 21.1°C$$

2. To find the Kelvin temperature we add 273:

$$T = t_C + 273 = 21.1 + 273 = \boxed{294 \text{ K}}$$

EXERCISE The "high-temperature" superconductor $YBa_2Cu_3O_7$ becomes superconducting when the temperature is lowered to 92 K. Find the superconducting threshold temperature in degrees Fahrenheit. (*Answer:* −294°F)

17-4 The Ideal-Gas Law

The properties of gases at low densities allow the definition of the ideal-gas temperature scale. If we compress such a gas while keeping its temperature constant, the pressure increases. Similarly, if a gas expands at constant temperature, its pressure decreases. To a good approximation, the product of the pressure and

---

[†] We write 1C° to indicate a *temperature change* of one Celsius degree, in contrast to 1°C, which means a temperature of one degree Celsius.

[‡] For most purposes, we can round off the temperature of absolute zero to −273°C.

volume of a low-density gas is constant at a constant temperature. This result was discovered experimentally by Robert Boyle (1627–1691), and is known as **Boyle's law:**

$$PV = \text{constant} \quad \text{(constant temperature)}$$

A more general law exists that reproduces Boyle's law as a special case. According to Equation 17-4, the absolute temperature of a low-density gas is proportional to its pressure at constant volume. In addition—a result discovered experimentally by Jacques Charles (1746–1823) and Joseph Gay-Lussac (1778–1850)—the absolute temperature of a low-density gas is proportional to its volume at constant pressure. We can combine these two results by stating

$$PV = CT \qquad\qquad 17\text{-}6$$

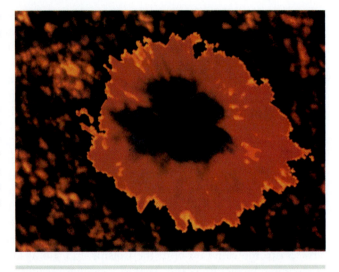

The atmosphere of Venus is almost entirely CO_2. However, measurements by the Pioneer Venus Orbiter show an atomic hydrogen cloud surrounding Venus. The crescent-shaped image shows atomic oxygen, and the bars show hydrogen extending far above the atmosphere. Since the escape speed on Venus is 10.3 km/s, slightly smaller than the escape speed on earth, and since the Venusian atmosphere is considerably warmer than earth's atmosphere, all of the hydrogen in the atmosphere at the time of the formation of Venus should have escaped by now.

where C is a constant of proportionality. We can see that this constant is proportional to the amount of gas by considering the following. Suppose that we have two containers with identical volumes, each holding the same amount of the same kind of gas at the same temperature and pressure. If we consider the two containers as one system, we have twice the amount of gas at twice the volume, but at the same temperature and pressure. We have thus doubled the quantity $PV/T = C$ by doubling the amount of gas. We can therefore write C as a constant k times the number of molecules in the gas N:

$$C = kN$$

Equation 17-6 then becomes

$$PV = NkT \qquad\qquad 17\text{-}7$$

The constant k is called **Boltzmann's constant.** It is found experimentally to have the same value for any kind of gas:

$$k = 1.381 \times 10^{-23}\,\text{J/K} = 8.617 \times 10^{-5}\,\text{eV/K} \qquad\qquad 17\text{-}8$$

An amount of gas is often expressed in moles. A **mole** (mol) of any substance is the amount of that substance that contains Avogadro's number N_A of atoms or molecules, defined as the number of carbon atoms in 12 g of ^{12}C:

$$N_A = 6.022 \times 10^{23} \qquad\qquad 17\text{-}9$$

AVOGADRO'S NUMBER

If we have n moles of a substance, then the number of molecules is

$$N = nN_A \qquad\qquad 17\text{-}10$$

Equation 17-7 is then

$$PV = nN_A kT = nRT \qquad\qquad 17\text{-}11$$

where $R = N_A k$ is called the **universal gas constant.** Its value, which is the same for all gases, is

$$R = N_A k = 8.314\,\text{J/(mol·K)} = 0.08206\,\text{L·atm/(mol·K)} \qquad\qquad 17\text{-}12$$

Figure 17-8 shows plots of $PV/(nT)$ versus the pressure P for several gases. For all gases, $PV/(nT)$ is nearly constant over a large range of pressures. Even oxygen, which varies the most in this graph, changes by only about 1 percent between 0 and 5 atm. An **ideal gas** is defined as one for which $PV/(nT)$ is constant for all pressures. The pressure, volume, and temperature of an ideal gas are related by

$$PV = nRT \qquad\qquad 17\text{-}13$$

<blockquote>IDEAL-GAS LAW</blockquote>

Equation 17-13, which relates the variables P, V, and T, is known as the ideal-gas law, and is an example of an **equation of state**. It describes the properties of real gases with low densities (and therefore low pressures). At higher densities, corrections must be made to this equation. In Chapter 20 we discuss another equation of state, the van der Waals equation, which includes such corrections. For any gas at any density, there is an equation of state relating P, V, and T for a given amount of gas. Thus the state of a given amount of gas is determined by any two of the three **state variables** P, V, and T.

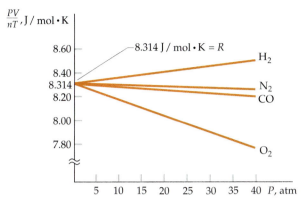

FIGURE 17-8 Plot of PV/nT versus P for real gases. In these plots, varying the amount of gas varies the pressure. The ratio PV/nT approaches the same value, 8.314 J/(mol·K), for all gases as we reduce the density, and thereby the pressure, of the gas. This value is the universal gas constant R.

VOLUME OF AN IDEAL GAS   **EXAMPLE 17-3**

What volume is occupied by 1 mol of an ideal gas at a temperature of 0°C and a pressure of 1 atm?

We can find the volume using the ideal-gas law, with $T = 273$ K:

$$V = \frac{nRT}{P}$$

$$= \frac{(1\ \text{mol})(0.0821\ \text{L·atm/[mol·K]})(273\ \text{K})}{1\ \text{atm}}$$

$$= \boxed{22.4\ \text{L}}$$

REMARKS Note that by writing R in L·atm/(mol·K), we could write P in atmospheres to get V in liters.

EXERCISE Find (*a*) the number of moles n and (*b*) the number of molecules N in 1 cm³ of a gas at 0°C and 1 atm. (*Answer* (*a*) $n = 4.46 \times 10^{-5}$ mol (*b*) $N = 2.68 \times 10^{19}$ molecules)

The temperature of $0°C = 273$ K and the pressure of 1 atm are often referred to as **standard conditions**. We see from Example 17-3 that under standard conditions, 1 mol of an ideal gas occupies a volume of 22.4 L.

Figure 17-9 shows plots of P versus V at several constant temperatures T. These curves are called **isotherms**. The isotherms for an ideal gas are hyperbolas. For a fixed amount of gas, we can see from Equation 17-13 that the quantity PV/T is constant. Using the subscripts 1 for the initial values and 2 for the final values, we have

$$\frac{P_2 V_2}{T_2} = \frac{P_1 V_1}{T_1} \qquad\qquad 17\text{-}14$$

<blockquote>IDEAL-GAS LAW FOR FIXED AMOUNT OF GAS</blockquote>

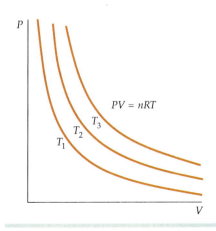

FIGURE 17-9 Isotherms on the PV diagram for a gas. For an ideal gas, these curves are hyperbolas given by $PV = nRT$.

HEATING AND COMPRESSING A GAS   **EXAMPLE 17-4**

A gas has a volume of 2 L, a temperature of 30°C, and a pressure of 1 atm. When the gas is heated to 60°C and compressed to a volume of 1.5 L, what is its new pressure?

PICTURE THE PROBLEM Since the amount of gas is fixed, the pressure can be found using Equation 17-14. Let subscripts 1 and 2 refer to the initial and final states, respectively.

1. Express the pressure P_2 in terms of P_1 and the initial and final volumes and temperatures:

$$\frac{P_1 V_1}{T_1} = \frac{P_2 V_2}{T_2}, \qquad P_2 = \frac{T_2 V_1}{T_1 V_2} P_1$$

2. Calculate the initial and final absolute temperatures:

$$T_1 = 273 + 30 = 303 \text{ K}$$
$$T_2 = 273 + 60 = 333 \text{ K}$$

3. Substitute numerical values in step 1 to find P_2:

$$P_2 = \frac{(333 \text{ K})(2 \text{ L})}{(303 \text{ K})(1.5 \text{ L})} (1 \text{ atm}) = \boxed{1.47 \text{ atm}}$$

EXERCISE How many moles of gas are in the system described in this example? (*Answer* $n = 0.0804$ mol)

The mass per mole of a substance is called its **molar mass** M. (The terms *molecular weight or molecular mass* are also sometimes used.) The molar mass of ^{12}C is, by definition, 12 g/mol or 12×10^{-3} kg/mol. Molar masses of the elements are given in the periodic table in Appendix E. The molar mass of a molecule such as CO_2 is the sum of the molar masses of the elements in the molecule. Because the molar mass of oxygen is 16 g/mol (actually 15.999 g/mol), the molar mass of O_2 is 32 g/mol and that of CO_2 is 12 + 32 = 44 g/mol.

The mass of n moles of a gas is given by

$$m = nM$$

and the density ρ of an ideal gas is

$$\rho = \frac{m}{V} = \frac{nM}{V}$$

Using $n/V = P/RT$ from Equation 17-13, we have

$$\rho = \frac{M}{RT} P \qquad\qquad 17\text{-}15$$

At a given temperature, the density of an ideal gas is proportional to its pressure.

THE MASS OF A HYDROGEN ATOM   **EXAMPLE 17-5**

The molar mass of hydrogen is 1.008 g/mol. What is the mass of one hydrogen atom?

PICTURE THE PROBLEM Let m be the mass of a hydrogen atom. Since there are N_A atoms in a mole, the molar mass M is given by $M = m N_A$. We can use this to solve for m.

The mass of a hydrogen atom is the molar mass divided by Avogadro's number:

$$m = \frac{M}{N_A} = \frac{1.008 \text{ g/mol}}{6.022 \times 10^{23} \text{ atoms/mol}}$$

$$= \boxed{1.67 \times 10^{-24} \text{ g/atom}}$$

REMARKS Note that Avogadro's number is essentially the reciprocal of the mass of the hydrogen atom measured in grams.

EXPANDING A GAS AT CONSTANT TEMPERATURE   **EXAMPLE 17-6**   **Try It Yourself**

A 100-g sample of CO_2 occupies a volume of 55 L at a pressure of 1 atm. (*a*) What is the temperature? (*b*) If the volume is increased to 80 L and the temperature is kept constant, what is the new pressure?

PICTURE THE PROBLEM Both questions can be answered using the ideal-gas law (Equation 17-13) if we first find the number of moles, n.

Cover the column to the right and try these on your own before looking at the answers.

Steps	Answers
(*a*) 1. The number of moles n is calculated from the mass of the sample m and the molar mass M of CO_2: The molar mass, from information in Appendix C, is 44 g/mol.	$n = \dfrac{m}{M} = 2.27 \text{ mol}$
2. Find the temperature T from the ideal-gas law.	$T = \dfrac{PV}{nR} = \boxed{295 \text{ K}}$
(*b*) Use $PV = $ constant to find the new pressure for $V = 80$ L.	$P_2 = \boxed{0.688 \text{ atm}}$

EXERCISE If the temperature is decreased at constant pressure, what happens to the volume? (*Answer*   It decreases)

17-5 The Kinetic Theory of Gases

The description of the behavior of a gas in terms of the macroscopic state variables P, V, and T can be related to simple averages of microscopic quantities such as the mass and speed of the molecules in the gas. The resulting theory is called **the kinetic theory of gases.**

From the point of view of kinetic theory, a gas consists of a large number of molecules making elastic collisions with each other and with the walls of a container. In the absence of external forces (we may neglect gravity), there is no preferred position for a molecule in the container,[†] and no preferred direction for its velocity vector. The molecules are separated, on the average, by distances that are large compared with their diameters, and they exert no forces on each other except when they collide. (This final assumption is equivalent to assuming a very low gas density, which, as we saw in the last section, is the same as assuming that the gas is an ideal gas. Because momentum is conserved, the collisions the

---

† Because of gravity, the density of molecules at the bottom of the container is slightly greater than at the top. As discussed in Chapter 13, the density of air decreases by half at a height of about 5.5 km, so the variation over a normal sized container is negligible.

molecules make with each other have no effect on the total momentum in any direction—thus such collisions can be neglected.)

Calculating the Pressure Exerted by a Gas

The pressure that a gas exerts on its container is due to collisions between gas molecules and the container walls. This pressure is a force per unit area and, by Newton's second law, this force is the rate of change of momentum of the gas molecules colliding with the wall.

Consider a rectangular container of volume V containing N gas molecules, each of mass m moving with a speed v. Let us calculate the force exerted by these molecules on the right-hand wall, which is perpendicular to the x axis and has area A. The molecules hitting this wall in a time interval Δt are those that are within distance $v_x \Delta t$ of the wall (Figure 17-10) and are moving to the right. Thus, the number of molecules hitting the wall during time Δt is the number per unit volume N/V times the volume $v_x \Delta t\, A$ times $\frac{1}{2}$ because, on average, only half the molecules are moving to the right. That is,

$$\text{Molecules that hit the wall} = \frac{1}{2}\frac{N}{V}v_x \Delta t\, A$$

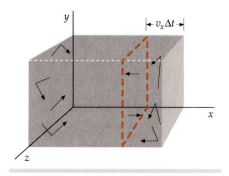

FIGURE 17-10 Gas molecules in a rectangular container. In a time interval Δt, the molecules closer to the right wall than the distance $v_x \Delta t$ will hit the right wall if they are moving to the right.

The x component of momentum of a molecule is $+mv_x$ before it hits the wall, and $-mv_x$ after an elastic collision with the wall. The change in momentum has the magnitude $2mv_x$. The magnitude of the total change in momentum $|\Delta \vec{p}|$ of all molecules during a time interval Δt is $2mv_x$ times the number of molecules that hit the wall during this interval:

$$|\Delta \vec{p}| = (2mv_x) \times \left(\frac{1}{2}\frac{N}{V}v_x \Delta t\, A\right) = \frac{N}{V}mv_x^2 A\, \Delta t \qquad \text{17-16}$$

The magnitude of the force exerted by the wall on the molecules and by the molecules on the wall is $|\Delta \vec{p}|/\Delta t$. The pressure is the magnitude of this force divided by the area A:

$$P = \frac{F}{A} = \frac{1}{A}\frac{|\Delta \vec{p}|}{\Delta t} = \frac{N}{V}mv_x^2$$

or

$$PV = Nmv_x^2 \qquad \text{17-17}$$

To allow for the fact that all the molecules in a container do not have the same speed, we merely replace v_x^2 with the average value $(v_x^2)_{av}$. Then, writing Equation 17-17 in terms of the kinetic energy $\frac{1}{2}mv_x^2$ associated with motion along the x axis, we have

$$PV = 2N(\tfrac{1}{2}mv_x^2)_{av} \qquad \text{17-18}$$

The Molecular Interpretation of Temperature

Comparing Equation 17-18 with Equation 17-7, which was obtained experimentally for any gas at very low densities, we can see that

$$PV = NkT = 2N(\tfrac{1}{2}mv_x^2)_{av}$$

or

$$\left(\tfrac{1}{2}mv_x^2\right)_{av} = \tfrac{1}{2}kT \qquad\qquad\qquad 17\text{-}19$$

THE AVERAGE ENERGY ASSOCIATED WITH MOTION IN THE x DIRECTION

Thus, the average kinetic energy associated with motion along the x axis is $\tfrac{1}{2}kT$. But there is nothing special about the x direction. On the average,

$$\left(v_x^2\right)_{av} = \left(v_y^2\right)_{av} = \left(v_z^2\right)_{av} \qquad\qquad\qquad 17\text{-}20$$

and

$$\left(v^2\right)_{av} = \left(v_x^2\right)_{av} + \left(v_y^2\right)_{av} + \left(v_z^2\right)_{av} = 3\left(v_x^2\right)_{av}$$

Writing $\left(v_x^2\right)_{av} = \tfrac{1}{3}\left(v^2\right)_{av}$ and K_{av} for the average translational kinetic energy of the molecules,[†] Equation 17-19 becomes

$$K_{av} = \left(\tfrac{1}{2}mv^2\right)_{av} = \tfrac{3}{2}kT \qquad\qquad\qquad 17\text{-}21$$

AVERAGE KINETIC ENERGY OF A MOLECULE

The absolute temperature is thus a measure of the average translational kinetic energy of the molecules. The total translational kinetic energy of n moles of a gas containing N molecules is

$$K = N\left(\tfrac{1}{2}mv^2\right)_{av} = \tfrac{3}{2}NkT = \tfrac{3}{2}nRT \qquad\qquad\qquad 17\text{-}22$$

KINETIC ENERGY OF TRANSLATION FOR n MOLES OF A GAS

where we've used $Nk = nN_Ak = nR$. Thus, the translational kinetic energy is $\tfrac{3}{2}kT$ per molecule and $\tfrac{3}{2}RT$ per mole.

We can use these results to estimate the order of magnitude of the speeds of the molecules in a gas. The average value of v^2 is, by Equation 17-21,

$$\left(v^2\right)_{av} = \frac{3kT}{m} = \frac{3N_AkT}{N_Am} = \frac{3RT}{M}$$

where $M = N_Am$ is the molar mass. The square root of $\left(v^2\right)_{av}$ is referred to as the **root mean square** (rms) speed:

$$v_{rms} = \sqrt{\left(v^2\right)_{av}} = \sqrt{\frac{3kT}{m}} = \sqrt{\frac{3RT}{M}} \qquad\qquad\qquad 17\text{-}23$$

Note that Equation 17-23 is similar to Equation 15-5 for the speed of sound in a gas:

$$v_{sound} = \sqrt{\frac{\gamma RT}{M}} \qquad\qquad\qquad 17\text{-}24$$

where $\gamma = 1.4$ for air. This is not surprising since a sound wave in air is a pressure disturbance propagated by collisions between air molecules.

[†] We include the word *translational* because the molecules may also have rotational or vibrational kinetic energy. Only the translational kinetic energy is relevant to the calculation of the pressure exerted by a gas on the walls of its container.

The rms Speed of Gas Molecules   **EXAMPLE 17-7**

Oxygen gas (O_2) has a molar mass of about 32 g/mol and hydrogen gas (H_2) has a molar mass of about 2 g/mol. Calculate (*a*) the rms speed of an oxygen molecule when the temperature is 300 K and (*b*) the rms speed of a hydrogen molecule at the same temperature.

PICTURE THE PROBLEM (*a*) We find v_{rms} using Equation 17-23. For the units to work out right, we use $R = 8.31$ J/(mol·K), and we express the molecular mass of O_2 in kg/mol. (*b*) Since v_{rms} is proportional to $1/\sqrt{M}$, and the molar mass of hydrogen is one-sixteenth that of oxygen, the rms speed of hydrogen is 4 times that of oxygen.

(*a*) Substitute the given values into Equation 17-23:

$$v_{rms} = \sqrt{\frac{3RT}{M}} = \sqrt{\frac{3(8.31\ \text{J/[mol·K]})(300\ \text{K})}{32 \times 10^{-3}\ \text{kg/mol}}}$$

$$= \boxed{483\ \text{m/s}}$$

(*b*) Use $v_{rms} \propto 1/\sqrt{M}$ to calculate v_{rms} for hydrogen:

$$\frac{v_{rms}(H_2)}{v_{rms}(O_2)} = \frac{\sqrt{M_{O_2}}}{\sqrt{M_{H_2}}}$$

so

$$v_{rms}(H_2) = \sqrt{\frac{M_{O_2}}{M_{H_2}}}\, v_{rms}(O_2) = \sqrt{\frac{32\ \text{g/mol}}{2\ \text{g/mol}}}\,(483\ \text{m/s})$$

$$= \boxed{1930\ \text{m/s}}$$

REMARKS The rms speed of oxygen molecules is 483 m/s = 1080 mi/h, about 1.4 times the speed of sound in air, which at 300 K is about 347 m/s.

EXERCISE Find the rms speed of a nitrogen molecule ($M = 28$ g/mol) at 300 K. (*Answer* 516 m/s)

The Equipartition Theorem

We have seen that the average kinetic energy associated with translational motion in any direction is $\frac{1}{2}kT$ per molecule (Equation 17-21) (or, equivalently, $\frac{1}{2}RT$ per mole), where k is Boltzmann's constant. If the energy of a molecule associated with its motion in one direction is momentarily increased, say, by a collision between the molecule and a moving piston during a compression, collisions between that molecule and other molecules will quickly redistribute the added energy. When the gas is again in equilibrium, energy will be equally partitioned among the translational kinetic energies associated with motion in the x, y, and z directions. This sharing of the energy equally among the three terms in the translational kinetic energy is a special case of the **equipartition theorem,** a result that follows from classical statistical mechanics. Each component of position and momentum (including angular position and angular momentum) that appears as a squared term in the expression for the energy of the system is called a **degree of freedom.** Typical degrees of freedom are associated with the kinetic energy of translation, rotation, and vibration, and with the potential energy of vibration. The equipartition theorem states that:

> When a substance is in equilibrium, there is an average energy of $\frac{1}{2}kT$ per molecule or $\frac{1}{2}RT$ per mole associated with each degree of freedom.

EQUIPARTITION THEOREM

In Chapter 18 we will use the equipartition theorem to relate the measured heat capacities of gases to their molecular structure.

Mean Free Path

The average speed of molecules in a gas at normal pressures is several hundred meters per second, yet if somebody across the room from you opens a perfume bottle, you don't detect the odor for several minutes. The reason for the time delay is that the perfume molecules do not travel directly toward you, but instead travel a zigzag path due to collisions with the air molecules. The average distance λ traveled by a molecule between collisions is called its **mean free path.** (The reason you smell the perfume at all is due to air currents (convection). The time for a perfume molecule to diffuse across a room is of the order of weeks.)

The mean free path of a gas molecule is related to its size, to the size of the surrounding gas molecules, and to the density of the gas. Consider one gas molecule of radius r_1 moving with speed v through a region of stationary molecules (Figure 17-11). The moving molecule will collide with another molecule of radius r_2 if the centers of the two molecules come within a distance $d = r_1 + r_2$ from each other. (If all the molecules are the same type, then d is the molecular diameter.) As the molecule moves, it will collide with any molecule whose center is in a circle of radius d (Figure 17-12). In some time t, the molecule moves a distance vt and collides with every molecule in the cylindrical volume $\pi d^2 vt$. The number of molecules in this volume is $n_V \pi d^2 vt$, where $n_V = N/V$ is the number of molecules per unit volume. (After each collision, the direction of the molecule changes, so the path actually zigs and zags.) The total path length divided by the number of collisions is the mean free path:

$$\lambda = \frac{vt}{n_V \pi d^2 vt} = \frac{1}{n_V \pi d^2}$$

This calculation of the mean free path assumes that all but one of the gas molecules are stationary, which is not a realistic situation. When the motion of all the molecules is taken into account, the correct expression for the mean free path is given by

$$\lambda = \frac{1}{\sqrt{2}\, n_V \pi d^2} \qquad\qquad 17\text{-}25$$

The average time between collisions is called the collision time τ. The reciprocal of the collision time, $1/\tau$, is equal to the average number of collisions per second, or the collision frequency. If v_{av} is the average speed, then the average distance traveled between collisions is

$$\lambda = v_{av}\tau \qquad\qquad 17\text{-}26$$

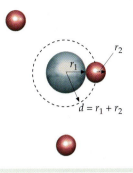

FIGURE 17-11 Model of a molecule (center sphere) moving in a gas. The molecule of radius r_1 will collide with any molecule of radius r_2 if their centers are a distance $d = r_1 + r_2$ apart, which is any molecule whose center is on a sphere of radius $d = r_1 + r_2$ centered about the molecule.

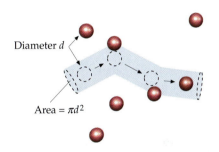

FIGURE 17-12 Model of a molecule moving with speed v in a gas of similar molecules. The motion is shown during time t. The molecule of diameter d will collide with any similar molecule whose center is in a cylinder of volume $\pi d^2 vt$. In this picture, all collisions are assumed to be elastic and all the molecules but one are assumed to be at rest.

MEAN FREE PATH OF A CO MOLECULE IN AIR   **E X A M P L E   1 7 - 8**   **Put It in Context**

The local poison control center wants to know more about carbon monoxide and how it spreads through a room. You are asked (a) to calculate the mean free path of a carbon monoxide molecule and (b) to estimate the mean time between collisions. The molecular mass of carbon monoxide is 28 g/mol. Assume that the CO molecule is traveling in air at 300K and 1 atm, and that the diameters of a CO molecule and air molecules are approximately 3.75×10^{-10} m.

PICTURE THE PROBLEM (a) Since d is given, we can find λ from $\lambda = 1/(\sqrt{2}\, n_V \pi d^2)$ using the ideal gas law to find $n_V = N/V$ (b) We can estimate the collision time by using v_{rms} for the average speed.

(a) 1. Write λ in terms of the number density n_V and the molecular diameter d:

$$\lambda = \frac{1}{\sqrt{2}\, n_V \pi d^2}$$

2. Use the equation $PV = NkT$ to calculate $n_V = N/V$:

$$n_V = \frac{N}{V} = \frac{P}{kT} = \frac{101.3 \times 10^3 \text{ Pa}}{(1.38 \times 10^{-23} \text{ J/K})(300 \text{ K})}$$

$$= 2.45 \times 10^{25} \text{ molecules/m}^3$$

3. Substitute this value of n_V and the given value of d to calculate λ:

$$\lambda = \frac{1}{\sqrt{2}\, n_V \pi d^2}$$

$$= \frac{1}{\sqrt{2}\,(2.45 \times 10^{25}/\text{m}^3)\,\pi(3.75 \times 10^{-10} \text{ m}^2)^2}$$

$$= \boxed{6.53 \times 10^{-8} \text{ m}}$$

(b) 1. Write τ in terms of the mean free path λ:

$$\tau = \frac{\lambda}{v_{av}}$$

2. Estimate v_{av} by calculating v_{rms}:

$$v_{rms} = \sqrt{\frac{3RT}{M}} = \sqrt{\frac{3(8.31 \text{ J/[mol·K]})(300 \text{ K})}{28 \times 10^{-3} \text{ kg/mol}}}$$

$$= 517 \text{ m/s}$$

3. Use $v_{av} \approx v_{rms}$ to estimate τ:

$$\tau = \frac{\lambda}{v_{av}} = \frac{6.53 \times 10^{-8} \text{ m}}{517 \text{ m/s}} = \boxed{1.26 \times 10^{-10} \text{ s}}$$

REMARKS Note that we put atmospheric pressure in pascals to get the proper units for λ. The mean free path is about 200 times the diameter of the molecule, and the collision frequency is about $1/\tau \approx 8 \times 10^9$ collisions per second.

*The Distribution of Molecular Speeds

We would not expect all of the molecules in a gas to have the same velocity. The calculation of the pressure of a gas allows us to calculate the square of the average speed and therefore the average energy of molecules in a gas, but it does not yield any details about the *distribution* of molecular velocities. Before we consider this problem, we will discuss the idea of distribution functions in general with some elementary examples from common experience.

Distribution Functions Suppose that a teacher gave a 25-point quiz to a large number N of students. To describe the results, the teacher might give the average score, but this would not be a complete description. If all the students received a score of 12.5, for example, that would be quite different from half the students receiving 25 and the other half zero, but the average score would be the same in both cases. A complete description of the results would be to give the number of students n_i who received a score s_i for all the scores received. Alternatively, one could give the fraction of the students $f_i = n_i/N$ who received the score s_i. Both n_i and f_i, which are functions of the variable s, are called **distribution functions.** The fractional distribution is somewhat more convenient to use. The probability that one of the N students selected at random received the score s_i equals the total number of students who received that score n_i divided by N, that is, the probability equals f_i. Note that

$$\sum_i f_i = \sum_i \frac{n_i}{N} = \frac{1}{N} \sum_i n_i$$

and since $\Sigma n_i = N$,

$$\sum_i f_i = 1 \qquad 17\text{-}27$$

Equation 17-27 is called the **normalization condition** for fractional distributions.

To find the average score, we add all the scores and divide by N. Since each score s_i was obtained by $n_i = Nf_i$ students, this is equivalent to

$$s_{av} = \frac{1}{N} \sum_i n_i s_i = \sum_i s_i f_i \qquad 17\text{-}28$$

Similarly, the average of any function $g(s)$ is defined by

$$g(s)_{av} = \frac{1}{N} \sum_i g(s_i) n_i = \sum_i g(s_i) f_i \qquad 17\text{-}29$$

In particular, the square of the average score of the square of the scores is

$$(s^2)_{av} = \frac{1}{N} \sum_i s_i^2 n_i = \sum_i s_i^2 f_i \qquad 17\text{-}30$$

The square root of $(s^2)_{av}$ is called the **root mean square score** or rms score. A possible distribution function is shown in Figure 17-13. For this distribution, the most probable score (that obtained by the most students) is 16, the average score is 14.2, and the rms score is 14.9.

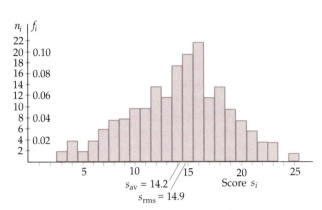

FIGURE 17-13 Grade distribution for a 25-point quiz given to 200 students. n_i is the number of students receiving grade s_i and $f_i = n_i/N$ is the fraction of students receiving grade s_i.

MAKING THE GRADE   **EXAMPLE 17-9**

Fifteen students took a 25-point quiz. Their scores were 25, 22, 22, 20, 20, 20, 18, 18, 18, 18, 18, 15, 15, 15, and 10. Find the average score and the rms score.

PICTURE THE PROBLEM The distribution function for this problem is $n_{25} = 1$, $n_{22} = 2$, $n_{20} = 3$, $n_{18} = 5$, $n_{15} = 3$, and $n_{10} = 1$. To find the average score, we use Equation 17-28. To find the rms score, we use Equation 17-30 and then take the square root.

1. By definition, s_{av} is:

$$s_{av} = \frac{1}{N} \sum_i n_i s_i$$

$$= \frac{1}{15}\big[1(25) + 2(22) + 3(20) + 5(18) + 3(15) + 1(10)\big]$$

$$= \frac{1}{15}(274) = 18.3$$

2. To calculate s_{rms}, first find the average of s^2:

$$(s^2)_{av} = \frac{1}{N} \sum_i n_i s_i^2$$

$$= \frac{1}{15}\big[1(25)^2 + 2(22)^2 + 3(20)^2 + 5(18)^2 + 3(15)^2 + 1(10)^2\big]$$

$$= \frac{1}{15}(5188) = 346$$

3. Take the square root of $(s^2)_{av}$:

$$s_{rms} = \sqrt{(s^2)_{av}} = \boxed{18.6}$$

Now consider the case of a continuous distribution, for example, the distribution of heights in a population. For any finite number N, the number of people who are *exactly* 2 m tall is zero. If we assume that height can be determined to any desired accuracy, there are an infinite number of possible heights, so the probability is zero that anybody has any one particular (exact) height. Therefore, we divide the heights into intervals Δh (for example, Δh might be 1 cm or 0.5 cm) and ask what fraction of people has heights that fall in any particular interval. For very large N, this number is proportional to the size of the interval, provided the interval is sufficiently small. We define the distribution function $f(h)$ as the fraction of the number of people with heights in the interval between h and $h + \Delta h$. Then for N people, $Nf(h)\,\Delta h$ is the number of people whose height is between h and $h + \Delta h$. Figure 17-14 shows a possible height distribution.

The fraction of people with heights in a given interval Δh is the area $f(h)\,\Delta h$. If N is very large, we can choose Δh to be very small, and the histogram will approximate a continuous curve. We can therefore consider the distribution function $f(h)$ to be a continuous function, write the interval as dh, and replace the sums in Equations 17-27 through 17-30 by integrals:

$$\int f(h)\, dh = 1$$

$$h_{av} = \int h f(h)\, dh \qquad\qquad \text{17-32}$$

$$[g(h)]_{av} = \int g(h)\, f(h)\, dh \qquad\qquad \text{17-33}$$

where $g(h)$ is an arbitrary function of h. Thus,

$$(h^2)_{av} = \int h^2 f(h)\, dh \qquad\qquad \text{17-34}$$

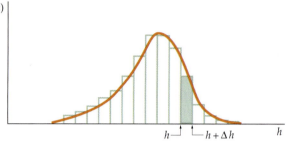

FIGURE 17-14 A possible height distribution function. The fraction of the number of heights between h and $h + \Delta h$ equals the shaded area $f(h)\,\Delta h$. The histogram can be approximated by a continuous curve as shown.

The probability of a person selected at random having a height between h and $h + dh$ is $f(h)dh$. A useful quantity characterizing a distribution is the **standard deviation** σ defined by

$$\sigma^2 = \left[(x - x_{av})^2\right]_{av} \qquad\qquad \text{17-35}a$$

Expanding the square on the right, we obtain

$$\sigma^2 = \left[x^2 - 2xx_{av} + x_{av}^2\right]_{av} = (x^2)_{av} - 2x_{av}x_{av} + x_{av}^2$$

or

$$\sigma^2 = (x^2)_{av} - x_{av}^2 \qquad\qquad \text{17-35}b$$

The standard deviation measures the spread of the values about the average value. For most distributions there will be few values that differ from x_{av} by more than a few multiples of σ. For the familiar bell-shaped distribution (called a normal distribution), about 68 percent of the values are expected to fall within $x_{av} \pm \sigma$.

In Example 17-7, we found that the rms value was greater than the average value. This is a general feature for any distribution (unless all the values are identical, in which case $x_{rms} = x_{av}$). We can see this from Equation 17-35b by noting that $x_{rms}^2 = (x^2)_{av}$. Then $\sigma^2 = (x^2)_{av} - x_{av}^2 = x_{rms}^2 - x_{av}^2$. Since σ^2 and x_{rms} are always positive, x_{rms} must always be greater than $|x_{av}|$.

For the familiar bell-shaped distribution (called a normal distribution), 68 percent of the values fall within $x_{av} \pm \sigma$, 95 percent fall within $x_{av} \pm 2\sigma$, and 99.7 percent fall within $x_{av} \pm 3\sigma$. This is known as the 68, 95, 99.7 rule.

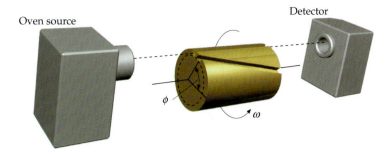

Oven source

Detector

ϕ

ω

FIGURE 17-15 Schematic diagram of the apparatus for determining the speed distribution of the molecules of a gas. A substance is vaporized in an oven and the vapor molecules are allowed to escape through a hole in the oven wall into a vacuum chamber. The molecules are collimated into a narrow beam by a series of slits (not shown). The beam is aimed at a detector that counts the number of molecules that are incident on it in a given period of time. A rotating cylinder stops most of the beam. Small slits in the cylinder (only one of which is depicted here) allow the passage of molecules that have a narrow range of speeds that is determined by the angular velocity of rotation of the cylinder. Varying the angular velocity of the cylinder and counting the number of molecules that reach the detector for each angular velocity give a measure of the number of molecules in each range of speeds.

The Maxwell–Boltzmann Distribution The distribution of the molecular speeds of a gas can be measured directly using the apparatus illustrated in Figure 17-15. In Figure 17-16, these speeds are shown for two different temperatures. The quantity $f(v)$ in Figure 17-16 is called the **Maxwell–Boltzmann speed distribution function**. In a gas of N molecules, the number with speeds in the range between v and $v + dv$ is dN, given by

$$dN = N f(v)\, dv \qquad\qquad 17\text{-}36$$

The fraction $dN/N = f(v)\, dv$ in a particular range dv is illustrated by the shaded region in the figure. The Maxwell–Boltzmann speed distribution function can be derived using statistical mechanics. The result is

$$f(v) = \frac{4}{\sqrt{\pi}}\left(\frac{m}{2kT}\right)^{3/2} v^2 e^{-mv^2/(2kT)} \qquad\qquad 17\text{-}37$$

MAXWELL–BOLTZMANN SPEED DISTRIBUTION FUNCTION

The most probable speed $v_{\max}$ is that speed for which $f(v)$ is maximum. It is left as a problem to show that

$$v_{\max} = \sqrt{\frac{2kT}{m}} = \sqrt{\frac{2RT}{M}} \qquad\qquad 17\text{-}38$$

Comparing Equation 17-38 with Equation 17-23, we see that the most probable speed is slightly less than the rms speed.

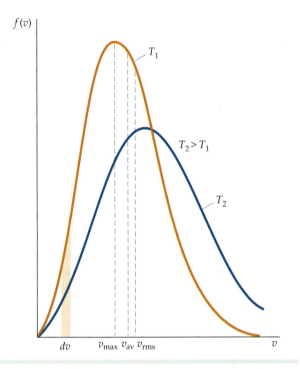

$f(v)$

T_1

$T_2 > T_1$

T_2

dv $v_{\max}$ v_{av} v_{rms} v

FIGURE 17-16 Distributions of molecular speeds in a gas at two temperatures, T_1 and $T_2 > T_1$. The shaded area $f(v)\, dv$ equals the fraction of the number of molecules having a particular speed in a narrow range of speeds dv. The mean speed v_{av} and the rms speed v_{rms} are both slightly greater than the most probable speed $v_{\max}$.

USING THE MAXWELL–BOLTZMANN DISTRIBUTION   **EXAMPLE 17-10**

Calculate the average value of v^2 for the molecules in a gas using the Maxwell–Boltzmann distribution function.

PICTURE THE PROBLEM The average value of v^2 is calculated from Equation 17-34 with v replacing h and $f(v)$ given by Equation 17-37.

1. By definition, $(v^2)_{av}$ is:

$$(v^2)_{av} = \int_0^\infty v^2 f(v)\, dv$$

2. Use Equation 17-37 for $f(v)$:

$$(v^2)_{av} = \int_0^\infty v^2 \frac{4}{\sqrt{\pi}} \left(\frac{m}{2kT}\right)^{3/2} v^2 e^{-mv^2/(2kT)} \, dv$$

$$= \frac{4}{\sqrt{\pi}} \left(\frac{m}{2kT}\right)^{3/2} \int_0^\infty v^4 e^{-mv^2/(2kT)} \, dv$$

3. The integral in step 2 can be found in standard integral tables:

$$\int_0^\infty v^4 e^{-mv^2/(2kT)} \, dv = \frac{3}{8} \sqrt{\pi} \left(\frac{2kT}{m}\right)^{5/2}$$

4. Use this result to calculate $(v^2)_{av}$:

$$(v^2)_{av} = \frac{4}{\sqrt{\pi}} \left(\frac{m}{2kT}\right)^{3/2} \frac{3}{8} \sqrt{\pi} \left(\frac{2kT}{m}\right)^{5/2} = \boxed{\frac{3kT}{m}}$$

■ **REMARKS** Note that our result agrees with $v_{rms} = \sqrt{3kT/m}$ from Equation 17-23.

In Example 17-6 we found that the rms speed of hydrogen molecules is about 1.93 km/s. This is about one-sixth of the escape speed at the surface of the earth, which we found to be 11.2 km/s in Section 11-3. So why is there no free hydrogen in the earth's atmosphere? As we can see from Figure 17-16, a considerable fraction of the molecules of a gas in equilibrium have speeds greater than the rms speed. When the rms speed of the molecules of a particular gas is as great as 15 to 20 percent of the escape speed for a planet, enough of the molecules have speeds greater than the escape speed so that the gas does not remain in the atmosphere of that planet very long before escaping. Thus, there is virtually no hydrogen gas in the earth's atmosphere. The rms speed of oxygen molecules, on the other hand, is about one-fourth that of hydrogen molecules, which makes it only about 4 percent of the escape speed at the surface of the earth. Therefore, only a negligible fraction of the oxygen molecules have speeds greater than the escape speed, and oxygen remains in the earth's atmosphere.

The Energy Distribution The Maxwell–Boltzmann speed distribution as given by Equation 17-37 can also be written as an energy distribution. We write the number of molecules with energy E in the range between E and $E + dE$ as

$$dN = NF(E) \, dE$$

where $F(E)$ is the energy distribution function. This will be the same number as given by Equation 17-37 if the energy E is related to the speed v by $E = \frac{1}{2}mv^2$. Then

$$dE = mv \, dv$$

and

$$Nf(v) \, dv = NF(E) \, dE$$

We can write

$$f(v) \, dv = Cv^2 e^{-mv^2/(2kT)} \, dv = Cve^{-E/(kT)} v \, dv = C\left(\frac{2E}{m}\right)^{1/2} e^{-E/(kT)} \frac{dE}{m}$$

where $C = (4/\sqrt{\pi})[m/(2kT)]^{3/2}$ (from Equation 17-37). The energy distribution function $F(E)$ is thus given by

$$F(E) = \frac{4}{\sqrt{\pi}} \left(\frac{m}{2kT}\right)^{3/2} \left(\frac{2}{m}\right)^{1/2} \frac{1}{m} E^{1/2} e^{-E/(kT)}$$

Simplifying, we obtain the **Maxwell–Boltzmann energy distribution function:**

$$F(E) = \frac{2}{\sqrt{\pi}}\left(\frac{1}{kT}\right)^{3/2} E^{1/2}e^{-E/(kT)} \qquad \text{17-39}$$

MAXWELL–BOLTZMANN ENERGY DISTRIBUTION FUNCTION

In the language of statistical mechanics, the energy distribution is considered to be the product of two factors: one, called the **density of states,** is proportional to $E^{1/2}$; the other is the probability of a state being occupied, which is $e^{-E/(kT)}$ and is called the **Boltzmann factor.**

SUMMARY

Topic	Relevant Equations and Remarks
1. Celsius and Fahrenheit Scales	On the Celsius scale, the ice point is defined to be 0°C and the steam point is 100°C. On the Fahrenheit scale, the ice point is 32°F and the steam point is 212°F. Temperatures on the Fahrenheit and Celsius scales are related by $$t_C = \tfrac{5}{9}(t_F - 32°) \qquad \text{17-2}$$
2. Gas Thermometers	Gas thermometers have the property that they all agree with each other in the measurement of any temperature as long as the density of the gas is very low. The ideal-gas temperature T is defined by $$T = \frac{273.16\ \text{K}}{P_3}P \qquad \text{17-4}$$ where P is the observed pressure of the gas in the thermometer and P_3 is the pressure when the thermometer is immersed in a water–ice–vapor bath at its triple point.
3. Kelvin Temperature Scale	The absolute temperature or temperature in kelvins is related to the Celsius temperature by $$T = t_C + 273.15\ \text{K} \qquad \text{17-5}$$
4. Ideal Gas	At low densities, all gases obey the ideal-gas law.
Equation of state	$PV = nRT$ 17-13
Universal gas constant	$R = N_Ak = 8.314\ \text{J/(mol·K)}$ $= 0.08206\ \text{L·atm/(mol·K)}$ 17-12
Boltzmann's constant	$k = 1.381 \times 10^{-23}\ \text{J/K} = 8.617 \times 10^{-5}\ \text{eV/K}$ 17-8
Avogadro's number	$N_A = 6.022 \times 10^{23}$ 17-9
Equation for a fixed amount of gas	A form of the ideal-gas law that is useful for solving problems involving a fixed amount of gas is $$\frac{P_2V_2}{T_2} = \frac{P_1V_1}{T_1} \qquad \text{17-14}$$

5. Kinetic Theory of Gases

Molecular interpretation of temperature	The absolute temperature T is a measure of the average molecular translational kinetic energy.
Equipartition theorem	When a system is in equilibrium, there is an average energy of $\frac{1}{2}kT$ per molecule (or $\frac{1}{2}RT$ per mole) associated with each degree of freedom.

Average kinetic energy   For an ideal gas, the average translational kinetic energy of the molecules is

$$K_{av} = (\tfrac{1}{2}mv^2)_{av} = \tfrac{3}{2}kT \qquad \textbf{17-21}$$

Total kinetic energy   The total translational kinetic energy of n moles of a gas containing N molecules is given by

$$K = N(\tfrac{1}{2}mv^2)_{av} = \tfrac{3}{2}NkT = \tfrac{3}{2}nRT \qquad \textbf{17-22}$$

rms speed of molecules   The rms speed of a molecule of a gas is related to the absolute temperature by

$$v_{rms} = \sqrt{(v^2)_{av}} = \sqrt{\frac{3kT}{m}} = \sqrt{\frac{3RT}{M}} \qquad \textbf{17-23}$$

where m is the mass of the molecule and M is the molar mass.

Mean free path   The mean free path of a molecule is related to its diameter d and the number of molecules per unit volume n_V by

$$\lambda = \frac{1}{\sqrt{2}\,n_V \pi d^2} \qquad \textbf{17-25}$$

*6. Maxwell–Boltzmann Distribution
$$f(v) = \frac{4}{\sqrt{\pi}}\left(\frac{m}{2kT}\right)^{3/2} v^2 e^{-mv^2/(2kT)} \qquad \textbf{17-37}$$

Energy distribution
$$F(E) = \frac{2}{\sqrt{\pi}}\left(\frac{1}{kT}\right)^{3/2} E^{1/2} e^{-E/(kT)} \qquad \textbf{17-39}$$

PROBLEMS

- • Single-concept, single-step, relatively easy
- •• Intermediate-level, may require synthesis of concepts
- ••• Challenging
- **SSM** Solution is in the *Student Solutions Manual*
- **iSOLVE** Problems available on iSOLVE online homework service
- **iSOLVE** ✓ These "Checkpoint" online homework service problems ask students additional questions about their confidence level, and how they arrived at their answer

In a few problems, you are given more data than you actually need; in a few other problems, you are required to supply data from your general knowledge, outside sources, or informed estimates.

Conceptual Problems

1 • **SSM** True or false:

(a) Two objects in thermal equilibrium with each other must be in thermal equilibrium with a third object.

(b) The Fahrenheit and Celsius temperature scales differ only in the choice of the zero temperature.

(c) The kelvin is the same size as the Celsius degree.

(d) All thermometers give the same result when measuring the temperature of a particular system.

2 • How can you determine if two bodies are in thermal equilibrium with each other if it is impossible to put them into thermal contact with each other?

3 • "One day I woke up and it was 20°F in my bedroom," said Mert to his old friend Mort. "That's nothing," replied Mort. "My room was once −5°C." Which room was colder?

4 •• Two identical vessels contain different ideal gases at the same pressure and temperature. It follows that (a) the number of gas molecules is the same in both vessels, (b) the total mass of gas is the same in both vessels, (c) the average speed of the gas molecules is the same in both vessels, (d) none of these answers is correct.

5 •• Figure 17-17 shows a plot of volume versus temperature for a process that takes an ideal gas from point A to point B. What happens to the pressure of the gas?

FIGURE 17-17
Problem 5

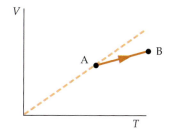

6 •• **SSM** Figure 17-18 shows a plot of pressure versus temperature for a process that takes an ideal gas from point A to point B. What happens to the volume of the gas?

FIGURE 17-18
Problem 6

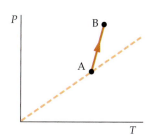

7 • True or false: The absolute temperature of a gas is a measure of the average translational kinetic energy of the gas molecules.

8 • By what factor must the absolute temperature of a gas be increased to double the rms speed of its molecules?

9 • How does the average translational kinetic energy of a molecule of a gas change if the pressure is doubled while the volume is kept constant? If the volume is doubled while the pressure is kept constant?

10 •• A vessel holds an equal number of moles of helium and methane, CH_4. The ratio of the rms speeds of the helium atoms to the CH_4 molecules is (a) 1, (b) 2, (c) 4, (d) 16.

11 • True or false: If the pressure of a gas increases, the temperature must increase.

12 • Why might the Celsius and Fahrenheit scales be more convenient than the absolute scale for ordinary, non-scientific purposes?

13 • **SSM** The temperature of the interior of the sun is said to be about 10^7 degrees. Do you think that this is degrees Celsius or kelvins, or does it matter?

14 • **i·SOLVE** ✓ If the temperature of an ideal gas is doubled while maintaining constant pressure, the average speed of the molecules (a) remains constant, (b) increases by a factor of 4, (c) increases by a factor of 2, (d) increases by a factor of $\sqrt{2}$.

15 • **i·SOLVE** ✓ If both temperature and volume of an ideal gas are halved, the pressure (a) diminishes by a factor of 2, (b) remains constant, (c) increases by a factor of 2, (d) diminishes by a factor of $\sqrt{2}$.

16 • The average translational kinetic energy of the molecules of an ideal gas depends on (a) the number of moles of the gas and its temperature, (b) the pressure of the gas and its temperature, (c) the pressure of the gas only, (d) the temperature of the gas only.

17 • If a vessel contains equal amounts, by weight, of helium and argon, which of the following are true?

(a) The pressure exerted by the two gases on the walls of the container is the same.
(b) The average speed of a helium atom is the same as that of an argon atom.
(c) The number of helium atoms and argon atoms in the vessel are equal.
(d) None of these statements is correct.

18 •• Two rooms, A and B, have equal volumes and are connected by an open door. Room A, which is air-conditioned, is 5C° cooler than room B. Which room has more air in it?

19 • Two different gases are at the same temperature. What can you say about the rms speeds of the gas molecules? What can you say about the average kinetic energies of the molecules?

20 •• Explain in terms of molecular motion why the pressure on the walls of a container increases when a gas is heated at constant volume.

21 •• **SSM** Explain in terms of molecular motion why the pressure on the walls of a container increases when the volume of a gas is reduced at constant temperature.

22 •• Oxygen has a molar mass of 32 g/mol and nitrogen has a molar mass of 28 g/mol. The oxygen and nitrogen molecules in a room have

(a) equal average kinetic energies, but the oxygen molecules are faster.
(b) equal average kinetic energies, but the oxygen molecules are slower.
(c) equal average kinetic energies and speeds.
(d) equal average speeds, but the oxygen molecules have a higher average kinetic energy.
(e) equal average speeds, but the oxygen molecules have a lower average kinetic energy.
(f) None of these answers is correct.

23 •• As any low-temperature physicist knows, liquid nitrogen is relatively cheap, while liquid helium is very expensive. One reason for this is that while nitrogen is the most common constituent of the atmosphere, helium is only found sealed in underground pockets of rock. Use ideas from this chapter to explain why this is true.

Estimation and Approximation

24 •• [SSM] A stoppered test tube that has a volume of 10 ml has 1 ml of water at its bottom and is at a temperature of 100°C and initially at a pressure of 1 atm (1.01×10^5 N/m²). Estimate the pressure inside the test tube when the water is completely boiled away.

25 ••• In Chapter 11, we found that the escape speed at the surface of a planet of radius R is $v_e = \sqrt{2gR}$, where g is the acceleration due to gravity at the surface of the planet. If the rms speed of a gas is greater than about 15 to 20 percent of the escape velocity of a planet, virtually all of the molecules of that gas will escape the atmosphere of the planet.

(a) At what temperature is v_{rms} for O_2 equal to 15 percent of the escape speed for the earth?
(b) At what temperature is v_{rms} for H_2 equal to 15 percent of the escape speed for the earth?
(c) Temperatures in the upper atmosphere reach 1000 K. How does this account for the low abundance of hydrogen in the earth's atmosphere?
(d) Compute the temperatures for which the rms speeds of O_2 and H_2 are equal to 15 percent of the escape velocity at the surface of the moon, where g is about one-sixth of its value on earth and $R = 1738$ km. How does this account for the absence of an atmosphere on the moon?

26 •• The escape velocity on Mars is 5.0 km/s and the surface temperature is typically 0°C. Calculate the rms speeds for (a) H_2, (b) O_2, and (c) CO_2 at this temperature. (d) Based on the criterion given in the chapter, are H_2, O_2, and CO_2 likely to be found in the atmosphere of Mars?

27 •• [SSM] Repeat Problem 26 for Jupiter, whose escape velocity is 60 km/s and whose temperature is typically −150°C.

Jupiter as seen from about twelve million miles. Because the escape speed at the surface of Jupiter is about 600 km/s, Jupiter easily retains hydrogen in its atmosphere.

Temperature Scales

28 • A certain ski wax is rated for use between −12 and −7°C. What is this temperature range on the Fahrenheit scale?

29 • [ISOLVE] The melting point of gold (Au) is 1945.4°F. Express this temperature in degrees Celsius.

30 • [SSM] [ISOLVE] What is the Celsius temperature corresponding to the normal temperature of the human body, 98.6°F?

31 • [ISOLVE✓] The length of the column of mercury in a thermometer is 4.0 cm when the thermometer is immersed in ice water and 24.0 cm when the thermometer is immersed in boiling water. (a) What should be the length at room temperature, 22.0°C? (b) If the mercury column is 25.4 cm long when the thermometer is immersed in a chemical solution, what is the temperature of the solution?

32 • The temperature of the interior of the sun is about 10^7 K. What is this temperature on (a) the Celsius scale and (b) the Fahrenheit scale?

33 • The boiling point of nitrogen N_2 is 77.35 K. Express this temperature in degrees Fahrenheit.

34 • [ISOLVE] The pressure of a constant-volume gas thermometer is 0.400 atm at the ice point and 0.546 atm at the steam point. (a) When the pressure is 0.100 atm, what is the temperature? (b) What is the pressure at 444.6°C, the boiling point of sulfur?

35 • [SSM] [ISOLVE] A constant-volume gas thermometer reads 50 torr at the triple point of water. (a) What will be the pressure when the thermometer measures a temperature of 300 K? (b) What ideal-gas temperature corresponds to a pressure of 678 torr?

36 • A constant-volume gas thermometer has a pressure of 30 torr when it reads a temperature of 373 K. (a) What is its triple-point pressure P_3? (b) What temperature corresponds to a pressure of 0.175 torr?

37 • At what temperature do the Fahrenheit and Celsius temperature scales give the same reading?

38 • [ISOLVE] Sodium melts at 371 K. What is the melting point of sodium on the Celsius and Fahrenheit temperature scales?

39 • The boiling point of oxygen at 1 atm is 90.2 K. What is the boiling point of oxygen on the Celsius and Fahrenheit scales?

40 •• On the Réaumur temperature scale, the melting point of ice is 0°R and the boiling point of water is 80°R. Derive expressions for converting temperatures on the Réaumur scale to the Celsius and Fahrenheit scales.

41 ••• [SSM] A thermistor is a solid-state device whose resistance varies greatly with temperature. Its temperature dependence is given approximately by $R = R_0 e^{B/T}$, where R is in ohms (Ω), T is in kelvins, and R_0 and B are constants that can be determined by measuring R at calibration points such as the ice point and the steam point. (a) If $R = 7360\ \Omega$ at the ice point and $153\ \Omega$ at the steam point, find R_0 and B. (b) What is the resistance of the thermistor at $t = 98.6°F$? (c) What is the rate of change of the resistance with temperature (dR/dT) at the ice point and the steam point? (d) At which temperature is the thermistor most sensitive?

The Ideal-Gas Law

42 • ✔ A gas is kept at constant pressure. If its temperature is changed from 50 to 100°C, by what factor does the volume change?

43 • ✔ A 10-L vessel contains gas at a temperature of 0°C and a pressure of 4 atm. How many moles of gas are in the vessel? How many molecules?

44 •• ✔ A pressure as low as 1×10^{-8} torr can be achieved using an oil diffusion pump. How many molecules are there in 1 cm^3 of a gas at this pressure if its temperature is 300 K?

45 •• SSM You copy the following paragraph from a Martian physics textbook: "1 *snorf* of an ideal gas occupies a volume of 1.35 *zaks*. At a temperature of 22 *glips*, the gas has a pressure of 12.5 *klads*. At a temperature of -10 *glips*, the same gas now has a pressure of 8.7 *klads*." Determine the temperature of absolute zero in *glips*.

46 •• A motorist inflates the tires of her car to a gauge pressure of 180 kPa on a day when the temperature is $-8.0°C$. When she arrives at her destination, the tire pressure has increased to 245 kPa. What is the temperature of the tires if we assume that (a) the tires do not expand or (b) that the tires expand by 7 percent?

47 •• A room is 6 m by 5 m by 3 m. (a) If the air pressure in the room is 1 atm and the temperature is 300 K, find the number of moles of air in the room. (b) If the temperature rises by 5 K and the pressure remains constant, how many moles of air leave the room?

48 •• SSM The boiling point of helium at 1 atm is 4.2 K. What is the volume occupied by helium gas due to evaporation of 10 g of liquid helium at 1 atm pressure and a temperature of (a) 4.2 K and (b) 293 K?

49 •• A container with a volume of 6.0 L holds 10 g of liquid helium. As the container warms to room temperature, what is the pressure exerted by the gas on its walls?

50 •• SSM An automobile tire is filled to a gauge pressure of 200 kPa when its temperature is 20°C. (Gauge pressure is the difference between the actual pressure and atmospheric pressure.) After the car has been driven at high speeds, the tire temperature increases to 50°C. (a) Assuming that the volume of the tire does not change and that air behaves as an ideal gas, find the gauge pressure of the air in the tire. (b) Calculate the gauge pressure if the volume of the tire expands by 10 percent.

51 •• Calculate the mass density of air at a temperature of 24°C and a pressure of 1 atm (1.01×10^5 N/m^2) using the ideal-gas law. Air is roughly 74 percent N$_2$ and 26 percent O$_2$.

52 •• ✔ A scuba diver is 40 m below the surface of a lake, where the temperature is 5°C. He releases an air bubble with a volume of 15 cm^3. The bubble rises to the surface, where the temperature is 25°C. What is the volume of the bubble right before it breaks the surface? *Hint: Remember that the pressure also changes.*

53 •• A hot-air balloon has a volume of 1.5 m^3 and is open at the bottom. If the air inside the balloon is at a temperature of 75°C, while the temperature of the air outside the balloon is 24°C, at a pressure of about 1 atm, what is the net force on the balloon and its contents? (Neglect the weight of the balloon itself.)

54 ••• A helium balloon is used to lift a load of 110 N. The weight of the balloon's skin is 50 N and the volume of the balloon when fully inflated is 32 m^3. The temperature of the air is 0°C and the atmospheric pressure is 1 atm. The balloon is inflated with sufficient helium gas that the net force on the balloon and its load is 30 N. Neglect changes of temperature with altitude. (a) How many moles of helium gas are contained in the balloon? (b) At what altitude will the balloon be fully inflated? (c) Does the balloon ever reach the altitude at which it is fully inflated? (d) If the answer to (c) is affirmative, what is the maximum altitude attained by the balloon?

Kinetic Theory of Gases

55 • SSM (a) Find v_{rms} for an argon atom if 1 mol of the gas is confined to a 1-L container at a pressure of 10 atm. (For argon, $M = 40 \times 10^{-3}$ kg/mol.) (b) Compare this with v_{rms} for a helium atom under the same conditions. (For helium, $M = 4 \times 10^{-3}$ kg/mol.)

56 • ✔ Find the total translational kinetic energy of 1 L of oxygen gas held at a temperature of 0°C and a pressure of 1 atm.

57 • Find the rms speed and the average kinetic energy of a hydrogen atom at a temperature of 10^7 K. (At this temperature, which is of the order of the temperature in the interior of a star, the hydrogen is ionized and consists of a single proton.)

58 • SSM In one model of a solid, the material is assumed to consist of a regular array of atoms in which each atom has a fixed equilibrium position and is connected by springs to its neighbors. Each atom can vibrate in the x, y, and z directions. The total energy of an atom in this model is

$$E = \tfrac{1}{2}mv_x^2 + \tfrac{1}{2}mv_y^2 + \tfrac{1}{2}mv_z^2 + \tfrac{1}{2}kx^2 + \tfrac{1}{2}ky^2 + kz^2$$

What is the average energy of an atom in the solid when the temperature is T? What is the total energy of 1 mol of such a solid?

59 • Show that the mean free path for a molecule in an ideal gas at temperature T and pressure P is given by

$$\lambda = \frac{kT}{\sqrt{2}P\pi d^2}$$

60 •• ✔ A pressure as low as $P = 7 \times 10^{-11}$ Pa has been obtained. Suppose that a chamber contains helium at this pressure and at room temperature (300 K). Estimate the mean free path λ and the collision time τ for helium in the chamber. Take the diameter of a helium molecule to be 10^{-10} m.

61 •• **SSM** Oxygen (O_2) is confined to a cubic container 15 cm on a side at a temperature of 300 K. Compare the average kinetic energy of a molecule of the gas to the change in its gravitational potential energy if it falls from the top of the container to the bottom.

*The Distribution of Molecular Speeds

62 •• Show that $f(v)$ given by Equation 17-37 is maximum when $v = \sqrt{2kT/m}$. Hint: Set $df/dv = 0$ and solve for v.

63 •• **SSM** $f(v)$ is defined in Equation 17-37. Because $f(v)\,dv$ gives the fraction of molecules that have speeds in the range dv, the integral of $f(v)\,dv$ over all the possible ranges of speeds must equal 1. Given the integral

$$\int_0^\infty v^2 e^{-av^2}\,dv = \frac{\sqrt{\pi}}{4}a^{-3/2}$$

show that $\int_0^\infty f(v)\,dv = 1$, where $f(v)$ is given by Equation 17-37.

64 •• Given the integral

$$\int_0^\infty v^3 e^{-av^2}\,dv = \frac{a^{-2}}{2}$$

calculate the average speed v_{av} of molecules in a gas using the Maxwell–Boltzmann distribution function.

65 •• **SSM** Current experiments in atomic trapping and cooling can create low-density gases of rubidium and other atoms with temperatures in the nanokelvin (10^{-9} K) range. These atoms are trapped and cooled using magnetic fields and lasers in ultrahigh vacuum chambers. One method that is used to measure the temperature of a trapped gas is to turn the trap off and measure the time it takes for molecules of the gas to fall a given distance! Consider a gas of rubidium atoms at a temperature of 120 nK. Calculate how long it would take an atom traveling at the rms speed of the gas to fall a distance of 10 cm if (*a*) it were initially moving directly downward and (*b*) if it were initially moving directly upward. Assume that the atom doesn't collide with any others along its trajectory.

General Problems

66 • At what temperature will the rms speed of an H_2 molecule equal 331 m/s?

67 •• (*a*) If 1 mol of a gas in a container occupies a volume of 10 L at a pressure of 1 atm, what is the temperature of the gas in kelvins? (*b*) The container is fitted with a piston so that the volume can change. When the gas is heated at constant pressure, it expands to a volume of 20 L. What is the temperature of the gas in kelvins? (*c*) The volume is fixed at 20 L, and the gas is heated at constant volume until its temperature is 350 K. What is the pressure of the gas?

68 •• **iSOLVE✓** A cubic metal box with sides of 20 cm contains air at a pressure of 1 atm and a temperature of 300 K. The box is sealed so that the volume is constant and it is heated to a temperature of 400 K. Find the net force on each wall of the box.

69 •• **SSM** Water, H_2O, can be converted into H_2 and O_2 gases by electrolysis. How many moles of these gases result from the electrolysis of 2 L of water?

70 •• A massless cylinder 40 cm long rests on a horizontal frictionless table. The cylinder is divided into two equal sections by a membrane. One section contains nitrogen and the other contains oxygen. The pressure of the nitrogen is twice that of the oxygen. How far will the cylinder move if the membrane is removed?

71 •• A cylinder contains a mixture of nitrogen gas (N_2) and hydrogen gas (H_2). At a temperature T_1 the nitrogen is completely dissociated but the hydrogen does not dissociate at all, and the pressure is P_1. If the temperature is doubled to $T_2 = 2T_1$, the pressure is tripled due to complete dissociation of hydrogen. If the mass of hydrogen is m_H, find the mass of nitrogen m_N.

72 •• **SSM** Three insulated vessels of equal volume V are connected by thin tubes that can transfer gas but do not transfer heat. Initially all vessels are filled with the same type of gas at a temperature T_0 and pressure P_0. Then the temperature in the first vessel is doubled and the temperature in the second vessel is tripled. The temperature in the third vessel remains unchanged. Find the final pressure P' in the system in terms of the initial pressure P_0.

73 •• A constant-volume gas thermometer with a triple-point pressure $P_3 = 500$ torr is used to measure the boiling point of some substance. When the thermometer is placed in thermal contact with the boiling substance, its pressure is 734 torr. Some of the gas in the thermometer is then allowed to escape so that its triple-point pressure is 200 torr. When it is again placed in thermal contact with the boiling substance, its pressure is 293.4 torr. Again, some of the gas is removed from the thermometer so that its triple-point pressure is 100 torr. When the thermometer is placed in thermal contact with the boiling substance once again, its pressure is 146.65 torr. Find the ideal-gas temperature of the boiling substance.

74 •• **SSM** The mean free path for O_2 molecules at a temperature of 300 K at 1 atm pressure ($p = 1.01 \times 10^5$ Pa) is $\lambda = 7.1 \times 10^{-8}$ m. Use this data to estimate the size of an O_2 molecule.

75 •• An experimental balloon contains hydrogen gas (H_2) at a temperature of 300 K and a pressure of 1 atm (1.01×10^5 N/m^2). (*a*) Calculate the mean-free path of a hydrogen molecule. Assume that a H_2 molecule is effectively spherical, with a mean diameter of 1.6×10^{-10} m. (*b*) Calculate the available volume per molecule (V/N), and find the average distance between each molecule and its nearest neighboring molecule (approximately the cube root of the available volume). Which is larger, the mean free path or the average nearest-neighbor distance between molecules?

76 ••• A cylinder 2.4 m tall is filled with 0.1 mol of an ideal gas at standard temperature and pressure (Figure 17-19). The top of the cylinder is then sealed with a piston whose mass is 1.4 kg and the piston is allowed to drop until it is in equilibrium. (*a*) Find the height of the piston, assuming that the temperature of the gas does not change as it is compressed. (*b*) Suppose that the piston is pushed down below its equilibrium position by a small amount and then released. Assuming that the temperature of the gas remains constant, find the frequency of vibration of the piston.

— 1.4 kg

FIGURE 17-19
Problem 76

77 ••• SSM The table below gives values of

$$\frac{4}{\sqrt{\pi}} \int_0^x z^2 e^{-z^2}\, dz$$

for different values of x. Use the table to answer the following questions: (*a*) For O_2 gas at 273 K, what fraction of molecules have speeds less than 400 m/s? (*b*) For the same gas, what percentage of molecules have speeds between 190 m/s and 565 m/s?

x	$\dfrac{4}{\sqrt{\pi}} \int_0^x z^2 e^{-z^2}\, dz$	x	$\dfrac{4}{\sqrt{\pi}} \int_0^x z^2 e^{-z^2}\, dz$
0.1	7.48×10^{-4}	0.7	0.194
0.2	5.88×10^{-3}	0.8	0.266
0.3	0.019	0.9	0.345
0.4	0.044	1.0	0.438
0.5	0.081	1.5	0.788
0.6	0.132	2.0	0.954

Heat and the First Law of Thermodynamics

THE WARM LEMONADE IN THIS PITCHER IS COOLED BY ADDING ICE. HEAT IS TRANSFERRED FROM THE LEMONADE TO THE ICE BECAUSE OF A DIFFERENCE IN TEMPERATURE.

? **How much ice should you add to a cup of lemonade to reduce the temperature of the lemonade from 20°C to 0°C? (See Example 18-4.)**

Heat is energy that is being transferred from one system to another because of a difference in temperature. In the seventeenth century, Galileo, Newton, and other scientists generally supported the theory of the ancient Greek atomists who considered thermal energy to be a manifestation of molecular motion. In the next century, methods were developed for making quantitative measurements of the amount of heat that leaves or enters an object, and it was found that if objects are in thermal contact, the amount of heat that leaves one object equals the amount that enters the other. This discovery led to the caloric theory of heat as a conserved material substance. In this theory, an invisible fluid called "caloric" flowed out of one object and into another and this "caloric" could be neither created nor destroyed.

The caloric theory reigned until the nineteenth century, when it was found that friction between objects could generate an unlimited amount of thermal energy, deposing of the idea that caloric was a substance present in a fixed amount.

The modern theory of heat did not emerge until the 1840s, when James Joule (1818–1889) demonstrated that the increase or decrease of a given amount of thermal energy was always accompanied by the decrease or increase of an equivalent quantity of mechanical energy. Thermal energy, therefore, is not itself conserved. Instead, thermal energy is a form of internal energy, and it is energy that is conserved. ➤ **In this chapter, we define heat capacity, and examine how heating a system can cause either a change in its temperature or a change in its phase. We then examine the relationship between heat conduction, work, and internal energy of a system and express the law of conservation of energy for the thermal systems as the first law of thermodynamics. Finally, we shall see how the heat capacity of a system is related to its molecular structure.**

18-1 Heat Capacity and Specific Heat

When energy is transferred to a substance by heating it, the temperature of the substance usually rises.[†] The amount of heat energy Q needed to raise the temperature of a substance is proportional to the temperature change and to the mass of the substance:

$$Q = C\Delta T = mc\Delta T \qquad \text{18-1}$$

where C is the **heat capacity,** which is defined as the amount of energy transferred via heating necessary to raise the temperature of a substance by one degree. The **specific heat** c is the heat capacity per unit mass:

$$c = \frac{C}{m} \qquad \text{18-2}$$

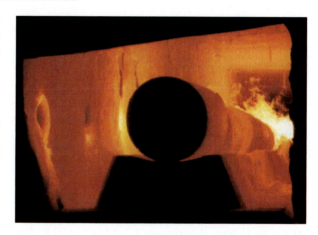

Steel ingots in a twin-tube tunnel furnace. The three 53-cm diameter carbon steel ingots seen here have been heated for about 7 hours to approximately 1340°C. Each 3200-kg ingot sits on a furnace car that transports it through the 81-m furnace, which is divided into twelve separate heating zones so that the temperature of the ingot is increased gradually to prevent cracking. The ingots, glowing a yellow-whitish color, exit the furnace to be milled into large, heavy-walled pipes.

The historical unit of heat energy, the **calorie,** was originally defined to be the amount of heat needed to raise the temperature of one gram of water one Celsius degree.[‡] Because we now recognize that heat is a measure of energy transfer, we can define the calorie in terms of the SI unit of energy, the joule:

$$1 \text{ cal} = 4.184 \text{ J} \qquad \text{18-3}$$

The U.S. customary unit of heat is the **Btu** (for British thermal unit), which was originally defined to be the amount of energy needed to raise the temperature of 1 pound of water by 1°F. The Btu is related to the calorie and to the joule by

$$1 \text{ Btu} = 252 \text{ cal} = 1.054 \text{ kJ} \qquad \text{18-4}$$

The original definition of the calorie implies that the specific heat of water is[§]

$$\begin{aligned} c_{\text{water}} &= 1 \text{ cal}/(\text{g}\cdot\text{C}°) = 1 \text{ kcal}/(\text{kg}\cdot\text{C}°) \\ &= 1 \text{ kcal}/(\text{kg}\cdot\text{K}) = 4.184 \text{ kJ}/(\text{kg}\cdot\text{K}) \end{aligned} \qquad \text{18-5}a$$

Similarly, from the definition of the Btu, the specific heat of water in U.S. customary units is

$$c_{\text{water}} = 1 \text{ Btu}/(\text{lb}\cdot\text{F}°) \qquad \text{18-5}b$$

---

† An exception occurs during a change in phase, as when water freezes or evaporates. Changes of phase are discussed in Section 18-2.

‡ The kilocalorie is then the amount of heat energy needed to raise the temperature of 1 kg of water by 1°C. The "calorie" used in measuring the energy equivalent of foods is actually the kilocalorie.

§ Careful measurement shows that the specific heat of water varies by about 1 percent over the temperature range from 0 to 100°C. We will usually neglect this small variation.

The heat capacity per mole is called the **molar specific heat** c',

$$c' = \frac{C}{n}$$

where n is the number of moles. Since $C = mc$, the molar specific heat c' and specific heat c are related by

$$c' = \frac{C}{n} = \frac{mc}{n} = Mc \qquad\qquad 18\text{-}6$$

where $M = m/n$ is the molar mass. Table 18-1 lists the specific heats and molar specific heats of some solids and liquids. Note that the molar heats of all the metals are about the same. We will discuss the significance of this in Section 18-7.

RAISING THE TEMPERATURE **EXAMPLE 18-1**

How much heat is needed to raise the temperature of 3 kg of copper by 20 C°?

PICTURE THE PROBLEM The amount of heat needed to raise the temperature of the substance (copper) is proportional to the temperature change (20 C°) and to the mass (3 kg) of the substance.

The required heat is given by Equation 18-1 with $c = 0.386$ kJ/kg·K from Table 18-1:

$$Q = mc\,\Delta T = (3 \text{ kg})(0.386 \text{ kJ/kg·K})(20 \text{ K})$$

$$= \boxed{23.2 \text{ kJ}}$$

REMARKS Note that we use $\Delta T = 20$ C° $= 20$ K. Alternatively, we could express the specific heat as 0.386 kJ/kg·C° and write the temperature change as 20 C°.

EXERCISE A 2-kg aluminum block is originally at 10°C. If 36 kJ of heat energy are added to the block, what is its final temperature? (*Answer* 30°C)

TABLE 18-1

Specific Heats and Molar Specific Heats of Some Solids and Liquids

Substance	c, kJ/kg·K	c, kcal/kg·K or Btu/lb·F°	c', J/mol·K
Aluminum	0.900	0.215	24.3
Bismuth	0.123	0.0294	25.7
Copper	0.386	0.0923	24.5
Glass	0.840	0.20	—
Gold	0.126	0.0301	25.6
Ice (−10°C)	2.05	0.49	36.9
Lead	0.128	0.0305	26.4
Silver	0.233	0.0558	24.9
Tungsten	0.134	0.0321	24.8
Zinc	0.387	0.0925	25.2
Alcohol (ethyl)	2.4	0.58	111
Mercury	0.140	0.033	28.3
Water	4.18	1.00	75.2

We see from Table 18-1 that the specific heat of water is considerably larger than that of other substances. Thus, water is an excellent material for storing thermal energy, as in a solar heating system. It is also an excellent coolant, as in a car engine.

Calorimetry

To measure the specific heat of an object we can first heat it to some known temperature, say the boiling point of water, then transfer it to a water bath of known mass and initial temperature, and, finally, measure the final equilibrium temperature of the object (and the bath). If the system is isolated from its surroundings (by insulating the container, for example), then the heat leaving the object will equal the heat entering the water and its container. This procedure is called **calorimetry**, and the insulated water container is called a **calorimeter**.

Large bodies of water, such as lakes or oceans, tend to moderate fluctuations of the air temperature nearby because the bodies of water can absorb or release large quantities of thermal energy while undergoing only very small changes in temperature.

Let m be the mass of the object, let c be its specific heat, and let T_{io} be its initial temperature. If T_f is the final temperature of the object in its water bath, the heat leaving the object is

$$Q_{out} = mc(T_{io} - T_f)$$

Similarly, if T_{iw} is the initial temperature of the water and container, and T_f is their final equilibrium temperature, then the heat absorbed by the water and container is

$$Q_{in} = m_w c_w (T_f - T_{iw}) + m_c c_c (T_f - T_{iw})$$

where m_w and $c_w = 4.18 \text{ kJ/kg·K}$ are the mass and specific heat of the water, and m_c and c_c are the mass and specific heat of the container. (Note that we have chosen the temperature differences so that the heat in and heat out are both positive quantities.) Setting these amounts of heat equal yields the specific heat c of the object:

$$Q_{out} = Q_{in} \qquad\qquad 18\text{-}7$$

$$mc\,(T_{io} - T_f) = m_w c_w (T_f - T_{iw}) + m_c c_c (T_f - T_{iw})$$

Because only temperature differences occur in Equation 18-7, and because the kelvin and Celsius degree are the same size, it doesn't matter whether we use kelvins or Celsius degrees.

MEASURING SPECIFIC HEAT   **E X A M P L E   1 8 - 2**

To measure the specific heat of lead, you heat 600 g of lead shot to 100°C and place it in an aluminum calorimeter of mass 200 g that contains 500 g of water initially at 17.3°C. If the final temperature of the mixture is 20.0°C, what is the specific heat of lead? [The specific heat of the aluminum container is 0.900 kJ/kg·K.]

PICTURE THE PROBLEM We set the heat leaving the lead equal to the heat entering the water and container and solve for the specific heat of lead c_{Pb}.

1. Write the heat leaving the lead in terms of its specific heat:

$$Q_{Pb} = m_{Pb}c_{Pb}|\Delta T_{Pb}|$$

2. Find the heat absorbed by the water:

$$Q_w = m_w c_w \Delta T_w$$

3. Find the heat absorbed by the container:

$$Q_c = m_c c_c \Delta T_c$$

4. Set the heat out equal to the heat in:

$$Q_{Pb} = Q_w + Q_c$$

$$m_{Pb}c_{Pb}|\Delta T_{Pb}| = m_w c_w \Delta T_w + m_c c_c \Delta T_c$$

where

$$\Delta T_c = \Delta T_w = 2.7 \text{ K and } |\Delta T_{Pb}| = 80 \text{ K}$$

5. Solve for c_{Pb}:

$$c_{Pb} = \frac{(m_w c_w + m_c c_c)\Delta T_w}{m_{Pb}|\Delta T_{Pb}|}$$

$$= \frac{[(0.5 \text{ kg})(4.18 \text{ kJ/kg·K}) + (0.2 \text{ kg})(0.9 \text{ kJ/kg·K})](2.7 \text{ K})}{(0.6 \text{ kg})(80 \text{ K})}$$

$$= \boxed{0.128 \text{ kJ/kg·K}}$$

REMARKS Note that the specific heat of lead is considerably less than that of water.

18-2 Change of Phase and Latent Heat

If heat is added to ice at 0°C, the temperature of the ice does not change. Instead, the ice melts. Melting is an example of a **phase change.** Common types of phase changes include fusion (liquid to solid), melting (solid to liquid), vaporization (liquid to vapor or gas), condensation (gas or vapor to liquid), and sublimation (solid directly to vapor, as when solid carbon dioxide [dry ice] changes to vapor). There are other types of phase changes as well, such as the change of a solid from one crystalline form to another. For example, carbon under intense pressure becomes a diamond.

Molecular theory can help us to understand why temperature remains constant during a phase change. The molecules in a liquid are close together and exert attractive forces on each other, whereas molecules in a gas are far apart. Because of this molecular attraction, it takes energy to remove molecules from a liquid to form a gas. Consider a pot of water sitting over a flame on the stove. At first, as the water is heated, the motion of its molecules increases and the temperature rises. When the temperature reaches the boiling point, the molecules can no longer increase their kinetic energy and remain in the liquid. As the liquid water vaporizes, the added heat energy is used to overcome the attractive forces between the water molecules as they spread farther apart in the gas phase. The added energy thus increases the potential energy of the molecules rather than their kinetic energy. Because temperature is a measure of the average translational *kinetic* energy of molecules, the temperature doesn't change.

For a pure substance, a change in phase at a given pressure occurs only at a particular temperature. For example, pure water at a pressure of 1 atm changes from solid to liquid at 0°C (the normal melting point of water) and from liquid to gas at 100°C (the normal boiling point of water).

The heat energy required to melt a substance of mass m with no change in its temperature is proportional to the mass of the substance:

$$Q_f = mL_f \qquad\qquad\qquad 18\text{-}8$$

Although melting indicates that the ice has experienced a change in phase, the temperature of the ice does not change.

where L_f is called the **latent heat of fusion** of the substance. At a pressure of 1 atm, the latent heat of fusion for water is 333.5 kJ/kg = 79.7 kcal/kg. If the phase change is from liquid to gas, the heat required is

$$Q_v = mL_v \qquad\qquad 18\text{-}9$$

where L_v is the **latent heat of vaporization.** For water at a pressure of 1 atm, the latent heat of vaporization is 2.26 MJ/kg = 540 kcal/kg. Table 18-2 gives the normal melting and boiling points, and the latent heats of fusion and vaporization at 1 atm, for various substances.

TABLE 18-2

Normal Melting Point (MP), Latent Heat of Fusion (L_f), Normal Boiling Point (BP), and Latent Heat of Vaporization (L_v) for Various Substances at 1 atm

Substance	MP, K	L_f, kJ/kg	BP, K	L_v, kJ/kg
Alcohol, ethyl	159	109	351	879
Bromine	266	67.4	332	369
Carbon dioxide	—	—	194.6[†]	573[†]
Copper	1356	205	2839	4726
Gold	1336	62.8	3081	1701
Helium	—	—	4.2	21
Lead	600	24.7	2023	858
Mercury	234	11.3	630	296
Nitrogen	63	25.7	77.35	199
Oxygen	54.4	13.8	90.2	213
Silver	1234	105	2436	2323
Sulfur	388	38.5	717.75	287
Water	273.15	333.5	373.15	2257
Zinc	692	102	1184	1768

† These values are for sublimation. Carbon dioxide does not have a liquid state at 1 atm.

CHANGING ICE INTO STEAM **EXAMPLE 18-3** Try It Yourself

How much heat is needed to change 1.5 kg of ice at −20 C° and 1 atm into steam?

PICTURE THE PROBLEM The heat required to change the ice into steam consists of four parts: Q_1, the heat needed to warm the ice from −20°C to 0°C; Q_2, the heat needed to melt the ice; Q_3, the heat needed to warm the water from 0°C to 100°C; and Q_4, the heat needed to vaporize the water. In calculating Q_1 and Q_3, we will assume that the specific heats are constant, with the values 2.05 kJ/kg·K for ice and 4.18 kJ/kg·K for water.

Cover the column to the right and try these on your own before looking at the answers.

Steps	Answers
1. Use $Q_1 = mc\,\Delta T$ to find the heat needed to warm the ice to 0°C.	$Q_1 = 61.5$ kJ $= 0.0615$ MJ

2. Use L_f from Table 18-2 to find the heat Q_2 needed to melt the ice.

$Q_2 = 500 \text{ kJ} = 0.500 \text{ MJ}$

3. Find the heat Q_3 needed to warm the water from 0°C to 100°C.

$Q_3 = 627 \text{ kJ} = 0.627 \text{ MJ}$

4. Use L_v from Table 18-2 to find the heat Q_4 needed to vaporize the water.

$Q_4 = 3.39 \text{ MJ}$

5. Sum your results to find the total heat Q.

$Q = Q_1 + Q_2 + Q_3 + Q_4 = \boxed{4.58 \text{ MJ}}$

REMARKS Notice that most of the heat was needed to vaporize the water, and that the amount needed to melt the ice was almost as much as that needed to raise the temperature of the water by 100 C°. A graph of temperature versus time for the case in which the heat is added at a constant rate of 1 kJ/s is shown in Figure 18-1. Note that it takes considerably longer to vaporize the water than it does to melt the ice or to raise the temperature of the water. When all of the water has vaporized, the temperature again rises as heat is added.

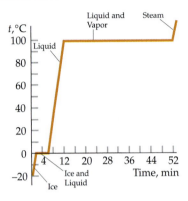

FIGURE 18-1

EXERCISE An 830-g piece of lead is heated to its melting point of 600 K. How much additional heat energy must be added to melt the lead? (*Answer* 20.5 kJ)

A Cool Drink　　　　　　**EXAMPLE 18-4**　**Put It in Context**

A 2-liter pitcher of lemonade has been sitting on the picnic table in the sun all day at 33°C. You pour 0.24 kg into a Styrofoam cup and add 2 ice cubes (each 0.025 kg at 0°C). (*a*) Assuming no heat lost to the surroundings, what is the final temperature of the lemonade? (*b*) What is the final temperature if you add 6 ice cubes?

PICTURE THE PROBLEM We set the heat lost by the lemonade equal to the heat gained by the ice cubes. Let T_f be the final temperature of the lemonade and water. We assume that lemonade has the same specific heat as water.

(*a*) 1. Write the heat lost by the lemonade in terms of the final temperature T_f:

$Q_{out} = m_L c |\Delta T| = m_L c (T_{Li} - T_f)$

2. Write the heat gained by the ice cubes and resulting water in terms of the final temperature:

$Q_{in} = m_{ice} L_f + m_{ice} c \Delta T_w = m_{ice} L_f + m_{ice} c T_f$

3. Set the heat lost equal to the heat gained and solve for T_f:

$$Q_{out} = Q_{in}$$

$$m_L c (T_{Li} - T_f) = m_{ice} L_f + m_{ice} c T_f$$

so

$$T_f = \frac{m_L c T_{Li} - m_{ice} L_f}{(m_L + m_{ice}) c}$$

$$= \frac{(0.24 \text{ kg})(4.18 \text{ kJ}/(\text{kg·C°}))(33°\text{C}) - (0.05 \text{ kg})(333.5 \text{ kJ/kg})}{(0.29 \text{ kg})(4.18 \text{ kJ}/(\text{kg·C°}))}$$

$$= \boxed{13.6°\text{C}}$$

(b) 1. For 6 ice cubes, $m_{ice} = 0.15$ kg. Find the final temperature as in step 3 of Part (a):

$$T_f = \frac{m_L c T_{Li} - m_{ice} L_f}{(m_L + m_{ice})c}$$

$$= \frac{(0.24 \text{ kg})(4.18 \text{ kJ/kg} \cdot \text{C}°)(33°\text{C}) - (0.15 \text{ kg})(333.5 \text{ kJ/kg})}{(0.39 \text{ kg})(4.18 \text{ kJ/kg} \cdot \text{C}°)}$$

$$= -10.4°\text{C}$$

2. This cannot be correct! No amount of ice at 0°C can lower the temperature of warm lemonade to below 0°C. Our calculation is wrong because our assumption in step 2 of Part (a) that all of the ice melts was wrong. Instead, the heat given off by the lemonade as it cools from 32°C to 0°C is not enough to melt all of the ice. The final temperature is thus:

$$T_f = \boxed{0°\text{C}}$$

 **PLAUSIBILITY CHECK** Let's calculate how much ice is melted. For the lemonade to cool from 33°C to 0°C, it must give off heat in the amount $Q_{out} = (0.24 \text{ kg})(4.18 \text{ kJ/kg} \cdot \text{C}°)(33°\text{C}) = 33.1$ kJ. The mass of ice that this amount of heat will melt is $m_{ice} = Q_{in}/L_f = 33.1 \text{ kJ}/(333.5 \text{ kJ/kg}) = 0.10$ kg. This is the mass of only 4 ice cubes. Adding more than 4 ice cubes does not lower the temperature below 0°C. It merely increases the amount of ice in the ice-lemonade mixture. In problems like this one, we should first find out how much ice must be melted to reduce the temperature of the liquid to 0°C. If less than that amount is added, we can proceed as in Part (a). If more ice is added, the final temperature is 0°C.

18-3 Joule's Experiment and the First Law of Thermodynamics

We can raise the temperature of a system by adding heat, but we can also raise its temperature by doing work on it.

Figure 18-2 is a diagram of the apparatus Joule used in a famous experiment in which he determined the amount of work needed to raise the temperature of 1 g of water by 1 C°. Here the system is a thermally insulated container of 1 g of water. Joule's apparatus converts the potential energy of falling weights into work done on the water by an attached paddle, as shown in the figure. Joule found that he could raise the temperature of his water sample by 1 F° by dropping 772 pounds of attached weights a distance of one foot. Converting to modern units, Joule found that it takes about 4.184 J (the energy units adopted by the scientific community in 1948) to raise the temperature of 1 g of water by 1 C°. The result that 4.184 J of mechanical energy is equivalent to 1 cal of heat energy is known as the **mechanical equivalence of heat**.

There are other ways of doing work on this system. For example, we could drop the insulated container of water from some height h, letting the system make an inelastic collision with the ground, or we could do mechanical work to generate electricity and then use the electricity to heat the water (Figure 18-3). In all such experiments, the same amount of work is required to produce a given temperature change. By the conservation of energy, the work done equals the increase in the internal energy of the system.

FIGURE 18-2 Schematic diagram for Joule's experiment. Insulating walls to prevent heat transfer enclose water. As the weights fall at constant speed, they turn a paddle wheel, which does work on the water. If friction is negligible, the work done by the paddle wheel on the water equals the loss of mechanical energy of the weights, which is determined by calculating the loss in the potential energy of the weights.

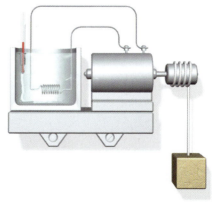

FIGURE 18-3 Another method of doing work on a thermally insulated container of water. Electrical work is done on the system by the generator, which is driven by the falling weight.

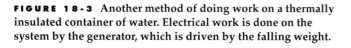

EXAMPLE 18-5

You drop a thermally insulated container of water from a height h to the ground. If the collision is perfectly inelastic and all of the mechanical energy lost goes into the internal energy of the water, what must h be for the temperature of the water to increase by 1 C°?

PICTURE THE PROBLEM The kinetic energy of the water just before it hits the ground equals its original potential energy mgh. During the collision, this energy is converted into thermal energy Q, which in turn causes a rise in temperature given by $Q = mc\Delta T$.

1. Set the potential energy equal to the thermal energy:

$$mgh = mc\Delta T$$

2. Solve for the height h:

$$h = \frac{c\Delta T}{g}$$

3. Substitute $c = 4.18$ kJ/kg·K and $\Delta T = 1$ C° = 1 K:

$$h = \frac{(4.18 \text{ kJ/kg·K})(1 \text{ K})}{9.81 \text{ N/kg}} = 0.426 \text{ km}$$

$$= \boxed{426 \text{ m}}$$

REMARKS Note that h is independent of the mass of the water. It is also rather large, which illustrates one of the difficulties with Joule's experiment—a large amount of work must be done to produce a measurable change in the temperature of the water.

Suppose that we perform Joule's experiment but replace the insulating walls of the container with conducting walls. We find that the work needed to produce a given change in the temperature of the system depends on how much heat is added to or subtracted from the system by conduction through the walls. However, if we sum the work done on the system and the net heat added to the system, the result is always the same for a given temperature change. That is, the sum of the heat transfer *into* the system and the work done *on* the system equals the change in the internal energy of the system. This is the **first law of thermodynamics,** which is simply a statement of the conservation of energy.

Let W_{on} stand for the work done by the surroundings *on* the system, and let W_{by} stand for the work done *by* the system on its surroundings. For example, suppose our system is a gas confined to a cylinder by a piston. If the piston compresses the gas, the surroundings do positive work on the gas, and W_{on} is positive. (However, if the gas expands against the piston, the surroundings do negative work on the gas, and W_{on} is negative.) Also, let Q_{in} stand for the heat transferred into the system. If heat is transferred into the system then Q_{in} is positive; if heat is transferred out of the system then Q_{in} is negative (Figure 18-4). Using these conventions, and denoting the internal energy by ΔE_{int}, the first law of thermodynamics is written

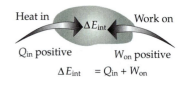

FIGURE 18-4 Sign convention for the first law of thermodynamics.

$$\Delta E_{int} = Q_{in} + W_{on} \qquad\qquad 18\text{-}10$$

The change in the internal energy of the system equals the heat transferred into the system plus the work done on the system.

FIRST LAW OF THERMODYNAMICS

Equation 18-10 is the same as the work energy theorem $W_{ext} = \Delta E_{sys}$ of Chapter 7 (Equation 7-9), except we have added the heat term Q_{in} and called the energy of the system ΔE_{int}.

E X A M P L E   1 8 - 6

You do 25 kJ of work on a system consisting of 3 kg of water by stirring it with a paddle wheel. During this time, 15 kcal of heat leaves the system due to poor insulation. What is the change in the internal energy of the system?

PICTURE THE PROBLEM We express all energies in joules and apply the first law of thermodynamics.

1. ΔE_{int} is found from the first law of thermodynamics:

$$\Delta E_{int} = Q_{in} + W_{on}$$

2. Heat is *removed* from the system, thus the heat *added* is negative:

$$Q_{in} = -15 \text{ kcal} = -(15 \text{ kcal})\left(\frac{4.18 \text{ kJ}}{1 \text{ kcal}}\right)$$

$$= -62.7 \text{ kJ}$$

3. The work done *on* the system is positive, thus:

$$W_{on} = 25 \text{ kJ}$$

4. Substitute these quantities and solve for ΔE_{int}:

$$\Delta E_{int} = Q_{in} + W_{on} = (-62.7 \text{ kJ}) + (25 \text{ kJ})$$

$$= \boxed{-37.7 \text{ kJ}}$$

REMARKS The internal energy decreases because the heat loss exceeds the work gain.

It is important to understand that the internal energy E_{int} is a function of the state of the system, just as P, V, and T are functions of the state of the system. Consider a gas in some initial state (P_i, V_i). The temperature T_i is determined by the equation of state. For example, if the gas is ideal, the $T_i = P_i V_i/(nR)$. The internal energy $E_{int,i}$ also depends only on the state of the gas, which is determined by any two state variables such as P and V, P and T, or V and T. If we compress the gas or let it expand, add to or remove heat from it, do work on it or let it do work, the gas will move through a sequence of states; that is, it will have different values of the state functions P, V, T, and E_{int}. If the gas is then returned to its original state (P_i, V_i), the temperature T and the internal energy E_{int} must equal their original values.

On the other hand, the net heat input Q and the work W done by the gas are not functions of the state of the system. There are no functions Q or W associated with any particular state of the gas. We could take the gas through a sequence of states beginning and ending at state (P_i, V_i) during which the gas did positive work and absorbed an equal amount of heat. Or we could take it though a different sequence of states such that work was done on the gas and heat was removed from the gas. It is correct to say that a system has a large amount of internal energy, but it is not correct to say that a system has a large amount of heat or a large amount of work. Heat is not something that is contained in a system. Rather, it is a measure of the energy that flows from one system to another because of a difference in temperature. Work is a measure of the energy that flows from one system to another because the point of contact of a force exerted by one system on the other undergoes a displacement not perpendicular to the force.

For very small amounts of heat added, work done, or changes in internal energy, it is customary to write Equation 18-10 as

$$dE_{int} = dQ_{in} + dW_{on} \qquad\qquad 18\text{-}11$$

In this equation, dE_{int} is the differential of the internal-energy function. However, neither dQ_{in} nor dW_{on} is a differential of any function. Instead, dQ_{in} merely represents a small amount of heat added to the system, and dW_{on} represents a small amount of work done by the system.

18-4 The Internal Energy of an Ideal Gas

The translational kinetic energy K of the molecules in an *ideal* gas is related to the absolute temperature T by Equation 17-22 in Chapter 17:

$$K = \tfrac{3}{2}nRT$$

where n is the number of moles of gas and R is the universal gas constant. If the internal energy of a gas is just this translational kinetic energy, then $E_{int} = K$, and

$$E_{int} = \tfrac{3}{2}nRT \qquad\qquad\qquad 18\text{-}12$$

Then the internal energy will depend only on the temperature of the gas, and not on its volume or pressure. If the molecules have other types of energy in addition to translational kinetic energy, such as energy of rotation, the internal energy will be greater than that given by Equation 18-12. But according to the equipartition theorem (Chapter 17, Section 5) the average energy associated with any degree of freedom will be $\tfrac{1}{2}RT$ per mole ($\tfrac{1}{2}kT$ per molecule), so again, the internal energy will depend only on the temperature and not on the volume or pressure.

We can imagine that the internal energy of a *real* gas might include other kinds of energy, which depend on the pressure and volume of the gas. Suppose, for example, that nearby gas molecules exert attractive forces on each other. Work is then required to increase the separation of the molecules. Then, if the average distance between the molecules is increased, the potential energy associated with the molecular attraction will increase. The internal energy of the gas will then depend on the volume of the gas as well as on its temperature.

Joule, using an apparatus like the one shown in Figure 18-5, performed a simple but interesting experiment to determine whether or not the internal energy of a gas depends on its volume. Initially, the compartment on the left in Figure 18-5 contains a gas and the compartment on the right has been evacuated. A stopcock that is initially closed connects the two compartments. The whole system is thermally insulated from its surroundings by rigid walls so that no heat can go into or out of the system *and* no work can be done. When the stopcock is opened, the gas rushes into the evacuated chamber. This process is called a **free expansion.** Eventually, the gas reaches thermal equilibrium with itself. Since no work has been done on the gas and no heat has been transferred to it, the final internal energy of the gas must equal its initial internal energy. If the gas molecules exert attractive forces on one another, the potential energy associated with these forces will increase as the volume increases. Since energy is conserved, the kinetic energy of translation must therefore decrease, which will result in a decrease in the temperature of the gas.

When Joule did this experiment, he found the final temperature to be equal to the initial temperature. Subsequent experiments verified this result for low gas densities. This implies that for a gas at low density—that is, for an ideal gas—the temperature depends only on the internal energy, or as we usually think of it, the internal energy depends only on the temperature. However, if the experiment is done with a large amount of gas initially in the left compartment so that the density is high, then the temperature after expansion is slightly lower than the temperature before the expansion. This indicates that there is a small attraction between the gas molecules of a real gas.

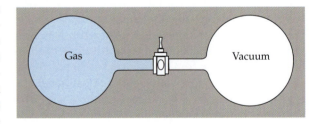

FIGURE 18-5 Free expansion of a gas. When the stopcock on the gas is opened, the gas expands rapidly into the evacuated chamber. Because no work is done on the gas and the whole system is thermally insulated, the initial and final internal energies of the gas are equal.

18-5 Work and the *PV* Diagram for a Gas

In many types of engines, a gas expanding against a moveable piston does work. For example, in a steam engine, water is heated in a boiler to produce steam. The steam then does work as it expands and drives a piston. In an automobile engine, a mixture of gasoline vapor and air is ignited, causing it to explode. The resulting high temperatures and pressures cause the gas to expand rapidly, driving a piston and doing work. In this section, we will see how we can mathematically describe the work done by an expanding gas.

Quasi-Static Processes

Figure 18-6 shows an ideal gas confined in a container with a tightly fitting piston that we assume to be frictionless. If the piston moves, the volume of the gas changes. The temperature or pressure or both must also change since these three variables are related by the equation of state $PV = nRT$. If we suddenly push in the piston to compress the gas, the pressure will initially be greater near the piston than far from it. Eventually the gas will settle down to a new equilibrium pressure and temperature. We cannot determine such macroscopic variables as T, P, or E_{int} for the entire gas system until equilibrium is restored in the gas. However, if we move the piston slowly in small steps and allow equilibrium to be reestablished after each step, we can compress or expand the gas in such a way that the gas is never far from an equilibrium state. In this kind of process, called a **quasi-static process,** the gas moves through a series of equilibrium states. In practice, it is possible to approximate quasi-static processes fairly well.

Let us begin with a gas at a high pressure, and let it expand quasi-statically. The magnitude of the force F exerted by the gas on the piston is PA, where A is the area of the piston and P is the gas pressure. As the piston moves a small distance dx, the work done *by* the gas on the piston is

$$dW_{\text{by gas}} = F_x\,dx = PA\,dx = P\,dV \qquad 18\text{-}13$$

where $dV = A\,dx$ is the increase in the volume of the gas. During the expansion the piston exerts a force of magnitude PA on the gas, but opposite in direction to the force of the gas on the piston. Thus, work done by the piston *on* the gas is just the negative of the work done *by* the gas

$$dW_{\text{on gas}} = -dW_{\text{by gas}} = -P\,dV \qquad 18\text{-}14$$

Note that for an expansion, dV is positive, the gas does work on the piston, so $dW_{\text{on gas}}$ is negative, and for a compression, dV is negative, work is done on the gas, so $dW_{\text{on gas}}$ is positive.

The work done on the gas during an expansion or compression from a volume of V_i to a volume of V_f is

$$W_{\text{on gas}} = -\int_{V_i}^{V_f} P\,dV \qquad 18\text{-}15$$

WORK DONE ON A GAS

To calculate this work we need to know how the pressure varies during the expansion or compression. The various possibilities can be illustrated most easily using a *PV* diagram.

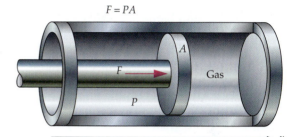

$F = PA$

FIGURE 18-6 Gas confined in a thermally insulated cylinder with a movable piston. If the piston moves a distance dx, the volume of the gas changes by $dV = A\,dx$. The work done by the gas is $PA\,dx = P\,dV$, where P is the pressure.

PV Diagrams

We can represent the states of a gas on a diagram of P versus V. Because by specifying both P and V we specify the state of the gas, each point on the PV diagram indicates a particular state of the gas. Figure 18-7 shows a PV diagram with a directed horizontal line representing a series of states that all have the same value of P. This line represents a *compression* at constant pressure. Such a process is called an **isobaric compression.** For a volume change of ΔV (ΔV is negative for a compression), we have

$$W_{\text{on}} = -\int_{V_i}^{V_f} P\,dV = -P\int_{V_i}^{V_f} dV = -P\Delta V$$

which is equal to the shaded area under the curve (directed line) in the figure. In general, for a compression the work done on the gas is equal to the area under the P-versus-V curve. (For an expansion the work done on the gas is equal to the negative of the area under the P-versus-V curve.) Because pressures are often given in atmospheres and volumes are often given in liters, it is convenient to have a conversion factor between liter-atmospheres and joules:

$$1\ \text{L·atm} = (10^{-3}\ \text{m}^3)(101.3 \times 10^3\ \text{N/m}^2) = 101.3\ \text{J} \qquad \text{18-16}$$

EXERCISE If 5 L of an ideal gas at a pressure of 2 atm is cooled so that it contracts at constant pressure until its volume is 3 L, what is the work done on the gas? (*Answer* 405.2 J)

Figure 18-8 shows three different possible paths on a PV diagram for a gas that is initially in state (P_i, V_i) and is finally in state (P_f, V_f). We assume that the gas is ideal and have chosen the original and final states to have the same temperature so that $P_iV_i = P_fV_f = nRT$. Since the internal energy depends only on the temperature, the initial and final internal energies are the same also.

In Figure 18-8a, the gas is heated at constant volume until its pressure is P_f, after which it is cooled at constant pressure until its volume is V_f. The work done on the gas along the constant-volume (vertical) part of path A is zero; along the constant pressure (horizontal) part of the path A, it is $P_f |V_f - V_i| = -P_f(V_f - V_i)$.

In Figure 18-8b, the gas is first cooled at constant pressure until its volume is V_f, after which it is heated at constant volume until its pressure is P_f. The work done on the gas along this path is $P_i|V_f - V_i| = -P_i(V_f - V_i)$, which is much less than that done along the path shown in Figure 18-8a as can be seen by comparing the shaded regions in Figure 18-8a and Figure 18-8b.

In Figure 18-8c, path C represents an **isothermal** compression, meaning that the temperature remains constant. (Keeping the temperature constant during the compression requires that heat be transferred out of the gas during the

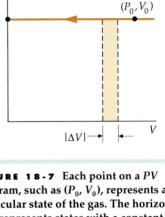

FIGURE 18-7 Each point on a PV diagram, such as (P_0, V_0), represents a particular state of the gas. The horizontal line represents states with a constant pressure P_0. The shaded area, $P_0 |\Delta V|$, represents the work done on the gas as it is compressed an amount $|\Delta V|$.

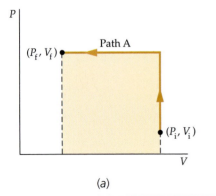

(a)

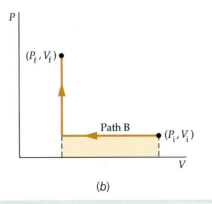

(b)

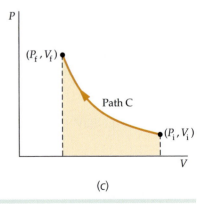

(c)

FIGURE 18-8 Three paths on PV diagrams connecting an initial state (P_i, V_i) and a final state (P_f, V_f). The corresponding shaded area indicates the work done on the gas along each path.

compression.) We can calculate the work done on the gas along path C by using $P = nRT / V$. Hence, the work done on the gas as it is compressed from V_i to V_f is

$$W_{on} = -\int_{V_i}^{V_f} P\,dV = -\int_{V_i}^{V_f} \frac{nRT}{V}\,dV$$

Since T is constant for an isothermal process, we can remove it from the integral. We then have

$$W_{isothermal} = -nRT \int_{V_i}^{V_f} \frac{dV}{V} = -nRT \ln \frac{V_f}{V_i} = nRT \ln \frac{V_i}{V_f} \qquad 18\text{-}17$$

WORK DONE ON GAS DURING ISOTHERMAL COMPRESSION

We see that the amount of work done on the gas is different for each process illustrated. Since for these states $E_{int,f} = E_{int,i}$, the net amount of heat added must also be different for each of the processes. This discussion illustrates the fact that both the work done and the heat added depend only on how a system moves from one state to another, but the change in the internal energy of the system does not.

WORK DONE ON AN IDEAL GAS   **EXAMPLE 18-7**

An ideal gas undergoes a cyclic process from point A to point B to point C to point D and back to point A as shown in Figure 18-9. The gas begins at a volume of 1 L and a pressure of 2 atm and expands at constant pressure until the volume is 2.5 L, after which it is cooled at constant volume until its pressure is 1 atm. It is then compressed at constant pressure until its volume is again 1 L, after which it is heated at constant volume until it is back in its original state. Find the total work done on the gas and the total heat added to it during the cycle.

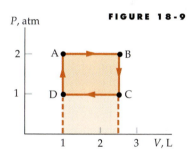
FIGURE 18-9

PICTURE THE PROBLEM We calculate the work done during each step. Since $\Delta E_{int} = 0$ for any complete cycle, the first law of thermodynamics implies that the total heat added to the gas plus the total work done on the gas equals zero.

1. From A to B the process is an isobaric (constant pressure) expansion, so the work done on the gas is negative. The work done on the gas equals the negative of the shaded area under the AB curve, shown in Figure 18-10*a*:

$$W_{AB} = -P\Delta V = -P(V_B - V_A)$$
$$= -(2\text{ atm})(2.5\text{ L} - 1\text{ L}) = -3\text{ L·atm}$$

2. Convert the units to joules:

$$W_{AB} = -3\text{ L·atm} \times \frac{101.3\text{ J}}{1\text{ L·atm}} = -304\text{ J}$$

3. From B to C (Figure 18-9) the gas cools at constant volume so the work done is zero:

$$W_{BC} = 0$$

4. As the gas undergoes an isobaric compression from C to D, the work done on it is positive. This work equals the area under the CD curve, shown in Figure 18-10*b*:

$$W_{CD} = -P\Delta V = -P(V_D - V_C)$$
$$= -(1\text{ atm})(1\text{ L} - 2.5\text{ L})$$
$$= 1.5\text{ L·atm} = \boxed{152\text{ J}}$$

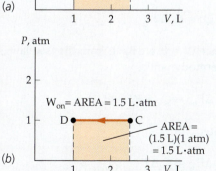

(a)

(b)

FIGURE 18-10 (*a*) The work done on the gas during the expansion from A to B is equal to the negative of the area under the curve. (*b*) The work done on the gas during the compression from C to D is equal to the area under the curve.

5. As the gas is heated back to its original state A, the volume is again constant (Figure 18-9), so no work is done:

$$W_{DA} = 0$$

6. The total work done by the gas is the sum of the work done along each step:

$$W_{total} = W_{AB} + W_{BC} + W_{CD} + W_{DA}$$

$$= (-304 \text{ J}) + 0 + 152 \text{ J} + 0 = \boxed{-152 \text{ J}}$$

7. Because the gas is back in its original state, the total change in internal energy is zero:

$$\Delta E_{int} = 0$$

8. The heat added is found from the first law:

$$\Delta E_{int} = Q_{in} + W_{on}$$

so

$$Q_{in} = \Delta E_{int} - W_{on} = 0 - (-152 \text{ J}) = \boxed{152 \text{ J}}$$

REMARKS The work done by the gas equals the negative of the work done on the gas, so the total work done by the gas during the cycle is $+152$ J. During the cycle the gas extracts 152 J of heat from its surroundings and does 152 J of work on its surroundings. This leaves the gas in its initial state. The total work done by the gas equals the area enclosed by the cycle in Figure 18-9. Such cyclic processes have important applications for heat engines, as we will see in Chapter 20.

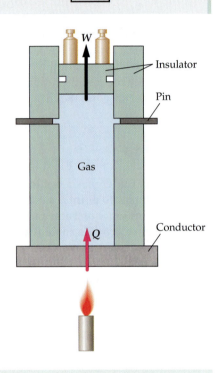

FIGURE 18-11 Heat is added and the pressure remains constant. The gas expands, thus doing positive work on the piston.

18-6 Heat Capacities of Gases

The determination of the heat capacity of a substance provides information about its internal energy, which is related to its molecular structure. For all substances that expand when heated, the heat capacity at constant pressure C_p is greater than the heat capacity at constant volume C_v. If heat is added at constant pressure, the substance expands and does positive work on its surroundings (Figure 18-11). Therefore, it takes more heat to obtain a given temperature change at constant pressure than to obtain the same temperature change when heated at constant volume. The expansion is usually negligible for solids and liquids, so for them $C_p \approx C_v$. But a gas heated at constant pressure readily expands and does a significant amount of work, so $C_p - C_v$ is not negligible.

If heat is added to a gas at constant volume, no work is done (Figure 18-12), so the heat added equals the increase in the internal energy of the gas. Writing Q_v for the heat added at constant volume, we have

$$Q_v = C_v \, \Delta T$$

Since $W = 0$, we have from the first law of thermodynamics

$$\Delta E_{int} = Q_v + W = Q_v$$

Thus,

$$\Delta E_{int} = C_v \Delta T$$

Taking the limit as ΔT approaches zero, we obtain

$$dE_{int} = C_v \, dT \qquad \qquad 18\text{-}18a$$

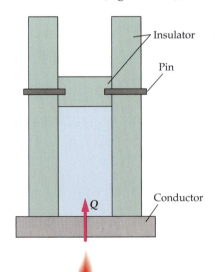

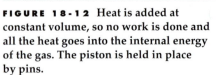

FIGURE 18-12 Heat is added at constant volume, so no work is done and all the heat goes into the internal energy of the gas. The piston is held in place by pins.

and

$$C_v = \frac{dE_{int}}{dT}$$ 18-18b

The heat capacity at constant volume is the rate of change of the internal energy with temperature. Since E_{int} and T are both state functions, Equations 18-18a and 18-18b hold for any process.

Now let's calculate the difference $C_p - C_v$ for an ideal gas. From the definition of C_p, the heat added at constant pressure is

$$Q_p = C_p \, \Delta T$$

From the first law of thermodynamics,

$$\Delta E_{int} = Q_p + W_{on} = Q_p - P\Delta V$$

Then

$$\Delta E_{int} = C_p \, \Delta T - P\Delta V \quad \text{or} \quad C_p \, \Delta T = \Delta E_{int} + P\Delta V$$

For infinitesimal changes, this becomes

$$C_p \, dT = dE_{int} + P \, dV$$

Using Equation 18-18a for dE_{int}, we obtain

$$C_p \, dT = C_v \, dT + P \, dV$$ 18-19

The pressure, volume, and temperature for an ideal gas are related by

$$PV = nRT$$

Taking the differentials of both sides of the ideal-gas law, we obtain

$$P \, dV + V \, dP = nR \, dT$$

For a constant-pressure process $dP = 0$, so

$$P \, dV = nR \, dT$$

Substituting this into Equation 18-19 gives

$$C_p \, dT = C_v \, dT + nR \, dT = (C_v + nR) \, dT$$

Therefore,

$$C_p = C_v + nR$$ 18-20

which shows that, for an ideal gas, the heat capacity at constant pressure is greater than the heat capacity at constant volume by the amount nR.

Table 18-3 lists measured molar heat capacities c'_p and c'_v for several gases. Note from this table that the ideal gas prediction, $c'_p - c'_v = R$, holds quite well for all gases. The table also shows that c_v' is approximately 1.5R for all monatomic gases, 2.5R for all diatomic gases, and greater than 2.5R for gases consisting of more complex molecules. We can understand these results by considering the molecular model of a gas (Chapter 17.) The total translational kinetic energy of n moles of a gas is $K = \frac{3}{2}nRT$ (Equation 17-22). Thus, if the internal energy of a gas consists of translational kinetic energy only, we have

$$E_{int} = \frac{3}{2}nRT \qquad\qquad 18\text{-}21$$

The heat capacities are then

$$C_v = \frac{dE_{int}}{dT} = \frac{3}{2}nR \qquad\qquad 18\text{-}22$$

C_v FOR AN IDEAL MONATOMIC GAS

and

$$C_p = C_v + nR = \frac{5}{2}nR \qquad\qquad 18\text{-}23$$

C_p FOR AN IDEAL MONATOMIC GAS

The results in Table 18-3 agree well with these predictions for monatomic gases, but for other gases, the heat capacities are greater than those predicted by Equations 18-22 and 18-23. The internal energy for a gas consisting of diatomic or more complicated molecules is evidently greater than $\frac{3}{2}nRT$. The reason is that such molecules can have other types of energy, such as rotational or vibrational energy, in addition to translational kinetic energy.

TABLE 18-3

Molar Heat Capacities in J/mol·K of Various Gases at 25°C

Gas	c_p'	c_v'	c_v'/R	$c_p' - c_v'$	$(c_p' - c_v')/R$
Monatomic					
He	20.79	12.52	1.51	8.27	0.99
Ne	20.79	12.68	1.52	8.11	0.98
Ar	20.79	12.45	1.50	8.34	1.00
Kr	20.79	12.45	1.50	8.34	1.00
Xe	20.79	12.52	1.51	8.27	0.99
Diatomic					
N_2	29.12	20.80	2.50	8.32	1.00
H_2	28.82	20.44	2.46	8.38	1.01
O_2	29.37	20.98	2.52	8.39	1.01
CO	29.04	20.74	2.49	8.30	1.00
Polyatomic					
CO_2	36.62	28.17	3.39	8.45	1.02
N_2O	36.90	28.39	3.41	8.51	1.02
H_2S	36.12	27.36	3.29	8.76	1.05

A system consisting of 0.32 mol of a monatomic ideal gas with $c'_v = \frac{3}{2}RT$ occupies a volume of 2.2 L at a pressure of 2.4 atm, as represented by point A in Figure 18-13. The system is carried through a cycle consisting of three processes:

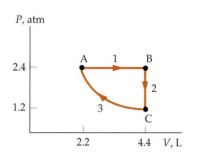

P, atm

FIGURE 18-13

1. The gas is heated at constant pressure until its volume is 4.4 L at point B.

2. The gas is cooled at constant volume until the pressure decreases to 1.2 atm (point C).

3. The gas undergoes an isothermal compression back to point A.

(a) What is the temperature at points A, B, and C? (b) Find W, Q, and ΔE_{int} for each process and for the entire cycle.

PICTURE THE PROBLEM You can find the temperatures at all points from the ideal-gas law. You can find the work for each process by finding the area under the curve, and the heat exchanged from the given heat capacity and the initial and final temperatures for each process. In process 3, T is constant, so $\Delta E_{\text{int}} = 0$ and the heat input plus the work done on the gas equals zero.

Cover the column to the right and try these on your own before looking at the answers.

Steps

(a) Find the temperatures at points A, B, and C using the ideal-gas law.

(b) 1. For process 1, use $W_1 = -P_c\Delta V$ to calculate the work, and $C_p = \frac{5}{2}nR$ to calculate the heat Q_1. Then use W_1 and Q_1 to calculate $\Delta E_{\text{int},1}$.

2. For process 2, use $C_v = \frac{3}{2}nR$ and $T_C - T_B$ from step 1 to find Q_2. Then, since $W_2 = 0$, $\Delta E_{\text{int},2} = Q_2$.

3. Calculate W_3 from $W = -nRT \ln(V_A/V_C)$ in the isothermal compression. Then, since $\Delta E_{\text{int},3} = 0$, $Q_3 = -W_3$.

4. Find the total work W, the total heat Q, and the total change ΔE_{int} by summing the quantities found in steps 2, 3, and 4.

Answers

$T_A = T_C = \boxed{201 \text{ K}}$, $T_B = \boxed{402 \text{ K}}$

$W_1 = -5.28 \text{ L·atm} = \boxed{-535 \text{ J}}$, $Q_1 = \boxed{1337 \text{ J}}$

$\Delta E_{\text{int},1} = Q_1 + W_1 = \boxed{802 \text{ J}}$

$W_2 = \boxed{0}$, $Q_2 = \boxed{-802 \text{ J}}$, $\Delta E_{\text{int},2} = \boxed{-802 \text{ J}}$

$W_3 = \boxed{371 \text{ J}}$, $Q_3 = \boxed{-371 \text{ J}}$, $\Delta E_{\text{int},3} = \boxed{0}$

$W_{\text{total}} = W_1 + W_2 + W_3$

$= (-535 \text{ J}) + 0 + 371 \text{ J} = \boxed{-164 \text{ J}}$

$Q_{\text{total}} = Q_1 + Q_2 + Q_3$

$= 1337 \text{ J} + (-802 \text{ J}) + (-371 \text{ J})$

$= \boxed{164 \text{ J}}$

$\Delta E_{\text{int,total}} = \Delta E_{\text{int},1} + \Delta E_{\text{int},2} + \Delta E_{\text{int},3}$

$= 802 \text{ J} + (-802 \text{ J}) + 0 = \boxed{0}$

REMARKS The total change in internal energy is zero, as it must be for a cyclic process. The total work done on the gas plus the total heat absorbed by the gas equals zero. The total work done on the gas equals the area under the CA curve minus the area under the AB curve, which equals the negative of the area enclosed by the three curves in Figure 18-13.

Heat Capacities and the Equipartition Theorem

According to the equipartition theorem stated in Section 5 of Chapter 17, the internal energy of n moles of a gas should equal $\frac{1}{2}nRT$ for each degree of freedom of the gas molecule. The heat capacity at constant volume of a gas should then be $\frac{1}{2}nR$ times the number of degrees of freedom of the molecule. From Table 18-2, nitrogen, oxygen, hydrogen, and carbon monoxide all have molar heat capacities at constant volume of about $\frac{5}{2}R$. Thus, the molecules in each of these gases have five degrees of freedom. About 1880, Rudolf Clausius speculated that these gases must consist of diatomic molecules that can rotate about two axes, giving them two additional degrees of freedom (Figure 18-14). The two degrees of freedom besides the three for translation are now known to be associated with their rotation about each of the two axes, x' and y', perpendicular to the line joining the atoms. The kinetic energy of a diatomic molecule is therefore

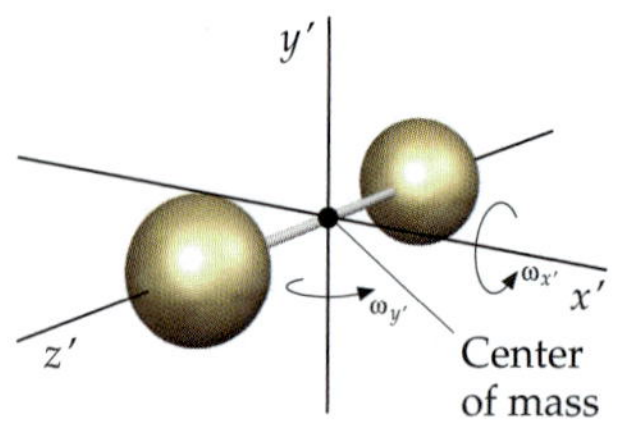

$$K = \tfrac{1}{2}mv_x^2 + \tfrac{1}{2}mv_y^2 + \tfrac{1}{2}mv_z^2 + \tfrac{1}{2}I_{x'}\omega_{x'}^2 + \tfrac{1}{2}I_{y'}\omega_{y'}^2$$

The total internal energy of n moles of such a gas is then

$$E_{int} = 5 \times \tfrac{1}{2}nRT = \tfrac{5}{2}nRT \qquad\qquad 18\text{-}24$$

and the heat capacity at constant volume is

$$C_v = \tfrac{5}{2}nR \qquad\qquad 18\text{-}25$$

Apparently, diatomic gases do not rotate about the line joining the two atoms—if they did, there would be six degrees of freedom and C_v would be $\frac{6}{2}nR = 3nR$, which is contrary to experimental results. Furthermore, monatomic gases apparently do not rotate at all. We will see in Section 18-8 that these puzzling facts are easily explained when we take into account the quantization of energy.

FIGURE 18-14 Rigid-dumbbell model of a diatomic molecule.

HEATING A DIATOMIC IDEAL GAS   **EXAMPLE 18-9**

Two moles of oxygen gas are heated from a temperature of 20°C and a pressure of 1 atm to a temperature of 100°C. Assume that oxygen is an ideal gas. (a) How much heat must be supplied if the volume is kept constant during the heating? (b) How much heat must be supplied if the pressure is kept constant? (c) How much work does the gas do in part (b)?

PICTURE THE PROBLEM The heat needed for constant-volume heating is $Q_v = C_v \Delta T$, where $C_v = \frac{5}{2}nR$ since oxygen is a diatomic gas. For constant-pressure heating, $Q_p = C_p \Delta T$, where $C_p = C_v + nR$. Finally, the amount of work done by the gas equals the negative of the work done on the gas, which can be found from $\Delta E_{int} = Q + W_{on}$. Alternatively, $W_{by} = P\Delta V$.

(a) 1. Write the heat needed for constant volume in terms of C_v and ΔT:

$$Q_v = C_v \Delta T$$

2. Calculate the heat for $\Delta T = 80\ C° = 80\ K$:

$$Q_V = C_V \Delta T = \tfrac{5}{2}nR\Delta T$$

$$= \frac{5}{2}(2\ \text{mol})\left(8.314\ \frac{\text{J}}{\text{mol}\cdot\text{K}}\right)(80\ \text{K})$$

$$= \boxed{3.33\ \text{kJ}}$$

(b) 1. Write the heat needed for constant pressure in terms of C_p and ΔT:

$$Q_p = C_p \Delta T$$

2. Calculate the heat capacity at constant pressure:

$$C_p = C_v + nR = \tfrac{5}{2}nR + nR = \tfrac{7}{2}nR$$

3. Calculate the heat added at constant pressure for $\Delta T = 80$ K:

$$Q_p = C_p \Delta T$$

$$= \frac{7}{2}(2\ \text{mol})\left(8.314\ \frac{\text{J}}{\text{mol·K}}\right)(80\ \text{K})$$

$$= \boxed{4.66\ \text{kJ}}$$

(c) 1. The work W_{on} can be found from the first law of thermodynamics:

$$\Delta E_{int} = Q_{in} + W_{on}, \quad \text{so} \quad W_{on} = \Delta E_{int} - Q_{in}$$

2. The internal energy change equals the heat added at constant volume which was calculated in Part (a):

$$\Delta E_{int} = Q_v = C_v \Delta T = \tfrac{5}{2}nR\Delta T$$

and

$$Q_p = C_p \Delta T = \tfrac{7}{2}nR\Delta T$$

so

$$W_{on} = \Delta E_{int} - Q_p = \tfrac{5}{2}nR\Delta T - \tfrac{7}{2}nR\Delta T = -nR\Delta T$$

$$= -(2\ \text{mol})\left(8.314\ \frac{\text{J}}{\text{mol·K}}\right)(80\ \text{K})$$

$$= -1.33\ \text{kJ}$$

3. The work done by the gas at constant pressure is then:

$$W_{by} = -W_{on} = \boxed{1.33\ \text{kJ}}$$

REMARKS Note that the change in internal energy is independent of the process. It depends only on the initial and final states.

EXERCISE Find the initial and final volumes of this gas from the ideal-gas law, and use them to calculate the work done by the gas if the heat is added at constant pressure from $W_{by} = P\Delta V$. (*Answer* $V_i = 48.0$ L, $V_f = 61.1$ L, $W = 13.1$ L·atm $= 1.33$ kJ)

18-7 Heat Capacities of Solids

In Section 18-1, we noted that all of the metals listed in Table 18-1 have approximately equal molar specific heats. Experimentally, most solids have molar heat capacities approximately equal to $3R$:

$$c' = 3R = 24.9\ \text{J/mol·K} \tag{18-26}$$

This result is known as the **Dulong–Petit law.** We can understand this law by applying the equipartition theorem to the simple model for a solid shown in Figure 18-15. According to this model, a solid consists of a regular array of atoms in which each of the atoms has a fixed equilibrium position and is connected by springs to its neighbors. Each atom can vibrate in the x, y, and z directions. The total energy of an atom in a solid is

$$E = \tfrac{1}{2}mv_x^2 + \tfrac{1}{2}mv_y^2 + \tfrac{1}{2}mv_z^2 + \tfrac{1}{2}k_{eff}x^2 + \tfrac{1}{2}k_{eff}y^2 + \tfrac{1}{2}k_{eff}z^2$$

where k_{eff} is the effective force constant of the hypothetical springs. Each atom thus has six degrees of freedom. The equipartition theorem states that a substance

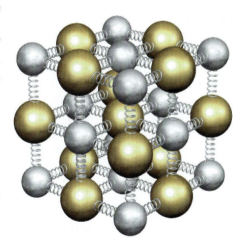

FIGURE 18-15 Model of a solid in which the atoms are connected to each other by springs. The internal energy of the molecule consists of the kinetic and potential energies of vibration.

in equilibrium has an average energy of $\frac{1}{2}RT$ per mole for each degree of freedom. Thus, the internal energy of a mole of a solid is

$$E_{\text{int,m}} = 6 \times \tfrac{1}{2}RT = 3RT \qquad\qquad 18\text{-}27$$

which means that c' is equal to $3R$.

EXAMPLE 18-10

The molar mass of copper is 63.5 g/mol. Use the Dulong–Petit law to calculate the specific heat of copper.

PICTURE THE PROBLEM The Dulong–Petit law gives the molar specific heat of a solid, c'. The specific heat is then $c = c'/M$ (Equation 18-6), where M is the molar mass.

1. The Dulong–Petit law gives c' in terms of R:

$$c' = 3R$$

2. Using $M = 63.5$ g/mol for copper, the specific heat is:

$$c = \frac{c'}{M} = \frac{3R}{M} = \frac{3(8.314\ \text{J/mol·K})}{63.5\ \text{g/mol}}$$

$$= 0.392\ \text{J/(g·K)} = \boxed{0.392\ \text{kJ/kg·K}}$$

REMARKS This differs from the measured value of 0.386 kJ/kg·K given in Table 18-1 by less than 2 percent.

EXERCISE The specific heat of a certain metal is measured to be 1.02 kJ/kg·K. (*a*) Calculate the molar mass of this metal, assuming that the metal obeys the Dulong–Petit law. (*b*) What is the metal? (*Answer* (*a*) $M = 24.4$ g/mol. (*b*) The metal must be magnesium, which has a molar mass of 24.31 g/mol)

18-8 Failure of the Equipartition Theorem

Although the equipartition theorem had spectacular successes in explaining the heat capacities of gases and solids, it had equally spectacular failures. For example, if a diatomic gas molecule like the one in Figure 18-14 rotates about the line joining the atoms, there should be an additional degree of freedom. Similarly, if a diatomic molecule is not rigid, the two atoms should vibrate along the line joining them. We would then have two more degrees of freedom corresponding to kinetic and potential energies of vibration. But according to the measured values of the molar heat capacities in Table 18-3, diatomic gases apparently do not rotate about the line joining them, nor do they vibrate. The equipartition theorem gives no explanation for this, nor for the fact that monatomic atoms apparently do not rotate about any of the three possible perpendicular axes in space. Furthermore, heat capacities are found to depend on temperature, contrary to the predictions of the equipartition theorem. The most spectacular case of the temperature dependence of heat capacity is that of H_2, as shown in Figure 18-16. At temperatures below 70 K, c'_v for H_2 is $\frac{3}{2}R$, which is the same as that for a gas of molecules that translate, but do not rotate or vibrate. At temperatures between

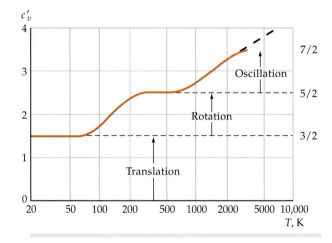

FIGURE 18-16 Temperature dependence of the molar heat capacity of H_2. (The curve is qualitative in those regions where c'_v is changing.)

250 K and 700 K, $c_v' = \frac{5}{2}R$, which is that for molecules that translate and rotate but do not vibrate. And at temperatures above 700 K, the H_2 molecules begin to vibrate. However, the molecules dissociate before c_v' reaches $\frac{7}{2}R$. Finally, the equipartition theorem predicts a constant value of $3R$ for the heat capacity of solids. While this result holds for many, although not all, solids at high temperatures, it does not hold at very low temperatures.

The equipartition theorem fails because the energy is **quantized.** That is, a molecule can have only certain values of internal energy, as illustrated schematically by the energy-level diagram in Figure 18-17. The molecule can gain or lose energy only if the gain or loss takes it to another allowed level. For example, the energy that can be transferred between colliding gas molecules is of the order of kT, the typical thermal energy of a molecule. The validity of the equipartition theorem depends on the relative size of kT and the spacing of the allowed energy levels.

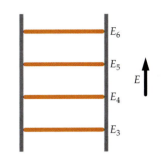

FIGURE 18-17 Energy-level diagram. **A system can have only certain discrete energies.**

> If the spacing of the allowed energy levels is large compared with kT, energy cannot be transferred by collisions and the classical equipartition theorem is not valid. If the spacing of the levels is much smaller than kT, energy quantization will not be noticed and the equipartition theorem will hold.

CONDITIONS FOR THE VALIDITY OF THE EQUIPARTITION THEOREM

Consider the rotation of a molecule. The energy of rotation is

$$E = \frac{1}{2}I\omega^2 = \frac{(I\omega)^2}{2I} = \frac{L^2}{2I} \qquad\qquad 18\text{-}28$$

where I is the moment of inertia of the molecule, ω is its angular velocity, and $L = I\omega$ is its angular momentum. In Section 10-5, we mentioned that angular momentum is quantized, and its magnitude is restricted to

$$L = \sqrt{\ell(\ell + 1)}\hbar \qquad \ell = 0, 1, 2, \ldots \qquad\qquad 18\text{-}29$$

where $\hbar = h/(2\pi)$, and h is Planck's constant. The energy of a rotating molecule is therefore quantized to the values

$$E = \frac{L^2}{2I} = \frac{\ell(\ell + 1)\hbar^2}{2I} = \ell(\ell + 1)\, E_{0r} \qquad\qquad 18\text{-}30$$

where

$$E_{0r} = \frac{\hbar^2}{2I} \qquad\qquad 18\text{-}31$$

is characteristic of the energy gap between levels. If this energy is much less than kT, we expect classical physics and the equipartition theorem to hold. Let us define a critical temperature T_c by

$$kT_c = E_{0r} = \frac{\hbar^2}{2I} \qquad\qquad 18\text{-}32$$

If T is much greater than this critical temperature, then kT will be much greater than the spacing of the energy levels, which is of the order of kT_c, and we expect classical physics and the equipartition theorem to be valid. If T is less than or of the order of T_c, then kT will not be much greater than the energy-level spacing, and we expect classical physics and the equipartition theorem to break down. Let's estimate T_c for some cases of interest.

1. *Rotation of H_2 about an axis perpendicular to the line joining the H atoms and through the center of mass* (Figure 18-18): The moment of inertia of H_2 about the axis is

$$I_H = 2M_H \left(\frac{r_s}{2} \right)^2 = \frac{1}{2} M_H r_s^2$$

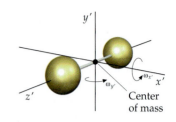

FIGURE 18-18 Rigid-dumbbell model of a diatomic molecule.

where M_H is the mass of an H atom and r_s is the separation distance. For hydrogen, $M_H = 1.67 \times 10^{-27}$ kg, and $r_s \approx 8 \times 10^{-11}$ m. The critical temperature is then

$$T_c = \frac{\hbar^2}{2kI} = \frac{\hbar^2}{kM_H r_s^2}$$

$$= \frac{(1.05 \times 10^{-34} \text{ J·s})^2}{(1.38 \times 10^{-23} \text{ J/K})(1.67 \times 10^{-27} \text{ kg})(8 \times 10^{-11} \text{ m})^2} \approx 75 \text{ K}$$

As we see from Figure 18-16, this is approximately the temperature below which the rotational energy does not contribute to the heat capacity.

2. O_2: Because the mass of O_2 is about 16 times that of H_2, and the separation is about the same, the critical temperature for O_2 should be about $(75/16) \approx$ 4.6 K. For all temperatures for which O_2 exists as a gas, $T \gg T_c$, so kT is much greater than the energy level spacing. Consequently, we expect the equipartition theorem of classical physics to apply.

3. *Rotation of a monatomic gas:* Consider the He atom, which consists of the He nucleus and two electrons. The mass of an electron is about 8000 times smaller than the mass of the He nucleus, but the radius of the nucleus is about 100,000 times smaller than the distance to the electron. Therefore, the moment of inertia of the He atom is almost entirely due to its two electrons. The distance from the He nucleus to its electrons is about half the separation distance of the H atoms in H_2, and the electron mass is about 2000 times smaller than that of the H nucleus. Thus, using $m_e = M_H/2000$ and $r = r_s/2$, we find the moment of inertia of the two electrons in He to be roughly

$$I_{He} = 2m_e r^2 \approx 2 \frac{M_H}{2000} \left(\frac{r_s}{2} \right)^2 = \frac{I_H}{2000}$$

The critical temperature for He is thus about 2000 times that of H_2 or about 150,000 K. This is much higher than the dissociation temperature (the temperature at which electrons are stripped from their nuclei) for helium. So, the gap between allowed energy levels is always much greater than kT and the He molecules cannot be induced to rotate by collisions occurring in the gas. Other monatomic gases have slightly greater moments of inertia because they have more electrons, but their critical temperatures are still tens of thousands of kelvins. Therefore, their molecules also cannot be induced to rotate by collisions occurring in the gas.

4. *Rotation of a diatomic gas about the axis joining the atoms:* We see from our discussion of monatomic gases that the moment of inertia for a diatomic gas molecule about this axis will also be due mainly to the electrons and will be of the same order of magnitude as for a monatomic gas. Again, the critical temperature, T_c, calculated in order for this rotation to occur due to collisions between molecules in the gas, exceeds the gas's dissociation temperature, making rotation under those circumstances impossible.

It is interesting to note that the successes of the equipartition theorem in explaining the measured heat capacities of gases and solids led to the first real understanding of molecular structure in the nineteenth century, whereas its failures played an important role in the development of quantum mechanics in the twentieth century.

ROTATIONAL ENERGY OF THE HYDROGEN ATOM   **EXAMPLE   18-11**

(*a*) **Estimate the lowest (nonzero) rotational energy for the hydrogen atom and compare it to** kT **at room temperature,** $T = 300$ **K.** (*b*) **Calculate the critical temperature** T_c.

PICTURE THE PROBLEM From Equation 18-30, the lowest rotational energy is for $\ell = 1$. We use Equation 18-30 to determine the energy in terms of the moment of inertia. We can neglect the moment of inertia of the nucleus because its radius is 100,000 times smaller than the radius of the atom. Therefore, the moment of inertia for the atom is essentially the moment of inertia of the electron about the nucleus. Then $I = m_e r^2$, where $r \approx 5 \times 10^{-11}$ m is the distance from the nucleus to the electron.

(*a*) 1. The lowest energy greater than zero occurs for $\ell = 1$:

$$E_\ell = \frac{\ell(\ell + 1)\hbar^2}{2I}, \qquad \ell = 0, 1, 2, \ldots$$

so

$$E_1 = \frac{1(1 + 1)\hbar^2}{2m_e r^2} = \frac{\hbar^2}{m_e r^2}$$

2. The numerical values are:

$$\hbar = 1.05 \times 10^{-34} \text{ J·s}$$
$$m_e = 9.11 \times 10^{-31} \text{ kg}$$
$$r = 5 \times 10^{-11} \text{ m}$$

3. Substitute the numerical values:

$$E_1 = \frac{\hbar^2}{m_e r^2} = \frac{(1.05 \times 10^{-34} \text{ J·s})^2}{(9.11 \times 10^{-31} \text{ kg})(5 \times 10^{-11} \text{ m})^2}$$

$$= \boxed{4.8 \times 10^{-18} \text{ J}}$$

4. The value of kT at $T = 300$ K is:

$$kT = (1.38 \times 10^{-23} \text{ J/K})(300 \text{ K}) = 4.1 \times 10^{-21} \text{ J}$$

5. Compare E_1 and kT:

$$\frac{E_1}{kT} = \frac{4.8 \times 10^{-18} \text{ J}}{4.1 \times 10^{-21} \text{ J}} \approx 10^3$$

$$\boxed{E_1 \text{ is about three orders of magnitude larger than } kT.}$$

(*b*) Set $kT_c = E_1$ and solve for T_c:

$$kT_c = E_1$$

$$T_c = \frac{E_1}{k} = \frac{4.8 \times 10^{-18} \text{ J}}{1.38 \times 10^{-23} \text{ J/K}} = \boxed{3.48 \times 10^5 \text{ K}}$$

REMARKS The critical temperature of a hydrogen atom is so high that the atom would be ionized well before the critical temperature could be reached.

18-9 The Quasi-Static Adiabatic Compression of a Gas

A process in which no heat flows into or out of a system is called an **adiabatic process.** Such a process can occur when the system is extremely well insulated, or when the process happens very quickly. Consider the quasi-static adiabatic compression of a gas in which the gas in a thermally insulated container is slowly compressed by a piston, which is thereby doing work on the gas. Because no heat enters or leaves the gas, the work done on the gas equals the increase in the internal energy of the gas, and the temperature of the gas increases. The curve representing this process on a PV diagram is shown in Figure 18-19.

We can find the equation for the adiabatic curve for an ideal gas by using the equation of state ($PV = nRT$) and the first law of thermodynamics ($dE_{int} = dQ_m + dW_{on}$). We have

$$C_v dT = 0 + (-PdV) \qquad \text{18-33}$$

where we have used $dE_{int} = C_v dT$ (Equation 18-18a), $dQ_m = 0$ (the process is adiabatic), and $dW_{on} = -PdV$ (Equation 18-15). Then, substituting for P using $P = nRT/V$, we obtain

$$C_v \, dT = -nRT \frac{dV}{V}$$

Rearranging,

$$\frac{dT}{T} + \frac{nR}{C_v} \frac{dV}{V} = 0$$

Integration gives

$$\ln T + \frac{nR}{C_v} \ln V = \text{constant}$$

Simplifying,

$$\ln T + \frac{nR}{C_v} \ln V = \ln T + \ln V^{nR/C_v} = \ln TV^{nR/C_v} = \text{constant}$$

Thus,

$$TV^{nR/C_v} = \text{constant} \qquad \text{18-34}$$

where the constants in the two preceding equations are not the same. Equation 18-34 can be rewritten by noting that $C_p - C_v = nR$, so

$$\frac{nR}{C_v} = \frac{C_p - C_v}{C_v} = \frac{C_p}{C_v} - 1 = \gamma - 1 \qquad \text{18-35}$$

where γ is the ratio of the heat capacities:

$$\gamma = \frac{C_p}{C_v} \qquad \text{18-36}$$

Therefore,

$$TV^{\gamma-1} = \text{constant} \qquad \text{18-37}$$

We can eliminate T from Equation 18-37 using $PV = nRT$. We then have

$$\frac{PV}{nR} V^{\gamma-1} = \text{constant}$$

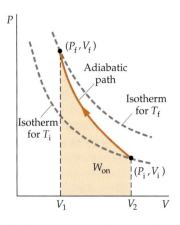

FIGURE 18-19 Quasi-static adiabatic compression of an ideal gas. The dashed lines are the isotherms for the initial and final temperatures. The curve connecting the initial and final states of the adiabatic compression is steeper than the isotherms because the temperature increases during the compression.

Clouds form if rising moist air cools due to adiabatic expansion of the air. Cooling causes water vapor to condense into liquid droplets.

or

$$PV^\gamma = \text{constant} \qquad\qquad 18\text{-}38$$

QUASI-STATIC ADIABATIC PROCESS

Equation 18-38 relates P and V for adiabatic expansions and compressions.

EXERCISE Show that for quasi-static adiabatic process $T^\gamma/P^{\gamma-1} = \text{constant}$.

The work done on the gas in an adiabatic compression can be calculated from the first law of thermodynamics:

$$dE_{int} = dQ_{in} + dW_{on} \quad \text{or} \quad dW_{on} = dE_{int} - dQ_{in}$$

Since $dE_{int} = C_v\, dT$ and $dQ_{in} = 0$, we have

$$dW_{on} = C_v\, dT$$

Then

$$W_{adiabatic} = \int dW_{on} = \int C_v dT = C_v \Delta T \qquad\qquad 18\text{-}39$$

ADIABATIC WORK ON GAS

where we have assumed that C_v is constant.[†] We note that the work done on the gas depends only on the change in the temperature of the gas. In an adiabatic compression, work is done on the gas, and its internal energy and temperature increase. In a quasi-static adiabatic *expansion*, work is done *by* the gas, and the internal energy and temperature decrease.

We can use the ideal-gas law to write Equation 18-39 in terms of the initial and final values of the pressure and volume. If T_i is the initial temperature and T_f is the final temperature, we have for the work done

$$W_{adiabatic} = C_v \Delta T = C_v(T_f - T_i)$$

Using $PV = nRT$, we obtain

$$W_{adiabatic} = C_v\left(\frac{P_f V_f}{nR} - \frac{P_i V_i}{nR}\right) = \frac{C_v}{nR}(P_f V_f - P_i V_i)$$

Using Equation 18-35 to simplify this expression, we have

$$W_{adiabatic} = \frac{P_f V_f - P_i V_i}{\gamma - 1} \qquad\qquad 18\text{-}40$$

ADIABATIC WORK ON GAS

QUASI-STATIC ADIABATIC COMPRESSION OF AIR   **EXAMPLE 18-12**

A quantity of air is compressed adiabatically and quasi-statically from an initial pressure of 1 atm and a volume of 4 L at temperature 20°C to half its original volume. Find (a) the final pressure, (b) the final temperature, and (c) the work done on the gas.

[†] For an ideal gas, E_{int} is proportional to the absolute temperature, and therefore $C_v = dE_{int}/dT$ is a constant.

PICTURE THE PROBLEM Because the process is both quasi-static and adiabatic, we know that PV^γ = constant, and $TV^{\gamma-1}$ = constant. These relations yield the final pressure and final temperature, respectively. Find γ using Equations 18-36, 18-20, and 18-25. The work done is found from $W_{adiabatic} = C_v\Delta T$ (Equation 18-39). For a diatomic gas, $C_v = \frac{5}{2}nR$. Let subscript 1 refer to initial values, and subscript 2 to final values. Then $P_1 = 1$ atm, $V_1 = 4$ L, $V_2 = 2$ L, $T_1 = 20°C = 293$ K.

(a) 1. Write PV^γ = constant in terms of initial and final values:

$$P_1 V_1^\gamma = P_2 V_2^\gamma$$

2. Find γ for a diatomic gas using Equations 18-36, 18-20, and 18-25:

$$\gamma = \frac{C_p}{C_v} = \frac{C_v + nR}{C_v} = 1 + \frac{nR}{C_v} = 1 + \frac{nR}{\frac{5}{2}nR}$$

3. Solve for P_2:

$$P_2 = P_1\left(\frac{V_1}{V_2}\right)^\gamma = (1\text{ atm})\left(\frac{4\text{ L}}{2\text{ L}}\right)^{1.4} = \boxed{2.64\text{ atm}}$$

(b) 1. Write $TV^{\gamma-1}$ = constant in terms of initial and final values:

$$T_1 V_1^{\gamma-1} = T_2 V_2^{\gamma-1}$$

2. Solve for T_2:

$$T_2 = T_1\left(\frac{V_1}{V_2}\right)^{\gamma-1} = (293\text{ K})\left(\frac{4\text{ L}}{2\text{ L}}\right)^{0.4}$$

$$= 387\text{ K} = \boxed{114°C}$$

(c) 1. Equation 18-39 gives the work done:

$$W_{adiabatic} = C_v\Delta T = \frac{5}{2}nR\Delta T$$

2. Using the ideal-gas law for the initial conditions, express nR in terms of P_1, V_1, and T_1:

$$W_{adiabatic} = \frac{5}{2}nR\Delta T = \frac{5}{2}\frac{P_1 V_1}{T_1}(T_2 - T_1)$$

$$= \frac{5}{2}\frac{(1\text{ atm})(4\text{ L})}{293\text{ K}}(387\text{ K} - 293\text{ K})$$

$$= 3.20\text{ L·atm} = \boxed{324\text{ J}}$$

REMARKS The work can also be calculated using Equation 18-40, but using $W_{adiabatic} = C_v\Delta T$ is preferable because it is more directly connected to a principle, the first law of thermodynamics, and thus is easier to recall.

Speed of Sound Waves

We can use Equation 18-38 to calculate the adiabatic bulk modulus of an ideal gas, which is related to the speed of sound waves in air. We first compute the differential of both sides of PV^γ = constant (Equation 18-38):

$$P\,d(V^\gamma) + V^\gamma\,dP = 0$$

or

$$\gamma PV^{\gamma-1}\,dV + V^\gamma\,dP = 0$$

Then

$$dP = -\frac{\gamma P\,dV}{V}$$

Referring to Equation 13-6, the adiabatic bulk modulus[†] is then:

$$B_{adiabatic} = -\frac{dP}{dV/V} = \gamma P \qquad\qquad 18\text{-}41$$

[†] The bulk modulus, discussed in Chapter 13, is the negative ratio of the pressure change to the fractional change in volume, $B = -\Delta P/(\Delta V/V)$. The isothermal bulk modulus, which describes changes that occur at constant temperature, differs from the adiabatic bulk modulus, which describes changes with no heat transfer. For sound waves at audible frequencies, the pressure changes occur too rapidly for appreciable heat flow, so the appropriate bulk modulus is the adiabatic bulk modulus.

The speed of sound (Equation 15-4) is given by

$$v = \sqrt{\frac{B_{adiabatic}}{\rho}}$$

where the mass density ρ is related to the number of moles n and the molecular mass M by $\rho = m/V = nM/V$. Using the ideal-gas law, $PV = nRT$, we can eliminate V from the density

$$\rho = \frac{nM}{V} = \frac{nM}{nRT/P} = \frac{MP}{RT}$$

Using this result and γP for $B_{adiabatic}$, we obtain

$$v = \sqrt{\frac{B_{adiabatic}}{\rho}} = \sqrt{\frac{\gamma P}{MP/(RT)}} = \sqrt{\frac{\gamma RT}{M}}$$

which is Equation 15-5, the speed of sound in a gas.

SUMMARY

1. The first law of thermodynamics, which is a statement of the conservation of energy, is a fundamental law of physics.
2. The equipartition theorem is a fundamental law of classical physics. It breaks down if the typical thermal energy kT is small compared to the spacing of quantized energy levels.

Topic	Relevant Equations and Remarks
1. Heat	Heat is energy that is transferred from one object to another because of a temperature difference.
Calorie	The calorie, originally defined as the heat necessary to raise the temperature of 1 g of water by 1°C, is now defined to be 4.184 joules.
2. Heat Capacity	Heat capacity is the amount of heat needed to raise the temperature of a substance by one degree. $$C = \frac{Q}{\Delta T} \qquad \text{18-1}$$
At constant volume	$$C_v = \frac{Q_v}{\Delta T}$$
At constant pressure	$$C_p = \frac{Q_p}{\Delta T}$$
Specific heat (heat capacity per unit mass)	$$c = \frac{C}{m} \qquad \text{18-2}$$
Molar specific heat (heat capacity per mole)	$$c' = \frac{C}{n} \qquad \text{18-6}$$

Heat capacity–internal energy relation	$C_v = \dfrac{dE_{int}}{dT}$	**18-18a**
Ideal gas	$C_p = C_v + nR$	**18-20**
Monatomic ideal gas	$C_v = \frac{3}{2}nR$	**18-22**
Diatomic ideal gas	$C_v = \frac{5}{2}nR$	**18-25**

3. Fusion and Vaporization   Both melting and vaporization occur at a constant temperature.

Latent heat of fusion	The heat needed to melt a substance is the product of the mass of the substance and its latent heat of fusion L_f:	
	$Q_f = mL_f$	**18-8**
L_f of water	$L_f = 333.5$ kJ/kg	
Latent heat of vaporization	The heat needed to vaporize a liquid is the product of the mass of the liquid and its latent heat of vaporization L_v:	
	$Q_v = mL_v$	**18-9**
L_v of water	$L_v = 2257$ kJ/kg	

4. First Law of Thermodynamics   The change in the internal energy of a system equals the heat transferred into the system plus the work done on the system:

$$\Delta E_{int} = Q_{in} + W_{on} \qquad \textbf{18-10}$$

5. Internal Energy E_{int}   The internal energy of a system is a property of the state of the system, as are the pressure, volume, and temperature. Heat and work are not properties of state.

Ideal gas	E_{int} depends only on the temperature T.	
Monatomic ideal gas	$E_{int} = \frac{3}{2}nRT$	**18-12**
Internal energy related to heat capacity	$dE_{int} = C_v\,dT$	**18-18b**

6. Quasi-Static Process   A quasi-static process is one that occurs slowly so that the system moves through a series of equilibrium states.

Isobaric	$P = $ constant	
Isothermal	$T = $ constant	
Adiabatic	$Q = 0$	
Adiabatic, ideal gas	$TV^{\gamma-1} = $ constant	**18-37**
	or	
	$PV^{\gamma} = $ constant	**18-38**
	where	
	$\gamma = \dfrac{C_p}{C_v}$	**18-36**

7. **Work Done on a Gas**	$W_{on} = -\int_{V_i}^{V_f} P\,dV = C_v\Delta T - Q_{in}$	18-10, 18-15, and 18-18

Constant volume	$W_{on} = -\int_{V_i}^{V_f} P\,dV = 0 \qquad V_f = V_i$	

Isobaric	$W_{on} = -\int_{V_i}^{V_f} P\,dV = -P\int_{V_i}^{V_f} dV = -P\Delta V$	

Isothermal	$W_{isothermal} = -\int_{V_i}^{V_f} P\,dV = -nRT\int_{V_i}^{V_f} \frac{dV}{V} = nRT\ln\frac{V_i}{V_f}$	18-17

Adiabatic	$W_{adiabatic} = C_v\Delta T$	18-39

8. **Equipartition Theorem**	The equipartition theorem states that if a system is in equilibrium, there is an average energy of $\frac{1}{2}kT$ per molecule or $\frac{1}{2}RT$ per mole associated with each degree of freedom.
Failure of the equipartition theorem	The equipartition theorem fails if the thermal energy ($\sim kT$) that can be transferred in collisions is smaller than the energy gap ΔE between quantized energy levels. For example, monatomic gas molecules cannot rotate because the first nonzero energy permitted is much greater than kT.
9. **Dulong–Petit Law**	The molar specific heat of most solids is $3R$. This is predicted by the equipartition theorem, assuming a solid atom has six degrees of freedom.

PROBLEMS

- • Single-concept, single-step, relatively easy
- •• Intermediate-level, may require synthesis of concepts
- ••• Challenging
- **SSM** Solution is in the *Student Solutions Manual*
- **iSOLVE** Problems available on iSOLVE online homework service
- **iSOLVE** ✓ These "Checkpoint" online homework service problems ask students additional questions about their confidence level, and how they arrived at their answer

In a few problems, you are given more data than you actually need; in a few other problems, you are required to supply data from your general knowledge, outside sources, or informed estimates.

Use $v = 340$ m/s for the speed of sound in air unless otherwise indicated.

Conceptual Problems

1 • Body A has twice the mass and twice the specific heat of body B. If they are supplied with equal amounts of heat, how do the subsequent changes in their temperatures compare?

2 • **SSM** The temperature change of two blocks of masses M_A and M_B is the same when they absorb equal amounts of heat. It follows that the specific heats are related by (a) $c_A = (M_A/M_B)c_B$, (b) $c_A = (M_B/M_A)c_B$, (c) $c_A = c_B$, (d) none of the above.

3 • The specific heat of aluminum is more than twice that of copper. Identical masses of copper and aluminum, both at 20°C, are dropped into a calorimeter containing water at 40°C. When thermal equilibrium is reached, (a) the aluminum is at a higher temperature than the copper, (b) the aluminum has absorbed less energy than the copper, (c) the aluminum has absorbed more energy than the copper, (d) both (a) and (c) are correct statements.

4 • Joule's experiment establishing the mechanical equivalence of heat involved the conversion of mechanical energy into internal energy. Give some examples of the internal energy of a system being converted into mechanical energy.

5 • SSM Can a system absorb heat with no change in its internal energy?

6 • In the equation $\Delta E_{int} = Q + W$ (the formal statement of the first law of thermodynamics), the quantities Q and W represent (a) the heat supplied to the system and the work done by the system, (b) the heat supplied to the system and the work done on the system, (c) the heat released by the system and the work done by the system, (d) the heat released by the system and the work done on the system.

7 • A real gas cools during a free expansion, though an ideal gas does not. Explain.

8 • An ideal gas at one atmosphere pressure and 300 K is confined to half of an insulated container by a thin partition. The partition is then removed and equilibrium is established. At that point of equilibrium, which of the following is correct? (a) The pressure is half an atmosphere and the temperature is 150 K, (b) the pressure is one atmosphere and the temperature is 150 K, (c) the pressure is half an atmosphere and the temperature is 300 K, (d) none of the above.

9 • A certain gas consists of ions that repel each other. The gas undergoes a free expansion with no heat exchange and no work done. How does the temperature of the gas change? Why?

10 •• SSM Two gas-filled rubber balloons of (initially) equal volume are at the bottom of a dark, cold lake. The top of the lake is warmer than the bottom. One balloon rises rapidly and expands adiabatically as it rises. The other balloon rises more slowly and expands isothermally. Which balloon is larger when it reaches the surface of the lake?

11 • A gas changes its state reversibly from A to C (Figure 18-20). The work done by the gas is (a) greatest for path A→B→C, (b) least for path A→C, (c) greatest for path A→D→C, (d) the same for all three paths.

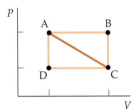

FIGURE 18-20 Problem 11

12 • When an ideal gas is subjected to an adiabatic process, (a) no work is done by the system, (b) no heat is supplied to the system, (c) the internal energy remains constant, (d) the heat supplied to the system equals the work done by the system.

13 • True or false:

(a) The heat capacity of a body is the amount of heat it can store at a given temperature.
(b) When a system goes from state 1 to state 2, the amount of heat added to the system is the same for all processes.
(c) When a system goes from state 1 to state 2, the work done on the system is the same for all processes.
(d) When a system goes from state 1 to state 2, the change in the internal energy of the system is the same for all processes.
(e) The internal energy of a given amount of an ideal gas depends only on its absolute temperature.
(f) A quasi-static process is one in which the system is never far from being in equilibrium.
(g) For any material that expands when heated, C_p is greater than C_v.

14 • SSM If a system's volume remains constant while undergoing changes in temperature and pressure, then (a) the internal energy of the system is unchanged, (b) the system does no work, (c) the system absorbs no heat, (d) the change in internal energy equals the heat absorbed by the system.

15 • When an ideal gas is subjected to an isothermal process, (a) no work is done by the system, (b) no heat is supplied to the system, (c) the heat supplied to the system equals the change in internal energy, (d) the heat supplied to the system equals the work done by the system.

16 •• The 1-L fuel tank of a gas grill contains 600 g of propane (C_3H_8) at a pressure of 2 MPa. What can you say about the phase state of the propane?

17 •• An ideal gas undergoes a process during which $P\sqrt{V}$ = constant and the volume of the gas decreases. What happens to the temperature?

18 •• SSM Which would you expect to have a higher heat capacity per *unit mass,* lead or copper? Why? (Don't look up the heat capacities to answer this question.)

19 •• Calculating the heat capacity of a liquid is very difficult, because of the strong intermolecular interactions and the random positions of the molecules in a liquid. However, simply based on counting degrees of freedom, would you expect a monatomic liquid to have a higher or lower heat capacity than the solid phase of the same substance it melted from? Assume that the melting temperature is high enough that you don't have to take quantum effects into consideration when calculating the heat capacity of the solid.

Estimation and Approximation

20 •• SSM A simple demonstration to show that heat is a form of energy is to take a bag of lead shot and drop it repeatedly onto a very rigid surface from a small height. The bag's temperature will increase, allowing an estimate of the heat capacity of lead. (a) Estimate the temperature increase of a bag filled with 1 kg of lead shot dropped 50 times from a height of 1 m. (b) In principle, the change in temperature is independent of the mass of the shot in the bag; in practice, it's better to use a larger mass than a smaller one. Why might this be true?

21 •• A "typical" microwave oven has a power consumption of about 1200 W. Estimate how long it should take to boil a cup of water in the microwave assuming that 50% of the power consumption goes into heating the water. Does this estimate mesh with everyday experience?

22 • A demonstration of the heating of a gas under adiabatic compression involves putting a few scraps of paper into a large glass test tube, which is then sealed off with a piston. If the piston compresses the trapped air very rapidly, the paper will catch fire. Assuming that the burning point of paper is 451°F, estimate the factor by which the volume of the air trapped by the piston must be reduced for this demonstration to work.

23 •• There is a small change in the volume of a liquid on heating at constant pressure. Use the following data for water to estimate the contribution this makes to the heat capacity of water between 4°C and 100°C:

Density of water at 4°C and 1 atm pressure: 1.000 g/cm^3

Density of liquid water at 100°C and 1 atm pressure: 0.9584 g/cm^3

24 •• **SSM** A certain molecule has vibrational energy levels that are equally spaced by 0.15 eV. Find the critical temperature T_c so for $T \gg T_c$ you would expect the equipartition theorem to hold and for $T \ll T_c$ you would expect the equipartition theorem to fail.

Heat Capacity; Specific Heat; Latent Heat

25 • **SSM** A "typical" adult male consumes about 2500 kcal of food in a day. (*a*) How many joules is this? (*b*) If this consumed energy is dissipated over the course of 24 hours, what is his average output power in watts?

26 • **SOLVE** A solar home contains 10^5 kg of concrete (specific heat = 1.00 kJ/kg·K). How much heat is given off by the concrete when it cools from 25 to 20°C?

27 • **SOLVE** How many calories must be supplied to 60 g of ice at −10°C to melt it and raise the temperature of the water to 40°C?

28 •• **SOLVE** How much heat must be removed when 100 g of steam at 150°C is cooled and frozen into 100 g of ice at 0°C? (Take the specific heat of steam to be 2.01 kJ/kg·K.)

29 •• **SOLVE** A 50-g piece of aluminum at 20°C is cooled to −196°C by placing it in a large container of liquid nitrogen at that temperature. How much nitrogen is vaporized? (Assume that the specific heat of aluminum is constant and is equal to 0.90 J/kg·K.)

30 •• **SOLVE** If 500 g of molten lead at 327°C is poured into a cavity in a large block of ice at 0°C, how much of the ice melts?

31 •• **SSM** **SOLVE** A 30-g lead bullet initially at 20°C comes to rest in the block of a ballistic pendulum. Assume that half the initial kinetic energy of the bullet is converted into thermal energy within the bullet. If the speed of the bullet was 420 m/s, what is the temperature of the bullet immediately after coming to rest in the block?

32 •• **SOLVE** A 1400-kg car traveling at 80 km/h is brought to rest by applying the brakes. If the specific heat of steel is 0.11 cal/g·K, what total mass of steel must be contained in the steel brake drums if the temperature of the brake drums is not to rise by more than 120 C°?

Calorimetry

33 • **SOLVE** A 200-g piece of lead is heated to 90°C and is then dropped into a calorimeter containing 500 g of water that is initially at 20°C. Neglecting the heat capacity of the container, find the final temperature of the lead and water.

34 • **SSM** **SOLVE** The specific heat of a certain metal can be determined by measuring the temperature change that occurs when a piece of the metal is heated and then placed in an insulated container made of the same material and containing water. Suppose a piece of metal has a mass of 100 g and is initially at 100°C. The container has a mass of 200 g and contains 500 g of water at an initial temperature of 20.0°C. The final temperature is 21.4°C. What is the specific heat of the metal?

35 •• In the 2002 Tour de France, champion bicyclist Lance Armstrong expended an average power of about 400 W, 5 hours a day for 20 days. What quantity of water, initially at 24°C, could be brought to the boiling point by harnessing all of that energy?

36 •• A 25-g glass tumbler contains 200 mL of water at 24°C. If two 15-g ice cubes, each at a temperature of −3°C, are dropped into the tumbler, what is the final temperature of the drink? Neglect thermal conduction between the tumbler and the room.

37 •• A 200-g piece of ice at 0°C is placed in 500 g of water at 20°C. The system is in a container of negligible heat capacity and is insulated from its surroundings. (*a*) What is the final equilibrium temperature of the system? (*b*) How much of the ice melts?

38 •• **SOLVE** ✓ A 3.5-kg block of copper at a temperature of 80°C is dropped into a bucket containing a mixture of ice and water whose total mass is 1.2 kg. When thermal equilibrium is reached, the temperature of the water is 8°C. How much ice was in the bucket before the copper block was placed in it? (Neglect the heat capacity of the bucket.)

39 •• **SOLVE** ✓ A well-insulated bucket contains 150 g of ice at 0°C. (*a*) If 20 g of steam at 100°C is injected into the bucket, what is the final equilibrium temperature of the system? (*b*) Is any ice left afterward?

40 •• **SOLVE** ✓ A calorimeter of negligible mass contains 1 kg of water at 303 K and 50 g of ice at 273 K. Find the final temperature T. Solve the same problem if the mass of ice is 500 g.

41 •• **SSM** A 200-g aluminum calorimeter contains 500 g of water at 20°C. A 100-g piece of ice cooled to −20°C is placed in the calorimeter. (*a*) Find the final temperature of the system, assuming no heat loss. (Assume that the specific heat of ice is 2.0 kJ/kg·K.) (*b*) A second 200-g piece of ice at −20°C is added. How much ice remains in the system after it reaches equilibrium? (*c*) Would you give a different answer for (*b*) if both pieces of ice were added at the same time?

42 •• **SOLVE** The specific heat of a 100-g block of material is to be determined. The block is placed in a 25-g copper calorimeter that also holds 60 g of water. The system is initially at 20°C. Then 120 mL of water at 80°C are added to the calorimeter vessel. When thermal equilibrium is attained, the temperature of the water is 54°C. Determine the specific heat of the block.

43 •• A 100-g piece of copper is heated in a furnace to a temperature t. The copper is then inserted into a 150-g copper calorimeter containing 200 g of water. The initial temperature of the water and calorimeter is 16°C, and the final temperature after equilibrium is established is 38°C. When the calorimeter and its contents are weighed, 1.2 g of water are found to have evaporated. What was the temperature t?

44 •• A 200-g aluminum calorimeter contains 500 g of water at 20°C. Aluminum shot with a mass 300 g is heated to 100°C and is then placed in the calorimeter. (*a*) Using the value of the specific heat of aluminum given in Table 18-1, find the final temperature of the system, assuming that no heat is lost to the surroundings. (*b*) The error due to heat transfer between the system and its surroundings can be minimized if the initial temperature of the water and calorimeter is chosen to be below room temperature, where Δt_w is the temperature change of the calorimeter and water during the measurement. Then the final temperature is $\frac{1}{2}\Delta t_w$ above room temperature. What should the initial temperature of the water and container be if the room temperature is 20°C?

First Law of Thermodynamics

45 • **SOLVE** A diatomic gas does 300 J of work and also absorbs 600 cal of heat. What is the change in internal energy of the gas?

46 • **SSM** **SOLVE** If 400 kcal is added to a gas that expands and does 800 kJ of work, what is the change in the internal energy of the gas?

47 • **SOLVE** A lead bullet moving at 200 m/s is stopped in a block of wood. Assuming that all of the energy change goes into heating the bullet, find the final temperature of the bullet if its initial temperature is 20°C.

48 • (*a*) At Niagara Falls, the water drops 50 m. If the change in potential energy goes into the internal energy of the water, compute the increase in its temperature. (*b*) Do the same for Yosemite Falls, where the water drops 740 m. (These temperature rises are not observed because the water cools by evaporation as it falls.)

49 • When 20 cal of heat are absorbed by a gas, the system performs 30 J of work. What is the change in the internal energy of the gas?

50 •• **SOLVE** A lead bullet initially at 30°C just melts upon striking a target. Assuming that all of the initial kinetic energy of the bullet goes into the internal energy of the bullet to raise its temperature and melt it, calculate the speed of the bullet upon impact.

51 •• **SSM** On a cold day you can warm your hands by rubbing them together. (*a*) Assume that the coefficient of friction between your hands is 0.5, that the normal force between your hands is 35 N, and that you rub them together at an average speed of 35 cm/s. What is the rate at which heat is generated? (*b*) Assume further that the mass of each of your hands is approximately 350 g, that the specific heat of your hands is about 4 kJ/kg·K, and that all the heat generated goes into raising the temperature of your hands. How long must you rub your hands together to produce a 5 C° increase in their temperature?

Work and the *PV* Diagram for a Gas

In Problems 52 through 55, the initial state of 1 mol of an ideal gas is $P_1 = 3$ atm, $V_1 = 1$ L, and $E_{int,1} = 456$ J, and its final state is $P_2 = 2$ atm, $V_2 = 3$ L, and $E_{int,2} = 912$ J.

52 • **SOLVE** The gas is allowed to expand at constant pressure to a volume of 3 L. It is then cooled at constant volume until its pressure is 2 atm. (*a*) Show this process on a *PV* diagram, and calculate the work done by the gas. (*b*) Find the heat added during this process.

53 • **SOLVE** The gas is first cooled at constant volume until its pressure is 2 atm. It is then allowed to expand at constant pressure until its volume is 3 L. (*a*) Show this process on a *PV* diagram, and calculate the work done by the gas. (*b*) Find the heat added during this process.

54 •• **SSM** The gas is allowed to expand isothermally until its volume is 3 L and its pressure is 1 atm. It is then heated at constant volume until its pressure is 2 atm. (*a*) Show this process on a *PV* diagram, and calculate the work done by the gas. (*b*) Find the heat added during this process.

55 •• The gas is heated and is allowed to expand such that it follows a straight-line path on a *PV* diagram from its initial state to its final state. (*a*) Show this process on a *PV* diagram, and calculate the work done by the gas. (*b*) Find the heat added during this process.

56 •• **SOLVE** One mole of the ideal gas is initially in the state $P_0 = 1$ atm, $V_0 = 25$ L. As the gas is slowly heated, the plot of its state on a *PV* diagram moves in a straight line to the state $P = 3$ atm, $V = 75$ L. Find the work done by the gas.

57 •• One mole of the ideal gas is heated while its volume changes, so that $T = AP^2$, where A is a constant. The temperature changes from T_0 to $4T_0$. Find the work done by the gas.

58 • **SSM** An *isobaric* expansion is one carried out at constant pressure. Draw several isobars for an ideal gas on a diagram showing volume as a function of temperature.

59 •• **SOLVE** An ideal gas initially at 20°C and 200 kPa has a volume of 4 L. It undergoes a quasi-static, isothermal expansion until its pressure is reduced to 100 kPa. Find (*a*) the work done by the gas, and (*b*) the heat added to the gas during the expansion.

Heat Capacities of Gases and the Equipartition Theorem

60 • The heat capacity at constant volume of a certain amount of a monatomic gas is 49.8 J/K. (*a*) Find the number of moles of the gas. (*b*) What is the internal energy of the gas at $T = 300$ K? (*c*) What is the heat capacity of the gas at constant pressure?

61 • The Dulong–Petit law was originally used to determine the molecular mass of a substance from its measured heat capacity. The specific heat of a certain solid is measured to be 0.447 kJ/kg·K. (*a*) Find the molecular mass of the substance. (*b*) What element is this?

62 •• **SSM** (*a*) Calculate the specific heats per unit mass of air at constant volume and constant pressure. Assume a temperature of 300 K and a pressure of 10^5 N/m². Assume that air is composed of 74% N_2 (molecular weight 28 g/mole) molecules and 26% O_2 molecules (molecular weight 32 g/mole) and that both components are ideal gases. (*b*) Compare your answer to the value listed in the *Handbook of Chemistry and Physics* for the heat capacity at constant pressure of 1.032 J/g·K.

63 •• One mole of an ideal diatomic gas is heated at constant volume from 300 to 600 K. (a) Find the increase in internal energy, the work done, and the heat added. (b) Find the same quantities if this gas is heated from 300 to 600 K at constant pressure. Use the first law of thermodynamics and your results for (a) to calculate the work done. (c) Calculate the work done in (b) directly from $dW = P \, dV$.

64 •• **SOLVE✓** A diatomic gas (molar mass M) is confined to a closed container of volume V at a pressure P_0. What amount of heat Q should be transferred to the gas in order to triple the pressure? (Express your answer in terms of P_0 and V.)

65 •• One mole of air ($c_v = 5R/2$) is confined at atmospheric pressure in a cylinder with a piston at 0°C. The initial volume, occupied by gas, is V. Find the volume of gas V' after the equivalent of 13,200 J of heat is transferred to it.

66 •• The heat capacity of a certain amount of a particular gas at constant pressure is greater than that at constant volume by 29.1 J/K. (a) How many moles of the gas are there? (b) If the gas is monatomic, what are C_v and C_p? (c) If the gas consists of diatomic molecules that rotate but do not vibrate, what are C_v and C_p?

67 •• **SSM** Carbon dioxide (CO_2) at 1 atm of pressure and a temperature of −78.5°C sublimates directly from a solid to a gaseous state, without going through a liquid phase. What is the change in the heat capacity (at constant pressure) per mole of CO_2 when it undergoes sublimation? Assume that the gas molecules can rotate but do not vibrate. Is the change in the heat capacity positive or negative on sublimation? The CO_2 molecule is pictured in Figure 18-21.

FIGURE 18-21 Problem 67

O   C   O

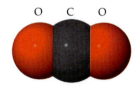

68 •• One mole of a monatomic ideal gas is initially at 273 K and 1 atm. (a) What is its initial internal energy? (b) Find its final internal energy and the work done by the gas when 500 J of heat are added at constant pressure. (c) Find the same quantities when 500 J of heat are added at constant volume.

69 •• List all of the degrees of freedom possible for a water molecule and estimate the heat capacity of water at a temperature very far above its boiling point. (Ignore the fact the molecule might dissociate at high temperatures.) Think carefully about all of the different ways in which a water molecule can vibrate.

Quasi-Static Adiabatic Expansion of a Gas

70 •• One mole of an ideal gas ($\gamma = \frac{5}{3}$) expands adiabatically and quasi-statically from a pressure of 10 atm and a temperature of 0°C to a pressure of 2 atm. Find (a) the initial and final volumes, (b) the final temperature, and (c) the work done by the gas.

71 • An ideal gas at a temperature of 20°C is compressed quasi-statically and adiabatically to half its original volume. Find its final temperature if (a) $C_v = \frac{3}{2}nR$ and (b) $C_v = \frac{5}{2}nR$.

72 • **SOLVE✓** Two moles of neon gas initially at 20°C and a pressure of 1 atm are compressed adiabatically to one-fourth of their initial volume. Determine the temperature and pressure following compression.

73 •• **SSM** Half a mole of an ideal monatomic gas at a pressure of 400 kPa and a temperature of 300 K expands until the pressure has diminished to 160 kPa. Find the final temperature and volume, the work done, and the heat absorbed by the gas if the expansion is (a) isothermal and (b) adiabatic.

74 •• Repeat Problem 73 for a diatomic gas.

75 •• One-half mole of helium is expanded adiabatically and quasi-statically from an initial pressure of 5 atm and temperature of 500 K to a final pressure of 1 atm. Find (a) the final temperature, (b) the final volume, (c) the work done by the gas, and (d) the change in the internal energy of the gas.

76 ••• **SSM** A hand pump is used to inflate a bicycle tire to a gauge pressure of 482 kPa (about 70 lb/in.²). How much work must be done if each stroke of the pump is an adiabatic process? Atmospheric pressure is 1 atm, the air temperature is initially 20°C, and the volume of the air in the tire remains constant at 1 L.

77 ••• An ideal gas at initial volume V_1 and pressure P_1 expands quasi-statically and adiabatically to volume V_2 and pressure P_2. Calculate the work done by the gas directly by integrating $P \, dV$, and show that your result is the same as that given by Equation 18-39.

Cyclic Processes

78 •• One mole of N_2 ($C_v = \frac{5}{2}nR$) gas is originally at room temperature (20°C) and a pressure of 5 atm. It is allowed to expand adiabatically and quasi-statically until its pressure equals the room pressure of 1 atm. It is then heated at constant pressure until its temperature is again 20°C. During this heating, the gas expands. After it reaches room temperature, it is heated at constant volume until its pressure is 5 atm. It is then compressed at constant pressure until it is back to its original state. (a) Construct an accurate PV diagram showing each process in the cycle. (b) From your graph, determine the work done by the gas during the complete cycle. (c) How much heat is added or subtracted from the gas during the complete cycle? (d) Check your graphical determination of the work done by the gas in (b) by calculating the work done during each part of the cycle.

79 •• **SSM** One mole of an ideal diatomic gas is allowed to expand along the straight line from 1 to 2 in the PV diagram (Figure 18-22). It is then compressed back isothermally from 2 to 1. Calculate the total work done on the gas during this cycle.

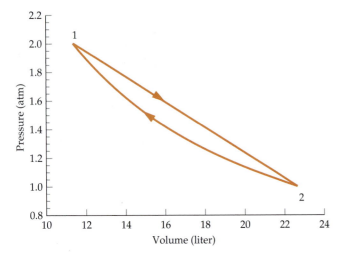

FIGURE 18-22 Problem 79

80 •• Two moles of an ideal monatomic gas have an initial pressure $P_1 = 2$ atm and an initial volume $V_1 = 2$ L. The gas is taken through the following quasi-static cycle: It is expanded isothermally until it has a volume $V_2 = 4$ L. It is then heated at constant volume until it has a pressure $P_3 = 2$ atm. It is then cooled at constant pressure until it is back to its initial state. (a) Show this cycle on a PV diagram. (b) Calculate the heat added and the work done by the gas during each part of the cycle. (c) Find the temperatures T_1, T_2, and T_3.

81 ••• At point D in Figure 18-23 the pressure and temperature of 2 mol of an ideal monatomic gas are 2 atm and 360 K. The volume of the gas at point B on the PV diagram is three times that at point D and its pressure is twice that at point C. Paths AB and CD represent isothermal processes. The gas is carried through a complete cycle along the path DABCD. Determine the total amount of work done by the gas and the heat supplied to the gas along each portion of the cycle.

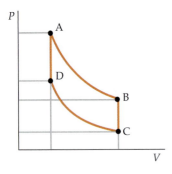

FIGURE 18-23 Problems 81 and 82

82 ••• **SSM** Repeat Problem 81 with a diatomic gas.

83 ••• An ideal gas of n mol is initially at pressure P_1, volume V_1, and temperature T_h. It expands isothermally until its pressure and volume are P_2 and V_2. It then expands adiabatically until its temperature is T_c and its pressure and volume are P_3 and V_3. It is then compressed isothermally until it is at a pressure P_4 and a volume V_4, which is related to its initial volume V_1 by $T_c V_4^{\gamma-1} = T_h V_1^{\gamma-1}$. The gas is then compressed adiabatically until it is back in its original state. (a) Assuming that each process is quasi-static, plot this cycle on a PV diagram. (This cycle is known as the Carnot cycle for an ideal gas.)

(b) Show that the heat Q_h absorbed during the isothermal expansion at T_h is $Q_h = nRT_h \ln(V_2/V_1)$. (c) Show that the heat Q_c given off by the gas during the isothermal compression at T_c is $Q_c = nRT_c \ln (V_3/V_4)$. (d) Using the result that $TV^{\gamma-1}$ is constant for an adiabatic expansion, show that $V_2/V_1 = V_3/V_4$. (e) The efficiency of a Carnot cycle is defined to be the net work done divided by the heat absorbed Q_h. Using the first law of thermodynamics, show that the efficiency is $1 - Q_c/Q_h$. (f) Using your results from the previous parts of this problem, show that $Q_c/Q_h = T_c/T_h$.

General Problems

84 • **ISOLVE✔** The volume of three moles of a monatomic gas is increased from 50 L to 200 L at constant pressure. The initial temperature of the gas is 300 K. How much heat must be supplied to the gas?

85 • In the process of compressing n moles of an ideal diatomic gas to one-fifth of its initial volume, 180 kJ of work is done on the gas. If this is accomplished isothermally at room temperature (293 K), how many calories of heat are removed from the gas?

86 • **SSM** What is the number of moles n of the gas in Problem 85?

87 • The PV diagram in Figure 18-24 represents 3 mol of an ideal monatomic gas. The gas is initially at point A. The paths AD and BC represent isothermal changes. If the system is brought to point C along the path AEC, find (a) the initial and final temperatures, (b) the work done by the gas, and (c) the heat absorbed by the gas.

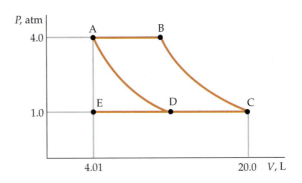

FIGURE 18-24 Problems 87-90

88 •• Repeat Problem 87 with the gas following path ABC.

89 •• **SSM** Repeat Problem 87 with the gas following path ADC.

90 •• Suppose that the paths AD and BC represent adiabatic processes. What then are the work done by the gas and the heat absorbed by the gas in following the path ABC?

91 •• At very low temperatures, the specific heat of a metal is given by $c = aT + bT^3$. For the metal copper, $a = 0.0108$ J/kg·K² and $b = 7.62 \times 10^{-4}$ J/kg·K⁴. (a) What is the specific heat of copper at 4 K? (b) How much heat is required to heat copper from 1 to 3 K?

92 •• Two moles of a diatomic ideal gas are compressed isothermally from 18 L to 8 L. In the process, 170 calories escape from the system. Determine the amount of work done by the gas, the change in internal energy, and the initial and final temperatures of the gas.

93 •• Suppose the two moles of a diatomic ideal gas in Problem 92 are compressed from 18 L to 8 L adiabatically. The work done on the gas is 820 J. Find the initial temperature and the initial and final pressures.

94 •• How much work must be done to 30 grams of CO at standard temperature and pressure to compress it to one-fifth of its initial volume if the process is (a) isothermal; (b) adiabatic?

95 •• Repeat Problem 94 if the gas is CO_2.

96 •• Repeat Problem 94 if the gas is argon.

97 •• **ISOLVE** A thermally insulated system consists of 1 mol of a diatomic ideal gas at 100 K and 2 mol of a solid at 200 K that are separated by a rigid insulating wall. Find the equilibrium temperature of the system after the insulating wall is removed, assuming that the solid obeys the Dulong–Petit law.

98 •• **SSM** When an ideal gas undergoes a temperature change at constant volume, its energy changes by $\Delta E_{int} = C_v \Delta T$. (a) Explain why this result holds for an ideal gas for any temperature change independent of the process. (b) Show explicitly that this result holds for the expansion of an ideal gas at constant pressure by first calculating the work done and showing that it can be written as $W = nR\Delta T$, and then by using $\Delta E_{int} = Q - W$, where $Q = C_p\Delta T$.

99 •• One mole of an ideal monatomic gas is heated at constant volume from 300 to 600 K. (a) Find the heat added, the work done by the gas, and the change in its internal energy. (b) Find these same quantities if the gas is heated from 300 to 600 K at constant pressure.

100 •• **SSM** Heat in the amount of 500 J is supplied to 2 mol of an ideal diatomic gas. (a) Find the change in temperature if the pressure is kept constant. (b) Find the work done by the gas. (c) Find the ratio of the final volume of the gas to the initial volume if the initial temperature is 20°C.

101 •• An insulated cylinder is fitted with a movable piston to maintain constant pressure. The cylinder initially contains 100 g of ice at −10°C. Heat is supplied to the contents at a constant rate by a 100-W heater. Make a graph showing the temperature of the cylinder contents as a function of time starting at $t = 0$, when the temperature is −10°C, and ending when the temperature is 110°C. (Use $c = 2.0$ kJ/kg·K for the average specific heat of ice from −10 to 0°C and of steam from 100 to 110°C.)

102 •• **SSM** Two moles of a diatomic ideal gas expand adiabatically. The initial temperature of the gas is 300 K. The work done by the gas during the expansion is 3.5 kJ. What is the final temperature of the gas?

103 •• One mole of monatomic gas, initially at temperature T, undergoes a process in which its temperature is quadrupled and its volume is halved. Find the amount of heat Q transferred to the gas. It is known that in this process the pressure was never less than the initial pressure, and the work done on the gas was the minimum possible.

104 •• If a small amount of a substance is dissolved into a liquid, the liquid pressure will rise slightly. For a dilute solution, the change in pressure follows the ideal-gas law: $PV = NkT$, where N is the number of solute molecules dissolved in the liquid, V is the liquid volume, and P is the increase in the liquid pressure. Calculate the increase in pressure when 20 g of table salt (NaCl) are dissolved in 1 L of water at a temperature of 24°C.

105 •• A vertical heat-insulated cylinder is divided into two parts by a movable piston of mass m. Initially the piston is held at rest. The top part is evacuated and the bottom part is filled with 1 mole of diatomic ideal gas at temperature 300 K. After the piston is released and the system comes to equilibrium, the volume, occupied by gas, is halved. Find the final temperature of the gas.

106 •• According to the Einstein model of a crystalline solid, the internal energy per mole is given by

$$U = \frac{3N_A kT_E}{e^{T_E/T} - 1}$$

where T_E is a characteristic temperature called the Einstein temperature, and T is the temperature of the solid in kelvins. Calculate the molar internal energy of diamond ($T_E = 1060$ K) at 300 K and 600 K, and thereby the increase in internal energy as diamond is heated from 300 K to 600 K.

107 ••• **SSM** In an isothermal expansion, an ideal gas at an initial pressure P_0 expands until its volume is twice its initial volume. (a) Find its pressure after the expansion. (b) The gas is then compressed adiabatically and quasi-statically back to its original volume, at which point its pressure is $1.32P_0$. Is the gas monatomic, diatomic, or polyatomic? (c) How does the translational kinetic energy of the gas change in these processes?

Note: Problems 108 and 109 involve nonquasi-static processes. Nevertheless, assuming that the gases participating in these processes approximate ideal gases, one can calculate the state functions of the end products of the reactions using the first law of thermodynamics and the ideal-gas law. For $T > 2000$ K, vibration of the atoms contributes to C_p of H_2O and CO_2 so that C_p of these gases is $7.5R$ at high temperatures. Also, assume the gases do not dissociate.

108 ••• The combustion of benzene is represented by the chemical reaction $2(C_6H_6) + 15(O_2) \rightarrow 12(CO_2) + 6(H_2O)$. The amount of energy released in the combustion of 2 mol of benzene is 1516 kcal. One mol of benzene and 7.5 mol of oxygen at 300 K are confined in an insulated enclosure at a pressure of 1 atm. (a) Find the temperature and volume following combustion if the pressure is maintained at 1 atm. (b) If, following combustion, the thermal insulation about the container is removed and the system is cooled to 300 K, what is the final pressure?

109 ••• **SSM** Carbon monoxide and oxygen combine to form carbon dioxide with an energy release of 280 kJ/mol of CO according to the reaction $2(CO) + O_2 \rightarrow 2(CO_2)$. Two mol of CO and one mol of O_2 at 300 K are confined in an 80-L container; the combustion reaction is initiated with a spark. (a) What is the pressure in the container prior to the reaction? (b) If the reaction proceeds adiabatically, what are the final temperature and pressure? (c) If the resulting CO_2 gas is cooled to 0°C, what is the pressure in the container?

110 ••• Use the expression given in Problem 106 for the internal energy per mole of a solid according to the Einstein model to show that the molar heat capacity at constant volume is given by

$$c_v' = 3R \left(\frac{T_E}{T}\right)^2 \frac{e^{T_E/T}}{(e^{T_E/T} - 1)^2}$$

111 ••• (a) Use the results of Problem 110 to show that the Dulong–Petit law, $c_v' \approx 3R$ holds for the Einstein model when $T > T_E$. (b) For diamond, T_E is approximately 1060 K. Numerically integrate $dE_{int} = c_v' \, dT$ to find the increase in the internal energy if 1 mol of diamond is heated from 300 to 600 K. Compare your result to that obtained in Problem 106.

112 ••• If a hole is punctured in a tire, the gas inside will gradually leak out of it. Let's assume the following: the area of the hole is A; the tire volume is V; and the time, τ, it takes for most of the air to leak out of the tire can be expressed in terms of the ratio A/V, the temperature T, the Boltzmann constant k, and the mass of the gas molecules inside the tire, m. (a) Under these assumptions, use dimensional analysis to find an estimate for τ. (b) Use the result of Part (a) to estimate the time it takes for a car tire with a nail hole punched in it to go flat.

The Second Law of Thermodynamics

THIS OLD-FASHIONED TRAIN ENGINE PRODUCES STEAM WHICH DOES WORK ON A PISTON THAT CAUSES THE TRAIN'S WHEELS TO MOVE. THE STEAM ENGINE'S EFFICIENCY IS LIMITED BY THE SECOND LAW OF THERMODYNAMICS.

? **What is the maximum possible efficiency of this engine? (See Example 19-4.)**

Solar energy is directed toward the solar oven at the center by this circular array of reflectors at Barstow, California.

We are often asked to conserve energy. But according to the first law of thermodynamics, energy is always conserved. What then does it mean to conserve energy if the total amount of energy in the universe does not change regardless of what we do? The first law of thermodynamics does not tell the whole story. Energy is always conserved, but some forms of energy are more useful than others. The possibility or impossibility of putting energy to *use* is the subject of the **second law of thermodynamics.** For example, it is easy to convert work into thermal energy, but it is impossible to remove energy as heat from a single reservoir and convert it entirely into work with no other changes. This experimental fact is one statement of the second law of thermodynamics.

No system can take energy as heat from a single reservoir and convert it entirely into work without additional net changes in the system or its surroundings.

SECOND LAW OF THERMODYNAMICS: KELVIN STATEMENT

➤ **In this chapter, we will encounter several other formulations of this law.**

A common example of the conversion of work into heat is movement with friction. For example, suppose you spend two minutes pushing a block this way and that way along a tabletop in a closed path, leaving the block in its initial position. Also, suppose that the block-table system is initially in thermal equilibrium with its surroundings. The work you do on the system is converted into internal energy of the system, and as a result the block-table system becomes warmer. Consequently, the system is no longer in thermal equilibrium with its surroundings. However, the system will transfer energy as heat to its surroundings until it returns to thermal equilibrium with those surroundings. Because the final and initial states of the system are the same, the first law of thermodynamics dictates that the energy transferred to the environment as heat equals the work done by you on the system. The reverse process never occurs—a block and table that are warm will never spontaneously cool by converting their internal energy into work that causes the block to push your hand around the table! Yet such an amazing occurrence would not violate the first law of thermodynamics or any other physical laws we have encountered so far. It does, however, violate the second law of thermodynamics. Thus, there is a lack of symmetry in the roles played by heat and work that is not evident from the first law. This lack of symmetry is related to the fact that some processes are *irreversible*.

There are many other irreversible processes, seemingly quite different from one another, but all related to the second law. For example, heat conduction is an irreversible process. If we place a hot body in contact with a cold body, heat will flow from the hot body to the cold body until they are at the same temperature. However, the reverse does not occur. Two bodies in contact at the same temperature remain at the same temperature; heat does not flow from one to the other leaving one colder and the other warmer. This experimental fact gives us a second statement of the second law of thermodynamics.

A process whose only net result is to transfer energy as heat from a cooler object to a hotter one is impossible.

SECOND LAW OF THERMODYNAMICS: CLAUSIUS STATEMENT

We will show in this chapter that the Kelvin and Clausius statements of the second law are equivalent.

19-1 Heat Engines and the Second Law of Thermodynamics

The study of the efficiency of heat engines gave rise to the first clear statements of the second law. A **heat engine** is a cyclic device whose purpose is to convert as much heat input into work as possible. Heat engines contain a **working substance** (water in a steam engine, air and gasoline vapor in an internal-combustion engine) that absorbs a quantity of heat Q_h from a high temperature reservoir, does work W on its surroundings, and gives off heat Q_c as it returns to its initial state, where

Q_h, W, and Q_c represent magnitudes and are positive.

The earliest heat engines were steam engines, invented in the eighteenth century for pumping water from coal mines. Today steam engines are used to generate electricity. In a typical steam engine, liquid water is heated under several hundred atmospheres of pressure until it vaporizes at about 500°C (Figure 19-1). This steam expands against a piston (or turbine blades), doing work, then exits at a much lower temperature and is cooled further in the condenser where heat is transferred from it, causing it to condense. The water is then pumped back into the boiler and heated again.

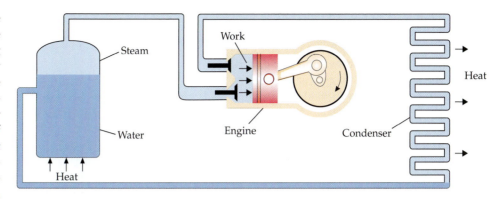

FIGURE 19-1 Schematic drawing of a steam engine. High-pressure steam does work on the piston.

Figure 19-2 is a schematic diagram of the heat engine used in many automobiles —the internal-combustion engine. With the exhaust valve closed, a mixture of gasoline vapor and air enters the combustion chamber as the piston moves down

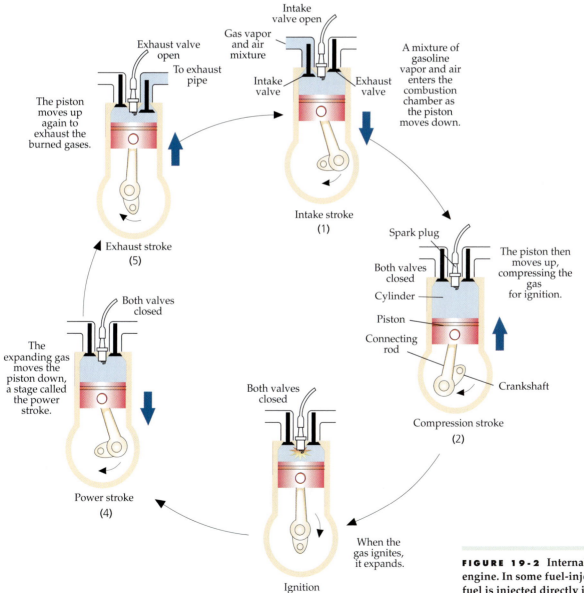

FIGURE 19-2 Internal-combustion engine. In some fuel-injected engines, the fuel is injected directly into the cylinder rather than into the air stream.

during the intake stroke. The mixture is then compressed, after which it is ignited by a spark from the spark plug. The hot gases then expand, driving the piston down and doing work on it in the stage called the power stroke. The gases are then exhausted through the exhaust valve, and the cycle repeats. An idealized model of the processes in the internal combustion engine is called the **Otto cycle** and is shown in Figure 19-3.

Figure 19-4 shows a schematic representation of a basic heat engine. The heat input is represented as coming from a **hot heat reservoir** at temperature T_h, and the exhaust goes into a **cold heat reservoir** at a lower temperature T_c. A hot or cold heat reservoir is an idealized body or system that has a very large heat capacity so that it can absorb or give off energy as heat with no noticeable change in its temperature. In practice, burning fossil fuel often acts as the high-temperature reservoir, and the surrounding atmosphere or a lake often acts as the low-temperature reservoir. Applying the first law of thermodynamics ($\Delta E_{int} = Q_{in} + W_{on}$) to the heat engine gives

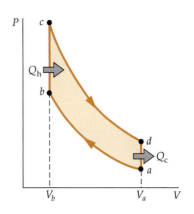

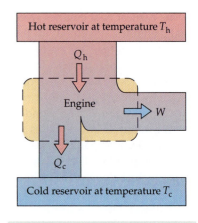

FIGURE 19-3 Otto cycle, representing the internal-combustion engine. The fuel-air mixture enters at *a* and is adiabatically compressed to *b*. It is then heated (by ignition from the spark plug) at constant volume to *c*. The power stroke is represented by the adiabatic expansion from *c* to *d*. The cooling at constant volume from *d* to *a* represents the exhausting of the combustion products and the intake of a fresh fuel-air mixture.

FIGURE 19-4 Schematic representation of a heat engine. The engine removes heat energy Q_h from a hot reservoir at a temperature T_h, does work W, and gives off heat Q_c to a cold reservoir at a temperature T_c.

$$W = Q_h - Q_c \qquad \text{19-1}$$

where W is the work done *by* the engine during one complete cycle, $Q_h - Q_c$ is the total energy transferred to the engine as heat during one cycle, and ΔE_{int} is the change in internal energy of the engine (including the working substance) during one cycle. Since the initial and final states of the engine for a complete cycle are the same, the initial and final internal energies of the engine are equal. Thus, $\Delta E_{int} = 0$.

The **efficiency** ε of a heat engine is defined as the ratio of the work done by the engine to the heat absorbed from the high temperature reservoir:

$$\varepsilon = \frac{W}{Q_h} = \frac{Q_h - Q_c}{Q_h} = 1 - \frac{Q_c}{Q_h} \qquad \text{19-2}$$

DEFINITION—EFFICIENCY OF A HEAT ENGINE

An exhaust manifold feeds the header pipes seen on this top-fuel dragster in order to carry heat away from the engine to reduce its temperature.

The heat Q_h is usually produced by burning some fuel like coal or oil that must be paid for, so it is desirable to get the most efficient use of the fuel as possible. The best steam engines operate near 40 percent efficiency; the best internal-combustion engines operate near 50 percent efficiency. At 100 percent efficiency ($\varepsilon = 1$), all the thermal energy absorbed from the hot reservoir would be converted into work and no thermal energy would be given off to the cold reservoir. However, *it is impossible to make a heat engine with an efficiency of 100 percent.* This experimental result is the **heat-engine statement of the second law of thermodynamics.** It is another way of expressing the Kelvin statement given earlier:

It is impossible for a heat engine working in a cycle to produce *only the effect* of extracting heat from a single reservoir and performing an equivalent amount of work.

SECOND LAW OF THERMODYNAMICS: HEAT-ENGINE STATEMENT

The word *cycle* in this statement is important because it *is* possible to convert heat completely into work in a noncyclic process. An ideal gas undergoing an isothermal expansion does just this. But after the expansion, the gas is not in its original state. To bring the gas back to its original state, work must be done on the gas, and some heat will be exhausted.

The second law tells us that to do work with energy extracted from a heat reservoir, we must have a colder reservoir available to receive part of the energy as exhaust. If this were not true, we could design a ship with a heat engine that was powered by simply extracting energy as heat from the ocean. Unfortunately, the lack of a colder reservoir for exhaust makes this enormous reservoir of energy unavailable for such use. (It is theoretically possible to run a heat engine between the warmer surface water of the ocean and the colder water at greater depths, but no practical scheme for using this temperature difference has yet emerged.) In order to convert completely disordered thermal energy at a single temperature into the completely ordered energy associated with work (with no other changes in the source or object), a separate cold reservoir must be used.

EFFICIENCY OF A HEAT ENGINE   **E X A M P L E   1 9 - 1**

During each cycle a heat engine absorbs 200 J of heat from a hot reservoir, does work, and exhausts 160 J to a cold reservoir. What is the efficiency of the engine?

PICTURE THE PROBLEM We use the definition of the efficiency of a heat engine (Equation 19-2).

1. The efficiency is the work done divided by the heat absorbed:

$$\varepsilon = \frac{W}{Q_h}$$

2. The heat absorbed is given:

$$Q_h = 200 \text{ J}$$

3. The work is found from the first law:

$$W = Q_h - Q_c = 200 \text{ J} - 160 \text{ J} = 40 \text{ J}$$

4. Substitute the values of Q_h and W to calculate the efficiency:

$$\varepsilon = \frac{W}{Q_h} = \frac{40 \text{ J}}{200 \text{ J}} = 0.20 = \boxed{20\%}$$

EXERCISE A heat engine has an efficiency of 35%. (*a*) How much work does it perform in a cycle if it extracts 150 J of energy as heat from a hot reservoir per cycle? (*b*) How much energy as heat is exhausted to the cold reservoir per cycle? (*Answer* (*a*) 52.5 J (*b*) 97.5 J)

EFFICIENCY OF AN IDEAL INTERNAL COMBUSTION ENGINE—   **E X A M P L E   1 9 - 2**   **Try It Yourself**
THE OTTO CYCLE

(*a*) Find the efficiency of the Otto cycle shown in Figure 19-3. (*b*) Express your answer in terms of the ratio of the volumes $r = V_a/V_b = V_d/V_c$.

PICTURE THE PROBLEM (*a*) To find ε, you need to find Q_h and Q_c. Heat transfer occurs only during the two constant-volume processes, *b* to *c* and *d* to *a*. You can thus find Q_h and Q_c and therefore ε in terms of the temperatures T_a, T_b, T_c, and T_d. (*b*) The temperatures can be related to the volumes using $TV^{\gamma-1} =$ constant for adiabatic processes.

Cover the column to the right and try these on your own before looking at the answers.

Steps

(a) 1. Write the efficiency in terms of Q_h and Q_c.

2. The heat out occurs at constant volume from d to a. Write Q_c in terms of C_v and the temperatures T_a and T_d.

3. The heat in occurs at constant volume from b to c. Write Q_h in terms of C_v and the temperatures T_c and T_b.

4. Substitute these values of Q_c and Q_h to find the efficiency in terms of the temperatures T_a, T_b, T_c, and T_d.

(b) 1. Relate T_c to T_d using $TV^{\gamma-1} = $ constant, and $V_a/V_c = r$.

2. Relate T_b to T_a as in step 1.

3. Use these relations to eliminate T_c and T_b from ε in Part (a) so that ε is expressed in terms of r.

Answers

$\varepsilon = 1 - \dfrac{Q_{cold}}{Q_{hot}} = 1 - \dfrac{Q_c}{Q_h}$

$Q_c = |Q_{d \to a}| = C_v|T_a - T_d| = C_v(T_d - T_a)$

$Q_h = Q_{b \to c} = C_v(T_c - T_b)$

$\varepsilon = \boxed{1 - \dfrac{T_d - T_a}{T_c - T_b}}$

$T_c V_c^{\gamma-1} = T_d V_d^{\gamma-1}$

$T_c = T_d \dfrac{V_d^{\gamma-1}}{V_c^{\gamma-1}} = T_d r^{\gamma-1}$

$T_b = T_a r^{\gamma-1}$

$\varepsilon = 1 - \dfrac{T_d - T_a}{T_d r^{\gamma-1} - T_a r^{\gamma-1}} = \boxed{1 - \dfrac{1}{r^{\gamma-1}}}$

REMARKS The ratio r (volume before compression/volume after compression) is called the compression ratio.

19-2 Refrigerators and the Second Law of Thermodynamics

A **refrigerator** is essentially a heat engine run backwards (Figure 19-5a.). The refrigerator's engine extracts thermal energy from the interior of the refrigerator (cold reservoir) and transfers it to the surroundings (hot reservoir) (Figure 19-5b). Experience shows that such a transfer always requires work—a result known as the **refrigerator statement of the second law of thermodynamics,** which is another way of expressing the Clausius statement:

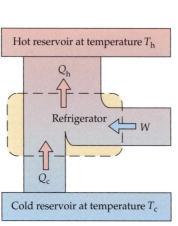

FIGURE 19-5
(a) Schematic representation of a refrigerator. Work W is done on the refrigerator and it removes heat energy Q_c from a cold reservoir and gives off heat Q_h.
(b) An actual refrigerator.

Hot reservoir at temperature T_h

Q_h

Refrigerator — W

Q_c

Cold reservoir at temperature T_c

(a)

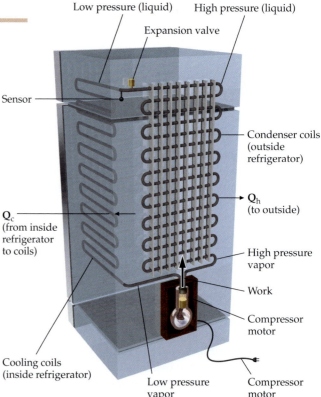

Low pressure (liquid)   High pressure (liquid)

Expansion valve

Sensor

Condenser coils (outside refrigerator)

$\mathbf{Q_h}$ (to outside)

Q_c (from inside refrigerator to coils)

High pressure vapor

Work

Compressor motor

Cooling coils (inside refrigerator)

Low pressure vapor

Compressor motor

(b)

> It is impossible for a refrigerator working in a cycle to produce *only the effect* of extracting heat from a cold object and reject the same amount of heat to a hot object.

SECOND LAW OF THERMODYNAMICS: REFRIGERATOR STATEMENT

Were the above statement not true, we could cool our homes in the summer with refrigerators that pumped thermal energy to the outside without using any electricity or any other energy.

A measure of a refrigerator's performance is the ratio Q_c/W of the heat removed from the low temperature reservoir to the work done on the refrigerator. (This work equals the electrical energy that comes from the wall outlet.) The ratio Q_c/W is called the **coefficient of performance** (COP):

$$COP = \frac{Q_c}{W} \qquad\qquad 19\text{-}3$$

DEFINITION—COEFFICIENT OF PERFORMANCE (REFRIGERATOR)

The greater the COP, the better the refrigerator. Typical refrigerators have coefficients of performance of about 5 or 6. In terms of this ratio, the refrigerator statement of the second law says that the COP of a refrigerator cannot be infinite.

MAKING ICE CUBES   **EXAMPLE 19-3**   **Put It in Context**

You have one hour before guests start arriving for your party when you suddenly realize that you forgot to buy ice for drinks. You quickly put one liter of water at 10°C into your ice cube trays and pop them into the freezer. Will you have ice in time for your guests? The label on your refrigerator states that the appliance has a coefficient of performance of 5.5 and a power rating of 550 W. You estimate that only 10 percent of the power goes to freezing the ice.

PICTURE THE PROBLEM Work equals power times time. We are given the power, so we need to find the work to determine the time. The work is related to Q_c by Equation 19-3. To find Q_c we calculate how much heat must be extracted from the water.

1. The time needed is related to the power available and the work required:

$$P = W/t$$
$$t = W/P$$

2. The work is related to the coefficient of performance and the heat extracted:

$$W = \frac{Q_c}{COP}$$

3. The heat Q_c removed from inside of the refrigerator equals the heat Q_{cool} to be removed from the water to cool it plus the heat Q_{freeze} to be removed from the water to freeze it:

$$Q_c = Q_{cool} + Q_{freeze}$$

4. The heat needed to cool 1 L of water (mass 1 kg) by 10°C is:

$$Q_{cool} = mc\Delta T$$
$$= (1\,\text{kg})(4.18\,\text{kJ}/(\text{kg}\cdot\text{K}))(10\,\text{K})$$
$$= 41.8\,\text{kJ}$$

5. The heat needed to freeze 1 L of water into ice cubes is:

$$Q_{freeze} = mL_f = (1\,\text{kg})(333.5\,\text{kJ/kg}) = 333.5\,\text{kJ}$$

6. Add these heats to obtain Q_c:

$$Q_c = 41.8 \text{ kJ} + 333.5 \text{ kJ} = 375.3 \text{ kJ} \approx 375 \text{ kJ}$$

7. Substitute Q_c into step 2 to find the work W:

$$W = \frac{Q_c}{\text{COP}} = \frac{375 \text{ kJ}}{5.5} = 68.2 \text{ kJ}$$

8. Use this value of W and 55 W for the available power to find the time t:

$$t = \frac{W}{P} = \frac{68.2 \text{ kJ}}{55 \text{ J/s}} = 1.24 \text{ ks} \times \frac{1 \text{ min}}{60 \text{ s}} = \boxed{20.7 \text{ min}}$$

REMARKS You won't make it in one hour. You should have done the calculation before you put the water into the refrigerator and then used only half as much water.

EXERCISE A refrigerator has a coefficient of performance of 4.0. How much heat is exhausted to the hot reservoir if 200 kJ of heat are removed from the cold reservoir? (*Answer* 250 kJ)

19-3 Equivalence of the Heat-Engine and Refrigerator Statements

The heat-engine and refrigerator statements (or the Kelvin and Clausius statements, respectively) of the second law of thermodynamics seem quite different, but they are actually equivalent. We can prove this by showing that if either statement is assumed to be false, then the other must also be false. We'll use a numerical example to show that if the heat-engine statement is false, then the refrigerator statement is false.

Figure 19-6a shows an ordinary refrigerator that uses 50 J of work to remove 100 J of energy as heat from a cold reservoir and rejects 150 J of energy as heat to a hot reservoir. Suppose the heat-engine statement of the second law were not true. Then a "perfect" heat engine could remove energy from the hot reservoir and convert it completely into work with 100 percent efficiency. We could use this perfect heat engine to remove 50 J of energy from the hot reservoir and do 50 J of work (Figure 19-6b) on the ordinary refrigerator. Then, the combination of the perfect heat engine and the ordinary refrigerator would be a perfect refrigerator, transferring 100 J of energy as heat from the cold reservoir to the hot reservoir without requiring any work, as illustrated in Figure 19-6c. This violates the refrigerator statement of the second law. Thus, if the heat-engine statement is false, the refrigerator statement is also false. Similarly, if a perfect refrigerator existed, it could be used in conjunction with an ordinary heat engine to construct a perfect heat engine. Thus, if the refrigerator statement is false, the heat-engine statement is also false. It then follows that if one statement is true, the other is also true. Therefore, the heat engine statement and the refrigerator statement are equivalent.

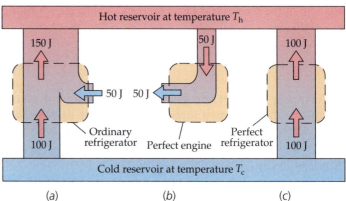

Hot reservoir at temperature T_h

150 J 50 J 100 J

50 J 50 J

Ordinary refrigerator Perfect engine Perfect refrigerator

100 J 100 J

Cold reservoir at temperature T_c

(a) (b) (c)

(a) An ordinary refrigerator removes 100 J from a cold reservoir, requiring the input of 50 J of work.

(b) A perfect heat engine violates the heat engine statement of the second law by removing 50 J from the hot reservoir and converting it completely into work.

(c) Putting the two together makes a perfect refrigerator that violates the refrigerator statement of the second law by transferring 100 J from the cold reservoir to the hot reservoir with no other effect.

FIGURE 19-6 Demonstration of the equivalence of the heat-engine and refrigerator statements of the second law of thermodynamics.

19-4 The Carnot Engine

According to the second law of thermodynamics, it is impossible for a heat engine working between two heat reservoirs to be 100% efficient. What, then, is the maximum possible efficiency for such an engine? A young French engineer,

Sadi Carnot answered this question in 1824, before either the first or the second law of thermodynamics had been established. Carnot found that a *reversible engine* is the most efficient engine that can operate between any two given reservoirs. This result is known as the Carnot theorem:

> No engine working between two given heat reservoirs can be more efficient than a reversible engine working between those two reservoirs.

CARNOT THEOREM

A reversible engine working in a cycle between two heat reservoirs is called a **Carnot engine,** and its cycle is called a **Carnot cycle.** Figure 19-7 illustrates the Carnot theorem with a numerical example.

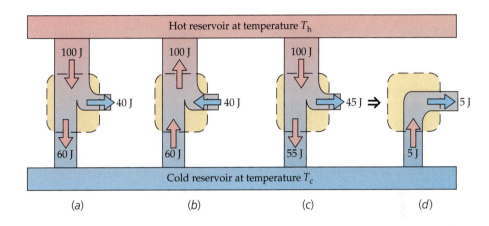

If no engine can have a greater efficiency than a Carnot engine, it follows that all Carnot engines working between the same two reservoirs have the same efficiency. This efficiency, called the **Carnot efficiency,** must be independent of the working substance of the engine and thus can depend only on the temperatures of the reservoirs.

Let us look at what makes a process reversible or irreversible. According to the second law, heat flows from hot objects to cold objects and never the other way around. Thus, the conduction of heat from a hot object to a cold one is *not* reversible. Also, friction can transform work into heat, but friction can never transform heat into work. The conversion of work into heat via friction is *not* reversible. Friction and other dissipative forces irreversibly transform mechanical energy into thermal energy. A third type of irreversibility occurs when a system passes through nonequilibrium states, such as when there is turbulence in a gas or when a gas explodes. For a process to be reversible, we must be able to move the system back through the same equilibrium states in the reverse order.

From these considerations and our statements of the second law of thermodynamics, we can list some conditions that are necessary for a process to be reversible:

FIGURE 19-7 Illustration of the Carnot theorem. (*a*) A reversible heat engine with 40 percent efficiency removes 100 J from a hot reservoir, does 40 J work, and exhausts 60 J to the cold reservoir. (*b*) If the same engine runs backwards as a refrigerator, 40 J of work are done to remove 60 J from the cold reservoir and exhaust 100 J to the hot reservoir. (*c*) An assumed heat engine working between the same two reservoirs with an efficiency of 45 percent which is greater than that of the reversible engine in (*a*). (*d*) The net effect of running the engine in (*c*) in conjunction with the refrigerator in (*b*) is the same as that of a perfect heat engine that removes 5 J from the cold reservoir and converts it completely into work with no other effect, violating the second law of thermodynamics. Thus, the reversible engine in (*a*) is the most efficient engine that can operate between these two reservoirs.

1. No mechanical energy is transformed into thermal energy by friction, viscous forces, or other dissipative forces.

2. Energy transfer as heat can only occur between objects at the same temperature (or infinitesimally near the same temperature).

3. The process must be quasi-static so that the system is always in an equilibrium state (or infinitesimally near an equilibrium state).

CONDITIONS FOR REVERSIBILITY

Any process that violates any of the above conditions is irreversible. Most processes in nature are irreversible. To have a reversible process, great care must be taken to eliminate frictional and other dissipative forces and to make the process quasi-static. Because this can never be completely accomplished, a reversible process is an idealization similar to the idealization of motion without friction in mechanics problems. Reversibility can, nevertheless, be closely approximated in practice.

We can now understand the features of a Carnot cycle, which is a reversible cycle between two reservoirs only. Because all heat transfer must be done isothermally in order for the process to be reversible, the heat absorbed from the hot reservoir must be absorbed isothermally. The next step is a quasi-static adiabatic expansion to the lower temperature of the cold reservoir. Next, heat is given off isothermally to the cold reservoir. Finally, there is a quasi-static, adiabatic compression to the higher temperature of the hot reservoir. The Carnot cycle thus consists of four reversible steps:

1. A quasi-static isothermal absorption of heat from a hot reservoir
2. A quasi-static adiabatic expansion to a lower temperature
3. A quasi-static isothermal exhaustion of heat to a cold reservoir
4. A quasi-static adiabatic compression back to the original state

STEPS IN A CARNOT CYCLE

One way to calculate the efficiency of a Carnot engine is to choose as the working substance a material of which we have some knowledge—an ideal gas, and then explicitly calculate the work done on it over a Carnot cycle (Figures 19-8a and 8b). Since all Carnot cycles have the same efficiency independent of the working substance, our result will be valid in general.

The efficiency of the Carnot cycle (Equation 19-2) is

$$\varepsilon = 1 - \frac{Q_c}{Q_h}$$

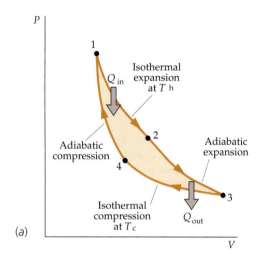

(a)

FIGURE 19-8 (a) Carnot cycle for an ideal gas:

Step 1: **Heat is absorbed from a hot reservoir at temperature T_h during an isothermal expansion from state 1 to state 2.**

Step 2: **The gas expands adiabatically from state 2 to state 3, reducing its temperature to T_c.**

Step 3: **The gas gives off heat to the cold reservoir as it is compressed isothermally at T_c from state 3 to state 4.**

Step 4: **The gas is compressed adiabatically until its temperature is again T_h.**

(b) Work is done on the gas or by the gas during each step. The net work done during the cycle is represented by the shaded area. All processes are reversible. All steps are quasi-static.

(b)

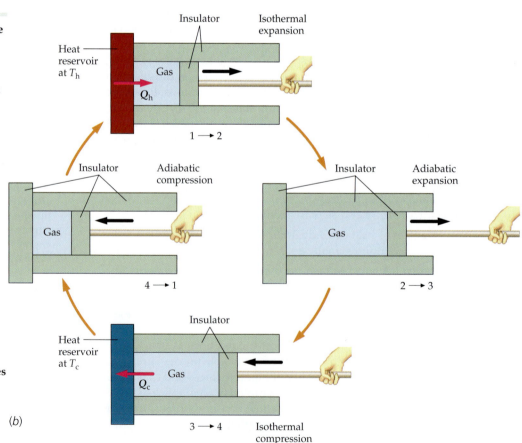

The heat Q_h is absorbed during the isothermal expansion from state 1 to state 2. The first law of thermodynamics is $\Delta E_{int} = Q_{in} + W_{on}$. For an isothermal expansion of an ideal gas $\Delta E_{int} = 0$. Applying the first law to the isothermal expansion from state 1 to state 2 we have $Q_h = Q_{in}$, so Q_h equals the work done by the gas.

$$Q_h = W_{by\ gas} = \int_{V_1}^{V_2} P\,dV = \int_{V_1}^{V_2} \frac{nRT_h}{V}\,dV = nRT_h \int_{V_1}^{V_2} \frac{dV}{V} = nRT_h \ln\frac{V_2}{V_1}$$

Similarly, the heat given off to the cold reservoir equals the work done on the gas during the isothermal compression at temperature T_c from state 3 to state 4. This work has the same magnitude as that done by the gas if it expands from state 4 to state 3. The heat rejected is thus

$$Q_c = W_{on\ gas} = nRT_c \ln\frac{V_3}{V_4}$$

The ratio of these heats is

$$\frac{Q_c}{Q_h} = \frac{T_c \ln\dfrac{V_3}{V_4}}{T_h \ln\dfrac{V_2}{V_1}} \qquad\qquad 19\text{-}4$$

We can relate the ratios V_2/V_1 and V_3/V_4 using Equation 18-37 for a quasi-static adiabatic expansion. For the expansion from state 2 to state 3, we have

$$T_h V_2^{\gamma-1} = T_c V_3^{\gamma-1}$$

Similarly, for the adiabatic compression from state 4 to state 1, we have

$$T_h V_1^{\gamma-1} = T_c V_4^{\gamma-1}$$

Dividing these two equations, we obtain

$$\left(\frac{V_2}{V_1}\right)^{\gamma-1} = \left(\frac{V_3}{V_4}\right)^{\gamma-1} \Rightarrow \frac{V_2}{V_1} = \frac{V_3}{V_4}$$

Coal-fueled electric generating plant at Four Corners, New Mexico.

Power plant at Wairakei, New Zealand, that converts geothermal energy into electricity.

Solar energy is focused and collected individually to produce electricity by these heliostats that are being tested at Sandia National Laboratory.

Control rods are inserted into this nuclear reactor at Tihange, Belgium.

Thus, Equation 19-4 gives

$$\frac{Q_c}{Q_h} = \frac{T_c \ln \dfrac{V_2}{V_1}}{T_h \ln \dfrac{V_2}{V_1}} = \frac{T_c}{T_h} \qquad\qquad 19\text{-}5$$

The Carnot efficiency ε_c is thus

$$\varepsilon_C = 1 - \frac{T_c}{T_h} \qquad\qquad 19\text{-}6$$

CARNOT EFFICIENCY

Equation 19-6 demonstrates that because the Carnot efficiency must be independent of the working substance of any particular engine, it depends only on the temperatures of the two reservoirs.

EFFICIENCY OF A STEAM ENGINE **EXAMPLE 19-4**

A steam engine works between a hot reservoir at 100°C (373 K) and a cold reservoir at 0°C (273 K). (a) What is the maximum possible efficiency of this engine? (b) If the engine is run backwards as a refrigerator, what is its maximum coefficient of performance?

An experimental wind-powered electric generator at Sandia National Laboratory. The propeller is designed for optimum transfer of wind energy to mechanical energy.

PICTURE THE PROBLEM The maximum efficiency is the Carnot efficiency given by Equation 19-6. To find the maximum COP, we use the definition of efficiency ($\varepsilon = W/Q_h$), the definition of COP (COP = Q_c/W), and Equation 19-5.

(a) The maximum efficiency is the Carnot efficiency:

$$\varepsilon_{max} = \varepsilon_C = 1 - \frac{T_c}{T_h} = 1 - \frac{273 \text{ K}}{373 \text{ K}}$$

$$= 0.268 = \boxed{26.8\%}$$

(b) 1. Write the expression for the COP if the engine is run in reverse for a single cycle:

$$COP = \frac{Q_c}{W}$$

2. Write the expression for the efficiency if the engine is run forward for a single cycle. (Since for maximum possible performance the engine is reversible, the values for Q_h, Q_c, and W are the same whether the engine is run backward or forward.):

$$\varepsilon = \frac{W}{Q_h}$$

3. Solve the step 2 result for the work and substitute it into the step 1 result:

$$COP = \frac{Q_c}{W} = \frac{Q_c}{\varepsilon Q_h}$$

4. Using Equation 19-5 and the Part (a) result, solve for the COP:

$$COP = \frac{Q_c}{\varepsilon Q_h} = \frac{T_c}{\varepsilon T_h} = \frac{273 \text{ K}}{0.268(373 \text{ K})}$$

$$COP = \boxed{2.73}$$

REMARKS Even though this maximum efficiency seems to be quite low, it is the greatest efficiency possible for any engine working between these temperatures. Real engines will have lower efficiencies because of friction, heat conduction, and other irreversible processes. Real refrigerators will have a lower coefficient of performance. It can be shown that the coefficient of performance of a Carnot refrigerator is $T_c/\Delta T$.

The Carnot efficiency gives us an upper limit on possible efficiencies, and is therefore useful to know. For example, we calculated in Example 19-4 that the Carnot efficiency is 26.8 percent. This means that, no matter how much we reduce friction and other irreversible losses, the best efficiency obtained between reservoirs at 373 K and 273 K is 26.8 percent. We would know, then, that an engine working between those two temperatures with an efficiency of 25 percent is a very good engine!

WORK LOST BY AN ENGINE   **EXAMPLE 19-5**

An engine removes 200 J from a hot reservoir at 373 K, does 48 J of work, and exhausts 152 J to a cold reservoir at 273 K. How much work is "lost" per cycle due to irreversible processes in this engine?

PICTURE THE PROBLEM The difference between maximum amount of work that could be done using a Carnot engine and 48 J is the work lost.

1. The work lost is the maximum amount of work that could be done minus the work actually done:

$$W_{lost} = W_{max} - W$$

2. The maximum amount of work that could be done is the work done using a Carnot engine:

$$W_{max} = \varepsilon_C Q_h$$

3. The work lost is then:

$$W_{lost} = \varepsilon_C Q_h - W$$

4. The Carnot efficiency can be expressed in terms of the temperatures:

$$\varepsilon_C = 1 - \frac{T_c}{T_h}$$

5. Substituting for ε_C gives:

$$W_{lost} = \left(1 - \frac{T_c}{T_h}\right)Q_h - W$$

$$= \left(1 - \frac{273 \text{ K}}{373 \text{ K}}\right)(200 \text{ J}) - 48 \text{ J}$$

$$= \boxed{5.6 \text{ J}}$$

REMARKS The 5.6 J of energy in the answer is not "lost" to the universe—total energy is conserved. That 5.6 J of energy exhausted into the cold reservoir by the non-ideal engine of the problem is only lost in that it would have been converted into useful work if an ideal (reversible) engine had been used.

WORK LOST BETWEEN HEAT RESERVOIRS **EXAMPLE 19-6**

If 200 J of heat are conducted from a heat reservoir at 373 K to one at 273 K, how much work capability is "lost" in this process?

We saw in the previous example that a Carnot engine working between these two reservoirs could do 53.6 J of work if it extracted 200 J from the 373-K reservoir and exhausted to a 273-K reservoir. Thus, if 200 J is conducted directly from the hot reservoir to the cold reservoir without any work being done, 53.6 J of this energy has been "lost" in the sense that it could have been converted into useful work.

EXERCISE A Carnot engine works between heat reservoirs at 500 K and 300 K. (*a*) What is its efficiency? (*b*) If it removes 200 kJ of heat from the hot reservoir, how much work does it do? (*Answer* (*a*) 40% (*b*) 80 kJ)

EXERCISE A real engine works between heat reservoirs at 500 K and 300 K. It removes 500 kJ of heat from the hot reservoir and does 150 kJ of work during each cycle. What is its efficiency? (*Answer* 30%)

The Thermodynamic or Absolute Temperature Scale

In Chapter 17, the ideal-gas temperature scale was defined in terms of the properties of gases at low densities. Because the Carnot efficiency depends only on the temperatures of the two heat reservoirs, it can be used to define the ratio of the temperatures of the reservoirs independent of the properties of any substance. We *define* the ratio of the thermodynamic temperatures of the hot and cold reservoirs to be

$$\frac{T_c}{T_h} = \frac{Q_c}{Q_h} \qquad \qquad 19\text{-}7$$

DEFINITION OF THERMODYNAMIC TEMPERATURE

where Q_h is the energy removed from the hot reservoir and Q_c is the energy exhausted to the cold reservoir by a Carnot engine working between the two reservoirs. Thus, to find the ratio of two reservoir temperatures, we set up a reversible engine operating between them and measure the energy transferred as heat to or from each reservoir during one cycle. The **thermodynamic temperature** is completely specified by Equation 19-7 *and* the choice of one fixed point. If the fixed point is defined to be 273.16 K for the triple point of water, then the thermodynamic temperature scale matches the ideal-gas temperature scale for the range of temperatures over which a gas thermometer can be used. Any temperature that reads zero at absolute zero is called an *absolute temperature scale.*

*19-5 Heat Pumps

A **heat pump** is a refrigerator with a different objective. Typically, the objective of a refrigerator is to cool an object or region of interest. The objective of a heat pump, however, is to heat an object or region of interest. For example, if you use a heat pump to heat your house you transfer heat from the cold air outside the house to the warmer air inside it. Your objective is to heat the region inside your house. If work W is done on a heat pump to remove heat Q_c from the cold reservoir and reject heat Q_h to the hot reservoir, the coefficient of performance for a heat pump is defined as

$$COP_{HP} = \frac{Q_h}{W}$$

19-8

DEFINITION—COEFFICIENT OF PERFORMANCE (HEAT PUMP)

This coefficient of performance differs from that for the refrigerator, which is Q_c/W (Equation 19-3). Using $W = Q_h - Q_c$, this can be written

$$COP_{HP} = \frac{Q_h}{Q_h - Q_c} = \frac{1}{1 - \dfrac{Q_c}{Q_h}}$$

19-9

The maximum coefficient of performance is obtained using a Carnot heat pump. Then Q_c and Q_h are related by Equation 19-5. Substituting $Q_c/Q_h = T_c/T_h$ into Equation 19-9, we obtain for the maximum coefficient of performance

$$COP_{HP\,max} = \frac{1}{1 - \dfrac{T_c}{T_h}} = \frac{T_h}{T_h - T_c} = \frac{T_h}{\Delta T}$$

19-10

where ΔT is the difference in temperature between the hot and cold reservoirs. Real heat pumps have coefficients of performance less than the $COP_{HP\,max}$ because of friction, heat conduction, and other irreversible processes.

The two coefficients are related. Using $Q_h = Q_c + W$, we can relate Equations 19-3 and 19-10:

$$COP_{HP} = \frac{Q_h}{W} = \frac{Q_c + W}{W} = 1 + \frac{Q_c}{W} = 1 + COP$$

19-11

AN IDEAL HEAT PUMP   **EXAMPLE 19-7** Try It Yourself

An ideal heat pump is used to pump heat from the outside air at −5°C to the hot-air supply for the heating fan in a house, which is at 40°C. How much work is required to pump 1 kJ of heat into the house?

PICTURE THE PROBLEM Use Equation 19-11 with $COP_{HP\,max}$ calculated from Equation 19-10 for $T_c = -5°C = 268$ K and $\Delta T = 45$ K.

Steps	Answers
1. Calculate the work from Equation 19-8:	$W = \dfrac{Q_h}{COP_{HP}}$
2. Calculate the COP_{HP} from Equation 19-10:	$COP_{HP} = COP_{HP\,max} = \dfrac{T_h}{\Delta T}$
3. Solve for the work:	$W = \dfrac{Q_h}{COP_{HP}} = Q_h\dfrac{\Delta T}{T_h} = (1\ \text{kJ})\dfrac{45\ \text{K}}{313\ \text{K}}$
	$W = \boxed{0.144\ \text{kJ}}$

REMARKS The $COP_{HP\,max} = T_h/\Delta T = 6.96$. That is, the energy transferred inside the house as heat is 6.96 times larger than the work done. (Only 0.144 kJ of work is needed to pump 1 kJ of heat into the hot-air supply in the house.)

19-6 Irreversibility and Disorder

There are many irreversible processes that cannot be described by the heat-engine or refrigerator statements of the second law, such as a glass falling to the floor and breaking or a balloon popping. However, all irreversible processes have one thing in common—the system plus its surroundings moves toward a less ordered state.

Suppose a box containing a gas of mass M at a temperature T is moving along a frictionless table with a velocity v_{cm} (Figure 19-9a). The total kinetic energy of the gas has two components: that associated with the movement of the center of mass $\frac{1}{2}Mv_{cm}^2$, and the energy of the motion of its molecules relative to its center of mass. The center of mass energy $\frac{1}{2}Mv_{cm}^2$ is ordered mechanical energy that could be converted entirely into work. (For example, if a weight were attached to the moving box by a string passing over a pulley, this energy could be used to lift the weight.) The relative energy is the internal thermal energy of the gas, which is related to its temperature T. It is random, non-ordered energy that cannot be converted entirely into work.

Now, suppose that the box hits a fixed wall and stops (Figure 19-9b). This inelastic collision is clearly an irreversible process. The ordered mechanical energy of the gas is converted into random internal energy and the temperature of the gas rises. The gas still has the same total energy, but now all of that energy is associated with the random motion of the gas molecules about the center of mass of the gas, which is now at rest. Thus, the gas has become less ordered (more disordered), and has lost some of its ability to do work.

19-7 Entropy

There is a thermodynamic function called **entropy** S that is a measure of the disorder of a system. Entropy S, like pressure P, volume V, temperature T, and

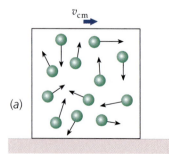

(a)

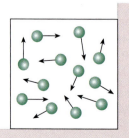

(b)

FIGURE 19-9

internal energy U, is a function of the state of a system. As with potential energy, it is the *change* in entropy that is important. The change in entropy dS of a system as it goes from one state to another is defined as

$$dS = \frac{dQ_{rev}}{T}$$ 19-12

where dQ_{rev} is the energy that must be transferred to the system as heat in a *reversible* process that brings the system from the initial state to the final state. If dQ_{rev} is negative, then the entropy change of the system is negative.

The term dQ_{rev} does not mean that a reversible heat transfer must take place in order for the entropy of a system to change. Indeed, there are many situations in which the entropy of a system changes when there is no transfer of heat whatsoever, for example, the box of gas colliding with the wall in Figure 19-9. Equation 19-12 simply gives us a method for *calculating* the entropy difference between two states of a system. Because entropy is a state function, the change in entropy when the system moves from one state to another depends only on the system's initial and final states, not on the process by which the change occurs.

Entropy of an Ideal Gas

We can illustrate that dQ_{rev}/T is in fact the differential of a state function for an ideal gas (even though dQ_{rev} is not). Consider an arbitrary reversible quasi-static process in which a system consisting of an ideal gas absorbs an amount of heat dQ_{rev}. According to the first law, dQ_{rev} is related to the change in the internal energy dE_{int} of the gas and the work done on the gas $(dW_{on} = -PdV)$ by

$$dE_{int} = dQ_{rev} + dW_{on} = dQ_{rev} - PdV$$

For an ideal gas, we can write dE_{int} in terms of the heat capacity, $dE_{int} = C_v dT$, and we can substitute nRT/V for P from the equation of state. Then

$$C_v dT = dQ_{rev} - nRT \frac{dV}{V}$$ 19-13

Equation 19-13 cannot be integrated unless we know how T depends on V. This is just another way of saying that dQ_{rev} is not a differential of a state function Q_{rev}. But if we divide each term by T, we obtain

$$C_v \frac{dT}{T} = \frac{dQ_{rev}}{T} - nR \frac{dV}{V}$$ 19-14

Since C_v depends only on T, the term on the left can be integrated as can the second term on the right.[†] Thus, dQ_{rev}/T is the differential of a function, the entropy function S.

$$dS = \frac{dQ_{rev}}{T} = \left(C_v \times \frac{dT}{T} \right) + nR \frac{dV}{V}$$ 19-15

For simplicity, we will assume that C_v is constant. Integrating Equation 19-15, we obtain

$$\Delta S = \int \frac{dQ}{T} = C_v \ln \frac{T_2}{T_1} + nR \ln \frac{V_2}{V_1}$$ 19-16

Equation 19-16 gives the entropy change of an ideal gas that undergoes a reversible expansion from an initial state of volume V_1 and temperature T_1 to a final state of volume V_2 and temperature T_2.

---

† Mathematically, the factor $1/T$ is called an integrating factor for Equation 19-13.

Entropy Changes for Various Processes

ΔS for an Isothermal Expansion of an Ideal Gas

If an ideal gas undergoes an isothermal expansion, then $T_2 = T_1$ and its entropy change is

$$\Delta S = \int \frac{dQ}{T} = nR \ln \frac{V_2}{V_1} \qquad \text{19-17}$$

The entropy change of the gas is positive because V_2 is greater than V_1. In this process, an amount of energy Q is transferred as heat from the reservoir to the gas. This heat equals the work done by the gas:

$$Q = W_{by} = \int_{V_1}^{V_2} P\,dV = nRT \int_{V_1}^{V_2} \frac{dV}{V} = nRT \ln \frac{V_2}{V_1} \qquad \text{19-18}$$

The entropy change of the gas is $+Q/T$. Because the same amount of heat leaves the reservoir at temperature T, the entropy change of the reservoir is $-Q/T$. The net entropy change of the gas plus the reservoir is zero. We will refer to the system under consideration plus its surroundings as the "universe." This example illustrates a general result:

In a reversible process, the entropy change of the universe is zero.

ΔS for a Free Expansion of an Ideal Gas

In the free expansion of a gas discussed in Section 18-4, a gas is initially confined in one compartment of a container, which is connected by a stopcock to another compartment that is evacuated. The whole system has rigid walls and is thermally insulated from its surroundings so that no heat can flow in or out, and no work can be done on (or by) the system (Figure 19-10). When the stopcock is opened, the gas rushes into the evacuated chamber. Eventually, the gas reaches thermal equilibrium with itself. Since there is no work done and no heat transferred, the final internal energy of the gas must equal its initial internal energy. If we assume that the gas is ideal, the final temperature T equals the initial temperature.

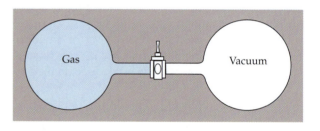

FIGURE 19-10 Free expansion of a gas. When the stopcock is opened, the gas expands rapidly into the evacuated chamber. Since no work is done on the gas and the whole system is thermally insulated, the initial and final internal energies of the gas are equal.

We might think that there is no entropy change of the gas because there is no heat transfer. But this process is not reversible, so we cannot use $\int dQ/T$ to find the change in entropy of the gas. However, the initial and final states of the gas in the free expansion are the same as those of the gas in the isothermal expansion just discussed. *Because the change in the entropy of a system for any process depends only on the initial and final states of the system, the entropy change of the gas for the free expansion is the same as that for the isothermal expansion.* If V_1 is the initial volume of the gas and V_2 is its final volume, the entropy change of the gas is given by Equation 19-17, or

$$\Delta S_{gas} = nR \ln \frac{V_2}{V_1}$$

In this case, there is no change in the surroundings, so the entropy change of the gas is also the entropy change of the universe:

$$\Delta S_u = nR \ln \frac{V_2}{V_1} \qquad \text{19-19}$$

Note that because V_2 is greater than V_1, the change in entropy of the universe for this irreversible process is positive; that is, the entropy of the universe increases. This is also a general result:

In an irreversible process, the entropy of the universe increases.

If the final volume in the free expansion were less than the initial volume, then the entropy of the universe would decrease—but this does not happen. A gas does not freely contract by itself into a smaller volume. This leads us to yet another statement of the second law of thermodynamics:

For any process, the entropy of the universe never decreases.

FREE EXPANSION OF AN IDEAL GAS   **E X A M P L E   1 9 - 8**

Find the entropy change for the free expansion of 0.75 mol of an ideal gas from $V_1 = 1.5$ L to $V_2 = 3$ L.

PICTURE THE PROBLEM For a free expansion of an ideal gas the initial and final temperatures are the same. Thus, the entropy change ΔS for a free expansion from V_1 to V_2 is the same as ΔS for an isothermal process from V_1 to V_2. For the isothermal process $\Delta E_{int} = 0$, so $Q = W_{by}$. First we calculate Q, then we set $\Delta S = Q/T$.

1. The entropy change is the same as for an isothermal expansion from V_1 to V_2:

$$\Delta S = \Delta S_{isothermal} = \frac{Q}{T}$$

2. The heat Q that would enter the gas during an isothermal expansion at temperature T equals the work done by the gas during the expansion:

$$Q = W_{by} = nRT \ln \frac{V_2}{V_1}$$

3. Substitute this value of Q to calculate ΔS:

$$\Delta S = \frac{Q}{T} = nR \ln \frac{V_2}{V_1}$$

$$\Delta S = (0.75 \text{ mol})(8.31 \text{ J/mol·K}) \ln 2$$

$$\Delta S = \boxed{4.32 \text{ J/K}}$$

ΔS for Constant-Pressure Processes   If a substance is heated from temperature T_1 to temperature T_2 at constant pressure, the heat absorbed dQ is related to its temperature change dT by

$$dQ = C_p \, dT$$

We can approximate reversible heat conduction if we have a large number of heat reservoirs with temperatures ranging from just slightly greater than T_1 to T_2 in very small steps. We could place the substance, with initial temperature T_1, in contact with the first reservoir at a temperature just slightly greater than T_1 and let the substance absorb a small amount of heat. Because the heat transfer is approximately isothermal, the process will be approximately reversible. We then place the substance in contact with the next reservoir at a slightly higher temperature, and so on, until the final temperature T_2 is reached. If heat dQ is absorbed reversibly, the entropy change of the substance is

$$dS = \frac{dQ}{T} = C_p \frac{dT}{T}$$

Integrating from T_1 to T_2, we obtain the total entropy change of the substance:

$$\Delta S = C_p \int_{T_1}^{T_2} \frac{dT}{T} = C_p \ln \frac{T_2}{T_1} \qquad\qquad 19\text{-}20$$

This result gives the entropy change of a substance that is heated from T_1 to T_2 by any process, reversible or irreversible, as long as the final pressure equals the initial pressure. It also gives the entropy change of a substance that is cooled. In the case of cooling, T_2 is less than T_1, and $\ln (T_2/T_1)$ is negative, giving a negative entropy change.

EXERCISE Find the change in entropy of 1 kg of water that is heated at constant pressure from 0°C to 100°C. (*Answer* $\Delta S = 1.31$ kJ/K)

ENTROPY CHANGES DURING HEAT TRANSFER   **EXAMPLE 19-9** **Try It Yourself**

Suppose 1 kg of water at temperature $T_1 = 30°C$ is added to 2 kg of water at $T_2 = 90°C$ in a calorimeter of negligible heat capacity at a constant pressure of 1 atm. (*a*) Find the change in entropy of the system. (*b*) Find the change in entropy of the universe.

PICTURE THE PROBLEM When the two amounts of water are combined, they eventually come to a final equilibrium temperature, T_f, that can be found by setting the heat lost equal to the heat gained. To calculate the entropy change of each mass of water, we consider a reversible isobaric heating of the 1-kg mass of water from 30°C to T_f and a reversible isobaric cooling of the 2-kg mass from 90°C to T_f using Equation 19-18. The entropy change of the system is the sum of the entropy changes of each part. The entropy change of the universe is the entropy change of the system plus the entropy change of its surroundings. To find the entropy change of the surroundings, assume no heat leaves the calorimeter during the time it takes the water to reach its final temperature.

Cover the column to the right and try these on your own before looking at the answers.

Steps	Answers
(*a*) 1. Calculate T_f by setting the heat lost equal to the heat gained:	$T_f = 70°C = 343$ K
2. Use your result for T_f and the data given to calculate ΔS_1 and ΔS_2:	$\Delta S_1 = 0.519$ kJ/K $\Delta S_2 = -0.474$ kJ/K
3. Add ΔS_1 and ΔS_2 to find the total entropy change of the system:	$\Delta S_{\text{system}} = \boxed{+0.0453 \text{ kJ/K}}$
(*b*) 1. Assuming no heat leaves the calorimeter, find the entropy change of the surroundings:	$\Delta S_{\text{surroundings}} = 0$
2. Add ΔS_{system} and $\Delta S_{\text{surroundings}}$ to find the entropy change of the universe:	$\Delta S_{\text{u}} = \boxed{+0.0453 \text{ kJ/K}}$

 REMARKS Note that we had to convert the temperatures to the absolute scale to calculate the entropy changes. The entropy change of the universe is positive, as expected.

ΔS for an Inelastic Collision Because mechanical energy is converted into thermal energy in an inelastic collision, such a process is clearly irreversible. The entropy of the universe must therefore increase. Consider a block of mass m

falling from a height h and making an inelastic collision with the ground. Let the block, ground, and atmosphere all be at a temperature T, which is not significantly changed by the process. If we consider the block, ground, and atmosphere as our isolated system, there is no heat conducted into or out of the system. The state of the system has been changed because its internal energy has been increased by an amount mgh. This change is the same as if we added heat $Q = mgh$ to the system at constant temperature T. To calculate the change in entropy of the system, we thus consider a reversible process in which heat $Q_{rev} = mgh$ is added at a constant temperature T. According to Equation 19-12, the change in entropy is then

$$\Delta S = \frac{Q_{rev}}{T} = \frac{mgh}{T}$$

This positive entropy change is also the entropy change of the universe.

ΔS for Heat Conduction from One Reservoir to Another

Heat conduction is also an irreversible process, and so we expect the entropy of the universe to increase when this occurs. Consider the simple case of heat Q conducted from a hot reservoir at a temperature T_h to a cold reservoir at a temperature T_c. The state of a heat reservoir is determined by its temperature and its internal energy only. The change in entropy of a heat reservoir due to a heat exchange is the same whether the heat exchange is reversible or not. If heat Q is put into a reservoir at temperature T, then the entropy of the reservoir increases by Q/T. If the heat is removed, then the entropy of the reservoir decreases by $-Q/T$. In the case of heat conduction, the hot reservoir loses heat, so its entropy change is

$$\Delta S_h = -\frac{Q}{T_h}$$

The cold reservoir absorbs heat, so its entropy change is

$$\Delta S_c = +\frac{Q}{T_c}$$

The net entropy change of the universe is

$$\Delta S_u = \Delta S_c + \Delta S_h = \frac{Q}{T_c} - \frac{Q}{T_h} \qquad\qquad 19\text{-}21$$

Note that, because heat flows from a hot reservoir to a cold reservoir, the change in entropy of the universe is positive.

ΔS for a Carnot Cycle

Because a Carnot cycle is by definition reversible, the entropy change of the universe after a cycle must be zero. We demonstrate this by showing that the entropy change of the reservoirs in a Carnot engine is zero. (Since a Carnot engine works in a cycle, the entropy change of the engine itself is zero, so the entropy change of the universe is just the sum of the entropy changes of the reservoirs.) The entropy change of the hot reservoir is

$\Delta S_h = -\dfrac{Q_h}{T_h}$ and the entropy change of the cold reservoir is $\Delta S_c = +\dfrac{Q_c}{T_c}$. These heats are related to the temperatures by the definition of thermodynamic temperature (Equation 19-7)

$$\frac{T_c}{T_h} = \frac{Q_c}{Q_h} \left(\text{or} \quad \frac{Q_h}{T_h} = \frac{Q_c}{T_c} \right)$$

The entropy change of the universe is thus

$$\Delta S_u = \Delta S_h + \Delta S_c = -\frac{Q_h}{T_h} + \frac{Q_c}{T_c} = -\frac{Q_h}{T_h} + \frac{Q_h}{T_h} = 0$$

The entropy change of the universe is zero as expected.

Notice that we have ignored any entropy change associated with the energy transferred via work from the Carnot engine to its surroundings. If this work is used to raise a weight, or some other ordered process, then there is no entropy change. However, if this work is used to push a block across a table top where friction is involved, then there is an additional entropy increase associated with this work.

ENTROPY CHANGES IN A CARNOT CYCLE **EXAMPLE 19-10**

During each cycle, a Carnot engine removes 100 J of energy from a reservoir at 400 K, does work, and exhausts heat to a reservoir at 300 K. Compute the entropy change of each reservoir for each cycle, and show explicitly that the entropy change of the universe is zero for this reversible process.

PICTURE THE PROBLEM Since the engine works in a cycle, its entropy change is zero. We therefore compute the entropy change of each reservoir and add them to obtain the entropy change of the universe.

1. The entropy change of the universe equals the sum of the entropy changes of the reservoirs:

$$\Delta S_u = \Delta S_{400} + \Delta S_{300}$$

2. Calculate the entropy change of the hot reservoir:

$$\Delta S_{400} = -\frac{Q_h}{T_h} = -\frac{100\ J}{400\ K} = \boxed{-0.250\ J/K}$$

3. The entropy change of the cold reservoir is Q_c divided by T_c, where $Q_c = Q_h - W$:

$$\Delta S_{300} = \frac{Q_c}{T_c} = \frac{Q_h - W}{T_c}$$

4. We use $W = \varepsilon_C Q_h$ (Equation 19-2) to relate W to Q_h. The efficiency is the Carnot efficiency (Equation 19-6):

$$W = \varepsilon Q_h, \text{ where } \varepsilon = \varepsilon_C = 1 - \frac{T_c}{T_h}$$

so

$$W = \left(1 - \frac{T_c}{T_h}\right)Q_h$$

5. Calculate the entropy change of the cold reservoir:

$$\Delta S_{300} = \frac{Q_h - W}{T_c} = \frac{Q_h - Q_h\left(1 - \frac{T_c}{T_h}\right)}{T_c} = \frac{Q_h}{T_h}$$

$$= \frac{100\ J}{400\ K} = \boxed{0.250\ J/K}$$

6. Substitute these results into step 1 to find the entropy change of the universe:

$$\Delta S_u = \Delta S_{400} + \Delta S_{300}$$

$$\Delta S_u = -0.250\ J/K + 0.250\ J/K = \boxed{0}$$

REMARKS Suppose that an ordinary, nonreversible engine removed 100 J from the hot reservoir. Because its efficiency must be less than that of a Carnot engine, it would do less work and exhaust more heat to the cold reservoir. Then the entropy increase of the cold reservoir would be greater than the entropy decrease of the hot reservoir, and the entropy change of the universe would be positive.

EXAMPLE   19-11

Because entropy is a state function, thermodynamic processes can be represented as *ST*, *SV*, or *SP* diagrams instead of the *PV* diagrams we have used so far. Make a sketch of the Carnot cycle on an *ST* plot.

PICTURE THE PROBLEM The Carnot cycle consists of a reversible isothermal expansion followed by a reversible adiabatic expansion, then a reversible isothermal compression followed by a reversible adiabatic compression. During the isothermal processes, heat is absorbed or expelled at constant temperature, so *S* increases or decreases at constant *T*. During the adiabatic processes, the temperature changes, but since $\Delta Q_{rev} = 0$, *S* is constant.

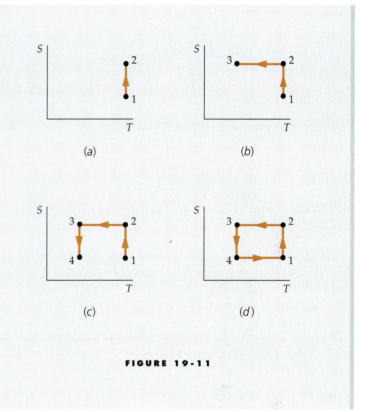

1. During the isothermal expansion (1 to 2 in Figure 19-11*a*), heat is absorbed reversibly so, *S* increases at constant *T*:

2. During the reversible adiabatic expansion (2 to 3 in Figure 19-11*b*), the temperature decreases while *S* is constant:

3. During the isothermal compression (3 to 4 in Figure 19-11*c*) heat is rejected reversibly, so *S* decreases at constant *T*:

4. During the reversible adiabatic compression (4 to 1 in Figure 19-11*d*) the temperature increases while *S* is constant:

FIGURE 19-11

■ **REMARKS** The Carnot cycle is a rectangle if plotted on an *S* versus *T* diagram.

19-8   Entropy and the Availability of Energy

If an irreversible process occurs, energy is conserved, but some of the energy becomes unavailable to do work and is "wasted." Consider a block falling to the ground. The entropy change of the universe for this process is mgh/T. When the block was at a height *h*, its potential energy mgh could have been used to do useful work. But after the inelastic collision of the block with the ground, this energy is no longer available because it has become the disordered internal energy of the block and its surroundings. The energy that has become unavailable (wasted) is equal to $mgh = T\Delta S_u$. This is a general result:

> In an irreversible process, energy equal to $T\Delta S_u$ becomes unavailable to do work, where *T* is the temperature of the coldest available reservoir.

For simplicity, we will call the energy that becomes unavailable to do work the "work lost":

$$W_{lost} = T\Delta S_u \qquad\qquad 19\text{-}22$$

A Sliding Box Revisited **EXAMPLE 19-12**

Suppose that the box shown in Figure 19-9*a* and *b* has a mass of 2.4 kg and slides with a speed of $v = 3$ m/s before crashing into a fixed wall and stopping. The temperature T of the box, table, and surroundings is 293 K and does not change appreciably as the box comes to rest. Find the entropy change of the universe.

PICTURE THE PROBLEM The initial mechanical energy of the box $\frac{1}{2}Mv^2$ is converted to the internal energy of the box-wall-surroundings system. The entropy change is equivalent to what would occur if the heat $Q = \frac{1}{2}Mv^2$ were added to the system reversibly.

The entropy change of the universe is Q/T:

$$\Delta S_u = \frac{Q}{T} = \frac{\frac{1}{2}Mv^2}{T} = \frac{\frac{1}{2}(2.4 \text{ kg})(3 \text{ m/s})^2}{293 \text{ K}}$$

$$\Delta S_u = \boxed{0.0369 \text{ J/K}}$$

REMARKS Energy is conserved, but the energy $T\Delta S_u = \frac{1}{2}Mv^2$ is no longer available to do work.

In the free expansion discussed earlier, the ability to do work was also lost. In that case, the entropy change of the universe was $nR \ln (V_2/V_1)$, so the work lost was $nRT \ln (V_2/V_1)$. This is the amount of work that could have been done if the gas had expanded quasi-statically and isothermally from V_1 to V_2, as given by Equation 19-17.

If heat is conducted from a hot reservoir to a cold reservoir, the change in entropy of the universe is given by Equation 19-21, and the work lost is

$$W_{lost} = T_c \, \Delta S_u = T_c \left(\frac{Q}{T_c} - \frac{Q}{T_h} \right) = Q \left(1 - \frac{T_c}{T_h} \right)$$

We can see that this is just the work that could have been done by a Carnot engine running between these reservoirs, removing heat Q from the hot reservoir and doing work $W = \varepsilon_C Q$, where $\varepsilon_C = 1 - T_c/T_h$.

19-9 Entropy and Probability

Entropy, which is a measure of the disorder of a system, is related to probability. Essentially, a state of high order has a low probability, whereas a state of low order has a high probability. Thus, in an irreversible process, the universe moves from a state of low probability to one of high probability.

Let us consider a free expansion in which a gas expands from an initial volume V_1 to a final volume $V_2 = 2V_1$. The entropy change of the universe for this process is given by Equation 19-19:

$$\Delta S = nR \ln \frac{V_2}{V_1} = nR \ln 2$$

Why is this process irreversible? Why can't the gas spontaneously compress back into its original volume? Such a compression would not violate the first law of thermodynamics, as there is no energy change involved. The reason that the gas does not compress to its original volume is merely that such a compression is extremely *improbable*. To see this, let's assume that the gas consists of only

10 molecules and that, initially, these molecules occupy the entire volume of their container. Then the chance that any one particular molecule will be in the left half of the container at any given time is $\frac{1}{2}$. The chance that any two particular molecules will both be in the left half is $\frac{1}{2} \times \frac{1}{2} = \frac{1}{4}$. (This is the same as the chance that a coin flipped twice will come up heads both times.) The chance that three particular molecules will be in the left half is $\frac{1}{2} \times \frac{1}{2} \times \frac{1}{2} = (\frac{1}{2})^3 = \frac{1}{8}$. The chance that all 10 molecules will be in the left half is $(\frac{1}{2})^{10} = \frac{1}{1024}$. That is, there is 1 chance in 1024 that all 10 molecules will be in the left half of the container at any given time.

Though the probability of all 10 molecules being on one side of the container is small, we would not be completely surprised to see it occur. If we look at the gas once each second, we could expect to see it happen once in every 1024 sec, or about once every 17 min. If we started with the 10 molecules randomly distributed and then found them all in the left half of the original volume, the entropy of the universe would have *decreased* by $nR \ln 2$. However, this decrease is extremely small, since the number of moles n corresponding to 10 molecules is only about 10^{-23}. Still, it would violate the entropy statement of the second law of thermodynamics, which says that for any process, the entropy of the universe never decreases. Therefore, if we wish to apply the second law of thermodynamics to microscopic systems such as a small number of molecules, we should consider the second law to be a statement of *probability.*

We can relate the probability of a gas spontaneously compressing itself into a smaller volume to the change in its entropy. If the original volume is V_1, the probability p of finding N molecules in a smaller volume V_2 is

$$p = \left(\frac{V_2}{V_1}\right)^N$$

Taking the natural logarithm of both sides of this equation, we obtain

$$\ln p = N \ln \frac{V_2}{V_1} = nN_A \ln \frac{V_2}{V_1} \qquad\qquad 19\text{-}23$$

where n is the number of moles and N_A is Avogadro's number. The entropy change of the gas is

$$\Delta S = nR \ln \frac{V_2}{V_1} \qquad\qquad 19\text{-}24$$

Comparing Equations 19-23 and 19-24, we see that

$$\Delta S = \frac{R}{N_A} \ln p = k \ln p \qquad\qquad 19\text{-}25$$

where k is Boltzmann's constant.

It may be disturbing to learn that irreversible processes, such as the spontaneous compression of a gas or the spontaneous conduction of heat from a cold body to a hot body, are not impossible—they are just improbable. As we have just seen, there is a reasonable chance that an irreversible process will occur in a system consisting of a very small number of molecules; however, *thermodynamics itself is applicable only to macroscopic systems*, that is, to systems that have a very large number of molecules. Consider trying to measure the pressure of a gas consisting of only 10 molecules. The pressure would vary wildly depending on whether no molecule, 2 molecules, or 10 molecules were colliding with the wall of the container at the time of measurement. The macroscopic variables of pressure and temperature are not applicable to a microscopic system with only 10 molecules.

As we increase the number of molecules in a system, the chance of an irreversible process occurring decreases dramatically. For example, if we have 50 molecules in a container, the chance that they will all be in the left half of the container is $(\frac{1}{2})^{50} \approx 10^{-15}$. Thus, if we look at the gas once each second, we could expect to see all 50 molecules in the left half of the volume about once in every 10^{15} seconds or once in every 36 million years! For 1 mole (6×10^{23} molecules), the chance that all will wind up in half of the volume is vanishingly small, essentially zero. For macroscopic systems, then, the probability of a process resulting in a decrease in the entropy of the universe is so extremely small that the distinction between improbable and impossible becomes blurred.

SUMMARY

The second law of thermodynamics is a fundamental law of nature.

Topic	Relevant Equations and Remarks
1. **Efficiency of a Heat Engine**	If the engine removes Q_h from a hot reservoir, does work W, and exhausts heat Q_c to a cold reservoir, its efficiency is $$\varepsilon = \frac{W}{Q_h} = \frac{Q_h - Q_c}{Q_h} = 1 - \frac{Q_c}{Q_h} \qquad \text{19-2}$$
2. **Coefficient of Performance of a Refrigerator**	$$\text{COP} = \frac{Q_c}{W} \qquad \text{19-3}$$
3. **Coefficient of Performance of a Heat Pump**	$$\text{COP}_{HP} = \frac{Q_h}{W} \qquad \text{19-8}$$
4. **Equivalent Statements of the Second Law of Thermodynamics**	
The Kelvin statement	No system can take energy as heat from a single reservoir and convert it entirely into work without additional net changes in the system or its surroundings.
The heat-engine statement	It is impossible for a heat engine working in a cycle to produce *only the effect* of extracting heat from a single reservoir and performing an equivalent amount of work.
The Clausius statement	A process whose only net result is to transfer energy as heat from a cooler object to a hotter one is impossible.
The refrigerator statement	It is impossible for a refrigerator working in a cycle to produce *only the effect* of extracting heat from a cold object and rejecting the same amount of heat to a hot object.
The entropy statement	The entropy of the universe (system plus surroundings) can never decrease.
5. **Conditions for a Reversible Process**	1. No mechanical energy is transformed into thermal energy by friction, viscous forces, or other dissipative forces. 2. Energy transfer as heat can only occur between objects at the same temperature (or infinitesimally near the same temperature). 3. The process must be quasi-static so that the system is always in an equilibrium state (or infinitesimally near an equilibrium state).

6. **Carnot Engine**

A Carnot engine is a reversible engine that works between two reservoirs. It uses a Carnot cycle, which consists of

Carnot cycle

1. A quasi-static isothermal absorption of heat at temperature T_h
2. A quasi-static adiabatic expansion
3. A quasi-static isothermal exhaustion of heat at temperature T_c
4. A quasi-static adiabatic compression back to the original state

Carnot efficiency

$$\varepsilon_C = 1 - \frac{Q_c}{Q_h} = 1 - \frac{T_c}{T_h}$$

19-6

7. **Thermodynamic Temperature**

The ratio of the thermodynamic temperatures of two reservoirs is defined to be the ratio of the heat exhausted to the heat intake of a Carnot engine running between the reservoirs.

$$\frac{T_c}{T_h} = \frac{Q_c}{Q_h}$$

19-7

8. **Entropy**

Entropy is a measure of the disorder of a system. The difference in entropy between two nearby states is given by

$$dS = \frac{dQ_{rev}}{T}$$

19-12

where dQ_{rev} is the heat added in a reversible process connecting the states. The entropy change of a system can be positive or negative.

Entropy and loss of work capability

During an irreversible process, the entropy of the universe S_u increases and an amount of energy

$$W_{lost} = T\Delta S_u$$

19-22

becomes unavailable for doing work.

Entropy and probability

Entropy is related to probability. A highly ordered system is one of low probability and low entropy. An isolated system moves towards a state of high probability, low order, and high entropy.

PROBLEMS

- Single-concept, single-step, relatively easy
- •• Intermediate-level, may require synthesis of concepts
- ••• Challenging
- **SSM** Solution is in the *Student Solutions Manual*
- Problems available on iSOLVE online homework service
- ✓ These "Checkpoint" online homework service problems ask students additional questions about their confidence level and how they arrived at their answer

In a few problems, you are given more data than you actually need; in a few other problems, you are required to supply data from your general knowledge, outside sources, or informed estimates.

Conceptual Problems

1 • How does kinetic friction in an engine affect its efficiency?

2 • **SSM** Explain why you can't just open your refrigerator to cool your kitchen on a hot day. Why is it that turning on a room air conditioner will cool down the room but opening a refrigerator door will not?

3 • Why do power-plant designers try to increase the temperature of the steam fed to engines as much as possible?

4 •• On a humid day, water vapor condenses on a cold surface. During condensation, the entropy of the water (a) increases, (b) remains constant, (c) decreases, (d) may decrease or remain unchanged.

5 • **SSM** In a reversible adiabatic process, (*a*) the internal energy of the system remains constant, (*b*) no work is done by the system, (*c*) the entropy of the system remains constant, (*d*) the temperature of the system remains constant.

6 •• True or false:

(*a*) Work can never be converted completely into heat.
(*b*) Heat can never be converted completely into work.
(*c*) All heat engines have the same efficiency.
(*d*) It is impossible to transfer a given quantity of heat from a cold reservoir to a hot reservoir.
(*e*) The coefficient of performance of a refrigerator cannot be greater than 1.
(*f*) All Carnot engines are reversible.
(*g*) The entropy of a system can never decrease.
(*h*) The entropy of the universe can never decrease.

7 •• An ideal gas is taken reversibly from an initial state P_i, V_i, T_i to the final state P_f, V_f, T_f. Two possible paths are (A) an isothermal expansion followed by an adiabatic compression, and (B) an adiabatic compression followed by an isothermal expansion. For these two paths, (*a*) $\Delta E_{int\,A} > \Delta E_{int\,B}$, (*b*) $\Delta S_A > \Delta S_B$, (*c*) $\Delta S_A < \Delta S_B$, (*d*) none of the above is correct.

8 •• **SSM** Figure 19-12 shows a thermodynamic cycle on an *ST* diagram. Identify this cycle and sketch it on a *PV* diagram.

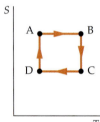

FIGURE 19-12
Problems 8 and 68

9 •• Figure 19-13 shows a thermodynamic cycle on an *SV* diagram. Identify the type of engine represented by this diagram.

FIGURE 19-13
Problem 9

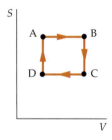

10 •• Sketch an *ST* diagram of the Otto cycle.

11 •• Sketch an *SV* diagram of the Carnot cycle.

12 •• Sketch an *SV* diagram of the Otto cycle.

13 •• Figure 19-14 shows a thermodynamic cycle on an *SP* diagram. Make a sketch of this cycle on a *PV* diagram.

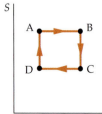

FIGURE 19-14 Problem 13

14 • **SSM** Which has a greater effect on increasing the efficiency of a Carnot engine, a 5-K increase in the temperature of the hot reservoir or a 5-K decrease in the temperature of the cold reservoir?

Estimation and Approximation

15 •• Estimate the maximum efficiency of an automobile engine with a compression ratio of 8:1. Assume the Otto cycle and assume $\gamma = 1.4$.

16 •• **SSM** (*a*) Estimate the highest COP possible for a "typical" household refrigerator. (*b*) If the refrigerator draws 600 W of electrical power, estimate the rate at which heat is being drawn from the refrigerator compartment.

17 •• The temperature of the sun is about 5400 K, the Earth's average temperature is about 290 K, and the solar constant (the intensity of sunlight reaching the Earth's orbit) is about 1.3 kW/m². (*a*) Calculate the total power of sunlight hitting the Earth. (*b*) Calculate the net rate at which the Earth's entropy is increasing due to the flow of solar radiation. (*c*) Calculate the net rate at which the sun's entropy is decreasing just due to the outflow of solar radiation *hitting the Earth*.

18 •• (*a*) Using the information given in Problem 17 and the known distance from the Earth to the sun (1.5×10^{11} m), calculate the total power that the sun radiates into space. (*b*) There are about 10^{11} stars like the sun in the Milky Way galaxy, and 10^{11} galaxies in the universe. Use this information to estimate the rate at which the entropy of the universe is increasing, assuming that the average temperature of the universe is 2.73 K.

19 •• A typical human body produces about 100 W of heat. Estimate the increase in entropy of the universe produced by a single human body over the course of a spring day where the temperature is 70°F during the day and 55°F at night.

20 ••• **SSM** How long, on average, should we have to wait until all of the air molecules in a room rush to one half of the room? (As a friend of mine put it, "Don't hold your breath. . . .") Assume that the air molecules are contained in a 1 m × 1 m × 1 m box and that they reshuffle their positions 100 times per second. Calculate the average time it should take for all the molecules to occupy only one half of the box if there are (*a*) 10 molecules, (*b*) 100 molecules, (*c*) 1000 molecules, and (*d*) 1 mole of molecules in the box. (*e*) The highest vacuums that have been created to date have pressures of about 10^{-12} torr. If a typical vacuum chamber has a capacity of about 1 liter, how long will a physicist have to wait before all of the gas molecules in the vacuum chamber occupy only one half of it? Compare that to the expected lifetime of the universe, which is about 10^{10} years.

Heat Engines and Refrigerators

21 • **ISOLVE** ✔ An engine with 20% efficiency does 100 J of work in each cycle. (*a*) How much heat is absorbed in each cycle? (*b*) How much heat is rejected in each cycle?

22 • **ISOLVE** ✔ An engine absorbs 400 J of heat and does 120 J of work in each cycle. (*a*) What is its efficiency? (*b*) How much heat is rejected in each cycle?

23 • █SOLVE✔ An engine absorbs 100 J and rejects 60 J in each cycle. (*a*) What is its efficiency? (*b*) If each cycle takes 0.5 s, find the power output of this engine in watts.

24 • SSM █SOLVE A refrigerator absorbs 5 kJ of energy from a cold reservoir and rejects 8 kJ to a hot reservoir. (*a*) Find the coefficient of performance of the refrigerator. (*b*) The refrigerator is reversible and is run backward as a heat engine between the same two reservoirs. What is its efficiency?

25 •• An engine operates with 1 mol of an ideal gas, for which $C_v = \frac{3}{2}R$ and $C_p = \frac{5}{2}R$, as its working substance. The cycle begins at $P_1 = 1$ atm and $V_1 = 24.6$ L. The gas is heated at constant volume to $P_2 = 2$ atm. It then expands at constant pressure until $V_2 = 49.2$ L. During these two steps, heat is absorbed by the gas. The gas is then cooled at constant volume until its pressure is again 1 atm. It is then compressed at constant pressure to its original state. During the last two steps, heat is rejected by the gas. All the steps are quasi-static and reversible. (*a*) Show this cycle on a *PV* diagram. Find the work done, the heat added, and the change in the internal energy of the gas for each step of the cycle. (*b*) Find the efficiency of the cycle.

26 •• An engine using 1 mol of a diatomic ideal gas performs a cycle consisting of three steps: (1) an adiabatic expansion from an initial pressure of 2.64 atm and an initial volume of 10 L to a pressure of 1 atm and a volume of 20 L, (2) a compression at constant pressure to its original volume of 10 L, and (3) heating at constant volume to its original pressure of 2.64 atm. Find the efficiency of this cycle.

27 •• █SOLVE An engine using 1 mol of an ideal gas initially at $V_1 = 24.6$ L and $T = 400$ K performs a cycle consisting of four steps: (1) an isothermal expansion at $T = 400$ K to twice its initial volume, (2) cooling at constant volume to $T = 300$ K, (3) an isothermal compression to its original volume, and (4) heating at constant volume to its original temperature of 400 K. Assume that $C_v = 21$ J/K. Sketch the cycle on a *PV* diagram and find its efficiency.

28 •• SSM One mole of an ideal monatomic gas at an initial volume $V_1 = 25$ L follows the cycle shown in Figure 19-15. All the processes are quasi-static. Find (*a*) the temperature of each state of the cycle, (*b*) the heat flow for each part of the cycle, and (*c*) the efficiency of the cycle.

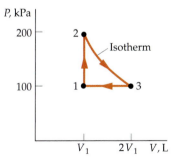

FIGURE 19-15 Problem 28

29 •• An ideal gas ($\gamma = 1.4$) follows the cycle shown in Figure 19-16. The temperature of state 1 is 200 K. Find (*a*) the temperatures of the other three states of the cycle and (*b*) the efficiency of the cycle.

FIGURE 19-16 Problem 29

30 ••• The *diesel cycle* shown in Figure 19-17 approximates the behavior of a diesel engine. Process *ab* is an adiabatic compression, process *bc* is an expansion at constant pressure, process *cd* is an adiabatic expansion, and process *da* is cooling at constant volume. Find the efficiency of this cycle in terms of the volumes V_a, V_b, V_c, and V_d.

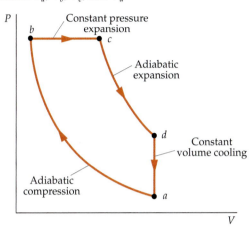

FIGURE 19-17 Diesel cycle for Problem 30

31 •• SSM "As far as we know, Nature has never evolved a heat engine"—Steven Vogel, *Life's Devices*, Princeton University Press (1988). (*a*) Calculate the efficiency of a heat engine operating between body temperature (98.6°F) and a typical outdoor temperature (70°F), and compare this to the human body's efficiency for converting chemical energy into work (approximately 20%). Does this contradict the Second Law of Thermodynamics? (*b*) From the result of Part (*a*), and a general knowledge of the conditions under which most warm-blooded animal life exists, explain why no warm-blooded animals have evolved heat engines to supply their internal energy.

32 ••• The Clausius equation of state is $P(V - bn) = nRT$, where b is a constant. Show that the efficiency of a Carnot cycle is the same for a gas that obeys this equation of state as it is for one that obeys the ideal-gas equation of state, $PV = nRT$.

Second Law of Thermodynamics

33 •• A refrigerator takes in 500 J of heat from a cold reservoir and gives off 800 J to a hot reservoir. Assume that the heat-engine statement of the second law of thermodynamics is false, and show how a perfect engine working with this refrigerator can violate the refrigerator statement of the second law.

34 •• SSM If two adiabatic curves intersect on a *PV* diagram, a cycle could be completed by an isothermal path between the two adiabatic curves shown in Figure 19-18. Show that such a cycle could violate the second law of thermodynamics.

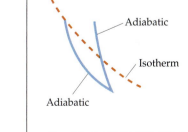

FIGURE 19-18 Problem 34

Carnot Engines

35 • **iSOLVE**✓ A Carnot engine works between two heat reservoirs at temperatures T_h = 300 K and T_c = 200 K. (a) What is its efficiency? (b) If it absorbs 100 J from the hot reservoir during each cycle, how much work does it do? (c) How much heat does it give off during each cycle? (d) What is the COP of this engine when it works as a refrigerator between the same two reservoirs?

36 • **iSOLVE** An engine removes 250 J from a reservoir at 300 K and exhausts 200 J to a reservoir at 200 K. (a) What is its efficiency? (b) How much more work could be done if the engine were reversible?

37 •• A reversible engine working between two reservoirs at temperatures T_h and T_c has an efficiency of 30%. Working as a heat engine, it gives off 140 J of heat to the cold reservoir. A second engine working between the same two reservoirs also gives off 140 J to the cold reservoir. Show that if the second engine has an efficiency greater than 30%, the two engines working together would violate the heat-engine statement of the second law.

38 •• A reversible engine working between two reservoirs at temperatures T_h and T_c has an efficiency of 20%. Working as a heat engine, it does 100 J of work in each cycle. A second engine working between the same two reservoirs also does 100 J of work in each cycle. Show that if the efficiency of the second engine is greater than 20%, the two engines working together would violate the refrigerator statement of the second law.

39 •• **SSM** A Carnot engine works between two heat reservoirs as a refrigerator. It does 50 J of work to remove 100 J from the cold reservoir and gives off 150 J to the hot reservoir during each cycle. Its coefficient of performance COP = Q_c/W = (100 J)/(50 J) = 2. (a) What is the efficiency of the Carnot engine when it works as a heat engine between the same two reservoirs? (b) Show that no other engine working as a refrigerator between the same two reservoirs can have a COP greater than 2.

40 •• **iSOLVE** A Carnot engine works between two heat reservoirs at temperatures T_h = 300 K and T_c = 77 K. (a) What is its efficiency? (b) If it absorbs 100 J from the hot reservoir during each cycle, how much work does it do? (c) How much heat does it give off in each cycle? (d) What is the coefficient of performance of this engine when it works as a refrigerator between these two reservoirs?

41 •• In the cycle shown in Figure 19-19, 1 mol of an ideal gas (γ = 1.4) is initially at a pressure of 1 atm and a temperature of 0°C. The gas is heated at constant volume to T_2 = 150°C and is then expanded adiabatically until its pressure is again 1 atm. It is then compressed at constant pressure back to its original state. Find (a) the temperature T_3 after the adiabatic expansion, (b) the heat entering or leaving the system during each process, (c) the efficiency of this cycle, and (d) the efficiency of a Carnot cycle operating between the temperature extremes of this cycle.

FIGURE 19-19
Problem 41

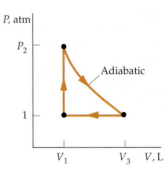

42 •• **iSOLVE** A steam engine takes in superheated steam at 270°C and discharges condensed steam from its cylinder at 50°C. Its efficiency is 30%. (a) How does this efficiency compare with the maximum possible efficiency for these temperatures? (b) If the useful power output of the engine is 200 kW, how much heat does the engine discharge to its surroundings in 1 h?

*Heat Pumps

43 • **SSM** **iSOLVE**✓ A heat pump delivers 20 kW to heat a house. The outside temperature is −10°C and the inside temperature of the hot-air supply for the heating fan is 40°C. (a) What is the coefficient of performance of a Carnot heat pump operating between these temperatures? (b) What must be the minimum power of the engine needed to run the heat pump? (c) If the COP of the heat pump is 60% of the efficiency of an ideal pump, what must be the minimum power of the engine?

44 • **iSOLVE**✓ A refrigerator is rated at 370 W. (a) What is the maximum amount of heat it can remove in 1 min if the inside temperature of the refrigerator is 0°C and it exhausts into a room at 20°C? (b) If the COP of the refrigerator is 70% of that of an ideal pump, how much heat can it remove in 1 min?

45 • Rework Problem 44 for a room temperature of 35°C.

Entropy Changes

46 • What is the change in entropy of 1 mol of water at 0°C that freezes?

47 •• Consider the freezing of 50 g of water by placing it in the freezer compartment of a refrigerator. Assume the walls of the freezer are maintained at −10°C. The water, initially liquid at 0°C, is frozen into ice and cooled to −10°C. Show that even though the entropy of the ice decreases, the net entropy of the universe increases.

48 • **iSOLVE** Two moles of an ideal gas at T = 400 K expand quasi-statically and isothermally from an initial volume of 40 L to a final volume of 80 L. (a) What is the entropy change of the gas? (b) What is the entropy change of the universe for this process?

49 • The gas in Problem 48 is taken from the same initial state ($T = 400$ K, $V_1 = 40$ L) to the same final state ($T = 400$ K, $V_2 = 80$ L) by a process that is not quasi-static. (*a*) What is the entropy change of the gas? (*b*) What can be said about the entropy change of the universe?

50 • What is the change in entropy of 1.0 kg of water when it changes to steam at 100°C and a pressure of 1 atm?

51 • ISOLVE✓ What is the change in entropy of 1.0 kg of ice when it changes to water at 0°C and a pressure of 1 atm?

52 •• A system absorbs 200 J of heat reversibly from a reservoir at 300 K and gives off 100 J reversibly to a reservoir at 200 K as it moves from state A to state B. During this process, the system does 50 J of work. (*a*) What is the change in the internal energy of the system? (*b*) What is the change in entropy of the system? (*c*) What is the change in entropy of the universe? (*d*) If the system goes from state A to state B by a nonreversible process, how would your answers for Parts (*a*), (*b*), and (*c*) differ?

53 •• SSM ISOLVE A system absorbs 300 J from a reservoir at 300 K and 200 J from a reservoir at 400 K. It then returns to its original state, doing 100 J of work and rejecting 400 J of heat to a reservoir at a temperature T. (*a*) What is the entropy change of the system for the complete cycle? (*b*) If the cycle is reversible, what is the temperature T?

54 •• ISOLVE Two moles of an ideal gas originally at $T = 400$ K and $V = 40$ L undergo a free expansion to twice their initial volume. What is (*a*) the entropy change of the gas and (*b*) the entropy change of the universe?

55 •• A 200-kg block of ice at 0°C is placed in a large lake. The temperature of the lake is just slightly higher than 0°C, and the ice melts. (*a*) What is the entropy change of the ice? (*b*) What is the entropy change of the lake? (*c*) What is the entropy change of the universe (the ice plus the lake)?

56 •• A 100-g piece of ice at 0°C is placed in an insulated container with 100 g of water at 100°C. (*a*) When thermal equilibrium is established, what is the final temperature of the water (Ignore the heat capacity of the container.) (*b*) Find the entropy change of the universe for this process.

57 •• SSM A 1-kg block of copper at 100°C is placed in a calorimeter of negligible heat capacity containing 4 L of water at 0°C. Find the entropy change of (*a*) the copper block, (*b*) the water, and (*c*) the universe.

58 •• If a 2-kg piece of lead at 100°C is dropped into a lake at 10°C, find the entropy change of the universe.

59 •• ISOLVE A 1500-kg car traveling at 100 km/h crashes into a concrete wall. If the temperature of the air is 20°C, calculate the entropy change of the universe.

60 •• SSM A box is divided into two identical halves by an impermeable partition through its middle. On one side is 1 mole of ideal gas A; on the other, 1 mole of ideal gas B (which is different from A). (*a*) Calculate the change in entropy when the partition is lifted, and the two gases mix together. (*b*) If we repeat the process with the same type of gas

in each side, should the entropy change when the partition is lifted? Explain. (Think carefully about this question!)

Entropy and Work Lost

61 •• SSM ISOLVE If 500 J of heat is conducted from a reservoir at 400 K to one at 300 K, (*a*) what is the change in entropy of the universe, and (*b*) how much of the 500 J of heat conducted could have been converted into work using a cold reservoir at 300 K?

62 •• One mole of an ideal gas undergoes a free expansion from $V_1 = 12.3$ L and $T_1 = 300$ K to $V_2 = 24.6$ L and $T_2 = 300$ K. It is then compressed isothermally and quasi-statically back to its original state. (*a*) What is the entropy change of the universe for the complete cycle? (*b*) How much work is wasted in this cycle? (*c*) Show that the work wasted is $T\Delta S_u$.

General Problems

63 • ISOLVE✓ An engine with an output of 200 W has an efficiency of 30%. It works at 10 cycles/s. (*a*) How much work is done in each cycle? (*b*) How much heat is absorbed and how much is given off in each cycle?

64 • ISOLVE✓ In each cycle, an engine removes 150 J from a reservoir at 100°C and gives off 125 J to a reservoir at 20°C. (*a*) What is the efficiency of this engine? (*b*) What is the ratio of its efficiency to that of a Carnot engine working between the same reservoirs? (This ratio is called the *second law efficiency*.)

65 • An engine removes 200 kJ of heat from a hot reservoir at 500 K in each cycle and exhausts heat to a cold reservoir at 200 K. Its efficiency is 85% of a Carnot engine working between the same reservoirs. (*a*) What is the efficiency of this engine? (*b*) How much work is done in each cycle? (*c*) How much heat is exhausted in each cycle?

66 •• (*a*) Calvin Cliffs Nuclear Power Plant, located on the Hobbes River, generates 1 GW of power. In this plant liquid sodium circulates between the reactor core and a heat exchanger located in the superheated steam that drives the turbine. Heat is transferred into the liquid sodium in the core, and out of the liquid sodium (and into the superheated steam) in the heat exchanger. The temperature of the superheated steam is 500 K. Waste heat is dumped into the river, which flows by at a temperature of 25°C. (*a*) What is the highest efficiency that this plant can have? (*b*) How much waste heat is dumped into the river every second? (*c*) How much heat must be generated to supply 1 GW of power? (*d*) Assume that new, tough environmental laws have been passed (to preserve the unique wildlife of the river). Because of this, the plant is not allowed to heat the river by more than 0.5°C. What is the minimum flow rate that the Hobbes river must have (in L/sec)?

67 • ISOLVE To maintain the temperature inside a house at 20°C, the power consumption of the electric baseboard heaters is 30 kW on a day when the outside temperature is −7°C. At what rate does this house contribute to the increase in the entropy of the universe?

68 •• The system represented in Figure 19-12 (Problem 8) is 1 mol of an ideal monatomic gas. The temperatures at points A and B are 300 and 750 K, respectively. What is the thermodynamic efficiency of the cyclic process ABCDA?

69 •• (a) Which process is more wasteful: (1) a block moving with 500 J of kinetic energy being slowed to rest by friction when the temperature of the atmosphere is 300 K, or (2) 1 kJ of heat being conducted from a reservoir at 400 K to one at 300 K? *Hint: How much of the 1 kJ of heat could be converted into work in an ideal situation?* (b) What is the change in entropy of the universe for each process?

70 •• Helium gas ($\gamma = 1.67$) is initially at a pressure of 16 atm, a volume of 1 L, and a temperature of 600 K. It is expanded isothermally until its volume is 4 L and is then compressed at constant pressure until its volume and temperature are such that an adiabatic compression will return the gas to its original state. (a) Sketch this cycle on a PV diagram. (b) Find the volume and temperature after the isobaric compression. (c) Find the work done during each cycle. (d) Find the efficiency of the cycle.

71 •• [SSM] A heat engine that does the work of blowing up a balloon at a pressure of 1 atm extracts 4 kJ from a hot reservoir at 120°C. The volume of the balloon increases by 4 L, and heat is exhausted to a cold reservoir at a temperature T_c. If the efficiency of the heat engine is 50% of the efficiency of a Carnot engine working between the same reservoirs, find the temperature T_c.

72 •• Show that the COP of a Carnot refrigerator is related to the efficiency of a Carnot engine by $\text{COP} = T_c/(\varepsilon_C T_h)$.

73 •• [ISOLVE ✓] A freezer has a temperature $T_c = -23°C$. The air in the kitchen has a temperature $T_h = 27°C$. Since the heat insulation is not perfect, some heat flows into the freezer at a rate of 50 W. Find the power of the motor that is needed to maintain the temperature in the freezer.

74 •• Two moles of a diatomic gas are taken through the cycle ABCA as shown on the PV diagram in Figure 19-20. At A the pressure and temperature are 5 atm and 600 K. The volume at B is twice that at A. The segment BC is an adiabatic expansion and the segment CA is an isothermal compression. (a) What is the volume of the gas at A? (b) What are the volume and temperature of the gas at B? (c) What is the temperature of the gas at C? (d) What is the volume of the gas at C? (e) How much work is done by the gas in each of the three segments of the cycle? (f) How much heat is absorbed by the gas in each segment of the cycle? (g) What is the thermodynamic efficiency of this cycle?

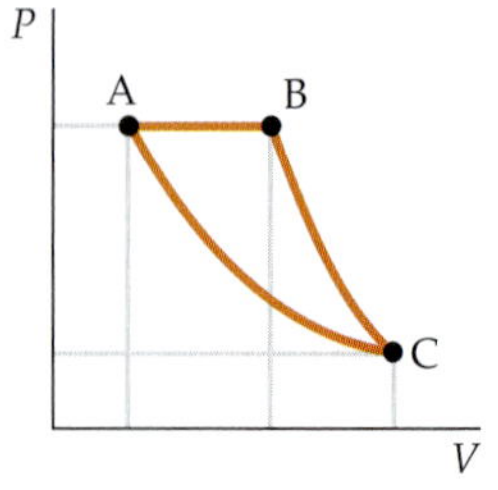

FIGURE 19-20
Problems 74 and 76

75 •• Two moles of a diatomic gas are carried through the cycle ABCDA shown in the PV diagram in Figure 19-21. The segment AB represents an isothermal expansion, the

segment BC an adiabatic expansion. The pressure and temperature at A are 5 atm and 600 K. The volume at B is twice that at A. The pressure at D is 1 atm. (a) What is the pressure at B? (b) What is the temperature at C? (c) Find the work done by the gas in one cycle and the thermodynamic efficiency of this cycle.

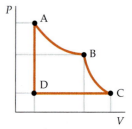

FIGURE 19-21
Problems 75 and 77

76 •• Repeat Problem 74 for a monatomic gas.

77 •• Repeat Problem 75 for a monatomic gas.

78 •• Compare the efficiency of the Otto engine and the Carnot engine operating between the same maximum and minimum temperatures.

79 ••• [SSM] Using the equation for the entropy change of an ideal gas when the volume and temperature change and $TV^{\gamma-1}$ is a constant, show explicitly that the entropy change is zero for a quasi-static adiabatic expansion from state (V_1, T_1) to state (V_2, T_2).

80 ••• (a) Show that if the refrigerator statement of the second law of thermodynamics were not true, then the entropy of the universe could decrease. (b) Show that if the heat-engine statement of the second law were not true, then the entropy of the universe could decrease. (c) An alternative statement of the second law is that the entropy of the universe cannot decrease. Have you just proved that this statement is equivalent to the refrigerator and heat-engine statements?

81 ••• Suppose that two heat engines are connected in series, such that the heat exhaust of the first engine is used as the heat input of the second engine as shown in Figure 19-22. The efficiencies of the engines are ε_1 and ε_2, respectively. Show that the net efficiency of the combination is given by

$$\varepsilon_{net} = \varepsilon_1 + (1 - \varepsilon_1)\varepsilon_2.$$

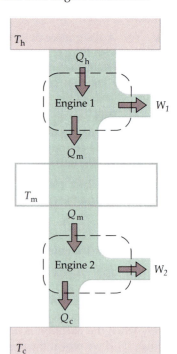

FIGURE 19-22
Problems 81 and 82

82 ••• **SSM** Suppose that each engine in Figure 19-22 is an ideal reversible heat engine. Engine 1 operates between temperatures T_h and T_m and Engine 2 operates between T_m and T_c, where $T_h > T_m > T_c$. Show that

$$\varepsilon_{net} = 1 - \frac{T_c}{T_h}$$

This means that two reversible heat engines in series are equivalent to one reversible heat engine operating between the hottest and coldest reservoirs.

83 ••• Bertrand Russell once said that if a million monkeys were given a million typewriters and typed away at random for a million years, they would produce all of Shakespeare's works. Let's limit ourselves to the following fragment of Shakespeare (*Julius Caesar* III:ii):

> *Friends, Romans, countrymen! Lend me your ears.*
> *I come to bury Caesar, not to praise him.*
> *The evil that men do lives on after them,*
> *The good is oft interred with the bones.*
> *So let it be with Caesar.*
> *The noble Brutus hath told you that Caesar was ambitious,*
> *And, if so, it were a grievous fault,*
> *And grievously hath Caesar answered it . . .*

Even with this small fragment, it will take a lot longer than a million years! By what factor (roughly speaking) was Russell in error? Make any reasonable assumptions you want. (You may even assume that the monkeys are immortal.)

Thermal Properties and Processes

? **What might eventually happen to this bridge if it did not have expansion joints? (See Example 20-1.)**

When an object absorbs thermal energy, various changes may occur in the physical properties of the object. For example, its temperature may rise, accompanied by an expansion or contraction of the object, or the object may liquefy or vaporize, during which its temperature remains constant.
➤ **In this chapter, we examine some of the thermal properties of matter and some important processes involving thermal energy.**

20-1 Thermal Expansion

When the temperature of an object increases, the object typically expands. (Consider that on concrete highways, expansion joints appear every 10 to 15 m, allowing the road to expand without cracking.) Suppose that we have a long rod of length L at a temperature T. When the temperature changes by ΔT, the fractional change in length ΔL is proportional to ΔT:

$$\frac{\Delta L}{L} = \alpha \Delta T \qquad \qquad 20\text{-}1$$

where α, called the **coefficient of linear expansion,** is the ratio of the fractional change in length to the change in temperature:

$$\alpha = \frac{\Delta L / L}{\Delta T}$$

20-2

The units for the coefficient of linear expansion are reciprocal Celsius degrees (1/°C), which are the same as reciprocal kelvins (1/K). The value of α for a solid or liquid doesn't vary much with pressure, but it may vary significantly with temperature. Equation 20-2 gives the average value over the temperature interval ΔT. The coefficient of linear expansion at a particular temperature T is found by taking the limit as ΔT approaches zero:

$$\alpha = \lim_{\Delta T \to 0} \frac{\Delta L / L}{\Delta T} = \frac{1}{L} \frac{dL}{dT}$$

20-3

The accuracy obtained by using the average value of α over a wide temperature range is sufficient for most purposes.

The **coefficient of volume expansion** β is similarly defined as the ratio of the fractional change in volume to the change in temperature (at constant pressure):

$$\beta = \lim_{\Delta T \to 0} \frac{\Delta V / V}{\Delta T} = \frac{1}{V} \frac{dV}{dT}$$

20-4

Like α, β does not usually vary with pressure for solids and liquids, but may vary with temperature. Average values for α and β for various substances are given in Table 20-1.

For a given material, $\beta = 3\alpha$. We can show this by considering a box of dimensions L_1, L_2, and L_3. Its volume at a temperature T is

$$V = L_1 L_2 L_3$$

The rate of change of the volume with respect to temperature is

$$\frac{dV}{dT} = L_1 L_2 \frac{dL_3}{dT} + L_1 L_3 \frac{dL_2}{dT} + L_2 L_3 \frac{dL_1}{dT}$$

Dividing each side of the equation by the volume, we obtain

$$\beta = \frac{1}{V} \frac{dV}{dT} = \frac{1}{L_3} \frac{dL_3}{dT} + \frac{1}{L_2} \frac{dL_2}{dT} + \frac{1}{L_1} \frac{dL_1}{dT}$$

We can see that each term on the right side of the above equation equals α, and so we have

$$\beta = 3\alpha$$

20-5

Similarly, the coefficient of area expansion is twice that of linear expansion.

The increase in size of any part of an object for a given temperature change is proportional to the original size of that part of the body. For example, if we increase the temperature of a steel ruler, the effect will be similar to that of a (very slight) photographic enlargement. That is, the dimensions of the ruler itself will be larger, as will the distance between the equally spaced lines. **If the ruler has a 1-cm-diameter hole in it, say between the 3-cm and 4-cm lines, the hole will get larger, just as the distance between the 3-cm and 4-cm lines does.**

TABLE 20-1

Approximate Values of the Coefficients of Thermal Expansion for Various Substances

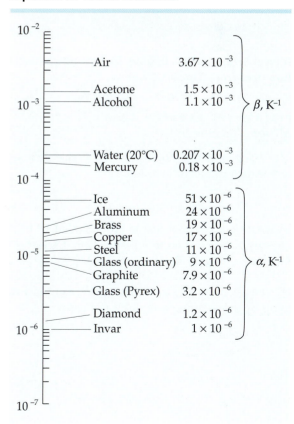

Air	3.67×10^{-3}
Acetone	1.5×10^{-3}
Alcohol	1.1×10^{-3}
Water (20°C)	0.207×10^{-3}
Mercury	0.18×10^{-3}

β, K^{-1}

Ice	51×10^{-6}
Aluminum	24×10^{-6}
Brass	19×10^{-6}
Copper	17×10^{-6}
Steel	11×10^{-6}
Glass (ordinary)	9×10^{-6}
Graphite	7.9×10^{-6}
Glass (Pyrex)	3.2×10^{-6}
Diamond	1.2×10^{-6}
Invar	1×10^{-6}

α, K^{-1}

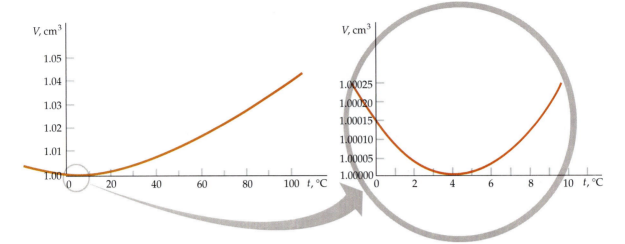

FIGURE 20-1 Volume of 1 g of water at atmospheric pressure versus temperature. The minimum volume, which corresponds to the maximum density, occurs at 4°C. At temperatures below 0°C, the curve shown is for supercooled water. (Supercooled water is water that is cooled below the normal freezing point without solidifying.)

Most materials expand when heated and contract when cooled. Water, however, presents an important exception. Figure 20-1 shows the volume occupied by 1 g of water as a function of temperature. The minimum volume, and therefore the maximum density, is at 4°C. Thus, when water below 4°C is cooled, it expands rather than contracts and vice versa. This property of water has important consequences for the ecology of lakes. At temperatures above 4°C, the water in a lake becomes denser as it cools and therefore sinks to the bottom. But as the water cools below 4°C, it becomes less dense and rises to the surface. This is the reason that ice forms first on the surface of a lake. Since ice is less dense than liquid water, it remains at the surface and acts as a thermal insulator for the water below. If water behaved like most substances and contracted when it froze, then ice would sink and expose more water at the surface that would then freeze. Lakes would fill with ice from the bottom up and would be much more likely to freeze completely in the winter, killing fish and other aquatic life.

An Expanding Bridge   **EXAMPLE 20-1**

A steel bridge is 1000 m long. By how much does it expand when the temperature rises from 0 to 30°C?

PICTURE THE PROBLEM Use $\alpha = 11 \times 10^{-6}\,\text{K}^{-1}$ from Table 20-1 and calculate ΔL from Equation 20-1.

The change in length for a 30 C° (30 K) change in temperature is the product of α, L, and ΔT:

$$\Delta L = \alpha L \Delta T = (11 \times 10^{-6}\,\text{K}^{-1})(1000\,\text{m})(30\,\text{K})$$

$$= \boxed{0.33\,\text{m} = 33\,\text{cm}}$$

REMARKS Expansion joints are included in bridges to relieve the enormous stresses that would occur without them. Excessive stress caused by temperature increases can cause the bridge to buckle.

We can calculate the stress that would result in a steel bridge without expansion joints by using Young's modulus (Equation 12-1):

$$Y = \frac{\text{stress}}{\text{strain}} = \frac{F/A}{\Delta L/L}$$

Then

$$\frac{F}{A} = Y\frac{\Delta L}{L} = Y\alpha\Delta T$$

For $\Delta T = 30$ K, $\Delta L/L = 0.33$ m$/1000$ m as found in Example 20-1. Then using $Y = 2 \times 10^{11}$ N/m^2 (from Table 12-1),

$$\frac{F}{A} = Y\frac{\Delta L}{L} = (2 \times 10^{11}\,\text{N/m}^2)\,\frac{0.33\,\text{m}}{1000\,\text{m}} = 6.6 \times 10^{7}\,\text{N/m}^2$$

This stress is about one-third of the breaking stress for steel under compression. A compression stress of this magnitude would cause a steel bridge to buckle and become permanently deformed.

A COMPLETELY FILLED GLASS **EXAMPLE 20-2**

While working in the laboratory, you fill a 1-L glass flask to the brim with water at 10°C. You heat the flask, raising the temperature of the water and flask to 30°C. How much water spills out of the flask?

PICTURE THE PROBLEM The glass flask and the water both expand when heated, but the water expands more, so some spills out. We calculate the amount spilled by finding the changes in volume for $\Delta T = 20$ K using $\Delta V_a = \beta V \Delta T$ with $\beta = 1.1 \times 10^{-3}$ K^{-1} for water (from Table 20-1), and $\Delta V_g = \beta V \Delta T = 3\alpha V \Delta T$ with $\alpha = 9 \times 10^{-6}$ K^{-1} for glass. The difference in these volume changes equals the volume spilled.

1. The volume of water spilled ΔV_s is the difference in the changes in volume of the water and glass:

$$\Delta V_s = \Delta V_a - \Delta V_g$$

2. Find the increase in the volume of the water:

$$\Delta V_a = \beta_a V \Delta T$$

3. Find the increase in the volume of the glass flask:

$$\Delta V_g = \beta_g V \Delta T = 3\alpha_g V \Delta T$$

4. Subtract to find the amount of water spilled:

$$\Delta V_s = \Delta V_a - \Delta V_g = \beta_a V \Delta T - \beta_g V \Delta T$$
$$= (\beta_a - \beta_g)\,V\Delta T = (\beta_a - 3\alpha_g)\,V\Delta T$$
$$= [0.207 \times 10^{-3}\,\text{K}^{-1} - 3(9 \times 10^{-6}\,\text{K}^{-1})](1\,\text{L})(20\,\text{K})$$
$$= 3.6 \times 10^{-3}\,\text{L} = \boxed{3.6\,\text{mL}}$$

BREAKING COPPER **EXAMPLE 20-3**

A copper bar is heated to 300°C. Then it is clamped rigidly between two fixed points so that it can neither expand nor contract. If the breaking stress of copper is 230 MN/m^2, at what temperature will the bar break as it cools?

PICTURE THE PROBLEM As the bar cools, the change in length ΔL that *would* occur if the bar contracted is offset by an equal stretching due to tensile stress in the bar. The stress F/A is related to the stretching ΔL by $Y = (F/A)/(\Delta L/L)$, where Young's modulus for copper is $Y = 110$ GN/m^2 (from Table 12-1). The maximum allowable stretching occurs when F/A equals 230 MN/m^2. Thus, we find the temperature change that would produce this maximum contraction.

1. Calculate the change in length ΔL_1 that would occur if the bar were unclamped and cooled by ΔT:

$$\Delta L_1 = \alpha L \Delta T$$

2. A tensile stress F/A stretches the bar by ΔL_2, where $L_1 + \Delta L_2 = 0$:

$$Y = \frac{F/A}{\Delta L_2/L}, \quad \text{so} \quad \Delta L_2 = L\frac{F/A}{Y}$$

3. Substitute the step 1 and step 2 results into $\Delta L_1 + \Delta L_2 = 0$ and solve for ΔT with the stress equal to the breaking value:

$$\Delta L_1 + \Delta L_2 = 0$$

$$\alpha L \Delta T + L\frac{F/A}{Y} = 0$$

so

$$\Delta T = -\frac{F/A}{\alpha Y}$$

$$= -\frac{230 \times 10^6 \text{ N/m}^2}{(17 \times 10^{-6} \text{ K}^{-1})(110 \times 10^9 \text{ N/m}^2)}$$

$$= -123 \text{ K} = -123 \text{ C}°$$

4. Add this result to the original temperature to find the final temperature at which the bar breaks:

$$T_f = T_1 + \Delta T = 300°\text{C} - 123 \text{ C}° = \boxed{177°\text{C}}$$

20-2 The van der Waals Equation and Liquid–Vapor Isotherms

At ordinary pressures most gases behave like an ideal gas. However, this ideal behavior breaks down when the pressure is high enough or the temperature is low enough such that the density of the gas is high and the molecules are, on average, closer together. An equation of state called the **van der Waals equation** describes the behavior of many real gases over a wide range of pressures more accurately than does the ideal-gas equation of state ($PV = nRT$). The van der Waals equation for n moles of gas is

$$\left(P + \frac{an^2}{V^2}\right)(V - bn) = nRT \qquad \text{20-6}$$

THE VAN DER WAALS EQUATION OF STATE

The constant b in this equation arises because the gas molecules are not point particles but objects that have a finite size; therefore, the volume available to each molecule is reduced. The magnitude of b is the volume of one mole of gas molecules. The term an^2/V^2 arises from the attraction of the gas molecules to each other. As a molecule approaches the wall of the container, it is pulled back by the molecules surrounding it with a force that is proportional to the density of those molecules n/V. Because the number of molecules that hit the wall in a given time is also proportional to the density of the molecules, the decrease in pressure due to the attraction of the molecules is proportional to the square of the density and therefore to n^2/V^2. The constant a depends on the gas and is small for inert gases, which have very weak chemical interactions. The terms bn and an^2/V^2 are both negligible when the volume V is large, so at low densities the van der Waals equation approaches the ideal-gas law. At high densities the van der Waals equation provides a much better description of the behavior of real gases than does the ideal-gas law.

Figure 20-2 shows PV isothermal curves for a substance at various temperatures. Except for the region where the liquid and vapor coexist, these curves are described quite accurately by the van der Waals equation and can be used to determine the constants a and b. For example, the values of these constants that give the best fit to the experimental curves for nitrogen are $a = 0.14$ Pa·m^6/mol^2 and $b = 39.1$ mL/mol. This volume of 39.1 mL per mole is about 0.2 percent of the

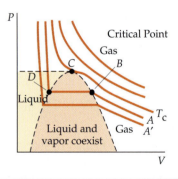

FIGURE 20-2 Isotherms on the PV diagram for a substance. For temperatures above the critical temperature T_c, the substance remains a gas at all pressures. Except for the region where the liquid and vapor coexist, these curves are described quite well by the van der Waals equation. The pressure for the horizontal portions of the curves in the shaded region is the vapor pressure, which is the pressure at which the vapor and liquid are in equilibrium. To the left of the shaded region for temperatures below the critical temperature, the substance is a liquid and is nearly incompressible.

volume of 22.4 L occupied by 1 mol of nitrogen under standard conditions. Since the molar mass of nitrogen is 28 g/mol, if 1 mol of nitrogen molecules were packed into a volume of 39.1 mL, then the density would be

$$\rho = \frac{M}{V} = \frac{28 \text{ g}}{39.1 \text{ mL}} = 0.72 \text{ g/mL} = 0.72 \text{ kg/L}$$

which is almost the same as the density of liquid nitrogen, 0.80 kg/L.

The value of the constant b can be used to estimate the size of a molecule. Since 1 mol (N_A molecules) of nitrogen has a volume of 39.1 cm³, the volume of one nitrogen molecule is

$$V = \frac{b}{N_A} = \frac{39.1 \text{ cm}^3/\text{mol}}{6.02 \times 10^{23} \text{ molecules/mol}} = 6.50 \times 10^{-23} \text{ cm}^3/\text{molecule}$$

If we assume that each molecule occupies a cube of side d, we obtain

$$d^3 = 6.50 \times 10^{-23} \text{ cm}^3$$

or

$$d = 4.0 \times 10^{-8} \text{ cm} = 0.4 \text{ nm}$$

which is a plausible estimate for the "diameter" of a nitrogen molecule.

At temperatures below T_c, the van der Waals equation describes those portions of the isotherms outside the shaded region in Figure 20-2 but not those portions inside the shaded region. Suppose we have a gas at a temperature below T_c that initially has a low pressure and a large volume. We begin to compress the gas while holding the temperature constant (isotherm A in the figure). At first the pressure rises, but when we reach point B on the dashed curve, the pressure ceases to rise and the gas begins to liquefy at constant pressure. Along the horizontal line BD in the figure, the gas and liquid are in equilibrium. As we continue to compress the gas, more and more gas liquefies until point D on the dashed curve, at which point we have only liquid. Then, if we try to compress the substance further, the pressure rises sharply because a liquid is nearly incompressible.

Now consider injecting a liquid such as water into a sealed evacuated container. As some of the water evaporates, water-vapor molecules fill the previously empty space in the container. Some of these molecules will hit the liquid surface and rejoin the liquid water in a process called condensation. Initially, the rate of evaporation will be greater than the rate of condensation, but eventually equilibrium will be reached. The pressure at which a liquid is in equilibrium with its own vapor is called the **vapor pressure.** If we now heat the container slightly, the liquid boils, more liquid evaporates, and a new equilibrium is established at a higher vapor pressure. Vapor pressure thus depends on the temperature. We can see this from Figure 20-2. If we had started compressing the gas at a lower temperature, as with isotherm A' in Figure 20-2, the vapor pressure would be lower, as is indicated by the horizontal constant-pressure line for A' at a lower value of pressure. The temperature for which the vapor pressure for a substance equals 1 atm is the **normal boiling point** of that substance. For example, the temperature at which the vapor pressure of water is 1 atm is 373 K (= 100°C), so this temperature is the normal boiling point of water. At high altitudes, such

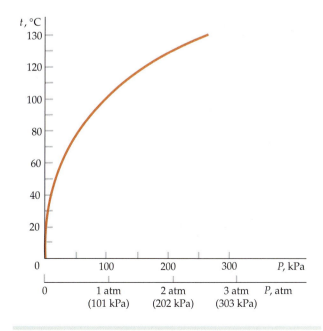

FIGURE 20-3 Boiling point of water versus temperature.

as on the top of a mountain, the pressure is less than 1 atm, therefore, water boils at a temperature lower than 373 K. Figure 20-3 gives the vapor pressures of water at various temperatures.

At temperatures greater than the critical temperature T_c, a gas will not liquefy at any pressure. The critical temperature for water vapor is 647 K (= 374°C). The point at which the critical isotherm intersects the dashed curve in Figure 20-2 (point C) is called the **critical point.**

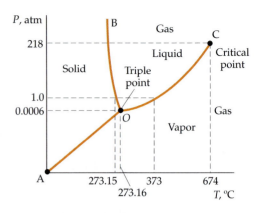

20-3 Phase Diagrams

Figure 20-4 is a plot of pressure versus temperature at a constant volume for water. Such a plot is called a **phase diagram.** The portion of the diagram between points O and C shows vapor pressure versus temperature. As we continue to heat the container, the density of the liquid decreases and the density of the vapor increases. At point C on the diagram, these densities are equal. Point C is called the **critical point.** At this point and above it, there is no distinction between the liquid and the gas. Critical-point temperatures T_c for various substances are listed in Table 20-2. At temperatures greater than the critical temperature a gas will not liquefy at any pressure.

If we now cool our container, some of the vapor condenses into a liquid as we move back down the curve OC until the substance reaches point O in Figure 20-4. At this point, the liquid begins to solidify. Point O is the **triple point,** that one point at which the vapor, liquid, and solid phases of a substance can coexist in equilibrium. Every substance has a unique triple point at a specific temperature and pressure. The triple-point temperature for water is 273.16 K (= 0.01°C) and the triple-point pressure is 4.58 mmHg.

At temperatures and pressures below the triple point, the liquid cannot exist. The curve OA in the phase diagram of Figure 20-4 is the locus of pressures and temperatures for which the solid and vapor coexist in equilibrium. The direct change from a solid to a vapor is called **sublimation.** You can observe sublimation by putting a few loose ice cubes in the freezer compartment of a no-frost (self-defrosting) refrigerator. Over time, the ice cubes will shrink and eventually disappear due to sublimation. This happens because the atmospheric pressure is well above the triple-point pressure of water, and therefore, equilibrium is never established between the ice and water vapor. The triple-point temperature and pressure of carbon dioxide (CO_2) are 216.55 K and 3880 mmHg, which means that liquid CO_2 can only exist at pressures above 3880 mmHG (= 5.1 atm). Thus, at ordinary atmospheric pressures, liquid carbon dioxide cannot exist at any temperature. When solid carbon dioxide "melts," it sublimates directly into gaseous CO_2 without going through the liquid phase, hence the name "dry ice."

The curve OB in Figure 20-4 is the melting curve separating the liquid and solid phases. For a substance like water for which the melting temperature decreases as the pressure increases, curve OB slopes upward to the left from the triple point, as in this figure. For most other substances, the melting temperature increases as the pressure increases. For such a substance, curve OB slopes upward to the right from the triple point.

For a molecule to escape (evaporate) from a substance in the liquid state, energy is required to break the molecular bonds at the liquid's surface. Vaporization cools the liquid left behind. If water is brought to a boil over heat, this cooling effect keeps the temperature of the liquid constant at the boiling point. This is the reason that the boiling point of a substance can be used to calibrate thermometers. However, water can also be made to boil without adding heat by evacuating the air above it, thereby lowering the applied pressure. The energy needed for vaporization is then taken from the water left behind. As a result, the water will cool down, even to the point that ice forms on top of the boiling water!

FIGURE 20-4 Phase diagram for water. The pressure and temperature scales are not linear but are compressed to show the interesting points. Curve OC is the curve of vapor pressure versus temperature. Curve OB is the melting curve, and curve OA is the sublimation curve.

TABLE 20-2

Critical Temperatures T_c for Various Substances

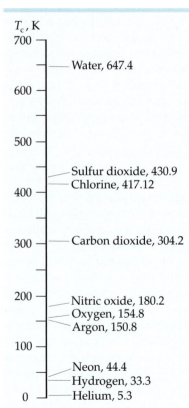

20-4 The Transfer of Thermal Energy

Thermal energy is transferred from one place to another by three processes: conduction, convection, and radiation.

In **conduction,** energy is transferred as heat by interactions among atoms or molecules, although there is no transport of the atoms or molecules themselves. For example, if one end of a solid bar is heated, the atoms in the heated end vibrate with greater energy than do those at the cooler end. The interaction of the more energetic atoms with their neighbors causes this energy to be transported along the bar.[†]

In **convection,** energy is transported as heat by direct mass transport. For example, warm air in a region of a room expands, its density decreases, and the buoyant force on it due to the surrounding air causes it to rise. Energy is thus transported upward along with the mass of warm air.

In **radiation,** energy is transported as heat through space in the form of electromagnetic waves that move at the speed of light. Thermal radiation, light waves, radio waves, television waves, and X rays are all forms of electromagnetic radiation that differ from one another in their wavelengths and frequencies.

In all mechanisms of heat transfer, the rate of cooling of a body is approximately proportional to the temperature difference between the body and its surroundings. This result is known as **Newton's law of cooling.**

In many real situations, all three mechanisms for heat transfer occur simultaneously, though one may be more dominant than the others. For example, an ordinary space heater uses both radiation and convection. If the heating element is quartz, then the main mechanism of heat transference is radiation. If the heating element is metal (which does not radiate as efficiently as quartz), then convection is the main mechanism by which heat is transmitted, with the heated air rising to be replaced by cooler air. Fans are often included in heaters to speed the convection process.

Conduction

Figure 20-5a shows an insulated uniform solid bar of cross-sectional area A. If we keep one end of the bar at a high temperature and the other end at a low temperature, energy is conducted down the bar from the hot end to the cold end. In the steady state, the temperature varies linearly from the hot end to the cold end. The rate of change of the temperature along the bar dT/dx is called the **temperature gradient.**

Let ΔT be the temperature difference across a small segment of length Δx (Figure 20-5b). If ΔQ is the amount of heat conducted through the segment in some time Δt, then the rate of conduction of heat $\Delta Q/\Delta t$ is called the thermal current I. Experimentally, it is found that the thermal current is proportional to the temperature gradient and to the cross-sectional area A:

$$I = \frac{\Delta Q}{\Delta t} = kA\frac{\Delta T}{\Delta x} \qquad 20\text{-}7$$

DEFINITION—THERMAL CURRENT

The proportionality constant k called the *thermal conductivity*, depends on the composition of the bar.[‡] In SI units, thermal current is expressed in watts, and the thermal conductivity has units of $W/(m \cdot K)$.[§] In practical calculations in the

---

[†] If the solid is a metal, the transport of thermal energy is helped by free electrons, which move throughout the metal.
[‡] Don't confuse the thermal conductivity with Boltzmann's constant, which is also designated by k.
[§] In some tables, the energy may be given in calories or kilocalories and the thickness in centimeters.

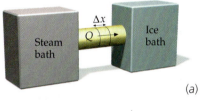

(a)

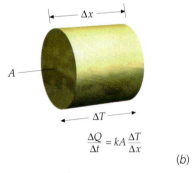

$$\frac{\Delta Q}{\Delta t} = kA\frac{\Delta T}{\Delta x}$$

(b)

FIGURE 20-5 (*a*) An insulated conducting bar with its ends at two different temperatures. (*b*) A segment of the bar of length Δx. The rate at which thermal energy is conducted across the segment is proportional to the cross-sectional area of the bar and the temperature difference across the segment, and it is inversely proportional to the length of the segment.

United States, the thermal current is usually expressed in Btu per hour, the area in square feet, the length (or thickness) in inches, and the temperature in degrees Fahrenheit. The thermal conductivity is then given in Btu·in./(h·ft²·F°). Table 20-3 gives the thermal conductivities of various materials.

If we solve Equation 20-7 for the temperature difference, we obtain

$$\Delta T = I \frac{\Delta x}{kA} \qquad 20\text{-}8$$

or

$$\Delta T = IR \qquad 20\text{-}9$$

TEMPERATURE CHANGE VERSUS CURRENT

where $\Delta x/(kA)$ is the **thermal resistance** R:

$$R = \frac{\Delta x}{kA} \qquad 20\text{-}10$$

DEFINITION—THERMAL RESISTANCE

EXERCISE Calculate the thermal resistance of an aluminum slab of cross-sectional area 15 cm² and thickness 2 cm. (*Answer* 0.0563 K/W = 56.3 mK/W)

EXERCISE What thickness of silver would be required to give the same thermal resistance as a 1-cm thickness of air of the same area? (*Answer* $\Delta x = (1 \text{ cm})(429)/(0.026) = 16{,}500 \text{ cm} = 165 \text{ m}$)

TABLE 20-3

Thermal Conductivities k for Various Materials

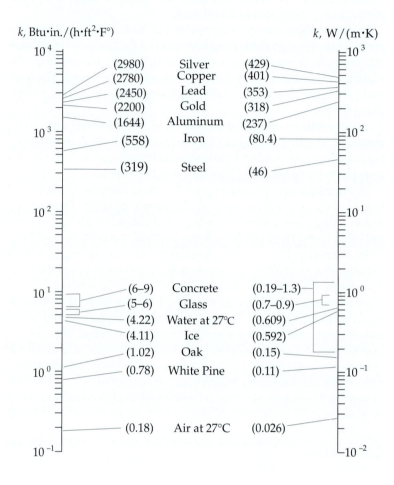

k, Btu·in./(h·ft²·F°)	Material	k, W/(m·K)
(2980)	Silver	(429)
(2780)	Copper	(401)
(2450)	Lead	(353)
(2200)	Gold	(318)
(1644)	Aluminum	(237)
(558)	Iron	(80.4)
(319)	Steel	(46)
(6–9)	Concrete	(0.19–1.3)
(5–6)	Glass	(0.7–0.9)
(4.22)	Water at 27°C	(0.609)
(4.11)	Ice	(0.592)
(1.02)	Oak	(0.15)
(0.78)	White Pine	(0.11)
(0.18)	Air at 27°C	(0.026)

In many practical problems, we are interested in the flow of heat through two or more conductors (or insulators) in series. For example, we may want to know the effect of adding insulating material of a certain thickness and thermal conductivity to the space between two layers of wallboard. Figure 20-6 shows two thermally conducting slabs of the same cross-sectional area but of different materials and of different thickness. Let T_1 be the temperature on the warm side, T_2 be the temperature at the interface between the slabs, and T_3 be the temperature on the cool side. Under the conditions of steady-state heat flow, the thermal current I must be the same through both slabs. This follows from energy conservation; for steady-state flow, the rate at which energy enters any region must equal the rate at which it exits that region.

If R_1 and R_2 are the thermal resistances of the two slabs, we have from Equation 20-9 for each slab

$$T_1 - T_2 = IR_1$$

and

$$T_2 - T_3 = IR_2$$

Adding these equations gives

$$\Delta T = T_1 - T_3 = I(R_1 + R_2) = IR_{eq}$$

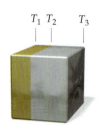

FIGURE 20-6 Two thermally conducting slabs of different materials in series. The equivalent thermal resistance of the slabs in series is the sum of their individual thermal resistances. The thermal current is the same through both slabs.

or

$$I = \frac{\Delta T}{R_{eq}}$$ 20-11

where R_{eq} is the **equivalent resistance.** Thus, for thermal resistances in series, the equivalent resistance is the sum of the individual resistances:

$$R_{eq} = R_1 + R_2 + \cdots$$ 20-12

THERMAL RESISTANCES IN SERIES

This result can be applied to any number of resistances in series. In Chapter 25, we will find that the same formula applies to electrical resistances in series.

To calculate the amount of heat leaving a room by conduction in a given time, we need to know how much heat leaves through the walls, the windows, the floor, and the ceiling. For this type of problem, in which there are several paths for heat flow, the resistances are said to be in parallel. The temperature difference is the same for each path, but the thermal current is different. The total thermal current is the sum of the thermal currents through each of the parallel paths:

$$I_{total} = I_1 + I_2 + \cdots = \frac{\Delta T}{R_1} + \frac{\Delta T}{R_2} + \cdots = \Delta T \left(\frac{1}{R_1} + \frac{1}{R_2} + \cdots \right)$$

or

$$I_{total} = \frac{\Delta T}{R_{eq}}$$ 20-13

This thermogram of a house shows the heat energy being radiated to its surroundings.

where the equivalent thermal resistance is given by

$$\frac{1}{R_{eq}} = \frac{1}{R_1} + \frac{1}{R_2} + \cdots$$ 20-14

THERMAL RESISTANCES IN PARALLEL

We will encounter this equation again in Chapter 25 when we study electric conduction through parallel resistances. Note that for both resistors in series (Equation 20-11) and resistors in parallel (Equation 20-13) I is proportional to ΔT, which is in agreement with Newton's law of cooling.

THERMAL CURRENT BETWEEN TWO METAL BARS   **E X A M P L E   2 0 - 4**

Try It Yourself

Two insulated metal bars, each of length 5 cm and rectangular cross section with sides 2 cm and 3 cm, are wedged between two walls, one held at 100°C and the other at 0°C. (Figure 20-7). The bars are lead and silver. Find (a) the total thermal current through the two-bar combination, and (b) the temperature at the interface.

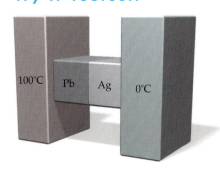

FIGURE 20-7 Two thermally conducting slabs of different materials in parallel.

PICTURE THE PROBLEM The bars are thermal resistors connected in series. (a) You can find the total thermal current from $I = R_{eq}/\Delta T$, where the equivalent resistance R_{eq} is the sum of the individual resistances. Using Equation 20-10 and the thermal conductivities given in Table 20-3, the individual resistances can be determined. (b) You can find the temperature at the interface by applying $I = R_1/\Delta T$ to the lead bar only, and solving for ΔT in terms of the value for I found in Part (a).

Cover the column to the right and try these on your own before looking at the answers.

Steps

Answers

(a) 1. Write the equivalent thermal resistance in terms of the thermal resistances of the two bars.

$R_{eq} = R_{Pb} + R_{Ag}$

2. Using Equation 20-10, write each resistance in terms of the individual thermal conductivities and geometric parameters:

$R_{Pb} = \dfrac{\Delta x_{Pb}}{k_{Pb}A_{Pb}}, R_{Ag} = \dfrac{\Delta x_{Ag}}{k_{Ag}A_{Ag}}$

3. Use Equation 20-13 to find the thermal current.

$I = \Delta T/R_{eq} = \boxed{232\ \text{W}}$

(b) 1. Calculate the temperature difference across the lead bar using the current and thermal resistance found in Part (a).

$\Delta T_{Pb} = IR_{Pb} = 54.9\ \text{K} = 54.9°C$

2. Use your result from the previous step to find the temperature at the interface.

$T_{if} = 100°C - \Delta T_{Pb} = \boxed{45.1°C}$

3. Check your result by finding the temperature difference across the silver bar.

$\Delta T_{Ag} = IR_{Ag} = 45.1°C$

THE METAL BARS REARRANGED

E X A M P L E   2 0 - 5

The metal bars in Example 20-4 are rearranged as shown in Figure 20-8. Find (a) the thermal current in each bar, (b) the total thermal current, and (c) the equivalent thermal resistance of the two-bar system.

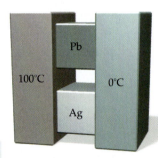

100°C   Pb   0°C   Ag

FIGURE 20-8

PICTURE THE PROBLEM The current in each bar is found from $I = \Delta T/R$, where R is the thermal resistance of the bar (found in Example 20-4). The total current is the sum of the currents. The equivalent resistance can be found from Equation 20-14 or from $I_{total} = \Delta T/R_{eq}$.

(a) Calculate the thermal current for each bar:

$I_{Pb} = \dfrac{\Delta T}{R_{Pb}} = \dfrac{100\ \text{K}}{0.236\ \text{K/W}} = \boxed{424\ \text{W}}$

$I_{Ag} = \dfrac{\Delta T}{R_{Ag}} = \dfrac{100\ \text{K}}{0.194\ \text{K/W}} = \boxed{515\ \text{W}}$

(b) Add your results to find the total thermal current:

$I_{total} = I_{Pb} + I_{Ag}$

$= 424\ \text{W} + 515\ \text{W} = \boxed{938\ \text{W}}$

(c) 1. Use Equation 20-14 to calculate the equivalent resistance of the two bars in parallel:

$\dfrac{1}{R_{eq}} = \dfrac{1}{R_{Pb}} + \dfrac{1}{R_{Ag}} = \dfrac{1}{0.236\ \text{W}} + \dfrac{1}{0.194\ \text{W}}$, so

$R_{eq} = \boxed{0.107\ \text{K/W}}$

2. Check your result using, $I_{\text{total}} = \Delta T / R_{\text{eq}}$:

$$I_{\text{total}} = \frac{\Delta T}{R_{\text{eq}}};$$

$$R_{\text{eq}} = \frac{\Delta T}{I_{\text{total}}} = \frac{100 \text{ K}}{938 \text{ W}} = 0.107 \text{ K/W}$$

REMARKS Note that the equivalent resistance is less than either of the individual resistances. This is always the case for parallel resistors.

In the building industry, the thermal resistance of a square foot of cross-sectional area of a material is called its **R factor** R_f. Consider a 32 ft² sheet of insulating material with thickness Δx and R factor R_f of 7.2. That is, each square foot (Figure 20-9) has a thermal resistance of 7.2 F°/(Btu/h). The 32 square feet are in parallel, so the net resistance R_{net} is calculated using Equation 20-14 giving

$$\frac{1}{R_{\text{net}}} = \frac{1}{R_f} + \frac{1}{R_f} + \cdots = \frac{32}{R_f}, \quad \text{so} \quad R_{\text{total}} = \frac{R_f}{32}$$

Thus, the total thermal resistance R in F°/(Btu/h) equals the R factor divided by the area A in square feet. That is

$$R_{\text{net}} = \frac{R_f}{A}$$

FIGURE 20-9 For a 1-in. thickness of this material, the $R_f = 7.2$.

Since the total resistance R_{total} is related to the conductivity by $R_{\text{net}} = \Delta x/(kA)$ (Equation 20-10), we can express the R factor by

$$R_f = R_{\text{net}} A = \frac{\Delta x}{k} \qquad \text{20-15}$$

DEFINITION—R FACTOR

where Δx is the thickness in inches and k is the conductivity in Btu·in./(h·ft²·F°). Table 20-4 lists R factors for several materials. In terms of the R factor, Equation 20-9 for the thermal current is

$$\Delta T = IR_{\text{net}} = \frac{I}{A} R_f \qquad \text{20-16}$$

For slabs of insulating material of the same area in series, R_f is replaced by the equivalent R factor $R_{f,\text{eq}}$

$$R_{f,\text{eq}} = R_{f1} + R_{f2} + \ldots$$

For parallel slabs, we calculate the thermal current through each slab and add all these currents together in order to obtain the total current.

TABLE 20-4

R Factors $\Delta x/k$ for Various Building Materials

Material	Thickness, in.	R_f, h·ft²·F°/Btu
Building board		
Gypsum or plasterboard	0.375	0.32
Plywood (Douglas fir)	0.5	0.62
Plywood or wood panels	0.75	0.93
Particle board, medium		
density	1.0	1.06
Finish flooring materials		
Carpet and fibrous pad	1.0	2.08
Tile		0.5
Wood, hardwood finish	0.75	0.68
Roof insulation	1.0	2.8
Roofing		
Asphalt roll roofing		0.15
Asphalt shingles		0.44
Windows		
Single-pane		0.9
Double pane		1.8

EXAMPLE 20-6 **Put It in Context**

You are helping your friend's family put new asphalt shingles on the roof of their winter cabin. The 60 ft × 20 ft roof is made of 1-in. pine board covered with asphalt shingles. There is room for 2 in. of roof insulation, and your friend's family is wondering how much of a difference it would make to their energy bill if they were to install the two inches of insulation. Knowing that you are studying physics, they ask for your opinion.

PICTURE THE PROBLEM To assess the situation, you first calculate the R factor for each layer of the roof. Since the layers are in series, the equivalent R factor is just the sum of the individual R factors. The aim is to calculate the equivalent R factor of the roof with and without the insulation. The R factors for asphalt shingles and for roof insulation are found in Table 20-4. The R factor for the pine board is calculated from its thermal conductivity, which is found in Table 20-3. Note that when you shingle a roof you have to overlap the shingles, so there are two layers of asphalt shingling on the roof.

1. The equivalent R factor is the sum of the individual R factors:

$$R_{f,eq} = R_{f,pine} + R_{f,asph} + R_{f,insul}$$

2. The R factor for the double layer of shingles is twice the R factor for one layer:

$$R_{f,asph} = 2(0.44 \text{ h·ft}^2\text{·F°/Btu})$$
$$= 0.88 \text{ h·ft}^2\text{·F°/Btu}$$

3. The R factor for 2 in. of roof insulation is twice that for 1 in.:

$$R_{f,insul} = 2(2.8 \text{ h·ft}^2\text{·F°/Btu})$$
$$= 5.6 \text{ h·ft}^2\text{·F°/Btu}$$

4. The R factor for 1-in.-thick pine is obtained from the conductivity:

$$R_{f,p} = \frac{\Delta x_p}{k_p} = \frac{1 \text{ in.}}{0.78 \text{ Btu·in./(h·ft}^2\text{·F°)}}$$
$$= 1.28 \text{ h·ft}^2\text{·F°/Btu}$$

5. The equivalent R factor without the insulation is:

$$R'_{f,eq} = R_{f,pine} + R_{f,asph}$$
$$= 1.28 \text{ h·ft}^2\text{·F°/Btu} + 0.88 \text{ h·ft}^2\text{·F°/Btu}$$
$$= 2.16 \text{ h·ft}^2\text{·F°/Btu}$$

6. The equivalent R factor with insulation is:

$$R_{f,eq} = R_{f,pine} + R_{f,asph} + R_{f,insul} = R'_{f,eq} + R_{f,insul}$$
$$= 2.16 \text{ h·ft}^2\text{·F°/Btu} + 5.6 \text{ h·ft}^2\text{·F°/Btu}$$
$$= 7.76 \text{ h·ft}^2\text{·F°/Btu}$$

7. One comparison of the two equivalent R factors is their ratio:

$$\frac{R'_{f,eq}}{R_{f,eq}} = \frac{2.16}{7.76} = 0.28$$

8. By adding the insulation the heat loss rate per square foot is reduced by 78%. Is it 78% of a large heat loss or a small heat loss? Using Equation 20-16 we calculate the thermal current I through the entire roof.

$$\Delta T = I R_{net} = \frac{I}{A} R_f$$

$$I' = \frac{A}{R'_{f,eq}} \Delta T = \frac{(60 \text{ ft})(20 \text{ ft})}{2.16 \text{ h·ft}^2\text{·F°/Btu}} \Delta T$$

$$= \left[556 \text{ (Btu/h)/F°}\right] \Delta T$$

9. To complete the calculation we estimate that the temperature inside the cabin is maintained at 70°F and the temperature outside the cabin during the winter is typically 40°F colder.

$$I' = \left[556 \text{ (Btu/h)/F°}\right]\Delta T$$
$$= \left[556 \text{ (Btu/h)/F°}\right](40°F) = 22,200 \text{ Btu/h}$$

and

$$I = 0.28 I' = 0.28(22,200 \text{ Btu/h}) = 6200 \text{ Btu/h}$$

so the reduction due to the insulation is

$$I - I' = 22,200 \text{ Btu/h} - 6200 \text{ Btu/h}$$

$$= 16,000 \text{ Btu/h}$$

10. Estimate the savings that would result from adding the 2 in. of insulation.

| See the following Remarks for an estimate of the cost. |

REMARKS Installing 2 in. of roof insulation reduces the heat loss through the roof by 22,200 Btu/h. The cabin is heated with propane, and the energy content of propane is about 92,000 Btu/gal. Insulating the roof reduces consumption by approximately 6 gal of propane every 24 h of use. Propane costs about $1.40/gal, so this amounts to a savings of approximately $8.40 per day, or $252 per month. Your friend's family is impressed by the potential savings (and by the benefits of your physics knowledge). They decide to install the 2 in. of roofing insulation.

EXERCISE How much additional savings can be had by adding even more insulation to the roof? (*Answer*   The maximum additional savings is 6200 Btu/h which would save $68 per month.)

REMARKS These cost estimates do not include the cost of purchasing and installing the insulation.

The thermal conductivity of air is very small compared with that of solid materials, which makes air a very good insulator. However, when there is a large air gap—say, between a storm window and the inside window—the insulating efficiency of air is greatly reduced because of convection. Whenever there is a temperature difference between different parts of the air space, convection currents act quickly to equalize the temperature, so the effective conductivity is greatly increased. For storm windows, air gaps of about 1 to 2 cm are optimal. Wider air gaps actually reduce the thermal resistance of a double-pane window due to convection.

The insulating properties of air are most effectively used when the air is trapped in small pockets that prevent convection from taking place. This is the principle underlying the excellent insulating properties of both goose down and Styrofoam.

If you touch the inside surface of a glass window when it is cold outside, you will observe that the surface is considerably colder than the inside air. The thermal resistance of windows is mainly due to thin films of insulating air that adhere to either side of the glass surface. The thickness of the glass has little effect on the overall thermal resistance. The air film on each side typically adds an R factor of about 0.45 per side. Thus, the R factor of a window with N separated glass layers is approximately $0.9N$ because of the two sides of each layer. Under windy conditions, the outside air film may be greatly decreased, leading to a smaller R factor for the window.

Convection

Convection is the transport of energy as heat by the transport of the material medium itself. This thermal property is responsible for the great ocean currents as well as the global circulation of the atmosphere. In the simplest case, convection arises when a fluid (gas or liquid) is heated from below. The warm fluid then expands and rises as the cooler fluid sinks. The mathematical description of convection is very complex because the flow depends on the temperature difference in different parts of the fluid, and this temperature difference is affected by the flow itself.

The heat transferred from an object to its surroundings by convection is approximately proportional to the area of the object and to the difference in temperature between the object and the surrounding fluid. It is possible to write an equation for the energy transported as heat by convection and to define a coefficient of convection, but the analyses of practical problems involving convection is quite complex and will not be discussed here.

Radiation

All objects emit and absorb electromagnetic radiation. When an object is in thermal equilibrium with its surroundings, it emits heat and absorbs heat at the same rate. The rate at which an object radiates energy is proportional to both the area of the object and to the fourth power of its absolute temperature. This result, found empirically by Josef Stefan in 1879 and derived theoretically by Ludwig Boltzmann about five years later, is called the **Stefan-Boltzmann law:**

$$P_r = e\sigma AT^4 \qquad\qquad 20\text{-}17$$

STEFAN–BOLTZMANN LAW

where P_r is the power radiated, A is the area, σ is a universal constant called Stefan's constant, which has the value

$$\sigma = 5.6703 \times 10^{-8}\,\text{W}/(\text{m}^2\cdot\text{K}^4) \qquad\qquad 20\text{-}18$$

and e is the **emissivity** of the object, a fractional quantity between 0 and 1 that is dependent upon the composition of the surface of the object.

When electromagnetic radiation falls on an opaque object, part of the radiation is reflected and part is absorbed. Light-colored objects reflect most visible radiation, whereas dark objects absorb most of it. The rate at which an object absorbs radiation is given by

$$P_a = e\sigma AT_0^4 \qquad\qquad 20\text{-}19$$

where T_0 is the temperature of the source of the radiation.

If an object emits more radiation than it absorbs, then it cools, while the object's surroundings absorb radiation from the object and become warmer. If the object absorbs more radiation than it emits, then the object warms and its surroundings cool. The net power radiated by an object at temperature T in an environment at temperature T_0 is

$$P_{net} = e\sigma A(T^4 - T_0^4) \qquad\qquad 20\text{-}20$$

When an object is in thermal equilibrium with its surroundings, $T = T_0$, and the object emits and absorbs radiation at the same rate.

An object that absorbs all the radiation incident upon it has an emissivity equal to 1, and it is called a **blackbody.** A blackbody is also an ideal radiator. The concept of a blackbody is important because the characteristics of the radiation emitted by such an ideal object can be calculated theoretically. Materials such as black velvet come close to being ideal blackbodies. The best practical approximation of an ideal blackbody is a small hole leading into a cavity, such as a keyhole in a closet door (Figure 20-10). Radiation incident on the hole has little chance of being reflected out the hole before the walls of the cavity absorb it. Thus, the radiation emitted out of the hole is characteristic of the temperature of the walls of the cavity.

The radiation emitted by an object at temperatures below approximately 600°C is not visible. Most radiation emissions are concentrated at wavelengths much longer than those of visible light.[†] As an object is heated, the rate of energy

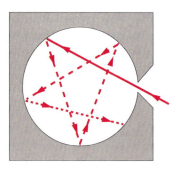

FIGURE 20-10 A hole in a cavity approximates an ideal blackbody. Radiation entering the cavity has little chance of leaving the cavity before it is completely absorbed. The radiation emitted through the hole (not shown) is therefore characteristic of the temperature of the walls of the cavity.

[†] When we study light, we will see that visible light is electromagnetic radiation with wavelengths between about 400 and 700 nm.

emission increases, and the energy radiated extends to higher frequencies (and shorter wavelengths). Between about 600 and 700°C, enough of the radiated energy is in the visible spectrum for the object to glow a dull red. At higher temperatures, it may become bright red or even "white hot." Figure 20-11 shows the power radiated by a blackbody as a function of wavelength for several different temperatures. The wavelength at which the power is a maximum varies inversely with the temperature, a result known as Wien's displacement law:

$$\lambda_{max} = \frac{2.898 \text{ mm·K}}{T} \qquad 20\text{-}21$$

WIEN'S DISPLACEMENT LAW

This law is used to determine the surface temperatures of stars by analyzing their radiation. It can also be used to map out the variation in temperature over different regions of the surface of an object. Such a map is called a thermograph. Thermographs can be used to detect cancer because cancerous tissue results in increased circulation which produces a slight increase in skin temperature.

The spectral-distribution curves shown in Figure 20-11 played an important role in the history of physics. It was the discrepancy between theoretical calculations (using classical thermodynamics) of what the blackbody spectral distribution should be and the actual experimental measurements of spectral distributions that led to Max Planck's first ideas about the quantization of energy in 1900.

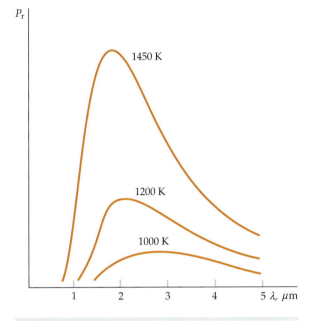

FIGURE 20-11 Radiated power versus wavelength for radiation emitted by a blackbody. The wavelength of the maximum power varies inversely with the absolute temperature of the blackbody.

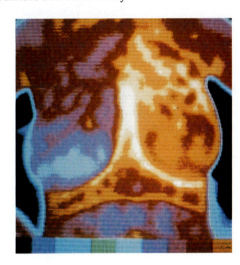

A thermograph was used to detect this cancerous tumor.

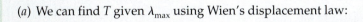

RADIATION FROM THE SUN   **EXAMPLE   20-7**

(a) The radiation emitted by the surface of the sun emits maximum power at a wavelength of about 500 nm. Assuming the sun to be a blackbody emitter, what is its surface temperature? (b) Calculate λ_{max} for a blackbody at room temperature, $T = 300$ K.

(a) We can find T given λ_{max} using Wien's displacement law:

$$\lambda_{max} = \frac{2.898 \text{ mm·K}}{T}$$

so

$$T = \frac{2.898 \text{ mm·K}}{\lambda_{max}} = \frac{2.898 \text{ mm·K}}{500 \text{ nm}}$$

$$= \boxed{5800 \text{ K}}$$

(b) We can find λ_{max} from Wien's displacement law for $T = 300$ K:

$$\lambda_{max} = \frac{2.898 \text{ mm·K}}{300 \text{ K}} = 9.66 \times 10^{-3} \text{ mm}$$

$$= \boxed{9.66 \text{ } \mu m}$$

REMARKS The peak wavelength from the sun is in the visible spectrum. The blackbody radiation spectrum describes the sun's radiation fairly well, so the sun is indeed a good example of a blackbody.

For $T = 300$ K, the spectrum peaks in the infrared at wavelengths much longer than the wavelengths visible to the eye. Surfaces that are not black to our eyes may act as blackbodies for infrared radiation and absorption. For example, it has been found experimentally that the skin of human beings of all races is black to infrared radiation; hence, the emissivity of skin is 1.00 for its own radiation process.

RADIATION FROM THE HUMAN BODY   **E X A M P L E   2 0 - 8**   **Try It Yourself**

Calculate the net rate of heat loss in radiated energy for a naked person in a room at 20°C, assuming the person to be a blackbody with a surface area of 1.4 m² and a surface temperature of 33°C (= 306 K). (The surface temperature of the human body is slightly less than the internal temperature of 37°C because of the thermal resistance of the skin.)

Cover the column to the right and try these on your own before looking at the answers.

Steps   **Answer**

Use $P_{net} = e\sigma A(T^4 - T_0^4)$ with $e = 1$, $T = 306$ K, and $P_{net} = 111$ W
$T_0 = 293$ K.

REMARKS This large energy loss is approximately equal to the basal metabolic rate of about 120 W. We protect ourselves from this great loss of energy by wearing clothing, which, because of its low thermal conductivity, has a much lower outside temperature and therefore a much lower rate of thermal radiation.

When the temperature of an object T is not too different from the surrounding temperature T_0, a radiating object obeys Newton's law of cooling. We can see this by writing Equation 20-20 as

$$P_{net} = e\sigma A(T^4 - T_0^4) = e\sigma A(T^2 + T_0^2)(T^2 - T_0^2)$$
$$= e\sigma A(T^2 + T_0^2)(T + T_0)(T - T_0)$$

When $T - T_0$ is small, we can replace T by T_0 in the sums with little change in the result. Then

$$P_{net} = e\sigma A(T^4 - T_0^4) \approx e\sigma A(T_0^2 + T_0^2)(T_0 + T_0)(T - T_0) = 4e\sigma AT_0^3 \, \Delta T$$

The net power radiated is approximately proportional to the temperature difference, in agreement with Newton's law of cooling. This result can also be obtained by using the differential approximation.

$$\Delta P_r \approx \left. \frac{dP_r}{dT} \right|_{T=T_0} (T - T_0)$$

where $P_r = e\sigma A(T^4 - T_0^4)$. For a small temperature difference $T - T_0$ we have

$$\Delta P_r \approx e\sigma A \, 4T^3 \big|_{T=T_0} (T - T_0) = 4e\sigma AT_0^3 \, \Delta T$$

Topic	Relevant Equations and Remarks	
1. Thermal Expansion		
Coefficient of linear expansion	$\alpha = \dfrac{\Delta L/L}{\Delta T}$	20-2
Coefficient of volume expansion	$\beta = \dfrac{\Delta V/V}{\Delta T} = 3\alpha$	20-4, 20-5
2. The van der Waals Equation of State	The van der Waals equation of state describes the behavior of real gases over a wide range of temperatures and pressures, taking into account the space occupied by the gas molecules themselves and the attraction of the molecules to one another.	
	$\left(P + \dfrac{an^2}{V^2}\right)(V - bn) = nRT$	20-6
3. Vapor Pressure	Vapor pressure is the pressure at which the liquid and gas phases of a substance are in equilibrium at a given temperature. The liquid boils at that temperature for which the external pressure equals the vapor pressure.	
4. The Triple Point	The triple point is the unique temperature and pressure at which the gas, liquid, and solid phases of a substance can coexist in equilibrium. At temperatures and pressures below the triple point, the liquid phase of a substance cannot exist.	
5. Heat Transfer	The three mechanisms by which thermal energy is transferred are radiation, conduction, and convection.	
Newton's law of cooling	For all mechanisms of heat transfer, if the temperature difference between the body and its surroundings is small, the rate of cooling of a body is approximately proportional to the temperature difference.	
6. Heat Conduction		
Current	The rate of conduction of thermal energy is given by	
	$I = \dfrac{\Delta Q}{\Delta t} = kA\dfrac{\Delta T}{\Delta x}$	20-7
	where I is the thermal current, k is the coefficient of thermal conductivity, and $\Delta T/\Delta x$ is the temperature gradient.	
Thermal resistance	$\Delta T = IR$	20-9
	where R is the thermal resistance:	
	$R = \dfrac{\Delta x}{kA}$	20-10
Equivalent resistance:		
series	$R_{eq} = R_1 + R_2 + \ldots$	20-12
parallel	$\dfrac{1}{R_{eq}} = \dfrac{1}{R_1} + \dfrac{1}{R_2} + \ldots$	20-14

R factor	The R factor is the thermal resistance in in.·ft²·°F/(Btu/h) for a square foot of a slab of material $$R_f = R_{net}A = \frac{\Delta x}{k}$$	**20-15**

7. Thermal Radiation

Rate of power radiated	$$P_r = e\sigma A T^4$$	**20-17**
	where $\sigma = 5.6703 \times 10^{-8}$ W/m²·K⁴ is Stefan's constant and e is the emissivity, which varies between 0 and 1 (depending on the composition of the surface of the object). Materials that are good heat absorbers are also good heat radiators.	
Net power radiated by an object at T to its environment at T_0	$$P_{net} = e\sigma A(T^4 - T_0^4)$$	**20-20**
Blackbody	A blackbody has an emissivity of 1. It is a perfect radiator, and it absorbs all radiation incident upon it.	
Wein's law	The power spectrum of electromagnetic energy radiated by a blackbody has a maximum at a wavelength λ_{max}, which varies inversely with the absolute temperature of the body: $$\lambda_{max} = \frac{2.898 \text{ mm·K}}{T}$$	**20-21**

PROBLEMS

- Single-concept, single-step, relatively easy
- •• Intermediate-level, may require synthesis of concepts
- ••• Challenging
- **SSM** Solution is in the *Student Solutions Manual*
- Problems available on iSOLVE online homework service
- ✓ These "Checkpoint" online homework service problems ask students additional questions about their confidence level, and how they arrived at their answer

In a few problems, you are given more data than you actually need; in a few other problems, you are required to supply data from your general knowledge, outside sources, or informed estimates.

Conceptual Problems

1 • **SSM** Why does the mercury level first decrease slightly when a thermometer is placed in warm water?

2 • A large sheet of metal has a hole cut in the middle of it. When the sheet is heated, the area of the hole will (*a*) not change, (*b*) always increase, (*c*) always decrease, (*d*) increase if the hole is not in the exact center of the sheet, (*e*) decrease only if the hole is in the exact center of the sheet.

3 • Mountaineers say that you cannot hard boil an egg on the top of Mount Rainier. This is true because (*a*) the air is too cold to boil water, (*b*) the air pressure is too low for stoves to burn, (*c*) boiling water is not hot enough to hard boil the egg, (*d*) the oxygen content of the air is too low, (*e*) eggs always break in their backpacks.

4 • Which gases in Table 20-2 cannot be liquefied by applying pressure at 20°C?

5 •• **SSM** The phase diagram in Figure 20-12 can be interpreted to yield information on how the boiling and melting points of water change with altitude. (*a*) Explain how this information can be obtained. (*b*) How might this information affect cooking procedures in the mountains?

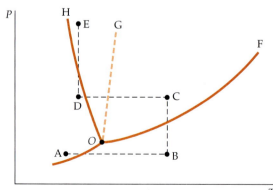

FIGURE 20-12 Problem 5

6 • If the absolute temperature of an object is tripled, the rate at which it radiates thermal energy (a) triples, (b) increases by a factor of 9, (c) increases by a factor of 27, (d) increases by a factor of 81, (e) depends on whether the absolute temperature is above or below zero.

7 • **SSM** In a cool room, a metal or marble table top feels much colder to the touch than does a wood surface even though they are at the same temperature. Why?

8 • True or false:

(a) During a phase change, the temperature of a substance remains constant.
(b) The rate of conduction of thermal energy is proportional to the temperature gradient.
(c) The rate at which an object radiates energy is proportional to the square of its absolute temperature.
(d) All materials expand when they are heated.
(e) The vapor pressure of a liquid depends on the temperature.

9 • The earth loses heat by (a) conduction, (b) convection, (c) radiation, (d) all of the above.

10 • Which heat-transfer mechanisms are the most important in the warming effect of a fire in a fireplace?

11 • Which heat-transfer mechanism is important in the transfer of energy from the sun to the earth?

12 •• Explain why turning down the temperature of a house at night in winter can save money on heating costs. Why doesn't the cost of the fuel consumed to heat the house back up in the morning equal the savings realized by cooling it down?

13 •• Two cylinders made of materials A and B have the same lengths; their diameters are related by $d_A = 2d_B$. When the same temperature difference is maintained between the ends of the cylinders they conduct heat at the same rate. Their thermal conductivities are related by (a) $k_A = k_B/4$, (b) $k_A = k_B/2$, (c) $k_A = k_B$, (d) $k_A = 2k_B$, (e) $k_A = 4k_B$.

14 • Infrared light is sometimes referred to as "heat waves." Explain why infrared light has received this label, and why the label is inaccurate.

15 • **SSM** In artistic nomenclature, blue is often referred to as a "cool" color, while red is referred to as a "warm" color. In physics, however, red is considered a "cooler" color than blue. Explain why.

Estimation and Approximation

16 ••• Liquid helium is stored in containers fitted with 7-cm-thick "superinsulation" consisting of numerous layers of very thin aluminized Mylar sheets. The rate of evaporation of liquid in a 200-L container is about 0.7 L per day. Assume the container is spherical and that the external temperature is 20°C. The specific gravity of liquid helium is 0.125 and the latent heat of vaporization is 21 kJ/kg. Estimate the thermal conductivity of superinsulation.

17 •• Estimate the thermal conductivity of the skin, given that the body of an "average" adult male has about 1.8 m^2 of skin area, and produces about 130 W of heat when resting. Use an internal temperature of 37°C (98.6°F) and an external skin temperature of 33°C. Assume that the skin has an average thickness of about 1 mm.

18 •• **SSM** Estimate the effective emissivity of the earth, given the following information: the solar constant (the intensity of light incident on the earth from the sun) is 1370 W/m^2, 70 percent of this light is absorbed by the earth, and the earth's average temperature is 288 K. (Assume that the effective area that is absorbing the light is πR^2, where R is the earth's radius, while the blackbody-emission area is $4\pi R^2$.)

19 •• Black holes in orbit around a normal star are detected from earth due to the frictional heating of infalling gas into the black hole, which can reach temperatures greater than 10^6 K. Assuming that the infalling gas can be modeled as a blackbody radiator, estimate λ_{max} for use in an astronomical detection of a black hole. (Remark: This is in the x-ray region of the electromagnetic spectrum.)

Thermal Expansion

20 • **iSOLVE** A steel ruler has a length of 30 cm at 20°C. What is its length at 100°C?

21 •• (a) Define a coefficient of area expansion. (b) Calculate it for a square and a circle, and show that it is two times the coefficient of linear expansion.

22 •• **iSOLVE** The density of aluminum is 2.70×10^3 kg/m^3 at 0°C. What is the density of aluminum at 200°C?

23 •• **iSOLVE** A copper collar is to fit tightly about a steel shaft that has a diameter of 6.0000 cm at 20°C. The inside diameter of the copper collar at that temperature is 5.9800 cm. To what temperature must the copper collar be raised so that it will just slip on the steel shaft, assuming the steel shaft remains at 20°C?

24 •• **SSM** **iSOLVE** Repeat Problem 23 when the temperature of both the steel shaft and copper collar are raised simultaneously.

25 •• A container is filled to the brim with 1.4 L of mercury at 20°C. When the temperature of container and mercury is raised to 60°C, 7.5 mL of mercury spill over the brim of the container. Determine the linear expansion coefficient of the container.

26 •• **iSOLVE** ✓ A hole is drilled in an aluminum plate with a steel drill bit whose diameter at 20°C is 6.245 cm. In the process of drilling, the temperature of the drill bit and of the aluminum plate rise to 168°C. What is the diameter of the hole in the aluminum plate when it has cooled to room temperature?

27 •• **SSM** A rookie crew was left to put in the final 1 km of rail for a stretch of railroad track. When they finished, the temperature was 20°C, and they headed to town for some refreshments. After an hour or two, one of the old-timers noticed that the temperature had gone up to 25°C, so he said, "I hope you left some gaps to allow for expansion." From the look on their faces, he knew that they had not, and they all rushed back to the work site. The rail had buckled into an isosceles triangle. How high was the buckle?

28 •• **iSOLVE** A car has a 60-L steel gas tank filled to the top with gasoline when the temperature is 10°C. The coefficient of volume expansion of gasoline is $\beta = 0.900 \times 10^{-3}$ K^{-1}. Taking the expansion of the steel tank into account, how much gasoline spills out of the tank when the car is parked in the sun and its temperature rises to 25°C?

29 •• A thermometer has an ordinary glass bulb and thin glass tube filled with 1 mL of mercury. A temperature change of 1°C changes the level of mercury in the thin tube by 3.0 mm. Find the inside diameter of the thin glass tube.

30 •• *iSOLVE* ✓ A mercury thermometer consists of a 0.4-mm capillary tube connected to a glass bulb. The mercury level rises 7.5 cm as the temperature of the thermometer increases from 35°C to 43°C. Find the volume of the thermometer bulb.

31 ••• A grandfather's clock is calibrated at a temperature of 20°C. (*a*) On a hot day, when the temperature is 30°C, does the clock run fast or slow? (*b*) How much does it gain or lose in a 24-h period? Assume that the pendulum is a thin brass rod of negligible mass with a heavy bob attached to the end.

32 ••• *iSOLVE* ✓ A steel tube has an outside diameter of 3.000 cm at room temperature (20°C). A brass tube has an inside diameter of 2.997 cm at the same temperature. To what temperature must the ends of the tubes be heated if the steel tube is to be inserted into the brass tube?

33 ••• **SSM** What is the tensile stress in the copper collar of Problem 23 when its temperature returns to 20°C?

The van der Waals Equation, Liquid–Vapor Isotherms, and Phase Diagrams

34 • *iSOLVE* ✓ (*a*) Calculate the volume of 1 mol of steam at 100°C and a pressure of 1 atm, assuming that it is an ideal gas. (*b*) Find the temperature at which the steam will occupy the volume found in Part (*a*) if it obeys the van der Waals equation with $a = 0.55$ Pa·m^6/mol^2 and $b = 30$ cm^3/mol.

35 •• From Figure 20-3, find (*a*) the temperature at which water boils on a mountain where the atmospheric pressure is 70 kPa, (*b*) the temperature at which water will boil in a container in which the pressure has been reduced to 0.5 atm, and (*c*) the pressure at which water will boil at 115°C.

36 •• **SSM** The van der Waals constants for helium are $a = 0.03412$ L^2·atm/mol^2 and $b = 0.0237$ L/mol. Use these data to find the volume in cubic centimeters occupied by one helium atom and to estimate the radius of the atom.

37 ••• (*a*) For a van der Waals gas, show that the critical temperature is $8a/27Rb$ and the critical pressure is $a/27b^2$. (*b*) Rewrite the van der Waals equation of state in terms of the reduced variable $V_r = V/V_c$, $P_r = P/P_c$, and $T_r = T/T_c$.

Heat Conduction

38 • *iSOLVE* A copper bar 2 m long has a circular cross section of radius 1 cm. One end is kept at 100°C and the other end is kept at 0°C. The surface of the bar is insulated so that there is negligible heat loss through it. Find (*a*) the thermal resistance of the bar, (*b*) the thermal current I, (*c*) the temperature gradient $\Delta T/\Delta x$, and (*d*) the temperature of the bar 25 cm from the hot end.

39 • *iSOLVE* ✓ A 20 × 30-ft slab of insulation has an R factor of 11. How much heat (in Btu/h) is conducted through the slab if the temperature on one side is 68°F and on the other side it is 30°F?

40 •• *iSOLVE* ✓ Two metal cubes with 3-cm edges, one copper (Cu) and one aluminum (Al), are arranged as shown in Figure 20-13. Find (*a*) the thermal resistance of each cube, (*b*) the thermal resistance of the two-cube system, (*c*) the thermal current I, and (*d*) the temperature at the interface of the two cubes.

FIGURE 20-13
Problem 40

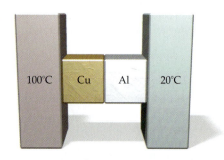

41 •• The cubes in Problem 40 are rearranged in parallel as shown in Figure 20-14. Find (*a*) the thermal current carried by each cube from one side to the other, (*b*) the total thermal current, and (*c*) the equivalent thermal resistance of the two-cube system.

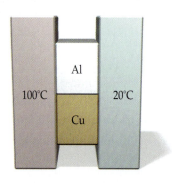

FIGURE 20-14 **Problem 41**

42 •• **SSM** The cost of air conditioning a house is proportional to the rate at which heat flows from the house and is inversely proportional to the coefficient of performance (COP) of the air conditioner. We denote the temperature difference between the house and its surroundings as ΔT. Assuming that the rate at which heat flows from a house is proportional to ΔT and that the air conditioner is operating ideally, show that the cost of air conditioning is proportional to $(\Delta T)^2$.

43 ••• A spherical shell of thermal conductivity k has inside radius r_1 and outside radius r_2 (Figure 20-15). The inside of the shell is held at a temperature T_1, and the outside at temperature T_2. In this problem, you are to show that the thermal current through the shell is given by

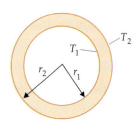

FIGURE 20-15 **Problem 43**

$$I = \frac{4\pi k r_1 r_2}{r_2 - r_1}(T_2 - T_1) \qquad \text{20-22}$$

Consider a spherical element of the shell of radius r and thickness dr. (*a*) Why must the thermal current through each such element be the same? (*b*) Write the thermal current I through such a shell element in terms of the area $A = 4\pi r^2$, the thickness dr, and the temperature difference dT across the element. (*c*) Solve for dT in terms of dr and integrate from $r = r_1$ to $r = r_2$. (*d*) Show that when r_1 and r_2 are much larger than $r_2 - r_1$, Equation 20-22 (shown above) is the same as Equation 20-7.

44 •• **SSM** **SOLVE** For a boiler at a power station, heat must be transferred to boiling water at the rate of 3 GW. The boiling water passes through copper pipes having a wall thickness of 4.0 mm and a surface area of 0.12 m² per meter length of pipe. Find the total length of pipe (actually there are many pipes in parallel) that must pass through the furnace if the steam temperature is 225°C and the external temperature of the pipes is 600°C.

45 ••• A steam pipe of length L is insulated with a layer of material of thermal conductivity k. Find the rate of heat transfer if the temperature outside the insulation is T_1, the temperature inside is T_2, the outside radius of the insulation is r_1, and the inside radius is r_2.

Radiation

46 • **SSM** **SOLVE**✓ Calculate λ_{max} for a human blackbody radiator, assuming the surface temperature of the skin to be 33°C.

47 • **SOLVE**✓ The heating wires of a 1-kW electric heater are red hot at a temperature of 900°C. Assuming that 100% of the heat output is due to radiation and that the wires act as blackbody radiators, what is the effective area of the radiating surface? (Assume a room temperature of 20°C.)

48 •• **SOLVE**✓ A blackened, solid copper sphere of radius 4.0 cm hangs in a vacuum in an enclosure whose walls have a temperature of 20°C. If the sphere is initially at 0°C, find the rate at which its temperature changes, assuming that heat is transferred by radiation only.

49 •• **SOLVE** The surface temperature of the filament of an incandescent lamp is 1300°C. If the electric power input is doubled, what will the temperature become? *Hint: Show that you can neglect the temperature of the surroundings.*

50 •• Liquid helium is stored at its boiling point (4.2 K) in a spherical can that is separated by a vacuum space from a surrounding shield that is maintained at the temperature of liquid nitrogen (77 K). If the can is 30 cm in diameter and is blackened on the outside so that it acts as a blackbody radiator, how much helium boils away per hour?

General Problems

51 • **SSM** **SOLVE** A steel tape is placed around the earth at the equator when the temperature is 0°C. What will the clearance between the tape and the ground (assumed to be uniform) be if the temperature of the tape rises to 30°C? Neglect the expansion of the earth.

52 •• Show that change in the density of an isotropic material due to an increase in temperature ΔT is given by $\Delta\rho = -\beta\rho\Delta T$.

53 •• **SOLVE** The solar constant is the power received from the sun per unit area perpendicular to the sun's rays at the mean distance of the earth from the sun. Its value at the upper atmosphere of the earth is about 1.35 kW/m². Calculate the effective temperature of the sun if it radiates like a blackbody. (The radius of the sun is 6.96×10^8 m.)

54 •• **SOLVE**✓ To determine the R value of insulating material that comes in sheets of $\frac{1}{2}$-in. thickness, you construct a cubical box of 12 in. per side and place a thermometer and a 100-W heater inside the box. After thermal equilibrium has been attained, the temperature inside the box is 90°C when the external temperature is 20°C. Determine the R value of this material.

55 •• A 2-cm-thick copper sheet is pressed against a sheet of aluminum. What should be the thickness of the aluminum sheet so that the temperature of the copper–aluminum interface is $(T_1 + T_2)/2$, where T_1 and T_2 are the temperatures at the copper–air and aluminum–air interfaces?

56 •• At a temperature of 20°C, a steel bar of radius 2.2 cm and length 60 cm is jammed horizontally perpendicular between two vertical concrete walls. With a blowtorch, the temperature of the bar is raised to 60°C. Find the force exerted by the bar on each wall.

57 •• (a) From the definition of β, the coefficient of volume expansion (at constant pressure), show that $\beta = 1/T$ for an ideal gas. (b) The experimentally determined value of β for N_2 gas at 0°C is 0.003673 K⁻¹. Compare this value with the theoretical value $\beta = 1/T$, assuming that N_2 is an ideal gas.

58 •• One way to construct a device with two points whose separation remains the same in spite of temperature changes is to bolt together one end of two rods, both of which have different coefficients of linear expansion as in the arrangement shown in Figure 20-16. (a) Show that the distance L will not change with temperature if the lengths L_A and L_B are chosen such that $L_A/L_B = \alpha_B/\alpha_A$. (b) If material B is steel, material A is brass, and $L_A = 250$ cm at 0°C, what is the value of L?

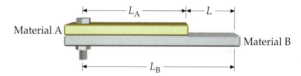

FIGURE 20-16 Problem 58

59 •• On the average, the temperature of the earth's crust increases 1.0 C° for every 30 m of depth. The average thermal conductivity of the earth's crust is 0.74 J/m·s·K. What is the heat loss of the earth per second due to conduction from the core? How does this heat loss compare with the average power received from the sun? (The solar constant is about 1.35 kW/m².)

60 •• **SOLVE** A copper-bottomed saucepan containing 0.8 L of boiling water boils dry in 10 min. Assuming that all the heat flows through the flat copper bottom, which has a diameter of 15 cm and a thickness of 3.0 mm, calculate the temperature of the outside of the copper bottom while some water is still in the pan.

61 •• **SSM** A hot-water tank of cylindrical shape has an inside diameter of 0.55 m and inside height of 1.2 m. The tank is enclosed with a 5-cm-thick insulating layer of glass wool whose thermal conductivity is 0.035 W/m·K. The metallic interior and exterior walls of the container have thermal conductivities that are much greater than that of the glass wool. How much power must be supplied to this tank in order to maintain the water temperature at 75°C when the external temperature is 1°C?

62 ••• The diameter of a rod is given by $d = d_0(1 + ax)$, where a is a constant and x is the distance from one end. If the thermal conductivity of the material is k what is the thermal resistance of the rod if its length is L?

63 ••• A solid disk of radius r and mass m is spinning in a frictionless environment with angular velocity ω_1 at temperature T_1. The temperature of the disk is then changed to T_2. Express the angular velocity ω_2, rotational kinetic energy E_2, and angular momentum L_2 in terms of their values at the temperature T_1 and the linear expansion coefficient α of the disk.

64 ••• Write a spreadsheet program to graph the temperature of the earth as a function of emissivity, using the results of Problem 18. How much does the emissivity have to change in order for the average temperature to rise by 1 K? This can be thought of as a model for the effect of increasing concentrations of greenhouse gases like methane and CO_2 in the earth's atmosphere.

65 ••• A small pond has a layer of ice 1 cm thick floating on its surface. (a) If the air temperature is $-10°C$, find the rate in centimeters per hour at which ice is added to the bottom of the layer. The density of ice is 0.917 g/cm^3. (b) How long does it take for a 20-cm layer to be built up?

66 ••• **SSM** A 200-g copper container holding 0.7 L of water is thermally isolated from its surroundings—except for a 10-cm-long copper rod of cross-sectional area 1.5 cm² connecting it to a second copper container filled with an ice and water mixture so its temperature remains at 0°C. The initial temperature of the first container is $T_0 = 60°C$. (Assume the heat capacity of the rod to be negligible.) (a) Show that

the temperature T of the first container changes over time t according to

$$T = T_0 e^{-t/RC}$$

where T is in degrees Celsius, R is the thermal resistance of the rod, and C is the total heat capacity of the container plus the water. (Neglect the heat capacity of the rod.) (b) Evaluate R, C, and the "time constant" RC. (c) Show that the total amount of heat Q conducted in time t is

$$Q = CT_0(1 - e^{-t/RC})$$

(d) Using a spreadsheet program, graph both $T(t)$ and $Q(t)$; from the graph, find the time it takes for the temperature of the first container to be reduced to 30°C.

67 ••• A blackened copper cube that is 1 cm along an edge is heated to a temperature of 300°C, and then is placed in a vacuum chamber whose walls are at a temperature of 0°C. In the vacuum chamber, the cube cools radiatively. (a) Show that the (absolute) temperature T of the cube follows the differential equation:

$$\frac{dT}{dt} = -\frac{e\sigma A}{C}(T^4 - T_0^4)$$

where C is the heat capacity of the cube, A is its surface area, e the emissivity, and T_0 the temperature of the vacuum chamber. (b) Using Euler's method, numerically solve the differential equation to find $T(t)$, and graph it. Assume $e = 1$. (The Euler method is discussed in Section 4 of Chapter 5.) How long does it take the cube to cool to a temperature of 15°C?

APPENDIX A

SI Units and Conversion Factors

Basic Units

Length	The *meter* (m) is the distance traveled by light in a vacuum in $1/299,792,458$ s.
Time	The *second* (s) is the duration of 9,192,631,770 periods of the radiation corresponding to the transition between the two hyperfine levels of the ground state of the ^{133}Cs atom.
Mass	The *kilogram* (kg) is the mass of the international standard body preserved at Sèvres, France.
Current	The *ampere* (A) is that current in two very long parallel wires 1 m apart that gives rise to a magnetic force per unit length of 2×10^{-7} N/m.
Temperature	The *kelvin* (K) is $1/273.16$ of the thermodynamic temperature of the triple point of water.
Luminous intensity	The *candela* (cd) is the luminous intensity, in the perpendicular direction, of a surface of area $1/600,000$ m^2 of a blackbody at the temperature of freezing platinum at a pressure of 1 atm.

Derived Units

Force	newton (N)	$1\,N = 1\,kg{\cdot}m/s^2$
Work, energy	joule (J)	$1\,J = 1\,N{\cdot}m$
Power	watt (W)	$1\,W = 1\,J/s$
Frequency	hertz (Hz)	$1\,Hz = cy/s$
Charge	coulomb (C)	$1\,C = 1\,A{\cdot}s$
Potential	volt (V)	$1\,V = 1\,J/C$
Resistance	ohm (Ω)	$1\,\Omega = 1\,V/A$
Capacitance	farad (F)	$1\,F = 1\,C/V$
Magnetic field	tesla (T)	$1\,T = 1\,N/(A{\cdot}m)$
Magnetic flux	weber (Wb)	$1\,Wb = 1\,T{\cdot}m^2$
Inductance	henry (H)	$1\,H = 1\,J/A^2$

Conversion Factors

Conversion factors are written as equations for simplicity;
relations marked with an asterisk are exact.

Length

1 km = 0.6215 mi

1 mi = 1.609 km

1 m = 1.0936 yd = 3.281 ft = 39.37 in.

*1 in. = 2.54 cm

*1 ft = 12 in. = 30.48 cm

*1 yd = 3 ft = 91.44 cm

1 lightyear = 1 $c \cdot$y = 9.461×10^{15} m

*1 Å = 0.1 nm

Area

*1 m^2 = 10^4 cm^2

1 km^2 = 0.3861 mi^2 = 247.1 acres

*1 $in.^2$ = 6.4516 cm^2

1 ft^2 = 9.29×10^{-2} m^2

1 m^2 = 10.76 ft^2

*1 acre = 43,560 ft^2

1 mi^2 = 640 acres = 2.590 km^2

Volume

*1 m^3 = 10^6 cm^3

*1 L = 1000 cm^3 = 10^{-3} m^3

1 gal = 3.786 L

1 gal = 4 qt = 8 pt = 128 oz = 231 in^3

1 in^3 = 16.39 cm^3

1 ft^3 = 1728 $in.^3$ = 28.32 L
       = 2.832×10^4 cm^3

Time

*1 h = 60 min = 3.6 ks

*1 d = 24 h = 1440 min = 86.4 ks

1 y = 365.24 d = 3.156×10^7 s

Speed

*1 m/s = 3.6 km/h

1 km/h = 0.2778 m/s = 0.6215 mi/h

1 mi/h = 0.4470 m/s = 1.609 km/h

1 mi/h = 1.467 ft/s

Angle and Angular Speed

*π rad = 180°

1 rad = 57.30°

1° = 1.745×10^{-2} rad

1 rev/min = 0.1047 rad/s

1 rad/s = 9.549 rev/min

Mass

*1 kg = 1000 g

*1 tonne = 1000 kg = 1 Mg

1 u = 1.6606×10^{-27} kg

1 kg = 6.022×10^{26} u

1 slug = 14.59 kg

1 kg = 6.852×10^{-2} slug

1 u = 931.50 MeV/c^2

Density

*1 g/cm^3 = 1000 kg/m^3 = 1 kg/L

(1 g/cm^3)g = 62.4 lb/ft^3

Force

1 N = 0.2248 lb = 10^5 dyn

*1 lb = 4.448222 N

(1 kg)g = 2.2046 lb

Pressure

*1 Pa = 1 N/m^2

*1 atm = 101.325 kPa = 1.01325 bars

1 atm = 14.7 lb/$in.^2$ = 760 mmHg
      = 29.9 in.Hg = 33.8 ftH_2O

1 lb/$in.^2$ = 6.895 kPa

1 torr = 1 mmHg = 133.32 Pa

1 bar = 100 kPa

Energy

*1 kW·h = 3.6 MJ

*1 cal = 4.1840 J

1 ft·lb = 1.356 J = 1.286×10^{-3} Btu

*1 L·atm = 101.325 J

1 L·atm = 24.217 cal

1 Btu = 778 ft·lb = 252 cal = 1054.35 J

1 eV = 1.602×10^{-19} J

1 u·c^2 = 931.50 MeV

*1 erg = 10^{-7} J

Power

1 horsepower = 550 ft·lb/s = 745.7 W

1 Btu/h = 1.055 kW

1 W = 1.341×10^{-3} horsepower
    = 0.7376 ft·lb/s

Magnetic Field

*1 T = 10^4 G

Thermal Conductivity

1 W/(m·K) = 6.938 Btu·in./(h·ft^2·F°)

1 Btu·in./(h·ft^2·F°) = 0.1441 W/(m·K)

APPENDIX B

Numerical Data

Terrestrial Data

Free-fall acceleration g	9.80665 m/s^2; 32.1740 ft/s^2
(Standard value at sea level at 45° latitude)[†]	
Standard value	
At sea level, at equator[†]	9.7804 m/s^2
At sea level, at poles[†]	9.8322 m/s^2
Mass of earth M_E	$5.98 \times 10^{24} \text{ kg}$
Radius of earth R_E, mean	$6.37 \times 10^6 \text{ m}$; 3960 mi
Escape speed $\sqrt{2R_E g}$	$1.12 \times 10^4 \text{ m/s}$; 6.95 mi/s
Solar constant[‡]	1.35 kW/m^2
Standard temperature and pressure (STP):	
Temperature	273.15 K
Pressure	101.325 kPa (1.00 atm)
Molar mass of air	28.97 g/mol
Density of air (STP), ρ_{air}	1.293 kg/m^3
Speed of sound (STP)	331 m/s
Heat of fusion of H_2O (0°C, 1 atm)	333.5 kJ/kg
Heat of vaporization of H_2O (100°C, 1 atm)	2.257 MJ/kg.

† Measured relative to the earth's surface.
‡ Average power incident normally on 1 m² outside the earth's atmosphere at the mean distance from the earth to the sun.

Astronomical Data[†]

Earth	
Distance to moon[‡]	$3.844 \times 10^8 \text{ m}$; $2.389 \times 10^5 \text{ mi}$
Distance to sun, mean[‡]	$1.496 \times 10^{11} \text{ m}$; $9.30 \times 10^7 \text{ mi}$; 1.00 AU
Orbital speed, mean	$2.98 \times 10^4 \text{ m/s}$
Moon	
Mass	$7.35 \times 10^{22} \text{ kg}$
Radius	$1.738 \times 10^6 \text{ m}$
Period	27.32 d
Acceleration of gravity at surface	1.62 m/s^2
Sun	
Mass	$1.99 \times 10^{30} \text{ kg}$
Radius	$6.96 \times 10^8 \text{ m}$

† Additional solar-system data is available from NASA at <http://nssdc.gsfc.nasa.gov/planetary/planetfact.html>.
‡ Center to center.

Physical Constants[†]

Gravitational constant	G	$6.673(10) \times 10^{-11}$ N·m²/kg²
Speed of light	c	$2.997\,924\,58 \times 10^8$ m/s
Fundamental charge	e	$1.602\,176\,462(63) \times 10^{-19}$ C
Avogadro's number	N_A	$6.022\,141\,99(47) \times 10^{23}$ particles/mol
Gas constant	R	$8.314\,472(15)$ J/(mol·K)
		$1.987\,2065(36)$ cal/(mol·K)
		$8.205\,746(15) \times 10^{-2}$ L·atm/(mol·K)
Boltzmann constant	$k = R/N_A$	$1.380\,6503(24) \times 10^{-23}$ J/K
		$8.617\,342(15) \times 10^{-5}$ eV/K
Stefan-Boltzmann constant	$\sigma = (\pi^2/60)k^4/(\hbar^3 c^2)$	$5.670\,400(40) \times 10^{-8}$ W/(m²k⁴)
Atomic mass constant	$m_u = \frac{1}{12} m(^{12}C)$	$1.660\,538\,73(13) \times 10^{-27}$ kg = 1u
Coulomb constant	$k = 1/(4\pi\epsilon_0)$	$8.987\,551\,788\ldots \times 10^9$ N·m²/C²
Permittivity of free space	ϵ_0	$8.854\,187\,817\ldots \times 10^{-12}$ C²/(N·m²)
Permeability of free space	μ_0	$4\pi \times 10^{-7}$ N/A²
		$1.256\,637 \times 10^{-6}$ N/A²
Planck's constant	h	$6.626\,068\,76(52) \times 10^{-34}$ J·s
		$4.135\,667\,27(16) \times 10^{-15}$ eV·s
	$\hbar = h/2\pi$	$1.054\,571\,596(82) \times 10^{-34}$ J·s
		$6.582\,118\,89(26) \times 10^{-16}$ eV·s
Mass of electron	m_e	$9.109\,381\,88(72) \times 10^{-31}$ kg
		$0.510\,998\,902(21)$ MeV/c^2
Mass of proton	m_p	$1.672\,621\,58(13) \times 10^{-27}$ kg
		$938.271\,998(38) \times$ MeV/c^2
Mass of neutron	m_n	$1.674\,927\,16(13) \times 10^{-27}$ kg
		$939.565\,330(38)$ MeV/c^2
Bohr magneton	$m_B = eh/2m_e$	$9.274\,0008\,99(37) \times 10^{-24}$ J/T
		$5.788\,381\,749(43) \times 10^{-5}$ eV/T
Nuclear magneton	$m_n = eh/2m_p$	$5.050\,783\,17(20) \times 10^{-27}$ J/T
		$3.152\,451\,238(24) \times 10^{-8}$ eV/T
Magnetic flux quantum	$\phi_0 = h/2e$	$2.067\,833\,636(81) \times 10^{-15}$ T·m²
Quantized Hall resistance	$R_K = h/e^2$	$2.581\,280\,7572(95) \times 10^4$ Ω
Rydberg constant	R_H	$1.097\,373\,156\,8549(83) \times 10^7$ m⁻¹
Josephson frequency-voltage quotient	$K_J = 2e/h$	$4.835\,978\,98(19) \times 10^{14}$ Hz/V
Compton wavelength	$\lambda_C = h/m_e c$	$2.426\,310\,215(18) \times 10^{-12}$ m

[†] The values for these and other constants may be found on the Internet at http://physics.nist.gov/cuu/Constants/index.html. The numbers in parentheses represent the uncertainties in the last two digits. (For example, 2.044 43(13) stands for 2.044 43 ± 0.000 13.) Values with without uncertainties are exact, including those values with ellipses (like the value of pi is exactly 3.1415...).

For additional data, see the following tables in the text.

Periodic Table of Elements

1	2	3	4	5	6	7	8	9	10	11	12	13	14	15	16	17	18
1 H 1.00797																	2 He 4.003
3 Li 6.941	4 Be 9.012											5 B 10.81	6 C 12.011	7 N 14.007	8 O 15.9994	9 F 19.00	10 Ne 20.179
11 Na 22.990	12 Mg 24.31											13 Al 26.98	14 Si 28.09	15 P 30.974	16 S 32.064	17 Cl 35.453	18 Ar 39.948
19 K 39.102	20 Ca 40.08	21 Sc 44.96	22 Ti 47.88	23 V 50.94	24 Cr 52.00	25 Mn 54.94	26 Fe 55.85	27 Co 58.93	28 Ni 58.69	29 Cu 63.55	30 Zn 65.38	31 Ga 69.72	32 Ge 72.59	33 As 74.92	34 Se 78.96	35 Br 79.90	36 Kr 83.80
37 Rb 85.47	38 Sr 87.62	39 Y 88.906	40 Zr 91.22	41 Nb 92.91	42 Mo 95.94	43 Tc (98)	44 Ru 101.1	45 Rh 102.905	46 Pd 106.4	47 Ag 107.870	48 Cd 112.41	49 In 114.82	50 Sn 118.69	51 Sb 121.75	52 Te 127.60	53 I 126.90	54 Xe 131.29
55 Cs 132.905	56 Ba 137.33	57–71 Rare Earths	72 Hf 178.49	73 Ta 180.95	74 W 183.85	75 Re 186.2	76 Os 190.2	77 Ir 192.2	78 Pt 195.09	79 Au 196.97	80 Hg 200.59	81 Tl 204.37	82 Pb 207.19	83 Bi 208.98	84 Po (210)	85 At (210)	86 Rn (222)
87 Fr (223)	88 Ra (226)	89–103 Actinides	104 Rf (261)	105 Ha (260)	106 (263)	107 (262)	108 (265)	109 (266)									

Rare Earths (Lanthanides)

57 La 138.91	58 Ce 140.12	59 Pr 140.91	60 Nd 144.24	61 Pm (147)	62 Sm 150.36	63 Eu 152.0	64 Gd 157.25	65 Tb 158.92	66 Dy 162.50	67 Ho 164.93	68 Er 167.26	69 Tm 168.93	70 Yb 173.04	71 Lu 174.97

Actinides

89 Ac 227.03	90 Th 232.04	91 Pa 231.04	92 U 238.03	93 Np 237.05	94 Pu (244)	95 Am (243)	96 Cm (247)	97 Bk (247)	98 Cf (251)	99 Es (252)	100 Fm (257)	101 Md (258)	102 No (259)	103 Lr (260)

The 1–18 group designation has been recommended by the International Union of Pure and Applied Chemistry (IUPAC).

Atomic Numbers and Atomic Masses†

Name	Symbol	Atomic Number	Mass	Name	Symbol	Atomic Number	Mass
Actinium	Ac	89	227.03	Mercury	Hg	80	200.59
Aluminum	Al	13	26.98	Molybdenum	Mo	42	95.94
Americium	Am	95	(243)	Neodymium	Nd	60	144.24
Antimony	Sb	51	121.75	Neon	Ne	10	20.179
Argon	Ar	18	39.948	Neptunium	Np	93	237.05
Arsenic	As	33	74.92	Nickel	Ni	28	58.69
Astatine	At	85	(210)	Niobium	Nb	41	92.91
Barium	Ba	56	137.3	Nitrogen	N	7	14.007
Berkelium	Bk	97	(247)	Nobelium	No	102	(259)
Beryllium	Be	4	9.012	Osmium	Os	76	190.2
Bismuth	Bi	83	208.98	Oxygen	O	8	15.9994
Boron	B	5	10.81	Palladium	Pd	46	106.4
Bromine	Br	35	79.90	Phosphorus	P	15	30.974
Cadmium	Cd	48	112.41	Platinum	Pt	78	195.09
Calcium	Ca	20	40.08	Plutonium	Pu	94	(244)
Californium	Cf	98	(251)	Polonium	Po	84	(210)
Carbon	C	6	12.011	Potassium	K	19	39.098
Cerium	Ce	58	140.12	Praseodymium	Pr	59	140.91
Cesium	Cs	55	132.905	Promethium	Pm	61	(147)
Chlorine	Cl	17	35.453	Protactinium	Pa	91	231.04
Chromium	Cr	24	52.00	Radium	Ra	88	(226)
Cobalt	Co	27	58.93	Radon	Rn	86	(222)
Copper	Cu	29	63.55	Rhenium	Re	75	186.2
Curium	Cm	96	(247)	Rhodium	Rh	45	102.905
Dysprosium	Dy	66	162.50	Rubidium	Rb	37	85.47
Einsteinium	Es	99	(252)	Ruthenium	Ru	44	101.1
Erbium	Er	68	167.26	Rutherfordium	Rf	104	(261)
Europium	Eu	63	152.0	Samarium	Sm	62	150.36
Fermium	Fm	100	(257)	Scandium	Sc	21	44.96
Fluorine	F	9	19.00	Selenium	Se	34	78.96
Francium	Fr	87	(223)	Silicon	Si	14	28.09
Gadolinium	Gd	64	157.25	Silver	Ag	47	107.870
Gallium	Ga	31	69.72	Sodium	Na	11	22.990
Germanium	Ge	32	72.59	Strontium	Sr	38	87.62
Gold	Au	79	196.97	Sulfur	S	16	32.064
Hafnium	Hf	72	178.49	Tantalum	Ta	73	180.95
Hahnium	Ha	105	(260)	Technetium	Tc	43	(98)
Helium	He	2	4.003	Tellurium	Te	52	127.60
Holmium	Ho	67	164.93	Terbium	Tb	65	158.92
Hydrogen	H	1	1.0079	Thallium	Tl	81	204.37
Indium	In	49	114.82	Thorium	Th	90	232.04
Iodine	I	53	126.90	Thulium	Tm	69	168.93
Iridium	Ir	77	192.2	Tin	Sn	50	118.69
Iron	Fe	26	55.85	Titanium	Ti	22	47.88
Krypton	Kr	36	83.80	Tungsten	W	74	183.85
Lanthanum	La	57	138.91	Uranium	U	92	238.03
Lawrencium	Lr	103	(260)	Vanadium	V	23	50.94
Lead	Pb	82	207.2	Xenon	Xe	54	131.29
Lithium	Li	3	6.941	Ytterbium	Yb	70	173.04
Lutetium	Lu	71	174.97	Yttrium	Y	39	88.906
Magnesium	Mg	12	24.31	Zinc	Zn	30	65.38
Manganese	Mn	25	54.94	Zirconium	Zr	40	91.22
Mendelevium	Md	101	(258)				

† More precise values for the atomic masses, along with the uncertainties in the masses, can be found at http://physics.nist.gov/PhysRefData/.

APPENDIX D

Review of Mathematics

In this appendix, we will review some of the basic results of algebra, geometry, trigonometry, and calculus. In many cases, we will merely state results without proof. Table D-1 lists some mathematical symbols.

Equations

The following operations can be performed on mathematical equations to facilitate their solution:

1. The same quantity can be added to or subtracted from each side of the equation.

2. Each side of the equation can be multiplied or divided by the same quantity.

3. Each side of the equation can be raised to the same power.

It is important to understand that the preceding rules apply to each *side* of the equation and not to each term in the equation.

EXAMPLE D-1

Solve the following equation for x: $(x - 3)^2 + 7 = 23$.

1. Subtract 7 from each side:	$(x - 3)^2 = 16$
2. Take the square root of each side:	$x - 3 = \pm 4$
3. Add 3 to each side:	$x = 4 + 3 = 7$
	or
	$x = -4 + 3 = -1$

REMARKS Note that in step 2 we do not need to write $-(x - 3) = \pm 4$ because all possibilities are included in

$x - 3 = \pm 4$.

CHECK THE RESULT We check our result by substituting each value into the original equation: $(7 - 3)^2 + 7 = 16 + 7 = 23$ and $(-1 - 3)^2 + 7 = 16 + 7 = 23$.

Mathematical Symbols

$=$	is equal to
$\neq$	is not equal to
$\approx$	is approximately equal to
$\sim$	is of the order of
$\propto$	is proportional to
$>$	is greater than
$\geq$	is greater than or equal to
$\gg$	is much greater than
$<$	is less than
$\leq$	is less than or equal to
$\ll$	is much less than
Δx	change in x
$\|x\|$	absolute value of x
$n!$	$n(n - 1)(n - 2) \ldots 1$
Σ	sum
$\lim$	limit
$\Delta t \to 0$	Δt approaches zero
$\dfrac{dx}{dt}$	derivative of x with respect to t
$\dfrac{\partial x}{\partial t}$	partial derivative of x with respect to t
$\int$	integral

EXAMPLE D-2

Solve the following equation for x:

$$\frac{1}{x} + \frac{1}{4} = \frac{1}{3}$$

1. Subtract from each side:

$$\frac{1}{x} = \frac{1}{3} - \frac{1}{4} = \frac{4}{12} - \frac{3}{12} = \frac{1}{12}$$

2. Multiply each side by $12x$:

$$x = 12$$

REMARKS This type of equation occurs both in geometric optics and in analyses of electric circuits. Although it is easy to solve, errors are often made. A typical mistake is to take the reciprocal of each *term*, obtaining $x + 4 = 3$. Taking the reciprocal of each term is not allowed; taking the reciprocal of each *side* of an equation is allowed. Note that multiplying each side by $12x$ in step 2 is equivalent to taking the reciprocal of each side of the equation.

Direct and Inverse Proportion

The relationships of direct proportion and inverse proportion are so important in physics that they deserve special consideration. Often much algebraic manipulation can be avoided through a simple knowledge of these relationships. Suppose, for example, that you work for 5 days at a certain pay rate and earn $400. How much would you earn at the same pay rate if you worked 8 days? In this problem, the money earned is *directly proportional* to the time worked. We can write an equation relating the money earned M to the time worked t using a constant of proportionality R:

$$M = Rt$$

The constant of proportionality in this case is the pay rate. We can express R in dollars per day. Since $400 was earned in 5 d, the value of R is $400/(5\text{ d}) = $80/\text{d}$. In 8 d, the amount earned is therefore

$$M = ($80/\text{d})(8\text{ d}) = $640$$

However, we do not have to find the pay rate explicitly to work the problem. Since the amount earned in 8 d is $\frac{8}{5}$ times that earned in 5 d, this amount is

$$M = \tfrac{8}{5}($400) = $640$$

We can use a similar example to illustrate inverse proportion. If you get a 25% raise, how long would you need to work to earn $400? Here we consider R to be a variable and we wish to solve for t:

$$t = \frac{M}{R}$$

In this equation, the time t is *inversely proportional* to the pay rate R. Thus, if the new rate is $\frac{5}{4}$ times the old rate, the new time will be $\frac{4}{5}$ times the old time or 4 d.

There are some situations in which one quantity varies as the square or some other power of another quantity where the ideas of proportionality are also very useful. Suppose, for example, that a 10-in. diameter pizza costs $8.50. How much would you expect a 12-in. diameter pizza to cost? We expect the cost of a pizza to be approximately proportional to the amount of its contents, which is proportional to the area of the pizza. Since the area is in turn proportional to the square of the diameter, the cost should be proportional to the square of the diameter. If we increase the diameter by a factor of $12/10$, the area increases by a factor of $(12/10)^2 = 1.44$, so we should expect the cost to be $(1.44)(\$8.50) = \$12.24$.

EXAMPLE D-3

The intensity of light from a point source varies inversely with the square of the distance from the source. If the intensity is 3.20 W/m² at 5 m from a source, what is it at 6 m from the source?

1. Write an equation expressing the fact that the intensity varies inversely with the square of the distance:

$$I = \frac{C}{r^2}$$

where C is some constant.

2. Let I_1 be the intensity at $r_1 = 5$ m and I_2 be the intensity at $r_2 = 6$ m, and express the ratio I_2/I_1 in terms of r_1 and r_2:

$$\frac{I_2}{I_1} = \frac{C/r_2^2}{C/r_1^2} = \frac{r_1^2}{r_2^2} = \left(\frac{r_1}{r_2}\right)^2 = \left(\frac{5}{6}\right)^2 = 0.694$$

3. Solve for I_2:

$$I_2 = 0.694 I_1 = (0.694)(3.20\ \text{W/m}^2) = 2.22\ \text{W/m}^2$$

Linear Equations

An equation in which the variables occur only to the first power is said to be linear. A linear equation relating y and x can always be put into the standard form

$$y = mx + b \qquad\qquad \text{D-1}$$

where m and b are constants that may be either positive or negative. Figure D-1 shows a graph of the values of x and y that satisfy (Equation D-1). The constant b, called the **intercept,** is the value of y at $x = 0$. The constant m is the **slope** of the line, which equals the ratio of the change in y to the corresponding change in x. In the figure, we have indicated two points on the line, x_1, y_1 and x_2, y_2, and the changes $\Delta x = x_2 - x_1$ and $\Delta y = y_2 - y_1$. The slope m is then

$$m = \frac{y_2 - y_1}{x_2 - x_1} = \frac{\Delta y}{\Delta x}$$

If x and y are both unknown, there is no unique solution for their values. Any pair of values x_1, y_1 on the line in Figure D-1 will satisfy the equation. If we have two equations, each with the same two unknowns x and y, the equations can be solved simultaneously for the unknowns.

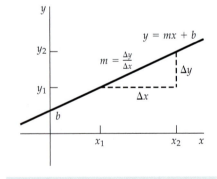

FIGURE D-1 Graph of the linear equation $y = mx + b$, where b is the intercept and $m = \Delta y/\Delta x$ is the slope.

EXAMPLE D-4

Find the values of x and y that satisfy

$$3x - 2y = 8 \qquad\qquad \text{D-2}$$

and

$$y - x = 2 \qquad\qquad \text{D-3}$$

PICTURE THE PROBLEM Figure D-2 shows a graph of each of these equations. At the point where the lines intersect, the values of x and y satisfy both equations. We can solve two simultaneous equations by first solving either equation for one variable in terms of the other variable and then substituting the result into the other equation. An alternative method is to multiply one equation by a constant such that one of the unknown terms is eliminated when the equations are added or subtracted.

1. Solve (Equation D-3) for y: $y = x + 2$

2. Substitute this value for y into (Equation D-2): $3x - 2(x + 2) = 8$

3. Simplify and solve for x: $3x - 2x - 4 = 8$

     $x = 12$

4. To solve these equations using the alternative method, we first multiply (Equation D-3) by 2: $2y - 2x = 4$

5. Add this equation to (Equation D-2): $2y - 2x = 4$

     $\dfrac{3x - 2y = 8}{3x - 2x = 12}$

     $x = 12$

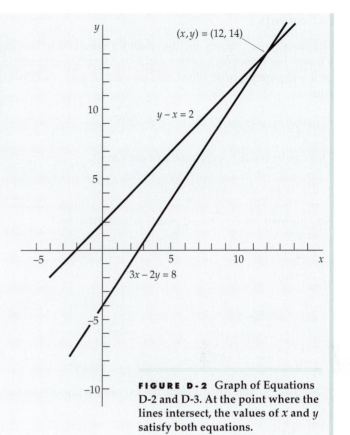

FIGURE D-2 Graph of Equations D-2 and D-3. At the point where the lines intersect, the values of x and y satisfy both equations.

Factoring

Equations can often be simplified by factoring. Three important examples are

1. Common factor: $2ax + 3ay = a(2x + 3y)$
2. Perfect square: $x^2 - 2xy + y^2 = (x - y)^2$
3. Difference of squares: $x^2 - y^2 = (x + y)(x - y)$

The Quadratic Formula

An equation that contains a variable to the second power is called a *quadratic equation*. The standard form for a quadratic equation is

     $ax^2 + bx + c = 0$ D-4

where a, b, and c are constants. The general solution of this equation is

     $x = -\dfrac{b}{2a} \pm \dfrac{1}{2a}\sqrt{b^2 - 4ac}$ D-5

When b^2 is greater than $4ac$, there are two solutions corresponding to the + and − signs. Figure D-3 shows a graph of y versus x where $y = ax^2 + bx + c$. The curve, called a **parabola,** crosses the x axis twice. The values of x for which $y = 0$ are the solutions to (Equation D-4). When $b^2 < 4ac$, the graph of y versus x does not intersect the x axis, as is shown in Figure D-4, and there are no real solutions to (Equation D-4). When $b^2 = 4ac$, the graph of y versus x is tangent to the x axis at the point $x = -b/2a$.

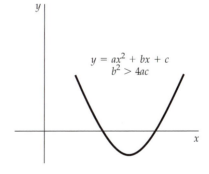

$y = ax^2 + bx + c$
$b^2 > 4ac$

FIGURE D-3 Graph of y versus x when $y = ax^2 + bx + c$ for the case $b^2 > 4ac$. The two values of x for which $y = 0$ satisfy the quadratic equation (Equation D-4).

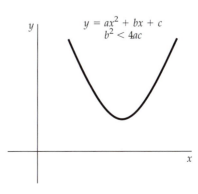

$y = ax^2 + bx + c$
$b^2 < 4ac$

FIGURE D-4 Graph of y versus x when $y = ax^2 + bx + c$ for the case $b^2 < 4ac$. In this case, there are no (real) values of x for which $y = 0$.

Exponents

The notation x^n stands for the quantity obtained by multiplying x times itself n times. For example, $x^2 = x \cdot x$ and $x^3 = x \cdot x \cdot x$. The quantity n is called the **power,** or the **exponent,** of x. When two powers of x are multiplied, the exponents are added:

$$(x^m)(x^n) = x^{m+n} \qquad\qquad \text{D-6}$$

This can be readily seen from an example:

$$x^2 x^3 = (x \cdot x)(x \cdot x \cdot x) = x^5$$

Any number raised to the 0 power is defined to be 1:

$$x^0 = 1 \qquad\qquad \text{D-7}$$

Then

$$x^n x^{-n} = x^0 = 1$$
$$x^{-n} = \frac{1}{x^n} \qquad\qquad \text{D-8}$$

When two powers are divided, the exponents are subtracted:

$$\frac{x^n}{x^m} = x^n x^{-m} = n^{n-m} \qquad\qquad \text{D-9}$$

Using these rules, we have

$$x^{1/2} \cdot x^{1/2} = x$$

so

$$x^{1/2} = \sqrt{x}$$

When a power is raised to another power, the exponents are multiplied:

$$(x^n)^m = x^{nm} \qquad\qquad \text{D-10}$$

Logarithms

If y is related to x by $y = a^x$, the number x is said to be the logarithm of y to the base a and is written

$$x = \log_a y$$

then if $x = 1$ then $y = a^1 = a$ and

$$\log_a a = 1 \qquad\qquad \text{D-11}$$

and, if $x = 0$ then $y = a^0 = 1$ and

$$\log_a 1 = 0 \qquad\qquad \text{D-12}$$

Also, if $y_1 = a^n$ and $y_2 = a^m$, then

$$y_1 y_2 = a^n a^m = a^{n+m}$$

and

$$\log_a y_1 = n, \log_a y_2 = m, \text{ and } \log_a y_1 y_2 = n + m$$

so

$$\log_a y_1 y_2 = \log_a y_1 + \log_a y_2 \qquad\qquad \text{D-13}$$

It immediately follows that

$$\log_a y^n = n \log_a y \qquad\qquad \text{D-14}$$

There are two bases in common use: base 10, called **common logarithms,** and base $e(e = 2.728\ldots)$, called **natural logarithms.** When no base is specified, the base is usually understood to be 10. Thus, $\log 100 = \log_{10} 100 = 2$ since $100 = 10^2$.

The symbol ln is used for natural logarithms. Thus,

$$\log_e x = \ln x \qquad\qquad \text{D-15}$$

and $y = \ln x$ implies

$$x = e^y \qquad\qquad \text{D-16}$$

Logarithms can be changed from one base to another. Suppose that

$$z = \log x \qquad\qquad \text{D-17}$$

Then

$$10^z = 10^{\log x} = x \qquad\qquad \text{D-18}$$

Taking the natural logarithm of both sides of (Equation D-18), we obtain

$$z \ln 10 = \ln x$$

Substituting $\log x$ for z (see Equation D-17) gives

$$\ln x = (\ln 10)\log x \qquad\qquad \text{D-19}$$

The Exponential Function

When the rate of change of a quantity is proportional to the quantity itself, the quantity increases or decreases exponentially. An example of *exponential decrease* is nuclear decay. If N is the number of radioactive nuclei at some time, then the change dN in some very small time interval dt will be proportional to N and to dt:

$$dN = -\lambda N \, dt$$

where the constant of proportionality λ is the decay rate. The function N satisfying this equation is

$$N = N_0 e^{-\lambda t} \qquad \text{D-20}$$

where N_0 is the number at time $t = 0$. Figure D-5 shows N versus t. A characteristic of exponential decay is that N decreases by a constant factor in a given time interval. The time interval for N to decrease to half its original value is its half-life $t_{1/2}$. The half life is obtained from (Equation D-20) by setting $N = \frac{1}{2}N_0$ and solving for the time. This gives

$$t_{1/2} = \frac{\ln 2}{\lambda} = \frac{0.693}{\lambda} \qquad \text{D-21}$$

An example of *exponential increase* is population growth. If the number of organisms is N, the change in N after a small time interval dt is given by

$$dN = +\lambda N \, dt$$

where λ is a constant that characterizes the rate of increase. The function N satisfying this equation is

$$N = N_0 e^{\lambda t} \qquad \text{D-22}$$

A graph of this function is shown in Figure D-6. An exponential increase is characterized by a doubling time T_2, which is related to λ by

$$T_2 = \frac{\ln 2}{\lambda} = \frac{0.693}{\lambda} \qquad \text{D-23}$$

If the rate of increase λ is expressed as a percentage, $r = \lambda/100\%$, the doubling time is

$$T_2 = \frac{69.3}{r} \qquad \text{D-24}$$

For example, if the population increases by 2 percent per year, the population will double every $69.3/2 \approx 35$ years. Table D-2 lists some useful relations for exponential and logarithmic functions.

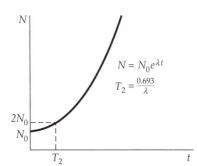

FIGURE D-6 Graph of N versus t when N increases exponentially. The time T_2 is the time it takes for N to double.

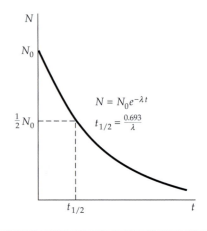

FIGURE D-5 Graph of N versus t when N decreases exponentially. The time $t_{1/2}$ is the time it takes for N to decrease by one-half.

TABLE D-2

Exponential and Logarithmic Functions

$e = 2.71828;$

$e^0 = 1$

If $y = e^x$, then $x = \ln y$.

$e^{\ln x} = x$

$e^x e^y = e^{(x+y)}$

$(e^x)^y = e^{xy} = (e^y)^x$

$\ln e = 1; \quad \ln 1 = 0$

$\ln xy = \ln x + \ln y$

$\ln \dfrac{x}{y} = \ln x - \ln y$

$\ln e^x = x; \quad \ln a^x = x \ln a$

$\ln x = (\ln 10) \log x$

$\quad = 2.3026 \log x$

$\log x = (\log e) \ln x = 0.43429 \ln x$

$e^x = 1 + x + \dfrac{x^2}{2!} + \dfrac{x^3}{3!} + \cdots$

$\ln(1 + x) = x \pm \dfrac{x^2}{2} + \dfrac{x^3}{3} \pm \dfrac{x^4}{4} + \cdots$

Geometry

The ratio of the circumference of a circle to its diameter is a natural number π, which has the approximate value

$$\pi = 3.141592$$

The circumference C of a circle is thus related to its diameter d and its radius r by

$$C = \pi d = 2\pi r \qquad \text{circumference of circle} \qquad \text{D-25}$$

The area of a circle is

$$A = \pi r^2 \qquad \text{area of circle} \qquad \text{D-26}$$

The area of a parallelogram is the base b times the height h (Figure D-7) and that of a triangle is one-half the base times the height (Figure D-8). A sphere of radius r (Figure D-9) has a surface area given by

$$A = 4\pi r^2 \qquad \text{spherical surface area} \qquad \text{D-27}$$

and a volume given by

$$V = \tfrac{4}{3}\pi r^3 \qquad \text{spherical volume} \qquad \text{D-28}$$

A cylinder of radius r and length L (Figure D-10) has surface area (not including the end faces) of

$$A = 2\pi r L \qquad \text{cylindrical surface} \qquad \text{D-29}$$

and volume of

$$V = \pi r^2 L \qquad \text{cylindrical volume} \qquad \text{D-30}$$

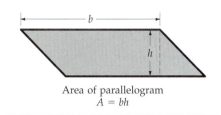

Area of parallelogram
$A = bh$

FIGURE D-7 Area of a parallelogram.

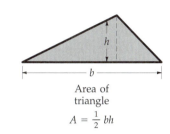

Area of triangle
$A = \frac{1}{2}\,bh$

FIGURE D-8 Area of a triangle.

Trigonometry

The angle between two intersecting straight lines is measured as follows. A circle is drawn with its center at the intersection of the lines, and the circular arc is divided into 360 parts called **degrees.** The number of degrees in the arc between the lines is the measure of the angle between the lines. For very small angles, the degree is divided into minutes (') and seconds (") with $1' = 1°/60$ and $1" = 1'/60 = 1°/3600$. For scientific work, a more useful measure of an angle is the radian (rad). In radian measure, the angle between two intersecting straight lines is found by again drawing a circle with its center at the intersection of the lines. The measure of the angle in radians is then defined as the length of the circular arc between the lines divided by the radius of the circle (Figure D-11). If s is the arc length and r is the radius of the circle, the angle θ measured in radians is

$$\theta = \frac{s}{r} \qquad \text{D-31}$$

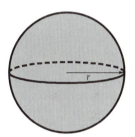

Spherical surface area
$A = 4\pi r^2$
Spherical volume
$V = \frac{4}{3}\pi r^3$

FIGURE D-9 Surface area and volume of a sphere.

Since the angle measured in radians is the ratio of two lengths, it is dimensionless. The relation between radians and degrees is

$$360° = 2\pi \text{ rad}$$

or

$$1 \text{ rad} = \frac{360°}{2\pi} = 57.3°$$

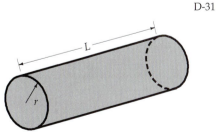

Cylindrical surface area
$A = 2\pi r L$
Cylindrical volume
$V = \pi r^2 L$

FIGURE D-10 Surface area (not including the end faces) and volume of a cylinder.

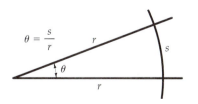

$\theta = \dfrac{s}{r}$

FIGURE D-11 The angle θ in radians is defined to be the ratio s/r, where s is the arc length intercepted on a circle of radius r.

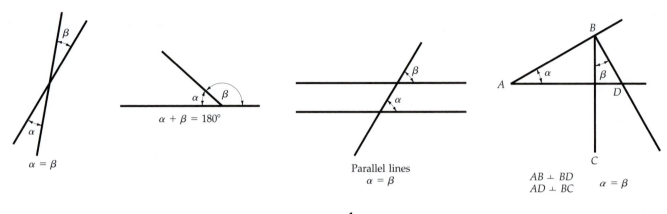

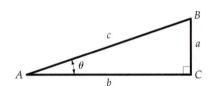

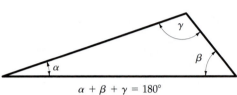

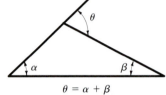

FIGURE D-12 Some useful relations for angles.

Figure D-12 shows some useful relations for angles.

Figure D-13 shows a right triangle formed by drawing the line BC perpendicular to AC. The lengths of the sides are labeled a, b, and c. The trigonometric functions $\sin\theta$, $\cos\theta$, and $\tan\theta$ for an acute angle θ are defined as

$$\sin\theta = \frac{a}{c} = \frac{\text{opposite side}}{\text{hypotenuse}} \qquad \text{D-33}$$

FIGURE D-13 A right triangle with sides of length a and b and a hypotenuse of length c.

$$\cos\theta = \frac{b}{c} = \frac{\text{adjacent side}}{\text{hypotenuse}} \qquad \text{D-34}$$

$$\tan\theta = \frac{a}{b} = \frac{\text{opposite side}}{\text{adjacent side}} = \frac{\sin\theta}{\cos\theta} \qquad \text{D-35}$$

Three other trigonometric functions, defined as the reciprocals of these functions, are

$$\sec\theta = \frac{c}{b} = \frac{1}{\cos\theta} \qquad \text{D-36}$$

$$\csc\theta = \frac{c}{a} = \frac{1}{\sin\theta} \qquad \text{D-37}$$

$$\cot\theta = \frac{b}{a} = \frac{1}{\tan\theta} = \frac{\cos\theta}{\sin\theta} \qquad \text{D-38}$$

The angle θ whose sine is x is called the arcsine of x, and is written $\sin^{-1} x$. That is, if

$$\sin\theta = x$$

then

$$\theta = \arcsin x = \sin^{-1} x \qquad \text{D-39}$$

The arcsine is the inverse of the sine. The inverse of the cosine and tangent are defined similarly. The angle whose cosine is y is the arccosine of y. That is, if

$$\cos \theta = y$$

then

$$\theta = \arccos y = \cos^{-1} y \qquad \text{D-40}$$

The angle whose tangent is z is the arctangent of z. That is, if

$$\tan \theta = z$$

$$\theta = \arctan z = \tan^{-1} z \qquad \text{D-41}$$

The Pythagorean theorem

$$a^2 + b^2 = c^2 \qquad \text{D-42}$$

gives some useful identities. If we divide each term in this equation by c^2, we obtain

$$\frac{a^2}{c^2} + \frac{b^2}{c^2} = 1$$

or, from the definitions of $\sin \theta$ and $\cos \theta$,

$$\sin^2 \theta + \cos^2 \theta = 1 \qquad \text{D-43}$$

Similarly, we can divide each term in (Equation D-42) by a^2 or b^2 and obtain

$$1 + \cot^2 \theta = \csc^2 \theta \qquad \text{D-44}$$

and

$$1 + \tan^2 \theta = \sec^2 \theta \qquad \text{D-45}$$

These and other useful trigonometric formulas are listed in Table D-3.

The following equation is actually two equations. For one of the equations the two upper signs are taken, and for the other equation the two lower signs are taken. The same sign rule applies to others of the following "equations":

$$\sin(A \pm B) = \sin A \cos B \pm \cos A \sin B$$

$$\cos(A \pm B) = \cos A \cos B \mp \sin A \sin B$$

$$\tan(A \pm B) = \frac{\tan A \pm \tan B}{1 \mp \tan A \tan B}$$

$$\sin A \pm \sin B = 2 \sin\left[\tfrac{1}{2}(A \pm B)\right] \cos\left[\tfrac{1}{2}(A \mp B)\right]$$

$$\cos A + \cos B = 2 \cos\left[\tfrac{1}{2}(A + B)\right] \cos\left[\tfrac{1}{2}(A \pm B)\right]$$

$$\cos A \pm \cos B = 2 \sin\left[\tfrac{1}{2}(A + B)\right] \sin\left[\tfrac{1}{2}(B \pm A)\right]$$

$$\tan A \pm \tan B = \frac{\sin(A \pm B)}{\cos A \cos B}$$

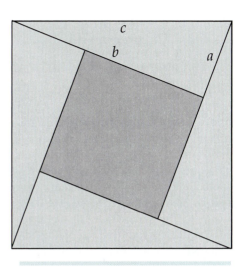

When this figure was first published the letters were absent and it was accompanied by the single word "Behold!" Using the drawing, establish the Pythagorean theorem ($a^2 + b^2 = c^2$).

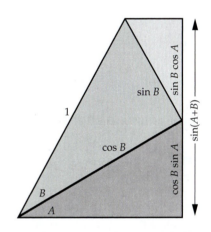

Using this drawing, establish the identity $\sin(A + B) = \sin A \cos B + \cos A \sin B$. You can also use it to establish the identity $\cos(A + B) = \cos A \cos B - \sin A \sin B$. Try it.

TABLE D-3

Trigonometric Formulas

$$\sin^2 \theta + \cos^2 \theta = 1; \quad \sec^2 \theta - \tan^2 \theta = 1; \quad \csc^2 \theta - \cot^2 \theta = 1$$

$$\sin 2\theta = 2 \sin \theta \cos \theta$$

$$\cos 2\theta = \cos^2 \theta - \sin^2 \theta = 2 \cos^2 \theta - 1 = 1 - 2 \sin^2 \theta$$

$$\tan 2\theta = \frac{2 \tan \theta}{1 - \tan^2 \theta}$$

$$\sin \tfrac{1}{2}\theta = \sqrt{\frac{1 - \cos \theta}{2}}; \quad \cos \tfrac{1}{2}\theta = \sqrt{\frac{1 + \cos \theta}{2}}; \quad \tan \tfrac{1}{2}\theta = \sqrt{\frac{1 - \cos \theta}{1 + \cos \theta}}$$

EXAMPLE D-5

Use the isosceles right triangle shown in Figure D-14 to find the sine, cosine, and tangent of 45°.

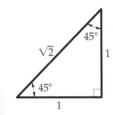

FIGURE D-14
An isosceles right triangle.

PICTURE THE PROBLEM It is clear from the figure that the two acute angles of this triangle are equal. Since the sum of the three angles in a triangle must equal 180°, and the right angle is 90°, each acute angle must be 45°. If we multiply each side of any triangle by a common factor, we obtain a similar triangle with the same angles as the first. We can therefore choose any convenient length for one side. We choose the equal sides to have a length of 1 unit.

1. Find the length of the hypotenuse from the Pythagorean theorem:

$$c = \sqrt{a^2 + b^2} = \sqrt{1^2 + 1^2} = \sqrt{2} \text{ units}$$

2. Calculate sin 45° from its definition:

$$\sin 45° = \frac{a}{c} = \frac{1}{\sqrt{2}} = 0.707$$

3. Calculate cos 45° from its definition:

$$\cos 45° = \frac{b}{c} = \frac{1}{\sqrt{2}} = 0.707$$

4. Calculate tan 45° from its definition:

$$\tan 45° = \frac{a}{b} = \frac{1}{1} = 1$$

EXAMPLE D-6

The sine of 30° is exactly 0.5. Find the ratios of the sides of a 30–60° right triangle.

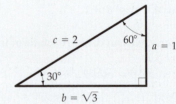

FIGURE D-15 A 30–60° right triangle.

PICTURE THE PROBLEM This common triangle is shown in Figure D-15. We choose a length of 1 unit for the side opposite the 30° angle.

1. Calculate the hypotenuse from the definition of the sine and the choice of 1 unit for the opposite side:

$$\sin 30° = \frac{a}{c} = \frac{1}{c} = 0.5$$

2. Use the Pythagorean theorem to find the length b of the side opposite the 60° angle:

$$c = \frac{1}{0.5} = 2$$

$$b = \sqrt{c^2 - a^2} = \sqrt{2^2 - 1^2} = \sqrt{3}$$

3. Use these results to calculate cos 30°, tan 30°, sin 60°, cos 60°, and tan 60°:

$$\cos 30° = \frac{b}{c} = \frac{\sqrt{3}}{2} = 0.866$$

$$\tan 30° = \frac{a}{b} = \frac{1}{\sqrt{3}} = 0.577$$

$$\sin 60° = \frac{b}{c} = \cos 30° = 0.866$$

$$\cos 60° = \frac{a}{c} = \sin 30° = 0.500$$

$$\tan 60° = \frac{b}{a} = \frac{\sqrt{3}}{1} = 1.732$$

For small angles, the length a is nearly equal to the arc length s, as can be seen in Figure D-16. The angle $\theta = s/c$ is therefore nearly equal to $\sin \theta = a/c$:

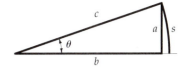

$$\sin \theta \approx \theta \qquad \text{for small values of } \theta \qquad\qquad\qquad \text{D-46}$$

Similarly, the lengths c and b are nearly equal, so $\tan \theta = a/b$ is nearly equal to both θ and $\sin \theta$ for small values of θ:

FIGURE D-16 For small angles, $\sin \theta = a/c$, $\tan \theta = a/b$, and the angle $\theta = s/c$ are all approximately equal.

$$\tan \theta \approx \sin \theta \approx \theta \qquad \text{for small values of } \theta \qquad\qquad \text{D-47}$$

Equations D-46 and D-47 hold only if θ is measured in radians. Since $\cos \theta = b/c$ and these lengths are nearly equal for small values of θ, we have

$$\cos \theta \approx 1 \qquad \text{for small values of } \theta \qquad\qquad\qquad \text{D-48}$$

EXAMPLE D-7

By how much do $\sin \theta$, $\tan \theta$, and θ differ when $\theta = 15°$?

1. Convert 15° to radians:

$$\theta = 15° \frac{2\pi \text{ rad}}{360°} = 0.262 \text{ rad}$$

2. Find $\sin 15°$ and $\tan 15°$ using a calculator:

$$\sin 15° = 0.259$$

$$\tan 15° = 0.268$$

3. Compute the percentage difference between θ and $\sin \theta$:

$$\frac{|\sin \theta - \theta|}{\theta} = \frac{|0.259 - 0.262|}{0.262} = \frac{0.003}{0.262} = 0.011 \approx 1\%$$

4. Compute the percentage difference between θ and $\tan \theta$:

$$\frac{|\tan \theta - \theta|}{\theta} = \frac{|0.268 - 0.262|}{0.262} = \frac{0.006}{0.262} = 0.023 \approx 2\%$$

REMARKS For smaller angles, the approximation $\theta < \sin \theta < \tan \theta$ is even more accurate.

Example D-7 shows that if accuracy of a few percent is needed, small angle approximations can be used only for angles of about 15° or less. Figure D-17 shows graphs of θ, $\sin \theta$, and $\tan \theta$ versus θ, for small values of θ.

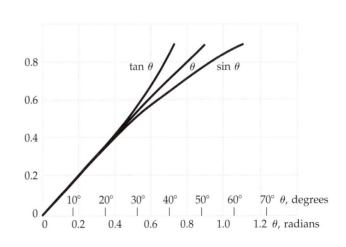

FIGURE D-17 Graphs of $\tan \theta$, θ, and $\sin \theta$ versus θ for small values of θ.

Figure D-18 shows an obtuse angle with its vertex at the origin and one side along the x axis. The trigonometric functions for a general angle such as this are defined by

$$\sin \theta = \frac{y}{c} \qquad \text{D-49}$$

$$\cos \theta = \frac{x}{c} \qquad \text{D-50}$$

$$\tan \theta = \frac{y}{x} \qquad \text{D-51}$$

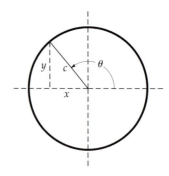

FIGURE D-18 Diagram for defining the trigonometric functions for an obtuse angle.

Figure D-19 shows plots of these functions versus θ. The sine function has a period of 2π rad. Thus, for any value of θ $\sin(\theta + 2\pi) = \sin \theta$ and so forth. That is, when an angle changes by 2π rad, the function returns to its original value. The tangent function has a period of π rad. Thus, $\sin(\theta + 2\pi) = \sin \theta$ and so forth. Some other useful relations are

$$\sin(\pi - \theta) = \sin \theta \qquad \text{D-52}$$

$$\cos(\pi - \theta) = -\cos \theta \qquad \text{D-53}$$

$$\sin(\pi/2 - \theta) = \cos \theta \qquad \text{D-54}$$

$$\cos(\pi/2 - \theta) = \sin \theta \qquad \text{D-55}$$

The trigonometric functions can be expressed as power series in θ. The series for $\sin \theta$ and $\cos \theta$ are

$$\sin \theta = \theta - \frac{\theta^3}{3!} + \frac{\theta^5}{5!} - \frac{\theta^7}{7!} + \dots \qquad \text{D-56}$$

$$\cos \theta = 1 - \frac{\theta^2}{2!} + \frac{\theta^4}{4!} - \frac{\theta^6}{6!} + \dots \qquad \text{D-57}$$

When θ is small, good approximations are obtained using only the first few terms in the series.

(a)
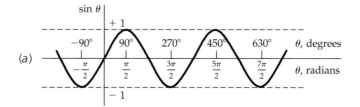

(b)
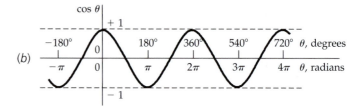

(c)
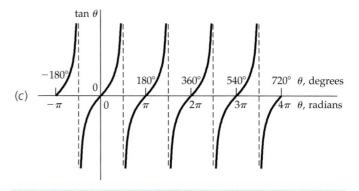

FIGURE D-19 The trigonometric functions $\sin \theta$, $\cos \theta$, and $\tan \theta$ versus θ.

The Binomial Expansion

The binomial theorem is very useful for making approximations. One form of this theorem is

$$(1 + x)^n = 1 + nx + \frac{n(n-1)}{2!}x^2 + \frac{n(n-1)(n-2)}{3!}x^3 + \dots \qquad \text{D-58}$$

If n is a positive integer, there are $n + 1$ terms in this series. If n is a real number other than a positive integer, there are an infinite number of terms. The series is valid for any value of n if x^2 is less than 1. It is also valid for $x^2 = 1$ if n is positive. The series is particularly useful if x^2 is small compared to 1. Then each term is much smaller than the previous term and we can drop all but the first two or three terms in the equation. If x^2 is much less compared to 1, we have

$$(1 + x)^n \approx 1 + nx, \qquad |x| \ll 1 \qquad \text{D-59}$$

EXAMPLE D-8

Use (Equation D-59) to find an approximate value for the square root of 101.

1. Write $(101)^{1/2}$ so it is in the form $(1 + x)^n$ with x much less than 1:

   $(101)^{1/2} = (100 + 1)^{1/2} = (100)^{1/2}(1 + 0.01)^{1/2} = 10(1 + 0.01)^{1/2}$

2. Use Equation D-59 with $n = \frac{1}{2}$ and $x = 0.1$ to expand $(1 + 0.01)^{1/2}$.

   $(1 + 0.01)^{1/2} \approx 1 + \frac{1}{2}(0.01) = 1.005$

3. Substitute this result into the equation in step 1:

   $(101)^{1/2} = 10(1 + 0.01)^{1/2} \approx \boxed{10.05}$

REMARKS We can assess the accuracy of this result by computing the first term in Equation D-58 that was neglected. This term is

$$\frac{n(n-1)}{2}x^2 = \frac{\frac{1}{2}(-\frac{1}{2})}{2}(0.01)^2 = -\frac{0.0001}{8} \approx -0.00001 = -0.001$$

We therefore expect our answer to be correct to within about 0.001%. The value of $(101)^{1/2}$ to eight significant figures is 10.049875, which differs from 10.05 by 0.000124 or about 0.001% of 10.05.

Complex Numbers

A general complex number z can be written

$$z = a + bi \qquad\qquad \text{D-60}$$

where a and b are real numbers and $i = \sqrt{-1}$. The quantity a is called the real part and the quantity ib is called the imaginary part of z. We can represent a complex number in a plane as shown in Figure D-20, where the x axis is the real axis and the y axis is the imaginary axis. We can use the relations $a = r\cos\theta$ and $b = r\sin\theta$ from Figure D-20 to write the complex number z in polar coordinates:

$$z = r\cos\theta + ir\sin\theta \qquad\qquad \text{D-61}$$

where $r = \sqrt{a^2 + b^2}$ is called the magnitude of z.

When complex numbers are added or subtracted, the real and imaginary parts are added or subtracted separately:

$$z_1 + z_2 = (a_1 + ib_1) + (a_2 + ib_2) = (a_1 + a_2) + i(b_1 + b_2) \qquad \text{D-62}$$

However, when two complex numbers are multiplied, each part of one number is multiplied by each part of the other number:

$$z_1 z_2 = (a_1 + ib_1)(a_2 + ib_2) = a_1 a_2 + i^2 b_1 b_2 + i(a_1 b_2 + a_2 b_1) \qquad \text{D-63}$$

$$= a_1 a_2 - b_1 b_2 + i(a_1 b_2 + a_2 b_1)$$

where we have used $i^2 = -1$.

The complex conjugate z^* of the complex number z is that number obtained by replacing i with $-i$. If $z = a + ib$ then

$$z^* = (a + ib)^* = a - ib \qquad\qquad \text{D-64}$$

The product of a complex number and its complex conjugate equals the square of the magnitude of the number:

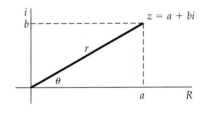

$z = a + bi$
$= r\cos\theta + (r\sin\theta)i$
$= r(\cos\theta + i\sin\theta)$

FIGURE D-20 Representation of a complex number in a plane. The real part of the complex number is plotted along the horizontal axis, and the imaginary part is plotted along the vertical axis.

$$zz^* = (a + ib)(a - ib) = a^2 + b^2$$

D-65

A particularly useful function of a complex number is the exponential $e^{i\theta}$. Using the expansion for e^x given in Table D-2, we have

$$e^{i\theta} = 1 + i\theta + \frac{(i\theta)^2}{2!} + \frac{(i\theta)^3}{3!} + \frac{(i\theta)^4}{4!} + \cdots$$

Using $i^2 = -1$, $i^3 = -i$, $i^4 = +1$, and so forth and separating the real parts from the imaginary parts, this expansion can be written

$$e^{i\theta} = \left(1 - \frac{\theta^2}{2!} + \frac{\theta^4}{4!} - \cdots\right) + i\left(\theta - \frac{\theta^3}{3!} + \cdots\right)$$

Comparing this result with Equations D-56 and D-57, we can see that

$$e^{i\theta} = \cos\theta + i\sin\theta$$

D-66

Using this result, we can express a general complex number as an exponential:

$$z = a + ib = r\cos\theta + ir\sin\theta = re^{i\theta}$$

D-67

where $r = \sqrt{a^2 + b^2}$.

Solving differential equations using complex numbers

Consider an equation of the form

$$a\frac{d^2x}{dt^2} + b\frac{dx}{dt} + cx = A\cos\omega t$$

D-68

that represents a physical process, such as a damped harmonic oscillator driven by a sinusoidal force, or a series RLC combination being driven by a sunusoidal potential drop. Each of the parameters in (Equation D-68) is a real number. We wish to obtain the steady-state solution to this equation using complex numbers. To do so, we first construct the equation:

$$a\frac{d^2y}{dt^2} + b\frac{dy}{dt} + cy = A\sin\omega t$$

D-69

Equation D-69 has no physical meaning and we have no interest in solving it. However, it is of use in solving Equation D-68. After multiplying both sides of (Equation D-69) by i we add it and (Equation D-68) to obtain

$$\left(a\frac{d^2x}{dt^2} + ai\frac{d^2y}{dt^2}\right) + \left(b\frac{dx}{dt} + bi\frac{dy}{dt}\right) + (cx + ciy) = A\cos\omega t + Ai\sin\omega t$$

We next combine terms to get

$$a\frac{d^2(x + iy)}{dt^2} + b\frac{d(x + iy)}{dt} + c(x + iy) = A(\cos\omega t + i\sin\omega t)$$

D-70

whose validity depends on the derivative of a sum being equal to the sum of the derivatives. We simplify our result by defining $z = x + iy$ and by using the identity $e^{i\omega t} = \cos\omega t + i\sin\omega t$. Substituting these into (Equation D-70) we obtain

$$a\frac{d^2z}{dt^2} + b\frac{dz}{dt} + cz = Ae^{i\omega t} \qquad\qquad \text{D-71}$$

which we now solve for z. Once z is obtained we can solve for x using $x = \text{Re}(z)$.

Since we are looking only for the steady state solution for (Equation D-68) we can assume its solution is of the form $x = x_0 \cos(\omega t - \phi)$, where ϕ is a constant. This is equivalent to assuming that the solution to (Equation D-71) is of the form $z = z_0 e^{i\omega t}$, where z_0 is a complex number. Then $dz/dt = i\omega z$ $d^2z/dt^2 = -\omega^2 z$, and $e^{i\omega t} = z/z_0$.

Substituting these into (Equation D-70) gives $-a\omega^2 z + i\omega b z + cz = A\dfrac{z}{z_0}$

Dividing both sides of this equation by z and solving for z_0 gives

$$z_0 = \frac{A}{-a\omega^2 + i\omega b + c}$$

Expressing the denominator in polar form gives $(-a\omega^2 + c) + i\omega b = \sqrt{(-a\omega^2 + c)^2 + \omega^2 b^2}\, e^{i\phi}$, where $\tan\phi = \omega^2 b^2/(-a\omega^2 + c)$. Thus,

$$z_0 = \frac{A}{\sqrt{(-a\omega^2 + c)^2 + \omega^2 b^2}}\, e^{-i\phi}$$

so

$$z = z_0 e^{i\omega t} = \frac{A}{\sqrt{(-a\omega^2 + c)^2 + \omega^2 b^2}}\, e^{i(\omega t - \phi)}$$

$$= \frac{A}{\sqrt{(-a\omega^2 + c)^2 + \omega^2 b^2}}(\cos(\omega t - \delta) + i\sin(\omega t - \delta)) \qquad \text{D-72}$$

It follows that

$$x = \text{Re}(z) = \frac{A}{\sqrt{(-a\omega^2 + c)^2 + \omega^2 b^2}}\cos(\omega t - \phi) \qquad \text{D-73}$$

Differential Calculus

When we say that x is a function of t, we mean that for each value of t there is a single corresponding value of x. An example is $x = At^2$, where A is a constant. To indicate that x is a function of t, we sometimes write $x(t)$ for x. Figure D-21 is a graph of x versus t for a typical function $x(t)$. At a particular value $t = t_1$, x has the value of x_1 as indicated. At another value t_2, x has the value x_2. The change in t, $t_2 - t_1$, is written $\Delta t = t_2 - t_1$ and the corresponding change in x is written $\Delta x = x_2 - x_1$. The ratio $\Delta x/\Delta t$ is the slope of the straight line connecting (x_1, t_1) and (x_2, t_2). If we make Δt smaller and smaller, the line connecting (x_1, t_1) and (x_2, t_2) approaches the line that is tangent to the curve at the point (x_1, t_1). The slope of this tangent line is called the derivative of x with respect to t and is written dx/dt:

$$\frac{dx}{dt} = \lim_{\Delta t \to 0} \frac{\Delta x}{\Delta t} \qquad\qquad \text{D-74}$$

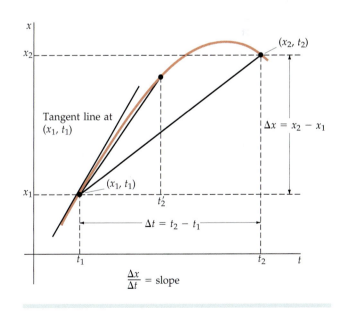

FIGURE D-21 Graph of a typical function $x(t)$. The points (x_1, t_1) and (x_2, t_2) are connected by a straight line. The slope of this line is $\Delta x/\Delta t$. As the time interval beginning at t_1 is decreased, the slope for that interval approaches the slope of the line tangent to the curve at time t_1, which is the derivative of x with respect to t.

The derivative of a function of t is generally another function of t. If x is a constant, the graph of x versus t is a horizontal line with zero slope. The derivative of a constant is thus zero. In Figure D-22, x is proportional to t:

$$x = Ct$$

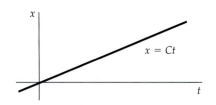

This function has a constant slope equal to C. Thus the derivative of Ct is C. Table D-4 lists some properties of derivatives and the derivatives of some particular functions that occur often in physics. It is followed by comments aimed at making these properties and rules clearer. More detailed discussion can be found in most calculus books.

FIGURE D-22 Graph of the linear function $x = Ct$. This function has a constant slope C.

TABLE D-4

Properties of Derivatives and Derivatives of Particular Functions

Linearity

1. The derivative of a constant times a function equals the constant times the derivative of the function:

$$\frac{d}{dt}[Cf(t)] = C\frac{df(t)}{dt}$$

2. The derivative of a sum of functions equals the sum of the derivatives of the functions:

$$\frac{d}{dt}[f(t) + g(t)] = \frac{df(t)}{dt} + \frac{dg(t)}{dt}$$

Chain rule

3. If f is a function of x and x is in turn a function of t, the derivative of f with respect to t equals the product of the derivative of f with respect to x and the derivative of x with respect to t:

$$\frac{d}{dt}f(t) = \frac{df}{dx}\frac{dx}{dt}$$

Derivative of a product

4. The derivative of a product of functions $f(t)g(t)$ equals the first function times the derivative of the second plus the second function times the derivative of the first:

$$\frac{d}{dt}[f(t)g(t)] = f(t)\frac{dg(t)}{dt} + \frac{df(t)}{dt}g(t)$$

Reciprocal derivative

5. The derivative of t with respect to x is the reciprocal of the derivative of x with respect to t, assuming that neither derivative is zero:

$$\frac{dx}{dt} = \left(\frac{dt}{dx}\right)^{-1} \quad \text{if} \quad \frac{dt}{dx} \neq 0$$

Derivatives of particular functions

6. $\dfrac{dC}{dt} = 0$ where C is a constant

7. $\dfrac{d(t^n)}{dt} = nt^{n-1}$

8. $\dfrac{d}{dt}\sin \omega t = \omega \cos \omega t$

9. $\dfrac{d}{dt}\cos \omega t = -\omega \sin \omega t$

10. $\dfrac{d}{dt}\tan \omega t = \omega \sec^2 \omega t$

11. $\dfrac{d}{dt}e^{bt} = be^{bt}$

12. $\dfrac{d}{dt}\ln bt = \dfrac{1}{t}$

EXAMPLE D-9

Find the derivative of $x = at^2 + bt + c$, where a, b, and c are constants.

PICTURE THE PROBLEM From rule 2, we can differentiate each term separately and add the results.

1. Use rules 1 and 7 to find the derivative of the first term:
$$\frac{d(at^2)}{dt} = 2at^1 = 2at$$

2. Compute the derivatives of the second and third terms:
$$\frac{d(bt)}{dt} = b, \quad \frac{d(c)}{dt} = 0$$

3. Add these results:
$$\frac{dx}{dt} = 2at + b$$

Comments on Rules 1 Through 5

Rules 1 and 2 follow from the fact that the limiting process is linear. We can understand rule 3, the chain rule, by multiplying $\Delta f/\Delta t$ by $\Delta x/\Delta x$ and noting that, since x is a function of t, both Δx and Δf approach zero as Δt approaches zero. Since the limit of a product of two functions equals the product of their limits, we have

$$\lim_{\Delta t \to 0} \frac{\Delta f}{\Delta t} = \lim_{\Delta t \to 0} \frac{\Delta f}{\Delta x} \frac{\Delta x}{\Delta t} = \left(\lim_{\Delta t \to 0} \frac{\Delta f}{\Delta x} \right)\left(\lim_{\Delta t \to 0} \frac{\Delta x}{\Delta t} \right) = \frac{df}{dx} \frac{dx}{dt}$$

Rule 4 is not immediately apparent. The derivative of a product of functions is the limit of the ratio

$$\frac{f(t + \Delta t)g(t + \Delta t) - f(t)g(t)}{\Delta t}$$

If we add and subtract the quantity $f(t + \Delta t)g(t)$ in the numerator, we can write this ratio as

$$\frac{f(t + \Delta t)g(t + \Delta t) - f(t + \Delta t)g(t) + f(t + \Delta t)g(t) - f(t)g(t)}{\Delta t}$$

$$= f(t + \Delta t)\left[\frac{g(t + \Delta t) - g(t)}{\Delta t} \right] + g(t)\left[\frac{f(t + \Delta t) - f(t)}{\Delta t} \right]$$

As Δt approaches zero, the terms in square brackets become $dg(t)/dt$ and $df(t)/dt$, respectively, and the limit of the expression is

$$f(t)\frac{dg(t)}{dt} + g(t)\frac{df(t)}{dt}$$

Rule 5 follows directly from the definition:

$$\frac{dx}{dt} = \lim_{\Delta t \to 0} \frac{\Delta x}{\Delta t} = \lim_{\Delta x \to 0} \left(\frac{\Delta t}{\Delta x} \right)^{-1} = \left(\frac{dt}{dx} \right)^{-1}$$

Comments on Rule 7

We can obtain this important result using the binomial expansion. We have

$$f(t) = t^n$$

$$f(t + \Delta t) = (t + \Delta t)^n = t^n \left(1 + \frac{\Delta t}{t}\right)^n$$

$$= t^n \left[1 + n\frac{\Delta t}{t} + \frac{n(n-1)}{2!}\left(\frac{\Delta t}{t}\right)^2 + \frac{n(n-1)(n-2)}{3!}\left(\frac{\Delta t}{t}\right)^3 + \cdots \right]$$

Then

$$f(t - \Delta t) - f(t) = t^n \left[n\frac{\Delta t}{t} + \frac{n(n-1)}{2!}\left(\frac{\Delta t}{t}\right)^2 + \cdots \right]$$

and

$$\frac{f(t - \Delta t) - f(t)}{\Delta t} = nt^{n-1} + \frac{n(n-1)}{2!}t^{n-2}\,\Delta t + \cdots$$

The next term omitted from the last sum is proportional to $(\Delta t)^2$, the following to $(\Delta t)^3$, and so on. Each term except the first approaches zero as Δt approaches zero. Thus

$$\frac{df}{dt} = \lim_{\Delta x \to 0} \frac{f(t + \Delta t) - f(t)}{\Delta t} = nt^{n\pm 1}$$

Comments on Rules 8 to 10

We first write $\sin \omega t = \sin \theta$ with $\theta = \omega t$ and use the chain rule,

$$\frac{d \sin \theta}{dt} = \frac{d \sin \theta}{d\theta}\frac{d\theta}{dt} = \omega \frac{d \sin \theta}{d\theta}$$

We then use the trigonometric formula for the sine of the sum of two angles θ and $\Delta \theta$:

$$\sin(\theta + \Delta \theta) = \sin \Delta \theta \cos \theta + \cos \Delta \theta \sin \theta$$

Since $\Delta \theta$ is to approach zero, we can use the small-angle approximations

$$\sin \Delta \theta \approx \Delta \theta \quad \text{and} \quad \cos \Delta \theta \approx 1$$

Then

$$\sin(\theta + \Delta \theta) \approx \Delta \theta \cos \theta + \sin \theta$$

and

$$\frac{\sin(\theta + \Delta \theta) - \sin \theta}{\Delta \theta} \approx \cos \theta$$

Similar reasoning can be applied to the cosine function to obtain rule 9.

Rule 10 is obtained by writing $\tan \theta = \sin \theta / \cos \theta$ and applying rule 4 along with rules 8 and 9.

$$\frac{d}{dt}(\tan\theta) = \frac{d}{dt}(\sin\theta)(\cos\theta)^{-1} = \sin\theta\frac{d}{dt}(\cos\theta)^{-1} + \frac{d(\sin\theta)}{dt}(\cos\theta)^{-1}$$

$$= \sin\theta(-1)(\cos\theta)^{-2}(-\sin\theta) + (\cos\theta)(\cos\theta)^{-1}$$

$$= \frac{\sin^2\theta}{\cos^2\theta} + 1 = \tan^2\theta + 1 = \sec^2\theta$$

Comments on Rule 11

Again we use the chain rule

$$\frac{de^\theta}{dt} = \frac{bde^\theta}{bdt} = b\frac{de^\theta}{d(bt)} = b\frac{de^\theta}{d\theta} \qquad \text{with} \qquad \theta = bt$$

and the series expansion for the exponential function:

$$e^{\theta+\Delta\theta} = e^\theta e^{\Delta\theta} = e^\theta\left[1 + \Delta\theta + \frac{(\Delta\theta)^2}{2!} + \frac{(\Delta\theta)^3}{3!} + \cdots\right]$$

Then

$$\frac{e^{\theta+\Delta\theta} - e^\theta}{\Delta\theta} = e^\theta + e^\theta\frac{\Delta\theta}{2!} + e^\theta\frac{(\Delta\theta)^2}{3!} + \cdots$$

As $\Delta\theta$ approaches zero, the right side of the equation above approaches e^θ.

Comments on Rule 12

Let

$$y = \ln bt$$

Then

$$e^y = bt \qquad \text{and} \qquad \frac{dt}{dy} = \frac{1}{b}e^y = t$$

Then using rule 5, we obtain

$$\frac{dy}{dt} = \left(\frac{dt}{dy}\right)^{-1} = \frac{1}{t}$$

Integral Calculus

Integration is related to the problem of finding the area under a curve. It is also the inverse of differentiation. Figure D-23 shows a function $f(t)$. The area of the shaded element is approximately $f_i\Delta t_i$, where f_i is evaluated anywhere in the interval Δt_i. This approximation improves if Δt_i is very small. The total area from t_1 to t_2 is found by summing all the area elements from t_1 to t_2 and taking the limit as each Δt_i approaches zero. This limit is called the integral of f over t and is written

$$\int_{t_1}^{t_2} f\,dt = \text{Area} = \lim_{\Delta t_i \to 0}\sum_i f_i\,\Delta t_i$$

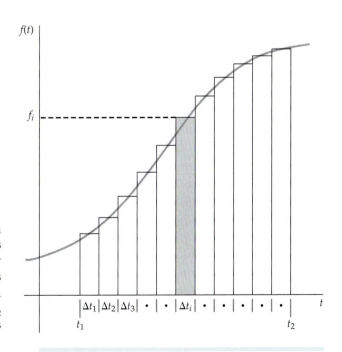

FIGURE D-23 A general function $f(t)$. The area of the shaded element is approximately $f_i\Delta t_i$, where f_i is evaluated anywhere in the interval.

If we integrate some function $f(t)$ from t_1 to some general value of t, we obtain another function of t. Let us call this function y:

$$y = \int_{t_1}^{t} f\, dt$$

The function y is the area under the f-versus-t curve from t_1 to a general value t. For a small interval Δt, the change in the area Δy is approximately $f\Delta t$.

$$\Delta y \approx f\, \Delta t$$

$$f \approx \frac{\Delta y}{\Delta t}$$

If we take the limit as Δt approaches 0, we can see that f is the derivative of y:

$$f = \frac{dy}{dt}$$

The relation between y and f is often written

$$y = \int f\, dt$$

where $\int f\, dt$ is called an **indefinite integral.** To evaluate an indefinite integral, we find the function y whose derivative is f. The definite integral of f from t_1 to t_2 is $y(t_1) - y(t_2)$, where $df/dt = y$:

$$\int_{t_1}^{t_2} f\, dt = y(t_2) - y(t_1)$$

EXAMPLE D-10

Find the indefinite integral of $f(t) = t$.

The function whose derivative is t is $\frac{1}{2}t^2$ plus any constant:

$$\int t\, dt = \tfrac{1}{2}t^2 + C$$

where C is any constant

Table D-5 lists some important integration formulas. More extensive lists of differentiation and integration formulas can be found on the Internet. They can be found by using a search engine and searching for *table of integrals*, and also in handbooks such as Herbert Dwight's *Tables of Integrals and Other Mathematical Data*, fourth edition, Macmillan Publishing Company, Inc., New York, 1961.

TABLE D-5

Integration Formulas[†]

1. $\displaystyle\int A\, dt = At$

2. $\displaystyle\int At\, dt = \tfrac{1}{2}At^2$

3. $\displaystyle\int At^n\, dt = A\frac{t^{n+1}}{n+1} \quad n \neq -1$

4. $\displaystyle\int At^{-1}\, dt = A\ln t$

5. $\displaystyle\int e^{bt}\, dt = \frac{1}{b}e^{bt}$

6. $\displaystyle\int \cos \omega t\, dt = \frac{1}{\omega}\sin \omega t$

7. $\displaystyle\int \sin \omega t\, dt = -\frac{1}{\omega}\cos \omega t$

8. $\displaystyle\int_0^\infty e^{-ax}\, dx = \frac{1}{a}$

9. $\displaystyle\int_0^\infty e^{-ax^2}\, dx = \frac{1}{2}\sqrt{\frac{\pi}{a}}$

10. $\displaystyle\int_0^\infty xe^{-ax^2}\, dx = \frac{2}{a}$

11. $\displaystyle\int_0^\infty x^2e^{-ax^2}\, dx = \frac{1}{4}\sqrt{\frac{\pi}{a^3}}$

12. $\displaystyle\int_0^\infty x^3e^{-ax^2}\, dx = \frac{4}{a^2}$

13. $\displaystyle\int_0^\infty x^4e^{-ax^2}\, dx = \frac{3}{8}\sqrt{\frac{\pi}{a^5}}$

[†] In these formulas, A, b, and ω are constants. In formulas 1 through 7 an arbitrary constant C can be added to the right side of each equation. The constant a is greater than zero.

ILLUSTRATION CREDITS

Chapter 1

Opener p. 1 Jeff Divine/FPG/Getty; **p. 4 (a)** The Granger Collection; **(b)** © 1999 Geoffrey Wheeler; **p. 6 (a)** McDonald Observatory; **(b)** Bruce Coleman; **p. 7** Eunice Harris/Photo Researchers, Inc.; **p. 11 (a)** IBM Almaden Research Center; **(b)** From *The Body Victorius,* The Delacourt Press/Lennart Nilsson; **(c)** Kent and Donnan Dannen/Photo Researchers; **(d)** NASA; **(e)** Smithsonian Institution.

Chapter 2

Opener p. 17 Corbis/Stock Market; **p. 23** Novastock/Dembinsky Photo Associates; **p. 24** Estate of Harold E. Edgerton/Palm Press Inc.; **p. 25** Gunter Ziesler/Peter Arnold Inc.; **p. 27** © Sydney Harris; **p. 28 Figure 2-9** James Sugar/Black Star; **p. 31** ©1994 General Motors Corporation. All rights reserved GM Media Archives; **p. 32 (left)** Stanford Linear Accelerator, U.S. Department of Energy; **(right)** Stanford Linear Accelerator Center, U.S. Department of Energy; **p. 40** Courtesy of Gene Mosca; **p. 46** Courtesy of Chuck Adler.

Chapter 3

Opener p. 53 Kevin Miller/Getty.

Chapter 4

Opener p. 85 John Neubauer/FPG/Getty; **p. 86** Jose Dupont/Explorer/Photo Researchers; **p. 88** NASA/Science Source/Photo Researchers; **p. 92 (a)** Cotton Coulson/Woodfin Camp and Assoc.; **(b)** Gary Ladd; **(c)** Los Alamos National Lab; **(d)** Science Photo Library/Photo Researchers; **p. 94 (b)** Fundamental Photographs; **Figure 4-7** David J. Phillip/AP Wide World.

Chapter 5

Opener p. 117 Courtesy of BMW; **p. 119 (top)** F.P. Bowden and D. Tabor, *Friction and Lubrication of Solids,* Oxford University Press, 2000; **(bottom)** Uzi Landman and David W. Leudtke/Georgia Institute of Technology; **p. 121** Nicole Villamora; **p. 122** Jean-Claude LeJeune/Stock, Boston; **p. 132** Courtesy of BMW; **p. 133** Sandia National Laboratory; **p. 134** NASA; **p. 135 (top)** Joe McBride/Stone; **(bottom)** Stuart Williams/Dembinsky Photo Associates; **p. 147 Figure 5-57** David de Lossy/The Image Bank.

Chapter 6

Opener p. 151 Getty; **p. 156** Courtesy of Dr. Roger Craig; **p. 168** Bill Bacjman/Photo Researchers; **p. 171 Figure 6-25** David J. Phillip/AP Wide World.

Chapter 7

Opener p. 183 Mark E. Gibson/Dembinsky Photo Associates; **p. 185** Loren Winters/Visuals Unlimited; **p. 194 (top left)** Visual Horizons/FPG International; **(top right)** New York State Commerce Department; **p. 195** Courtesy of Blyth Offshore Wind Limited; **p. 197** The Photo Works/Photo Researchers, Inc.; **p. 198** Stan Sholik/FPG International; **p. 205** Leicester University/Science Photo Library/Photo Researchers; **p. 215** Courtesy of PASCO.

Chapter 8

Opener p. 217 Jerry Wachter/Photo Researchers, Inc.; p. 223 **Figure 8-14** Estate of Harold E. Edgerton/Palm Press Inc.; **p. 228** Courtesy of Daedalon Corporation; **p. 229** NASA; **p. 234 (bottom)** Robert R. Edwards/BOB-E Photography; **p. 235 Figure 8-26** Romilly Lockyer/The Image Bank; **p. 237** Courtesy of Mercedes-Benz of N.A., Montvale, NJ; **p. 240** Estate of Harold E. Edgerton/Palm Press Inc.; **p. 244 (top)** Joe Strunk/Visuals Unlimited; **(bottom)** M. Hans/Vandystadtl/Photo Researchers, Inc.; **p. 246 (bottom right)** Brookhaven National Laboratory; **p. 251 Figure 8-47** NASA/Superstock.

Chapter 9

Opener p. 267 Photodisk; **p. 268** Fred Habegger/Grant Heilman Photography, Inc.; **p. 269** David Malin/Anglo-Australian Telescope Board; **p. 271** David Malin/Anglo-Australian Telescope Board; **p. 278** D.S. Kerr/Visuals Unlimited, Inc.; **p. 280** Richard Menga/Fundamental Photographs, Inc.; **p. 284** Fundamental Photographs, Inc.; **p. 288** Loren Winters/Visuals Unlimited; **p. 290** Scott Goldsmith/Stone/Getty; **p. 298 Figure 9-41** ©Treë.

Chapter 10

Opener p. 309 © Michael Newman/PhotoEdit; **p. 314** Dick Luria/Science Source/Photo Researchers; **p. 317** Courtesy of Segway; **p. 318 (left)** © The Harold E. Edgerton 1992 Trust; **(right)** Mike Powell/Getty; **p. 319** Dick Luria/FPG International; **p. 321** NASA/Goddard Space Flight Center; **p. 330** Chris Sorenson/The Stock Market; **p. 331** Chris Trotman/DUOMO/Corbis; **p. 337 Figure 10-56** Courtesy of Tangent Toy Co.

Relativity

Opener p. R-1 Courtesy of NASA.

Chapter 11

Opener p. 339 Stocktrek/Corbis; **p. 340** Collection of Historical Scientific Instruments, Harvard University; **p. 344 (top)** NASA; **(bottom)** NASA; **p. 345** Courtesy Central Scientific Company; **p. 353** NASA.

Chapter 12

Opener p. 370 Courtesy of Department of Physics, Purdue University; **p. 372** © 2002 Estate of Alexander Calder/Artists Rights Society (ARS), New York; **p. 379** Photodisk.

Chapter 13

Opener p. 395 Andy Pernick/Bureau of Reclamation; **p. 401** Vanessa Vick/Photo Researchers, Inc.; **p. 403** Chuck O'Rear/Woodfin Camp and Assoc.; **p. 405** David Burnett/Woodfin Camp and Assoc.; **p. 407 (top)** Estate of Harold E. Edgerton; **(bottom)** Takeski Takahara/Photo Researchers, Inc.; **p. 408** P. Motta/Photo Researchers, Inc.; **p. 412** Michael Dunn/The Stock Market; **p. 415** Picker International.

Chapter 14

Opener p. 425 Barry Slaven/Visuals Unlimited; **p. 427** Citibank; **p. 429** NASA; **p. 433** Institute for Marine Dynamics; **p. 442** Richard Menga/Fundamental Photographers; **p. 447 (top)** Monroe Auto Equipment; **(bottom)** David Wrobel/Visuals Unlimited; **p. 449** Eye Wire/Getty; **p. 451** Royal Swedish Academy of Music.

Chapter 15

Opener p. 465 John Cetrino/Check Six/Picture Quest; **p. 466 Figure 15-1** Richard Menga/Fundamental Photographs; **Figure 15-2** Richard Menga/Fundamental Photographs; **p. 478 (top right)** David Sacks/The Image Bank/Getty; **(left)** Maynard and Boucher/Visuals Unlimited; **p. 480** From Winston E. Cock, *Lasers and Holography,* Dover Publications, New York, 1981; **p. 484** Courtesy of Davies Symphony Hall; **p. 485 (top and bottom)** Fundamental Photographs; **p. 486 (top)** Bernard Benoit/Photo Researchers, Inc.; **(a)** Education Development Center; **p. 491 (top, a)** Sandia National Laboratory; **(top, b)** Robert de Gast/Photo Researchers; **(c)** Estate of Harold E. Edgerton/Palm Press Inc.; **(bottom, b)** Education Development Center; **p. 496 Figure 15-30** Estate of Harold E. Edgerton/Palm Press Inc.

Chapter 16

Opener p. 503 David Yost/Steinway & Sons; **p. 507** Rubberball Productions; **p. 510 (a)** Berenice Abbott (8J 1328)/Photo Researchers; **p. 513 (left)** University of Washington; **(center)** University of Washington; **(right)** University of Washington; **p. 519** Professor Thomas D. Rossing, Northern Illinois University, DeKalb; **p. 525** Courtesy of Chuck Adler.

Chapter 17

Opener p. 532 Hoby Finn/PhotoDisk/Getty; **p. 535 (a)** Courtesy of Taylor Precision Products; **(b)** Courtesy Honeywell, Inc.; **p. 536** Richard Menga/Fundamental Photographs; **p. 538** NASA; **p. 554** Jet Propulsion Laboratory/NASA.

Chapter 18

Opener p. 558 Donna Day/PhotoDisk/Getty; **p. 559** Phoenix Pipe & Tube/Lana Berkovich; **p. 561** From Frank Press and Raymond Sievert, *Understanding Earth,* 3rd ed., W.H. Freeman and Co., 2001; **p. 562** From Donald Wink, Sharon Gislason, and Sheila McNicholas, *The Practice of Chemistry,* W.H. Freeman and Co., 2002; **p. 582** Will and Deni McIntyre/Photo Researchers.

Chapter 19

Opener p. 595 (top) Paul Chesley/National Geographic/Getty; **p. 595 (bottom)** Sandia National Laboratory; **p. 598** © 2002 Robert Briggs; **p. 601** Anderson Ross/PhotoDisk/Getty; **p. 605 (right)** Michael Collier/Stock, Boston; **(left)** Jean-Pierre Horlin/The Image Bank; **p. 606 (top left)** Sandia National Laboratory; **(top right)** Peter Miller/The Image Bank; **(bottom)** Sandia National Laboratory.

Chapter 20

Opener p. 628 Frank Siteman/Stock Boston, Inc./PictureQuest; **p. 637** Alfred Pasieka/Photo Researchers, Inc.; **p. 639** Courtesy of Eugene Mosca; **p. 643** Science Photo Library/Photo Researchers, Inc.

ANSWERS

Problem answers are calculated using $g = 9.81$ m/s² unless otherwise specified in the Problem. Differences in the last figure can easily result from differences in rounding the input data and are not important.

ANSWERS

1. (c)
3. (c)
5. (a)
7. (e)
9. (a) True
   (b) False
   (c) True
11. 1.19×10^{57}
13. (a) 10^{-8} m
    (b) 20 atoms
15. (a) 3×10^{10} diapers
    (b) 1.5×10^7 m^3
    (c) 0.6 mi^2
17. $177 M
19. (a) 0.000040 W
    (b) 0.000000004 s
    (c) 3,000,000 W
    (d) 25,000 m
21. (a) C_1 is in m; C_2 is in m/s
    (b) C_1 is in m/s^2
    (c) C_1 is in m/s^2
    (d) C_1 is in m; C_2 is in s^{-1}
    (e) C_1 is in m/s; C_2 is in s^{-1}
23. (a) 4×10^7 m
    (b) 6.37×10^6 m
    (c) 2.48×10^4 mi; 3.96×10^3 mi
25. 210 cm
27. 1.28 km
29. (a) 36.0 km/h·s
    (b) 10.0 m/s^2
    (c) 88.0 ft/s
    (d) 26.8 m/s
31. 4050 m^2
33. (a) m/s^2
    (b) s
    (c) m
35. T^1
37. mv^2/r
41. M/L^3
43. (a) 30,000

(b) 0.0062
(c) 0.000004
(d) 217,000
45. (a) 1.44×10^5
    (b) 2.55×10^{-8}
    (c) 8.27×10^3
    (d) 6.27×10^2
47. 4×10^6
49. (a) 1.69×10^3
    (b) 4.8
    (c) 5.6
    (d) 10
51. 31.7 y
53. 2.0×10^{23}
55. (a) 1.41×10^{17} kg/m^3
    (b) 216 m
57. (a) 4.85×10^{-6} parsec
    (b) 3.08×10^6 m
    (c) 9.47×10^{15} m
    (d) 6.33×10^4 AU
    (e) 3.25 c·y
59. The claim is conservative as the actual weight of water used is closer to 55,000 tons.
61. (a) $n = 3/2$; $C = 17.0$ y/(Gm)$^{3/2}$
    (b) 0.510 Gm
63. 1.16×10^{19} lb

1. 0
3. It is safer to land against the wind.
5. (a) Negative
   (b) During the last five steps, gradually slow the speed of walking, until the wall is reached.
   (c)

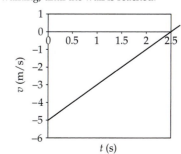

7. (a) True

    (b) True in one dimension

9. False

11. (a)

13. (b)

15. Yes. In any round-trip, A to B, and back to A, the average velocity is zero.

17. No. If the velocity is constant, a graph of position as a function of time is linear with a constant slope equal to the velocity.

19. (b)

21. (a) False

    (b) True

23. $v_{\text{top of flight}} = 0; a_{\text{top of flight}} = -g$

25. (b)

27. (c)

29. (c)

31. (c)

33. (a)

35. (d)

37. (d)

39. Velocity: (a) negative at t_0 and t_1; (b) positive at t_3, t_4, t_6, and t_7; (c) zero at t_2 and t_5

    Acceleration: (a) negative at t_4; (b) positive at t_2 and t_6; (c) zero at t_0, t_1, t_3, t_5, and t_7

41. (a) Graphs (a), (f), and (i)

    (b) Graphs (c) and (d)

    (c) Graphs (a), (d), (e), (f), (h), and (i)

    (d) Graphs (b), (c), and (g)

    (e) Graphs (a) and (i) are mutually consistent. Graphs (d) and (h) are mutually consistent. Graphs (f) and (i) are also mutually consistent.

43. (a) 54.2 m/s; (b) $-123g$

45. 4.02 m/s²

47. 14.2 ms

49. (a) 0.278 km/min

    (b) −0.0833 km/min

    (c) 0

    (d) 0.128 km/min

51. (a) 2.25 h

    (b) 4.99 h

    (c) 880 km/h

    (d) 611 km/h

53. (a) 4.33 y

    (b) 4.33×10^6 y; No

55. 35.8 m

57. (a) 0

    (b) 0.333 m/s

    (c) −2.00 m/s

    (d) 1.00 m/s

59. 122 km/h; $1.04v_{\text{av}}$

61. (a)

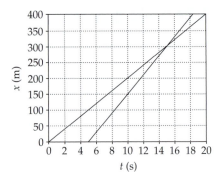

    (b) 15 s

    (c) 300 m

    (d) 100 m

63. 6 h

65. −2.00 m/s²

67. (a) 2 m

    (b) $\Delta x = (2t - 5)\Delta t + (\Delta t)^2$, where x is in meters if t is in seconds.

    (c) $v = 2t - 5$, where v is in meters per second if t is in seconds.

69. (a) $a_{\text{av,AB}} = 3.33$ m/s²; $a_{\text{av,BC}} = 0; a_{\text{av,CE}} = -7.50$ m/s²

    (b) 75.0 m

    (c)

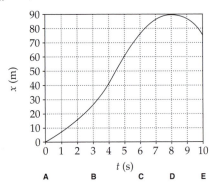

    (d) At point D, $t = 8$ s, the graph crosses the time axis and so $v = 0$.

71. (a) 80.0 m/s

    (b) 400 m

    (c) 40.0 m/s

73. 15.6 m/s²

75. (a) 4.68 s

    (b) 20.4 m

    (c) 0.991 s and 3.09 s

77. (a)

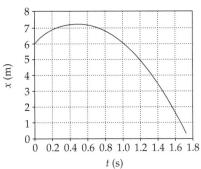

    (b) 7.27 m

    (c) 1.73 s

    (d) 11.9 m/s

79. 43.6 m

81. 68.0 m/s

83. (a) 666 m

    (b) 13.6 m/s

85. (a) 10.4 s

    (b) 27.4 s

    (c) 12.8 s

87. (a) 19.0 km

    (b) 2 min 18 s

    (c) 610 m/s

89. 40.0 cm/s; -6.88 cm/s^2

91. (a) 4.76 m/s or 10.7 mi/h

    (b) 0.595

93. 10.9 m

95. 27.6 m

97. 4.59 km

99. 2.40 m; 1.40 s

103. (a) -25.7 m/s^2

    (b) 2.33 s

105. (a) $1.03 \, c \cdot y/y^2 = 1.03 \, c/y$

    (b) ≈ 2 d

107. 4.80 m/s

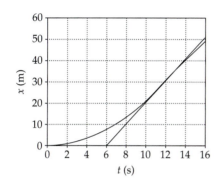

109. $\frac{1}{3} h$

111. (a) 34.7 s

    (b) 1.20 km

    (c)

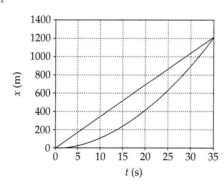

113. (a) $L_1 = \frac{2}{3} L$

    (b) $t = \frac{2}{3} t_{fin}$

115. (a) $A = 90$ m

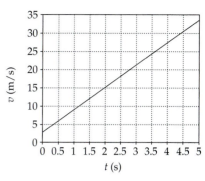

    (b) $x(t) = (3 \text{ m/s}^2)t^2 + (3 \text{ m/s})t$; $\Delta x = 90.0$ m

117. $x(t) = (\frac{7}{3} \text{ m/s}^3)t^3 - (5 \text{ m/s})t$

119. (a) 0.250 m/s per box

    (b) $v(1 \text{ s}) = 0.930$ m/s; $v(2 \text{ s}) = 3.20$ m/s; $v(3 \text{ s}) = 6.20$ m/s

    (c) $x(3 \text{ s}) = 7.00$ m

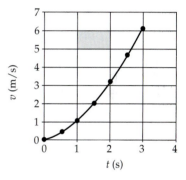

121.

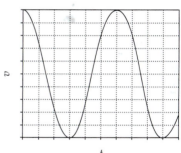

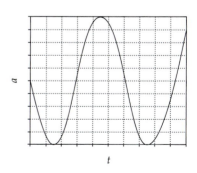

123. (a) $v(t) = (0.1 \text{ m/s}^3)t^2$

    (b) 2.23 m/s

125. 12.8 m/s^2; 30.5%

127. (a) $a = \omega v_{max}\cos(\omega t)$; Because a varies sinusoidally with time, it is *not* constant.

    (b) $x = x_0 + \dfrac{v_{max}}{\omega}[1 - \cos(\omega t)]$

129. (a) $x(t) = x_0 e^{(t - t_0)/b}$ $(b = 1\,\text{s})$

(b) Because the numerical value of b, expressed in SI units, is one, the numerical values of a, v, and x are the same at each instant in time.

Chapter 3

1. The magnitude of the displacement of a particle is less than or equal to the distance it travels along its path.

3. The displacement for any trip around the track is ZERO. Thus we see that no matter how fast the race car travels, the average velocity is always ZERO at the end of each complete circuit.

5. No. The magnitude of a component of a vector must be less than or equal to the magnitude of the vector. If the angle θ shown in the figure is equal to 0° or multiples of 90°, then the magnitude of the vector and its component are equal.

7. No

9. (e)

11. (c)

13. (a) The velocity vector, as a consequence of always being in the direction of motion, is tangent to the path.

(b)

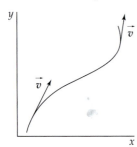

15. (a) A car moving along a straight road while braking

(b) A car moving along a straight road while speeding up

(c) A particle moving around a circular track at constant speed

17. (a)

(b)

(c)

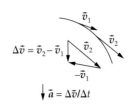

19.

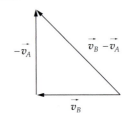

21. True

23. (d)

25. (a) False

(b) True

27.

29. (a)

| | Direction of velocity |
Path	vector
AB	north
BC	northeast
CD	east
DE	southeast
EF	south

(b)

| | Direction of acceleration |
Path	vector
AB	north
BC	southeast
CD	0
DE	southwest
EF	north

(c) The magnitudes are approximately equal.

31. The droplet leaving the bottle has the same horizontal velocity as the ship. During the time the droplet is in the air, it is also moving horizontally with the same velocity as the rest of the ship. Because of this, it falls into the vessel, which has the same horizontal velocity. Because you have the same horizontal velocity as the ship, you see the same thing as if the ship were standing still.

33. True

35. The principal reason is aerodynamic drag; when moving through a fluid such as the atmosphere, the ball's acceleration will depend strongly on its velocity.

37. 14.8 m/s

39. $R = 22.2\,\text{m}; \alpha = 22.5°$

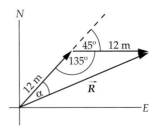

41. (a)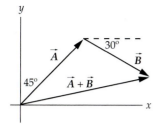

(b)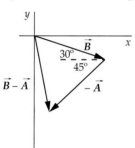

(c)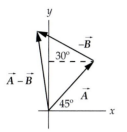

(d)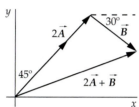

(e)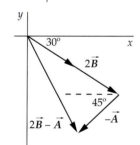

43. (b)

45.

	A	θ	A_x	A_y
(a)	10 m	30°	8.66 m	5 m
(b)	5 m	45°	3.54 m	3.54 m
(c)	7 km	60°	3.50 km	6.06 km
(d)	5 km	90°	0	5 km
(e)	15 km/s	150°	−13.0 km/s	7.50 km/s
(f)	10 m/s	240°	−5.00 m/s	−8.66 m/s
(g)	8 m/s²	270°	0	−8.00 m/s²

47. (a) 5.83; 31.0°

(b) 122; −35.0°

(c) 5.39; $\theta = 42.1°$; $\phi = 236°$

49. (a) $\vec{v} = (5 \text{ m/s})\hat{i} + (8.66 \text{ m/s})\hat{j}$

(b) $\vec{A} = (-3.54 \text{ m})\hat{i} + (-3.54 \text{ m})\hat{j}$

(c) $\vec{r} = (14 \text{ m})\hat{i} - (6 \text{ m})\hat{j}$

51. $\vec{D} = (3 \text{ m})\hat{i} + (3 \text{ m})\hat{j} + (3 \text{ m})\hat{k}$; $D = 5.20$ m

53. $\vec{v}_{av} = (14.1 \text{ km/h})\hat{i} + (-4.1 \text{ km/h})\hat{j}$

55. (b)

57. (a) $\vec{v}_{av} = (33.3 \text{ m/s})\hat{i} + (26.7 \text{ m/s})\hat{j}$

(b) $\vec{a}_{av} = (-3.00 \text{ m/s}^2)\hat{i} + (-1.77 \text{ m/s}^2)\hat{j}$

59. $\vec{v} = (30 \text{ m/s})\hat{i} + [40 \text{ m/s} - (10 \text{ m/s}^2)t]\hat{j}$; $\vec{a} = (-10 \text{ m/s}^2)\hat{j}$

61. (a) $\vec{v}_{av} = (20 \text{ m/s})(-\hat{i} + \hat{j})$

(b) $\vec{a}_{av} = (-2 \text{ m/s}^2)\hat{i}$

(c) $\Delta\vec{r} = (600 \text{ m})(-\hat{i} + \hat{j})$

63. (a) 13.1° west of north

(b) 300 km/h

65. 8.47°; 2.57 h

67. You should fly your plane across the wind.

69. (a) $\vec{r}_{AB} (6 \text{ s}) = (120 \text{ m})\hat{i} + (4 \text{ m})\hat{j}$

(b) $\vec{v}_{AB} (6 \text{ s}) = (-20 \text{ m/s})\hat{i} - (12 \text{ m/s})\hat{j}$

(c) $\vec{a}_{AB} = (-2 \text{ m/s}^2)\hat{j}$

71. $1.52 \times 10^{-6} \text{ m/s}^2$; $1.55 \times 10^{-7}g$

73. $3.44 \times 10^{-3}g$; $6.07 \times 10^{-4}g$

75. 33.4 min^{-1}

77. $h = \dfrac{(v_0 \sin \theta_0)^2}{2g}$

79. 33.8 m/s

81. 20.3 m/s; 36.2°

83. 69.3°

85. (a) 18.0 m/s

(b) 14.0°

87. (a) 8.14 m/s

(b) 23.2 m/s

89. −63.4°

91. 209 m

93. (a) 0.452 s

(b) 22.6 m

95. (a) 485 km

(b) 1.70 km/s

101. $L = \dfrac{2v_0^2 \tan \theta}{g \cos \theta}$

103. 10.8 m/s; $v = (6.50 \text{ m/s})\hat{i} + (-21.6 \text{ m/s})\hat{j}$

105. 40.5 m/s; 0.994 s

107. 7.41 m/s; 0.756 s; 15.9 m/s; 17.5 m/s; 25.0°

109. 0.785 m

111. 4.91 m/s²; 8.50 m/s²

113. (a)

(b) $\vec{v} = (5 \text{ m/s})\hat{i} + (10 \text{ m/s})\hat{j}$; 11.2 m/s

115. 31.3°; 8.06 m

117. Fourth step

119. (a) $v_{min} = \dfrac{x}{\cos\theta}\sqrt{\dfrac{g}{2(x\tan\theta - h)}}$

   (b) $v_{min} > 26.0$ m/s $= 58.0$ mi/h

   (c) $h_{max} < x\tan\theta$

Chapter 4

1. If an object with no net force acting on it is at rest or is moving with a constant speed in a straight line (i.e., with constant velocity) relative to the reference frame, then the reference frame is an inertial reference frame.

3. No. If the net force acting on an object is zero, its acceleration is zero. The only conclusion one can draw is that the *net* force acting on the object is zero.

5. No. Correctly predicting the direction of the subsequent motion requires knowledge of the initial velocity as well as the acceleration.

7. The mass of an object is an intrinsic property of the object whereas the weight of an object depends directly on the local gravitational field. Therefore, the mass of the object would not change and $w_{grav} = mg_{local}$. Note that if the gravitational field is zero then the gravitational force is also zero.

9. Your apparent weight would be greater than your true weight when observed from a reference frame that is accelerating upward. That is, when the surface on which you are standing has an acceleration a such that a_y is positive.

11. (a) $F_{n21} = m_1 g$

    (b) $F_{n12} = m_1 g$

    (c) $F_{nT2} = (m_1 + m_2)g$

    (d) $F_{n2T} = (m_1 + m_2)g$

13. (b)

15. (c)

17. (a)   (b)

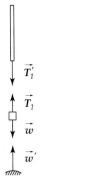

19. (a) True

    (b) False

    (c) False

    (d) False

21. (d)

23. The velocity of the elevator has no effect on the person's apparent weight.

25. (a) 782 N; 62.6 N

    (b) Because there is no acceleration, the forces are the same going up and going down the incline.

27. (a) 6.00 m/s²

    (b) 1/3

    (c) 2.25 m/s²

29. -3.75 kN

31. (a) 4.24 m/s² @ 45.0°

    (b) 8.40 m/s² @ 105°

33. 12.0 kg

35. (a) 4.00 m/s²

    (b) 2.40 m/s²

37. (a) $\vec{a} = (1.50\text{ m/s}^2)\hat{i} + (-3.50\text{ m/s}^2)\hat{j}$

    (b) $\vec{v} = (4.50\text{ m/s})\hat{i} + (-10.5\text{ m/s})\hat{j}$

    (c) $\vec{r} = (6.75\text{ m})\hat{i} + (-15.8\text{ m})\hat{j}$

39. (a) 530 N

    (b) 119 lb

41. (a) 60.0 N

    (b) 57.7 N

43. $T_2 > T_1$

45. (a) 36.9°

    (b) 4.08 N

    (c) 3.43 N; 2.40 N; 3.43 N

47. (a) $\vec{a} = (0.500\text{ m/s}^2)\hat{i} + (2.60\text{ m/s}^2)\hat{j}$

    (b) $\vec{F}_3 = (-5.00\text{ N})\hat{i} + (-26.0\text{ N})\hat{j}$

49. (a) $T = \dfrac{w}{2\sin\theta}$; $\theta = 90°$; $T \to T_{max}$ as $\theta \to 0°$

    (b) 19.6 N

51. (a) 11.8 kN

    (b) 9.81 kN

    (c) 7.81 kN

53. (a) 3.82 kN

    (b) 4.30 kN

55. 56.0 N

57. (d)

59. (a) 508 N; 508 N

    (b) mg; 0

61. 552 N

63. (a)

65. (e)

67. (a) 19.6 N

    (b) 19.6 N

    (c) 25.6 N

    (d) 14.6 N

69. (a) 1.31 m/s²

    (b) 16.7 N; 21.3 N

71. (a) $a = \dfrac{F}{m_1 + m_2}$; $F_{2,1} = \dfrac{Fm_1}{m_1 + m_2}$

    (b) 0.400 m/s²; 0.800 N

75. (a)

77. (a) $a = \dfrac{g(m_2 - m_1\sin\theta)}{m_1 + m_2}$; $T = \dfrac{gm_1 m_2(1 + \sin\theta)}{m_1 + m_2}$

    (b) 2.45 m/s²; 36.8 N

79. (a) 1.37 m/s²; 61.4 N

    (b) 1.19; The answer is the ratio of two quantities with the same units and so has no units.

81. (a) 398 N

    (b) 368 N

83. (a) 5.00 cm

    (b) $a_{5kg} = 4.91$ m/s²; $a_{20kg} = 2.45$ m/s²; $T = 24.5$ N

85. 1.36 kg or 1.06 kg

87. $F = 2T = \dfrac{4m_1 m_2 g}{m_1 + m_2}$

91. (a) -100 m/s^2

    (b) 6.13 cm

    (c) 35.0 ms

93. 305 N; 1.55 kN

95. (a) 1.50 m/s

    (b) 1.50 m

    (c) 0.500 m/s

    (d) 12.0 N

97. (a) $a = \dfrac{F}{m_1 + m_2}$

    (b) $F_{\text{net}} = \dfrac{F m_2}{m_1 + m_2}$

    (c) $T = \dfrac{F m_1}{m_1 + m_2}$

    (d)

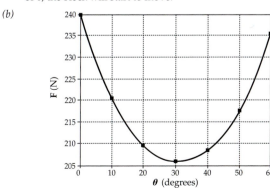

Yes . . . correct answers appear above.

99. (a) 55.0 g

    (b) 2.45 m/s²; 2.03 N

101. (a) $\frac{1}{3}(F_2 + 2F_1)$

     (b) $\dfrac{3T_0}{4C}$

Chapter 5

1. The force of friction between the object and the floor of the truck must be the force that causes the object to accelerate.

3. (d)

5. (b)

7. As the spring is extended, the force exerted by the spring on the block increases. Once that force is greater than the maximum value of the force of static friction on the block, the block will begin to move. However, as it accelerates, it will shorten the length of the spring, decreasing the force that the spring exerts on the block. As this happens, the force of kinetic friction can then slow the block to a stop, which starts the cycle over again. One interesting application of this to the real world is the bowing of a violin string: the string under tension acts like the spring, while the bow acts as the block, so as the bow is dragged across the string, the string periodically sticks and frees itself from the bow.

9. (e)

11. Block 1 will hit the pulley before block 2 hits the wall.

13. (d)

15. (d)

17. For a rock, which has a relatively small surface area compared to its mass, the terminal speed will be relatively high; for a lightweight, spread-out object like a feather, the opposite is true.

    Another issue is that the higher the terminal velocity is, the longer it takes for a falling object to reach terminal velocity: from this, the feather will reach its terminal velocity quickly, and fall at an almost constant speed very soon after being dropped; a rock, if not dropped from a great height, will have almost the same acceleration as if it were in free-fall for the duration of its fall, and thus be continually speeding up as it falls.

19. (a) M/T; kg/s

    (b) M/L; kg/m

    (c) ML/T²

    (d) 56.9 m/s

    (e) 86.9 m/s

21. (b)

23. (a) 15.0 N

    (b) 12.0 N

25. 500 N

27. (a) -5.89 m/s^2

    (b) 76.4 m

29. (a) 49.1 N

    (b) 123 N

31. 4.57°

33. (a) 0.667

    (b) 2.16 m/s²; 1.36 s

35. (a)

37. 2.36 m/s²; 37.2 N

39. (a) 0.599

    (b) 9.25 m

    (c) 4.73 m/s

41. (a) 2.75 m/s²

    (b) 10.1 s

43. (a) 0.965 m/s²

    (b) 0.184 N

45. (a) 25.0°

    (b) 0.118 N

47. (a) The static-frictional force opposes the motion of the object, and the maximum value of the static-frictional force is proportional to the normal force F_N. The normal force is equal to the weight minus the vertical component F_V of the force F. Keeping the magnitude F constant while increasing θ from 0 results in a decrease in F_v and thus a corresponding decrease in the maximum static-frictional force f_{max}. The object will begin to move if the horizontal component F_H of the force F exceeds f_{max}. An increase in θ results in a decrease in F_H. As θ increases from 0, the decrease in F_N is larger than the decrease in F_H, so the object is more and more likely to slip. However, as θ approaches 90°, F_H approaches 0, and no movement will be initiated. If F is large enough and if θ increases from 0, then at some value of θ, the block will start to move.

    (b)

49. *(b)*

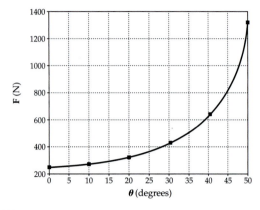

51. *(a)* 0.238

 (b) 1.40 m/s²

53. *(a)* 17.7 N

 (b) 1.47 m/s²; 5.88 N

 (c) 1.96 m/s²; 7.87 m/s²

55. *(a)* 0.163 m/s²

 (b) 0.0381 m

 (c) −0.254 m/s²

57. −8.41 glapp/plipp²; 0.191

59. *(a)* −1.57 N; 83.8 N

 (b) 6.49 N; 37.5 N

61. *(a)* −2.60 m/s²; 19.2 m

 (b) −2.11 m/s²; 23.7 m

63. *(a)* 0.297

 (b) 2.82 m/s

65. *(c)*

67. *(a)* 1.41 m/s

 (b) 8.50 N

69. *(a)* 8.33 m/s²; upward

 (b) 542 N; upward

 (c) 1.18 kN; upward

71. $T_2 = [m_2(L_1 + L_2)]\left(\dfrac{2\pi}{T}\right)^2; T_1 = [m_2(L_1 + L_2) + m_1 L_1]\left(\dfrac{2\pi}{T}\right)^2$

73. 53.3°; 410 N

75. *(a)* 0.395 N

 (b) 0.644

77. $3.44 \times 10^{-3}g; 6.07 \times 10^{-4}g$

79. 51.6°

81. *(a)* $a_c = \dfrac{v_0^2}{r}\left(\dfrac{1}{1 + \left(\dfrac{\mu_k v_0}{r}\right)t}\right)^2$

 (b) $a_t = -\mu_k a_c$

 (c) $a = a_c\sqrt{1 + \mu_k^2}$

83. 12.8 m/s

85. *(a)* 7.25 m/s

 (b) 0.536

87. 21.7°

89. *(a)* 7.832 kN

 (b) −766 kN

91. $v_{min} = 20.1$ km/h; $v_{max} = 56.1$ km/h

93. 2.79×10^{-4} kg/m

95. 88.2 km/h

97. 3.31 s; 100

99. $y(3.5 \text{ s}) \approx 60.4$ m; $y_{max} \approx 60.6$ m @ $t = 3.3$ s; $t_{flight} \approx 7$ s; The ball spends a little longer coming down than it does going up.

101. 0.511

103. *(a)* 0.289

 (b) 600 N

105. 1.49 kN

107. $a = g(\sin \theta_1 - \tan \theta_0 \cos \theta_1)$

109. *(a)* 49.4 m/s²

 (b) 4.49 s

111. *(a)* 193 N

 (b) 51.8 N

 (c) The sled does not move.

 (d) μ_k is undetermined.

 (e) 536 N

113. 0.433

115. 23.6 rev/min

117. *(a)* Toward the earth's axis.

 (b) A stone dropped from a hand at a location on the earth. The effective weight of the stone is equal to $m\vec{a}_{st, surf}$, where $\vec{a}_{st, surf}$ is the acceleration of the falling stone (neglecting air resistance) relative to the local surface of the earth. The gravitational force on the stone is equal to $m\vec{a}_{st, iner}$, where $\vec{a}_{st, iner}$ is the acceleration of the stone relative to an inertial reference frame. These accelerations are related by $\vec{a}_{st, surf} + \vec{a}_{surf, iner} = \vec{a}_{st, iner}$, where $\vec{a}_{surf, iner}$ is the acceleration of the local surface of the earth relative to the inertial frame (the acceleration of the surface due to the rotation of the earth). Multiplying through this equation by m and rearranging gives $m\vec{a}_{st, surf} = m\vec{a}_{st, iner} - m\vec{a}_{surf, iner}$, which relates the apparent weight to the acceleration due to gravity and the acceleration due to the earth's rotation. A vector addition diagram can be used to show that the magnitude of $m\vec{a}_{st, surf}$ is slightly less than that of $m\vec{a}_{st, iner}$.

 (c) 983 cm/s²

Chapter 6

1. *(a)* False

 (b) True

 (c) True

3. False

5. No. The work done on any object by any force $\vec{F}$ is defined as $dW \equiv \vec{F} \cdot d\vec{r}$. The direction of $\vec{F}_{net}$ is toward the center of the circle in which the object is traveling, and $d\vec{r}$ is tangent to the circle. No work is done by the net force because $\vec{F}_{net}$ and $d\vec{r}$ are perpendicular, so the dot product is zero.

7. Because $W \propto x^2$, doubling the distance the spring is stretched will require four times as much work.

9. *(d)*

11. *(a)* False

 (b) False

 (c) True

13. *(a)* False

 (b) False

15. *(a)* 0.245 m

 (b) 120 J

17. $\approx 1\%$

19. 20.8 kJ

21. (a) 147 J

    (b) 266 J

23. 10.6 kJ

25. (a) 6.00 J

    (b) 12.0 J

    (c) 3.46 m/s

27. $W = -\frac{1}{2}kx_1^2 - \frac{1}{3}ax_1^3$

29. (a) $m(y) = 40\,\text{kg} - (1\,\text{kg/m})y$

    (b) 5.89 kJ

31. (a) 4.17 N

    (b) $\vec{T}$, $\vec{F}_g$, and $\vec{F}_n$; Because all of these forces act perpendicularly to the direction of motion of the object, none of them do any work.

35. 180°

37. (a) −24

    (b) −10

    (c) 0

39. (a) 1.00 J

    (b) 0.213 N

43. No. Let $\vec{A} = \hat{i}$, $\vec{B} = 3\hat{i} + 4\hat{j}$ and $\vec{C} = 3\hat{i} - 4\hat{j}$ and form $\vec{A} \cdot \vec{B}$ and $\vec{A} \cdot \vec{C}$.

45. (b) The results of (a) and (b) tell us that $\vec{a}$ is perpendicular to $\vec{v}$ and parallel (or antiparallel) to $\vec{r}$.

47. (a) 98.1 W

    (b) 392 J

49. (a) $v = (\frac{5}{8}\,\text{m/s}^2)t$

    (b) $P = 3.13t$ W/s

    (c) 9.38 W

51. 445 W

53. $v = \sqrt{v_0^2 + 2gH}$

55. 4.71 kJ

57. (a) 392 J

    (b) 2.45 m; 4.91 m/s

    (c) 24.1 J; 368 J

    (d) 392 J; 19.8 m/s

59. (a) 0.100 m

    (b) 0.141 m

61. (a) $U(\theta) = (m_2\ell_2 - m_1\ell_1)g \sin\theta$

    (b) U is a minimum at $\theta = -\pi/2$ and a maximum at $\theta = \pi/2$

    (c) $U = 0$ independently of θ

63. (a) $F_x = \dfrac{C}{x^2}$

    (b) F_x is positive for $x \neq 0$ and therefore $\vec{F}$ is directed away from the origin.

    (c) $U(x)$ decreases with increasing x.

    (d) F_x is negative for $x \neq 0$ and therefore $\vec{F}$ is directed toward the origin. $U(x)$ increases with increasing x.

65. $U(x) = \dfrac{a}{x} + U_0$

67. (a) $F_x = 4x(x + 2)(x - 2)$

    (b) −2 m, 0, 2 m

    (c) Unstable equilibrium at $x = -2$ m; stable equilibrium at $x = 0$; unstable equilibrium at $x = 2$ m

69. (a) 0 and 2 m; neutral equilibrium for $x > 3$ m

    (b)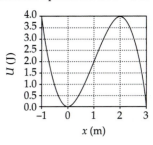

    (c) Stable equilibrium at $x = 0$; unstable equilibrium at $x = 2$ m

    (d) 2.00 m/s

71. (a) $U(y) = -mgy - 2Mg\left(L - \sqrt{y^2 + d^2}\right)$

    (b) $y = d\sqrt{\dfrac{m^2}{4M^2 - m^2}}$

    (c) Stable equilibrium

73. (a) 706 MJ

    (b) 11.8 MW

75. 0.500 m

77. (a) 34.4 N

    (b)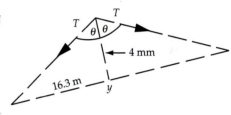

    1.68 N

    (c) 3.38 mJ

79. (a) $F(x) = mC^2x$

    (b) $W = \frac{1}{2}mC^2x_1^2$

81. In the following, if t is in seconds and m is in kilograms, then v is in m/s, a is in m/s², P is in W, and W is in J.

    (a) $v = (6t^2 - 8t)$; $a = (12t - 8)$

    (b) $P = 8mt(9t^2 - 18t + 8)$

    (c) $W = 2mt_1^2(3t_1 - 4)^2$

83. 5.74 km

85. (a)

x	W
(m)	(J)
−4	6
−3	4
−2	2
−1	0.5
0	0
1	0.5
2	1.5
3	2.5
4	3

    (b)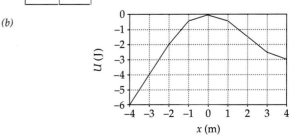

87. (b) $W = (10\pi\,\text{m})F_0$ if the rotation is clockwise; $-(10\pi\,\text{m})F_0$ if the rotation is counterclockwise. Because $W \neq 0$ for a complete circuit, $\vec{F}$ is not conservative.

89. (a)

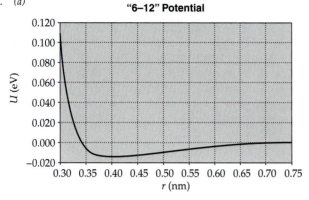

"6–12" Potential

(b) The minimum value is about -0.0107 eV, occurring at a separation of approximately 0.380 nm. Because the function is concave upward at this separation, this separation is one of stable equilibrium, although very shallow.

(c) -6.69×10^{-12} N; 7.49×10^{-11} N

Chapter 7

1. (a)
3. (a) False
   (b) False
5. As she starts pedaling, chemical energy inside her body is converted into kinetic energy as the bike picks up speed. As she rides it up the hill, chemical energy is converted into gravitational potential energy. While freewheeling down the hill, potential energy is converted to kinetic energy, and while braking to a stop, kinetic energy is converted into thermal energy (a more random form of kinetic energy) by the frictional forces acting on the bike.
7. (d)
9. No. From the work–kinetic energy theorem, no total work is being done on the rock, as its kinetic energy is constant. However, the rod must exert a tangential force on the rock to keep the speed constant. The effect of this force is to cancel the component of the force of gravity that is tangential to the trajectory of the rock.
11. 33.6 s
13. 3.04×10^{19} J/y; $\approx 6\%$
15. 1.10×10^6 L/s
17. (c)
19. 3.89 m
21. 5.05 m
23. 25.6°
25. $U = \dfrac{[mg(\sin\theta + \mu_s \cos\theta)]^2}{2k}$
27. $6mg$
29. (c)
31. $6mg$
33. (a) 31.0 m
    (b) -31.7 J
    (c) 33.7 m/s

35. (a) 151 m
    (b) 45.3 m/s
37. (a) $\frac{5}{2}mgL$
    (b) $6mg$
39. (a) 20.2°
    (b) 6.39 m/s
41. $v = L\sqrt{2\dfrac{g}{L}(1 - \cos\theta) + \dfrac{k}{m}\left(\sqrt{\frac{13}{4} - 3\cos\theta} - \frac{1}{2}\right)^2}$
43. (a) 94.2 kJ
    (b) The energy required to do this work comes from chemical energy stored in the body.
    (c) 471 kJ
45. (a) 104 J
    (b) 70.2 J
    (c) 33.8 J
    (d) 2.91 m/s
47. (a) 7.67 m/s
    (b) 58.9 J
    (c) 0.333
49. (a) $W_f = (13.7\,\text{N})y$
    (b) $E_{mech} = -(13.7\,\text{N})y$
    (c) 1.98 m/s
51. 0.875 m; 2.49 m/s
53. (a) 9.00×10^{13} J
    (b) $\$2.5 \times 10^6$
    (c) 28,400 y
55. 1.88×10^{-28} kg
57. 3.56×10^{14} reactions
59. 0.782 MeV
61. (a) 3.16 kg
    (b) 8.04×10^9 kg
63. (b)
65. 57.6 MJ
67. (a) 0.208
    (b) 3.45 MJ
69. (a) From the FBD we can see that the forces acting on the box are the normal force exerted by the inclined plane, kinetic friction force, and the gravitational force (the weight of the box) exerted by the earth.
    (b) 0.451 m
    (c) 1.33 J
    (d) 2.52 m/s
71. 11.3 kW; -6.77 kW
73. (a) 1.60 kJ
    (b) 619 J
    (c) 23.4 m/s
75. (a) 147 J
    (b) The energy is transferred to the girder from its surroundings, which are warmer than the girder. As the temperature of the girder rises, the atoms in the girder vibrate with a greater average kinetic energy, leading to a larger average separation, which causes the girder's expansion.

77. (a)

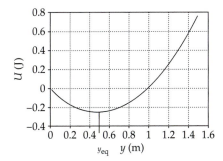

(b) $F = -ky + mg$

(c) $y_{max} = \dfrac{2mg}{k}$

(d) $y_{eq} = \dfrac{mg}{k}$

(e) $W_f = \dfrac{m^2 g^2}{2k}$

79. (a) 17.3 m

(b) 4.91 kN

(c) 4.91 m/s²

(d) 13.4 kN, upward

(e) 5.46 kN; 63.9°

(f) 1.44 kN

81. (a) 491 N; 981 N

(b) 9.82 kW; 29.4 kW

(c) 8.85°

(d) 6.36 km/L

83. (a) 17.4 MJ

(b) 1.39×10^{10} J

(c) 9.73×10^9 J

(d) 1.59 MW

87. (a) $v = \sqrt{\dfrac{2mgY}{M + m}}$

(b) $v = \sqrt{\dfrac{2mgY}{M + m}}$

89. $D = 2\sqrt{HL(1 - \cos\theta)}$

91. (a) $K_{max} = mgh + \dfrac{m^2 g^2}{2k}$

(b) $x_{max} = \dfrac{mg}{k} + \sqrt{\dfrac{m^2 g^2}{k^2} + \dfrac{2mgh}{k}}$

(c) $x = \dfrac{mg}{k} + \sqrt{\dfrac{2m^2 g^2}{k^2} + \dfrac{4mgh}{k}}$

93. (a)

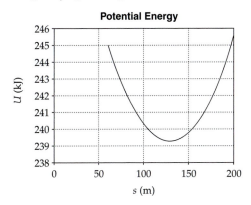

Potential Energy

(b) 5.39 kJ

Chapter 8

1. A doughnut.

3. (b)

5. No. Consider a 1-kg block with a speed of 1 m/s and a 2-kg block with a speed of 0.707 m/s. The blocks have equal kinetic energies but momenta of magnitude 1 kg·m/s and 1.414 kg·m/s, respectively.

7. $\vec{p}_{recoil} = \vec{p}_{rifle} = -\vec{p}_{bullet}$ or $\vec{p}_{rifle} + \vec{p}_{bullet} = 0$

9. Conservation of momentum requires only that the net external force acting on the system be zero. It does not require the presence of a medium such as air.

11. Think of someone pushing a box across a floor. Her push on the box is equal but opposite to the push of the box on her, but the action and reaction forces act on *different objects*. You can only add forces when they act on the same object.

13. The problem is that the comic situations violate the conservation of momentum! To move forward requires pushing something backward, which Superman doesn't appear to be doing when flying around. In a similar manner, if Superman picks up a train and throws it at Lex Luthor, he (Superman) ought to be tossed backward at a pretty high speed to satisfy the conservation of momentum.

15. The friction of the tire against the road causes the car to slow down. This is rather subtle, as the tire is in contact with the ground without slipping at all times, so as you push on the brakes harder, the force of static friction of the road against the tires must increase. Also, of course, the brakes heat up, and not the tires.

17. Assume that the ball travels at 80 mi/h = 35 m/s. The ball stops in a distance of about 1 cm, so the distance traveled is about 2 cm at an average speed of about 18 m/s. The collision time is $\dfrac{0.02\ m}{18\ m/s} \approx 1$ ms.

19. (a) False

(b) True

(c) True

21. (a) The loss of kinetic energy is the same in both cases.

(b) The percentage loss is greatest for the case in which the two objects have oppositely directed velocities of magnitude $\frac{1}{2}v$.

23. (b) is correct because all of 1's kinetic energy is transferred to 2 when $m_2 = m_1$.

25. The water is changing direction when it rounds the corner in the nozzle. Therefore, the nozzle must exert a force on the stream of water to change its direction, and, from Newton's 3rd law, the water exerts an equal but opposite force on the nozzle.

27. No. $\vec{F}_{ext,net} = d\vec{p}/dt$ defines the relationship between the net force acting on a system and the rate at which its momentum changes. The net external force acting on the pendulum bob is the sum of the force of gravity and the tension in the string and these forces do not add to zero.

29. Think of the stream of air molecules hitting the sail. Imagine that they bounce off the sail elastically–their net change in momentum is then roughly twice the change in momentum that they experienced going through the fan. Another way of looking at it: initially, the air is at rest, but after passing through the fan and bouncing off the sail, it is moving backward; therefore, the boat must exert a net force on the air pushing it backward, and there must be a force on the boat pushing it forward.

31. (a) 2.33 s
    (b) 6.74 m/s
33. (0.233 m, 0)
35. (2.00 m, 1.40 m)
37. (1.50 m, 1.36 m)
41. $z_{cm} = \frac{1}{2} R$
43. $\vec{v}_{cm} = (3 \text{ m/s})\hat{i} - (1.5 \text{ m/s})\hat{j}$
45. $\vec{a}_{cm} = (2.4 \text{ m/s}^2)\hat{i}$
47. (a) $F_n = (m_p + m_b)g$
    (b) $F_n = (m_p + 2m_b)g$
    (c) $F_n = m_p g$
49. (a) $F_n = (m_p + m_b)g$
    (b) $F_n = m_p g + m_b g \left(1 + \sqrt{1 + \frac{2kh}{m_b g}} \right)$
51. $\vec{v}_{10} = (4 \text{ m/s})\hat{i}$
53. $\vec{v}' = 2v\hat{i} - v\hat{j}$
55. $-\sqrt{\frac{gh}{3}} \hat{i}$
57. (a) 43.5 J
    (b) $\vec{v}_{cm} = (1.50 \text{ m/s})\hat{i}$
    (c) $\vec{v}_{1,rel} = (3.50 \text{ m/s})\hat{i}$ and $\vec{v}_{2,rel} = (-3.50 \text{ m/s})\hat{i}$
    (d) 36.8 J
    (e) $K_{cm} = 6.75 \text{ J} = K - K_{rel}$
59. (a) 10.8 N·s
    (b) 1.34 kN
61. 1.81 MN·s; 10.602 MN
63. 230 N
65. (a) $\vec{I} = (1.08 \text{ N·s})\hat{i}$ (directed into wall)
    (b) 360 N, into wall
    (c) 0.480 N·s, away from wall
    (d) 3.84 N, away from wall
67. (a) 20.0 m/s
    (b) 20% of the initial kinetic energy is transformed into thermal energy, sound, and the deformation of metal.
69. (a) −2.00 m/s
    (b) The collision was inelastic.
71. (a) $\vec{v}_{cm} = (23.1 \text{ m/s})\hat{i}$
    (b) −254 m/s
73. (a) 5.00 m/s
    (b) 0.250 m
    (c) $v_{1f} = 0$; $v_{2f} = 7.00 \text{ m/s}$
75. (a) $0.200v_0$
    (b) $0.400v_0$
77. 450 m/s
81. $h = \frac{v^2}{8g}\left(\frac{m_1}{m_2}\right)^2$
83. 0.0529
85. $1.50 \times 10^6 \text{ m/s}$
87. (a) $\vec{v}_1 = (312 \text{ m/s})\hat{i} + (66.6 \text{ m/s})\hat{j}$
    (b) 5.61 km
    (c) 35.8 kJ
89. 0.913
91. (a) 20% of its mechanical energy is lost.
    (b) 0.894

93. (a) 1.70 m/s
    (b) 0.833
95. (a) 60°
    (b) 2.50 m/s; 4.33 m/s
97. (a) 1.00 m/s; 1.73 m/s
    (b) The collision was elastic.
99. $v_1 = 8.66 \text{ m/s}$; $v_2 = 5.00 \text{ m/s}$
101. In an elastic collision

$$K_i = K_f = \frac{p_1^2}{2}\left[\frac{m_1^2 + 6m_1 m_2 + m_2^2}{m_1^2 m_2 + m_1 m_2^2}\right] = \frac{p_1'^2}{2}\left[\frac{m_1^2 + 6m_1 m_2 + m_2^2}{m_1^2 m_2 + m_1 m_2^2}\right]$$

If $p_1' = +p_1$, the particles do not collide.
103. (a) $\vec{v}_{cm} = 0$
    (b) $\vec{u}_3 = (-5 \text{ m/s})\hat{i}$; $\vec{u}_5 = (3 \text{ m/s})\hat{i}$
    (c) $\vec{u}_3' = (5 \text{ m/s})\hat{i}$; $u_5' = 0.75 \text{ m/s}$
    (d) $\vec{v}_3' = (5 \text{ m/s})\hat{i}$; $\vec{v}_5' = (-3 \text{ m/s})\hat{i}$
    (e) 60.0 J; 60.0 J
105. (a) 360 kN
    (b) 120 s
    (c) 1.72 km/s
107. (a) $\tau_0 = 1 + \frac{a_0}{g}$
    (b) $v_f = gI_{sp}\left(\ln\frac{m_0}{m_f} - \frac{1}{\tau_0}\left(1 - \frac{m_f}{m_0}\right)\right)$
    (c)

    (d) 28.1
109. 0.192 m/s; 31.3 mJ; 12.0 mJ
111. 0.462 m/s
113. (a) $\vec{p} = -(1.10 \times 10^5 \text{ kg·km/h})\hat{i} + (1.05 \times 10^5 \text{ kg·km/h})\hat{j}$
    (b) 43.4 km/h; 46.3° west of north
115. (a) 6.26 m/s
    (b) 20.0 m
117. 3.72 m
119. (a) The velocity of the basketball will be equal in magnitude but opposite in direction to the velocity of the baseball.
    (b) $v_{1f} = 0$
    (c) $v_{2f} = 2v$
121. (a) 29.6 km/s
    (b) 8.10; The energy comes from an immeasurably small slowing of Saturn.
123. $3.00 \times 10^5 \text{ m/s}$
125. (a) 0.600 m/s^2
    (b) 960 N
127. No. The driver was traveling at 23.3 km/h.
129. 8.85 kg
131. $\frac{1}{14}r$
133. (a) $0.716^N E_0$
    (b) 55

135. (a) $y_{cm} = \dfrac{v^2}{2L}t^2$

    (b) $a_{cm} = \dfrac{v^2}{L}$

    (c) $F = \dfrac{vt}{L}\left(\dfrac{v}{gt} + 1\right)Mg$

137. $v_{2f} = \left(\dfrac{m_b}{m_2 + m_b}\right)\left[1 + \dfrac{m_1}{m_1 + m_b}\right]v;$

    $v_{1f} = -\dfrac{m_2 m_b(2m_1 + m_b)}{(m_1 + m_b)^2 (m_2 + m_b)}v$

139. -0.960 m/s^2

141. $v = (1.70 \text{ m}^{1/2}/\text{s})\sqrt{L}$

Chapter 9

1. Because r is greater for the point on the rim, it moves the greater distance. Both turn through the same angle. Because r is greater for the point on the rim, it has the greater speed. Both have the same angular velocity. Both have zero tangential acceleration. Both have zero angular acceleration. Because r is greater for the point on the rim, it has the greater centripetal acceleration.

3. (c)

5. (d)

7. No. A net torque is required to *change* the rotational state of an object. A net torque may decrease the angular speed of an object. All we can say for sure is that a net torque will *change* the angular speed of an object.

9. (b)

11. (b)

13. For a given applied force, this increases the torque about the hinges of the door, which increases the door's angular acceleration, leading to the door being opened more quickly. It is clear that putting the knob far from the hinges means that the door can be opened with less effort (force). However, it also means that the hand on the knob must move through the greatest distance to open the door, so it may not be the quickest way to open the door. Also, if the knob were at the center of the door, you would have to walk around the door after opening it, assuming the door is opening toward you.

15. (b)

17. (b)

19. (a)

21. True. If the sphere is slipping, then there is kinetic friction that dissipates the mechanical energy of the sphere.

23. 10.3%

25. 6.42

27. (a) 15.6 rad/s

    (b) 46.8 rad

    (c) 7.45 rev

    (d) 73.0 m/s^2

29. (a) 40.0 rad/s

    (b) 0.960 m/s^2; 192 m/s^2

31. (a) 0.589 rad/s^2

    (b) 4.71 rad

33. (d)

35. 1.04 rad/s; 9.92 rev/min

37. (a) 1.87 N·m

    (b) 124 rad/s^2

    (c) 620 rad/s

39. (a) $g \sin\theta$

    (b) Because the line-of-action of the tension passes through the pendulum's pivot point, its lever arm is zero and it causes no torque.

    (c) $g \sin\theta$

41. (a) $d\tau_f = \dfrac{2\mu_k Mg}{R^2}r^2 dr$

    (b) $\tau_f = \frac{2}{3}MR\mu_k g$

    (c) $\Delta t = \dfrac{3R\omega}{4\mu_k g}$

43. 56.0 kg·m^2

45. (a) 28.0 kg·m^2

    (b) 28.0 kg·m^2

47. 2.60 kg·m^2

49. (b) $I_{cm} = \frac{1}{12}m(a^2 + b^2)$

51. $5.41 \times 10^{-47} \text{ kg·m}^2$

55. $I = \frac{3}{10}MR^2$

57. $I_x = 3M\left(\dfrac{H^2}{5} + \dfrac{R^2}{20}\right)$

59. (a) 84.6 mJ

    (b) 347 rev/min

63. (a) 19.6 kN

    (b) 5.89 kN·m

    (c) 0.267 rad/s

    (d) 1.57 kN

65. (a) 3.62 rad/s

    (b) 3.62 rad/s

67. Unless M, the mass of the ladder, is zero, $v_r > v_f$. It is better to let go and fall to the ground.

69. 3.11 m/s^2; $T_1 = 12.5 \text{ N}$; $T_2 = 13.4 \text{ N}$

71. 8.23 m/s

73. (a) $a = \dfrac{g}{1 + \dfrac{2M}{5m}}$

    (b) $T = \dfrac{2mMg}{5m + 2M}$

75. (a) 72.0 kg

    (b) 1.37 rad/s^2; $T_1 = 294 \text{ N}$; $T_2 = 746 \text{ N}$

77. (a) $a = \dfrac{g \sin\theta}{1 + \dfrac{m_1}{2m_2}}$

    (b) $T = \dfrac{\frac{1}{2}m_1 g \sin\theta}{1 + \dfrac{m_1}{2m_2}}$

    (c) $E = m_2 gh$

    (d) $E_{bottom} = m_2 gh$

    (e) $v = \sqrt{\dfrac{2gh}{1 + \dfrac{m_1}{2m_2}}}$

    (f) For $\theta = 0$: $a = T = 0$

    For $\theta = 90°$: $a = \dfrac{g}{1 + \dfrac{m_1}{2m_2}}$, $T = \frac{1}{2}m_1 a$, and $v = \sqrt{\dfrac{2gh}{1 + \dfrac{m_1}{2m_2}}}$

    For $m_1 = 0$: $a = g \sin\theta$, $T = 0$, and $v = \sqrt{2gh}$

79. 0.0864 m/s^2; 3.14 m/s

81. 0.192 m/s^2; 0.962 N

83. 1.13 kJ

85. 45.9 m

87. $19.5°$

89. (a) $a = \frac{2}{3} g \sin \theta$

  (b) $f_s = \frac{1}{3} mg \sin \theta$

  (c) $\theta_{max} = \tan^{-1}(3\mu_s)$

91. $v' = \sqrt{\frac{4}{3}} v$

93. 223 J

97. (a) $\alpha = \dfrac{2F}{R(M + 3m)}$; counterclockwise

  (b) $a_c = \dfrac{F}{M + 3m}$

  (c) $a_{CB} = -\dfrac{2F}{M + 3m}$

99. (a) 0.400 rad/s^2; 0.200 rad/s^2

  (b) 4.00 N

101. (a) $s_1 = \dfrac{12}{49} \dfrac{v_0^2}{\mu_k g}$, $t_1 = \dfrac{2}{7} \dfrac{v_0}{\mu_k g}$, and $v_1 = \dfrac{5}{7} v_0$

  (b) $5/7$

  (c) 26.6 m; 3.88 s; 5.71 m/s

103. $v = \dfrac{2r\omega_0}{7}$

105. (a) 360 kN

  (b) 120 s

  (c) 1.72 km/s

107. (a) $v = 1.57 v_0$

  (b) $\Delta t = \dfrac{4}{7} \dfrac{v_0}{\mu_k g}$

  (c) $\Delta x = 0.735 \dfrac{v_0^2}{\mu_k g}$

111. $I = 2mR^2$

113. 0.134 m

115. (a) 7.36 m/s^2

  (b) 14.7 m/s^2

  (c) 2.43 m/s

117. (a) 780 kJ

  (b) 90.3 N·m; 150 N

  (c) 1380 rev

119. (a) 15.0 m

  (b) 15.4 rad/s

121. (a) S^2

  (b) S^3

  (c) S^5

123. (a) $\omega = \sqrt{\dfrac{4g}{3r}}$

  (b) $F = \frac{7}{3} Mg$

125. (a) $v = \sqrt{\dfrac{2MgD \sin \theta}{M + \dfrac{I}{r^2}}}$

  (b) $f_s = \dfrac{Mg \sin \theta}{1 + \dfrac{R}{r}}$

127. (a) 14.7 m/s^2

  (b) 66.7 cm

129. 41.7 J

131. The solid line on the graph shown below shows the position y of the bucket when it is in free fall and the dashed line shows y under the conditions modeled in this problem.

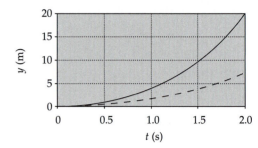

133. (a) 25.7 N

  (b) 3.21 kg

  (c) 1.10 m/s^2

Chapter 10

1. (a) True

  (b) True

  (c) True

3. $90°$

5. (a) Doubling $\vec{p}$ doubles $\vec{L}$.

  (b) Doubling $\vec{r}$ doubles $\vec{L}$.

7. False

9. (e)

11. It is easier to crawl radially outward. In fact, a radially inward force is required just to prevent you from sliding outward.

13. The hardboiled egg is solid inside, so everything rotates with a uniform velocity. By contrast, it is difficult to get the viscous fluid inside a raw egg to start rotating; however, once it is rotating, stopping the shell will not stop the motion of the interior fluid, and, for this reason, the egg may start rotating again after momentarily stopping.

15. (b)

17. (b)

19. (a) The lifting of the nose of the plane rotates the angular momentum vector upward. It veers to the right in response to the torque associated with the lifting of the nose.

  (b) The angular momentum vector is rotated to the right when the plane turns to the right. In turning to the right, the torque points down. The nose will move downward.

21. (b)

23. The center of mass of the rod-and-putty system moves in a straight line, and the system rotates about its center of mass.

25. 4.17 rev/s

27. (a) $2.40 \times 10^{-8} \text{ kg·m}^2/\text{s}$

  (b) 5.22×10^{52}; 2.29×10^{26}

  (c) The quantization of angular momentum is not noticed in macroscopic physics because no experiment can differentiate between $\ell = 2 \times 10^{26}$ and $\ell = 2 \times 10^{26} + 1$.

29. (a) 0.331

    (b) Because experimentally $C < 2/5 = 0.4$, the mass density must be greater near the center of the earth.

31. 10.1 rad/s

33. $\vec{\tau} = FR\hat{k}$

35. (a) $24\hat{k}$

    (b) $-24\hat{j}$

    (c) $-5\hat{k}$

39. $\vec{B} = 4\hat{j} + 3\hat{k}$

45. (a) 54.0 kg·m²/s

    (b) ω increases as the particle approaches the point and decreases as it recedes.

47. (a) 1.33×10^{-5} kg·m²/s

    (b) 1.33×10^{-5} kg·m²/s

    (c) 1.33×10^{-5} kg·m²/s

    (d) 8.83×10^{-5} kg·m²/s; -6.17×10^{-5} kg·m²/s

49. (a) 4.00 N·m

    (b) $(0.192 \text{ rad/s}^2)t$

51. (a) $\tau_{\text{net}} = Rg(m_2 \sin\theta - m_1)$

    (b) $L = vR\left(\dfrac{I}{R^2} + m_1 + m_2\right)$

    (c) $a = \dfrac{g(m_2 \sin\theta - m_1)}{\dfrac{I}{R_2} + m_1 + m_2}$

55. (a) 5.00 rev/s

    (b) 622 J

    (c) The energy comes from your internal energy.

57. 9.67 mm/s

59. (a) $L_0 = r_0 m v_0$

    (b) $K_0 = \frac{1}{2}mv_0^2$

    (c) $T = F_c = m\dfrac{v_0^2}{r_0}$; $W = -\dfrac{3}{2}mv_0^2$

61. 54.7°

63. (a) 3.46×10^{-47} kg·m²

    (b) 1.99 meV; 5.98 meV; 12.0 meV

65. 82.5 m/s

67. $v_{\text{cm}} = \dfrac{mv}{M+m}$; $\omega = \dfrac{mMvd}{\frac{1}{12}ML^2(M+m) + Mmd^2}$

69. $v = \sqrt{\dfrac{(0.5M + 0.8m)(\frac{1}{3}Md^2 + 0.64md^2)g}{0.32dm^2}}$

71. (a) $v_{\text{cm}} = \dfrac{K}{M}$

    (b) $\dfrac{4K}{M}$

    (c) $-\dfrac{2K}{M}$

    (d) $x = \frac{1}{6}\ell$

73. 0.349

75. 12 rad/s; 10.8 J

77. (a) 18.1 J·s

    (b) 0.414 rad/s

    (c) 15.2 s

    (d) 0.0791 J·s

79. (a) $\vec{L} = -(47.7 \text{ kg·m}^2/\text{s})\hat{k}$

    (b) $\vec{\tau} = (15.9 \text{ N·m})\hat{k}$

81. (a) 243 J·s

    (b) 306 J

83. (a) No, L decreases.

    (b) Its kinetic energy is constant.

    (c) v_0 (The kinetic energy remains constant.)

85. Yes.

87. $v_r = \dfrac{\ell\omega}{2L}\sqrt{(L^2 - \ell^2)}$

91. (a) 0.228 rad/s

    (b) 0.192 rad/s

93. 4.47×10^{22} N·m

95. 12.5 rad/s

97. (a) 26.5 rad/s

    (b) $\vec{L} = (0.303 \text{ kg·m}^2/\text{s})e^{(1.41\text{s}^{-1})t}$

Chapter R

1. Neither of you has determined the proper time interval. By definition, the proper time interval is measured by a clock in the rest frame of the car, that is, by a clock in the car.

3. Yes. If two events occur at the same time *and* place in one reference frame they occur at the same time *and* place in all reference frames. (Any pair of events that occur at the same time *and* at the same place in one reference frame are called a space-time coincidence.)

5. (a)

7. $1 + (8.61 \times 10^{-11})$

9. 6.00 ns

11. (a) 6.63 m

    (b) 12.6 m

13. (a) 599 m

    (b) 13.4 km

15. (a) 129 y

    (b) 87.6 y

17. (a) 0.600 m

    (b) 2.50 ns

19. 0.800c

21. (a) $4.50 \times 10^{-10}\%$

    (b) 0.142 ms

23. 25.0 min; 25.0 min

25. 60.0 min

27. 0.400c; event B can precede event A provided $\nu > 0.400c$

29. (a) 11.3 y

    (b) 40.0 y

31. (a) 1.005

    (b) 1.155

    (c) 1.667

    (d) 7.089

33. (a) $0.155E_0$

    (b) $1.29E_0$

    (c) $6.09E_0$

35. 2.97 GeV

39. (b) 0.866c

    (c) 0.999c

41. (a) 0.794%

    (b) 68.7%

43. (a) 0.943

45. (a) 617 eV

    (b) 79.6 eV

    (c) 7.96 eV

47. (a) 0.745

    (b) 5.00 ft

    (c) No. In Keisha's rest frame, the back end of the ladder will clear the door before the front end hits the wall of the shed, while in Ernie's rest frame, the front end will hit the wall of the shed while the back end has yet to reach the door.

Chapter 11

1. (a) False

    (b) True

3. (d)

5. (a)

7. The gravitational field is proportional to the mass within the sphere of radius r and inversely proportional to the square of r, i.e., proportional to $r^3/r^2 = r$.

9. (d)

11. $1.08 \times 10^{11} M_s$

13. (a) 2.78 h

    (b) 19.3×10^{42} kg·m²/s; 7.85×10^{42} kg·m²/s; 0.703%

    (c) 4.80×10^{-4} rad/s

15. 84.0 y

17. (a) 1.59×10^{11} m

    (b) 2.71×10^{10} m; 2.91×10^{11} m

19. (a) 90°

    (b) 0.731 AU

21. (a) 1.90×10^{27} kg

    (b) 0.282 m/s²; 0.0356 m/s²

23. (a) 8.18×10^4 s

    (b) 1.22×10^9 m

25. 1.99×10^{30} kg

27. $10w$, where w is your weight on earth.

29. 2.27×10^4 m/s

31. 1.43

33. (a) 7.37 m

    (b) 0.0319 mm

35. 0.605

37. (a) 2.27 kg

    (b) It is the *inertial* mass of m_2.

39. 10^9 m

41. $W = \dfrac{GMm_0}{R}$

43. 6.94 km/s

45. (a) $\vec{F}_{outside} = -\dfrac{GMm_0}{r^2} \hat{r}$

    (b) $U(r) = -\dfrac{GMm_0}{r}$; $U(R) = -\dfrac{GMm_0}{R}$

    (d) $U(r) = U(R) = -\dfrac{GMm_0}{R}$

(e)

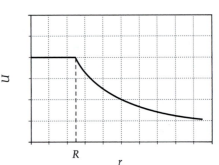

47. 2.38 km/s

49. 19.4 km/s

51. (a) 62.7 MJ

    (b) 17.4 kW·h

    (c) \$139

53. (a) 7.31 h

    (b) 1.04 GJ

    (c) 8.72×10^{12} J·s

55. 11.1 GJ

57. $\vec{g} = (4 \text{ N/kg})\hat{i}$

59. (a) $\vec{g} = \dfrac{Gm}{L^2}\hat{i} + \dfrac{Gm}{L^2}\hat{j}$

    (b) $g = \sqrt{2}\dfrac{Gm}{L^2}$

61. (a) $\vec{g} = (-1.67 \times 10^{-11} \text{ N/kg})\hat{i}$

    (b) $\vec{g} = (-8.34 \times 10^{-12} \text{ N/kg})\hat{i}$

    (c) 2.48 m

63. (a) $M = \frac{1}{2}CL^2$

    (b) $\vec{g} = \dfrac{2GM}{L^2}\left[\ln\left(\dfrac{x_0}{x_0 - L}\right) - \left(\dfrac{L}{x_0 - L}\right)\right]\hat{i}$

65. (a) 0

    (b) 0

    (c) 3.20×10^{-9} N/kg

67. $g_1 = g_2$

69. (a) $F = \dfrac{Gm(M_1 + M_2)}{9a^2}$

    (b) $F = \dfrac{GmM_1}{3.61a^2}$

    (c) 0

71. (a) $F_g = \dfrac{mg}{R}r$

    (b) $F_N = \left(\dfrac{mg}{R} - m\omega^2\right)r$

    (c) The change in mass between you and the center of the earth as you move away from the center is more important than the rotational effect.

73. $g(x) = G\left(\dfrac{4\pi\rho_0 R^3}{3}\right)\left[\dfrac{1}{x^2} - \dfrac{1}{8(x - \frac{1}{2}R)^2}\right]$

75. $\omega = \sqrt{\dfrac{4\pi\rho_0 G}{3}}$

77. 0.104 mm/s

79. (a) $\vec{F} = -\dfrac{GMm}{d^2}\left[1 - \dfrac{\frac{d^3}{4}}{\left\{d^2 + \frac{R^2}{4}\right\}^{3/2}}\right]\hat{i}$

    (b) $\vec{F}(R) = -0.821\dfrac{GMm}{R^2}\hat{i}$

81. 249 y

83. (a) $W = GM_E m \left(\dfrac{1}{r_1} - \dfrac{1}{r_2} \right)$

(b) $W = mgR_E^2 \left(\dfrac{1}{R_E} - \dfrac{1}{R_E + h} \right)$

85. 8.96×10^7 m

87. 1.70 Mm

91. $v = 1.64\sqrt{\dfrac{GM}{a}}$

93. For $r < R_1$, $g(r) = 0$; For $r > R_2$, $g(r) = \dfrac{GM}{r^2}$;

For $R_1 < r < R_2$, $g(r) = \dfrac{GM(r^3 - R_1^3)}{r^2(R_2^3 - R_1^3)}$

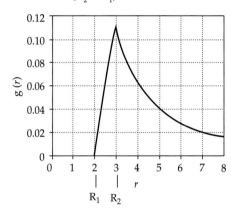

95. $g = \dfrac{2G\lambda}{r}$

97. (b) $U = -\dfrac{GMm_0}{L} \ln\left(\dfrac{x_0 + L/2}{x_0 - L/2} \right)$

99. 33.5 pN

101. (a) The gravitational force is greater on the lower robot, so if it were not for the cable its acceleration would be greater than that of the upper robot, and they would separate. In opposing this separation the cable is stressed.

(b) 220 km

Chapter 12

1. (a) False
   (b) True
   (c) True
   (d) False

3. No. The definition of the center of gravity does not require that there be any material at its location.

5. This technique works because the center of mass must be directly under the balance point. Hence the intersection of the two lines must be at the center of mass.

7. (b)

9. (c)

11. The tensile strengths of stone and concrete are at least an order of magnitude lower than their compressive strengths, so you want to build compressive structures to match their properties.

13. (b) 200 N/m

15. 318 N

17. (b) Taking long strides requires a larger coefficient of static friction because θ is then large.

(c) If μ_s is small, that is, there is ice on the surface, θ must be small to avoid slipping.

19.

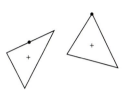

21. $(x_{cg}, y_{cg}) = \left(\dfrac{\frac{1}{2}a^2b - \pi aR^2 + \pi R^3}{ab - \pi R^2}, \frac{1}{2}b \right)$

23. 692 N; 90°; 2.54 kN; No block is required to prevent the mast from moving.

25. 0.728 m

27. $F_2 = \dfrac{1}{2}W$; $F_1 = \dfrac{\sqrt{3}}{2}W$

29. (a) 5.00 m
    (b) 4.87 m

31. $\vec{F}_1 = \dfrac{Mg\sqrt{h(2R - h)}}{h - R}\hat{i} + Mg\hat{j}$

33. (a) $\vec{F} = (30.0\text{ N})\hat{i} + (30.0\text{ N})\hat{j}$
    (b) $\vec{F} = (35.0\text{ N})\hat{i} + (45.0\text{ N})\hat{j}$

35. (a) $F_n = Mg - F\sqrt{\dfrac{2R - h}{h}}$
    (b) $F_{c,h} = F$
    (c) $F_{c,v} = F\sqrt{\dfrac{2R - h}{h}}$

37. (a) 6.87 N
    (b) 1.65 N·m
    (c) −8.26 N; 15.1 N

39. 636 N; 21.5°

41. (a) 70.7 N
    (b) 1.77 m
    (c) 3.54 m
    (d) 497 N

43. $\tau_{net} = (69.3\text{ N})b - (40.0\text{ N})a$

45. $D = \frac{1}{2}(\sqrt{3}b - a)$

47. $y = \dfrac{35.7\text{ m} - 30.4x}{3.57\text{ m} - (294\text{m}^{-1})x}$

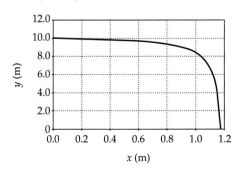

49. $h = \mu_{\mathrm{s}} L \tan \theta \sin \theta$

51. $\mu_{\mathrm{s}} = \dfrac{2h}{L \tan \theta \sin \theta}$

53. 59.0°

55. (a) 41.6 N

    (b) 0.136%

57. 5.01°

61. (a) $1.82 \times 10^6 \, \mathrm{N/m^2}$

    (b) 6.62 mJ

63. 0.686

65. It will not support the elevator.

69. $F_L = 117 \, \mathrm{N}$; $F_R = 333 \, \mathrm{N}$

71. $w_1 = 1.50 \, \mathrm{N}$; $w_2 = 7.00 \, \mathrm{N}$; $w_3 = 3.50 \, \mathrm{N}$

73. 0.148

75. $\mu_{\mathrm{s}} < 0.500$

77. $\mu_{\mathrm{s}} = \frac{1}{2}(\cot \theta - 1)$

79. (a) 147 N

    (b) 3.62 m

81. The block will tip before it slides.

83. $\mu_{\mathrm{s}} < 0.500$

85. (a) The stick remains balanced as long as the center of mass is between the two fingers. For a balanced stick the normal force exerted by the finger nearest the center of mass is greater than that exerted by the other finger. Consequently, a larger static-frictional force can be exerted by the finger closer to the center of mass, which means the slipping occurs at the other finger.

    (b) The finger farthest from the center of mass will slide inward until the normal force it exerts on the stick is sufficiently large to produce a kinetic-frictional force exceeding the maximum static-frictional force exerted by the other finger. At that point the finger that was not sliding begins to slide, the finger that was sliding stops sliding, and the process is reversed. When one finger is slipping the other is not.

87. (a) 23.0 m/s

    (b) 29.1 m/s

89. (c) $\ell_5 = 1.142 \, \mathrm{m}$; $\ell_{10} = 1.464 \, \mathrm{m}$; $\ell_{100} = 2.594 \, \mathrm{m}$

    (d) Increasing N in the spreadsheet solution suggests that the sum of the individual offsets continues to grow as N increases without bound. The series is, in fact, divergent and the stack of bricks has no maximum offset or length.

91. 566 N

93. $F_n = 2mg$; $F = \dfrac{mg}{\cos \theta}$; $F_W = mg \dfrac{R - r}{\sqrt{R(2r - R)}}$

Chapter 13

1. (e)

3. (d)

5. Nothing. The fish is in neutral buoyancy, so the upward acceleration of the fish is balanced by the downward acceleration of the displaced water.

7. (b)

9. It blows over the ball, reducing the pressure above the ball to below atmospheric pressure.

11. False

13. The buoyant force acting on the ice cubes equals the weight of the water they displace (i.e., $B = w_f = \rho_f V_{fg}$). When the ice melts, the volume of water displaced by the ice cubes will occupy the space previously occupied by the submerged part of the ice cubes. Therefore the water level remains constant.

15. Because the pressure increases with depth, the object will be compressed and its density will increase. Thus it will sink to the bottom.

17. The drawing shows the beaker and a strip within the water. As is readily established by a simple demonstration, the surface of the water is not level while the beaker is accelerated, showing that there is a pressure gradient. That pressure gradient results in a net force on the small element shown in the figure.

19. From Bernoulli's principle, the opening above which the air flows faster will be at a lower pressure than the other one, which will cause a circulation of air in the tunnel from opening 1 toward opening 2. It has been shown that enough air will circulate inside the tunnel even with the slightest breeze outside.

21. 0.673 kg

23. 103 kg

25. 29.8 inHg

27. 230 N

29. 198 atm

31. (a) 14.8 kN

    (b) 0.339 kg

33. 0.453 m

35. $F = \dfrac{\rho g a^3}{8}$

37. 4.36 N

39. (a) $11.1 \times 10^3 \, \mathrm{kg/m^3}$

    (b) lead

41. $800 \, \mathrm{kg/m^3}$; 1.11

43. $250 \, \mathrm{kg/m^3}$

45. 3.89 kg

47. $2.46 \times 10^7 \, \mathrm{kg}$

49. 491 kN

51. (a) 9.28 cm/s

    (b) 0.331 cm

    (c) 8.31 cm, in reasonable agreement with everyday experience.

53. (a) 12.0 m/s

    (b) 133 kPa

    (c) The volume flow rates are equal.

55. (a) 4.58 L/min

    (b) 763 cm²

57. 144 kPa

59. (a) 21.2 kg/s

    (b) 636 kg·m/s

    (c) 899 kg·m/s; 899 N

61. (a) $x = 2\sqrt{h(H - h)}$

    (b) $h = \frac{1}{2}H \pm \frac{1}{2}\sqrt{H^2 - x^2}$

63. (b) $P_{\mathrm{top}} = P_{\mathrm{atm}} - \rho g d$

65. 1.43 mm

67. 93.4 mi/h; Since most major league pitchers can throw a fastball in the low-to-mid-90s, this drag crisis may very well play a role in the game.

69. 0.0137; 0.0115

71. The net force is zero. Neglecting the thickness of the table, the atmospheric pressure is the same above and below the surface of the table.

73. 1061 kg/m^3

75. 65.7%

77. If you are floating, the density (or specific gravity) of the liquid in which you are floating is immaterial as you are in translational equilibrium under the influence of your weight and the buoyant force on your body. Thus the buoyant force on your body is your weight in both (a) and (b).

79. $V = \dfrac{m}{0.96\rho_{\text{w}}}$

81. 11.8 cm

83. 1 m is a reasonable diameter for the pipeline.

85. $h_{\text{A}} = 12.6 \text{ m}; h_{\text{B}} = 9.78 \text{ m}$

87. (a) 64.6%

    (b) 10.7 kN

    (c) 17.9 m/s^2

89. $3.31 \times 10^{-3} \text{ mmHg or } 3.31 \text{ } \mu\text{mHg}$

91. 1.37

93. (a) 70.0 m^3

    (b) 7.47 m/s^2

95. (c) 0.126 km^{-1}

97. (a) 33.9 kN

    (b) 39.8 kN; 36.1 kN

99. (c) $h = \left(\sqrt{H} - \dfrac{A_2}{2A_1} \sqrt{2g}\, t \right)^2$

    (d) 1 h 46 min

Chapter 14

1. $0; 4\pi^2 f^2 A$

3. (a) False
   (b) True
   (c) True

5. (a)

7. False

9. Assume that the first cart is given an initial velocity v by the blow. After the initial blow, there are no external forces acting on the carts, so their center of mass moves at a constant velocity $v/2$. The two carts will oscillate about their center of mass in simple harmonic motion where the amplitude of their velocity is $v/2$. Therefore, when one cart has velocity $v/2$ with respect to the center of mass, the other will have velocity $-v/2$. So, the velocity with respect to the laboratory frame of reference will be $+v$ and 0, respectively. Half a period later, the situation is reversed; so, one will appear to move as the other stops, and vice-versa.

11. True

13. Examples of driven oscillators include the pendulum of a clock, a bowed violin string, and the membrane of any loudspeaker.

15. Because f' varies inversely with the square root of m, taking into account the effective mass of the spring predicts that the frequency will be reduced.

17. (d)

19. (b)

21. 8π

23. (a) 3.00 Hz
    (b) 0.333 s
    (c) 7.00 cm
    (d) 0.0833 s; Because $v < 0$, the particle is moving in the negative direction at $t = 0.0833$ s.

25. (a) $x = (25\ \text{cm})\cos[(4.19\ \text{s}^{-1})t]$
    (b) $v = -(105\ \text{cm/s})\sin[(4.19\ \text{s}^{-1})t]$
    (c) $a = -(439\ \text{cm/s}^2)\cos[(4.19\ \text{s}^{-1})t]$

27. (a) $x = (27.7\ \text{cm})\cos[(4.19\ \text{s}^{-1})t - 0.445]$
    (b) $v = -(116\ \text{cm/s})\sin[(4.19\ \text{s}^{-1})t - 0.445]$
    (c) $a = -(486\ \text{cm/s}^2)\cos[(4.19\ \text{s}^{-1})t - 0.445]$

29. (a)

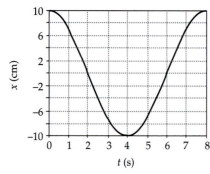

(b)

t_f	t_i	Δx
(s)	(s)	(cm)
1	0	2.93
2	1	7.07
3	2	7.07
4	3	2.93

31. (a) 7.85 m/s; 24.7 m/s²
    (b) −6.28 m/s; −14.8 m/s²

33. (a) 0.313 Hz
    (b) 3.14 s
    (c) $x = (40\ \text{cm})\cos[(2\ \text{s}^{-1})t + \delta]$

35. 22.5 J

37. (a) 0.368 J
    (b) 3.84 cm

39. 1.38 kN/m

41. (a) 6.89 Hz
    (b) 0.145 s
    (c) 0.100 m
    (d) 4.33 m/s
    (e) 187 m/s²
    (f) 36.3 ms; 0

43. (a) 682 N/m
    (b) 0.417 s
    (c) 1.51 m/s
    (d) 22.7 m/s²

45. (a) 3.08 kN/m
    (b) 4.16 Hz
    (c) 0.240 s

47. (a) 0.438 m/s
    (b) 0.379 m/s; 120 m/s²
    (c) 95.5 ms

49. 0.262 s

51. 10.1 kJ

53. (a) 0.997 Hz
    (b) 0.502 s
    (c) 0.294 N

55. (a) 46.66 cm
    (b) 0.261 s
    (c) 0.767 m/s

57. (a) 0.270 J
    (b) −0.736 J
    (c) 1.01 J
    (d) 0.270 J

59. (a) 1.90 cm
    (b) 0.0542 J
    (c) ±0.224 J
    (d) 0.334 J

61. 12.2 s

63. 11.7 s

65. $T = 2\pi\sqrt{\dfrac{L}{g(1 - \sin\theta)}}$

67. 1.10 s

69. 0.504 kg·m²

71. (b) 3.17 s

73. 21.1 cm from the center of the meter stick

77. (a) 1.63572 m
    (b) 14.5 mm, upward

79. 13.5°

81. 3.14%

85. (a) 0.314

    (b) -3.13×10^{-2} percent

87. (a) 1.57%

    (c) $0.430E_0$

89. (a) 1.01 Hz

    (b) 2.01 Hz

    (c) 0.352 Hz

91. (a) 4.98 cm

    (b) 14.1 rad/s

    (c) 35.4 cm

    (d) 1.00 rad/s

93. (a) 0

    (b) 4.00 m/s

95. (a) 14.1 cm; 0.444 s

    (b) 23.1 cm; 0.363 s

    (c) $(14.1 \text{ cm})\sin\left[(14.1 \text{ s}^{-1})t\right]$; $(23.1 \text{ cm})\sin\left[(17.3 \text{ s}^{-1})t\right]$

97. (a) $v = -(1.2 \text{ m/s})\sin\left[(3 \text{ rad/s})t + \dfrac{\pi}{4}\right]$

    (b) -0.849 m/s

    (c) 1.20 m/s

    (d) 1.31 s

99. (a) The normal force is identical to the tension in a string of length r that keeps the particle moving in a circular path and a component of mg provides, for small displacements θ_0 or s_2, the linear restoring force required for oscillatory motion.

    (b) The particles meet at the bottom. Because s_1 and s_2 are both much smaller than r, the particles behave like the bobs of simple pendulums of equal length and, therefore, have the same periods.

101. 1.62 s

103. 3.86×10^{-7} N·m/rad

105. g' is closer to g than is g''. Thus the error is greater if the clock is elevated.

107. (a) $\mu_s = \dfrac{Ak}{(m_1 + m_2)g}$

     (b) A is unchanged. E is unchanged since $E = \frac{1}{2}kA^2$. ω is reduced by increasing the total mass of the system and T is increased.

109. (b) 2.04 cm/s^2

113. (a) $x = 0$

     (b) $v_s = x_0\sqrt{\dfrac{k}{m_b + m_p}}$

     (c) $x_f = x_0\sqrt{\dfrac{m_p}{m_b + m_p}}$

115. (a)

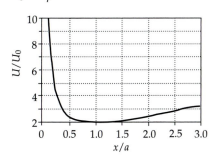

(b) $x_0 = a$ or $\alpha_0 = 1$

(c) $U(x_0 + \varepsilon) = U_0\left[1 + \beta + (1 + \beta)^{-1}\right]$

(d) $U(x_0 + \varepsilon) = \text{constant} + U_0\dfrac{\varepsilon^2}{a^2}$

119. 6.44×10^{13} rad/s

121. $7.78\sqrt{\dfrac{R}{g}}$

123. (a) 0.0478

     (b) 0.00228

127. (a)

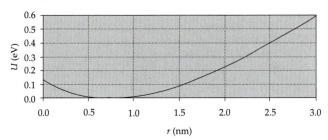

(b) $r = r_0$; $k = 2\beta^2 D$

(c) $\omega = 2\beta\sqrt{\dfrac{D}{m}}$

Chapter 15

1. The speed of a transverse wave on a rope is given by $v = \sqrt{F/\mu}$ where F is the tension in the rope and μ is its linear density. The waves on the rope move faster as they move up because the tension increases due to the weight of the rope below.

3. True

5. The speed of the wave v on the bullwhip varies with the tension F in the whip and its linear density μ according to $v = \sqrt{F/\mu}$. As the whip tapers, the wave speed in the tapered end increases due to the decrease in the mass density, so the wave travels faster.

7. No; Because the source and receiver are at rest relative to each other, there is no relative motion of the source and receiver and there will be no Doppler shift in frequency.

9. The light from the companion star will be shifted about its mean frequency periodically due to the relative approach to and recession from the earth of the companion star as it revolves about the black hole.

11. (a) True

    (b) False

    (c) False

13. There was only one explosion. Sound travels faster in water than air. Abel heard the sound wave in the water first, then, surfacing, heard the sound wave traveling through the air, which took longer to reach him.

15.

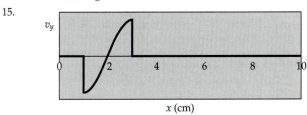

17. Path C. Because the wave speed is highest in the water, and more of path C is underwater than A or B, the sound wave will spend the least time on path C.

19. (a) 78.5 m

    (b) 69.7 m

    (c) 70.5 m . . . about 1% larger than our result in part (b) and 11% smaller than our first approximation in (a).

21. 270 m/s; 20.6%

23. 1.32 km/s

25. 19.6 g

27. (a) 265 m/s

    (b) 15.0 g

29. (b) 40.0 N

33. The lightning struck 680 m from the ball park, 58.4° W (or E) of north.

39. (a) $y(x,t) = A \sin k(x - vt)$

    (b) $y(x,t) = A \sin 2\pi\left(\dfrac{x}{\lambda} - ft\right)$

    (c) $y(x,t) = A \sin 2\pi\left(\dfrac{x}{\lambda} - \dfrac{1}{T}t\right)$

    (d) $y(x,t) = A \sin\dfrac{2\pi}{\lambda}(x - vt)$

    (e) $y(x,t) = A \sin 2\pi f\left(\dfrac{x}{v} - t\right)$

41. 9.87 W

43. (a) The wave is traveling in the $-x$ direction.; 5.00 m/s

    (b) 10.0 cm; 50.0 Hz; 0.0200 s

    (c) 0.314 m/s

45. (a) 6.82 J

    (b) 44.0 W

47. (a) 79.0 mW

    (b) Increasing f by a factor of 10 would increase P_{av} by a factor of 100. Increasing A by a factor of 10 would increase P_{av} by a factor of 100. Increasing F by a factor of 10^4 would increase v by a factor of 100 and P_{av} by a factor of 100.

    (c) Depending on the adjustability of the power source, increasing f or A would be the easiest.

49. (a) 0.750 Pa

    (b) 4.00 m

    (c) 85.0 Hz

    (d) 340 m/s

51. (a) 3.68×10^{-5} m

    (b) 8.27×10^{-2} Pa

53. (a) The displacement s is zero.

    (b) 3.68 μm

55. (a) 138 Pa

    (b) 21.7 W/m²

    (c) 0.217 W

57. (a) 50.3 W

    (b) 2.00 m

    (c) 4.45×10^{-3} W/m²

59. (a) 20.0 dB

    (b) 100 dB

61. 90.0 dB

65. (a) 100 m

    (b) 0.126 W

67. (a) 100 dB

    (b) 50.3 W

    (c) 2.00 m

    (d) 96.5 dB

69. (a) 81.1 dB

    (b) 80.0 dB; Eliminating the two least intense sources does not reduce the intensity level significantly.

71. 87.8 dB

73. 57.0 dB

75. (a) 260 m/s

    (b) 1.30 m

    (c) 262 Hz

77. (a) 1.70 m

    (b) 247 Hz

79. 153 Hz

81. 1021 Hz or a fraction increase of 2.06%; Because this fractional change in frequency is less than the 3% criterion for recognition of a change in frequency, it would be *impossible* to use your sense of pitch to estimate your running speed.

83. 349 mi/h

85. 7.78 kHz

87. 15.0 km west of P

89. (a) $f' = (1 - u_r/v)(1 - u_s/v)^{-1} f_0$

91. 1.33 m/s

93. (a) 824 Hz

    (b) 849 Hz

95. 184 m

97. -2.07×10^{-5} nm; 99 2.25×10^8 m/s

99. $2.25 \times 10^{\wedge}8$ m/s . . . where the upper arrow means the 8 is an exponent.

101. 20.8 cm

103. 3.42 m/s

105. 529 Hz; 474 Hz

107. 7.99 m

109. (a) 55.1 N/m²

    (b) 3.46 W/m²

    (c) 0.109 W

111. 77.0 kN

113. 204 m

115. 24.0 cm

117. (b) $v_0 = \sqrt{\dfrac{F}{\mu}}$

    (c) As seen by an observer at rest, the pulse remains at the same position because its speed along the chain is the same as the speed of the chain. With respect to a fixed point on the chain, the pulse travels through 360°.

119. (b) 2.21 s

Chapter 16

1.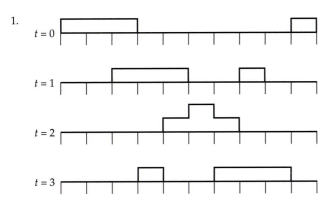

3. (c)

5. (b)

7. (a)

9. since $v \propto T$, increasing the temperature increases resonant frequencies.

11. No; the wavelength of a wave is related to its frequency and speed of propagation ($\lambda = v/f$). The frequency of the plucked string will be the same as the wave it produces in air, but the speeds of the waves depend on the media in which they are propagating. Since the velocities of propagation differ, the wavelengths will not be the same.

13. When the edges of the glass vibrate, sound waves are produced in the air in the glass. The resonance frequency of the air columns depends on the length of the air column, which depends on how much water is in the glass.

15. (b)

17. The pitch is determined mostly by the resonant cavity of the mouth, and the frequency of sounds he makes is directly proportional to their speed. Since $v_{He} > v_{air}$ (see Equation 15-5), the resonance frequency is higher if helium is the gas in the cavity.

19. Pianos are tuned by ringing the tuning fork and the piano note simultaneously and tuning the piano string until the beats are far apart (i.e., the time between beats is very long). If we assume that 2 s is the maximum detectable period for the beats, then one should be able to tune the piano string to at least 0.5 Hz.

21. 34.0 Hz; Because $v \propto T$, the frequency will be somewhat higher in the summer.

23. 7.07 cm

25. (a) 90.0°

    (b) $\sqrt{2}A$

27. (a) 0

    (b) $2I_0$

    (c) $4I_0$

29. (a) $\frac{1}{4}\lambda$

    (b) $\frac{1}{4}\lambda$

31. (a) 60.0 cm

    (b) $\dfrac{2\pi}{5}$

    (c) 24.0 m/s

33. 4726 Hz; 9452 Hz

35. (b)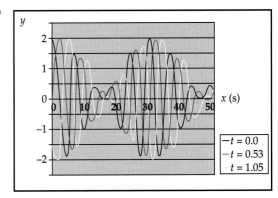

    (c) 0.500 m/s

37. 1.81; 51.5°

39. (a) 0.279 m

    (b) 1.22 kHz

    (c)

m	θ_m
	(rad)
3	0.432
4	0.592
5	0.772
6	0.992
7	1.354
8	undefined

    (d) 0.0698 rad

41. 1.98 rad or 113°

43. (a) 70.5 Hz

    (b) The person on the street hears no beat frequency as the sirens of both ambulances are Doppler shifted up by the same amount (approximately 35 Hz).

45. (a) 2.00 m; 25.0 Hz

    (b) $y_3(x,t) = (4 \text{ mm})\sin kx \cos \omega t$, where $k = \pi\text{m}^{-1}$ and $\omega = 50\pi\text{s}^{-1}$.

47. (a) 521 m/s

    (b) 2.80 m; 186 Hz

    (c) 372 Hz; 558 Hz

49. 141 Hz

51. (a) 31.4 cm; 47.7 Hz

    (b) 15.0 m/s

    (c) 62.8 cm

53. (a)

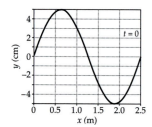

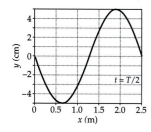

(b) 12.6 ms

(c) Since the string is moving either upward or downward when $y(x) = 0$ for all x, the energy of the wave is entirely kinetic energy.

55. (a) 70.8 Hz

(b) 4.89 Hz

(c) 35

57. 452 Hz; It would be better to have the pipe expand so that v/L, where L is the length of the pipe, is independent of temperature.

59. (a) 80 cm

(b) 480 N

(c) You should place your finger 9.23 cm from the scroll bridge.

61. (a) 75.0 Hz

(b) The harmonics are the 5th and 6th.

(c) 2.00 m

63. (a) 0.574 g/m

(b) 1.29 g/m; 2.91 g/m; 6.55 g/m

65. (a) The two sounds produce a beat because the third harmonic of the A string equals the second harmonic of the E string, and the original frequency of the E string is slightly greater than 660 Hz. If $f_E = (660 + \Delta f)$Hz, a beat of $2\Delta f$ will be heard.

(b) 661.5 Hz

(c) 79.6 N

69. 76.8 N; 19.2 N; 8.53 N

71. (a) N/f_0

(b) $\Delta x/N$

(c) $2\pi N/\Delta x$

(d) N is uncertain because the waveform dies out gradually rather than stopping abruptly at some time; hence, where the pulse starts and stops is not well defined.

73. (a) 3.40 kHz; 10.2 kHz; 17.0 kHz

(b) Frequencies near 3400 Hz will be most readily perceived.

75. $\frac{1}{3}\lambda$

77. 6.62 m

79. (a) 1.90 cm; 3.59 m/s

(b) 0; 0

(c) 1.18 cm; 2.22 m/s

(d) 0; 0

81. (a) At resonance, standing waves are set up in the tube. At a displacement antinode, the powder is moved about; at a node the powder is stationary, and so it collects at the nodes.

(b) $2fD$

(c) If we let the length L of the tube be 1.2 m and assume that $v_{air} = 344$ m/s (the speed of sound in air at 20°C), then the 10th harmonic corresponds to $D = 25.3$ cm and a driving frequency of 680 Hz.

(d) If $f = 2$ kHz and $v_{He} = 1008$ m/s (the speed of sound in helium at 20°), then D for the 10th harmonic in helium would be 25.3 cm, and D for the 10th harmonic in air would be 8.60 cm. Hence, neglecting end effects at the driven end, a tube whose length is the least common multiple of 8.60 cm and 25.3 cm (218 cm) would work well for the measurement of the speed of sound in either air or helium.

83. (a) The pipe is closed at one end.

(b) 262 Hz

(c) 32.4 cm

85. (a) $y_1(x,t) = (0.01 \text{ m})\sin\left[\left(\dfrac{\pi}{2}\text{ m}^{-1}\right)x - (40\pi \text{ s}^{-1})t\right]$;

$y_2(x,t) = (0.01 \text{ m})\sin\left[\left(\dfrac{\pi}{2}\text{ m}^{-1}\right)x + (40\pi \text{ s}^{-1})t\right]$;

(b) 2.00 m

(c) $v_y(1 \text{ m},t) = -(2.51 \text{ m/s})\sin(40\pi \text{ s}^{-1})t$

(d) $a_y(1 \text{ m},t) = -(316 \text{ m/s}^2)\cos(40\pi \text{ s}^{-1})t$

87. $y_{res}(x,t) = 0.1 \sin(kx - \omega t)$

89. (b) 203 Hz

91. (a) What you hear is the fundamental mode of the tube and its overtones. A more physical explanation is that the echo of the finger snap moves back and forth along the tube with a characteristic time of $2L/c$, leading to a series of clicks from each echo. Since the clicks happen with a frequency of $c/2L$, the ear interprets this as a musical note of that frequency.

(b) 38.6 cm

93. (a) Since no conditions were placed on its derivation, this expression is valid for all harmonics.

(b) 1.54%

95. (a) $v_y(x,t) = -\omega_1 A_1 \sin \omega_1 t \sin k_1 x - \omega_2 A_2 \sin \omega_2 t \sin k_2 x$

(b) $dK = \frac{1}{2}\mu[\omega_1^2 A_1^2 \sin^2 \omega_1 t \sin^2 k_1 x + 2\omega_1\omega_2 A_1 A_2 \sin \omega_1 t$ $\sin k_1 x \sin \omega_2 t \sin k_2 x + \omega_2^2 A_2^2 \sin^2 \omega_2 t \sin^2 k_2 x]dx$

(c) $K = \frac{1}{4}m\omega_1^2 A_1^2 \sin^2 \omega_1 t + \frac{1}{4}m\omega_2^2 A_2^2 \sin^2 \omega_2 t$

97. (a)

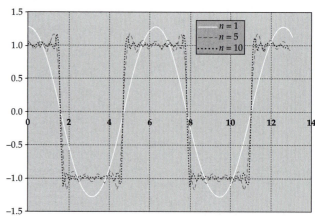

(b) $f(2\pi) = 1$ which is equivalent to the Liebnitz formula.

99. (b)

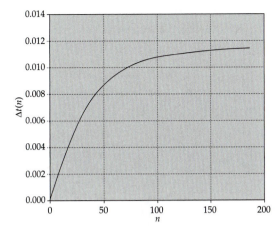

(c) The frequency heard at any time is $1/\Delta t_n$, so because Δt_n increases over time, the frequency of the culvert whistler decreases.; 7.65 kHz

Chapter 17

1. (a) False
   (b) False
   (c) True
   (d) False

3. Mert's room was colder.

5. From the ideal-gas law we have $P = nRT/V$. In the process depicted, both the temperature and the volume increase but the temperature increases faster than does the volume. Hence the pressure increases.

7. True

9. K_{av} increases by a factor of 2; K_{av} is reduced by a factor of $\frac{1}{2}$.

11. False

13. Since $10^7 \gg 273$, it does not matter.

15. (b)

17. (d)

19. The ratio of the rms speeds is inversely proportional to the square root of the ratio of the molecular masses. The kinetic energies of the molecules are the same.

21. Because the temperature remains constant, the average speed of the molecules remains constant. When the volume decreases, the molecules travel less distance between collisions, so the pressure increases because the frequency of collisions increases.

23. The average molecular speed of He gas at 300 K is about 1.4 km/s, so a significant fraction of He molecules have speeds in excess of earth's escape velocity (11.2 km/s), and thus "leak" away into space. Over time, the He content of the atmosphere decreases to almost nothing.

25. (a) 3.61×10^3 K
    (b) 225 K
    (c) If $v_{rms} > \frac{1}{5}v_e$ or $T \ge 25T_{atm}$, H_2 molecules escape. Therefore, the more energetic H_2 molecules escape from the upper atmosphere.
    (d) 164 K; 10.3 K; If we assume that the temperature on the moon with an atmosphere would have been approximately 1000 K, then all O_2 and H_2 would have escaped during the time since the formation of the moon to the present.

27. (a) 1.24 km/s
    (b) 310 m/s

(c) 264 m/s
(d) O_2, CO_2, and H_2 should be found on Jupiter.

29. 1063°C

31. (a) 8.40 cm
    (b) 107°C

33. −319°F

35. (a) 54.9 torr
    (b) 3704 K

37. −40°C = −40°F

39. −183°C; −297°F

41. (a) $B = 3.94 \times 10^3$ K; $R_0 = 3.97 \times 10^{-3}$ Ω
    (b) 1.31 kΩ
    (c) −389 Ω/K; −433 Ω/K
    (d) The thermistor is more sensitive (i.e., has greater sensitivity, at lower temperatures).

43. 1.79 mol; 1.08×10^{24} molecules

45. −83.2 glips

47. (a) 3.66×10^3 mol
    (b) 60.0 mol

49. 10.0 atm

51. 1.19 kg/m³

53. 2.56 N

55. (a) 276 m/s
    (b) 872 m/s

57. 499 km/s; 2.07×10^{-16} J

61. $K/\Delta U = 7.95 \times 10^4$

65. (a) 0.142 s
    (b) 0.143 s

67. (a) 122 K
    (b) 244 K
    (c) 1.43 atm

69. 111 mol; 55.5 mol

71. $7m_H$

73. 400.49 K

75. (a) 4.10×10^{-26} m
    (b) 4.28 nm; The mean free path is larger by approximately a factor of 1000.

77. (a) 48.9%
    (b) 70.6%

Chapter 18

1. $\Delta T_B = 4\Delta T_A$

3. (c)

5. Yes, if the heat absorbed by the system is equal to the work done by the system.

7. $W_m + Q_m = \Delta E_{int}$; For an ideal gas, ΔE_{int} is a function of T only. Since $W = 0$ and $Q = 0$ in a free expansion, $\Delta E_{int} = 0$ and T is constant. For a real gas, ΔE_{int} depends on the density of the gas because the molecules exert weak attractive forces on each other. In a free expansion, these forces reduce the average kinetic energy of the molecules and, consequently, the temperature.

9. The temperature of the gas increases. The average kinetic energy increases with increasing volume due to the repulsive interaction between the ions.

11. (a)

13. (a) False

    (b) False

    (c) False

    (d) True

    (e) True

    (f) True

    (g) True

15. (d)

17. If V decreases, the temperature decreases.

19. The heat capacity of a substance is proportional to the number of degrees of freedom per molecule associated with the molecule. Since there are 6 degrees of freedom per molecule in a solid, and only 3 per molecule (translational) for a monatomic liquid, you would expect the solid to have the higher heat capacity.

21. 1.63 min, an elapsed time that seems to be consistent with experience.

23. $c_p = (1.01\%)c_{water}$

25. (a) 10.5 MJ

    (b) 121 W

27. 7.48 kcal

29. 48.8 mg

31. 365°C

33. 20.8°C

35. 453 kg

37. (a) 0°C

    (b) 125 g

39. (a) 4.94°C

    (b) No ice is left.

41. (a) 2.99°C

    (b) 199.8 g

    (c) The answer would be the same.

43. 618°C

45. 2.21 kJ

47. 176°C

49. 53.7 J

51. (a) 6.13 W

    (b) 19.0 min

53. (a) 405 J

    (b) 861 J

55. (a) 507 J

    (b) 963 J

57. $\frac{3}{2}P_0V_0$

59. (a) 555 J

    (b) 555 J

61. (a) 55.7 g/mol

    (b) Fe

63. (a) 0; 6.24 kJ; 6.24 kJ

    (b) 8.73 kJ; 6.24 kJ; 2.49 kJ

    (c) 2.49 kJ

65. 59.6 L

67. $\Delta C_P = -\frac{13}{2}Nk$

69. $C_{V,water} = 5Nk$

71. (a) 465 K

    (b) 387 K

73. (a) 300 K; 7.80 L; 1.14 kJ; 1.14 kJ

    (b) 208 K; 5.41 L; 574 J; 0

75. (a) 263 K

    (b) 10.8 L

    (c) −1.48 kJ

    (d) 1.48 kJ

79. −142 J

81. $Q_{D \to A} = 8.98$ kJ; $Q_{A \to B} = 13.2$ kJ; $Q_{B \to C} = -8.98$ kJ; $Q_{C \to D} = -6.56$ kJ; $W_{cycle} = 6.62$ kJ

83. (a)

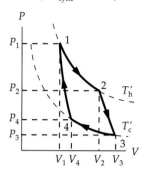

85. 180 kJ

87. (a) 65.2 K; 81.2 K

    (b) 1.62 kJ

    (c) 2.22 kJ

89. (a) 65.2 K; 81.2 K

    (b) 2.65 kJ

    (c) 3.25 kJ

91. (a) 9.20×10^{-2} J/kg·K

    (b) 0.0584 J/kg

93. 47.6 kPa; 51.5 K; 71.2 K; 148 kPa

95. (a) 2.49 kJ

    (b) 3.20 kJ

97. 171 K

99. (a) $W = 0$; $Q = 3.74$ kJ

    (b) $\Delta U = 3.74$ kJ; $Q = 6.24$ kJ; $W = 2.50$ kJ

101.

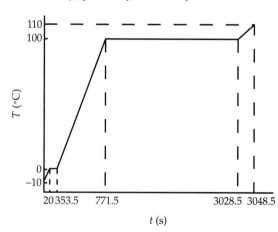

103. $4RT$

105. 396 K

107. (a) $\frac{1}{2}P_0$

    (b) diatomic

    (c) In the isothermal process, T is constant, and the translational kinetic energy is unchanged. In the adiabatic process, $T_3 = 1.32T_0$, and the translational kinetic energy increases by a factor of 1.32.

109. (a) 93.5 kPa

(b) 6266 K; 1.30 MPa

(c) 56.7 kPa

111. (b) $\Delta U = 4621$ J, a result in good agreement with the result of Problem 106.

Chapter 19

1. Friction reduces the efficiency of the engine.

3. Increasing the temperature of the steam increases the Carnot efficiency, and generally increases the efficiency of any heat engine.

5. (c)

7. (d)

9. Note that A→B is an adiabatic expansion. B→C is a constant volume process in which the entropy decreases; therefore heat is released. C→D is an adiabatic compression. D→A is a constant volume process that returns the gas to its original state. The cycle is that of the Otto engine (see Figure 19-3).

11.

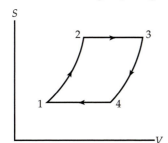

13.

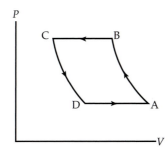

15. 56.5%

17. (a) 1.66×10^{17} W

(b) 5.66×10^{14} J/K·s

(c) 3.09×10^{13} J/K·s

19. 29.8 kJ/K

21. (a) 500 J

(b) 400 J

23. (a) 40.0%

(b) 80.0 W

25. (a)

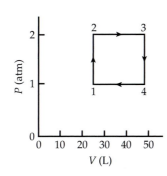

$W_{1\rightarrow2} = 0; Q_{1\rightarrow2} = 3.74$ kJ

$W_{2\rightarrow3} = 4.99$ kJ; $Q_{2\rightarrow2} = 12.5$ kJ

$W_{3\rightarrow4} = 0; Q_{3\rightarrow4} = -7.48$ kJ

$W_{4\rightarrow1} = 2.49$ kJ; $Q_{4\rightarrow1} = -6.24$ kJ

(b) 15.4%

27.

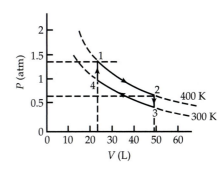

13.1%

29. (a) 600 K; 1800 K; 600 K

(b) 15.4%

31. (a) 5.16%; The fact that this efficiency is considerably less than the actual efficiency of a human body does not contradict the Second Law of Thermodynamics. The application of the second law to chemical reactions such as the ones that supply the body with energy have not been discussed in the text.

(b) Most warm-blooded animals survive under roughly the same conditions as humans. To make a heat engine work with appreciable efficiency, internal body temperatures would have to be maintained at an unreasonably high level.

35. (a) 33.3%

(b) 33.3 J

(c) 66.7 J

(d) 2.00

37. Let the first engine be run as a refrigerator. Then it will remove 140 J from the cold reservoir, deliver 200 J to the hot reservoir, and require 60 J of energy to operate. Now take the second engine and run it between the same reservoirs, and let it eject 140 J into the cold reservoir, thus replacing the heat removed by the refrigerator. If ε_2, the efficiency of this engine, is greater than 30%, then Q_{h2}, the heat removed from the hot reservoir by this engine, is $140 \text{ J}/(1 - \varepsilon_2) > 200$ J, and the work done by this engine is $W = \varepsilon_2 Q_{h2} > 200$ J. The end result of all this is that the second engine can run the refrigerator, replacing the heat taken from the cold reservoir, and do additional mechanical work. The two systems working together then convert heat into mechanical energy without rejecting any heat to a cold reservoir, in violation of the second law.

39. (a) 33.3%

(b) If COP > 2, then 50 J of work will remove more than 100 J of heat from the cold reservoir and put more than 150 J of heat into the hot reservoir. So running engine (a) to operate the refrigerator with a COP > 2 will result in the transfer of heat from the cold to the hot reservoir without doing any net mechanical work in violation of the second law.

41. (a) 100°C

(b) $Q_{1\rightarrow2} = 3.12$ kJ; $Q_{2\rightarrow3} = 0; Q_{3\rightarrow1} = -2.91$ kJ

(c) 6.73%

(d) 35.5%

43. (a) 5.26

(b) 3.19 kW

(c) 4.81 kW

45. (a) 173 kJ
    (b) 121 kJ
47. $\Delta S_u = 2.40 \, \text{J/K}$
49. (a) 11.5 J/K
    (b) Since the process is not quasi-static, it is nonreversible and the entropy of the universe must increase.
51. 1.22 kJ/K
53. (a) 0
    (b) 267 K
55. (a) 244 kJ/K
    (b) −244 kJ/K
    (c) $\Delta S_u > 0$
57. (a) −117 J/K
    (b) 137 J/K
    (c) 20.3 J/K
59. 1.97 kJ/K
61. (a) 0.417 J/K
    (b) 125 J
63. (a) 20.0 J
    (b) 66.7 J; 46.7 J
65. (a) 51.0%
    (b) 102 kJ
    (c) 98.0 kJ
67. 113 W/K
69. (a) Process (1) is more wasteful of *mechanical* energy. Process (2) is more wasteful of *total* energy.
    (b) 1.67 J/K; 0.833 J/K
71. 313 K
73. 10.0 W
75. (a) 253 kPa
    (b) 462 K
    (c) 6.96 kJ; 25.9%
77. (a) 253 kPa
    (b) 416 K
    (c) 6.58 kJ; 34.8%
79. 180 J
83. $\approx 10^{478}$

Chapter 20

1. The glass bulb warms and expands first, before the mercury warms and expands.
3. (c)
5. (a) With increasing altitude P decreases; from curve OF, T of the liquid-gas interface diminishes, so the boiling temperature decreases. Likewise, from curve OH, the melting temperature increases with increasing altitude.
    (b) Boiling at a lower temperature means that the cooking time will have to be increased.
7. The thermal conductivity of metal and marble is much greater than that of wood; consequently, heat transfer from the hand is more rapid.
9. (c)
11. In the absence of matter to support conduction and convection, radiation is the only mechanism.

13. (a)
15. The temperature of an object is inversely proportional to the maximum wavelength at which the object radiates (Wein's displacement law). Since blue light has a shorter wavelength than red light, an object for which the wavelength of the peak of thermal emission is blue is hotter than one which is red.
17. 18.1 mW/(m·K)
19. 2.90 nm
21. (a) $\gamma \equiv \dfrac{\Delta A/A}{\Delta T}$
    (b) $\gamma \approx 2\alpha\Delta T$
23. 217°C
25. $15.4 \times 10^{-6} \, \text{K}^{-1}$
27. 5.24 m
29. 0.255 mm
31. (a) The clock runs slow.
    (b) 8.21 s
33. $3.68 \times 10^{-12} \, \text{N/m}^2$
35. (a) 90°C
    (b) 82°C
    (c) 170 kPa
37. (b) $\left(P_r + \dfrac{3}{V_r^2}\right)(3V_r - 1) = 8T_r$
39. 2.07 kBtu/h
41. (a) $I_{Cu} = 962 \, \text{W}; I_{A1} = 569 \, \text{W}$
    (b) 1.53 kW
    (c) 0.0523 K/W
43. (a) Conservation of energy requires that the thermal current through each shell be the same.
45. $I = \dfrac{2\pi kL}{\ln(r_1/r_2)}(T_2 - T_1)$
47. $9.35 \times 10^{-3} \, \text{m}^2$
49. 1598°C
51. 2.10 km
53. 5767 K
55. 1.18 cm
57. (b) $\dfrac{\beta_{exp} - \beta_{th}}{\beta_{th}} < \boxed{0.3\%}$
59. $1.26 \times 10^{10} \, \text{kW}; <0.002\%$
61. 132 W ignoring the cylindrical insulation; 142 W taking the insulation into account.
63. $L_2 = L_1; \omega_2 \approx (1 - 2\alpha\Delta T)\omega_1; E_2 = E_1(1 - 2\alpha\Delta T)$
65. (a) 0.698 cm/h
    (b) 11.9 d
67. (b) 40.5 min

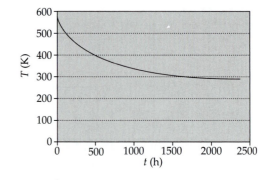

INDEX

Physical Constants[†]

Atomic mass constant	$m_u = \frac{1}{12}m(^{12}C)$	$1\,u = 1.660\,538\,73(13) \times 10^{-27}$ kg
Avogadro's number	N_A	$6.022\,141\,99(47) \times 10^{23}$ particles/mol
Boltzmann constant	$k = R/N_A$	$1.380\,6503(24) \times 10^{-23}$ J/K $8.617\,342(15) \times 10^{-5}$ eV/K
Bohr magneton	$m_B = e\hbar/(2m_e)$	$9.274\,008\,99(37) \times 10^{-24}$ J/T = $5.788\,381\,749(43) \times 10^{-5}$ eV/T
Coulomb constant	$k = 1/(4\pi\epsilon_0)$	$8.987\,551\,788\ldots \times 10^9$ N·m²/C²
Compton wavelength	$\lambda_C = h/(m_e c)$	$2.426\,310\,215(18) \times 10^{-12}$ m
Fundamental charge	e	$1.602\,176\,462(63) \times 10^{-19}$ C
Gas constant	R	$8.314\,472(15)$ J/(mol·K) = $1.987\,2065(36)$ cal/(mol·K) = $8.205\,746(15) \times 10^{-2}$ L·atm/(mol·K)
Gravitational constant	G	$6.673(10) \times 10^{-11}$ N·m²/kg²
Mass of electron	m_e	$9.109\,381\,88(72) \times 10^{-31}$ kg = $0.510\,998\,902(21)$ MeV/c^2
Mass of proton	m_p	$1.672\,621\,58(13) \times 10^{-27}$ kg = $938.271\,998(38)$ MeV/c^2
Mass of neutron	m_n	$1.674\,927\,16(13) \times 10^{-27}$ kg = $939.565\,330(38)$ MeV/c^2
Permittivity of free space	ϵ_0	$8.854\,187\,817\ldots \times 10^{-12}$ C²/(N·m²)
Permeability of free space	μ_0	$4\pi \times 10^{-7}$ N/A²
Planck's constant	h	$6.626\,068\,76(52) \times 10^{-34}$ J·s = $4.135\,667\,27(16) \times 10^{-15}$ eV·s
	$\hbar = h/(2\pi)$	$1.054\,571\,596(82) \times 10^{-34}$ J·s = $6.582\,118\,89(26) \times 10^{-16}$ eV·s
Speed of light	c	$2.997\,924\,58 \times 10^8$ m/s
Stefan-Boltzmann constant	σ	$5.670\,400(40) \times 10^{-8}$ W/(m²·K⁴)

† The values for these and other constants can be found in Appendix B as well as on the Internet at http://physics.nist.gov/cuu/Constants/index.html. The numbers in parentheses represent the uncertainties in the last two digits. (For example, 2.044 43(13) stands for 2.044 43 ± 0.000 13.) Values without uncertainties are exact. Values with ellipses are exact (like the number $\pi = 3.1415\ldots$).

Derivatives and Definite Integrals

$$\frac{d}{dx}\sin ax = a\cos ax$$

$$\int_0^\infty e^{-ax}\,dx = \frac{1}{a}$$

$$\int_0^\infty x^2 e^{-ax^2}\,dx = \frac{1}{4}\sqrt{\frac{\pi}{a^3}}$$

$$\frac{d}{dx}\cos ax = -a\sin ax$$

$$\int_0^\infty e^{-ax^2}\,dx = \frac{1}{2}\sqrt{\frac{\pi}{a}}$$

$$\int_0^\infty x^3 e^{-ax^2}\,dx = \frac{4}{a^2}$$

$$\frac{d}{dx}e^{ax} = ae^{ax}$$

$$\int_0^\infty xe^{-ax^2}\,dx = \frac{2}{a}$$

$$\int_0^\infty x^4 e^{-ax^2}\,dx = \frac{3}{8}\sqrt{\frac{\pi}{a^5}}$$

The a in the six integrals is a positive constant.

Vector Products

$$\vec{A} \cdot \vec{B} = AB\cos\theta \qquad \vec{A} \times \vec{B} = AB\sin\theta\,\hat{n} \quad (\hat{n}\text{ obtained using right-hand rule})$$

For additional data, see the following tables in the text.

Geometry and Trigonometry

$C = \pi d = 2\pi r$ — definition of π

$A = \pi r^2$ — area of circle

$V = \frac{4}{3}\pi r^3$ — spherical volume

$A = dV/dr = 4\pi r^2$ — spherical surface area

$V = A_{base}L = \pi r^2 L$ — cylindrical volume

$A = dV/dr = 2\pi rL$ — cylindrical surface area

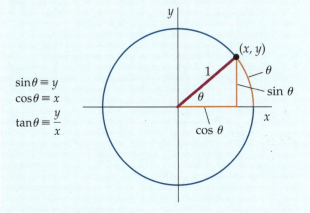

$o = h\sin\theta$
$a = h\cos\theta$

$\sin^2\theta + \cos^2\theta = 1$

$\sin(A \pm B) = \sin A \cos B \pm \cos A \sin B$

$\cos(A \pm B) = \cos A \cos B \mp \sin A \sin B$

$\sin A \pm \sin B = 2\sin[\frac{1}{2}(A \pm B)]\cos[\frac{1}{2}(A \mp B)]$

$\sin\theta \equiv y$
$\cos\theta \equiv x$
$\tan\theta \equiv \dfrac{y}{x}$

IF $|\theta| \ll 1$, THEN
$\cos\theta \approx 1$ AND $\tan\theta \approx \sin\theta \approx \theta$ (θ in radians)

Quadratic Formula

If $ax^2 + bx + c = 0$, then $x = \dfrac{-b \pm \sqrt{b^2 - 4ac}}{2a}$

Binomial Expansion

If $|x| < 1$, then $(1 + x)^n =$
$$1 + nx + \frac{n(n-1)}{2!}x^2 + \frac{n(n-1)(n-2)}{3!}x^3 + \ldots$$

If $|x| \ll 1$, then $(1 + x)^n \approx 1 + nx$

If $|\Delta x|$ is small, then $\Delta F \approx \dfrac{dF}{dx}\Delta x$